CAMBRIDGE LIBRARY COLLECTION

Books of enduring scholarly value

British and Irish History, Nineteenth Century

This series comprises contemporary or near-contemporary accounts of the political, economic and social history of the British Isles during the nineteenth century. It includes material on international diplomacy and trade, labour relations and the women's movement, developments in education and social welfare, religious emancipation, the justice system, and special events including the Great Exhibition of 1851.

Official Catalogue of the Industrial Department

This official catalogue, reissued here in its updated third edition, appeared in 1862 to accompany London's International Exhibition of that year. Held from May to November in South Kensington, on a site now occupied by the Natural History Museum and the Science Museum, the exhibition served to showcase the progress that had been made in a diverse range of crafts, trades and industries since the Great Exhibition of 1851. Over 6 million visitors came to view the wares of more than 28,000 exhibitors from Britain, her empire and beyond. The catalogue contains brief entries for participants, giving details of their name, location and products. The coverage includes mining, engineering, textiles, ceramics, metals, printing, photography, musical instruments, and pharmaceuticals. Containing a ground plan of the exhibition space as well as many contemporary advertisements, this publication remains an instructive resource for social and economic historians.

Cambridge University Press has long been a pioneer in the reissuing of out-of-print titles from its own backlist, producing digital reprints of books that are still sought after by scholars and students but could not be reprinted economically using traditional technology. The Cambridge Library Collection extends this activity to a wider range of books which are still of importance to researchers and professionals, either for the source material they contain, or as landmarks in the history of their academic discipline.

Drawing from the world-renowned collections in the Cambridge University Library and other partner libraries, and guided by the advice of experts in each subject area, Cambridge University Press is using state-of-the-art scanning machines in its own Printing House to capture the content of each book selected for inclusion. The files are processed to give a consistently clear, crisp image, and the books finished to the high quality standard for which the Press is recognised around the world. The latest print-on-demand technology ensures that the books will remain available indefinitely, and that orders for single or multiple copies can quickly be supplied.

The Cambridge Library Collection brings back to life books of enduring scholarly value (including out-of-copyright works originally issued by other publishers) across a wide range of disciplines in the humanities and social sciences and in science and technology.

Official Catalogue
of the
Industrial Department

Anonymous

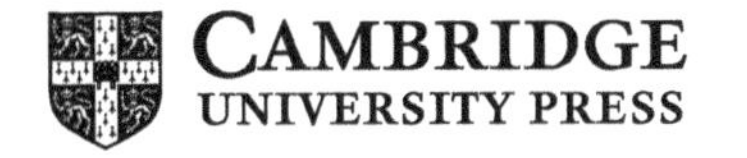

CAMBRIDGE
UNIVERSITY PRESS

University Printing House, Cambridge, CB2 8BS, United Kingdom

Published in the United States of America by Cambridge University Press, New York

Cambridge University Press is part of the University of Cambridge.
It furthers the University's mission by disseminating knowledge in the pursuit of education, learning and research at the highest international levels of excellence.

www.cambridge.org
Information on this title: www.cambridge.org/9781108067157

This edition first published 1862
This digitally printed version 2013

ISBN 978-1-108-06715-7 Paperback

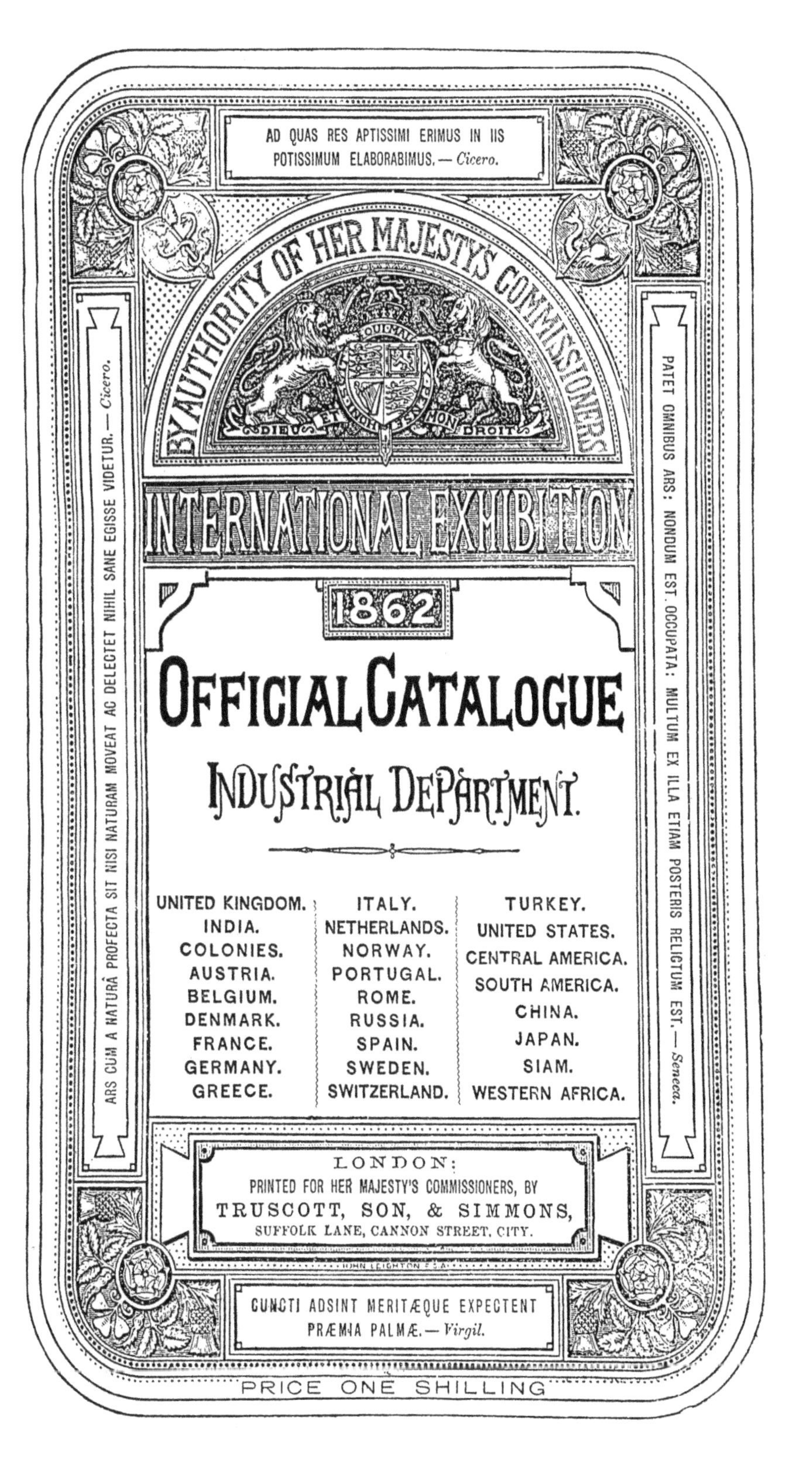
AD QUAS RES APTISSIMI ERIMUS IN IIS POTISSIMUM ELABORABIMUS. — Cicero.
BY AUTHORITY OF HER MAJESTY'S COMMISSIONERS
DIEU ET MON DROIT
HONI SOIT QUI MAL Y PENSE
ARS CUM A NATURÂ PROFECTA SIT NISI NATURAM MOVEAT AC DELECTET NIHIL SANE EGISSE VIDETUR. — Cicero.
PATET OMNIBUS ARS: NONDUM EST OCCUPATA: MULTUM EX ILLA ETIAM POSTERIS RELICTUM EST. — Seneca.
INTERNATIONAL EXHIBITION
1862
OFFICIAL CATALOGUE
INDUSTRIAL DEPARTMENT.
UNITED KINGDOM. INDIA. COLONIES. AUSTRIA. BELGIUM. DENMARK. FRANCE. GERMANY. GREECE.
ITALY. NETHERLANDS. NORWAY. PORTUGAL. ROME. RUSSIA. SPAIN. SWEDEN. SWITZERLAND.
TURKEY. UNITED STATES. CENTRAL AMERICA. SOUTH AMERICA. CHINA. JAPAN. SIAM. WESTERN AFRICA.
LONDON:
PRINTED FOR HER MAJESTY'S COMMISSIONERS, BY
TRUSCOTT, SON, & SIMMONS,
SUFFOLK LANE, CANNON STREET, CITY.
CUNCTI ADSINT MERITÆQUE EXPECTENT PRÆMIA PALMÆ. — Virgil.
PRICE ONE SHILLING

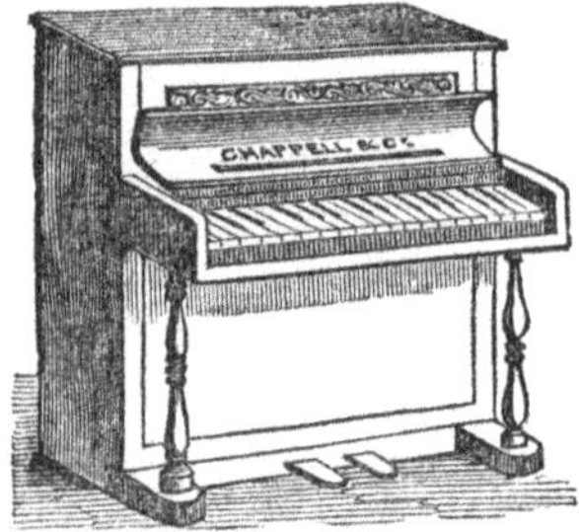

INTERNATIONAL EXHIBITION, 1862.

Regulations with respect to the Admission of Visitors to the Exhibition.

1. The Exhibition will open at Ten in the morning (except on Saturdays, when it will open at Twelve), and will close at Seven in the evening in May, June, and July, and half an hour before sunset after that date. Bells will be rung a quarter of an hour before closing.

2. The Royal Horticultural Society having arranged a new entrance to their Gardens from Kensington Road, the Commissioners have agreed with the Council of the Society to establish entrances to the Exhibition from the Gardens, and to issue a joint ticket, giving the owner the privilege of admission both to the Gardens and to the Exhibition on all occasions when they are open to visitors, including the Flower Shows and Fêtes held in the Gardens, up to the 18th of October, 1862.

3. There are two principal entrances for visitors :—

(1.) In Prince Albert's Road. (2.) In Exhibition Road.

And four secondary entrances. Three of these are from the Horticultural Gardens, for the owners of the joint tickets, Fellows of the Society, and other visitors to the Gardens, and one in Cromwell Road for the Picture Galleries. There are several exit doors.

4. The regulations necessary for preventing obstructions and danger at the several entrances will be issued from time to time.

5. Admittance to the Exhibition will be given only to the owners of Season Tickets, and to visitors paying at the doors.

REFRESHMENT AND RETIRING ROOMS.

12. Refreshments are provided according to an authorised scale of charges hung up in the rooms. The First and Second Class Rooms are on the North side of the Building, looking into the Horticultural Gardens. Other rooms are in the Eastern and Western Annexes; and light Refreshments are served at Buffets near the middle entrance to the Picture Galleries, and under the Western Dome.

13. There are Retiring Rooms, Lavatories, &c., in the North-East Transept and the South-East Transept, adjoining Exhibition Road, and in the North-West Transept and South-West Transept, adjoining Prince Albert's Road. There are also two Retiring Rooms, for Ladies only, in the Galleries on the South side of the Building, adjoining Cromwell Road. A moderate charge is made for the use of them.

LOST ARTICLES.

14. Inquiries respecting articles lost or found should be made at the Police Office, in the South Central Court.

POST OFFICE, RAILWAY INQUIRY OFFICE, TELEGRAPH OFFICE.

15. The Post Office, for the use of Visitors, is at the end of the North-East Transept on the right hand. Letters for the Country may be posted till 5 P.M. o'clock. There is also a Pillar Post at the Western Entrance. The Railway Inquiry Office is in the North-East Tower. The Telegraph Office is at the Central Entrance in Cromwell Road.

FOREIGN MONEY EXCHANGE OFFICES.

16. Foreign Money Exchange Offices, established by Messrs. Adam Spielmann & Co., will be found outside the doors at the Eastern and Western Dome, and the Cromwell Road; and in the Building near the Greek Court, at the entrance to the Horticultural Gardens.

June 2nd, 1862. (By Order) F. R. SANDFORD, *Secretary*.

DESCRIPTION OF THE BUILDING.

THE Exhibition Building—including picture-galleries and annexes—covers twenty-four acres and a-half of the land purchased by the Commissioners of 1851 out of the surplus from the first International Exhibition. The line of picture-galleries forms the upper part of the long southern front in the Cromwell-road, and there are also two wings, east and west; the north-east and south-east transepts form the eastern side in the Exhibition-road—intersected midway by the eastern dome; and the north-west and south-west transepts in Prince Albert's-road, form the western side—intersected midway by the western dome. The northern front consists of the upper and lower refreshment rooms, built in and over the arcades at the southern end of the Horticultural Gardens. The nave runs from the east to the west dome, through the building—cutting it into two unequal parts. The open, or glass courts, between the refreshment-rooms and the nave, are called the North Courts; and the courts between the picture-galleries and the nave, are called the South Courts. The eastern annex is a long continuation of the north-east transept—running north; and the western annex is a long continuation of the north-west transept, also running north.

GROUND PLAN OF INTERNATIONAL EXHIBITION.

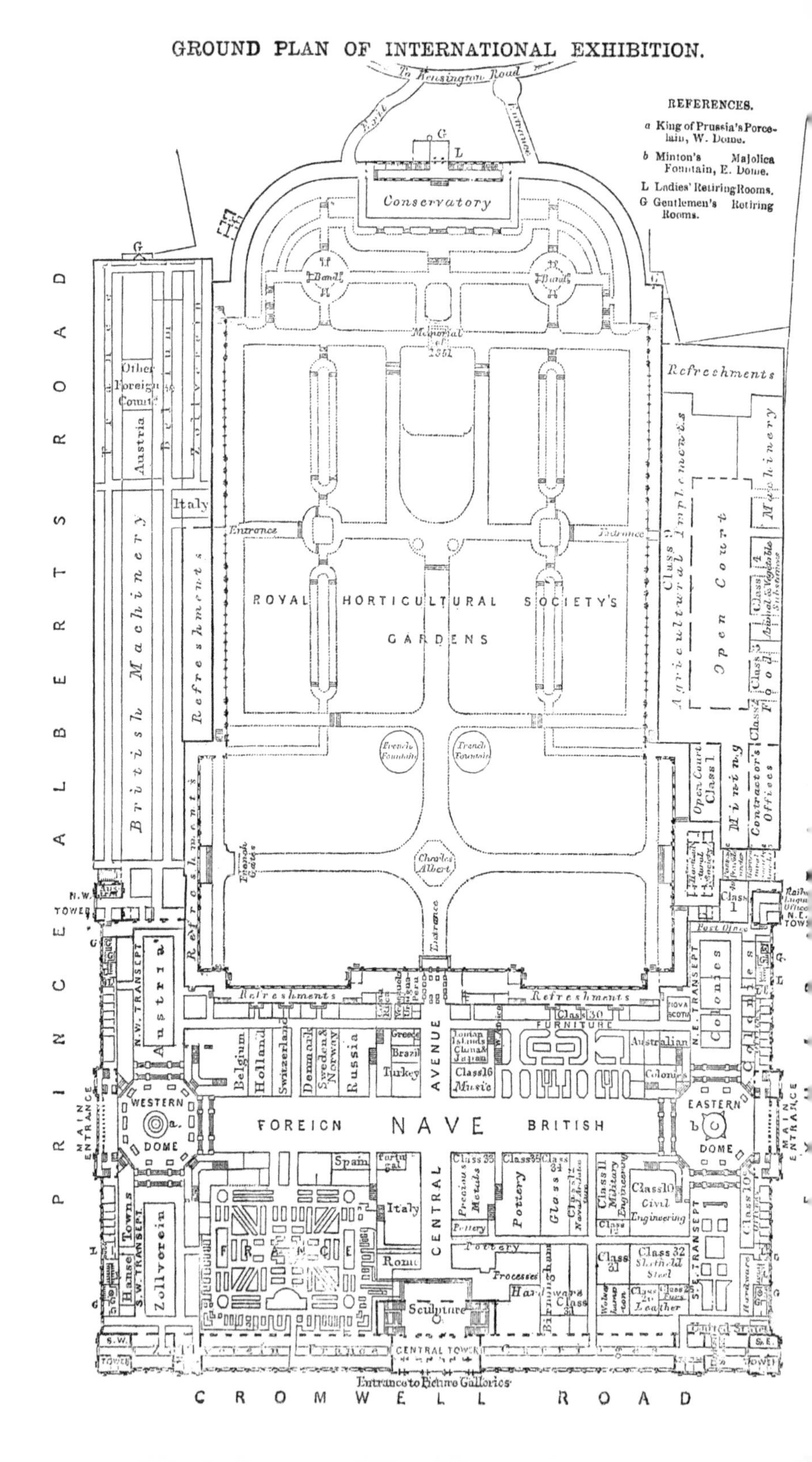

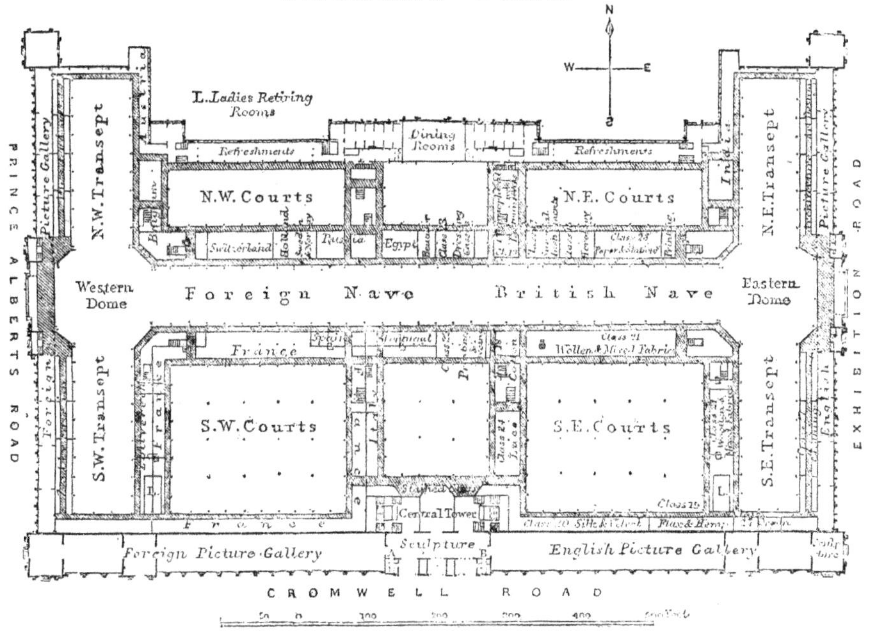

The Crimson Cards with white figures distributed through the Building are those which correspond with the numbers in this Catalogue.

NOTICE.

All corrections forwarded to the Office of the Official Catalogue up to the 1st of September have been made in this (the third) edition, excepting in those cases where the alterations would have necessitated a departure from the prescribed regulations relating to the description of exhibitors' articles. Some few names will be found out of alphabetical order, which is owing to the necessity of preserving the general sequence of numbers; and the names of fresh exhibitors, or those which, from any cause, have been removed from their original position, appear at the end of each class or sub-class.

BOOKS, &c., SOLD IN THE BUILDING.

No articles are sold under the authority of Her Majesty's Commissioners except the following:—

1. THE INDUSTRIAL CATALOGUE. Containing the Name and Address of every Exhibitor, with a Brief Description of his Goods. Price 1*s*.

2. THE FINE ARTS CATALOGUE. Containing the Name of the Artist, and the Title of every Work of Art exhibited, and the Name of the Exhibitor. Price 1*s*.

3. A CONCISE HISTORY OF THE INTERNATIONAL EXHIBITION OF 1862: Its Rise and Progress, its Building and Features, and a Summary of all former Exhibitions. By JOHN HOLLINGSHEAD. With Numerous Illustrations and Plans. Price 2*s*. 6*d*., or 3*s*. 6*d*. bound in cloth.

4. THE OFFICIAL ILLUSTRATED CATALOGUE. Containing the Name and Address of every British Exhibitor, and detailed Descriptions of Articles exhibited. Illustrated with many hundred Engravings. In Thirteen Parts, price 1*s*. each.

CLASS

PART 1.

1. Mining, Quarrying, Metallurgy, and Mineral Products.
2. Chemical Substances and Products, and Pharmaceutical Processes.
3. Substances used for Food, including Wines.
4. Animal and Vegetable Substances used in Manufactures.

PART 2.

5. Railway Plant, including Locomotive Engines and Carriages.
6. Carriages not connected with Railroads.

PART 3.

7. Manufacturing Machines and Tools.

PART 4.

8. Machinery in general.

PART 5.

9. Agricultural and Horticultural Machines and Implements.

PART 6.

10. Civil Engineering, Architectural, and Building Contrivances.
11. Military Engineering, Armour, and Accoutrements, Ordnance, and Small Arms.
12. Naval Architecture and Ship's Tackle.

PART 7.

13. Philosophical Instruments and Processes depending upon their use.
14. Photographic Apparatus and Photography.
15. Horological Instruments.
16. Musical Instruments.
17. Surgical Instruments and Appliances.

CLASS

PART 8.

18. Cotton.
19. Flax and Hemp.
20. Silk and Velvet.
21. Woollen and Worsted, including Mixed Fabrics generally.
22. Carpets.
23. Woven, Spun, Felted, and Laid Fabrics, when shown as specimens of Printing or Dyeing.
24. Tapestry, Lace, and Embroidery.

PART 9.

25. Skins, Fur, Feathers, and Hair.
26. Leather, including Saddlery and Harness.
27. Articles of Clothing.

PART 10.

28. Paper, Stationery, Printing, and Bookbinding.
29. Educational Works and Appliances.

PART 11.

30. Furniture and Upholstery, including Paper-hangings and Papier-mache.

PART 12.

31. Iron and General Hardware.
32. Steel and Cutlery.

PART 13.

33. Works in Precious Metals, and their imitations, and Jewellery.
34. Glass.
35. Pottery.
36. Manufactures not included in previous classes.

5. HUNT'S HAND-BOOK TO THE OFFICIAL CATALOGUES. An Explanatory Guide to the Natural Productions and Manufactures of the International Exhibition, 1862. By ROBERT HUNT, Esq., F.R.S., F.S.S., &c., Author of the Synopsis and Hand-Book to the Official Catalogues of 1851. Price 6*d*. each Part; or, bound in Two Volumes, price 7*s*.

Part 1. Raw Materials and Manufactures directly from them Chemicals, Food, &c.
" 2. Machinery, Tools, Implements.
" 3. Engineering, Civil and Military. Naval Architecture, &c.
" 4. Iron and Steel. Metal Manufactures, Precious Metals.
Part 5. Jewellery. Glass. Fictile Manufactures.
" 6 Textile Materials and Manufactures, &c.
" 7. General Manufactures (Handicraft).
" 8. Philosophical Instruments, &c. Paper, Printing, &c. Education.
" 9. The British Colonial Possessions.
" 10. Foreign Countries.

And of E. STANFORD, 6, Charing Cross, London, S.W.

6. SYNOPSIS OF THE CONTENTS OF THE INDUSTRIAL DEPARTMENT OF THE EXHIBITION. By ROBERT HUNT, Esq., F.R.S., F.S.S., &c. Price 6*d*.

And of E. STANFORD, 6, Charing Cross, London, S.W.

7. GROUND PLANS OF THE BUILDING, THE GALLERIES, THE ANNEXES, AND THE HORTICULTURAL GARDENS. Price 3*d*.

8. VIEW OF THE INTERNATIONAL EXHIBITION BUILDING, SOUTH FRONT. Engraved by J. Le Keux, from a Drawing by Godfrey Sykes. Price 6*d*.

9. VIEW OF THE INTERNATIONAL EXHIBITION BUILDING, WEST FRONT. Engraved by J. Le Keux, from a Drawing by Godfrey Sykes. Price 6*d*.

10. PHOTOGRAPHIC VIEWS OF THE EXHIBITION. By the London Stereoscopic Company. *At various prices.*

11. MEDALS STRUCK IN THE BUILDING. *At various prices.*

12. OPERA AND OTHER GLASSES, *on Sale or Hire, at moderate charges.*

*** *Visitors are strictly forbidden to touch any of the articles exhibited.*

International Exhibition
1862.

OFFICIAL
CATALOGUE
OF THE
INDUSTRIAL DEPARTMENT.

THIRD EDITION.

LONDON:
Printed for Her Majesty's Commissioners, by
TRUSCOTT, SON, & SIMMONS,
SUFFOLK LANE, CANNON STREET, CITY.

INTRODUCTION.

A BRIEF explanation of the system pursued in respect to the compilation of the Official Catalogue, containing, as it probably does, an enumeration of many hundred thousand articles, may be deemed necessary; more, however, as a record of facts than from any desire to unduly expatiate upon the difficulties encountered.

A printed form, by order of Her Majesty's Commissioners, was forwarded in December last to each British Exhibitor, requesting him to fill in his name, address, and the briefest possible description of his goods, the whole to be included in sixteen words. These instructions were generally followed; but owing, in many instances, to the illegibility of the writing in the returns, and the confusion arising from the recurrence of names capable of being used either as a Christian or a Surname, a perpetual correspondence was, at the onset, involved, which, however, threatened to become so incessant that the Compiler was obliged, without further delay, to accept—*cum grano salis*—the descriptions as they were offered, which, in several instances, were rather a statement of the Exhibitor's calling than a list of his articles. A rigorous system of condensation, beyond even that at first required, was deemed necessary, owing to the probable bulk of the Foreign list of Exhibitors; for though, at that period, the exact dimensions of this portion of the Catalogue could only be surmised, it had, nevertheless, to be provided for; but while names and addresses were abbreviated, and redundances of all kinds expunged—especially those prejudging the merits of the articles—care was taken not to injure, by curtailment, any speciality of the Exhibitor, and pains were especially bestowed upon scientific nomenclature, in whatever department it occurred. The assistance rendered by the Superintendents of the various classes, and by the Foreign Commissioners, has been found of great value; and it is to be hoped that technicalities have not suffered beyond that which is unavoidable when enumerating a list of articles, dispersed (at the period of compilation) over the whole world, with nothing but the *lex scripta*, in a Babel of languages, as a guide.

As the Official Catalogue, owing to the great expense of its production, arising principally from the incessant necessity for revises, appears in a very condensed form,—its dimensions being, in fact, in an inverse ratio to the labour bestowed,—it was deemed right, in the interests of the Exhibitors, to give a notice of their articles, together with names and addresses, in the Illustrated Catalogue; and the visitor will do well to provide himself with a copy of Mr. Robert Hunt's Synopsis, and also with his Hand-book; for these works will, there is little doubt, fill up any hiatus in the information required.

The difficulties have naturally been greater in the Foreign division than in the British; for, added to the same impossibility of *vivâ voce* communications, has been the almost hopeless task, not of translating foreign languages into the vernacular, but of rendering the foreign translator's English into idiomatic expression, and of finding correct English equivalents in respect to many articles which have no technical synonyme in this country.

From this brief explanation it will probably be acknowledged that an Official Catalogue, while a necessary production, is somewhat an unpromising one, enjoining a larger amount of consideration from those who detect flaws in its contents than any other work in the form of an epitome. It is to be hoped, however, the labour and great anxiety involved in its construction will prevent its value from being materially affected; and as the utmost possible amount of correction will be made in later editions, it is requested that all information in respect to important errata be sent to the Office of the Catalogue, with the class, number, and page where they occur.

SYDNEY WHITING.

EXHIBITION BUILDING,
September 1st, 1862.

CONTENTS.

UNITED KINGDOM.

COLONIAL POSSESSIONS.

FOREIGN STATES.

EXPLANATION AS TO NUMBERS.

The numbers referring to the English Exhibitors are on Crimson Cards, and run from 1 to 6965 without any duplicate, but at the end of each Class and Sub-Class a break in the numerical order occurs, for the purpose of inserting Exhibitors' names during new editions without disturbing the general sequence. A number missing in its usual place is owing either to the erasion of an Exhibitor's name after the numbers were fixed, or to a transposition.

The numbers referring to the Foreign Exhibitors in most instances occur according to the arrangements of the Foreign Commissioners, as it was found impossible to alter the system they adopted in reference both to numbers and to alphabetical order.

Thus, in the English Department, the numbers in the Catalogue will correspond with the numbers on the cards of the Exhibitors, whatever may be their position in the Building; but in the Foreign Courts it will be necessary to refer, not only to the number in the Catalogue, but also to the name of the country, which will be found prominently displayed.

INTERNATIONAL EXHIBITION, 1862.

OFFICIAL DIRECTORY.

HER MAJESTY'S COMMISSIONERS.

THE EARL GRANVILLE, K.G., LORD PRESIDENT OF THE COUNCIL.
THE DUKE OF BUCKINGHAM AND CHANDOS.
SIR C. WENTWORTH DILKE, BART.
THOMAS BARING, ESQ., M.P.
THOMAS FAIRBAIRN, ESQ.

F. R. SANDFORD, *Secretary and General Manager.*

Special Commissioner for Jury Department.—DR. LYON PLAYFAIR, C.B., F.R.S.

OFFICES OF HER MAJESTY'S COMMISSIONERS:—
EXHIBITION BUILDING, SOUTH KENSINGTON, LONDON, W.

STAFF.

SECRETARIAT.

Assistant Secretary—LOUIS LINDON.
Correspondence—HON. EDWIN B. PORTMAN. *Assistants*—M. KINSEY, R. J. S. SMITH.
Board Clerk and Shorthand Writer—H. S. KEWLEY.
Registrar—C. MARTYN. *Assistants*—H. J. GIBBS, E. TOMPSON, C. H. G. PEASE.
Storekeeper—J. LINCOLN. *Clerk*—JOHN SYMONS.
Post-Office Clerk—C. R. BIGLAND. *Assistant*—T. W. CHURCH.
Clerks—J. EVANS, A. FRASER.
Office Keeper—S. MILLIE.

FINANCE.

Principal Financial Officer—J. J. MAYO. *Assistants*—D. C. MAUNSELL, S. J. NICOLLE. *Clerks*—PERCY JACKSON, JOHN REID.

BUILDING AND VENTILATION.

Engineer and Architect—CAPTAIN FRANCIS FOWKE, R.E. *Assistants*—CAPTAIN PHILLPOTTS, R.E.; LIEUTENANTS HARRISON and BUCKLE, R.E.
Superintendent of Decoration—J. G. CRACE.

CATALOGUES.

Superintendent of Official Catalogue—SYDNEY WHITING.
Superintendent of Fine Art Catalogue—CHARLES W. FRANKS.
Superintendent of Illustrated Catalogue—J. CUNDALL. *Assistant*—J. W. MCGAULEY.
Superintendent for Sale of Catalogues—W. H. FOSTER.

JURY DEPARTMENT.

Deputy Commissioners—

LIEUTENANT BROOKE, R.E.	For Classes 12 and 13.
DR. DALZELL, F.R.S.E.	„ 28, 29, and 35.
R. HUNT, F.R.S.	„ 1, 2, and 31.
O. JONES, V P.R.I.B.A.	„ 22, 30, 33, and 34.
CAPTAIN PHILLPOTTS, R.E.	„ 8 and 10.
J. O. PLAYFAIR	„ 25, 26, and 27.
HON. E. PORTMAN	„ 15, 16, 32, and 36.
H. SANDHAM	„ 5, 6, and 7.
G. WALLIS	„ 18, 19, 20, 21, 23, and 24.
PROFESSOR J. WILSON, F.R.S.E.	„ 3, 4, and 9.

Honorary Deputy Commissioners—

LORD FREDERICK CAVENDISH	For Class 11.
DR. W. S. PLAYFAIR, F.R.C.S. H.M. Indian Army	„ Classes 14 and 17.

Secretary of Juries—J. F. ISELIN, M.A.
Clerk for Summons—W. A. WELSH.
Clerk for Awards—
Postal Clerk—W. J. RAMEL.
General Clerk—
Office Attendant—SERJEANT BARNARD.

EXECUTIVE.

Assistant Manager (English Side)—RICHARD A. THOMPSON. *Clerks*—W. E. STREATFEILD, H. BLACKWELL.

Assistant Manager (Foreign Side)—PHILIP CUNLIFFE OWEN. *Deputy*—LIEUTENANT E. T. BROOKE, R.E. *Clerks*—J. W. APPELL, A. S. COLE.

Chief Superintendent—A. N. SHERSON. *Assistant*—R. LESTER. *Clerk*—H. MOTTLEY.

In command of Royal Engineers attached to the Exhibition—MAJOR EDWARDS, R.E.

Superintendents of Districts—

No. 1. Eastern Annexe (Classes 1, 2, 3, 4, 9, and parts of 5, 7, 8)—C. W. QUIN, F.C.S. *Deputy*—E. V. LINDON.

No. 2. Colonial Department—DR. LINDLEY, F.R.S. *Deputy*—P. L. SIMMONDS, F.S.S.

No. 3. East Dome (Classes 10, 11, 12)—MAJOR A. K. MOFFAT, H.M.B.S. Corps. *Deputy*—CAPTAIN WALLACE BARROW.

No. 4. South Transept (Classes 6, 25, 26, 31, 32), and America—T. A. WRIGHT. *Deputy*—J. BOULT.

No. 5. Classes 16, 30, 33, 34, 35—J. B. WARING, F.R.I.B.A. *Deputies*—J. C. GAPPER, W. TURNER.

No. 6. Galleries North of the Nave (Classes 13, 15, 17, 28, 36, India), and Staircases—C. R. WELD. *Deputy*—W. H. JORDAN.

No. 7. Galleries South of the Nave (Classes 18, 19, 20, 21, 22, 23, 24, 27), and Staircases—G. WALLIS. *Deputies*—J. TAYLOR, E. J. ALBO, J. LIDFORD.

No. 8. Western Annexe (Classes 5, 7, 8)—D. K. CLARKE. *Deputy*—JOHN CUNDY. *Assistants*—E. ROSE, JOSIAH EXLEY. *Clerk*—W. W. GREEN.

No. 9. Fine Art Galleries (Classes 37, 38, 38A, 40), Central Tower (Classes 14, 29), and Supervision of Statuary—R. F. SKETCHLEY. *Deputies*—J. S. RIPLEY, JUN., C. JOHNSON. *Inspector of Statuary*—J. MITCHELL.

No. 10. Foreign Side (Ground Floor)—J. C. FOX.

No. 11 Foreign Side (Galleries)—H. SLATER.

Fire, Water, and Gas Arrangements—CAPTAIN BENT.

Refreshments—CAPTAIN DONNELLY, R.E.

General Cleanliness, Retiring Rooms, and Circulation.—CAPTAIN MAYNARD, R.N.R.

Indian Department—DR. FORBES WATSON.

Musical Arrangements—LIEUTENANT CAUTLEY, R.E.

CUSTOMS.

Surveyor General—F. St. John.
Examining Officers—W. F. Bramall, *Principal;* G. H. Scrivenor. *Assistant ditto and Clerk*—F. Cockburn.
Out-door Officers—A. Cobb, C. Buchanan, J. Atkinson.
Messenger—W. Thomas.

ACTING COLONIAL COMMISSIONERS.

CANADA.—Sir W. E. Logan, Brown Chamberlin, J. B. Hurlbert, and Professor H. H. Miles, M.A.

VANCOUVER.—The Hon. A. J. Langley.

BRITISH COLUMBIA.—Richard Charles Mayne, R.N.

NEW BRUNSWICK.—Thos. Daniel.

NOVA SCOTIA.—Andrew M. Uniacke.

PRINCE EDWARD'S ISLAND.—H. Haszard, 25, Old Basinghall Street.

NEWFOUNDLAND.—F. N. Gisborne.

BERMUDA.—W. C. Fahie Tucker.

WEST INDIAN COLONIES, &c.

BAHAMAS.—Samuel Harris.

BARBADOS.—Stephen Cave, M.P.

BRITISH GUIANA.—Sir Wm. H. Holmes; and A. F. Ridgway.

JAMAICA.—Edward Chitty, Lucas Barret, and Alex. F. Ridgway.

ST. VINCENT.—George C. Stacpoole, M.D.

TRINIDAD.—Sir W. H. Holmes.

AFRICAN COLONIES AND COUNTRIES.

NATAL.—W. C. Sargeaunt.

ST. HELENA.—N. Solomon.

LIBERIA.—Gerard Ralston.

AUSTRALASIAN COLONIES.

NEW SOUTH WALES.—Edward Hamilton, and Sir Daniel Cooper; Sedgwick Cowper, *Secretary*.

QUEENSLAND.—M. H. Marsh, M.P., Alfred Denison, and Arthur Hodgson.

VICTORIA.—Sir Redmond Barry and C. E. Bright; J. G. Knight, F.R.I.B.A., *Secretary*.

SOUTH AUSTRALIA.—Francis S. Dutton.

WESTERN AUSTRALIA.—Alex. Andrews.

TASMANIA.—Joseph Milligan, M.D.; F. A. Ducroz; and J. A. Youl.

NEW ZEALAND.—John Morrison; M. Holmes for Otago.

EASTERN COLONIES.

CEYLON.—E. Rawdon Power.

MAURITIUS.—James Morris.

ACTING FOREIGN COMMISSIONERS.

BELGIUM.—F. FORTAMPS, Senator, 27, Ovington Square, Brompton; O. DELEPIERRE, Secretary of Legation and Consul, 3, Howley Place, Maida Hill; and E. C. DE GRELLE, 56, Cannon Street, E.C.

DENMARK.—A. WESTENHOLZ, Consul-General, 26, Mark Lane, E.C.

FRANCE.—M. LE PLAY, Conseiller d'Etat, Commissaire Gènèral de l'Empire Français, Hôtel de la Commission Impérial, Cromwell Road.

AUSTRIA.—CHEVALIER DE SCHWARZ, 6, Onslow Crescent, Brompton.

PRUSSIA.—GEHEIMER OBER-REGIERUNGS-RATH HOENE; REGIERUNGS-RATH A. ALTGELT; and C. HEIDMAN, 10, Hereford Square, Brompton.

BADEN.—DR. DIETZ, 40, Brompton Square, S.W.

BAVARIA.—DR. C. BEEG, 48, Brompton Square, S.W.

HANOVER.—DR. C. KARMARSCH, 9, Sussex Place, South Kensington, W.

HESSE-DARMSTADT.—GEHEIM-RATH ECKHARDT, 45, Brompton Square, S.W.; and COMMERCIEN-RATH F. FINK, 48, Brompton Square, S.W.

SAXONY.—REGIERUNGS-RATH M. L. WIESSNER, 13, Alfred Place, W.

WURTEMBERG.—DR. VON STEINBEIS, 53, Brompton Square, S.W.

HANSE TOWNS and MECKLENBURG.—A. L. J. MEIER, 26, Gloucester Grove, Old Brompton, W.

GREECE.—A. C. IONIDES, Consul-General, 17, Gracechurch Street, City.

IONIAN ISLANDS.—H. DRUMMOND WOLFF, C.M.G., London.

ITALY.—MARCHESE G. B. DI CAVOUR, M.P.; and Commendatore G. DEVINCENZI, M.P., 44, Thurloe Square, S.W.

NETHERLANDS.—T. A. VAN EYK, LL.D.; and Dr. E. H. VAN BAUMHAUER, 56, Brompton Crescent, S.W.

NORWAY.—E. TIDEMAND, of the Royal Norwegian Department for the Interior. 7, South Street, Thurloe Square.

SWEDEN.—C. F. WAERN, 66, Sussex Lodge, Addison Road, South Kensington.

PORTUGAL.—LE VISCOUNT DE VILLA MAIOR, 91, Sloane Street, S.W.

ROME.—H. DOYLE, 17, Cambridge Terrace, Hyde Park, W.

RUSSIA.—A. LEVSHIN, Privy Councillor and Senator, 2, Sydney Place, Fulham Road, S.W.; GABRIEL DE KAMENSKY, 33, Victoria Road, W.; GEORGE PETERSON, 56, Brompton Square, S.W.; and THEODORE JORDAN, 35, Grove Place, Brompton.

SPAIN.—G. E. BALLERAS, 14, Cromwell Road.

SWITZERLAND.—G. VOGT; and F. BUCHSER, 1, Ovington Terrace, Brompton.

TURKEY.—WORTLEY DE LA MORE, 55, Gloucester Street, South Belgravia.

EGYPT.—LUDOVIC LORIA, 52, Pelham Street, Brompton, S.W.

UNITED STATES.—J. E. HOLMES, 17, Norfolk Street, Strand.

COSTA RICA.—G. W. EWEN, 4, Adam Court, Old Broad Street, E.C.

BRAZIL.—F. I. DE CARVALHO MORIERA, Brazilian Minister; Agent, W. H. CLARK, 6, Leinster Terrace, Hyde Park.

PERU.—H. KENDALL, Consul, 11, New Broad Street, E.C.

URUGUAY.—T. QUESTED FINNIS, Alderman of the City of London, Tower Street.

VENEZUELA.—F. H. HEMMING, Consul, 25, Moorgate Street, E.C.

ECUADOR.—F. G. GASTON, 75, Mark Lane, E.C.; Z. GERSTENBERG, 11, Warnford Court, E.C.; and L. LEVINSOHN, 7, Finsbury Square, E.C.

ARGENTINE REPUBLIC.—M. B. SAMPSON, Consul, 1, George Street, Mansion House, E.C.

NICARAGUA.—J. J. DE ARRIETA, 101, Sloane Street, S.W.

GUATEMALA.—J. SAMUEL, Consul, 1, Coleman Street Buildings, Moorgate Street E.C.

LIST OF OBJECTS OF INTEREST PLACED IN THE MAIN AVENUES OF THE BUILDING.

The Statues, and Works pertaining to the Fine Arts, will be found in the Fine Arts Catalogue.

EASTERN DOME.

Dent's clock round stained-glass window.
Gilt pyramid, representing the quantity of gold received from Australia.
Collection of dressing-cases, by various contributors.
Hunt and Roskell's collection of plate.
Minton's great fountain, in modern majolica ware.
Asprey's collection of dressing-cases.
Crystal candelabra.

NAVE.

Drinking fountain, by Earp, of Lambeth, sculptured stone, enriched with Swiss, Devonshire, and other coloured marbles.
Granite pyramid.
An assortment of the various kinds of leather and skins, with a machine for testing the strength of leather, paper, &c.
Collection of alpaca and other fabrics.
A collection of Birmingham small arms, consisting of rifles, pistols, swords, bayonets, &c.; exhibited by Bentley and Playfair, Joseph Bourne, Cook & Son, Cooper & Goodman, Hollis & Sheath, King & Phillips, Pryse & Redman, W. L. Sargant, W. Scott & Son, Joseph Smith, Swinburn & Son, Tipping & Lawden, Wm. Tranter, Thomas Turner, James Webley, Joseph Wilson, and B. Woodward & Sons.
Model of "Warrior" Armstrong gun.
Nicholay's fur collection.
Fur collection by various Exhibitors.
Equatorial.
Mersey Steel Company's gun.
Crace's specimens of furniture.
Collection of small equatorials.
The Norwich gates: their massive portions are in cast iron, and their ornamental in forged iron; the panels, and wreaths round the bars, &c., consist of leaves and flowers, executed with the hammer; heraldic ornaments surmount the principal gates and pillars.
Art Union trophy.
Gun carriage.
Animal and vegetable substances placed in the compartments of an hexagonal pile, standing upon steps, and surmounted by a dome.
Collection of gutta percha and other articles.
Cooke & Sons' collection of astronomical instruments.
Philosophical instruments from Kew Observatory, Richmond, consisting of self-registering magnetometers and meteorological apparatus, exhibited by the British Association for the Advancement of Science.
Lighthouse apparatus, consisting of lanterns for light-ship, lenses, &c.
Buckingham's large equatorial.
Copeland's porcelain.
China, consisting of Her Majesty's dessert service, made at the Worcester Porcelain Works, and exhibited by Messrs. Kerr & Binns.
Peters & Son's drag, with powerful screw safety-break.
Silver-grey granite obelisk.
Osler's case of cut crystal glass, showing prismatic colours.
Toy collection.
Elkington's plated goods.
Trophy containing elaborate works in gold, silver, and jewels (Emanuel).
Model of Milan Cathedral.

Here Middle Avenue intersects.

Roman collection.
Jackson & Graham's furniture.
United States collection.
Trollope & Son's sideboard and mantelpiece.
A lighthouse lantern, and revolving dioptric apparatus, by Chance Bros.
A collection of porcelain.
Case with Venetian mosaics. Case with vases and other articles in Venetian glass, &c., being imitations of Oriental stones.
Turkish glass case.
Miscellaneous collection from Venezuela, Uruguay, and Peru.
Italian furniture.
Articles in Siberian graphite.
Bronzes, porphyry, porcelain, mosaics, &c. belonging to the Emperor of Russia.
Spanish furniture, arms, porcelain, &c.
Stuffed Siberian animals, graphite, &c.
Norway anchors, chain cables, &c.
Swedish terra-cotta articles, porphyry, porcelain, cannon, artistic castings, &c.
French furniture.
Danish porcelain.
Swiss carved wood articles.
Armour exhibited by the French Emperor.
Swiss musical boxes, carved wood and ivory, &c.
Stearic candle collection.
Dutch plate, glass, &c.
Continuation of French furniture.
Amsterdam Candle Company's collection.
"Star of the South" and other Dutch diamonds.
Candle collection.
Gold lace.
Belgian plate-glass.
Memorial of various Exhibitions.
Candle collection.
Model of Berlin Exchange.
Sheets of Belgian and French plate-glass.
Plate presented by Berlin to Princess Royal.

WESTERN DOME.

Prussian breach-loading steel gun.
Arms trophy.
Telegraph apparatus.
Dressing-cases.

Plate, including shield, presented to Crown Prince and Princess of Prussia.
Dressing-cases.
Lamp collection.
Walking-stick collection.
Leather mosaic case.
Shawl collection.
Muslin collection.
Plate presented to Princess Royal by Berlin.
King of Prussia's porcelain.
Music platform.
Cabinet and casket collections.
Carved wood pulpit.
Shawl collection.
Hedgland's & Walker's organs are at the sides of the western stained glass window.

EASTERN TRANSEPT.

NORTH END.

Case containing articles in aluminium, minerals, &c.
Canada Inclosure.
New Brunswick Inclosure.
Prince Edward's Island Inclosure.
Passage.
Vancouver's Island Inclosure.
British Columbia Inclosure.
Nova Scotia coal pyramid.
Tasmania Inclosure.
Glass cases.
An organ, by Forster and Andrews, Hull. It has 46 registers, 2,475 pipes, 6 composition pedals, 2 sforzando pedals, and 1 pneumatic combination pedal; it is 30 feet high, 22 feet wide, and 14 feet deep; the pneumatic movement is applied to the great and pedal organs, and there are four different pressures of wind. The bellows is blown by two of Jay's hydraulic engines, having a water pressure of 35lbs. to the square inch.
Glass cases containing samples of gold from Victoria, manufactures, &c.

Here Eastern Dome intersects.

Brass candelabra.
A screen, for Hereford Cathedral, designed by G. G. Scott, and executed by Skidmore & Co. It is 36 feet long, and 35 feet high; consists of wrought iron, and hammered copper foliage. The central figure represents Our Lord rising from the tomb; other bronze figures are at each side. The panels are filled in with mosaics of various coloured stones, &c.
A Gothic bell-tower, with nine cast-steel bells; under and about it are grouped specimens of steel, showing its application to the manufacture of ordnance and every kind of arms, the rolling stock of railways, &c.
Hart & Son's collection of ornamental brass and iron work.
Specimens of steel, made directly from cast-iron, by Bessemer's process.
Eight brass bells, by Warner & Sons, Cripplegate, fitted up with chiming apparatus, so as to be rung by one man. The largest weighs about one ton.
A Turret Clock, by Dent; striking the hours on a bell weighing between three and four tons, and the quarters on four smaller bells. The wheels are of gun-metal: each of the four dials is seven feet in diameter, and might be twelve.
The Coalbrookdale Company's iron-work gates, 54 feet wide and 25 feet high; attached is a fountain, in iron-work, 25 feet high, with a statue of Oliver Cromwell, and figures of Peace and War, &c.

SOUTH END.

MIDDLE AVENUE.

NORTH END.

Statue of Shakespere.
Small fountain.
Group of figures.
Liverpool import case.
Lady Godiva—by Fuller. Candelabrum.
Smith & Son's fur case.
Irish linen case.

Here the Nave intersects.

Venetian billiard table.
Various models in cork, &c.
Cannon.
A Clock, by Benson; striking the hours and quarters on five bells, by Mears, the largest weighing 22 cwt. The works are 300 feet from the dial, which is situated in the Great Central Tower, the connections being carried under ground. The weights exceed a ton, and are 200 feet from the works. A new remontoir escapement allows the use of the great weight required to drive a clock of such size, so distant from the dial, and moving hands of such magnitude—sometimes in opposition to the wind. A two-seconds compensation pendulum is employed.
Model of an ancient temple.
Equestrian statue of Viscount Harding.
Marble fountains.
Bas-reliefs, statues, &c.

SOUTH END.

WESTERN TRANSEPT.

NORTH END.

Austria.

Hardware Inclosure.
Passage containing toys, pipes, caskets, &c.
Furniture and glass Inclosure.
Passage containing philosophical instruments, telegraph apparatus, and mechanical toys.
Musical instrument, and photograph Inclosure.
Passage containing books, &c.
Porcelain Inclosure.
Small candelabra.
Large candelabra, vases, &c.
Small fountain.
Large crystal candelabra, &c.
Diamond and pearl jewellery.
Rich bindings and albums.
Gold jewellery.

Here Western Dome intersects.

Zollverein.

Armour; wire and elaborate cast-iron work.
Cutlery trophy.
Artistic cast-iron articles.
Collection of tea-urns, lamps, &c.
Space containing lamps, bronzes, &c.
Inclosure containing toys, perfumery, &c.

Passage containing furniture, paper-hangings, boots and shoes, and fancy articles.
—— containing jewellery, and horological instruments.
Inclosure containing pictures, toys, preserves, &c.
—— containing cutlery and wines.
—— containing drawing instruments, paper, &c.
Inclosure containing cutlery, plate, albums, &c.
—— containing toys, drawing requisites.
—— containing tools, clocks, balances, stamped tin articles.
—— containing paintings, vases of gilt crystal, &c.
—— containing toys and plaster casts.
—— containing furniture of richly carved oak, &c.
—— containing tools, soap, paper, &c.
—— containing surgical and philosophical instruments, electric and other clocks, photographs.
Automatic organ.

South End.

WINDOWS.

North End of East Transept.

Centre Window: from Doncaster, by Hardman, of Birmingham. Right side: the Robin Hood Window, by Messrs. Chance, of Birmingham. Left side: Scriptural subject, by the same.

North End of West Transept.

Centre Window: from Worcester Cathedral, by Hardman, of Birmingham. Right side; Ecclesiastical subject, by Heatons, Butters, and Bayne. Left side: by Holland, Warwick.

HORTICULTURAL SOCIETY'S GARDENS.

Baron Marochetti's model of the monument erected to Charles Albert at Turin.
East fountain, by Durene, of cast-iron, 45 feet high and 53 feet in diameter.
West fountain, by Barbazate, of cast-iron, 30 feet high, and 53 feet in diameter.

ARTICLES BELONGING TO HER MAJESTY THE QUEEN, EXHIBITED BY PERMISSION OF HER MAJESTY.

(By Messrs. R. & S. Garrard and Co; Class 33, South Court, facing the Nave).

The Koh-i-noor, exhibited as a specimen of diamond-cutting.

Three large and fine Rubies, from the Treasury at Lahore, mounted as a necklace in gold and enamel, in the Indian style, with large diamonds pendant.

An Ornament for the centre of the table, representing a covered fountain, designed in the style of the Palace of the Alhambra, executed in silver, silver-gilt, and enamel; round the base is a group of horses, portraits of favourite animals, the property of Her Majesty; on the lower portion of the base, which is designed to represent a ruin, are introduced the flamingo and vulture, and also various plants, natives of Arabia.

(By Messrs. Elkington & Co., Class 33, in the Nave.)

A pair of Candelabra, modelled from a design by H.R.H. the late Prince Consort being a portion of a set retained for use at Balmoral.

A silver equestrian Statuette, "Godiva," presented by H.M. the Queen to H.R.H. the late Prince Consort.

(By Messrs. Hunt and Roskell, Class 33, under Gallery, facing the Nave).

A Vase, by Antoine Vechte, in oxidized silver, damaskened. Subject: "The Centaurs and Lapithæ." On the pedestal are groups and entablatures illustrative of the same subjects.

A Vase and Pedestal, by Antoine Vechte, in oxidized silver, Marine composition. The bassi relievi represent Venus and Adonis, and Thetis presenting to her son Achilles the armour forged by Vulcan.

(Exhibited by permission of the Royal Personages to whom the Articles belong, by Messrs. R. & S. Garrard; Class 33, South Court, facing the Nave).

A Jewelled and Enamelled Cup, in silver-gilt, the gift of Her Majesty and the Prince Consort to their grandchild and godson, the Prince Frederic of Prussia, on the occasion of his christening. The cup is designed with emblems and figures typical of baptism; on the stem are the arms of England and Prussia, and on the base a group of St. George and the Dragon.

A Tazza-form Cup, presented by Her Majesty the Queen to the Hereditary Duke of Baden, on the occasion of his christening. The cup is treated in the Renaissance style, with emblematical wreaths of wheat and vine, symbolical of the sacrament, surmounted with a group typical of baptism; on the body are introduced the arms of England and Baden.

A richly chased Christening Bowl, in Renaissance style, with winged figures supporting coronet and wreaths of flowers; on the edge of the Bowl is an emblematic figure pouring water. This Cup was presented by H.M. the Queen to the daughter of His Excellency Count Bernsdorf, on the occasion of her christening.

SECTION I.

CLASS 1.

MINING, QUARRYING, METALLURGY, AND MINERAL PRODUCTS.

South Court, Eastern Annex.

1 AARON, E. & W. *Liverpool.*—Halkyn hydraulic limestone, Halkyn Chirt stone, and Holywell Roman cement stone.

2 ABERDARE COAL CO. *Cardiff, Glamorganshire.*—Merthyr steam coal from 4 feet and 9 feet seam.

3 ABERDARE IRON CO. *Aberdare, Glamorganshire.*—Coal. Iron ore, pigs, refined metal, and railway iron.

4 ABERDARE STEAM FUEL CO. *London, Cardiff, and Aberdare.*—Patent steam fuel.

5 ADAIR, J. G. *Bellgrove, Ballybrittas.*—Coal and mineral productions of Co. Donegal.

6 AYTOUN, R. 3, *Fettes-row, Edinburgh.*—Safety cage for miners, and hoist.

7 BANKART, F. & SONS, *Briton Ferry, Glamorganshire.*—Copper, and ores.

8 BARBER, WALKER & CO. *Eastwood, Nottinghamshire.*—Coals.

9 BARKER, R. *Wood Bank, Egremont, Cumberland.*—Hematite iron ores, and spar.

10 BARKER, RAWSON, & CO. *Sheffield.*—Leads: white, red, and refined.

11 BARNES, T. A. *Whitby, Yorkshire.*—Grosmont ironstone, and pig iron, Whitby.

12 BARRINGER & CARTER, *Mansfield, Nottinghamshire.*—Fine red casting sand, from Mansfield.

13 BARROW, B. *Ryde, Isle of Wight.*—Mineral products of the Isle of Wight.

14 BARROW, R. *Stavely Works, near Chesterfield.*—Coal, ironstone, and iron.

15 BATSON, A. *Ramsbury, Wilts.*—Devonshire madrepore inlaid table.

16 BAYLY, J. *Plymouth.*—Ores of copper, tin, and lead.

17 BEADON, W. *Otterhead, Honiton, Devon.*—Siliceous sands for stuccoing, plastering, &c. Mineral black, natural pigment. Fine clays, iron ores, &c.

18 BELL BROS. *Newcastle-on-Tyne.*—Aluminium and its alloys. Pig iron and iron ores from Cleveland.

19 BENNETT, T. 11, *Woodbridge-st. Clerkenwell.*—Leaf gold.

20 BENNETTS, W. *Camborne.*—Safety skip, for passing men, &c. from mines.

21 BENTHOLL. H. 14, *Chatham-pl. Blackfriars.*—Porphyry.

22 BENTLEY, J. F. *Stamford.*—Designs in stone from Stamford, &c.

23 BICKFORD, SMITH, & CO. *Tuckingmill, Cornwall*—Safety fuse for blasting.

24 BIDDULPH, J. & CO. *Swansea.*—Minerals, iron-ore, coal.

25 BIDEFORD ANTHRACITE MINING CO. *Bideford, Devon.*—Mineral black paint and culm.

26 BIRD, E. *Matlock, Bath.*—Obelisk, paper-weights in black marble, &c.

27 BIRD, W. & CO. 2, *Laurence Pountney-hill, E.C.*—Iron, steel, and tin plates.

28 BIRLEY, S. *Ashford, Derbyshire.*—Inlaid marble table, with arabesques, &c.

29 BLAENAVON IRON AND COAL CO. *Monmouthshire, and London.*—Iron bars, weldless tyres, girders, &c.

30 BLAENCLYDACH COAL CO. *Neath.*—Coal.

31 BOLCKOW & VAUGHAN, *Middlesbro', Tees.*—Coal, coke, ironstone, pig-iron, &c.

32 BOUNDY, T. *Swansea.*—Arsenic.

33 BOWLING IRON CO. *near Bradford, Yorks. and London.*—Boiler plates, tyres, bars, angles, &c.

34 BOXALL, J. J. *Pulborough, Sussex.*—Green Sand stone.

35 BRADLEY, C. L. *Richmond, Yorks.*—Copper and lead ore.

36 BREWER, R. *Corsham, Chippenham.*—Stone vase and cubes of Bath stone.

37 BRIGHT, S. & CO. *Buxton.*—Black marble vases and Mosaic work.

38 BRISTOL AND FOREST OF DEAN CO. *Princess Royal Colliery, near Lydney.*—Coal, from Yorkley and Whittington seams.

39 BROWN, J. & CO. *Sheffield.*—Rolled iron armour plates, and steel rails.

40 BROWN & JEFFCOCK, *Barnsley.*—Coals and ironstones from S. Yorkshire field, and geological sections.

41 BROWN & RENNIE, *Kilsyth, by Glasgow.*—Coal and coke.

42 BROWNE, W. *St. Austell, Cornwall.*—China clay, china stone, and iron ore.

43 BRUNTON, J. D. *Barge-yard, Bucklersbury.*—Condensed peat and peat charcoal.

44 BRUNTON, W. & CO. *Penhellick Safety Fuse Works, near Camborne.*—Fuse for blasting.

45 BUDD, J. P. *Swansea.*—Iron, and tin plates.

46 BULL, G., D.D. *Redhall, Co. Antrim.*—Large quartz crystal, or Irish diamond.

47 BUTLIN, T. & CO. *East End Iron Works, Wellingborough.*—Iron and iron ores.

48 BUTTERLEY CO. *Butterley Iron Works, Alfreton.*—Section of coal pit. Armour plates, deck beams, rolled girders, joists, &c.

49 BWLCH Y GROES SLATE CO. *Llanberis, Carnarvon.*—Roofing slate.

50 BYERS, J. & SON *Stockton-on-Tees, Durham.*—Lead ores, lead, silver, litharge, and lead manufactures.

51 CAITHNESS, EARL OF, 17, *Hill-st.*—Caithness flags.

52 CALOW, J. T. *Staveley, Derbyshire.*—Safety apparatus for shafts of mines, &c.

53 CAMPBELL, BROS. *William-st. Blackfriars.*—Pig and bar iron from Calder and Govan works.

54 CANNAMANNING CHINA CLAY CO. *Newton Abbot.* — China, pipe, and paper clays.

55 CASE & MORRIS, *Ince, Wigan.*—Section of actual strata of Rose Bridge and Ince Hall Collieries, coal, &c.

56 CHAFFER, T. *Burnley, Lancashire.*—Flag and paving stones.

57 CHAMBERS, J. *Alfreton.*—Coal.

58 CHEESEWRING GRANITE CO. 6, *Cannon-st. Cornwall.*—Design by John Bell for memorial of Exhibition, 1851. (*Nave.*)

59 CHILD, W. J. & T. *Hull.*—Millstones.

60 CLAY CROSS CO. *Clay Cross, Chesterfield.*—Coal, lime, limestone, ironstone, and pig iron.

61 COAL OWNERS OF NORTHUMBERLAND AND DURHAM, *Newcastle-on-Tyne.*—Map and section of coal field.

62 COCHRANE & CO. *Woodside, Dudley.*—Iron pipes and pig iron.

64 COLLEY, G. 8, *Upper Dorset-st. Pimlico.*—Vase in free stone.

65 CONNORREE MINING CO. *Ovoca, Ireland.*—Minerals and ores.

66 COPELAND, G. A. *Carwythenack House, Constantine.*—Waterproof blasting cartridges.

67 CORBETT, W. F. *Gt. Charles-st. Birmingham.*—Apparatus to prevent overwinding at pits.

68 CORBETT, J. *Bromsgrove, Worcestershire.*—Table and provision salt.

69 COURAGE, A. & CO. *Bagillt, Flintshire.*—Lead, sanitary pipes, zinc, &c.

70 COWPEN COAL CO. *Blyth.*—Block of Cowpen Hartley steam coal.

71 COX, BROS. & CO. *Derby.*—Lead, shot, plate of Derbyshire silver, &c.

72 CRAIG, G. & SON, *Pavement Works, Thurso, Caithness.*—Caithness flags for tables, shelving, and pavement.

73 CRAWLEY, C. E. 17, *Gracechurch-st.*—Miners' safety-lamp.

74 CRAWLEY, G. B. *Neath.*—Coal.

75 CRAWSHAY, H. & CO. *Lightmoor Collieries, Cinderford.*—Rock-Vein coal.

76 CRAWSHAY, H. & CO. *Abbot's-wood Mines, Cinderford.*—Black Brush iron ore.

77 CROWN PRESERVED COAL CO. 62, *Moorgate-st.*—Preserved coal.

78 CWMORTHIN SLATE CO. *Merionethshire.*—Slates and slabs.

79 DABBS, J. *Stamford.*—Freestone, from Ketton quarries.

80 DAGLISH, J. *Hetton Collieries, Durham.*—Model of ventilating furnace. Self-registering water-guage.

81 DAVIS, D. *Bute-cres. Cardiff.*—Sample of Merthyr steam coals.

82 DAVIS, J. *Ulverstone*—Iron pyrites from Millom.

83 DAWES, W. H. & G. *Denby Iron Works. Derby.*—Coal and ironstone.

84 DENBY, W. 3, *Denby-pl. Sidmouth.*—Mosaic table composed of silicious pebbles found at Sidmouth.

85 DENMAN, LORD, *Stoney, Middleton.*—Grit stone from the district.

86 DEVON AND COURTENAY CLAY CO. *Newton Abbot.*—Pipe and potters' clay.

87 DEVON GREAT CONSOLS, MINE, *Tavistock.*—Copper ores.

88 DEVONSHIRE, DUKE OF.—Slate from Burlington quarries, Lancashire.

89 DOVE, D. *Nutshill Quarries, Glasgow.*—Grindstones.

90 DOWLAIS IRON CO. *Dowlais, Merthyr Tydvil.*—Samples of manufactured iron.

91 DUNCAN, FALCONER, & WHITTON, *Carmyllie, by Arbroath*—Step and pavement slabs.

92 DYBALL, T. *Kirton Lindsay.*—Iron ore, (for Sir Culling Eardley.)

93 EAST CORNWALL ARSENIC CO. 9, *Parade, Plymouth.*—Arsenical mundic and arsenic.

94 EASTWOOD & SONS, *Derby.*—Samples of iron.

95 EBBW VALE & PONTYPOOL COS. *Newport, Monmouthshire.*—Minerals; tin plates; iron manufactures.

97 EDWARDS, WOOD, & GREENWOOD, *Tame Valley Colliery, Tamworth.* — Iron pyrites and fire clay.

98 ELLAM, JONES, & CO. *Maskeaton Mills, Derby.*—Emery, and oxide of iron paint, made from the ore, expressly for iron-work.

99 ELLIS & EVERARD, *Markfield Granite Quarries, Leicestershire.*—Paving setts, macadamizing and building stones, &c:

100 EVANS & ASKIN, *Birmingham.*—Nickel, cobalt, and German silver.

101 FARNLEY IRON CO. *near Leeds.*—Coal, boiler-plate, tyres, fire-clay goods, &c.

102 FAYLE & CO. 31, *George-st. Hanover-sq.*—Blue clay for earthenware.

103 FINNIE, A. & SON, *Kilmarnock.*—Steam and house coal.

104 FIRTH, BARBER, & CO. *Oak's Colliery, Barnsley.*—Steam and house coal from Barnsley seam.

105 FITZGERALD, R. *Clare View, Tarbert, Co. Kerry.*—Peat.

106 FORSTER, G. B. *Cowpen Colliery, Blyth.*—Model of coal pit, with cages and apparatus.

107 FORSTER, R. *Gateshead.* — Grindstones.

108 FOWLER, W. & CO. *Sheepbridge Iron Works, Chesterfield.*—Coal and ironstone, of which armour-plate iron is made.

109 FRANKLIN, F. *Galway.* — Polished marble and frame.

110 FREEMAN, W. & J. *Millbank, Westminster.*—Granites and stones.

111 FRYAR, M *School of Mines, Glasgow.*—Plans and drawings relating to mining.

112 GAMMIE, G. *Shotover House, Oxfordshire.*—Native Oxford ochre.

113 GARDNER, R. *Sansaw, Shrewsbury.*—Grimshill stone, copper ore, and barytes.

114 GARLAND, T *Fairfield, Redruth.*—Arsenic.

115 GENERAL MINING CO. FOR IRELAND, *Dublin.*—Zinc ores, spelter, fire-clays, ochres from Tipperary.

116 GEOLOGICAL SURVEY OF THE UNITED KINGDOM, 28, *Jermyn-st.*—Published maps and sections, 1 inch and 6 inch scales.

117 GIBBS & CANNING, *Tamworth.* — Stoneware sewerage pipes, fire-bricks, &c.

118 GILBERTSON, W. & Co. *Swansea.*—Tin-plates, black plates, and taggers.

119 GILKES, WILSON, PEASE, & Co. *Middlesbro'.*—Pig iron, test bars, and iron ores.

120 GODDARD, E. for E. BLAKE, *Newton Abbot.*—Pipe, potter's clay, &c.

121 GOLDSWORTHY, T. & SONS, *Hulme, Manchester.*—Whetstones, knife-cleaning machines, emery, &c.

122 GOVERNOR & Co. OF COPPER MINERS, *Cwm Avon Works, Glamorganshire.*—Coal, iron ores, and metals.

123 GOWANS, J. *Rochville, Merchiston-pk. Edinburgh.*—Boring machine, galvanic apparatus for blasting, and wedge.

124 GRAHAM, A. *Huddersfield.*—Building stones and hard-paving stones.

125 GRANVILLE, EARL, *Shelton, Staffordshire.*—Minerals and pig iron.

126 GRAY, J., M.D. *Glasgow.*—Modification of Davy's lamp.

127 GREAVES, J. W. *Port Madoc, North Wales.*—Roofing slates.

128 GREAVES & KIRSHAW, *Warwick, and Paddington.*—Hydraulic Lias lime and cement. Polished lias stone.

129 GREENWELL, G. C. *Radstock.*—Sections and specimens of Somersetshire coal-field.

130 GREENWELL, G. C. for WESTBURY IRON CO. *Wilts.*—Section of ironstone and furnace products.

131 GREGORY, J. R. 25, *Golden-sq. London.*—Minerals, fossils, and rocks; Devonian fossil fishes from Scotland.

132 HALIFAX CORPORATION, *Halifax, Yorks.*—Coal and products from Halifax.

133 HALL, J. & T. *Derby.*—Marble and spar ornaments, &c.

134 HALL, J. & Co. *Stourbridge.*—Fire clays, gas retorts, bricks, crucibles, &c.

135 HALLIDAY, T. C. *Greetham, Rutland.*—Clipham stone blocks.

136 HAMPSHIRE, J. K. *Whittington Collieries, Chesterfield.*—Safety apparatus, for raising and lowering persons in shafts.

137 HAMPSHIRE, M. & Co. *Spring-st. Huddersfield.*—Building stones.

139 HARPER & MOORES, *Lower Delph Clay Works, Stourbridge.*—Glass house pot clay, retorts, fire bricks, lumps, &c.

140 HARRIS, J. *Newton Abbot, Devonshire.*—Minerals from district.

141 HARRISON, AINSLIE, & Co. *Newland Furnace, Ulverston.*—Hematite and puddling ores, and Lorn pig iron.

142 HARRY, G. *Swansea.*—Copper, silver, iron, zinc, and nickel ores, and metals.

143 HAWKSWORTH, W. & Co. *Linlithgow.*—Cast steel, engraver's plate, steel rifle barrels, &c.

144 HEATH, EVANS, & Co. *Aberdare.*—Steam coal.

145 HEAVEN, W. H. *Lundy Island, Clovelly, North Devon.*—Specimens of district granite.

146 HEGINBOTHAM, P. & SON, *Shallcross Mills, Whaley Bridge, Stockport.*—Sulphate of barytes.

147 HENDERSON, G. W. M. *Fordell, Fifeshire.*—Carved block of sandstone.

148 HENDERSON, J., C.E. *Truro, Cornwall.*—Plan and section of a mine.

149 HENGISTBURY IRON MINING CO.—*Christchurch, Hants.*—Iron ore.

150 HENSON, R. 113A, *Strand.*—Ornaments in marble, and minerals.

151 HEWLETT, A. for the EARL OF CRAWFORD AND BALCARRES, *Wigan*—Cannel gas coal.

152 HIGGS, S. & SON, *Penzance.*—Safety-lamp, tin and copper ores, and plans.

153 HILL, F. *Helston.*—Ores, metals, minerals, clay, marle, and stone from Helston district.

154 HIRD, DAWSON, & HARDY, *Low Moor Iron Works.*—Iron in various stages.

155 HOLLAND, S. & Co. *Port Madoc, Carnarvonshire.*—Roofing slates.

156 HOLMES, J. *Bolton Wood Quarries, Bradford.*—Monument (Open Court).

157 HOLROYD, J. & SONS, *Brighouse.*—Flag and building stones.

158 HOOPER & MOORE, *Stourbridge.*—Fire bricks, &c.

159 HOWARD, HON. J. 1, *Whitehall-place.*—Forest of Dean stone, coal, iron ore, clay, pottery.

160 HOWARD, RAVENHILL, & Co. *Rotherhithe, London.*—Bridge links, rolled entire.

162 HOWIE, J. *Hurtford Colliery, Kilmarnock, Scotland.*—Coal.

163 HUNT, J. *Porthleven, Helston, Cornwall.*—Model of ore-separator, portable gold-washer, and lead ores.

164 HYNAM, J. 7, *Princes-sq. Finsbury.*—Purified Fuller's earth for various manufactures; crucibles.

165 IBBERSON, J. *Lockwood, Huddersfield.*—Building stones.

166 IRVING, G. V. *Newton, Lanarkshire N.B.*—Minerals from the Lead Hills.

168 JENKINS, W. H. & Co. *Truro, Cornwall.*—Ochres, fluor spar, &c. and glaze for porcelain.

169 JENNINGS & Co. *Swansea.*—Arsenical ore, arsenic unrefined, refined crystals, powdered and lump arsenic.

170 JENNINGS, W. *Victoria-st. Hereford.*—"Three Elms Quarry" stone.

171 JOHNSON, MATTHEY, & Co. 79, *Hatton-garden.*—Platinum, and preparations of precious metals.

172 JOHNSON, W. W. & R. & SONS, *Limehouse.*—Wetterstedt's metal for roofing, &c.

173 JONES & CHARLTON, *Duckinfield, Manchester.*—Self-extinguishing detector safety-lamp.

174 JONES, DANIEL, *Bradford-on-Avon, Wilts.*—Stone from Bath Farleigh Downs.

175 JONES, DAVID, *Hay, South Wales.*—Grey sandstone from Pontvain quarries.

176 JONES, DUNNING, & Co. *Middlesborough.*—Pig iron.

177 JONES, I. *Swansea.*—Flat chain.

178 JONES, W. *Port Tennant, Swansea.*—Fuel for steam purposes.

179 JORDAN, H. K. 2, *Clifton-wood-ter. Clifton.*—Minerals.

180 JORDAN, J. B. *Museum of Practical Geology, London, S.W.*—Models of mineral forms constructed of cardboard.

182 JORDAN, W. H. 14, *Langham-st. Regent-st.*—Model of pit frame and safety cage.

183 JULEFF, J. *Pednandrea, Redruth, Cornwall.*—Cornish crucibles, various.

184 KAY, W. *Hayhill, Ochiltree, Ayrshire.*—Pair of curling-stones.

185 KAYE, G. *Ryecroft-edge, Huddersfield.*—Building stones.

186 KELL, R. & Co. *Newcastle-on-Tyne.*—Grindstones.

187 KING BROS. *Stourbridge Fire-Clay Works.*—Clay retorts, bricks, &c.

188 KINSMAN, Rev. R. B. *Tintagel, Camelford.*—Specimens of slates from Tintagel quarries, &c.

189 KIRKSTALL FORGE Co. *Leeds.*—Samples of manufactured malleable iron.

190 KNIGHT, F. WINN Esq. M.P. *Exmoor, South Molton.*—Iron ores.

191 KNOWLES, A. *Highbank, Pendlebury.*—Safety cage for coal mines.

192 LAMACRAFT, W. *Newton Abbot.*—Clays.

193 LAW LIFE ASSURANCE SOCIETY, *Fleet-st. E.C.*—Green and black marble, from Connemara, County Galway.

194 LAYCOCK, J. *Newcastle.*—Iron jointed prop used in working pillars in coal mines.

195 LEE MOOR PORCELAIN CLAY Co. *Plympton.*—Porcelain clay, architectural bricks, &c.

196 LEESWOOD GREEN COLLIERY Co. *near Mold.*—Section of Leeswood-green cannel coal.

197 LEETCH, J. 68, *Margaret-st. W.*—Preparation of fluor spar for polishing.

198 LEISS, F. 30, *Southampton-st. Strand.*—Articles made of mica.

199 LEVER, ELLIS, *West Gorton Works, Manchester.*—Flexible tubing, fly-doors, and brattice.

200 LEVICK & SIMPSON, *Newport.*—Specimens of rails, bars, and cold blast pig-iron.

201 LILLESHALL Co. *Shiffnal, Shropshire.*—Minerals, castings, malleable specimens.

202 LIVINGSTONE, A. S. *Llanelly, Carmarthenshire.*—Patent fuel.

203 LIZARD SERPENTINE Co. 20, *Surrey-st. Strand.*—Sundry works manufactured from serpentine.

204 LLANGOLLEN SLAB & SLATE Co. 4, *South Wharf-rd. Paddington.*—Enamelled slate work; large slate slab.

205 LLETTY SHENKIN COAL Co. *Cardiff.*—Smokeless Welsh steam coal.

206 LOMAS, J. & SONS, *Marble Works, Bakewell, Derbyshire.*—Polished Derbyshire marble.

207 LONDONDERRY, MARCHIONESS OF, *Seaham Hall.*—Pensher sand stone. A model of Seaham harbour and town.

209 LOWES & ROBINSON, *Stanhope, Darlington.*—Minerals. Section of Weardale strata, by Robinson & Son.

210 LOWRY, J. W. 45, *Robert-st. Hampstead-rd. N.W.*—Engravings of fossils for the Geological Survey of United Kingdom.

211 LUCAS & BARRATT, *Stockton-on-Tees.*—Pig-iron.

212 LUMBY, J. *Stamford.*—Ironstone pyrites-coal fire, and terra cotta clay.

213 LUND HILL COAL Co. *Barnsley.*—Barnsley steam and house coal.

214 MACDONALD, A. *Aberdeen.*—Specimens of granite used in buildings, decorations, memorials, and general purposes.

215 MAGNUS, L. S. 3, *Adelaide-pl. London-bridge.*—Coals and products, and iron ores.

216 MARGAM TIN PLATE Co. *Taibach, Glamorgan.*—Tin and terne plates, sheet, bar, and rod iron.

217 MARLBOROUGH, DUKE OF, *Blenheim Palace, Woodstock.*—Iron ore from Fawler mines, Charlbury, Oxfordshire.

218 MARSHALL, E. S. 31, *John-st. Tottenham-ct.-rd.*—Gold and silver leaf.

219 MARTIN, E. & SON, *St. Austell.*—China clay, and China stone.

220 MARTIN, R. *Higher Blowing House, St. Austell.*—China clay, and stone.

221 MATTHEWS, J. *Royston.*—Coprolites.

222 M'CALL, R. *near Limerick.*—Magnetic iron ore.

223 MEESON & Co. *Grays, Essex.*—Products of Grays chalk pits.

224 MEIK, T. *Sunderland.*—Model of the mode of shipping coals.

225 MERSEY STEEL AND IRON Co. *Liverpool.*—Crank shaft and forgings.

226 MICHELL, R. R. & Co. *Marazion, Cornwall.*—Model of tin-smelting furnace. Moulds, tools, kettles, &c.

227 MICHELL, S. *St. Austell, Cornwall.*—Porcelain clay, washed and unwashed.

228 MICKLETHWAIT, R. *Ardsley House, Barnsley, Yorkshire.*—Grindstones.

229 MITCHELL, W. B. 16, *Broom Hill, Sheffield.*—Minerals and products of South Yorkshire.

230 MITCHELL, W. B. *Sheffield.*—Ores and minerals of the High Peak, Derbyshire.

231 MONA MINE Co. *Amlwch, Anglesey.*—Produce of copper mining and smelting.

232 MONK BRIDGE IRON Co. *Leeds.*—Combined cast-steel and iron tyres, minerals, &c.

233 MONTEIRO, L. A. 51, *Manchester-st. W.*—Stalagmite.

234 MOORE & MANBY, *Billiter-sq. London.*—Iron for engineers, &c.

235 MORCOM, J. *St. Austell, Cornwall.*—Manganese and iron ores.

236 MORE, F. *Linley Hall, Shropshire.*—Lead ores. Ancient lead, spades, &c.

237 MOREWOOD & ROGERS, *Stratford, Essex.*—Articles coated with lead, tin and lead, tin and zinc, &c.

238 MORGAN, R. & SONS, *Llanelly, Carmarthenshire.*—Anthracite malting coal.

239 MOSER & SONS, *Southwark.*—Sections of rolled iron.

240 MOULDED PEAT CHARCOAL CO. *Fenchurch-st. London.*—Charcoal, foundry blacking, iron and tin plates, and peat products.

241 MUCKLESTON, E. THE REV. *Stoke Cobham.*—Stone from Whitesbourne quarry, Shropshire.

242 MURPHY, J. *Penzance.*—Serpentine ornaments.

243 MURRAY, A. 24, *New-st. Spring-gardens.*—Anthracite from Broadmoor and Landshipping.

244 MURRAY, T. *Chester-le-Street, Durham.*—Working model of underground steam-engine.

245 MUSEUM OF PRACTICAL GEOLOGY, *Jermyn-st.*—Model of Holmbush Mine, constructed by T. B. Jordan, Clapham.

246 MYLNE, R. W. 21, *Whitehall-pl.*—Map,—tertiary and cretaceous districts,—France, England, &c. with contoured seas.

247 NEWALL, D. H. & J. *Granite Works, Dalbeattie.*—Stone monument and a fountain.

248 NEWCASTLE, DUKE OF, *Shireoak Colliery, Worksop.*—Steam coal, ironstone, and views of colliery.

249 NICHOLLS, J. *Trekenning House, near St. Columb.*—Copper and lead ores, slates, and porphyry.

250 NICHOLSON, M. *Whittington Collieries, near Chesterfield.*—Stone kerb and arching for coal shaft bottom.

251 NIXON, TAYLOR, & CORY, *Cardiff.*—Coal used by Royal yacht, Warrior, Black Prince; section of strata.

252 NORTH GUNBARROW CHINA CLAY CO. *Newton Abbot.*—Porcelain clay, &c.

253 NORTHUMBERLAND, DUKE OF, *Alnwick Castle.*—Freestone, from quarries in Northumberland.

254 NOWELL & ROBSON, *Idle, near Leeds.*—York landings, paving and block stone.

255 OAKES & CO. *Alfreton, Derby.*—Coal and iron.

256 OKEY, S. F. & CO. *Castleford Iron Works.*—Coal, iron ore, iron, &c.

257 ORD & MADDISON, *Darlington.*—Limestones and marbles.

258 OXLAND, R. 42, *Park-st. Plymouth.*—Model of furnace for dressing tin ores containing wolfram. Tungsten and compounds.

259 PACKARD, E. & CO. *Ipswich.*—Coprolites from Green Sand and Craig.

260 PALMER, C. M. *Newcastle-on-Tyne.*—Specimens of coke.

261 PARK-END COAL CO. *New Fancy Pit.*—Park-end high delf and smith coal.

262 PARKINSON, J. 81, *Cheapside.*—Minerals: Ashburton and Ilsington. Bethell's Anthracite coke. Dr. Smith's peat fuel. Fire igniters, &c.

263 PARKSIDE MINING CO. *Whitehaven.*—Hematite iron ore, with section.

264 PATENT METALLIC FUSE CO. *Wadebridge.*—Waterproof fuses for blasting.

265 PATENT PLUMBAGO CRUCIBLE CO. *Battersea Works, S.W.*—Crucibles for melting metals, portable furnaces, &c.

266 PAULL, J. M. *Alston, Cumberland.*—Cage for the ascent and descent of workmen, &c. in mines. Ores, &c.

267 PEAKE, S. *Minera, Wrexham.*—Stone.

269 PEARCE, W. JUN. *Boscawen-bridge, Truro.*—Inlaid Serpentine and Steatite tables, column, mausoleum, and dolphin tazza.

270 PEARSON, E. 11, *South Quay, Great Yarmouth.*—Amber, jet, agates, petrifactions, &c.

271 PEARSON, W. *Heddon Quarry, Northumberland.*—Freestone for building.

272 PEASE, I. & I. W. *Darlington.*—Model of Upleatham ironstone mines, limestone, coals, coke, fire-bricks, vases, &c.

273 PERRENS & HARRISON, *Stourbridge.*—Clays, retort, and fire-bricks.

274 PHILLIPS & DARLINGTON, 26, *Moorgate-st.-chambers, Moorgate.*—Fuel from partially coked or torrefied coal.

275 PHILLPOTTS, I. *Newport, Monmouthshire.*—Risca black vein steam coal, and miners' tools.

276 PHIPPARD, T. *the Priory, Wareham, Dorset.*—Pottery clays and sands, &c.

277 PIKE, W. & J. *Wareham, Dorsetshire.*—Clays for fine pottery, &c.

278 PIRNIE COAL CO. *Leven Fife. N.B.*—Cannel coal for oil or gas.

279 POLGLAZE & VICTOR, *Wadebridge.*—Metallic safety fuses.

280 POLKINGHORNE, W. *Tywardreath, Cornwall.*—A synopsis of the Cornwall ticketings for copper ores, from 1800 to 1860.

281 PORT NANT GRANITE CO. *Pwlheli, North Wales.*—Piers, and permanently-rough pavement for bridges.

282 PORTER, W. 21, *Pitt-st. Old Kent-rd.*—Millstones.

283 POTTER, A. *Newcastle-on-Tyne.*—Gas retorts, lumps, fire-bricks, &c.

284 POWELL, T. & SONS, *Cardiff.*—Coal and plan of colliery.

285 POWELL, W. J. *Tisbury, Wilts.*—Coralline flint, &c. from Tisbury oolites.

286 PRICE, DR. D. S. 26, *Great George-st. Westminster.*—Illustrations of the smelting of iron in Great Britain. Analysis of specimens.

288 PURIFIED FUEL CO. 16, *George-st. Mansion-house.*—Block fuel from coal.

289 QUEENSGATE WHITING CO. *Beverley, Yorkshire.*—Fine, hard, and soft Paris white.

290 QUILLIAM, T. *Castletown, Isle of Man.*—Manx marble and stone.

291 RAMSAY, G. H. & SONS, *Derwenthaugh, Newcastle-on-Tyne.*—Retorts, fire-clay, cannel coal, coke, &c.

292 RAY, J. *Kilburne Colliery, Derby.*—Kilburne coal and ironstone.

293 RAY, J. *Ulverstone.*—Slates.

294 RAYNES, LUPTON, & CO. *Liverpool.*—Penmaenmawr stone for paving streets.

295 READWIN, T. A., F.G.S. *Stretford, Manchester.*—British gold ores.

296 REDRUTH LOCAL COMMITTEE, *Redruth, Cornwall.*—Mineral produce of West Cornwall.

297 REID, P. S. *Chester-le-Street.*—Boreing tools.

298 RENWICK & NICHOLSON, *Newcastle-on-Tyne.*—Coal and coke.

299 RHIWBRYFDIG SLATE CO. *Portmadoc.*—Roofing slates.

300 RHORYDD SLATE CO. *Port Madoc.*—Roofing slates and ridges.

301 RHOS COLLIERY CO. *Llanelly.*—Anthracite coal.

302 ROBINSON & SON, *Stanhope, Darlington.*—Section of Weardale strata.

303 ROBINSON, W. & CO. *Gospel Oak Works, Tipton.*—Flat and corrugated iron, variously prepared and treated.

304 ROBSON, R. *Newcastle-on-Tyne.*—Specimens of freestone.

305 ROGERS & RAWLINGS, *Bradford-on-Avon.*—Font in Bath stone.

306 ROGERS, E. *Abercarn.*—Iron ores, plans and description.

307 ROGERS, P. *Swansea.* — Enamelled slates and marble.

308 ROSS OF MULL GRANITE CO. 35, *Parliament-st.*—Polished red granite, &c.

309 RUDDOCK, S. 22, *Bloomfield-ter. Pimlico.*—Statuette of St. Agnes.

310 RUSSELL, J. *Newport, Monmouthshire.*—Steam, cooking, and household coal.

311 SALT CHAMBER OF COMMERCE, *Northwich.*—Rock salt, marine salt, and other manufactured salt of various countries.

312 SALTER, J. W. 28, *Jermyn-st.*—Geological map, coloured on a new principle.

314 SCARTH, W. T. *Raby-castle, Darlington.*—Stone, basalt, ironstone, lead ore, &c. from Teesdale.

315 SCHLESINGER, J. *George-st. Birmingham.*—Emery-cloth and paper, &c.

316 SCHNEIDER, HANNAY, & CO. *Barrow-in-Furness.*—Model of blast furnaces.

317 SCHULL BAY COPPER MINING CO. 33, *Great Winchester-st. E.C.*—Malachite and copper ores.

318 SCOTTISH IRONMASTERS: Baird, W. & Co., Merry and Cunningham, Dixon, W. S., Houldsworth & Co., Wilson's trustees, Addie, R., Wilsons & Co., Dunlop & Co., and others, *Glasgow.* — Ironstones from which Scotch pig-iron is made; also, specimens of pig-iron.

319 SEAFIELD, EARL OF, *Cullen-house.*—Serpentine and specimens from Portsoy. Cairngorm crystals from Strathspey.

320 SECCOMBE, J. *Liskeard, Cornwall.*—Crystallized oxide of copper.

321 SEWELL, E. *Fulneck, Leeds.*—Model of Tong Ironstonefield, West of Leeds.

322 SHELTON BAR IRON CO. *Stoke, Staffordshire.*—Boiler-plate and iron.

323 SHEPHERD, T. *Bath.*—Crossway stone.

324 SHEPHERD & EVANS, *Aberdare.*—Smokeless steam coal.

325 SHIELD & DINNING, *Langley Lead Works, Haydon Bridge.*—Lead ores, lead smelting and refining.

326 SIM, W. *Glasgow.*—Silver grey granite Obelisk, decorated with incised ornaments inlaid with gold. *Nave.*)

327 SIMON, L. *Nottingham.*—Bronze powder, varnish, and printing ink.

328 SIMPSON, O. N. *Little Casterton Quarry, near Stamford.*—Oolite building stones.

330 SMAILE, R. & CO. *Newcastle-on-Tyne.*—Pressed crucibles.

331 SMITH E. J. *Gateshead.*—Stone.

332 SMITH, R. *The Priory, Dudley.*—Hot and cold blast iron, and minerals.

333 SMITH, S. *Ashford, Derbyshire.*—Vases of black Derbyshire marble,

334 SOPWITH, T. 43, *Cleveland-st. W.*—Illustrations of Allenhead's lead mining.

335 SOWERBY & PHILLIPS, *Newcastle-on-Tyne.*— Waldridge, Hutton seam, gas, and smiths' coal.

336 SPARK, H. K. *Darlington.*—Coal, coke, ironstone, iron, and fire-bricks.

337 SPARKS, W. *Crewkerne, Somerset.*—Stone and limestone.

338 SQUIRES, C. & SONS, *Stourbridge.*—Model of glass house furnace, &c.

339 STANIER & SON, *Silverdale, Newcastle, Staffordshire.*—Bars for ships' knees.

340 STARK, J. C. *Torquay.*—Devonshire marbles.

341 STICK, H. & CO. *Swansea.* — Tin plates and iron.

342 STICKLEY, J. *Cross-st. Hatton-garden.*—Leaf gold and other beaten metals.

343 STRATON & CARGILL, *Arbroath, Forfarshire.* — Polished and dressed pavement.

344 SUNDERLAND LOCAL COMMITTEE, 13, *Bridge-st.*—Model of docks and harbour entrance.

345 SUTTON & ASH, *Snow-hill, Birmingham.*—Patterns of sections of rolled iron.

346 SWANSEA LOCAL COMMITTEE. — Copper, silver, iron, zinc, and nickel ores and metals.

347 SWEETLAND, TUTTLE, & CO. 55, *Old Broad-st.*—Copper in various stages.

348 TASKER, T. *Billinge-hill Quarry, near Wigan.*—Scythestones and grindstones.

349 TAVISTOCK COMMITTEE, *Tavistock.*—Copper and tin ores, stones, clay, &c.

350 TAYLOR, BROS. & CO. *Leeds.*—Tyres, cranks, axles, &c.

351 TAYLOR, H. *Coal Trade Office, Newcastle-on-Tyne.*—Plans and sections of coal-fields of Durham and Northumberland.

352 TEAGUE, M. *St. Paul, Penzance.*—Obelisk.

353 TERRET, J. *Coleford, Gloucestershire.*—Brick tiles and pipes.

354 THOMAS, H. *Lisburne mines, Cardiganshire.*—Lead ores.

355 ———Silver-lead ore from Glogfach mine, Cardiganshire.

356 ———Silver-lead ore from Log-y-las mine, Cardiganshire.

357 ———Silver-lead ore from Frongoch mine, Cardiganshire.

358 ———Silver-lead ore from Cwmystwith mine, Cardiganshire.

359 ———Silver-lead ore from Cefn-cwm-brwyns mine, Cardiganshire.

360 ———Silver-lead ore from Goginan mine, Cardiganshire.

361 THOMAS, H. *Lisburne mines, Cardiganshire.*—Silver-lead ore from Cwm-Erfin mine, Cardiganshire.
362 ———Silver-lead ore from East Darren mine, Cardiganshire.
363 ———Silver-lead ore from Nanty mine, Montgomeryshire.
364 THOMPSON, HATTON, & CO. *Bilston.* —Iron and tin-plate articles.
365 THOMPSON, W. *Elmer-st. Grantham.* —Ancaster free-stone and rag.
367 TOMLINSON, A. *Bakewell, Derbyshire.* —Inlaid tables, &c. of Derbyshire marble.
368 TONKIN, J. *Pool, Cornwall.*—Transverse section of Dolcoath mine.
369 TOWNSHEND, WOOD, & CO. *Swansea.* —Bar and sheet iron, tin, terne, and black plates.
370 TRASK, C. *Norton-sub-Hamdon, Somerset.*—Ham stones.
371 TRICKETT, G. 21, *Cannon-st. E.C.*—A burnt iron column.
372 TRICKETT, S. *Isle of Dogs.*—Stones, granite, and marble.
373 TRICKETT AND HOLDSWORTH, *Horsforth, near Leeds.*—Bramley Fall stone, for docks, bridges, and basement-course of large buildings.
374 TROTTER, THOMAS, & CO. *Winnallshill; near Coleford.*—Lathe-turned columns. Ashlar, sawn and planed, from Brixdale quarries. Coal.
375 TRUSCOTT, C. & CO. *St. Austell, Cornwall.*—China clays, China stones, &c.
376 TUFFLEY, E. *Avening, near Stroud, Gloucestershire.* — Model of a staircase in Painswick stone.
377 TURNBULL, M. JUN. *Tranwell, Morpeth, Northumberland.*—Specimen of freestone.
378 TURNER, CASSONS, & CO. *Portmadoc.* —Slates and slate slabs.
379 TURNER, J. 1, *Hall Bank, Buxton.*—Derbyshire marble vases and tables, inlaid.
380 TYLER, J. W. 4, *Wood-st. Westminster.* —Improvements in laying of zinc.
381 TYM, J. *Castleton, Derbyshire.*—Fluor spar vases, candlesticks, chalices, &c.
382 USWORTH COLLIERY, OWNERS OF, *Usworth, Washington, Durham.*—Freestone.
383 VIGRA AND CLOGAU COPPER MINING CO. 28, *Bucklersbury.*—Gold and gold ores from Merionethshire.
384 VINT, G. & BROS. *Idle, near Leeds.*—Obelisk in Gazeby stone.
385 VOSS, J. *Woodyhide, Corfe Castle, Dorset.*—Purbeck marble.
386 WAGSTAFFE & CO. *Fremator Quarries, near Tavistock.*—Pedestal of granite, 4 ft. 8 in. by 16 in.
387 WALCOTT, GEORGE, 24, *Abchurch-lane.*—Gas retort bed, full size set of six.
388 WALKER, D. *Turin-hill Quarries near Arbroath.*—Flagstones.
389 WARING, C. H. *Neath.*—Miners' safety lamps; cannot be opened without extinguishing the light.
390 WARNER, A. 31, *Threadneedle-st.*—Iron treated by chemical processes.
391 WARNERS, LUCAS, & BARRETT, *Norton Furnam, Stockton-on-Tees.*—Samples of pig iron and railway chairs.
392 WATSON, H. *Newcastle-on-Tyne.*—Safety lamps for coal mines; and underground pump.
393 WAYNE, T. & CO. *Aberdare.*—Coal and iron.
394 WEARDALE IRON CO. *Ferryhill, Durham.*—Iron, steel, and their minerals.
395 WELSH SLATE CO. *Port Madoc, Carnarvon.* — Slabs and building slates from quarries near Festiniog, Carnarvon.
396 WESCOMB, C. *Exeter.*—Spiral fluted nails.
397 WESTON & GRICE, *West Bromwich, near Birmingham.* — Bar-iron and railway fastenings.
398 WHARNCLIFFE SILKSTONE COLLIERY. *Sheffield.*—Coal.
399 WHEELER, P. & CO. *St. Austell, Cornwall.*—China clay (Kaolin), and stone.
400 WHITELAW, J. *Preston Grange Colliery, Prestonpans, Scotland.* — Model of miner's safety cage.
401 WHITEWAY & CO. *Kingsteignton, Devonshire.*—Tobacco pipe, potter's clays, &c.
402 WICKLOW COPPER MINE CO. 43, *Dame-st. Dublin.*—Iron pyrites rich in sulphur.
403 WILLIAMS, R. & CO. *Portmadoc, Carnarvonshire.*—Slate ridges for finishing roofs.
404 WILLIAMSON, C. *Plas-yn-Morfu, Holywell, Flintshire.*—Ores of lead and zinc.
405 WILLIAMSON, J. *Kerridge, Macclesfield.*—Building stones, granite paving setts, &c.
406 WILLIAMSON, R. 18, *Lothian-road, Camberwell.*—Model to illustrate system of ventilating collieries.
407 WILSON, G. B. *Forest-hall, Newcastle-on-Tyne.*—Specimen of freestone for building.
408 WILSON, J. *Grantham.*—Carved font of Ancaster stone.
409 WILSON, SIR T. M. *Charlton House.* —Founders', and other sands.
410 WIMSHURST'S PATENT METAL FOIL CO. 20, *Cannon-st. E.C.*—Sheet of cut lead, one mile long.
411 WOLSTON, R. W. *Brixham, Devon.*—Iron paints and composition for coating materials under water.
412 WOMBWELL MAIN COAL CO. *New Barnsley.*—Coal and sections.
413 WOOD & DAGLISH, *Hetton Colliery, Durham*—Models of plan for ventilating collieries, and conveyance of coal under ground.
414 WOOD, T. & CO. *Cliff Wood, & Spinkwell Quarries, Bradford.*—Ashlar stone.
415 WOODHOUSE & JEFFCOCK, *Derby.* —Coals from Shipley collieries, Derbyshire.
416 ———Coals and ironstones from Victoria colliery, Warwickshire.
417 ———Coals from Cinderhill colliery, Nottingham.
418 ———Coals from Granville colliery Derbyshire.
419 ———Coals from Wyken colliery, Warwickshire.
420 ———Coals from Moira colliery, Leicestershire.
421 ———Coals and ironstone from Oakerthorpe and Highfield collieries, Derby.
422 ———Coals from Baddesley collieries.

423 WOODHOUSE & JEFFCOCK, *Derby*.—Coals from Gresley colliery.
424 ———Coals from Swanwick collieries.
425 ———Model of Shipley colliery. Coals and ironstones.
426 WOODRUFF, T. 4, *Quadrant, Buxton*.—Derbyshire tables, vases, &c.
427 WOODWARD BROS. *Ruabon*.—Building, grinding, and scythe stones.
428 WRIGHT, J. & SON, *John st. Granite Works, Aberdeen*.—Polished granite vase, &c.
429 YNISCEDWYN IRON CO. *Swansea*.—Refined metal, anthracite, coal, and iron, iron ores, &c.
431 YSTALYFERA IRON CO. *Swansea*.—Anthracite pig, bar, and sheet iron, boiler plates, tin, terne plates, and nails.

———

432 ABEL.—Fuse for blasting by magneto-electricity.
433 COSHAM, H. *Parkfield Colliery*.—Coals from Bristol coal field.
434 FISHER, G. *Taff Vale Railway Co*.—Model of coal-shipping apparatus.
435 FLETCHER & CO. *Marston, Cheshire*.—Rock salt for cattle.
436 GADLY'S IRON CO. *Aberdare, Glamorganshire*.—Iron, and steam coal.
437 GAPPER, J. C. 65, *Great Russell-st*.—Geological survey of a portion of the Lincolnshire iron field.
438 HADLEY, J. *London*.—Lead and quartz from mines in Wales.
439 HUNT, COL. *Eaton-place*.—Coke.
440 LAYCOCK, J. & CO. *Seghill Colliery, Newcastle-on-Tyne*.—"Carr's Hartley" steam coal.
441 MURCHISON, S. *London*.—Copper ores, lead ores, &c.
442 SERPENTINE MARBLE CO. *Waterloo-place*.—Serpentine and Devonshire marble.
443 TENNANT, J. *Strand, London*.—Septaria table.
444 WRIGHT, S. *Buttermere, near Cockermouth*.—Slates.

———

CLASS 2.

CHEMICAL SUBSTANCES AND PRODUCTS, AND PHARMACEUTICAL PROCESSES.

Eastern Annex, South-East Passage.

Sub-Class A.—Chemical Products.

458 ADAMS, J. *Victoria-pk. Sheffield*.—Chemicals.
459 ALLEN, F. *Bow-common*.—Aniline and fine chemicals.
460 ALBRIGHT & WILSON, *Oldbury*.—Amorphous phosphorus, and other chemical products.
461 ALLHUSEN, C. & SONS, *Newcastle-on-Tyne*.—Alkali and bleaching powder.
462 ANDREW, F. W. 3, *Neville-ter. Brompton*.—Cement for repairing ceramic articles, &c.
463 AVRIL, J. 12, *Castle-st. Holborn*.—Insect killing powder.
464 BAILEY, J. *Shooters-hill, Staffordshire*.—Colors for porcelain, glass, &c.
465 BAILEY, W. & SON, *Horseley-fields Chemical Works, Wolverhampton*.—Chemicals.
466 BAKER, E. & SONS, *Birmingham*.—Paste blacking, black lead, &c.
467 BAKER, F. B. *Hampton-ct*.—Crystals of sulphate of magnesia, copper, and alum.
468 BALKWILL & CO. *Plymouth*.—Metallic and white arsenic, &c.
469 BARNES, J. B. 1, *Trevor-ter. Knightsbridge*.—Volatile organic acids and their ethers, &c.
470 BARRELL, J. 26, *Upper Eaton-st*.—Plate powder.
471 BARTLETT, BROS. & CO. *Camden-town*.—Soluble silicates and aluminates, insoluble made without heat, and artificial stone.
472 BELL & BLACK, 15, *Bow-lane, Cheapside*.—Wax vesta wire fusees, and Congreve matches.
473 BELL, I. L. *Newcastle-on-Tyne*.—Aluminate of soda. Oxichloride of lead.
474 BERGER, S. & CO. *Bromley-by-Bow*.—Rice starch.
475 BETTS, A. 41, *North Bar-st. Banbury*.—Blackings, &c.
476 BLAYDON CHEMICAL CO. *Newcastle-on-Tyne*.—Chemical manures.
477 BLINKHORN, SHUTTLEWORTH, & CO. *Spalding*.—Composition for preventing incrustations in steam-boilers.
478 BLUNDELL, SPENCE, & CO. *Hull and London*.—Varnishes, paints, oils, &c.
479 BOLTON & BARNITT, 146, *Holborn-bars*.—Chemical products.
480 BORWICK, G. *Little Moorfields*.—Baking powder.
481 BOUCK, J. T. & CO. 32, *Dickenson-st. Manchester*.—Tar products, and other chemicals.
482 BOWDITCH, Rev. W. R. *Wakefield*.—Illustrations of the purification of gas from sulphur: safety lamps.
484 BRAMWELL & CO. *Newcastle-on-Tyne*.—Prussiate of potash.
485 BRAY & THOMPSON, *Alum Works, Chatterley, near Tunstall, Staffordshire*.—Alum.
486 BRODIE, B. C. *Oxford*.—Graphite, chemically disintegrated and purified.

487 BROOMHALL, J. *London.*—Starch from rice, wheat, potatoes, and sago.

488 BRYANT & MAY, *London.* — Safety matches and chemical lights.

489 BUCKLEY, J. (THE TRUSTEES OF THE LATE) *Manchester.*—Sulphate of iron.

490 BUSH, W. J. 30, *Liverpool-st. E.C.*—Essences and essential oils.

491 CAHN, D. 12, *North-bdgs. Finsbury-circus.*—Blacks for printers.

492 CALLEY, S. *Brixham, Devon.*—Metallic paints, iron ochres.

493 CARR, T. *Birkenhead.*—Soluble superphosphate.

494 CATTELL, DR. *Euston-sq.* — Purified gutta percha, varnishes, carbons, inks, &c.

495 CHANCE, BROS. & CO. *Alkali Works, near Birmingham.*—Artificial manures and chemical products.

496 CHICK, G. B. *Bristol.*—Indigo, stone blue, and black lead.

497 CHURCH, A. H. 170, *Great Portland-st. W.*—Rare chemical products.

498 COLLINGS, H. A. 48, *Whiskin-st. London, E.C.* — Jeweller's rouge, polishing powders, &c.

499 COLMAN, J. & J. 26, *Cannon-st. E.C.*—Starch and blue.

500 CONDY, H. B. *Battersea.*—Disinfecting fluid, and other preparations.

501 COWAN & SONS, *Hammersmith-bridge Works, Barnes.*—Bones, animal charcoal, &c.

502 COX & GOULD, *Chicksand-st. Whitechapel.*—Chemicals, various.

503 CRISP, E., M.D. *Chelsea.*—Colours from the biles of 500 animals: photographs.

504 DAVIS, A. 30, *Union-st. Bishopsgate.*—Polishing paste.

505 DAVY, MACKMURDO, & CO. 100, *Upper Thames-st.*—Mercurial preparations, and other chemicals.

506 DAWSON, D. *Miln's-bdg. Huddersfield.*—Benzole, aniline, Magenta-powder, &c.

508 DOUBLEDAY, H. *Coggeshall, Essex.*—Dextrine for glazing textile fabrics.

509 DUNELL, R. G. *Ratcliff-highway.*—Colours for decorators, &c.

510 DUNN, A. *Dalston.* — Marking ink pencils, &c.

511 DUNN, HEATHFIELD, & CO. *Princes-sq. Finsbury.*—Chemical and pharmaceutical products.

512 EMERY, F. & SON, *Cobridge, Staffordshire.*—Porcelain, glass, and earthenware colours, &c.

513 ESCHWEGE, H. *Mincing-lane.*—Potable wood naphtha, and pure spirit.

514 EVANS, T. 18, *Newland-st. Pimlico.*—Composition for harness, polish, &c.

515 EVERETT & CO. 51, *Fetter-lane.*—Blacking and varnish for boots.

516 FENN, J. 4, *North-ter. Grosvenor-sq.*—Blacking, varnish, &c.

517 FLEMING, A. B. & CO. *Chemical Works, Leith, Edinburgh.*—Vegetable carbon.

518 FOOT, C. & CO. *Battersea.* — Acetic, nitric, and other acids.

519 FOULKES & WALLWORTH, *Birkenhead.*—Cement for glass, wood, &c.

520 GASKELL, DEACON, & CO. *Widnes Dock, Warrington.* — Bleaching powder, alkalies, and colours.

521 GILES & BARRINGER, *Hackney-wick. London.*—Starch, polish, and blacking.

522 GREATOREX, F. 281, *King's-rd. Chelsea.*—Blacking.

523 GRIMWADE, RIDLEY, & CO. 31, *Gt. St. Helen's.*—Anti-corrosion paint.

524 HAAS & CO. *Leeds.*—Dyes.

525 HALLETT, G. & CO. 52, *Broadwall, Blackfriars.* — Antimony, and preparations therefrom.

526 HARE, J. & CO. *Temple-gate, Bristol.*—Colours.

527 HAWORTH & BROOKE, 33, *Lower King-st. Manchester.*—Oxides of tin.

528 HIRST, BROOKE, & TOMLINSON, *Leeds*—Chemicals, pharmaceutical preparations, varnishes, &c.

529 HOLLIDAY, R. *Chemical and Lamp Works, Huddersfield.*—Coal tar products and dyes.

530 HOPKIN & WILLIAMS, 5, *New Cavendish-st. W.*—Chemical and pharmaceutical products.

532 HOWARDS & SONS, *Stratford, Essex.*—Quinine and other cinchona alkaloids and chemicals.

533 HULLE, J. *Lombard-rd. Battersea.*—Vegetable alkaloids and their salts.

535 HURLET & CAMPSIE ALUM CO. *Glasgow.*—Alum, red and yellow prussiates.

536 HUSKISSON, W. & SONS, 77, *Swinton st. W.C.*—Chemical products.

537 HUTCHINSON & EARLE, *Widnes Docks, near Warrington.*—Specimens illustrating the process of alkali manufacture.

538 HYNAM, J. *Princes-sq. Finsbury.*—Matches, vestas, and fusees.

539 JAMES, E. *Sutton-rd. Plymouth.*—Starches, blues, black leads, &c.

540 JARROW CHEMICAL CO. *South Shields*—Soda, alkali, bicarbonate of soda, bleaching powder, &c.

541 JOHNSON & SONS, *Basinghall-st. London.*—Lunar caustic and other chemicals.

542 JOHNSON, W. W. & R. & SONS, *Limehouse.*—White lead.

543 JOHNSTONE, R. *Black Works, Agar-twn. St. Pancras.*—Vegetable and spirit blacks.

544 JONES, J. M. *Gloucester.*—Composition for preserving leather.

545 JONES, O. & CO. *York-rd. Battersea.*—Rice starch.

546 JONES, W. J. 12, *Victoria-rd. Belgravia.*—Chemical products for dyeing and cleaning.

547 JUDSON, D. & SON, 10, *Scott's-yd. City.*—Dyes and dye stuffs.

548 KANE, W. J. *Chemical Works, Dublin*—Chemicals.

549 KINGSTON, S. *Spalding.*—New pain for iron and external work.

550 KLABER, H. *Albion-pl. E.C.*—Wax Vestas, flaming fusees, matches, &c.

551 KUKLA & CO. *Pentonville-rd.*—Artificial salt.

552 LAMBERT, W. T. 9, *Tabernacle-row, Finsbury.*—Mercury, potee powders, refined black lead, &c.

553 LANGDALE, E. F. 72, *Hatton-garden.*—Essential oils, fruit essences, &c.

554 LANGLEY, W. 3, *Salters' Hall-ct. City.*—Fine colours and bronze-powders.

555 LEATHART, C. 19A, *High-st. Newington Butts.*—Green colour without arsenic, &c.

556 LETCHFORD & Co. *Whitechapel.*—Wax vestas and matches.

557 LEWIS, J. & SON, *Pontardawe Chemical Works, Swansea.*—Acetate of soda.

558 LONDON MANURE Co. 116, *Fenchurch-st.*—Artificial manures, &c.

559 LONGBOTTOM, J. & Co. *Belgrave Foundry, Leeds.*—Substances carbonized by patent process.

560 LUCAS, G. 44, *Kennedy-st. Manchester.*—Mineral filling for metallic plates.

561 MACKAY & Co. *Inverness.*—Chemical manure.

562 MANDER BROS. *Wolverhampton and London.*—Varnish and japan.

563 MARSHALL, J. SON, & Co. *London and Leeds.*—Cudbear, orchil, indigo carmine, lac dye, dyewoods, &c.

564 MASON, C. F. A. 13, *Walcot-pl. S.*—Blacking.

565 MAY & BAKER, *Garden Wharf, Battersea.*—Mercurial and other chemical products.

566 MELINCRYTHAN CHEMICAL Co. *Neath.*—Acetates and other products derived from the dry distillation of wood.

567 METROPOLITAN ALUM Co. *Bow-common.*—Alum.

568 MILLER, G. & Co. *Glasgow.*—Products of Boghead mineral and of coal tar.

569 MOCKFORD & Co. 7, *Mincing-lane.*—Chemicals, colours.

570 MORSON, T. & SON, *Southampton-row.*—Chemical and pharmaceutical products.

571 MUSPRATT, BROS. & HUNTLEY, *Liverpool, and Flint, N. Wales.*—Soda, chloride of lime, sulphate of alumina, &c.

572 NAYLOR, W. 4A, *James-st. Oxford-st.*—Samples of varnish, &c.

573 NEWMAN, J. *Soho-sq.*—Pigments.

574 ODLING, A. 30, *Glasshouse-st. Vauxhall.*—Ammonia.

575 PALING & Co. *Newark, Nottinghamshire.*—Starch, printers' flour, manures, and turnip-fly preventative, &c.

577 PATENT NITRO-PHOSPHATE Co. 109, *Fenchurch-st. E.C.*—Manures, and materials used in their manufacture.

578 PATENT PLUMBAGO CRUCIBLE Co. *Battersea Works, S.W.*—Plumbago, black lead, graphite, &c.

579 PEACOCK & BUCHAN, *Southampton.*—Composition for preserving ship's bottoms, &c.

580 PEGG, HARPER, & Co. *Derby.*—Painters' colours, plaster of Paris, barytes, &c.

581 PERKIN & SONS, *Greenford-green, Middlesex.*—Specimens illustrating the manufacture and application of Perkin's aniline purple.

582 PINCOFFS & Co. *Manchester.*—Alizarine and garancine.

583 POTTER, W. H. 23, *Clapham-rd.-pl. Surrey.*—Manures.

584 REA, J. 115, *Wardour-st.*—Shellacs, resins, and varnishes.

585 RECKITT, J. & SONS, *Hull.*—Starches, blues, and black-leads.

586 REEVES & SONS, 113, *Cheapside.*—Fine pigments.

587 RICHARDSON BROS. & Co. 17, *St. Helen's-pl.*—Refined saltpetre.

588 ROBERTS, DALE, & Co. *Manchester.*—Chemical products, pigments, aniline colours, &c.

589 ROOTH, J. S. *Chesterfield.*—Chemicals.

590 ROSE, W. A. 66, *Upper Thames-st.*—Varnishes, colours, oils, &c.

591 ROWNEY & Co. 51, *Rathbone-pl.*—Fine pigments.

592 RUMNEY, R. *Ardwick Chemical Works, Manchester.*—Illustrations of new chemicals for dyeing and calico printing; introduced since 1851.

593 ——— Silicate of soda and potash, uric acid and compounds.

594 RUMSEY, W. S. 3, *Clapham-rise.*—Chemical productions for polishing metals.

595 SAVORY & MOORE, 143, *New Bond-st.*—Chemicals.

596 SCOTT, W. L. *Westbourne-pk. W.*—Fabrics dyed with dianthine and aniline green.

597 SHAND, G. *Stirling.*—Tar and chemical products.

598 SHANKS, J. *St. Helen's, Lancashire.*—Specimens illustrative of the manufacture of chlorine.

599 SIDEBOTTOM, A. *Camberwell.*—Painting that will resist water and air: copying fluid: Lakes, &c.

600 SIMPSON, MAULE, & NICHOLSON, 1 & 2, *Kennington-rd.*—Coal-tar products and dyes.

601 SMITH, B. & SON, *Spitalfields.*—Archil, cudbear, and orchelline: and lichens from which they are produced.

602 SMITH, T. L. & Co. *St. James'-rd. Holloway.*—Starch.

603 SMITH, T. W. *Lower-st. Islington.*—Magenta lake, and other pigments.

604 SMITH, T. & H. *London & Edinburgh.*—Products from opium, alöin, caffeine, &c.

605 SPENCE, P. *Pendleton Alum Works, Manchester.*—Alum. Raw and calcined shale.

606 SPRINGFIELD STARCH Co. 104, *Upper Thames-st.*—Starch and British gum.

607 STANFORD, E. C. C. *Worthing, Sussex.*—New products distilled from seaweeds.

608 STENHOUSE, J. &c. *Rodney-st. Islington.*—Rare chemicals.

609 STIFF & FRY, *Redcliff-st. Bristol.*—Starch, and other products from rice and wheat.

610 STRUVE & Co. *German Spa, Brighton.*—Artificial mineral waters.

611 SYMONS, T. *Derby.*—Oil of vitriol, sulphate of ammonia, colcothar.

612 TUDOR, S. & W. *London, and Lead Works, Hull.*—Carbonate of lead and white lead.

613 VERSMANN, F. 7, *Bury-ct. St. Mary-axe.*—Wolfram ores, tungstate of soda, and antiflammable fabrics, &c.

614 VINCENT, C. W. 2, *Greyhound-ct. Milford-lane.*—Varnishes for making black and coloured printing inks.

615 WALKER ALKALI Co. *Newcastle.*—Hyposulphite of soda, alkalies, &c.

616 WALLIS, G. & T. 64, *Long-acre.*—Resins, oils, varnishes, &c.

617 WARD, F. O. *Hertford-st. Mayfair.*—Series illustrating new process for extracting alkali from natural alkaliferous silicates.

618 ——— Series illustrating new process for separating the animal and vegetable ingredients of mixed rags.

619 WARD, J. & Co. 452, *Garscube-rd. Glasgow.*—Kelp and its products.

620 WHAITE, H. 24, *Bridge-st. Manchester.*—Composition for painting flags.

621 WHITE, J. & J. *Shawfield Works, Glasgow.*—Bichromate of potash.

622 WHITWORTH, G. & Co. *Jamaica-row, Bermondsey.*—Concentrated fish manure.

623 WILKINSON, HEYWOODS, & CLARK, *Battle-bridge, London, N.*—Varnishes, colours, oxidized oils, &c.

624 WILSHERE & RABBETH, *Gt. Western-rd. Paddington.*—Varnishes and colours.

625 WILSON & FLETCHER, *Jubilee-st. Mile-end, E.*—Aniline colours and other pigments, &c.

626 WILSON, J. & SONS, *Hurlet, near Glasgow.*—Alum, and products from bones, &c.

627 WINSOR & NEWTON, 38, *Rathbone-pl. W.*—Fine colours.

628 WOOD, E. *Port-hill, Stoke-on-Trent.*—Borax, boracic acid, and china glaze.

629 WOOD & BEDFORD, *Leeds.*—Orchil and cudbear.

630 WOTHERSPOON, W. *Glenfield Starch Works, Paisley.*—Glenfield starch from sago flour.

631 WRIGHT, FRANCIS, & Co. 11, *Old Fish-st. E.C.*—Pharmaceutical preparations and chemicals.

632 YOUNG, J. *Bathgate.*—Paraffine, and paraffine oil from different kinds of coal.

633 YOUNG, J. W. *Neath, Glamorganshire.*—Paint and pigments.

---

634 CROOKES, W. 20, *Mornington-road, N.W.*—Thallium, a new elementary body.

635 DAY & MARTIN, 97, *High Holborn.*—Liquid and paste blacking.

636 KEATING, 79, *St. Paul's Church-yard.*—Medicinal plants from the west coast of Africa.

637 HOULDER, W. *London and Norwood, Middlesex.*—Sulphate of copper cross, &c.

Sub-Class B.—Medical & Pharmaceutical Processes.

644 ALLEN & HANBURYS, *Plough-ct. Lombard-st.*—Drugs and pharmaceutical preparations.

645 BASS, J. 81, *Hatton-garden.*—Pharmaceutical products.

646 BASTICK, W. *Brook-st. London.*—Pharmaceutical preparations.

647 BROWN, T. B. 103, *Icknield-st. Birmingham.*—Cantharidine blistering tissue &c.

648 BULLOCK & REYNOLDS, 3, *Hanover-st. W.*—Chemical and pharmaceutical products.

649 CURTIS & Co. 48, *Baker-st. W.*—Pharmaceutical preparations.

650 DARBY & GOSDEN, 140, *Leadenhall-st.*—Pharmaceutical products.

651 DENOUAL, J. 1, *Walpole-st. New Cross, S.E.*—Nauseous and alterable drugs enclosed in gelatine.

652 DICKINSON, W. *Queen's-gdns. London.*—Medicinal preparations.

653 DUNCAN, FLOCKHART, & Co. *Edinburgh.*—Chloroform from pure alcohol and methylated alcohol. Chloric ether.

654 GARDNER, J., M.D. 23, *Montague-st.*—Pharmaceutical chemicals.

655 HOLLAND, W. *Market Deeping.*—Essential oils, extracts, dried plants, and roots.

656 HOOPER, W. 7, *Pall Mall East.*—Chemical and pharmaceutical preparations.

657 LAMACRAFT & Co. 6, *Upper Rathbone-place.*—Medical plaisters, &c. &c.

658 LAURENCE, W. H. 163, *Sloane-st. S.W.*—Cod liver oil.

659 LE MAOUT, *Princes-st. Soho.*—Gelatine capsules enclosing nauseous drugs.

660 MACFARLAN, J. F. & Co. *North-bridge, Edinburgh.*—Chemical preparations from opium, green-heart bark, galls, and methylated spirit.

661 MAJOR, J. 5, *Park-lane, Piccadilly, W.*—Horse and cattle medicines.

662 MOFFATT, G. D. *Dundas-st. Edinburgh*—Cod liver oil.

663 MURRAY, SIR J., M.D. *Anatomy Office, Temple-st. Dublin.*—Fluid magnesia, camphor, and aërated extract of bark.

664 PHARMACEUTICAL SOCIETY OF GT. BRITAIN, COMMITTEE OF THE, 17, *Bloomsbury-sq.*—Systematic collection of drugs and preparations used in medicine.

665 RANSOM, W. *Hitchin.*—Medicinal extracts and essential oils.

666 SQUIRE, P. 277, *Oxford-st.*—Chemical and pharmaceutical products.

667 TUSTIAN, J. *Milcombe, near Banbury. Oxon.*—*Conf. Rosæ*, and other pharmaceutical preparations.

668 USHER, R. *Bodicot, near Banbury.*—Rhubarb and other medicinal herbs.

669 WATERS, R. 2, *Martin's-lane, Cannon-st. London, E.C.*—Quinine wine.

670 WATTS, J. & Co. 107, *Edgware-rd. London, W.*—Extracts and pharmaceutical preparations.

CLASS 3.

SUBSTANCES USED FOR FOOD.

Eastern Annex, East Side.

Sub-Class A.—Agricultural Produce.

700 ADKINS, T. K. *Wallingford.*—English flour made by Callard's process.

701 ASPREY, J. *Sandleford, Berks.*—Wheat, barley, and oats.

702 BAKERS, WHITE, & MORGAN, *Hibernian-chambers, London.*—British and foreign hops.

703 BARRY, DYKES, & Co. *Type-st, Finsbury.*—Chicory, cocoa, and mustard.

704 BROWN & POLSON, *Paisley.*—Corn flour and starch.

705 BUTLER & MCCULLOCH, *Covent Garden-market.*—Dried plants, flowers, roots, and seeds.

706 CAHILL, M. *Ballyconra, Kilkenny.*—Wheat and oats.

707 CARTER & Co. 238, *High Holborn.*—Seeds, flowers, and floral designs.

708 CHAMBERS, W. E. *Corn-market, Mark-lane.*—Cereals.

709 CHITTY, E. *Guildford.* — Wheaten flour.

710 CHRISTIE, W. *Steam Flour Mills, Chelsea.*—Wheaten flour.

711 DAVIS, E. J. *Globe Wharf, Mile-end-rd.*—Compressed hay and other forage.

712 FORDHAM, T. *Snelsmore Hill, Newbury, Berks.*—Wheat, beans, oats, &c.

713 FULLER, C. *Newnham-farm, Wallingford, Essex.*—Wheat.

714 HALLETT, F. F. *Brighton.*—Pedigree nursery wheat.

715 HENRI'S HORSE AND CATTLE FEED Co. *London Bridge.*—Cattle condiments.

716 IRWIN, E. *Ballymore, Boyle, Ireland.*—Black oats.

717 KIRK & SWALES, *New Wortley, near Leeds.*—Grain, flour, and malt.

718 KITCHIN, J. *Dunsdale, Westerham, Kent.*—Pocket of Golding hops.

719 LIVERPOOL COMMITTEE OF THE INTERNATIONAL EXHIBITION, 1862, *Liverpool.*—Imports and their appliances.

720 MACKEAN, W. *St. Mirren's, Paisley.*—Corn flour and starches.

721 PACK, T. H. *Ditton-ct. Maidstone, Kent.*—Pocket of hops.

722 PAINE, CAROLINE, *Farnham, Surrey.*—Pocket of Farnham hops.

723 PALING, W. & E. *Newark, Nottinghamshire.*—Cattle food and condiment.

724 POLSON, W. & Co. *Paisley.*—Indian corn and sago flour.

725 RAYNBIRD, CALDECOTT, & BAWTREE, *Basingstoke.*—Seed-corn and seeds.

726 ROBINSON, BELLVILLE, & Co. 64, *Red Lion-st. Holborn.*—Prepared barley and other cereals.

727 SIMPSON, A. *Snow-hill, Birmingham.*—Food for cattle.

728 STEVENS, R. *Collyweston, Northamptonshire.*—Cereals, &c.

729 STRANGE, W. *Banbury, Oxon.* — Wheat and beans.

730 STYLES, T. 148, *Upper Thames-st.*—Ashby's groats.

731 SUTTON & SONS, *Reading.*—Seeds, grasses, &c.

732 TAUNTON, W. *Redlynch, Salisbury.*—Corn and seeds.

733 TAYLOR, J. & SONS, *Bishop's Stortford. Herts.*—Malt (various).

734 THORLEY, J. *Newgate-st. City.*—Food for cattle.

735 WEBB, R. *Calcot, Reading.*—Mummy wheat, cobnuts, and filberts.

736 WELLSMAN, J. *Moulton, Newmarket.*—Malt, barley, barley grown from oats, &c.

737 WOOLLOTON, C. & SONS, 246, *Borough.*—British and foreign hops.

738 WRENCH, J. & SONS, *London-bridge.*—Favourite English cereals of the London Corn Market.

739 WRIGHT, I. & SON, *Gt. Bentley, Essex.*—Grass, ferns, and agricultural seeds.

---

740 CORRY, J. R. *Compton-pl. Canonbury-sq.*—Cereals.

741 ENGLAND, G. J. *Dudley.*—Specimens of malt.

Sub-Class B.—Drysaltery, Grocery, &c.

752 BAKER, SIMPSON, & Co. *Cork and London.*—Machine-made biscuits.

753 BARNES, MORGAN, & Co. 156A, *Upper Thames-st.*—Bottled fruits, jams, and pickles.

754 BATTY & Co. *Pavement, Finsbury.*—Pickles, sauces, jams, bottled fruits, &c.

755 BEATTIE, J. & Co. *Virginia-st. Glasgow.*—Raw, refined, and crushed sugars.

756 BEXFIELD & WOOD, 110, *Long Acre.*—Wedding cake ornaments.

757 BOLLAND, R. *Chester.* — Wedding cake.

758 BOVILL, F. A. 24, *Park-ter. Regent's-pk.* — Jellies, fruit syrups, and culinary essences.

759 BROUGHTON, T. A. B. & Co. *Bristol.*—Treble refined salt.

760 CADBURY BROS. *Birmingham and London.*—Chocolate and cocoas.

761 CLARENCE, T. 2, *Church-pl. Piccadilly.*—Cayenne sauce.

762 CLERIHEW, W. *Richmond-hill, Aberdeen.*—Drawing of process for curing coffee.

763 CLYDE SUGAR REFINERS' ASSOCIATION, *Greenock.*—Refined sugar.

764 COCKS, C. *Reading.*—Reading sauce pickles, &c.

765 Collier & Son, *Foster-st. Bishopsgate.*—Cocoa, chocolate, chicory, and coffee.
766 Colman, J. & J. 26, *Cannon-st, E.C.*—Mustard.
767 Copland & Co. 30, *Bury-st. St. Mary-Axe.*—Preserved meats, fruits, vegetables, &c.
768 Coxshall, J. *Waltham Abbey, Essex.*—Gingerbread.
769 Crosse & Blackwell, *Soho-sq. W.*—Preserved provisions and condiments.
770 Dakin & Co. 1, *St. Paul's-churchyard.*—Collection of teas.
771 Dawson & Morris, 96, *Fenchurch-st. E.C.*—Isinglass.
772 Dewar, T. *Newcastle-on-Tyne.*—Mustard and process of manufacture.
773 Dodson, H. 98, *Blackman-st. Southwark.*—Unfermented bread, biscuit, powder, cakes, &c.
774 Dorguin, E. 9, *Baker-st. Portman-sq.*—Cho-ca, chocolate, and bonbons.
775 Duncan, A. M'E. & Co. *Gorey, Jersey.*—Preserved animal and vegetable substances.
776 Dunn & Hewett, *Pentonville-hill.*—Iceland moss cocoa, &c.
777 Du Parcq, C. *Jersey.*—Cocoas and cider.
778 Elder, A. *Edinburgh.*—Holyrood sauce.
779 Fadeuilhe, V. B. 29 *&* 30, *Botolph-lane.*—Dry milk in powder.
780 Fahrmbacher, M. 4, *Sion-sq. E.*—Artificial confectionery.
781 Farmer, J. & Co. *Edgeware-rd. W.*—Cocoa: cocoa fat refined.
782 Fortnum, Mason, & Co. 180, *Piccadilly.*—Collection of preserved fruits.
783 Fry, J. S. & Sons, *Bristol and London.*—Chocolate, cocoa, chicory.
784 Gamble, Power, & Co. 78, *Frenchchurch-st. Cork.*—Preserved provisions in tin cases.
785 Garrard, J. T. *Needham-market, Suffolk.*—Hams, bacon, chaps, and ox tongues.
786 Harrison, R. & J. *Leeds.*—Mustard, seeds, and chicory.
787 Hart, J. W. *St. Mary-Axe, E.C.*—Isinglass.
788 Hassall, A. H., M.D. 74, *Wimpole-st.*—Specimens illustrating the adulteration of food.
789 Hay, 6, *North Audley-st. W.*—Dutch rusks.
790 Hexter, H. *Eccleston-st. Pimlico.*—Currie powder, vanilla, and essences.
791 Hill & Jones, *Jewry-st. Aldgate.*—Sweetmeats, syrups, jams, &c.
792 Howard & Co. *Scott-st. Hull.*—Starch and Indian confection flour.
793 Huntley & Palmer, *Reading.*—Biscuits and cakes.
794 James, J. E. *Birnam, Scotland.*—Sauces.
795 Jones, R. & Trevithick, F. H. 30, *Botolph-lane, London, E.C.*—Azotised raw meat, &c.
796 Keiller, J. & Son, *Dundee.*—Confections, marmalade, and preserves.
797 Langdale, E. F. 72, *Hatton-garden, E.C.*—Essences, syrups, &c.
798 Lebaigue, H. 9, *Langham-st. W.*—Chocolate.
799 Lewis, J. R. 16, *Gould-sq. City.*—Liquorice root and extract.
800 Liverpool Preserved Provision Co.—Preserved provisions.
801 McCall & Stephen, *Adelphi Biscuit Factory, Glasgow.*—Machine-made biscuits.
802 McCall, J. & Co. 137, *Houndsditch.*—Preserved provisions.
803 McClelland, G. *Wigtown, N.B.*—Preserved potato and extract of hops.
804 McCraw, E. C. *Winsford, Cheshire.*—Steam-made salt.
805 Mackay, J. 121, *George-st. Edinburgh.*—Essences from spices and herbs, &c.
806 Mackie, J. W. 108, *Princes-st. Edinburgh.*—Rusks and biscuits.
807 Makepeace, S. *Merton, Surrey.*—Preserved herb mixtures, flavouring essences, &c.
808 Marshall & Son, *Tavistock-house, Covent-garden.*—Condiments, pickles, sauces, &c.
809 Marshall, T. W. 2, *Richmond-ter. Camberwell.*—Crystallized liqueurs and creams.
810 Martineau, D. & Sons, *London.*—Illustrations of sugar refining.
811 Moore, E. D. & Co. *Wood's-eaves, Newport, Salop.*—Concentrated milk: its combination with cocoa and chocolate. Concentrated wort.
812 Morton, J. T. 106, *Leadenhall-st.*—Preserved provisions and jams.
813 Myzoule, J. H. 72, *Southampton-st. N.*—Confectionery.
814 Nelson, Dale, & Co. *Bucklersbury.*—Isinglass, gelatine, and lozenges.
815 Parsons, Fletcher, & Co. *Bread-st.*—Corn flour, dietetic cocoas, &c.
816 Partridge, E. 22, *Leadenhall-st.*—Pickles, sauces, preserved fruits and meats.
817 Peek, Frean, & Co. *London.*—Biscuits made by steam power.
818 Phillips & Co. 8, *King William-st. City.*—Collection of teas.
819 Reckitt & Co. *Hull.*—Machine made fancy biscuits.
820 Robb, A. 79, *St. Martin's-lane, W.C.*—Infants' food, cakes, and biscuits.
821 Schooling & Co. 14, *Gt. Garden-st. Whitechapel.*—Confectionery in packets, &c.
822 Scott, W. L. *Westbourne-pk. W.*—Table showing chief articles of food and drink, and their adulterants.
823 Shackle, Maria, & R. W. 1, *Jeffery's-ter. Kentish-town-road, Camden-town.*—Ornamental confectionery.
824 Smith, Suttie, & Co. *Arbroath.*—Sweetmeats and confections.
825 Smith, G. & Co. 23, *Little Portland-st, W.*—Isinglass, gelatine, and jelly.
826 Spratt, J. 118, *Camden-road-villas, N.W.*—Cakes for cats, poultry, pigs, and dogs.
827 Stanes, J. 4, *Cullum-st. City.*—Coffee branches in various stages of growth.
828 Thomas, F. *Ealing-lane, Brentford.*—Flowers in sugar.

829 TURNER, G. & R. H. 111, *High-st. Borough.*—Wedding cakes.

830 VICKERS, J. 23, *Little Britain.*—Specimens of isinglass.

831 WARE, G. R. 11, *Marchmont-st. London, W.C.*—Confectionery.

832 WARRINER, G. *Aldershot.*—Preparations to facilitate cookery.

833 WEBSTER, J. M. 58, *Pall Mall.*—"Old English Sauce."

834 WESTON & WESTALL, 115, *Lower Thames-st.*—Refined salt.

835 WIGNALL, R. H. 98, *London-rd. Liverpool.*—Toffee and cocoa-nut ice.

836 WOOD, G. 15, *Commercial-st. Leeds.*—Wedding cakes, &c.

837 WOTHERSPOON, J. & Co. *Glasgow and London.*—Lozenges, comfits, marmalade, jams, &c.

838 WOTHERSPOON, R. & Co. *Glasgow and London.*—Victoria colourless lozenges, &c.

839 WRIGHT, F. *Kensington.*—Fruit essences for summer beverages.

---

840 SANSON, DU FAVILLE, & Co. *Broughton House, Islington, N.*—Crystal fish-gelatine.

841 GARTON & Co. *Bristol.*—Saccharum, or grape-sugar, and its products in brewing, wine making, and distilling.

Sub-Class C.—Wines, Spirits, Beer, and other Drinks, and Tobacco.

851 ARCHER, J. A. *Broadway, Westminster.*—Tobaccos.

852 BAKER, F. *Virginia Mills, Stockport.*—Manufactured tobacco and cigars.

853 BASS, RATCLIFF, & GRETTON, *Burton-on-Trent.*—Pale, Australian, and strong ales.

854 BIGGS, A. *Birmingham.*—Manufactured tobacco.

855 BOLLMANN, CONDY, & Co. 48, *Half-moon-st. Bishopsgate.*—Malt vinegar.

856 DYER, W. *Littlehampton.*—British champagne.

858 EVANS & STAFFORD, *Leicester.*—Cigars and tobacco; Leicester, Stilton, and Derby cheese.

859 FOWLER, J. & Co. *Prestonpans, N.B.*—Beer and India pale ale.

860 FRYER, D. *Epney, Stonehouse, Gloucestershire.*—Cyder and perry.

861 GARRETT, N. *Aldeburgh, Suffolk.*—Crystallized malt.

862 GOODES, G. & S. 51, *Newgate-st. E.C.*—Cigars, tobacco, and snuff.

863 HEATLEY, J. *Alnwick, Northumberland.*—Manufactured tobaccos.

864 HICKS, J. R. *East Bergholt, Suffolk.*—English wines.

865 HILTON, A. *Barnard Castle.*—Rum shrub.

866 HOOPER, W. 7, *Pall-mall East.*—Artificial mineral waters.

867 HUGGINS, E. S. 2, *Albert-st. Derby.*—Liqueur orange brandy.

868 HYAMS, M. *Bath-st. London.*—Illustrations of the manufacture of cigars, &c.

869 JONAS, E. BROS. 78, *High Holborn.*—Cigars and tobacco.

870 KENT, W. & S. & SONS, *Upton-on-Severn.*—Cordials, cyder, perry, spirits, and vinegar.

871 MART & Co. 130, *Oxford-st.*—Wines, preserved fruits, &c.

872 PITT & Co. 28, *Wharf-rd. City-rd. N.*—Aërated quinine water.

873 RICHARDSON, SANDERS, & Co. *Hope Brewery, near Notting-hill Gate.*—New description of beer.

874 SALT, T. & Co. *Burton-upon-Trent.*—Pale and Burton ales.

875 SHARMAN, A. *Walham-green.*—New beverage from fruit of the carob tree.

876 SILICATED CARBON FILTER Co.—*Bolingbroke-gardens, Battersea.* — Filtered liquors.

877 TAYLOR, HUMPHREY & Co. *Shawfield-st. Chelsea.*—English liqueurs and cordials, &c.

878 WALKER, A. & W. 3, *Peartree-st. E.C.*—British wines.

879 WILLS, W. D. & H. O. & SONS, *Bristol.*—Bird's-eye and other tobaccos.

CLASS 4.

ANIMAL AND VEGETABLE SUBSTANCES USED IN MANUFACTURES.

Eastern Annex, East Side.

Sub-Class A.—Oils, Fats, and Wax, and their Products.

910 BARCLAY & SON, 170, *Regent-st.*—Bleached wax, candles of various materials.

911 BAUWENS, F. L. *Oil Works*, 15, *St. Anne-st. Westminster.*—Candles, soap, and oils.

912 BRECKNELL, TURNER, & SONS, *Haymarket.*—Candles, soaps, oils, &c. Hand-candlesticks, with peculiar glass shades.

913 CANTRILL & CO. *King's-cross, N.*—Railway and other greases.

914 CATTELL, DR. *Euston-sq.*—Oils, fats, tracing cloths, anti-attrition compounds, &c.

915 CLARKE, S. 55, *Albany-st. N.W.*—Night lights and lamps.

916 COOK, E. & CO. *East London Soap Works, Bow.*—Soaps.

917 COWAN & SONS, *Hammersmith-bridge Works, Barnes.*—Soaps, and model of works.

918 FIELD, J. C. & J. *Upper Marsh, Lambeth.*—Paraffin and stearine candles, wax, &c.

919 GIBBS, D. & W. *City Soap Works, E.C.*—Specimens of hard, soft, and scented soaps.

920 GOSSAGE, W. & SONS, *Warrington.*—Soap and silicate of soda.

921 HALE, W. S. & SONS, 71, *Queen-st. London.*—Stearic acid. British sperm and composite candles.

923 AUSTIN, R. T. 95, *Shrubland-grove, Dalston.*—Flowers in wax.

924 KNIGHT, J. & SONS, *Soap and Candle Works, Old Gravel-lane, E.*—Primrose soap.

925 LAMBERT, ELIZABETH B. *Spring Villa, Tunbridge.*—A Kentish bank of flowers in July, modelled in wax.

926 LANGTON, BICKNELL, & SONS, *Newington Butts, S.*—Sperm oil and spermaceti, in various stages of manufacture.

927 LUMSDEN, ISABELLA, 8, *Trevor-ter. Rutland-gate.*—Bouquet of wax flowers, in frame.

928 MACKEAN, W. *St. Mirren's, Paisley.*—Soaps.

929 MAKEPEACE, ELIZA, *Merton, Surrey.*—Orchidæ in wax and anatomized leaves.

931 MEECH, H. J. 3, *North-pl. Kennington-rd. Lambeth.*—Wax figures.

934 MITTON, T. *Old-sq. Blackburn.*—Candles.

935 NEIGHBOUR, G. & SONS, 127, *Holborn.*—Machine oil, &c.

936 OGLEBY, C. & CO. 58, *Paradise-st. Lambeth.*—Refined spermaceti, paraffine, and stearic acid; with candles made of them.

937 MARTIN & PENFOLD, *Tenison-st. Lambeth.*—Tubular candles.

938 PIERSON, J. 66, *Mortimer-st. W.*—Flowers modelled in wax.

939 PRICE'S PATENT CANDLE CO. *Belmont, Vauxhall.*—Specimens showing improvements in making candles, oils, and glycerine.

940 RICH, W. 14, *Great Russell-st. Bloomsbury.*—Wax figures and flowers.

941 ROBIN & HOUSTON, *Paisley.*—Soap.

942 ROSE, W. A. *Upper Thames-st.*—Lubricating compounds, oils, &c.

943 ROWE, T. B. & CO. *Brentford, Middlx. W.*—Soaps.

944 SENTIS, J. *Abercorn-st. Paisley.*—Stearine, oil and soap from grease recovered from soap suds.

945 SHIPLEY, MISS J. 34, *Carter-st. Manchester.*—Wax flowers.

946 SYMONS, MRS. 9, *Devonshire-ter. Notting Hill-gate.*—Wax flowers.

947 TAYLOR, W. & CO. *Leith.*—Soaps, stearine, and candles.

948 TREWOLLA, MRS. R. *Halesowen.*—Group of wax flowers.

949 TUCKER, F. & CO. *Kensington.*—Wax, stearine, and bleached tallow candles. Stearine and bleached tallow.

950 WEST OF ENGLAND SOAP CO. *Plymouth.*—Soaps, paraffin, and composite candles.

951 WILKINS, PRISCILLA, 49, *St. Paul's-rd. Kennington, S.*—Wax flowers and fruit.

952 WILLES, MARGARET H. *Marshside, Lower Edmonton.*—Wax flowers, ornamental leather flowers, &c.

953 WILLIAMS, J. & SON, *Clerkenwell.*—Soaps, with illustrative processes.

---

954 ROGERS, E. S. & CO. *Manchester.*—Oils, from vegetable substances, and fatty matters; and grease used in manufacturing and locomotion.

Sub-Class B.—Other Animal Substances used in Manufactures.

965 AZÈMAR, J. C. *The Waldrons, Croydon.*—Ivory turnings by an amateur.

966 BARNES, S. & T. 3, *Shouldham-st. W.*—Ivory, wood, and bone hair, tooth, and nail-brushes.

967 BARRY, BROS. *Meriton's Wharf, S.E.*—English sheep skins.

968 BERENDT & LEVY, *Leeds.*—Low wools.

969 BERTHOLD & PHILLIPS. 31, *Gloucester-ter New-rd. Commercial-rd. East.*—Tortoise-shell combs.

970 BILLINGTON, MISSES, *Lord-st. Southport.*—Group of shell-flowers.

971 BUXTON, W. *Limetree-lodge. Rotherhithe.*—Wools grown in the United Kingdom.

972 CANTOR & Co. 6, *Houndsditch, N.E.*—Turkey sponges.

973 COPE, R. & SONS, *Uttoxeter.*—Cabinet-maker's glue.

974 COX, J. & G. *Gorgie Mills, Edinburgh.*—Gelatine and glue.

975 DARNEY, J. & SONS, *Kinghorn, Scotland; and Drury-lane, W.C.*—Scotch glue, sizing, &c.

976 DOBSON, J. *Joseph-st. Bradford.*—Combs.

977 DORRIEN, C. *Ashdean, near Chichester.*—Sussex merino wools.

978 DUTTON, T. R. 19, *Holywell-row, Shoreditch.*—Wood and ivory carvings and turnings.

979 FENTUM, M. 85, *New Bond-st. W.*—Works in ivory and hardwoods.

980 FISHER, W. & SONS, *Orchard-pl. Sheffield.*—Umbrella, matchet, and knife handles of pressed horn.

981 FOX, T. B. 37, *St. John-st. Devizes.*—Fleeces.

982 GLASS, G. M. *Brandon-st. Walworth.*—Gelatine.

983 GREEN, J. 7, *Sherborne-st. Islington.*—Sheet gelatine for tracing, wrappers for confectioners, &c.

985 HASTILOW, C. 3, *Queen-st. E.C.*—Chessmen, ivory balls, and fancy goods.

986 HEINRICH, J. *Lower Kennington-lane.*—Tortoiseshell combs.

987 HITCH, M. *Evesham, Worcestershire.*—Imitation tortoiseshell combs.

988 JACOB, B. 68, *Leadenhall-st.*—Shells and shellwork.

989 JAQUES, J. & SON, 102, *Hatton-garden.*—Fancy ivory goods.

990 JEWESBURY, H. W. & Co. 1 & 2, *Mincing-la. E.C.*—Varieties of cochineal.

991 JOHNSON, P. *Wigan.*—Specimens of concentric turning.

992 JOWITT & SONS, *Leeds.*—Wools.

993 LAMMLER, G. 2, *South-st. Finsbury-market.*—Meershaum pipes and amber mouth pieces.

994 LUBLINSKI, R. 185, *City-rd.*—Umbrella and parasol handles, &c.

995 MANNINGS, G. *Wedhampton, near Devizes.*—Fleeces of wool from Wilts.

996 MARLBOROUGH, DUKE OF, *Blenheim, Oxon.*—Oxfordshire Down wool, and blankets manufactured therefrom.

997 MASON, G. *Yateley, Hants.*—British silk and flax.

998 MILLER, H. *Bury St. Edmunds.*—Spiral turning by a patent lathe.

999 MOORE, W. S. 47, *Percival-st. E.C.*—Ivory, wood, and bone, hair, tooth and nail brushes.

1000 NIMMO, T. & Co. *Rivald's-green Works, Linlithgow, N.B.*—Glues and gelatine.

1001 NUPPNAU, E. 27, *Norfolk-st. Strand.*—Vases, cups, &c. turned in ivory.

1002 OLLEY, T. G. 98, *Bolsover-st. London, W.*—Specimens of ornamental turnery.

1003 PLAYNE, C. *Nailsworth, Stroud.*—Turnings in ivory.

1004 PROCKTER & BEVINGTON, 124, *Grange-rd. Bermondsey.*—Glues.

1005 PUCKRIDGE, F. 56 & 57, *Kingsland-rd.*—Goldbeaters' skin.

1006 RICHARDSON, E. & J. *Newcastle-on-Tyne.*—Glues and gelatines.

1007 ROYAL AGRICULTURAL SOCIETY OF ENGLAND, 12, *Hanover-sq. W.*—Wool.

1008 RYLEY, E. C. *Great Prescot-st. E.*—Specimens of amateur turnings in turnery and hard wood.

1009 SALOMONS, A. *Old Change, E.C.*—Turned ivory.

1010 SAMUEL, M. 7, *East Smithfield.*—Shells, matting, canes, &c.

1011 SANDS, T. C. *Mortimer-st. Leeds.*—Wool cleaned by machinery.

1012 SASSÉ, P. C. 53, *Wynyatt-st. Clerkenwell.*—Looking-glasses, paper-knives, card-cases, chessmen, &c. in ivory.

1013 SISSON, J. & SON, *Kendal.*—Various comb manufactures.

1015 STEWART, R. S. & Co. *Aberdeen Comb Works, and London.*—Combs.

1016 TUCKER, E. & Co. *Belfast, Ireland.*—Starch and glue.

1017 TUCKER, H. *Fleet-lane, Farringdon-st. E.C.*—Gold-beaters' moulds; and skin, for scientific and other purposes.

1019 WRIGHT, F. *Needham Market, Suffolk.*—Glues made from hides and skins.

1020 YOUNG, B. & Co. *Spa-rd. Bermondsey.*—Size, glue, and gelatine.

1021 REBOW, J. GURDON, *Wyvenhoe-park, Colchester.*—Sheep's wool.

1022 PEEL, J. *Pudsey.*—Medallion and spiral turnings in wood and marble.

Sub-Class C.—Vegetable Substances used in Manufactures, &c.

1033 ADAMSON, R. Gardener, *Balcarres, Fifeshire.*—Baskets for fruits and cut flowers.

1034 AGAVA PATENT HAIR CO. *Newlay, near Leeds.*—Fibre from aloe plant, a substitute for horsehair.

1035 ALDRED, T. 126, *Oxford-st.*—Bows, arrows, fishing-rods, and tackle.

1036 ALLEN, MATILDA, 17, *Percy-st. Bedford-sq.*—Models of plants, showing the blossoms, seed vessels, &c.

1037 ANDERSON, R. *Dunkeld, Perthshire.*—Salmon and trout flies.

1038 BAILEY, J. *King's Cliffe, Northamptonshire.*—Butter prints, taps, spoons, spice-boxes, &c.

1039 BAZIN, G. 9, *Denmark-pl. Wells-st. Hackney.*—Taper swan-floats, &c.

1040 BELOE, W. L. *Home-pl. Coldstream, Berwickshire.*—Fishing-tackle, &c.

1041 BERNARD, J. 4, *Church-pl. Piccadilly.*—Fishing-rods, tackle, flies, &c.

1042 BLACHE & Co. 21, *Wilson-st. Finsbury-sq.*—Knife-cut veneers from various woods.

1044 BOLLANS, W. *King's Cliffe, Northamptonshire.*—Wood turnings and carvings.

1045 BURLEY R. & Co. *Glasgow.*—Pt. steel core handles for hammers, &c.

1046 CAMP, W. 81, *Tottenham Court-road.*

—Arm clubs, skittles, and other specimens of turning.

1047 CHEVALIER, BOWNESS & SON, 12, *Bell-yd. Temple-bar.*—Fishing-rods and tackle.

1048 CLARK, G. F. H. & Co. *Camomile-st. E.C.*—Resinous gums for varnish and hat manufacturing.

1049 CLARK & Co. 79, *Cannon-st. Westminster.*—India-rubber fabrics and felt.

1050 CLARKE, J. R. 26, *Trafalgar-st. Walworth.*—Mosaic Tunbridge ware.

1051 CLARKSON, T. C. 56, *Stamford-st. Blackfriars.*—Articles made in cork.

1052 CLEMENCE, H. 55, *Upper Stamford-st. S.*—Corks.

1053 COHEN, C. 18, *Bury-st. City.*—Sticks and canes for umbrellas, &c.

1054 COLES, W. F. 52, *Aldermanbury.*—Patent cork linings, and cork carpeting.

1055 COLLYER, R. H., M.D. *Alpha-rd. N.W.* Paper materials, raw to completed states, with machinery.

1056 COSSENS, E. J. 15, *Little Queen-st. Holborn.*—Figure with basket, carved in elder pith.

1057 COSSER, R. 13, *Stucley-ter. Hampstead-rd. N.*—Fancy baskets.

1058 COTTON SUPPLY ASSOCIATION, *Manchester.*—Cotton samples and cotton tree.

1059 COW, P. B. & Co. 46, *Cheapside, London.*—India-rubber waterproof fabrics, and vulcanized India-rubber goods.

1060 DAHMEN, MARSHALL, & Co. *Ditton Works, Thames Ditton.*—Fibre and vegetable substances connected with textile fabrics and paper.

1061 DANKS, J. 56½, *Webber-row, S.*—New invented door mats, made of cocoa-nut fibre and wool.

1062 DEED, J. S. & SONS, 451, *Oxford-st.*—Cocoa mats and matting.

1064 DUFFIELD, J. 12, *Gt. Chapel-st. Oxford-st.*—Dairy utensils, moulds, &c.

1066 EVERARD, H. W. *Union Mills, Manchester.*—Vulcanized india-rubber brace, surgical and other webs, &c.

1067 FARLOW, C. 191, *Strand.*—Fishing rods and tackle.

1068 FARRANT, R. E. 16, *Queen's-row, Buckingham-gate.*—Carved plates and potatoe bowls.

1069 FAUNTLEROY, R. & Co. *Bunhill-row, E.C.*—Foreign hardwoods, dyewoods, fancy-woods, &c.

1070 FAUNTLEROY, R. & SONS. *Potter's-fields, Tooley-st.*—Foreign hardwoods and ivory and mother-o'-pearl shells.

1071 FORSTER, T. *Streatham, Surrey, S.*—Articles in vulcanite (ebonite), made from India-rubber vulcanized waste.

1073 GIEHR, R. 4, *George's-row, City-row.*—Chairs and fancy baskets.

1075 GOUGH & BOYCE, 12, *Bush-lane, London, E.C.*—Kamptulicon or elastic floor-cloth.

1076 GOULD, A. 268, *Oxford-st.*—Fishing-rods, eel-traps, &c.

1077 GOWLAND & Co. 3, *Crooked-lane, London-bridge.*—Every description of fishing tackle.

1078 GUTTA PERCHA Co. *Wharf-rd. City-rd.*—Articles in gutta percha; insulated telegraph wires.

1079 HANCOCK, J. L. 266, *Goswell-st. E.C.*—Manufactures in vulcanized India-rubber.

1080 HAWES, J. 7. *Adelphi-ter.*—Preserved natural flowers.

1081 HEEKS, M. H. 61, *White Lion-st. Pentonville.*—Wicker baskets and balloon-car.

1082 HEINRICH, J. 36, *Lower Kennington-lane, S.*—Tortoiseshell combs.

1083 HINKS, J. 64, *George-st. Birmingham.*—Wood turnings.

1084 HODGES, R. E. 44, *Southampton-row, Russell-sq.*—Applications of India-rubber accumulators, or springs. Measuring instruments.

1085 HOLLINGSWORTH & WILLOUGHBY, 3, *Wenlock-rd. N.*—Veneers cut by knife machinery.

1086 HOOPER, W. 7, *Pall Mall East.*—Vulcanite and vulcanized India-rubber goods.

1087 HORSEY, J. 36A, *Belvidere-rd. S.*—Articles in India-rubber for personal use, &c.

1088 HOWARD, J. *Luton, Beds.*—Blocks for ladies' hats and bonnets.

1089 HYAMS, M. *Bath-st. London.*—Thistle-down, as substitute for silk, &c.

1090 HYDE, E. *Kingston-on-Thames.*—Cocoa-nut fibre brushes and mats.

1091 JAMES, J. JUN. 1, *Cleveland-ter. Bath.*—Models of basket-work.

1092 JONES & Co. 111, *Jermyn-st. S.W.*—Fishing rods, reels, lines, flies, &c.

1093 KING, F. 56, *Wells-st. Oxford-st.*—Brooms and brushes manufactured from bass.

1094 KOLLE, H. & SON, *Glemsford, Suffolk, and Queen-st. Cheapside, London.*—Cocoa nut fibre manufactures.

1095 LATARCHE, P. 18, *Coldbath-sq. Clerkenwell.*—Wickered flasks and baskets.

1096 LEATHER CLOTH Co. 56, *Cannon-st. West.*—Leather cloth.

1097 LEE, T. 33, *Old-st. London, E.C.*—Life-preserving swimming vest.

1098 LENTON, R. 7, *Bartholomew-st. Exeter.*—Wicker flower stands and bird-cages.

1099 LUDBROOK, S. *Bancroft-pl. Mile End.*—Dressed piassava or bass. Brooms and brushes made of the same.

1100 MACKAY, A. 107, *High-st. Edinburgh.*—Wicker work articles.

1101 MACINTOSH, C. & Co. *Cannon-st. City.*—India rubber manufactures.

1102 MACNEILL, F. & Co. *Bunhill-row, E.C.*—Asphalted roofing, kamptulicon floor-cloth, and other goods.

1103 MADDEN, SUSANNA, 56, *Long-lane, E.C.*—Skittles and balls.

1104 MASON, G. ESQ. *Yately, Hants.*—Specimens of flax and silk cultivated at Yately, Hants.

1105 MEYERS, B. *Mill-lane, Tooley-st.*—Canes, sticks, whips, &c.

1106 MORLEY, J. 12, *Carrington-st. Nottingham.*—Artificial salmon and trout flies.

1107 MORRIS, C. 4, *Mountnod-sq. Lewisham.*—Combs, tools, and fancy baskets.

1108 NOBLE, G. & J. A. 4, *George-yard, Lombard-st.*—Textile fibres.

1109 NORTH BRITISH RUBBER Co. *Edinburgh.*—India-rubber manufactures.

1110 OLIVER, W. & SONS, 120, *Bunhill-row, E.C.*—Fine mahogany and rare foreign woods.

1111 PACKER, R. L. 38, *Union-st. Lambeth-walk, S.* — Glove-stretchers and powder-boxes.

1112 PARKES, A. *Birmingham.*—"Parkesine," of all colours.

1113 PEACH, J. & SONS, *Derby.*—Silk lines and salmon line.

1114 PETERS, W. & SON, 71, *Long-acre.*—Fishing-tackle.

1115 PILLINER, S. A. 4, *Hatfield-pl. Blackfriars.*—Anatomized leaves.

1117 RAYNBIRD, H. *Basingstoke.*—Timber, bark, hoops, &c.

1118 RECKITT & Co. *Eureka Works, Hulme, Manchester.*—American leather-cloth, &c.

1119 ROBERTSON, A. *Holloway Mills, N.*—Barrel package of wood, as substitute for canisters.

1120 ROUTLEDGE, T. *Eynsham-mills, Oxford.*—Esparto, or alfa, and half-stuff for paper manufacture.

1121 SCOTTISH VULCANITE Co. *Edinburgh.*—Vulcanite combs, &c.

1122 SCOTT, W. L. *Westbourne-park, W.*—Specimens of cotton, in "fasciculæ," showing length of staple.

1123 SEITHEN, A. B. 1, *Wharf-rd. City-rd.*—Corks.

1124 SHEPHERD, BRIGGS, & Co. *Portobello Mills, Wakefield.*—Cocoa fibre mats, &c.

1125 SILVER, S. W. & Co. 66 & 67, *Cornhill.*—Articles in Indiarubber and ebonite.

1126 SIMMONDS, P. L. 8, *Winchester-st. Pimlico.*—Nuts, seeds, fibres, &c., scientifically named and their applications.

1127 SKILBECK, J. *Upper Thames-st.* — Woods and articles used in dyeing.

1128 SMEE, W. & SONS, 6, *Finsbury-pavement.*—Woods used for furniture.

1129 SMITH, T. & SONS, *Herstmonceux, Sussex.*—Basket manufactures.

1130 SMITH, W. & A. *Mauchline, Ayrshire.*—Scottish fancy wood work.

1131 SPILL, G. & Co. 149, *Cheapside.*—Vegetable leather, leather cloths, waterproof fabrics.

1133 STEVENS, M. *Royal Mews, Pimlico.*—Anatomized leaves.

1134 STEVENS, W. 2, *Rock-pl. Tottenham-road, Kingsland.*—Preserved natural flowers.

1135 SWAAB, S. L. 9, *Hunter-st. Brunswick-sq.*—Prepared India fibres, &c.

1136 TAYLER, HARRY, & Co. 19, *Gutter-lane, Cheapside.* — Kamptulicon for floors, knife-boards, &c.

1137 TAYLOR, B. 169, *St. John-st.-road.*—Vegetable ivory turnings.

1138 TOPLIS, T. & J. *Ashby-de-la-Zouch.*—Flower stands, work-baskets, &c.

1139 TRELOAR, T. 42, *Ludgate-hill, E.C.*—Brushes, cocoa-nut fibre articles.

1140 TRESTRAIL, F. G. & Co. 19 & 20, *Walbrook.*—Kamptulicon, or india-rubber and cork floor cloth.

1141 TUCK, T. H. & Co. 35, *Cannon-st. E.C.*—Elastic packing and rubber manufactures for mechanical purposes.

1142 TURNBULL, T. *William-st. Portland Town.*—Specimens of sawn wood.

1143 WALDEN, S. J. *Whitefriars, E.C.*—Articles in wicker work.

1144 WALKER & STEMBRIDGE, *Ducksfoot-lane, London.*—Gums and gum-resins.

1145 WANSBROUGH, J. *Guildford-st. S.*—Waterproof flocked cloth and india-rubber goods.

1146 WARNE, W. & Co. 9, *Gresham-st. west, E.C.*—Manufactures of india-rubber.

1147 WELLS & HALL, 60, *Aldermanbury.*—Elastic braids and fabrics.

1148 WEST HAM GUTTA-PERCHA Co. 18, *West-st. Smithfield.* — Gutta-percha, gutta-rubber, telegraph wire.

1149 WHITEHEAD, T. 37, *Eastcheap, E.C.*—Straw envelopes for packing bottles.

1150 WILDEY & Co. 7, *Holland-st. Blackfriars-road.*—Mats, matting, &c. of cocoa-nut fibre.

1151 WILSON, A. & G. 19, *Waterloo-place, Edinburgh.*—Fishing tackle.

1152 WRIGHT, C. 376, *Strand, W.C.*—Fishing-rod, tackle, and archery.

1153 WRIGHT, J. *Kelso, Scotland.*—Artificial flies and casting lines.

1154 CRESSY, T. S. *Burton-on-Trent.* — Corks made by machinery.

1155 HERMANN, A. 4, *Oxenden-st. Coventry-st.*—Landscapes in cork.

1156 WATTON, F. & Co. *British Grove Works, Chiswick.*—Camplicon, or India rubber substitute, made from oxidized oils.

Sub-Class D.—Perfumery.

1163 ATKINSON, J. & E. 24, *Old Bond-st.*—Perfumery and toilet articles.

1164 BAYLEY & Co. 17, *Cockspur-st.*—Perfumery, toilet articles, distilled waters, &c.

1165 BENBOW & SON, 12, *Little Britain, E.C.*—Perfumery and toilet articles.

1166 BONUS, WM. E. 9, *Charles-st. Manchester-sq.*—Essential oils, &c.

1167 BREIDENBACH, F. H. 157B, *New Bond-st.*—Perfumery.

1168 CLEAVER, F. S. 32 & 33, *Red Lion-st. Holborn, W.C.* — Fancy soap and perfumery.

1169 CONDY BROS. & Co. 15, *Garlick-hill, London, E.C.*—Oils and extracts, fruit essences, &c.

1170 DELCROIX & SON, 39, *Great Castle-st. W.*—Perfumery.

1171 EDE, R. B. & Co. 21, *Bow-la. E.C.*—Perfumery and domestic requisites.

1172 EWEN, J. 17, *Garlick-hill, London, E.C.*—Clarified fats.

1173 GOSNELL, J. & Co. 12, *Three King-ct. Lombard-st.*—Perfumery, soaps, hair-brushes, &c.

1174 HIRST, BROOKE, & TOMLINSON, *Leeds.*—Perfumed toilet soap, and perfumery.

Eastern Annex, East Side.

1175 KEITH, G. 55, *Gt. Russell-st. Bloomsbury.*—Perfumery.

1176 LANGDALE, E. F. 72, *Hatton-garden, E.C.*—Essential oils.

1177 LEWIS, J. 6, *Bartlett's-buildings, E.C.* —Perfumes extracted by cold process. Toilet and iodine soaps, oil, &c.

1178 LLOYD, W. A. 19, *Portland-rd. Regents-pk. W.*—Aquarium.

1179 LOW, R. SON, & CO. 330, *Strand.*—Perfumery and toilet articles.

1180 MOREAU, J. 88, *Regent-st. W.*—Perfumery.

1181 PEARS. A. & F. 91, *Great Russell-st. Bloomsbury.*—Transparent soap.

1182 PERKS, S. *Hitchin, Herts.*—Essential oil of lavender, &c.

1184 PIESSE & LUBIN, 2, *New Bond-st.*—Frangipani, violet, patchouly, and other perfumes.

1187 RIMMEL, E. 96, *Strand, London.*—Perfumery, perfumery materials, toilet soaps, and perfume vaporizer.

1188 ROBSON, J. M. 32, *Laurence-lane, Cheapside.*—Fancy soaps and perfumery.

1189 SAUNDERS, J. T. 148, *Oxford-st.*—Perfumery.

1190 THOMPSON, J. 6, *King-st. Holborn, W.C.*—Toilet soaps and distilled perfumes.

1191 VICKERS, S. 13, *Boat-lane, Leeds.*—Perfumery, &c.

1192 WARRICK BROS. *Garlick-hill, London.* —Essential oils, perfumes, pomades, &c.

1193 WHARRY, J. *Chippenham, Wilts.*—Lavender water.

1194 WHITAKER & GROSSMITH, 120, *Fore-st. Cripplegate.* — Perfumery and toilet soaps.

1195 YARDLEY & STATHAM, 7, *Vine-st. Bloomsbury.*—Fancy soaps and perfumery.

SECTION II.

CLASS 5.

RAILWAY PLANT, INCLUDING LOCOMOTIVE ENGINES AND CARRIAGES.

Western and Eastern Annexes.

1227 ADAMS, W. B. *Holly-mount, London.*—Wheels, springs, and rail joints.

1228 ALLAN, A. *Perth.* — Straight link valve motion; pressure gauges; and compound buffer.

1229 ANDERSTON FOUNDRY CO. *Glasgow.*—Permanent-way materials.

1230 ARMSTONG, SIR W. G. & CO. *Elswick Engine Works, Newcastle-upon-Tyne.*—Locomotive engine and tender.

1231 ASHBURY, J. *Openshaw, Manchester.*—Saloon carriage, goods waggon, wheels and axles, axles, tyres, and bar iron. (E.A.)

1232 AYTOUN, R. 3, *Fettes-row, Edinbro'.*—A railway brake.

1233 BAIN, MCNICOL, & YOUNG, *Edinbro'.*— Simultaneous-acting level-crossing gates. (E.A.)

1234 BAINES, W. & CO. *London Works, Smethwick, Birmingham.*—Switches, crossings, turn-tables (model of), and girders.

1236 BATESON, S. S. 17, *Bolton-st. London, W.*—Feed-water heating apparatus.

1237 BAYLISS, SIMPSON, & JONES. 43, *Fish-st.-hill.*—Iron hurdles, fencing, cable chains, anchors, &c.

1238 BEYER, PEACOCK, & CO. *Gorton Foundry, Manchester.* — Locomotive-engine and tender.

1239 BIDDELL, G. A. *Ipswich.*—Chilled railway crossings.

1240 BROWN, G. & I. & CO. *Rotherham Iron Works.*—Solid iron tyre, and steel-faced ditto. (E.A.)

1241 BUTTERLEY IRON CO. *Derby.*—Rail.

1242 CLARK, G. 30, *Craven-st. Strand.*—Gas-signals for railways, light-houses, &c.

1243 COPLING, J. *Hackney.*—Railway-signals, and guards' communication with passengers, &c.

1244 CORLETT, H. L. *Inchicore, Dublin.*—Rails, brackets, joint chairs, buffing springs, &c. (E.A.)

1245 DAVIDSON, J. *Leek, Staffordshire.*—System of communication on railway trains.

1246 DERING, G. E. *Lockleys, Welwyn, Herts.*—Permanent-way.

1247 DIXON & CLAYTON, *Bradford.*—Rolled spoke-iron, railway wheels, and tyre fasteners. (E.A.)

1248 DUNN, T. & CO. *Windsor Bridge Iron Works, Manchester.*—Turn-tables, traversers, engines, and hydraulic machinery. (W. & E.A.)

1249 EDINGTON, T. & SONS, *Phœnix Iron Works, Glasgow.*—Railway chairs and sleepers.

1250 ENGLAND, G. & CO. *Hatcham Iron Works, London, S.E.*—Locomotive engine with tender. Traversing screw-jack, &c. (W. & E.A.)

1251 FAIRBAIRN, W. & SONS, *Manchester.*—Locomotive engine.

1252 FAY, C. *Manchester.*—Self-adjusting continuous railway carriage-brake, which wears out the block without any regulating. (E.A.)

1253 GARDNER, S. *Neath.*—Axle-box and truck-buffer. (E.A.)

1254 GLOUCESTER WAGGON CO. *Gloucester.*—Waggon for discharging coal into ships. (E.A.)

1255 GOVERNOR & CO. OF COPPER MINES IN ENGLAND, 10, *New Broad-st.-mews, London.*—Rail.

1256 GRANT, W. 6, *Alice-st. Liverpool.*—Mirrors and signals in railway trains.

1258 HATTERSLEY, W. 135, *St. George-st. E.*—Passengers signal for railway carriages.

1259 HENSON, W. F. 15, *New Cavendish-st. Portland-pl. London, W.*—Railway buffer and bearing springs. (E.A.)

1260 HOY, J. 6, *Pickering-pl. W.*—Railway signal.

1261 HUGHES, H. *Falcon Works, Loughborough.*—Models of plant used by railway contractors. (E.A.)

1262 ISCA FOUNDRY CO. *Newport, Monmouthshire.*—Switches, crossings, chairs, axle-boxes, wheels, &c.

1263 KINGSTON, W. H. 68, *Upper Stamford-st. S.*—Means of verbal communication on railway trains.

1264 KITCHIN, R. *Warrington.*—Weighing machinery, cranes, &c.

1267 LITTLE, C. *Bradford.*—Safety coupling for railway waggons. (E.A.)

1268 LLOYDS, FOSTERS, & CO. *Old Park Iron Works, Wednesbury.*—Wheels, axles, turn-tables, cranes, tyres, and iron. (E.A.)

1269 LONDON AND NORTH WESTERN RAILWAY, *Works, Crewe and Wolverton.*—Locomotive engines and tenders.

1270 ——— Apparatus for supplying water to tenders whilst in motion.

1272 M'CONNELL, J. *West Houghton, Bolton-le-Moors.*—Self-acting railway signal for day and night.

1273 MACINTOSH C. & CO. 3, *Cannon-st. London, and Cambridge-st. Manchester.*—Vulcanised rubber, buffers, springs, &c. (E.A.)

1274 MANNING, WARDLE, & CO. *Boyne Engine Works, Hunslet, Leeds.* — Colliery locomotive engine.

Eastern and Western Annexes.

1275 MORRIS, E. 8, *Albert-sq. Clapham-rd. S.*—Iron wedge for railway chairs.

1276 MOULTON & Co. *Bradford, Wilts.*—Buffers.

1277 MURPHY, J. *Railway Works, Newport, Monmouthshire.*—Railway wheels; safety bolts and nuts. (E.A.)

1278 NEATH ABBEY IRON CO. *Neath.*—Colliery locomotive engine, &c.

1279 NEILD & Co. *Dallam Iron Works, Warrington, Lancashire.* — Railway wheels, axles, tyres, and bar iron, &c. (E.A.)

1280 NEILSON & Co. *Hyde-park Locomotive Works, Glasgow.*—Eight feet wheel express engine.

1281 NETHERSOLE, W. E. *Swansea.*—Model of the frame of a railway-waggon, showing exhibitor's side-chain arrangement. (E.A.)

1282 NETHERSOLE, W. E. *Swansea.*—Model of improvements in draw gear and end tipping waggon flaps.

1283 NEWALL, J. *Bury, Lancashire.*—Continuous railway breaks, signal and gas apparatus for lighting trains.

1284 ORDISH & LE FEUVRE, 18, *Gt. George-st. S.W.*—Railway chairs and sleepers.

1285 OWEN, W. *Phœnix Works, Rotherham.*—Wrought iron wheels, axles, and solid tyres. (E.A.)

1286 PARSONS, P. M. 9, *Arthur-st. West, London Bridge.*—Railway switch; samples of white brass, for railway purposes, &c.

1287 PATENT SHAFT AND AXLETREE Co. *Brunswick Iron Works, Wednesbury.*—Wheels, axles, tyres, &c. (E.A.)

1288 PERMANENT WAY Co. 26, *Gt. George-st. Westminster.*—Rail joints and preserved timber for sleepers.

1289 PERRY, H. J. Jun. 3, *Greenwich-rd. Greenwich.*—Working model of atmospheric railway. (E.A.)

1290 POOLEY, H. & SON, *Liverpool.*—Railway, commercial, and mining weighing apparatus.

1291 RANSOMES & SIMS, *Ipswich.*—Station pumping-engine and boiler, compressed railway fastenings, &c.

1292 REAY & USHER, *South Hylton Forge, Sunderland.*—Locomotive crank axle of cast-steel, &c. (E.A.)

1294 RICHARDSON G. & CHATTAWAY, E. 1, *New Broad-st.*—Communication between guard and engine-man; railway-break.

1295 RICHARDSON, R. 26, *Gt. George-st. Westminster.*—Railway switches, and rail fastenings.

1296 SCOTT, S. T. 23, *Charterhouse-st. E.C.*—Models of safety couplings for railway-carriages. (E.A.)

1298 SEATON, W. 44, *Albemarle-st.*—Safety saddle-rail.

1299 SHARP, STEWART, & Co. *Atlas Works, Manchester.*—Coal-burning goods-engine.

1300 SIMONS, W. & Co. *London Works, Renfrew.*—Railway-chairs, sleepers, &c.

1301 SPENCER & SONS, *Newcastle-on-Tyne.*—Cast-steel tyres, volute spring-buffers, springs, &c. (E.A.)

1302 STEVENS & SON, *Darlington Works, Southwark.*—Semaphore signals.

1303 STAFFORD, CAPT. P. P. *St. James-sq.*—Self-acting railway signal, for preventing accidents in tunnels, &c.

1304 TIZARD, W. L. 12, *Mark-lane.*—Washer for railway fish-plates, &c.

1305 TRUSS, T. S., C.E. 53, *Gracechurch-st. E.C.*—Railway chairs and packing. (E.A.)

1306 VICKERS, A. *Bristol.*—Method of opening, shutting, and fastening, four gates simultaneously, applicable to railway crossings. (E.A.)

1307 WALKER, W. 3, *Atholl-lane, Edinbro'.*—Invoice-box and ticket-keeper.

1308 WESTON & GRICE, *Stour Valley Iron Works, West Bromwich.*—Bar-iron and railway fastenings.

1309 WISE, F. 22, *Buckingham-st. Adelphi.*—Ramie's railway chairs, without wedge or bolt.

1310 WRIGHT, J. & SONS, *Saltley Works, Birmingham.*—First-class carriage, for the Egyptian railway. (E.A.)

1311 WRIGHT, P. *Constitution Hill Works, Dudley.*—Railway-wheels and axles. (E.A.)

1312 WRIGHT, T. *George-yard, Lombard-st.*—Permanent way.

1314 LEIGH, W. *Golden-ter. Richmond-rd. Dalston.*—Indicators, to show the names of stations inside carriage.

1315 SHAW, H. *Dublin.*—Railway-brake, by which a train may be brought to rest in 60 yards.

1316 SINCLAIR, R. *Stratford, E.*—A locomotive engine, which has run 44950 miles, without repairs.

1317 ————Carriage wheels and roller axle box; model of a truss bridge.

1318 BRADLEY, W. & Co. *Sheffield.*—Railway buffer, and draw springs; fish plate, with square bolt and expansion.

CLASS 6.

CARRIAGES NOT CONNECTED WITH RAIL OR TRAM ROADS.

South-East Court.

1338 ALDEBERT, I. 57, *Long-acre.* — A barouche-landau, constructed with steel instead of iron.

1339 ANDREWS, A. 14, *Above-bar-st. Southampton.*—"Eugenie" park phaeton.

1340 ANGUS, H. *Westgate-st. Newcastle-on-Tyne.* — Double-seated brougham, with improved break.

1341 BENNION & HEALEY, 16, *Berry-st. Liverpool.*—Brougham, weighing 8 cwt.

1342 BLACK, H. & SON, *Berners'-st. Oxford-st.*—A light C under-spring coach.

1343 BOOKER & SONS, 14, *Mount-st. Grosvenor-sq.*—A "sociable."

1344 BOYALL, R. J. *Grantham Carriage Manufactory.*—Park phæton on inverted double C springs.

1345 BRABY, J. & SON, *Newington-causeway, Southwark.*—A spring-waggon, with improved wheels and break.

1346 BRIGGS, G. & Co. 45, *Wigmore-st. W.* —A carriage.

1347 BURNETT, E. *High-st. Ashford, Kent.* —Cart, to form either cart or sleigh.

1348 BURTON, H. L. 12, *Nowell's-bdgs. Islington, N.* — Perambulators and propellers.

1349 CAMPBELL, F. 33, *English-st. Dumfries.*—Scotch elm varnished sporting-cart.

1350 CAMPBELL, R. F. 8, *Brook-st. Hyde-pk.*—Model of a plan for prevention of accident to carters, &c.

1351 CASE, C. J. 36, *Jamaica-st. E.*—Model of an omnibus, made of brass and steel.

1352 CHANTLER, J. D. *St. Mary-st. Manchester.*—Light four-wheeled carriage.

1353 CLARKE BROS. *Shiffnal, Salop.*—Tubular iron carriage-shafts.

1354 COCKSHOOT, J. *New Bridge-st. Manchester.*—Extremely light brougham.

1355 COLE, W. *Craven-pl. Kensington.*—Brougham, C and under springs.

1356 COOK & HOLDWAY, 12, *Mount-st. Grosvenor-sq.*—Sociable landau, with registered head.

1357 COOPER, BLACKFORD, & SON, 140, *Long-acre.*—Carriage laces and trimmings.

1358 CORBEN & SONS, 30, *Gt. Queen-st. W.C.*—" A Diorопha."

1359 COUSINS, E. *Alfred-st. High-st. Oxford.*—A pony carriage.

1360 CROSS, T. W. & CO. *Hunslet-rd. Leeds.*—Bath chair and perambulators.

1361 DART & SON, 12, *Bedford-st. Covent-garden.*—Coach lace.

1362 DAVIES & SON, 15, *Wigmore-st.*—Landau with concealed self-acting steps opening with the doors.

1363 DAVIES & SONS, *Abingdon-st. Northampton.*—A trotting phaeton.

1364 EDWARDS, SON, & CHAMBERLAYNE, 21, *Newman-st. W.*—A light four-wheel carriage.

1365 ELL G. & CO. 366, *Euston-road, London.*—Van, and models of heavy vehicles.

1366 EVANS, J. 32, *Tarlton-st. Liverpool.*—An improved Hansom cab.

1367 FELTON, W. J. & C. 2, *Halkin-pl. Belgrave-sq.*—New brougham.

1368 FINDLATER, W. *Gas-st. Birmingham.* —Light brougham.

1369 FULLER, J. & SONS, *College-st. Bristol.*—Stanhope phæton waggonette with screw break.

1370 FULLER, S. & A. *Kingsmead-st. Bath.* —Brougham under seven cwt.

1371 GITTINS, R. 28, *New-st. Dorset-sq.*—New carriage axletree.

1372 GLOVER, J. T. *East-gate, Warwick.*—Light waggonette.

1373 HADLEY, C. 37, *Lower Hurst-st. Birmingham.* — Drawings of omnibuses, cabs, broughams, carts, &c.

1374 HALE, S. W. 27, *Park-lane, W.*—Elcho sociable.

1375 HALL & SONS, 98, *Long-acre.*—A barouche on elliptic springs, with self-acting body-steps.

1376 HARVEY, J. *Richmond, Surrey.*—A drawing of a two-wheel closed carriage.

1377 HAWKINS, J. *Hatfield-st. Blackfriars-rd.*—Arms, axletrees, and springs.

1378 HAZELDINE, G. 5, *Lant-st. Borough.*—Road van.

1379 HIGGINSON, C. JUN. 65, *George-st. Portman-sq.*—Carriage heraldry.

1381 HOLMES, H. & A. *London-rd. Derby.* —Park sociable, with landau head, on C and under springs.

1382 HOLROYD, NOBLE, & COLLIER, *Halifax.*—Machine-made wheels. Imitation wicker paneling, and carved wood mouldings.

1383 HOOPER & CO. 28, *Haymarket.*—"Sefton" landau with flat falling head, and a "Craven" barouche on C and under springs.

1384 HORSLEY, C. & SON, *Hungate-st. Beccles.*—A light brougham.

1385 HOULGATE, F. 8, *Westbrough-st. Scarborough.*—Circular-fronted brougham.

1386 HOWITT, W. J. 25, *Denmark-pl. Soho.* —C springs, and coach-smith's work.

1387 HUTLEY, F. 11, *Long Acre.*—Carriage laces.

1388 HUTTON, J. & SONS, *Summer-hill,*

Dublin.—Round-fronted brougham and Irish car.

1389 IVALL & LARGE, 56, *South Audley-st.*—Four-in-hand coach, with drag.

1390 JONES, W. 70, *Upper Seymour-st. N.W.*—Paintings used in manufacture of carriages.

1391 KESTERTON, E. 94, *Long Acre.*—The "Amempton," an open or close carriage.

1392 KINROSS, W. *Stirling.*—Two-wheel buggy.

1394 LA ROCHE, J. & MEHEW, J. 5, *James-pl. Chelsea.*—Velocipede: all the iron-work of tubular iron.

1395 LENNY, C. & Co. 9, *Park-lane, W.*—Sociable landau.

1396 MCDOUGALL, A. & SON, 36, *Rupert-st. W.*—Horse van.

1397 MCNAUGHT & SMITH, 9, *Tything, Worcester.* — Waggonette, with movable head.

1398 MACNEE, J. & Co. 106, *Princes-st. Edinburgh.*—Landau Clarence.

1399 MANN, J. H. & Co. *Twickenham.*—Park phaeton, with improved fore-carriage.

1400 MASON, H. H. *Carriage Works, Kingsland-bridge.* — Waggonette for eight persons.

1401 MILFORD, T. & SON, *Wheel Works, Thorverton, Devon.*—A cart.

1402 MULLINER, F. *Bridge-st. Northampton.*—Fitzroy phaeton, constructed with malleable steel; wheels of hickory.

1403 MULLINER, H. *Leamington.*—Four-wheel dog-cart: folds open and forms waggonette; head drops on.

1404 NEWHAM, E. *Market Harboro'.*—Light sociable phaeton, with seats and dash removable.

1405 NEWNHAM & SON, *Bath.* — Light landau waggonette, with folding leather head and improved arrangement of seats.

1406 NEWTON, J. 10, *Werrington-st. N.W.*—Folding double-seated perambulator.

1407 NURSE & Co. 200, *Regent-st. W.*—Sociable landau on elliptic springs.

1408 OFFORD, R. & J. 79, *Wells-st. W.*—Carriage.

1409 PARKER, F. 75, *Regent-st. Cambridge.* — Registered family cart with improved springs.

1410 PARSONS, G. *Martock, Somerset.*—Wheels for common roads.

1411 PARTRIDGE, E. *Smethwick, near Birmingham.*—Collinge and mail carriage axletrees.

1413 PEARCE & COUNTZE, 103, *Long-acre.*—Sociable landau.

1414 PETERS & SONS, *Park-st. London, W.*—Park barouche and a brougham.

1415 REAY & USHER, *South Hylton Iron Works, Sunderland.*—Axle block forgings, finished under the forge hammer.

1416 RENDALL, J. & W. *High-st. Stoke Newington, N.*—A two ton coal van.

1417 RIDGES, J. E. *Tudor Coach Factory, Cleveland-rd. Wolverhampton.* — Miniature landau, open or close.

1418 RIGBY & ROBINSON, 7, *Park-lane, Piccadilly.*—Elcho landau.

1420 ROCK & SON, *Hastings.*—A dioropha with improvements.

1421 ROGERS, I. *North Audley-st.* — Sociable landau, complete whether open or close.

1422 ROGERS, W. & Co. *Bristol.*—The "Clifton Waggonette."

1423 SAWYER, W. *St. James-st. Dover.*—Drawings of velocipedes.

1424 SEADON & JONES, 60, *Whitechapel.*—Sociable, with removable inclosure, to carry six persons.

1425 SELLERS, J. A. 313, *Oxford-st. W.*—Model of a carriage.

1426 SHANKS, R. H. & F. 4, *Gt. Queen-st. W.C.*—Light step-piece landau, opening very low.

1427 SHEPHERD, J. 1, *Cheapside, Birmingham.*—Brougham on noiseless springs.

1428 SHERWIN, J. *Tabernacle-walk, Finsbury.*—Axletrees, waggon and cart arms.

1429 SHILLIBEER, G. 40, *City-rd.*—Vis-à-vis omnibus, separate seats inside; outside seats reached from interior.

1430 SHORT, J. 23, *Cleveland-st. Fitzroy-sq.*—Heraldic mountings for carriages, harness, &c.

1431 SILK & SONS, *Long-acre.*—Landau on horizontal springs.

1432 SIMPSON, H. C. *Barker-st. Shrewsbury.*—Car with extra luggage accommodation.

1433 SMITH, J. B. 17, *Green-st. Bath.*—Silver mounted perambulator.

1434 STAREY, T. R. *Nottingham.*—Landau, with flat fall of head, elastic springs and silent wheels, chain tyres, &c.

1435 STARTIN & MACKENZIE, *Benacre-st. Birmingham.*—Headed phaeton.

1436 STEVENSON & ELLIOT, *Stirling, Scotland.*—Light phaeton with moveable side glasses.

1437 STOCKEN, F. 5A, *Halkin-st. Belgrave-sq.*—Carriage.

1438 STRICKLAND, H. 9, *Macclesfield-st. Soho.*—Specimens of carriage herald painting.

1439 THOMSON, G. *Stirling, N.B.*—Light waggonette, close or open, with reversible seat.

1440 THOMSON, W. 40, *Canal-st. Perth.*—Four-wheel dog cart, and a waggonette convertible into a dog cart.

1441 THORN, W. & F. 19, *Gt. Portland-st. W.*—Summer and winter carriage, close or open.

1442 THORNTON, E. M. 6, *Brooke-st. Holborn.*—Rein clip, for the reins when out of hand.

1443 THRUPP & MABERLY, 269, *Oxford-st.*—A light elliptic spring coach.

1444 TURRILL, H. L. 67, *South Audley-st.*—Carriage.

1445 VEZEY, R. & E. *Long-acre, and Bath.*—Brougham with concealed step, noiseless springs: model of Her Majesty's state coach.

1446 WARD, J. 5, *Leicester-sq.*—Invalid chairs.

1447 WATERS, G. & SON, 72, *North-end, Croydon.*—Open carriage.

1448 WATKINS & HORNSBY, *Duke-st. Birmingham.*—Carriage axles.

1449 WHITTINGHAM, T. & WILKIN, 136, *Long-acre.*—Carriage laces, and imitation cane-work.

1450 WICKSTEED, F. 18, *Upper St. Martins-lane.*—Carriage drawings.

1451 WINDOVER, C. S. *Huntingdon.*—A carriage forming a barouche, sociable, coach, and landau.

1452 WOODALL & SON, 28 *Orchard-st. W.*—Side light coach, with improvements in ventilation.

1454 WYBURN & CO. 121, *Long-acre.*—Landau and a brougham.

CLASS 7.

MANUFACTURING MACHINES AND TOOLS.

Western and Eastern Annexes.

Sub-Class A.—Machinery employed in Spinning and Weaving.

1486 ANDERSTON FOUNDRY CO. *Glasgow.*—Fancy looms.

1487 APPERLY, J. & CO. *Dudbridge, Stroud.*—Woollen machinery, &c.

1488 BOOTH & CHAMBERS, *Bury, Lancashire.*—Loom for weaving, various motions; and reed-holder.

1489 CLARKE, I. P. *Leicester.*—Reels, spools, and mill-bobbins.

1490 CLARKE, T. A. W. *Leicester.*—Machine for covering India rubber rings.

1491 COMBE, J. & CO. *Belfast.*—Flax machinery.

1492 COOK & HACKING, *California Iron Works, Bury, Lancashire.*—Self-acting Heald knitting-machine.

1493 COTTON SUPPLY ASSOCIATION, *Manchester.*—Indian native churka, and roller-gin for cleaning cotton.

1494 CRABTREE, T. *Halifax, Yorkshire.*—Card-setting machinery.

1495 DAVIS, E. & J. *Leeds and Derby.*—Yarn-tester.

1496 DE BERGUE, S. *Manchester.*—Reeds and dents for weaving. Steel-wire for crinolines.

1497 DICKINSON, W. & SONS, *Phœnix Iron Works, Blackburn.*—Loom for fancy weaving, and fast loom, Taylor's pt. Power-looms, &c.

1498 DIXON, J. & SONS, *Steeton, via Leeds.*—Bobbins, rollers, keys, treenails, drawer knobs, and boxes.

1499 DOBSON & BARLOW, *Bolton, Lancashire.*—Machinery for preparing and spinning cotton.

1500 DUGDALE, J. & SONS, *Blackburn.*—Looms for twilled and plain cloth, and cop-winding machine; roving and throstle spindle collars.

1501 FAIRBAIRN, P. & CO, *Leeds.*—Rope-spinning machinery.

1502 FERRABEE, J. & CO. *Stroud, Gloucestershire.*—Woollen machinery for forming bats of fleece, &c.

1503 GATENBY & PASS, *Manchester.*—Reeds, dent, dent-wires, &c.

1504 GORDON, J. 3, *Billiter-sq. London.*—Roller gin for cleaning cotton, worked by the foot.

1505 HALEY, J. & SONS, *Cloth Hall Machine Works, Dewsbury.*—Rag-grinding machine.

1506 HARDING, T. R. *Leeds.*—Wool-combing implements.

1508 HARRISON, J. & SONS, *Blackburn, Lancashire.*—Series of machinery for preparing and weaving light and heavy fabrics.

1509 HATTERSLEY, G. & SON, *North Brook Works, Keighley, near Bradford.*—Looms for fancy goods.

1510 HENDERSON & CO. *Durham.*—Power-loom, for Brussels and velvet carpets.

1511 HETHERINGTON & SONS, *Vulcan Works, Manchester.*—Cotton-preparing and spinning machinery.

1512 HEWKIN, H. *Oldham, Lancashire.*—Model of cotton mill.

1513 HIGGINS, W. & SONS, *Manchester.*—Pt. cotton machinery: carding-engines, frames, throstle, &c.

1514 HINE, R. E. & CO. *Manchester.*—Machine for spinning, doubling, and twisting, silk, &c.

1515 HODGSON, G. *Bradford.*—Looms, with latest improvements.

1516 IRWIN & SELLERS, *Preston.*—Box-wood logs, cuttings, shuttles, bobbins, pickers.

1517 JACKSON & GRAHAM, *Oxford-st.*—Jacquard carpet power-loom.

1519 KERR, J. & CO. *Douglas Foundry, Dundee.*—Five-bowled cheating calender; calender for finishing cloth.

1520 LAWSON, S. & SONS, *Hope Foundry, Leeds.*—Flax machinery and self-acting tools.

1521 MACLEA & MARCH, *Leeds.*—System of spiral gill cone preparings for short hosiery wool.

1522 MASON, J. *Rochdale.*—Machinery for preparing and spinning cotton and woollen.

1523 MORRISON, T. & G. *Paisley.*—Jacquard-machine.

1524 NIGHTINGALE, W & C. *Old-st. E.C.* —Horse-hair curling machine.

1525 OLDHAM, J. C. *Heywood, near Manchester.*—Power loom shuttles.

1526 PARKER, C. & SONS, *Dundee.*—Flax, hemp and jute, power looms, &c.

1527 PERRY, J. *Shipley Field Mills, near Bradford.*—Machinery for preparing and combing wool; circular combs, gills, and fallers, &c.

1528 PLATT BROS. & Co. *Hartford Ironworks, Oldham.*—Machinery for preparing and spinning cotton and woollen.

1530 ROBINSON, J. & R. & Co. 30, *Milk-st. Cheapside.*—Silk velvet loom. (Process Court.)

1531 ROWAN, J. & SONS, 152, *York-st. Belfast.*—Scutching machine.

1532 SHARP, STEWART, & Co. *Atlas Works, Manchester.*—Reel winding-machine, for silk, linen, or cotton sewing-thread.

1533 SMITH, W. & BROS. *Heywood, Lancashire.*—Woollen and other looms, shuttles, shafts.

1535 STUART, J. & W. *Musselburgh, near Edinbro'.*—Fishing net loom.

1537 TUER & HALL, *Bury, near Manchester.*—Shearing machine, looms.

1538 WALKER & HACKING, *Bury, Lancashire.*—Machinery for preparing and spinning cotton yarn.

1539 WARD, G. 77, *Darwen-st. Blackburn.* —Heald knitting machine.

1541 WILSON, L. & SONS, *Todmorden.*—Bobbins.

1542 WHITESMITH, I. 29, *Govan-st. Glasgow.*—Power loom with six shuttles and twilling combined.

1543 WREN & HOPKINSON, *London-rd. Manchester.*—Machinery for making cotton sewing-thread and spinning silk.

1544 TAYLOR, E. *Kinghorn, Scotland.*—Handsorting and intersecting machine heckles.

Sub-Class B.—Machines and Tools employed in various Manufactures.

1552 ANNABLE & BLENCH, 28, *St. John-st. E.C.*—Printing-machine.

1553 ARMITAGE, M. & H. & Co. *Mousehole Forge, near Sheffield.*—Anvils, vices, hammers, &c. (E.A.)

1554 BARRETT, EXALL, & ANDREWES, *Reading.*—Dauglish's bread and biscuit machinery. (E.A.)

1555 BERTRAM, G. *Sciennes, Edinbro'.*—80-inch paper-making and cutting machine.

1556 BESLEY, R. & Co. *Fann-st. Aldersgate-st.*—Type-casting machine in operation.

1557 BEYER, PEACOCK, & Co. *Gorton Foundry, Manchester.*—Wheel-lathe, to turn and bore, up to 7 feet diameter; and triple-headed drilling machine.

1558 BISSELL, W. *Union-st. Wolverhampton.*—Flooring and bench-cramp, machine, lifting-jack.

1560 BRADBURY & Co. *Rhodes Bank Foundry, Oldham.*—Sewing-machines. (Process Court.)

1561 BRADLEY & CRAVEN, *Westgate-common Foundry, Wakefield.* — Brick-making machine. (E.A.)

1562 BRUNTON, J. D. *Barge-yd. Bucklersbury.*—Peat-fuel, and machinery for preparation of same.

1563 BUCKTON, J. & Co. *Well House Foundry, Leeds.*—Self-acting engineers' tools.

1564 BUNNETT & Co. *Deptford, Kent.*—Brick-making machine.

1565 BURN, R. *Lochrin Engine Works, Edinburgh.*—Envelope label dies, and roller-gins, &c.

1566 CARVER, W. 5, *Todd-st. Manchester.* —Sewing-machines. (Process Court.)

1567 CASSON, J. *Wellington-st. Woolwich.* —Machines for dressing dried fruits. (E.A.)

1569 CLARKE, T. A. W. *Leicester.*—Machine for covering elastic thread, by a new method.

1570 CLAYTON, H. & Co. *Atlas Works, near Dorset-sq. N.W.*—Brick, tile, and pipe-machines.

1571 COHEN, B. S. 9, *Magdalen-row, Gt. Prescott-st.*—Pencil manufacturing. (Process Court.)

1572 COLLEY, E. E. 5, *West-cottages, West-st. Walworth.*—Working model of Hoe's printing-machine.

1573 COLLIER, L. *River-st. Rochdale.*—Confectioners' and biscuit-bakers' machines. Sugar-mills, &c. (E.A.)

1574 CONISBEE, W. 40, *Herbert-bldgs. Waterloo-rd.*—A Main's printing-machine.

1575 COOK D. & Co. *Glasgow.*—Steam riveting-machine; bour pan, for evaporating sugar-cane juice.

1576 COOKE, S. & SONS, *York.* — Fancy lathes.

1577 CORYTON, J. 89, *Chancery-lane.*—Type composer. (Process Court.)

1578 COWAN, T. W. *Kent Iron Works, Greenwich.*—Compressed air hammer.

1579 COX & SON, 28 & 29, *Southampton-st. Strand.*—Wood-carving machine.

1581 CRAWHALL & CAMPBELL, *Glasgow.* Horizontal boring machine.

1582 DAY & SON, *Gate-st. Lincoln's Inn Fields.*—Lithographic and copper plate presses. (Process Court.)

1583 DEANE & DAVIES, 19, *Blackfriars-st. Manchester.*—Sewing machines, presses, gas apparatus, &c.

1584 DE BERGUE, C. & Co. *Manchester and London.* — Steam hammers, punching machines, &c.

1585 DONKIN, B. & Co. *near Grange-road, Bermondsey.*—Paper-making and cutting machines.

1586 DOULTON & Co. *Lambeth Pottery, S.* —Potter's wheel, worked by steam, showing the process of manufacture.

1587 DUPPA, T. D. *Longville, Westanstow, Shropshire.* — Vice-bench, for carpenters, coopers, &c. (E.A.)

1589 EASTERBROOK & ALLCARD, *Sheffield.* —Engineering and railway tools, machines, &c. (E.A.)

1590 Eastwood, C. *Virginia-pl. Leeds.*—A cutting and measuring machine for brushes. (Process Court.)

1591 Eastwood & Sons, *Railway Iron Works, Derby.*—Steam hammer, iron, &c.

1592 Effertz, P. 71, *Coupland-st. Manchester.*—A brick machine and model, drain-pipe machine, and drawings.

1594 Fairbairn, P. & Co. *Leeds.*—Engineering tools.

1595 Fentum, M. 85, *New Bond-st.*—Lathe and saw for working in ivory.

1596 Ferrabee, H. 75, *High Holborn.*—British sewing machine. (Process Court.)

1597 Ferrabee, J. & Co. *Stroud, Glo'stershire.*—Adjusting spanners or screw wrenches.

1598 Forrest & Barr, *Glasgow.*—Wood planing and moulding machine, for ship and other builders.

1599 Fox, Bros. *Derby.*—Lathe, drilling machine, and planing ditto.

1600 Gadd, W. & Son, *Fishergate, Nottingham.*—A screwing machine, on a new principle.

1601 Garrett, B. 5, *Cumberland-st. Shoreditch.*—Imperial printing presses, and book-binding press. (E.A.)

1602 Garside, H. *Coupland-st. Manchester.*—Electrograph machine for engraving the copper cylinders used in calico printing.

1603 Geeves, W. *Caledonian Mills, New Wharf-rd. Islington, N.*—Saw frame.

1604 Gerish, F. W. *East-rd. City-rd.*—A platen press, with rotary motive power.

1605 Gehrling, J. 15, *William-st. North, Caledonian-rd.* — Eyelet machines, various tools and steelyards.

1606 Gibbs, D. & W. *City Soap Works.*—Machinery for grinding and compressing soap.

1607 Glasgow, J. *Trafford-st. Manchester.*—Screwing machine.

1608 Glen & Ross, *Greenhead Engine Works, Glasgow.* — Rigby's double-acting steam hammers.

1609 Grafton, H. 80, *Chancery-lane.*—Machine for making solvable paper tubes.

1610 Greenwood & Batley, *Albion Works, Leeds.*—Machinery for working in wood and metals, and making boots, &c.

1611 Greig, D. & J. *Edinburgh.*—Paper cutting machine, presses, and case of plates.

1612 Guinness & Co. *Cheapside.*—Sewing machine. (Process Court.)

1613 Harrild & Sons, 25, *Farringdon-st.*—Newspaper addressing machine, and other new printing materials. (Process Court.)

1614 Harrison, —, 16, *Bishopsgate-st. Within.*—Magnetic printing press. (Process Court.)

1615 Harrison, C. W. *Lorrimore-road, Walworth.*—Electro-magnetic printing press. (Process Court.)

1616 Harvey, G. & A. *Albion Works, Glasgow.*—Machine tools.

1617 Hawkins & Co. 38, *Lisle-st. London, W.*—Self-acting steam fly press.

1618 Hetherington & Sons, *Manchester.*—Tools.

1619 Hill, P. *Bertram House, Hampstead.* — Post-office stamping machine. (Process Court.)

1621 Holtzapffel, L. & Co. 64, *Charing-cross.* — Lathes; sawing, cutting, grinding, measuring, drawing, and printing apparatus.

1622 Hudswell & Clarke, *Jack-lane, Engine Works, Leeds.*—Hammers for smithing.

1623 Hughes, H. *Homerton.*—Chain goffering machines. Specimens of embossing in relief. (Process Court.)

1624 Hughes & Kimber, *Red Lion-passage, Fleet-st.*—Lithographic and copper plate press, &c.

1625 Hulse, J. S. *Manchester.*—Machine tools.

1626 Hunt, J. & Co. *Clay Hall Iron Works, Old Ford, Bow.*—Machine for cutting the teeth of wood or metal wheels.

1627 Hunt & Roskell, 156, *New Bond-st.*—Process of cutting and polishing diamonds.

1628 Imray, J. *Bridge-rd. Lambeth.*—Steam hammer, hydraulic anvil, and striker.

1630 Jaques, J. *Prescot.*—Spring dividers and compasses. (Process Court.)

1631 Jarrett, G. 37, *Poultry, City.*-Stamping and embossing presses. (Process Court.)

1632 Johnson & Atkinson, 31, *Red Lion-sq.*—Machinery for casting and finishing type.

1634 Jones, Lavinia, *Bow-st. Covent-garden.* — Miniature Albion printing-press, cases of type, and furniture, with appliances. (Process Court.)

1635 Jones, W. 246, *High Holborn.*—Embossing and stamping presses. (Process Court.)

1636 Keiller, W. *Perth.*—Cinnamon and cassia cutting machine, &c. (.E.A.)

1637 Keith, W. 11, *Three Crown-sq.*—Sewing machine.

1638 Kendall & Gent, *Salford, Manchester.*— Self-acting machine for cutting tubes for engineers and boiler makers.

1639 Kennan & Sons, *Fishamble-st. Dublin.*—Sculpturing machine, amateur's lathes. (Process Court.)

1640 Kershaw, J. & J. *Store-st. Works, Manchester.*—Lathe-shaping machine.

1641 Kirkstall Forge Co. *Leeds.*—Naylor's single or double action steam hammer.

1642 Lamb, J. *Holborn Paper Mills, Newcastle, Staffordshire.*—Laying apparatus, attached to paper cutting machine; felt not required.

1644 Lee, H. C. 11, *Laurence-Pountney-lane, E.C.* — Knitting machine. (Process Court.)

1645 Legg, R. 14, *Owen's-row, Clerkenwell.*—Tobacco cutting machine and four-horse steam engine.

1646 Leigh, E. & Son, *Manchester.*—Top rollers; model of steam-ship.

1647 Lely, A. *Redditch.* — Machine for grooving needles, and ditto for polishing the eyes. (Process Court.)

1648 Lewis, J. 51, *High-st. Bloomsbury.*—Machine for boring and fret cutting. (Process Court.)

1649 Lockett, J. Sons, & Leake, *Manchester.*—Pentagraph engraving machine.

1650 Lyons, M. 143, *Suffolk-st. Birmingham.*—Apparatus for depositing metals.

1651 MACLEA & MARCH, *Leeds.*—Lathes: planing, shaping, and slotting machines.

1652 MCDOWALL, J. & SONS, *Walkinshaw Foundry, Johnstone, Glasgow.*—Wood planing and moulding machine.

1653 MCKENZIE, A. & Co. 32, *St. Enoch's-sq. Glasgow.*—Sewing machine, with specimens of work. (Process Court.)

1654 MCKERNAN, L. 98, *Cheapside.*—Sewing machines. (Process Court.)

1656 MARSHALL, T. J. 80½, *Bishopsgate-without.*—Paper-making machines, &c.

1657 MATHIESON, A. & SON, *Tool Works, East Campbell-st. Glasgow.*—Tools. (E.A.)

1658 MIERS, W. J. 15, *Lamb's Conduit-passage, W.C.*—Machine for cutting and drawing ovals.

1659 MILLER & RICHARD, *Edinburgh, and London.*—Printing press.

1660 MILLS, J. *Stockport.*—Tapered pins, and keys, made by machinery.

1661 MILWARD, H. & SONS, *Redditch, Worcestershire.*—Processes in needle-making machinery.

1662 MITCHEL, W. H. 16, *Newton-st. W.C.*—Type composing and distributing machines.

1663 MORGAN & Co. *Paisley.*—Block cutting-machine. (Process Court.)

1665 MORRALL, A. *Studley Mills, London and Manchester.*—Needles and thimbles in process of manufacture.

1666 MORRISON, R. & Co. *Newcastle-on-Tyne.*—Steam hammer, with piston and bar forged solid.

1667 M'QUEEN, BROS. 184, *Tottenham-court-road.*—Copper plate printing machine. (Process Court.)

1668 MUIR & Co. *Britannia Works, Manchester.*—Machine tools.

1669 NAPIER, D. & SONS, 51, *York-road, Lambeth.*—Printing press; machine for making rifle bullets with cold lead, by compression.

1670 NASMYTH, J. & Co. *Bridgewater Foundry, Patricroft, near Manchester.*—Differential, dividing, punching, and other machines. Steam hammers, &c.

1674 NEWBERY, R. C. & Co 5, *President-st. west, E.C.*—Machine for making enamelled cloth collars. (Process Court.)

1675 NEWTON WILSON & Co. 144, *High Holborn.*—Sewing, and other machines. (Process Court.)

1677 OATES, J. P. *Erdington, Birmingham.*—Photograph of machine for making solid bricks. (E.A.)

1678 PAGE, E. & Co. *Victoria Iron Works, Bedford.*—Brick and pipe machinery. (E.A.)

1679 PALMER, H. R. 308, *Albany-rd. S.*—Parallel motion stamping, printing, endorsing, and paging machines. (Process Court.)

1680 PARKER, W. & SONS, *Northampton.*—Boot and shoe-making machine.

1681 PATENT FILE MACHINE AND FILE MANUFACTURING CO. *Manchester.*—Self-acting machines for cutting files.

1682 PEARSON, W. & Co. *Leeds.*—Cut-nail and sewing machines. (Process Court).

1683 PERRY, T. & SON, *Highfields, Bilston.*—Case-hardened rolls, for rolling metals. (E.A.)

1684 PETO, BRASSEY, & BETTS, *Birkenhead.*—Drilling machine, and machine for punching holes at one operation.

1685 PETTER & GALPIN, *Belle Sauvage Works, E.C.*—Printing machine.

1686 PINCHES, T. R. & Co. 27, *Oxendon-st. Haymarket, S.W.*—Medal press. (Process Court.)

1687 PORTER & Co. *Carlisle.*—Lozenge and biscuit machine.

1688 POWIS, JAMES, & Co. *Victoria Works, Blackfriars-rd.*—Wood-working machines and steam-engines.

1689 PRENTIS & GARDNER, *Steam-engine and Paper Machine Works, Maidstone, Kent.*—Knotter or paper-strainer.

1690 PRESTON, F. & Co. *Manchester.*—Copying and stamping machines, presses, &c. (Process Court.)

1691 REYNOLDS, J. G. 33, *Wharf-road, City-road, London.*—Machine for making tobacco-pipes. (Process Court.)

1692 RHODES, J. *Morley, near Leeds.*—Rag-machine.

1693 RHODES, J. *Grove Works, Wakefield.*—Steam hammer; punching and shearing machine.

1694 ROBERTS, R. & Co. 10, *Adam-st. Adelphi.*—Drawings of Jacquard punching machine, and angle iron punching machine.

1695 ROBINSON, T. & SON, *Rochdale.*—Sawing, planing, moulding, morticing, tenoning, and sharpening machines, for woodwork.

1696 ROSS, J. *Leith.*—Double-cylinder printing machine, with self-acting set-off sheet apparatus.

1697 RYDER, W. *Bolton, Lancashire.*—Machines for forging and fluting rollers.

1698 SALISBURY, S. C. *Coventry.*—Knot-stitch sewing machine. (Process Court.)

1699 SEGGIE, A. *Edinburgh.*—Lithographic press for finest work.

1700 SERVICE, W. *Mitcham, Surrey.*—Sewing machines, with double-feed action. (Process Court.)

1701 SHANKS & CO. engineers, 6, *Robert-st. Adelphi.*—Drilling, bolt-screwing, and other machines. Steam hammer. Models, &c.

1702 SHARP & BULMER, *Middlesboro'.*—Hand brick or tile machine, 5,000 bricks per day. (E.A.)

1703 SHARP, STEWART, & Co. *Atlas Works Manchester.*—Workshop tools, wheel lathe Giffard's injectors.

1704 SHARRATT & NEWTH, *Clerkenwell.*—Glaziers' diamonds. (E.A.)

1705 SHEPHERD, HILL, & Co. *Union Foundry, Leeds.*—Machinery.

1706 SIEBE, A. 5, *Denmark-st. Soho.*—Paper knotting machine.

1707 SIEMENS, HALSKE, & Co. 3, *Gt. George-st.*—India rubber covering machines. Submarine cable.

1708 SIMPSON, R. E. & Co. *Glasgow.*—Single and double action shuttle sewing machine. (Process Court.)

1709 SINCLAIR, J. 541, *Castle-hill, Edinburgh.*—Dandy roll, for water-mark on paper.

Western and Eastern Annexes.

1710 SINIBALDI, MADAME C. *South-villas, South-st. Greenwich.*—Chain machine, cranks, pistons, printing-press, &c.

1711 SMITH, A. *Princes-st. Leicester-sq. W.* —Machinery for making submarine cables and wire ropes.

1712 SMITH & COVENTRY, *Ordsal-lane, Manchester.*—Radial drill, screwing lathe, and other tools for cutting metal.

1714 SMITH & HAWKES, *Eagle Foundry, Birmingham.*—Chilled-cast rolls, testing-machine, diagrams, &c. (E.A.)

1715 SMITH, BEACOCK, & TANNETT, *Victoria Foundry, Leeds.*—Self-acting machine tools for shaping, slotting, turning, and rifling.

1716 SMITH, C. 30, *White-st. Hulme, Manchester.*—Soap-cutting machine.

1717 SMITH, E. *Cemetery-rd. Sheffield.*—Pointing and carving machine, for objects in relievo (Process Court).

1718 SMITH, J. & Co. 62, *London-wall.*—Continuous motion shuttle sewing machines (Process Court).

1720 SMITH, J. & SON, 8, *Up. Fountain-pl. City-rd.*—Model moulds and rollers for paper making, &c.

1722 STEVENS, G. *Cambridge-rd. London.* —Machinery for kneading dough. (E.A.)

1723 STONE, J. *Deptford.*—Machine for making 1,000 cast-metal nails at one time.

1724 STOTHERT & PITT, *Bath.*—Machine for striking and scraping leather hides; model ditto for rolling leather.

1725 SWEET, A. 20, *St. James-place, Hampstead-road.*—Case of locks, and machine.

1726 THOMAS, W. F. & Co. 66, *Newgate-st.*—Sewing machines. (Process Court.)

1727 THOMPSON, R. H. *Dockyard, Woolwich.*—Sawing and other machines.

1728 THWAITES & CARBUTT, *Vulcan Iron Works, Bradford, Yorkshire.*—Steam hammers and engineering tools.

1729 TIDCOMBE, G. & SON, *Watford, Herts.* —A continuous-sheet paper-cutting machine.

1730 ULLMER, E. & W. *Castle-st. Holborn.* —Cylindrical printing machine, and diagonal paper cutting ditto.

1732 VICARS, T. & T. & Co. *Wheat Sheaf Foundry, Liverpool.*—Bread and biscuit machinery. (E.A.)

1733 VICTORIA SEWING MACHINE Co. 97, *Cheapside.*—Sewing machines.

1734 WATERLOW & SONS, *London.*—Railway-ticket printing machine, may be worked by hand or power.

1735 WATKINS, T. 89, *Bridge-st. Bradford.* —Porcelain guides, washers, and other articles used in machinery.

1736 WATSON, H. *High Bridge Works, Newcastle-on-Tyne.*—Brass and copper rolls for paper mills; Jullious's self-regulating pulp elevator, &c.

1737 WEATHERLEY, H. 54, *Theobald's-rd.* —Confectioner's machines, for hand or steam power.

1738 WESTON & HORNER, 80, *White Cross-st. London, E.C.*—Self-feeding mortising machine.

1739 WHIGHT & MANN, *Gipping Works, Ipswich.*—"Excelsior" sewing machine. (Process Court.)

1740 WHITFIELD, H. *Rainhill, near Prescot.*—Lancashire files.

1741 WHITMEE, J. & Co. 70, *St. John-st. E.C.*—Mills, weighing machines; Tice's patent gas regulators. (E.A.)

1742 WHITWORTH, J. & Co. *Chorlton-st. Manchester.*—Machinery for cutting metals and timber.

1743 WILSON, W. *Campbellfield, Glasgow.* —Semi dry pulverised clay brick making machine.

1744 WOOD, J. & R. M. 89, *West Smithfield, E.C.*—Printing and stereotyping machinery and type.

1745 WORSSAM, S. & Co. 304, *King's-rd. Chelsea.*—Wood working machines.

1746 WRIGHT, J. *Pathend, Kirkcaldy.*—Mould making machines for producing printing surfaces, &c. (Process Court).

1747 WYLIE, A. C. 8, *Cannon-st. London.* —Condie's steam hammers.

1748 YATES, W. S. *Stamford-st. Leeds.*—Machine to assort bristles.

1749 YOUNG, J. & T. *Ayr.*—Vertical saw-frame, to cut from twenty-four inches broad, and from five inches thick.

1750 YOUNG'S PT. TYPE Co. 77, *Fleet-st.* —Type composing machine, and type composing and distributing machines. (Process Court.)

1751 YOUNGMAN, C. T. 25, *West-st. E.C.*—Paper bags, made by steam machinery (Process Court).

1752 COTTRILL, E. 52, *St. Paul's-sq. Birmingham.*—Die-sinker, stamper, and piercer.

1753 SPARKES HALL, J. 308, *Regent-st.*—Shoe-makers' upright bench.

CLASS 8.

MACHINERY IN GENERAL.

Western and Eastern Annexes.

1780 ADAMSON, D. & Co. *Newton Moor Iron Works, near Manchester.* — Twenty-tons hydraulic lifting-jack: and steam boiler.

1781 ADCOCK, J. *Marlborough-rd. Dalston.* — Distance indicator for wheel carriages. (E.A.)

1782 ALLEN, HARRISON, & Co. *Cambridge-st. Mills, Manchester.*—Gun-metal fittings for steam engines. (E.A.)

1783 APPLEBY BROS. 69, *King William-st. City.*—Cranes, engines, wheels, pumps, &c. (E.A.)

1784 ARMSTRONG, R. *North Woolwich, E.* —Steam boiler, model, and drawing. (E.A.)

1785 ARMSTRONG, SIR W. G. & Co. *Elswick Engine Works, Newcastle-on-Tyne.*—Models exhibiting "Armstrong's hydraulic system."

1786 ASHTON, J. P. 2, *Upper Holland-st. Kensington.*—Steam engine and hoist.

1787 ASKEW, C. *Charles-st. Hampstead-rd.* —Brewer's refrigerator, boiler, and cowl. (E.A.)

1788 BAINES & DRAKE, *Glasgow.*—Engine and boiler mountings. (E.A.)

1789 BALFOUR, H. T. 16, *Adam-st. Strand.* —Quartz crushing machine. (E.A.)

1790 BARNETT, S. 23, *Forston-st. Hoxton, N.*—Soda-water machinery.

1791 BARRETT, EXALL, & ANDREWES, *Reading.* — 30 horse-power double-cylinder horizontal high-pressure expansive condensing engine.

1792 BASTIER, J. U. 19, *Manchester-bdgs. Westminster.*—Chain pump.

1793 BATE, J. & Co. 18, *Crescent, Birmingham.*—Bottle corking and washing machines. (E.A.)

1794 BAYMAN, H. *Johnson-st. Old Gravel-lane, E.*—Lifting jacks, iron blocks, ship's hearth, and winch. (E.A.)

1795 BEAUMONT, F. W. *Clapham.* — Self-acting steam boiler-feeding and general meter. (E.A.)

1796 BECK, J. 133A, *Gt. Suffolk-st. Southwark.*—Valves for gas, water, and steam: fire cocks, &c.

1797 BELLHOUSE, E. T. & Co. *Eagle Foundry, Manchester.*—Steam engine, hydraulic pumps and cocks, models of presses, mills, engine boiler, &c.

1798 BELLIS & SEEKINGS (late Bach & Co.) *Broad-st. Birmingham.*—Two-and-a-half horse-power vertical steam engine.

1799 BENSON, W. *Robin Hood-st. Nottingham.*—Three horse-power steam engine on pillar.

1800 BLINKHORN, SHUTTLEWORTH, & Co. *Spalding.*—Fire engines. (E.A.)

1801 BODMER, R. & L. R. 2, *Thavies-inn, Holborn.*—Safety valves. (E.A.)

1802 BOTHAMS, J. C. *Salisbury.*—Water meters, tap to check waste, &c.

1803 BOWSER & CAMERON, *Glasgow.* — Five-ton derrick crane.

1804 BRADFORD, T. *Manchester and Fleet-st. London.*—Washing, wringing, drying, mangling, and knife-cleaning machinery, &c. (E.A.)

1805 BRAY'S TRACTION ENGINE Co. 17, *Pall Mall east.*—Traction engine, for common roads. (E.A.)

1806 BRIDLE, H. *Bridport, Dorset.* — Double-action refrigerator, for brewing and distilling. (E.A.)

1807 BRIGGS & STARKEY, *Leeds and Liverpool.*—Washing, wringing, and mangling machines. (E.A.)

1808 BROUGHTON COPPER Co. *Manchester.* —Varieties of copper and brass work; fire-engines. (E.A.)

1809 BRYANT & COGAN, 55, *Broadmead, Bristol.*—Edge-laid leather mill-band. (E.A.)

1810 BUNNETT & Co. *Deptford, Kent.*—Concentric steam engine, working without fly-wheel; brick-making machine.

1812 CARR, T. *New Ferry, near Birkenhead.*—Disintegrated mills and fan blower. (E.A.)

1813 CARRETT, MARSHALL, & Co. *Sun Foundry, Leeds.*—Hydraulic and other engines, pumps, steam hammer, and steam carriage. (W. & E.A.)

1814 CATER, H. 9, *Anchor-terrace, Southwark-bridge.*—Model of multitubular steam boiler. (E.A.)

1815 CHADBURN BROS. *Nursery, Sheffield.* —Pressure guages, tallow feeders, &c. (E.A).

1816 CHALMERS, D. 43, *Holmhead-st. Glasgow.*—Hot air engine.

1817 CHANDLER, J. *Mark-lane, E.C.*—Flat glass water guages, for steam boilers, &c.

1818 CHANTRELL, G. F. 6, *Hatton-garden, Liverpool.*—Model of animal charcoal revivifying furnace for sugar refineries.

1819 CHAPLIN, A. & Co. *Glasgow.*—Drawing of traction-engine for common roads, steam-crane used in Exhibition.

1820 CHEDGEY, J. *Grove, Southwark.*—Mangle, with glass bed and rollers, glass pump, and glass pipes. (E.A.)

1821 CHESHIRE SALT Co. *Winsford, Cheshire.*—Apparatus for the manufacture of salt.

1822 CLARK, D. K. 11, *Adam-st. Adelphi* —Smoke consumer and feed-water heater.

1823 CLARK, J. L. 2, *Sambrook-ct. E.C.*—Smoke consuming fire bars. (E.A.)

1824 CLAYTON, SHUTTLEWORTH, & Co.—78, *Lombard-st.*—High-pressure steam engines. (E.A.)

1826 COFFEY, J. A. *Finsbury.*—Pharmaceutical and other apparatus. (E.A).

1827 COLQUHOUN & THOMSON, 1, *Lawrence*

Pountney-hill.—Moveable girder fire bars. (E.A.)

1828 COOMBE & Co. 30, *Mark-lane.*—French mill stones, flour machines, wire brushes, &c. (E.A.)

1829 CORCORAN, B. & Co. *Mark-lane.*—Metallic cloth; model of malt kiln; millstones. (E.A.)

1830 COWAN, T. W. *Kent Iron Works, Greenwich.*—Sixteen horse-power trunk engines. (W. & E.A.)

1831 CROSS, T. W. & Co. *Leeds.*—Fire engines.

1833 DAVIES, J. & G. *Limerick Foundry, Tipton.*—Rotary engine.

1834 DAVIS, J. *Ulverston.*—Steam engine, with fixed valve adapted.

1835 DAWSON, C. S. *Thames Ditton, Surrey.*—Hydrostatic engine. (E.A.)

1836 DAWSON, J. *Greenpark, Scotland.*—A machine for protecting the revenue derived from the manufacture of spirits. (E.A.)

1837 DEACON, H. *Appleton, near Warrington.*—100 millions four-wheeled counter: three-wheeled electric clock.

1838 DINGWALL, W. 4, *Idvies-st. Dundee.*—Water meter. (E.A.)

1839 DIXON, E. *Wolverhampton.*—Wrought iron gas tubes and connections. (E.A.)

1840 DONKIN, B. & Co. *near Grange-rd. Bermondsey.*—Turbine water wheel, and gas valve. Drilling apparatus for mains.

1841 DORWARD, W. L. 15, *Camden-sq. Camberwell.*—Rotary engine.

1842 DUNCAN, T. 44, *West Derby-st. Liverpool.*—A water meter, from which power may be obtained.

1843 EADIE & SPENCER, *Glasgow.*—Iron tubes for boilers. (E.A.)

1844 EASTON, AMOS, & SONS, *Grove, Southwark.*—Centrifugal Appold pump; turbine; hydraulic ram, &c.

1845 EDWARDS, C. J. & SON, 32, *Gt. Sutton-st. London, E.C.*—Leather bands, hose, and fire buckets. (E.A.)

1846 EDWARDS, R. 12, *Fairfield-place, Bow.*—Models of machinery for pulverizing.

1847 ENGLAND, G. & Co. *Hatcham Iron Works, Hatcham.*—Screw jack. (E.A.)

1848 EVERITT, A. & SONS, *Birmingham.*—Brass, copper, and iron articles. (E.A.)

1849 FARROW & JACKSON, 18, *Gt. Tower-st. E.C.*—Machines, &c. for wines, spirits, oil. (E.A.)

1850 FAWCETT, PRESTON, & Co. *Liverpool.*—Cane mill and engine.

1851 FERRABEE, H. 75, *High Holborn.*—Steam and water pressure gauges. (E.A.)

1852 FERRABEE, J. & Co. *Stroud, Gloucestershire.*—Direct-action steam engine, with cut-off valve.

1853 FLEET, B. *East-st. Walworth. S.*—Steam soda-water machine, with bottling apparatus.

1854 FORREST & BARR, *Glasgow.*—Derrick crane. (E.A.)

1855 FORRESTER, G. & Co. *Vauxhall Foundry, Liverpool.*—Triple effect vacuum pan apparatus and air pumps.

1856 FOWLER, B. & Co. *Whitefriars-st. Fleet-st.*—Force pumps, fire engines, and hydraulic rams.

1857 FRIEAKE & GATHERCOLE, 81, *Mark-lane, City.*—Salinometers, telegraphs, indicators, &c. (E.A.)

1858 GALLAGHER, J. *Wolverhampton.*—Self-acting bottling machine. (E.A.)

1859 GALLOWAY, W. & J. *Manchester.*—Models of boilers. Safety valve and lifting jack. (E.A.)

1860 GERARDIN & WATSON, 43, *Poland-st. Oxford-st.*—Beer engine and tavern bar fittings. (E.A.)

1861 GIBBON, R. *Royal Brewery, Brentford.*—Malt-crushing machine. (E.A.)

1862 GODWIN, R. A. 151, *Newport-st. Lambeth.*—Flood pump, double-actioned, &c.; all valves accessible by raising the outlet valve.

1863 GOODALL, H. *Derby.*—Machine for grinding and sifting, and a machine for making bread, &c. (E.A.)

1864 GOUGH & NICHOLS, *Back Quay-st. Works, Manchester.*—Vertical portable engine.

1865 GRAUTOFF, B. A. & Co. 4, *Lime-st. sq. E.C.*—Steam and vacuum gauges and salometers. (E.A.)

1866 GRAY, J. W. & SON, 114, *Fenchurch-st. City.*—Spherical steam engine.

1867 GREENING & Co. *Victoria Iron Works, Manchester.*—Oscillating steam engine, with surface valve.

1868 GREW, N. 8, *New Broad-st. City.*—Model of a locomotive engine for running on the ice.

1869 GRIMALDI, F. & Co. 30, *Bucklersbury, City.*—Rotatory boilers.

1870 GWYNNE & Co. *Essex-st. Wharves, Strand.*—Double-acting centrifugal pump, worked by horizontal engines.

1871 HACKWORTH, J. W. *Darlington.*—High pressure engine and model of locomotive.

1872 HANCOCK, J. & F. & Co. *Tipton-green Furnaces, Staffordshire.*—Drawing condensing engine, with improved vacuum in the cylinder. (E.A.)

1873 HANDS, J. *Cardigan-st. Birmingham.*—Horizontal steam engine, two horse-power.

1874 HANDYSIDE, A. & Co. *Derby.*—Brewing machinery.

1875 HARGREAVES, W. *Crawshaw-booth, Manchester.*—Waterfall washing machine. (E.A.)

1876 HARLOW, R. *Stockport.*—Multitubular fire bridge and heat generator shown in section of steam boiler.

1877 HARRISON, J. 8, *New Broad-st.*—Cast iron boiler. (E.A.)

1878 HARRISS & RISSE, *New Oxford-st.*—Pressure guages.

1879 HART, D. *Whitechapel-rd. London.*—Weighing crane, weigh-bridge, and weighing machines. (E.A.)

1880 HARVEY & Co. *Hayle, Cornwall.*—Model of pumping engine for London water companies, and a model of safety apparatus.

1882 HEPBURN & SONS, 25, *Long-lane, Bermondsey.*—Machine belts and leather. (E.A.)

1883 HERKLESS, W. *Broad-close, Glasgow.*—Machine for grinding tanner's bark. (E.A.)

1884 HILL, J. *Ashford, Kent.* — Flour dressing machine, with silent feed, &c. (E.A.)

1885 HOLGATE, J. & Co. 33, *Dover-rd. Southwark.*—Leather mill bands and hose pipes. (E.A.)

1886 HOLMES, F. H. *Northfleet, Kent.* — Magneto-electric machine and light; light-house regulators.

1888 HORTON, SON, & KENDRICK, *Southwark.*—Models of steam engine boilers. (E.A.)

1889 HOWORTH, J. *Victoria Works, Farnworth, near Bolton.*—Revolving Archimedean screw ventilators, and chimney-tops. (E.A.)

1890 HUGHES, J. & SONS, 91, *Dover-st. Borough.*—Mill stones.

1891 HUMPHREYS & TENNANT, *Deptford-pier.*—Marine engine.

1892 HURRY, H. C. *Rookswood Villa, Worcester.*—An electro-magnetic motive engine.

1893 HUXHAMS & BROWN, *Exeter.*—Bark mill, hydraulic lift, and French burr millstone, for wheat.

1894 IMPERIAL IRON TUBE CO. *Birmingham.*—Iron, brass, and copper tubes, and fittings. (E.A.)

1895 IMRAY, J. *Bridge-rd. Lambeth.*—Horizontal and vertical steam engines.

1897 KEY, J. *Whitebank, Kirkcaldy.*—Horizontal direct acting screw engines, of 80 horse power.

1898 KING, C. B. 20, *Abingdon-st. Westminster.*—Design for traction engine and steam carriage. (E.A.)

1899 KING, J. C. 12, *Portland-rd. W.*—Tubular carriage axle, and wood washers. (E.A.)

1900 KIRKALDY, J. & SONS, 166, *Wapping.*—Ship's portable fire engine. (E.A.)

1901 KNOWELDEN & Co. *Park-st. Southwark.*—Pumps, hydraulic motive engines, cranes, &c.

1902 LAIRD, BROS. *Birkenhead.*—A pair of 40 horse-power horizontal direct acting engines.

1903 LAMBERT, T. & SONS, *Lambeth.*—Hydraulic press pumps, lift ditto; steam engine fittings.

1904 LANSDALE, R. *Pendleton, Manchester.*—Compound rotary washing, wringing, and mangling machine. (E.A.)

1905 LA ROCHE, P. 6, *Blacklands-ter. Chelsea.*—Beer engine and appliances.

1906 LAWRENCE, H. M. & Co. *London Works, Liverpool.*—Machine for making ice by steam.

1907 LAWRENCE, J. 5, *Formosa-ter. Maida-hill, W.*—Refrigerator, mash tun, and other brewing machinery. (E.A.)

1908 LEADBETTER, T. & Co. 13, *Gordon-st. Glasgow.*—Force pump, fire plug, hydraulic ram, and water closet.

1909 LEONI, S. *St. Paul-st. N.*—Taps, steam cocks, &c. of adamas.

1910 LILLESHALL Co. *Shiffnal, Shropshire.*—Pair of blast engines.

1912 LLOYD & LLOYD, *Albion Tube Works, Birmingham.*—Wrought iron tubes and fittings. (E.A.)

1913 LLOYD, G. 70, *Gt. Guildford-st. Southwark.*—Noiseless centrifugal fan blowing machines, mine ventilators, &c.

1914 LOUGH, J. & Co. 69, *Fenchurch-st.*—Union joints and pipe fittings. (E.A.)

1915 LUMLEY & WATSON, 50, *Lower Shadwell, E.*—Steam crane, blocks, and winch.

1916 MCCALLUM, D. 1, *Octagon, Plymouth.*—Electro-magnetic engine.

1917 MCFARLANE, W. 39, *Stockwell-st. Glasgow.*—Cylinder mangle, washing, and wringing machines. (E.A.)

1919 MCGLASHAN & MERRYWEATHER, *Coal-yard, Drury-lane.*—Boiler fittings, plumbers' brass work, pumps, &c. (E.A.)

1920 MACINTOSH & Co. *London and Manchester.*—Vulcanised rubber mechanical articles and appliances. (E.A.)

1921 MCONIE, W. & A. *Scotland-st. Engineer Works, Glasgow.* — 30 horse-power steam engine and sugar mill.

1922 MACORD, R. H. 63, *Lower Thames-st.*—Bottling machines. (E.A.)

1923 MANCHESTER WATER-METER CO. *Ardwick, Manchester.*—High pressure water-meters.

1924 MANLOVE, ALLIOTT, & Co. *Bloomsgrove Works, Nottingham.*—Engines, centrifugal sugar machines, &c.

1925 MARTIN, W. A. 55, *Great Sutton-st. E.C.*—Rocking furnace bars. (E.A.)

1926 MAUDSLAY, SONS, & FIELD, *Lambeth.*—Marine engine.

1927 MAY, W. & Co. *Birmingham.*—Double cylinder steam engine and surface condenser, corn mill, &c.

1928 MERRYWEATHER & SON, *Long-acre.*—Fire-engines, hose, buckets, fire-escapes, &c.

1929 MICKELTHWATE, A. *Sheffield.*—Metallic, hemp, and leather belting, metallic and leather soles. (E.A.)

1930 MIDDLETON, T. *Loman-st. Southwark.*—Murray's chain pump.

1931 MILLER & PIERCE, *Glasgow.*—Fire pump for ships.

1932 MIRRLEES & TAIT, *Glasgow.*—Steam engine and sugar mill in motion.

1933 MONCTON, E. H. C. *Wansford.*—Model of a steam generater. (E.A.)

1934 MOORE, E. 55, *Upper Marylebone-st. W.*—Gauges, steam fittings, &c. (E.A.)

1935 MORGAN, J. & Co. *Stafford-st. Birmingham.*—Beer machines.

1936 MORRISON, R. & Co. *Newcastle-on-Tyne.*—High pressure surface condensing expansive marine engine. cut-off variable.

1937 MURRAY, E. & Co. 2, *Walbrook-buildings, City.*—Moving Argand fire bars; metallic lubricant. (E.A.)

1938 NAPIER, D. & SON, 51, *York-rd. Lambeth.*—Centrifugal machine for curing sugar; automaton mint weighing machine.

1939 NAPIER, R. & SONS, *Glasgow.*—Drawings of marine engines.

1940 NEAL, T. 45, *St. John-st. Smithfield.*—Grinding mills. (E.A.)

1942 NEEDHAM & KITE, *Phœnix Iron Works, Vauxhall.*—Filter for semi-fluids. (E.A.)

1943 NEILL, E. B. 11, *Parliament-st. W.C.* Ericson's caloric air engine (2 h-p.)

1944 NEWTON, KEATES, & Co. *Liverpool.* —Copper and brass articles for engineers, &c. (E.A.)

1945 NOBES & HUNTER, 16, *St. Andrew's-rd. Borough.*—Leather bands, hose, buckets. (E.A.)

1946 NORMANDY & Co. *London.*—Apparatus for obtaining aërated fresh water from sea water.

1947 NORTH BRITISH RUBBER Co. *Edinbro'.*—India rubber belting for machinery.

1948 NORTH MOOR FOUNDRY Co. *Oldham.* —Turbines, fans, blast machines, ship ventilators, steam engines, hydraulic governors, &c.

1949 NORTON, L. 38, *Belle Sauvage-yard, Ludgate-hill.*—Model pumps; cloth tentering and wool-drying machine.

1950 ORKNEY, EARL OF, 3, *Ennismore-pl. Hyde-park.*—Rotary engine.

1951 OXLEY & Co. *Parsonage, Manchester.* —Mill furnishings, lubricators, syphon boxes, air valves, leather strappings, sliver cans. (E.A.)

1952 PARKIN, W. 13½, *Lovell-st. Attercliffe-rd. Sheffield.*—Metallic railway key; cast-steel piston head. (E.A.)

1953 PATENT FRICTIONAL GEARING Co. *Glasgow.*—Robertson's frictional gearing.

1954 PEEL, WILLIAMS, & PEEL, *Soho Iron Works, Manchester.*—Steam engine, hydraulic press, and pumps, for beetroot sugar works.

1955 PENN, J. & SONS, *Greenwich.*—Marine engines.

1956 PERREAUX & Co. 5, *Jeffrey's-sq. London, E.C.*—India rubber mechanical appliances.

1957 POTIER, W. *Green-st. Blackfriars-rd.* —Gut wheel bands. (E.A.)

1958 POTTS, J. *Derby-lane, Burton-on-Trent.*—Working model of steam engine, made of glass, showing movement of piston, valves, &c.

1959 PRELLER, C. A. 4, *Lant-st. S.E.*—Leather machine-driving belts. (E.A.)

1960 RANDOLPH, ELDER, & Co. *Glasgow.* —Drawing of marine engine.

1961 RANSOMES & SIMS, *Ipswich.* — Portable steam engine, 20 horse-power; portable steam crab, 5 horse-power. (E.A.)

1962 RAVENHILL, SALKELD, & Co. *London.*—Models of marine steam engines.

1963 RAWLINGS, J. 10, *Carlton-hill East, N.W.*—Machine for cleaning boots. (E.A.)

1964 RENNIE, G. & SONS, 6, *Holland-st. Blackfriars.* — Marine screw single-trunk engines, of 200 horse-power, of H.M. Ship Reindeer.

1965 RICHARDSON, T. & SONS, *Hartlepool.* —Inverted-cylinder marine engines.

1966 RICHMOND, J. *Hackney-wick Works, Victoria-park, N.E.* — Counting machines. (W. & E.A.)

1968 ROBERTS, R. & Co. 10, *Adam-st. Adelphi.*—Drawing and model of Turbine.

1969 ROBERTS, W. *Millwall, Poplar.*—Fire engines. (E.A.)

1970 ROBINSON, W. *Bridgwater.*—Machine for cleaning casks without unheading. (E.A.)

1971 ROSE, W. 37, *Victoria-st. Manchester.* —Fire engines, &c. (E.A.)

1972 ROUTLEDGE & OMMANNEY, *Salford, Manchester.*—Engines; pumps; boiler feeder; machine for cleaning brass turnings, &c.

1973 RUSE, C. 24, *Hereford-pl. Commercial-rd. East, London, E.*—Beer machines. (E.A.)

1974 RUSSELL. J. & Co. *London and Manchester.*—Wrought iron tubes for boilers, &c. (E.A.)

1975 RUSSELL, J. & SONS. *Wednesbury.*—Tubes and fittings. (E.A.)

1977 RUSTON, PROCTOR, & Co. *Lincoln.*—Traction steam engines. (E.A.)

1978 SALTER, G. & Co. *West Bromwich.*—Spring balances, dynamometers, pressure-gauges, &c. (E.A.)

1979 SAMUELSON, M. & Co. *Hull.*—Oil mill.

1980 SANDERS, F. 473, *Oxford-st.*—Improvements in beer engines. Spirit pumps. (E.A.)

1981 SANDYS, VIVIAN, & Co. *Copper House Foundry, Hayle, Cornwall.*—16 horse-power horizontal steam engine.

1982 SCOTT, G. 35, *Page's-walk, Bermondsey.*—Portable engine; oscillating engine; surface condenser. (E.A.)

1983 SEARBY, G. 2, *Crown-ct. Threadneedle-st.*—Steam guage.

1984 SHAND & MASON, *Blackfriars-rd.*—Various fire-engines, implements, &c. (E.A.)

1985 SHEPARD, E. C. *Victoria-st. Westminster.*—Magneto-electric machine, for electric light. Street-lamp carburator.

1986 SIÈBE, D. 17, *Mason-st. Lambeth.*—"Harrison's" ice making machine.

1987 SIEMENS, C. W. 3, *Gt. George-st. Westminster.*—Regenerative gas engine and furnaces. Fluid meters.

1988 SIMPSON, G. *Glasgow.*—Pumps.

1989 SIMPSON, G. 315, *Oxford-st.*—Freezing machines, refrigerators, seltzogenes, &c.

1990 SISSONS & WHITE, *Hull.* — Steam pile driver. (E.A.)

1991 SMITH, BROS. &. Co. *Hyson Green Works, Nottingham.*—Pressure and vacuum gauges. (E.A.)

1992 SOUL, M. A. 3, *Leadenhall-st.*—Salinometer for steam-boilers using salt water (Long's patent). (E.A.)

1993 STEER, W. *Crossland-st. Nottingham.* —Electro-magnetic engine.

1994 STEPHENSON TUBE Co. *Birmingham.* Seamless metal tubes, calico printing rollers &c. (E.A.)

1995 STONE, J. *Deptford, Kent, S.E.*—Three-throw ships' pump, and fire engines.

1996 STONES, SETTLE, & WILKINSON, *King-st. Hull.*—Brass works for engineers. (E.A.)

1997 STRATFORD, W. 6, *Edward-st. Mile End-rd.*—Furnaces and bars. (E.A.)

1999 STUBBS, W. 1, *Union-st. Mile End.*—Coopering in wood, bone, and ivory. (E.A.)

2000 SUMMERSCALES, W. & SON, *Coney-lane Mills, Keighley, Yorkshire.*—Brush and dash wheel washing wringing, and mangling machine. (E.A.)

2001 SYMONS, C. 2, *George-st. S.E.*—Sewing machine.

2002 TANGYE, BROS. & PRICE, *Birmingham.*—Working model of hydraulic wool and cotton press, and an hydraulic ship jack.

2003 TAPLIN, B. D. & Co. *Traction Engine Works, Lincoln.*—Traction engine. (E.A.)

2004 TAYLOR, J. & Co. *Britannia Engine Works, Birkenhead.*—Traction engine; steam winch with deck pumps. Model of steam crane, and of Stephenson's first locomotive. (W. & E.A.)

2005 TENNANT, T. M. & Co. *Newington Works, Edinburgh.*—8 horse-power upright portable steam engine; 6 horse-power horizontal steam engine. (W. & E.A.)

2006 THOMPSON & STATHER, *Green-lane, Sculcoates, Hull.*—Hydraulic presses, with improved pumps and stops.

2007 THORNEWILL & WARHAM, *Burton-upon-Trent.*—Winding engine for collieries, &c. (E.A.)

2008 TIZARD, W. L. 12, *Mark-lane.*—Surface refrigerator, fermenting apparatus, and washing machine. (E.A.)

2009 TOD & M'GREGOR, *Glasgow.*—Inverted cylinder marine engines.

2010 TOPHAM, C. 31, *Bush-lane, E.C.*—Smith's self-expanding apparatus for cleaning tubular boilers. (E.A.)

2011 TRUSS, T. S. 53, *Gracechurch-st.*—Elastic joint for gas and water pipes. (E.A.)

2012 TYLER, H. & Co. 85, *Up. Whitecross-st.*—Soda water machine, presses, well-engine, lift-pump, &c.

2013 TYLOR, J. & SONS, *Warwick-lane, Newgate-st.*—Pumps, fire-engines, steam fittings. (E.A.)

2014 WALKER & SON, 58, *Oxford-st. Birmingham.*—Steam boiler alarm water guages, &c. (E.A.)

2015 WARD, F. O. *Hertford-st. Mayfair.*—Horizontal steam engine, combined with double-acting hydraulic power pumps, on cistern bed—new principle.

2016 WARNER & SONS, *Cres. Cripplegate.*—Water wheels, irrigators, pumps, &c.

2017 WEBB & SON, *Comb's Tannery, Stowmarket.*—Leather: machine bands, buckets, and hose, &c. (E.A.)

2018 WEIR, E. 142, *High Holborn.*—Washing, wringing, and mangling machines: cinder lifter, bread kneader. (E.A.)

2019 WENHAM, F. H. 1, *Union-rd. Clapham-rise, S.*—A 10 horse-power thermo-expansion steam engine. (E.A.)

2020 WESTON, T. A. 31, *Essex-st. Strand.*—Pulley block and lifting apparatus. (E.A.)

2021 WHITE, J. 7, *Trinity-st. Borough.*—Engine oil feeders, cistern, filters, bands, &c. (E.A.)

2022 WHITMEE, J. & Co. 70, *St. John-st. Clerkenwell.*—Crushers, cutters, mills, and machines: Jolley's American provision safes and refrigerators. (E.A.)

2023 WHITMORE & SONS, *Wickham-market, Suffolk.*—Steam engine, corn mill, &c.

2024 WILKINS, W. P. *Ipswich.*—Condensing steam engine of 20 horse-power, pumps, and stop valve. (E.A.)

2025 WILLIAMSON, W. 133, *High Holborn.*—Washing, wringing, and mangling machine. (E.A.)

2026 WILLIAMSON, BROS. *Canal Iron Works, Kendal, Westmoreland.*—Vortex turbines, fan, pumps, &c.

2027 WILSON, J. C. & Co. 14A, *Cannon-st. E.C.*—Portable steam sugar-cane mill.

2028 WISE, F. 22, *Buckingham-st. Adelphi.*—Feed water regulator. indicator, and alarm for steam boilers. (E.A.)

2029 WOOD, R. & SONS, *Leeds.*—20 horse double cylinder engine, steam pump, &c.

2030 WOODCOCK & LEE, 33, *Old-st. London, E.C.*—Machine for measuring and rolling cloth. (E.A.)

2031 WORSDELL, T. *Berkley-st. Birmingham.*—Steam crane; hydraulic and screw lifting jacks; hydraulic wire testing machines. (E.A.)

2032 WRIGHT, E. T. *Goscote Iron Works, near Walsall.*—Model of diagonal seam steam boiler. (E.A.)

2033 YARROW, A. F. *Barnsbury.*—Locomotive steam carriage. (E-A.)

2035 ZANNI, G. 51, *Lambs' Conduit-st.*—Self-basting roasting apparatus. (E.A.)

2036 NORRIS & Co. *Shadwell.*—Leather hose and belts.

2037 MUYGRIDGE, E. J. 16, *Southampton-st. Strand.*—A washing machine, on the principle of a fulling mill.

2038 WHITEHEAD, J. *Preston.*—Drain pipe, and brick machines; wire netting.

CLASS 9.

AGRICULTURAL AND HORTICULTURAL MACHINES AND IMPLEMENTS.

West Side of Eastern Annex.

2071 AMIES & BARFORD, *Peterborough*.—Portable steaming apparatus, sack elevator, water ballast roller, and clod crusher.

2072 ASHBY, T. W. & Co. *Stamford*.—Hay-maker, rotary harrows, mills, oil-cake breaker, chaff-cutters, steam-engine, &c.

2073 AVELING, J. *Rochester*.—Agricultural locomotive engine.

2074 BAIN, M'NICOL, & YOUNG, 29, *Cross-causeway, Edinburgh*.—Wire netting, fencing, and iron gates.

2075 BALL, W. *Rothwell, Northampton*.—Agricultural cart, ploughs, &c.

2076 BAMLETT, A. C. *Middleton Tyas, Yorks*.—Manual delivery reaper.

2077 BARNARD, BISHOP, & BARNARDS, *Norwich*.—Root pulpers, Norfolk pig troughs, iron chairs, and wire netting.

2078 BARRETT, EXALL, & ANDREWES, *Reading*.—Thrashing machines, steam engines, mills, and agricultural machinery.

2079 BENTALL, E. H. *Heybridge, Essex*.—Chaff cutters, corn and seed crushers, root pulpers, and oil-cake mills.

2080 BEGBIE, J. *Haddington, N.B.*—Sack holder and lifter; machine for sowing seed.

2081 BELL, G. *Inchmichael, by Errol, N.B.*—Reaping machine with sheaffer.

2082 BOBY, R. *Bury St. Edmunds, Suffolk*.—Machines for cleaning and separating grain; wort pump.

2083 BOOTHMAN, J. *Gisburn Coates, near Skipton*.—Observatory bee hive and feeding box.

2084 BOYD, J. *Lewisham*.—Brush lawn mower, shaft roller, and tubular scythe handles.

2085 BROWN & MAY, *North Wilts Foundry, Devizes*.—Portable steam engine, and patent sluice cock.

2086 BURGESS & KEY, *Newgate-st. London, E.C.*—Reaping and mowing machines, and agricultural implements, &c.

2087 BURRELL, C. *St. Nicholas Works, Thetford, Norfolk*.—Traction engine; combined thrashing and dressing machine, &c.

2088 BUSBY IMPLEMENT CO. *Bedale, Yorks*.—Ploughs, horse hoes, carts, and turnip tailers, &c.

2089 CAMBRIDGE, W. C. *Bristol*.—Clod crushers, chain harrows, &c.

2090 CARSON & TOONE, *Warminster, Wilts*.—Chaff-engines, turnip cutters, horse-hoes, cheese presses, &c.

2091 CHANDLER, R. *Old Ford, Bow, Middlesex*.—Models of steam cultivating apparatus.

2092 CHILDS & OWEN, 481, *New Oxford-st.*—Grain separater.

2093 CLAY, C. *Stennard Iron Works, Wakefield, Yorks*.—Clay's cultivator and horse hoe.

2094 CLAYTON, SHUTTLEWORTH, & Co. *Lincoln and London*.—Steam engines, thrashing machinery, &c.

2095 COLEMAN & SONS, *Chelmsford*.—8 h.-p. steam engine and apparatus for cultivation; cultivator and potato digger.

2096 COMINS, J. *South Molton*.—Self-cleaning clodcrusher, set of drags, horse-hoe, paring-plough.

2097 CORNES, J. *Barbridge Works, Nantwich*.—Chaff cutters.

2098 COULTAS, J. JUN. *Spittlegate, Grantham*.—General purpose and corn and seed drills and horse hoe.

2099 CRANSTON, W. M. 58, *King William-st.*—Wood's grass mowing machine.

2100 CROSS, T. W. & Co. *Washington Works, Leeds*.—Garden engines.

2101 CROSSKILL, A. & E. *Beverley*.—Clod crusher, farm railway and trucks, wheels and axles, models, &c.

2102 CROSSKILL, W., TRUSTEES OF, *The Beverley Iron Works, Yorkshire*.—Agricultural implements.

2103 CROWLEY & SONS, *Newport Pagnell, Bucks*.—Model of a steam plough.

2104 CUTHBERT, R. & Co. *Newton-le-Willows, Bedale*.—Reaping machine.

2105 DENNIS, T. H. P. *Chelmsford, Essex*.—Metallic horticultural building.

2106 DORE, J. 17, *Exmouth-st. Clerkenwell*.—Garden, watering, rolling, and syringeing machine.

2107 DOWNIE, R. SEN. *Barnet, Hertfordshire*.—Open beehive and unicomb case.

2108 DRAY, TAYLOR, & Co. 4, *Adelaide-pl. London-bridge*.—Tubular iron gates.

2109 DRAY, W. & Co. *Farningham, Kent*.—Reaping machine, with drop platform.

2110 DRUMMOND, P. R. *Perth*.—Land cleanser—gathers, lifts, and carts stones, &c.

2111 EATON, J. *Thrapstone*.—Turnip thinner and horse hoe, &c.

2112 FENN, R. *Rectory, Woodstock*.—Cottager's bee-hive.

2113 FERRABEE, J. & Co. *Stroud, Gloucestershire*.—Lawn mowing and sweeping machines.

2114 FERRYMAN, E. *Mendep-place, Oundle, Northamptonshire*.—Patent self-kneading lever churn.

2115 FOWLER, J. JUN. 28, *Cornhill, E.C.*—Steam ploughs.

2116 FRY, A. & T. *Temple-gate, Bristol*.—Cart, and American horse-rake.

2117 GARRETT & SON, *Leiston-works, Suffolk*.—Agricultural machinery.

2118 GIBBONS, P. & H. P. *Wantage, Berkshire*.—Portable combined double-blower thrashing machine.

West Side of Eastern Annex.

2119 GRAY, J. *Danvers-st. Chelsea.*—Conservatory, tubular boiler, patent valve.

2120 GRAY & SONS, *Belfast.*—Agricultural machinery.

2121 GRAY, J. & CO. *Uddingston, near Glasgow.*—Agricultural implements, machines, and engine.

2122 GREEN, T. *Smithfield Iron Works, Leeds, and London.*—Lawn-mowing machines.

2123 HALKETT, P. 142, *High Holborn.*—Guideways; entire steam agriculture.

2124 HANCOCK, J. & F. & CO. *Tipton Green Furnaces, Staffordshire.*—Pulverizing-plough; butter machines; windlass.

2125 HAYWOOD, J. JUN. *Phœnix Foundry, Derby.*—Cast-iron vases and chairs.

2126 HENSMAN, W. & SON, *Linslade Works, Leighton Buzzard, Beds.* — Steam plough, drills, &c.

2127 HEREMAN, S. 7, *Pall-mall, East.*—Sir Joseph Paxton's patent hothouses.

2128 HILL & SMITH, *Brierly-hill, Staffordshire.* — Iron-fencing and hurdles, and land rollers.

2129 HOLMES & SONS, *Norwich.*—Corn-seed and manure drills, portable engine, thrashing machine, seed-sheller.

2130 HORNSBY & SONS, *Grantham.*—Steam engines, thrashing machines, drills, winnowing machines, ploughs, machines, &c.

2131 HOWARD, J. & F. *Bedford.*—Steam ploughs and steam cultivators, harrows, horse rakes, haymaking machines.

2132 HUGHES & SONS, 29, *Mark-lane.*—Model machines.

2133 HUGHES, H. *Regent-st. Loughborough.*—Bee-hive.

2134 HUMPHRIES, E. *Pershore, Worcestershire.*—Portable steam engine and combined finishing thrashing machines.

2135 HUNT & PICKERING, *Leicester.* — Corn crusher mills, root pulpers, oil-cake breakers, ploughs, rakes, whippletrees, &c.

2136 HUNT, T. R. & R. *Earls Colne, Essex.* — Steam-power machine for hulling clover and trefoil seed.

2137 HUNTER, P. 64, *Nicolson-st. Edinburgh.*—Churns, dairy utensils, Scotch cooper work, &c.

2138 JAMES, I. *Tivoli, Cheltenham.*—Liquid manure distributor and pump.

2139 KAY, T. *Holbeck Moor Pottery, near Leeds.*—Pots, fern cases, &c.

2140 KEMP, MURRAY, & NICHOLSON, *Stirling, N.B.*—Combined reaper and mower; 2-horse grubber.

2141 KENNAN & SONS, *Dublin.* — Iron fences, log saws, lawn mowers, and root blasters.

2142 KINGSTON, S. *Spalding.* — Rotary cupola bee-hive on Nutt's principle.

2143 LEACH, G. *Leeds.*—Models of patent "steam mole" for cultivating land, &c.

2144 LEE, C. 12, *Warwick-cresc. W.* — Water barrow, runner, greenhouse ventilator.

2145 LESLIE, B. 2, *Abercorn-pl. N.W.*—Model of a pump, for irrigation in India, worked by wind.

2146 LIPSCOMBE & CO. 233, *Strand.* — Improved fountain jets.

2147 LOVEY, E. *Ponsnooth, Perran-wharf, Cornwall.*—Bee-hives.

2148 MAGGS & HINDLEY, *Bourton, Dorset.* —Agricultural machinery.

2149 MAPPLEBECK & LOWE, *Birmingham.* —Draining tools and agricultural implements.

2150 MARRIOTT, J. *Gracechurch-st.* — Apiary, hives, &c.

2151 MESSENGER, T. G. *Loughborough.*—Triangular tubular boiler, hinged valve and indicator.

2152 MILFORD & SON, *Wheel Works, Thorverton, Devon.*—Carts and waggons.

2153 MOODY, C. P. *Holway, Sherborne, Dorset.*—Field gate, formed of machine-made duplicate ports.

2154 MORTON, H. J. & CO. *Basinghall-buildings, Leeds.*—Corrugated iron roofs and buildings, and cable-strained fences.

2155 MUNN, MAJOR, *Throwley, Kent.*—Model of a bee hive.

2156 MUSGRAVE BROS. *Ann-st. Belfast.*—Iron stalls for cattle.

2157 NEIGHBOUR, G. & SON, 149, *Regent-st.* —Bee hive: bees at work.

2158 NICHOLSON, W. N. *Trent Iron Works, Newark.*—Hay machines, horse rakes, oil cake crushers, sack lifters, rollers, &c.

2159 NIXEY, W. G. 12, *Soho-sq.*—Garden labels.

2160 ORMSON, H. *Stanley-bdge. Chelsea.*—Conservatory, hot water tubular boilers, &c.

2161 PAGE, E. & CO. *Victoria Iron Works, Bedford.*—Ploughs, horse hoes, horse rakes, chaff cutters, harrows, &c. &c.

2162 PETTITT, W. J. *Dover.*—Maj. Munn's bar-frame hive.

2163 PHILLIPS, G. *Harrow.*—Bee hives.

2164 PICKSLEY, SIMS, & CO. *Leigh, near Manchester.*—Agricultural implements.

2165 PRIEST & WOOLNOUGH, *Kingston-on-Thames.*—Horse hoes, turnip, manure, and corn drills.

2166 PRINCE & CO. 4, *Trafalgar-sq.*—Mechanical models of inventions.

2167 RANKIN, R. & J. *Liverpool.*—Corn cleaner, for removing smut, &c. from grain.

2168 RANSOMES & SIMS, *Ipswich.*—Steam engines, thrashing machines, screens, mills ploughs, and agricultural machinery.

2169 READ, R. 35, *Regent-circus.*—Horticultural engines, machines, and syringes.

2170 REEVES, R. & J. *Bratton, Westbury, Wilts.*—Liquid manure drills, &c.; manure distributors, &c.

2171 RICHMOND & CHANDLER, *Salford, Manchester.*—Chaff cutters, and machinery for the preparation of food for cattle, &c.

2172 ROBEY & CO. *Lincoln and London.* —Steam engine, thrashing machine, &c.

2173 ROWLEY, J. J. *Rowthorne, Chesterfield.*—Hedge clipping machine, grass mower and reaper combined.

2174 ROWSELL, S. *Buckland St. Mary, Somerset.*—Tubular iron horse rake; gates.

2175 RUSTON, PROCTOR, & CO. *Lincoln.*—Eight h.-p. portable engine, and combined finishing and threshing machine.

2176 ST. PANCRAS IRONWORK CO. *Old St. Pancras-rd.*—Conservatory, and glass walls.

2177 SAMSON & JEWELL, *St. Helier's, Jersey.*—Combined paring and breaking cultivator, in lieu of Skim plough.

2178 SAMUELSON, B. *Banbury and London.* — Harvesting and food preparing machinery.

2179 SCOTT, T. 18, *Parliament-st.*—Self-regulating drinking trough for cattle, &c.

2180 SCOTT, T. *Newcastle, Co. Down, Ireland.*—Machine and apparatus for seed separating, dressing, &c.

2181 SELLAR, G. & SON, *Huntley, Aberdeenshire.*—Ploughs and ridging ploughs.

2182 SHANKS & SON, 18, *Cannon-st. City.* —Lawn mowers.

2183 SHARPE, B. *Hanwell-park, Middlesex.* —Grass harrows.

2184 SMITH, G. 31, *St. John's-sq. Clerkenwell.*—Enamelled garden labels.

2185 SMITH, W. *Kettering, Northamptonshire.*—Horse hoe. Winnowing, sugar, and currant machines.

2186 SMYTH, J. & SONS, *Peasenhall, Suffolk.* —Drilling and sowing machines.

2187 SNOWDEN, W. *Gloucester.*—Paring plough.

2188 STANLEY, J. M. & Co. *Midland Works, Sheffield.*—Octagon conservatory.

2189 STEEVENS, W. 6, *Godolphin-rd. Hammersmith.*—Steam-plough, for ploughing, cultivating, and tilling.

2190 TASKER & SONS, *Andover.*—Combined thrashing and winnowing machine.

2191 TAYLOR, J. & SONS, *Kensall-green, London, W.*—Conservatory, tubular boiler furnace doors, &c.

2192 TEGETMEIER, W. B. *Muswell-hill, N.* —Bee hives.

2193 THOMPSON, H. A. *Lewes, Sussex.*—Entrance and other gates. Iron piers, post caps, drainage instruments, &c.

2194 TURNER, E. R. & F. *Ipswich.*—Portable steam engine, thrashing and dressing machine, mills, &c.

2195 TUXFORD & SONS, *Boston, Lincolnshire.*—Portable steam engines, road locomotives, thrashing, stacking, grinding, and sawing machinery.

2196 TYE, J. *Lincoln.* — Double-mill, French stones, and governors.

2197 TYLER, H. & Co. 85, *Up. Whitecross-st. London, E.C.*—Garden engines, conservatory pump, syringes, fountain-jets.

2198 UNDERHILL, W. S. *Newport, Salop.* —Corn elevator, thrashing machine, field and barn implements, and game and poultry fences.

2199 WALLIS & HASLAM, *Basingstoke.*—Two-horse and four-horse portable thrashing machines, &c.

2200 WARNER, J. & SONS, *Crescent, Cripplegate.*—Garden engines, pumps, syringes, fountains, and fumigator for graperies.

2201 WEEKS, J. & Co. *King's-rd. Chelsea.*—Boiler, heating stacks, models of conservatories, &c.

2202 WEIR, E. 142, *High Holborn.*—Spirit draining levels, with French and English scales, churns, and irrigating pumps.

2203 WILKINSON, WRIGHT, & Co. *Boston, Lincolnshire.*—Stacking machines, or straw carriers, &c.

2204 WILLISON, R. *Alloa, N.B.*—Ventilator for vineries, lift and force pump, &c.

2205 WOODBOURNE, J. *Park Iron Works, Kingsley, near Alton, Hampshire.*—Machine for packing hops.

2206 WOODS & COCKSEDGE, *Stowmarket, Suffolk.*—Prize horse powers, grinding mills, root-pulpers, weed extirpators, models.

2207 YOUNG, J. & T. *Ayr.*—Drill for mangold-wurzel and turnip seed.

2208 ARCHER, E. *New North-rd. Islington.* —Churn on a new principle: and high-pressure liquid manure tap.

2209 HAYES, E. *Watling Works, Stony Stratford.*—Portable steam engine: windlass for steam ploughing.

CLASS 10.

CIVIL ENGINEERING, ARCHITECTURAL, AND BUILDING CONTRIVANCES.

South Court.

Sub-Class A.—Civil Engineering and Building Contrivances.

2226 ALGER'S FURNACE Co. 4, *Victoria-st. Westminster.*—Elliptical blast furnace.

2228 ALLEN, H. 17, *Percy-st. W.*—Double model.

2229 ARBUCKLE, G. B. & SCOTT, T. *Charlton, S.E.*—Models of iron armour for ships and forts.

2230 ARCHITECTURAL POTTERY Co. *Poole, Dorset.*—Mosaic and glazed tiles and bricks, and tile tubs.

2231 ASHBY, R. 34, *Smith-st. Chelsea.*—Model for fire-proof building.

2233 BALE, T S *Mount Pleasant, Newcastle, Staffordshire.* — Mosaic and ornamental floor and wall tiles and glazed bricks.

2234 BARNETT, S. 23, *Forston-st. Hoxton.* —Diving apparatus.

2235 BARRETT, H. 12, *York-bdgs. Adelphi.* —Model of fire-proof flooring.

2237 BEART'S PATENT BRICK Co. *King's-cross.*—Bricks and agricultural drain pipes.

2238 BELLMAN & IVEY, 14, *Buckingham-st. Fitzroy-sq. W.*—Scagliola.

2239 BETHELL, J. 38, *King William-st. E.C.*—Creosoted woods.

2240 BOWER, G. *St. Neots, Huntingdonshire.*—Gas apparatus and purifier.

2241 BROOKE, E. *Field House Fire Clay*

Works, Huddersfield.—Glazed sewer tubes, fire-bricks, furnaces, &c.

2242 BROWN, J. *Chapel-field, Norwich.*—Models for rendering windows tight.

2243 BROWN, R. *Surbiton, Surrey.*—Tiles, bricks, in colours, ornamental ridge, &c.

2244 BROWN-WESTHEAD, M. *Manchester.*—Hoist-governor, to prevent accidents occurring from excessive speed, &c.

2245 BRUNEL, I. *Duke-st. Westminster.*—Models of Saltash and Chepstow bridges, designed by the late Mr. Brunel.

2246 BUNNETT & CO. *Deptford, Kent.*—Revolving iron shutters, brass sashes, &c.

2247 BURGESS, T. H. 4, *Upper Marsh, Lambeth.*—Stand, for boots being made without sitting.

2248 BURT & POTTS, 65, *York-st. Westminster.*—Water-tight iron window.

2249 CARTWRIGHT, J. M. & Co. *Burton-on-Trent.*—Fire-bricks and arches for locomotive engines.

2250 CHALMERS, J. *London.*—Drawings of proposed railway to connect England and France

2252 CHAPPUIS, P. E. 69, *Fleet-st.*—Reflectors for diffusing daylight, and reflecting artificial light.

2253 CHRISTMAS, R. & JONES, 28, *Lord-st. Birkenhead.*—Castellated circular turret, random rubbed; white quartz.

2254 CLARK & CO. *Gate-st. Lincoln's-inn-fields.*—Revolving shutters in steel, iron, and wood, fire-places, &c.

2255 CLARK, E. 24, *Great George-st. Westm.*—Model of hydraulic graving docks.

2256 CLARKE, G. *South-cres.-mews. Burton-cres.*—Fire escape.

2257 CLARKE, J. V. 251, *High Holborn.*—Gas regulators and apparatus.

2258 CLERK, F. N. *Mitre Works, Wolverhampton.*—Metal roofing and galvanized fittings.

2259 CLIFF, J. & SON, *Wortley, Leeds.*—Clay retorts, fire-bricks, &c.

2261 COCKEY, E. & SONS, *Frome, Selwood.*—Valves for regulating the flow of gas in gas manufactories.

2262 CORY, W. & SON, *Commercial-road, Lambeth.*—Machinery for discharging colliers and other vessels.

2263 COSSER, F. C. 145, *York-rd. Lambeth.*—Models of railway carriage signal, steam-engine, chimney pot and ventilator.

2265 COWEN, J. & Co. *Newcastle-on-Tyne.*—Gas retorts, fire-bricks, tiles, &c.

2266 CRESSWELL, J. 100, *Islington, Birmingham.*—Self-folding shutters.

2267 DODMAN, G. 4, *Back South Parade, Manchester.*—Hoist safe, on eccentric principles.

2268 DOULTON, H. & Co. *Lambeth.*—Stoneware pipes and other articles.

2269 DOWNING, G. F. 122, *King's-ra. Chelsea.*—Model of double straining frame for floor cloth.

2270 DUNCAN, R. 174, *Trongate, Glasgow.*—Self-acting time and tide guage.

2271 EASTWOOD, J. & W. *Belvidere-rd. Lambeth.*—Bricks, tiles, and other manufactures.

2272 EDINGTON, T. & SONS, *Phœnix Iron Works, Glasgow.*—Cast iron pipes.

2273 EDMUNDSON, J. & CO. *Dublin.*—Wigham's portable gas apparatus, and for cooking, &c.

2274 EDWARDS, G. H. 90, *Aldersgate-st. E.C.*—Fastening for sash lines, instantaneously connected and disconnected.

2276 ELKIN, W. H. 27, *Belvedere-rd. S.*—A window; can be cleaned without danger.

2277 ERRINGTON, J. E., V.P. Inst. C.E. 13, *Duke-st. Westminster.*—Viaduct across valley of the Lune.

2278 FAIRFAX, BRYSON, & CO. *Birmingham, &c.*—Model of Maillefert's aerostatic tubular diving bell.

2279 FAYLE & CO. 31, *George-st. Hanover-sq.*—Building blocks, bricks, &c.

2281 FISHE, BROS. & CO. *Stourbridge.*—Fire bricks, gas retorts, and glass-house pots.

2282 FISON, C. O. *Stowmarket, Suffolk.*—Kiln tiles for drying malt, &c.; white bricks pantiles.

2285 FREEN & CO.—Specimens of cement.

2286 GARRETT, BROS. *Paddington and Staffordshire*—Tiles, pavements, &c.

2288 GIBBS & CANNING, *Tamworth.*—Glazed stoneware, fire-bricks, &c.

2289 GIBSON & TURNER, *Ball's-bridge, Dublin.*—Models of bridges.

2290 GILKES, WILSON, & CO. *Middlesbro'-on-Tees.*—Model of railway viaduct over the river Beelah, Westmoreland.

2291 GLOVER, G. & CO. *Ranelagh Works, Pimlico.*—Standard gasometers.

2292 GRAY, J., M.D. *Glasgow.*—Coating to preserve iron, wood, and stone.

2293 GREENWOOD, J. 10, *Arthur-st. West, London-bridge.*—India rubber stops for air-tight joints.

2294 HARTLEY, T. *Esher-st. Westminster.*—Specimens of marble.

2295 HAWKSHAW, J. & BARLOW, W. H. 33, *Gt. George-st. Westminster.*—Model of proposed suspension bridge at Clifton.

2296 HEINKE, MESSRS. 79, *Gt. Portland-st.*—Diving apparatus and submarine lamp, &c.

2298 HEMMING, S. C. & CO. 21, *Moorgate-st.*—Iron buildings and iron roofing.

2299 HOLLAND, W. *St. John's, Warwick.*—Apparatus for raising and lowering window sashes.

2300 HOOD, S. & SON, 68, *Up. Thames-st.*—Wrought iron sashes, staircase, and ballusters.

2301 HOWIE, J. *Hurlford Fire Clay Works, N.B.*—Fire bricks, vases, fountain, &c.

2302 INGHAM, W. & SONS, *Wortley, Leeds.*—Fire-bricks, gas retorts, terra cotta, &c.

2303 JACKSON, R. W. *Greatham Hall, Durham.*—Model of harbour and docks of West Hartlepool.

2304 JAMIESON, R. *Glasgow.*—Permeating timber, coating stone, wood, iron, &c.

2306 JONES, W. *Springfield Tileries, Newcastle, Staffordshire.*—Terra metallic ridging, roofing, and paving tiles.

2307 KENNEDY, LIEUT.-COL. 54, *Torrington-sq.*—Financial illustrations of railways, and other public works.

2308 KNIGHT, BEVAN & STURGE, 155, *Fenchurch-st.*—Portland cement.

2309 LAIDLAW & SON, *Edinburgh and Glasgow.*—Gas meters and fittings.

2310 LAWRENCE, BROS. *City Iron Works, London.*—Working models of patent sluices; lifts, cranes, and diving bell apparatus.

2311 LEE, SON, & SMITH, 16, *Up. Ground-st. Blackfriars.*—Limestone, lime, cement, plaster.

2313 MCFARLANE, W. & Co. *Saracen Foundry, Glasgow.*—Architectural cast-iron appliances.

2314 MACINTOSH, J. 40, *North Bank, Regent's-park.*—Telegraphic cables.

2316 MACNEILL, SIR J. 23, *Cockspur-st.*—Model of bridge over the river Boyne, at Drogheda.

2318 SHROPSHIRE COLLECTIVE SERIES OF ARCHITECTURAL CLAY MANUFACTURES OF SHROPSHIRE COAL FIELDS, classified and arranged by G. Maw, F.S.A., on behalf of the following Exhibitors:—

Burton, J. & E.
Coalbrook Dale Co.
Davis, G. & Co.
Doughty, —.
Evans, R.
Exley, W.
Lewis, G. W.
Madeley Wood Co.
Maw & Co.
Simpson, W. B. & Sons.
Thorn, Mrs.

2320 MEARS, G. & Co. 267, *Whitechapel-rd.*—Bell, 15 cwt. and self-acting striking apparatus: models.

2321 MOREWOOD & Co. *Dowgate Dock, London, E.C.*—Galvanized iron roofs, farm buildings, &c.

2322 MORTON, F. *Liverpool.*—Railway fences, iron telegraph poles, galvanized corrugated roofs, &c.

2323 MURRAY, J. 7, *Whitehall-pl.*—Cellular wine-bin.

2324 NORMAN, R. & N. *Burgess-hill, Sussex.*—Ornamental bricks, tiles, &c.

2325 PAINE, MRS. *Farnham, Surrey.*—Stone, &c. made from soluble silica.

2326 PART, J. C. 186, *Drury-lane.*—Martin's cement and plaster of Paris.

2327 PATENT BITUMENIZED WATER GAS & DRAINAGE PIPE CO. 14A, *Cannon-st. E.C.*—Inoxidable pipes, one-fourth weight of iron pipes.

2328 PEAKE, T. *Stoke-upon-Trent, and City-rd. Basin, London, E.C.*—Terro-metallic bricks, tiles, &c.

2330 PERKINS, A. M. 6, *Francis-st. Regent-sq. London.*—Warming and ventilating buildings by hot water.

2331 PITTAR, S. J. 91, *Newman-st. Oxford-st.*—Model bridges, pocket umbrella, &c.

2332 PORTER & Co. *Lincoln.*—Gas-works for private use, with drawings.

2333 RAMAGE, R. 55A, *Holywell-st. Millbank.*—Glass, metal, and other ventilators.

2334 RENNIE, G. & SON, 6, *Holland-st. Blackfriars.*—Model of system of docking, in connection with floating graving docks.

2335 REYNOLDS, W. *Sheffield.*—Artificial stone, metallic mortar.

2336 ROSHER, F. & G. *Blackfriars, London, E.C.*—Garden ornaments of artificial stone, &c.

2337 SCOTT, M. *Parliament-st.* — Models of timber breakwater; submarine foundations; and a diving apparatus.

2338 SIEBE, A. 5, *Denmark-st. Soho.*—Diving apparatus.

2340 SIMMONS, G. 7, *New Palace-yd. Westminster.*—Gas and water connector.

2341 SLACK & BROWNLOW, *Manchester.*—Self-acting cistern filter.

2342 SMITH, A. *Princes-st. Haymarket.*—Door spring, casement fastening, &c.

2343 SPARKES-HALL, J. 308, *Regent-st.*—Upright bench, for a working shoemaker; self-acting prismatic ventilator.

2344 STEPHENSON, W. & SONS, *Newcastle-on-Tyne.*—Fire-clay gas retorts, fire bricks, &c.

2345 STEWART, D. Y. & Co. *Glasgow.*—Cast-iron pipes.

2346 STUTTER, C. *Woolpit, Suffolk.*—White and red facing bricks, stable clinkers, and other kiln goods.

2347 SZERELMEY, N. C. *Laboratory, New Palace, S.W.*—Zopissa; granitic preserving and indurating compounds.

2349 THORN & Co. *Grosvenor-rd. S.W.*—Atmospheric bells, Trinidad asphalte, specimen stone from Old Westminster Bridge, &c.

2350 TOD & MACGREGOR, *Glasgow.*—Model of graving dock and basin on the Clyde.

2351 TUPPER & Co. 61A, *Moorgate-st.*—Galvanized iron.

2352 TURNER, W. & GIBSON, J. W. *Dublin.*—Balance rolling bridges, for railways over water and public roads, iron roofs, &c.

2353 VAVASSEUR & Co. *Sumner-st. Southwark.* — Galvanized, corrugated, and plain sheet iron, &c.

2354 VIGNOLES, C. 21, *Duke-st. Westminster.*—Models and drawings of Bilboa railway, Spain.

2355 VIEILLE MONTAGNE ZINC CO. *Manchester-bldngs. Westminster.*—Models showing the use of zinc.

2357 WALKER, C. & SONS, *Little Sutton-st.*—Gas valves, water valves, hydrants, regulating columns, &c.

2358 DIXON, DIXON, *Croydon, Surrey.*—Model of a locomotive engine.

2359 GRAY, J. & SON, 114, *Fenchurch-st.*—Ship pumping apparatus.

2360 HARRIS, J. *Hanwell.* — Pneumatic locks, and atmospheric telegraphs for mines, &c.

2361 HEMANS, G. W. 13, *Queen-st. Westminster.*—Drawing of a railway bridge over the river Shannon.

2362 MACKENZIE, H. *Ardross and Dundonnel.*—Model for separating weirs in salmon weirs.

2363 BARLOW, P. W. 26, *St. George's-st.*—Model of iron bridge over the Thames at Lambeth.

2364 WHITE, J. B. & BROS. 17, *Millbank-st. Westminster.*—Cements.

2365 SPURGEON, B. W. *Derby.* — Ornamental cement works.

Sub-Class B.—Sanitary Improvements and Constructions.

2368 ASKEW, J. 27, *Charles-st. Hampstead-rd.*—Reversible ventilated window, for cleaning from inside.

2369 BAZALGETTE, J. W. *Spring-gardens*, Drawings of, and models connected with the Metropolitan Main Drainage.

2370 BEAGLE & Co. 71, *Cannon-st. west.*—Ventilators.

2371 BEAUMONT, E. B. *Darfield Pottery, Barnsley, Yorkshire.*—Sanitary tubes, terra-cotta, gas retorts, fire-bricks, filters, &c.

2372 BODMER BROS. *Newport, Monmouthshire.*—Bricks, &c. of unburnt artificial stone.

2374 BROOKS, B. & SMITH, R. & J. 154, *Goswell-st.*—New invented sash-bars for windows.

2375 BURTON & WALLER, *Holland-st. Southwark.*—Drainage apparatus.

2376 CHANTRELL, G. F. 6, *Hatton-garden, Liverpool.*—Water closet, with flushing apparatus.

2377 CHEAVIN, S. *Pen-st. Boston.*—Rapid belt water purifier, damp-proof paints and cement, &c.

2378 CLIFF, J. & Co. *Lambeth.*—Vitrified stoneware drain-pipes, &c.

2379 COOKE, W., C.E. 26, *Spring-gardens, S.W.*—Sanitary appliances.

2380 DALE, T. *Gt. Yarmouth Water Works.*—Service box, for supplying water closets and preventing waste.

2381 DANCHELL, T. H. & Co. 38, *Red Lion-sq.*—Filtering, water softening, and water testing apparatus.

2382 EDWARDS, F. & SON, 49, *Gt. Marlboro'-st. W*—Model of chimneys and ventilating houses.

2383 FIELD & ALLEN, 27, *Frednesh-st. Edinburgh.*—Articles for house-building purposes.

2384 FINCH, J. 11, *Adam-st. Adelphi.*—Porcelain bath designed by his late R. H. Prince Albert.

2385 GOTTO, F. *Leighton Buzzard, Bedfordshire.*—Self-discharging effluvia trap.

2386 JENNINGS, G. *Holland-st. Blackfriars-rd.*—Domestic, sanitary, and building appliances.

2387 KEY, E. *Sharrington, via Thetford, Norfolk.*—Models of country cottages.

2388 KEYNSHAM BLUE LIAS LIME AND CEMENT CO. 6, *Martin's-lane, Cannon-st.*—Samples of blue lias lime.

2389 KITE, C. 20, *Liverpool-st. King's-cross.*—Chimney tops, ventilators, and stable requisites.

2390 LIPSCOMBE & Co. 233, *Strand.*—Self-cleansing charcoal water filters.

2391 LOVEGROVE, J. *Town Hall, Hackney.*—Effluvia traps.

2392 MCKINNELL, J. 15, *Langham-st.*—Ventilator for buildings, &c.

2393 MOORE, J. 81, *Fleet-st.*—Ventilators for houses.

2394 NIXON, T. *Kettering, Northampton.*—Greenhouse.

2395 PIERCE, W. 5, *Jermyn-st. London, S.W.*—Huthnance's heating apparatus for drying rooms, &c. Improved stone grates.

2396 PRITCHARD, W. 3, *Ware-st. Kingsland-rd.*—Life protector for window cleaning.

2397 RIDDELL, J. H. 155, *Cheapside.*—Portable cooking stove, and slow combustion heating stove.

2399 ROSSER, S. E. *Northumberland-st. Strand.*—Warming, ventilating, and desiccating apparatus.

2400 SILICATED CARBON FILTER CO. *Bolingbroke-gardens, Battersea.* — Dahlke's silicated carbon filters.

2401 SMITH, G. & Co. *Glasgow.*—Sanitary structures, such as baths and dry-deodorising closets and stable fittings.

2402 SOWOOD, T. *Blue Boar-ct. Manchester.*—Models in reference to ventilating, heating and smoke curing.

2403 SPENCER, T. 32, *Euston-sq.*—Filters for purifying water with magnetic iron oxide.

2406 TENWICK, J. *Albion Foundry, Clarendon-st. Landport.*—Ventilators for sewers, &c.

2407 TYE & ANDREW, *Brixton-rd.*—Effluvia trap, &c.

2408 UNDERHAY, F. G. *Clerkenwell.*—Regulator water-closets, valves, and basin apparatus.

2409 WARNER, J. & SONS, *Crescent, Cripplegate.*—Ships', and portable water-closets, sanitary contrivances, &c.

2411 WOODCOCK, W. 26, *Great George-st. Westminster.*—Close stoves, open fire-places.

2412 WOODWARD, J. *Swadlincote, Burton-on-Trent.*—Terra-cotta chimney-tops, sewerage pipes, &c.

2413 CENTRAL COTTAGE IMPROVEMENT SOCIETY, 37, *Arundel-st. Strand.*—Models and plans of labourers' cottages.

2414 HAGAN, C. *Office of Works, Tower of London.*—Ventilating chimney tops.

2415 MUNRO, W. *Inverness.*—Models of labourers' cottages, &c.

2416 TAYLOR, J. JUN. 53, *Parliament-st.*—Sanitary building appliances.

Sub-Class C.—Objects shown for Architectural Beauty.

2423 BLANCHARD, M. H. 74, *Blackfriars-rd.*—Articles in terra cotta.

2424 BOUCNEAU, A. 48, *Warren-st.*—French and Italian marble chimney-pieces.

2425 THE COUNCIL OF THE ARCHITECTURAL MUSEUM, 18, *Stratford-pl. W.*—Architectural and decorative carvings in stone and wood.

2426 CLAY, C. 21, *Sidmouth-st. Regent-sq*—Inlaid marble table.

2427 EARP, T. 1, *Kennington-rd. Lambeth*—Architectural and decorative carvings.

2428 EDWARDES BROS. & BURKE, 144, *Regent-st.*—Sculptured staty. chimney pieces; mediæval monument.

2429 FIELD, W. 13, *Parliament-st.* — A marble column.

2430 FORSYTH, J. 8, *Edward-st. Hampstead-rd.*—Model of marble font and cover, for Lord Dudley.

2431 GEORGI, G. 18, *Homer-st. Lambeth.*—Scagliola imitation of Florentine mosaics: ornamental models for marble chimney-pieces, &c.

2432 GOMM, H. J. 18, *Royal-st. Lambeth.*—A Caen stone chimney piece.

South Court.

2433 HARMER, J. M. 10, *Thornhill-bridge-pl. Caledonian-rd.*—Models of architectural ornaments.

2434 JACKSON, G. & SONS, 49, *Rathbone-pl.*—*Carton pierre* enrichments.

2435 MAGNUS, G. E. 40, *Upper Belgrave-pl. S.W.*—Enamelled slate bath, billiard table, chimney pieces, &c.

2436 MITCHELL, G. *Walton-st. Brompton.*—A marble chimney-piece.

2437 EARP, T. 1, *Kennington-rd.Lambeth.*—A drinking-fountain, mediæval style (see No. 2427).

2438 PALMER, J. E. *Guildhall, Swansea.*—Model of a font in Maltese stone.

2439 PALMER, J. E. *Swansea.*—Model of a font.

2440 POOLE H. & SON, 11, *Gt. Smith-st. Westminster.*—Mosaic and inlaid pavement for Chichester cathedral, from designs by W. Slater.

2441 PULHAM, J. *Broxbourne.* — Architectural and garden decorations in terra-cotta.

2442 RICHARDSON, E. *Horewood-sq. London.*—Mural monuments, &c.

2443 ROBERTSON & HUNTER, *Granite Works, Wellington-rd. Aberdeen.*—Drinking fountains of Aberdeenshire granites.

2444 SERPENTINE MARBLE CO. *No. 5, Waterloo-pl.*—Vases, font, and pedestals.

2445 STANLEY, W. *Brighton Cottages, Camberwell.* — Enamelled slate, stone, and marbled glass.

2446 THOMAS, J. 32, *Alpha-rd. Regent's-pk.*—Carved chimney piece, with room and cornice decoration for H.M. the Queen.

2447 HARTLEY, F. H. & CO. *Earl-st. Horseferry-rd. Westminster.*—An Elizabethan baronial chimney piece in various marbles.

2448 HARTLEY, T. H. *Westminster.*—An Elizabethan baronial carved bardilla marble chimney piece, with niches for figures.

2449 POOLE & SON, *Gt. Smith-st. Westminster.*—Specimen of marble, mosaic, and inlaid pavement.

2450 SPURGEON, —. *Derby.* — Chimney piece surmounted by frame, illustrating the capabilities of Martin's cement.

CLASS 11.

MILITARY ENGINEERING, ARMOUR AND ACCOUTREMENTS, ORDNANCE AND SMALL ARMS.

South Court.

Sub-Class A. — Clothing and Accoutrements.

2466 CARTER, LT.-COL. *Monmouth.*—Military accoutrements, boots, &c.

2467 CATTANACH, W. *Bankfoot, Perth.*—Highland-dress purses.

2468 FIRMIN & SONS, 153, *Strand.*—Metal buttons, and military ornaments.

2469 HOLMES, T. 15, *Princess'-ter. Regent's-park.*—Cartouch box.

2470 MACKENZIE, CAPT. J. D. *R.E. Office, Devonport.*—Light volunteer knapsack.

2471 MITCHELL, H. 39, *Charing-cross.*—Photographs of British war medals, &c.

2472 MUNN, MAJOR, *Throwley, Kent.*—Improved cartouch box.

2473 TROUBRIDGE, COL. SIR T., Bt. C.B. 8, *Queens's-gate, W.*—Volunteer or tourist valise, suspended by a metal yoke.

2474 O'HALLORAN, COL. 50, *Elgin-crescent.*—Experimental knapsack, for infantry; and a sporan.

Sub-Class B.—Tents and Camp Equipages.

2486 CLARKE, W. H. 3, *Vernon-pl. Bloomsbury.*—Models of ambulance, waggons, &c.

2487 COTTON, C. P. 8, *Lower Pembroke-st. Dublin.*—Model of improved tent.

2488 EDGINGTON, B. *Duke-st. London-bdg.*—Model of military tent, with stove, &c.

2489 EDGINGTON, F. *Thomas-st. Old Kent-rd.*—Models of tents, &c.

2490 EDGINGTON, J. & CO. 17, *Smithfield-bars, E.C.*—Marquee; rick cloth; models of tents, &c.

2491 PICHLER, F. 162, *Gt. Portland-st.*—Folding and self-supporting tent.

2492 RHODES, MAJOR G. (S. W. SILVER & CO. 34, *Bishopsgate*). — Portable waterproof tents.

2493 TURNER, G. *Northfleet, Kent.*—Models of improved tents and marquees.

2494 UNITE, J. 130, *Edgeware-rd.*—Models of tents. Rick cloth, &c.

Sub-Class C.—Arms, Ordnance, &c.

2505 ADAIR, COL. S. 7, *Audley-sq.*—Military model of London and adjacent country.

2506 ADAMS, R. 76, *King William-st. City.*—Breech-loading guns. Rifles, and revolvers.

2507 AKRILL, E. *Beverley, Yorkshire.*—Breech-loaders—Gilby's patent.

2508 ALLEN, J. W. 22, *& 31, West Strand.*—Camp bucket, canteen, &c.

2509 ARMSTRONG, SIR W. G. *Newcastle-on-Tyne.* — Breech-loading and muzzle-loading ordnance, with projectiles.

2510 BAKER, F. T. 88, *Fleet-st.*—Sporting guns and rifle.

2511 BAYLISS & SON, *St. Mary's-sq. Birmingham.*—Military implements.

2512 BIELEFELD, C. 21, *Wellington-st. Strand.*—Gun wads and cartridges.

2513 BIRMINGHAM MILITARY ARMS TRADE, *Birmingham.*—Military rifles, guns, and pistols, in the form of a Trophy.

2514 BLAKELY, CAPT. 34, *Montpelier-sq. London.*—Blakely's cannon.

2515 BRAZIER, J. *The Ashes, Wolverhampton.*—Gun locks, implements, breech-loader barrels, actions, &c.

2516 BREECH-LOADING GUN CO. *Great Portland-st. London, W.*—Guns, rifle rests, slings, stadiometers, &c.

2517 BRIDER, G. 30, *Bow-st. Covent-gard.*—Implements for fire arms.

2518 BRINE, LIEUT. R.N. *Army & Navy Club.*—Model of Crimean monument; specimens, &c.

2519 BRITTEN, BASHLEY, *Redhill.*—Rifled cannon; projectiles for rifled cannon.

2520 BROWN, CAPT. *Romsey.*—Artificial parchment; self-lubricating ramrod; solid paper tubes; and compressed gunpowder cartridges, &c.

2521 BROWN, J. 8, *Shelley-ter. Stoke Newington.*—Repeating pistol, to fire 14 times.

2522 BURNETT, C. J. *Edinburgh.*—Various fire-arms and projectiles.

2523 CALISHER & TERRY, 24, *Whittall-st. Birmingham.*—Breech-loading rifles and pistols.

2524 CLINTON, LORD, R.N. & HART, G. W. *Southsea.*—Shot-proof embrasure.

2525 SECRETARY OF WAR, *Chatham.*—Model of proposed land defences for Plymouth, and models of parts of Preston and Colchester Barracks.

2526 CULLING, C. *Downham Market, Norfolk.*—Safety gun.

2527 DAW, G. H. 57, *Threadneedle-st.*—Breech-loading arms.

2528 DONNELLY, CAPT. R.E. *South Kensington Museum.*—Rolling drawbridge, without counterbalance weight, or extra length.

2529 DOUGALL, J. D. 59, *St. James'-st. London, S.W.*—Breech-loading fire-arms, the mechanism of which interlocks the barrels and stock.

2530 DRYDEN, C. 10, *Denmark-st. Soho.*—Gun locks.

2531 DU CANE, CAPT. E. F. *Victoria-rd. London, W.*—Iron forts, &c.

2532 EBRALL, S. *Shrewsbury.*—Guns and rifles.

2533 FAIRMAN, J. 23, *Jermyn-st.*—Breech-loading guns.

2534 FAWCETT, PRESTON, & CO. *Liverpool.*—Gun on carriage and boat-slide combined, for land and sea service.

2535 FAWCUS, G. *Alma-pl. North Shields.*—Scaling ladders.

2536 FOWKE, CAPT. F., R.E. *Park-house, South Kensington.*—Collapsing canvas boat pontoons; fire engine for military purposes.

2537 FOX, LT.-COL. A. LANE, GR. GDS. *Clapham.*—Model illustrating parabolic theory, for the range of projectiles in vacuo.

2538 GARDEN, R. S. 29, *Piccadilly.*—Punt gun on Prince's breech-loading principle.

2539 GIBBS, G. *Corn-st. Bristol.*—Breech and muzzle-loading double guns; sporting and target rifles.

2540 GLADSTONE, H. & CO. 22, *Lawrence Pountney-lane.*—Capt. Hayes' seamless skin cartridge.

2541 GISBORNE, F. N. 3, *Adelaide-pl. E.C.*—Electrographic target and signal apparatus.

2542 GRAINGER, J. 60, *Vyse-st. Birmingham.*—Guns, pistols, rifles, &c.

2543 GREENFIELD, J. & SON, 10, *Broad-st. W.*—Minié bullet compressing machine.

2544 GREENER, W. *Birmingham.*—Rifle artillery, double guns, rifles, &c.

2545 HALE, W. 6, *John-st. Adelphi.*—Rockets, comet shells, and apparatus for directing their flight.

2546 HARRINGTON, J. 6, *Lansdowne-ter. West Brixton.*—A self-capping rifle.

2547 HEMMING & CO. 21, *Moorgate-st.*—Electric targets, &c.

2548 HODGES, E. C. 6, *Florence-st. Islington.*—Breech-loading action.

2549 HOLLAND, H. J. 98, *New Bond-st.*—Breech-loading guns and rifles.

2550 JACKSON, R. 30, *Portman-pl. W.*—Rifle, muzzle loader.

2551 JAMES, COL. SIR H. *Southampton.*—Maps, books and instruments, and specimens of photo-zincography.

2552 JEFFERY, A. *Limehouse, E.*—Muzzle-loading rifled ordnance projectiles, showing application of Minié principle to cannon.

2553 JEFFRIES, G. *Golden Ball-st. Norwich.*—Portable cartridge machine and breech-loading gun.

2554 JONES, S.-MAJOR J. *Chatham.*—Iron band gabion, field and floating suspension bridges, &c.

2555 JOYCE & CO. *Upper Thames-st.*—Percussion caps, gun wadding, cartridges, &c.

2557 LANCASTER, C. W. 151, *New Bond-st. W.*—Breech-loading guns and rifles, military rifles (oval bore), &c.

2558 LANG, J. 22, *Cockspur-st.*—Guns, rifles, revolvers and other pistols, &c.

2559 LEETCH, J. 68, *Margaret-st. W.*—Breech-loading fire-arms, &c.

2560 LEWIS, G. E. 32, *Lower Loveday-st. Birmingham.*—Sporting guns, pistols, rifles, &c.

2561 LONDON ARMOURY CO. *Bermondsey.*—Machine-made Enfield rifles, revolving pistols, &c.

2562 LOVELL, J. W. *Chatham.*—Sap shield.

2563 LUCAS, W. H. 109, *Victoria-st.*—Model of a self-adjusting rolling bridge for forts.

2564 MACINTOSH, J. 40, *North-bank, Regent's-pk.*—Breech-loaders, ordnance, and cartridges.

2565 MANTON & SONS, 4, *Dover-st. Piccadilly.*—Guns and rifles.

2566 MARRISON, R. *Great Oxford-st. Norwich.*—Self-extracting breech-loaders, cartridge chargers, &c.

2567 MERSEY STEEL & IRON CO. *Liverpool.*—Guns, armour plates, &c.

2568 MILLER & PEARER, *Glasgow.*—Brass cannon.

2569 MONT STORM, WM. 3, *Rood-lane, E.C.*—Breech-loading fire-arms.

2570 MOORE & HARRIS, *Birmingham.*—Sporting guns, breech-loaders, &c.

2571 MORTIMER & SON, *Edinburgh.*—Breech-loading guns, rifles, &c.

2572 MURCOTT & HANSON, 68, *Haymarket.*—Breech-loaders.

2573 NEWTON, W. E. 66, *Chancery-lane.*—Gabion; new revêtment; fire-place for tents.

2574 PAPE, W. R. 36, *Westgate-st. Newcastle-on-Tyne.*—Sporting guns and rifles.

2575 PARFREY, G. *Victoria-rd. Pimlico.*—Breech-loading double-barrel gun, sliding action, and draws its own cartridges.

2576 PARSON, W. *Swaffham, Norfolk.*—Double guns.

2577 PARSONS, P. M. 9, *Arthur-st. west, London-bridge.*—Breech-loading fire-arms.

2578 PATON, E. *Perth.*—Breech-loading rifles, &c.

2579 POTTER, J. *Lynn, Norfolk.*—Breech and muzzle-loading double guns, and machine for compressing rifle bullets.

2580 PRINCE, F. W. 15, *Wellington-st. London-bridge.*—Breech-loading cannon and small arms.

2581 PURSALL, W. & Co. 45, *Hampton-st. Birmingham.*—Percussion caps. Military and sporting ammunition.

2582 REEVES, C. *Charlotte-st. Birmingham.*—Swords. Rifles to load at breech or muzzle, &c.

2583 REILLY, E. M. & Co. 502, *New Oxford-st.*—Guns, breech-loaders, rifles, revolvers, &c.

2584 RESTELL, T. 43, *Broad-st. Birmingham.*—Breech-loading rifles and small cannon.

2585 RIGBY, W. J. *Dublin.*—Rifles, guns, breech-loading stanchion gun, &c.

2586 RICHARDS, W. & Co. *Birmingham and London.*—Breech-loading rifles, guns, cannon, &c.

2587 SCHLESINGER, J. *George-st. Birmingham.*—Needle fire-arm breech-loader.

2588 SCOTT, M. 26, *Parliament-st. Westminster.*—Sunken, but moveable barrier, to exclude enemies' ships from ports.

2589 SCOTT, W. & C. & SONS, 95, *Bath-st. Birmingham.*—Muzzle- and breech-loading uns, rifles, &c.

2590 SCOTT & ARBUCKLE, 18, *Parliament-st.*—Hythe position trigger, discharged by pressure instead of pull.

SECRETARY OF STATE FOR WAR, *War-Office, London.*

2591 ———Models of works of fortifications.

2592 ———Model of proposed barrack at Colchester.

2593 ———Plan of Netley Hospital.

2594 ———Plan for Herbert Hospital (Woolwich).

2595 ———Plan for Regimental Hospital, for 60 and 120 men.

2596 ———Models of ambulances.

2597 ———Models of 12-pounder Armstrong gun, carriage and limber, forge waggon, rocket carriage, &c.

2598 ———Model of ballistic and gun pendulum.

2599 ———Armstrong rifled 110-pndr. cannon.

2600 ———*Royal Carriage Department* — Armstrong guns-carriages and slide for naval service, travelling, boat, and field carriages.

SECRETARY OF STATE FOR WAR, *War-Office, London.*

2601 ———*Royal Gun Factories*—Rifled ordnance and their details, tools, guages, and drawings of machinery.

2602 ———*Royal Laboratory*—Cartridges, fuzes, projectiles, tubes, &c.

2603 ———*Royal Small Arms Factory*—Specimens illustrating the manufacture of Enfield rifles, &c.

2604 ———*Chemist*—Fuzes to be fired by magneto-electricity and electro-magnetism.

2605 SMITH, G. 40, *Davies-st. Berkeley-sq.*—Sporting guns and rifles.

2606 SYLVEN, T. 33, *Leicester-sq.*—A muzzle-loading gun, altered to breech-loader, &c.

2607 TREEBY, T. 1, *Westbourne-ter.-villas, W.*—Self-indicating target.

2608 TRULOCK & HARRISS, 9, *Dawson-st. Dublin.*—Breech and muzzle-loading sporting guns and rifles.

2609 TYLER, CAPT. R.E. *Hampton Court.*—Sheet-iron gabion.

2610 VALLANCE, P. 4, *Bolton-rd. St. John's-wood.*—Telescopic rifle sights.

2611 WHITE, T. *Portsmouth.*—Model of portable barracks and fittings, &c.

2612 WHITWORTH RIFLE & ORDNANCE Co. *Sackville-st. Manchester.*—Rifled ordnance, gun carriages &c.

2613 WILKINSON & SON, 27, *Pall-mall.*—Guns, rifles, and swords for real service. Defensive armour.

2614 WILLIAMS, A. H. & BOCCIUS, J. G. 135, *Fenchurch-st.*—Percussion cap-holder.

2615 WOODWARD, J. 64, *St. James's-st. S.W*—Guns and rifles, muzzle and breech-loading.

2616 WOOLLCOMBE, R. W. 14, *St. Jean d'Acre-ter. Stoke, Devonport.*—Cannon and projectiles, for projection with cycloidal rotation.

2617 WYLEY, A. *Rose-lodge, Belfast.*—Self-cocking, self-capping breech-loader.

2618 BLAND, CAPT. 23, *Dorset-sq.*—Marking apparatus for rifle practice, requiring no machinery.

2619 BLAND, CAPT. & GODFREY, CAPT. 23, *Dorset-sq.*—Snapping caps and aiming target.

2620 CLINTON, LORD R.—Stadiometer.

2621 MCNEIL, CAPT.—Self-acting targets.

2622 MINTON, S. *Isleworth, London, S.W.*—Rotary battery for coast defences.

2623 MOFFAT, MAJOR A. 5, *Porchester-ter. North.*—Models of an hospital for 80, and of a lavatory for 100 men.

2624 REID, COL. C. 45, *Pall-mall.*—Weapons with which H.M. Goorkah regiments are armed.

2625 THE GUN BARREL Co. *Lawrence Pountney-hill, Cannon-st. City.*—Gun barrels, and iron for them.

2625 SHEDDEN, T. *Ardgartan House, Arrochar.*—Military breach-loading rifle, on a new principle.

2626 VIVIAN, MAJ. *Torquay.*—Trajectory target, self-acting.

CLASS 12.

NAVAL ARCHITECTURE—SHIPS' TACKLE.

South Court.

Sub-Class A.—Ship Building for purposes of War and Commerce.

2646 ASTON, J. J. 4, *Middle Temple-lane.* —Working model boat, with disc propeller.

2647 AYLEN, CAPT. J., R.N. *Welton, near Brough, Yorkshire.* —Wedge-armed anchor, shackle, &c.

2648 BASIRE, J. 4, *King-st. Westminster.*— Models of Brown & Harfield's capstans, and C. Langley's unsinkable ship.

2649 BETHELL, J. 38, *King William-st. E.C.*—Models of a new mode of building ships.

2650 BROWNING, H. *Clifton-wood, Bristol.* —Composition for the preservation of ship's bottoms.

2651 BURDEN, W. *Hay Well, Great Malvern.*—Oblique paddle wheel, illustrating new theory of motion.

2652 BURNETT, C. J. *Edinbro'.* — Fan propellers, with shields and accompaniments.

2653 CAIRD & CO. *Greenock.*—Models of steam-ships and marine engines.

2654 CALLEY, S. *Brixham, Devon.*—Ship's sheathing, compositions for metals and wood, metallic paints, &c.

2655 CAMPBELL, R. F. 8, *Brook-st. Hyde-park.*—Apparatus for management of vessels: new mechanical motion.

2656 CARR, T. *New Ferry, near Birkenhead.*—Models of steering apparatuses.

2657 CLAY, J. 82, *Castle-st. Edgeley.*— Models of ships and propellers.

2658 CLIBBETT, W. *Appledore, Devon.*— Half model of barque.

2659 CLIFFORD, C. 49, *Fenchurch-st.*— Systems of unlashing, lowering, and releasing ships' boats.

2660 COMMISSIONERS IRISH LIGHTS.— Fastnet Rock Light-house, off Cape Clear, S.C. Ireland.

2661 COMMISSIONERS OF NORTHERN LIGHT HOUSES, *Edinburgh.*—Light-house apparatus and models.

2662 CORPORATION OF TRINITY HOUSE, *London.*—Models of light-houses and of a light-vessel.

2663 COUCHMAN, J. W. *Tottenham-green.* —Models of sheet-making, iron bridge, and floating battery.

2664 COULSON, J. & CO 11 & 12, *Clement's-lane, E.C.*—General iron-work for ships.

2665 CUNNINGHAM, H. D. P. *near Gosport, Hants.*—Self-reefing topsail, &c.

2666 DANDO, W. E. 29, *Percy-st. Bedford-sq.* —Apparatus for lowering boats.

2667 DAY W. & CO. *Bow-rd. E.*—Marine cements and compositions.

2668 DENNY BROS. W. *Dumbarton.* — Sectional model of screw steamer, models of two screw steamers.

2669 DUNCAN, R. *Trongate, Glasgow.*— Slip cradle or carriage, and self-acting time and tide guage.

2670 DUNLOP, D. *Hurlet, Glasgow.*—Steam ram: sweeping enemies from decks by machinery.

2671 EDDY, C. W. *Sutton, Loughborough.* —Naval inventions.

2672 ELIOT, E. J. 7, *Southampton-row, W.C.*—Hydraulic apparatus for raising sunken vessels.

2673 ELLIS, G. 4, *Collier-st. North.*— War ships' safety ports.

2675 FORMBY, R. *Liverpool.*—Apparatus for working ships' pumps by water-power.

2676 FRYER, 3, *Leadenhall-st.*—Model of direct acting steam engines and propellers.

2677 FULLER, G. L. 69, *Lombard-st.*— Model of floating ship lift for open water, Mackelcan's pt.

2678 FYFE, T. 46, *Leicester-sq.*—Submarine ship, ship appliances, &c.

2679 GITTINS, R. 28, *New-st. Dorset-sq.*— Model of a new propelling power for steam-ships.

2680 GRANTHAM, J. 31, *Nicholas-lane, E.C.* —Plan for preserving iron ships from concussions, &c.

2681 GRAY, J. W. & SON, 114, *Fenchurch-st.*—Appliances connected with ships.

2682 GREEN, MESSRS. R. & H. *Blackwall.* —Models of ships, iron beams in wooden ships, boat-lowering apparatus.

2683 GRIFFITHS, R. 69, *Mornington-rd. London, N.W.*—Two-screw propellers, and model of armour-plated frigate.

2685 HALL, R. 37, *Princes-stairs, Rotherhithe.*—Ship's figure-head.

2686 HALL, J. & J. *Arbroath and Dundee* —Half model of vessel.

2687 HEWITT, W. 3, *Brislington-crescent, Bristol.*—Feathering screw-propeller.

2688 HIGGINS, A. 10, *St. Vincent's-Parade, Clifton.*—A trader for narrow rivers with new arrangement of rudder.

2689 HORNSEY, W. *West-front, Southampton.*—Marine engine-room. Telegraphs and gongs.

2690 IMRAY, J. *Bridge-rd. Lambeth.*— Hirsch's propeller, with boss for altering pitch.

2691 JAMIESON, R. *Glasgow.*—Compositions for marine architectural purposes.

2692 JECKS, I. *Gt. Yarmouth.*—Ship, with passage to allow missiles to pass through her.

2693 JONES, J. JUN. *Liverpool.*—Models of angulated iron-cased ships, &c.

2694 KING, J. C. 12, *Portland-rd. W.*— Design of a ship of steel, cast in sections.

2695 KIRKALDY, D. 4, *Corunna-st. Glasgow.*—Coloured engineering drawing, R.M.S. Persia; drawings, &c.

2696 LAIRD, J. SONS, & CO. *Birkenhead.*— Models of ships.

2697 LORDS OF THE ADMIRALTY, *Whitehall.*—Models, showing progress made in naval architecture.

2698 MCGREGOR, J. *Beechwood, Partick, near Glasgow.*—Model of screw steamer.

2699 MARE, MESSRS. C. J. & Co. *Millwall.*—Models of the armour-plated war-frigate Northumberland," and of the "Himalaya."

2700 MITCHELL, C. & Co. *Newcastle-on-Tyne.*—Steamers for rivers and sea.

2701 MULLEY, W. R. *Lockyer-st. Plymouth.*—Model of an auxiliary reserve rudder.

2702 PALMER & SWIFT, *Langbourn Chambers, Fenchurch-st.*—Hydraulic marine propellers, regulated independent of the engine.

2703 PATTERSON, W. *Bristol.*—Models of steam ships, yachts, &c.

2704 PEARSE, M. & Co. *Stockton-on-Tees.*—Model of Government troop steamer for the Lower Indus.

2705 PERETTE, A. 25, *Curzon-st. Mayfair.*—Centrifugal and centripetal propeller (combined).

2706 PILE, SPENCE, & Co. *Dockyard, West Hartlepool.*—Models of steam vessels for graving dock, &c.

2707 PORT OF DUBLIN CORPORATION, *Ballast Office, Dublin.*—Light-house models.

2708 PROCTER, S. *Churwell, Leeds.*—Model of auxiliary screw three-masted schooner: inverted cylinders, enclosing slides for screw, &c.

2709 RANDOLPH, ELDER, & Co. *Glasgow.*—Models of vessels.

2710 RENNIE, G. & SONS, 6, *Holland-st. Blackfriars.*—Models of ships; model of a floating graving dock.

2711 RICHARDS, J. *Iron Works*, 27, *Hill-st. Milford.*—Model of iron ship, anchors, and cable.

2712 RICHARDSON, C. J. 34, *Kensington-sq.*—Drawings of projecting shields for ships.

2713 RICHARDSON, DUCK, & Co. *South Stockton Iron Ship Yard, Stockton-on-Tees.*—Models of iron screw steamers.

2714 RICHARDSON, J. W. & Co. *Low Walker, Newcastle-on-Tyne.*—Models of ships and steamers.

2715 ROBERTS R. & Co. 10, *Adam-st. Adelphi.*—Models of screw steamers, windlass and screw.

2716 ROBERTSON, A. J. *Hattonburn, Kinross.*—Models of ships.

2717 ROGERSON, J. & Co. *Newcastle-on-Tyne.*—Models of floating dock and a ferry steamer.

2718 ROSE & CROWDER, *Wapping.*—Lift dock for repairing ships in tideless waters.

2719 RUSSELL, J. S. *London.*—Models of ships built on the wave principle since 1851.

2720 SADLER, W. *Tredegar-pl. Bow-rd.*—Models of frigates, floating batteries, and gun-boats.

2721 SAMUDA BROS. *London.*—Models of steam-vessels—Leinster, Victoria, and Tamar, armour-cased frigates, &c.

2722 SHARPE, B. *Hanwell-pk. Middlesex.*—Shot proofing, gunnery instruments, naval architecture, &c.

2723 SIMONS, W. & Co. *London Works, Renfrew, near Glasgow.*—Models and plans of marine architecture.

2724 SIMPSON, R. *Dundee.*—Models of clipper ship and steamers.

2725 THAMES IRONWORKS & SHIPBUILDING Co. *Blackwall.*—Models of iron-cased frigate Warrior, and other vessels.

2726 THOMPSON, H. L. 47, *Parliament-st.*—Models and drawings of iron ships.

2727 TOD & M'GREGOR, *Glasgow.*—Model of a screw steamer.

2728 TOVELL, G. R. *Ramsey, I. of Man.*—Models of ships and vessels.

2729 TRUSS, T. S. 53, *Gracechurch-st.*—Swift propeller.

2730 VERNON, T. & SON, *Liverpool.*—Models: Woodside landing stage; Caisson, Malta dockyard; and plans of ships.

2731 VINES, R. 3, *Gt. College-st. Camden Town.*—Steamer's floats, for propelling steam vessels without backwater.

2732 WALKER, W. H. *Liverpool.*—Floating hydraulic ship lift.

2733 WATSON, G. 50, *Lower Shadwell, E.*—Boat lowering and disengaging apparatus.

2734 WRIGHT, J. W. 4, *Cumberland-pl. Old Kent-rd. S.E.*—Paddle wheels of an improved construction.

LORDS OF THE ADMIRALTY, *Whitehall:*

2735 Two models, representing the old and modern methods of launching ships.

2736 A series of sixteen models, representing the progress of naval architecture from the first ship of the Royal Navy (1488) to the present time.

2737 The Queen's yachts "Victoria and Albert," "Fairy," and "Osborne."

2738 A series of four models, representing the principal details of different modes of construction for ships of the Royal Navy, from the earliest of which there is any record, to that now adopted in iron-cased ships.

2739 On the wall, 36 half-block models, showing the lines of the different classes of the ships of the Royal Navy from the year 1765, to the iron-cased ships of the present day.

2740 H.M.S. "Queen," representing a full-rigged vessel of war at the time of the last Exhibition.

2740A BELL, E. 31, *Fleet-st. City, E.C.*—Model of "Euryalus" frigate.

2741 COLES, CAPT. R.N.—Models of cupola ships of war.

2741A HENDERSON, T. *Clifton College, Dartmouth, Devon.*—Small models of ships of war.

2742 JERNINGHAM, CAPT. A. W., R.N. *H.M.S. Cambridge, Devonport.*—A port-hole in an iron-plated ship, allowing great elevation, depression, &c.

2742A KYNASTON, MRS. A. 31, *Eastcheap, E.C.*—Patent slip or life hooks for lowering boats.

2742B NASMYTH, J.—Steam-hammer.

Sub-Class B.—Boat and Barge Building, and Vessels for Amusement, &c.

2743 AYLING, E. 50, *Lower Fore-st. Lambeth.*—Racing oars and sculls.

2744 BIFFEN, W. *Middle-wall, Hammersmith.*—Models of boats.

2745 CANNON, H. 14, *Blackwall*.—Model of "Peterboat" for whitebait fishery.

2746 CORYTON, J. 89, *Chancery-lane*.—Ship and life-boat, exemplifying vertical wave-line system.

2747 HALKETT, P. A. 142, *High Holborn*.—Life and other boats.

2748 HAMLEY, J. I. 16, *Capland-st. Lisson-grove*.—Model of life-boat.

2749 HAWKESWORTH, A. & ANNESLEY, G. 65, *Lincoln's Inn-fields*.—Model of the Hartlepool Seamen's Association life-boat in use since 1853.

2750 HUTCHINS, W. *Crooms-hill, Greenwich*.—Pneumatic life-boat and shot-proof ship's cutter.

2751 JORDESON, T. P. *Eastcheap*.—Apparatus for converting ships' boats into life-boats.

2752 LEARWOOD, T. *Truro, Cornwall*.—Life-boat propelled without oars.

2753 PRESTON, LIEUT. T., R.N. *Lowestoft*.—Double rudders and propeller.

2754 PYM, J. 4, *Laurence Pountney-hill*.—Hulk for raising sunken vessels (Pym's.)

2755 RICHARDSON, H. T. *Aberhirnant Bala, N.W.*—Model of Richardson's iron tubular life-boat.

2756 ROYAL NATIONAL LIFE-BOAT INSTITUTION, 14, *John-st. Adelphi*.—Models of life-boats and of other life-saving apparatus, &c.

2757 SEARLE, E. *Stangate, Lambeth*.—Model of state barge.

2758 STEVENS, W. *Trinity-sq. Tower-hill*.—Model, ship's boat, and rigging.

2759 THOMPSON, N. 21, *Rochester-rd. N.W.*—Models of machinery for building boats.

2760 TWYMAN, H. 26, *Hardres-st. Ramsgate*.—Lugger life-boat.

2761 WATSON & DAVISON, 5, *Munster-sq. Regent's-pk*.—Safety rowlocks.

2762 WENTZELL, A. *Lambeth*.—Models of boats.

2763 CHUBBS & CO. *St. Paul's-churchyard*.—Model of "Fairy" yacht.

Sub-Class C.—Ships' Tackle and Rigging.

2774 ADCOCK, J. *Marlborough-rd. Dalston*.—A marine "odometer," or improved ship's log.

2775 BERREY, CAPT. G. A. 32, *Fenchurch-st*.—A "sphereometer," for facilitating great circle sailing.

2776 BIRT, J. 5, *Wellclose-sq*.—Model of the mortar and rocket life-saving apparatus.

2777 BLAKENEY, J. W. & CO. *Hull*.—Ship's compass and binnacle.

2778 BROWN, J. H. *Adelaide-pl. London-bridge*.—A floating buoy for saving ship's papers.

2779 BROWN, LENOX, & CO. *Billiter-sq. E.C.*—Malleable cast-iron blocks and sheaves.

2780 DANBY, J. F. 11, *Cantelowes-rd. N.W.*—Model of "Danby's anchor."

2781 GIFFORD, W. J. 19, *Princes-st. London, W.C.*—Models gaffyard rig.

2782 GLADSTONE, T. M. 30, *Parliament-st*.—Models of an anchor.

2783 GODDARD, J. M. 9, *Ship-st.-lane*.—Specimens in the manufacture of ships' blocks.

2784 HAWKS, CRAWSHAY, & SONS, *Newcastle-on-Tyne*.—Model of Trotman's anchor, 95 cwt., supplied to H.M. frigate Warrior.

2785 HERBERT, G. *Dartford*.—Models of floating bodies moored from centre of gravity. Ships' motion-metre self-registering.

2786 HOLSGROVE & REED, *Sunderland*.—Model of Reed's anchor. Iron ship knees and forgings.

2787 HOLTUM, W. *Church-st. Walmer*.—Model of a balista for communication with stranded vessel.

2788 HUNTER, S. 22, *Grey-st. Newcastle-on-Tyne*.—Model of a new anchor.

2789 JEULA, H. *Lloyd's, E.C.*—Martin's anchor.

2790 LAING, J. 2, *McVicar's-lane, Dundee*.—Helixometer, pump, and ventilator for ships.

2791 LONGHURST, J. *Ticehurst, Sussex*.—Breakless cable chain.

2792 MACDONALD, J. 13, *Henry-st. Vauxhall*.—Compass, ships' lamps, &c.

2793 MARTIN, C. 10, *Bath-place, Hatcham*.—Improved Porter's anchors, and a bombe-mitraille (unloaded).

2794 MOORE, C. *Swansea*.—A spherical indicator, for nautical and astronomical purposes.

2795 PARKES, H. P. *Chain and Anchor Works, Tipton, Staffordshire*.—Chains and anchors.

2796 PEACOCK, G., F.R.G.S. *Starcross, Devon*.—Refuge buoy-beacon, granulated cork poncho-mattress, life and treasure preserver, and anchor.

2797 RETTIG, E. *London*.—Martin's anchor.

2798 RICH, W. 14, *Great Russell-st. Bloomsbury*.—Kite to carry a line, man, &c. from stranded vessels.

2799 ROGERS, *St. Leonard's-rd. Poplar*.—Models of boat-lowering gear, chain cable, stopper, controller, windlass, &c.

2800 ROGERS, W. *Waterloo-st. Swansea*.—A ship's steering apparatus.

2801 ROYAL HUMANE SOCIETY, 4, *Trafalgar-sq. W.C.*—Models of apparatus used for rescuing the seemingly drowned or dead.

2802 SAMUEL, D. A. 3, *Cedar-pl. West Ham, Essex*.—Steering apparatus.

2803 SMITH, R. 23, *Fish-st.-hill*.—Cork life buoy, jacket, belt, &c.

2804 SOLOMON, J. 22, *Red Lion-sq*.—Spheriometers for nautical calculations, in reference to great circle sailing.

2805 STONE, J. *Deptford*.—Side lights, blocks, boat fittings, &c.

2806 TENWICK, J. *Clarendon-st. Landport*.—Steering apparatus.

2807 TROTMAN, J. 42, *Cornhill*.—Model of Trotman's anchor used on board H.M. yacht Victoria and Albert.

2808 TYLOR, J. & SONS, *Warwick-lane, Newgate-st*.—Apparatus for distilling fresh from sea water.

2809 WALKER & SON, *Oxford-st. Birmingham*.—Ship log and sounding machines.

2810 WARD, CAPT. J. R., R.N. *New Brentford*.—Cork life-buoy.

2811 WATSON, T. 49, *Rupert-st. W.*—Application of friction break for ships' capstans; a model.

2812 WEST, J. G. & Co. 93, *Fleet-st.*—Ships' and boats' binnacles, and liquid compasses.

2813 WIGHT, A. 14, *Lansdowne-cres. W.*—Compound iron cable.

2814 WOOD & Co. *Liverpool, London, and Stourbridge.*—Chains, cables, and anchors.

LORDS OF THE ADMIRALTY, *Whitehall:*

2815 Specimens of Admiralty charts.

2816 Specimens of compasses in use in the Royal Navy.

2817 Model of the main-deck compasses of the "Warrior."

2818 A self-registering barometer, by Vice-Admiral Sir A. Milne, K.C.B.

2819 TAYLOR, J.—New sextant, and compass.

2820 MITCHELL, A. & SON, 19, *Adam-st. Adelphi, London.*—Screw-piles and moorings.

2821 WARDEN, G. & Co. 12, *London-st. London, E.C.*—Windlass: heavy and quick purchases. cables always ready, and given by one man.

2822 MOORE, T. 33, *Regent-circus, Piccadilly.*—Windlass worked by endless screws, self-acting chain-stopper, &c.

2823 EMERSON & WALKERS, 21, *Bridge-st. Snnderland.*—Patent ship's windlass.

2824 SELWYN, CAPT. J. H. R.N. *Checquers Court, Tring, Herts.*—Cylinder for laying, under-running, and raising telegraphic submarine cable.

2825 RANKINE, A. *Castle-st. Kircudbright.*—Working model of an iron cased cupola screw propeller war steamer.

CLASS 13.

PHILOSOPHICAL INSTRUMENTS, AND PROCESSES DEPENDING UPON THEIR USE.

Gallery, North Court.

2845 ACKLAND, W. 19, *Church-row, Newington Butts.*—Dividing engine and instruments divided thereby.

2846 ADIE, P. 395, *Strand.*—Philosophical and scientific instruments.

2847 ADIE, R. 55, *Bold-st. Liverpool.*—Philosophical and scientific instruments.

2848 ALDOUS, W. L. 47, *Liverpool-st. King's-cross.*—Microscopic drawings of the human breath and other curiosities.

2849 ALISON, S. DR. 80, *Park-st. W.*—Differential double stethoscope; sphygmoscopes; stethogomiometer; and hydrophone, used in chest diseases.

2850 ALLAN, T. 1, *Adelphi-ter.*—Automatic telegraph; electro magnetic engine; submarine cables, &c.

2851 BAGOT, C. E., M.D. *Dublin.*—Nephelescope, for viewing the upper strata of clouds.

2852 BAILEY, J. W. 162, *Fenchurch-st.*—Optical instruments.

2853 BAKER, C. 244, *High Holborn.*—Microscopes and their appliances: various instruments.

2854 BARNETT, J. 3, *Whitehall-st. Tottenham.*—Microscopic preparations.

2855 BEALE, PROF. *London.*—Microscopes and preparations for ditto.

2856 BELLHOUSE, W. D. 1, *Park-st. Leeds.*—Medical galvanic apparatus.

2857 BENHAMS & FROUD, 42, *Chandos-st. Charing-cross.*—Chemical apparatus in platina.

2858 BESTALL, W. 1, *Victoria-cot. Kennington-pk.*—Apparatus for exhibiting polarised light.

2860 BLIGH, J. 30, *Charles-st. Berkeley-sq.*—Thermometer and self-acting ventilator.

2861 BOLTON & BARNITT, 146, *Holborn-bars.*—Chemical, galvanic, and pneumatic apparatus.

2862 BRAHAM, J. *Bristol.*—Spectacles, from their earliest invention to the present day.

2863 BRETT, J. W. 2, *Hanover-sq.*—Submarine telegraph cables; Roman type printing telegraph.

2864 BRITISH AND IRISH MAGNETIC TELEGRAPH CO. *Liverpool.*—Telegraphic instruments, submarine cables, &c.

2865 BRITISH ASSOCIATION, *Kew Observatory.*—Philosophical instruments.

2866 BROWN, D. S. *Eton Lodge, Ashby-rd. Islington.*—Meteorological instruments.

2867 BUCKINGHAM, J. *Walworth-common.*—Refracting and portable telescopes, equatorially mounted.

2868 BURROW, W. & J. *Great Malvern.*—Landscape glasses and target telescopes.

2869 BURTON, E. 47, *Church-st. Minories.*—Optical and mathematical instruments.

2870 BUSS, T. O. 3, *Upper East Smithfield.*—Hydrometers and saccharometers for fluids.

2871 BUTTERS, T. E. 4, *Belvedere-cres. Lambeth.*—Parallel glass, for optical purposes.

2872 CAMERON, P. *Glasgow.*—Marine compass; marine barometer; instrument for determining ship's position at sea.

2873 CASARTELLI, J. 43, *Market-st. Manchester.*—Microscopes, telescopes, mining and surveying instruments.

2874 CASELLA, L. P. 23, *Hatton-garden.*—Mathematical, philosophical, and meteorological instruments.

2875 CHADBURN, BROS. *Nursery, Sheffield.*—Optical instruments.

2876 CHANCE BROS. *Birmingham.*—Dioptric sea lights and lanterns.

2877 CHATTERTON, T. 14, *King's-ter. Bagnigge Wells-rd.*—Barometers.

2878 CLARK, G. 30, *Craven-st. Strand.*—Improvements in electro telegraphic cables.

2879 COOK, J. E. 22, *Mearns-st. Greenock.*—Damp and water resisting mirrors.

2880 COOKE, T. & SONS, *Buckingham Works, York.* — Telescopes, equatorials, transit, and philosophical instruments, &c.

2881 CHEYNE & MOSELEY, J. B. .—Recording apparatus, to serve as a check upon signal-men, engine-drivers, and others.

2882 COX, F. B. 50, *Camden-st. Birmingham.*—Engineer's and other rules.

2883 COX, F. J. 22, *Skinner-st.*—Dissolving views, apparatus, &c.

2884 CRABBE, REV. G. *Merton Rectory, Thetford.*—Instrument showing solar time at noon within one second.

2885 CRONMIRE, J. M. & H. 10, *Bromehead-st. Commercial-rd. East.*—Mathematical instruments.

2886 CUTTER, W. G. *Crystal Palace.*—Stereoscopes, debuscopes, &c.

2888 DALLMEYER, J. H. 19, *Bloomsbury-st.*—Telescopes, microscopes, lenses, apparatus, &c.

2889 DANCER, J. B. *Manchester.*—Microscopes, microscopic photographs, telescope, &c.

2890 DARKER, W. H. 9, *Paradise-st. Lambeth.*—Illustrations of action of polarized light on crystalline and other bodies.

2891 DAVIS, E. & J. *Leeds and Derby.*—Coal dials, anemometers, gauges, steam thermometers, &c.

2892 DE GRAVE, SHORT, & FANNER, 59, *St. Martin's-le-Grand.* — Hydrostatic assay balances, scales and weights for diamonds, bullion, &c.

2893 DE LA RUE, W. *Observatory, Cranford.* — Astronomical photographs. Photographs of eclipse, 1861.

2894 DENT & CO. 61, *Strand.*—Ship and azimuth compasses and dipleidoscopes.

2895 DESVIGNES, MR. *Lewisham, S.E.*—Mimoscope, a philosophical toy.

2896 DUNCAN, C. S. *Monmouth-rd. Bayswater.*—Rattan deep sea electric telegraph cable.

2897 ELLIOTT BROS. 30, *Strand.*—Philosophical and surveying instruments, &c.

2898 FORBES, R. C. 95, *Warwick-st. Liverpool.*—Artificial horizon for observing altitudes in hazy weather.

2899 FRITH, P. & CO. *Sheffield and London.*—Optical instruments.

2900 GLASS, ELLIOTT, & CO. 10, *Cannon-st. London.*—Electric submarine telegraph cables.

2901 GODDARD, J. T. *Whitton, near Hounslow.* — Cloud mirror and sunshine recorder.

2902 GOWLAND, G. *Liverpool.* — Compasses, sextants, and binocular glasses, &c.

2903 GREEN, S. & SON 7, *Helmet-row, Old-st. London, E.C.*—Compasses and sun dials.

2904 GRIFFIN, J. J. 119, *Bunhill-row.* — Apparatus for experiments in chemistry and physics.

2905 GRUBB, T. *Dublin.*—Great Equatorial (achromatic) 12 inches aperture, equipoised throughout.

2906 HART, W. D. 7, *North College-st Edinburgh.*—Electrical apparatus.

2907 HELY, A. A. 26, *Up. Albany-st. Regent's-park.*—Pocket telescope.

2908 HENLEY, W. S. 46, *St. John-st.-rd.*—Magnets and magnetic instruments, &c.

2910 HETT, A. 4, *Albion-grove, Islington.*—Injected microscopic preparations.

2911 HICKS, J. 8, *Hatton-garden.*—Meteorological instruments.

2912 HIGHLEY, S. 70, *Dean-st. Soho.*—Educational miscroscopes, apparatus, &c.

2913 HINTON, W. 21, *Greville-st.*—Barometers.

2915 HOOPER, W. 7, *Pall-mall East.*—Submarine telegraph cables.

2916 HORNE & THORNTHWAITE, 123, *Newgate-st.*—Meteorological instruments, chemical apparatus, microscopes, polariscopes, &c.

2917 HUDSON & SONS, *Greenwich.* — Animal, vegetable, and fossil tissues and structures, &c.

2918 HUGHES, J. *Queen-st. Ratcliff, E.*—Nautical, optical, and surveying instruments.

2919 JACKSON & TOWNSON, 89, *Bishopsgate Within.*—Chemical and scientific apparatus for general and special purposes.

2920 JOHNSON, H. 39, *Crutched Friars.*—Volutors for tracing spiral curves: Deep-sea pressure-guages: Deep-sea thermometers.

2921 JOHNSON, W. 188, *Tottenham-court-rd.*—Spectacle frames cut from solid steel, &c.

2922 KIESSLER & NEU, *Spencer-st. Goswell-rd.*—Analytical balances, &c.

2923 KNIGHT, G. & SONS, 2, *Foster-lane, E.C.*—Philosophical apparatus.

2924 KULLBERG, V. 12, *Cloudesley-ter. N.*—Self-registering mariners' compass.

2925 LADD, W. 12, *Beak-st. W.*—Focimeter for lighthouses: Induction Coils and apparatus: Microscope with magnetic stage; Air pump.

2926 LADD & OERTLING, 192, *Bishopsgate-st. Without.*—Bullion, chemical, and assay balances; Metal hydrometers and saccharometers.

2927 LAING, J. *Perth-rd. Dundee.*—Instrument giving motion to objects of the stereoscope.

2928 LANKESTER, DR. E. 8, *Saville-row.*—An ozonometer, for registering the hourly variations of ozone.

2929 LEWIS, J. *Dublin.* — Automaton register and pentagraph.

2930 LOWE, RIGHT HONOURABLE R. 34, *Lowndes-sq.* — Spectacles which magnify without glass or any other refracting medium.

2931 MACDONALD, DR. 4, *Coburg-pl. Kennington-lane.* — Instrument for finding the longitude at sea.

2934 MINCHIN, H. *Lower Dominick-st. Dublin.* — Galactoscope, for measuring the transparency of milk.

2935 MOORE, C. *Quay-parade, Swansea.*—Indicator for ascertaining problems, and magnetic variation of compasses, &c.

2936 MORTIMER, J. *Mansfield.* — Instrument for determining inclination and declination of magnetic needle

2937 MURRAY & HEATH, 43, *Piccadilly.*

—Various scientific and philosophical apparatus.

2938 MUSSELWHITE, J. *Devizes.*—An improved syphon.

2939 NEGRETTI & ZAMBRA, 1, *Hatton-garden, E.C.*—Philosophical and meteorological instruments.

2940 NEWTON & Co. 3, *Fleet-st. London.*—Mathematical, surveying, and philosophical instruments, &c.

2941 NORMAN, J. 178, *City-road.*—Microscopic objects.

2942 ORCHARD, J. 2, *Phillimore-pl. Kensington.*—Barometers; Optical and philosophical instruments.

2943 PARKES, J. & SON, *St. Mary's-row, Birmingham.*—Microscopes, telescopes, and philosophical instruments.

2944 PASTORELLI, F. & Co. 208, *Piccadilly.*—Metford's theodolite; new level with micrometer for distances, and other philosophical instruments.

2945 PILLISCHER, M. 88, *New Bond-st. W.*—Microscopes and other optical instruments.

2946 POWELL & LEALAND, 170, *Euston-rd.*—Binocular microscopes with rotating thin stage, object glasses 2¼ in to 1-25th.

2947 PULVERMACHER, J. L. 73, *Oxford-st.*—Galvano-piline, a self-supplying constant battery.

2948 READE, REV. J. B. *Ellesborough, Tring.*—Hemispherical condenser for microscopes, illustrating a new principle in illumination.

2949 REID, BROS. 25, *University-st. W.C.*—Electric telegraph materials.

2950 ROGERS, J. 216, *Gresham House, City.*—Telegraph wires and cables.

2951 RONCHETTI, J. B. 9, *Cambridge-st. Golden-sq.* — Hydrometers, and thermometers.

2952 ROSS, T. 2, *Featherstone-bldngs.*—Optical instruments.

2953 SALMON, W. J. 100, *Fenchurch-st.*—Microscopes.

2954 SAX, J. 8, *Hatton-garden.*—Chemical and bullion balances and weights.

2955 SCOTT, W. E. *Westbourne-park, London, W.*—Self-registering maximum thermometer for deep sea, &c. and microscopic specimens.

2956 SHARP, E. B. *Hanwell-park, Middlesex.*—Submarine electric telegraphs, paying out machinery, &c.

2957 SHARP, H. 38, *Bowden-st. Sheffield.*—Achromatic microscope objectives.

2958 SHAW, W. T. 6, *Park-villas, Dalston.*—Stereotrope, or stereoscopic thaumatrope.

2959 SIEMENS, HALSKÉ, & Co. 3, *Gt. George-st. Westminster.*—Telegraphic apparatus, for land and submarine lines.

2960 SILVER, S. W. & Co. 3, *Bishopsgate-st.*—Electrical machine with ebonite plate.

2962 SPENCER, BROWNING, & Co. 111, *Minories.*—Telescopes; Crooke's spectroscopes; aneroid barometers; and nautical instruments.

2963 SMITH, E. 16, *Queen Anne-st. London, W.*—Potash-box to abstract carbonic acid during expiration.

2964 SMITH, BECK, & BECK, 6, *Coleman-st. E.C.*—Optical instruments.

2965 SMYTH, C. P. *Edinburgh.*—Rotary ship clinometer; model of compound rotary apparatus; electric registering anemometer.

2966 SPRATT, A. 118, *Camden-rd.-villas.*—Weather indicator.

2967 SPRATT, J. 118, *Camden-rd.-villas.*—Lightning conductors, reproducing points, lock isolator, &c.

2968 STANLEY, W. F. 3, *Great Turnstile, W.C.*—Mathematical and surveying instruments.

2969 STEVENSON, P. *Edinburgh.*—Scientific apparatus and brewers' implements.

2970 STEWART, B. *Kew Observatory, Richmond.*—Philosophical instruments.

2971 SUB-MARINE ELECTRIC TELEGRAPH Co. 58, *Threadneedle-st.*—Samples of their sub-marine cables.

2972 SUFFELL, 132, *Long Acre.*—Surveying and drawing implements.

2973 SUGG, W. *Marsham-st. Westminster.*—Photometer and apparatus for determining illuminating power, specific gravity, &c. of coal gas.

2974 SWIFT, J. 3, *Matson's-ter. Kingsland-rd.*—Improvements in construction of microscope.

2976 TREE, J. & Co. 22, *Charlotte-st. Black-friars-rd. S.*—Rules, scales for scientific and other purposes, and levels.

2977 TYER, E. 15, *Old Jewry-chambers.* Train-signalling telegraphs, &c.

2978 UNIVERSAL PRIVATE TELEGRAPH Co. 448, *Strand.*—Wheatstone's magneto-alphabetic telegraphs for railways or private use.

2979 VARLEY, A. 1, *Raglan-ter. Highbury.*—Apparatus for heating greenhouses.

2980 VARLEY, C. 7, *York-pl. Kentish-town.*—Philosophical instruments, microscope, &c.

2981 VARLEY, C. F. 4, *Fortess-ter. Kentish-town.*—Telegraph apparatus, insulators, batteries, electrometer, &c.

2982 VULLIAMY, L. L. & H. P. *Clapham-common.*—Model of electro-magnetic motive-engine.

2983 WALKER, C. V. *Fernside, Redhill.*—Telegraph train signals.

2984 WALTER, J. 17, *Water-st. Liverpool.* Barometer and weather indicator.

2985 WARNER, J. 72, *Fleet-street.*—Philosophical and other instruments.

2986 WATSON, H. *Newcastle-on-Tyne.*—Armstrong's hydro-electric machine.

2987 WEBB, H. *George-st. Balsall-heath.*—Microscopic objects.

2988 WELLS & HALL, *Mansfield-st. Southwark.*—Telegraph conductors, submarine cables, and wires for magnetic coils.

2989 WENHAM, F. H. *Effra-vale-lodge, Brixton.*—A binocular microscope.

2990 WEST, F. L. 31, *Cockspur-st. Charing-cross.*—Barometers.

2991 WHITEHOUSE, N. 2, *Cranbourn-st.*—Opera glasses, spectacles, &c.

2992 WILDE, H. *St. Anne's-sq. Manchester.*—Alphabetic dial telegraphs, worked by magneto-electricity.

2993 WILKINS & Co. *Long-acre, London*—Lighthouse apparatus; lantern for light. ship, lenses, &c.

2994 WOOD, E. G. 74, *Cheapside.*—Philosophical and other instruments.

2995 YEATES, A. 12, *Brighton-pl. New Kent-rd.*—Astronomical, geodulical, and nautical instruments.

2996 YEATES & SON, 2, *Grafton-st. Dublin.*—Philosophical and mathematical instruments.

2997 YOUNG, J. *Dalkeith.*—Carbon for electrical batteries, and electrodes for electric light.

2998 INTERNATIONAL DECIMAL ASSOCIATION, PROFESSOR LEVI, *Farrar's-bdgs. Temple.*—Illustrations of the decimal and metric system of all nations.

2999 FIELD, R. & SON, *New-st. Birmingham.*—Microscopes, telescopes, and surveying instruments.

3000 GUTTA PERCHA COMPANY, 18, *Wharf-rd. City-rd.*—Submarine telegraph cables, and other insulated wire.

3001 HALL, A. J. 2, *William-st. Clerkenwell.*—Machine for describing elipses and other oval curves.

3002 HOLLAND, J. V. 22, *New-st. Spring-gardens.*—Steam expansion gauges.

3003 HUSBANDS & CLARKE, *Denmark-st. Bristol.*—Optical instruments.

3004 MICROSCOPICAL SOCIETY, *London.*—Peters' machine, for microscopic writing.

3005 NICHOLL & FOWLER, 16, *Aldersgate-st. E.C.*—Weighing and measuring apparatus.

3006 REGISTRAR-GENERAL, *Somerset House.*—Tables calculated and stereoglyphed by the Swedish calculating machine.

3007 TENNANT, PROF. J. 149, *Strand, W.C.*—Models of crystals, in glass.

3008 TREMLETT, R. 7, *Guildford-pl. Clerkenwell.*—Barometers and air-pumps.

3009 HUSBANDS & CLARK, 1, *Denmark-st. Bristol.*—Traversing theodolite and portable equatorial.

3009 BECKLEY, R. *Kew Observatory.*—Anemometer, and original photographs of the sun.

3010 HARRIS, T. & SONS, 52, *Great Russell-st. W.C.*—Telescopes, spectacles, and opera-glasses.

3011 COWPER, RT. HON. W. (on the part of Her Majesty's Government).—Babbage's difference engine.

3012 WHITE, J. *Renfield-st. Glasgow.*—Portable atmospheric electrometer, water-dropping collector, marine galvanometer, tangent-galvanometer, static-reaction governor.

3013 GRIESBACH, J. H. 19, *Carlton-rd. Maida-vale.*—Apparatus for ascertaining the pitch, by printing the vibrations, per second, of strings.

CLASS 14.

PHOTOGRAPHIC APPARATUS AND PHOTOGRAPHY.

Central Tower and Gallery, North Court.

3029 ADAMS, A. 26, *Bread-st. Aberdeen.*—Carte de visite, stereoscopic views.

3030 ALFIERI, C. *Northwood, Hanley, Staffordshire.*—Illustrations of Welsh scenery, &c.; negatives made in field camera.

3031 AMATEUR PHOTOGRAPHIC ASSOCIATION, 26, *Haymarket, London*—Photographs by the members of the Association.

3032 ANGEL, O. *High-st. Exeter.*—Photographs, enlarged by the solar camera from collodion negatives.

3033 AUSTEN, W. 5, *Buxton-pl. Lambeth-rd.*—Presses, camera stands, head rests, &c.

3034 BARNES, R. F. 64A, *New Bond-st.*—Photographs.

3036 BASSANO, A. 122, *Regent-st. W.*—Coloured, crayon, and plain photographic portraits.

3037 BEARD, R. 31, *King William-st. London-bridge.*—Coloured and plain photographs and microscopic portraits.

3039 BEDFORD, F. 23, *Rochester-rd. Camden-rd. Villas.*—Photographs: landscape and architecture by the wet collodion process.

3040 BENNETT, A. W. 5, *Bishopsgate Without, London.*—Photographs: application of photography to illustration of books.

3041 BIRD, P. H. 1, *Norfolk-sq. W.*—Photographs of views.

3042 BIRNSTINGL, L. & Co. 7, *Coleman-st. E.C.*—Photographs.

3043 BLAND & Co. 153, *Fleet-st. London, E.C.*—Photographic cameras, materials, and apparatus.

3045 BOOTH, H. C. *Harrogate, Yorkshire.*—Portraits, photographed from life, on paper and ivory, plain and coloured.

3046 BOURNE, S. *Moore & Robinson's Bank, Nottingham.*—Photographic landscapes, by the Fothergill dry process.

3047 BOURQUIN & Co. 13, *Newman-st. Oxford-st.*—Photographic materials, albums, &c.

3048 BOWERS, H. T. *Gloucester.*—Photographic views, collodion and wax papers, enlarged copy of ancient print, &c.

3049 BREESE, C. S. *Acock's-green, near Birmingham.*—Instantaneous transparent stereographs on glass.

3051 BROTHERS, A. *St. Ann's-sq. Manchester.*—Group finished in water colours; portrait on ivory; portraits untouched.

3052 BROWNRIGG, S. W. *Eblana-ter. Phœnix-park, Dublin.*—Photographs.

Central Tower and Gallery, North Court.

3053 BULL, J. T. & G. *Gt. Queen-st. Lincoln's-inn.*—Photographic profiled accessories, and artistic backgrounds.

3054 BURNETT, C. J. 21, *Ainslie-pl. Edinburgh.*—Photograpic prints with uranium, copper, palladium, platinum, &c.

3055 BURTON, J. & PATESON, R. 28, *Avenham-lane, Preston, Lancashire.*—Landscapes and buildings.

3056 CADE, R. 10, *Orwell-pl. Ipswich.*—Machinery and architecture illustrated; also views and portraits.

3057 CAMPBELL, D. *Cromwell-pl. Ayr.*—Large views: Land of Burns.

3058 CAITHNESS, EARL OF, 17, *Hill-st. W.*—Photographic views.

3060 CLAUDET, A. 107, *Regent-st.*—Photographic portraits: stereoscopic, and visiting cards, enlarged to the natural size.

3061 COLNAGHI, P. & D., SCOTT & Co. 13 *&* 14, *Pall Mall East.* — Photographs from ancient and modern pictures, portraits, &c.

3062 CONTENCIN, J. 4, *White Cottages, Grosvenor-st. Camberwell.*—Various photographs from drawings, &c.

3063 CORDINGLEY, W. 14, *Wells-st. St. Helen's, Ipswich.*—Camera stand.

3064 COX, F. J. 22, *Skinner-st. London.*—Lenses, cameras, portable field apparatus, and instantaneous shutters.

3065 CRAMB BROS. *Glasgow.* — Photographs on ivory; views in Palestine; half life-size portraits, not enlarged.

3066 CRITCHETT, C. 11, *Woburn-sq.* — Photographs.

3067 CRUTTENDEN, J. *Week-st. Maidstone.*—Photographs.

3068 CUNDALL, DOWNES, & Co. 168, *New Bond-st.*—Photographs from nature and from drawings.

3069 DALLMEYER, T. H. 19, *Bloomsbury-st. W.C.*—Photographic lenses, cameras, apparatus, &c.

3070 DANCER, J. B. 43, *Cross-st. Manchester.*—Microscopic photographs.

3071 DAVIS, T. S. 3, *Stanley-ter. Stockwell, S.*—Photographic manipulating camera.

3072 DOLAMORE & BULLOCK, 30, *Regent-st. Waterloo-pl. S.W.*—Photographs.

3073 EASTHAM, J. 22, *St. Ann's-sq. Manchester.*—French and English Treaty of Commerce, opal portraits.

3074 FENTON, R. 2, *Albert-ter.*—Photographs.

3075 FIELD, J. *Dornden, Tonbridge Wells.*—Specimens of photolithography; plates engraved on stone by the sun.

3077 GANDY, T. 40, *South-st. Grosvenor-sq.*—Portraits.

3078 GORDON, R. M. 38, *Alpha-rd. St. John's-wood.*—Photographs of Madeira.

3079 GORDON, R. *Bembridge, Isle of Wight.*—Isle of Wight scenery.

3080 GRAHAM, J. *Surrey Lodge, Lambeth.*—Photographic panoramic views of Jerusalem, Syria, Naples, and Pompeii.

3081 GREEN, B. R. 41, *Fitzroy-square.*—Coloured photographs.

3082 GRIFFITHS, J. & BARBER, 2, *Reeves-ter. Canal-rd. Mile End-rd.*—Daguerreotypes, with electrotype copies therefrom; and other photographs.

3083 GRISDALE, J. E. 73, *Oxford-st. W.*—Photographic camera.

3084 GUSH & FERGUSON, 179, *Regent-st.*—Photographic miniatures, collodion process.

3085 HAMILTON, A. R. *Maple-rd. Surbiton, S.W.*—Photographs of the Waterloo medal, by B. Pistrucci.

3086 HARE, G. 140, *Pentonville-road, N.*—Photographic portrait, landscape, stereoscopic, and carte de visite cameras.

3087 HARMER, R. 131, *Shoreditch.*—Photographs illustrating a new method of printing, adapted for book illustration.

3088 HART, F. W. 13, *Newman-st. Oxford-st. London.*—Views.

3090 HEATH & BEAU, 283, *Regent-st. W.*—Miniatures and photographs.

3091 HEATH, VERNON, 43, *Piccadilly.*—Various portraits. English and Scottish landscapes.

3092 HEMPHILL, W. D., M.D. *Clonmel.*—Photographs of antiquitics, &c. at Cashel and Cahir, Co. Tipperary, Ireland.

3093 HENNAH, T. H. 108, *King's-rd. Brighton.*—Collodion photographs.

3094 HERING, H. 137, *Regent-st. London.*—Frames of plain and coloured photographs, portraits and views.

3095 HIGHLEY, S. 70, *Dean-st. Soho.*—Operators' actinometer, micrographic apparatus, dropping bottle, and photographers' travelling lamp.

3096 HILL, D. O. *Edinburgh.* — Photographs.

3097 HOCKIN & WILSON, 38, *Duke-st. Manchester-sq. W.*—Photographic set, and tent; collodion, &c. in hermetically sealed tubes.

3098 HOLDEN, REV. DR. *Durham.* — Photographs of cathedrals and abbeys.

3099 HOPKIN & WILLIAMS, 5, *New Cavendish-st*—Photographic chemicals.

3100 HORNE & THORNETHWAITE, 123, *Newgate-st.*—Photographic lenses, cameras, apparatus, and chemicals.

3101 JAMES, COLONEL SIR H., R.E. *Ordnance Survey Office, Southampton.*—Plans reduced by photography, photozincographs, and photopapyrographs.

3102 JEFFREY, W. 114, *Gt. Russell-st. Bloomsbury, W.C.*—Photographs from busts of Alfred Tennyson, William Fairbairn, &c.

3103 JEFFERY, W. *Shepherd & Co.* 97, *Farringdon-st. E.C.* — Photographic tent, 14 lbs. weight.

3104 JONES, B. *Selkirk-villa, Cheltenham.*—Photographic pictures from glass negatives.

3105 JOUBERT, F. 36, *Porchester-ter. W.*—Photographs in vitrifiable colour, burnt in on glass; collodion photographs, and phototypes.

3106 KATER, E. 46, *Sussex-gardens, Hyde-pk.*—Photographs of ancient armour from Mr. Meyrick's collection.

3107 KEENE, R. *All Saints, Derby.* — Photographs illustrating scenery and antiquities of Derbyshire.

3108 KILBURN, W. E. 222, *Regent-st.*—Photographic portraits.

3109 KING, H. N. 42½, *Milsom-st. Bath.*—Cartes de visite; portraits of celebrities; views and stereoscopic slides.

3110 LAMB, J. 191, *George-st. Aberdeen.*—Views or portraits, or both.

3111 LEAKE, J. C. *Poplar, London, E.*—Photographic operating tent.

3113 LICKLEY, A. *Allhallowgate, Ripon Yorkshire.*—Camera with shade and shutter; positive collodion photographs.

3115 LOCK & WHITFIELD, 178, *Regent-st.*—Photographic miniatures.

3116 LONDON SCHOOL OF PHOTOGRAPHY, 103, *Newgate-st. &c.*—Photographs.

3117 LONDON STEREOSCOPIC CO. 54, *Cheapside, E.C.*—Instantaneous stereoscopic views, large views, and portraits.

3118 MACDONALD, SIR A. K. BART. *Woolmer, Liphook, Hants.*—Photographic views.

3119 MACKENZIE, W *Paternoster-row.*—Photographic illustrations for the Queen's bible, by Frith.

3120 M'LEAN, MELHUISH, & HAES, 26, *Haymarket.*—Photographic apparatus; untouched and coloured photographs.

3121 MARRIOTT, M. *Montpelier-sq. London, S.W.*—Panoramic camera; portable stereoscopic cameras for dry processes.

3122 MAULL & POLYBLANK, 187A, *Piccadilly.*—Photographs.

3123 MAYALL, J. E. 226, *Regent-st.*—Portraits of eminent personages, studies from life. A crayon machine and daguerreotypes.

3124 MAYER BROS. 133, *Regent-st.*—Photographic portraits.

3125 MAYLAND, W. *Cambridge.*—Views of the University and its vicinity.

3126 MOENS, W. J. C. *Lewisham.*—Views of water supply of ancient Carthage; temples in Greece, and others.

3127 MUDD, J. 10, *St. Ann's-sq. Manchester.*—Landscape photographs.

3128 MURRAY & HEATH, 43, *Piccadilly.*—Cameras, tent, baths, draining frames, plate-holders, and other photographic apparatus.

3129 NEGRETTI & ZAMBRA, *Hatton-garden.*—Transparent glass pictures and apparatus.

3130 NEWCOMBE, C. T. 135, *Fenchurch-st. E.C.*—Photographs.

3131 NICHOLSON, A. 23, *St. Augustine-rd. Camden Town.*—Photographs from plates prepared by Fothergill's process.

3132 OLLEY, W. H. 2, *Bolingbroke-ter. Stoke Newington.*—Photographs from the microscope, by reflecting process.

3133 OTTEWILL, T. & CO. *Charlotte-ter. Islington.*—Photographic apparatus.

3134 PENNY, G. S. 14, *Rodney-ter. Cheltenham.*—Photographs by various processes.

3135 PIPER, J. D. *Ipswich.*—Landscapes, &c. by collodion process.

3136 PONTING, T. C. 32, *High-st. Bristol.*—Photographs enlarged from small negatives. Iodized negative collodion, sensitive for years.

3138 POULTON, S. 352, *Strand, W.C.*—Stereoscopic slides. Photographs, untouched and coloured.

3139 POUNCY, J. *Dorchester, Dorset.*—Photographs printed in carbon.

3140 PRETSCH, P. 3, *Guildford-pl. Foundling.*—Photographic engraving, and printing with ordinary printer's ink, &c.

3141 PROUT, V. 15, *Baker-st. Portman-sq. W.*—Reproductions of pictures—various subjects.

3142 PYNE, J. B. JUN. 40, *Roxburgh-ter. Haverstock-hill, N.W.*—Photographic copies of pictures, sculpture, portraits from life, &c.

3143 RAMAGE, J. *Edinburgh.*—Photo-lithographs.

3144 REEVES, A. 257, *Tottenham Court-rd.*—Microscopic photographs and microscope.

3145 REJLANDER, O. G. 42, *Darlington-st. Wolverhampton.*—Various photographs.

3146 RICHARDSON, T. W. *Brede, Sussex, and Staplehurst.*—A reflecting camera.

3147 ROBINSON, H. P. 15, *Upper Parade, Leamington.*—Photographs.

3148 ROSS & THOMSON 90, *Princes-st. Edinburgh.*—Photographs by the collodion process.

3149 ROSS, T. 2 & 3, *Featherstone-buildings, Holborn.*—Photographic lenses, cameras, stands, and apparatus.

3150 ROUCH, W. W. 180, *Strand.*—Apparatus and chemicals; photographs, taken with new binocular camera and Hardwich's brom-iodized collodion.

3151 RUSSELL, J. *East-st. Chichester.*—Ruins of Chichester Cathedral after the fall of the spire.

3152 SHEPHERD & CO. 97, *Farringdon-st.*—Photographic lenses, cameras, apparatus, &c.

3153 SIDEBOTHAM, J. 19, *George-st. Manchester.*—Photographic landscapes, by the collodio-albumen process.

3154 SIMPSON, H. 1, *Savile-pl. Lambeth.*—Photographic cabinets, forming complete operating rooms.

3155 SKAIFE, T. 47, *Baker-st. W.*—Pistolgraph, with a selection of its productions called pistolgrams.

3156 SMITH, L. *Cookridge-st. Leeds.*—Photographic views.

3157 SMYTH & BLANCHARD, *George-st. Euston-sq.*—Instantaneous photographs and life-size photographs.

3158 SOLOMON, J. 22, *Red Lion-sq.*—Photographic apparatus, &c.

3160 SPACKMAN, B. L. *Kensington Museum.*—Photographs of the gardens of Horticultural Society. Various art re-productions. Exhibition building.

3161 SPENCER, J. A. 7, *Gold Hawk-ter. Shepherd's Bush, W.*—Albumenized and other prepared photographic papers.

3162 SPODE, J. *Hawkesyard-park, near Rugeley.*—Proofs from collodion negatives.

3163 STOVIN & CO. *Whitehead's-grove, Chelsea.*—Principal buildings, London; microscopic photographs.

3164 STUART-WORTLEY, LIEUT.-COL. A. H. P. *Carlton Club, Pall Mall.*—Photographs of Vesuvius, during the eruption of 1861-2.

3165 SUTTON, E. 204, *Regent-st. W.*—Miniature photographs, plain and coloured.

3166 SWAN, H. 5, *Bishopsgate Without, London.*—Large (and apparently single) pictures rendered stereoscopic. New stereoscopes.

3167 TALBOT, W. H. FOX, *Lacock Abbey, Wiltshire.*—Photoglyphic engravings, produced by the action of light alone.

3168 TELFER, W. 194, *Regent-st.*—Untouched and coloured photographs.

3169 THOMPSON, C. THURSTON, *South Kensington Museum.*—Photographs from the Raphael cartoons, and pictures by J. M. W. Turner.

3170 THOMPSON, S. 20, *Portland-rd. Notting-hill, W.*—Photographs, landscapes, architectural subjects, &c.

3171 TRAER, J. R. 47, *Hans-pl. S.W.*—Photographs of microscopic objects.

3172 TURNER, B. B. *Haymarket.*—Photographs from paper negatives taken by the Talbot process.

3173 VERSCHOYLE, LT. COL. 23, *Chapel-st. Belgrave-sq.*—Photographs, by wet, and collodion-albumen processes.

3174 WALKER, C. & SON, *Windsor-rd. Lower Norwood.*—Carbotype photographs, unchangeable; silver printed duplicates, changeable.

3175 WARDLEY, G. 10, *St. Ann's-sq. Manchester.*—Photographic landscapes: negatives produced by the Taupenot process.

3176 WARNER, W. H. *Ross, Herefordshire.*—Architectural and miscellaneous photographs.

3177 WATKINS, H. 215, *Regent-st.*—Photographic portraits.

3178 WATKINS, J. & C. 34, *Parliament-st. S.W.*—Portraits, plain and coloured.

3179 WHITE, H. 7, *Southampton-st. Bloomsbury.*—Photographic landscapes.

3180 WHITING, W. & SONS, *Camden Town.*—Portable developing cameras for working wet collodion in the open air.

3181 WILDING, W. H. 2, *Chesterfield-st. King's-cross.*—Universal eccentric camera front; instantaneous camera.

3182 WILLIAMS, T. R. 236, *Regent-st. W.*—Untouched and coloured photographic portraits, vignettes, cartes de visite, &c.

3183 WILSON, G. W. *Aberdeen.*—Views by the wet collodion process.

3184 WILSON, SIR T. M. *Charlton House.*—The Geysers, Iceland.

3186 WRIGHT, C. 235, *High Holborn.*—Photographic portraits and copies of paintings.

3187 WRIGHT, DR. H. G. *London.*—Portable photographic apparatus, including tent, &c.

3188 MULLINS, H. *Jersey.*—Photographic portraits.

3189 MOULE, J. 15, *Seabright-pl. Hackney-rd. N.E.*—Photographic apparatus for producing portraits by artificial light; and photographs taken at night.

CLASS 15.

HOROLOGICAL INSTRUMENTS.

Gallery, North Court.

3218 ADAMS, F. B. & SONS, 21, *St. John-sq. E.C.*—Reversible chronometer, watches, &c.

3219 AGAR, W. *Bury, Lancashire.*—Working men's watches.

3220 ARMSTRONG, T. *Manchester.*—Watchmen's clocks; steam or speed clock, &c.

3221 AUBERT & LINTON, 252, *Regent-st.*—Watches and ornamental clocks.

3222 BAILEY, J. & CO. *Albion Works, Manchester.*—Turret clocks.

3223 BARRAUD & LUND, 41, *Cornhill.*—Chronometers and watches.

3224 BAYLISS, W. *Finmere, Oxfordshire.*—Model of new remontoire escapement.

3225 BENNETT, J. 65, *Cheapside.*—Marine and pocket chronometers; clocks, gold and silver watches, &c.

3226 BENSON, J. W. *Ludgate-hill.*—Gold and silver watches and clocks.

3227 BLACKIE, G. 24, *Amwell-st. E.C.*—Balance for chronometers, and new auxiliary.

3228 BROCK, J. 21, *George-st. Portman-sq.*—Marine chronometers.

3229 BROOKS, S. A. *Northampton-sq. E.C.*—Watch jewels and materials.

3230 CAMERER, KUSS, & CO. 2, *Broad-st. Bloomsbury.*—Part quarter skeleton on ten bells, trumpeter, and cuckoo clocks.

3231 CAMPBELL, A. 63, *Cheapside.*—Gold and silver watches.

3232 CHEVALIER, B. 4, *Red Lion-st. E.C.*—Chronometer and watch cases.

3233 CLARK, DR. *Finmere-house, Oxfordshire.*—Astronomical clock; impelled by gravitation, requires no oil to the escapement.

3234 COATHUPE, CAPT. H. B. 1, *Abingdon-ter. Kensington.*—"Silent clocks," &c.

3235 COLE, J. F. 5, *Queen-sq. Bloomsbury.*—Chronometers, watches, new horological models, &c.

3236 COLE, T. 6, *Castle-st. Holborn.*—Clocks of novel construction.

3237 CONDLIFF, J. 4, *Fraser-st. Liverpool.*—A skeleton clock.

3238 COOKE, T. & SONS, *Buckingham Works, York.*—Astronomical and other clocks, &c.

3239 CRISP, W. B. 81, *St. John-st.-rd.*—Chronometers.

3240 DAVIES, C. W. *Notting-hill.*—Clock showing time and longitude at important places.

3241 DAVIS, W. & SONS, 84, *King William-st. City.*—Chronometers, watches, clocks, &c.

3242 DELOLME, H. 48, *Rathbone-pl. Oxford-st.*—Chronometers, clocks, and watches.

3243 DENT & CO. 61, *Strand.*—Chronometers, regulators, watches, &c.

3244 DENT, M. F. & CO. 34, *Cockspur-st. S.W.*—Chronometers, watches, and clocks; new auxiliary compensation balance, &c.

3245 DE SOLLA, J. & SON, 34, *Southampton-ter. Waterloo-bridge.*—Lilliputian alarm clocks.

3246 DETTMANN, T. *Minories.*—Astronomical clock, with constant ball escapement; electro-magnetic clock, &c.

3247 EHRHARDT, W. 26, *Augusta-st. Birmingham.*—Watches, and instruments connected with them.

3248 FAIRER, J. 188, *St. George-st. E.*—Turret clocks, watches, &c.

3249 FORREST, J. 29, *Myddelton-st. E.C.*—Pocket watches.

3250 FRODSHAM, C. 84, *Strand.*—Watches and astronomical clocks; new equation double compensation balances, &c.

3251 FRODSHAM & BAKER, 31, *Gracechurch-st.* — Chronometers, watches, and clocks.

3252 GANEVAL & CALLARD, 27, *Alfred-st. Islington.*—Watch pendulum, spring, and wire manufactures.

3253 GREENWOOD, J. & SONS, 6, *St. John's-sq. E.C.*—Quarter and bracket clocks, &c.

3254 GUIBLET & RAMBAL, 11, *Wilmington-sq. Clerkenwell.*—Keyless fuzee watches. Pocket chronometers.

3255 GUILLAUME, E. & C. 16, *Myddelton-sq. E.C.*—Watches and repeaters.

3256 GUMPEL, C. G. 2, *Gordon-cottages, Brixton.*—A system of electric clocks.

3259 HAWLEYS, 287, *High Holborn.* — Regulator, to be wound once in 12 months.

3260 HIGHFIELD, BROS. 5, *King Edward-ter. Islington.*—Marine and pocket chronometers, watches, and regulator.

3261 HILL, C. J. *Chapel-fields, Coventry.*—Watches and pearl dials.

3262 HISLOP, W. 108, *St. John-st.-rd. London, E.C.*—Standard or observatory clock, showing mean and sidereal time.

3263 HOLDSWORTH, S. 220, *Upper-st. Islington.*—Chronometer and watch jewels.

3264 HOLL, F. R. 284, *City-rd.*—Keyless chronometers and watches.

3265 HOLLIDAY, T. 108, *Goswell-rd.*—Watch cases and dials.

3266 HOLLOWAY & CO. 128, *Minories.*—Clocks of simple construction.

3267 HOWARD, R. 29, *King-sq. Goswell-rd.*—Sunk seconds dials, &c.

3268 HOWELL, JAMES, & CO. *Regent-st.*—Clocks, watches, &c.

3270 HUTTON, J. 10, *Mark-lane.*—Marine and other chronometers, &c.

3271 JACKSON, W. H. & S. 66, *Red Lion-st. Clerkenwell.*—Chronometers; day of month, keyless, and other watches.

3272 JOHNSON, E. D. 9, *Wilmington-sq. Clerkenwell.*—Chronometers, watches, pendulums, &c.

3273 JONES, J. 338, *Strand.*—Watches.

3275 KLAFTENBERGER, C. J. 157, *Regent-st.*—Minute repeaters, chronometers, lever and duplex watches.

3276 KULLBERG, V. 12, *Cloudesley-ter. N.*—Chronometers, watches, and clocks.

3277 LANGE, C. 9, *Salisbury-st. Strand.*—Watches and timepieces.

3278 LEONARD, G. W. *Cloudesley-ter. Liverpool-rd.*—Compensation balances.

3279 LOSADA, J. R. 105, *Regent-st.*—Chronometers, clocks, and astronomical pendulums, &c.

3282 MARRIOTT, B. 38, *Upper-st. Islington.*—Watches, &c.

3283 MERCER, T. 45, *Spencer-st. Clerkenwell.*—Marine chronometers.

3284 MOORE, B. R. & J. 38, *Clerkenwell-close.*—Turret and other clocks.

3285 MORRIS, W. *Blackheath, S.E.*—Electric regulator, with centre seconds; and clocks beating simultaneously.

3286 MUIRHEAD, J. & SON, *Glasgow.* — Various clocks, engine counters, ship chronometers, &c.

3287 MURRAY, J. 30, *Cornhill, London, E.C.* — Chronometers, keyless and other watches, clocks, &c.

3288 NEAL, J. 18, *Edgware-rd. W.*—Onyx clocks, duplex lever, and novel chronometer watches.

3289 NICOLE & CAPT, 14, *Soho-sq.*—Keyless watch and conteur.

3290 ORAM, G. J. 19, *Wilmington-sq. Clerkenwell.*—Watches and chronometers.

3291 PARKINSON & FRODSHAM, 4, *Change-alley, E.C.*—Chronometers, watches, regulators, astronomical clocks, &c.

3292 PLASKETT, W. & SON, 12A, *Globe-rd. N.E.*—Improved marine chronometers.

3293 POOLE, J. 57, *Fenchurch-st.*—Chronometers and watches.

3294 PORTHOUSE AND FRENCH, 16, *Northampton-sq.*—Chronometers and watches.

3295 THE PRESCOTT COMMITTEE FOR THE EXHIBITION OF TOOLS, HOROLOGICAL INSTRUMENTS, &c.:—

Preston, J.	Stockley, Jas.
Hewett, S. & J.	Taylor, Richard.
Wycherley, J.	Preston, Wm.
Copple, J. & W.	Saggerson, E.
Scarisbruk, C.	Molyneux, Wm.
Hunt, J. & Co.	Whitfield, J. J.
Ford, R.	Alcock, J.
Welsby, J.	Jacques, J.
Brown, Ann.	Smith, J.
Johnson, C. B.	Naylor, Thos.
Houghton, S.	Heyes, Thos.
Pendleton, P.	

3297 QUAIFE, T. *Hawkhurst. Kent.*—Chime clock, 50 changes; and chronometer.

3300 ROTHERHAM & SONS, *Coventry.*—Gold and silver watches and parts of watches.

3302 RUSSELL, T. & SON, *Liverpool.*—Watches: hard-tempered nickel movements.

Gallery. North Court.

3303 SAMUEL, A. & SON, 29, *Charterhouse-sq.*—English watches.

3304 SANDERS, J. 15, *West-bar, Sheffield.*—Regulator timepiece and keyless watches.

3305 SCHOOF, W. G. 9, *Ashby-st. Northampton-sq.*—Regulator with detached escape ment, and mercurial pendulum.

3306 SEWILL, J. 61, *S. Castle-st. Liverpool.*—Watches, and pocket and marine chronometers.

3307 SHEPHERD, C. 53, *Leadenhall-st. City.*—Galvano-magnetic clocks.

3309 SMITH, J. & SONS, *St. John-sq. Clerkenwell.* — Turret and house clocks, &c.; illuminated and other dials.

3311 STRAM, N. *Ashby-st. Northampton-sq.*—Reversible and self-winding watch.

3312 STRATH, BROS. 7, *Park-ter. Camden-town.*—Models of watches.

3313 TANNER & SON, *Lewes.*—Clock with perpetual register of day, week, and month.

3316 THOMSON & PROFAZE, 25, *New Bond-st. W.*—Watches and clocks.

3318 VIVIER, O. 21, *Sekford-st. Clerkenwell.*—Fusee keyless watches, with various movements.

3319 WALES & M'CULLOCH, 56, *Cheapside.*—Gold and silver watches.

3320 WALKER, J. 68, *Cornhill.*—Watches and clocks.

3321 WALSH, A. P. 46, *Wilmington-sq. Clerkenwell.*—Watches and chronometers.

3322 WATKINS, A. 67, *Strand, London.*—Models of direct action time-keeper; watches and movements.

3324 WEBSTER, R. 74, *Cornhill.*—Watches, chronometers, keyless watches, touch watches, &c.

3325 WHITE, E. 20, *Cockspur-st. S.W.*—Chronometers, watches, and clocks.

3326 WHITTAKER, R. 7, *Gt. Sutton-st. Clerkenwell.*—Dome-capped lever watch.

3327 WOOD, T. J. 12, *Long-lane, City.*—English finished "Black Forest Clocks."

3329 YOUNG, J. *Knaresboro'.*—Improvements in lever watches to save time in repairing, &c.

3330 MCLENNAN, J. 6, *Park-pl.*—Pocket chronometers.

3331 PETIT, S. A. 69, *Princes-st. Leicester-sq.*—Regulators, watches, &c.

CLASS 16.

MUSICAL INSTRUMENTS.

North Court.

3360 ALLISON, R. & SONS, *Wardour-st. W.*—Oak piano, temp. Charles I. and model.

3362 BATES, T. C. & SON, 6, *Ludgate-hill.*—Cottage-pianofortes; a village church organ, with *ad libitum* automaton player.

3364 BESSON, F. 198, *Euston-rd.*—Musical instruments, "transposition register." Prototypes, &c.

3365 BETTS, A. 27, *Royal Exchange.*—Violins.

3366 BEVINGTON & SONS, *Greek-st. Soho, London.*—An organ, of three manuals and pedals: chancel organs, two, and five stops.

3367 BOND, W. & J. 44, *Norton-st. Liverpool.* — Piano: novel construction of wrist plank.

3368 BOOSEY & CHING, 24, *Holles-st. London.*—Harmonium, with self-blowing machine, and others.

3369 BOOSEY & SONS, 28, *Holles-st. London.*—Military and other musical instruments.

3370 BRINSMEAD, J. 15, *Charlotte-st. Fitzroy-sq.*—Perfect check repeating grand and upright piano on a new principle, and models.

3371 BRINSMEAD, J. 15, *Charlotte-st. Fitzroy-sq.*—Tubulated equipollent boudoir pianos, with check-repeater actions; and models.

3372 BROADWOOD, J. & SONS, 33, *Gt. Pulteney-st.*—Grand pianos, with parts and models illustrative of construction.

3373 BROOKS, H. & Co. *London.*—Pianos, hammer rails, keys, actions, mouldings, frett carvings, &c.

3374 BUTLER, G. *Greek-st. Soho.*—Cornets, saxhorns, flutes, and drums.

3375 CADBY, C. 39, *Liquorpond-st.* — Pianos and harmoniums.

3376 CARD, E. J. 29, *St. James'-st.*—Semi-metallic and metal flutes.

3377 CHALLEN, C. & SON, 3, *Berners-st. Oxford-st.*—Oblique, and cottage pianos.

3378 CHAPPELL, A. 214, *Regent-st.*—Military clarionets, bassoons, flutes, azemars, silent practice drum, &c.

3379 CHAPPELL & Co. 50, *New Bond-st.*—Pianos and harmoniums, with and without pedals.

3380 CHIDLEY, E. 28, *Store-st. W.C.*—Treble and baritone concertinas.

3381 CHIDLEY, R. 135, *High Holborn.*—Harmoniums and concertinas.

3382 CLINTON & Co. 35, *Percy-st. W.*—Wood and metal flutes.

3383 COLLARD & COLLARD, 16, *Grosvenor-st. W.*—Models and actions of pianos.

3384 CONS, F. & F. 81, *John-st. Tottenham court-road.*—The interior action of a piano.

3385 COOK, C. & H. E. *Tavistock-pl.*—Piano silk fronts.

3386 Corfe, E. 28, *Bedford-ter. Old Ford-rd. Victoria-park.*—Musical strings.

3387 Coxhead, C. J. *Castle-st. Shrewsbury.*—Oblique piano, with new action.

3388 Croger, T. 483, *Oxford-st.*—Æolian harps, and other musical instruments.

3389 Davis, J. M. 40, *Esher-st. Kennington-lane.*—Valves to musical instruments; action enclosed.

3390 Dearlove, M. W. 156, *North-st. Leeds.*—Violins, viola, &c.

3392 Distin, H. 10, *Gt. Newport-st. St. Martin's-lane.*—Military musical instruments.

3393 Dodd, J. *Image-cottage, Holloway-rd. Islington, N.*—Violin-Tenor, and Violoncello bows; silvered music string: specimens, &c.

3394 Duff & Hodgson (late Towns), 20, *Oxford-st.*—Pianos.

3395 Eavestaff, W. 17, *Sloane-st.*—A trichord walnut wood piano, 7 octaves.

3396 Eavestaff, W. G. 60, *Gt. Russell-st. Bloomsbury.*—Piano.

3397 Fincham, J. 110, *Euston-rd. London.*—Stops of organ metal pipes shown in skeleton organ.

3398 Forster and Andrews, *Hull.*—A grand church organ, and a chancel organ.

3399 French, J. M. 67, *Bull-st. Birmingham.*—Cottage grand piano, with tubular braced back.

3400 Geary, J. *Prince of Wales-rd. Kentish Town.*—Piccolo piano and semi-cottage ditto.

3401 Glassbarrow, C. 104, *Gt. Russell-st.*—New and improved piano.

3402 Glen, T. 2, *North Bank-st. Edinburgh.*—Bagpipes in metal, for tropical climates.

3403 Greaves, E. 76, *Milton-st. Sheffield.*—Æolian pitch pipes, tuning forks, hammers, metronomes, &c.

3404 Greiner & Sandilands, 1, *Golden-sq. London.*—Pianofortes, with choir tuning.

3405 Hampton, C. 31, *Charlotte-st. Fitzroy-sq.*—Improvements in pianos.

3406 Harrison, & Co. 65, *John-st. Fitzroy-sq.*—Piano, with iron clipper plates, and rustless gilt steel wire.

3407 Higham, J. *Victoria-bridge, Manchester.*—Brass musical instruments.

3408 Hill, W. E. 192, *Waterloo-bridge-rd.*—Gold and silver mounted violin, &c. Bows. Viola, &c.

3409 Holdernesse, W. 444, *Oxford-st. W.C.*—Cottage piano.

3410 Holman, J. N. & E. 43, *London-st. Fitzroy-sq.*—Patent model action of piano.

3411 Hopkins, T. M. *Worcester.*—Double bass, with apparatus for producing enharmonic scales of harmonics.

3412 Hopkinson, J. & J. 235, *Regent-st.*—Piano fortes and models.

3413 Hughes, W. & Co. 148, *Drury-lane.*—Covered strings for piano; wires, &c.

3414 Imhof & Mukle, 547, *Oxford-st.*—Orchestrion, or self-acting organ.

3415 Ivory & Prangley, 275, *Euston-rd. London.*—Semi-cottage pianoforte with grand action and keys.

3416 Jackson & Paine, *Store-st. W.C.*—Anti-blocking hopper pianofortes.

3417 Kind, C. 50, *George's-grove, Holloway.*—Model of a grand pianoforte action: new invention.

3418 Kirkman J. & Son, 3, *Soho-sq.*—Pianofortes.

3419 Knoll, C. & Co. 187, *Tottenham-ct.-rd. W.*—Grand and other pianofortes.

3420 Köhler, J. 35, *Henrietta-st. W.C.*—Military brass musical instruments.

3421 Lachenal, L. 8, *Little James-st. W.C.*—Full compass English concertinas.

3422 Locke, E. C. 7, *Gt. Ducie-st. Manchester.*—The Peri Campanula, or Fairy Bells.

3423 Luff, G. & Son, 103, *Gt. Russell-st. Bloomsbury, W.C.*—Model piccolo piano.

3424 Matthews, W. & Sons, 5, *St. James-st. Nottingham.*—Piano, with propeller action.

3425 Metzler, G. & Co. *Gt. Marlborough-st. W.*—Military instruments, &c.

3426 Minasi, C. 3, *St. James-ter. Kentish Town-rd.*—Music stool; harmonium.

3427 Moore, J. & H. 104, *Bishopsgate-st. Within.*—Microchordon grand pianoforte.

3428 Murphy, G. 28, *Cheapside.*—Piano.

3429 Oates, J. P. *Erdington, Birmingham.*—Cornet with "Champion" pistons, and water exit.

3430 Oetzmann & Plumb, 151, *Regent-st.*—Pianos.

3431 Peachey, G. 73, *Bishopsgate-st. Within.*—Tri-chord piccolo pianos.

3432 Potter, H. 36, *Charing-cross.*—Flutes, valve brass instruments, and drums.

3433 Priestley, F. 15, *Berners-st. W.*—Syren pianos.

3435 Rudall, Rose, Carte, & Co. 20, *Charing-cross.*—Military brass, and other instruments.

3436 Russell, G. 35, *Brook-st. Euston-rd.*—Patent improved semi-cottage piano.

3437 Rüst & Co. 34, *Gt. Marlborough-st. W.*—Pianoforte, with tubular sounding-board, and newly-constructed case.

3438 Scowen, T. L. *Allen-rd. Stoke Newington.*—Music time-keeper and accentor; compass for drawing circles without centre marks.

3440 Simpson, J. 266, *Regent-st.*—Concertinas, flutes, and flageolets.

3441 Sparks, W. G. 13, *Eversholt-st. Oakley-sq.*—Pianos.

3442 Starck, J. E. 25, *Old-st. St. Luke's, E.C.*—Musical instruments.

3443 Thompson, H. 322, *Regent-st.*—Orchestral piano, extra pedal, producing chords and octaves.

3445 Walker, J. W. 27, *Francis-st. Bedford-sq. W.C.*—Church and chamber organs.

3446 Ward, H. 100, *Gt. Russell-st. Bloomsbury.*—Piano.

3448 Willis, H. *Albany-st. Regent's-pk.*—An organ with four manual and pedal organ and 60 stops.

3450 Wornum, R. & Sons, *Store-st. Bedford-sq.*—Pianos.

3452 Pohlmann & Son, *Halifax.*—Grand upright action, oblique, and pianos, with three unisons, &c.; piano made in 1773.

3453 NUTTING & ADDISON, 210, *Regent-st.*—A piano.

3454 STIDOLPH, G. F. & J. *Woodbridge, Suffolk.*—Mimina organ, the whole of the pedal notes produced from a single pipe.

3455 NUTTING & ADDISON, 210, *Regent-st. W.*—A piano-forte.

3456 KELLY, C. 11, *Charles-st. Middlesex Hospital, W.*—Harmonium, with forty stops and two rows of keys.

3457 HEDGELAND, W.—Church organ, with three sets of keys, the movement reversed; Gothic case.

3458 JONES, H. *Fulham-rd. Brompton.*—An organ.

CLASS 17.

SURGICAL INSTRUMENTS AND APPLIANCES.

Gallery, North Court.

3482 ARBUCKLE, J. *South-bridge, Edinbro'.*—Truss, &c.

3483 ASH, C. & SONS, 9, *Broad-st. Golden-sq.*—Artificial teeth and dental materials.

3484 ATKINSON, B. F. 3, *Hemmings-row, Charing-cross.*—Trusses and splints.

3485 BAILEY, W. H. 418, *Oxford-st.*—Trusses, elastic stockings, and surgical instruments.

3486 BARLING, J. 7, *High-st. Maidstone, Kent.*—Crystal gold, in sponge and leaf, for dentists.

3487 BASSINGHAM, B. 5, *Ruby-st. Wisbech.*—Artificial leg.

3488 BIGG, H. H. 29, *Leicester-sq.*—Orthopœdic and anatomical appliances.

3489 BLACKWELL, W. & Co. *Cranbourn-st. and Bedford-ct. W.*—Surgical instruments, crutches, trusses, &c.

3490 BLUNDELL, W. 3, *Holles-st. Cavendish-sq.*—Artificial teeth.

3491 BROWN, S. S. *Ellesmere Works, Runcorn.*—Surgical appliances.

3492 BROWNING, E. 38, *Montague-sq. W.*—Artificial teeth, &c.

3493 CALKIN, J. 12, *Oakley-sq. N.W.*—Transparent ventilating eye protector.

3494 CAPLIN, 9, *York-pl. Baker-st.*—Electro-chemical bath for invalids.

3495 CAPPIE, J., M.D. *Edinburgh.*—Obstetric forceps.

3496 CARTE, A., M.D. *Royal Hospital.*—nstruments for aneurism, by compression.

3497 CLELAND & HILL, 146, *George-st. Glasgow.*—Artificial limbs.

3498 CLOVER, J. 3, *Cavendish-pl.*—Inhaler, chloroform, &c. Gives chloroform vapour any strength required, under 4½ per cent.

3499 COGHLAN, J., M.D. *Wexford.*—Surgical instruments.

3500 COLES, W. & Co. 3, *Charing-cross.*—Spiral spring trusses.

3501 COLLINS, D. J. 48, *Foley-st. W.*—Surgical appliances and instruments.

3502 COXETER, J. 24, *Grafton-st. east, Tottenham-ct.-rd.*—Surgeons' instruments.

3503 CRAPPER & BRIERLEY, *Hanley, Staffordshire.*—Porcelain trays, mineral teeth, &c.

3504 DIXON, T. 4 & 7, *St. James's-pl.*—Nightingale cradle and bed.

3505 DURROCH, W. F. 28, *St. Thomas-st. east.*—Surgical instruments, &c.

3506 ERNST, F. G. 19, *Calthorpe-st. W.C.*—Anatomical appliances and surgical instruments.

3507 EVANS & STEVENS, 12, *Old Fish-st. St. Paul's.*—Collection of surgical instruments.

3508 EVANS, C. *The Hospital, Birkenhead.*—Arm splint.

3509 EVRARD, J. 35, *Charles-st. Middlesex Hospital.*—Surgical and dental instruments.

3510 FAULKNER, H. 24, *Keppel-st. Russell-sq.*—Artificial teeth in vulcanite.

3511 FAULKNER, J. 2, *Mornington-cres. Hampstead-rd. N.W.*—Vulcanite base for artificial teeth.

3512 FERGUSON, J. & J. 21, *Giltspur-st. E.C.*—Surgical instruments.

3513 FITKIN, W. 88, *Fleet-st.*—Elevator for extracting teeth; and other dental instruments.

3514 FRANCOIS, H. 42, *Judd-st. Euston-rd.*—Artificial teeth.

3515 FRESCO, A. 7, *Grosvenor-st. W.*—Artificial teeth.

3516 GABRIEL, M. & A. 34, *Ludgate-hill.*—Artificial teeth with air cells and soft gums.

3517 GANNON, T. *Liquorpond-st. London, E.C.*—Self-adjusting leg and foot rest.

3518 GORDON, DR. *Edinburgh.*—New forceps and elevator, adopted for the extraction of all kinds of teeth and stumps.

3519 GARRETT, J. A. 38, *Wardour-st. W.*—Trusses and surgical bandages.

3520 GILL, T. D. 84, *John-st. Tottenham Court-rd.*—Gas vulcanizer for dentists.

3521 GRAY & HALFORD, 171, *Goswell-rd. E.C.*—Artificial human eyes.

3522 GRAY, J. & Co. 154, *Fitzwilliam-st. Sheffield.*—Surgical appliances and apparatus.

3523 GRIFFITHS, R. 2, *Duke-st. Smithfield.*—Medicine chests and sample cases.

3524 GROSSMITH, W. R. 175, *Fleet-st.*—Artificial eyes, legs, arms, hands, &c.

3525 HALLAM, F. H. 9, *Endell-st. Long-acre.*—Dental instruments.

3526 HARNETT, W. 12, *Panton-sq. Coventry-st.*—Mineral teeth in vulcanite and gold, &c.

3528 HAYES, G., M.D. 66, *Conduit-st. London, W.*—Mechanical dentistry.

3530 HILLIARD, W. B. *Renfield-st. Glasgow.* —Surgical instruments, artificial leg, trusses, &c.
3531 HOOPER, W. *Pall-mall East.*—Hydrostatic beds, cushions, and lift for invalids.
3532 HOY, J. 6, *Pickering-pl. W.*—Truss for hernia.
3533 HUDSON, T. *South Shields.*—Tooth stump inst. &c.
3534 HUXLEY, E. 12, *Old Cavendish-st.*—Surgical bandages and moc-main trusses.
3535 JOHNSON, T. *Commercial-rd. E.*—Model of apparatus for slinging horses.
3536 LAMBERT, P. 18, *Charlotte-st. W.C.*—Artificial teeth.
3537 LAURENCE & CO. *Islington.*—Horse-hair and bath gloves.
3538 LAWSON, BUXTON, & CO. *Shales Moor Works, Sheffield.*—Surgical, dental, and veterinary instruments.
3539 LEARWOOD, T. *Fairmantle-st. Truro.* —Artificial limbs and trusses.
3540 LEMALE, T. & CO. 62, *Chandos-st. W.C.*—Artificial teeth and gums.
3541 LINDSEY, M. J. 37, *Ludgate-st. City.* —Trusses.
3542 LONGDON, F. & CO. *Derby.*—Surgical appliances.
3543 LOWS, A. 19, *Lowther-st. Carlisle.*—Artificial teeth, with continuous gums.
3544 MACINTOSH, C. & CO. *Cannon-st. London.*—Vulcanised surgical and chemical apparatus.
3545 MACINTOSH, J. 40, *North-bank, Regent's-park.*—Collodion as setting for artificial teeth.
3546 MARSDEN, W. J. *Upper Thorpe-rd. Sheffield.*—Respirators, chest protectors, eye-shade, &c.
3547 MASTERS, M. 1, *Paragon-pl. New Kent-rd.*—Artificial hands, arms, legs, &c.
3548 MATTHEWS, W. 8, *Portugal-st. W.C.* —Surgical instruments and appliances.
3549 MAURICE, J. 3, *Langham-pl. W.*—Artificial teeth, and applications of vulcanized India-rubber.
3550 MAW, S. & SON, 11, *Aldersgate-st.*—Surgical instruments.
3551 MILLER, C. M., M.D. *Claremont-villa, Stoke Newington-rd.*—Spectacles for conical cornea.
3552 MILLIKIN, J. 9, *St. Thomas's-st. Borough.*—Surgeons' instruments and appliances.
3554 MOGGRIDGE & DAVIS, 18, *George-st. Hanover-sq.*—Specimens in dentistry.
3555 MORRISON, J. D. *Edinburgh.* —Dental appliances, processes, and products.
3556 MOSELEY, MESSRS. 30, *Berners-st. W.*—Artificial teeth and dental appliances.
3557 NORMAN, S. JUN. 1, *Cheltenham-pl. Westminster-rd.*—Lift for a short leg, and shell for boot.
3559 O'CONNELL, E. *Bury, Lancashire.*—Apparatus for giving nourishment to infants, &c.
3561. PARSONS, MESSRS. J. 15, *Manor-row, Bradford.*—Artificial teeth.
3562 PATRICK, H. W. 18, *Broad-st. Golden-sq. W.*—Artificial palates, teeth, and dental applications.
3563 PAUL, A. 27, *Mecklenburgh-sq.*—Douche bath (2 models).
3564 PEARCE, W. & CO. *Bridge-st. Bristol.* —Surgical appliances.
3565 PINDAR, C. 19, *John-st. Blackfriars-rd.*—Pill and plaster machine, &c.
3566 POLLARD, C. & E. *Alfred-pl. Thurloe-sq.*—Turkish bath, and bathing sandal.
3567 POWELL, S. 2, *Surrey-cottages, Old Kent-rd.*—Breast drawers, glass syringes, tube bottles, &c.
3568 PRATT, J. F. 420, *Oxford-st. W.*—Apparatus for various deformities; surgical instruments.
3569 PUCKRIDGE, F. L. 4, *York-pl. Walworth.*—Liston's court plasters, gold beater's skins.
3570 PULVERMACHER, T. L. 73, *Oxford-st.*—Piline and galvanic battery, &c.
3571 REDFORD, G. *Cricklewood.*—Stretcher, in halves, fitting universally, and medicine pouch.
3572 REIN, F. C. 108, *Strand.*—Surgical instruments, and acoustic appliances.
3573 REIN, MRS. S. 108, *Strand.*—Elastic supports, stockings, kneecaps, &c.
3574 REYNOLDS, J. 20, *St. Anne-st. Liverpool.*—Artificial leg. Trusses and appliances for deformity.
3575 RIMMEL, E. 96, *Strand.*—Aromatic disinfector, for hospitals, sick-rooms, &c.
3576 ROGERS, C. 40, *Gt. Tindell-st. Birmingham.*—Either side double-lever truss.
3577 ROGERS, M. 18, *New Burlington-st. W.*—Artificial teeth.
3578 ROOFF, W. B. 7, *Willow-walk, Kentish Town.*—Respirators, acoustic and medical instruments, &c.
3579 RUSSELL, CAPT. G. *Swan-hill, Shrewsbury.*—Hospital beds and appliances.
3580 SANSOM, DR. A. E. *Lower-rd. Islington.*—Apparatus for administering chloroform.
3581 SAVORY & MOORE, 143, *New Bond-st.* —Medicine chests, panniers, and cases for the use of troops.
3583 SILLIS, F. 2, *George-st. Euston-sq.*—Artificial legs, hands, and arms, spring crutches, &c.
3584 SIMPSON, H. 55, *Strand.*—Surgical instruments.
3585 SMALE, BROS. 19, *Gt. Marlborough-st.*—Mineral teeth, dental implements, and appliances.
3586 SMITH, J. C. *Week-st. Maidstone.*—Tooth instruments.
3587 SMITH, W. & F. 253, *Tottenham-court-rd.*—Water bed.
3588 SPARKS & SON, 28, *Conduit-st. W.*—Surgical bandages and appliances.
3589 SPRATT, W. H. 2, *Brook-st. Hanover-sq.*—Trusses and orthopœdic instruments.
3590 SYKES, MARY, 280, *Regent-st.*—Surgical bandages.
3591 THRING, C. 3, *Little Randolph-st. Camden-town.*—Arm sling, &c.
3592 TOMPSON, W. A. 18, *Cecil-st. Strand.* —Inhaler for applying caustic solution, internally, in throat diseases.
3593 TUFNELL, J. *Mount-st. Dublin.*—Tubular bougies for surgical purposes.

Gallery, North Court.

3594 TWEEDIE, W. 337, *Strand.*—Respirator, of gold wire.

3596 WALTERS, F. 16, *Moorgate-st.*—Surgical instruments.

3597 WEEDON, T. *Hart-st. Bloomsbury.*—Instruments for microscopical preparations, &c.

3598 WEISS, J. & SON, 62, *Strand.*—Surgical instruments.

3599 WELLS, G. S. 59, *Euston-sq.*—Artificial teeth and gums.

3600 WELTON, T. 13, *Grafton-st. Fitzroy-sq.*—Jointed pin leg, and other instruments.

3601 WELTON & MONCKTON, 13, *Grafton-st. Fitzroy-sq.*—Magnetic chair and battery.

3602 WESTBURY, R. 26, *Old Millgate, Manchester.*—Trusses and deformity instruments.

3603 WETHERFIELD, J. *Henrietta-st. Covent-gard.*— Amadou plaster, a surgica appliance.

3604 WHIBLEY, E. 41, *Radnor-st. Chelsea*—Surgical operating table.

3605 WHICKER & BLAISE, 67, *St. James-st.*—Surgical instruments and appliances.

3606 WHITE, J. 228, *Piccadilly.*—Truss, surgical appliances, &c.

3607 WHITING, W. & SONS, *High-st. Camden-town.*—Spinal supports for curvatures.

3608 WILLIAMS, G. J. 17, *Cavendish-pl. Cavendish-sq. W.* — Artificial palates and teeth.

3609 WOOD, W. R. *Carlisle-house, Brighton.*—Models and general dentistry.

3611 YOUNG, J. A. 47, *Bath-st. Glasgow.*—Dental forceps, &c.

3612 READ, MESSRS. 8, *Holles-st. Cavendish-sq.*—Artificial teeth.

3613 NUNN, R. M. *Grays, Essex.*—Medica inspirator.

SECTION III.

CLASS 18.

COTTON.

Gallery, South.

3640 ASHWORTH, E. & SONS, *Egerton Mills, Bolton.*—Cotton of various kinds.

3641 AULD, BERRIE, & MATHIESON, *Union-st. Glasgow.*—Scotch muslins.

3642 BARLOW, GOODY, & JONES, *Manchester.*—Quilts, counterpanes, cotton blankets, dimities, &c.

3643 BRITTAIN, T. *Manchester.*—Sponge cloths, for cleaning machinery, &c.

3644 BROOK, J. & BROS. *Meltham Mills, Huddersfield.*—Sewing cottons, crochet, and embroidery.

3645 BROWN, SHARPS, & TYAS, 18, *Watling-st.*—Embroidered muslins.

3646 CARLILE, J., SONS, & CO. *Barkend Mills, Paisley.*—Cotton and linen threads.

3647 CHRISTIE, H. *Blackfriars Mills, Manchester.*—Yarns, various.

3648 CHRISTY, W. M. & SONS, *Fairfield, near Manchester.*—Towels, blankets, and counterpanes.

3649 CLARK & CO. *Seedhill and Cumberland Mills, Paisley.*—Sewing, crochet, and embroidering cottons.

3650 CLARK, J. JUN. & CO. *Mile-end, Glasgow.* –Sewing thread.

3651 CLARKE, I. P. *Leicester.*—Sewing cottons, reels, spooling and mill bobbins.

3652 COATS, J. & P. *Fergushe Works, Paisley.*—Sewing cottons.

3653 COATS, NEILSON, & CO. *Thorn Mill, Johnstone, Renfrewshire.* — Yarns for embroidering muslins and other yarns.

3654 COPESTAKE, MOORE, CRAMPTON, & CO. 5, *Bow Church-yard.*—Sewed muslin manufactures.

3655 CREWDSON & WORTHINGTON, *Manchester.*—Shirtings.

3656 DICKINS & CO. *Spring Vale Works, Middleton.*—Dyed and polished yarns and sewings.

3657 ERMEN & ENGELS, *Manchester.*—Cotton for sewing, crochet, knitting, and embroidery.

3658 EVANS, W. & CO. *Derby.*—Cotton threads for embroidery, &c.

3659 FAULKNER, H. 6, *Castle-ct. Lawrence-lane.*—Cotton twines: run 30 per cent. longer length than hemp, same weight.

3660 FORD, F. *Stanley-st. Mills, Manchester.* —Cottons, various; spools, balls, skeins, &c.

3661 GILLS & HARTLEY *Wood-st.*—Coutils for stays.

3662 GOODAIR, SLATER, & SMITH, *Kent-st. Mills, Preston.*—Longcloths, and shirtings.

3663 GREENWOOD & WHITTAKER, *Manchester.*—Water twist, shirting calicoes.

3665 HASTINGS, W. *Huddersfield.*—Cotton yarns for warps and wefts.

3666 HAWKINS, J. & SONS, *Green Bank Mills, Preston, Lancashire.*—Cloths for shirts, &c.

3667 HAWORTH, R. & CO. *Manchester.*—Jeannetts, India twills, &c.

3668 HOLLINS, E. *Sovreign Mills, Preston.* —Cotton shirtings and sheetings.

3669 HOPWOOD, R. & SON, *Nova Scotia Mills, Blackburn.*—Calicoes in process of manufacture.

3670 HORROCKSES, MILLER, & CO. 9, *Bread-st. London, E.C.*—Long cloths and twilled shirtings.

3671 HOULDSWORTH, T. & CO. *Manchester.* —Cotton yarn.

3672 HUDSON, J. & SONS, *Leicester.*—Sewing cotton.

3673 JACK, J. R. 37, *Virginia-st. Glasgow.* —Jacquard muslin window curtains.

3674 JOHNSON, J. & FILDES, *Manchester.*—Quilts, counterpanes, toilette covers, &c.

3675 JOHNSON, J. M. *Britannia Mill, Mirfield.*—Yarns, Teviots, and fancy warps.

3676 KENYON, J. T. & CO. *White Hall Mill, Over Darwen, near Blackburn.*—Printing cloths and India and China sheetings.

3677 KERR & CLARK, *Linside Thread Works, Paisley.*—Spool cotton, enamelled and six cord

3678 KESSELMEYER & MELLODEW, *Manchester.*—Fast pile silk, imitation silk, and cotton velvets, velveteens, cords, beaverteens, moleskins, &c.

3679 LOWTHIAN, FAIRLIE, & CO. *Carlisle.* —Ginghams, checks, stripes, drills, &c.

3680 MANCHESTER COTTON TWINE CO. *Manchester.*—Cotton twine and cotton mill bands.

3681 MANLOVE, S. *Holy Moor Mills Chesterfield.*—Sewing cotton, embroidery, &c.

3683 MARTIN, JOHNSON, & JOULE, *Manchester.*—Dimities and damasks.

3684 MOORE, J. 33, *Piccadilly, Manchester.* —Velvet ribbons, with patent edges.

3635 MORGAN, J. *Ducie Works, Manchester.*—Candle wicks.

3686 NORTHCOTE, S. & CO. 29, *St. Paul's Churchyard.*—Embroidered collars, sleeves, and other goods.

3687 OUTRAM, R. & CO. 13, *Watling-st.*—Plain and figured muslins, counterpane, quilt, &c.

3688 PHILLIPS, J. 8, *Lawrence-lane, Cheapside*—Woven and other fancy quiltings.

Gallery, South.

3689 RAWORTH, J. T. *Leicester*.—Nine-cord, six-cord, and glacè sewing cotton.

3690 SHAW, JARDINE, & Co. *Manchester*.—Spinning and doubling lace, yarns, &c.

3691 SMITH, W. J. & Co. 40, *Faulkener-st. Manchester*.—Drills, quiltings, corset ribs, &c.

3693 SWAINSON, BIRLEY, & Co. *Preston*.—Calicoes for shirts, &c.

3694 SYMINGTON, R. B. & Co. 9, *Cochrane-st. Glasgow*.—Harness figured muslin curtains, lappets and linings.

3695 TOWNSEND, T. & SON, *Coventry*.—Cotton yarns.

3696 WILSON, T. & D. & Co. 145, *Ingram-st. Glasgow*.—Plain and fancy muslins.

3697 WRIGLEY, H. & E. *Huddersfield*.—Single and double cotton yarns, grey, gassed, bleached, and coloured granderelle.

3698 YATES, BROWN, & HOWAT, *Spring-field-ct. Glasgow*.—Muslins, Jacquard muslin curtains, &c.

3700 HALLE & UDALLE, *Manchester*.—Cotton velveteens.

CLASS 19.

FLAX AND HEMP.

Gallery, South-East Transept.

3728 AINSWORTH, T. *Chator Mills, White-haven*.—Sewing machine linen threads, flax yarns.

3729 AUSTIN, J. *Princes-st. Finsbury*.—Sash, blind, and picture lines, &c.

3730 BARBOUR, W. & SONS, *Lisburn*.—Thread, yarns, flax, and linens.

3732 BAXTER, BROS. & Co. *Dundee*.—Yarns and linen.

3733 BELFAST LOCAL COMMITTEE, *Belfast*.—Trophy: flax, and flaxen manufactures of Ireland.

3734 BELL, R. & Co. 13, *Donegal-st. Belfast*.—Damask goods.

3735 BENNETT & THORN, 190, *High-st. Borough*.—Hemp, flax, ropes, mats, &c.

3736 BIRD, R. *Crewkerne, Somerset*.—Linen and woollen saddlery webs, straining webs, &c.

3737 BIRREL BROS. *Dunfermline*.—Damask table-cloths, napkins, &c.

3738 BOOTH, W. & Co. *Leeds*.—Bed-ticks, sheetings, and drills.

3739 BROWN & LIDDELL, *Belfast*.—Table-linen, birds-eye diaper, sheetings, &c.

3740 BROWNE, W. *Patent Rope Works, Wivenhoe, Colchester*.—Cordage, rope, lines, and twines.

3741 BUCKINGHAM, J. 33, *Broad-st. Bloomsbury, W.C.*—Rope web matting, sackings, &c.

3742 CARTER BROS. *Oak Mills, Barnsley*.—Sheetings, towelings, damasks, drills, &c.

3743 CHARLEY, J. & W. & Co. *Seymour-hill, Belfast*.—Irish linens.

3744 CLEUGH, A. *Imperial Mills, Bromley, London, E.*—Jute, hemp, and flax yarns, rug weft, twines, &c.

3745 CLIBBORN, HILL, & Co. *Banbridge, Co. Down, Ireland*.—Bird-eyed diapers, bleached.

3746 CONNOR, F. *Linen Hall, Belfast*.—Linen drills.

3747 COSTERTON & NAYLER, *Flax Works, Scole, Norfolk*.—Prepared flax, tows, yarn, and waste for paper-making.

3748 CROGGON & Co. 2, *Dowgate-hill, E.C.*—Asphalte roofing, inodorous ship sheathing, and dry hair felts.

3749 DAGNALL & TILBURY, *Farm-lane, Walham-green, S.W.*—Mats, matting lines, twines, coir yarn and fibre.

3750 DEWAR, D., SON, & SONS, *Wood-st. London*.—Table linens, cambric handkerchiefs, sheetings, Irish linens, &c.

3751 DUNBAR, DICKSONS, & Co. *Belfast*.—Linens, sheetings, cambric, &c.

3752 DUNBAR, MCMASTER, & Co. *Gilford, Co. Down, Ireland*.—Flax, linen, yarns, threads, &c.

3753 EDGINGTON, F. *Thomas-st. Old Kent-rd*.—Marquee, flag, rick-cloth, sack, and tarpaulin manufactures.

3754 EDWARD, A. & D. & Co. *Logie Works, Dundee*.—Linens, yarns, and jute fabrics.

3755 ELSTOB & BLINKHORN, *Spalding, Lincolnshire*.—Canvass folding buckets, portable water cisterns, and seamless hose piping.

3756 FAULDING, STRATTON, & BROUGH 13, *Coventry-st*.—Table linen.

3757 FENTON, SON, & Co. *Belfast*.—Bleached linens and damasks.

3758 FINLAYSON, BOUSFIELD, & Co. *John-stone, near Glasgow*.—Threads.

3759 FLEMING, W. & J. & Co. *Baltic Works, Glasgow*.—Jute yarns, carpetings, baggings, sacks, &c.

3760 FOX, C. J. *Doncaster*.—Wool sheets, canvas, tarpaulings, sacks, &c.

3761 FRASER, D. & SON, *Arbroath*.—Sail canvas, duck, tarpauling, &c. &c.

3762 GAVIN, P. & SONS, *Leith Ropery, Leith*.—Power-loom sailcloth.

3763 GILL, J. *Headingley, near Leeds*.—Grey, bleached, and dyed linen yarns and twines.

3765 GRIMOND, J. & A. D. *Dundee.*—Jute carpeting, Hessians, sacking, jute yarns.

3766 GRIMSTON, R. & T. & Co. *Clifford Mills, near Tadcaster.*—Shoe threads, &c.

3767 GUNDRY, J. & Co., & PYMORE MILL Co. *Bridport, Dorsetshire.*—Shoe threads and yarns, twines, nets, seines, &c.

3768 HARFORD, G. *Newcastle-on-Tyne.*—Sail cloth.

3769 HARRIS, J. & SONS, *Derwent Mills, Cockermouth.*—Linen threads, dyed, bleached, and variously finished and made up.

3770 HAWKE, E. H. & SON, *Scorrier, Cornwall.*—Various descriptions of rope, safety fuse, &c.

3771 HIND, J. & SONS, *Durham-st. Mills, Belfast.*—Brown and bleached linens; linen and cambric yarns.

3772 HOLDSWORTH, W. B. & Co. *Leeds.*—Hemp and flax yarns, threads, and twines.

3773 JAFFE BROS. *Belfast.*—Linens; linen and cambric handkerchiefs.

3774 JOHNSTON & CARLISLE, *Brookfield Mills, Belfast.*—Yarns and linens.

3775 KINNIS, W. & Co. *Dunfermline, N.B.*—Linen, damasks, sheetings, &c.

3776 LOCKHART, N. & N. *Kirkcaldy, Scotland.*—Fishing nets.

3777 LOCKHART, N. & SONS, *Kirkcaldy, Fife.*—Ticks, sheetings, towellings, sackings, and sacks.

3778 M INTYRE & PATTERSON, *Belfast.*—Flax and yarns.

3779 MARSHALL & Co. *Leeds.*—Yarns, threads, and woven fabrics of linen.

3780 MATHEWSON, J. & SON, *Dunfermline, N.B.*—Damask table-cloths, napkins, &c.

3781 MATIER, H. & Co. *Belfast.*—Linens and handkerchiefs.

3782 MILLER, O. G. *Dundee.*—Line, tow, and jute yarns.

3783 MOIR, J. & SON, *Dundee.*—Ducks, linens, sheetings, &c.

3784 MONCUR, A. & SON, *Dundee.*—Linens, salt, and grain sacks, and sacking.

3785 MOORE, W. F. *Cronkbourne, Douglas, Isle of Man.*—Sail cloth and fishing nets.

3786 MOORE & WEINBERG, *Belfast.*—Linen manufactures.

3787 MORISON, J. *Norton Folgate.*—Marquee, tent, tarpaulin, rick cloth, &c.

3788 NORMAND, J. & SONS, *Dysart, Fifeshire.*—Damasks, diapers, and hucks.

3789 NUTT, R. 31, *Trippett, Hull.*—Oil press, hairs, and press bagging.

3790 PATERSON, J. *Dundee.*—Hemp carpeting, manilla and coir mattings, rugs, &c.

3791 PRESTON, SMYTH, & Co. *Belfast.*—Flax, yarns, linens, cambrics, &c.

3792 RICHARDSON, J. N., SONS, & OWDEN *Belfast.*—Flax, linen and damask, &c.

3793 ROBERTSON, J. *Middle Hills, neat Coupar Angus, Perthshire.*—Flax, tow, sheetings, hessians, sackings, and various goods.

3794 RUSSELL, J. N. & SONS, *Lansdowne Mills, Limerick.*—Munster flax, yarns, linens &c.

3795 SAMSON, H. & SONS, *Hill-bank, Dundee.*—Linens, various.

3796 SMIETON, J. & SON, *Dundee.*—Linen and union dowlas, Osnaburgs, chequelas, coletas, Russias, &c.

3797 SMITH, T. & W. *Royal Exchange-bdgs., and Newcastle-on-Tyne.*—Wire and hemp ropes.

3798 STEPHENS, J. P. & Co. *Bridport* Twines, lines, nets, seines, sail canvas, &c.

3799 STUART, J. & W. *Musselburgh, Scotland.*—Fishing nets and twines made by machinery.

3800 TERRELL, W. & SONS, 6, *Welsh-back, Bristol.*—Manilla and other cordage, twine, cord, line, &c.

3801 THOMSON, D. & J. & Co. *Seafield Works, Dundee.*—Jute carpeting, sacking, pocketing, tarpaulin, bagging, yarns, &c.

3802 UNITE, J. 130, *Edgware-rd.*—Hemp and flax, raw and manufactured.

3803 WALKER, J. & Co. *Arbroath.*—Sail cloth and sail twine.

3804 WALKER, J. & H. *Dundee.*—Jute yarns, guano bags, wool packs, &c.

3805 WILFORD, J. & SONS, *Brompton, Northallerton, Yorkshire.*—Linen drills.

3806 WILKS, BROS. & SEATON, 80, *Watling-st. London, E.C.*—White, coloured, fancy, and silk and wool flannels.

3807 WILSON BROS. 29, *Lowther-st. Whitehaven.*—Twilled and plain double sail cloth.

3808 WILSON, G. *Hutton Rudby, Cleveland, Yorkshire.*—Long cloth, and flax sail cloth.

3809 YEOMAN & Co. *Osmotherly, Northallerton.*—Drills, ducks, huckabacks, and yarns.

3810 HOUNSELL, W. & Co. *Bridport.*—Twines, lines, nets, canvas, &c.

CLASS 20.

SILK AND VELVET.

South-East Gallery.

3840 ADSHEAD, W. & Co. *Higher Fence, Macclesfield.*—Dyed silks.

3841 ALLEN, J. *Spa-mills, Derby.*—Elastic gusset webs, braids, &c.

3842 ALSOP, DOWNES, SPILSBURY, & Co. *Leek, Staffordshire.*—Braids, bindings, serges, buttons, sewings, twist.

3843 BALLANCE, T. & SON, 13, *Spital-sq.*—Silks and velvets.

3845 BICKHAM, POWNALL, & Co. 2, *York-st. Manchester.*—Broad silk goods.

3846 BIRCHENOUGH, J. *Macclesfield.*—Scarfs, handkerchiefs, sarsnetts, &c.

3848 BROCKLEHURST, J. T. & SONS, *Macclesfield.*—Thrown silk, and waste silk goods.

3849 BROWETT, F. *Coventry.*—Ladies' dress trimmings, &c.

3851 CAMPBELL, HARRISON, & LLOYD, 19, *Friday-st. E.C.*—Silk and velvet fabrics.

3852 CARR, T. & Co. *Leek.*—Silk throwsters, bindings, machine twist sewings, serges, &c.

3853 CARTER & PHILLIPS, *Coventry.*—Ribbons.

3854 CASH, J. & J. *Coventry.*—Ribbons and cambric frillings.

3855 CHADWICK, J. 12A, *Mosley-st. Manchester.*—Silks.

3856 CHADWICK, J. 33, *Fountain-st. Manchester.*—Silks.

3857 CHRISTY & Co. *Stockport.*—Power-made hat plushes, velvets, and piled fabrics.

3858 CORNELL, LYELL, & WEBSTER, 15, *St. Paul's Church-yard.*—Ribbons and moire antiques.

3859 CORNS, W. W. & Co. *Macclesfield.*—Scarfs, dress silks, ties, shawls, &c.

3860 COURTAULD, S. & Co. 19, *Aldermanbury.*—Crapes and aerophanes.

3861 COX, R. S. & Co. *Coventry, & 7, St. Paul's-churchyard.*—Ribbons, silks, and handkerchiefs, &c.

3862 CRITCHLEY, BRINSLEY, & Co. *Macclesfield.* — Silk handkerchiefs and scarfs, &c.

3863 CUNLIFFE, PIGGOTT, & Co. *Spring-gardens, Manchester.*—Silks.

3864 DALTON & BARTON, *Coventry.* — Ribbon, carriage lace, and upholstery trimmings.

3865 DAVIDSON & MYATT, *Leek, Staffordshire.*—Machine twist for sewing machines, &c.

3866 ELISE, L. M. & ISAACSON, F. W. 170, *Regent-st.*—English silks.

3868 FRANKLIN, W. & SON, *Coventry.*—Ribbons, &c.

3869 GIBSON, S. JUN. *Leek.*—Silk twist, sewing silks, buttons, &c.

3870 GRANT & GASK, 59 *to* 62, *Oxford-st.*—Tissue de Verre.

3871 GROUT & Co. 12, *Foster-lane, London.*—Crapes, aerophanes, and lisses.

3872 HADWEN, J. *Kelroyd Mills, Halifax.*—Silk waste yarns and tissues: silk waste and mohair yarns and tissues.

3873 HART, J. *Coventry.*—Silk ribbons.

3874 HENNELL & ELD, *Coventry.*—Raw and thrown silk.

3875 HOULDSWORTH, J. & Co. 23, *Portland-st. Manchester.*—Furniture fabrics, machine embroideries, &c.

3876 KEITH & Co. 124, *Wood-st. London.*—Furniture silks; silks for carriage lining.

3877 KEMP, STONE, & Co. 35, *Spital-sq. N.E.*—Broad silks and velvets.

3878 LE MARE, E. R. *Manchester.*—Silk goods.

3879 NEWSOME, C. *Coventry.*—Ribbons.

3880 PAYN, J. J. *Aldermanbury.*—Silk reps, tissues, brocatelles, borders, &c.

3881 PEEL, GREENHALGH, & Co. & G. WHYATT & SON, *Bury and Manchester.*—Silk union velvets by steam power.

3882 POTTS & WRIGHT, *Macclesfield.*—Sarcenet and handkerchiefs.

3883 POWNALL, STUBBS, & Co. *Leek, Staffordshire.*—Sewing silks, twists, buttons, whip-lashes, tassels, military ornaments, &c.

3884 RATLIFF, J. & SON, *Coventry.*—Plain and fancy ribbons.

3885 RUSSELL, DALGLISH, & Co. *Blackhall Factory, Paisley.* — Thrown silks, fringes, sewings, &c.

3886 SALKELD, J. & Co. *Dalton, near Huddersfield.*—Illustrations of silk throwing and spinning. Sewings.

3887 SEAMER, T. 5, *Milk-st. Cheapside.*—Silks and velvets.

3888 SLATER, BUCKINGHAM, & SLATER, 35, *Wood-st. London.* — Cravats, scarfs, and ties.

3889 SIMPSON, M. & W. *Leek.*—Machine sewing silk without knots; trimmings, &c.

3890 SLINGSBY, H. *Park-st. Coventry.*—Silk scarfs, neck-ties, trimmings, &c.

3891 SMALL, W. *Macclesfield.* — Sarsnetts and scarfs.

3893 SUDBURY LOCAL COMMITTEE, *Sudbury.*—Silks: velvets and brocades.

3894 TAYLOR, S. 45, *Friday-st. London, E.C.*—Silks: velvets, satins, &c.

3895 THOMPSON, W. & Co. *Galgate, near Lancaster.*—Material in waste and yarn.

3896 THORP, J. & S. 20, *Piccadilly, Manchester.*—Galloons, doulles, bindings, ribbons, &c.

3897 VAVASSEUR, TAYLOR, & Co. 3, *Watling-st.*—Ladies' silks, scarves and handkerchiefs.

3898 WALTERS, D. & SONS, 45, *Newgate-st.*—Furniture silks.

3899 WANKLYN, W. 42, *Cheapside.*—Silks, handkerchiefs, &c.

3900 WATSON & HEALEY, *Rochdale.*—Velvet and plush, from spun silk waste.

3901 WINKWORTH & PROCTER, *Manchester.*—Coloured glacè, plain and figured chenè, &c.

3902 WREFORD & Co. 18, *Aldermanbury, London, E.C.*—Sewing silks and twists.

3903 CROSS, P. R. *Sudbury, Suffolk.*—Respirator scarf.

3904 CLABBURN, SONS, & CRISP, *Norwich.*—Silk shawls.

CLASS 21.

WOOLLEN AND WORSTED, INCLUDING MIXED FABRICS.

South-East Gallery.

3934 ADIE, S. 115, *Regent-st.*—Tartans, shaw s, cloakings, linsey woolseys, &c.

3935 AKROYD, J. & SON, *Halifax.*—Worsted yarns and fabrics, furniture and dress goods, &c.

3936 ANDERSON, J. & A. *Princes-sq. Glasgow.*—Gingham and fancy dresses.

3937 ANDREWS, H. & Co. *Albion-st. Leeds.*—Cloths, meltons, and tweeds.

3938 ARMITAGE, BROS. *Huddersfield.*—Fancy coatings and doeskins.

3939 ARMITAGE, S. & B. *Shepley, near Huddersfield.*—Woollen and other goods.

3940 BARBER, J. & SONS, *Holmbridge Mills, near Huddersfield.*—Fancy trouserings and coatings.

3941 BARKER, B. & SON, *Cookridge-st. Leeds.*—Woollen cloths, beavers, and cassimeres.

3942 BARRON, R. *Gillroyd Mills, Morley and Leeds.*—Union cloths, made from Sydney wool and mungo.

3944 BENNETT, S. & SON, *Winsham, near Chard, Somerset.*—Woollen cloths.

3945 BIRCHALL, J. D. *Wellington-st. Leeds.*—Woollen cloths and fancy coatings.

3946 BIRD, O. *Stroud, Gloucestershire.*—Billiard cloth, doeskins, &c.

3947 WALKER, BIRRELL, & Co. *Glasgow.*—Fancy dress fabrics.

3948 BISHOP, SON, & HEWITT, *Leeds.*—Waterproof tweeds, &c.

3949 BLAKELEY, BROS. *Dewsbury.*—Shoddy and mungo, from woollen rags.

3950 BLISS, W. & Co. 26, *Basinghall-st. London, and Chipping Norton, Oxon.*—Shawls, tweeds, serges, rifle cloth, &c.

3952 BOLINGBROKE, C. & T. & JONES, *Norwich.*—Paramattas, poplins, shawls, dresses, &c.

3953 BOWMAN, J. & SON, *Langholm, N.B.*—Scotch tweeds.

3954 BRADFORD LOCAL COMMITTEE.—Wools.

3955 ———Yarns.

3956 BRADFORD LOCAL COMMITTEE.—Alpacas and mohair goods, plain and figured.

3957 ———Orleans cloths, plain and figured.

3958 ———Cobourg, paramatta, barathea, reps, cords, cloths.

3959 ———Lastings, serge de Berri, crapes, stockinetts, gambroons, camlets.

3960 ———Italian summer cloths, Russell, and mottled cords.

3961 ———Umbrella cloths.

3962 ———Mixed and mottled worsted and alpaca goods, and winseys.

3963 ———Fancy goods: alpaca, mohair, worsted, silk, &c.

3964 ———Worsted goods: merinos, says, shalloons, &c.

3965 ———Moreens.

3966 ———Damasks, reps, and table covers.

3967 ———Wool shawls, delaines, and shawl cloths.

3968 BRAITHWAITE & Co. *Kendal.*—Woollens, coatings, coat linings, checks, &c.

3969 BREWIN & WHETSTONE, *Leicester.*—Worsted, lambs' wool, and merino yarns.

3970 BRIGGS & SONS, *Carlton Cross Mills, Leeds.*—Woollen shawls.

3971 BROWN BROS. *Buckholm Mills, Galashiels, Scotland.*—Fancy Scotch tweeds.

3972 BROWN & COLLANDER, *Yeadon, near Leeds.*—Tweeds, meltons, and fancy cloakings.

3973 BROWN, J. & H. & Co. *Selkirk, N.B.*—Scotch tweeds, shawls, and mauds.

3974 BULL & WILSON, 52, *St. Martin's-lane.*—Woollen cloths and vestings.

3975 BURGESS, A. *Leicester.*—English, foreign, and colonial wool.

3977 BUTTERWORTH, J. & SON, *Grunbooth-mills, near Rochdale.*—Flannels, and processes of making them.

3978 CALEY BROS. *Windsor.*—Furniture,

silk damasks, and satins. Silk and cotton diaphano for transparent window blinds.

3979 CARR, I. & Co. *Tiverton, Bath.*—Fur, beavers, meltons, &c.

3980 CARTER, W. & GEISSLER, H. *Kirkburton.*—Cloakings, coatings, checks, &c.

3981 CHEETHAM, C. G. & W. *Woodbottom Mills, Horsforth, near Leeds.*—Volunteer army cloths.

3982 CHILD, J. & J. *Shelley, near Huddersfield.*—Waistcoats, ladies' skirts, and mantles.

3983 CLABBURN, SONS, & CRISP, *Norwich.*—Poplins, fancy dresses, &c.

3984 CLARK, J. & T. *Trowbridge, Wilts.*—Woollen cloths.

3985 CLAY, J. T. *Rastrick, near Huddersfield.*—Woollen and worsted fancy goods.

3988 COCHRANE, J. & W. *Mid Mill, and Netherdale, Galashiels.*—Scotch tweeds.

3989 COGSWELL, J. & Co. *Trowbridge, Wiltshire.*—Woollen doeskins, ribbs, deerskins, Venetians, &c.

3990 COLLIER, H. SEN. *Crawley Mills, Witney, Oxon.*—Witney blankets.

3991 COMYNS, A. SON & Co. 10, *College-green, Dublin.*—Irish friezes and tweeds.

3992 COOKE, A. M. 115, *Cheapside.*—Alpacas, mohairs, &c.

3993 COOK, T. SON, & WORMALD, *Dewsbury Mills.*—Blankets, rugs, and cloths.

3994 COOPER, A. & Co. *Leeds.*—Cloth caps.

3995 COOPER, D. & J. *Leeds.*—Woollen and union cloths.

3996 CRAVEN, J. *Thornton, near Bradford.*—Superfine llama, d'Ecosse, rep, and other shawls.

3997 CROMBIE, J. & J. *Grandholme Works, Aberdeen.*—Woollen goods.

3998 CROSLAND, B. *Oaks Mill, near Huddersfield.*—Mohairs, sealskins, furs, and velvet cloths.

3999 CROSLAND, W. & H. *Huddersfield.*—Woollen and angola goods.

4000 CROSS, W. 62, *Queen-st. Glasgow.*—Tartan and fancy woollen shawls and piece goods.

4001 CROWTHER, B. 60, *Albion-st. Leeds.*—Blankets and woollens.

4002 CUBITT, WILSON, & RANDALL, 36, *King-st. Cheapside.*—Tapestry, Utrecht velvets, and plushes.

4003 DALRYMPLE, W. *Union Mills, I. of Man.*—Woollen goods.

4004 DAVIES R. S. & SONS, *Stonehouse Mills, Gloucestershire.*—Cloths, cassimeres, and doeskins.

4006 DAY, NEPHEW, & Co. *Dewsbury.*—Pilots, cheviots, velvet piles, and woollens.

4007 DAY & WATKINSON, *Huddersfield.*—Drab kerseys and Bedford cords.

4008 DICKSONS & LAINGS, *Wilton Mills, Hawick.*- Wool tweeds and plaids.

4009 DIXON, T. D. *Morley, near Leeds.*—Union cloth, made from Sidney wool and waste.

4010 DOBSON, J. & A. *Innerleithen, N.B.*—Tweeds, shirtings, shawls, and woollen goods.

4011 DOBSON & RILEY, *Fenay Mills, Huddersfield.*—Fancy woollen manufactures.

4012 DOLAN, J. C. *Leeds.*—Army, police, and other cloths.

4013 DRINKWATER, W. *Salford Woollen Mills, Manchester.*—Woollen cords, worsted, and cotton tweeds.

4014 EARLY, E. & SON, *West-end, Witney.*— Witney blankets, tiltings, yarns, rugging.

4015 EARLY, J. & Co. *Witney.*—Blankets, pilot and collar cloths, and tweeds.

4016 EARLY, R. JUN. *Witney.*—Princes' checks, webs, kerseys, tiltings, mop yarns.

4017 ECROYD, W. & SONS, *Lomeshaye Mills, Burnley.*—Worsted and mixed stuffs, dyed, &c.

4018 EDMONDS & Co. *Bradford, Wiltshire.*—Blue cloth, and wool dyed black.

4019 ELWORTHY, W. & T. *Wellington, Somersetshire.*—Serges, blanket, yarns, &c.

4023 FIELD, R. *Skelmanthorpe, Huddersfield.*—Vestings, Balmoral skirts, &c.

4024 FIELDING & JOHNSON, *Leicester.*—Worsted for hosiery, &c.

4025 FIRTH, E. & SONS, *Heckmondwike.*—Blankets, cloths, sealskins, mohairs, rugs, &c.

4026 FORBES & HUTCHESON, 5, *Forbes-pl. Paisley.*—Shawls and dresses.

4027 FOX, J. J. & SON, *Devizes.*—Cloth of English wool.

4028 FRANCIS & FLINT, *Nailsworth Mills, Stroud.*—Wool-dyed black and blue cloths, doeskins, &c.

4029 FRY, W. & Co. 31, *Westmoreland-st. Dublin.*—Irish poplins, or tabinets, and brocatelles.

4030 FYFE, A. & Co. 77, *Queen-st. Glasgow.*—Fancy dress fabrics.

4031 FYFE, H. & SON, Manufacturers, *Glasgow.*—Ginghams, drugget skirts, fancy dresses.

4032 GARVIE & DEAS, *Perth.*—Linen and cotton fabrics, hosiery, &c.

4033 GILL, R. & SON, *Inverleithen.*—Scotch tweeds, wool tartans, &c.

4034 GLOYNE, C. G. *Bradford-rd. Dewsbury.*—Pilots, witneys, velvet piles, mantle cloths, &c.

4035 BINNS, G. & SONS, *Deighton, near Huddersfield.*—Tweeds, diagonals, and coatings.

4036 GOTT, B. & SONS, *Leeds.*—Cloths, blankets, woollen yarns.

4037 GOW, BUTLER, & Co. 55, *Wilson-st. Glasgow.*—Shawls and fancy dress fabrics.

4038 GREENWOOD & CARTWRIGHT, *Rawfolds, near Leeds.*—Woollen cloths for Eastern markets.

4039 GREENWOOD, J. & SONS, *Dewsbury.*—Pilots, petershams, witneys, cheviots, &c.

4040 GRIST, SONS, & Co. *Brimscombe, Glo'stershire.*—Mattress wools, flocks, and shoddy, made from woollen rags.

4041 GRIST, SON, & TABRAM, *Nailsworth, Glo'stershire.*—Shoddy, from woollen rags; flocks.

4043 HARGREAVE & NUSSEY'S, *Farnley Low Mills, near Leeds.*—Woollen cloths.

4044 HARTLEY, J. & Co. *Low Fold Mills, Leeds.*—Cloths and beavers.

4045 HARTLEYS & HARDWICK, *Aire-st. Leeds.*—Cloths, unions, tweeds, and coatings.

4046 HASTINGS, R. *The Votch Mills, near Stroud.*—Woollen cloths and doeskins.

4047 HATTERSLEY, G. & SON, *Quarmby, Clough Mills, near Huddersfield.*—Fancy union cloths.

4048 HEMINGWAY, J. *Watergate Mill, Dewsbury.*—Woollen yarns for carpets.

4049 HEWITT, H. & ST. J. *Heytesbury, Wilts.*—Fine woollen fabrics.

4050 HEY, G. *Kirkbarton, near Huddersfield.*—Fancy woollen cloths.

4051 HILL, J. & Co. *Banbury.*—Mohair and worsted plush, worsted webbing, &c.

4052 HINCHCLIFF, D. *Providence Mill, Morley, Leeds.*—Union cloths, of wool and other materials.

4053 HINCHLIFFE, J. & J. *Morley and Leeds.*—Union cloth, made from Sydney wool waste and mungo.

4054 HINDE, F. & SON, *Norwich.*—Paramattas, tamatioes, grenadines, poplins, &c.

4055 HITCHCOCK, G. & Co. 74, *St. Paul's Churchyard.*—Ladies' fancy dresses.

4056 HODGES, T. W. & SONS, *Leicester.*—Elastic webs.

4057 HOLDSWORTH, J. & Co. *Halifax and Bradford.*—Textile fabrics, worsted yarns, &c.

4058 HOLLINS, W. & Co. *Pleasby Works, near Mansfield, Nottingham.*—Cotton, merino, silk, cashmere, yarns.

4059 HOLROYD, S. & Co. *Huddersfield.*—Cotton warp and doeskin.

4060 HOLT, RUSSELL, & BATES, 114, *St. Martin's-lane.*—Woollen goods.

4061 HOOPER, C. & Co. *Eastington Mills, Stonehouse, Glo'ster.*—Superfine cloth, doeskins, and elastics.

4062 HOWGATE, HOLT, & Co. *Dewsbury.*—Woollen and union cloths.

4063 HOWSE, MEAD, & SONS, 18, *St. Paul's Churchyard.*—Woollen cloths.

4064 HUDSON & BOUSFIELD, *Leeds.*—Woollen and union cloths.

4065 HUNT & Co. *Lodgemore and Frome Hall Mills, Stroud, Glo'stershire.*—Superfine black cloths and doeskins, scarlets, billiards, &c.

4066 HUNT, WINTERBOTHAM, & Co. *Cam and Dursley Mills, Glo'stershire.*—Woollen cloths.

4067 IRELAND, J. & Co. *Kendal.*—Railway rugs, kerseys, linseys, &c.

4068 IRWIN, E. *Leeds.*—Double milled woollen cloths, &c.

4070 JAY, G. & SON, *Albion Mills, Norwich.*—Mohair and alpaca yarns.

4071 JEBSON, J. & J. *Skelmanthorpe, near Huddersfield.*—Fancy waistcoatings and skirtings.

4072 JENKINS, W. *Carmarthen-rd. Swansea.*—Welsh cloth and flannel.

4073 JOHNSTON, J. *New Mill, Elgin.*—Tweeds and mauds, of Cheviot, Australian, and Vicugua wool.

4074 JONAS, SIMONSEN, & Co. *Huddersfield.*—Woollen and union goods, and felt carpets.

4075 JONES, RANDALL, & WAY, 127, *Cheapside.*—Fancy woollen goods.

4076 JORDAN, J. *Huddersfield.*—Vestings, coatings, mantle cloths.

4077 KENYON, J. & T. *Huddersfield.*—Fancy woollens.

4078 KELSALL & KEMP, *Rochdale, Lancashire.*—Flannels, swanskins, and dometts.

4079 KERR, SCOTT, & KILNER, 58, *Cannon-st. west.*—Shawls.

4080 KOHNSTAMM, H. 33, *Dowgate-hill, Cannon-st. E.C.*—Leather cloths, with materials.

4081 LAIDLAW, W. & SONS, *Hawick.*—Hosiery, plaids, tweeds, &c.

4082 LAIRD & THOMSON, 69, *Ingram-st. Glasgow.*—Mixed fabrics, woollen, and poncho cloth.

4083 LAVERTON, A. *Westbury, Wiltshire.*—Woollen cloths, silk mixtures, arctic fur, and sable beavers.

4084 LEACH, J. & SONS, 83, *Wood-pl. Cheapside.*—Flannels, baize, &c.

4085 LEES, G. *Gala-bank Mill, Galashiels.*—Saxony shawls and cloakings, angolas, tweeds.

4087 LIDDELL, BENNETT, & MARTIN, *Upperhead Mills, Huddersfield.*—Fancy woollens.

4088 LIEBMANN, M. *Huddersfield.*—New felted carpet, called "Airdale Felt."

4091 LITTLES, LEACH, & Co. *Britannia Mills, Leeds.*—Superfine wool and piece-dyed cloths, doeskins, unions, &c.

4092 LOCKE, CROSIER, & EDWARDS, 64, *Friday-st. E.C.*—Shawls, mantle cloths.

4094 LOCKWOOD & KEIGHLEY, *Huddersfield.*—Woollen cords, velveteens, &c.

4095 LUPTON, W. & Co. *Leeds.*—Woollen cloths.

4096 MACDOUGALL & Co. *Inverness.*—Tweeds, tartans, linseys, plaids, &c.

4097 MALLINSON, D. *Lepton, near Huddersfield.*—Vestings, skirts, and trouserings.

4098 MARLING, STRACHAN, & Co. *Ebley and Stanley Mills, Stroud, Glo'stershire.*—Wool-dyed cloths, doeskins, and cassimeres.

4099 MARRIOTT, T. & SON, *Wakefield, Yorkshire.*—Worsted and woollen yarns.

4100 MELLOR, J. & SON, *Huddersfield.*—Doeskins, sataras, and fancy unions.

4101 MIDDLETON & ANSWORTH, *Norwich, and* 92, *Watling-st. E.C.*—Fabrics for dresses, shawls, and crinoline.

4102 MILLMAN & FOXWELL, *Nind Mills, Wotton-under-Edge.*—Superfine woollen cloths and doeskins.

4103 MILNER & HALE, *Huddersfield.*—Coatings and mantle cloths.

4104 MILNER & NOKES, *Thurlstone, Yorkshire.*—Fancy hairlines.

4106 MITCHELL & WHYTLAW, *Glasgow.*—Fancy dress fabrics.

4107 MORGAN, J. & Co. *Paisley.*—Shawls and woollen tartans.

4108 NEWBY & WOODHOUSE, *Bookfoot Brighouse, Yorkshire.*—Fancy unions.

1100 NOLDA, C. & Co. 2, *Church-ct. Old Jewry.*—Fancy woollens.

4111 NORTON, J. *Clayton West, near Huddersfield.*—Shawls, mantle cloths, coatings, rugs, and dress goods.
4112 OATES & BLAKELEY, *Dewsbury.*—Pilot and velvet cloths, tweeds, and cloakings.
4113 OATES, H. & SON, *Heckmondwike, near Leeds.*—Blankets, woollens, army goods, &c.
4114 O'REILLY, DUNNE, & Co. 30, *College-green, Dublin.*—Irish poplin and tabinets.
4115 PATON, J. & D. & Co. *Tillicoultry, Scotland.*—Woollen shawls and cloakings.
4116 PATON, J., SON, & Co. *Kilncraig's Factory, Alloa, N.B.*—Woollen hosiery, yarns.
4118 PEACE, D. *Shelley, near Huddersfield.*—Waistcoatings.
4121 PEACE, W. *Shipley, near Huddersfield.*—Vestings.
4122 PEASE, H. & Co. *Darlington and Bradford.*—Worsted mixed fabrics and yarns.
4123 PIM, BROS. & Co. *South Gt. George's-st. Dublin.*—Irish poplins.
4124 PLAYNE, P. P. & C. *Nailsworth, near Stroud.*—Superfine woollen cloths.
4125 PODD, T. & Co. *Leicester.*—Worsted, woollen, and Berlin yarns.
4127 RATCLIFFE & SONS, *Staley, Huddersfield.*—Plain and fancy flannels.
4128 REID & TAYLOR, *Langholm, Scotland.*—Tweeds and mauds.
4129 RHODES, D. & SONS, *Dewsbury.*—Seal skins, velvet piles, cheviots, travelling rugs, &c.
4130 RIPLEY, R. *Isle Mill, Holbeck, near Leeds.*—Woollen yarn.
4131 ROBERTS, JOWLINGS, & Co. *Lightpill Mills, Stroud.*—Woollen cloths and doeskins.
4132 ROBERTSON, J. *Coupar Angus, Perthshire.*—Fast coloured mixtures of flax, &c. and flax volunteer uniforms.
4133 SALTER, S. & Co. *Trowbridge, Wilts.*—Plain and fancy woollens.
4134 SCHOFIELD J. & SON, *Commercial Mills, Huddersfield.*—Tweeds, wool and angola yarns, knickerbocker yarns.
4135 SCHWANN, KELL, & Co. *Huddersfield.*—Cloakings and coatings.
4136 SCOTT, A. & SON, *Valley Mills, Morley, near Leeds.* — Woollen and union cloths.
4138 SHAW, J. *Huddersfield.*—Trouserings.
4139 SHAW & BEAUMONT, *Kirkheaton and Huddersfield.*—Fancy woollen trouserings and coatings.
4140 SHEARD, M. & SONS, *Batley, Yorkshire.*—Pilots, velvets, and reversible cloths.
4141 SHEPPARD, W. B. & G. *Frome.*—Cloths and woollen goods.
4142 SIBBALD, J. & Co. *Abbot's Mill, Galashiels.*—Scotch woollen trouserings.
4143 SMITH, W. JUN. *Morley and Leeds.*—Union cloths made from Sydney wool, machine waste, and mungo.
4144 SMITH, R. & SON, *Hayford Mills, near Stirling.*—Wincey, linsey, and woollens.
4146 SMITH, W. SON, & Co. *Leeds.*—Woollen cloths, meltons, naps, witneys.
4147 SPEIRS, D. & Co. *Paisley.*—Paisley, Scotch, and woollen shawls.
4148 SPENCE, J. & Co. 78, *St. Paul's-churchyard.*—Shawls, poplins, mohairs, challies, and alpacas.
4149 STANCOMB, W. & J. *Trowbridge Wilts.*—Fancy woollens.
4150 STANDEN & Co. 112, *Jermyn-st. St. James's.*—Hand-knit Shetland goods.
4151 STANTON & SON, *Stafford Mills, near Stroud.*—Woollen cloth.
4152 STARKEY, J. & A. *Sheepridge, near Huddersfield.*—Woollen cords, Bedford cords and velveteens.
4155 STOCKDALE, W. *High Burton, near Huddersfield.*—Fancy cloakings and coatings.
4156 SYKES, D. *Brookfield Mills, Hunslet.*—Woollen cloths.
4157 SYKES, G. *Dalton, Huddersfield.*—All wool and mixed fancy coatings, and Balmoral skirts.
4159 TAYLOR, J. & SONS, *Newsome, Huddersfield.*—Fancy woollen and silk trouserings and coatings, &c.
4161 THORPE, W. *Almondbury, near Huddersfield.* — Trouserings, coatings, and tweeds.
4162 THRESHER & GLENNY, 152, *Strand.*—Silk, wool, flannels, cloths, &c.
4163 TOLSON, BROS. *Dalton, near Huddersfield.*—Quiltings, dresses, &c.
4164 TOWLER, ROWLING, & ALLEN, 15, *Watling-st. E.C.*—Norwich paramattas, poplins, fancy dress, shawl.
4165 TURNBULL, W. & Co. 21, *Glassford-st. Glasgow.*—Scotch tweeds
4166 TURNER & MUTER, 14, *West Nile-st. Glasgow.*—Dresses in mixed fabrics of cotton, mohair, and silk.
4168 VICKERMANN, B. & SONS, *Huddersfield.*—Broad and narrow cloths.
4169 WADE, J. & SON, *Morley, near Leeds.*—Union cloths.
4170 HOLROYD, J. & Co. *Leeds.*—Plain and fancy woollen and union cloths.
4171 WALKER, G. *Lindley, near Huddersfield.*—Coatings.
4172 WALL & Co. *Welshpool, N. Wales.*—Flannels, tweeds, and clothing.
4173 WANDLE FELT Co. *Manchester.*—Cloths and felts for mechanical operations.
4174 WATSON & NAYLOR, *Pike-mills, Kidderminster.* — Worsted yarns, carded and combed.
4175 WATSON, W. & SONS, *Dangerfield-mills, Hawick, N.B.*—Tweeds and plaids.
4176 WEBB, T. R. & Co. *Huddersfield.*—Elastic stockingetts.
4177 WHEELER, T. & Co. *Abbey-mills, Leicester.*—Elastic webs.
4178 WHEELER, W. S. & SONS, 4, *Ludgate-st.*—Woollen trouserings and coatings.
4179 WHITE, J. 12, *Frederick-st. Edinburgh.* — Hand-knitted Shetland woollen articles.
4180 WHITEHEAD, E. 21, *Rook-st. Manchester.*—Poplins and poplinettes.
4181 WHITEHILL, M. & Co. *Paisley*—Embroidered shawls and table covers.
4182 WHITELEY, T. & SON, *Stainland.*—Tweeds.
4185 WILKS, BROS. & SEATON, 80, *Watling-st. London, E.C.*—Irish linens.

4186 WILSON, J. J. & W. *Kendal.*—Railway wrappers, tweeds, horse clothing, &c.
4187 WILSON, W. *Allars-cres. Hawick, N.B.*—Tweeds and cloakings.
4188 WILSON & ARMSTRONG, *Hawick, N.B.*—Tweeds and wool shawls.
4189 WISE & LEONARD, *Nailsworth, and Holcombe Mills, near Stroud.*—Superfine black and blue cloths, kerseymere, and double beaver.
4190 WOODHOUSE, J. & Co. *Leeds, and New Mills, Holbeck.*—Tweeds, meltons, and union cloths.
4193 WRIGLEY, J. & T. C. & Co. *Dungeon Mills, near Huddersfield.*—Woollen goods.
4194 WRIGLEY, J. & SONS, *Huddersfield.*—Liveries and meltons and carriage linings.
4195 WURTZBURG, E. & Co. *Leeds.*—Woollen and union cloths.
4196 YUILL, J. & Co. *Paisley.*—Shawls, mufflers, and dresses.

4197 WALKER, J. & SONS, *Lindley, near Huddersfield.*—Mohairs, Hudson-bays, sealskins, cashmere, furs, brennas, shells, rugs, &c.

CLASS 22.

CARPETS.

Under North-East Gallery, and on Gallery Walls.

4227 BOYLE, J. W. 9, *Gt. Marlborough-st.*—Carpets.
4228 BRINTON & LEWIS, *Kidderminster.*—Various carpets and rugs.
4229 CAWLEY, J. 28, *Red Lion-st. Clerkenwell.*—Adelaide mats and rugs.
4230 COOKE, HINDLEY, & LAW, 12, *Friday-st. E.C.*—Various descriptions of carpeting.
4231 CROSSLEY, J. & SONS, *Halifax.*—Carpets, rugs, &c.
4232 DIXON, H. J. & SONS, *Long Meadow Mills, Kidderminster.*—A Brussels velvet-pile carpet.
4233 DOWNING, G. F. *Knightsbridge.*—Seamless floor cloth, ten yards wide.
4234 FAIRFAX, KELLEY, & SONS, *Heckmondwike.*—Carpets, blankets, seal skins, and carriage rugs.
4235 FILMER, J. H. & SON, *Berners-st. W.*—Carpets, designed in competition by students of the S. Kensington School of Art.
4237 FIRTH, E. & SONS, *Heckmondwike.*—Various descriptions of carpeting.
4238 GOATLEY & CHORLEY, 39, *Westminster-bridge-rd.*—Floor cloth manufactures.
4239 GREGORY, C. 212, *Regent-st.*—Axminster and other carpets, &c.
4240 GREGORY, THOMSON, & Co. *Kilmarnock.*—Specimens of carpeting.
4241 HARE & Co. *Bristol.*—Floor cloths: Corinium pavement, tiles, and marbles, &c.
4242 HARRISON, C. *Stourport.*—Brussels and other carpets.
4243 HAWKSWORTH, S. *Baker-st. Doncaster.*—Floor cloth in paint imitation of mosaic pavement.
4244 HEAD, J. & Co. *Stourport, Worcestershire.*—Carpets and rugs.
4245 HENDERSON & Co. *Durham.*—Samples of various carpeting.
4246 JACKSON & GRAHAM, *Oxford-st.*—Brussels, Axminster, and other carpets.
4247 KINDON & POWELL, *Swan-st. Old Kent-rd. S.E.*—Floor cloths, 'table covering, stair cloths.
4248 LAPWORTH, BROS. 22, *Old Bond-st.*—Axminster, Brussels, and Wilton pile carpets.
4249 MORTON & SONS, *Kidderminster and London.*—Velvet and Brussels carpets.
4250 NAIRN, M. & Co. *Scottish Floor Cloth Manufactory, Kirkaldy.*—Printed floor cloth.
4251 PALMER BROS. *Kidderminster.*—Hand-loom Brussels carpet.
4252 ROLLS J. & SONS, *Kennington-lane, Lambeth, S.*—Floor cloth.
4253 SEWELL, HUBBARD, & BACON, *Old Compton-rd. Soho.*—Axminster carpet.
4254 SMITH, T. & Co. 9, *Gt. Marlborough-st. W.*—Carpets.
4255 SMITH & BABER, 1, *South-place Knightsbridge.*—Floor cloth.
4256 SOUTHWELL, H. & M. *Bridgnorth.*—Bordered Wilton carpet.
4257 STEVENSON, W. 16, *Piccadilly.*—Bordered pile carpet.
4258 SWALLOW, M. & SONS, *Heckmondwike.*—Carpets, alpaca coat linings, &c.
4259 TAPLING, T. & Co. 1 *to* 8, *Gresham-st. West, City.*—Axminster carpets.
4260 TAYLER, HARRY & Co. 19, *Gutter-lane, Cheapside.*—Kamptulicon floor cloth.
4261 TEMPLETON, J. & Co. *Glasgow and London.*—Carpets without seam, and in breadths, rugs, mats, &c.
4262 TEMPLETON, J. & J. S. *Glasgow.*—Axminster carpeting, and silk and wool curtains and covers.
4263 WATSON, BONTOR, & Co. 36, *Old Bond-st.*—Axminster, Brussels, and velvet carpets.

Under North-East Gallery, and on Gallery Walls.

4264 WAUGH & SONS, 3 & 4, *Goodge-st.*—Axminster and other carpets.

4265 WHITWELL & CO. *Dockray Hall Mills, Kendal.*—Carpeting of various descriptions.

4266 WHYTOCK, R. & Co, 9, *George-st. Edinburgh.*—Wilton, and Scoto-Axminster carpet.

4268 WOODWARD BROS. & CO. *Kidderminster.*—Carpeting and rugs.

4269 WOODWARD, H. & SONS, *Stour Vale Mills, Kidderminster.*—Brussels and Tournay velvet carpets.

4270 HUMPHRIES, J. & SONS, *Mile-st. Kidderminster.*—Velvet-pile carpets.

4271 HARVEY, NICHOLS, & CO. *Knightsbridge.*—Carpets, &c.

CLASS 23.

WOVEN, SPUN, FELTED, AND LAID FABRICS, AS SPECIMENS OF PRINTING OR DYEING.

South-East Gallery.

4300 BARLOW S. & CO. *Stakehill, Chadderton, Manchester.*—Cotton goods bleached by Barlow's process.

4301 BAYNES & SON, *Queen's-rd. Bayswater.*—Dyed furniture hangings, &c.

4302 BEAULIEU, J. & Co. 7, *Sloane-st. S.W.*—Dyed goods.

4304 BERRIE, J. 13, *Oldham-st. Manchester.*—Old silk and merino cleaned.

4305 BLACK & WINGATE, *Glasgow.* — Imitation cambrics, handkerchiefs, embroideries, &c.

4306 BOTTERILL, J. *Leeds.*—Specimens of dyed articles.

4307 BRADSHAW, HAMMOND, & CO. *Levenshulme Works, and* 35, *Mosley-st. Manchester.*—Calico prints.

4308 BUTTERWORTH & BROOKS, *Manchester.*—Calico prints and muslin-de-laines.

4309 CALDER VALE PRINTING CO. *Manchester.*—Silks and Orleans printed by machine.

4310 CLARKSON, T. & CO. 17, *Coventry-st. Haymarket.*—Block and machine printed chintz furnitures.

4311 DAILY & CO. *St. James's-pl. Hampstead-rd.*—Specimens of dyeing and cleaning.

4312 DEWHURST, S. & CO. *Broughton Works, Manchester.* — Bookbinder's cloth, beetled twills, &c.

4313 DONOVAN, R. 2, *Gt. Pulteney-st.*—Tapestry part cleaned.

4314 GEORGE, T. W. & CO. *Spring-gardens Dye Works, Leeds.* — Worsted and mixed stuff goods, in various colours, and black.

4315 GIRDWOOD, W. & CO. *Old-park, Belfast.*—Printed linens, lawn dresses, &c.

4316 GRAFTON, F. W. & CO. *Broad-oak, Accrington.*—Printed calicoes.

4317 HANDS, SON, & CO. *Coventry.*—Silks dyed in the skein.

4318 HINE, R. E. & CO. 36, *York-st. Manchester.*—Silk handkerchiefs printed, and other silks.

4319 HOYLE, T. & SONS, *Mayfield, Manchester.* — Cambrics, challis, and delaine prints.

4320 KEYMER, J. & CO. *Dartford, Kent.*—Printed silk handkerchiefs.

4321 LITTLEWOOD, WILSON, & CO. *Manchester.*—Muslin, calico, and delaine prints.

4322 LOCKETT, J., SONS, & LEAKE, *Manchester.*—Engravings for calico printers.

4323 MCKINNELL & CO. *Ancoats-bridge Works, Manchester.*—Blue and madder printed cottons.

4324 MACNAB, J. 145, *Ingram-st. Glasgow.*—Printed calicoes and muslins.

4325 MCNAUGHTON & THOM, 80, *Mosley-st. Manchester.*—Printed calicoes.

4326 MONTEITH, H. & CO. *Glasgow.*—Dyed and printed goods.

4327 MUIR, BROWN, & CO. 29, *West George-st. Glasgow.* — Cotton and woollen fabrics.

4328 NEWTON BANK PRINTING CO. 51, *Mosley-st. Manchester.*—Printed cottons.

4329 EWING, J. O. & CO. 32, *St. Vincent-place, Glasgow.*—Turkey red goods.

4330 ORMEROD, R. & CO. 50, *Mosley-st. Manchester.*—Patent printed ribbons.

4331 OXFORD, J. & CO. *Bury-ct. St. Mary Axe.*—Printed corahs, China crape shawls, &c.

4332 PALMER & CO. *Holme Works, Carlisle.*—Taffata silk and cotton, umbrella cloths, &c.

4333 PULLAR, R. & SONS, *Perth.*—Umbrella cloths and cotton goods.

4334 RICHARDSON, B. S. *Priory-fields, Coventry.*—Silk samples of dyeing.

4335 SALOMONS, A. *Old Change, London E.C.*—British printed cottons and muslins.

4338 SMITH, S. & Co. *Horton Dyeworks, Bradford.*—Dyed and finished stuff.

4339 STEAD, MCALPINE, & Co. *Cummersdale Print-works, Carlisle.*—Calico and cotton damask, chintz furnitures.

4340 STIRLING, W. & SONS, *Glasgow.*—Plain and printed cottons.

4341 TATTON, S. *Mill-st. Leek.*—Sewing silk and twist.

4342 TURNER, C. *Leeds, and* 18, *Lawrence-lane, London, E.C.*—Felt carpets and goods.

4343 TURNER, NORRIS, & TURNER, *Manchester, and Hayfield, Derbyshire.*—Printed cottons.

4345 VICTORIA FELT CARPET Co. *Leeds and London.*—Felt carpets, table covers, tablings, &c. &c.

4346 WALFORD, FAIRER, & HARRISON, *London.*—Printed bandannas and silks.

4347 WATSON & STARK, 51, *George-st. Manchester.*—Printed trouserings, coatings, and cambrics.

4348 WELCH, T. *Merton, Surrey.*—Woollen cloth, printed patterns.

4349 WILKINSON, J. SON, & Co. *Leeds.*—Felt carpets, rugs, squaes, numnahs, boot felt, &c.

4350 YATES, M. W. 2, *Wood-st. London, E.C.*—Table covers.

4351 ROSENDALE PRINTING Co, 8, *Nicholas-st. Manchester.*—Prints.

4352 WHITWELL, BUSHER, & Co. *Kendal*—Dyed worsted and woollen yarns.

4353 THWAITES, J. *Kendal.*—Chromatic arrangement of woollen carpet yarns.

CLASS 24.

TAPESTRY, LACE, AND EMBROIDERY.

South-East Gallery.

4381 ABRAHAM, R. & SONS, 5, *Lisle-st. W.*—Cover for articles used in Jewish synagogues, embroidery, &c.

4382 ADAMS, T. & Co. *Nottingham.*—Curtains and other lace.

4383 ALLEN, C. 108, *Grafton-st. Dublin.*—Irish, point, and other lace.

4384 AUSTIN, J. *Princes-st. Finsbury.*—Picture and sash lines, chandelier rope, &c.

4385 BAGLEY, J. W. *Nottingham.*—Valenciennces, Maltese, and Honiton laces, &c.

4386 BARNETT, MALTBY, & Co. *Stoney-st. Nottingham.*—Silk laces, nets, shawls, &c.

4387 BATES, J. MISS, *Newington, Surrey.*—A lace fall and other articles worked by hand.

4388 BLACKBORNE, A. 35, *South Audley-st. W.*—Foreign and English laces.

4389 BORWICK, MISS, 2, *Henstridge-villas, St. John's-wood.*—A raised crotchet counterpane.

4390 BRADBURY, CULLEN, & FISHER, *Broadway, Nottingham.*—Lace shawls, falls, &c.

4392 CARDWELL, C. *Northampton.*—Articles of pillow lace.

4393 CATT, 198, *Sloane-st.*—Dress trimmings and fancy needlework.

4394 CHAMBERS, J. & Co. 4, *Upper Sackville-st. Dublin.*—Silk embroidery on muslin cloth, handkerchiefs, &c.; sewed muslin and under-clothing.

4395 CLARKE, E. 18A, *Margaret-st. Cavendish-sq.*—Specimens of Honiton lace, &c.

4396 COPESTAKE, MOORE, CRAMPTON, & Co. 5, *Bow Church-yard.*—Lace.

4397 COWAN & Co. 24, *St. Vincent-pl. Glasgow.*—Embroidered muslins. Frillings, &c.

4398 CROOME, MISS, *Middleton Cheney, Banbury.*—Egyptian-work dress and local-made lace.

4399 DART & SON, 12, *Bedford-st. Covent-gard.*—Lace for carriages.

4400 DEBENHAM, SON, & FREEBODY, 44, *Wigmore-st.*—Various articles in Honiton, Buckinghamshire and Nottingham lace.

4401 RIEGO DE LA BRANCHARDIERE, MLLE. 1, *Princes-st. W.*—Crochet and modern point lace.

4402 DIXON, G. 13, *Goldsmith-st. Cheapside.*—General upholstery, trimmings, &c.

4403 DUNNICLIFF & SMITH, *Nottingham.*—Valenciennes and other lace.

4404 EHRENZELLER, F. 35, *Cannon-st. west, City.*—Needlework lace articles, &c.

4406 ELLINGTON & RIDLEY, 89, *Watling-st. E.C.*—Trimmings, &c. for upholstery purposes.

4407 ERNE, COUNTESS OF, 95, *Eaton-sq.*—Irish valenciennes lace.

4408 EVANS, R. & Co. 24, *Watling-st. E.C.*—Upholstery and dress trimmings, &c.

4409 FORREST, J. & SONS, 101, *Grafton-st. Dublin.*—Irish point lace tunics.

4410 GILBERT, T. *High Wycombe.*—Pillow lace goods.

4411 GOBLET, H. F. 20, *Milk-st. City.*—Irish crochet, tatting, lace, and needlework.

4412 GODFROY, E. *Buckingham.*—Guipure tunic flounce, veils and lappets, &c.

4413 GRAHAM, A. 34, *York-st. Glasgow.*—

Patterns for needlework, cut velvet, silk, &c.

4414 GREEN, A. 136, *Buchanan-st. Glasgow.*—Fancy needlework, and materials used.

4415 HAYES, MISS E. J. 24, *Richmond-pl. East-st. Walworth.*—Infant's embroidered cloak and bead embroidery.

4416 HAYWARDS (D. BIDDLE), 81, *Oxford-st.*—Honiton and other British laces. Guipure, &c.

4417 HERBERT, T. & Co. *Houndsgate, Nottingham.*—Tattings, crochets, muslins, laces, fringes, &c.

4418 HEUGH, WIGHT, & Co. 56, *Friday-st.*—Arnold's stitched frilling.

4419 HEYMANN & ALEXANDER, *Nottingham.*—Nets, laces, and curtains.

4420 HIGGINS, EAGLE, & HUTCHINSON, 57, *Cannon-st. West.*—British lace goods.

4421 HORNSEY, J. *Bedford.*—Maltese and other laces, and lace goods.

4422 HOWELL, JAMES, & Co. *Regent-st.*—Lace.

4423 HYDE, MRS. 7, *Finsbury-pl. South.*—Needlework in Berlin wool, and imitations of natural flowers and ferns.

4424 HYDE, ARCHER & Co. 7, *Finsbury-pl. South, London.*—Upholstery trimmings, &c.

4425 INDUSTRIAL SOCIETY, 76, *Grafton-st. Dublin.*—Irish point, guipure, and crochet lace.

4426 JACOBY, M. & Co. *Nottingham.*—Lace goods.

4427 JONES, W. & Co. 236, *Regent-st. London.*—Gold lace embroidery. Accoutrements, &c.

4428 KEITH & Co. 124, *Wood-st. London.*—Upholstery fringe.

4429 LECHÊNE, A. 2, *Foley-st. Portland-pl.*—Chromo on velvet, for decorative tapestry, &c.

4430 MERY, L. & Co. Draftsmen, 87, *Up. Ground-st. Blackfriars.*—Patent velvet application.

4431 LESTER & SONS, *Bedford.*—Laces and lace articles.

4432 LONG, G. *Loudwater, Wycombe, Bucks.*—Hats, &c. made on the pillow lace principle, and straw embroidery.

4433 MACARTHUR, D. & Co. 26, *Bothwell-st. Glasgow.*—Muslin, lace, fancy linen, and crape goods.

4434 MACDONALD, HELEN J. 1, *Stafford-st. Edinburgh.*—Silk patchwork table-cover.

4436 MADDERS, W. & Co. *Leamington-pl. Manchester.*—Anti-macassar, and other articles embroidered by machinery.

4437 MALLET, H. *Nottingham.*—Various laces, guipures, blonde, shawls, mantillas, &c.

4438 MANLOVE, ALIOTT, & LIVESEY, *Bloomsgrove Works, Nottingham.*—Nets, laces, and trimmings.

4439 MANLY, G. N. 43, *New Finchley-rd.*—Irish lace.

4440 MEE, C. 71, *Brook-st. Grosvenor-sq.*—Embroidery and ornamental needlework.

4441 NEWMAN & PURNEY, 118, *Oxford-st. W.*—Fringes and trimmings.

4442 NORTHCOTE, S. & Co. 29, *St. Paul's Churchyard, E.C.*—Lace shawls, flounces parasol covers, &c.

4443 PALMER, H. *Dunse, Berwickshire, Scotland.*—Silk embroidery, &c.

4444 PALMER & COTTRELL, 87, *Blackfriars-rd. S.*—Painted velvets and cloth work, &c.

4445 PARSONS & SON, 82, *Long-acre.*—Carriage and other laces.

4447 PULLING & MOODY, 39, *Gresham-st. E.C.*—Crape goods.

4448 RADLEY, E. 20, *Lambs Conduit-st. W.C.*—Upholsterers' trimmings.

4449 RECKLESS & HICKLING, *Nottingham.*—Lace shawls, flounces, &c.

4450 ROBINSON, H. *Watling-st.*—Real and imitation British lace.

4451 SARJEANT, J. *Sandy, Bedfordshire.*—Lace articles.

4455 STANDRING, J. & BRO. *Manchester.*—Braids, cords, laces, fringes, &c.

4457 STILLWELL, SON, & LEDGER, 25, *Barbican, City.*—Gold lace, embroidery, and army and navy fittings.

4458 URLING, G. F. 224, *Regent-st.*—Articles in fine Honiton lace.

4459 VERKRUZEN & Co. 96, *Hatton-garden.*—Traced embroideries, gold drawings on velvet, &c.

4460 VICCARS, R. *Padbury, Buckingham.*—Lace and lace articles.

4461 VICKERS, W. *Nottingham.*—Various articles of black silk lace.

4462 VOKES, F. S. T. *Royal Surrey Theatre.*—Braiding for clothes and regimentals.

4463 WELSTED, H. *Ballywalter, Castletownroche, Co. Cork.*—Pocket handkerchiefs and other articles.

4464 WILLS, S. & Co. *Broadway, Nottingham.*—Machine lace, nets, curtains, &c.

4465 WILSON, MISS C. G. *Guildhall, Broad Sanctuary, S.W.*—Chessboard. Needlework.

4466 WOOLCOCK, C. M. & A. 13, *Old Quebec-st. Oxford-st.*—Tapestry drapery, portière, folding screen, &c.

4467 FETLEY, C. *Trafalgar-st. Bradford.*—Embroidered quilt on satin, &c.

4468 DEAN, MISS H. L. 55, *Manchester-road, Bradford.*—Machine quilted skirts.

4469 CLARKE, J. *Eastbrook Mills, Bradford.*—Netted table-cover.

CLASS 25.

SKINS, FUR, FEATHERS, AND HAIR.

Transept, South Court.

Sub-Class A.—Skins and Feathers.

4499 BEVINGTON & MORRIS, 67, *Cannon-st. West.*—Furs; angora, and other rugs.

4500 CLARK, C. & J. *Street, near Glastonbury.*—Rugs, boots, and shoes.

4501 DRAKE, R. 25, *Piccadilly.*—Manufactured furs, various.

4502 HOLDEN, J. T. *Collis Works, Birmingham.*—Self-fasteners for mantles.

4503 INCE, T. & C. 75, *Oxford-st.*—Furs.

4504 JEFFS, R. 244, *Regent-st.*—Fur skins and manufactured skins.

4505 LILLICRAPP, W. 19A, *Davies-st. Berkeley-sq.*—Sealskin cloaks, and furrier's goods.

4506 MEYER, S. M. 71, *Cannon-st. West, E.C.*—Furs, various.

4507 POLAND, G. & SON, 90, *Oxford-st. W.*—Furs and skins.

4508 ROBERTS, E. B. 239, *Regent-st.*—Manufactured furs.

4509 SENGER, A. H. 3, *Lamb-alley, Bishopsgate-st.*—Fur sealskin in different stages of manufacture.

4510 SMITH, G. & SONS, 9 *to* 11, *Watling-st.*—Furs.

4511 TUSSAUD BROS. 105, *Marylebone-rd.*—Method of saving skins from furs and applying artificial pelts.

4512 NICHOLAY, E. J. 82, *Oxford-st.*—Furs, fur ornaments, carpets, rugs, &c.

Sub-Class B.—Feathers.

4522 DE COSTA ANDRADE & CO. 7 & 8, *Cripplegate-bldgs. E.C.*—Ostrich feathers, raw and dressed.

4523 SUGDEN, SON, & NEPHEW, 16, *Aldermanbury.*—Case of ostrich and fancy feathers.

Sub-Class C.—Manufactures from Hair.

4534 ASTON, J. 20, *Dale End, Birmingham*—Household and saddlery brushes.

4535 BARRETT, A. 63, *Piccadilly.*—Brushes.

4536 BLYTH & SONS, 4, *Chiswell-st. London.*—Feathers, horse hair, and wools.

4537 BOOTH & FOX, 80, *Hatton-garden.*—Feather purifiers, feathers, down quilts, and petticoats.

4538 BROWNE, F. 47, *Fenchurch-st.*—Perukes, head dresses, &c.

4539 CARLES, H. R. 45, *New Bond-st. W.*—Imperceptible capillamenta.

4540 CHILD, W. H. 21, *Providence-row, Finsbury.*—Electro-galvanic, metallic, and other brushes.

4541 CLIFFT, J. *Bristol.*—Ornamental hair in wigs, &c. Specimens of human hair dyed.

4542 CONDRON, T. & R. 51, *Bingfield-st. Caledonian-rd. N.*—Fancy brushes.

4543 COOPER & HOLT, 50 & 51, *Bunhill-row, Finsbury.*—Curled horse hair bed feathers and wool.

4544 DICKINSON, J. & SON, *South market, Meadow-lane, Leeds.*—Machine brushes for manufacturing trades.

4545 DOHERTY, THE MISSES, *Sligo.*—Horsehair ornaments, made by peasant girls.

4546 DOUBBLE, T. 18, *Bartlett's-bldgs.*—Brooms, brushes, and combs.

4547 DOUGLAS, R. 21 & 23, *New Bond-st.*—Wigs, &c.

4548 DOW, A. 1, *Hardwick-st. Liverpool.*—Brushes for plate and jewellery.

4549 ELLINGTON & RIDLEY, 89, *Watling-st. E.C.*—Bed feathers, eider down, bed quilts.

4550 ESSEX, F. 53, *Percival-st.*—Wool rugs, foot muffs, boots, furs, &c.

4551 FARRANT, R. E. 16, *Queen's-row, Buckingham-gate.*—Various brushes.

4552 FORSTER, G. 9, *Hatton-wall, London, E.C.*—Shaving brushes, ivory and bone.

4553 GOSNELL, J. & CO. 12, *Three King-ct. Lombard-st.*—Perfumery. Hair and other brushes.

4554 GRAY, E. M. 44, *Ebury-st. Pimlico.*—Hair flowers.

4555 GRAY, L. 44, *Ebury-st. Pimlico.*—Hair coronet.

4556 GREENWOOD, B. *Bond-st. Bradford.*—Brooms and brushes.

4557 HASTINGS, S. *Limerick.*—Brushes made of oak taken from the old cathedral of Limerick.

4558 HERRMANN, A. 4, *Oxenden-st.*—Hair work, scissor lace work, &c.

4559 HEWLETT, A. H. 5, *Burlington-arcade.*—Artificial hair, hair dye, scalps, &c.

4560 HOPEKIRK, W. 88, *Westminster-bridge-rd.*—Wigs, partings, and scalps.

4561 HOVENDEN, R. & SONS, *Gt. Marlborough-st. W.*—Human hair, raw and manufactured.

4562 HOWARD, W. 23, *Gt. Russell-st. Bloomsbury.*—Introduction of gutta-percha for securing brush hairs.

4563 JEFFCOAT, J. 9, *Middle Queen's-bdgs Brompton.*—Painters' brushes, the bristles being tied on self-tightening principles.

4564 King, G. & Son, 116, *Bunhill-row.*—Brushes for manufacturing purposes.

4565 Kolle, H. & Son, 65, *Queen-st. Cheapside.*—Crinolines, &c.

4566 Loyer & Son, 33, *Gracechurch-st.*—Self-supplying water-brush for carriages.

4567 Marsh, J. 175, *Piccadilly.*—Hair, perfumery, and brushes.

4568 Mason, T. 40, *Portland-st. Leeds.*—Hair front and parting, hair-dye, &c.

4569 Metherell, J. K. 47, *Carey-st. W.C.*—Full-buttomed wig.

4570 Nash, T. Jun. 134, *Gt. Dover-st. Boro'.*—Paint and other brushes.

4571 Nightingale, W. & C. *Wardour-st.*—Feathers and downs. Horsehair.

4573 Pemberton, A. 15, *Broad-st. Worcester.*—Saddlery and stable brushes.

4574 Saville, H. *Leeds.*—Scalp with mechanical movement, &c.

4575 Smith, A. *Wentworth-st. Whitechapel.*—Brushes and piassara.

4576 Taylor, R. 3, *Brunswick-pl. South Kensington.*—Perukes and designs in human hair.

4577 Truefitt, H. P. 20 & 21, *Burlington-arcade.*—Specimens of wig making, &c.

4578 Truefitt, W. 1, *New Bond-st.*—Perukes, ladies' head-dresses, and articles in hair.

4579 Unwin & Albert, 24, *Piccadilly.* Perukes and ornamental hair.

4580 Vickers, S. 13, *Boar-lane, Leeds.*—Perukes, scalps, &c.

4581 Wall, T. 3, *Upper-arcade, Bristol.*—Artificial flowers worked in human hair.

4582 Watkins, C. A. 10, *Greek-st. Soho.*—Painting brushes, graining tools, shaving brushes, &c.

4583 Webb, E. *Worcester.*—Hair seating, curled hair, cider and hop cloth crinoline, &c.

4584 Whitfield, S. & Sons, *Birmingham.*—Purified and unpurified bed feathers.

4585 Williams, J. 46, *Westminster-rd.*—Brooms and brushes.

4586 Winter, W. 205, *Oxford-st. London.*—Ornamental hair.

4587 Wyatt, C. 1, *Conduit-st. W.*—Shades of dyed hair.

CLASS 26.

LEATHER, INCLUDING SADDLERY AND HARNESS.

Transept, South Court.

Sub Class A.—Leather.

4618 Batty, J. *Tottenham.*—East India and English sheepskins and seal fleshers.

4619 Bevington & Morris, 67, *Cannon-st. West, E.C.*—Leather.

4620 Bevington & Sons, 2, *Cannon-st. West, E.C.*—Leather trophy.

4621 Boak, A. 59, *West-port, Edinburgh.*—Hog skins tanned, and other leather.

4622 Bowen & Atkins, *Shipston-on-Stour, Worcestershire.* — Oak bark, tanned leather, and glove leather.

4623 British and Foreign Tanning Co. 37, *Gt. George-st. Bermondsey.*—Sole leather.

4625 Clark, J. & Sons, 76, *Dean-st. Soho.*—Leather manufactures.

4626 Cooper, F. E. *Brunswick-ct. Bermondsey.*—Leather for bookbinding and upholsterers.

4627 Deed, J. S. & Sons, 451, *Oxford-st.*—Leather, skins, hides, rugs, &c.

4628 Drake, R. *Bristol.*—Oak bark and valonia tanned leather.

4629 Draper, H. *Kenilworth.*—2 butts, tanned two years, soles from 3 butts.

4630 Essex, W. & Sons, 28, *Stanhope-st. Strand.*—Leather for carriages, harness, &c.

4631 Fisher, N. & Sons, 31, *Maze-pond, Southwark, S.E.*—Enamelled, curried, and coloured leather.

4632 Flitch, J. J. & Co. *Leeds.*—Fancy leather.

4633 Franklin, W. & J. *Walsall.*—Leather for bridles, saddles, and harness.

4634 George, C. 102, *Dean-st. Soho-sq.*—Morocco and Russia leather.

4635 Hemsworth, Linley, & Wilks, 30, *West Smithfield.*—Boot and shoe leather.

4636 Hepburn & Sons, *Bermondsey.*—Rough, dressed, and machine leather.

4637 Holden, E. T. *Walsall.*—Coach, saddle, harness, and other leather.

4638 Holmes, T. & Son, *Antaby-rd. Tannery, Hull.*—Walrus hide belting.

4639 Hudson, S. 65, *Dawson-st. Dublin.*—Saddlery, harness, &c.

4640 Hyde, Archer, & Co. 7, *Finsbury-pl. S.*—Coach, harness, and saddle leather.

4641 Jones, W. H. & Son, 179, *High-st. Boro'.*—Enamelled and curried leather.

4642 Lambert, Blakey, & Mowbray, *Bermondsey New-rd.*—Blocked boot-fronts, tops, jockey legs, Spanish cordovan, kip butts, &c. &c.

4643 Lever, J. *Neate-st. Old Kent-rd.*—Vellums and parchments.

4644 Liddell, W. H. 135, *West-port, Edinburgh.*—Hog skins, harness and bridle leather.

4645 LLOYD, T. 16, *Newcastle-st. Strand.*—Parchment, vellum, forrel, &c.

4646 M'RAE, J. & J. 43 & 46, *Bermondsey-st.*—Various specimens of leather.

4647 MARSHALL, W. & SON, *Ladyburn, near Greenock.*—Saddlers' basils.

4648 MATHEWS, G. *Market-st. Bermondsey.*—Goat, calf, sheep, and seal skins, and horse hides.

4649 MATTHEWS, W. *Spa-rd. Bermondsey.*—Enamelled and patent leather.

4650 MOFFAT, J. *Musselburgh.*—Hides and leather.

4651 MONTGOMERY, G. F. *Dowgate-hill, London, E.C.*—Hides preserved, prepared, and tanned by Lapervuse's process.

4652 MUNDY, W. P. *Tyers'-gateway, Bermondsey.* — Tanned East India kips. and English calf skins.

4653 NORRIS & CO. *Shadwell, London, E.*—Leather for machinery, belting, hose pipes, and fire buckets.

4654 POOLE, J. & C. *Walworth-common, London, S.*—Boot tops, legs, and fronts.

4655 PULLMAN, R. & J. 17, *Greek-st. Soho.*—Chamois, deer, and buff leather.

4656 RICHARDSON, E. & J. *Newcastle-on-Tyne.*—Enamelled and various leathers.

4657 ROBERTS, D. & E. *Pages-walk, Bermondsey.*—Leather manufactures.

4658 SAXTON, WADDINGTON, & CAREY, *Bartholomew-close.*—Skins, kips, fronts, shoe legs, jockey legs, cordovan, grained calf, &c.

4659 SHAW & MORRIS, *Wyld's-rents, Bermondsey.*—Harness leather.

4660 SIDEBOTTOM, A. *Crown-st. Camberwell.*—Kid skins for gloves.

4661 SMITH, E. *Camomile-st. London, E.C.*—Boot fronts, tops, jockey legs, hides, &c.

4662 SOMERVILLE, BROS. *Netherfield, Kendal.*—Shoe and harness leather.

4663 SOUTHEY & CO. 16, *Little Queen-st. W.C.*—Hides and skins.

4664 STOCKIL, W. 37, *Long-lane.*—Curried and blocked calf leather.

4665 SUTTON, W. *Scotby Works, near Carlisle.*—Boot and shoe leather in Spanish cordovan. Shoe hides and kips.

4666 TOMLINS, W. *Black Swan-yd. Bermondsey-st.*—Leather, parchment, and vellum.

4667 WILSON, WALKER, & CO. *Sheepscar, Leeds.*—Fancy sheep leather and coloured calf.

4668 WINSOR, G. & SON, 58, *Russell-st. Bermondsey.*—Sheepskin wool rugs.

4669 YORKSHIRE LEATHER CO. 482, *New Oxford-st.*—Harness and leather.

4670 FOORD & MOIR, *Pontsburgh Tanworks, Edinburgh.* — Tanned and curried leather; grained shoe-butts, and calf-skin, &c.

4671 MEGSON, E. *Dean-gate, Manchester.*—Hose-pipes.

Sub Class B.—Saddlery, Harness, &c.

4680 ANGUS, J. & CO. 131, *Trongate, Glasgow.*—Saddle trees, hames, &c.

4682 BANTON, E. *Walsall.*—Saddlers' ironmongery, &c.

4683 BARTLEY, C. A. 20B, *Portman-st. Portman-sq.*—Harness and saddlery.

4684 BLACKWELL, S. 259, *Oxford-st.*—Saddlery and harness.

4685 BLYTH, R. & SONS, 4, *Park-lane, W.*—Saddles, harness pad, &c.

4686 BOURNE, T. 5, *College-road, Cork.* — Carriage harness, with improved tug buckles.

4687 BRACE, H. *Walsall.*—Saddles, bridles harness, and horse appointments.

4688 BRILEFOLD, C. 21, *Wellington-st.*—Saddle-trees.

4689 BROWN & SON, 7, *Moat-row, Birmingham.*—Saddle-trees.

4690 CALLOW, T. & SON, 8, *Park-lane.*—Whips.

4691 CAMPBELL, J. & CO. *Adams-row, Walsall.*—Deer and other saddles, harness, and saddlers' ironmongery.

4692 CARTER, LT.-COL. *Monmouth.*—Harness on new principles.

4693 CLARK, W. & SON, *Leeds.*—Saddlery, harness, and horse clothing.

4694 COOPER, M. 2 & 3, *Railway-st. York.*—Saddles, horse clothing, &c.

4695 COWAN, J. *Union-st. Glasgow.*—Scotch cart and van harness.

4696 CUFF & SON, 18, *Cockspur-st.*—Saddlery.

4697 DAVIES, A. 33, *Strand.*—Saddles harness, bridles, &c.

4698 DEER, F. A. *Neath, Glamorganshire.*—Saddles and harness, &c.

4699 DOVEY, F. 30, *Brownlow-st. Drury-lane.*—Saddle-trees.

4700 DUFFY, J. *Market Harboro', Leicestershire.*—Safety collar.

4701 DUNLOP, J. *Haddington.*—Farm car harness.

4702 ELLAM, R. 213, *Piccadilly.*—Whips saddles, harness, &c.

4703 GARDEN & SON, 200, *Piccadilly.*—Saddles and harness.

4704 GARNETT, W. 4, *Bridgeman-pl. Walsall.*—Saddles and gig harness.

4705 GIBSON & CO. 6, *New Coventry-st.*—Saddlery, &c.

4706 GORDON, A. 99, *Piccadilly.*—Harness and saddles.

4707 GRAY, E. 44, *High-st. Sheffield.*—Saddles and harness.

4708 GREATREX, C. & SON, *Walsall.*—Coach and saddlery ironmongery, harness, whips, &c.

4709 GREEN, R. 8, *Edwards-st. Portman-sq.*—Saddlery and harness.

4710 HARGRAVES, J. & SON, *Carlisle.*—New styles of horse clothing, girth, and other webs.

4711 HAWKINS, J. V. 26, *Francis-st. Tottenham-court-rd. W.C.*—Various saddles.

4712 HAYNES & SON, 27 & 28, *Brownlow-st. Long-acre.*—Saddle-trees.

4713 HENTON & SON, 7, *Bridge-st. Lambeth.*—Saddle and harness manufacture.

4714 HESKETH, J. *Ashton-upon-Mersey Cheshire.*—Harness.

4715 HINKSON, J. 76, *Dame-st. Dublin.*—Saddlery and harness.

4716 HODDER, C. 8, *Nelson-st. Bristol.*—Park and hunting saddles, &c.

4717 HOLGATE, J. & Co. 33, *Dover-rd. Southwark.*—Saddlery and harness.

4718 HOLMES, *Derby.*—Harness.

4719 HOLMES, H. M. JUN. *London-rd. Derby.*—Pillar rein, and a saddle drier.

4720 HOOD & STEPHENSON, *Dunse, Berwickshire, N.B.*—Agricultural and other harness; saddle and bridles.

4721 HOUGHTON, G. *Tewkesbury.*—Spring seat saddle and tree, showing action of spring.

4722 JACKMAN, J. 110, *Wardour-st.*—Saddlery, bridles, holsters, and saddle bags.

4723 LANE, H. W. 3, *Little Compton-st. W.*—Saddle trees.

4724 LANGDON, MESSRS. 9, *Duke-st. Manchester-sq.*—Saddles and harness.

4725 LEA, CORPORAL MAJOR, *Royal Horse Guards.*—Collar for prevention of crib-biting.

4726 LENNAN, W. 29, *Dawson-st. Dublin.*—Saddlery, harness, &c.

4727 M'DOUGALL, A. *Upper Thames-st.*—Van and cart harness.

4728 M'NAUGHT & SMITH, *Worcester.*—West India mule and other harness.

4729 MARTIN, W. H. 65, *Burlington-arcade.*—Whips, canes, sticks, &c.

4730 MIDDLEMORE, W. *Birmingham.*—Saddlery, harness, and saddlers' ironmongery.

4731 MERRY, S. 21, *St. James-st.*—Harness, bridles, saddles, and horse clothing.

4732 MORE, J. & SON, *Market-st. Finsbury.*—Harness, pads, collars, round reins, and pole-pieces.

4733 NANSON, R. *English-st. Carlisle.*—Saddles and portmanteaus.

4734 NICHOLSON, W. H. JUN. 57, *Market-st. Manchester.*—Side saddle and hunting saddle.

4735 NICKOLLS, G. A. 1, *Oxford-market.*—Harness crupper.

4736 OERTON, F. B. *Walsall.*—Saddlery, harness, and saddlers' ironmongery.

4737 OLDFIELD & SON, 1, *Motcomb-st. Belgrave-sq.*—Saddlery and harness.

4738 OWEN, J. A. 7, *Lisle-st. Leicester-sq.*—Saddles and harness.

4739 PEARL, J. J. 2, *Friendly-pl. Old Kent-rd.*—Pad cloths, fronts, and rosettes.

4740 PEAT, H. 14, *Old Bond-st.*—Saddles and harness.

4741 PETCH, T. 40, *Albert-st. N.W.*—Gig harness, saddles, pads, tops, and trees.

4742 RAND & BECKLEY, 297, *Oxford-st. London.*—Saddlery and harness.

4743 SHATTOCK, J. M. & Co. *Bristol.*—Harness, saddles, and saddle-trees.

4744 SHIPLEY, J. G. 181, *Regent-st.*—Whips, saddlery, harness.

4745 SMITH, R. & Co. 1, *Beech-st. City.*—Whips, hunting crops, walking canes.

4746 SWAINE & ADENEY, 185, *Piccadilly.*—Whips, thongs, canes, and sporting apparatus.

4747 TIBBITS, J. & SON, *Walsall.*—Harness mountings, saddlers' ironmongery, &c.

4748 URCH & Co. 84, *Long Acre.*—Army and hunting saddlery, harness, and horse clothing, &c.

4749 WEIR, J. *Dumfries.*—Riding saddles, cart and gig harness.

4750 WHILLOCK, D. 24, *Tabernacle-row, St. Luke's.*—Saddles.

4751 WHIPPY, STEGGALL, & Co. *North Audley-st.*—Saddles and harness.

4752 WHITE, J. C. *Liverpool-st. City.*—Harness with improved tugs, and saddlery.

4753 WILKINSON & KIDD, 257, *Oxford-st.*—Saddles and harness.

4754 WILLIAMS, W. E. *High-st. Wandsworth, Surrey.*—Horse collars, harness, and saddlery.

4755 WRIGHT, S. *Stowmarket, Suffolk.*—Set of gig harness, plated furniture.

Sub Class C.—Manufactures generally made of Leather.

4766 CROSBIE, A. W. G. *Dumfries.*—Leather snuff-boxes.

4767 EARRATT, J. & R. 25, *Henrietta-st. Covent-garden.*—Military officers' accoutrements, &c.

4768 GEORGE, J. 81, *Dean-st. Soho.*—Ornamental leather for walls, screens, and covering furniture.

4769 HENDERSON, H. & SONS, *Dundee.*—Grained leather, machine-closed uppers, hose-pipes, and belting.

4770 NICHOLLS, H. 52, *Regent-st.*—Leather habiliments, prepared skins.

4771 REVELL, J. 272, *Oxford-st.*—Ornamental leather work and potichomanie.

4772 STAGG, T. 37, *Devonshire-st. Bloomsbury.*—Ornamental hand-gilding on leather, velvet, silk, &c.

4773 TURNER, P. 31, *Dean-st. Soho.*—Leather prepared for embroidery.

CLASS 27.

ARTICLES OF CLOTHING.

South-East Angle.

Sub-Class A.—Hats and Caps.

4804 ASHTON, J. & SONS, *Cornwall-rd. Lambeth.*—Gentlemen's hats.
4805 BLAIR, J. & Co. *Glasgow.*—Satin hats.
4806 BOOTH & PIKE, *Manchester.*—Materials used for hats.
4807 BRIGGS & PREEDY, 98, *Gracechurch-st.*—Hats, caps, felt hats, and umbrellas.
4808 CARRINGTON, S. & T. *Stockport, Cheshire.*—Felt and silk hats.
4809 CHRISTY'S, *Gracechurch-st.* — Illustrations of the manufacture of felted and silk hats.
4810 DOUGLAS & URE, *Glasgow.*—Hand-knitted Scotch caps.
4811 ELLWOOD & SONS, *Great Charlotte-st. Blackfriars-rd.*—Air-chamber hats, helmets, &c.
4812 GAIMES, SANDERS, & NICOL, 22, *Birchin-lane, Cornhill.* — Light ventilating hats.
4813 GARRARD, R. & J. *Loman-st. Southwark.*—Japanned leather and felt hats, &c.
4814 HEATH, R. 25, *St. George's-pl. Knightsbridge.*—Inventions in hats and umbrellas.
4815 HUSBAND, R. *Parsonage, Manchester.*—Hats, in silks and felts.
4816 JACOBS, W. *Dorchester, Dorset.*—Police and other hats.
4817 LINCOLN & BENNETT, 1, *Sackville-st. Piccadilly.*—Hats, &c.
4818 MELTON, H. 194, *Regent-st.*—Ladies' and gentlemen's hats.
4819 MOLLADY, J. & E. E. *Denton, near Manchester.*—Felt and silk hats.
4820 PRITCHARD, A. & F. 31, *Stamford-st. London, S.*—Hats, and combined cork and felt hats for India.
4821 SIMMONS & WOODROW, *Oldham.*—Silk and felt hats, and children's felt hats.
4822 SIMPSON, J. A. 6, *St. George's-cresc. Liverpool.*—Hats.
4823 SMITH, J. 8, *Merchant-st. Bristol.*—Hats and caps.
4825 TRESS & Co. 27, *Blackfriars-rd.*—Silk, felt, and beaver hats.
4826 WESTLANDS, LAIDLAW, & Co. *Glasgow.*—Patent expanding and other hats, helmets, Scotch bonnets, &c.
4827 WILSON, W. & Co. *Newcastle-on-Tyne.*—Hatters' furs, silk and felt hats and caps.
4828 ZOX, L. 85, *Long-acre.*—Fancy hats and caps.

Sub-Class B.—Bonnets and General Millinery.

4840 BORNE, C. & SON, 11, *Queen-st. Oxford-circus.*—Bonnet shapes and crowns.
4841 EMES, MISS, 31, *St. John's-villas, Adelaide-rd. N.W.*—Patent dress fastenings.
4842 FOSTER, SON, & DUNCUM, 16, *Wigmore-st. London, W.*—Artificial flowers, of various materials.
4843 FRANCIS, MISS E. 26, *Wellington-rd. Stoke Newington.*—Ladies' night caps.
4844 JONES, W. 85, *Chapel-st. Pentonville.*—Artificial May-tree, made of muslin and cambric.
4845 SHERRIN, S. H. 24, *Well-st. Cripplegate.*—Gold and silver flowers. Bridal ornaments.
4846 STUART & TAYLOR, 37, *Old Change.*—Millinery, flowers, wreaths, &c.
4847 TRESOLDI & BAKER, 20 *&* 21, *Coppice-row, Clerkenwell.*—Materials for artificial flowers.
4848 VALLI, D. *St. John's-lane, Smithfield.*—Artificial florists' materials.
4849 VYSE & SONS, *London.*—General millinery, mantles, shawls, &c.
4850 WHITE, W. 21, *Nassau-st. W.*—Artificial flowers.

Sub-Class C.—Hosiery, Gloves, and Clothing in General.

4861 ALLEN & SOLLY, *Nottingham.*—Hosiery and samples showing cotton spinning from earliest date.
4862 ASHWELL, T. & Co. *Nottingham and London.*—Hosiery.
4863 AUSTIN, J. *Princes-st. Finsbury, E.C*—Crinoline steel, cords, blind and picture lines
4864 NEWLAND, W. B. 24, *Gutter-lane E.C.*—Collars and wrists.
4865 BARRS, W. & SON, 7, *Edmund-st Birmingham.*—Umbrella furniture.
4866 BIGGS & SONS, *Leicester.*—Hosiery gloves, boots, shoes, &c.
4867 BINYON A. 37, *Eastcheap, E.C.* — Chest expander.
4868 BLYTH, C. & Co. 4, *Cripplegate-bdgs. London.*—Shirts, clothing, &c.
4869 BOSS, J. A. 1, *Little Love-lane, Cheapside.*—Umbrellas, sunshades, parasols, &c.
4870 BOWEN, B. *Chipping Norton, Oxfordshire.*—Leggings and gloves.
4871 BRIDE, J. H. 68, *Grange-rd. Bermondsey.*—Shirts, collars, and fronts.
4872 BRIE, J. & Co. 43, *Conduit-st.* —

Shirts, collars, dressing gowns, handkerchiefs, &c

4873 BROCKSOPP, T. 114, *Wood-st. London.* —Hosiery, &c.

4874 CARPENTER & Co. 43, *Temple-st. Birmingham.*—Webs, braces, belts, &c.

4875 CARTWRIGHT & WARNERS, *Loughborough.*—Merino hosiery. Skirts, &c.

4876 COLES, W. F. 5, *Aldermanbury-postern, E.C.*—Cork soles and fleecy hosiery.

4877 COOPER, T. & Co. 2, *South-bridge, Edinburgh.*—Braces.

4878 CORAH, N. & SONS, *Leicester.*—Hosiery and hosiery yarns.

4879 DESBOROUGH, S. 24, *Noble-st. E.C.* — Patent umbrella and parasol ribs and stretchers.

4880 DICKSONS & LAINGS, *Hawick.*—Cheviot and lambs-wool hosiery and under-clothing.

4881 ELLIS, J. & Co. 79, *Castle-st. Bristol.* —Ladies' stays and corsets.

4882 ELSTOB, W. 19, *Woodstock-st. Oxford-st.*—Belt, breeches, and trousers.

4883 ENSOR, T. & SONS, *Milborne-port, Somerset.*—Gloves.

4884 EWEN, R. *Hawick, Scotland.*—Lambs-wool, hosiery, and under-clothing.

4885 FIRKINS, J. & Co. *Worcester.*—Kid and other gloves.

4886 FOSTER & Co. *Oxford.*—Ecclesiastical and academical robes, &c.

4887 FOSTER, PORTER, & Co. 47, *Wood-st. London.*— Gloves, hosiery, shirts, collars, ties, and shawls.

4888 FOWKE, CAPT. F. *London.*—Patent umbrella.

4889 FOWNES, BROS. & Co. 41, *Cheapside, E.C.*—Every description of gloves, &c.

4890 GRANGER, A. 308, *High Holborn.*—Wearing apparel manufactured of paper.

4892 HALLIDAY, T. W. *Dundee.*—Gentlemen's seamless felt wearing apparel.

4893 HARRIS, R. & SONS, *Leicester.*—Hosiery, gloves, braces, &c.

4894 HARRISON, C. H. 13, *Wood-st. Cheapside.*—Umbrellas and parasols.

4895 HEPPLE, J. *Wenlaton, Newcastle-on-Tyne.*—Cutting mensurators.

4896 HUDSON, J. & SONS, *Leicester.*—Hosiery.

4897 JOHNSON, W. G. *Nottingham.*—Gloves, hair netts, lace, scarfs, hosiery, and under shirts.

4898 JOHNSTON, J. *St. Ninians, Stirling.*—Checked tartan hose.

4899 JONES, F. J. 10, *Aldermanbury, E.C.*—Shirts, collars, belts, braces, &c.

4901 JOY, STANDEN, & Co. *Oxford.*—Academical, ecclesiastical, and civil robes, &c.

4902 LAING, J. *Hawick, Scotland.*--Hosiery, &c.

4903 LAURENCE, F. R. 21, *Southampton-st.* — Shirt collars, shirts, and other white goods.

4904 LAWRENCE, W. 2, *St. Paul's-villas, Ball's-pond.*—Chamois leather, under-clothing, and travelling sheets.

4905 LINKLATER, R. *Lerwick, Shetland Isles.*—Shetland knitted shawls, veils, and hose, in wool and silk.

4906 MACINTOSH, C. & Co. *Cannon-st. London.*—Waterproof and elastic articles of clothing.

4907 MCINTYRE, HOGG, & Co. *Addle-st. E.C.*—Shirts and collars.

4908 MARION & MAITLAND, MESDAMES, 238, *Oxford-st.*—Corsets.

4911 MEYER, S. & M. 71, *Cannon-st. West.* —Umbrellas, various.

4912 MEYERS, M. 9, *Gt. Alie-st. E.*—Umbrellas and parasols.

4913 MIDDLEMASS, J. *South-bridge, Edinboro'.*—Presbyterian pulpit gowns. Clothing and shirts.

4914 MONEY, A. K. *Woodstock, Oxfordshire.*—Leather gloves.

4915 MORLEY, J. & R. 18, *Wood-st. London.*—Hosiery.

4916 MUNDELLA, HINE, B. H. & Co. *Nottingham.*—Hosiery principally by power machines.

4917 NEVILL, J. B. & W. & Co. 13, *Gresham-st. West. E.C.*—Hosiery.

4918 NEVILL, W. & Co. *Langham Factory, Godalming.*—Hosiery.

4920 PAYNE, T. *Hinckley, Leicestershire.* —Hosiery.

4923 REYNOLDS, G. W. & Co. 12, *Cheapside.*—Crinoline-skirts, stays, &c.

4924 SALOMONS, A. *Old Change, E.C.*—Stays, crinolines, and corsets.

4925 SANGSTER, W. & J. 140, *Regent-st.*—Umbrellas and parasols.

4926 SCOTT, P. & Co. *Edinburgh.*—Shirt; seamless coat.

4927 SHARP, PERRIN, & Co. *Old Change, E.C.*—Ladies' under-clothing, baby-linen, &c.

4928 SILVER, S. W. & Co. *Cornhill.*—Shirts, caps, and waterproof garments.

4929 SINCLAIR, R. & Co. 80, *Wood-st. E.C.* — Shirts, collars, and ladies' under-clothing.

4930 SMYTH & Co. 37, *Lower Abbey-st. Dublin.*—Hosiery.

4931 STEARS, S. 36, *Briggate, Leeds.*—Parasol.

4932 SWEARS & WELLS, 192, *Regent-st.*—Lilliputian hosiery.

4933 TAYLOR, B. *Birmingham.*—Braces, belts, leggings, webs, girths, bridles, &c.

4934 THOMSON, W. T. & C. H. *Fore-st. London.*—Crown crinolines.

4935 TILLIE & HENDERSON, *Glasgow.*—Shirts and under-clothing.

4936 WADDINGTON & SONS, 1, *Coleman-st. London.*—Umbrellas and parasols.

4937 WARD, STURT, & SHARP, *Wood-st. London.*—Hosiery and gloves.

4939 WELCH & SONS, 44, *Gutter-lane, London.*—Straw hats, millinery, and flowers.

4940 WELLS, J. S. *Mount-st. Nottingham.* —Cotton and merino hose, vests, pantaloons, &c.

4941 WHITE, F. & W. E. *Loughborough.*—Hosiery.

4942 WHITBY BROS. *Yeovil, Somerset.*—Foreign lambskins, gloves, and leather.

4944 WHITEHEAD, W. & SON, 63, *North-*

bridge, Edinburgh.—Tartan hosiery, and Vicunia under-clothing.

4945 WILSON, W. *Allan-cres. Hawick, N.B.*—Hosiery and under-clothing.

4946 WILSON & ARMSTRONG, *Nassau-st. Dublin.*—Balbriggan hosiery.

4947 WILSON & Co. 18, *Southampton-row, Russell-sq.*—Baby's clothes protector.

4948 WILSON & MATHESON, *Glassford-st. Glasgow.*—Umbrellas, caps, elastic and other belts, &c.

4949 WYATT, J. W. & Co. 65, *Bunhill-row.*—Crinoline skirts.

4950 WILSON, C. E. & Co. *Monkwell-st. Falcon-sq. E.C.*—Leglet.

4951 BLANKLEY, W. & F. 10, *Silver-st. Wood-st. E.C.*—Belts, braces, and crinolines.

Sub-Class D.—Boots and Shoes.

4960 ALDRED, W. *Manchester.*—Boots and shoes with adjustable heels. Detached heels and fastenings for same.

4961 ALLEN, C. E. *High-st. Haverfordwest.*—Boots and shoes.

4962 ATLOFF, J. G. 69, *New Bond-st.*—New style of boots.

4963 BALL, W. & Co. *New Weston-st. Bermondsey.*—Machine and hand-closed boot uppers.

4964 BARRON, W. J. & Co. 66 *&* 67, *Aldermanbury.*—Elastic webs and shoe mercery.

4965 BAULCH, C. *Bristol.*—Rivet sole boots; cork clumps, and self-adjusting leather clogs.

4968 BIRD, W. 86, *Oxford-st.*—Ladies' boots.

4969 BOWLER, J. 19, *Blandford-st. Manchester-sq.*—Mechanical boot stretcher, lasts, &c.

4970 BOWLEY & Co. *Charing-cross.*—Boots and shoes.

4971 BRISON, R. 1, *St. Augustine's-parade.*—Anatomical lasts.

4972 BROWN, E. 67, *Prince's-st. W.*—Boots, blacking, and polishes.

4973 CARTER, LIEUT.-COL. *Monmouth.*—The Hythe boot.

4974 CHAPPELL, J. 388, *Strand.*—The Pulvinar boot.

4975 CHARLESWORTH, W. *Stamford-st. Leicester.*—Riveted and sewn boots and shoes.

4976 CHRISTMAS, G. 12, *Mount-st. Westminster-rd.*—Boot to prevent splashing.

4977 CLARK, C. & J. *Street, near Glastonbury.*—Boots, shoes, and slippers.

4978 CLARKE, E. W. 12, *Southampton-row, Russell-sq.*—Boots for lame and tender feet; volunteer gaiter boots, and others.

4979 CREAK, J. *Wisbeach.*—Ladies' and gentlemen's side spring Bluchers and button boots.

4980 CREMER & Co. 126, *New Bond-st. W.*—Boot with elastic spring in arch of foot.

4981 CROSSDALE, J. 2, *Rotherfield-st. Islington.*—Ventilating boots and shoes.

4982 DAVIES, J. 46, *Gt. Queen-st. Lincoln's-inn-fields.*—Gentlemen's boots.

4983 DENNANT, J. 6, *Bedford-pl. Vauxhall-rd.*—Machine closed boot uppers, &c.

4984 DERHAM BROS. *Bristol and London.*—Machine-made boots and shoes.

4985 DOWIE, J. 455, *Strand.*—Elasticated leather sole boots.

4986 DUTTON, W. H. & SONS, *Knightsbridge, S.W.*—Boots and shoes.

4987 EAST, S. 103, *Fore-st. Exeter.*—Boots, shoes, rifle leggings, &c.

4988 EVANS, R. *Newtown, Montgomeryshire.*—Boots and specimens of workmanship.

4989 FRAMPTON, S. 79, *Regent-st. London.*—Ladies' and children's boots and shoes.

4990 GARNER, D. 23, *Clarence-rd. Bristol.*—Anatomical and riveting lasts; portable boot trees.

4991 GLEW, J. H. 19, *Howland-st. Fitzroy-sq.*—Ladies' boots and shoes.

4992 GORDON, E. 6A, *Princes-st. Leicester-sq.*—Boots, various.

4993 GRUNDY, T. 44, *St. Martin's-lane.*—Easy boots.

4994 GULLICK, *Pall Mall.*—Boots, and spur box.

4995 GUNDRY & SONS, 1, *Soho-sq.*—Variety of boots and shoes.

4996 HALL, C. G. 89, *Regent-st.*—Boots and shoes.

4997 SPARKES HALL, J. 308, *Regent-st.*—Elastic boots and shoes.

4998 HALL & Co. *Wellington-st. Strand.*—Pannus-corium or leather-cloth boots and shoes.

4999 HALLAM, J. & E. 149, *Waterloo-rd.*—Clogs.

5000 HAMILTON, J. 4, *Diana-place, Euston-rd.*—Ladies' boots.

5001 HARTLEY, J. 11, *King-st. St. James's.*—Top boots of English leather.

5002 HEATH, AUSTIN, & MYCOCK, *Browning-st. Stafford.*—Ladies' boots and shoes.

5003 HICKSON, W. & SONS, *Smithfield.*—Boots and shoes.

5004 HOOK & KNOWLES, 66, *New Bond-st.*—Boots and shoes, over shoes, brogues, &c.

5005 HUDSON, A. *Cranbrook.*—Boots and shoes with inner sole, for tender feet.

5006 HUTCHINGS, J. T. 5, *Inkerman-ter. Woolwich.*—Boots and shoes, with composition soles.

5007 JAMES, A. 2, *Trevor-sq. Knightsbridge.*—Boots and shoes.

5008 JENNETT, J. 44, *Whitcomb-st. Leicester-sq.*—Boot trees and lasts.

5009 JOSEPH, J. & SONS, 13, *Skinner-st. Snow-hill.*—Ladies' and children's boots and shoes.

5010 JUDGE, C. 6, *Sion-pl. Walworth.*—Leather buttons, laces, and leggings.

5012 KNIGHT & MAY, *Tewkesbury.*—Riveted boots and shoes, goloshes, &c.

5013 LANAGAN, 9, *Brownlow-st. Bedford-row.*—Principle in bootmaking illustrated. Instrument for measuring distorted feet.

5014 LANGDALE, H. 57, *Mount-st. Grosvenor-sq.*—Boots and shoes and needlework.

South-East Angle.

5015 LATHAM, J. 214, *Oxford-st.*—Ladies' boots and shoes.

5016 LEPRINCE, 261, *Regent-st.*—Chameleon shoes, changing colour.

5017 LINE, W. & J. *Daventry.* — Gentlemen's boots and shoes.

5018 LOWLEY, J. 71, *Briggate, Leeds.* — Boots and shoes.

5019 MABANE, J. & W. 3, *Templer-st. Leeds.*—Boot tops and dressed leather.

5021 MEDWIN, J. & Co. 86, *Regent-st.*—Elastic boots and shoes.

5022 MURRAY, J. F. 34, *Great Russell-st. Bloomsbury.*—Morning slippers.

5023 NEALE, G. 4, *Albert-pl. Holloway.*—Boots for persons with contracted hips, ankles, &c.

5024 NORMAN, S. W. & E. G. 3 *&* 4, *Oakley-st. S.*—Boot without stitch in sole, cork heel boots and shoes.

5025 OGDEN, F. 66, *Princes-st. Leicester-sq.* — Boot, breeches, and glove trees, &c.

5026 PALEY, R. *Leeds.*—Boots and shoes.

5027 PANZETTA & ANDREW, 141, *New Bond-st.*—Shooting boots.

5028 PARKER, W. & SONS, *Wood-st. Northampton.* — Boots and shoes, a small machine, &c.

5029 PATENT PLASTIQUE LEATHER CO. *Quay, Ipswich.*—Boots, with sole and heel moulded solid.

5030 PEAL, N. *Duke-st. Grosvenor-sq.*—Boots, shoes, and materials.

5031 PHIPPS, BARKER, & Co. *Sloane-st. S.W.*—Boots and shoes.

5032 POCOCK BROS. *Southwark Bridge-rd. S.E.*—Boots and shoes.

5033 REID, J. 99, *Regent-st.*—Ladies' and gentlemen's boots and shoes.

5034 ROBERT, A. 26, *Change-alley, Cornhill.*—Boots.

5035 ROBERTS, D. 9, *New Bond-st.*—Hunting and military boots.

5036 SCARD, A. 8, *Bow-lane, Cheapside.*—Specimens of the process of bootmaking.

5037 SEAGER, *Ipswich.*—Boots and shoes.

5038 SOMERVELL BROS. *Netherfield, Kendal.*—Boot and shoe uppers, and leggings.

5039 STAGG, A. 34, *Little South-st. Wisbeach.*—Last, boot and glove trees.

5040 STOKES, H. 27A, *Coventry-st. Haymarket.*—Gaiter boots.

5041 SURRIDGE, W. H. 275, *Regent-st. London, W.* — Boots and their appliances, with illustrations of boot making.

5042 TALLERMAN, R. & SON, 131, *Bishopsgate Without.*—Ladies' and children's waterproof boots and shoes.

5043 TODD, T. 24, *Colliergate, York.*—Fancy boots.

5044 WALKER & KEMPSON, *Leicester.*—Riveted boots.

5045 WALSH, W. 44, *Bolsover-st. Portland-place.*—Ladies' and gentlemen's boots.

5047 WINTER, C. *Norwich.*—Ladies' boots.

5048 YAPP, P. S. 200, *Sloane-st. London.*—Boots and shoes.

5049 NORTH BRITISH RUBBER CO. *Castle Mills, Edinburgh.*—Rubber boots and shoes.

5050 SCZERELMEY, N. C. *Park-road, Clapham.*—Pannonia leather cloth boots and shoes.

CLASS 28.

PAPER, STATIONERY, PRINTING, AND BOOKBINDING.

Gallery, North Court.

Sub-Class A.—Paper, Card, and Millboard.

5081 BARLING, J. *Park Mill, East Malling, near Maidstone.*—Paper and mill-board made from hop-bine.

5082 BURGESS & WARD, *Mendip Paper Mills, near Wells, Somerset.*—Straw paper, with illustrations of its manufacture and applications.

5084 GREER, A. & Co. *Dripsey and Glenville Mill, Cork.*—Writing papers, &c.

5085 LAMB, J. *Newcastle-under-Lyne.* — Pottery, tissue, and other papers.

5087 ROUTLEDGE, T. *Eynsham Mills, Oxford.*—Paper from esparto fibre; half stuff from same.

5088 SAUNDERS, T. H. *Queenhithe.*—Hand and machine-made paper.

5089 SZERELMEY, N. C. *Park-rd. Clapham.* — Arabian zopissa waterproof, and paper boards processes.

5090 HOOK, TOWNSEND, C., & Co. *Snodland Paper Works, near Rochester, Kent.*—Writing papers, webs for envelopes, &c.

5091 TURNBULL, J. L. & J. *Holywell-mount, Shoreditch.* — Drawing and other boards, card, &c.

5092 WOOLLEY & Co. 210, *High Holborn*—Paste boards, cards, &c.

5093 MORLEY, W. W. *Wooburn, Bucks.*—Mill-boards.

5094 TOWLE & JEFFERY, *Oxford.*—Paper boards, and pipes made of straw.

Sub-Class B.—Stationery.

5102 ARNOLD, P. & J. 135, *Aldersgate-st.*—Writing inks.

5104 BANKS & CO. *Keswick.*—Pencils, pen-holders, leads, &c.

5105 BARCLAY, R. 29, *Bucklersbury, E.C.*—Indelible paper, bank-cheques, account-books, &c.

5106 BAUERRICHTER & CO. 41, *Charterhouse-sq. London.*—Ornamental boxes.

5107 BAUS, H. 59, *Hatton-garden, London.*—Impressions from seals, models, &c.

5108 BENNETT, C. W. & CO. 14, *Smith-st. E.C.*—Fancy boxes.

5109 BLACKWOOD & CO. 18, *Bread-st-hill.*—Inks, sealing-wax, &c.

5110 BOUSQUET, I. 28, *Barbican.*—Gold, silver, and foil paper.

5111 BOWDEN, G 314, *Oxford-st. W.*—Elastic and other bands.

5112 BRETNALL, 24, *Huntley-st. Tottenham-court-rd.*—Tracing-papers, cloths, &c.

5113 BROOKMAN & LANGDON, 28, *Gt. Russell-st. W.C.*—Pencils and holders.

5114 BROOKS, H. & CO. *Cumberland-market, London, N.W.*—Various stationery articles.

5116 BROWN, W. & CO. 41, *Old Broad-st. E.C.*—Banking and commercial stationery.

5117 CALDWELL BROS. 15, *Waterloo-pl. Edinbro'.*—Stamped paper and envelopes.

5118 CANTON R. 27, *College-st. E.C.*—Fancy stationery.

5119 CARLYLE, G. 28, *Bold-st. Liverpool.*—Manifold writer and carbonic paper.

5120 CLEMENTS & NEWLING, 96, *Wood-st. London.*—Stationery for drapers, &c.

5121 COCHRAN, P. *Liverpool.*—Inks.

5122 COHEN, B. S. 9, *Magdalen-row, Gt. Prescott-st.* — Compressed Cumberland lead artists' and account-book pencils.

5123 COLLYER, R. H., M.D. 8, *Alpha-rd. St. John's-wood.*—Chemical ink pencils.

5124 CORFIELD, J. & SON, 7, *Farringdon-st. E.C.*—Marble papers, head-bands, book-edges, &c.

5125 COWAN, A. & SONS, 77, *Cannon-st. West.*—Writing paper, account books, &c.

5126 CREESE, J. & CO. *Birmingham.*—Illuminated and other tablets, mouldings, &c.

5127 DOBBS, KIDD, & CO. *London.*—Fancy stationery, &c.

5129 EDWARDS, E. *Birmingham.*—Ink, inkstands, paper weights, &c.

5130 ELLIOTT, D. *Park-rd. Old Kent-rd.*—Marking ink, &c.

5131 FASE & SON, *Edward-ter. Kensington.*—Piccolo writing cases, puzzles, &c.

5132 FETHERSTON, J. J. 18, *Suffolk-st. Dublin.*—Official dies, seals, stamps, &c.

5133 GILL, MISS, 9, *High-st. Hastings.*—Artificial paper flowers.

5134 GOODALL, C. & SON, *Camden Town, N.W.*—Playing cards.

5135 GOODHALL & DINSDALE, *Pancras-lane, E.C.*—Commercial stationery.

5136 GRAHAM, T. & R. 10, *High-st. Paisley.*—Spool tickets.

5137 HARRINGTON, J. *Lansdowne-ter. Brixton.*— Apparatus, with 12 blades for pointing pencils.

5138 HIGGINSON, MRS. *Uxbridge.*—Paper flowers.

5140 HOWARD, W. 23, *Great Russell-st. Bloomsbury.*—Tracing papers, vellums, and mounting linens.

5141 HYDE & CO. 61, *Fleet-st.*—Writing inks, sealing-wax, and clamp copying-book.

5142 IBBOTSON & LANGFORD, *Manchester.*—Fancy paper and pasteboard.

5144 JARROLD & SONS, 47, *St. Paul's-churchyard.*—Bookbinding.

5145 JOB, BROS. & CO. 75, *Cannon-st. West.*—Drawing and tracing papers.

5146 JOHNS, G. E. 9, *Bath-st. Newgate-st.*—Decorated boxes.

5147 JOHNSON & ROWE, 17, *Warwick-sq. E.C.*—Pocket-books, purses, &c.

5148 JONES & CAUSTON, 47, *Eastcheap, City.*—Account books, stationery, and printing.

5149 KING, J. 56, *Seymour-st. N.W.*—Fancy stationery.

5150 LAW & SONS, 37, *Monkwell-st. London, E.C.*—Bookbinders' cloths.

5151 LETTS, *Royal Exchange.*— Account-books, diaries, and works for MS. purposes; leather-work; maps and engravings.

5152 LUNTLEY, J. 42, *Bishopsgate-st. Without.*—Ticket receipt and till protector.

5153 LYONS, W. *Fennel-st. Manchester.*—Inks.

5154 MCGLASHAN, H. 8, *Helmet-row, Old-st. St. Luke's.*—Gold, silver, and foil paper.

5156 MARTIN, T. *Newton Abbot, Devon.*—Wax impressions of seals engraved by machinery.

5157 MATTHEWS, W. 1, *Wigmore-st.*—Waxed papers, for wrapping oily substances, &c.

5158 MEAD AND POWELL, 73, *Cheapside.*—Account books and stationery.

5159 MEEK, G. 2, *Crane-ct. Fleet-st.*—Embossed and laced papers and envelopes.

5161 MORDAN, F. 326, *City-rd.*—Gold pens, pencil cases, purses, inks, &c.

5162 NICHOLSON, J. 45, *Leeds-rd. Bradford.*—Account books, fancy stationery, &c.

5163 ORTNER & HOULE, 3, *St. James's-st. S.W.*—Heraldic seal and die engraving. Embossing and designing.

5164 PATERSON, BROS. *Peel-grove, Old Ford-rd.*—Mechanical guard books for filing, &c.

5165 PERRY, J. & CO. 37, *Red Lion-sq.*—Pencils, elastic bands, and inkstands.

5166 POLLARD, G. 10, *Walbrook, City.*—Envelopes, and paper for do.

5167 REYNOLDS, J. & SONS, *Vere-st. Lincoln's Inn-fields.*—Playing cards, and cards for the blind.

5168 RIDDIFORD, J. 14, *Cowley-st. Westminster.*—Rice-paper flowers, &c.

5169 ROBINSON, J. B. & SON, 17, *Bouverie-st. Fleet-st.*—Paper boxes.

5170 ROWSELL, S. W. & SON, 31, *Cheapside, E.C.*—Account books.

5171 SHOLL, J. 5A, *Chapel-st. Spital-sq.*—Writing paper and copying fluid.

5172 SMITH, J. 42, *Rathbone-pl.*—Tracing cloth and envelopes.

5174 STEAD, C. & SON, *Dalton, near Huddersfield.*—Design or point paper.

5175 STODART, M. 31, *Cloudesley-ter. Islington.*—Rice paper flowers, &c.

5176 STRAKER, S. & SONS, 26, *Leadenhall-st.*—Stationery.

5177 SUTCLIFFE, W. 101, *Bunhill-row, E.C.*—Labels and glove bands.

5178 TANNER BROS. *Welsh-back, Bristol.*—Account books, bookbindings, &c.

5179 THOMPSON, H. *Weybridge-heath, Surrey.*—Imitative cameo wafers.

5180 UNWIN, G. 31, *Bucklersbury, E.C.*—Specimens of printing, bookbinding, &c.

5181 WARNER, R. 18, *Newman-st. W.*—Specimens of seal-engraving.

5182 WATERSTON, G. *Edinburgh and London.*—Sealing wax.

5183 WEBSTER, H. 23, *Litchfield-st. Soho.*—Travelling inkstands and writing cases.

5184 WEDGWOOD, R. & SONS, 9, *Cornhill.*—Manifold writers.

5185 WETHERFIELD, R. 1, *Henrietta-st. Covent-garden.*—Paper flowers.

5186 WILLIAMS, COOPERS, & CO. 85, *West Smithfield.*—Commercial stationery, &c.

5187 WILSON, J. L. 128, *St. John-st.*—Cloth for bookbinding.

5188 WILSON, R. *Keswick, Cumberland.*—Lead pencils, &c.

5189 WOOD, J. T. 278, *Strand.*—Engravings, stationery, &c.

5190 HOUGHTON, W. *New Bond-st.*—The Queen's note-paper.

5191 WARD, M. *Belfast.*—Ledgers, bookbinding, and illuminating.

5192 WARNER, G. E. *Poland-st.*—Seals.

5193 WYON, J. S. —Impressions from seals.

5194 MURDOCK, C. & SON, 4, *George-st. Edinburgh.* — Specimens of seal engraving, die cutting, and chromolithography, as applied to heraldry.

Sub-Class C.—Plate, Letter Press, and other Modes of Printing.

5200 ADAMS & GEE, 23, *Middle-st. Smithfield.*—Specimens of printing on metal.

5201 ASHBY & CO. 79, *King William-st. E.C.*—Ornamental engraving for bank-notes, certificates, and bonds.

5202 AUSTIN, S. *Hertford, Herts.*—Printed books in oriental and other languages; and bookbinding.

5203 BAGSTER, S. & SONS, 15, *Paternoster-row.*—Polyglot typography; binding, illumination, &c.

5204 BANK OF ENGLAND, *Threadneedle-st.*—Specimens of surface printing.

5206 BELL & DALDY, *Fleet-st.*—Books.

5207 BEMROSE, & SONS, *Derby, and Matlock Bath.* — Bound books, letterpress and lithographic printing.

5208 BESLEY, R. & CO. *Fann-st. Aldersgate-st.*—Metal types and specimens.

5209 BISHOP, J. 4, *North Audley-st.*—Engravings by clockwork to prevent forgery.

5210 BLACK, A. & C. *Edinburgh.*—Books.

5212 BOOTH, L. *Regent-st.*—Reprint of first edition of Shakspeare.

5213 BONNEWELL, W. H. & CO.—76, *Smithfield, E.C.*—Wood types, &c.

5214 BRADBURY, WILKINSON, & CO. 13, *Fetter-lane, E.C.*—Bank notes, bonds, bills of exchange, &c.

5215 BRANSTON F. W. *The Grove, Southwark.*—Advertising tablets.

5216 BROOKS, V. 1, *Chandos-st. Charing-cross.*—Lithographic printing.

5217 CASLON, H. W. & CO. 23, *Chiswell-st.*—Printing types and printed specimens.

5218 COATHUPE, CAPT. H. B. 1, *Abingdon-lane, Kensington.*—Printing on metals.

5219 COLLINGRIDGE, W. H. *Aldersgate-st. London.* — Printing, engraving; materials used, &c.

5220 COLLINS, W. *Glasgow.*—Bindings, bibles, testament, &c.

5221 COLVILL, H. J. M. 52, *Queen-st. Camden-town.*—Printing in pure silver.

5222 COX, G. J. 46, *Stanhope-st. Hampstead-rd.*—Impressions taken from ferns and leaves, lace, &c.

5223 CROSS, J. & SON, 18, *Holborn-hill.*— Engravings, lithographs, &c. cut by machinery.

5224 DAY & SON, 6, *Gate-st. Lincoln's-inn-fields.*—Specimens of lithography, chromo-lithography, and plate printing.

5225 DE LACY, G. 38, *Seckforde-st. Clerkenwell.*—Tools, letters, blocks, &c. for bookbinding.

5226 DICKES, W. 5, *Old Fish-st. Doctors' Commons.*—Engraving and oil-colour printing.

5227 DULAU & CO. 37, *Soho-sq.*—Bound books, various.

5228 ELECTRO PRINTING BLOCK COMPANY, *Burleigh-st. Strand.*—Enlargements and reductions from copper plates, wood blocks, &c.

5229 EYRE & SPOTTISWOODE, 43, *Fleet-st.*—Bibles, prayer books, &c. in various bindings.

5230 FAITHFULL, MISS, *Victoria Press, Gt. Coram-st.*—A specimen of printing by women.

5231 FALKNER, G. *King-st. Manchester.*—Engravings on stone and specimens therefrom.

5232 FIGGINS, V. & J. *West-st. Smithfield.*—Newspaper and book founts.

5233 FONTANNE, A. 5, *Bunhill-row, E.C.*—Cast brass type.

5234 GABALL, J. H. 3, *Russell-ct. Brydges-st.*—Letter-press fac-simile of ancient MS.

5235 GARDNER, T. B. 45, *Greek-st. Soho.*—Stencil plates, &c.

5236 GAUCI, P. *London.*—Chromolithographs.

5237 GEORGE, B. *Hatton-garden.*—Show tablets and frames.

5238 GILMOUR & DEAN, *Roy. Exchange-pl.*

Glasgow.—Lithography, engraving, die-cutting, embossing, and ornamental printing.

5239 GRANT & CO. *Broadway, Ludgate-hill.*—Colour printing.

5240 GRIFFITH & FARRAN, *St. Paul's-churchyard.*—Illustrated gift books.

5241 GROOM, WILKINSON, & CO. *Queen's Head-passage, Paternoster-row.* — Manufacturers' patterns and show-cards.

5242 GUITTON & MENUEL, 2, *Bartlett's-passage, Fetter-lane.*—Brass types, &c.

5243 HANHART, M. & N. 64, *Charlotte-st. Fitzroy-sq.*—Lithography and chromo-lithography.

5245 HAYMAN, BROS. 13, *Gough-sq. Fleet-st.*—Ornamental letter-press.

5246 HOME, MESSRS. R. & CO. *Edinburgh.* —Printed music.

5247 HUGHES & KIMBER, *Red Lion-passage, Fleet-st.*—Plates for engraving.

5248 JEWELL, J. H. 104, *Great Russell-st. Bloomsbury.*—Music engraving and printing.

5249 JOHNSON, J. M. & SON, *Castle-st. Holborn.*—Show cards and tablets.

5250 KENT, W. & CO. 23, *Paternoster-row.* —Books.

5251 LAVARS, T. *Broad-st. Hall, Bristol.*—Chromo-lithography.

5253 LEFEVRE, C. 12, *Red Lion-st. Clerkenwell.*—Colours printed on leather.

5254 LEIGHTON BROS. *Milford-house, Strand.* — Surface colour printing by machinery.

5255 LEIGHTON & LEIGHTON, 9, *Buckingham-st. W.C.* — Wood-engravings, and processes connected with the production of printing surfaces.

5256 LINTON, W. J. 85, *Hatton-garden, E.C.*—Engraving for surface-printing, &c.

5257 LONGMAN, GREEN, LONGMAN, & ROBERTS, *Paternoster-row.* — Printing and bookbinding.

5258 LOW, S., SON, & CO. *Ludgate-hill.*—Illustrated books.

5259 MACKENZIE, W. *Paternoster-row.*—Bible, illustrated by photographs.

5260 MACLURE, MACDONALD, & MACGREGOR, *London, Glasgow, Liverpool, and Manchester.*—Lithography and engraving.

5261 MACPHERSON, D. 28, *Salisbury-st. Edinburgh.* — Stereotype plates; moulding material used for an indefinite period.

5262 MCQUEEN, BROS. 184, *Tottenham-court-rd.*—Plate printing.

5263 MANSELL, J. *Red Lion-sq.* — Oil coloured prints. Ornamental papers and cards.

5264 MELVILLE & CO. 68, *Marylebone-rd.* —Solid indelible ink.

5265 MILLER & RICHARD, *Edinburgh and London.* — Newspaper, book, old style, and other printing types.

5266 MOREL, V. & CO. 3, *Playhouse-yard. E.C.*—Electrotypes and surface plates, produced by a new process.

5267 MOTTRAM, J. 35A, *Ludgate-hill.* — Medallion and engine-turned ruling.

5268 MUNRO, F. P. 4, *Gibson-st. S.*—Hand stamps for letters.

5269 MURRAY, J. *Albermarle-st.*—Books.

5270 NAPIER, J. 13, *East Sciennes-st. Edinburgh.*—New method of stereotyping.

5271 PARSONS, FLETCHER, & CO. *Bread-st.* —Printing and lithographic inks.

5272 PARTRIDGE, S. W. 9, *Paternoster-row.* —Illustrated books, &c.

5273 PATENT TYPE-FOUNDING CO. 31, *Red Lion-sq.*—Type cast and dressed by machinery.

5275 ROWNEY, G. & CO. *Rathbone-pl.*—Chromo-lithography.

5276 SCHENK, F. R. 50, *George-st. Edinburgh.*—Lithography.

5277 SCOTT, R. J. 8, *Whitefriars-st.* — Blocks for wood-engraving.

5278 SEARBY, W. 2, *Crown-court, City.*—Relief engraving for bank notes, &c.

5279 SIDEY, C. 5, *Stephen-st. Tottenham Court-rd.*—Engraved music plates, plans, &c.

5280 SILVERLOCK, H. *Doctors' Commons.* —Engraving, electrotyping, stereotyping, &c.

5281 SKINNER, J. 47, *Whitecross-st. E.C.* —Glyphography.

5282 SMITH, B. & SON, 7, *Wine Office-court, Fleet-st.* -Printing ink, &c.

5283 SMITH, ELDER, & CO. 65, *Cornhill, London.*—Books.

5284 SPIERS & SON, *Oxford.*—Book of arms of the colleges.

5285 SPRAGUE, R. W. & CO. 5, *Ave Maria-lane.*—Lithography, ornamental writing on vellum, and new lithotype process.

5286 STANDIDGE & CO. 36, *Old Jewry.*—Fac similes of documents, and lithographic printing.

5287 STANESBY, S. 6, *St. George's-ter. Kensington.*—Illuminated books, &c.

5288 STEPHENSON, BLAKE, & CO. *Sheffield.* —Printing types.

5289 TERRY, C. 183, *High Holborn.*—Chromo-lithographs.

5290 TRÜBNER & CO. 60, *Paternoster-row.* —Books.

5291 ULLMER, F. *Old Bailey.*—Wood type.

5292 UNDERWOOD, T. *Birmingham.* — Chromo-lithographs.

5293 WALLER, 18, *Hatton-garden.*—Lithographs.

5294 WALLIS, GEO. 16, *Victoria-grove, Brompton.*—Specimens of the new art of autotypography.

5295 WATTS, W. M. *Crown-ct. Temple-bar.* —Printing and embossing for the blind, &c.

5296 WESTWOOD, PROFESSOR J. O. *Oxford.* —Volume of fac-similes of Anglo-Saxon MSS.; with carved wooden covers.

5297 WHITEMAN, F. J. 19, *Little Queen-st. Lincoln's-inn.*—Specimens of writing engraving, and marking plates.

5299 WINSTONE, B. 100, *Shoe-lane, E.C.*—Printers' ink, specimens, &c.

5300 WYATT, S. 22, *Gerrard-st. W.*—Lithographs.

5301 CLAY, R., SON, & TAYLOR, 7, *Bread-st.-hill.*—Specimens of printing.

5302 BELL & DALDY, *Fleet-st.*—Books.

5303 BERRI, D. C. 36, *High Holborn.* —Letter-stamps, as used in the Post-office.

5304 BRADBURY & EVANS, *Fleet-st.* — Books.

5305 CHAMBERS, W. & R. *Paternoster-row.* —Books.

5306 CIVIL ENGINEER & ARCHITECTS' JOURNAL, *Warwick-ct. Gray's Inn.*—Specimens of architectural lithography.

5307 FRY, J. *Cosham, Bristol.*—Books.

5308 HOGG, J. & SONS, *London.*—Books.

5309 JONES, O. 9, *Argyle-pl. London.* — Books.

5310 MCMILLAN & CO. *Henrietta-st. Covent-garden.*—Books.

5310A MUYGRIDGE. E. J. 16, *Southampton-st. Strand.*—Specimens of his plate-printing, the ink being introduced from underneath by perforations.

Sub-Class D.—Bookbinding.

5310B BEDFORD, F. 9, *Gloucester-st. Warwick-sq.*—Ornamental bookbinding.

5311 BONE W. & SON, 76, *Fleet-st.*—Bookbindings.

5312 CHATELIN, A. 15, *Newman-st. W.*—Bookbindings.

5313 CLARK, W. *Dunfermline.* — Bookbindings.

5315 HOLLOWAY, M. M. 25, *Bedford-st. Strand.*—Bookbinding.

5316 JEFFREY, J. *Charlotte-st. Portland-pl.*—Bookbinding, &c.

5317 LEIGHTON, J. 12, *Ormonde-ter. Regent's-park.*—Designs principally executed for British publishers. Illustrated books, &c.

5318 LEIGHTON, J. & J. 40, *Brewer-st. Golden-sq.*—Specimens of bookbinding, &c.

5319 LEIGHTON, SON, & HODGE, 13, *Shoe-lane, London.*—Publishers' bookbinding.

5320 NELHAM, W. 48, *Liverpool-st. King's-cross.*—Bookbinding.

5321 POTTS, WATSON, & BOLTON, *Garter-court, Barbican.*—Designs on leather for upholstery, &c.

5322 RAINES, T. 24, *Great Ormond-st.*—Specimens of bookbinding.

5323 RAMAGE, J. *North-bridge. Edinbro'.*—Palæography, inlaid with leather, hand-worked; other bookbindings.

5324 REYNOLDS, W. 6, *Eldon-st. Finsbury.* —Leather backs for binding volumes of music.

5325 RIVIÈRE, R. 196, *Piccadilly.*—Bookbindings.

5326 SETON & MACKENZIE, 80, *George-st. Edinburgh.*—Bookbindings.

5327 TONKINSON, J. 16, *St. John-st. Clerkenwell.*—Book-clasping, edging, and mounting in metals.

5328 WESTLEYS & CO. 10, *Friar-st. Doctors'-commons.*—Bookbindings.

5329 WRIGHT, J. TRUSTEES OF THE LATE, 16, *Noel-st. Soho, W.C.*—Bookbinding.

5330 ZAEHNSDORF, J. 30, *Brydges-st. Covent-garden.*—Bookbindings.

CLASS 29.

EDUCATIONAL WORKS AND APPLIANCES.

Central Tower.

Sub-Class A.—Books, Maps, Diagrams, and Globes.

5361 ALBITES, A., Paris, *Edgbaston Proprietary School.*—French instruction books.

5362 ALLMAN, T. 42, *Holborn-hill.* — Educational works and appliances.

5363 BEAN, J. W. *Leeds.*—Copy books.

5364 BELL, W. 30, *Burton-st. W.C.*—Figurative representations of universal history and chronology.

5365 BELL & DALDY, *Fleet-st.*—Educational works.

5366 BERTHON, REV. E. L. *Romsey.* — Large globe of new construction for educational purposes.

5367 BETTS, J. 115, *Strand.*—Globe, geographical slates, maps, &c.

5368 BEVAN, H. *St. Mary's-st. Shrewsbury.* —Tablets for arithmetical operations.

5369 BISHOP, T. B. *Wimbledon, Surrey.*—Chronological charts.

5370 BLACK, A. & C. *Edinburgh.*—Books, atlases, and maps.

5371 BLACKWOOD, W. & SONS, 45, *George-st. Edinburgh.*—Geological maps, &c.

5372 BLANCHARD, REV. H. D. *Lund, Beverley.*—A school register.

5373 BOUVERIE, J. *Maida-hill W.*—Books and illustrative drawings.

5374 BOWER, B. *Cheadle.*—Dial map of Alderley Edge.

5375 BRITISH AND FOREIGN BIBLE SOCIETY, 10, *Earl-st. Blackfriars.*—One hundred and ninety-one versions of the Holy Scriptures, in various languages.

5376 CASSELL, PETTER, & GALPIN, *La Belle Sauvage-yd. Ludgate-hill.*—Educational works.

5377 CHRISTIAN VERNACULAR EDUCA-

TIONAL SOCIETY, 5, *Robert-st. Adelphi.* — Publications in the Indian languages.

5379 CRAMPTON, T. *The Butts, Brentford, W.*—School books, and various school appliances.

5380 CRUCHLEY, G. F. 81, *Fleet-st.*—Maps, atlases, globes, &c.

5381 CUIPERS, P. 32, *Castle-st. Holborn. E.C.*—Maps of Central America, &c.

5382 CURWEN, J. *Plaistow, E.*—Diagrams and publications on the Tonic Sol-fa method.

5383 DARTON & HODGE, *Holborn-hill.*—Educational books, maps, prints, &c.

5384 DAY & SON, 6, *Gate-st. Lincoln's Inn-fields.*—Coloured diagrams for educational purposes.

5385 DEAN & SON, 11, *Ludgate-hill.*—Maps, educational works, &c.

5386 EASTON, W. *Scudamore-schools, Hereford.*—Arithmetic and mensuration.

5387 FLETCHER, P. *Clyde-st. Edinburgh.*—Globe for use of the blind.

5388 FYFE, W. *Dorchester.*—Text-books, diagrams, calendars, &c.

5389 GALL & INGLIS, *Edinburgh.*—Charts, maps, atlas, and educational works.

5390 GILBERT, J. 2, *Devonshire-grove, Old Kent-rd.*—Educational works.

5391 GORDON, J. 51, *Hanover-st. Edinburgh.*—Educational books, cards with objects, &c.

5392 GOVER, E. *Princes-st. Bedford-row.* — Atlases and other educational publications.

5393 GRIFFITH & FARRAN, *St. Paul's Church-yd.*—Educational books.

5394 HALL, A., VIRTUE, & CO. 25, *Paternoster-row.*—Educational books.

5396 HOGG, J. & SONS, 9, *St. Bride's Avenue, Fleet-st.*—Educational books.

5397 HOPPER, A. *Edgbaston, near Birmingham.*—Lessons for the deaf and dumb, &c.

5398 JARROLD & SONS, 47, *St. Paul's Churchyard.*—Educational works, copy-books, &c.

5399 JONES, A. *Shakspere-ter. Stoke Newington.*—Publications of the United Association of Schoolmasters.

5400 JONES, REV. C. W. *Pakenham, Suffolk.*—Adults' reading books.

5401 KNIPE, J. A. *Moorville, near Carlisle.*—Geological maps of the British isles, and England and Wales, and Scotland.

5402 LONGMAN, GREEN, LONGMAN, & ROBERTS, *Paternoster-row.* — Educational books and appliances.

5403 LUCAS, G. 44, *Kennedy-st. Manchester.*—Globes, for schools, &c.

5404 MACKIE, S. J. 25, *Golden-sq. W.*—Scientific and educational diagrams and books.

5405 MACMILLAN & CO. *Cambridge.*—Educational books.

5406 MAIR & CO. 34, *Bedford-st. Strand.*—School Directory.

5407 MARTIN, G. W. *Exeter Hall.*—Books, instruments, and apparatus for musical education.

5408 MORRISON, LIEUT. R.N. 17, *Surrey-st. Strand.*—Astronomical diagrams.

5409 NEWTON & SON, 66, *Chancery-lane.*—Mode of mounting globes and orreries.

5410 NUTT, D. 270, *Strand.*—Educational books.

5411 OAKEY, H. 20, *Newman-st. Oxford-st.*—Musical atlas.

5412 OLIVER & BOYD, *Edinburgh.*—Educational class books.

5413 PARKER, SON, & BOURN, 445, *West Strand.*—Hullah's musical works, &c.

5415 POTTS, R. *Trin. Coll. Cambridge.*—Educational works.

5416 PROPRIETORS OF "WEEKLY DISPATCH" *Fleet-st.*—Maps of London, &c.

5417 RELIGIOUS TRACT SOCIETY, 56, *Paternoster-row, London.*—Printed books, &c.

5418 REYNOLDS, J. 174, *Strand.*—Specimens of school diagrams.

5419 SALMON, E. W. *Nottingham.* — An engraved plan of Nottingham.

5420 SCOTT, DR. W. R. *Exeter.*—Works for the deaf and dumb, &c.

5421 SHEAN, W. 1, *Liverpool-st. City.*—Attendance register book.

5422 SIMPKIN, MARSHALL, & CO. *Stationers'-hall-ct.*—Dr. Cornwell's educational works.

5423 SMITH & SON, 172, *Strand.*—Map of London, and globes.

5424 STANFORD, E. 6, *Charing-cross, London.*—Maps, atlases, globes, books, &c.

5425 STEPHENSON, THE REV. N. *Shirley, near Birmingham.*—A Church catechism.

5426 SUNDAY SCHOOL UNION, 56, *Old Bailey, E.C.*—Books and educational appliances.

5427 THELWALL, S. M. 9, *Stanhope-st. Bath.*—Syllabic book and tables.

5428 THIMM, F. 3, *Brook-st. Grosvenor-sq.*—Grammars of the European languages.

5429 TILLEARD, J. *Council Office, London.*—Works on music.

5430 WALLIS, G. 16, *Victoria-grove, West Brompton.*—Drawing book for designers, &c.

5431 WALTON & MABERLY, *Upper Gower-st.*—Educational and scientific works.

5432 WATKINS, J. *The College, Dulwich, S.*—Educational diagrams, to illustrate the doctrine of fractions.

5433 WHARTON, J. 42, *Queen-sq.*—Logical arithmetic, algebra, &c.

5434 WHITE, G. *Abbey-st. Schools, N.E.*—Educational works.

5435 WILKINS, E. P., M.D. *Newport, I. W.*—Maps in modelling instead of linear shading.

5436 WYLD, J. 13, *Charing-cross.*—Geographical and topographical novelties.

5437 ARUNDEL SOCIETY, 24, *Bond-st.* — Publications of the Society.

5438 BORSCHITZKY, J. F. 32, *Tavistock-pl. London, W.C.* — Educational music and diagrams, on the international system.

5439 BRUCCIANI, D. 39, *Russell-st. Covent-garden.*—Art studies for schools.

Sub-Class B.—School Fittings and Furniture.

5447 Abbatt, R. *Stoke Newington.*—Inventions for educational purposes.
5448 Agar, W. T. *Lymington, Hants.*—Pedal music desk—foot turning the leaves.
5449 Andrews, R. 144, *Oxford-st. Manchester.*—Guida Mano, for giving a correct position to children while learning the piano.
5450 Association for Promoting Welfare of the Blind, 127, *Euston-rd.*—Manufactures and educational apparatus.
5451 Atkins, R. 27, *Bethel-st. Norwich.*—School desks, easels, and general fittings.
5452 Barnard, J. & Son, 339, *Oxford-st. W.*—New chemical colours, sketching apparatus, educational diagrams, &c.
5453 Biggs & Sons, 24, *Guildford-st. East, W.C.*—Educational works and appliances.
5454 Blind Asylum, *Bristol.*—Mats, baskets, brushes, embossed books, &c.
5456 Bowden, G. 314, *Oxford-st. W.*—Sketch books and appliances for artists and schools.
5457 Briggs, R. & Son, 1, *Welbeck-st. Cavendish-sq.*—Drawing materials, &c.
5458 British and Foreign School Society, *Borough-rd. London, S.E.*—Educational appliances.
5459 Brockedon, W. & Co. 34, *Gt. Ormond-yd. London.*—Pure Cumberland lead, for official documents, solid ink, &c.
5460 Campbell, T. 24, *South Richmond-st. Edinburgh.*—Plan for making rugs in Blind Asylum.
5461 Colson, J. *St. Swithin-st. Winchester.*—Drawings of College at Winchester.
5462 Congregational Board of Education, *Homerton, N.E.*—Model of schools. School apparatus.
5463 Croger, T. 483, *Oxford-st.*—Automaton musical instrument.
5464 Cronmire, J. M. & H. 10, *Bromehead-st. Commercial-rd. East.*—Mathematical instruments.
5465 Croydon, W. J. 9, *York-pl. Windsor.*—Drawing models.
5466 Denton, A. 8, *South-pl. Finsbury.*—Miniature models for gold and silver, &c.
5467 Fetherston, J. I. 18, *Suffolk-st. Dublin.*—Alphabets; chromatic colour charts, and boxes.
5468 Ford, R. D. 32, *Great Carter-lane, Doctors'-commons, E.C.*—Desks and school fittings.
5469 Gall, J. *Myrtle-bank, Trinity, Edinburgh.*—Alphabet for the blind.
5470 Glover, S. A. *Great Malvern.*—Sol-fa harmonicon.
5471 Gray, J. 33, *Richmond-pl. Edinburgh.*—Raised maps for the blind.
5472 Green, B. R. 41, *Fitzroy-sq.*—Rustic drawing models.
5473 Grew, T. *Plaistow-park, West Ham, Essex.*—Mathematical drawing instruments.
5474 Hammer, G. M. 44, *Harrington-st. N.W.*—Models of school fittings, and educational apparatus.
5475 Hancock, C. *Quadrant-rd. Highbury New-park.*—Animals and figures modelled in paper.
5477 Haskins, J. F. 14, *Victoria-st. E.C.*—Class-room desks, and musical works.
5478 Hay, J. H. *Kennington Oval.*—Class register, portfolios, &c.
5479 Herdman, W. G. *West-villa, Liverpool.*—A scene in the Midland Counties.
5480 Holmes, C. *London-rd. Derby.*—School desk and form.
5482 Home and Colonial Training Institution, *King's Cross, London.*—Models of an infant school, of a school desk, &c.
5483 Houghton, P. & Co. 25, *Stamford-villas, Fulham.*—Drawing stumps, &c.
5484 Hughes, G. A. 11, 47a, *Edgware-rd.*—Books, &c. for the blind.
5486 Jackson, E. S. 3, *Sheffield-ter. Kensington.*—Floreated motto.
5487 Johnston, W. & A. K. 4, *St. Andrew-sq. Edinburgh.*—Illustrations of physiology—colours.
5488 Joseph, Myers, & Co. 144, *Leadenhall-st. E.C.*—Educational models and appliances.
5489 Leathes, Major H. M. *St. Margaret's, Herringfleet, Suffolk.*—Model picture.
5490 London Society for Teaching the Blind to Read, &c. *Upper Avenue-rd. N.W.*—Baskets, brushes, knitting.
5491 Macintosh, C. & Co. *Cannon-st. London.*—Educational requisites.
5492 Martin, J. *Alfreton-rd. Nottingham.*—Writing machine for the blind.
5493 Mill, J. 1, *Foundling-ter. W.C.*—Educational articles for children.
5494 Mitford, B. *Cheltenham.*—Conversational and other tablets.
5495 Moon, W., F.R.G.S. *Brighton.*—Diagrams, &c. for the blind.
5496 Moore, G. B. 9, *Lansdowne-ter. N.W.*—Crayon drawing, fixed for portfolios.
5497 Musselwhite, J. *Devizes.*—Moveable note music board.
5498 National Society for the Education of the Poor, *Sanctuary, Westminster.*—Educational appliances.
5499 Newman, J. 24, *Soho-sq.*—Artists' colours, varnishes, brushes, &c.
5500 Pearce, T. B. 93, *Newman-st.*—The musical octave dissected, &c.
5501 Pemberton, R. *Euston-sq.*—An organ for the use of schools.
5502 Philanthropic Society's Farm School, *Redhill.*—Model of a school-house.
5503 Pitman, I. *Bath.*—Phonetic shorthand and printing alphabets.
5504 Rahles, Dr. F. 13, *Albert-st. N.W.*—Alphabet and spelling game.
5505 Reeves & Sons 113, *Cheapside, E.C.*—Series of drawing materials. Artists' materials.
5506 Reformatory & Refuge Union, 118, *Pall-mall.*—Apparatus used in reformatories, and illustrative models.
5507 Ridley, Rev. N. J. 10, *Paternoster-*

row.—Articles, apparatus, and models illustrating "Book-Hawking."

5508 ROBERSON & CO. *Long Acre.*—Drawing materials, &c.

5509 ROWNEY, G. & CO. *Rathbone-pl.*—Artists' materials.

5510 RUSSELL & BUGLER, *Ashford, Kent.*—Improved school desks.

5511 RYFFEL, T. E. 5, *Upper Stamford-st. Blackfriars.*—Calculating cubes and multiplication table represented without figures.

5512 SCHOOL FOR THE INDIGENT BLIND, *St. George's-fields, Southwark.*—Books, &c. for the blind.

5513 SCIENCE & ART DEPARTMENT, *South Kensington.*—Illustrations of the course of instruction in schools of art.

5514 SEATON, J. L. 3, *Frederick-pl. Hampstead-rd.*—Desks, forming tables and seats also.

5515 SHARP, G. 16, *Wentworth-pl. Dublin.*—Models for elementary drawing.

5516 SHERRATT, T. JUN. 5, *Westmoreland-pl. W.*—Planetary clock.

5517 SOCIETY FOR PROMOTING CHRISTIAN KNOWLEDGE, *Gt. Queen-st. Lincoln's-inn-fields.*—Books, maps, prints, &c.

5518 SPENCER, W. *Beverley.*—Arithmetical exercises on cards, &c.

5519 STANLEY, W. F. 8, *Gt. Turnstile, Holborn.*—Mathematical instruments.

5520 STEPHENS, H. 18, *St. Martin's-le-Grand.*—Inks, papier-mache, slates, &c.

5521 SUNDAY-SCHOOL INSTITUTE, 41, *Ludgate-hill.*—Materials for organization and management of Sunday-schools.

5522 WEDGWOOD & SONS, 9, *Cornhill.*—Writing machines for the blind.

5523 WEDLAKE, 58, *Warren-st.* — Doll-houses.

5525 WOLFF, E. & SON, 23, *Church-st. Spitalfields.*—Pencils, crayons, &c.

5526 CRANBOURNE, THE VISCOUNT 20, *Arlington-st.*—Books for the blind.

5527 HOWARD, *Colchester.* — Plans of school-buildings.

5528 JOHNSON, E. C. *Savile-row.*—Books, for the blind.

5529 WILLIAMS, A. *Windsor.*—Model of improved school desk and form.

Sub-Class C.—Toys and Games.

5538 ASSER & SHERWIN, 81, *Strand.*—Games and evening amusements.

5539 BAER, J. 399A, *Oxford-st.*—Scientific pegs; wood carvings.

5541 BURLEY, G. 28, *George-st. Blackfriars-rd.*—Dolls.

5542 CAMP, W. 81, *Tottenham-ct.-rd.*—Cricket stumps, skittle, lawn billiard, and bowling green balls.

5543 CREMER, 27, *New Bond-st.*; *and* CREMER, JUN. 210, *Regent-st.* — Toys and games.

5544 DARK, M. & SONS, *Lord's Cricket-ground, N.W.*—Cricket-bats and wickets.

5545 DARK, R. *Lord's Cricket-ground.*—Balls, leg-guards, gauntlets, &c.

5546 DUKE & SON, *Penshurst, Kent.*—Cricket-balls and cricket appliances.

5547 FELTHAM, J. & CO. 2, *Barbican.*—Cricketing, archery, fencing requisites.

5548 GILBERT, W. *Rugby.*—Foot-balls and foot-ball shoes.

5549 HOFFMANN, H. *Norland-sq. Notting-hill.*—The "Kindergarten," illustrated.

5550 JAQUES, J. & SON, 102, *Hatton-garden.*—Games, chessmen, &c.

5551 JEFFERIES & MALINGS, *Wood-st. Woolwich.*—Bats, balls, racket shoes, &c.

5552 JOHNSON, S. 6, *Heathpole-st. W.*—Model of doll's house.

5553 KEEN, T. E. 144, *Leadenhall-st.*—A scientific toy.

5554 KENNEDY & CO. 15, *Westbourne-grove-ter. Bayswater.*—Mechanical "trotting pony," and spring bassinette.

5555 LILLYWHITE, J. 5, *Seymour-st. Euston-sq.*—Articles connected with cricket.

5556 LOYSEL, E. 92, *Cannon-st. E.C.*—Game of Tournoy.

5557 MATHEWS, C. E. *Oatlands Park, Surrey.*—A miscellaneous game.

5558 MEAD & POWELL, 73, *Cheapside.*—Rocking horses, perambulators, baby-jumpers, &c.

5559 MONTANARI, A. 198, *Oxford-st.*—Wax dolls.

5560 MOORE, J. L. *West-st. Dorking.*—Set of cricket stumps.

5561 NICHOLSON, H. *Rochdale.*—Cricket balls.

5562 NOVRA, H. 95, *Regent-st.*—Conjuring tricks and puzzles.

5563 NORMAND, G. B. 54, *Old Compton-st. Soho.*—India rubber balls and balloons.

5564 PAGE, E. J. 6, *Kennington-row, S.*—Articles for cricket.

5565 PEACOCK, T. 515, *New Oxford-st.*—Dolls.

5566 PIEROTTI, C. 13, *Mortimer-st. Oxford-st.*—Foreign and English toys.

5567 PIEROTTI, H. 13, *Mortimer-st. Oxford-st.*—Dolls and wax figures.

5568 PRAETORIUS & WARNER, 32, *Tavistock-pl. W.C.*—Kindergarten materials and diagrams.

5569 PRINCE, MISS A. *Norfolk-cres. W.*—The English "Pinakothek."

5570 PRITCHARD, W H. 29, *Peartree-st. St. Luke's.*—Nursery friend and walking assistant.

5571 RICH, W. 14, *Gt. Russell-st. Bloomsbury.*—Kites.

5572 ROTH, M. *Old Cavendish-st. London.*—Gymnastic figures, diagrams, &c.

5573 SPRATT, I. 1, *Brook-st. W.*—Games and toys.

5574 VAN NOORDEN, P. E. 115, *Gt. Russell-st. W.C.*—Musical games.

5575 WILSON, G. *Castle-st. Shrewsbury.*—Chess tables, nacre and pearl specimens, &c.

5576 WOODMAN, W. 13, *Three Colt-ct. Worship-st.*—Backgammon table.

5577 ZIMMERMAN, W. 18, *Bishopsgate-st. Without.*—Toys, games, &c.

Sub-Class D.—Illustrations of Elementary Science.

5588 ASHMEAD, G. B. 10, *Duke-st. Grosvenor-sq.*—British small birds. Nests and eggs.

5589 BARTLETT, A. D. & SON, *Zoological Gardens, Regent's-pk.*—Preserved and mounted birds, &c.

5590 BOHN, J. 45, *Essex-st. Strand.*—Vivaria and implements.

5591 CUTLER, H. G. 8, *Earl-st. S.E.*—Aquarium.

5592 DAMON, R. *Weymouth.*—Foreign and British shells, fossils, &c.

5594 ELLIOTT BROS. 30, *Strand.*—Sectional and other models.

5595 GARDNER, J. 292, *Oxford-st.*—Stuffed birds, &c.

5596 GARDNER, J. 426, *Oxford-st.*—Stuffed birds.

5597 GREGORY, J. R. 25, *Golden-sq.*—Minerals, fossils, educational collections, &c.

5598 GRIFFIN, J. J. 119, *Bunhill-row.*—Apparatus for elementary instruction in science.

5599 HENSON, E. M. 113A, *Strand.*—Educational collections, &c.

5600 HIGHLEY, S. 70, *Dean-st. Soho.*—Educational apparatus, specimens, &c. for Natural History purposes.

5601 HOLT, E. 24, *White-rock, Hastings.*—Sea-weeds, with names.

5602 LA TOUCHE, THE REV. J. D. *Stokesay, Newton, Salop.*—An orrery for schools.

5603 LAWRENCE, E. 10, *King-st. Lambeth-walk.*—Magic lantern.

5604 LLOYD, W. A. *Portland-rd. W.*—Vivarium.

5605 MAJOR, R, 2, *Sussex-ter. Old Brompton.*—Stuffed birds and animals.

5606 PINNELL, T. 30A, *Thomas-st. Oxford-st.*—Fern frame and aquarium combined.

5607 RICKMAN & HOBBS, 17, *Grove-pl. Brixton-rd.*—Fossils discovered at Peckham and Dulwich, 1860—61.

5608 RIGG, A. & J. *Chester.*—Apparatus for teaching motions and principles of machinery, &c.

5609 ROBERTSON, C. 13, *Queen-st. Oxford.*—Elementary zoological series.

5610 SHARP, C. *Sheffield.*—Educational series of microscopic objects.

5611 SHORT, W. 50, *Praed-st. W.*—Stuffed birds.

5612 STATHAM, W. E. 111, *Strand, W.C.*—Educational sets of scientific apparatus.

5613 SUTTON, C. 2, *Hampstead-st. Fitzroy-sq.*—Dissolving view lanterns, and oxyhydrogen apparatus.

5615 WARD, E. H. *Vere-st. W.*—Specimens of natural history.

5616 WILKINSON, C. W. 8, *Lihton-st. Islington.*—Aquarium,

5617 WILSON, F. W. 1, *Myrtle-ter. Sydenham.*—Mounted animals, birds, &c.

5618 WRIGHT, B. 36, *Gt. Russell-st.*—Elementary geological collections.

CLASS 30.

FURNITURE, PAPER-HANGING, AND DECORATION.

North Court.

Sub Class A.—Furniture and Upholstery.

5651 ANDREWS, W. S. 6, *King's-row, Walworth.*—Flower vases and ornaments, in wood, &c.

5652 ANNOOT, C. 16, *Old Bond-st.*—Cabinet, gilt chairs, and carved centre table.

5653 ARROWSMITH, A. J. 80, *New Bond-st. London, W.*—Solid parqueterie for borders, panelling, altar floors, &c.

5654 ASPINWALL, W. 70, *Grosvenor-st.*—Furniture.

5655 ASSER & SHERWIN, 81, *Strand.*—Bagatelle table for conversion into billiard table.

5656 AVERY, J. 81, *Gt. Portland-st.*—Window blinds.

5657 AYCKBOURN, F. 17, *Bond-st. Vaux-hall-cross.*—Bed, with air-tubes in tick case.

5658 AYRES, W M. 59, *St. Anne-st. Liverpool.*—Models of furniture and appliances.

5659 BAKER, REV. R. S. *Hargrave Rectory, Kimbolton.*—Carved oak eagle lectern.

5660 BALDWIN, C. *Amersham, Bucks.*—Ornamental wood work.

5662 BAYLIS, W. H. 69, *Judd-st. London, W.C.*—Jewel casket and frame, carved in wood.

North Court.

5663 BEARD, J. *Stonehouse, near Stroud.*—Model of sofa bed.

5664 BEDFORD, E. 42, *King-st.-ter. Islington, N.*—Looking glass and carved frame.

5665 BELLERBY, W. 10, *Bootham, York.*—Oak reading stands, with carved panels on burnt wood.

5666 BETTRIDGE, J. & Co. *Birmingham.*—Papier-mâché and japanned tea trays, tables, chairs, caddies, inkstands, writing desks.

5668 BIELEFELD, C. 21, *Wellington-st. Strand.*—Papier maché decorations and works of art.

5669 BIGGS & SON, 31, *Conduit-st. Bond-st.*—Picture frame.

5670 BIRD & HULL, 106, *King-st. Manchester.*—Bed-room furniture and an ebony cabinet.

5671 BORNEMANN, A. F. G. *Bath.*—Chairs.

5672 BRADLEY, J. 129, *Fore-st. Exeter.*—Tables painted in imitation of Devonshire marbles, &c.

5673 BRIDGES, H. 406, *Oxford-st.*—Furniture and dairy articles.

5674 BROOKS, H. & Co. *London.*—Music stools and what-not.

5675 BROWN, BROS. 165, *Piccadilly.*—Easy chairs and folding bedstead.

5676 BROWN, G. & A. 25, *Newman-st. W.*—Composition furniture for gilding and interior decorations.

5677 BRUNSWICK, M. 26, *Newman-st. W.*—Marqueterie and buhl furniture.

5678 BRYER, W. *Southampton.*—Crucifixion, &c.; Moment of Victory, &c.

5679 BURGES, W. 15, *Buckingham-st. Strand.*—Painted and decorative furniture.

5680 BURRIDGE, H. 15, *Grenville-st. Brunswick-sq.*—Granite and marble papers.

5681 BURROUGHES & WATTS, *Soho-sq.*—Oak billiard table.

5683 CALDECOTT, W. 54, *Great Russell-st. Bloomsbury.*—A carved English oak sideboard, tables, chairs, &c.

5684 CASEY, *Academy-st. Cork.*—Carved wood plate-chest.

5685 CHANCE, J. H. 28, *London-st. Fitzroy-sq.*—Gilding, in various coloured golds.

5686 CHAPLIN, R. P. 52, *Frith-st. Soho.*—Couches and chairs for invalids.

5687 CHARLTON, T. & Co. 128, *Mount-st. Grosvenor-sq.*—Suite of bed-room furniture.

5688 COLLMANN, L. W. 53, *George-st. Portman-sq.*—Sideboard in oak, &c.

5689 COOKE, REV. R. H. *Cheltenham.*—Lectern for church.

5690 COX & SON, 28 & 29, *Southampton-st. Strand.*—Church furniture, embroidery, wood and stone carvings.

5691 CRACE, J. G. 14, *Wigmore-st. W.*—Decorations, and cabinet furniture.

5692 CRIMMIN, J. *Killarney.*—Fancy cabinet articles, in bog oak, &c.

5693 CRISWICK & DOLMAN, 6, *New Compton-st. Soho.*—Imitation carved bedstead; book-cases; candelabrum.

5694 DEAR, W. 30 & 31, *St George's-pl. Hyde-park.*—Marqueterie and Mosaic tables. Draperies, &c.

5695 DEXHEIMER, C. 27, *Connaught-ter.*—Marqueterie table, buhl cabinet, &c.

5696 DRAPER, F. 70, *Great Titchfield-st. London, W.*—Picture frames, &c.

5697 DREW, J. 2, *Great Warner-st. Clerkenwell.*—Portable house furniture.

5698 DYER & WATTS, 1, *Northampton-st. Islington.*—Suite of bed-room furniture.

ECCLESIOLOGICAL SOCIETY, 78, *New Bond-st.*

5699 BLENCOWE, MISS, *West Walton, Wisbeach.*—Frontals for Clehonger Church, and Peterborough Cathedral.

5701 BRANDON, R. 17, *Clements Inn, W.C.*—Model of roof.

5702 NICHOLLS, T. 68, *Hercules-bldngs, Lambeth.*—Effigy and tomb; Waltham reredos figure in alabaster.

5703 CLAYTON & BELL, 311, *Regent-st. W.*—Incised group, in stone, from the pavement of Lichfield Cathedral.

5704 EARP, T. *Kennington-rd. Lambeth.*—Reredos and lectern.

5705 FORSYTH, J. 8, *Edward-st. Hampstead-rd. N.W.*—Chichester Cathedral: stall ends and ornamental framing, panel from pulpit, mural tablets, monument.

5707 NORTON, J. 24, *Old Bond-st.*—Statue of Edward III.; font, in alabaster and marble; casts of Bedminster reredos, &c.

5708 PARRY, T. G. *Highnan-ct. Gloucester.*—Frescoes.

5709 PHILIP, J. B. 1, *Roehampton-pl. Vauxhall Bridge-rd.*—Effigy of Dr. Mill, and chimney piece.

5711 REDFERN, F. 29, *Clipstone-st. Fitzroy-sq. W.*—Cast of Resurrection, from Sherborne Mortuary Chapel.

5712 SHAW, N. *London.*—Book-case.

5713 SLATER, W. 4, *Carlton Chambers, S.W.*—Westrop monument, from Limerick Cathedral.

5714 STREET, G. E. 33, *Montague-pl. Bedford-sq. W.*—Iron font cover.

5715 TEULON, *Craig's-ct. S.W.*—Reredos.

5716 WHITE, W. 30A, *Wimpole-st.*—Church plate; reredos.

5716A GODWIN, W. *Lugwardine, Hereford*—Encaustic tiles.

5716B GRAYS, DAVISON, 370, *Euston-rd.*—An organ.

5717 EGAN, J. *Arbutus Factory, Killarney.*—Fancy cabinet table, &c.

5718 ELLIOTT, H. 6, *Vere-st. Oxford-st. London.*—Satin-wood wardrobe, and revolving pedestal.

5719 ELLIS, C. 21, *Bedford-st. Covent-garden.*—Dining-room, and other furniture.

5720 FAULDING, J. 338, *Euston-rd.*—Specimens of fret-cutting and curvilinear sawing.

5721 FILMER, T. H. & SON, 32, *Berners-st. W.*—Articles of furniture and decoration.

5722 FOLSCH, F. W. *Long Buckby, Northamptonshire.*—A jewel-box and dressing-case, inlaid wood.

5723 FORSYTH, J. 8, *Edward-st. Hampstead-rd.*—Book-case, wood carvings, &c.

5724 FOX, T. 93, *Bishopsgate-st. Within.*—Suite of drawing-room furniture, in walnut and holly

5725 FREYBERG, J. *Grosvenor-st. West.*—Furniture of a lady's boudoir.

5726 GANN, MARY C. 32, *Dorset-sq.*—Drawing-room tree-stand. New tile, natural, with colours burnt and unburnt.

5727 GARDNER. J. H. *Poppin's-court, Fleet-st.*—Dressing-table and glass.

5728 GARROOD, R. E. *Chelmsford, Essex.* Mitre box, for mitreing picture frame mouldings.

5730 GEORGE, CLEMENT, & SON, 16, *Berners-st.*—Oak sideboard, in the Italian style, and chairs.

5731 GILLOW & Co. 176, *Oxford-st.*—Sideboard, cabinet, and carved chairs.

5733 GOW, J. 13, *Argyle-st. King's-cross.*—Moulds for papier mache ornaments.

5734 GRIFFITHS, J. 89A, *Ratcliffe-ter. Liverpool.*—Carved boxwood miniature frame.

5735 HARDING, MADDOX, & BIRD, 70, *Fore-st. Finsbury.*—Louis XVI. bedstead and silk hangings.

5736 HARLAND & FISHER, 33, *Southampton-st. Strand.*—Specimens of ecclesiastic and domestic decorative art.

5737 HARROLD, C. & G. *Hinckley.*—Figured oak table and panels.

5738 HATCHWELL, H. & S. B. *Newton Abbot.*—Revolving church stool.

5739 HAWKINS, S. 54, *Bishopsgate-st.*—Model of dining-table.

5740 HEAL & SON, 196, *Tottenham-ct.-rd.*—Bed-room furniture.

5742 HERMANN, F. 54, *Devonshire-st. Portland-pl.*—Library cabinet in buhl.

5743 HERRING, SON, & CLARK, 109, *Fleet-st. E.C.*—Carved sideboard, easy chair, and dining-room chairs, en suite.

5744 HINDLEY, C. & SONS, 134, *Oxford-st.*—Cabinet and other furniture, silk hangings, &c.

5745 HOLLAMBY, H. *Tunbridge Wells.*—Mosaics in natural coloured woods.

5746 HOLLAND, W. *Stained Glass and Decorative Works, St. John's, Warwick.*—Wall decorations and patent sash.

5747 HOLLAND & SONS, 23, *Mount-st. Grosvenor-sq.*—Furniture, inlaid table, gilt cabinet, &c.

5748 HOWARD & SONS, 26, *Berners-st.*—Book-case fittings, library table, and seats.

5749 HUMBLE, G. & J. *Kelso, N.B.*—Arm-chair of antique pattern.

5750 INGLEDEW, C. 7, *Market-row, Oxford-st.*—Dining-room and library chairs.

5751 JACKSON & GRAHAM, *Oxford-st.*—Decorative furniture, &c.

5752 JACKSON, G. & SONS, 49, *Rathbone-pl. Oxford-st.*—Furniture in carton-pierre and papier mache.

5753 JEFFERSON, J. H. 46, *College-green, Bristol.*—Gilt frames.

5754 JENNER & KNEWSTUB, 33, *St. James's-st.*—Fancy cabinets, dressing cases, and work table.

5755 JOHNSTON, MRS. *Ashley, Newmarket.*—Oval frames in leather, &c.

5756 JOHNSTONE & JEANES, 67, *New Bond-st.*—Fancy cabinet.

5758 JONES & Co. 214, *Piccadilly, W.*—Decoration and furniture.

5759 JONES & WILLIS, *Temple-row, Birmingham.*—Church furniture and decoration.

5760 JOUBERT, A. 18, *Maddox-st. W.*—Cabinet, music chair, and an easy chair.

5761 KENDALL, T. H. *Chapel-st. Warwick.*—Articles of furniture and specimens of wood carving.

5762 KIRK & PARRY, *Sleaford.*—Ancaster stone font, with English oak cover.

5763 KNIGHT, T. 7, *George-st. Bath.*—Library table, and ebony chair.

5764 LAKE & SON, *Old Kent-rd.*—English oak bedstead-foot, iron bedstead, and bedstead-pillar, for secret and secure fastening.

5765 LAMB, J. *Manchester.*—Sideboard in oak, &c.

5766 LAUNSPACH, L. 9, *Upper Berkeley-st. Hyde-park.*—Cabinets and tables.

5767 LAWFORD, 89, *Newman-st.*—Reclining arm-chairs, table-bedsteads; self-adjusting flap dining-table.

5769 LECAND, S. 246, *Tottenham-court-rd.*—Console table and glass, &c.

5770 LENZBERG & WALTON, 492, *New Oxford-st.*—Cornices, window blinds, &c.

5771 LEVIEN, J. M. 10, *Davies-st. Grosvenor-sq.*—Cabinet in Pompeian style, and other decorative furniture.

5772 LITCHFIELD & RADCLYFFE, 30, *Hanway-st. Oxford-st.*—Ebony and ivory furniture.

5773 LOTH, J. T. *Carlisle.*—The Victoria casket, a pencil-case, urn stands, mosaic marble chess-table, &c.

5774 LOVEGROVE, J. I. *Isleworth.*—Decorative church writing.

5775 LOWSON, G. *Broughty-ferry, near Dundee.*—Inlaid chess table.

5776 MCCALLUM & HODSON, *Birmingham.*—Papier-machè and japan goods.

5777 MCDONALD, D. *Melrose, Roxburghshire.*—Carved Davenport desk, &c.

5778 MCFARLANE, W. *Saracen Foundry, Glasgow.*—Oak bookcase, with bronze mountings.

5779 MCLAUCHLAN, D. J. & SON, *Water-lane, Blackfriars.*—Gilt carved console table and glasses.

5780 MADDOX, G. 21, *Baker-st.*—Bed-room furniture in deal, inlaid and French polished.

5781 MARGETTS & EYLES, 127, *High-st. Oxford.*—Chimney-glass, style Louis XIV.

5782 MASSEY, T. *City-walls, Chester.*—Table inlaid with foreign woods.

5783 MORRIS, MARSHALL, FAULKNER, & Co. 8, *Red Lion-sq.*—Decorative furniture, tapestries, &c.

5786 NORTH, B. *West Wycombe, Bucks.*—Fancy chairs.

5787 NOSOTTI, C. 399, *Oxford-st.*—Looking glasses, gilt furniture, and decorations.

5788 NUTCHEY, J., D.M. 5, *West-st. Soho.*—Reading table, executed by turning lathes.

5789 OGDEN, H. *Manchester.*—Sideboard, in oak: drawing-room settee and chairs.

5790 PAGE, H. M. 23, *Coventry-st.*—Gilt console table, glass, girandole, cheval screen, &c.

5791 PALMER, H. 7, *St. Michael's-pl. Bath.*—Tables, &c.

5792 PARKER, J. *Woodstock, Oxon.*—Buck-horn hall furniture.

5794 PASHLEY, J. 19, *Red Lion-sq. Holborn.*—Louis XVI. console table and glass.

5795 PATERSON, T. 15, *Rupert-st. Haymarket.*—Carvings and cabinets.

5796 PERRY, W. 5, *North Audley-st.*—Specimens of wood carving.

5797 PHILLIPS, T. 10, *Park-st. Bristol.*—Looking glass.

5798 POOLE & MACGILLIVRAY, 25, *Princes-st. Cavendish-sq.*—Jewel stand and two chairs.

5800 RICHARDSON, T. 9, *Swift's-row, Carlisle.*—Desk of oak, from Carlisle Cathedral.

5801 RIVETT, W. & S. 50, *Crown-st. Finsbury-sq.*—A mahogany sideboard.

5802 ROGERS, G. A. 21, *Soho-sq.*—Wood carvings, &c.

5803 ROGERS, W. G. 21B, *Soho-sq. W.*—Wood carvings.

5804 RORKE, J. 75, *Oakley-st. Lambeth.*—Projecting letters for shop fronts, &c.

5805 ROSS & CO. *Ellis's-quay, Dublin.*—Portable furniture.

5806 ROWLEY, C. *Bond-st. Manchester.*—Patterns of picture and other frames.

5807 SANDERS, W. C. 59, *Queen Anne-st. London, W.*—Leather carving.

5808 SANDEMAN, R. *Edinburgh.*—Mirror tables, in walnut and Quebec ash.

5809 SANDFIELD, J. 38, *Newman-st. Oxford-st.*—Imitation ormulu metal miniature frames.

5812 SCOTT, H. D. *Boston, Lincolnshire.*—Frame and birds carved in wood.

5813 SCOTT, J. & T. 10, *George-st. Edinburgh.*—Cabinet in the style of Louis XVI.

5814 SCOWEN, T. L. *Allen-rd. Stoke Newington.*—Canopies for carriages, boats, &c.

5815 SEDDON, T. *London.*—Articles of furniture.

5816 SEDLEY, A. *Regent-st.*—Furniture.

5817 SHARP, D. F. *Ingram-ter. Sleaford.*—Carved wood bracket.

5818 SILVER, S. W. & CO. 4, *Bishopsgate Within.*—Portable furniture.

5819 SKIDMORE ART MANUFACTURES CO. *Coventry.*—Furniture in mediæval style.

5820 SMEE, W. & SONS, 6, *Finsbury-pavement.*—Modern household furniture and bedding.

5821 SOUTHGATE, J. 76, *Watling-st.*—Camp, barrack, and military equipàge, and cabin furniture.

5822 SPIERS & SON, *Oxford.*—Oxford Cyclopean washstands. Ornamental furniture.

5823 STANTON, T. 22, *Davies-st. Berkeley-sq.*—Decorative door, chandelier, candelabra, &c.

5824 STATHER, J. *Hull.*—Photographic oak paper hangings, washable. Granite column imitated with machine painted paper.

5825 STEVENS, J. 64, *East-st. Taunton.*—Carved mahogany sideboard, representing game, fish, and fruits.

5826 STEVENS, G. H. 14, *Smith-sq. Westminster, W.C.*—Glass mosaic jardinieres, table tops, &c.

5828 STRAHAN, R. & CO. *Dublin.*—Carved cabinet.

5829 STRAPPS, M. G. *Wisbeach, Cambridgeshire.*—Carved oak chair, in alto-relievo.

5830 STRONG, W. 137½, *New Bond-st.*—Glass carved frame. A clock case.

5831 SUTER, A. 65, *Fenchurch-st.*—New castor for furniture with specimen on chair.

5832 TAYLOR, H. J. *Daisy-hill, Dewsbury.*—Table painted in imitation of inlaid woods.

5833 TAYLOR & SONS, 167, *Gt. Dover-st. S.E.*—Expanding dining tables for ships' use, and foot of bedstead.

5834 TAYLOR, J. & SON, 109, *Princes-st. Edinburgh.*—Sideboard, cabinet, &c.

5835 THOMAS, J. 32, *Alpha-rd.*—Marble figure; portion of drawing-room.

5836 THURSTON & CO. *Catherine-st. Strand.*—Billiard table and model.

5837 TOLLNER, J. *John-st. Tottenham-court-rd.*—Inlaid wood table.

5838 TOMLINSON, M. *Hulme, Manchester.*—Specimens of medical shop-fittings; new dentists' operating chairs.

5839 TOMS & LUSCOMBE, 103, *New Bond-st. W.*—Buhl cabinets and tables.

5840 TRAPNELL, C. & W. 2, *St. James's, Barton, Bristol.*—Oak sideboard, ebonised candelabrum, and a cabinet.

5841 TROLLOPE, G. & SONS, 15, *Parliament-st.*—Carved chimney piece, decorations, and cabinet furniture.

5842 TUCKER, J. *North-st. Finsbury.*—Fancy writing-table with escretoire, &c.

5843 TUDSBURY, R. J. *Edwinstowe, Ollerton, Notts.*—Carvings from nature in lime-wood.

5844 TURLEY, R. 381, *Summer-lane, Birmingham.*—Japanned and papier-maché articles.

5845 TUTILL, G. 83, *City-rd.*—Banner's india-rubber preparations to prevent cracking, &c.

5846 TWEEDY, T. H. *Newcastle-on-Tyne.*—Sideboards.

5847 VOKINS, J. & W. 14, *Gt. Portland-st.*—Registered folio frames, for displaying drawings or engravings.

5848 WALKER, J. *Kensington-pl. Notting-hill.*—Carvings in lime-trees—Spring.

5849 WALLIS, T. W. *Louth, Lincolnshire.*—Wood-carvings from nature.

5850 WARD, J. *Leicester-sq.*—Self-propelling and recumbent invalid's chairs.

5851 WEBB, J. 22, *Cork-st. W.*—An inlaid table top.

5852 WERTHEIMER, 154, *New Bond-st. W.*—Louis XVI. console table, surface, steel in silver, mounts in the style of Goutière.

5853 WESTRUP, C. 83, *Old-st.-rd. E.C.*—Fancy occasional chairs.

5854 WHITE, J. *Shrewsbury.*—Gilt fire screen.

5855 WHYTOCK, R. & Co. 9, *George-st. Edinburgh.*—Pollard oak sideboard.

5856 WILKIE, J. 1, *Addington-pl. Lambeth.*—Carved oak figure of our Lord.

5857 WILKINSON, C. & SON, 8, *Old Bond-st.*—Carved and decorative furniture.

5858 WILSON J. W. & Co. 18A, *Wigmore-st. W.*—Bedstead, bagatelle-board, couch and chair.

5859 WINFIELD, W. 22, *Upper Charlton-st. Fitzroy-sq.*—Flowers carved in walnut.

5860 WOOLVERTON, C. *South-quay, Gt. Yarmouth.*—Window-blind.

5861 WRIGHT, W. 27, *Smith-st. St. George's, Birmingham.*—A pearl-table and chimney-glasses, with pearl frames, &c.

5862 WRIGHT & MANSFIELD, 3, *Gt. Portland-st. W.*—Sideboard with Wedgewood plaques, and other decorative furniture.

5863 FOXLEY, G. *Welwyn, Herts.*—Carved wood spoons, forks, &c.

5864 HALSTEAD & SON, *Chichester.*—Grille, for choir of Chichester Cathedral.

5865 HOPE, BERESFORD.—Model of a monument to the Viscountess Beresford.

5866 OWEN, J. *Sheffield.*—Iron bedstead, with patent mattress.

5867 RICHARDSON, *Stamford.*—Carved chairs.

Sub Class B.—Paper Hangings and General Decoration.

5877 ARTHUR, T. 3, *Sackville-st. W.*—Ornamental panel and pilaster.

5879 BUCHAN & SON, *Southampton.*—Renaissance decorations, illustrating the story of Undine.

5880 CARLISLE & CLEGG, 81, *Queen-st. Cheapside.*—Decorative paper-hangings.

5881 COOKE, W. *Grove Works, Leeds.*—Paper-hangings.

5882 COTTERELL, BROS. *Bristol.*—Panel decorations.

5883 COULTON, I. L. 7, *Robert-st. Hampstead-rd.*—Allegorical arabesque decoration, style of Louis XVI.

5884 DOW, R. 59, *George-st. Perth.*—Imitations of woods and mouldings.

5885 EARLE, J. H. 28, *Howland-st. Fitzroy-sq.*—Drawing-room decoration.

5886 GIRARDET, F. 4, *Charles-st. Manchester-sq.*—Specimens of graining and marbling.

5887 GODDARD, W. E. *Hull, Yorkshire.*—Pyrography, or carving upon charred wood, for decorations, &c.

5888 GRANT, W. A. 81, *King-st. Camden-town.*—Imitations of woods and marbles.

5889 GREEN & KING, 23, *Baker-st. W.*—Painted washable wall decorations.

5890 GRIFFIN, J. 7, *Nankin-st. East India-rd. Poplar.*—Imitation of woods.

5891 HASWELL, D. O. 49, *Greek-st. Soho.*—Specimens of writing for signs and tablets.

5892 HAWTHORNE, J. 98, *St. John-st. Clerkenwell.*—Wood coloured by ink.

5893 HAYWARD & SON, 88, *Newgate-st.*—Church and domestic decorations.

5894 HEYWOOD, HIGGINBOTTOM, SMITH, & Co. *Manchester.*—Machine-made paper-hangings.

5896 HORNE, R. 41, *Gracechurch-st.*—Block-printed paper-hangings.

5897 HUMMERSTON BROS. *Leeds, Yorkshire.*—Painted imitations of woods and marbles.

5898 HUNT, C. 40, *Spring-st. Paddington.*—Imitation of woods and marbles, &c.

5899 HURWITZ, B. 9, *Southampton-st. Strand.*—Interior decorations, furniture, &c.

5900 JEFFREY & Co. 500, *Oxford-st.*—Paper-hangings.

5901 JONES & Co. *Arlington-st. Islington.*—Paper-hangings.

5902 KENSETT, J. 18, *Southampton-st. Strand.*—Imitatons of wood and marble.

5903 KERSHAW, T. 38, *Baker-st.*—Decorations for walls, &c.

5904 LAINSON, G. 1, *Henry-pl. Clapham.*—Wall decorations. Pilasters painted on satin.

5905 LAMAMY, A. 3, *Percy-st. Bedford-sq.*—Tableaux in marqueterie and wood mosaic.

5906 LEA, C. J. *High-st. Lutterworth.*—Decoration for wall of a church.

5907 MCLACHLAN, J. 35, *St. James's-st. Piccadilly.*—Artistic drawing and dining-room decorations, and imitations of woods and marbles.

5908 MASLIN & Co. 32, *Foley-st. W.*—Imitations of British and foreign marbles and serpentines on paper.

5909 MORANT, BOYD, & MORANT, 91, *New Bond-st.*—Decoration and articles of furnitnre.

5910 NAYLOR, W. 4A, *James-st. Oxford-st.*—Patterns of staining deal, to imitate all kinds of woods, &c.

5912 OWEN, A. I. 249, *Oxford-st. London.*—Interior decorations and furniture, &c.

5913 PEARSE, J. S. 8, *Barnsbury-st. Islington.*—White and coloured enamel centre, as applied to walls, &c.

5914 PITMAN, W. 210, *Euston-rd. N.W.*—Mediæval paintings and designs.

5915 PURDIE, BONNAR, & CARFRAE, 77, *George-st. Edinburgh.*—Decoration for a drawing-room in French style.

5916 PURDIE, COWTAN, & Co. 314, *Oxford-st.*—Dining-room and other decorative imitation woods and paintings, in water, glass, &c.

5917 READ, W. 153, *Marylebone-rd.*—Imitations of woods and marbles.

5918 RODGERS, J. & J. *Sheffield.*—Painted wall and wood-work decorations.

5919 SCHISCHKAR, E. *Leightcliffe, Yorkshire.*—Marmography, produced by chemical means on transparent articles.

5920 SCOTT, CUTHBERTSON, & Co. *Whitelands, Chelsea.*—Block printed paper-hangings.

5921 SIBTHORPE, R. & SON, *Dublin.*—Specimens of internal decoration.

5922 SIMPSON, W. B. & SONS, 456, *West Strand.*—Painted wall decoration.

North Court.

5923 SMITH, C. 43, *Upper Baker-st.*—Decoration in painted imitations of inlaid marbles, &c.

5924 SMITH, G. T. 1, *Wenlock-rd. City-rd.*—Ornamental wood work printed by agency of heat.

5925 SOUTHALL, C. & Co. 157, *Kingsland-rd.*—Grained woods and marble on paper.

5927 STEINITZ, C. *London Parquetry Works, Grove-lane, Camberwell, S.*—Parquetry floors, veneered wall panellings, and ceilings.

5928 STEPHENS, H. 18, *St. Martin's-le-Grand, London.*—Wood, stained, as a substitute for paint.

5929 TAYLOR, J. 5, *Compton-st.*—Imitations of woods and marbles.

5930 TURNER & OWST, *Elizabeth-st. Pimlico.*—Block-printed paper-hangings with frieze and pilaster.

5931 WARNE, S. 4, *Bruton-st. Berkeley-sq.*—Furniture, decoration, and paper-hangings.

5933 WHITE & PARLBY, 50, *Gt. Marylebone-st.*—Architectural decorations in relief. Decorative furniture for gilding.

5934 WILLIAMS, COOPER'S, & Co. 85, *West Smithfield.*—Wall decorations in Italian and other styles.

5935 WILSHERE & RABBETH, *Gt. Western-rd. Paddington.*—Varnish panels, colours, and wood-stains.

5936 WOOLLAMS, H. 110, *High-st. Manchester-sq.*—Paper-hangings.

5937 WOOLLAMS, J. & Co. 69, *Marylebone-lane.*—Paper-hangings, decorations, &c.

5938 ASPREY, C. 166, *New Bond-st.*—A Davenport.

5939 BURNS & LAMBERT, *Gt. Portland-st.*—Mediæval paper-hangings.

5940 OWEN, J. *Sheffield.*—Springs and attachments for mattress frames, bedsteads, sofas, &c.

5941 GUSHLOW, G. 34, *Newman-st. Oxford-st.*—Plaster casts and composition imitation of bronzes, &c.

5942 TAYLOR, J. 5, *Compton-st. Brunswick-sq. W.C.*—Specimens of graining.

CLASS 31.

IRON AND GENERAL HARDWARE.

South Court.

Sub-Class A.—Iron Manufactures.

5969 ADAMS, W. S. & SON, 57, *Haymarket.*—Cooking apparatus.

5970 ADCOCK, R. C. 4, *Halkin-st. West.*—Indicating door bolt.

5971 ADDIS, W. 6, *Leicester-st. Leicester-sq.*—Kitchen range, stoves, &c.

5973 ALLEN, T. *Clifton.*—Metallic tubular bedsteads.

5975 AVERY, W. & T. *Digbeth, Birmingham.*—Scales and weighing machines.

5976 BACKHOUSE, W. N. 46, *Westgate-st. Ipswich.*—Kitchen ranges.

5977 BAILY, W. & SONS, 71, *Gracechurch-st.*—Iron-work, gates, staircase, &c.

5978 BAMBER, W. C. 12, *Little College-st. S.W.*—Mortice, balance, night latch, &c.

5979 BARLOW, J. 14, *King William-st. City.*—Cask tilt and jack screen, &c.

5980 BARNARD, BISHOP, & BARNARDS, *Norwich.*—Park gates in ornamental wrought iron.

5981 BARRETT, R. & SON, *Beech-st. Barbican.*—Chimney-sweeping and drain machinery.

5982 BARTLETT, J. & SON, *Welsh Back, Bristol.*—Weighing-machines and balances.

5983 BARTON, J. 370, *Oxford-st.*—Stable harness room fittings, &c.

5984 BAYLISS, SIMPSON, & JONES, 43, *Fish-st.-hill.*—Iron hurdles, fencing, gates, cable-chains, &c.

5985 BAYMAN, H. 1, *Johnson-st. E.*—Lifting jacks, iron blocks, and single winch.

5986 BENHAM & SONS, 19, *Wigmore-st. W.*—Metal work, stoves, fenders, ranges, &c.

5987 BENNETT, W. *Liverpool.*—Ranges, smoke-jacks, stoves, and grates.

5988 BERRY, G. 19, *Buttesland-st. N.*—Locks, with crypted guards.

5989 BILLING, J. *Westminster.*—Stove for reducing smoke, &c.

5990 BILLINGE, J. *Ashton, near Wigan.*—Hinges.

5991 BINKS BROS. *Mill Wall, Poplar.*—Wire ropes, conductors, &c.

5992 BISSELL, W. *Union-st. Wolverhampton.*—Rim and mortice locks.

5993 BLACKETT, F. W. 31, *West Smithfield.*—Inaccessible lock.

5995 BOLTON, T. & SONS, *Birmingham.*—Rolled metals, wire and tubes, &c.

5996 BOOBBYER, J. H. 14, *Stanhope-st. Strand.*—Locks, bolts, and hinges.

5997 BRACHER & GRIPPER, 11, *Cannon-st. West.*—Iron safes, boxes, and locks.

5998 BRAMAH & Co. 124, *Piccadilly.*—Iron safes, despatch boxes, and locks.

5999 BRIERLEY & GEERING, *Birmingham.*—Bedsteads.

6000 BROWN, J. & Co. *Glasgow.*—Gill air-warmers, stoves, &c.

6001 BROWN, BROS. *Lyme Regis.*—Cooking ranges, various.

6002 BROWN & GREEN, *George-st. Luton.*—Kitchen range for cure of smoky chimneys, &c.

6003 BROWN, LENNOX, & Co. *Millwall, Poplar.*—Screw bench.

6004 BRYON, T. *Salop-st. Wolverhampton.*—Bedstead.

6005 BUIST, G. 70, *St. Mary's-wynd, Edinburgh.*—Lightning conductors and metallic cords.

6006 BULLOCK, T. & SON, *Cleveland-st. Birmingham.*—Ivory, bone, wood, and horn buttons.

6007 BURCHFIELD, T. & SON, 8, *West Smithfield.*—Improved meat-chopping machine.

6008 BURNEY & BELLAMY, *Millwall, Poplar.*—Iron tanks and cisterns, &c.

6009 BUTLER, J. & SONS, 4, *Elm-st. Gray's Inn-lane.*—Brass, copper, and iron wove wire.

6010 CARPENTER & TILDESLEY, *Somerford Works, Willenhall.*—Rim, hall door locks, curry-combs, &c.

6011 CARRINGTON, J. 4, *Queen's-mews, Queen's-gate, Kensington.*—Model of a horse stall, &c.

6012 CARRON Co. 15, *Upper Thames-st.*—Sugar pan, bright range stoves, &c.

6013 CASEY, W. F. 10, *Raven-row, Stepney.*—Models of scales for weighing bullion.

6014 CHAMBERS, W. *Oozell-st. Birmingham.*—Metallic bedstead, pillars, and rails.

6015 CHATWOOD & DAWS, *Bow-st. Bolton.*—Locks, gunpowder escapement, bankers' safes.

6016 CHILLINGTON IRON Co. *Wolverhampton.*—Machine-made horse shoes.

6017 CHUBB & SON, 57, *St. Paul's-churchyard.*—Detector locks, fire-proof and thief-proof safes, strong-room doors.

6018 CLARK, T. & C. & Co. *Wolverhampton.*—Hollow ware and general casting.

6019 COALBROOKDALE Co. *Coalbrookdale, Shropshire.*—Plain and ornamental iron-work.

6020 COLLINS & GREEN, 7, *Albion-pl. Blackfriars.*—Stoves and grates.

6021 COOLEY & FOWKE, *Castle-st. Wolverhampton.*—Hardware.

6023 CORMELL, J. *Lansdowne Iron Works, Cheltenham.*—Wrought iron tanks, cisterns, cattle troughs, &c., coated inside.

6024 CORNFORTH, J. *Berkley-st. Mills, Birmingham.*—Steel and iron wire and wire nails.

6025 COTTAM & Co. 2, *Winsley-st. London.*—Stable fittings, ironwork, &c.

6026 COTTERILL, E. *Vittoria-st. Birmingham.*—Detector locks, &c.

6027 COX, S. *Walsall.*—Saddlers' iron-mongery, and harness mountings.

6028 CRICHLEY H. *Sheffield-pl. Birmingham.*—Stove grates, mantel pieces, &c.

6029 DAVIES, E. *Galvanized Iron Works, Snow-hill, Wolverhampton.*—Galvanised iron, for various uses, &c.

6030 DAWBARN, R. *The Brink, Wisbech.*—Clamp for instantaneous stoppage of leaks in flexible hose.

6031 DAY & MILLWARD, *Birmingham.*—Weighing machines, scales, steelyards, &c.

6032 DEANE, E. 1, *Arthur-st. East, E.C.*—Range, steel ovens, steel boiler, roasting apparatus, &c.

6033 DEELEY, A. S. *Brasshouse Passage, Birmingham.*—Wrought iron shoe heels and toe tips.

6034 DEELEY, G. H. & Co. *Campbell-st. Dudley.*—Chains for mining, &c.

6035 DIXON, A. *Birmingham.*—Knife cleaners, &c.

6036 DOBSON, E. & W. 24, *Fieldgate-st. Whitechapel.*—Branding irons.

6037 DOCKER & ONIONS, *Thorp-st. Birmingham.*—Smith's bellows, portable forges, &c.

6038 DOLLAR, T. A. 56, *New Bond-st.*—Methods of horse-shoeing.

6039 DOWLER, G. *Great Charles-st. Birmingham.*—Inkstands, corkscrews, candle shades, &c.

6040 DOWLING, E. 2, *Little Queen-st. Holborn.*—Scales, weights, and mills.

6042 DULEY & SONS, *Northampton.*—Ranges and patent bushes for axles.

6043 DYKE & Co. 15, *Aston-pl. N.*—Ice closet and chests.

6044 EASTHOPE, W. *Wyle Cop, Shrewsbury.*—Cooking apparatus.

6045 EDELSTEN & WILLIAMS, *Birmingham.*—Iron wire, buttons, solid-headed pins.

6046 EDGE & SON, *Coalport, Ironbridge, Shropshire.*—Flat gatten chains, cables, and wire ropes.

6047 EDWARDS, E. *Birmingham.*—Glass finger plates, lock furniture, bell pulls, &c.

6048 EDWARDS, F. & SON, 49, *Gt. Marlborough-st. W.*—Porcelain tile grates, fire-brick grates, &c.

6049 EDWARDS, W. *Edgbaston, Birmingham.*—Crinoline fire protectors.

6051 ELLIOTT, J. 67, *Division-st. Sheffield.*—Quadrant weighing machines.

6052 ELLIOTT'S PATENT SHEATHING AND METAL Co. *Newhall, Birmingham.*—Rolled metals, wire, bolts, spikes, nails, &c.

6054 ELLIS, G. H. *Grantham, Lincolnshire.*—Boot, knife and fork cleaners; traps, &c.

6056 EVANS, J. & Co. 34, *King William-st. E.C.*—Stoves, and cooking apparatus.

6057 EYLAND, M. & SONS, *Walsall.*—Spectacles, eye-glasses, buckles, &c.

6058 FEETHAM, M. & Co. 9, *Clifford-st. London.*—Iron and brass work.

6059 FIELD, W. & SON, 224, *Oxford-st.*—Horse shoes.

6060 FIELDHOUSE, G. & Co. *Wolverhampton.*—Steel coffee and other mills.

6061 FINCH, J. *Priory-st. Works, Dudley.* —Fenders, fire-irons, bedstead-castings, &c.

6062 FINLAY, J. *Glasgow.* — Powerful radiating grates.

6063 FIRMAN & SONS, 153, *Strand.*— Military ornaments.

6064 FITZWYGRAM, LT.-COL. 15*th Hussars, Dublin.*—Improved horse shoes.

6065 FLAVEL, S. & Co. *Leamington.*— Improved kitchener.

6066 FRANCIS, E. *Camden-pl. Dublin.*— Horse shoes, shod hoofs, &c.

6067 FREARSON, J. *Clement-st. Birmingham.*—Hooks and eyes.

6068 FULLER, W. 60, *Jermyn-st. London.* —Freezer for making ices.

6069 GALE, S. 320, *Oxford-st. W.*—Inventions for domestic uses.

6070 GEDDES, J. 4, *Cateaton-st. Manchester.* —Ornamental wire plant stands, model rosery, and verandah.

6071 GENERAL IRON FOUNDRY CO. *Upper Thames-st.*—Stoves, mantels, bronzes, &c.

6072 GIBBONS, J. *St. John's Lock Manufactory, Wolverhampton.*—Ornamental locks and general ironmongery.

6073 GIBBONS & WHITE, 345, *Oxford-st.*— Weather-tight casements, and lock furniture.

6074 GIBSON, T. *Cape Works, Birmingham.*—Springs, axletrees, and carriage iron work.

6075 GILLETT, W. 18, *Back-st. Bristol.*— Bottling machines.

6076 GINGELL, W. J. *Bristol.*—Model of corn meter.

6077 GLASS, ELLIOTT, & CO. 10, *Cannon-st.* —Iron and steel wire ropes.

6078 GODDARD, *Nottingham.* — Cooking apparatus for close or open fires.

6079 GOLLOP, EMILIA, *Charles-st. City-rd.* —Floor springs, hinges, &c.

6080 GRAY, A. & SONS, *Weaman-st. Birmingham.*—Fire-irons, &c.

6081 GRAY, J. & SON, 85, *George-st. Edinburgh.*—Stove and range.

6082 GREEN, J. *Irving-st. Birmingham.*— Builders' iron-work and other articles.

6083 GREENING & CO. *Manchester.*—Machine-made wire park fencing.

6084 GREENING, N. & SONS, *Warrington, Lancashire.*—Wire cloth woven by steam power.

6085 GRIFFITHS & BROWETT, *Birmingham.*—Wrought iron, tinned, japanned, and enamelled wares.

6086 GROUT, A. 8, *Shephard-st. Spitalfields.*—Various models and wire fences.

6087 GUY, S. 3, *Haunch of Venison-yd. Brook-st. New Bond-st.*—Specimens of horse shoeing.

6088 HAGUE, T. *Bridge-st. Sheffield.*— Fire-irons.

6089 HALE J. *Hatherton Works, Walsall.* —Spring hooks, curb-chains, bitts, &c.

6090 HALL, R. *Leith, Scotland.* — Iron branding stamps and their impressions.

6092 HAMILTON & CO. 3, *Royal Exchange.* —Locks and safes.

6093 HAMMOND, TURNER, & SONS, *Birmingham.*—Buttons, military ornaments, &c.

6094 HANDYSIDE, A. & CO. *Britannia Foundry, Derby.*—Fountain and vases.

6095 HARLEY, G. *Warwick-st, Wolverhampton.*—Lock and night latches.

6096 HARLOW & CO. *Smethwick, Birmingham.*—Metallic bedsteads.

6097 HAWKINS, J. & CO. 38, *Lisle-st. Leicester-sq.*—Bits, stirrups, spurs, &c.

6098 HAYWARD BROS. 117, *Union-st. Southwark.*—Ranges, coal-hole plates, and lock furniture.

6099 HEATON, R. & SONS, *The Mint, Birmingham.*—Coins complete and in progress.

6100 HENN, I. *Rea-st. Works, Birmingham.*—Taper pointed wood screws in iron and brass, &c.

6101 HEWENS, R. 120, *Warwick-st. Leamington-priors.*—Kitchener with regulator.

6102 HIATT & CO. *Masshouse-lane, Birmingham.*—Police handcuffs, leg irons, dog collars, &c.

6103 HILL & SMITH, *Brierly-hill, Staffordshire.*—Forged iron work, axletrees, &c.

6104 HILLIARD & CHAPMAN, 56, *Buchanan-st. Glasgow.* — Knife cleaners and sharpeners.

6105 HOBBS, ASHLEY, & CO. 76, *Cheapside.*—Locks and door fastenings.

6106 HOOD, S. & SON, 68, *Upper Thames-st.*—Stable fittings.

6107 HOOD, W. 12, *Upper Thames-st.*— Drinking fountains, lamp posts, and specimen castings.

6108 HOOLE, H. E. *Green-lane Works, Sheffield.*—Grates, fenders, fire-irons.

6109 HOPKINS, J. H. & SONS, *Granville-st. Birmingham.*—Block-tin, stamped tinned iron, and japanned articles.

6110 HULSE & HAINES, *Ichnield-st. Birmingham.*—Brass and iron bedsteads.

6111 HURST, C. H. *Royal-rd. Kennington-park.*—Tap wrench, mallets, and cement for repairing glass and china, &c.

6112 ILES, C. *Peel Works, Birmingham.*— Thimbles, pins, needles, &c.

6113 ILIFFE, & PLAYER BROS. *Birmingham.* — Buttons, medals, and military ornaments.

6114 INGRAM, G. W. *Lombard-st. Birmingham.*—Powder flasks, shot pouches, Crimping and goffering machines.

6115 ISMAY, T. & CO. *Dover.*—Ranges for large kitchens.

6116 JAMES FOUNDRY CO. *Walsall.*—Iron and brass, and builders' ironmongery.

6117 JAMES & SONS, *Bradford-st. Birmingham.*—Self-boring wood-screws.

6118 JAMES, J. & SONS, *Victoria Works, Redditch.*—Needles and fish hooks.

6119 JEAKES, C. & CO. 5, *Gt. Russell-st. London.*—Kitchen ranges and fittings, grates, brass, and iron-work.

6120 JEAVONS, I. & D. *Petit-st. Works, Wolverhampton.*—Wrought-iron hollow ware, &c.

6121 JEFFREY & JAFFRAY, 2, *Allen's-court, Oxford-st.*—Wire work.

6122 JENKINS, HILL, & JENKINS, *Milton Works, Birmingham.* — Wire-iron, iron and steel wires, &c.

6124 JONES & ROWE, *Worcester.*—Range, with steam closet, &c.

6126 KEITH, G. 55, *Gt. Russell-st. Bloomsbury.*—Ice-machines, &c.

6127 KENNARD, R. W. & Co. 67, *Upper Thames-st.*—Cast iron park gates and railing, verandahs, garden seats, &c.

6128 KENRICK, A. & SONS, *West Bromwich.*—Cast-iron, tinned and enamelled hollow ware, &c.

6129 KENT, G. 199, *High Holborn.*—Knife-cleaning machines, and other inventions for promoting domestic economy.

6131 KNIGHT, MERRY, & Co. *Bradford-st. Birmingham.*—General tin-plate articles.

6132 LAMBERT, BROS. *Walsall, Staffordshire.*—Tubes, iron and brass fittings, bedsteads, &c.

6133 LANE, H. *Wednesfield, near Wolverhampton.*—Wild beast, game, and vermin traps.

6134 LEADBEATER, J. & Co. 125, *Aldersgate-st.*—Fire and thief-proof safes.

6135 LEIGHTON, J. 40, *Brewer-st. Golden-sq.*—Stoves to prevent smoke. Maltese chimney-caps.

6137 LEWIS, W. 6, *New Westgate-buildings, Bath.*—Gas cooking-stove, &c.

6138 LINES, W. D. & PALMER, W. 1, *Marlborough-rd. St. John's-wood.*—Horse shoes.

6139 LINLEY, T. & SONS, *Stanley-st. Sheffield.*—Blast bellows, forges, vice-benches, &c.

6140 LLOYD, M. *Charles Henry-st. Birmingham.*—Malleable nails.

6141 LLOYD, T. & SONS, 15, *Old-st.-rd. Shoreditch.*—Steel mills.

6142 LONGDEN & Co. *Sheffield.*—Cooking apparatus, stove, stair balusters, &c.

6143 LYON, A. 32, *Windmill-st. Finsbury.*—Mincing machines and digesters.

6144 McCONNEL, R. *Glasgow.*—Locks, latches, and fastenings.

6146 MANDER, WEAVER, & Co. *Wolverhampton.*—Aluminium casket, and aluminium in various forms.

6147 MANGER, J. *Russell-st. Liverpool.*—Suction bellows, with frame.

6148 MAPPLEBECK & LOWE, *Birmingham.*—Ranges, stove grates, fenders, &c.

6149 MARTINEAU, F. E. & Co. *Cliveland-st. Birmingham.*—Hinges.

6150 MATHEWS, W. *Mount-st. Berkeley-sq.*—Horse shoes and shoeing hammer.

6151 MAXWELL, H. & Co. 161, *Piccadilly, W.*—Spurs and sockets.

6152 MAY, A. 259, *High Holborn.*—Gas oven, ranges, drying closet, &c.

6153 MEDHURST, T. 465, *Oxford-st.*—Weighing machines.

6154 MILLS, J. 40, *Gt. Russell-st. W.C.*—Stove and range, with special improvements.

6155 MORETON, J. & Co. 22, *Bush-lane, E.C.*—Foreign and colonial hardware.

6156 MOREWOOD & Co. *Dowgate-dock, London, E.C. and Birmingham.*—Galvanized iron.

6157 MORRISON, D. & Co. *Birmingham.*—Metallic furniture.

6158 MORTON, J. & ON, *Bellfield Works, Sheffield.*—Stove grates, fenders, and fire irons.

6159 MUSGRAVE BROS. *High-st. Belfast.*—Slow combustion stoves; grates; iron fittings for stables, &c.

6160 NASH, R. *Ludgate-hill-passage, Birmingham.*—Presses, dies, &c.

6161 NASH, S. 253, *Oxford-st.*—Stoves, &c.

6162 NASH & HULL, 202, *Holborn, W.C.*—Glass, wood, and brass letters, and stencil plate.

6163 NETTLEFOLD & CHAMBERLAIN, *Broad-st. Birmingham.*—Screws, &c.

6164 NETTLETON, J. 4, *Sloane-sq. Chelsea.*—Open fired ventilating stove and pan.

6165 NEVE, J. & Co. *Horseley-fields, Wolverhampton.*—Cut nails, shoe bills, heel and toe tips, washers, &c.

6166 NEWTON, T. *Walsall.*—Saddle harness and carriage ironmongery.

6167 NICHOLAS, R. 32, *Water-st. Birmingham.*—Roasting jack.

6168 NICHOLSON, W. N. *Trent Iron Works, Newark.*—Cooking range, cottage stoves and fittings, &c.

6169 NOCK & PRICE, *Union-pas. Birmingham.*—Gas cooking range.

6170 NYE, S. & Co. *Wardour-st. Soho.*—Mincing machines, masticators, and coffee mills.

6171 ONIONS, J. C. *Bradford-st. Birmingham.*—Smiths' bellows, portable forges, anvils, vices, &c.

6172 OTTLEY, T. *Spencer-st. Birmingham.*—Gold, silver, and bronze medals.

6173 OWEN, W. *Phœnix Works, Rotherham.*—Bradley's kitchener.

6174 PALMER, J. & SONS, *Beech-lanes, near Birmingham.*—Screw railway wrenches, hammers, &c.

6175 PATENT ENAMEL Co. *Birmingham.*—Glass enamelled hollow-ware, tablets, &c.

6176 PERRY & SON, *Bilston.*—Bedsteads; safes.

6177 PEYTON & PEYTON, *Bordesley Works, Birmingham.*—Metallic bedsteads, hat stands, &c. of wrought and cast iron combined.

6178 PHILLIPS, T. 55, *Skinner-st. Snow-hill.*—Gas bath, gas cooking apparatus, &c.

6180 PIERCE, W. 5, *Jermyn-st. London.*—Grates, fenders, fire lumps, &c.

6181 PIGOTT & Co. *St. Paul's-sq. Birmingham.*—Buttons, ornaments, medals, shirt studs, sleeve links, &c.

6182 PLIMLEY, J. T. & Co. *Wolverhampton.*—General ironmongery.

6183 POTTER, T. 44, *South Molton-st.*—Bronzed casting and wrought iron work.

6184 POUPARD, W. *Blackfriars-rd.*—Weighing machine, spirometer balance, &c.

6185 PRICE, C. & Co. *Wolverhampton.*—Double-action detector locks; fire proof safes.

6186 PRICE, G. *Cleveland Works, Wolverhampton.*—Fire safes, chests, and doors; mortise and other locks.

6187 PULLINGER, C. *Selsey, Sussex.*—Rat traps, &c.; each one caught re-setting the trap.

6188 RADCLYFFE, T. *Leamington.* — Ranges and smokeless feeding screw.

6189 RAWLINS, E. 27, *Whittall-st. Birmingham.* — Stampings and pressings of iron and steel.

6190 REDMAYNE & Co. *Wheathill Foundry, Rotherham.*—Stove grates, umbrella stands, &c.

6191 REYNOLDS, J. *Crown and Phœnix Works, Birmingham.*—Nails, bills, washers, brackets, &c.

6192 REYNOLDS, J. 57, *New Compton-st. W.C.*—Wire work and metallic netting.

6193 RHODER, W. *Westgate, Bradford.*—Indestructible fire-proof safe.

6195 RICHARDS, W. & Co. *Imperial Wire Works,* 370, *Oxford-st.*—Wire work.

6196 RICKETS & HAMMOND, 5, *Agar-st. Strand.* — Gas-range, and gas-stove, globe-light chandeliers, and ventilating globe lights.

6197 RIDDELL, J. H. 155, *Cheapside.*—Slow combustion boiler.

6198 RITCHIE, J. *Canongate, Edinburgh.* —Metallic cord, picture and sash lines, &c.

6199 RITCHIE, WATSON, & Co. *Etna Foundry, Glasgow.* — Kitcheners, cabooses, grates, plumbers' goods, &c.

6200 ROBERTS, W. *Lion Foundry, Northampton.* — Stoves, ranges, cast-iron tables, chairs, &c.

6201 ROBERTSON & CARR, *Chantry Works, Sheffield.*—Grates, stoves, fenders, &c.

6202 ROBOTHAM, S. *Bradford-st. Birmingham.*—Wire fencing and general wire work.

6203 ROCKE, W. *Phœnix Foundry, Wolverhampton.*—Machinery from pig and wrought iron, refined.

6204 ROGERS, P. & Co. 106, *Digbeth, Birmingham.* —Steelyards, scale beams, scales, weighing machines, stocks and dies.

6205 ROLLASON, A. & SONS, *Bromford Mills, Erdington, near Birmingham.*—Steel music wire and metals.

6206 ROWLEY, C. & Co. *Newhall-st. Birmingham.*—Fancy goods.

6207 ROWLEY, SARAH A. *Clement-st. Birmingham.*—Buttons and studs.

6208 RYFFEL, I. E. 5, *Upper Stamford-st. Blackfriars.*—Hygeian stove.

6209 ST. PANCRAS IRON WORK Co. *Old St. Pancras-rd.*—Stable fittings and ornamental gates.

6210 SCOTT, J. W. *Sidbury Works, Worcester.*—Solid leather buttons, wads, washers, &c.

6211 SHERWIN, J. *Tabernacle-walk, Finsbury.*—Kitchen range.

6212 SMITH, F. & Co. *Halifax.*—Bar iron, and wire in various stages.

6213 SMITH, T. *St. John's-sq. Wolverhampton.*—Hardware and cutlery.

6214 SMITH, T. & W. *Newcastle-on-Tyne.* —Wire ropes.

6215 SMITH & WELLSTOOD, *Columbian Stove Works, Glasgow.* — Stoves, ranges, boilers, &c.

6216 SPOKES, J. *North-st-mews, Fitroy-sq.* —Meat screens and refrigerators.

6217 STANDING, T. *Preston, Lancashire.*—Galvanized wire netting, fencing staples, liquid agitator, &c.

6218 STANDLEY, W. 38, *Park-st. Walsall.* —Bits, spurs, stirrups, bridles, cruppers, &c.

6219 STANLEY, J. M. & Co. *Sheffield.*—Ranges, cooking apparatus, grates, &c.

6220 STARK, J. C. 13, *Strand, Torquay.*—Grates and kitchen-ranges.

6221 STEEL & GARLAND, *Sheffield.* — Stoves, grates, and fenders.

6222 STEPHENSON, J. & Co. *Poplar.*—Wire rope for ships' standing rigging.

6223 STUART & SMITH, *Roscoe-pl. Sheffield.* —Stoves, grates, fenders, &c.

6224 STUBBS, W. H. *Park-cres. London, N.W.*—Instruments for cleaning forks.

6225 TALBOT, C. & S. *Gt. Titchfield-st. Marylebone.*—Cooking apparatus and utensils.

6226 TANN, J. 30, *Walbrook, E.C.*—Locks, safes, iron doors, and boxes, &c.

6227 TAYLOR, J. JUN. 53, *Parliament-st.* —Smoke consuming grates, &c.

6228 TAYLOR, W. *Sheepcote-st. Birmingham.*—Shutter-bars, door-spring, &c.

6229 THOMAS, W. H. 6, *Sloane-st.*—Fire screen for dining room; door porters.

6230 TITFORD, R. V. & Co. 117, *Leadenhall-st. E.C.*—Scales, weighing machines, weights, &c.

6231 TONKS, S. *Gt. Hampton-st. Birmingham.*—Galvanised iron and japanned goods.

6232 TOOVEY, E. & C. *Wolverhampton.*—Ironmongery and hardware.

6233 TUCKER & REEVES, 181, *Fleet-st.*—Locks, tin boxes, and fire-proof safes.

6234 TURNER, G. 13, *Rose-ter. Fulham-rd. Brompton.*—Machinery and articles for domestic uses.

6235 TYLOR & PACE, 5, *Queen-st. Cheapside.*—Bedsteads, window blinds, &c.

6237 VINCENT, R. *St. George's-place, Camberwell.*—Smoke resisting stove door.

6238 WALKER & CLARK, 24, *Denmark-st. Soho, W.C.*—Wire cloth and work.

6239 WALKER, T. & SON, *Oxford-st. Birmingham.*—Stoves for buildings.

6241 WARDEN, J. & SONS, *Edgbaston-st. Birmingham.*—Railway screw bolts, nuts, and railway appliances.

6242 WATKIN, W. & Co. *High-st. Stourbridge.*—Spades, anvils, vices, &c.

6243 WATKINS & KEENE, *London Works, Birmingham.*—Bolts, nuts, couplings, tie-rods, &c.

6244 WEBSTER & HORSFALL, *Birmingham.* —Steel music wires.

6245 WELDON, C. & J. *Cheapside.*—Covered and other buttons.

6246 WENHAM LAKE ICE Co. 140, *Strand.* —Refrigerators, freezing powders, &c.

6247 WEST, HARRIET, 344, *Euston-rd.*—Iron and wire work.

6248 WHALEY, BURROWS, & FENTON, *Queen's Ferry Wire and Wire Rope Works, near Flint.*—Winding drums, wire-ropes, and cables.

6249 WHITE, T. *Thorpe, Hesley, near Rotherham.*—Bolts, nails, and various articles.

6250 WHITFIELD, S. & SONS, *Birmingham.* —Iron bedsteads. Fire and thief-proof safes.

6251 WHITFIELD, T. & Co. *Birmingham.*—General iron goods.

6252 WHITLEY, J. *Ashton, near Warrington.*—Wrought-iron hinges and handles.

6253 WILDS, W. *Hertford, Herts.*—Ceiling ventilator. Model of stove.

6254 WILKINS & WEATHERLY, 39, *Wapping.*—Rope of steel, iron, and copper wire.

6255 WILLS BROS. 12, *Euston-rd.*—Art drinking fountains.

6256 WINCHESTER, GRAVELEY, & SAGER, *Upper East Smithfield.*—Sea-water distilling and cooking apparatus.

6257 WINFIELD, R. W. & SON, *Birmingham.*—Tubes, bedsteads, gas-fittings, brass-foundry, &c.

6258 WINTER, H. 3, *Paragon-rd. Hackney.*—Weighing-machine.

6259 WITHERS, G. & SONS, *Park Works, West Bromwich.*—Safes and money chests.

6260 WOOD BROS. *Stourbridge.*—Chain cables, anchors, anvils, vices, &c.

6261 WOODIN, D. 2, *Upper Park-pl. Dorset-sq.*—Horse shoes, which prevent slipping.

6262 WRIGHT, G. & Co. *Burton Weir, Sheffield.*—Iron-foundry articles.

6263 WRIGHT, P. *Constitution-hill Works, Dudley.*—Vices, anvils, cramps, &c.

6264 WRIGHT & NORTH, *Wolverhampton.*—Iron and steel, and combined metal tyre bars.

6265 YATES, HAYWOOD, & DRABBLE, *London.*—Ornamental furniture in cast-iron.

6267 YORK, S. & Co. *Wolverhampton.*—Hardware goods.

6268 YOUNG, W. 34, *Queen-st. Cheapside.*—Lamps, gas-burners, and smokeless grates.

6269 AUBIN, C. *Wolverhampton.*—Nettlefold's guardian locks, and fancy keys.

6270 COTTRILL, E. *St. Paul's-sq. Birmingham.*—Copying-presses and dies.

6272 MOORE, J. *Birmingham.*—Medals.

BETTRIDGE, J. & Co. have been removed to Class 30.

Sub-Class B.—Manufactures in Brass and Copper.

6277 ALDER, H. *Grange Works, Edinburgh.*—Gas meters.

6278 ALDRED, W. 28, *Pall Mall, Manchester.*—Gas-burners, made from silver and sheet metals.

6280 BENHAMS & FROUD, 42, *Chandos-st. Charing-cross.*—Copper, zinc, and brass manufactures.

6282 BIDDLE, E. EMILY, *Victoria-st. Birmingham.*—Gilt rims and ornaments for books and cabinets.

6283 BISCHOFF, BROWN, & Co. *Langham Works, George-st. Great Portland-st. London.* — Gas meter, with floating measuring chamber.

6284 BLEWS & SONS, *Birmingham.* — Bells, and brass-foundry articles; chandeliers, weights and measures.

6285 BRIGHT, R. (successor to Argand & Co.) 37, *Bruton-st.*—Lamps and wicks.

6286 CARTWRIGHT, SAMBIDGE, & KNIGHT, 39, *Lombard-st. Birmingham.* — Chandeliers, brackets, gas fittings, &c.

6287 CHAMBERS & Co. 216, *Bradford-st. Birmingham.*—Railway, ship, and carriage lamps, coach furniture, &c.

6288 COWAN, W. & B. *Buccleuch-st. Works, Edinburgh.*—Wet and dry gas meters.

6289 CROLL, RAIT, & Co. *Kingsland-rd. N.E.*—Dry gas meter and apparatus.

6290 DALE, R. & SON, 195, *Upper Thames-st.*—Copper kitchen furniture.

6291 DEFRIES, J. & SONS, 147, *Houndsditch.*—Brass chandeliers, bronzes, hall lanterns, &c.

6292 DICKIE, C. *Dundee.*—Wire work, working models, bell hanging.

6293 DRURY, F. 10, *Duke-st. W.*—Musical inventions applicable to clocks.

6294 DUCKHAM, H. A. F. 44, *Clerkenwell-green.*—Gas meters and gas regulators.

6295 EDGE, T. *Gt. Peter-st. Westminster.*—Gas meters.

6296 EVERED, R. & SON, *Bartholomew-st. Birmingham.*—Brass foundry articles.

6297 FELE, J. & Co. *St. James'-sq. Wolverhampton.* — Brass chandeliers, gas fittings, &c.

6298 FORREST, G. & SON, *Nevill's-ct. New-st.-sq.*—Candelabrum and chandelier.

6299 GARDNER, H. & J. 453, *Strand.*—Lamps, chandeliers, candelabra, &c.

6300 GLOVER, G. & Co. *Ranelagh Works, Pimlico.*—Dry gas-meters, pneumatometers, and photometers.

6301 GLOVER, T. *Suffolk-st. Clerkenwell-green.*—Dry gas meters and gas holders.

6302 GRAY, BAILEY, & BARTLET, *Berkley-st. Birmingham.* — Gaseliers, tea trays, coal vases, &c.

6303 GREENWAY, W. *Princess-st. Birmingham.*—General brass-foundry articles.

6304 GUEST & CHRIMES, *Rotherham.*—Water-works articles.

6305 HARDMAN, J. & Co. *Gt. Charles-st. Birmingham.* — Mediæval metal manufactures.

6306 HARROW, W. & SON, 14, *Portland-st. Soho.* — Chandeliers, tripods, brackets, &c.

6307 HART & SON, *Cockspur-st.*—Ecclesiastical metal work.

6308 HARVEY, T. 13, *Bradford-st. Walsall.*—Brass and plated harness furniture.

6309 HICKLING & COX, *Birmingham.*—Copper and iron boat nails, rivets, and washers, &c.

6310 HICKMAN, J. *William-st. Birmingham, N.*—Brass cocks.

6311 HILL, J. *Broad-st. Birmingham.*—Chandeliers, gas-fittings, stampings for metallic bedsteads, &c.

6312 HIND, J. 118, *Kingsland-rd.*—Engraved and inlayed metals.

6313 HINKS, J. & SON, *Crystal Lamp Works, Birmingham.* — Lamps for burning hydro-carbon oils.

6315 HORSEY & BAKER, *Worcester-st. Southwark.*—Tea urns, taps, with steel protectors, &c.

6316 HUGHES, R. H. *Atlas Works, Hatton-garden.*—Safety indicating chandeliers.

6317 HULETT, D. & Co. 56, *High Holborn.* —Gas-fittings.

6319 JOHNSTON BROS. 190, *High Holborn.* —Gaselier for a cathedral.

6320 LAMBERT, T. & SON, *Lambeth.*—Plumbers', gasfitters', ironmongers', and engineers' furnishings.

6321 LEALE, A. 4, *Litchfield-st. W.C.*—Vanes, weathercocks, moulds, &c.

6322 LEONI, S. 34, *St. Paul-st. N.*—Taps, gas-burners, &c.

6323 LOYSEL, E., C.E. 92, *Cannon-st. E.C.* —Percolator, keyless locks.

6324 MACKEY, C. *Gt. Hampton-row, Birmingham.*—Brass knobs, vases, furniture ornaments, &c.

6325 MARRIAN, J. P. *Birmingham.*—Goods in brass for ships.

6326 MATTHEWS, E. 377, *Oxford-st.*—Engraving in metal for ecclesiastical decoration and other purposes.

6327 MESSENGER & SONS, *Broad-st. Birmingham.*—Chandeliers, and general gas fittings, &c.

6328 MIDWINTER, E. & Co. 68, *Snow-hill.* —Tea urns, tea kettles, coal scoops, &c.

6329 NAYLOR, J. *Hulme, Manchester.*—Lamps, various.

6330 NORTH, E. P. *Exeter-row, Birmingham.*—Ornamental metallic panelling.

6331 NUNN, W. 179, *St. George-st. E.*—Signal lanterns.

6332 OERTON, F. B. *Walsall.*—Carriage lamps, axles, springs, &c.

6333 OLIVER, G. & J. 286, *Wapping, E.*—Dioptric ship's signal lamps.

6335 PHILLP, C. J. *Caroline-st. Birmingham.* — Gaseliers, brackets, and gas fixtures.

6336 PONTIFEX, H. & SONS, 55, *Shoe-la. London.*—Brewing and distilling apparatus.

6337 PONTIFEX, R. & SON, 14, *Upper St. Martin's-la.* — Copper, brass, and steel plates.

6338 PROSSER, W., & H. J. STANDLY, 24, *Dorset-pl. Dorset-sq.*—Lamps for lime light.

6339 PYRKE, J. S. & SONS, *Dorrington-st. London.*—Urns and swing kettles.

6340 REID, J. *Edinburgh.*—Gas saturator, for preventing evaporation of water from meter.

6341 RENNIE & ADCOCK, *Easy-row Works, Birmingham.* — Chandeliers, candelabra, bronzes, &c.

6342 RICHARDS, W. *Crawford-passage, Clerkenwell.*—Gas meters.

6343 SARSON, T. F. *Leicester.*—Lamp that can be repaired in a few minutes.

6344 SINGER, J. W. *Frome, Somerset.*—Brass lectern, altar rails, and mediæval ornaments.

6345 SKIDMORE ART MANUFACTURES Co. *Coventry.*—Screen for Hereford Cathedral; gas corona, pendants, standards, &c.

6346 SOUTTER, W. *New Market-st. Birmingham.*—Tea urns and kettles.

6347 STEER, J. *Weaman-st. Birmingham.* —Cornices, poles, rings, &c.

6348 STONE, J. *Deptford.*—Boat and ship nails, &c.

6349 STRODE, W. 16, *St. Martin's-le-Grand. E.C.*—Sun burner with valve, and candelabra.

6350 SUGG, W. *Marsham-st. Westminster.* —Gas meters, governors and pressure gauges, lava burners, and public lamp governors.

6351 TAYLOR, J. & Co. *Loughborough.*—Bells hung in frame.

6352 THOMASON, T. & Co. *St. Paul's-sq. Birmingham.*—Ecclesiastical and domestic Gothic metal work.

6354 TONKS, W. & SONS, *Moseley-st. Birmingham.*—Brasswork.

6355 UNDERHAY, F. G. *Crawford-passage, Clerkenwell.*—Gas meter.

6356 VERITY, B. & SONS, 32 *King-st. Covent-garden.*—Gaseliers and brass works.

6358 WARNER, J. & SONS, *Crescent, Cripplegate.*—Bells, urns, baths, lamps, braziery, &c.

6359 WEST & GREGSON, *Oldham.*—Model gas (station) meter, with its appurtenances.

6360 WOOTTON & POWELL, *Parade Works, Birmingham.* — Gas chandeliers and wall brackets.

6361 WYATT, A. 22, *Gerrard-st. Soho.*—State carriage lamps.

6362 YOUNG, J. & SON, 46, *Cranbourn-st. London, W.C.*—Weighing-machines, for persons or goods.

Sub-Class C.—Manufactures in Tin, Lead, Zinc, Pewter, and General Braziery.

6373 AZULAY, B. *Rotherhithe, Surrey.*—Heat-retaining vessels for boiling water, &c.

6374 BEARD & DENT, 21, *Newcastle-st. Strand.*—Plumbers' appliances.

6375 BRABY, F. & Co. *Euston-rd. London.* —Galvanized zinc, iron, perforated metals.

6376 CHATTERTON, J. *Wharf-rd. City-rd.* —Lead, block tin, and composition pipe.

6377 COOKSEY, H. R. *Bordesley, Birmingham.*—Coffin plates, handles, and ornaments.

6378 DIXON, J. & SONS, *Sheffield.*—Britannia metal wares.

6379 ELLIS, J. 136, *King's-rd. Brighton.*—The Elutriator, for decanting wine.

6380 EWART, HENRIETTA, 346, *Euston-rd. London.*—Various zinc goods.

6381 FOXALL, S. 52, *William-st. Regent's-pk.*—Confectioner's moulds, piecer, &c.

6382 GILBERT, J. A. & Co. *Clerkenwell, London.*—Grocers' fittings and appliances.

6383 HICKMAN & CLIVE, *William-st. N. Birmingham.*—Coffin furniture.

6384 LOVEGROVE, J. J. 6, *Pembroke-pl. Isleworth.*—Specimens of plumbing, from 14th century to present time.

6385 LOVERIDGE, H. & Co. *Wolverhampton.*—Papier maché trays, wares, &c.

6386 MARSTON, J. *London Works, Bilston, Staffordshire.*—Various japanned and other goods.

6387 PERRY, E. *Jeddo Works, Wolverhampton.*—Japan and tin wares.

6388 TYLOR, J. & SONS, *Warwick-lane, Newgate-st. E.C.*—Baths for private dwellings.

6390 WATTS & HARTON, 61, *Shoe-lane, London.*—Pewter articles.

6391 WILSON, R. & W. *London.*—Baths, and plate-warmer.

6392 WOLVERHAMPTON ELECTRO-PLATE Co. *Peel Works, Wolverhampton.*—Silver-plated wares.

6393 ZOBEL, J. 139, *Euston-rd.*—Ornaments for gas and water, &c.

CLASS 32.

STEEL CUTLERY AND EDGE TOOLS.

Transept, South Court.

Sub-Class A.—Steel Manufactures.

6425 ACADIAN CHARCOAL IRON Co. (Limited), *Sheffield.*—Pig and bar iron, steel, and steel tools and cutlery.

6426 ALLCOCK, S. & Co. *Unicorn Works, Redditch.*—Needles, fishing tackle, &c.

6427 BESSEMER, H. 4, *Queen-st.-pl. New Cannon-st.*—Specimens of Bessemer iron and steel.

6428 BOULTON, W. & SON, *Redditch.*—Needles, fish-hooks.

6429 BRANDAUER, C. & Co. 407, *New John-st. West*—Steel pens and pen holders.

6430 BROWN, J. & Co. *Atlas Steel and Iron Works, Sheffield.*—Steel springs, buffers, files, rails, axles, and forgings.

6431 CALDWELL BROS. 15, *Waterloo-pl. Edinburgh.*—Serpentine pen.

6432 CAMMELL, C. & Co. *Cyclop Works, Sheffield.*—Iron, steel, files, springs, forgings, railway materials.

6433 DEWSNAP, J. 10, *St. Thomas-st. Sheffield.*—Leather and cabinet goods, dressing-cases, &c.

6434 GILLOTT, J. *London and Birmingham.*—Metallic pens, and pen holders.

6435 GOODMAN, G. 82, *Caroline-st. Birmingham.*—Elastic pins and needles.

6436 HADFIELD & SHIPMAN, *Attercliffe Steel Wire Mills, Sheffield.*—Steel wire for crinoline, umbrella ribs, ropes, fish-hooks, springs, &c.

6437 HINKS, WELLS, & Co. *Birmingham.*—Steel pens and holders.

6438 HOEY, T. & Co. 25, *New-row West, Dublin.*—Pins and hair-pins.

6439 HUTCHINSON, P. & SON, *Kendal.*—Fish-hooks, and fishing tackle.

6440 KIRBY, BEARD, & Co. 62, *Cannon-st. West, E.C.*—Pins, needles, fish-hooks, sewing cotton, &c.

6441 KNIGHTS, W. & Co. *Shaksperean Works, Stratford-on-Avon.*—Pins and needles.

6442 LEWIS, H. & SON, *Queen-st. Redditch.*—Sewing needles and fish-hooks.

6443 MILWARD, H & SONS, *Redditch.*—Needles and fish-hooks.

6444 MITCHELL, W. *Birmingham; and 74, Cannon-st. West, E.C.*—Metallic pens and holders.

6445 MITCHELL, W. 41, *London-st. Fitzroy-sq.*—Springs.

6446 MOGG, J. & Co. *Adelaide Works, Redditch.*—Needles and fishing tackle.

6447 MYERS & SON, *Birmingham.*—Steel pens, letter clips, paper knives, &c.

6448 NAYLOR, VICKERS, & Co. *Sheffield.*—Cast-steel disc wheels, tyres, axles, &c.

6450 PAGE, W. & J. 71, *Mott-st. Birmingham.*—Corkscrews, and steel toys.

6451 PEACE, J. & Co. *Sheffield.*—Bright rolled steel, and saws from same; Crinoline; Steel busks.

6452 PERRY, J. & Co. 37, *Red Lion-sq.*—Metal pens.

6453 REYNOLDS, G. W. & Co. 12, *Cheapside.*—Steel wire, crinoline steel, umbrella frames, &c.

6454 ROWELL, J. 7, *St. Albans-row, Carlisle.*—Artificial flies and fish-hooks.

6456 SHORTRIDGE, HOWELL, & Co. *Hart-*

ford Steel Works, Sheffield.—Steel, files, and general articles.

6457 SMITH, J. W. *High Cross-st. Leicester.*—Self-acting, and other hosiery needles.

6458 SMITH & HOUGHTON, *Warrington.*—Piano, pinion, and round steel wires.

6459 SOMMERVILLE, A. & Co. *Birmingham.*—Gilt pointed, and other steel pens.

6460 SPENCER, J. & SONS, *Newcastle-on-Tyne.*—Cast-steel tyres, volute spring buffers, springs, steel, and files.

6461 THOMAS, S. & SONS, *British Needle Mills, Redditch.*—Needles and fish-hooks.

6462 TOWNSEND, G. & Co. 12, *Walbrook, London.*—Sewing machine and other needles, and tools for making them.

6463 TURNER, R. & Co. *London and Redditch.*—Needles, and fish-hooks.

6464 TURNEY, G. L. 20, *Lawrence-lane, Cheapside.*—Needles and pins.

6465 TURNOR, M. & Co. *Icknieldport-rd. Birmingham.*—Metal pens, holders, and pen boxes.

6466 TURTON BROS. *Phœnix Steel Works, Sheffield.*—Steel, files, saws, engineering tools, &c.

6467 TURTON, T. & SONS, *Sheaf and Spring Works, Sheffield.*—Steel, files, edge-tools, railway springs, &c.

6468 WALKER, H. 47, *Gresham-st.*—Ridged and other needles, fish-hooks, &c.

Sub-Class B.—Cutlery and Edge Tools.

6480 ADDIS, J. B. 159, *Waterloo-rd.*—Carvers' tools.

6481 ADDIS, S. J. 50, *Worship-st. Shoreditch.*—Carvers, and general edge tools.

6482 ALLARTON, T. & POWELL, *Birmingham.*—Awls, and sewing machine needles.

6483 BADGER, C. 1, *Stangate, Lambeth, S.*—Planes in iron and gun metal.

6484 BAKER, W. *Pembroke-st. London, N.*—Awls, trade bodkins, and needles.

6485 BARKER, R. & SON, *Easingwold.*—Butchers' and table steels, made from the best refined cast steel.

6486 BEACH, W. *Salisbury.*—Case of cutlery.

6487 BEARDSHAW, G. *Tomcrop-lane, Sheffield.*—Cutlery, &c.

6488 BOLSOVER, T. *Ford, Ridgeway, Sheffield.*—Sickles, and reaping-hooks.

6490 BOOTH, H. E. & Co. *Norfolk Works, Norfolk-lane, Sheffield.*—Table knives and forks; spear; butcher, bowie, and dagger knives.

6491 BROOKES & CROOKES, *Atlantic Works, Sheffield.*—Knives, razors, and dressing case fittings.

6492 BROWN, H. & SONS, 108, *Rockingham-st. Sheffield.*—Braces, bits, joiners' tools, augers, gimblets, skates, tool chests.

6493 BUCK, J. *Newgate-st. and Waterloo-rd.*—Tools.

6495 CHAMPION & Co. 169, *Broad-lane, Sheffield.*—Fine scissors.

6496 COCKBAIN, J. *Portland-pl. Carlisle.*—Joiners' and cabinet tools.

6497 DIGGINS, G. 20, *Bessborough-pl. Pimlico.*—Iron and metal planes, &c.

6498 DRABBLE, J. & Co. *Orchard Works, Sheffield.*—Table and other knives, forks, &c.

6499 EADON, M. & SONS, *President Works, Sheffield.*—Steel, saws, files, machine knives, &c.

6500 EASTWOOD, G. 31, *Walmgate, York.*—Planes for joiners, cabinet makers, &c.

6502 FLETCHER, J. C. *Crown Works, Sheffield.*—Tools, &c.

6503 FULLER, J. H. 70, *Hatton-garden.*—Tube cutters, stocks and dies, taps, &c.

6504 GALLIENNE, G. 138, *Goswell-st.*—General cutlery.

6505 GIBBINS, J. & SONS, *Sheffield.*—Cutlery, &c.

6506 GILBERT BROS. *Sheffield.*—Razors, pen, pocket, and sportsmen's knives.

6507 GILPIN, W. SEN. & Co. *Wedges' Mills, Cannock, Staffordshire.*—Edge tools, augers, matchets, iron and steel.

6508 GORRILL, R. & SON, 159, *Eyre-st. Sheffield.*—Fine scissors.

6509 GRAY, J. H. *Pelham-st. Nottingham.*—Improved skates.

6510 GREENSLADE, E. A. & W. *Thomas-st. Bristol.*—Planes.

6511 GREER, J. 90, *Newgate-st.*—Cutlery, and knives for various trades.

6512 HANNAH, A. *Catton, Glasgow.*—Tools for boring wood.

6513 HARDY, T. 44, *Milton-st. Sheffield.*—Stilettoes, corkscrews, and articles in steel, &c.

6514 HARGREAVES, SMITH, & Co. *Eyre-lane, Sheffield.*—Cutlery and hardware.

6515 HASLAM, J. & SONS, *Ridgeway, near Sheffield.*—Scythes, sickles, and reaping hooks.

6516 HAWCROFT, W. & SONS, *Bath Works, 53, Bath-st. Sheffield.*—Variety of razors.

6517 HAYWOOD, J. & Co. *Sheffield.*—Pruning knife, and cutlery in general.

6518 HEATH, S. *Union-pl. Paddock, Walsall.*—Spring splitting machine and saddlers' tools.

6519 HILL, J. V. 5, *Grays-inn-rd. King's-cross.*—Saws.

6520 HOWARTH, J. *Broomspring Works, Sheffield.*—Edge and general trade tools.

6521 JACKSON, NEWTON, & Co. *Sheaf Island Works, Sheffield.*—Edge tools, cutlery, &c.

6522 JOLLEY, J. & T. *Excelsior Works, Warrington.*—Files, railway ticket nippers, engineers' tools, &c.

6523 JOWITT, T. & SON, *Sheffield.*—Steel and files for engineers.

6525 KING & PEACH, *Hull.*—Planes, various.

6526 KINGSBURY, T. 9, *New Bond-st.*—Dressing-case cutlery.

6527 LINNEKER, R. & J. *Cobnar Works, Sheffield.* — Scythes, sickles, chaff-machine knives, and straw knives.

6528 MAPPIN BROS. 222, *Regent-st.*—Cutlery.

6529 MAPPIN & Co. *Oxford-st.*—Cutlery.

6530 MARSH, BROS. & Co. *Pond's Works, Sheffield.*—Tools and cutlery.

6532 MATTHEWMAN, B. JUN. 80, *Milton-st. Sheffield.*—Scissors.

6533 MECHI & BAZIN, 4, *Leadenhall-st.*—Cutlery.

6534 MITCHELL, J. W. 1, *Bridge-house-pl. Newington-causeway.*—Saws and tools.

6535 MITCHELL, W. H. 3, *Britannia-pl. Limehouse.* — Hand, panel, and tenon saws.

6537 MONK, T. 74, *Edward-st. Birmingham.*—Moulders', plasterers', and stonemasons' tools.

6538 MOSELEY, J. & SON, 54, *Broad-st. Bloomsbury.*—Planes and joiners' tools.

6539 MOSELEY, J. & SON, *King-st. Covent-garden.*—Tools, cutlery, and steel.

6540 MUSHET, R. & Co. *Coleford and Sheffield.*—Cutlery, edge tools, and samples of steel.

6541 NURSE, C. *Mill-st. Maidstone.*—Carpenters' planes.

6542 PARKES, F. & Co. *Sutton Works, Birmingham.*—Cast-steel forks, spades, draining edge, and plantation tools.

6543 PARKIN, J. *Steel Works, Harvest-lane, Sheffield.*—Tools, &c.

6544 PEACE, WARD, & Co. *Agenoria Steel Works, Sheffield.*—Tools and cutlery.

6545 RODGERS, J. & SONS, 6, *Norfolk-st. Sheffield.*—Cutlery.

6546 RUSSELL, T. & Co. *Canada Works,* 38, *Charles-st. Sheffield.*—Saws.

6548 SAYNOR & COOKE, *Paxton Works, Sheffield.*—Pruning, budding knives, scissors, &c.

6550 SHIRLEY, W. *Crescent Works,* 19, *Carver-st. Sheffield.*—Bowie, and other knives, &c.

6551 STEER & WEBSTER, *Castle Hill, Works, Sheffield.* — Scissors, razors, knives, table cutlery, &c.

6552 SUTTON, W. & SONS, *New Town-row, Birmingham.*—Awl blades.

6553 TAYLOR, H. 105, *Fitzwilliam-st. Sheffield.*—Various trades' tools.

6554 THOMAS, R. *Icknield Edge Tool Works, Birmingham.*—Edge tools.

6555 TUTON, M. *Scarboro'-rd. Driffield.*—Stand forks, and hedge tools.

6556 UNWIN & ROGERS, *Rockingham-st. Sheffield.*—General cutlery.

6557 WALDRON, W. & SONS, *Bellbroughton, Stourbridge.*—Scythe, hay, and chaff-knife. Hook, and edge-tools, &c. &c.

6558 WARD, G. 171, *Eyre-st. Sheffield.*—Knives, lancets, &c.

6559 WARD, T. 31, *Brightmore-st. Sheffield.*—Improved pen-knives.

6560 WILKINSON, T. & SON, *Sheffield.*—Cutlery, scissors, tailors' shears, &c.

6561 WILKINSON, W. & SONS, *Spring Works, Grimesthorpe, near Sheffield.*—Various shears.

6562 WINKS, B. & SONS, *Sheffield.*—Razors table-knives, and scalping blades.

6563 WOSTENHOLM, G. & SON, *Washington Works, Sheffield.* — Knives, razors, and scissors.

CLASS 33.

WORKS IN PRECIOUS METALS, AND THEIR IMITATIONS AND JEWELLERY.

South Court, Central Division.

6595 ADAMS, G. W. *Hosier-lane, London.*—Knives, forks, spoons, and various articles.

6596 ADKINS, H. & SONS, 22, *Weaman's-row, Birmingham.*—Electro-plated goods.

6597 ANGELL, J. 10, *Strand.*—Jewellery; gold and silver plate.

6598 ASTON, T. & SON, *Regent-place, Birmingham.*—Jewellery, &c.

6599 ATKIN BROS. *Sheffield and London.*—Electro-plate, Britannia metal, silver and plated cutlery.

6600 ATTENBOROUGH, R. 19, *Piccadilly.*—Silver cups, ebony and silver casket jewellery; watches.

6601 BALLENY, J. 74, *Hatton-garden.*—Jewellery.

6602 BARKER, W. 43, *Paradise-st. Birmingham*—Silver, plated, and metal wares.

6603 BARRY, W. E. *Egyptian Hall, Piccadilly.*—Gilt metal work mountings for artistic productions.

6604 BELL, J. & Co. *Newcastle-on-Tyne.*—Groups in aluminium.

6605 BENSON, J. M. 46, *Cornhill, E.C.*—Argentine and electro-plate.

6606 BIRMINGHAM COMMITTEE, *Birmingham.*—Gold and silver jewellery; gold and silver plated jewellery, chains, &c. Exhibitors:

C. Green.	H. Manton.
G. Hazleton.	J. Russell.
F. A. Harrison.	Shaw & Co.
Hilliard & Thomas.	B. W. Westwood.
J. Lees.	W. Spencer.
R. A. Loach.	

6607 BIDEN, J. 37, *Cheapside.*—Stone and other engravings. Signet rings, &c.

6608 BRAGG, T. & J. *Vittoria-st. Birmingham.*—Gold bracelets, brooches, earrings, pins, studs, and links.

6609 BRYAN, C. *West Cliff, Whitby.*—Set brooches, bracelets, earrings, coronets, &c.

6610 BRYDONE & SONS, 29, *Princes-st. Edinburgh.*—Devices in hair.

6611 COLLIS, G. R. & Co. 130, *Regent-st.*—Silver and electro-plated services.

6612 DERRY & JONES, *Great Hampton-st. Birmingham.*—Plated table service, and general articles in plate.

6613 DIXON, J. 95, *Lillington-st. Pimlico, S.W.*—Bronze medals.

6614 DIXON, J. & SONS, *Sheffield.*—Best Sheffield and electro plate.

6615 DODD, P. G. & SON, 45, *Cornhill.*—Artistic works in the precious metals, &c.

6616 DONNE, W. & SONS, 51, *Cheapside.*—Engraving on precious metals.

6617 DUCLOS, L. D. 19, *Whiskin-st. Clerkenwell.*—Cameos, &c.

6618 DUNCAN, J. 4, *St. Nicholas-st. Aberdeen.*—Granite jewellery, &c.

6619 ELKINGTON & Co. 22, *Regent-st.*—Manufactures in silver, electro-plate, bronze, and enamelled metal.

6620 ELLIS BROS. *Exeter.*—Brooch, and bracelet of Sidmouth pebbles.

6621 EMANUEL, E. & E. 101, *High-st. Portsmouth.*—Works of art, &c. in the precious metals; jewels; horological machinery.

6622 EMANUEL, H. 70, *Brook-st. Hanover-sq.*—Works of art in precious metals, jewels, plate, &c.

6623 FORRER, A. 32, *Baker-st.*—Hair, jewellery, brooches, bracelets, &c.

6624 FRANCIS, W. 13, *Hemingford-rd. Islington.*—Pencils for any sized leads.

6625 GARRARD, R. & S. & Co. 25, *Haymarket.*—Works of art in silver, plate, and jewellery.

6626 GOGGIN, J. 74, *Grafton-st. Dublin.*—Ornamental jewellery in bog oak, &c.

6627 GREEN, C. 48, *Augusta-st. Birmingham.*—Signet rings.

6628 GREEN, R. A. 82, *Strand.*—Jewellery.

6629 HANCOCK, C. F. *Bruton-st. Bond-st.*—Jewellery, and works of art in precious metals, Victoria crosses, &c.

6630 HARRISON, W. W. *Montgomery Works, Fargate, Sheffield.*—Electro-silver plate.

6631 HAZLETON, 45, *Northampton-st. Birmingham.*—Filagree jewellery.

6632 HILLIARD & THOMASON, *Spencer-st. Birmingham.*—Silver fancy goods.

6633 HOWELL, JAMES, & Co. 9, *Regent-st.*—Goldsmith's work, jewellery, &c.

6634 HUNT & ROSKELL, 156, *New Bond-st.*—Artistic works in silver and gold, jewellery, watches, &c.

6635 JAMIESON, G. 107, *Union-st. Aberdeen.*—Granite and pebble ornaments.

6636 JENNER & KNEWSTUB, 33, *St. James-st.*—Gold and silversmiths' work, &c.

6637 JOHNSON, J. 22, *Suffolk-st. Dublin.*—Bog-oak ornaments.

6638 KEITH, J. 41, *Westmorland-pl. City-rd.*—Church plate.

6639 KONINGH, H. DE, 79, *Dean-st. Soho.*—Clock, and specimens of enamelling.

6640 LAMBERT & Co. *Coventry-st. London, W.*—Chased shield, cistern, beakers, chalices, &c.

6641 LA ROCHE, MISS E. 21, *Noel-st. St. James's.*—Pierced ornaments for workers in metal.

6642 LAW, J. 3, *North-side, Bethnal-green.*—Gold and silver leaf, &c.

South Court, Central Division.

6643 LEE, B. 41, *Rathbone-pl.*—Hair jewellery.

6644 LEES, J. 37, *Spencer-st. Birmingham.*—Chains, hooks, swivels, rings, &c.

6645 LISTER, W. & SONS, 12, *Mosley-st. Newcastle-on-Tyne.* — Silver-plate and jewellery.

6646 LOACH, A. M. 5, *Regent-parade, Birmingham.*—Plated gold brooches, bracelets, and lockets.

6647 LOEWENSTARK, A. D. & SON, 1, *Devereux-ct. Strand, W.C.*—Masonic jewels and paraphernalia, medals, and filagree work.

6648 LONDON & RYDER, 17, *New Bond-st.*—Jewellery, diamond work, and silver plate.

6649 MANTON, H. 110, *Gt. Charles-st. Birmingham.*—Fancy silver goods.

6650 MAPPIN BROS. 222, *Regent-st.*—Electro silver plate.

6651 MAPPIN & Co. 78, *Oxford-st.*—Dressing bags and cases, &c.

6652 MARSHALL, W. & Co. 24, *Princess-st. Edinburgh.*—Gold and silver enamelled jewellery in antique style.

6653 MARTIN, HALL, & Co. *Sheffield.*—Silver and electro-plate and silver-plated cutlery.

6654 MUIRHEAD, J. & SON, *Glasgow.*—Silver and electro services, covers, dishes, &c.

6655 NELIS, J. *Omagh, Ireland.*—Pearls found in the river Strule, Omagh.

6656 PARKER & STONE, 7, *Myddelton-st. Clerkenwell.*—Gold chains and jewellery.

6657 PAYNE, E. R. 21, *Old Bond-st. Bath.*—Two vases in silver from the antique.

6658 PHILLIPS, R. 23, *Cockspur-st. London, N.W.*—Works in gold, silver, coral, and precious stones.

6659 PORTLAND Co. 6, *Ridinghouse-st. London.*—Silver and electro-plated goods.

6660 PRIME, T. & SON, *Magneto-Plate Works, Birmingham.*—Services in silver and electro-plate.

6661 READING, J. 82, *Spencer-st. Birmingham.*—Key rings, hooks, swivels, &c.

6662 REID & SONS, *Newcastle-on-Tyne.*—Silver and electro-plated services, bronze, &c.

6663 RETTIE, MIDDLETON & SONS, *Union-st. Aberdeen.*—Granite jewellery, and silver crest brooches.

6664 ROBINSON, H. 61, *Bolsover-st. Euston-rd.*—Inkstands, caskets, vases, &c.

6665 RUSSELL, J. *Warstone-lane, Birmingham.*—Silver and gold jewellery.

6666 SHAW, C. T. *Gt. Hampton-st. Birmingham.*—Gold brooches, bracelets, rings, and chains.

6667 SMITH & NICHOLSON, 12, *Duke-st. Lincoln's-inn-fields.*—Silver and electro-plated articles.

6668 SPENCER, W. 33, *Regent-place, Birmingham.*—Jewellery.

6669 SPURRIER, W. 5, *New Hall-st. Birmingham.*—Electro-silver tea services, entreè dishes, &c.

6670 TATNELL, H. *Salisbury-sq. Fleet-st.*—Electro silver-plated wares.

6671 TENNANT, J. 149, *Strand, London.*—Stones used in jewellery, &c.

6672 THOMAS, 153, *New Bond-st.*—Silver plate.

6674 WESTWOOD, B. W. 20, *Warstone-lane, Birmingham.*—Rings, brooches, pins, &c.

6675 WHEATLEY, J. A. 31, *English-st. Carlisle.*—Cumberland silver and lead; the Cumbrian cup; jewellery, &c.

6676 WIDDOWSON & VEALE, 73, *Strand.*—Silver-plate and jewellery.

6677 WILEY, W. E. *Graham-st. Birmingham.*—Gold pens, pencil cases, &c.

6678 WILKINSON, H. & Co. *Sheffield.*—Silver and electro-plate, cutlery, &c.

6679 WILKINSON, T. & Co. *Birmingham.*—Electro-plate centre piece, &c.

6680 WILLS BROS. 12, *Euston-rd. N.W.*—Works in electro deposit, silver, &c.

6681 MACCARTHY, H. 21, *Lower Grosvenor-st.*—Groups, in electro-silver and bronze.

6682 RESTELL, R. 35, *High-st. Croydon.*— Double-lock jewellery, a new mode of fastening.

6683 THWAITES, J. H. B. *Bristol.*—Specimens illustrative of a new method of cutting precious stones.

6684 HARRISON, F. A. *Birmingham.*—Jewellery.

6685 GOGGIN, C. 13, *Nassau-st. Dublin.*—Bog oak ornaments.

CLASS 34.

GLASS, FOR DECORATIVE AND HOUSEHOLD PURPOSES.

South Court, Central Division.

Sub-Class A.—Stained Glass, and Glass used in Buildings and Decorations.

6710 BAILLIE, T. & Co. 118, *Wardour-st.* —Stained glass windows.
6711 BALLANTINE, J. & SON, 42, *George-st. Edinburgh.*—Stained glass windows.
6713 BARNETT, H. M. *Newcastle-on-Tyne.* —Stained glass medallion and foliage window.
6716 CHANCE, BROS. & Co. *Glass Works, near Birmingham.* — Stained glass window, (East Transept.) Crown, sheet, plate, painted, and optical glass, shades, &c.
6717 CLAUDET & HOUGHTON, 89, *High Holborn.*—Glass shades; stained glass windows, &c.
6718 CLAYTON, J. R. & BELL, A. 311, *Regent-st.*—Stained and painted glass windows.
6719 COX & SON, 28 & 29, *Southampton-st.* Stained glass windows.
6721 FIELD & ALLAN, *Edinburgh & Leith.* —Stained glass window.
6722 FORREST, J. A. 58, *Lime-st. Liverpool.* —Two stained glass windows.
6723 GIBBS, A. 38, *Bedford-sq.*—Stained glass windows.
6724 GIBBS, C. 148, *Marylebone-rd. Regent's-park.*—Stained glass windows.
6725 HARDMAN, J. & Co. 166, *Gt. Charles-st. Birmingham.*—Stained glass windows.
6726 HARTLEY, J. & Co. *Sunderland.*—Specimens of stained glass, in the two windows terminating the nave.
6727 HEATON, BUTLER, & BAYNE, *Cardington-st. Euston-rd.* — Stained glass windows.
6730 HOLLAND, & SON, *St. John's, Warwick.*—Stained glass. (In Transept.)
6731 JAMES, W. H. 37, *High-st. Camden-town.*—Enamelled window glass, &c.
6732 LONG, C. 17, *Queen's-rd. Bayswater.* —Ornamental, embossed, and painted window glass.
6775 MOORE, J. 81, *Fleet-st.*—Ventilators, self-shadowed glass for windows, tablets, &c.
6734 MORRIS, MARSHALL, FAULKNER, & Co. 8, *Red Lion-sq. London.*—Stained glass windows.
6735 O'CONNOR, M. & A. & W. H. 4, *Berners-st. Oxford-st.*—Stained glass windows.
6737 POWELL, J. & SONS, *Glass Works, Whitefriars, E.C.*—Stained glass for windows.
6738 PREEDY, F. 13, *York-pl. Portman-sq.*—Stained glass windows.
6739 PRINCE, A. & Co. 4, *Trafalgar-sq.*—Specimens of illuminated glass.
6740 REES & BAKER, 175, *Goswell-rd. E.C.* —Stained glass windows.
6741 WARD & HUGHES, 67, *Frith-st. Soho-sq.*—A painted glass window for St. Anne's Ch. Westminster.
6742 WARRINGTON, J. P. 43, *Hart-st. Bloomsbury-sq.*—Stained glass windows.
6743 WARRINGTON, W. JUN. 17, *North-umberland-pl. W.*—Specimens of stained glass.
6744 WARRINGTON, W. SEN. 35, *Connaught-ter. W.*—Progressive examples of stained glass, from the twelfth century.

6745 LAVERS & BARRAUD, *Endell-st. W.C.* —Stained glass windows.
6746 HERBERT, MRS. F. 20, *Royal Avenue-ter. Chelsea.*—Paper transparencies.

Sub-Class B.—Glass for Household Use and Fancy Purposes.

6756 AIRE & CALDER GLASS BOTTLE Co. 61, *King William-st.*—Glass bottles.
6757 ALEXANDER, AUSTIN, & POOLE, *Earl-st. Blackfriars.*—Glass bottles, jars, and insulators.
6712 BARKER, S. & Co. *Sloane-st. S.W.*—Ornamental window glass, &c.
6758 BOWRON, BAILY, & Co. *Stockton-on-Tees.*—Glass bottles, glass for architectural and various uses.
6759 BROCKWELL, F. H. 80, *Leather-lane.* —Ink glasses, flasks, mounted bottles, &c.
6760 BROWN, M. L. 47, *St. Martin's-lane.* —Table glass.
6761 CANDLISH, J. 224, *High-st. Wapping.* —Wine bottles.
6762 COPELAND, MR. 160, *New Bond-st.*—English crystal glass.
6763 DEFRIES, J. & SONS, 147, *Houndsditch.* —Chandeliers, lustres, and table glass.
6764 DOBSON & PEARCE, 19, *St. James'-st. S.W.*—Table glass, lustres, gaseliers, and modern Venetian glass.
6765 DOWLING, E. 2, *Little Queen-st. Holborn.*—Glass weights and rosettes.
6720 EDMETT, B. 10, *Long-acre, W.C.*—Illuminated and other writing on glass.
6766 GARRETT, J. *Arundel-pl. Haymarket.* —Drinking goblet.
6767 GREEN, J. 35, *Upper Thames-st.*—Tables glass, chandeliers, and lustres.
6728 HETLEY, H. 13, *Wigmore-st.*—Horticultural and window glass, &c.
6729 HETLEY, J. & Co. 35, *Soho-sq. W.*—Glass shades, stained and other window glass.
6768 HODGETTS, W. J. *Wordsley, near Stourbridge.*—Table and toilet glass.

6769 HOOMAN & MALISKESKI, 490, *Oxford-st. London.*—Photographic portraiture for the interior of glass vases, &c.

6770 JENNINGS, G. 263, *High Holborn.*—Gilding and writing on glass, ornamentings, &c.

6771 KILNER BROS. *Upper Thames-st.*—Glass bottles.

6772 LLOYD & SUMMERFIELD, *Park Glass Works, Birmingham.* — Cut and engraved glass, and glass window bars.

6773 MARCH, T. C. *St. James's-palace.*—Table decorations.

6774 MILLAR, J. & Co. *Edinburgh.*—Engraved glass.

6733 MOORE, E. & Co. *South Shields.*—Pressed flint glass table ware.

6776 MOTT & SONS, *Liverpool.*—Vessels for vivifying draught beer.

6777 NAYLOR & Co. 7, *Princes-st. Cavendish-sq.*—Classical designs, shapes, and engraving.

6778 NORTHUMBERLAND GLASS CO. *Newcastle-on-Tyne.*—Various specimens of British flint glass.

6779 OSLER, F. & C. 45, *Oxford-st. W.*—Pair of colossal candelabra, in crystal glass; table glass, and lapidary cutting.

6780 PEARCE, W. & Co. 9, *Brooke-st. Holborn.*—Toilets and smelling bottles.

6781 PELLATT & Co. 59, *Baker-st. London.*—Table glass and chandeliers.

6782 PHILLIPS, E. *Shelton, Staffordshire.*—Glass gaseliers, chandeliers, candelabra, and table glass.

6783 PHILLIPS, W. P. & G. 359, *Oxford-st.*—Table and dessert pieces.

6785 PRICE, J. 41, *Castle-st. W.C.*—Embossed and burnished gold writing on glass.

6786 READWIN, W. R. 44, *Warwick-st. Pimlico.*—Writing on glass.

6787 ROYSTON, *Jesus-lane, Cambridge.*—Illuminated alphabets, gilded on glass.

6788 RUST, J. & Co. *Lambeth.*—Lamp glasses, globes and perfumers' bottles, soluble glass for soap makers.

6789 SHARPUS, T. & CULLUM, W. 13, *Cockspur-st. Pall Mall.*—Crystal table glass, &c.

6790 SINCLAIR, C. 5, *City-rd. Finsbury-sq.*—Glass chandeliers.

6791 SPIERS & SON, *Oxford.*—Specimens of table glass.

6792 STOREY & SON, 19, *King William-st. City.*—Table and ornamental flint glass.

6793 TOOGOOD, W. 37, *Mount-st. Grosvenor-sq. W.*—Glass bottles, show jars, &c.

6794 WESTWOOD & MOORE, *Brierley Hill.*—Glass bottles.

6795 WHEELER, J. J. 1, *Henry-pl. Chelsea.*—Medical glass, glass surgical instruments, and chemical apparatus.

6796 WOOD, J. H. *Camberwell, S.*—Anglo-Venetian silvered mirrors.

6797 CHANCE, BROS. & Co. *Glass Works, near Birmingham.* — Crown, sheet, plate, painters', and optical glass, shades, &c.

6798 CLAUDET & HOUGHTON, 89, *High Holborn.*—Glass shades.

6799 LAVERS & BARRAUD, *Endell-st. Bloomsbury.*—Painted glafs.

6800 POWELL, J. & SONS, *Whitefriars.*—Glass for all purposes.

CLASS 35.

POTTERY.

North Court, Central Division.

6827 ASHWORTH, G. L. & BROS. *Hanley, Staffordshire.*—Services, &c. in china and earthenware, &c.

6828 ATKINS, T. & SON, 62, *Fleet-st. E.C.*—Hydro-pneumatic glass fountains, with carbon filters.

6829 BATTAM & SON, *Gough-sq. Fleet-st. E.C.*—Ceramic works.

6830 BELL, J. & M. P. & Co. *Glasgow Pottery, Glasgow.* — Porcelain, earthenware, parian and terra cotta.

6831 BEVINGTON, MESSRS. & SON, *Hanley, Staffordshire.*—China, porcelain, parian, &c.

6832 BLANCHARD, M. *Blackfriars-rd. Southwark.*—Articles in terra-cotta.

6833 BLASHFIELD, J. M. *Stamford, Lincolnshire.*—Terra cotta, and other pottery.

6834 BOOTE, MESSRS. *Burslem, Staffordshire.*—Tiles and pottery.

6835 BOURNE, J. & SON, *Macclesfield-st. City Basin.*—Vitrified stoneware, &c.

6836 BROWN, M. L. 47, *St. Martin's-lane.*—Dinner, dessert, and other services.

6837 BROWN, W. *Trent Pottery, Burslem,*—Earthenware dinner, tea, and toilet services.

6838 BROWN-WESTHEAD, MOORE, & Co. *Hanley.*—Parian, china, earthenware, &c.

6839 BROWNFIELD, W. *Cobridge, Staffordshire Potteries.*—Earthenware dinner, dessert, and toilet services, &c.

6840 BULLOCK, C. *Britannia China Works, Longton.*—China tea, breakfast, and dessert sets.

6841 CLIFF, J. & SON, *Wortley, near Leeds.*—Ornamental and useful works in terra-cotta.

6842 COLE, H., C.B. *South Kensington.*—Use of earthenware pressed Mosaics for exterior decoration of buildings.

6843 COMER, R. *Thorpe Hamlet, near Norwich.*—Pottery.

6844 COPELAND, W. T. 160, *New Bond-st.*—Examples of ceramic manufactures.

6845 CRYSTAL PALACE ART UNION.—Presentation pieces in ceramic statuary.

6846 DANIELL & Co. (See Rose & Co.)

6847 DAVENPORT, BANKS, & Co. *Castle Field Pottery, Etruria, Stoke-on-Trent.*—Jet, jasper, antique wares, &c.

6848 DIMMOCK, J. & Co. *Hanley.*--Earthenware.

6849 DOULTON & WATTS, *Lambeth Pottery.*—Glazed stoneware filters, &c.

6850 DUDSON, J. *Hanley.*—Stoneware and china figures.

6851 DUKE, SIR J. & NEPHEWS, *Hill Pottery, Burslem.*—Ceramic productions; Limoges enamels; majolica and Palissy ware, &c.

6852 FELL, T. & Co. *Newcastle-on-Tyne.*—Dinner and chamber ware, lamps, &c.

6853 GOODE, T. & Co. 19, *South Audley-st. W.*—Porcelain dessert service, &c.

6854 GOODWIN, E. & SON, *St. Clement's, Ipswich.*—Clay pipes.

6855 GOSS, W. H. *Stoke-on-Trent.*—Statuettes, vases, pazzi, &c.

6856 GRANGER, G. & Co. *Worcester.*—China, chemical porcelain services, parian, &c.

6857 GRIMSLEY, T. *Oxford.*—Works in terra cotta.

6858 HOLLAND, W. T. *South Wales Pottery, Llanelly.*—Printed earthenware.

6859 HYAMS, M. *Bath-st. London.*—Designs for smoking pipes, &c.

6860 INDERWICK, J. 58, *Prince's-st. Soho.*—Tobacco pipes.

6861 JAMIESON, J. & Co. *Borrowstounness, N.B.*—Fine earthenware.

6862 JENNINGS, G. *Holland-st. Blackfriars-rd.*—Jars, and bottles with patent capsules.

6863 KERR, W. H. & Co. *Royal Porcelain Works, Worcester.*—Porcelain, parian, and stone china.

6864 KEYS & BRIGGS, *Stoke-upon-Trent.*—Parian, majolica, mosaic, jet, and porous ware.

6865 KOSCH, F. *Hanley.*—China and earthenware, printed by new process in gold and colours.

6866 LIDDLE, ELLIOT, & SON, *Dalehall Pottery, Longport, Staffordshire.*—Earthenware services, parian, &c.

6867 LIVESLEY, POWELL, & Co. *Hanley.*—Parian, china, and earthenware.

6868 LOCKETT, J. *Longton.*—China and earthenware of all kinds.

6869 MELI, G. *Stoke-upon-Trent.*—Parian statuettes, vases, ornaments, jugs, butter-tubs, dessert-pieces, &c.

6870 MID-LOTHIAN POTTERY Co. *Portobello, near Edinburgh.*—Bottles, filters, jars, &c.

6871 MILLAR, J. & Co 5, *St. Andrew's-st. Edinburgh.*—Ornamental pottery, &c.

6872 MILLICHAMP, H. *Princes-st. Lambeth.*—Terra-cotta vases, &c.

6873 MINTON & Co. *Stoke-upon-Trent.*—China, earthenware, majolica, parian, tiles, &c.

6874 NORTHEN, W. *Union Pottery, Vauxhall.*—Specimens of earthenware of all kinds.

6875 OLD HALL EARTHENWARE, Co. *Hanley, Staffordshire.*—Various specimens of earthenware.

6876 PELLATT & Co. 59, *Baker-st. Portman-sq. W.*—Specimens of ceramic manufacture.

6877 PHILLIPS, W. P. & G. 359, *Oxford-st. W.*—China dessert service, vases, &c.

6878 POWELL, W. & SONS, *Temple-gate, Pottery, Bristol.*—Various articles in stoneware.

6879 PRICE, C. & J. R. *Thomas-st. Bristol.*—Stoneware vases, filters, &c.

6881 ROBERTS, J. *Upnor, Rochester, Kent.*—Terra-cotta filters, stoves, &c.

6882 ROSE & Co. DANIELL & Co. *New Bond-st. London.*—China manufacturers.

6883 SHARPE BROS. *Swadlincote, Burton-on-Trent.*—Closet basins and wares.

6884 SHARPUS, T., & CULLUM, W. 13, *Cockspur-st. Pall Mall.*—China services, &c.

6885 SHERWIN, H. *Wolstanton, Stoke-on-Trent.*—Dinner plates of parsley-leaf patterns, &c.

6886 SOUTHORN, E. *Broseley, Salop.*—Glazed tobacco pipes.

6887 SOUTHORN, W. & Co. *Broseley, Shropshire.*—Glazed tobacco pipes.

6888 STIFF, J. *London Pottery, High-st. Lambeth.*—Water-filters, chemical apparatus, &c.

6889 STOREY & SON, 19, *King William-st. City.*—China, porcelain, and earthenware.

6890 TEMPLE, E. 184, *Regent-st.*—Ceramic statuary and specimens.

6891 TURNER, BROMLEY, & HASSALL, *Stoke-upon-Trent.*—Parian groups, statuettes, vases, &c.

6892 WATHEN & LICHFIELD, *Fenton, Staffordshire.*—Earthenware dinner, and tea jugs, and toilette sets.

6893 WEDGWOOD, J. & SONS, *Etruria, Staffordshire.*—Various descriptions of wares and vases.

6894 WILKINSON & RICKHUSS, *Hanley, Staffordshire.*—China services, parian vases, jugs, &c.

6895 YNISMEDEW BRICK & PIPE Co. *Pontardawe, Swansea.*—Sewerage pipes, vases, tazzas, chimney pots, and faced bricks.

6896 ROBERTSON, MRS. J.—Service of tartan-plaid pattern.

6897 GROVE, R. H. *Barlaston, near Stone, Staffordshire.*—Lustre ware.

CLASS 36.

TOILET, TRAVELLING, AND MISCELLANEOUS ARTICLES.

Gallery, North Court.

Sub-Class A.—Dressing Cases and Toilet Articles.

6926 ASPREY, C. 166, *New Bond-st.*—Dressing cases, travelling bags, and despatch boxes.

6927 AUSTIN, T. & G. 39, *Westmoreland-st. Dublin.*—Dressing cases, despatch boxes, carriage bags.

6928 BETJEMENN, G. & SONS, 36, *Pentonville-road, London.*—Dressing cases, writing sets, and book slides.

6929 BOURN, E. *College-st. Bristol.*—Russia leather travelling desks, stationery, writing and dressing cases.

6930 GEBHARDT, ROTTMANN, & CO. 24, *Lawrence-lane, Cheapside.*—Dressing cases and bags, writing desks, and photograph albums.

6931 HOWELL, JAMES, & CO. 9, *Regent-st. Pall Mall.*—Dressing cases, travelling bags, &c.

6932 JENNER & KNEWSTUB, 33, *St. James's-st.*—Dressing cases, travelling bags, and despatch boxes.

6934 LEUCHARS, W. 38, *Piccadilly.*—Dressing and writing case, travelling bag, &c.

6935 MAPPIN BROS. 222, *Regent-st.*—Dressing cases and bags.

6936 MAPPIN & CO. 77, *Oxford-st.*—Dressing cases and bags.

6937 MECHI & BAZIN, 4, *Leadenhall-st.*—Travelling dressing bags and cases, despatch boxes, &c.

6938 PARKINS & GOTTO, 25, *Oxford-st.*—Writing and dressing cases, bags, despatch boxes, &c.

6939 TOULMIN & GALE, 7, *New Bond-st.*—Despatch boxes, writing desks, and dressing cases.

6940 WEST, F. 1, *St. James's-st. Pall Mall.*—Dressing and writing cases, travelling bags.

Sub-Class B.—Trunks and Travelling Apparatus.

6951 ALLEN, J. W. 31, *Strand.*—Portmanteaus, trunks, dressing cases, &c.

6952 BARRETT, B. BROS. 184, *Oxford-st.*—Travelling goods.

6954 CAVE, H. JANE, 1, *Edward-st. Portman-sq.*—Waterproof dress and bonnet baskets.

6955 DAY, W. & SON, 353, *Strand.*—Portmanteaus and travelling requisites generally.

6956 FISHER, S. 211, *Strand.*—Dressing cases and portmanteaus.

6957 HARROW & SON, 38, *Old Bond-st.*—Light travelling basket and imperial trunk.

6958 KANE, G. 70, *Dame-st. Dublin.*—Portmanteaus, &c.

6959 LAST, J. 38, *Haymarket.*—Trunks, hat boxes, bags, portmanteaus, and imperials.

6960 LAST, S. 256, *Oxford-st.*—Solid and other portmanteaus; new invented leather bag, &c.

6961 PRATT, H. *Chester-ter. Eaton-sq.*—Travelling wardrobe and compendium portmanteau.

6962 SILVER, S. W. & CO. 4, *Bishopsgate within.*—Portmanteaus, bags, &c.

6963 SOUTHGATE, J. 76, *Watling-st.*—Solid leather portmanteau, travelling trunks and bags.

6964 WATSON, C. J. 162, *Piccadilly.*—Portmanteau, with collapsing fittings for a hat.

6965 WILKS, E. *Cheltenham.*—Portmanteaus, &c.

6966 BROWN, E. 4, *St. Augustine's-parade, Bristol.*—Dressing cases, tourist's cases, portfolios, &c.

COLONIAL POSSESSIONS.

(GENERALLY UNDER THE N.E. TRANSEPT.)

AUSTRALIA, SOUTH.

N.E. Transept, West-side.

1 MACDONNELL, HIS EXCELLENCY SIR R., C.B.—A case of insects; a specimen of malachite.

2 GENERAL COMMITTEE.—Native woods; a collection of stuffed birds; photographs of public buildings in Adelaide.

3 DUTTON, F. J.—Malachite, and other minerals; volumes of debates, votes, and proceedings of the Parliament, and Acts of the Legislature of the Colony.

WINES.

4 EVANS, H.—Shiraz, 1857-8; Riesling, 1857; Espanoir, 1857 & 1860; Muscatel, 1860.

5 GILBERT, J.—Shiraz, 1861; Vordeilho and Riesling, 1852.

6 GREEN, W.—Pineau, 1858.

7 HECTOR, E.—Montura, 1858.

8 AULD, P.—Palomino Blanco, 1859; Verdeilho and Donzelinho, 1860.

9 WARK, DR.—Black Portugal, 1859.

10 KINGSTON, G. S.—Molle negro, 1861, &c.

11 BEASELEY, F.—Black Portugal and Verdeilho, 1860.

12 WYMAN, J.—Shiraz, 1860.

13 HILLS, C.—Verdeilho, 1858.

14 BARNARD, G. H.—Mixed, 1860.

15 RANDALL, D.—Shiraz and carbonet, 1860.

16 DAVIS, A. H.—Moore Farm, 1859.

17 OVERBURY, T.—Muscat, of Alexandria, 1860.

18 IND, G. F.—19 WARNE, J. B.—20 DAVIS, A. H.—Wines.

---

21 CANT, GRIFFIN.—22 MCDOUGALL, J.—23 DUNN, J.—24 HAY, J.—25 BELL, A.—26 BUTTFIELD, J. B.—27 WEDD, W.—28 STEVENS, J.—29 WEHL, E.—30 WADDELL, J.—Wheat.

31 WEHL, DR.—Small-grained wheat.

32 CANT, G.—Barley.

33 WHITE, S.—34 STEVENS, J.—35 MAGAREY, T.—36 DUFFIELD, W.—37 DUNN, J.—38 HARRISON BROS.—39 HART, J. & Co.—40 BEEBY & DUNSTAN.—Flour.

41 MURRAY, A.—Biscuits.

42 CAMPBELL, J.—Muscatel raisins, and soft shell almonds.

43 GRAVES, T.—Dried apricots.

44 DAVIS, F. C.—Jams.

45 SKIPPER, —.—A nautilus shell.

46 NEALES, J. B.—A box of gold specimens.

47 PRIEST, T.—Slabs of slate.

48 ENGLISH & AUSTRALIAN COPPER CO.—Refined copper in cake and tile.

49 S. AUSTRALIAN MINING ASSOCIATION.—Copper ore and other minerals from the Burra Burra and Karkulta mines.

50 WHEAL ELLEN MINING CO.—51 CUMBERLAND MINING CO.—Copper and lead ores.

52 GT. NORTHERN MINING CO.—53 WALLAROO Do.—54 WIRRA WILKA Do.—55 WORTHING Do.—56 PREAMINNA Do.—57 DUREYA Do.—58 KAPUNDA Do.—59 MOUNT ROSE MINE.—Copper ores.

60 CORNWALL MINING CO.—A block of copper ore, weighing 6 tons.

61 ROLLISON, W.—62 ENGLEHART, DR.—63 MAURAU, DR.—Collections of minerals.

64 RODDA, R. V.—Copper ore smelted by his patent process.

65 KELLETT, J.—Polished marble and slate.

66 CRABB, R. S.—Malachite.

67 MELLOR, J.—A reaping machine; model of do.; cart-wheels and felloes; samples of wood; a plank of blue gum timber.

68 BENDA, A.—Native woods.

69 HAIGH, J. F.—Alpaca and angora wool.

70 PEACOCK, W. & SONS.—Wool in fleeces.

71 ANDERSON, A.—72 HAWKER, HON. G.—73 KELLY, P.—74 BOWMAN, BROS.—75 MURRAY, J.—Wool.

76 BURFORD, W. H.—Soap.

77 CHAMBERS, J.—Curiosities brought by J. M. STUART from the centre of Australia.

AUSTRALIA, WESTERN.

Near the Nave, E. Transept.

1 SAMSON, L. *Fremantle.*—Copper and lead ore from Wheal Fortune Mine.

2 SCOTT, D. *Fremantle.*—Ditto.

3 W. A. MINING ASSOCIATION, *Perth.*—Copper ore from Wanernooka Mine.

4 DRUMMOND, J. N. *Champion Bay.*—Copper ore from Gillireah Mine.

5 HORROCKS, J. L. *Champion Bay.*—Copper from Gwalla Mine.

6 SHENTON, G. *Perth.*—Surface copper ore from Wheal Arrino Mine.

7 SHENTON, A. AND CENTRAL COMMITTEE, *Perth.*—Fossils from the Greenough, York, and Kojenup districts.

8 MEADE, THE REV. W. *Albany.*—Black metallic sand.

9 ELLIOT, G. *Bunbury.*—Surface iron ore and black sand from the Bunbury district.

10 SHENTON, A.—Specimens from the Northam district.

11 BARKER, S. A. *Guildford.*—Magnetic iron ore from Cotes.

12 CARSON, J. *Perth.*—Porcelain clay, from

the Darling Range, and crucible made therefrom.

13 HABGOOD, R. M. *London.*—Lead ore from the Geraldine Mine.

14 WALSH, J. *Perth.*—Modelling clay.

15 CENTRAL COMMITTEE, *Perth.*—"Wilgi" clay, with which the Aborigines rub themselves.

16 MILLARD, J. *Toodyay.*—Stone used instead of Turkey-stone.

17 R. ENGINEER DEPARTMENT, *Fremantle.*—A bar reduced from magnetic iron ore.

18 DRUMMOND, J. *Toodyay.*—Mineral, containing asbestos.

19 NEWMAN, E. *Fremantle.*—A pile of Jarrah wood.

20 CARR, J. G. C. *Perth.*—A log of sandal wood, weighing 454 lbs.; fibrous rush.

21 KENWORTHY, J. *York.*—A log of sandal wood.

22 KING, C. *Perth.*—A wheel of native gum wood; planks of Jarrah wood.

23 SMITH, T. *Perth.*—Jarrah wood.

24 JOYCE, W. *Perth.*—Shea oak shingles for roofing.

25 CENTRAL COMMITTEE, *Perth.*—Sections of excrescences from mahogany trees; cabinet-work and turnery; bark of the tea-tree, reducible to pulp, for paper, &c.; wool from York district.

26 R. ENGINEER DEPARTMENT, *Fremantle.*—Mahogany posts, 17 years immersed in water and mud; a cabinet and slab.

27 CARSON, J. *Perth.*—A cask made of casuarina.

28 CLIFTON, W. P. *Leschenault.*—Logs, a slab, and parts of wheels.

29 JOHNSON, H. *Perth.*—An inlaid dumb waiter.

30 DURLACHER, A. *Newcastle.*—A sandal wood pedestal.

31 SLOANE, W. *Perth.*—White and red gum, for shipwrights' work.

32 LOCAL COMMITTEE, *York.*—Native woods, and parts of wheels.

33 WHITFIELD, G. *Toodyay.*—Varieties of eucalyptus wood.

34 RANFORD, B. B. *Perth.*—Colonial leathers: sole leather, kip, calf, kangaroo, &c.; barks used for tanning.

35 HOMFRAY, R. *Perth.*—Native spears, shields, kylies, head-dresses, nose bones, hair girdles, opossum wool, &c.

36 YELVERTON & CO. *Vasse.*—A collection of timbers.

37 SAMSON, L.—A beam of Jarrah wood.

38 PADBURY, W. *Perth.*—Silver wattle bark.

39 MCKNOE, MRS.—Feather flowers.

40 O'NEIL, MISS.—Feather screens.

41 LUKIN & KNIGHT, MISSES, *Perth.*—A muff of parrot feathers.

42 BURGES, S. *Tipperary.*—Skins of native animals.

43 FARMANER, F. *Perth.*—Emu skin.

43A GREGORY, F. T.—Pearls and pearl oysters from Nicol Bay.

44 HILLMAN, A. *Perth.*—Native silk.

45 FLEAY, J. *Beverley.*—Wool.

46 CLIFTON, G. *Fremantle.*—A collection of shells, pearls; a fishing net, rope, &c. from Nichol Bay.

47 JACKSON & CO. *Perth.*—An inlaid work-table.

48 LEAK, MRS. G. M. *London.*—An inlaid chess table.

49 DU CANE, MRS. F. F. *London.*—Pressed flowers from Swan River.

50 DU CANE, THE REV. A.—Hortus siccus of flowers, from Swan River.

51 LITTLE, T. *Dardanup.*—Frontignan wine, of 1860 and 1861.

52 MACGUIRE, J. *Dardanup.*—Frontignan wine of 1861, from the Wellington district.

53 CENTRAL COMMITTEE.—Wine of 1860, from Pyrton.

54 CLIFTON, W. P. *Leschenault.*—Pedro Ximenes, white Frontignan, and black St. Peter's wines; olive oil; wheat reported to yield 27 to 28 bushels per acre, each weighing nearly 70 lb.

55 CLIFTON, MRS.—Australian Madeira, of 1860.

56 BARLEO, F. P. *Crawley.*—Black cluster grape wine.

57 JECKS, T. *Guildford.*—Olive oil.

58 PARKER, S. S. *York.*—3 bush. of wheat.

59 BAIN, J. *Fremantle.*—Preserved meat and fish.

60 HARDEY, J. *Perth.*—Preserved fruits.

61 CLIFTON, G.—Jelly seaweed, "euchemia speciosa," for culinary purposes.

62 CARR, J. G. C.—Gum from the manna eucalyptus, and blackboy, or xanthorhœa.

63 BARKER, S. A. *Guildford.*—Gum from the eucalyptus resinifera.

64 DRUMMOND, J. *Toodyay.*—Gum from the "boro blackboy" tree; and wheat.

65 SHENTON, A. *Perth.*—Native hops.

65A SANFORD, H. A. *London.*—Furs and dressed skins; native weapons, from the Murchison, taken in action; baskets of blackboy, or xanthorhœa; emu eggs; minerals from Victoria district.

66 BOSTOCK, C. *Freemantle.*—Wheat in ear.

67 SHENTON, G. *Perth.*—Wheat in ear.

68 MUIR & SONS, *Forest Hill.*—Wheat, 66 lb. the bushel.

BAHAMAS.

Northern Courts, West-side of N.E. Transept.

1 LOCAL COMMISSIONERS.—Native woods, viz.: Yellow wood, green ebony, iron wood, naked wood, Braziletto, cedar, Madura, and horseflesh mahogany.

Fibre of the pita plant, pine-apple leaf and forest pine, and specimens of cordage; indigenous cotton and seed.

Coarse grain and blown salt, produced by solar evaporation.

Arrow-root and starch.

Shell-work, palmetto hats, fans, and walking sticks; baskets and ornaments from the Jumbu Bean.

2 GEORGE, J. S., M.C.P.—Cascarilla bark, turtle shells, conch pearls, ambergris, &c.

3 HARRIS, THE HON. G. D., M.C.P.—Several kinds of sponges, and a collection of shells.

4 General Apothecaries' Co.—Cascarilla bark, canella alba, surgical sponges.
5 Harris, Samuel. — Conch pearls, sponges, shells, and shell-work.

BARBADOS.

Eastern-side of N.E. Transept.

S. Cave, M.P. (*Commissioner for the Colony*).
1 Sugar made by common process on Drax Hall, the first estate on which sugar was grown in Barbados.
2 Sugar from Hannay's Estate, made by the oscillating process—a new method of stirring the sugar after it has been poured from the copper into the cooler. The sugar crystallizes in larger grains, and sells for nearly 2*s*. per cwt. more than sugar made by the old process.
3 Sugar from Maxwell's Estate, made by same process.
4—15 Various specimens of sugar, made by the ordinary process.
16 Two samples of rum from Sunbury and Hampton Estates.
17 Vine cotton (Gossypium vitifolium).
18 Arrowroot.
19 Starch made from the sweet potato (Batatas edulis).
20 Indigo.
21 Barbados tar (Petroleum), with two samples of lubricating oil made from it; these last sent by L. R. Valpy, Esq.
22 Fibre of the ochro plant.
23 Silk cotton (Eriodendron anfractuosum).
24 The flower of the sugar cane arrow.
25 Basket of fruit, &c. of the island in wax.
26 Vase of flowers of the island, made of feathers.
27 Box of flowers, made of shells.
28 Two gourds of aloes, one split.
29 Guinea corn (Sorghum vulgare).
30 Simmons, Mrs. J. A.—Flowers of the island, in wax.
31 ———— Another case of the same.
32 ———— Two cases of fruits of the island, in wax.
33 ———— The flower of the night blowing cereus (Cereus grandiflorus) in wax.
34 St. John, Mrs.—Case of shells, moss, and seaweed, arranged in vase.
35 Chambers, G. H.—Specimen of white coral, which forms the basis of the island of Barbados.

BERMUDA.

Eastern-side of N.E. Transept.

Tucker, W. C. Fahie, Commissioner for Bermuda.—Cedar furniture, &c.; specimens of woods; samples of work in palmetto, straw, grasses, and flowers; cotton; fibre; preserves; lime-juice; seeds; pepper; honey; beeswax; tannic acid; starch; models; pumice and brain stones; petrefactions; sponges; marine specimens; coral; turtle-shell; limestone; lime, &c.

BORNEO.

Northern Courts, near the Horticultural Gardens Entrance.

1 St. John, Spencer.—Native arms, mats, block of antimony ore, antimony paint, coal, native cloths, vegetable tallow, sago, dammar, rubber.
2 Grant, C. T. C.—Gold and silver ornaments of native manufacture.

BRITISH COLUMBIA.

Centre of N.E. Transept.

1 Executive Committee at New Westminster.—Model of a log hut.
2 ———— Shingles from Cedar Wood.
3 ———— Pales of ditto.
4 ———— Fence post of ditto.
5 ———— Stakes of ditto.
6 ———— Horizontal section of Douglas fir.
7 ———— A rent slab of ditto.
8 ———— Log of yew.
9 ———— Ditto dogwood.
10 ———— Ditto cherry tree.
11 ———— Specimens of birds stuffed.
11a ———— Indian curiosities.
12 ———— Sample of wheat.
13 ———— Ditto of peas.
14 ———— Ditto of hops.
15 ———— Pickled fish.
16 ———— Drawing of a tree of Douglas fir, 309 feet high.
17 ———— Ten horizontal sections of Douglas fir.
18 Dr. J. W. Gilbert, Esq. F.R.S.—A walking-stick of curled maple, with engraved head of Cariboo gold.
19 Bank of British North America.—A case of specimens of native gold.
20 Hydrographer's Office, Admiralty.—Map of the south coast, and of Vancouver.
21 Commander Mayne, R.N.—Photographs of New Westminster.

CANADA.

Centre and Eastern-side of N.E. Transept.

1 Larue, A. & Co. *Three Rivers*.—Bog iron ore, with samples of wrought and cast iron obtained from the same at Radnor Forges.
2 Seymour, G. *Madoc*.—Magnetic iron ore.
3 Cowan, A. *Kingston*.—Magnetic iron ore; phosphate of lime; mica; plumbago; friable sandstone; feldspar.
4 Orton, J. *Hastings Road*.—Magnetic iron ore.
5 Closter, C. C. *Gaspé Basin*.—Lead ore from Indian Cove, Gaspé.
6 Wright, J. & Co. *Montreal*.—Lead ore from Upton.
7 Foley & Co. *Montreal*.—Lead ore and pig lead from Ramsay Mine, with a plan of the mine.
8 Montreal Mining Company, *Montreal*.—Copper ore, undressed and dressed, from

Bruce Mine, Lake Huron, with plans of the mine; copper from Mamainse, Lake Superior.

9 WEST CANADA MINING COMPANY, *Wellington Mine, Lake Huron.*—Copper ore, undressed and dressed, from Wellington Mine, Lake Huron, with a plan of the mine.

10 DAVIES, W. H. A., & C. DUNKIN, *Montreal.*—Copper ore, undressed and dressed, from Acton Mine, Actonvale, with plans of the mine.

11 MOORE, G. B. & Co. *Montreal.*—Copper ore from Upton Mine.

12 POMROY, ADAMS, & Co. *Sherbrooke.*—Copper ore from Wickham & Yale's Durham Mines, with plans of the mines.

13 SHAW, BIGNOL, & HUNT. *Quebec.*—Copper ore from Black River Mine, St. Flavien.

14 ENGLISH AND CANADIAN MINING COMPANY, *Quebec.*—Copper ore, undressed and dressed, from Harvey's Hill Mine, with plans of the mine.

15 FLOWERS, MACKIE, & Co. *Montreal.*—Copper ore from St. Francis Mine, Cleveland, and Coldspring Mine, Melbourne, with plans of the mines.

16 GRIFFITH & BROTHERS, *Cleveland.*—Copper ore from Jackson's Mine.

17 SWEET, S. & Co. *North Sutton.*—Copper ore from Sweet's Mine, N. Sutton, with a plan of the mine.

18 ROBERTSON, G. D. & Co. *St. Hyacinthe.*—Copper ore from Craig's Range Mine, Chester.

19 MCCAW, T. *Montreal.*—Copper ore from Ascot or Haskell Hill Mine, Ascot.

20 FLETCHER, R. H. *Bruce Mines.*—Smelted copper.

21 WALTON, B. *Montreal.*—Roofing slates, serpentine, chromic iron.

22 BROWN, A. S. *Brockville.*—Cobaltiferous pyrites from Elizabethtown, and phosphate of lime from North Elmsley.

23 CANADIAN OIL COMPANY, *Hamilton.*—Crude and refined rock oil from Enniskillen.

24 WATKINS & INGLIS, *Hamilton.*—Crude and refined rock oil.

25 RUSSELL & Co. *Kingston.*—Plumbago from Pointe du Chêne Graphite Mine, county of Argenteuil.

26 FINLAY, M. *Quebec.*—Fire clay.

27 CHEESEMAN, C. R. *Phillipsburg.*—Building stones and marble.

28 O'DONNELL, H., C.E. *Quebec.*—Building stone (gneiss) and sewerage pipe tile.

29 BROWN, T. *Thorold.*—Hydraulic cement, crude and prepared.

30 PEEL & COMPTE, *Montreal.*—Bricks.

31 BULMER & SHEPPARD, *Montreal.*—Bricks.

32 TREADWELL, P. C. *L'Original.*—Drain tiles.

33 MISSISQUOI DRAIN-TILE COMPANY. *Missisquoi.*—Drain tiles.

34 MARTINDALE, T. *Oneida.*—Crude and prepared plaster (gypsum).

35 DONALDSON, J. *Oneida.*—Crude and prepared plaster (gypsum).

36 TAYLOR, A., *York, Grand River.*—Crude and prepared plaster (gypsum), with a plan of the mine.

37 CARON, E. *St. Anne de Montmorenci.*—Iron ochre.

38 GIBB, T. *Toronto.*—White bricks and drain tiles.

39 BELL, R., M.P. *Ottawa.*—Building stones used in the construction of the Parliament House, Ottawa.

40 KNOWLES, W. *Arnprior.*—Marble from Arnprior.

41 GEOLOGICAL SURVEY OF CANADA, *Montreal.*—Collection of ores of iron, lead, copper, nickel, silver, gold, platinum, indormine, chromic iron, molybdenite, dolomite magnesite, bituminous shale, soapstone, potstone, mica rock, mica, plumbago, asbestus, fire clay, building stones, marbles, serpentines, slates, bricks, flagstones, hydraulic limestone, common limestones, whetstones, grindstones, millstone, buhrstone, freshwater shell marl, iron ochres, sulphate of barytes, lithographic stones, agates, albite, orthoclase, jasper conglomerate, epidosite sandstone for glass making, moulding sand, peat, a collection of the crystalline rocks, with a geological map of Canada.

42 PROVANCHER, ABBÉ, *St. Joachim.*—Specimens of woods, branches, leaves, and flowers.

43 PRIEUR, F. X.—Specimens wood, obtained in County St. John, southern extremity Lower Canada.

44 LEPAGE, J. B.—Specimens woods, obtained at Rimouski.

45 DUBORD, DR.—Specimens of woods, collected in County St. Maurice.

46 COULTÉE, L. M.—Specimens of woods, obtained in the county of Ottawa.

47 PRICE, D.—Specimens of woods, collected in the County of Chicoutimi.

48 PATTON, DUNCAN, & Co.—Collection of commercial woods.

49 GINGRAS, G. *Quebec.*—Nine pieces of sawn wood.

50 GIROUX, O. *Quebec.*—Vegetable extracts.

51 TURGEON & OUELLET, *Quebec.*—Preserved fish.

52 TÊTU, C. H. *River Ouelle.*—Skins of white porpoise and seal; oils of white porpoise, shark, and cod liver.

53 COTÉ, O. *Quebec.*—Collection of furs.

54 VAN ALLEN, D. R. *Chatham.*—Sections of trees and planks.

55 SHARP, S., of the G. W. R. *Hamilton.*—Sections of trees, planks, and specimens polished woods.

56 SKEAD, J. *Ottawa.*—Sections of trees and foliage; and planks.

57 LAURIE, J. *Markham.*—Sections of trees and planks.

58 DICKSON, A. *Pakenham.*—Specimens polished woods.

59 MCKEE, H. *Norwich.*—Specimens of shrubs, twigs, and leaves.

60 CHOATE, J. *Ingersoll.*—Planks.

61 MCCRACKEN, A. *London.*—Planks.

62 MCKELLAR, A. *Chatham.*—Sections trees.

63 BURROWS, *Simcoe.*—Sections trees.

63½ TREMBICKE, A. L. Engineer, *G. T. Railway.*—Sections of trees.

64 BRONSON, A. *Bayham.*—Plank.

65 FRUIT GROWERS' ASSOCIATION, U.C.

—Coloured plates of fruit grown in Upper Canada (open air).

66 PASSMORE, S. W. *Toronto.*—Ducks birds, fishes.

67 THOMPSON, J. *Montreal.*—Case of 103 birds.

68 CROOKS, MISS K. *Hamilton.*—Native plants, flowers, leaves of trees, &c. from vicinity of Hamilton.

69 FLEMING, J. *Toronto, U.C.*—Wheat, oats, barley; seeds of tare, millet, peas, turnips, flax, red onion, &c.

70 THE AGRICULTURAL SOCIETY OF BEAUHARNOIS, LOWER CANADA.—Barley, oats, peas, rye, wheat, flax, and grass seed.

71 THE AGRICULTURAL SOCIETY OF HUNTINGDON, L.C.—Samples of barley, Indian corn or maize, oats, peas, wheat.

72 BOA, W. *of St. Laurent, Island of Montreal, L.C.*—Barley, beans, Indian corn, meal, oats, peas, wheat, buckwheat, and oat straw.

73 BEAUDRY, P. *St. Damase, L.C.*—Barley and wheat.

74 LOGAN, J. *Petite Côte, Island of Montreal, L.C.*—Barley, beans, butter, maize or Indian corn, oats, and wheat.

75 MALO, P. *St. Damase, L.C.*—Barley, wheat.

76 MCKINNON, D. *Somerset, Megantic, L.C.*—Samples of barley and wheat.

77 ROCHELEAU, A. *St. Bruno, L.C.*—Barley, flax.

78 WILKINS, C. *Rougemont, L.C.*—Barley, flax, Indian corn, maple sugar.

79 EVANS, W. *Montreal, L.C.*—Beans, Indian corn, peas, Timothy seed.

80 BROWN, D. *Nelsonville, L.C.*—Cheese, maple sugar.

81 LYMANS, CLARE, & CO. *Montreal, L.C.*—Clover seed, Timothy seed, flax seed.

82 PEEL (COUNTY) AGRICULTURAL SOCIETY, U.C.—Barley, wheat, peas.

83 DENISON, R. L. *Toronto, U.C.*—Indian corn stalks.

84 SHAW, A. *Toronto, U.C.*—Indian corn, rye, peas.

85 MARTIN, P. DIT LADOUCEUR, *St. Laurent, L.C.*—Indian corn.

86 DAWES & SONS, *Lachine, Island of Montreal, L.C.*—Hops.

87 MCKEE, H. *Norwich, U.C.*—Honey.

88 THE AGRICULTURAL SOCIETY OF WENTWORTH AND HAMILTON, U.C.—Oats, wheat.

89 BADHAM, T. *Drummondville, L.C.*—Oats.

90 MATTHIEU, H. *St. Hyacinthe, L.C.*—Oats.

91 CUMMING, H. *Megantic, L.C.*—Peas.

92 L'HEUREUX, REV. F. *Contrecœur, L.C.*—Maple sugar.

93 ALIX, J. B. *St. Cesaire, L.C.*—Maple sugar.

94 SHARON, H. *Southwick, U.C.*—Maple sugar.

95 THE AGRICULTURAL BOARD OF UPPER CANADA.—Samples of wheat.

96 ROBERTSON J. *Nepean, U.C.*—Wheat.

97 BEARDMAN, J. *Nepean, U.C.*—Wheat.

98 BRUNELLE, L. *St. Hyacinthe, L.C.*—Buck wheat.

99 DRUMMOND, J. *Petite Côte, Island of Montreal, L.C.*—Wheat.

100 LAMONDE, J. *St. Damase, L.C.*—Wheat.

101 STEWART, D. *Inverness, L.C.*—Wheat.

102 PILOTE, REV. F. *St. Anne's College, L.C.*—Wheat.

103 EAST BRANT AGRICULTURAL SOCIETY.—Wheat.

104 BLAIKIE & ALEXANDER, *Toronto, U.C.*—One sample of flax, and four samples of straw.

105 MCNAUGHTON, E. A. *Newcastle, U.C.*—Prepared arrowroot, flax seed, &c.

106 REINHARDT, G. *Montreal, L.C.*—Smoked meats and sausages.

107 PIGEON, N. *Montreal, L.C.*—Wine from native Canadian grapes.

108 PAULET, MDME. *Montreal, L.C.*—Native wine.

109 CRAWFORD, D. *Toronto, U.C.*—Canadian mustard.

110 LAIDLEY & TORREY, *Toronto, U.C.*—Box of wool.

111 LYMAN & CO. *Montreal, L.C.*—Arctusine; Canadian yellow wax.

112 WHEELER, J. *Montreal, L.C.*—Toilet soap.

113 LARUE & CO. *Three Rivers, L.C.*—Railway wheels from Radnor forges, St. Maurice.

114 LOWE, J. *Grand Trunk Railway, L.C.*—Model of direct action, self-balanced, oscillating cylinder for steam-engine, and a steam gauge.

115 MARTIN, J. *Toronto, U.C.*—Model of steam superheater for locomotive.

116 SHARP, S., *G. W. Railway of Canada, Hamilton, U.C.*—Models of sleeping and freight cars.

117 LEDUC, C. *Montreal, L.C.*—A four-wheeled open carriage.

118 BAWDEN, W. *Hochelaga, Montreal, L.C.*—A brick and tile making machine, with model of pug mill.

119 RICHARD, E. O. *Quebec, L.C.*—Model of an improved water wheel.

120 MOORE, T. *Etobicoke.*—Wooden handles for tools, &c.

121 TONGUE & CO. *Ottawa, U.C.*—A large collection of tools.

122 WASHBURN, S. *Ottawa, U.C.*—Axes.

123 JEFFERY, J. *Côte des Neiges, Montreal, L.C.*—An iron plough.

124 PATERSON, J. *Montreal, L.C.*—An iron swing plough.

125 COLLARD, H. *Gananoque, U.C.*—A cultivator, with wheels.

126 COMER, L. *Hinchinbrooke, Frontenac, U.C.*—Model of an improved beehive.

127 GASKIN, CAPT. R. *Kingston, U.C.*—Agricultural implements, &c.

128 MCSHERRY, J. *St. David's, U.C.*—An iron plough.

129 MORLEY, J. *Thorold, U.C.*—An iron swing plough.

130 MYERS & SON, *Toronto, U.C.*—A patent churn.

131 WHITING & CO. *Oshawa, U.C.*—A collection of agricultural implements.

132 MAYNARD, REV. MR. *Toronto, U.C.*—Model of fish-tail submarine propeller.

133 OATS, R. H. *Toronto, U.C.*—Model of patent instantaneous reefers.

134 KING, T. D. *Montreal, L.C.*—Diagram of the mean diurnal changes of temperature of air and water of the St. Lawrence, Montreal.

135 THOMPSON, J. E. *Toronto, U.C.*—Heating and other apparatus.

136 NOTMAN, W. *Montreal, L.C.* — Two portfolios, and a collection in frames of photographs.

137 BONALD, G. S. D. *McGill University, Montreal, L.C.*—An apparatus for detecting consumption, &c.

138 PALMER, DR. H. *London, U.C.*—A medical magnetic instrument.

139 DUNPHY, MRS. P. *St. Malachi, L.C.*—Woollen yarn.

140 STEPHEN & CO. *Montreal, L.C.*—Woollen cloths.

141 FAIRBANK, E. *Clifton, U.C.*—Screens, mats, plumes, &c.

142 THOMPSON, T. *Toronto, U.C.* — A Shaftoe saddle.

143 ANGUS & LOGAN, *Montreal, L.C.*—Paper.

144 COUNCIL OF PUBLIC EDUCATION IN LOWER CANADA: Superintendent—Hon P. J. O. CHAUVEAU.—A large collection of educational works.

145 ALLEN, W. *Montreal.*—School furniture.

146 EDWARDS, J. *Toronto.*—Specimens of penmanship.

147 NELSON & WOOD, *Montreal, L.C.*—Brooms and brushes.

148 EDDY, E. B. *Ottawa, U.C.*—Tubs, pails, washboards.

149 MCILROY, T. *Brampton, Peel County, U.C.*—A walnut invalid bedstead.

150 SNELL, W. H. *Victoria Iron Works, Montreal, L.C.*—Nail plates, &c.

151 BULLOCK, W. *Toronto, U.C.*—Stained glass.

152 MILLS, J. *Hamilton, U.C.*—Ornamental tiles.

153 GIBB, T. *Toronto, U.C.*—Drain tiles.

154 MISSISQUOI DRAIN TILE COMPANY, *L.C.*—Drain tiles.

155 PALSGRAVE, C. T. *Montreal.*—Type and impressions.

156 HENRY, P. *Montreal, L.C.*—Cigars, &c.

157 BRIDGE, A. *Westbrook, Kingston, U.C.*—Small fancy tub of Canadian woods.

158 HAYCOCK & CO. *Ottawa, U.C.* — Canadian walnut box, containing specimens of building stones used for New Parliament Houses, Canada.

159 LEWIS, C. *Ingersoll, U.C.*—A fancy keg of Canadian woods.

160 ROBERTSON, G. *Kingston, U.C.*—A case of blacking.

161 BENSON & ASPDEN, *Edwardsburgh, U.C.*—Samples of Indian corn starch.

162 MCNAUGHTON, E. A. *Newcastle, Durham County, U.C.*—Flour and potato starch.

163 HOPKINS, J. W. Architect, *Montreal, L.C.*—Architectural drawings.

164 LAWFORD & NELSON, *Montreal, L.C.*—Interior view of a building for skating during the winter.

165 HOPKINS, LAWFORD, & NELSON, *Montreal, L.C.* — Photograph of building erected by them.

166 JACOBI, O. R. *Montreal, L.C.*—Views in oil of local scenery.

167 WESTMACOTT, S. *Toronto, U.C.*—Two landscapes—Canadian scenery.

168 WHALE, R. *Burford, U.C.*—Landscapes, &c.

169 RODDEN, W. *Montreal, L.C.*—Plantagenet water.

170 SOVEREIGN, L. L. *Simcoe, C.W.*—A combined plough, drill, and harrow. A garden drill.

171 THOMPSON, MISS. — Wreath of Canadian autumn leaves.

172 HODGES, MESSRS. — Pictures of Canadian scenery and life, by Kreighoff.

172 ARMSTRONG, D. *Owen Sound.* — Spring wheat.

173 BENNING, D. *Beauharnois.*—English oats.

174 AGRICULTURAL SOCIETY, *Beauharnois.*—Flax seed.

175 BRODIE, J. *Beauharnois.*—Late peas.

175 CAREY, J. *Flamboro West, U.C.*—Spring wheat.

176 CARROLL, AMBRIDGE, & CO. *Hamilton, U.C.*—Crude and refined kerosene.

177 DRUMMOND, J. *Petite Côte, L.C.*—Spring wheat.

178 FILIATREAU, J. B. *Beauharnois, L.C.*—Late rye.

179 GALBRAITH, J. *Beauharnois.* — Canadian barley.

180 GENDRON, J. *Beauharnois.* — Early peas.

181 GERIE, A. *Ancaster, U.C.* — Soule's wheat.

182 MALO, G. *St. Damase, L.C.*—Black Sea wheat.

183 CANADA OIL WORKS, *Hamilton.* — Rock oil.

184 M'DONALD, *Beauharnois, L.C.* — Wheat.

185 M'GAW, T. *East Whitby, U.C.* — Spring wheat.

186 MUIR, J. *Huntingdon, L.C.* — Oats, 80 bush. to acre.

187 PERCIL, J. *Huntingdon, L.C.*—Peas.

188 ROSE, E. H. *Chatham, U.C.*—Walnut veneers.

189 SAUNDERS, W. *London, U.C.*—Medicinal herbs and fluids; perfumery.

190 SCHUYLER, S. *Huntingdon, L.C.* — Indian corn.

191 TAIT, C. *Beauharnois, L.C.*—Black Sea wheat.

192 THOMPSON, D. *Beauharnois.*—Two-row barley.

193 TREMBICKI, A. L. *Montreal.*—Sections of wood.

194 WILSON, J. *Wellington, U.C.*—Barrel oatmeal.

---

195 BILLINGS, E, *Montreal.*—Figures and descriptions of Canadian organic remains.

196 HUNT, T. STERRY, *Montreal.* — A collection of the crystalline rocks of Canada.

CAPE OF GOOD HOPE.

Northern Courts, near West-side of N.E. Transept.

GHISLIN, T. G. — Novel applications of South African algae, as a substitute for horn, as handles in cutlery, whips, &c. A collection of African vegetable fibres, adapted to textile fabrics, brushes, rope, paper, &c. Various kinds of wood. Curiosities, natural and artificial; and specimens of aboriginal industry.

BOWLER, J. W. *Cape Town.* — View of Cape Town from Blue Berg (water-colour).

———View of Keeskamma river, British Kaffraria.

CEYLON.

Under Gallery, East-side of N.E. Transept.

1 CEYLON COMMITTEE :—

Sponge (*N. Prov.*); oyster shells (*E. Prov.*); chenaub (*W. Prov.*).

Oils of gingelley, margosa, elleppey, and dugong (*N. Prov.*); cocoa-nut oil (*N. S. W. & N.W. Provs.*).

Medicinal substances (*N. Prov.*).

Green, black, and gingelley, gram, rice, honey, vinegar, fruits, and spices (*N. Prov.*); rice, &c. (*W. & Cent. Provs.*); cocoa-nuts (*E. W. Provs.*); palmyra flour (*E. & N. Provs*).

Sunn fibres, and a variety of gums (*N. Prov.*); fibres (*Cent. Prov.*); Tagoolacoody sap and root (*E. Prov.*); hemp (*E. Prov.*); cotton (*N. E. & Cent. Provs.*).

Stuffing moss lichen (*Maritime Provs.*).

Tobacco (*N. & N.W. Provs.*).

Bees' wax, vegetable wax, and birds' nests (*N. Prov.*).

Dye stuffs (*N. S. E. W. N.W. Central & Maritime Provs.*).

Tanning substances (*N. E. W. N.W. & Maritime Provs.*).

Monkey skins (*N. & E. Provs.*); sheep, goat, and bullock skins (*W. N. & E. Provs.*); cheetah and tiger skins (*N. Prov.*).

A musical instrument called "Nagoola" (*E. Prov.*).

Lace and gold embroidery (*W. Prov.*).

Carved cocoa-nut shells; manufactured articles in wood (*W. Prov.*).

Articles made of brass (*E. Prov.*).

Mats, baskets, &c. of date and screw pine leaves, pooswell and coir mats (*W. Prov.*); coir mats, native figures, cigars and cheroots (*N. Prov.*); talipot leaves, a mat and tent of talipot (*Cent. Prov.*); towels, napkins, and a table cloth (*E. Prov.*).

Rope and cordage made of a variety of substances (*N. & Cent. Prov.*); deerskin rope (*E. Prov.*).

A bullock-pack, a casting-net, a peacock-feather fan, a pingo, a harpoon, bows and arrows, yokes, a scoop, a pitchfork, and ploughs (*E. Prov.*).

A dalada shrine (*W. Prov.*).

2 RATTEMAHATMEYA OF LOWER DOOMBERA.—Iron, rice, kittool fibres, Singhalese axes, knives.

3 RATTEMAHATMEYA OF UPPER DOOMBERA.—Iron, rice, and cotton.

4 RATTEMAHATMEYA OF UDUNUWERA.—Rice, kittool fibres, cotton, Singhalese axes, bill-hooks, walking sticks.

5 RATTEMAHATMEYA OF YATINUWERA.—Rice, Singhalese axes, cocoa-nut fibre ropes, a mammoty, bill-hooks, walking-sticks, Kandyan whips.

6 RATTEMAHATMEYA OF LOWER HEWAHETTE.—Rice, bill-hooks, ploughs.

7 POWER, T. C.—Plumbago and iron, cotton, areca-nut cutters, Singhalese axes, knives, a sickle, a mammoty, and a tavelum pack, a cotton-cleaning machine (*W. Prov.*).

8 STEELE, T.—Plumbago, red and white coral (*S. Prov.*); kittool fibre (*N. S. & Cent. Provs.*).

9 ILLANGAKOON, J. D. S. *Modliar.*—Plumbago, gums (*S. Prov.*).

10 BRODIE, A. O.—Iron (*Cent. Prov.*); neganda fibre (*N. & Cent. Prov.*); neganda mats (*Cent. Prov.*).

11 WIJESINHE, J. D. S. *Modliar.*—Iron, painted boards, Singhalese axes, knives (*S. Prov.*).

12 ISMAIL, C. L. M. *Lebbe, Marikan.*—Oils of citron and lemon grass (*S. Prov.*); citronella (*N. Prov.*).

13 BREARD.—Cinnamon-oil (*W. Prov.*).

14 COREA, D. C. *Modliar.*—Oils of cinnamon and camphor (*S. Prov.*); kittool fibre (*N. S. & Cent. Provs*).

15 OBEYESEKERE, T. F. S. *Modliar.*—A large collection of oils and medicinal substances (*S. Prov.*).

16 PIERIS, T. A.—Paddy and rice, fibres, neganda mats, ropes of various substances, two inlaid daggers (*N. W. Prov.*); Cadjie gum (*N.W. & W. Provs.*).

17 KARUMARATINE, A. R. *Modliar.* — Paddy and rice, spices (*S. Prov.*).

18 SMITH, D.—Cinnamon (*W. Prov.*).

19 DANIEL, J. B.—Coffee (*W. Prov.*).

20 WRIGHT, W. H. — Fruits, &c. (*Cent. Prov.*)

21 SHAND & Co.—Coir fibre (*W. Prov.*).

22 THWAITES, G. H. U.—Vanilla (*Cent. Prov.*).

23 TISSA, SIRISUMANA. — Jackwood, for dyeing (*S. Prov.*).

24 FORBES, W. G.—Lace (*S. Prov.*).

25 DISSANAIKE, A. T. *Modliar.*—Carved cocoa-nut shell lamp and shells (*S. Prov.*).

26 WICKREMERATINE, A. B. *Modliar.*—Carved cocoa-nut shells; porcupine-quill desk (*S. Prov.*).

27 WEIRALASIRINAYANA ABEYARATINE, DON C. & I.—Carved ebony Davenport and jewel-case (*S. Prov.*).

28 DE COSTA, DON A.—Ebony table and chairs (*S. Prov.*).

29 KARUNANAIKE, DON F.—Ebony foot-stool (*S. Prov.*).

30 WIMALASIRIRIAYANA, A. DE S.—Pair of lyre tables (*S. Prov.*).

31 LAYARD, C. P.—Pair of ebony flower-stands (*W. Prov.*)

32 SIM, CAPT. — Pair of tamarind card tables (*W. Prov.*).

33 DE SILVA, DON S.—Singhalese axes.

34 Ratnaike, W. B. *Modliar.* — Rush mats (*S. Prov.*).

35 De Soyza, I. *Modliar.*—Manufactured articles from the cocoa-nut tree (*W. Prov.*).

36 De Soyza, L. *Modliar.*—Table napkins (*S. Prov.*).

37 Mendis, A. *Modliar.*—Walking-sticks (*W. Prov.*).

38 Templer, Mrs.—Caltura baskets (*W. Prov.*).

39 Giboore, Jos.—Models of Ceylon boats.

40 Croft. J. McGrigor, M.D.—Dugong oil.

41 D'Oyly, Rev. C. T.—Dagger worn by the last of the kings of Kandy.

CHANNEL ISLANDS.

North Central Courts, near the Staircase.

JERSEY.

1 Bénezit, Mme.—Artificial flowers.

2 Fothergill, Mrs.—Algæ.

3 Labalastière, P.—Eau de Cologne.

4 Mullins, H.—Photographs.

5 Peacock, R. A.—Model of patent dock gates.

GUERNSEY.

1 Arnold, A.—Iodine, and chemical products obtained from sea-weed.

2 Bishop, A.—Model of an improved paddle-wheel steam-boat.

DOMINICA.

West-side of N.E. Transept.

Dominica Commission: Hon. J. Imray, M.D., Chairman. — Cotton; coffee; cocoa; Indian corn meal; arrowroot and other starches; sugar; spices; preserves; gums; honey; beeswax; specimens of sugar-cane, guavas, bananas, &c.; cork and other woods; about 170 specimens of indigenous woods polished, and their foliage; seeds; grasses; tortoiseshell and ornaments; specimens of natural history; building stone.

HONDURAS, BRITISH.

Wall near Horticultural Gardens, Northern Entrance.

Collection of woods; aloe fibre; hammock made of ditto; samples of sugar; rum, coloured and uncoloured; tobacco, rice, pimento, black beans, turmeric, arnatto, pepitá, or cucurbit seed.

INDIA.

N.E. Gallery.

1 India Museum, *Fife-house, Whitehall.* —Topographical model of India, constructed by R. Montgomery Martin, Esq.—Model of India, recast by F. Pulman, from the original by R. Montgomery Martin, Esq. drawn and coloured by Edward Stanford.

CLASS 1.

2 Fifty-four samples of soils with analyses, including types of all the principal soils of India.

3 East India Iron Co. by E. J. Burgess.—Samples of iron ores, charcoals, iron, steel; and various articles of manufacture therefrom.

4 Riddell, Robert, Esq. *London.*—Large cat's-eye stone from Ceylon.

5 Government of India.—Samples of deposits containing gold, from the rivers of Rangoon; samples of gold washings, from other localities in India.

6 D'Almeida, Jose, *Singapore.*—Samples of tin ore from the Malay Peninsula.

7 Burn, Hon. Capt. *Singapore.*—Samples of tin ore and tin from Malacca.

8 Geach, F. & Moniot, J. *Singapore.*—Tin ore from Kassang.

9 Government of India.—Tin ore from Province Wellesley, &c.; antimony ore from Candahar; galena from Beloochistan and Rangoon; red and yellow sulphuret of arsenic from Rangoon.

10 Mitchell, Capt. J.—Magnetic iron sand from various localities in Tinnevelly; copper ore from Nellore.

11 Madras Central Committee.—Iron and chrome ores, from Salem and elsewhere.

12 Beckett, Mr.—Iron ores from Ghologat.

13 Government of India.—Iron ores from Jubbulpore, Kumaon, Cuttack, and numerous other localities.

14 The Maharajah of Gwalior.—Iron ore and pig iron.

15 Kumaon Iron Co.—Iron and iron castings.

16 Government of India.—Sixteen kinds of pig iron from the Kymore range; iron and steel from Chittledroog.

17 Oldham, Professor T. — Series of coals, representing the principal coal fields of India.

18 Bivar, Major H. S. *Assam.*—Coal from Assam.

19 Government of India.—Coal from Cuttack, Chota Nagpore, Chittagong, and other localities.

20 Martin, J. N.—Petroleum from Assam.

21 Government of India. — Petroleum and naphtha from Burmah and Akyab.

22 Burn, Hon. Capt. *Singapore.*—Plumbago from Malacca.

23 Rajah of Vizianagarum.—Plumbago from Vizagapatam.

24 Government of India. — Sulphur from the Mountains of Beloochistan.

25 Collyer, Col. *Singapore.*—Brick clays from Singapore.

26 Hunter, Dr. A. *Madras.*—Samples of pottery clays.

27 Government of India.—Pottery clays

from Mysore; pottery clays from other localities.

28 SANDYS, T. — Mortars, cements, and concretes.

29 GOVERNMENT OF INDIA.—Samples of indurated potstone, from different localities; felspathic and porphyritic granites; hornblende and disintegrated granites; slates and hone-stones from Ulwar, Chittledroog, &c.

30 RAJAH OF VIZIANAGARUM.—Black and white talc.

31 BECKETT, MR. of *Assam.*—Black and white earth from Assam.

32 GOVERNMENT OF INDIA.—Fullers earth from Chittledroog, limestones, &c.; corundrum and emery from Mysore; gervoo and suttee from Shahabad.

33 CAMPBELL, DR. *Darjeeling.*—Flexible and other sandstones.

34 COLLYER, COL.—Sandstone for building purposes, from Singapore.

35 MCPHERSON, DR.—Samples of sands from Cape Comorin.

36 GOVERNMENT OF INDIA.—Rock crystal, cornelians, agates, bloodstones, jaspers, and pebbles, from numerous localities; Cambay agates, and manufactures therefrom; pebbles, jaspers, and stones from Assam.

37 GUTHRIE, COL. *Calcutta.*—Collection of manufactures in jade and crystal.

38 CALCUTTA LOCAL COMMITTEE.—Collection of manufactures in soapstone, from Agra.

39 GOVERNMENT OF INDIA. — Culinary vessels and domestic articles of stone; geological specimens from Assam, and rock formations.

40 OLDHAM, PROF. AND OTHERS.—Complete series of fossil cephalopodæ, belemnitidæ, nautilidæ, and other paleontological specimens.

41 HALLIDAY, FOX, & Co. *London.*—Minerals from Rangoon.

CLASS 2.

43 GOVERNMENT OF INDIA.—Salts, saltpetres, and mineral waters.

44 THE RAO OF KUTCH.—Alum shale and alum.

45 RAJAH OF VIZIANAGARUM. — Alum stone and kayoo royee.

46 CHEE YAM CHUAN, *Malacca.*—Sixty-eight medicinal roots.

47 SHORTT, DR. JOHN, *Chingleput.*—Thirteen medicinal substances, and twenty-one seeds from Chingleput.

48 GOVERNMENT OF INDIA.—Thirty medicinal substances from Moulmein; ninety medicinal substances from Moulmein; eighty medicinal substances from the Punjaub.

49 KOONEY LALL DEY.—Forty-four medicinal substances.

50 PATNA OPIUM AGENCY.—Series representing the manufacture of opium.

51 BENARES OPIUM AGENCY.—Series representing the manufacture of opium.

52 GOVERNMENT OF INDIA.—Opium from Lucknow; gunjah from Cuttack.

53 CALCUTTA LOCAL COMMITTEE.—Thirty-three medicinal substances from Calcutta.

54 TAN KIM SENG, *Singapore.*—Medical arrack, &c.

55 D'ALMEIDA, J. *Singapore.* — Cubebs and castor oil seeds.

56 COELHO, V. P.—Medicinal substances from South Canara.

57 KOONEY LALL DEY.—Nearly 200 medicinal substances.

58 GOVERNMENT OF INDIA.—Rose water, keora water, and other chemical products.

59 HALLIDAY, FOX, & Co. *London.*—Eighty-three medicinal substances from Rangoon.

59A BABOO GOPAUL CHUNDER GOOPTE. —Camphor cup from Calcutta.

CLASS 3.

60 INDIA MUSEUM, *Whitehall.*—Specimens of Indian grains and pulses, with analyses by Dr. J. Forbes Watson.

61 BALFOUR, SURGEON-MAJOR E. *Madras.* —Different sorts of grain in the ear.

62 HUNTER, DR. A. *Madras.*— Grains, pulses, &c.

63 GOVERNMENT OF INDIA.—Paddy and rice from Bangalore; grains from various parts of Mysore; pulut and other rice from Province Wellesley.

64 COELHO, V. P. — Paddy and other grains from South Canara.

65 MADRAS CENTRAL COMMITTEE.—Series of grains, pulses, &c.

66 CALCUTTA LOCAL COMMITTEE.—Pulses and cereals from Calcutta.

67 BULLOCK, MR. *Arracan.* — Series of paddy and rice.

68 GOVERNMENT OF INDIA.—Large series of paddy & rice from Chota Nagpore, & Ulwar.

69 PHAIRE, LIEUT. W. *Assam.*—Fifty-seven samples of paddy and rice, from Assam.

70 GOVERNMENT OF INDIA.—Large collection of grains from Akyab, Moulmein, Rangoon, &c.

71 HALLIDAY, FOX, & Co. *London.*—Collection of paddy and rice from Rangoon; series of native fruits in bottles, from Penang, Malacca, &c.

72 GOVERNMENT OF INDIA.—Three collections of models of Indian fruits; preserves and pickles from Lucknow.

73 D'ALMEIDA, JOSE, *Singapore.*—White and black pepper from Singapore and Rhio, long pepper from Java, and cinnamon from Singapore.

74 COELHO, V. P.—Pepper, kavàte, an cardamoms, from South Canara.

75 PHAIRE, LIEUT. W. *Assam.* — Long pepper.

76 CALCUTTA LOCAL COMMITTEE.—Series of spices and condiments.

77 SINGAPORE LOCAL COMMITTEE.—Nut megs, mace, cloves, &c. from Penang; peppe from Soosoo and Trang, Sumatra.

78 HALLIDAY, FOX, & Co.—Spices from Rangoon.

79 GEORGE, MESSRS. *Calcutta.* — Speed's steam-made arrowroot and tapioca of the crop of 1862.

80 GALOPIN.—Arrowroot from Burdwan.

81 COELHO, V. P. *South Canara.*—Bynee palm flour, amroota bally, raggy flour, and arrowroot.

82 LEIBERT, M.—Arrowroot from a wild plant growing in the jungles, and ordinary arrowroot.

83 SINGAPORE LOCAL COMMITTEE.—Sago flour from Singapore, Luigee, and Sarawak; pearl sago and arrowroot from Singapore; sweet potato flour, sago, and tapioca, from Malacca; tapioca from Province Wellesley; and sago from Penang.

84 BENGAL CENTRAL COMMITTEE.—Arrowroot from Akyab and Cuttack; wild arrowroot from Cuttack.

85 CAREW & CO. *Shahjehanpore.*—Cane-juice rum.

86 THOMPSON, DR.—Mango spirit.

87 CALCUTTA LOCAL COMMITTEE.—Six intoxicating liquors from Calcutta; rice arrack from Cuttack.

88 COELHO, V. P. *South Canara.*—Palmyra arrack, Cashew arrack, cocoa nut, and sugar-cane arracks.

89 CAREW & CO. *Shahjehanpore,* AND OTHER CONTRIBUTORS.—Sugar and sugar-candy from different localities.

90 GOVERNMENT OF INDIA.—Honey from Ulwar, Beerbhoom, Coorg, and Chittledroog.

91 ONSLOW, COL. W. CAMPBELL.—Coffee from Mysore.

92 SINGAPORE LOCAL COMMITTEE.—Coffee from Singapore, Malacca, Province Wellesley, Sumatra, Penang Hill, &c.; cocoa grown at Singapore.

93 FISCHER & CO. MESSRS. *Salem.*—Coffee.

94 LEIBERT, M.—Coffee grown at Seetagurah tea and coffee plantation.

95 MCPHERSON, DR.—First mandarin tea, at 20*s.* per lb., from Malacca.

96 CAMPBELL, DR. A. *Darjeeling.*—Specimens of teas, from various localities of India.

97 LIEBERT, M. *Seetagurrah Tea Plantation, Hazareebaugh.*—Pekoe, 3 sorts, orange Pekoe, and gunpowder teas.

98 HIGGS, SEVENOAKS, & MELANY.—Pekoe teas.

99 HANNAY, H. E. S.—Orange Pekoe tea.

100 BARRY & WAGENTRIEBER.—Varieties of teas.

101 WHATOORAM, JEMADUR, *Assam.*—Green tea.

102 CHACHAR TEA CO.—Varieties of tea.

103 MORGAN, T. *Debrooghur, Upper Assam.*—Pekoe, Souchong, and Congou teas.

104 LLOYD, CAPT. *Assam.*—Orange and flowery Pekoe teas.

105 BAINBRIDGE, H. G. *Assam.*—Flowery Pekoe, from Assam, and China leaf.

106 MORGAN, C. H. *Luckimpore, Upper Assam.*—Pekoe, Souchong, and Congou teas.

107 STRAFFORD, W. *Assam.*—Varieties of teas.

108 TYDD, FORBES, & CO. *Cachar Victoria Tea Plantation.*—Souchong and Pekoe teas.

109 BORRADAILE, JOHN, & CO.—Varieties of teas.

110 HOPE TOWN TEA ASSOCIATION, *Darjeeling.*—Varieties of teas.

111 WARRAND, T. *Warrand Field Tea Plantation.*—Souchong tea.

112 RICHARDS, G. *Willow Bank Tea Plantation.*—Souchong teas.

113 MEGRIE, C. R. TROAP.—Souchong, Pouchong, and Bohea teas.

114 MCIVER, K. *Konsannie Tea Plantation.*—Hyson and other teas.

115 CAMPBELL, DR. *Darjeeling.*—Brick tea.

116 WILLIAMSON, G. *Bonganakooah Tea Plantution, &c.*—Pekoe teas.

117 TARIKOLLALIS, MOONSHEE MAHOMED, *Darjeeling.*—Souchong tea.

118 SCANLAN, P. H. *Kursing Plantation.*—Varieties of teas.

119 MASSON, CAPT. *Tackvor Tea Co.*—Specimens of tea.

120 WOOD, OLLIFFE, & CO. *Darjeeling, &c.*—Pekoe teas.

121 GOVERNMENT PLANTATIONS OF *Hurbuns Wala, Kowlaghur, Gurhwal, Howalbagh, Agartala, Bhurtpore, Bheim Tal, and Kanjea.*—Specimens of teas.

122 COELHO, V. P. *South Canara.*—Betelnuts, and Byne-seeds, used as a substitute for them.

123 GOVERNMENT OF INDIA.—Tobacco from Mysore, Akyab, Rangoon, Cuttack, Lucknow, and Moulmein.

124 HALLIDAY, FOX, & CO. *London.*—Tobacco, twist, and other substances, from Rangoon.

125 SINGAPORE LOCAL COMMITTEE.—Trepang, sharks'-fins, birds'-nests, fish-maws, agar-agar, dried mushrooms, dried shell-fish, &c.

CLASS 4.

126 INDIA MUSEUM, *Whitehall.*—A collection of Asiatic silk-producing moths, with illustrations of their transformations, and samples of silks, prepared by F. Moore.

126A BASHFORD, T.—Bengal silk in its different stages.

127 PHAIRE, LIEUT. W. *Assam.*—White Erie silk cocoons.

128 GOVERNMENT OF INDIA, AND OTHERS.—Raw silk from Shikarpoor, &c. Tusser silk and cocoons from Cuttack, Chota Nagpore, &c. Raw wool from Sindh, Astagram, &c. Camels'-hair, from Sindh. Goats'-wool, from Thibet, Cashmere, &c. Series of shawl-wools, from Cashmere, Thibet, &c.

129 CAMPBELL, DR. A. *Darjeeling.*—Thibetian wool.

130 HALLIDAY, FOX, & CO. *London.*—Burmese raw silk, buffalo horns and ivory tusks, from Rangoon.

131 PHAIRE, LIEUT. W. *Assam.*—Deer-horns.

132 LOCAL COMMITTEES.—Bees'-wax, from Mysore, Burmah, &c.

133 SINGAPORE LOCAL COMMITTEE.—Tye-klawa, or bat guano, from the coast of Penang.

134 COELHO, V. P. *South Canara.*—Series of oil-seeds.

135 CALCUTTA LOCAL COMMITTEE.—Thirty-six specimens of oil-seeds.

136 GOVERNMENT OF INDIA.—Oil-seeds, from Chota Nagpore, &c.

137 COELHO, V. P. *South Canara.*—Collection of vegetable oils and fats.

138 SINGAPORE LOCAL COMMITTEE. — Macassar and other oils; vegetable tallows and fats.

139 CALCUTTA LOCAL COMMITTEE.—Oils and fats, from numerous localities.

140 WOOD, C. B. — Safflower and other oils.

141 THOMPSON, DR. — Oil of *Argemone Mexicana.*

142 CHETUMBARA PILLAY, *Madras.* — Peacock ointment and vegetable oils.

143 SAINTE BROS. *Cossipore.* — Refined cocoa-nut oils, stearine candles, and carriage candles.

144 SMITH & Co. *Calcutta.*—Cod-liver oil.

145 LODER, H. E.—Vegetable-wax.

146 SCOTT, G. *Penang.*—Essential oils.

147 KOONEY LALL DEY.—Attars, essences, and oils.

148 COELHO, V. P.—Sandal-wood oil, from South Canara.

149 BAULOO MOODLY, C. *Madras.* — Orange-oil.

150 CLEGHORN, DR. *Madras.* — Citronelle oil.

151 GOVERNMENT OF INDIA.—Samples of fixed and essential oils, from Lucknow and elsewhere.

152 HALLIDAY, FOX, & Co. *London.*— Gingelly oils, from Rangoon.

153 PHAIRE, LIEUT. W. *Assam.*—Stick lac.

154 KOONEY LALL DEY.—Shell lacs.

155 LOCAL COMMITTEES.—Lac from Jhansee, the forests of the Kymore range, &c.; wood oil from Akyab, &c.; turpentine of *Pinus longifolia* from Calcutta; and varnishes from Burmah.

156 HALLIDAY, FOX, & Co. *London.*— Gums and resins from Rangoon.

157 NEUBRONNER, T.—Gutta percha from Malacca; gutta gree grip and gutta babee.

158 DE WIND, A.—Gutta terbole, used to adulterate gutta percha.

159 MARTIN, W. C.—White caoutchouc from the Cossia Hills.

160 OSBORNE, MR. — White caoutchouc from Gorruckpore.

161 SHORTT, DR. JOHN, *Chingleput.* — Caoutchouc from *Calotropis gigantea*, *Euphorbia antiquorum*, and *Euphorbia tortilis.*

162 LOCAL COMMITTEES. — Caoutchouc from other localities.

163 H.H. THE SULTAN OF ZANZIBAR. — Specimens of copal.

164 SINGAPORE LOCAL COMMITTEE. — Dammars from Malacca and the Malay Peninsula.

165 COELHO, V. P.—Gamboge from South Canara.

166 SHORTT, DR. JOHN, *Chingleput.*— Collection of gums and resins.

167 GOVERNMENT OF INDIA.—Gums and resins from Cuttack, Calcutta, &c.

168 AUERBACK, RUDOLPH, & Co. *London.* —Samples of indigo.

169 FISCHER & Co. *Salem.*—Samples of indigo.

170 ARBUTHNOT, W. R. *Cuddapah.*—Kurpah indigo.

171 CRAKE, W. H.—Dry-leaf indigo.

172 RAMASHESHIA, *Vellore.*—Indigo from North Arcot.

173 STEEL, H. *Hooghly.* — Samples of indigo.

174 THE RAJAH OF ULWAR.—Sample of indigo.

175 JARDINE & Co.—Samples of indigo.

176 PHAIRE, LIEUT. W. *Assam.*—Indigo and madder from Assam.

177 WAGENTRIEBER, W. G. *Assam.* — Samples of roam dye.

178 THOMPSON, DR. R. F. — Vegetable green and yellow dyes.

179 GOVERNMENT OF INDIA.—Indigo from Sindh, &c.

180 HALLIDAY, FOX, & Co.—Dye stuffs and tanning materials from Rangoon.

181 CAMPBELL, DR. A. *Darjeeling.*—Sixteen specimens of lichens.

182 JUNG BAHADOOR, SIR.—Nepaul madder.

183 BIVAR, MAJOR H. S. & WAGENTRIEBER, W. G.—Madder from Assam.

184 PHAIRE, LIEUT. W. & MICHEL, H. L. —Madders from Assam.

185 MARTIN, J. N. — Madder from Assam.

186 CALCUTTA LOCAL COMMITTEE.—Series of dye stuffs.

187 MADRAS CENTRAL COMMITTEE.—Lac, kapila, and other dyes, from Salem.

188 GOVERNMENT OF INDIA.—Dye stuffs and tanning materials from Cuttack, Arracan, &c.

189 SHORTT, DR. *Chingleput.*—Series of dye stuffs and tanning materials.

190 SINGAPORE LOCAL COMMITTEE. — Gambier, barks, and other tanning substances.

191 RIDDELL, R. *London.*—Samples of cotton grown at Hyderabad.

192 BINGHAM, R. W. *Chynepore, Bengal.* —Brown nankeen cotton.

193 FISCHER & Co. *Salem.* — Bourbon cotton.

194 BLAND, REV. R. *Gowhatty.* — Raw cotton.

195 HUNTER, DR. A. *Madras.* — Raw cotton.

196 RAJAH OF VIZIANAGARUM. — Booroogа cotton.

197 PHAIRE, LIEUT. W. *Assam.*—Cotton and cotton seed.

198 BIVAR, MAJOR H. S. — Raw cotton from Luckimpore.

199 WAGENTRIEBER, W. G. *Assam.* — Assamese cotton from Mutiak.

200 MORTON, CAPTAIN. — Cotton from Assam.

201 THE MAHARAJAH OF ULWAR.—Raw cotton.

202 THE MAHARAJAH OF GWALIOR. — Raw and cleaned cotton.

202A HUTCHINSON, A.—Sample of cotton grown in Province Wellesley.

203 GOVERNMENT OF INDIA AND LOCAL COMMITTEES.—Above two hundred samples of cotton, from various localities.

204 HALLIDAY, FOX, & Co. *London.* — Burmese cotton.

204 A SMITH, FLEMING, & Co. *London.*—Eighteen samples of cotton.

205 SHORT, DR. *Chingleput.*—Silk cottons from *Asclepias volubilis* and *Dæmia extensa.*

206 MADRAS CENTRAL COMMITTEE, AND OTHERS.—Mudar and Bombax silk cottons.

207 BELFAST INDIAN FLAX COMPANY.—Flax grown at Sealkote, in the Punjaub.

208 AHMUTY & Co. — Samples of fibres, from Nepaul.

209 JADOORAM RURROA SUDDER, AMEER OF LUCKIMPORE. — Fibre of *Urtica tenacissima.*

210 MICHEL, H. L.—Hemp from Assam, and fibre of *Sterculia urens.*

211 PHAIRE, LIEUT. W. — Jute, from Assam.

212 HIGGS, REV. E. H.—Fibre of *Urtica tenacissima*, from Debroghur.

213 EAST INDIA FLAX COMPANY.—Samples of flaxes.

214 SHORTT, DR. *Chingleput.*—Substances employed for basket-making.

215 CAMPBELL, DR. *Darjeeling.*—Twenty specimens of reeds.

216 COELHO, V. P.—Series of fibres, from South Canara.

217 SECRETARY TO BOARD OF REVENUE, *Chingleput.*—Twenty samples of fibres.

218 RAJAH OF VIZIANAGARUM.—Samples of fibres.

219 MADRAS CENTRAL COMMITTEE. — Palmyra and other fibres from Salem.

220 HUNTER, DR. A. *Madras.*—Series of fibres.

221 SHORTT, DR. *Chingleput.* — Series of mat-making materials and fibres, from Chingleput.

222 EVANS, C.—Bark of gum-tree, used as oakum.

223 LOCAL COMMITTEES.—Numerous specimens of fibres, from Lucknow.

224 RIDDELL, R. *London.*—Specimens of fibres from India, and articles of manufacture therefrom.

224A STAUFEN, W. *London*, inventor and patentee.—Ejou fibre, curled and imitation hair; brushes made therefrom. *Agave Americana*, and dyed for weaving and military purposes; hair-cloth made therefrom.

225 MAITLAND, COL. *Madras.* — Fifteen specimens of woods used at the gun-carriage manufactory, Madras.

226 COELHO, V. P. *South Canara.*—Forty-nine woods of South Canara.

227 MAN, HON. MAJOR, *Singapore.* — Thirty-four woods of Penang.

228 RAJAH OF VIZIANAGARUM. — Seven woods of Vizagapatam.

229 BAKAR, H. H. INCHE WAN ABOO.—Twenty-seven woods of the Malayan Peninsula.

230 EVANS, C. — Forty-one woods from Malacca.

231 SHORTT, DR. J. *Chingleput.*—Seventy-eight woods from Chingleput.

232 BENGAL CENTRAL COMMITTEE AND OTHERS.—115 woods from Chota Nagpore; 18 woods from Lahore; 28 woods from Umritsur; 40 woods from Jubbulpore; 18 woods from Lucknow; 4 woods from Midnapore; 10 woods from Chittagong; 115 timber-woods from Moulmein; 113 woods from Rangoon; 14 woods from Cuttack; 42 woods from Akyab; 6 woods from Philibeet.

232A MYSORE GOVERNMENT — 96 woods from Mysore, partly tested for strength, with details of results.

232B LOCAL COMMITTEES. — Pak putta turnery-ware from Lahore; lac-turnery from Patna, &c.

233 REID, COL.—Seventy-three specimens of timber from Assam.

234 PHAIRE, LIEUT. W.—Three specimens of woods, and twelve canes, from Assam.

235 LAZARUS, C.—Mahogany plank, from a tree grown in Calcutta.

236 CAMPBELL, DR. *Darjeeling.*—Seventeen specimens of timber.

237 HALLIDAY, FOX, & Co. *London.* — Forty-two timbers of Rangoon.

238 BIVAR, MAJOR H. S. *Luckimpore.*—Weaver-bird's nest, and tinder from sago-palm tree.

239 MCPHERSON, DR.—Bark-tinder from Malacca.

240 SHORTT, DR. *Chingleput.*—Mudar-root charcoal.

241 COELHO, V. P. — Areca-nut boiled juice, used by the lower classes instead of varnish for wood work.

CLASS 6.

242 GOVERNNENT OF INDIA.—Carts and palanquins.

CLASS 7.

243 BOMBAY LOCAL COMMITTEE.—Raw materials and tools for inlaid work.

244 PHAIRE, LIEUT. W.—Weaving apparatus from Assam.

245 COLLECTOR OF DHARWAR.—Machines for cleaning cotton, &c.

246 GOVERNMENT OF INDIA. — Cotton-cleaning machines from Chittledroog, &c.; rice machines from Burmah; sugar and oil mills; cashmere and carpet looms.

247 HALLIDAY, FOX, & Co. *London.*—Burmese loom.

CLASS 8.

248 TEMPLEMORE, E. *Madras.*—Model of diving-bell raft.

249 FORBES, DR. *Bombay.*—Model of pile-driving machine.

CLASS 9.

250 COLLECTOR OF DHARWAR.—Models of agricultural implements.

251 CAMPBELL, DR. *Darjeeling.*—Agricultural implements.

252 GOVERNMENT OF INDIA.— Agricultural machines and implements from Bangalore, &c.

253 RAJAH OF VIZIANAGARUM. — Corn fan from Vizagapatam.

CLASS 11.

254 HIS HIGHNESS THE RAO OF KUTCH.—Fourteen rich inlaid arms, and complete chain armour.

255 CAVENAGH, HON. COL. — Six krises from Singapore.
256 THE RAJAH OF TRINGANU. — Six spears with gold ferules.
257 MADRAS CENTRAL COMMITTEE.—Four war knives from Malabar.
258 MCPHERSON, DR.—Upas poison, upas arrows, and other weapons.
259 NAWAB OF BHAUWULPORE.—Dagger and musket gold-mounted.
260 HARRISON, CAPT.—Dagger, silver-mounted with ivory handle.
261 MAHOMED ZAMA, *Peshawar.*—War knife.
262 CAMPBELL, DR. *Darjeeling.* — Arms from Darjeeling, &c.
263 RAJAH OF PUTTIALA.—Swords and daggers, gold and silver-mounted.
264 NAWAB OF RAMPORE.—Daggers.
265 MARTIN, JOHN, *Calcutta.*—Daggers in scabbards.
266 GOVERNMENT OF INDIA.—Arms, armour, and weapons, from numerous localities.

CLASS 12.

267 PAYN, W. H. *Dover.*—Templemore's recovery buoy for saving submerged property, from Madras.
268 GOVERNMENT OF INDIA.—Models of boats.
269 HALLIDAY, FOX, & CO. *London.*—Models of boats from Burmah.

CLASS 13.

270 BENGAL CENTRAL COMMITTEE.—Trisector from Agra.
271 CAUTLEY, SIR PROBY, K.C.B. — Troughton's improved level and prismatic compass, made by native workmen in India.

CLASS 14.

272 PLAYFAIR, CAPT. & ARMSTRONG, LIEUT. *Malacca.*—Photographs of natives.
273 HUNTER, DR. A. *Madras.*—Photographs.
274 STEVENSON, COL.—Stereographs.
275 RAJAH OF PUTTIALA.—Photographs.
275A GOVERNMENT OF INDIA.—A large collection of photographs of the different tribes of India, chiefly taken by amateurs.

CLASS 16.

276 BAYLY, H. S. *Northampton.*—Curious Burmese gong.
277 CAMPBELL, DR. *Darjeeling.*—Musical instruments from Thibet, &c.
278 HALLIDAY, FOX, & CO. *London.*—Burmese bells and gongs.

CLASS 18.

278 A INDIA MUSEUM, *London.*—Set of eighteen books, containing working specimens of all the principal textile manufactures of India.
279 THE RAO OF KUTCH.—Cotton cloth, dhotees, loonghees, &c.
280 COLLECTOR FOR BROACH, BOMBAY CENTRAL COMMITTEE. — Towels, napkins, table-cloths, canvas, checks, and ticking; cotton yarns spun by the aid of native mills.
281 BUCKLE, MAJOR, *Rewa Kanta.*—Cotton cloths manufactured in Rewa Kanta.
282 ANDERSON, MAJOR W. C. *Dharwar.*—Cotton fabrics of various kinds.
283 COLLECTOR OF AHMEDABAD.—Dhootees, doputtas, sadees, loongees, &c.
284 COLLECTOR FOR KURRACHEE, *Upper Sindh.*—Loossees and sheets.
285 GOVERNMENT OF MYSORE.—Cotton cloths from Bangalore; cotton ropes from Astagram; table-cloths and towels from Chittledroog.
286 H.H. INCHE WAN ABOO BAKAR.—Bugis cotton sarongs from Macassar; sarongs, salendongs, and handkerchiefs, from Java.
287 MADRAS CENTRAL COMMITTEE.—Arnee muslins; cotton cloths and napkins from Madras.
288 A. RAB RAO *Wallajahpatta, North Arcot.*—Chintz cloth.
289 BASETH RAO, *North Arcot.*—Chintz cloth.
290 GOVERNMENT OF INDIA.—Chogas, pushminees, and loongees, from Peshwar; cotton cloths from Goojrat, Ferozepore, Agra, &c.; table-cloth, &c. made at the jails in Lahore and the Punjab; patterns of cotton cloths used by agricultural tribes of the Punjab.
291 PHAIRE, LIEUT. W. *Assam.*—Cotton thread and cotton cloths.
292 CAMPBELL, DR. *Darjeeling.*—Striped and checked cotton cloths, scarfs, girdles, and other cotton goods.
293 RAJAH OF PUTTIALA.—Five pieces of glazed calico.
294 CHEKE, DR. N. H.—Cotton goods made at the Central Jail, Benares.
295 LLOYD, CAPT. E. — Muslin from Assam.
296 NAWAB OF RAMPORE.—Loongees and dhooties.
297 PETTIANGERI CHETTI.—Fine Maderpak muslins from North Arcot.
298 HUNTER, DR. A. *Madras.*—Table-cloths, towels, napkins, d'oyleys, checks, and other cloths, manufactured at the jail at Chingleput.
299 CALCUTTA LOCAL COMMITTEE.—Cotton fabrics from Lucknow; muslin and dhooties from Hooghly; table-cloths and towels from Beerbhoom; table-covers and dusters from Patna; napkins and towels from Dinapore; checked cloths from Bhaugulpore men's cloths from Burmah; country cloths from Cuttack; table-cloths and napkins from Jubbulpore; towels from Allahabad; ducks and tent cloths made at the jail of Meerut cotton fabrics from various localities.
300 COTTON SUPPLY ASSOCIATION. *Manchester.*—Twenty pieces of cloth made exclusively from Indian cotton, dyed and printed.
301 BINGHAM, R. W. *Chynepore, Bengal*—Coarse calico of brown cotton, spun and woven at Cheynepore.
302 H.H. MEER ALI MOORAD OF KHYRPOOR.—Chintz and other cloths.

302A India Government.—Mulmul, and other fine muslins, from Dacca.

CLASS 19.

303 Phaire, Lieut. W. *Assam.*—Four specimens of rope from the jail at Tejpore.

304 Ahmuty & Co.—Ropes.

305 Bainbridge, H. *Assam.*—Twine of rhea fibre, and net made thereof.

306 Borneo Co.—Sacks, bags, and cloth.

307 Campbell, Dr. *Darjeeling.*—Four specimens of jute cloth.

308 The Ameer of Luckimpore.—Rope of *Urtica tenacissima.*

309 Government of India.—Agave ropes from Balasore; kodal and scalie rope from Cuttack; red and white fibre and cordage from Moulmein; hemp and coir ropes from Calcutta; twine from various fibres, from Lucknow; rope and twine from Hooghly; sunn and other ropes from Ulwar; aloe fibre sacking, cord, and twine, from Chota Nagpore; aloe fibre rug from Allahabad; Moong and other twines from Lahore; ropes, &c. from various localities.

498 Rajah of Vizianagarum.—Three goldfinch nets.

CLASS 20.

310 H.H. Meer Ali Moorad of Khyrpoor.—Silks in thirteen colours.

311 Collector of Shikarpoor, *Upper Sindh.*—Silk cloths, &c. from Bokhara, Persia, &c.

312 The Rao of Kutch.—Specimens of silk cloth.

313 Collector of Ahmedabad.—Silk fabrics and kincaubs.

314 Mysore Government.—Silk shawls and cloths from Bangalore and Chittledroog.

315 Campbell, Dr. *Darjeeling.*—Silk fabrics.

316 Maharajah of Puttiala.—Coloured silk fabrics.

317 Cheke, Dr. *Benares.*—Silk turban and handkerchief, made at Central Jail.

318 Nawab Khan of Peshawar.—Plain and shot silks.

319 Nawab of Fareedkote.—Silk loongees from Lahore.

320 Baboo Kasseram.—Eria silk cloth from Assam.

321 Baboo Jeoram Deka, *Barovah, Peskar.*—Eria silk cloth from Assam,

322 Choro Chunder, *Mouzador.*—Eria silk cloth from Assam.

323 Beckett, W. A. O.—Cloth of Moonga silk.

324 Baboo Mohun Chundere, *Barovah Moonsiff.*—Mosquito curtain of Moonga silk.

325 Major, H. S. *Bivar.*—Silk scarfs from Assam.

326 Michel, H. L. Esq. *Luckimpore.*—Eria silk cloth.

327 Saccaram, Sahib, *Tanjore.*—Women's silk cloth.

328 Nawab of Bhawulpore.—Doputtas, &c. of Bhawulpore silk.

329 Bainbridge, H.—Silk cloths from Assam.

330 Phaire, Lieut. W. *Assam.*—Moonga and Eria silk shawls.

331 Veitch, Lieut.-Col. Hamilton, *Assam.*—Mezankoree, Eria, and Moonga silk cloths.

332 Thompson, R. F. *Malda.*—Flowered silk.

333 Lewis, J. M.—Flowered silk.

334 Baboo Huns Geer, *Gossain.*—Flowered silks, &c.

335 Government of India.—Silk fabrics from Berhampore, Umritsur, Mooltan, Lahore, Peshawar, Hooghly, Akyab, Rangoon, Cuttack, Midnapore, Burdwan, Bhaugulpore, &c.

336 H.H. Inche Wan Aboo Baker, *Singapore.*—Silk sarongs, cloths, and handkerchiefs.

CLASS 21.

337 The Rao of Kutch.—Woollen cloths of Kutch wool, and mixed fabrics.

338 Marayadu, Kuntadu, & Paoudu, *North Arcot.*—Grey and black cumblies.

339 Nawab of Rampore.—Plain shawls.

340 Campbell, Dr. *Darjeeling.*—Purse of Thibetian wool.

341 Government of India.—Coloured wools for needlework from Calcutta; blankets from Lahore, Allahabad, Mysore, &c.; mixed cotton and silk fabrics from Umritsur; gambrooms from Loodiana, &c.; silk and cotton fabrics from Ahmedabad; woollen cloths and cloaks from Shikarpoor.

371 Rai Loll Chund, Bahadoor of Umritsur.—Umritsur and Cashmere shawls.

372 Rai Hive, *Dyab of Umritsur.*—Jamewar shawls.

373 Mohammed Shaw, *Umritsur.*—Cashmere, Umritsur, and Noorpoor shawls and scarfs.

374 Bhai Kullian Singh, *Umritsur.*—Umritsur shawls.

375 Mohun Loll, *Umritsur.*—Cashmere shawls.

376 Gunpat, *Brahmin of Umritsur.*—Shawls from Umritsur and Cashmere.

377 Munsa Ram, *of Rampore.*—Shawl from Rampore.

378 Kashi Ram, *of Loodiana.*—Chuddars and shawls from Loodiana.

379 Missar, Churd.—Chuddars and scarfs from Cashmere.

380 Davee Sahai & Chumba Mull.—Shawls from Cashmere and Umritsur, Rampore and Pudderowhah; shawl borders, caps, chogas, waistcoats, &c., and other embroidered work.

381 Missar Gema Chund.—Cashmere, chuddahs, &c.

382 Rai Nursing Doss.—Half-shawl from Umritsur.

394 Government of India.—A large collection of shawls and roomals, from Delhi Benares.

395 McPherson, Dr. *Madras.*—Cashmere shawls, and other embroidered articles.

396 Farmer & Rogers, *Regent-st. London.*—Long black India shawl and square gold-worked shawl.

CLASS 22.

342 WATSON, BONTOR & Co. *London.*—Masulipatam carpets.

343 SUBBI CHETTI, PARNAPA CHETTI, & VEMKATACHELLI CHETTI.—Carpets made at Wallajahpetta, North Arcot.

344 SACCARAM SAHIB.—Silk rug made at Tanjore.

345 DALLAS, M. *Lahore.*—Carpet made at the Central Jail.

346 JUBBULPORE SCHOOL OF INDUSTRY.—Woollen carpets of Persian pattern.

347 MCANDREW, MAJOR.—Carpets made by Thugs at Lahore.

348 THE NAWAB OF BHAWULPORE.—Small carpet and silk rug.

349 NAJIR KHARWULL, KHAN OF PESHWAR.—Woollen Bokhara carpet.

350 CHEKE, DR. N. H. *Central Jail, Benares.*—Cotton and imitation Kidderminster carpets.

351 BABOO KISSORAN DAROGAH.—*Debroghur.*—Cotton carpet.

352 GOVERNMENT OF INDIA. Carpets from Sindh, Bangalore, Patna, Meerut, Shahpore, Lahore, &c.

353 ROBINSON, J. R. *Welbeck-st. London.*—Woollen carpet from Madras; woollen rugs from Lahore; and mat from Travancore.

354 COLLYER, COL. *Singapore.*—Coir and rattan matting, made by prisoners.

355 H.H. INCH WAN ABOO BAKAR.—Mats from pandang leaf, from Java.

356 SINGAPORE LOCAL COMMITTEE.—Mat of Mungkwang palm leaf from Malacca.

357 MODIN SAIB, & OTHERS.—Wandawash mats, made in North Arcot.

358 RAJAH OF TIPPERAH.—Ivory mat from Chittagong.

359 TURNBULL, R. D. *Calcutta.*—Setal puttee mats.

360 BABOO, GHOSE.—Mats from Jessore.

361 GOVERNMENT OF INDIA.—Straw mats from Pulghaut; matting of Daib grass; mat from Midnapore; mushnud mats from Midnapore, and the rushes of which they are made; mats from Hooghly; matting of wild date-palm leaf, from Chotanagpore.

CLASS 23.

362 WAGENTRIEBER, W. G. *Assam.*—Cloths dyed with the roam dye.

363 MADRAS CENTRAL COMMITTEE.—Silks dyed at Salem; cotton cloth, dyed with pomegranate rind.

364 RAO VENKATA & RAO PAPANA.—Arcot dyed cloths; printed chintz cloths.

365 GOVERNMENT OF INDIA:—Six roomals from Ulwar; seven silks dyed in Burmah; cloth dyed with *Butea frondosa, Cæsalpinia, sappan, koosum,* and *kamba goonda*; book of examples of native dyes, from Nepaul; eleven cotton yarns, dyed by the Burmese; printed or stamped cotton from Seetapore; series of cloths dyed with different substances, from Allahabad; cloths dyed with al. and indigo, from Jhansee.

CLASS 24.

366 H.H. MEER ALI MORAD OF BHYRPOOR.—Embroidered muslins, skirts for dresses, cloths, caps, slippers, &c.

367 THE RAO OF KUTCH.—Embroidered scarf and table cloth.

368 H.H. THE RAJAH OF TRINGANU.—Embroidered sarongs.

369 KHAJAH ABDOOL GUNNY, *Dacca.*—Dacca muslins.

370 RAMOHUN ROY, *Dacca.*—Dacca muslins, shawls, and other embroidered tissues.

383 NAWAB OF FAREEDKOTE.—Phulkarees of embroidered cotton.

384 RAJAH OF PUTTIALA.—Embroidered quilts, shawls, and saddles.

385 MCPHERSON, DR.—Calcutta work petticoat embroidery.

386 RAJA OF VIZIANAGARUM.—Lace-bordered bunjees, &c. from Vizagapatam.

387 RAJA GOREE, *of Oude.*—Gold embroidered parasol.

388 AGA ALLEE KHAN, *of Lucknow.*—Gold embroidered Cashmere gown.

389 RAJAH GOREE SHUNKER, *of Moram, Oude.*—Gold-embroidered shawl, roomals, slippers, and topees.

390 NAWAB SHURF-OOD DOOLAB OF LUCKNOW.—Gold-embroidered purses, caps, and shoes.

391 TYKISHEN, RAJAH OF LUCKNOW.—Tag-topee and mundeel-topee, embroidered in gold.

392 H.H. THE MAHARAJAH OF ULWAR.—Gold-embroidered crown and hat; gold-embroidered shoes and turbans.

393 H.H. JUNG-BAHADOOR OF NEPAUL.—Gold-embroidered belt.

397 BURZORJIE, SONS, & Co. *London.*—Embroidered table covers, from Sindh.

397A GOVERNMENT OF INDIA.—Embroidered work, from Surat, Agra, and Benares; embroidered muslins, from Sindh and Kutch; lace, lappets, veils, berthas, sleeves, insertions, &c. from Tinnevelly; Calcutta chickun work; embroidered caps, from Peshawar, Benares, and Lahore; embroidered muslin-work, from Madras.

CLASS 25.

398 REID, COL. C., C.B.—Two stuffed Royal Bengal tigers.

399 MCPHERSON, DR. *Hyderabad.*—Tiger skin and peacock feather fan.

400 RAJAH OF VIZIANAGARUM.—Red bulbul feathers.

401 VETCH, COL. *Assam.*—Bundle of peacock feathers.

402 DOWLEANS, A. M. ESQ. *Calcutta.*—Seven peacock feather fans.

403 CAMPBELL, DR. *Darjeeling.*—Yak's tail, skins, feathers, and horns, from Thibet.

404 GOVERNMENT OF INDIA.—Fur coat and waistcoat, from Bokhara and Musher; chamois' skin and dyed leather, from Nepaul; peacocks' feathers and fans, from Nepaul, Kurnool, &c.; collection of feathers, tipped

and coloured, with articles manufactured therefrom, from Calcutta.

CLASS 26.

405 H.H. MEER ALI MOORAD, OF KHYRPOOR, *Sindh.* — Horse trappings, housings, saddles, and saddle-cloths.

406 CAMPBELL, DR. *Darjeeling.*—Thibetan saddle-cloth.

407 GOVERNMENT OF INDIA.—Saddle and bridle, from Shikarpoor; set of harness artillery harness; buffalo and cow hide, tanned in the English manner.

408 HALLIDAY, FOX, & CO. *London.*—Burmese saddle.

CLASS 27.

409 H.H. MEER ALI MORAD, OF KHYRPOOR.—Sindh caps.

410 H.H. THE RAO OF KUTCH.—Turbans, caps, &c. from Kutch.

411 COLLECTOR OF BELGAUM.—Collection, illustrating the clothing of all classes.

412 COLLECTOR OF DHARWAR.—Collection, illustrating the ordinary and holiday clothing of all classes and both sexes; Malay fisherman's hat of palm leaf.

413 MCPHERSON, DR.—Gloves, stockings, slippers, &c.

414 RAJA OF VIZIANAGARUM. — Native shoes and wooden sandals.

415 CAMPBELL, DR. *Darjeeling.*—Caps, boots, &c., from Thibet, &c.

416 COELHO, V. P.—Leaf caps worn by the Sudra classes of South Canara.

417 EMAUM BUCKET KHAN.—Shoes worn by Mugaree men.

418 GOVERNMENT OF INDIA. — Bandela shoes, from Jhansee; wooden sandals, or clogs, from Umritsur and Bareilly; killat, or dress of honour, from Benares; clothing, from Ahmedabad; gloves, socks, and stockings, from Serinuggur.

419 HALLIDAY, FOX, & CO.—Articles of clothing, from Burmah.

CLASS 28.

420 HUNTER, DR. A. *Madras.*—Samples of paper.

421 BABOO DUMBROO DHUR OLKA.—Bark used by former kings of Assam to write upon.

422 CAMPBELL, DR. *Darjeeling.*—Thibetan work in 12 volumes, and printing blocks in the Thibetan character.

423 THUILLIER, COL.—Atlas of the Himalaya mountains.

424 OLDHAM, PROFESSOR T. — Memoirs of the geological surveys of India. Paleontologica Indica, and geological maps.

425 RAMSAY, COL.—Fine Nepaul paper.

426 GOVERNMENT OF INDIA.—Twenty samples of paper from Madras; country paper made at Ulwar; paper made from bamboo, old rags, old hemp, aloes fibre, plantain fibre: old records, &c.; Nepaul paper of *Daphne cannabina*; arsenical paper from Hooghly; country-made and other papers from Lucknow, Meerut, Jhansee, &c.

CLASS 29.

427 MCPHERSON, DR. *Madras.* — Malay leaf letter.

428 RAJAH OF PUTIALA.—Specimens of caligraphy.

429 BARU DASSENA DHU SINGH, *of Loodiana.*—Specimens of caligraphy.

430 NAWAB SHURFOD DOWLAH, *Lucknow.* —Three tables of Persian phrases.

431 GOVERNMENT OF INDIA.—Specimens of Sheekista and Nagree writing.

CLASS 30.

432 THE RAO OF KUTCH.—Carved box of ebony.

433 DESCHAMPS, J. *Madras.*—Rosewood arm-chairs and small chairs, and carved looking-glass frames.

434 COMARECK, W. *Madras.*—Articlesin papier mâché.

435 H.H. THE RAJAH OF VIZIANAGARUM. —Writing-desks, jewel-boxes, &c. of inlaid work.

436 MCPHERSON, DR. *Madras.* — Carved inlaid and ornamental work.

437 BHIMJEE BYRAMJEE, *Bombay.* — Carved blackwood furniture.

438 GOVERNMENT OF INDIA. — Sandal-wood writing-desks, jewel-boxes and work-boxes from Surat; inlaid tables, writing-desks, &c. from Bombay; collection of ivory inlaid work from Bombay; collection of articles in carved sandal-wood from Coompta, Surat, and Mysore; carved black-wood furniture from Bombay; lac-ware boxes, &c. from Singapore; lac-ware work from Banganpully and Madras; horn-work from Vizagapatam; carved and inlaid work from Bareilly; papier mâché articles from Cashmere and Lahore; lac-ware from Beerbhoom; carved work from Shahjehanpore.

439 GLADDING, J. 39, *City-rd. London.*—Carved sandal-wood boxes from India.

440 HALLIDAY, FOX, & CO. *London.*—Burmese boxes, &c.

441 KIRKMAN, J. & SONS, 3, *Soho-sq. London.*—Grand piano in rosewood case elaborately carved by natives at Madras.

442 NEAL, J. 18, *Edgware-rd. London.*—Carved and inlaid sandal work-boxes, &c.

443 FARMER & ROGERS, *Regent-st. London.*—Carved and inlaid sandal-wood envelope case, &c.

CLASS 31.

444 MCPHERSON, DR. — Collection of domestic utensils from British Burmah.

445 SHAIK FAKEER HOSSEIN. — Eight domestic utensils of brass from Sewan.

446 BABOO KUSSUBRAM BARROWAH.—Brass dish from Luckimpore.

447 BAHADOOR, SIR JUNG. — Banker's padlocks from Nepaul.

448 ROWLATT, CAPT. E.—Iron kodali or mattock.

449 BINDASHURA PURSORD, *Sarun, Patna.* —Iron umbrella, with contents, and brass water engine.

450 OLDHAM, PROF. — Two castings in brass from Manbhoom.

451 CAMPBELL, DR. *Darjeeling.*—Iron spoon, with Dorje head, &c.

452 GOVERNMENT OF INDIA.—Brass and copper vessels from Nepaul and Cuttack; fifteen domestic utensils of brass from Hooghly; brass hookahs from Mooradabah, &c.

453 HALLIDAY, FOX, & CO. *London.*—Burmese pestle and mortar.

CLASS 32.

454 H.H. MEER ALI MOORAD OF KHYRPOOR.—Gold and silver mounted sword, knife, and betel cutter.

455 MCPHERSON, DR.—Elaborately carved knife handles from British Burmah.

456 PHAIRE, LIEUT. W. *Assam.*—Specimens of Assamese cutlery.

457 MARTIN, J. W. — Two Assamese knives.

458 GOVERNMENT OF INDIA.—Steel wire from Cuttack and Bangalore; Persian knife and dagger from Bokhara; hunting knife from Cuttack; collection of manufactures in steel from Salem.

459 HALLIDAY, FOX, & CO. *London.*—Burmese razors, tweezers, and areca-nut cutter.

CLASS 33.

460 RIDDELL, R. *London.*—Ornament of pearls taken at the siege of Seringapatam.

461 H.H. THE RAO OF KUTCH.—A collection of manufactures in gold and silver.

462 CHOCALINGUM, *Madras.*—Gold and silver bracelets, &c.

463 RASALINGUM ASSARY, *District of Trichinopoly.*—Bangles, bracelets, and other articles of jewellery.

464 RAJAH DEONARIAN SINGH OF BENARES.—Brooches, lockets, and jewellery.

465 CAMPBELL, DR. *Darjeeling.* — Ear-rings, rosaries, ornaments, and coins.

466 H.H. THE MAHARAJAH OF ULWAR. —Set of jewels worn by ladies of rank in Ulwar.

467 MARTIN, J. *Calcutta.* — Necklace of 122 pearls, 9 emeralds, topaz, and diamond, and gold rings set with emeralds and diamonds, and other jewellery.

468 BAHADOOR, SIR JUNG.—Silver filagree boxes.

469 H.H. THE RAJA OF ULWAR.—Coins of the States, and silver hookah.

470 RAO, G. N. GAJAPATI. — Casket, bracelets, &c. from Vizagapatam.

471 MCPHERSON, DR.—Silver and seed bracelets.

472 RAJAH OF VIZIANAGARUM.—Tangled gold ring.

473 SAH MAKKHUN LALL MAHAJAUN OF LUCKNOW.—Silver spice box, &c.

474 H.H. THE RAJA OF PUTIALA.—Silver tea-pot, butter-pot, and mug.

475 SHEIK KOTULOODON HUSSEIN KHAN, *Lucknow.*—Fancy tinsel boxes, bracelets, &c.

476 RAI NURSING DOSS, OF UMRITSUR.—Box of agate and crystal set in gold.

477 PLOWDEN, G.—Calcutta Trades' Plate for 1860-61.

478 GOVERNMENT OF INDIA.—Gold and filagree work, brooches, bracelets, and jewellery, from Bauda, Benares, Allahabad, Calcutta, &c.; sandal-wood bracelets, mounted in gold and silver, from Mysore; coins from Ahmedabad; miscellaneous articles of jewellery.

479 NEAL, J. *Edgware-rd. London.*—Gold and silver filagree bracelets, and other articles of Indian jewellery.

480 HENCKELL, DU BUISSON, & CO. *London.*—Gold bracelets and articles in silver, made by Mootianassary, of Trichinopoly.

481 FLETCHER C. NORTON, *London.*—Cup of purest silver of Burmese manufacture.

482 HALLIDAY, FOX, & CO. *London.*—Ruby ring, ear-rings, and bracelets, from Burmah.

483 DOWLEANS, A. M. *Calcutta.*—Tusks mounted in silver as tablets.

CLASS 34.

484 GOVERNMENT OF INDIA.—Ink and pen tray of glass, and glass bangles and rings, from Meerut.

CLASS 35.

485 HUNTER, DR. A. *School of Industrial Arts, Madras.*—Collection of pottery made at the above institution.

486 MCPHERSON, DR.—Card trays from Java.

487 RAJAH OF VIZIANAGARUM.—Pottery from Vizagapatam.

488 SHAIK FAKEER HOSSEIN.—Nineteen specimens of pottery.

489 THATHNA LINGA CHARRI.—Stoneware from Akkèr.

490 ARMUGA UDAGAR, *North Arcot.*—Pottery made in the Gudiatam Talug.

491 GOVERNMENT OF INDIA.—Collections of pottery from Allahabad, Gyah, Patna, Moradabad, Bellary, and Lucknow; collection of pottery made at the Bangalore Industrial School.

492 CAMPBELL, DR. *Darjeeling.*—Thirty-three specimens of pottery.

493 HALLIDAY, FOX, & CO. *London.*—Burmese water-pots.

CLASS 36.

495 MADRAS CENTRAL COMMITTEE.—Four dozen figures from Condapully, Guntoor.

496 MCPHERSON, DR. — Miscellaneous wares from Malacca and Java.

497 WEST, W. E. L. *Madras.*—Moharum festival cage.

499 PHAIRE, LIEUT. W. *Assam.*—Wooden comb, ivory back-scratcher, and necklaces.

500 LOTHOW DHUR PHOOKER, *Assam.*—Ivory chessmen.

501 CAMPBELL, DR. J. *of Darjeeling.*—Spice-boxes, bowls, baskets, &c.

502 MAHARAJAH OF PUTIALA.—Chess board, chessmen, walking-stick, and sandalwood spoon.

503 GOMES, J. M.—Modelling in wax.

504 CHEE YAM CHUAN.—Articles manufactured from cocoa-nut sheels, &c.

505 ANGUS, G. *Singapore.*—Two fishing lines, weight, and hook, for trolling in deep water, from Singapore.

506 MOONSHEE KAFAITOLLAH MOONSIFF. —Ivory comb and stick, from Assam.

507 BABOO POORUANUND BOORUAH. — Ivory comb, &c. from Assam.

508 RAJAH OF TIPPERAH.—Ivory fan, from Chittagong.

509 GOWHATTY, P. B. — Ivory back-scratcher.

510 BABOO GOBINDRAM SHUMAH, *Luckimpore.*—Ivory comb, &c. from Assam.

511 SUTHERLAND, DR. *Moonghyr.* — Bangles.

513 DOWLEANS, A. M. *Calcutta.*—Fans.

515 SINGAPORE LOCAL COMMITTEE. — Mat bags, palm leaf boxes, brooms, cage, &c. from Province Wellesley.

516 GOVERNMENT OF INDIA.—Children's toys from Mysore; punkahs, &c. from Nepaul; chess and draughtsmen from Cuttagh; fifty-seven figures, representing different trades, from Kishnaghur; figures from Lucknow; chess and draughtsmen from Burhampore; inlaid turnery-ware, from Umritsur; walking-sticks from Lahore, Umritsur, Ulwar, &c.; shell bracelets from Dacca; manufactures of small wares in ivory, from Lahore and Umritsur; Burmese figures, from Assam; horns and ivory wares, from various localities; table-mats, &c. from Moonghyr; chowries and hookahs, from Hooghly; fans, from Jessore, Moonghyr, &c.

517 HALLIDAY, FOX, & CO. *London.*— Burmese cards and chessmen.

CLASS 37.

518 BURN, HON. CAPT.—Model of Malay house.

519 MCPHERSON, DR.—Elaborately carved Malacca temple.

520 GOVERNMENT OF INDIA.—Model of pagoda, in Burmese alabaster; wooden model of the Tasaung at Rangoon,: also of the bazaar, at Kishnaghur.

521 H.H. RAO OF KUTCH. — Model of Hindoo Temple of Shiva; model of Bunian chutree.

522 HALLIDAY, FOX, & CO. *London.*— Burmese pagoda.

CLASS 38.

523 MCPHERSON, DR. *Madras.*—Packet of illustrations of native art pictures of various places; carved sandal-wood boxes, with illustrations of native art, on ivory, from Delhi, Agra, Benares, &c.

524 HUNTER, DR. *Madras.*—Twenty-seven specimens of geometrical drawings, as taught among the Hindoos.

525 GOVERNMENT OF INDIA.—Large collection of examples of native paintings on ivory, &c. with portraits, from Ulwar, Lucknow. Delhi, Umritsur, &c.

526 HALLIDAY, FOX, & CO. *London.*— Burmese historical pictures.

CLASS 39.

527 SEVARAMMUDU, VENGABATHUDU, & KALAPA, *North Arcot.*—Collection of images, &c. in copper.

528 MCPHERSON, DR. *Madras.*—Bronze, gilt, and silver figures of Godama, of ancient Hindoo art; elephant with howdah, &c.

529 GOVERNMENT OF INDIA.—Pithwork figure of the Rajah of Tanjore, and other pith-work figures; statuettes, from Berhampore.

CLASS 40.

530 HUNTER, DR. A. *Madras.*—Collection of engravings.

531 GOVERNMENT OF INDIA.—Books of representations of various trades, from Umritsur, &c.

532 COOKE, M. C. *London.*—Wood-cut block of sandal-wood, with impressions of engraving (experimental).

[The articles for which space cannot be given have been removed to the India Museum, Fife House, Whitehall-yard, there to constitute a supplementary collection.]

JAMAICA.

East-side of N.E. Transept, under Gallery.

1 MCCRINDLE, J.—Animated nature.

2 HARRIS, R.—Animated nature; liqueurs; chemical preparations; meals; manufactures.

3 BOWERBANK, DR. L. Q. — Pottery; sticks; chemical preparations; meals; oils; manufactures; ladies' ornamental work.

4 POTTS, DR.—Animated nature; woods; basts; sticks; botanical specimens; seeds; coffee; chemical preparations; starches; oils; manufactures.

5 BARRETT, L.—6 FYFE, HON. A. G.— 7 SAWKINS, J. G.—8 BELL, J. C.—Minerals.

9 R. SOC. OF ARTS, JAMAICA.—Minerals; woods; botanical specimens; rums; liqueurs; sugars; chemical preparations; starches; oils; fine arts.

10 COOPER, CAPT. R.N.—Minerals; chemical preparations; starches.

11 SAILMAN, J.—Manures.

12 SOC. INDUSTRY, HANOVER.—Woods; sticks; rums; liqueurs,; sugars; chemical preparations; manufactures.

13 GORDON, R.M.—Woods; basts; coffee; chemical preparations; manufactures, &c.

14 HAUGHTON, ——Woods.

15 CAMPBELL, REV. J.—Woods; basts; botanical specimens; seeds; coffee; chemical preparations; starches; manufactures.

16 WILSON, N.—Fibres and basts; fruit in spirits.

17 NASH, MRS. J.—Basts; botanical specimens; seeds; ladies' ornamental work.

18 CAMPBELL, DR. C. — Basts; seeds; oils.

19 MAXWELL, DR. J.—Basts; botanical specimens; seeds; chemical preparations; starches; oils; manufactures.

20 BROWNE, W.—Basts; coffee; manufactures.

21 WILSON, E. F.—Basts; seeds; coffee; oils.

22 HYAMS, REV. —.—Sticks.

23 MCINTYRE, —.—24 HEPBURN, R.—Botanical specimens.

25 PAINE, W. S.—Botanical specimens; coffee.

26 CLARK,—GENERAL PENITENTIARY.—Botanical specimens; manufactures.

27 VIAN, THE MISSES.—Wax models.

28 VICKARS, HON. B.—29 PHILLIPS, H.—30 JAMES, MISS E.—Seeds.

31 TRENCH, J. L.—Seeds; chemical preparations.

32 DARLING, HIS EXCEL. THE GOVERNOR.—Fruits in spirits; manufactures.

33 GORDON, R.—34 "SHEPHERDS' HALL."—35 MILLER, W.—Coffee.

EXHIBITORS OF RUM —36 ROBERTS & GRIFFITHS.—37 DINGWELL, J.—38 SHORTRIDGE, S.—39 CALLAGHAN, D. & HARRISON, J.—40. MITCHELL, J. W.—41 ESPEUT, P.—42 SHOUBORG, A. A.—43 EAST, HON. H.—44 MCKAY, W.—45 RUSSELL, R.—46 SOLOMON, HON. G.—47 LAWRENCE, J. F.—48 HAMMETT, J.—49 "HALL-HEAD ESTATE,"—50 WESTMORLAND, HON. H.—51 BERRY, W. SEN., & 52 WALLACE, J.—53 "SEVEN PLANTATION ESTATE."—54 MITCHELL, J. H.—55 CHILD, W. D.—56 MARTIN, G. L.—57 WESTMORELAND, W.—58 "GIBRALTAR ESTATE."—59 CAMPBELL, T.—60 SINCLAIR, D.—61 VICKARS, W.—62 FOGARTY, D.—63 VICKARS, HON. B.—64 "FONTABELLE ESTATE."—65 COLVILLE, E.—66 CAMPBELL, MISS E.—67 THARPE, J.—68 FISHER, J. W.—69 WETZLER, D. B., & CO.—70 WETZLER, D. N., & CO.—71 SEWELL, W.—72 GORDON, J. W.—73 HOSTOP, L.—74 BODDINGTON, & CO.—75 TYSON, H.—76 BARRETT, C. G.—77 CLARKE, E. H.—78 GRANT, MISS.—79 CLARKE, T. O.—80 FLETCHER, J. & PERYER, J.—81 HIND, R. & STIRLING, W.—82 MILNER, T. H.—83 ELMSLIE, J. & SHORTRIDGE, S.—84 HOSSACK, HON. W.—85 "HAYWOOD HALL."—86 DAWKINS, LIEUT.-COL. W. G.—87 "RAYMOND ESTATE."—88 "NEW RAMBLE ESTATE."—89 MELVILLE, J. C.—90 PENNANT, COL. THE HON. E. G. D.—91 WRAY, J. & NEPHEW.—92 HARVEY, J.—93 GEORGES, W. P.—94 VAZ, I. N.—95 BARCLAY, A.—96 PAINE, W. S.—97 "BLUE MOUNTAIN VALLEY."—98 "DOVOR ESTATE."—99 GARRIGUES, H. L.—100 JARRETT, J.—101 DEWAR, R.—102 LAWSON, G. M.—103 ATKINS, G. W.—104 SHAWE, R. F.—105 MCGRATH, G.—106 JARRETT, H. N.—107 WILLIAMS, J. L.—108 FAYERMORE, G.—109 DOD, F.—110 HOLT & ALLAN.—111 BODDINGTON, DAVIS, & CO.—112 HIND, R.—113 STEPHEN, A.—114 HOSKINS, J. A.—115 HUTTON, W.—116 PARKINS, DAVIS, & SELBY.—117 WHITELOCK, HON. H. A.—118 HEAVEN, W. A.—119 TOD, F.—120 VAZ, I. N.—121 SHARP, E.—122 HAWTHORN & WATSON.—123 WRAY, J., & CO.—124 DENOES, P.—125 GADPAILE, C.—126 ARNABOLDI, G.

127 DERBYSHIRE, J.—128 MELVILLE, J. C.—129 COOKE, A.—130 WRAY & CO.—131 DENOES, P.—132 DAVISON, MRS. J.—133 HENRY, H. G.—134 TRENCH, J.—135 DICKSON, R.—136 GRANT, C.—Liqueurs, &c.

137 GARRIGUES, H. L.—138 RUSSELL, R.—139 MCKENNON, HON. L. M.—140 PAINE, W. S.—141 HOLT & ALLAN.—142 WILLIAMS, J. L.—143 HIGHINGTON, J.—144 WHITELOCK, HON. H. A.—145 PARKINS, DAVIS, & SELBY.—146 HARVEY, J.—147 SHARP, E.—148 HENRY, W.—149 DOD, F.—150 WHITE, R.—151 GRANT, J.—152 BODDINGTON & CO.—153 VAZ, I. N.—154 STEPHEN, A.—155 HOSKINS, J. A.—156 HIND, R.—Sugar.

157 VICKARS, HON. B.—Coffee, &c.

158 SAWKINS, MRS.—159 AARON.—160 A'COURT, WM.—161 SEGUEIRA, E. G.—162 GREY, C.—163 BALL, T.—Chemical preparations.

164 EDWARDS, E. B.—165 BRASS, J.—Chemical preparations. Meals; manufactures.

166 GALL, J.—Chemical preparations. Starches; oils; manufactures.

167 BROWN, MRS. W.—Chemical preparations; starches.

168 NUGENT.—Chemical preparations.

169 KENTISH, MRS. S.—170 JOHNSON MISS M.—Starches.

171 BATLEY, D. W.—Meals.

172 KEMBLE, HON. H. J.—Oils.

173 ARNABOLDI, G.—Oils, seeds.

174 "BROWNSVILLE."—Oils; manufactures.

175 O'HALLORAN, J.—Manufactures; ladies' ornamental work.

176 CHITTY, E.—Manufactures; ladies' ornamental work; fine arts.

177 BELL, J.—178 LAWES & CO.—180 CAREY, W. O.—181 FINGZIES, J. K.—182 "AN AFRICAN."—183 THOMPSON, R.—Manufactures.

184 RANKINE, MRS.—185 HARRISON, MRS. & MISS.—186 JAMES, MISS.—187 ARNABOLDI, MISS.—188 POOLE, MRS. P.—189 TEAP, MISS C.—190 POTTS, MRS. DR.—Ladies' ornamental work.

191 SAVAGE, J. A.—192 MAULL & POLYBLANK.—Fine arts.

193 JAMAICA COTTON CO.—Basts, &c.

194 CHITTY, E. & RIDGEWAY, A. F.—Botanical specimens; manufactures.

195 RIDGWAY, MRS. G. & SISTERS.—The banner,—"Arms of Jamaica."

ADDENDA.

196 BOWERBANK, DR.—Woods.

197 COOPER, CAPT., R.N.—Woods.

198 CAMPBELL, DR. J.—Woods.

199 PAINE, W. S.—Woods.

200 HANKEY, THOMSON.—Rum.

201 MOWATT, W.—Cotton.

202 HAMILTON, HON. DR.—203 CASSON, J.—Coffee.

204 ORGILL, HERBERT.—Cotton.

205 HARMAN, REV. J.—Rum.

206 SCOTT, A.—Manufactures.

207 GALL, J.—Cotton.

208 LEVIEN.—209 CASELEY, R.—210

ADONIO, MARCUS.—211 HARVEY, MRS. WM.—Chemical preparations.
212 JAMAICA COTTON CO.—Manufactures.
213 DUBUISSON.—Liqueurs, &c.
214 ABRAHAMS, J.—Engraving on copper.
215 SIMONS.—Animated nature.
216 WILSON, N.—Woods; cotton.

MALTA.

East-side of N.E. Transept, under Gallery.

1 BORG, P. P. & CO.—Black silk lace articles.
2 MUNNERO, V.—Black silk lace articles.
3 GRECH, G.—Embroidered handkerchiefs.
4 AZZOPARDI, F.—Black silk lace articles.
5 MEILACH, A.—Black silk lace articles.
6 MICALLEF, S.—Black and white lace articles.
7 LEONARDIS & BELLIA.—Black lace articles.
8 BELLIA, M.—Lace collar, and a specimen of broad white lace.
9 BELLIA, P.—Black lace articles, silver filigree work.
10 MUIR, G.—Silver and silver gilt filigree work; stone vases; flower pots, &c.
11 VELLA, BROS.—Cotton counterpanes, quilts and table covers, and straw hats.
12 AGUIS, P. P.—Cottonina cloth.
13 MUSCAT, A.—Cottonina cloth.
14 ZAMMIT, G. B.—Cottonina cloth.
15 VELLA, A.—Woollen quilts and table cloths.
16 SCHEMBRI, DR. S.—Specimens of paste; seeds; honey; wax; and orange flower water.
17 FRANCALANZA, L.—Stone vases.
18 TESTA, F.—Stone vase.
19 SEGOND & BROTHER.—Carved frame.
20 MELI, G.—Samples of leather, hides, &c.
21 DARMANIN & SONS.—Specimens of marble mosaic work; and stone work.
22 DIMECH, C.—Stone work.
23 TONNA, G.—A counterbass.
24 CATANIA, A.—Carved wood-work, and paper pattern.
25 POLITO, REV. CANON.—Statuettes of saints and knights of Malta.

MAURITIUS.

Near Central Entrance from Horticultural Gardens.

1 WIEHE & CO. *Labourdonnas Estate.*—Raw and manufactured sugar, candy, and syrup.
2 ICERY, E. & CO. *La Gaité Estate.*—Manufactured sugar, &c.
3 BARLOW, H. *Lucia Estate.*—Pure and impure sugar.
4 BULLEN, R. *Gros Cailloux.*—Ordinary sugar.
5 BELZIM & HAREL, *Trianon.*—Sugar for British market, and for Australian market.
6 GUTHRIE, MESSRS. *Beauchamp.*—Sugar.
7 CURRIE, J. *Beau Séjour.*—Sugar to meet English duty, and turbined with water.
8 STEIN & CO. *Gros Bois.*—Sugar.
9 BREARD, F. *Savannah Estate.*—Sugar.
10 DUNCAN, J. *Director of the Botanic Gardens, Pamplemousses.*—A collection of fibres prepared at the gardens from indigenous plants. Arrow-root, spices, and litchi fruit.
11 D'ESMERY, P.—Manure as active as guano.
12 WICHÉ, P. A.—Coral lime, reef and inland coral, and a case of colonial spirits.
13 BERGICOURT & CO.—Cigars and snuff.
14 MURPHY, W. *Long Mountain.*—Twenty-five specimens of native woods; cotton.
15 FRESQUET, A.—Vacoa tree, the roots made into various articles; hats of date tree leaves, &c.
16 MARTINDALE, E.—Finest honey; a tree-fern walking-stick.
17 BARCLAY, LADY.—Arrow-root.
18 BONIEUX, *Arsenal.*—Specimens of lalo fibre *(Abelmoschus esculentus)*, biscuits, crystalized manioc, and arrow-root flour.
19 CLOSETS, M. DE.—Iron ore.
20 MAUGENDRE, C. A.—A barrel of disinfecting powder.
21 DUMONT, MDLLE.—Embroidered handkerchief.
22 BROUSSE, M.—23 LANGLOIS, C.—Vanilla.
23 BEDINGFIELD, HON. F.—A walking-stick.
24 MORRIS, MRS. J.—Bag made of acacia seeds; articles from "Seychelles Isles," of cocoa-nut leaves and grasses.
25 BEDINGFIELD, HON. F.—A walking stick.

NATAL.

Under Gallery, Northern Courts, West-side of N.E. Transept.

NATAL COMMISSIONERS (Hon. Sec. R. J. MANN, M.D.)—Extensive collections of—
1 Food substances: Sugar, arrow-root, coffee, cereals, pepper, roots, fruits, and preserves; tea, cheese, spirits, honey, cured meat, &c.
2 Horns, skins, carosses, tusks, &c. of native animals, the produce of the chase; feathers of the ostrich, crane, &c.; samples of wools, fleeces, &c., and sponge from the Umgeni.
3 Textile and other materials: Cotton, Kafir cotton, flax seed, hemp, &c.; specimens of a variety of woods, bark, fibres, &c.
4 Colonial manufactures: Tanned skins and articles of leather, horns, soap, candles, tallow, cigars, tobacco, bricks, tiles, models, &c.
5 Kafir manufactures, illustrating native industry and domestic economy: Shields, assegais, clubs, musical instruments, ornaments, implements, models, &c.
6 Mineral substances: Freestone, granite, plumbago, quartz, limestone, fossils, sulphuret of lead, mineral water, &c.
7 Specimens of natural history: Snakes, birds, insects, shells, &c.; Kafir medicines.
8 A counter and frame made of native woods a map of Natal; charts illustrating

the climate of Natal, from observations taken by Dr. MANN; water-colour drawings of colonial scenery, with photographs of scenery, portraits of natives, &c.

NEW BRUNSWICK.

Centre of N.E. Transept.

1 NEW BRUNSWICK COMMISSIONERS.—Specimens of wheat, oats, buckwheat, rye, barley, beans, Indian corn, wheat flour, barley meal, buckwheat meal, rye flour, oatmeal, and hulled barley.

Albertite coal, freestone, and granite.

Native woods, with twigs and leaves.

Native woods, unmanufactured, and manufactured into doors, ballusters, and articles of furniture.

A drill harrow, a mould board plough, a horse-rake, a saw frame, a double sleigh, a travelling waggon, ships' blocks, &c.

A sample of leather engine-hose and discharge pipe.

Models of steering apparatus; of capstan and windlass; of suspension-bridge over falls of St. John's River; of road and railway bridge over Hammond River; of railway bridge over Salmon River; of a N. B. railway train; snow plough and flange cleaner; of railway engine-house, St. John; of passenger locomotive and tender; and of a saw-mill.

Chilled locomotive car-wheels.

A collection of edge-tools, hammers, &c.

Homespun cloth, rug, socks and mitts Indian bead-work and dress, basket-work transparent shop window-blinds; dried grasses.

Preserved salmon and lobster.

Photograph views in the Colony.

2 SCRYMGEOUR, J.—Horse-shoes.

3 HEGAN, J. & J.—Sattinet; union tweed and flannel.

4 SCOVIL, N. H.—Nails, ship-spikes, &c.

5 PHILPS BROS.—Bunting, wrapping and sheathing paper.

7 ADAMS, W. H.—Carriage and railroad springs.

8 MAGEE, A.—Beaver cap, gauntlets, and coat; bear and lynx robe; silk hats, &c.

9 RANKINE, A.—Biscuits.

10 MCMILLAN, J. & A.—Bookbinding.

11 PEARCE, C.—Stand of brass castings.

13 FOSTER, T. A. D.—Case of dentistry.

14 PRICE & SHAW.—Single horse sleigh.

15 MATTHEWS, G. F.—Case of minerals.

16 SPURR, DR. WOLFE.—Oils from coal.

17 CHAPMAN, J.—Hearth-rug.

18 WESTMORELAND AND ALBERT MINING AND MANUFACTURING CO.—Fossil-fish from their mine.

19 POTTER, E.—Specimen frames; boxwood moulds.

20 BLACKTIN, C.—Circular saws, variety of saws, knives, &c.

21 BOWREN & COX.—Photographic views.

22 FLOOD & WOODBURN.—Ditto.

23 JARDINE, MISSES. — Model summer-house, made from cones.

24 THOMSON, MRS. R — Watch-pocket, made from cones.

25 STEVENS, MRS. D B.—What-not, made from cones.

27 FLEMING & HUMBERT, *Phœnix Foundry*.—Oscillating steam-engine.

30 MCFARLANE, P.—Forks and hoes.

31 REID, J. H.—Pair of moose horns.

32 GILBERT, S. H.—Model stone picker.

33 LAMONT, M.—Indian dress.

35 HARRIS, J.—Cast-iron enamelled mantle-piece.

36 PROVINCIAL PENITENTIARY.—Wooden ware, brushes, &c.

NEWFOUNDLAND.

Under Eastern Gallery, N.E. Transept.

1 NEWFOUNDLAND GOVERNMENT.—Copper, galena, silver lead, silver ores, marble, and iron; skins of the silver grey, patch, and red fox, the martin, otter, beaver, weasel, and bear; stuffed birds, &c.

A model screw propeller.

2 GISBORNE, F. N.—A map of the colony; a Carriboo stag's head; Esquimaux carvings, and feather bags; patent fire-damp indicator; patent ship steering signals, and a patent railway ticket date cutter.

3 O'BRIEN, HON. L.—Wheat, barley, and oats.

4 DOYLES, E.—Fish oils

5 FOX, C.—Cod-liver oil, seal and cod manure, deodorised.

6 O'DWYER, HON. R.—7 ANGEL.—Cold drawn seal oil.

8 PUNTON & MUNN.—Seal stearine, seal and cod oils.

9 MCBRIDE & KERR.—10 STABB, E.—11 DEARIN, J. J.—12 MOORE.—13 MORRIS, L.—14 MCPHERSON.—Cod-liver oil.

15 KNIGHT, S.—Preserved salmon and lobsters.

16 NORMAN, N.—Preserved curlew.

17 TILLY, MESSRS.—Preserved salmon.

18 MCMURDO, T.—Cochineal colouring.

19 DEARIN & FOX.—Advertising plates.

20 PAGE & CO.—Silver ores.

21 LEAMON, J.—Flax.

22 A LADY.—Snake root, and poplar blossoms.

NEW SOUTH WALES.

Under Stairs of N. Gallery, near the East Dome and Nave,

MACARTHUR, SIR W.—Woods in variety, of Southern districts. 193 specimens.

MOORE, C. *Sydney*.—Woods in variety, of Northern districts. 115 specimens.

1 THOMPSON, HENRY, *Camden*.—21 specimens of woods adapted for posts, spokes, felloes, &c.

2 HOLDSWORTH, J. B. *George-st.*—1-in. board of Wellington pine.

3 LENEHAN, A. *Castlereagh-st.*—Specimen of rosewood.

4 TRICKETT, J. *Royal Mint*.—Specimen of Wellington pine, with bark on.

5 CUTHBERT, — —Twenty-one woods for ship building; iron-bark knee weighing 19 cwt.

6 RILEY J. *Glenmore.*—Log of Myall timber.

7 JOLLY, W. & Co.—Bundle of forest oak shingles.

8 FOUNTAIN, — *Newtown.*—Three specimens of tree fern stems from Lane Cove.

9 MURRAY, HON. T. A.—Specimens of Myall and Boree woods.

10 HALL, L. *Sharwood.*—Small specimen of pine from the Billybong.

11 WARD, MR. *Maitland.*—Specimen of iron-bark wood that has been under ground for 25 years.

12 WILLIAMS, J. *Pitt-st. Sydney.*—Staves of mountain ash.

13 SAMUEL, S. *Pitt-st. Sydney.*—Specimen of Myall.

14 DAWSON, A. *Sydney.*—Twenty-three woods adapted for building purposes.

15 CHAPMAN, C. *Sydney.*—Undressed staves.

16 CASEY, J. B. *for McLeay River Committee.*—Specimens of woods.

17 RUDDER, E. W. *Kempsey, McLeay River.*—Iron-bark timber cut in 1836, constantly exposed to the atmosphere for 25 years.

18 ——— Two specimens of ash exposed to all weathers for 15 years.

19 ——— Blood wood that has been under ground 23 years.

20 ——— Turpentine wood that has been exposed for 29 years.

21 ——— Blackbutt that has been exposed to weather for 15 years.

22 ——— Mahogany that has been exposed for 17 years.

23 ——— Pine that has been used in a mill for 15 years.

24 ——— Three specimens of forest oak.

25 ——— Three specimens of fustic.

26 ——— Palm tree.

27 ——— Spotted gum wood.

28 ——— White blossoming acacia, has been exposed to weather five years.

29 ——— Monthly rose.

30 ——— Cherry-tree wood.

31 ——— Yellow wood.

32 ——— Bastard Myall.

33 ——— Gigantic American aloe.

34 & 35 ——— Yellow cedar (two specimens).

36 ——— Tulip wood.

37 MACARTHUR, SIR W.—Two posts 46 years old.

38 ——— Narrow leaved iron-bark posts 46 years old; section of post 40 years old; butt of post 38 years old.

39 MANNING, EDYE.—Timber from hull of vessel built in N.S. Wales in 1830; vessel still plying.

40 MORT, T. S.—Log of Brigalow timber.

41 CLARKE, S.—Six specimens of fancy cedar.

42 DEAN, A. *Liverpool-st.*—Two pieces of blackbutt timber, one of blue gum.

43 MACARTHUR, SIR W.—Specimen of blue gum of Camden.

44 HILL, E. S.—Specimen of iron bark of Paramatta, and tea tree of Brisbane Water.

44A DAWSON, A.—20 specimens of building wood.

44B HOWARD & Co. *Berner-st.*—Tables and cabinet work from woods of New South Wales.

45 GOSPER, J. *Colo, Hawkesbury River.*—Maize in cob.

46 ——— Maize shelled.

47 MACARTHUR, J. & W. *Camden-park.*—Six kinds of maize in cob.

48 ——— Four kinds of shelled maize.

49 ——— Four bundles of maize.

50 PECK, M. *Hunter River.*—Yellow maize, perfected in 100 days, 70 bushels pér acre.

51 BOWDEN, C. *Hunter River.*—Yellow maize in cob.

52 BATTEN, C. *Frederickton.* Maize in cob.

53 ANDERSON, CAMPBELL, & Co. *Sydney.*—Two samples of maize in cob.

54 MAYTOM & BOURNE, *near Stroud.*—Maize in cob.

55 DOYLE, A. J. *Midlorn, Hunter River.* Maize in cob.

56 PATERSON, T. *Lorn, Hunter River.*—Maize in cob.

57 OAKS, MONTAGUE C. *Seven Oaks, McLeay River.*—Maize in cob, yielding 100 bushels an acre.

58 CLEMENTS, INGHAM S. *Bathurst.*—Wheat.

59 MACARTHUR, J. & W. *Camden-park.*—Wheat.

61 CHAPPELL, T. *Mudgee.*—Wheat.

62 FUTTER, MR. *Lumley, Bungonia.*—Wheat.

63 THOMPSON, H. *Camden.*—Wheat.

64 MACARTHUR, J. & W. *Camden-park.*—White wheat, three samples, 68 lbs., 66 lbs., and 64 lbs. to the bushel.

65 THOMPSON, H. *Camden.*—Wheat flour.

66 SOLOMON, VINDEN, & Co. *West Maitland.*—Wheat flour.

67 SPRING, J. & SONS, *Surrey Hills Mills.*—Wheat flour.

68 ANDERSON, CAMPBELL, & Co. *Sydney.*—Wheat flour.

69 PEMELL, J. & SONS, *Paramatta-st.*—Wheat flour.

70 HAYS, J. *Burrowa.*—Wheat flour.

71 THOMPSON, H. *Camden.*—Wheat flour.

72 ——— Maize flour.

73 SOLOMON, VINDEN, & Co. *West Maitland.*—Maize flour.

74 ANDERSON, CAMPBELL, & Co. *Sydney.*—Two samples of maize flour.

75 STROUD MILLS.—Maize flour.

76 HAVENDEN, J. *Grafton.*—Cotton.

77 PECK, M. *Hunter River.*—Cotton.

78 FOGWELL, R. *Hunter River.*—Cotton.

79 HICKEY, E. *Osterley, Hunter River.*—Cotton.

80 MOSS, H. *Shoalhaven.*—Cotton.

81 NOWLAN, J. B. *Hunter River.*—Cotton, various samples.

82 VINDEN, G. *West Maitland.*—Cotton grown by J. M. Ireland, Williams River.

83 GOODES, C. W. *Sydney.*—Cotton grown at Clarence River.

84 CALDWELL, J. *Pitt-st.*—Feejee Island cotton.

85 PECK, M. *Hunter River.*—Arrowroot from Dalwood roots.

86 NEALE, MRS. *Elizabeth Farm, Paramatta.*—Arrowroot.
87 HASSALL, MISS, *Camden.*—Arrowroot.
88 ROBERTSON, W. *Hermitage, Grafton, Clarence River.*—Arrowroot.
89 FILMER, W. *West Maitland.*—Arrowroot.
90 ———Potato arrowroot.
91 BEAUMONT, W. *Botany.*—Arrowroot.
92 THORNTON, W. *McLeay River.*—Arrowroot pulp for paper stuff.
92A GUNST, DR. J. W. *Clarence River.*—Arrowroot.
93 LARDNER, H. *Clarence District.*—Nettle fibre ("urtica gigas")
94 GOSPER, J. *Colo, Hawksbury River.*—Kurrajong bark.
95 BOWLER, C. E. *Newtown.*—Fibre of Miranda reed.
96 BLACKET, E. T.—Kurrajong bark.
97 BAWDEN, T. *Lawrence, Clarence River.*—Nettletree and sycamore barks.
98 MOSS, H. *Shoalhaven.*—Burrawang fibre.
99 CALVERT & CASTLE, *Cavan, near Yass.*—Kangaroo grass, and another grass.
100 GREAVES, W. A. B. *Dovedale, Clarence River.*—Sycamore fibre.
101 LARDNER, A. *Grafton, Clarence River.* Nettletree fibre and Kurrajong bark.
102 ———Dilly bag of brown Kurrajong, made by aborigines of Clarence River.
103 MOORE, C. *Botanic Gardens.*—Kurrajong bark.
104 GARRARD, *Richmond River.*—Six bags made by aborigines, of native fibres.
105 DE MESTRE, A. *Terrara, Shoalhaven.*—Two kinds of bark used as fish poisons.
106 HILL, E. S. *Wollahra.*—Tea-tree bark, "phormium tenax," and fish wattle bark.
107 MACARTHUR, J. & W. *Camden-park.*—Native flax.
108 SCOTT, T. W. *Brisbane Water.*—Tea-tree bark.
109 SNAPE, P. *Stroud.*—Wattle bark.
110 THOMPSON, *Camden.*—Wattle bark.
111 KREFFT, G. *Australian Museum.*—Grass-tree gum.
112 SIMMONS, C. *Waverly.*—Grass-tree gum.
113 RUDDER, E. W. *McLeay River.*—Collection of gums, dyes, varnish, &c., and 200 specimens of dyeing.
114, 115 BOSCH, J.—Varnishes and gums.
116, 117 SMITH, CAPT.—White wine and red wine.
118 MACARTHUR, J. & W. *Camden.*—White wine of 1858.—119 White wine of 1858.—120 White wine of 1856.—121 White wine of 1849.—122 White wine of 1848.—123 Muscat of 1845.—124 Muscat of 1853.—125 G. Muscat of 1851.—126 Red wine of 1853.—127 Red wine of 1851, bottled in 1855.—128 White wine of 1851, bottled in 1855.—129 White wine of 1853, bottled in 1856.—130 Muscat, 1853, bottled in 1856.—131 White, 1858, bottled in 1862.—132 White, 1858, bottled in 1862.—133 White, 1858, bottled in 1862.—134 Red, 1854, bottled in 1862.—135 Red, 1854, bottled in 1862.—136 Muscat, 1854, bottled in 1862.—136A Red, 1849, bottled in 1861.
137 CARMICHAEL, H.—White wine.
138 IRELAND, J. M.—White wine.
139 PILE, G.—Amontillado.
140, 141 BETTINGTON, MRS.—White, vintages 1858 and 1859.
142 ———Claret, 1860.
143 BLAKE, J. E.—Red wine.
144, 145 MC DOUGALL, A. L.—White wine, 1860; red wine, Malbec, 1860.
146, 147 WINDEYER, A.—Red Hermitage and white Madeira, 1858.
148, 149 LINDEMANN, H. L.—Red and white Cawarra.
150 MC DOUGALL, A. L.—White wine, 1851.
151 COWPER, HON. C.—Wivenhoe Madeira, 1853.
152 MORT, T. S.—Irrawang white, bottled 1850.
153—158 COOPER, SIR D.—Dalwood red and white Lindemann's Cawarra (two kinds), Docker's Burgundy, and Lawson's Burgundy.
159 ASPINALL, T.—Australian wine.
160, 161, SANGAR, J. M.—Rosenberg Shiraz and Brown Muscat of 1860.
162—164 SANGAR, J. M.—Riesling, Tokay, and Aucarot wines of 1860.
165 SMITH, J.—Red Kyamba, 1859.
166 FRAUENSFILDER, J. P.—White wine, 1858.
167, 168 MOYSE, V.—Australian wines.
169, 170 RODD, B. C.—Australian wines.
171 ASPINALL, T.—White wines.
172 FARQUHAR, H. M.—Muscat of Camden.
173 SMITH, J.—White Kyamba, 1859.
174 BLAKE, J. E.—White Kaludah, 1856.
175 ———Red Kaludah, 1856.
176 ———White Kaludah, 1858.
177 WRIGHT, J. *Balgowrie, Wollongong.*—Cayenne from chilies of Nepaul kind.
178 MACARTHUR, J. & W. *Camden-park.*—Cayenne from chilies of Nepaul kind.
179 BOLLAND, MR. *West Maitland.*—Cayenne from chilies of the Nepaul kind.
180 THORNTON, CAPT. W.—Cayenne from chilies grown on the McLeay river.
181 CHURCH, J. *West Maitland.*—Tobacco, leaf and manufactured.
182 MC CORMACK, J. *West Maitland.*—Tobacco-leaf.
183 HAMILTON, & CO. G. & J. *Hunter-st.*—Biscuits.
184 WILKIE, G. & CO. *George-st.*—Biscuits.
185 JACQUES, MISS L. *Balmain.*—Four samples of native currant jelly.
186 SKILLMAN, MR. *near Stroud.*—Box of dried peaches.
187 CAPORN, W. G. *Port-st. Sydney.*—Orange wine.
188 MILLER, P. *Paramatta.*—Raspberry and mulberry wines.
189 LAVERS, J. V. *Sydney.*—Ginger wine.
190—195 MACARTHUR, J. & W. *Camden-park.*—Capers, sorghum, broom-millet, imphee, beans and carob pods.
196 MOSS, H. *Shoalhaven.*—Sponge from Jerrimgong Beach.

197 PRESCOTT, H. *Sydney.*—Rye-grass seed.

198 THORNTON, CAPT. — Sarsaparilla grown on McLeay River.

199 MACARTHUR, J. *Camden-park.*—Walnuts.

200 SCOTT, J. W. *Point-clare, Brisbane.*—Water sugar-cane.

201 LEDGER. C. *Sydney.*—Alpacas.

202 JONES, D. & Co. *George-st. Sydney.*—Native cat-skin rug.

203, 204 HARBOTTLE, W. *George-st.*—Whales' teeth, tortoiseshell, and bees'-wax.

205 NORRIE, J. S. *Pitt-st.*—Bees'-wax.

206 ROSE, MRS. *Campbell-town.*—Honey.

206A NORRIE, J. S.—Honey.

207 ROBB, MRS. JAS. *Kiama.*—Feathers.

208 CRAWLEY, T. W. *Market, Sydney.*—Skins of platypus.

209 CHILD, W. *Mount Vincent, West Maitland.*—Cochineal from the acacia.

210 BELL, H. *Pitt-st. Sydney.*—Bone manure.

211—213 HARBOTTLE, W. *George-st. Sydney.*—Sperm, southern whale, and Dugong oils.

214 KIRCHNER, W. *Grafton, Clarence-river.*—Oleine or tallow oil.

215 YOUDALE, J. *West Maitland.*—Neat's-foot oil.

216 BELL, H. *Pitt-st. Sydney.*—Neat's-foot oil.

217 RUDDER, E. W. *Kempsey, McLeay's River.*—Purified neat's-foot oil.

218 BATTLEY, J. *Castlereagh-st.*—Shank bones.

219 SKINNER, T. *Darling-point.*—Silk.

220 TURNER, G. *West Maitland.*—Silk.

221 SANDROCK, G. F.—Silk.

222 LORD, MRS. *Double Bay.*—Silk.

223 KELLICK, J. JUN. *Philip-st. Sydney.*—Silk.

224 LEE, MRS. SEN. *Paramatta.*—Silk.

225 WHITING, J. *Hanley-st. Woolloomooloo.*—Silk.

226 LORD, MRS. *South Head-road.*—Silk.

227 BELL, H. *Pitt-st.*—Beef.

228 MANNING, J. *Kamareuka.*—Beef.

229 BATTLEY, J. *Castlereagh-st.*—Ox tongues.

230 MYERS, P. *Pitt-st. Sydney.*—Fish.

231 NOTT, J. *West Maitland.*—Tallow.

232 ——— Alpaca tallow.

233 COMMISSIONERS OF SOUTH WALES EXHIBITION OF 1862.—Alpaca pomade.

234, 235 BATTLEY, J. *Sydney.*—Beef and mutton tallow.

236 CHILD, W. *Mount Vincent, West Maitland.*—Cheese.

237 HOLDEN, A. *Gresford, Paterson-river.*—Cheese.

238, 239 RILEY & BLOOMFIELD, MESSRS.—Fleece wool, washed on sheep's back.

240 COX, E.—Fleece wool, washed on sheep's back.

241 MARLAY E.—Fleece wool, washed on sheep's back.

242 LORD & RAMSAY.—Fleece wool, washed on sheep's back; from sheep the progeny of merino stock of Messrs. Macarthur.

243 MACANSH, J. D.—Wool, washed on sheep's back.

244 EBSWORTH & Co. *Sydney.*—Wool, washed on sheep's back.

244A DAUGAR & Co.—Four fleeces washed on sheep's back.

245 HAYES, T.—Scoured wool.

246 RILEY & BLOOMFIELD, MESSRS.—Scoured wool.

247 COX, E. K.—Scoured wool.

248 CLIVE, HAMILTON, & ROWLAND J. TRAIL.—Scoured wool, 1st, 2nd, 3rd, and 4th quality, and lambswool.

249 BETTINGTON, MRS.—Scoured wool.

250 LORD & RAMSAY.—Scoured wool.

251 COX, G. H. & A. B.—Scoured wool.

252 EBSWORTH & Co. *Sydney.*—Scoured wool.

253 BELL, H. *Sydney.*—Scoured wool.

254 RILEY & BLOOMFIELD, MESSRS.—Wool in grease.

255 COX, E. K.—Wool in grease.

256 CLIVE, HAMILTON, & ROWLAND J. TRAIL.—Wool in grease.

257 MACANSH, J. D.—Wool in grease.

258—260 COX, G. H. & A. B.—Wool in grease.

261 LEDGER, C.—Fleece of Cotswold merino ram, 3 years old, 10 months wool.

262 ——— Fleece Leicester merino ram, 2 years old, 10 months wool.

263 DONALDSON, SIR S. A.—Six fleeces in grease.

264 LEDGER, C. *Sydney.*—Alpaca wool.

265 COOPER, LADY.—Case of specimens of gold.

266 WILSON, D. *Adelong.*—Seven specimens of quartz and mundic.

267 THOMAS, J., C.E. *Railway Department.*—Iron ore, coal, and quartz,

268 BAWDEN, T. *Clarence-river.*—Carbonate of iron.

269 LYNCH, R. ESQ.—Auriferous quartz.

270 GORDON, MRS.—Lead, silver, and copper ores.

271 LARKIN, E. ESQ.—Stalagmitic deposit from Wianamatta rocks.

272 SAMUEL, S. ESQ. & CHRISTOE, J. P.—Copper ores and copper.

273 LODER, A. ESQ.—Combustible schist.

274 WILLIAMS, CAPT. D.—Silicate of magnesia.

275 LEVIEN, A. ESQ.—Specimens illustrative of strata encountered in working a rich claim.

276 BLACKET, T. ESQ.—Two kinds of building stone.

277 MACARTHUR, SIR W.—Nine specimens of building stones.

278 CROAKER, C. W.—Limestone, sulphate of barytes, and green carbonate of copper from Bathurst.

279 AUSTRALIAN AGRICULTURAL CO.—Iron ore from Stroud.

280 MOSS, H. ESQ.—Native alum from Shoalhaven.

281 LAIDLAW, T. ESQ.—Lead ore from Jobbin's mine.

282 HUME, H.—Iron, copper, and lead ores.

283 BROWN, MRS. W.—Silicified rock, with leaves, &c.

284 PEARSON, R. W.—Copper ore from Good Hope mines.

285 PATTEN, W. ESQ.—Polished Devonian marble.

286 BROWN, THOS. — Bituminous schist, granite, ironstone, sandstone, alum, sulphate of magnesia, slate, and limestone.

287 CHRISTOE, J. P.—Copper and its ores.

288 DAWSON, A.—Stones used in building.

289 WILSON, A.—Silicate of magnesia.

290 BLAXLAND, A. — Gypsum, emery powder, coal, and geological specimens.

291 THOMPSON, H. *Camden.*—Ironstone and iron ore of Camden district.

292 MOREHEAD & YOUNG, MESSRS.—Copper ores.

293 HOLDEN, A.—Sulphide of antimony.

294 DANGAR, T.—Auriferous quartz.

295 CLARKE, REV. W. B.—White porcelain clay.

296 KEENE, W. inspector of coal fields.—Coal from eleven seams in the colony.

297 JOPLIN, C.—Tooth and bones of fossil kangaroo.

298 SOLOMON, S. & H. *Eden, Twofold Bay.* Copper and lead ores, and auriferous sand.

299 VYNER, CAPT. A.—Polished marble.

300 MCMURRICK, MR.—Auriferous quartz from new diggings.

301 SNAPE, P.—Iron ore, coal, clay, and limestone, from Stroud.

302 HOWELL, MISS.—Two specimens of stone.

303 RUDDER, E. W. *McLeay River.*—Earthy cobalt.

304 SAMUEL, S. *Pitt-st.*—Marble, &c.

305 SNAPE, P. *Stroud.*—Ironstone, &c.

306 MCCULLUM, A. *Woolgarlow.*—Copper ore.

307, 308 SAMUEL, S.—Rock killas from Ophir mine; and iron ore.

309 AUSTRALIAN AGRICULTURAL CO.—Coal from bore-hole seam, Newcastle.

310 ——— Coal from Bellambi.

311 DALTON, F.—Specimens of minerals from Rocky River district.

312 ROYAL MINT, *Sydney.*—Samples of gold, characteristic of the gold fields of the Colony.

313 ——— Auriferous quartz from some of the veins in N.S. Wales, now being worked, or capable of being worked, with profit.

314 ROYAL MINT, *Sydney.*—Two cases, illustrating the various deposits encountered in sinking for gold in this colony, and the character of gold thus obtained.

316 KEANE, W. *Government Examiner of Coal Fields.*—Series of palæozoic fossils, in 40 compartments.

317 IRONSIDE, ADELAIDE E.—The marriage in Cana of Galilee, painted at Rome in 1861 by the Exhibitor.

318 INGELOW, G. K.—Water colour drawings of Entrance to Sydney Harbour, and Manly Beach.

319 MARTENS, C.—Sydney Head (water colour).

320 COOPER, LADY.—Two drawings by C. Martens, and one by Thomas, water colour.

321 COMMISSIONERS OF N.S. WALES EXHIBITION. — Six drawings, &c., by E. B. Boulton.

322 HAMILTON, E. *First Provost of University.*—The University of Sydney.

322A DENISON, A.—Senate House of the University.

323 NICHOLL, W. G.—Plaster bust, allegorical of Australia.

324 TENERANI.—Photograph of marble statute of W. C. Wentworth, by Tenerani, of Rome.

324A BLACKET, E. T.—Photographs in variety.

325 JOLLY & CO. *Sydney.*—Five photographic views.

326 PATERSON, J.—A. S. W. Co.'s patent ship Pyrmont—photograph.

327 FREEMAN, BROS. *George-st.* — Collection of photographs.

328 PLOMLEY, JENNER, *Parramatta River.* —Fifty stereographic views.

329 BLACKWOOD & GOODES, *George-st.*—Eight photographic views.

330 GALE, F. B. *Queanbyan.*—Photographic portraits of aborigines and half castes.

331 DALTON, E. *George-st.* — Fifteen frames of photographs.

332 WINGATE, MAJOR.—Panoramic view from Pott's Point—photograph.

333 MORT, T. S.—Five photographs.

334 CUTHBERT, J.—Ship building yard—photograph.

335 WILLIAMS, J. *Pitt-st.*—Masonic officers.

336 HOBBS, J. T.—School of Arts by an amateur—photograph.

337 JOLLY & CO. Messrs.—Five saw mills —photograph.

338 YOUNG, RT. HON. SIR J. and LADY.—Kangaroo and emu of Australian gold, by Hogarth.

339 FLAVELLE BROS. *George-st.*—Mounted inkstand.

340 FINCK & BACKEMANN, *Market-st.*—Bracelet and brooch of Australian gold.

341 HOGARTH, J. *New South Wales.*—Natives in precious metals.

342 BRUSH & MACDONNELL, *George-st.*—Two emu eggs mounted.

342A M'LEAY, MRS. G.—Brooch of white topaz from the Murrumbidgee.

342B M'LEAY, MRS. G.—Table ornaments made from the seed vessels of plants near Paramatta, by the nuns of Subiaca, New South Wales.

343 RIDLEY, REV. W. *Rushcutter's Bay.*—Two primers of aboriginal language.

344 COWPER, MRS.—Key to aboriginal language.

345 BERNICKE, C. L. *Kent-st.* — Bookbinding.

346, 347 SHERRIFF & DOWNING.—Account books and bookbinding.

348 RICHARDS, T. *Government Printing Office, Sydney.*—Bookbinding and publishing.

349 FURBER, A. G.—Bookbinding.

350 SANDS & KENNY, *George-st.*—Account books and printed books.

351 REES, G. H. *Castlereagh-st.*—Books illustrative of colonial binding.

352 WAUGH, J. W. *George-st.*—Bookbinding and publishing.

353 DEGOTARDI, G.—Specimens illustrative of the advancement of printing in Australia.

354 MOSS, L. *Hunter-st.*—Five specimens music publishing.

355 CLARKE, J. R. *George-st.*—Book of printed music, &c.

356 ANDERSON, J. R. *George-st. Sydney.*—Two specimens of printed music.

357 CLARKE, H. T. *Castlereagh-st.*—Two gig whips, four stock whips, and piece of hide.

358 HINTON, BROS. *George-st.*—Gentleman's saddle.

359 M'CALL, D. *Hunter-st.*—Gentleman's saddle, &c.

360 BOVIS, C. *King-st. Sydney.*—Three pairs of boots and pair of slippers.

361 VICKERY, J. 375, *George-st. Sydney.*—Four pairs of rivet boots; two pairs of slippers.

362 HALL & ANDERSON, *Pitt-st.*—Harness leather, sole and kip leather, saddles, &c., &c.

363 BRUSH, J. *George-st.*—Lady's side saddle.

364 LOBB, J. *Pitt-st.*—Boots.

365 BEGG, J. E. *Glenmore Tannery.*—Sole leather.

366 ROW, J. *Camden.*—Kip leather.

367 GOODLUCK, J. G. *Camden.*—Leather.

368 SMITH, J. *Botany Tannery.*—Skins and leather; improved graining board.

369 VINDEN, G. *West Maitland.*—Case of Colonial rosewood.

370 REYNOLDS, A. *Balmain.*—Cedar boat.

371 RENNEY, W. *Pitt-st.*—Work table.

372 TUCKEY, W.—Chest of drawers.

373 ————Library door.

374 ————Carved cedar font.

375 ————Carved Elizabethan frame.

376 ORAM, E. *Liverpool-st.*—Carved cedar truss.

377 MORT, T. S.—Carving of Moreton Bay staghorn fern.

378 FULLER.—Window blind.

379 MILGROVE, H. *Park-st.*—Two spiral-turned candlesticks.

380 JONES, W. & SON, 21, *Surrey-st.*—Cabinet of cypress and other woods.

381 COOPER, LADY.—Set of drawers, table, book-stand, &c.

382 GILLMAN, MRS.—Table top of cypress pine from Northern districts.

383 COMMISSIONERS OF N. S. W. EXHIBITION OF 1862.—Fourteen pieces of cabinet work from colonial woods.

384 EBSWORTH, F. E. & Co. *Bridge-st.*—Reel cotton manufactured from cotton of N. S. Wales.

385 PATERSON, MRS. *Manton Creek.*—Socks from opossum wool.

386 ZIONS, H. *Castlereagh-st.*—Coats and vests made from a colonial invention.

387 RUSSELL, CAPT. *Regentville.*—Colonial tweed.

388 FARMER & PAINTER.—Two suits of Colonial tweed.

389 CAMPBELL, M. M. *Sussex-st.*—Colonial tweed.

390 LONDON COMMISSIONERS OF N. S. WALES.—(A.) Merino wool manufactures: series under ten consecutive numbers of light and heavy fabrics manufactured by Joseph Craven, Benj. Gott & Co., Barker & Co., Verity of Bramley, Paton & Co., Pease & Co. Carr & Co., of Twerton; also sundry goods from the general committee of Bradford. (B.) Alpaca wool manufactures: series of cloths manufactured at Bradford from Australian alpaca wool.

390A BARKER & CO. *Leeds.*—Case showing various stages of manufacture between raw wool and cloth.

390B BRADFORD COMMITTEE.—Case showing various stages of manufacture in merino and alpaca wool.

391 DIRECTORS OF RANDWICK ASYLUM.—Cabbage tree plait.

392 DUFFIN, J. *Sussex-st.*—Cabbage tree manufactures.

393 ENGLISH, MISS KATE, *Jamberoo.*—Cabbage tree and plait.

394 PRESCOTT, H. *Sydney.*—Cabbage tree 100 hands.

395 NEW SOUTH WALES COMMISSIONERS.—Plait work, cap, belt, and mat, of cabbage-tree.

396 FARMER & PAINTER, *Pitt-st. Sydney.*—Two hats of cabbage tree.

397 HASSALL, REV. T. *Berrima.*—Cabbage-tree hat.

398 GREGORY & CUBITT, *Aldermanbury.*—Cabbage-tree hats.

399 BOUSFIELD, F. *Crystal Palace.*—Cabbage-tree hat.

400 BIDDELL BROS. *George-st.*—Confectionery.

401 CATES, W. G. *George-st.*—Confectionery.

402 AUSTRALIAN SUGAR CO. *Sydney.*—Sugar and spirits, manufactured in the colony.

403 SACHER & JOSSELIN, *Sydney.*—Confectionery.

404 WELLAM, N. *Burwood.*—Pottery.

405 ENEVIS, W. *Bathurst.*—Soap and candles.

406 KIRCHNER, W. *Grafton.*—Candles and soap.

407 KENSETT BROS. *Campbell-st.*—Blacking.

408 MONK, D. J. *Pitt-st. Redfern.*—Blacking.

409 BLAND, DR. *Sydney.*—Model of ship, showing mode of extinguishing fire in hold, &c.

410 BLAND, DR.—Model of atmotic ship.

411 DAWSON, R. *Lower George-st.*—Model of coffer dam.

412 COWPER, MISS.—Model of St. Philip's Church.

413 BLACKET, E. T. *Sydney.*—Model of St. Mary's church, Maitland.

414 THACKERAY, B.—Model of horse railroad, invented by P. Brawen.

415 WOORE, T. *Sydney.*—Model of supporting rails.

416 KIRKWOOD, D. S. *Bega.*—Model of bridge.

417 FRANCIS, H. *Balmain.*—Model gas retort.

418 COOKE, A. *Randwich.*—Model of St. John's church, Darlinghurst.

419 WOORE, T.—Model of bridge.

420 Low, J. C. *Pitt-st.*—Model of Sofala gold diggings.

420A Crystal Palace Co. —Model of Government dry dock in Sydney Harbour: length, 300 feet; depth, 25 feet.

420B ——— Model of Mort dry dock in Sydney Harbour: length, 350 feet; depth, 26 feet. The "Simla," one of the largest steamships in the fleet of the P. & O. Co. has been repaired in this dock.

421 Cowper, S. S., H. Moss, E. Herborn, and Miss Macarthur.—Aboriginal implements and weapons, and work-box.

422 Hall, J. B. *Richmond-ter.*—Case of birds.

423 Caporn, W. G. *Fort-st.*—Two mats; rock sea-weed.

424 Hensley, Mrs. *Hunter River.*—Seaweed.

425 Bate, J. E. *Merrimbula.*—Collection of sea weeds.

426 Krefft, G. *Australian Museum.*—Seven cases of reptiles.

427 Becker, A. *Australian Museum.*—Fishes, &c.

428 Rudder, E. W. *Kempsey.*—Sixty-eight specimens of birds.

429 Crawley, T. W. *Market, Sydney.*—Stuffed birds (120), sea-weeds, &c.

430 Sawyer, H. *Derwent-st. Glebe.*—Insects.

431 Barnes, H. *Australian Museum.*—Casts of reptiles.

432 Cooper, Sir D.—Various geological specimens.

433 Krefft, G. *Australian Museum.*—Reptiles in variety.

434 Gipps, Lady.—Case of birds.

435 Lavers, G. V. *George-st.*—Noyeau, and orange wine.

436 Monk, D. J. *Pitt-st. Redfern.*—Vinegar.

437 Leycester, A. A. *Singleton.*—Fishing rod.

438 Fuller, *Pitt-st.* —Dwarf Venetian blinds.

439 Chapman, C. *Sydney.* — Colonial staves.

440 Fletcher, D. *Wynyard-sq.*—Dental work.

441 Cooper, J. *Woolloomooloo.*—Church glazing.

442 Jennings, W. *George-st. Sydney.*—Cutlery.

443 Wacey, G. *William-st. Woolloomooloo.*—Intersection ornaments, &c.

444 The Surveyor-General. — Two maps of Victoria and N.S. Wales.

445 Bryce, J. *Laurence, Clarence River.*—Casks of different woods.

446 Hill, E. S. *Woolhara.*—Peg tops of various woods.

447 Macarthur, Sir W.—Cherry brandy.

447A ——— Axe handles.

448 Mann, G. K.—Hone stones.

449 Norrie, J. S. *Pitt-st.*—Fossil cedar.

450 Dalgleish, D. *Sydney.*—Fossil.

451 Manning, Sir W. M.—Vase made of grass tree.

452 Paterson, F. *Market-st.*—Twenty-six herbals.

453 Selfe, H. *Pitt-st.*—Broom of cabbage tree.

454 Hertzhaumer, C.—Surgical instruments.

455 Lackerstein, A.—Cayenne pepper.

456 Hayes, T.—Sheep-skin rugs.

457 Norrie, J. S. *Pitt-st.*—Cajeput oil, and a variety of ottos and essential oils.

458 Thomson, H. *Camden.*—Lime, bricks, bullock yoke, &c.

459 Schulte, R. *Woolloomaloo.*—Dyes and dyeing.

460 Shaw, G. B.—Two engravings.

461 Wainwright, J.—Flute of myall wood.

462 Jolly, W. & Co.—Bullock yoke, &c.

463 Macarthur, Sir W.—Walking sticks.

464 Enever, W. *Sydney.*—Coach wheel illustrating woods used in the trade.

465 Goodsell, F. J. *Newtown.*—Bricks.

466 Hale, T. *Bellambi.*—Quartz gold.

467 Hertzhaumer, C. *King-st.*—Surgical instruments, invented by Dr. Bland.

468 Bousfield, F. *Crystal Palace.*—Letters, medals, &c.

469 Macarthur, General. — Medal struck in 1856 to commemorate the establishment of Constitutional Government in Victoria.

NEW ZEALAND.

Under Gallery fronting Nave, near Eastern Dome.

PROVINCE OF AUCKLAND.

1 New Zealand, Bank of.—Otago gold.

2 Hunter.—Gold.

3 Reading, J. B.—Sample of gold from Terawiti.

4 Heaphy, C. & Ewen.—Gold and auriferous quartz.

5 Heaphy, C.—Minerals, ores, auriferous deposits, fossils, views, maps, frames, shells, magnetic sand.

6 Jones, A.—Copper ores.

7 Great Barrier Copper Mining Co.—Copper ore.

8 Holman, J.—Stalactite, building stone.

9 Smales, Rev. G.—Stalactite, lignite, lava and quartz in ditto, war weapons, paddles, obsidian.

10 Hancock, J.—Specimens of limestone, trachytic stone.

11 Cadman, J.—Coloring pigment, slab of mottled kauri; volcanic and building stones.

12 Gilbert, H.—Pumice stone.

13 Pollock, T.—Iron stone; fire clays.

14 Arrowsmith, W.—Iron sand.

15 Burnett, J.—Marble.

16 Buchanan, F.—Moss agate; spar.

17 White, J.—Agates, cornelians; native adzes; sinkers for fishing.

18 Brighton, W.—Silicate.

19 Elliott, G. E.—Sulphur, silicious incrustations of.

20 ANDREWS, H. F.—Sulphur; shells.
21 WAIHOIHOI COAL CO.—Coal.
22 COLE, G.—Petrified wood; soap stone.
23 PREECE, REV. G.—Petrified rimu (wood).
24 SCOTT, A.—Kauri gum.
25 THIERRY, E. D.—Petrified Kauri gum.
26 WELLS, S.—Blue obsidian; petrified wood.
27 OTAMA-YEA.—Specimens of petrifaction.
28 ROE & SHALDERS.—Slabs of Kauri.
29 RING, C.—Slab mottled Kauri, rimu, matai, auriferous earth.
30 BROWN, T. W.—Aleake wood.
31 GIBBONS, MESSRS.—Woods (Rewarewa and Hinau).
32 COMMISSIONERS INTERNATIONAL EXHIBITION FOR AUCKLAND. — Woods from museum at Auckland.
33 ELLIS.—Vase, cotton stand, with specimens of New Zealand woods.
34 NINNES, J.—Taraire wood.
35 MASON, J.—Loo table, made from New Zealand wood; Pohutuhawa wood.
36 MANAKAN SAW MILLS.—Specimens of woods.
37 REID, REV. A.—Mixed breed wool, dried apples.
38 MORGAN, REV. J.—Mixed breed wools, bark; war weapons, native garment.
39 RUNCIMAN, J.—Ewe Hoggett wool, cross (Leicester and Merino) wool; perennial rye grass seed.
40 SELLERS, MRS.—Wool-cross, Merino and Leicester.
41 WEST, J.—Long wool, mixed breed wool (Merino, Cotswold, and Leicester).
42 CHURCH MISSION SCHOOL, *Otaki*. — 43 BARTON.—44 TAYLOR, WATT, & CO.— 45 HUNTER.—46 LUDLAM.—Various samples of wool.
47 MOORE.—Samples of wether fleece.
48 SHEPHERD. T. JUN. — Wool (Hogg, Leicester).
49 LLOYD, NEIL.—Flax, ropes, lines.
50 PURCHAS, REV. A. & NINNES, J.— Patent flax.
51 MATTHEWS, W.—Door-mats, fibres.
52 PROBERT, J.—Flax, Pikiareo plant.
53 SMITH, J. S.—Flax basket.
54 TURNBULL, T.—Dressed flax and tow.
55 THERRY, BARON DE.—Flax and New Zealand fibre.
56 WEBSTER, G.—Flax, Kiwi egg, vegetable caterpillars, land shells, fishing-hooks.
57 TAYLOR, REV. R.—Textile materials, warlike and domestic implements.
58 INNES, J. & PURCHAS, G. A.—Coil rope, dressed by patent machinery.
59 HOLT, C.—Netting, wove by loom and shuttle.
60 JAMES, CAPT.—Rigging, made from New Zealand flax.
61 MCEWEN, A.—Californian Prairie grass.
62 HORNE, DR.—Ferns.
63 NEW ZEALAND SOCIETY.—Native robe.
64 OWEN, G. B.—Native garment.
65 ——— One cabinet made entirely from New Zealand woods, height 10 ft., 5 ft. broad, 3 ft. deep.
66 WHITE, W.—Carved tigér and alligator.
67 KING, E.—Stuffed birds, dried apples, gold, limestone, cotton.
68 GOODFELLOW, J.—Soap, candles.
69 BLEARZARD, R.—Buckets.
70 VOLCKNER, REV. C.—War canoe, mats, garments, baskets.
71 CHAMBERLIN, H.—Huni-Huni.
72 ——— Four native New Zealand fish-hooks.
73 COMBES, DALDY, & BURTT.—Guano from South Sea Islands.
74 BURTT.—Guano.
75 FOX, MRS.—Drawing of New Zealand flora.
76 COMBES & DALDY.—Coffee; cotton from S. Sea Islands; Kauri gum; iron sand.
77 CROMBIE, J. N.—Photographic views of local scenery, groups of members of House of Representatives.
78 JONES, F. L.—79 BARRAUD, C. F.— 80 MARTYN, A.—Views of local scenery.

PROVINCE OF NELSON.

1 NELSON COMMISSIONERS. — A library table, cloth, photographs and stereoscopes.
2 NELSON PROVINCIAL GOVERNMENT.— A collection of gold specimens, each weighing 50 ounces, coal, and maps.
3 NELSON INSTITUTE COMMITTEE.—Native copper, rock with scales of gold, red hæmatite and chrome ore.
4 NELSON CHAMBER OF COMMERCE.— Dressed flax.
5 DUN MOUNTAIN CO.—Copper ore and chrome ore.
6 MORSE, N. G.—Fleece of wool, long Leicester.
7 BLICK BROS.—Scoured wool, Hinau bark for dyeing cloth.
8 NATTRASS, L. New Zealand flax for paper, and blue colour from Nelson chrome.
9 CURTIS BROS.—Plumbago, from Pakawau.
10 HACKET, T. R.—Coal from Buller river and Waimangaroha, W. coast.
11 LEWTHWAITE, J.—Coal from Pakawau.
12 MCGEE, C.—A stick of Rata.
13 EVERETT, E.—Chrome ore from Marsden's sett.
14 WIESENHAVEN, C.—Iron sand, from Blind Bay.
15 REDWOOD, H. JUN.—Wheat.
16 MONRO, D.—Oats.
17 BAIGENT, SEN.—Timber, native woods, foliage of trees.
18 HARLEY, C.—Nelson hops.
19 ELLIOT, C.—Tea-poy of native wood.
20 ANDREWS, T.—Chrome ore from Ben Nevis.

PROVINCE OF OTAGO.

1 HOLMES, M.—500 ozs. of gold specimens from mines in Otago; views of local scenery; provincial newspaper printed on satin; samples of grasses, corn, and wool.

PROVINCE OF WELLINGTON.

1 HUNTER.—2 TAYLOR, WATT, & CO.—3 LUDLAM. — 4 SELLERS. — 5 MOORE. — 6 BARTON.—7 CHURCH MISSION SCHOOL.—Various samples of wool.

8 BARRAUD, C. F.—Sketches showing the growth of the rata tree, &c.

9 ——— Photographs.

10 TAYLOR, REV. E.—Maori implements and numerous textile materials.

11 HUNTER.—Terawiti gold.

12 RENDING, J. B.—Wairiki gold.

13 NEW ZEALAND SOCIETY. — Basket made of "flax."

14 ELLIS.—Cotton-stand of New Zealand woods.

NOVA SCOTIA.

West-side of N.E. Transept.

1 PROVINCIAL GOVERNMENT.—A collection of mineral specimens:—Gold from the quartz workings at Tangier, Sherbrooke, Wine Harbour, Allan's Mill, the Ovens, &c.; washings from the auriferous sands at the Ovens; gold bars, &c.

Iron and iron ore from the Londonderry Mines, and other localities.

Coal from the Sydney Mines, the Glass Bay Mines, and the Joggins; and oil-coal from Fraser's Mine.

2 SCOTT, JAMES, ESQ.—A column of coal, 34 feet in height, from the Albion Mines.

3 HOWE, PROFESSOR.—224 specimens of minerals, including barytes, copper, manganese, &c., freestone, granite, ironstone, &c., marbles, clay, slate, anhydrite, clays and mineral paints, infusorial earth and cements, iron and garnet sand, amethysts, jaspers, agates, stilbite, calc spar, ankerite, selenite, and topaz.

4 HONEYMAN, THE REV.—A large collection of specimens illustrating the geology of the colony.

5 DOWNS, A. — A stuffed bull-moose; a case of game birds and wild ducks.

6 FALES, A. JUN.—84 varieties of polished woods, leaves, cones, &c., and a collection of native plants.

6A HOWE, DR. — Medicinal and other plants.

7 BESSONET, MISS.—Water-colour paintings of native flowers.

8 HODGES, MISS.—Baskets made of cones.

9 LAWSON & PILLSBURY, MISSES.—Forest leaves, varnished.

10 BLACK, MRS. W.—Wax fruits & flowers.

11 CHASE, W.—Photograph of Nova Scotia vegetables.

12 COLEMAN, W.—Nova Scotian furs.

13 HALIBURTON, R. G. — Bayberry, or myrtle wax.

14 JONES, J. M.—Native fish, prepared in large glass jars, under direction of PROFESSOR AGASSIZ.

15 WILLIS, J. W.—A collection of edible shell-fish.

16 NOVA SCOTIA COMMISSIONERS.—Dried, pickled and preserved fish, as prepared for export.

17 FRASER, R. G.—Fish oils of the province.

18 ——— A large collection of fruits and vegetables preserved in alcohol; grain, garden, and field seeds.

19 HARRIS & MCKAY.—Flower seeds.

20 CURRY & CO.—*Windsor*—Patent axles.

21 DONALD & WATSON, *Halifax.*—Castings in brass, gaseliers, sleigh bells, &c.

22 CONNELLY, G. *Picton.*—Axes.

23 GRANT, P.—*St. Croix.*—Horse-shoes.

24 SULLIVAN, J.—*Halifax.*—Horse-shoes and a curd chopper.

25 LONDONDERRY MINING CO.—Bar iron.

26 BILL & KERRY, S. *Liverpool.*—Edge-tools, hay and manure forks, and skates.

27 CORNELIUS, J.—Jewellery, manufactured of native gold, pearls, amethysts, &c.

28 SCARFE, F. — Common and pressed bricks, and drain tiles.

29 MALCOLM, R.—Fire-bricks, drain pipes, and pottery.

30 WALLACE.—Grindstone.

31 PICTON.—Grindstone.

32 JOHNSTONE, W.—Carving in Wallace freestone.

33 HOLLOWAY, T.—Purchase blocks.

34 MOSHER, J.—Purchase blocks.

35 WILSON, W.—Purchase blocks, log reel, dead eyes, and belaying pins.

36 MCEWEN & REED, *Halifax.*—Sofas, chairs, and a cabinet, &c. of native woods.

37 GORDON & KEITH, *Halifax.*—Furniture, and a ship's wheel.

38 MOORE, J. *Truro.*—Ox yokes.

39 DICKIE, J.—Patent harrow.

40 FRASER, W. & SON, *Halifax.*—A piano of native wood.

41 BROCKLEY, MEISNER, & BROCKLEY.—A piano of native wood.

42 WYMAN & FREEMAN, *Milton.*—Laths.

43 O'BRIEN, G. L. (late).—A pony phaeton.

44 CURRY, E. & CO. *Windsor.*—A sleigh.

45 CAMERON, J. *New Glasgow.*—Model of a steamer.

46 MOSELEY, E.—Two working models on a new system.

47 MCCURDY, MISS E. *Onslow.*—Woollen cloth, frilled, and sewing thread.

48 DUNLOP, J. *Stewincke.*—Home-spun cotton and wool.

49 LAQUILLE MILLS.—Black and grey satinet.

50 CREED, G.—Grey homespun.

51 BEALS, MRS. *Bedford.* — Women's hose.

52 COWIE & SONS, *Liverpool.*—Skiding leather, hogskins, sole and harness leather.

53 SCOTT, MISS.—A leather picture frame.

54 PHILLIPS, N.—Bookbinding.

55 BLAIR, MRS. J. F. *Onslow.*—Sewing thread.

56 BEGG, MISS E.—A bonnet and hat.

57 CAMPBELL & MCLEAN.—Tobacco, and maple sugar.

58 LYTTLETON, CAPT.—Three water-colour drawings.

59 WOODS, J.—A pencil drawing.
60 HARDING, C.—A pen and ink drawing.
61 COGSWELL, DR.—A set of artificial teeth; two bottles of silex.
62 O'CONNELL, J.—Salmon and trout flies.
63 SARRE, N.—Hair tonic.
64 CROSSKILL, J.—Bear's grease, eau de Cologne, and native cordials.
65 DUPÉ, G.—Cider and bitters.

PRINCE EDWARD'S ISLAND

Centre of N.E. Transept.

LOCAL COMMITTEE (H. HASZARD, *Commissioner in London*).—Corn, pulse, agricultural seeds; flour, meal and pearl barley; pork and dairy produce; linen and woollen manufactures; furniture and screens of native wood; agricultural machine and implements; harness and leather work; native canoes and baskets; patent ship's tackle; horse shoes; preserved fish; samples of textile materials; osiers for basket work; oil painting; hats, &c.; bay-tree tallow; honey, &c.

QUEENSLAND.

Northern Courts, near N.E. Transept, adjoining N.S. Wales.

1 ATTORNEY-GENERAL, THE HON.—Arrow-root and walking-canes.
2 ALDRIDGE, MRS.—Arrow-root; Rosella and pine-apple jams, Granadilla jelly, citron marmalade, and seeds of the cycas media.
3 AUSTIN.—Wood fossils.
4 ARCHER, W.—Sandal-wood, Leichardt tree; and geological specimens.
5 BALFOUR, J.—Twenty-two fleeces of wool.
6, 7 BAZLEY, T. M.P.—Cashmere, manufactured from 250 lace-thread warp, spun from very fine cotton and wool, native produce.
8, 9 BIGGE, F. & F.—Wool and cloth manufactured by A. Laverton, from wool grown by F. & F. Bigge.
10 ——— Log of the Bunya Bunya.
11 BARTLET, N.—Mineral specimens; pictures; and engravings.
12 COXEN, C.—Specimens of Myall wood.
13 COXEN, MRS.—Arrow-root, bees'-wax, honey, and candied lemon-peel.
14 CAIRNCROSS, W.—Sea Island cotton.
15 CAMPBELL, J.—Mess beef, beef tallow; coal.
16 CADDEN, W.—Rosella jam.
17 CANNAN.—Aboriginal implements.
18 CURPHY, MR.—Cyprus pine-root.
19 CHILDS, T.—Sea-island cotton, Rosella jam and vinegar.
20 COMMISSIONERS FOR EXHIBITION.—Wheat, dugong oil, colonial rum, cotton and leather.
21 CHAPMAN, T. T.—Arrow-root.
22 COSTIN, W. J.—Dugong oil, bees'-wax, and arrow-root.
23 COSTIN, T.—Colonial saddle, and stock-whip.
24 CARMODY, W.—Maize.
25 CHALLINOR, G.—Photographs.
26 COCKBURN.—Specimens of silk.
27 COOPER, LADY.—Moreton-bay pearl, set in Australian gold.
28 CLARKE, J. & T.—Broadcloth manufactured at Troubridge from Queensland wool.
29 DAVIDSON, G. & W.—Fleeces of Queensland wool.
29A DAY, S.—Arrow-root.
30 DUDGEON, S. V.—Sample of silk.
31 DOUGLAS, R.—Colonial soap.
32 DENISON, A.—Two cones of the Bunya Bunya, or *Araucaria Bidwilli.*
33 FITZALLEN.—Maize.
34 FLEMING, J.—Flour.
35 FAIRFAX, W.—Specimens of printing.
36 GREGORY, C.—Aboriginal weapon.
37 GREGORY, A. C.—Fibre from pine-apple leaves.
38 GRAY, T.—Colonial leather and boots.
39 GAMMIE, G.—Plaid manufactured from Queensland wool.
40 HARTENSTEIN, A. T.—Arrow-root.
41 HOLDSWORTH, W. A. EXECUTORS OF.—Arrow-root.
42 HAYNES, M.—Geological specimens, cotton fibre from indigenous plants, and aboriginal decorations.
43 HOPE, HON. L.—Sugar-cane, rice, varieties of Banana fibre, sea-island cotton, and flax.
44 HOCKINGS, A. J.—Maize and preserves.
45, 46 HILL, W. *Botanical Gardens.*—120 specimens of woods, water-lily seeds, arrow-root, preserved tamarinds and ginger, medicinal and tanning barks, sarsaparilla, dye-woods, tobacco, rice, cotton, sugar cane and rattans, walking canes, fibre, gum, stock-whip handles, and aboriginal weapons, implements, and ornaments; framed collection of the foliage of the indigenous woods.
47 HOLMES.—Wool in grease.
48 HODGSON & WATTS.—Sixteen fleeces of washed wool.
49 HODGSON, MRS.—Bracelet of quandong seeds set in gold, wine labels ditto.
50 HODGSON, A.—Case of stuffed birds; stock-whip; gold Australian nugget; and map of Queensland.
51 ——— J.—Ivory; cayenne pepper; specimens of natural history.
52 ILLIDGE, R.—Scented soap.
53 JOHNSON, R. & J.—Arrow-root.
54 JOHNSON, J.—Honey, bees'-wax, and orange marmalade.
55 LOVE, G.—Arrow-root.
56 LOVE, E.—Arrow-root.
57 LADE, T.—Grape and pine-apple wine.
58 LAIDLEY, J.—Opossum rugs; and photographs.
59 LUTWYCHE, MR. JUSTICE—Table and chessmen made of Moreton Bay woods.
60 MARSHALL, W. H.—Arrowroot, bees'-wax, and honey.
61 MARSHALL & DEUCHAR.—Twelve fleeces of wool.

62 MARVONEY, M.—Arrow-root.

63 MARSH, M. H.—Wool in fleece; cloth manufactured at Leeds from his flocks; stock-whip; Moreton Bay chestnut (*castanospermum*); Queensland flag.

64 MARSH, MRS.—Sachét made of Queensland woods; brooches of white topaz found at Moreton bay, and bracelets of Myall.

65 MARSH, MISS.—Manna.

66 MACDONALD, C.—Wool in fleece, three cases of stuffed birds, opossum rug, aboriginal weapons, and stuffed native animals.

67 NORTH BRITISH AUSTRALIAN CO.—Wool in fleece.

68 MEEKS, N.—Black wattle bark.

69 O'CONNEL, CAPT.—One ton of copper ore, and specimens of copper.

70 PRATTEN, J.—Sea Island cotton.

71 PASHEN, MRS.—Arrow-root.

72 PATTERSON, S.—Cypress pine board.

73 PUGH, T.—Almanack, Queensland.

73A PAULEY, W.—Specimens of wood, and two turned bed-posts.

74 RODÈ.—Sea Island and upland cotton.

75 PETRIE, J.—Building stones.

76 PETTIGREW.—Model of a ship.

77 STEWART, J.—Arrow-root and maize.

78 STEWART, H.—Two varieties of arrow-root and Sea Island cotton.

79 SUTHERLAND, MRS.—Sea Island cotton, silk.

80 SHOLL, CAPT.—Natural curiosities.

81 SLAUGHTER, A. SEN.—Arrow-root, bees'-wax, honey.

82 SHEEHAN, N.—Specimen of silk.

83 THOZÈT.—Bitter spice, and cascarilla bark, tobacco-leaf, cigars, Sea Island and N.O. cotton, ebony-fibre, and plum-wine.

84 THOMPSON, P. W.—Sea Island cotton, maize, and banana-fibre.

85 THORNTON, W.—Stone tomahawk and calabash.

86 VOWLES, G.—Specimens of silk.

87 WILDER, J. W.—Photographs.

88 WHITE, J. C.—Gum from Myall tree, &c. &c. &c.

89 WAY, E.—Walking canes, rosella jams, bees'-wax, and ginger root.

90 WARNER, J.—Banana fibre, and map of Brisbane.

91 QUEENSLAND GOVERNMENT.—400 lbs. cleaned Sea Island cotton; grown in Queensland.

92 WILDASH, J.—*Ornithorhynchus* (water mole); *Echidna* (Australian hedgehog).

---

93 NORTH BRITISH AUSTRALASIAN CO. 49, *Moorgate-st. E.C.*—Wool.

ST. HELENA.

Close to Mauritius and China, near Horticultural Garden Entrance.

1 ST. HELENA COMMITTEE.—Cotton; coffee; specimens of woods with foliage, bark, &c.; native birds; and specimens of building stones.

ST. VINCENT.

Eastern-side of N.E. Transept.

1 CROPPER, R.—Specimens of the *bulimus rosaceus*, from the egg to the adult shell; sloughs of the common rock crab (*grapsus*); inner and outer bark of the mountain mahoe, *hibiscus elatus*, and rope made of the inner bark.

2 HAWTAYNE, G. H.—Arrow-root, guava jelly, coffee, cacao, plantain meal; gum from the G (?) tree; ginger; oil of the occoa nut, benna, canole nut, castor and ground nut; spices.

3 ANDERSON, F.—A cask of pozzolano.

4 STEWART, C. D. & CLOKE, E. J.—Arrow-root from the Fancy estate.

TASMANIA.

Centre of N.E. Transept.

1—14 ABBOTT, J.—Coal, ores, fancy woods, palings, staves, tanning bark, vegetable fibre, music.

15—17 ALLISON, N. P.—Wool, skins, shell necklace.

18 ALLISON, W. R.—Wedge-tailed eagle.

19, 20 ALLPORT, MR.—Walnuts, filberts.

21—30 ALLPORT, MRS. —— Preserves, vinegar, Tasmanette, water-colour painting, topaz brooch.

31—34 ALLPORT, MR. MORTON.—Shells, Huon pine, stereographs.

35 ALLPORT, MRS. M.—Fancy plait of rush pith.

36 BACKHOUSE, R.—Flax.

37 BAKER, I.—Coal, New Town.

38 BALF, J. D.—Platypus skin.

39 BARCLAY.—Freestone, Glenorchy.

40, 41 BARTLEZ, T. B.—Manna, native bread.

42 BARNET, G.—Bituminous coal, Mersy.

43 BLYTHE, W. L.—Wool.

44—46 BOUTCHER, W. R.—Wheat, wine, and vinegar.

47—113 BOYD, J.—Clay, bricks, pottery, fruit, vinegar, shells, fancy woods, ship-building and railway timbers, palings, skins, fern-trees, 230 feet spar.

114 BRYANT, MISS S.—Flying opossum skin.

115, 116 BURDON, MR.—Carriage wheels, blue gum plank.

117—120 BURGESS, MRS.—Embroidery.

121 BUTCHER, MRS.—Potter's clay.

122 BUTTON, W. S.—Glue.

123—184 CALDER, J. E.—Cubes of sand-stones, marble, limestone, sea-weed, fancy woods, deer horns, photographs, opossum-mouse in spirit.

185, 186 CAMERON, A. L.—Skins.

187 CARTER, W.—Bituminous coal, 12-ft seam, Fingal.

188 CHATFIELD, W.—Chesnut-faced owl.

189 CHILTON, R.—Strong bituminous coal, 4½-ft. seam.

190 CLARK, G. C.—Wools.

191 CLARK, MISS C. A. C.—Railway rug, fur; workbox of fancy woods.

192 CLIFFORD, S.—Stereoscopic views in Tasmania.

193 COLLINS, MISS.—Fancy basket.

194—330 COMMISSIONERS FOR TASMANIA. —Coal (bituminous and anthracite), ores, marble, freestone, grindstones, hones, gold-dust, whalebone, timber-trophy, whale-boats, casks, implements, vegetable fibre for manufacture of paper, barks for tanning and medicinal uses, models of fruits, furs, skins, furniture, guano.

331 COOK, MRS. H.—Myrtle-wood vase.

332 COX, E.—Native bread.

333, 334 COX, F. — Peppermint-wood 25 years cut, cantharides-beetle.

335—337 CRESSWELL, C. F. — Wheat, Talavera and Tuscan; preserves.

338—341 CROUCH, MRS. S. — Preserved meat, hams, Penguin skin.

342—347 CROWTHER, MRS. B.—Sinews, opossum fur, ornamental feather-work, bird-skins.

348—379 CROWTHER, W. L.—Spermaceti, oils, whale's jaws, split timber, sawn timber, railway sleepers, ship planking, blue gum 45 years in use.

380 CRUTTENDEN, T. — Opossum-wool gloves.

381 DALGETY, F. G. — Oil-painting of Hobart Town.

382 DENNY, H.—Bituminous coal.

383 DOBSON, A.—Opalized wood, Syndal.

384 DOOLEY, M.—Native bread.

385, 386 DOUGLAS.—Opalized wood, "La Perouse's" tree.

387—391 DOYLE.—Leather of sorts.

392 DU CROZ, MRS. — Couch rug, black native cat.

393 DU CROZ, F. A.—Rug of grey and black opossum fur.

394 DYSODILE CO. — Resiniferous shale: Mersey river.

395 EMMETT, S. — Gold dust: Hellyer river, Tasmania, N.W.

396 FAWNS, J. A.—Table of fancy woods of Tasmania.

397 FENCHIER, H.—Iron ore.

398 FINLAYSON, A. H.—Cabin bread in Huon pine case.

399 FLETCHER, D. S.—Wheat.

400 GERRARD, REV. T.—Wool.

401—404 GELL, P. H.—Clay, wheat, wools.

405 GIBSON, J.—Oats.

406 GILLON. — Granite from Clarks's Island.

407—409 GLEDHILL.—Boots, sorts, hones.

410—436 GOULD, C. — Ores, dysodile, alum, porphyry, granites, limestones, topazes, bituminous coals, marbles, skins, geological maps, table of Huon pine and muskwood.

437, 438 GOURLAY, F. R. — Porpoise oil, heart of fern tree.

439 GOWLAND & STANARD.—Freestone.

440 GRAY.—Hickory-wood knee.

441 GREENBALGH, M.—Rug.

442 GROOM, F.—Strong bituminous coal, Mount Nicholas, 12 feet seam.

443—445 GRUBB AND TYSON. — Bench screws, office rulers, shingles.

446 GUNN, RONALD.—Aromatic wood.

447, 448 HAES, FREDERICK.—Ornamental plat of rush-pith, and fern-tree vase.

449—452 HALL, R. — Granite, kangaroo, and platypus skins.

453 HAWKINS.—Palings.

454—456 HILL, R.—Blackwood-log, palings, and shingles.

457—460 HORNE, A. J.—Skins; wool.

461 HULL, J. F. — Leather of elephant seal skin.

462 HULL, H. M.—Manna, insects.

463 HULL, H. — Skins of tiger-cat of Tasmania.

464 IRWIN, D.—Muskwood.

465 JUDD, H.—Native bread.

466—468 JOHNSON, T.—Model of apparatus for conveying salmon over to Tasmania.

469 KERMODE, R. Q.—Fine wools.

470—476 LETTE, R. L.—Minerals, shells, woods, Xerotes fibre.

477—479 LEWES, J. L. — Lavender, she-oak, bronze pigeon.

480 LLOYD, MAJOR.—Wattle gum.

481 MACCRACKEN, R.—Beef in canisters.

482 MACCRACKEN, MISS. — Bronze-wing beetles.

483 MACDONALD, W.—Asbestos in serpentine.

484 MACFARQUHAR.—Rug of native furs.

485—488 MCGREGOR, J. — Ship timbers of blue gum.

489 MACLANACHAN, J.—Fine wool.

490—492 MARSHALL, J. — Wheat, oats, barley.

493—495 MEREDITH, C.—Bituminous coal, fibre, native bread.

496—507 MEREDITH, MRS.—Water-colour paintings of flowers of Tasmanian trees, shrubs, and plants, framed in muskwood.

508 MEREDITH, J.—Wool.

509—546 MILLIGAN, J.—Topazes, jacinths, beryl, cairngorm, rock crystal, opal, carnelian, garnet, schorl, hornstone, auriferous quartz, iron ores, galena, obsidian, pumice, alum, Epsom salt, aboriginal baskets and necklaces, shells, loo-table of muskwood inlaid.

547--549 MOORE, DR.—Palings, staff, gate.

550 MORRISON, A.—Busts (2) of Aborigines of Tasmania.

551 MORRISON, J. A.—Grindstone, kangaroo point.

552 NIXON, RIGHT REV. DR. Bishop of Tasmania. — Photograph of groups of Tasmanian aborigines.

553 NOAKE, E.—Flour.

554 OFFICER, R. — Cajeput oil, distilled from the leaves of blue gum.

555—558 OLDHAM, T.—Dray wheels, piles, and ship planking of blue gum.

559 PINK.—Beef.

560—566 POWEL, W. — Table of myrtle, and black-wood, vases; walking sticks, office rulers, turned of jaw and teeth of sperm whale.

567 PROCTOR, W.—Peppermint wood, 29 years in use.

568—570 PYBUS, R.—Arrow-root, grass-tree gum; peppermint-wood, 35 years cut and exposed.

571 RANSOM, J.—Bituminous coal, Mount Nicholas 12-ft. seam.

572, 573 RITCHIE, R.—Oatmeal, groats.

574—576 ROGERS.—Vegetable fibre-barks.

577, 578 ROSS, J.—Ship-timber of blue-gum.

579, 580 SANDERSON, M.—Teeth of sperm-whale. and walking-stick of jaw of the same.

581, 582 SCOTT, J.—Flours; she-oak timber.

583 SEARLE.—Flour.

584, 585 SHARLAND, F. W.—Wattle-tree gum, grindstone.

586 SHARLAND, W. S.—Golding hops.

587 SHAW, M.—Native bread.

588 SHOOLRIDGE, R.—Fossiliferous lime-stone.

589 SMITH, J. L.—Oats.

590 SMITH, P. T.—Wools.

591 STEVENSON.—Lard.

592—604 STUART, J. W.—Ink, basket-work, knee-caps, carriage-mats, ladies' boots.

605—615 STUTTARD, J.—Views in water-colours of scenery on the north coast of Tasmania.

616, 617 SWIFT, A. H.—Bituminous coal, east coast of Tasmania, seam 5 ft. to 6 ft. 10 in.

618 TUFFIELD, MISS.—Feathers of Tasmanian emu.

619 THOMAS, MR.—Skin of penguin, Bass's Straits.

620, 624 THOMPSON, MR. H.—Minerals, gloves of opossum fur, skins, emu's egg.

625 THORNE, J.—Red ochre.

626—628 TULLY, W A.—Paraffin oil, native bread, backgammon board.

629 WADE, MR.—Products of botanic garden.

630 WALCH & SONS.—Bookbinding,

631 WALKER, J. C.—Almonds grown by exhibitor.

632, 633 WEAVER W. G.—Alcohol, skin of musk duck.

634 WEDGE, J.—Paraffin oil.

635 WEDGE, J.H.—Vegetable fibre: a grass.

636, 637 WHITING, G.—Opossum-skin rugs.

638 WHITING, J.—Stalactites.

639 WHYTE, J.—Fresh-water limestone.

640 WILKINSON.—"More pork," skins, &c.

641, 642 WILSON, G.—Wheat, oats.

643 WILSON, J. J.—Preserved fruits of 1861.

644—648 WRIGHT, I.—Wheat, leather of sorts, "Cape Barren" goose.

649 YOUL, J. A.—Busts of two aborigines of Tasmania.

650—654 YOUNG, LADY.—Book-stand, writing ditto, paper-knife, paper-weight, casket.

655 SMITH, L.—Wheat.

656 COOPER, A. H.—Sketches of South Sea Whale fisheries.

657 MARSHALL, G.—Wheat.

658 SMITH, J.—Wheat.

659, 660 LINDLEY, G. H.—Wheat, barley.

661 WILLIAMS, W.—Flour.

662, 663 NUTT, R. W.—Fine wools from Malahide.

664—668 ARCHER, W.—Wools.

TRINIDAD.

Under Eastern Gallery, N.E. Transept.

TRINIDAD EXHIBITION COMMITTEE.—A collection of minerals: Asphalte from the Pitch Lake, glance pitch, iron ore from Maracas, gypsum from St. Joseph, tertiary coal from the eastern coast, and lignite from the Irois coast.

Chemical and pharmaceutical products.

A collection of food substances: Rice, in the husk and cleaned; ground nuts; gingilli or sesamum; varieties of cacao or cocoa; coffee, nutmegs, akee seeds, Brazil nuts, tea; flour of the bread fruit, plantain, yams, tania, sweet potatoe, cushcush (a kind of yam), bitter cassada, sweet cassada; ochro; starch from the cassada, arrow-root. and toloman (canna); cloves, nutmegs, black pepper, and vanilla.

Among substances used in manufactures: Oil from the cocoa-nut, pressed and boiled; whale, castor, Avocado-pear, and carap oils; balsam of copaiva; a collection of ornamental seeds, timite fruit and seed, vegetable ivory, grugru-nuts, rough leaves of curatella, skins of sharks, cauto bark, sponges, mamure; timite, raw and prepared; arnatto, from fresh and fermented seeds.

Textile materials: Wild and cultivated cotton; ochroma or corkwood cotton; leaf and fibres of œnocarpus batawa, of carata, macerated and unmacerated, of langue de bœuf or agave vivipara, sanseviera, of wild cane or heliconia, of musas, plantains, and various other plants.

141 specimens of native woods.

Rope from sterculia caribea and malachra radiata.

Indian jugs, pots, and garglets made of clay mixed with the ashes of cauto bark.

Plantains stewed in syrup, and fruit preserves.

Indian wicker-work made of tirite, a species of calathea; Indian impermeable baskets of the same. Plain and ornamented calabashes; razor strops made of various vegetable substances; fancy baskets of luffa fruit; Indian fans; cocayes, and ornamented articles made of seeds.

Patent fuel, manufactured from pitch, by HAMILTON WARNER, *San Fernando*.

Walking-sticks of native woods.

Photographs made by MR. WILLIAM TUCKER, *Port of Spain*.

VANCOUVER.

Centre of N.E. Transept.

1 EXECUTIVE COMMITTEE, *Victoria.*
Gold; copper and iron ore; coal, limestone, cement stone, slate, sandstone, granite.
A spar of Douglas fir for the International Exhibition flag-staff, 220 feet long.
White and Douglas pine, silver fir, spruce, yellow cypress, cedar, oak, yew, hemlock, maple, dog-wood, alder, white pine and cypress cones and twigs.
Wheat, barley, oats, peas, timothy, and potatoes of field growth. A bunch of barley, of timothy, and of hemp nettle. Garden vegetables.
A bundle of kelp. A specimen of the rock crab.
Oils of whale, seal, dog fish, and oulachan.
A sample of wool.
A pair of antlers. A buck.
Indian manufactures: hemp and net, from the hemp nettle; rope and mantle, from the bark of yellow cypress; hats, a basket, whaling tackle, a harpoon, float and line, halibut fish-hooks.
Models of a stern-wheel steamer, of a side-wheel steamer, of a centre-board schooner.
Specimens of red bricks, manufactured near Victoria.
Three small kegs, a claret-jug, and a drinking-cup, made of native oak.

2 HENLEY, *Clover Point.*—Fifty-two varieties of kitchen-garden seeds.

3 DRIARD, S.—Prepared meat, concentrated soup; apples, and other fruit; sardines and anchovies.

4 FOUCAULT.—Halibut and salmon.

5 FARDON.—Photograph views and portraits.

6 A COLONIAL AMATEUR.—Sketches of scenery near and in the town of Victoria, and wild flowers from Fern-wood.

VICTORIA.

Centre of N.E. Transept, and West-side of Do.

1 ABEL, PROF. J. *Ballarat.*—Meteorite found at Cranbourne; collection of minerals; wine.

2 ALBION QUARTZ MINING CO.—Five cwt. of auriferous quartz.

3 BANK OF AUSTRALASIA.—Samples of gold, 50 in number, from the various gold fields in Victoria.

4 BANK OF VICTORIA DIRECTORS.—Specimens of gold, for the most part alluvial.

5 BANK OF NEW SOUTH WALES, DIRECTORS OF VICTORIA BRANCH.—Specimens of the occurrence of gold in the matrix.

6 BANK ORIENTAL.—Two specimens of quartz rich in gold.

7 BISTOL REEF MINING CO.—Section showing the strata cut through in reaching the quartz reef; bag of quartz.

8 BACK CREEK LOCAL COMMITTEE.—Sections, showing strata cut through in reaching Cornubian Reef, All Nations Reef, and alluvial sinking.

9 BAILLIE & BUTTERS.—Large specimens of quartz, studded with gold; samples of wash-dirt; copper ore, tin ore, &c.

10 BLUCHER'S REEF CO. *Maryborough.*—Sample bag of quartz.

11 BLAND, W. H. CLUNES.—Twenty ounces of silver taken from Victorian gold; gold in various forms; sample of arsenic.

12 BENDIGO GOLD MINING CO.—Section, showing strata and workings of their reef, Bendigo.

13 BUCHANNAN'S REEF, *Inglewood.*—Piece of sandstone, showing quartz veins therein; piece of slate, &c.

14 BURKITT, A. H. *Beechworth.*—Sample of analysis of black sand, gold, &c. from the Middle Woolshed, Ovens District.

15 BENYON, J. *Tarnagulla.*—Alluvial specimen from Doctor's-gully, quartz studded with gold, various specimens of quartz, rich in gold.

16 BLIGH & HARBOTTLE, *Melbourne.*—Two samples of antimony.

17 BLACK HILL QUARTZ MINING CO.—Twenty-five tons of ordinary quartz for crushing.

18 BEECHWORTH LOCAL COMMITTEE.—Samples of granite and other building stones, wheat, Indian corn, and flour.

19 BARKLY, SIR H.—Specimen of meteoric iron from Western Port, and horseshoe made therefrom.

20 BREADING, P. G. *Castlemaine.*—Graptolite, found in forming Barker-street.

21 BANNERMAN, A. *Sandhurst.*—Twenty-seven specimens of auriferous quartz; sundry small specimens from the Eagle Hawk Reef.

22 BRIGHT, BROS. *Melbourne.*—Sample of iron ore from the Ovens District.

23 CLARK & SONS, *Melbourne.*—Tin ore; six casks of auriferous quartz, from Ajax Mine, Castlemaine.

24 CASTLEMAINE LOCAL COMMITTEE.—Sandstone, slate, fossils, &c.

25 CAMPBELL, *Back-creek.*—Fourteen fossils; two precious stones; quartz, with mundic and small crystal.

26 COTOWORTH & WOOD, *Morse's-creek.*—Specimens of gold in quartz, from Oriental Reef.

27 CHAMBERS & GITCHELL, *Beechworth.*—Gold in quartz, from reefs in the Ovens Districts.

28 COLLES, J. *Back-creek.*—Volcanic specimens from Mount Greenock, an extinct volcano.

29 CLUNES ALLIANCE MINING CO.—Section, showing strata in sinking shaft.

30 CLUNES MINING CO.—Twenty-five tons of quartz, to be crushed by machine in Exhibition.

31 CATHERINE REEF MINING CO.—Three samples of auriferous quartz; one sample of quartz tailings.

32 CAIRNS, WILSON, & AMOS, *Melbourne.*—Antimony, reduced from the ore; bar iron rolled from scraps.

33 CAKEBREAD, G. *Geelong.* — Block of limestone, polished.

34 GREAT REPUBLIC GOLD MINING CO.—Cask of auriferous wash-dirt, from a depth of 290 feet, through two layers of basaltic rock.

35 ROYAL SAXON GOLD MINING CO. *Inkerman-lead.* — Auriferous wash-dirt from a depth of 300 feet through two layers of basaltic rock.

36 PRINCE OF WALES GOLD MINING CO.—*Cobbler's-lead, Ballarat.*—One cask of wash-dirt.

37 NELSON GOLD MINING CO. *Sebastopol-hill.* — Wash-dirt, Auriferous cement, and lignite found at a depth of 378 feet through four layers basaltic rock.

38 TEMPERANCE GOLD MINING CO. *Band of Hope Reef, Lit. Bendigo, Ballarat.*—One cask quartz (Schist Reef.)

39 RED JACKET GOLD MINING CO.—Auriferous wash-dirt.

40 CLIFFORD, G. P. *Melbourne.*—Twenty-five small surface stones containing gold, from Fryers-town; quartz road metal containing gold.

41 COGDON, J. *Ballarat.*—Small specimen from Hiscocks prospecting claim, Buningong, where gold was first discovered.

42 MAJESTIC MINING CO. *Black-hill, Ballarat.*—Sample of quartz (Schist Reef).

43 INDEPENDENT MINING CO. *Little Bendigo, Ballarat.*—Sample of quartz (Schist Reef).

44 CAMP CO. *Cobbler's-lead, Ballarat.*—Wash-dirt 400 feet from surface through three layers of basaltic rock.

45 COMMISSIONERS OF VICTORIA EXHIBITION.—Three specimens of auriferous quartz, five specimens of quartz with gold.

46 CASTLEMAINE LOCAL COMMITTEE.—Two quartz crystals.

47 DARTMOUTH REEF, *Inglewood.*—Quartz stone with crystals, taken ten feet from surface.

48 DALY'S REEF, *Inglewood.*—Specimens of quartz.

49 DARCY, *Heathcote.*—Two quartz crystals found at 120 feet from surface.

50 DOWDING & CO. *Sandhurst.*—Specimens of quartz gold, and other metals, from Johnson's Reef, Bendigo.

51 DYER & CO. *Melbourne.*—Limestone and lime from Geelong and Point Nepean.

52 DALGETTY & CO.—Sample of tin ore.

53 EASTWOOD CAPT. *Sandhurst.*—Seven specimens of auriferous quartz, sample of conglomerate from the White Hills.

54 FOORD, G. *Melbourne.*—Collection of minerals associated with gold; meteoric iron etched to exhibit the structure; titanic iron sand; sample of coal and coke from New Caledonia.

55 GETHING, G. *Ballarat.*—Specimen of basaltic rock containing zeolithes.

56 GUILFORD, MR. *Loddon.*—One bag upper or below wash-dirt, one bag alluvial wash-dirt.

57 HART, G. H. *Sandhurst.*—Five samples of auriferous drift from the neighbourhood of Huntly.

58 HALL, J. *Emerald-hill.*—Sample of iron ore from Sandhurst, crude and reduced to pig iron.

59 HEFFERNAN, J. *Sandhurst.* — Iron ore from Sandhurst.

60 INDEPENDENT GOLD MINING CO. *Amherst, Back-creek.*—Section of the Company's claim at Rocky-flat.

61 JOSKE, PAUL, *Melbourne.*—A collection of specimens of quartz rich in gold, from the exhibitor's claim at Sandhurst.

62 JOSEPH, H. *Sandhurst.*—Specimen of quartz from Wellington Claim, Golden-gully; two pieces of quartz, road metal containing gold; specimen of gold in cement.

63 KNIGHT, J. G. *Melbourne* (Architect secretary to Victoria department of Exhibition).—Specimen of building stones at present known in Victoria, and treatise thereon; drain pipes, bricks, tiles, &c.

64 KIDD, P. R. *Fryer's-creek.*—Iron ore producing 70 per cent. of metal.

65 KER, R. *Melbourne.*—Sample of red granite from Western Port.

66 LEWIS, J. *Whroo.*—Specimens of gold-bearing quartz and other minerals, from Balaclava-hill, Whroo District.

67 LEVIATHAN REEF MINING CO. *Maryborough.*—Section, showing strata cut through in reaching quartz reef.

68 LEVY & SONS, *Melbourne.*—Sample of coal from Cape Patterson.

69 LEICESTER, C. *Melbourne.* — Case of minerals; specimens and illustrations of various methods of extracting gold.

70 MARINER'S REEF, *Maryborough.* — Section, showing strata cut through in getting to the reef; samples of quartz.

71 MAXWELL'S REEF (Laidlaw and Party, *Inglewood*).—Eighteen specimens of quartz containing gold, five containing sulphurets.

72 MEADS, R. G.—Specimens of galena, found in the Tullarook Ranges, near Goulburn River.

73 MCNAIR, J. — Clunes nugget, from Clunes, weighing 16 oz.

74 MITCHELL, A. *Avoca.*—Specimens of gold in calcined quartz.

75 MITCHELL, M. *Melbourne.* — Quartz, from the first opening of McIvor Caledonia Reef, McIvor.

76 MARYBOROUGH LOCAL COMMITTEE.—Samples of quartz, various stones, and quartz crystals.

77 MALAKOFF REEF CO. *Steglitz.* — Sample of auriferous sulphides.

78 NUGGETY MINING CO. *Campbell's-creek, Castlemaine.*—Specimens of quartz with gold; yield of reef, 25 oz to the ton.

79 NIXON WM. *Geelong.*—Sample of coal found at the surface, about 10 miles from Geelong.

80 POLKINGHORNE, J. *McIvor.*—Samples of tin ore, bar tin, antimony, oxide of calcium.

81 PARKINS, H. *Sandhurst.* — Two sections, showing strata in deep sinking at Huntly.

82 PRESHAW, W. J. *Castlemaine.*—Boulder taken from a freestone quarry; three quartz crystals.

83 POOLE, A. *Castlemaine.*—Fossils found at Talbot Quarry, Tarradale.

84 ROBERTSON, J. S. *Inglewood.*—Three casks auriferous wash-dirt, and quartz from Inglewood District.

85 ROBERTS & JONES, *Castlemaine.*—Slate flag; sample of granite.

86 RODDA, R. N.—Case of minerals and metals, operated upon by a patent process.

87 RICHARDS, A. *Scotchman's-gully, Bendigo.*—Section, showing distribution of alluvial deposits in connexion with gold.

88 RIGBY, E. *McIvor.*—Washing of gold and black sand from McIvor Creek.

89 SPECIMEN HILL MINING CO. *Eagle Hawk, Bendigo.*—Bottle containing quicksilver and alluvial gold; rough gold, fine gold, &c.

90 SANDHURST (BENDIGO) LOCAL COMMITTEE.—Quartz with gold, flagging building stones, slate.

91 ST. MUNGO GOLD MINING CO. *Bendigo.*—Large quartz stone, containing gold; two small ditto.

92 SMYTH, BROUGH, *Melbourne.*—Collection of rocks and fossils relating to the geology of Victoria.

93 SELWYN, A. C. R. Government Geologist.—Six cases of minerals, rocks, and fossils, relating to the geology of Victoria; gypsum, coal, &c.

94 STIELING, G. F. *Richmond.*—Sample of modelling clay, fine clay, &c.

95 TURNER, W. J. *Beechworth.*—Gold, alluvial gold in slate, bar tin gold in crystallized quartz; precious stones, jewellery.

96 TRIUMPHANT GOLD MINING CO. *Rocky-flat, Back-creek.*—Petrified wood found at 120 feet from surface.

97 VICTORIA GOVERNMENT, per HON. C. HAINES, Minister of Finance.—8,000 ounces of alluvial gold.

98 VICTORIA KAOLIN CO. *Bulla Bulla.*—Block of kaolin, and specimens of its manufactures in various forms.

99 WATSON, J. F. *Back-creek.*—Iron ore from the ranges between the Bet Bet and Adelaide lead.

100 WALL, DR.—Specimen of magnesian limestone.

101 WALTERS & WRIGHT, *St. Arnauds.*—Two quartz stones from a cross spar 90 feet deep; specimens containing gold and silver.

102 WRIGHT, G. E. *Inglewood.*—Large quartz boulder, and numerous specimens of gold in quartz.

103 WELLINGTON CLAIM, *Maryborough.*—Section showing strata in reaching the great reef.

104 WILKINSON, R. *Back-creek.*—Fossils, and three precious stones.

105 WILSON, E. *Back-creek.*—Sample of blue stone.

106 ASKUNAS & CO. *Melbourne.*—Guano from Flat Island.

107 BARNARD, J. *Kew.*—Hyoscyamus leaves, extract and tincture.

108 BOSISTO, J.—Oils, tinctures, varnishes drugs, &c.

109 CONNOR, D. *Bunyip-creek.*—Resin palm nuts, extracts, &c.

110 COLE, Mr. *Murray-river.*—Resins of the *Eucalypti.*

111 CAULFIELD, E. *Toorak.*—Olive oil.

112 DAINTREE, H. *Melbourne.*—Resins.

113 DENNY, E. & J. *Geelong.*—Meat manure.

114 FLETCHER, G.—Resins of various kinds.

115 GRAY, H. *Ballarat.*—Essential oils, pyroligneous acid.

116 HARRIS, *South Yarra.*—Resin of *Eucalyptus Viminalis.*

117 HOLDSWORTH, *Sandhurst.*—Sample pyroxylic spirit.

118 JOHNSON, W. *St. Kilda.*—Oils, resins, &c.

119 KRUSE, J. *Melbourne.*—Fluid magnesia, mineral waters, bees'-wax, leeches from Murray River.

120 MECKMERKAN & CO. *Flemington.*—Superphosphate of lime.

121 MACDONALD, MR. *Wickliffe.*—Samples of salts and crystals from Lake Bolac.

122 MORTON, W. L. *Melbourne.*—Sandarac from *Callitris Verrucosa.*

123 MULLER, DR.—Resins and oils from various indigenous trees and plants.

124 PRAGST, G. *Williamstown.*—Charcoal, tar, and the residue from wood leaves in the manufacture of vegetable gas.

125 ROBERTSON, DR. *Queen's-cliff.*—Resins and essential oils.

126 WOODWARD, G. *Kew.*—Two samples of Victorian guano.

127 AITKEN, T. *Melbourne.*—One barrel ale.

128 BEECHWORTH LOCAL COMMITTEE.—Small samples of wheat, Indian corn, and flour.

129 BENCRAFT, G. *Melbourne.*—Two barrels oatmeal.

130 BAYLES, & CO. *Melbourne.*—Two bags of wheat.

131 BOWLES, J. B. *Back-creek.*—Two small cases biscuits.

132 CASTLEMAINE LOCAL COMMITTEE.—Small sample bag barley.

133 CLARK, R. *Benalla.*—Sample of wheat, barrel of flour from Oven's District.

134 DOEPPER, H. *Richmond.*—Samples of maccaroni and vermicelli.

135 DANELLI, S. *Richmond.*—Samples of maccaroni and vermicelli.

136 DOCKER, REV. J. *Wangaruta.*—One bag wheat.

137 DENNYS, C. & J. *Geelong.*—Charqui; preserved meats.

138 DEWAR, J. *Gisborne.*—One bag wheat.

139 ELLIOT & FAWNS, *Sandhurst.*—One barrel ale.

140 FINLAY, J. *Emerald-hill.*—Two small sample bags of oats.

141 FALLON, J. F.—One bag wheat.

142 FRY, J. *Ascot-mills.*—Flour.

143 FORDHAM.—Bottles of fruit; assorted jams.

144 GREEN, RAWDON.—Mess beef in tierces.

145 GRANT, T. *Melton.*—Victorian prize barley.

146 GREEN, *Warnambool.*—Bag of wheat; barrel of flour.

147 GIRAUD, L. *Collingwood.*—Liqueur, confectionery.

148 HODGKINSON, W. *Prahran.* — Two bottles of honey; one bottle of mead; bees'-wax.

149 HADLEY, T. H. & Co. *Melbourne.*—Flour from wheat weighing 69 lbs. per bushel.

150 JOHNSON, J. *Newburn-park, Port Albert.*—Four tierces of mess beef and pork.

151 KRUSE, J. *Melbourne.*—Sample of Sorghum sugar.

152 KINNERSLEY, D. *Burrambeet.*—Victorian prize oats, 49 lbs. per bushel.

153 LAWRENCE, W. *Merri Creek.*—Three Stilton cheeses.

154 MCKENZIE & Co. *Melbourne.*—Oatmeal.

155 MUELLER, DR.—Tea; ginger; bark.

156 RAMSDEN, S. *Carlton-mills.*—Flour, bran, and wheat.

157 REYNOLDS & Co. *Melbourne.*—Seeds of agricultural produce.

158 RICHARDS, MR. *Albert River.*—Arrowroot grown at South Gipps Land.

159 SWALLOW & Co.—Two cases of biscuits.

160 SANDHURST LOCAL COMMITTEE.—Samples of wheat, Indian corn, and tobacco leaf.

161 STEWART, R. *Geelong.*—Three tins of biscuits; jam, and marmalade.

162 SMITH, T. *Collingwood.*—Two bags of wheat.

163 VICTORIA EXPLORATION COMMITTEE.—Dried beef and meat; Nardoo flour, on which Burke, Wills, and King, the explorers, for a long time subsisted.

164 WILKIE, J. & Co. *Melbourne.*—Prize wheat, weighing 69 lbs. 4 ozs., grown by William Thomson, Gisborne.

165 ABEL, A. T. *Ballarat.*—Wine: colonial.

166 ALBURY & MURRAY RIVER AGRICULTURAL SOCIETY.—Wine: colonial.

167 BRYDEN & HENDRICK, *Geelong.*—Wine: colonial.

168 BLAKE, J. A. *Melbourne.* — Wine: five cases.

169 BREQUET, F. *Geelong.*—Wine: Australian, Sauterne, Burgundy, Claret, white Sauterne.

170 COOPER, R. *Melbourne.*—Wine: red Victoria, white Victoria.

171 DUNOYER, J. *Geelong.*—Wine: white Pineau, Gris.

172 DIXON, P. G. *Melbourne.* — Orange bitters, ginger wine, ginger brandy, soda-water.

173 EVERIST, T. J.—Wine: white Carignan, white Gouais.

174 FALLON, J. F.—Wine: Aucarot, Carbeitrel, Sauvignon, Muscat, Riesling, red Scyras.

175 GROSMANN, *Melbourne.*—Wine: Burgundy.

176 HIRSCHI, F. *Castlemaine.* — Mount Alexander (red); ditto (white).

177 LEMME & Co. *Castlemaine.* — Red Castlemaine wine.

178 MCMULLEN, W. *Geelong.*—Wine: Hermitage; brandy.

179 MATE & Co.—Wine: Aucarot, white Muscat of Alexandria, white Tokay, Riesling (white).

180 NIFFENECKER BROS. *Barabool Hills.*—Wine: Auvernat, black cluster, Burgundy, sparkling Chasselas; brandy.

181 PASSELAIGUE.—Wine: Hermitage.

182 SIDEL, B. *Barabool Hills.*—Burgundy.

183 WEBER BROS. *Batesford, Geelong.*—Wines: Chasselas, Burgundy, sweetwater.

184 WALSH, H. S. *Hawthorne.*—Wine white Longfield.

185 ZORNE, E. *Oakleigh.*—Five bottles of tomato sauce.

186 VICTORIA EXHIBITION COMMISSIONERS.—Coloured plaster casts of fruits grown in Victoria, comprising 57 varieties of apples, 45 pears, 10 cherries, plums, strawberries, figs, oranges, melons, and a large assortment of vegetables.

187 ACCLIMITIZATION SOCIETY OF VICTORIA.—Hair from llamas, alpacas, camels, Angora goats, &c.

188 BARKER, J. & R. *Melbourne.*—Raw silk from worms fed on the black mulberry.

189 CASTLEMAINE LOCAL COMMITTEE.—Native cochineal.

190 CROPPER, W. H. *Melbourne.*—Raw silk.

191 CHUCK, T. *Melbourne.*—Small sample of native cotton and fibres.

192 CROFTS, MR. *Melbourne.*—Samples of raw silk.

193 DARDANELLI, SIG. *Melbourne.*—Raw silk.

194 DOWNIE & MURPHY, *Melbourne.*—Mixed, purified, and common tallow

195 GASKELL, J. *Melbourne.*—Pure emu oil; raw silk.

196 GOUGH & Co. *Richmond.* — Bag of malt from Victorian barley; one bag of malt from Californian barley.

197 HAYTER, H. H. *Melbourne.* — Specimens of *Cryptostemma Calendulaceum.*

198 LAMBERT, T. *Richmond.*—Four samples of bark wood for tanning.

199 LOUGHMAN & Co. *Melbourne.*—Tobacco leaf.

200 MEARS, J. & A. *Collingwood.*—Medical herbs and roots (14 varieties).

201 MACKMEIKAN & Co. *Flemington.* — Glue pieces and bone dust.

202 MURPHY, F. M. *Castlemaine.*—Native flax.

203 MUELLER, DR. — Fibres; various plants; lichens.

204 QUIRK, H. B. *Maryborough.*—Native silk.

205 REED, J. *Collingwood.* — Rope, &c. made of New Zealand flax, grown at the Botanical Gardens, Melbourne.

206 RIDGE, MRS. *Melbourne.*—Hair of the first cross with the Angora and common goat.

207 STABER, F. *Collingwood.* — Fibre of the *Yucca gloriosa*, from leaves grown at the Botanical Gardens, Melbourne.

208 SADDLER, T. *St. Kilda.*—Silk, from worms reared at Caulfield.

209 WILSON, E. *Melbourne.*—Hair of the Poiteau ass.

210 BAYLDON & GRAHAM, *Geelong.*—Scoured wool.

211 CLOUGH & Co. *Melbourne.*—Sixteen bales of choice wool of various brands; ninety-six fleeces.

212 CORRIGAN, S. B. *Geelong.*—Combing, clothing, and lambs'-wool.

213 COMMISSIONERS OF VICTORIAN EXHIBITION.—Two samples of wool.

214 DOUGLASS, A. & Co. *Geelong.*—Scoured combing, clothing, and lambs'-wool.

215 DEGRAVES, WM. *Coliban Park.*—Spanish merino wool.

216 ELDER & SON, *Kuruc Kuruc.*—Merino fleece wool.

217 GOLDSBOROUGH & Co. *Melbourne.*—Thirteen bales choice wool of various brands, forming one half of the trophy in conjunction with Clough and Co.

218 LEARMONTH, MESSRS. *Ercildoun.*—Washed fleece.

219 MARSHALL, T. *Geelong.*—Scoured wool.

220 RUSSELL, P. *Carngham.*—Cross-bred wool.

221 RUSSELL, T. *Wanook.*—Fleece wool.

222 ROWE, E. *Melbourne.*—First cross between merino and Cotswold wool.

223 SIMSON, R. *Langi Kal Kal.*—Beaufort fleece wool.

224 SPIRO, F. *Melbourne.*—Three bales scoured wool.

225 TONDEUR, O. & Co. *Melbourne.*—Trophy, containing 70 samples of wool, various brands.

226 VICTORIAN EXHIBITION COMMISSIONERS.—Specimens of Victorian timber, in all 447 pieces, the greater portion being in slabs, 8 feet in length and 4 inches thick; collected under the direction of Dr. Mueller, Government Botanist. The collection also comprises specimens contributed by Messrs. Beveridge, Allitt, Kidd, McHaffie, Levy Bros. Williams & Little, Rodgers, Weatherhead, and Dr. Backhaus.

227 PURCHAS, A. *Melbourne.*—Working models of a railway carriage, with self-acting brake, and of gas tender, for lighting railway trains.

228 RANDALL, WM. *Melbourne.*—Working model of a locomotive engine and tender.

229 HACKETT & Co. *Collingwood.*—An "Albert" street car.

230 WILLIAMS, WM. *Melbourne.*—Pair-horse carriage.

231 NICOL, D.—An improved saw-set.

232 KAY, J. A. *Melbourne.*—Sewing machine.

233 MACINTOSH, *Melbourne.*—Mining picks, hammers, drills, and gadges.

234 ROBARDT, O.—Model of puddling machine, prismatic cross tramel, parallel ruler, and beam compass.

235 BROWN, W. *Fitzroy.*—Model of a road-scraper by horse-power; model of a battery of stampers.

236 COMMISSIONERS OF VICTORIAN EXHIBITION.—Working battery of 12 stampers for crushing quartz and amalgamating gold with ripples, and amalgamation complete; manufactured by the Port Philip Gold Mining Company, at Clunes.

237 GROLEY, W. *East Collingwood.*—Model of a quartz-grinding and amalgamating mill.

238 MERIDETH, J. *Castlemaine.*—Mode of an improved gold amalgamator.

239 HARPER, R.—Working model of an automatic coffee roaster.

240 MCNAUGHT, *Chewton.*—Models of horse-puddling machines, &c.

241 STRACHAN, W. *Melbourne.*—Model of engine for extinguishing bush-fires.

242 THOMSON, R. & W.—Mercurial filter, for separating gold amalgam from the liquid mercury.

243 HENDERSON & BETT, *South Yarra.*—Swing plough.

244 ROBINSON & Co. *Melbourne.*—Victorian prize reaping machine with side delivery.

245 CHAMP, WM. *Pentridge.*—Model of a pump made by a Chinese prisoner at the penal establishment.

246 LAMBERT & CURTIS, *Collingwood.*—Perforated gratings, for stamper boxes connected with quartz crushing machinery.

247 LOVE, R. A. *Sandhurst.*—Model of a new compound truss suspension-bridge.

248 WHITE, J. H. *Melbourne.*—An improved fire-hose director, with revolving nozzles of different sizes; an improved lever hose.

249 HENSON, A. W. *Melbourne.*—Fowling-piece, bullet-moulds, ramrods.

250 FERGUSON, CAPT. *Williamstown.*—Model of a life-boat.

251 HEATH & JACKSON, *Geelong.*—Model of yacht, "Southern Cross."

252 HIDDLE, J. *Melbourne.*—Model of an improved shackle for heavy chains.

253 SKINNER, MR. *Melbourne.*—Model of the iron steamer, "Phantom."

254 VAIL, MR. *Melbourne.*—Model of wreck escape.

255 WHITE, MESSRS. *Williamstown.*—Working models of vessels built by exhibitors.

256 WILKIE, HON. D. *Melbourne.*—Model of a new form of propeller for steam navigation.

257 BOLTON, J. *Williamstown.*—A gravitating dial,

258 ROBARDT, OTTO, *Melbourne.*—Prismatic cross-trammel parallel ruler and beam-compass.

259 GRIMOLDI, J. *Melbourne.*—Barometer and thermometers.

260 AMHERST MUNICIPAL COUNCIL.—Photographs of views and buildings in the municipality and suburbs.

261 BALLARAT MUNICIPAL COUNCIL.—Views and buildings in the town and district of Ballarat.

262 BELFAST MUNICIPALITY.—Views of Belfast.

263 BEECHWORTH MUNICIPALITY.—Views.

264 CASTLEMAINE MUNICIPAL COUNCIL.—Photographs of views and buildings.

265 COX & LUCKEN, *Melbourne.*—Photographs of stores and buildings in Melbourne, &c.

266 CARLTON MUNICIPAL COUNCIL. — Panoramic view of Carlton.

267 DUNOLLY MUNICIPAL COUNCIL. — Photographs of views and buildings.

268 DAINTREE, R.—Photographs of panoramic views of Ballarat, Castlemaine, &c.; geological sections and views.

269 DAVIS.—Photographs of buildings in Melbourne and Fitzroy.

270 GEELONG CORPORATION. — Photographs of public buildings in Geelong.

271 GEELONG. — Photographs of banks and private buildings, presented by the owners of the property.

272 HAIGH, E. — Photographs of views and buildings in and around Melbourne.

273 JOHNSON, MESSRS.—A collection of photographic views.

274 KILMORE MUNICIPAL COUNCIL. — Views and buildings in the district.

275 KYNETON MUNICIPAL COUNCIL. — Views of Kyneton.

276 MOONAMBEL MUNICIPAL COUNCIL.— Photographs of views of the district.

277 MELBOURNE CITY COUNCIL.—General views of the city, photographed by Nettleton.

278 NETTLETON, *Melbourne.* — Photographs of buildings.

279 PUBLIC WORKS DEPARTMENT.—Photographs of public buildings in the neighbourhood of Melbourne.

300 RICHMOND MUNICIPAL COUNCIL. — Views, &c., in the municipality.

301 SMYTHESDALE MUNICIPAL COUNCIL. —Photographs of general views in the districts.

302 SANDRIDGE MUNICIPAL COUNCIL.— Photographs of views and buildings in Sandridge.

303 ST. KILDA MUNICIPAL COUNCIL. — Photographs of views and buildings in the municipality.

304 SPIERS & POND, MESSRS.—Photograph of racket ground, showing the "All England" match.

305 SANDHURST MUNICIPAL COUNCIL.— Views and buildings at Sandhurst.

306 VICTORIA VOLUNTEERS. — Photograph by Batchelder & O'Neil, Melbourne.

307 WILLIAMSTOWN MUNICIPAL COUNCIL.—Photographs of views and buildings in Williamstown.

308 OSBORNE, S. W.—Specimens of photolithography, the process invented and patented by exhibitor.

309 MATTHIAS, J. R. *Melbourne.*—A bass drum, constructed on a new principle.

310 THORNE, J. *Melbourne.*—New kind of silver strings for violins, tenors, violoncellos, &c.

311 WITTON, H. *Collingwood.* — Case of clarionet reeds.

312 BEANEY, J. G., F.R.C.S. — An improved fracture apparatus.

313 CHRINSIDE, T. *Werribee.*—Basket and nets made by natives on the Grampians; baskets and nets made of reeds.

314 HOPWOOD, H. *Echuca.*—Nets made of Victoria flax.

315 MACKENZIE, J. *Swan Hill.*—Fishing net made by natives of the Murray, from fibre cyperus vaginatus.

316 CHAMP, W. *Pentridge.* — Woollen door mats.

317 HOLLINGS & CHAMBERS, *Melbourne.* —Wool flock for upholsterers.

318 M'LENNAN & Co. *Castlemaine.*—Six pair of socks; one pair of gloves.

319 POTTS, MRS. R. *Melbourne.*—Point lace.

320 BEECHWORTH LOCAL COMMITTEE.— An opossum skin rug.

321 CLARK, J. *Melbourne.* — Kangaroo skins; opossum skins; flying squirrel; native cat skins.

322 FITZGERALD. RYAN, *Portland.*— Kangaroo skins; opossum skins.

323 GRAY, MRS. *Portland.*—Emu feathers; wool and skin of native cat, dyed with sea weed.

324 HART, J. *Melbourne.*—Large rug of native cat skins.

325 ROBERTSON, J. *Melbourne.*—Dressed and dyed feathers of Australian birds.

326 WILLIAMSON, J. *Collingwood.* — Sample of curled horse hair.

327 BREARLEY, BROS. *Geelong.* — Two crop butts, four crop sides, four sides dressed curried shoe leather.

328 CLARK, J. *Melbourne.*—Dressed and curried hides, shoe leather, harness leather, waxed and brown kangaroo dressed saddle leather curried.

329 CHIRNSIDE, T.—Stock whips, saddle girths, hide rope.

330 COMMISSIONERS OF VICTORIAN EXHIBITION.—Pack saddle, as made for the Burke and Wills exploring expedition.

331 CHRISTIAN, H. *Kew.*—Halters.

332 DOCKER, REV. J. *Wangaratta.*— Stock whip, with handle of Myall wood.

333 FORD, BROS. *Melbourne.* — Came shoes, made for the Burke and Wills exploring expedition.

334 LADE & SANDERS, *Melbourne.*—Ladies and gentlemen's riding saddles.

335 MCFARLANE, *Melbourne.* — Stock whip, 17 ft. long.

336 MUELLER, DR.—Pair of saddle bags, with wire and leather covers, as used by Exhibitor for drying plants when travelling in the bush.

337 CHAMP, W. *Pentridge.*—Cabbage tree hats; boots; uniform; prisoners' clothes, made at penal establishment.

338 FORD, BROS. *Melbourne.* — Patent washing hats.

339 GALVIN, J. *Melbourne.*—Case of hats.

340 KING, MRS. *Collingwood.*—Bonnets made of colonial straw.

341 OXLEY, G. W. *Talbot.*—Back Creek volunteer uniform, in Sydney tweed.

342 THOMAS & MURPHY, *Melbourne.*— Pair of jockey boots of kangaroo leather.

343 WALLWORTH, *Melbourne.*—Military hats (Busbies.)

344 ANGUS & ELLERAY, *Castlemaine.*— Specimen of printing.

345 BURNIE, J. D. *Wornambool.*—Copy of Warnambool Sentinel.

346 CLARSON, SHALLARD, & Co.—Printing in colours.

347 COOK & FOX, *Melbourne.* — Super

royal folio ledger, bound in vellum, double Russia bands.

348 CLOUGH & Co. *Melbourne.* — Vol. Clough's Circular, and loose copies of ditto.

349 DETMOLD, WM. *Melbourne.*—Specimens of book binding—3 vol . Shakespeare.

350 EVANS & SOMMERTON, MESSRS.—Vol. Maryborough and Dunolly Advertiser.

351 FRANKLYN, F. B. *Melbourne.*—Vols. Melbourne Herald, Bell's Life, and Illustrated Post.

352 FERRES, JOHN, Government Printer. —Copies of official books and acts of parliament; vellum binding; stereotyping.

353 HEATH & CORDELL, *Geelong.*—Almanack for 1862.

354 LEVEY, WM. *Melbourne.*—Copies of Victorian Ruff, Stud Book, Fistiana.

355 MELBOURNE PUBLIC LIBRARY, TRUSTEES.—Progress catalogue; Melbourne Benevolent Asylum ; Reports.

356 MELBOURNE UNIVERSITY.—Copy of Library Catalogue Calendar, 1861-1862.

357 MASON & FIRTH, MESSRS. *Melbourne.* —Copies of various pamphlets and päpers published by them.

358 PROPRIETORS OF ARARAT ADVERTISER.—Copy of journal.

359 —— OF BALLARAT STAR.—Volume of journal.

360 —— OF BUNINGONG TELEGRAPH.—Copy of journal.

361 —— OF CASTLEMAINE ADVERTISER. —Copy of journal.

362 —— OF CRESWICK & CLUNES ADVERTISER.—Copies of journal.

363 —— OF ECONOMIST.—Volume of journal.

364 —— OF CHRISTIAN TIMES.—Volume of journal.

365 —— OF FEDERAL STANDARD.—Copies of journal.

366 —— OF ILLUSTRATED AUSTRALIAN MAIL.—Copies of journal.

367 —— OF MOUNT ALEXANDER MAIL.—Volume of journal.

368 —— OF OVENS AND MURRAY ADVERTISER.—Copies of journal.

369 —— OF REVIVAL RECORD.—Copies of journal.

370 —— OF SOUTH BOURKE STANDARD. —Volume of journal.

371 —— OF TALBOT LEADER.—Copies of journal.

372 —— OF WARWAMBOOL EXAMINER. —Copies of journal.

373 ROYAL SOCIETY OF VICTORIA. — Copies of transactions.

374 SANDS & KENNY.—Copies of Melbourne Illustrated, Kelly on the Vine Cricketer's Guide, map of Australia.

375 SMYTH, R. B.—Mining Surveyors' Reports, 2 vols.

376 SUPREME COURT OF VICTORIA.—Copy of Catalogue of Law Library.

377 SYME & Co. *Melbourne.*—Vol. Daily Age, Weekly Age, Leader, and Farmers' Journal.

378 STATISTICS IN ILLUMINATED WRITING, showing the rise and progress of the corporations and municipalities in Victoria.

379 TURNER, J. *Melbourne.*—Letter dispatch box, pocket-books, card cases.

380 VICTORIAN COMMISSIONERS.—Copies of Melbourne Exhibition Certificates, and season tickets.

381 WILSON, E.—Vols. of Examiner and Yeoman newspapers.

382 WILSON & MACKINNON.—2 vols. Daily Argus, Weekly Argus.

383 BERNARD, W. H. *Beechworth.*—An improved scale diagram.

384 MARSH, S. D. *Melbourne.*—Portion of MS. Opera, "The Gentleman in Black."

385 O'BRIEN, J. *Hotham.*— Ornamental writing, in frame of the oak.

386 TOLHURST, G. P. *Melbourne.*—Music compositions.

387 WILKIE & Co. *Melbourne.*—Music published by Exhibitors.

388 ALCOCK & Co. *Melbourne.*—Billiard table made of myrtle wood, specimens of turned work, in wood and ivory.

389 BARRY, SIR REDMOND. — Portable wardrobe, of Colonial woods, by Thwaites & Son.

390 BOWIE, DR. *Yarra Bend Lunatic Asylum.*—A loo table, made of colonial woods.

391 CHAMP, W. *Pendridge.*—2 fire-screens, 1 work-table, 1 flower-stand, chess-board and men, solitaire board.

392 CARR & SON.—Model of Venetian window-blind.

393 MCKENZIE, D. *South Yarra.*—Rustic seat.

394 PASLEY, CAPT. R.E.—Inlaid desk of Australian woods, with gold mountings.

395 PIETOICHE, F.—Potichiomanie chimney-piece and ornament.

396 VICTORIAN EXHIBITION COMMISSIONERS.—A Gothic case, 9 feet by 9 feet, and 15 feet high for exhibiting gold and precious stones, writing-desks, work-boxes of inlaid wood.

397 WHITEHEAD, J. *Melbourne.*—Picture frames.

398 CAIRNS, WILSON, & AMOS, *Melbourne.* —Bar-iron, rolled out of scraps and refuse.

399 HUGHES & HARVEY, *Melbourne.*—Fletcher s anti-agitator or milk preserving cans.

400 JENKINSON, W. *Fitzroy.*— Portable oven for use in the bush.

401 LAMBLE, S. *Geelong.*—A set of horse shoes mounted on hoofs.

402 WARD, *Melbourne.*—A small sample of plain cutlery.

403 BRUCE, J. V.—Gold inkstand, with granite pedestal, presented to Exhibitor by the workmen on Melbourne and Murray River Railway.

404 COMMISSIONERS OF VICTORIAN EXHIBITION.—A pyramid, 44 feet 9¼ inches high, and 10 feet square at the base, representing the quantity of gold exported from Victoria from 1st October, 1851, to 1st October, 1861, viz., 26,162,432 ounces troy, equal to 1,793,995 lbs. avordupois, or 800 tons 17 cwt. 3 qrs. 7 lbs., equal in solid measurement to 1,492½ cubic feet of gold, of the value of 104,649,728*l.* sterling. Designed by J. G. Knight, Fel. R.I.B.A., Secretary to the Commissioners of Victorian Department.

405 RHODES, MRS. *Melbourne.*—Topaz found at Kangaroo Flat, near River Loddon.

406 TURNER, W. J. *Beechworth.*—7 brooches, 2 bracelets, 1 diamond ring, gold and precious stones.

407 WIPER, J. *Melbourne.*—Specimens of cut and bent glass, frame of coloured cut glass.

409 HIRSCHI, F. *Castlemaine.*—Pottery.

410 KELLY, T. *Brunswick.*—Stone ware drain-pipes.

411 STIELING, G. F. *Richmond.*—Pottery.

412 ARNOLD, C. *Melbourne.*—Five cases of myall wood pipes, case of myall wood.

413 ANDERSON, SHARP, & WRIGHT.—Machine made window sash and frame.

414 ADAMSON, H. A. *Queen's-cliff.*—Two cases of sea-weeds.

415 BEALE, MRS. *Emerald-hill.*—Marine bouquet made of sea-weeds and shells.

416 BEAL, W. J. *Prahran.*—Three cases of fancy soap.

417 BOEHM, J. *East Collingwood.*—Soap.

418 BALLARAT COMMITTEE.—Three pieces of steam bent timber.

419 BROWN, W. *Melbourne.*—Sample gold and produce bags.

420 CHATFIELD, C. M. *Melbourne.*—Australian earth soap.

421 CHAMP, W. *Pentridge.*—Coir matting, door-mats, paper-knives, walking-sticks, imitation books, made of Victoria woods, granite fountain.

422 CROMPTON, *Beechworth.*—Cigars made from tobacco grown at Beechworth.

423 COMMISSIONERS OF VICTORIAN EXHIBITION.—Fifty-one paper-knives of different Victorian woods.

424 DOWNIE & MURPHY, *Hobson's-bay Candle Works.*—Sample boxes of candles, soap, refined tallow, &c.

425 DOTHERS, E. *Castlemaine.*—Sample of prepared and common bricks.

426 EWING, J. A. *Fitzroy.*—Yeast powder.

427 FRASER, MR. *Back-creek.*—Six bottles of hair dye and atmospheric oil.

428 FITTS, C. *Melbourne.*—Sample of Victorian glue.

429 FOORD, G. *Melbourne.*—Two samples of hydraulic cement made from septaria.

430 GRAY, W. *Phillipstown.*—Bricks and tiles.

431 GRAY & WARING, *Melbourne.*—Butter churn.

432 GOERNAMANN, F. *Melbourne.*—Steam bent timber.

433 HEATH, R. *Geelong.*—Case of artificial teeth in gold and vulcanite.

434 HODGKINS, W. *Emerald-hills.*—Brushes.

435 HODGKINSON, G. *Prahran.*—Bricks, tiles, and terra-cotta.

436 HEWETH, *Beechworth.*—Soap.

437 KELSALL, J. *Buningong.*—Case of soap.

438 KNIGHT, J. G. *Melbourne.*—Bricks and tiles from various parts of the colony.

439 LEVY BROS. *Melbourne.*—Myall wood pipes mounted in gold.

440 LEE, P. *St. Kilda.*—Cigars.

441 MERCER, MRS. *Geelong.*—Frame of fancy leather work.

442 MILLER, F. McD. *Melbourne.*—Cartridge and compressed bullets of various kinds.

443 MONTGOMERY, R. *Collingwood.*—Samples of cork cuttings.

444 MEMMOTT, W. *Collingwood.*—Specimen of comb making, made on the gold-fields.

445 MELBOURNE ASPHALT CO.—Eleven small specimens of asphalt.

446 MOURANT, G. W. *Collingwood.*—Wooden taps and spiles.

447 MCILWRAITH, *Melbourne.*—Sheet lead and piping.

448 MCCAPE, A. *Chiltern.*—Patent powder-proof lock.

449 NIGHTINGALE, E. *Melbourne.*—Milliners' boxes.

450 PERRY, J. *Melbourne.*—Bent timber for carriages.

451 QUELCH, BROS. *Prahran.*—Case containing various kinds of candles.

452 SKEATS & SWINBOURNE, *Melbourne.*—Machine wrought mouldings for joiners' work.

453 SHIPP & CO. *Melbourne.*—Samples of wire work.

454 SHIEBLACH, C. *Back-creek.*—Candles and soap.

455 SANDHURST LOCAL COMMITTEE.—Bricks and tiles.

456 STIELING, G. F. *Richmond.*—Fire bricks.

457 SANSOM, H. *St. Kilda.*—A curious chain cut in wood.

458 SHECKEL, T. *Geelong.*—Fancy basket and willow work.

459 TALLERMANN, W. *Collingwood.*—Samples of india rubber manufacture.

460 WILKIE, MRS.—Two cases of sea-weeds.

461 WILLIAMS, W. *Melbourne.*—Machine-wrought timber for carriage building.

462 WHITELAW, G. *Ballarat.*—Four specimens of graining in imitation of wood.

463 WARD & CO. *Melbourne.*—Soap powder.

464 WOOD, W. J. *Toorak.*—Blacking.

465 WESTALL, MRS. *Melbourne.*—Case of wax flowers.

466 CROUCH & WILSON, *Melbourne.*—Design for Town-hall, Prahran; design for Wesleyan chapel, Collingwood.

467 BILLING, N. *Melbourne.*—Design for Presbyterian church.

468 KNIGHT, J. G. *Melbourne.*—Design for Government House.

469 PURCHAS & SAWYER.—Designs for Melbourne Savings' Bank, Bank of Australasia, Temple-court, and premises, Bourke-street.

470 TERRY, L. *Melbourne.*—Design for Flower, Macdonald & Co.'s warehouse.

471 BATEMAN, *Melbourne.*—Designs for woollen fabrics, introducing indigenous flowers and foliage.

472 CALDER, J. *Ballarat.*—Oil painting: View in the Pyrenees, Victoria.

473 DE GRUCHY & LEIGH, *Melbourne.*—Specimen of chromo-lithography, "The Three Marys."

474 EATON, MISS, *Melbourne.*—Three water-colour paintings of indigenous flowers.

475 GUERARD, E. VON, *Melbourne.*—Oil paintings of Victorian scenery. No. 1. Fern Tree Gully. No. 2. Stony Rises. No. 3. View of Geelong. No. 4. Mount William. No. 5. Forest scene. No. 6. Sydney Heads.

476 ROWE, G.—Six water-colour paintings of scenery in Victoria.

477 STRUTT, W. *Melbourne.*—Portrait of the late Colonel Neil, portrait of Major-General McArthur, Maoris driving off Settler's Cattle, and six other subjects.

478 TAYLOR, T. G.—Painting, "View at Fryer's Creek, in 1852;" "Road Making in Black Forest"—a sketch taken in 1852.

479 ARNOLDI, X. *Melbourne.*—Seal of the commoners of the Victorian Exhibition.

480 POOLE, MRS. G. H. *Melbourne.*—Small medallion in wax of Burke and Wills, the explorers.

481 SCURRY & MACKENNAL, *Melbourne.*—Two designs for fountains; design for chimney piece.

482 SUMMERS, C. *Melbourne.*—Collection of medallion portraits; design for the seal for Victoria Exhibition; cast of the head of Wonga Wonga, a native chief.

483 THOMAS, MISS, *Richmond.*—Bust of Dr. Barnett; figure, "Napea."

484 COSMOPOLITAN GOLD MINING CO. *Ballarat.*—Drawing showing the workings of the Company's claim.

485 DAVIDSON, MR.—Large geological map of Ballarat, showing the principal leads of gold in that district.

486 MCCOY, PROFESSOR, *Melbourne University.*—Three plates of Decades of the Memoirs of the Museum of Victoria.

487 SELWYN, A. R. C., GOVERNMENT GEOLOGIST.—Progress map of the geology of Victoria.

488 TOCKNELL, W. *Melbourne.*—Steel-plate engraving.

489 ————View of Melbourne in 1839.

491 BACKHAUS, DR. *Sandhurst.*—Platypus.

492 BEECHWORTH COMMITTEE.—Two small Murray River turtles, porcupine, small platypus.

493 CHAMP, W. INSPECTOR-GENERAL OF PENAL ESTABLISHMENTS, *Pendridge.*—Native weapons.

494 DUNN, E. H. *Beechworth.*—Seventeen bottles containing snakes, lizards, &c.

495 DENNYS, M. L. *Geelong.*—A collection of Victorian quails.

496 HOWITT, W., LEADER OF CONTINGENT EXPLORING PARTY.—Native bag containing pitcherry.

497 HOWARD, THE REV. W. C. AND (498) BROKEY, P. LE P.—A collection of birds shot in the Ovens District.

499 HENSON, W. A. *Melbourne.*—Collection of native birds.

500 JAMIESON.—Native weapons used by the Murray tribe.

501 JOHNSON, T. *Back Creek.*—Native tomahawk.

502 MURRAY RIVER FISHING CO.—Preserved specimens of Murray cod and other fish.

503 MCCOY, *Melbourne University.*—A collection of insects (six cases).

504 MUELLER, DR.—Native weapons.

505 MARTINDALE, MRS. *Back Creek.*—Two bottles containing snakes.

506 PRESHAW, W. J. *Castlemaine.*—Native tomahawk.

507 PAULL, J. M. *Back Creek.*—Two bottles containing reptiles.

508 SANBRIDGE, H. A.—Native peaches.

509 THOMAS, W. *Melbourne.*—A kurbur-er, or native bear; native te work, used by natives to produce fire; native basket.

510 COOP.—Lead piping.

511 MCILWRAITH.—Milled lead and piping.

512 CRISP.—Diamond and precious stones.

513 SMYTH, W.—A collection of quartz specimens, rich in gold, from Inglewood.

514 COMMISSIONERS OF VICTORIAN EXHIBITION.—Sixteen cases of cereals.

515 ATTWOOD, S.—Sample of wheat.

516 BUTCHER, BENJ. N.—Sample of wheat.

517 BUCHANAN, R. & J.—Two samples of wheat.

518 M'CASKILL.—Two samples of wheat.

519 COCKRANE.—Sample of wheat.

520 CONNOR, J. H.—Sample of wheat.

521 DYER, R.—Oats.

522 DARCY, M.—Sample of wheat.

523 FORREST, J.—Sample of wheat.

524 GAUL & LUMSDEN.—Sample of wheat.

525 HANCOCK, A. & B.—Sample of wheat.

526 HADLEY & CO.—Bag of flour.

527 JOHNSON, D.—Sample of wheat.

528 KINNERSLEY, D.—Sample of wheat.

529 KITSON, S.—Sample of wheat.

530 MORGAN, J.—Sample of wheat.

531 M'ALPINE, W.—Sample of wheat.

532 MORRISON, W. J.—Sample of wheat.

533 HALL, MR.—Box of beans.

534 PATTERSON, A.—Sample of wheat.

535 PATTEN, A.—Sample of wheat.

536 PORTER, B. C.—Sample of wheat.

537 SIMPSON, G. H.—Sample of wheat.

538 THOMSON, W.—Three samples of wheat.

539 VEARING, T.—Sample of wheat.

540 WESTLAKE, A.—Sample of wheat.

541 DALGETTY, F. G.—A collection of Victorian birds in three cases.

542 BENJAMIN, D.—Presentation gold cup, gold brooch.

FOREIGN STATES.

[*In some instances the Address, and Names of Articles of Exhibitors in the Foreign Departments were not known at the period of going to press.*]

AFRICA, CENTRAL.

Under Staircase, near Central Entrance to Horticultural Gardens.

BAIKIE, DR. W. BALFOUR, R.N.—

1-2 Striped men's cloth, from Hausa.
3-4 Cloth made of fibres of the wine-palm and cotton, from the right bank of Kwarra.
5 A tobe, poorest quality, made in Nupe.
6 A tobe of finer quality.
7 A white tobe with plaits, from Nupe.
8 Striped trowsers, Nupe or Hausa make.
9-10 Common cloth, for women from Bonu.
11 A woman's wrapper, made in Nupe.
12 A woman's wrapper, from Nupe.
13 A woman's wrapper, not made up, called "Locusts' tooth."
14 A wrapper containing red silk, called Maizha'n baki, or "red mouth."
15 An inferior wrapper, from Nupe.
16 Blue and white cloth, from Nupe.
17-18 Cloth made in Yoruba.
19-20 Cloths from Nupe.
21-25 Cloths from Yoruba.
26 Small cloth for girls, from Nupe.
27 Bag from Onitsha.
28 Mat from right bank of Kwarra.
29 Tozoli (sulphuret of lead), applied to the eyelids.
31 Man's wrapper, from Ki, in Bonu.
32 Woman's head-tie, or alfuta, from Nupe.
33 Bags for gunpowder, from Onitsha.
34-35 A calabash and ladle.
37 Red silk, or "Al harini," of Hausa.
38 Sword hangings, or "Amila," made at Kano, in Hausa.
39 Siliya, or red silk cord, from Kano.
40 Rope, from Onitsha.
41-42 Bags.
43 White cloth, or fari, made in Nupe and Hausa.
44 White cloth, from below the Confluence.
45 A white tobe, from Nupe.
46 Four calabashes, for pepper, &c.
47 A small calabash and lid, for food.
48-49 Pinnæ of leaves of the wine-palm, dried and used for thatching.
50 Fruit of a leguminous plant, which buries its fruit like Arachis hypogæa.
51 Grass cloth, of wine palm.
52 Two cloths, from Okwani.
53 White cloth, from below the Confluence.
54 White perforated cloth, from the Ibo country.
55 Mats from Onitsha.
56 Large man's wrapper, from Nupe.
1 A white mat of leaves of the fan-palm, from Bonu.
2 Mats of the fan-palm, from Bonu. Fan-palm mats, called guva, or "Elephant mats."
3 Fine mats and hats, of leaves of the Phœnix spinosa, dyed. Circular mats of the same material, used by chiefs, from Nupe.

AFRICA, WESTERN.

Northern Courts, under Staircase, near Central Entrance to Horticultural Gardens.

COMMERCIAL ASSOCIATION OF ABEOKUTA. —1 Oils: Of beni seed, obtained by fermentation and boiling. 2 Of Egusi, from wild melon seed. 3 Of palm, for home consumption; 4 for exportation, obtained by beating, pressing, and boiling the fruit. 5, 7 Of palm-nut, for home consumption; 6 for exportation. 11 Shea butter. 10 Egusi, or wild melon, fruit. 8 Beni seed. 9 Fruit of the Shea butter tree.

1 White cotton thread; 2 Dyed; 3 Blue. 4 Fine spun cotton. 5 Coarse strong spun cotton, called "Akase." 6 Akase cotton, cleaned and bowed; 7 In seed. 8 Seed itself of Akase cotton. 9, 10 Ordinary native cotton. 11, 12, 13 Ordinary green, black, and brown seeded cottons. 14 Silk cotton. 16 Country rope of bark. 17 Palm fibre. 18 Red dyed native silk, from Illorin. 20 Fibre used for native sponge. 23, 24, 25, Native silk, from a hairy silk-worm at Abeokuta. 26 Leaves of the cotton tree. 27 Pine-apple fibre. 29 Bow-string fibre. 30 Jute.

15 Long black pepper. 22 Senna. 21 A sample of native antimony, from Illorin.

Sundry native manufactures.

N.B.—Cotton is obtainable in any quantity, and is now grown extensively throughout the Yoruba country, especially to the east and north. Great quantities of cotton cloths, of a strong texture, are annually made, finding their way to the Brazils, and into the far interior. To obtain a largely increased supply of cotton, it is only necessary to open roads, and bring money to the market. Upwards of 2,000 bales have been exported this year, and the quantity would have been doubled or trebled if the country had been at peace. The present price is 4½d. per lb. The other fibres are not at present made for exportation, though, doubtless, some of them—jute, for instance—would be, if in demand. Of the native manufactures, the grass cloths, made from palm fibre, and the cotton cloths, are most prominent. Very nice leather work is done. The art of dyeing Morocco leather different colours has been introduced from the interior. Indigo is almost the only dye which can be obtained in considerable quantities. The natives manufacture all their own iron implements, and the quality of the metal is considered good.

2 McWilliam, The Late Dr. C. B.—1 Cloth, from the confluence of the Niger and Tchadda. 2 Raw silk from Egga. 3 Cotton from the confluence. 4 Fishing spear, used by the natives of Kakunda. 5 Spoons, from Gori market. 6 A curved horn for holding galena, used to paint the eyelids. 7 Cloths, from towns on the Gambia. 8 Grass mat, from Angola. 9 Grass mat, from Binguela.

3 Walker, R. B. *Gaboon.*—A collection of mats, fibres, commercial products, skins, native arms, musical instruments, &c. of the Ba Fan tribes; tusks, scrivelloes, and hippopotamus teeth; ebony and bar wood.

4 Barnard, John A. L. 8, *Alfred-villas, Dalston.*—Tallicoonah or Kundah oil, from *Carapa Tallicoonah*, and a bundle of ground nuts (*Arachis hypogœa*) in the haulm.

BELGIUM.

N.W. Court, No. 1, and N.W. Gallery, No. 1.

CLASS 1.

1 The Ministry of Public Works, *Brussels.*—Mineral products of Belgium, collected by M. Jules Van Scherpenzeel-Thim.

---

2 Amand, E. *Mettet, Namur.*—Iron ores, charcoal castings.

3 Brincourt-André, L. *Herbeumont, Luxemburgh.*—Various kinds of slate.

4 Coupery de Saint-Georges, E. *Dinant, Namur.*—Black marble in polished slabs and blocks.

5 De Jaer & Co. *Antwerp.*—Alluvial pyrites.

6 Dejaiffe-Devroye, T. *Saint-Martin-Bâlatre, Namur.*—Black marble from Golzinnes.

7 De Merckx de Corbais, Mrs.—Lead ores and potter's clay.

8 Dassonville de Saint Hubert, L. *Namur.*—Belgian mill-stones (siliceous).

9 Descamps, J. & Co. *Saint-Josse-ten-Noode, near Brussels.*—Grit-stone pavement.

10 Desmanet de Biesme, Viscount, *Biesme, Namur.*—Black marble from Golzinnes.

11 De Thier, A. *Theux.*—Black marble from Theux.

12 De Villers & Co. *Brussels.*—Sainte-Anne marbles, polished.

13 De Wyndt & Co. *Antwerp.*—Sulphur in rolls and flowers.

14 Dupierry, *Viel-Salm, Luxemburgh.*—Whet and grinding-stones.

Exhibitors' Committee for the District of Verviers:—

16 Société Anonyme de Corphalie, *Antheit, near Huy.*—Ores of zinc, lead, pyrites; refined lead; crude zinc.

17 Société Anonyme de la Nouvelle-Montagne, *Verviers.*—Ores of lead, zinc, iron; pyrites; metals; sulphur.

18 Société Anonyme de Rocheux et d'Oneux, *Theux.*—Pyrites; ores of zinc, lead, iron.

19 Société Anonyme des Hauts-Fourneaux et Laminoirs de Montigny-sur-Sambre, *Montigny-sur-Sambre, Hainault.*—Iron-ores, coke-castings, puddled steel.

20 Société Anonyme des Hauts-Fourneaux, Usines, et Charbonnages de Châtelineau, *Châtelineau, Hainault.*—Coke castings.

21 Société Anonyme des Hauts-Fourneaux, Usines et Charbonnages de Marcinelle et Couillet, *Couillet, Hainault.*—Pit-coal, iron ores, coke castings, puddled steel.

22 Société Amonyme des Hauts-Fourneaux et Charbonnages de Sclessin, *Sclessin, Liege.*—Iron ores, coke castings, pyrites.

23 Société des Minières de Honthem, *Dolhain.*—Ores of lead; pyrites.

24 Société Anonyme de Vezin-Aulnoye, *Huy.*—Iron ores, coke castings.

25 Société Anonyme du Bleyberg, *Bleyberg ès-Montzen.*—Zinc and lead ores, lead-pigs, zinc ingots, regulus of silver, glazed pottery, crystals.

---

26 La Plume-Rouxhe, J. N. *Salm-Château, Luxemburgh.*—Whet and grinding stones.

27 Lebens, E. *Brussels.*—Quartzite paving-stones from Hal.

28 Marchal, D. *Brussels.*—Grit-stone pavement, specimens of marble.

29 Muller, A. & Co. *Berg-Gladbach, near Cologne, Prusse.*—Belgian zinc in pigs.

30 Offergeld, P. J. *Viel-Salm, Luxemburgh.*—Hones.

32 Pierlot-Quarré, *Forrieres, Luxemburgh.*—Quartzite pavement; specimens of marble.

33 Sacqueleu, F. *Tournay, Hainault.*—Specimens of marble; flags, slabs; mangers, manger-fronts.

34 Tacquenier, A. C. & Bros. *Lessines Hainault.*—Chlorophyre pavement.

35 Verbist-Lamal, R. *Brussels.*—Black marble from Basècles, in a rough and finished state, and in small blocks.

36 Watrisse, L. *Dinant, Namur.*—Polished black-marble slabs.

CLASS 2.

37 Barbanson, P. *Brussels.*—Animal-black and bone-dust.

38 Bortier, P. *Ghistelles, West Flanders.*—Preparation of nitrate of lime for manure.

39 Brasseur, E. *Ghent.*—White lead, ultramarine.

40 Bruneel, *Ghent.*—Chemical products

adapted for the fine arts; acids and acetates; vinegar, oil, and alcohol from wood.

41 CAPPELLEMANS, J. B. SEN. DEBY, A. & Co. *Brussels.*—Chemical products.

42 COOSEMANS & *Berchem, Antwerp.*—Naphtha, photogene; lubricating, paraffine, and other oils.

43 DE CARTIER, A. *Auderghem, near Brussels.*—"*Minium de fer d'Auderghem,*" a preservative paint for iron and wood.

44 DELMOTTE-HOOREMAN, C. *Mariakerke, near Ghent.*—White lead.

45 DELTENRE-WALKER, *Brussels.*—Fine varnishes for various purposes: collodion.

46 DE MOOR, A. *Brussels.*—Elastic copal, and other varnishes.

47 DE SAEGHER, H. *Brussels.*—Composition for removing incrustations from steam-boilers.

48 GENNOTTE, L. *Brussels.*—Vegetable powder for the destruction of insects.

49 MATHYS, M. *Brussels.*—Varnishes.

50 MERTENS, BALTHAZAR, & Co. *Lessines, Hainault.*—Preservative blacking, lucifer matches.

51 MERTENS, G. *Overboelaere, East Flanders.*—Inodorous lucifer matches, blacking.

52 KAYSER, A. & POPELEMON, J. *Brussels.*—Pulp for straw paper, cellulose, collodion.

53 RAVE & Co. *Court-Saint-Etienne, Brabant.*—Alkal-oxide; powder for mining purposes, inexplosive in ordinary circumstances.

54 SEGHERS, B. *Ghent.*—Bone black and ivory black.

55 VANDER ELST, P. D. *Brussels.*—Sulphuric and nitric acids, sulphate of soda, copperas, bleaching powder, crystals of tin.

56 VANSETTER-CONINCKX & Co. *Neder-Overheembeek, near Brussels.*—Turpentine, animal black.

57 VERSTRAETEN, E. *Ghent.*—Animal black.

CLASS 3.

AGRICULTURAL ASSOCIATION OF THE DISTRICT OF YPRES:—

58 AGRICULTURAL ASSOCIATION.—Hops, wheat, rye, Indian wheat, pease, colza; *œillette,* a variety of poppy; leaf tobacco.

59 COEVOET, L. F. *Poperinghe.*—Hops, grown in 1861.

60 DE GRYSE, W. *Poperinghe.*—Hops.

61 DELBAERE, P. *Poperinghe.*—Wheat and pease.

62 DEMOOR, B. *Passchendaele.*—Kidney-beans.

63 GOMBERT & CAMERLYNCK, *Reninghelst.*—Hops.

64 LEBBE-BEERNAERT, B. *Poperinghe.*—Hops, wheat, oats, blue peas.

65 LESAFFRE A. *Gheluwe.*—Tobacco.

66 MALOU, J. B. *Dickebusch.*—Hops grown in 1861.

67 PATTYN, C.—Rye.

68 PEENE, BROS. *Elverdinghe.*—Hops.

69 QUAGHEBEUR-VERDONCK, P. *Poperinghe.*—Hops.

70 RICQUIER, L. *Warneton.*—Œillette, a variety of poppy; colza.

71 ROMMENS, F. *Poperinghe.*—Hops grown in 1861.

72 VANDERGHOTE, E. *Elverdinghe.*—Hops grown in 1861.

73 VANDERMEERSCH, J. B. *Bas-Warneton.*—Tobacco.

74 VANDROMME, P. *Westoutré.*—Hops grown in 1861.

AGRICULTURAL SOCIETY OF EAST FLANDERS:—

75 DE BERLAERE, KN. *Vinderhaute.*—Hops.

76 DE CROESER, BARON ED. *Mooreghem.*—Leaf tobacco; wheat, oats, kidney beans, pease.

77 DEMEULDER, J. F. *Poesele.*—Rye, Australian white wheat.

78 DERORE, J. *Mooreghem.*—Wheat and oats.

79 GHENT COMMITTEE.—Cereals, hops.

80 GUEQUIER, J. *Wachtebeke.*—Buck-wheat, sorgho.

81 LATEUR, L. *Mooreghem.*—Wheat, Polish oats.

82 VAN BUTSELE, G. *Nuckerke.*—Australian wheat in ears.

83 VAN PELT, J. F. *Tamise.*—Rye and barley.

83A MONTON & AUTHOMPEN, *Herstal, Liege.*—Starch, &c.

---

84 BALCAEN, P. *Peteghem.*—Chicory.

85 BEERNAERT, L. *Thourout, West Flanders.*—Wheat, rye, oats, buck-wheat, kidney-beans; tobacco.

86 BELPAIRE & OOMEN, *Antwerp.*—Cigars.

87 BENOIT, A. *Chermont, sous St. Hubert, Luxemburgh.*—Summer wheat, winter rye, summer barley and black oats.

88 BLAESS, C. B. *Borgerhout, near Antwerp.*—Vinegar from grain.

89 BORGHS, *Turnhout.*—Wheat and oats.

90 BORTIER, P. *Ghistelles.*—Giant wheat, in ear.

92 CAPOUILLET, P. *Brussels.*—Sugar-loaves, raw beet-root sugar.

93 DE BISEAU, T. *Entre-Monts, sous Buvrinnes, Hainault.*—Wheat.

94 DELANNOY, N. *Tournay, Hainault.*—Chocolate, cocoa, racahout and fancy articles.

95 D'ELPIER, C. *Castle of Mielen, Saint-Trond.*—Giant wheat, potatoes.

96 DE MARNEFFE-VAN PETEGHEM, *Alost, East Flanders.*—Alost hops, grown in 1861.

97 DE NOTER, R. *Laerne, East Flanders.*—Leaf tobacco.

98 DEWYNDT, J. & Co. *Antwerp.*—Samples of sugar candy.

99 DEYMANN, J. H. *Charleroi.*—Deymann-bitter, a liqueur.

100 DIERT DE KERKWERVE, BARON, *Castle of Hemixem, Antwerp.*—Wheat, rye, and oats.

101 DUBUS & DESCAMPS, *Brussels.*—Extract of tobacco, for the manufacture of cigars.

102 HANSSENS, B. & SON (TRITHART, Director), *Vilvorde, Brabant.*—Fecula, starch, gommeline, dextrine, leiogomme, gum arabic various pastes.

103 HEIDT-CUITIS, J. *Chokier, Liége.*—Starch, &c.

104 JORISSEN, L. *Liége.*—Alcohol.

105 JOVENEAUX, A. *Tournay, Hainault.*—Chocolate, hygienic preparations.

106 LE HON, F. SEN. *Brussels.*—Curaçao, and other liqueurs.

107 MERKEL, G. *Tournay, Hainault.*—Vinegar.

108 MIRLAND & Co. *Pecy, near Tournay.*—Apple paste.

109 NORTHERN AGRICULTURAL SOCIETY, *Antwerp.*—Wheat, rye, oats, hops, &c.

110 PAILLET-JONEAU, A. *Ville-en-Hesbaye, Liége.*—Syrups, prepared from fruit, and beet-root.

111 PATRON-JOLY, *Huy.*—Sparkling wines of Huy.

112 PENITENTIARY OF SAINT HUBERT (MARINUS, Director), *Luxemburgh.*—Wheat, rye, barley, oats, trefoil and grass.

114 REMY, E. & Co. *Wygmael, Louvain, Brabant.*—Rice starch.

115 SCHALTIN-DUPLAIS & Co. *Spa, Liége.*—Liqueurs, Elixir of Spa.

116 SCHOOFFS, J. B. *Brussels.*—Extracts used in the manufacture of liqueurs.

117 SERRÉ, L. *Hal, near Brussels.*—Beer.

119 SOCIÉTÉ ANONYME DES MOULINS A VAPEUR DE BRUXELLES, *Molenbeek, near Brussels.*—Starch.

120 STEENS, H. *Schooten.*—Wheat, rye oats, buck-wheat, colza, pease.

121 STEIN, A. & Co. *Antwerp.*—Cigars made with Havanah and other tobaccos.

122 TINCHANT, L. *Antwerp.*—Cigars of different kinds.

123 ULLENS, C. F. *Schooten.* — Wheat, barley, rye, and oats.

125 VAN BERCHEM & Co. *Brussels.*—Cigars.

126 VANDEN WYNGAERT, *Wilmarsdonck.*—Barley.

127 VANDEVELDE, N. *Ghent.*—Liqueurs, champagne beer, rectified gin.

128 VAN GEETERUYEN EVERAERT, *Hamme, East Flanders.*—Starch produced by steam power.

129 VAN PUT, *Antwerp.*—Hay.

130 VANSTRAELEN, H. *Hasselt, Limburgh.*—Gin from grain.

131 VAN VOLSEM, P. *Hal, Brabant.*—Rye and oats grown on heaths.

132 VERGOUTS, F. *Lillo, Antwerp.*—Australian white wheat, Polders oats.

133 VERHEYDEN, DILBECK, *Brabant.*—Wheat, rye, oats, hops.

134 VERHEYEN, P. J. *Turnhout, Antwerp.*—Hops grown on the heaths of the Antwerp Campine.

135 VERTONGEN, BROS. *Raegels.*—Wheat, rye, oats, trefoil.

136 WINCQ, J. B. *Ochamps, Luxemburgh.*—Oats.

CLASS 4.

AGRICULTURAL ASSOCIATION OF THE DISTRICT OF YPRES:—

137 AGRICULTURAL ASSOCIATION. — Steeped and unsteeped flax, madder.

138 BEERNAERT, L. *Thourout.*—Raw and unsteeped flax, peeled flax.

139 HERMAN, J. *Becelaere.*—Steeped flax, peeled flax.

140 VAN LEENE, D. *Dickebusch.*—Steeped flax.

141 VAN WALLEGHEM, C. *Zannebeke.*—Unsteeped flax.

142 VERMEULEN, A. *Becelaere.* — Unsteeped flax.

AGRICULTURAL SOCIETY OF EAST FLANDERS:—

143 REYNIERS, J. A. *Seveneeke.*—Flax and hemp in different stages of preparation.

144 VAN PELT, J. F. *Tamise.*—Hemp, raw, steeped, in filaments, carded.

145 VAN RENTERGHEM, L. *Drongen.*—Flax.

---

146 BERTOU, BROS. *Liége.* — Waterproof grease for shoes.

147 BIHET, H. *Huy.*—Glue.

148 BISSÉ, E. & Co. *Cureghem, near Brussels.*—Lubricating oils for steam-engines and clock-work.

149 BRUGELMAN & HALSTEAD, *Cureghem near Brussels.*—Artificial wools.

150 CLAUDE, L. *Brussels.*—Pure Colza oil.

151 DAVID, C. *Antwerp.*—Flemish flax.

152 DE BEHAULT, *Buggenhout, East Flanders.*—Raw flax.

153 DE BRUYN, J. *Termonde, East Flanders.*—Peeled flax, raw flax and hemp.

154 DE CATERS, BARON, *Antwerp.* — Flax.

155 DE COCK, BROS. *Brussels.*—Watchmakers' oil.

156 DE CONINCK, BROS. *Brussels.*—Household soap, purified oil for carcel lamps.

157 DE CURTE, V. *Ghentbrugge, near Ghent.*—Distilled stearine and candles.

158 DE MOOR, E. & Co. *Antwerp.*—Stearine, candles, and oleine, produced from the waters in which wool had been washed.

159 DENAEYER, P. *Lebbeke, East Flanders.*—Artificial wools (mungo-shoddy) of all colours and degrees of fineness.

160 DENS, *Putte.*—Flax.

161 DE ROUBAIX-JENAR & Co. *Brussels.*—Stearine, candles, raw materials, oleic acid, products of distillation.

162 DE ROUBAIX-OEDENKOVEN & Co. *Borgerhout, near Antwerp.*—Stearine candles, fat-acids.

163 DE SAINT-HUBERT BROS. *Warnant-Moulins, Namur.*—Flax steeped by a manufacturing process, and peeled mechanically.

164 DES CRESSONNIÈRES, *Brussels.*—Soaps for Turkey red dyeing, household and toilet soaps, "huile tournante."

166 DUBOIS-CREPY, *Mons.*—Perfumed and household soaps.

167 EECKELAERS, L. *Saint-Josse-ten-Noode.*—Toilet and household soaps.

168 FELHOEN, BROS. (FELHOEN-PECQUERIAU), *Courtrai.*—Flax in the different stages of the dressing and heckling processes.

169 GAUCHEZ, L. *Brussels.*—Oil for carding engines and for felting threads.

170 GENNOTTE, T. *Brussels.*—Night lights, porcelain floats, waxed wicks.

171 HANSOTTE-DELLOYE, V. G. *Huy, Liége.*—Glue.

173 LEFEBURE, J. *Brussels.*—Flax and hemp prepared in six hours without steeping or creaming, by a new process.

174 MARTIN, *Saint-Josse-ten-Noode.*—Watchmakers' oil prepared without acid.

175 MECHANT, H. *Hamme-Sainte-Anne, East Flanders.*—Specimens of flax in all stages of preparation.

176 MULLENDORFF, *Ixelles, near Brussels.*—Machinery oil freed from acids.

177 NORTHERN AGRICULTURAL SOCIETY, *Antwerp.*—Specimens of flax.

178 PEERS, BARON E. *Oostcamp, West Flanders.*—Raw flax grown in a heathy soil.

179 QUANONNE, C. & MIDDAGH, P. *Molenbeek-St.-Jean, near Brussels.*—Distilled stearic acid candles.

180 REYNAERT, CH. *Reninghe, West Flanders.*—Raw flax.

181 ROMBOUTS-VREVEN, *Hasselt, Limburgh.*—Bleached wax, tapers.

182 STEENS, H. *Schooten.*—Flax.

183 TAULEZ-BOTTELIER, C. *Bruges.*—Samples of peeled flax.

184 VAN DEN PUT, V. *Brussels.*—Soaps, and perfumery.

185 VANDERPLASSE, BROS. *Brussels.*—Watchmakers' oil.

186 VANDERSCHRIECK, BROS. *Antwerp (succur-saal at Saint-Denis, near Paris).*—Woollen rags.

187 VAN ROYE, G. & H. BROS. *Brussels.*—Purified Colza oil.

188 VAN-SETTER-CONINCKX & Co. *Brussels.*—Oil for lubricating machinery.

189 VERBESSEM, C. *Ghent.*—Glue, size, and gelatine.

190 VERCRUYSSE-BRACQ, F. *Deerlyk, near Courtrai.*—Flax, raw and peeled.

191 VERPOORTEN, *Bruges.*—Grains; oils, common and refined; oil cakes.

192 VERSCHEURE, J. *Oyghem, West Flanders.*—Raw, steeped, and peeled flax.

193 VERTONGEN, BROS. *Raevels.*—Flax.

194 WINNEN, BROS. *Brussels.*—Shoddy; artificial whale-bone.

195 ZOUDE, L. *Val de Poix, Luxemburgh.*—Beech-wood gun-stocks, and fellies.

CLASS 5.

196 ARNOULD, G. *Mons.*—Forged iron railway chairs; new kind of fish-plates.

197 BLONDIAUX & Co. *Thy-le-Château, Hainault.*—Rails, splints, and breakings of rails.

198 CEURVORST, S. P. *Antwerp.*—Patent railway break, J. Briere's principle.

199 COMPAGNIE GÉNÉRALE DE MATÉRIELS DE CHEMINS DE FER, *Brussels.*—Railway carriage; trophy of wheels and iron fittings for waggons, &c.

200 COFFIN, C. & J. *Brussels.*—Axle and cylinder for locomotive engine.

201 HEINDRYCKX, *Ixelles, near Brussels.*—Wrought iron chairs; tramway system; model of crossing.

202 SOCIÉTÉ DES HAUTS-FOURNEAUX, USINES ET CHARBONNAGES DE CHATELINEAU, *Châtelineau, Hainault.*—Rails.

203 SOCIÉTÉ ANONYME DES HAUTS-FOURNEAUX, USINES ET CHARBONNAGES DE MARCINELLE ET COUILLET, *Couillet, Hainault.*—A locomotive; rails.

204 SOCIÉTÉ ANONYME DE LA FABRIQUE DE FER D'OUGRÉE, *Seraing, Liége.*—Unwelded wheel tires for waggons and steam-engines, axles and wheels for waggons.

204A SOCIÉTÉ ANONYME DES HAUTS-FOURNEAUX ET LAMINOIRS DE MONTIGNY SUR SAMBRE, *Hainault.*—Rails, &c.

205 SOCIÉTÉ DES FORGES ET LAMINOIRS DE L'HEURE, *Marchienne-au-Pont, Hainault.*—Axles.

206 SOCIÉTÉ ANONYME DES FORGES DE LA PROVIDENCE, *Marchienne-au-Pont, Hainault.*—Waggon wheels forged in one piece.

207 THIRION, *Aische-en-Refail, Namur.*—Model of a waggon.

208 USINE VANDENBRANDE, *Schaerbeek, near Brussels.*—Patent double-acting excentric; railway crossings, as used in Belgium.

209 VANDER ELST, L. & Co. *Braine-le-Comte, Hainault.*—Weigh-bridge for railways.

CLASS 6.

210 DE RUYTTER, J. *Bruges.*—A clarence.

211 JONES, BROS. *Brussels.*—Various carriages.

212 VAN AKEN, BROS. *Antwerp.*—A calash with double suspension.

213 VAN AKEN, C. B.—Calash.

CLASS 7.

214 CAIL, J. F. HALOT, A. & Co. *Brussels.*—Various apparatus for sugar manufacturers and refiners.

215 COMPAGNIE GÉNÉRALE DE MATÉRIELS DE CHEMINS DE FER, *Brussels.*—A mortising machine for wood.

216 DAUTREBANDE, H. *Huy.*—Machine for manufacturing endless paper.

217 DE BRUYNE, E. *Hamme, East Flanders.*—Hair-cloth bags for oil presses.

218 DE BRUYNE & SON, *Waesmunster, East Flanders.*—Hair-cloth bags for the extraction of oil from seeds; also, one to be used with an hydraulic press.

219 DE GROOTE, C. *Brussels.*—Bottle corker with glass tube, with or without needle.

220 DEHAYNIN, F. *Gosselies et Marcinelle, Hainault.*—Two drawings of a machine for agglomerating coal. Coal-bricks.

221 DE KEYSER-DUMORTIER, *Eecloo, East Flanders.*—Apparatus for the combination of oil seeds into cakes.

221 DUMONT, E. *Liege.*—Endless mechanical sieve for preparing ores, coals, &c.

EXHIBITORS' COMMITTEE FOR THE DISTRICT OF VERVIERS.—Machinery for the manufacture of wool:—

225 BORG, J. D. & VANDERMAESEN, L. C. *Verviers.*—Machine for making velvet.

226 HOUGET, J. D. & TESTON, C. *Verviers.*—Spinning machines, &c.

229 MARTIN, C. *Pepinster.*—Machine for oil-pressing wool; carding machine; articulated pads.

230 MARTIN, T. *Verviers.*—Cards and ribbons of cards for wool and cotton.

231 NEUBARTH & LONGTAIN, *Verviers.*—Longitudinal shearing implements.

232 TROUPIN, J. P. *Verviers.*—Blades, tables and rulers in cast steel for shearing cloths, shawls, and stuffs.

233 WANKENNE ET DEBIAL, *Verviers.*—Shearing or smoothing implements.

---

235 FETU, A. & DELIÈGE, *Liége.*—Matters used in making cards for spinning wool and cotton.

236 GENNOTTE, L. *Brussels.*—Apparatus for making gaseous beverages.

237 GÉRARD, D. *Charleroi.*—The working of mines simplified.

239 LAROCHE & Co. *Brussels.*—Machine for the manufacture of paper.

239A LELONG, C. & BISCOP. J. B. *Wiers, Hainault.*—Waggon for coal mine.

240 LEROY, A. *Brussels.*—Sewing, embroidering, button-hole making, and overcasting machines.

241 LIBOTTE, N. *Gilly, Hainault.*—Two miners' cages with parachutes and waggons.

242 MEERENS, *Brussels.*—Flower bleaching apparatus, preserving the lace makers from the effects of white lead.

243 MERTENS, *Gheel, Antwerp.*—Flax and hemp scutching and peeling machines.

245 NYST, F. *Liege.*—Friction parachute for miners' cages.

246 PERRIN, N. *Brussels.*—Drill stock with an Archimedean screw, worked with one hand only.

247 PREVOT, C. *Haine St. Pierre, Hainault.*—Pegs for carpenters' work, vessels, &c.

248 RYCX, A. & SON, *Ghent.*—Patent cards for cotton.

249 SACRÉ, A. *Brussels.*—Flax-drawing apparatus, with double spiral system: winding and spreading machines.

250 VALLÉE, F. *Molenbeek-St.-Jean, near Brussels.*—Small working model for spinning flax, wool, and cotton (the invention of the exhibitor).

251 VAN DER ELST, L. & Co. *Braine-le-Comte, Hainault.*—Drawing for a paper manufacturing machine.

252 VANGINDERTAELEN & Co. *Brussels.*—Hygienic and economic apparatus for breweries, taverns, distilleries, &c.

253 VAN GOETHEM, C. & Co. *Brussels.*—Centrifugal machines for purifying sugar.

254 VERMEULEN, C. *Roulers.*—Shuttles for weaving.

255 VINCENT, J. *Alost.*—A Jacquard machine with 756 hooks.

256 WERGIFOSSE, *Brussels.*—Liege mangle, washing and calendering machines.

257 WINNEN, *Brussels.*—Mill for unravelling rags.

258 WISSAERT, J. *Brussels.*—Embossing and gilding plates for bookbinding purposes.

259 WYNANTS & MACKINTOSH, *Brussels.*—Frames for locking up printing formes without wedges or feather edges.

CLASS 8.

260 ARNOULD, G. *Mons.*—A water-level; free-air manometer; miner's lamp.

261 BERTIEAUX, H. *Antwerp.*—Steam-engine.

262 CAIL, J. F. HALOT, A. & Co. *Brussels*—Giffard-injectors; tubular steam-generator portable engine.

COMMITTEE OF THE EXHIBITORS OF VERVIERS:—

263 HOUGET, J. D. & TESTON, C. *Verviers*—Portable engine.

---

264 CUNGNE, U. *Langhemarcq, West Flanders.*—Weighing scales, with pans above.

265 DELANDTSHEER, *Brussels.*—Drawings and description of a new application of Woolf's principle to horizontal steam-engines.

268 FONDU, J. B. *Lodelinsart, Hainault.*—Economic bars perforated horizontally, forming tubes under the ignited fuel.

270 GOUTEAUX, P. J. *Gilly, Hainault.*—Check chains and safety apparatus, applicable to mining engines.

271 LIBOTTE, N. *Gilly, Hainault.*—A fire-grate.

272 OBACH, N. *Brussels.*—Weighing-machine with double mechanism and square platform.

273 PERARD, L. *Liége.*—Horizontal blowing machine with two cylinders, Fossey's principle, 200-horse-power.

274 PETIT, H. J. & Co. *Brussels.*—Level, with one air bulb and check screw.

275 PIROTTE, L. & SISTERS, *Brussels.*—Weighing machine, Roberval's principle.

277 SACRÉ, C. *Brussels.*—Hydrometer for alcohol, giving the corrected density, and applicable to water.

278 SCRIBE, G. *Ghent.*—Patent horizontal engine (Woolf's principle), with connected cylinders, 30 horse-power.

279 THIRION, *Aische-en-Refail, Namur.*—Model of a windmill, driven by a helicoid, without wheel and pinion gear.

280 VANDERHECHT, E. *Brussels.*—Model of an anti-collision apparatus, applicable to mines and traction.

281 WINAND, F. *Goffontaine, Liege.*—Patent safety screw-jack.

CLASS 9.

283 BORTIER, P. *Ghistelles, West Flanders.*—Plan, in relief, of Britannia farm, at Ghistelles.

284 DAMS, *Tilleur, Liége.* — Unalterable enamelled tickets, for flowers, fruits, shrubs, &c.

285 D'AUXY, MARQUIS, *Frasnes near Leuze, Hainault.*—A granary.

286 DE GREEF, E. *Hal, Brabant.*—Agricultural implements.

287 DELSTANCHE, *Marbais, Brabant.* — Improved plough.

288 DESOER, O. *Ben-Ahin, Liége.*—A skeleton roller.

291 LECOMTE, *Pont-a-Celles, Hainault.*—Iron plough with double mould-board.

292 MARIE, L. J. *Marchienne-au-Pont, Hainault.*—Apparatus for cleaning grain.

293 ODEURS J. M. *Marlinne, Limburgh.*—Common plough; sub-soil plough, with or without balance beam.

294 PAS, P. A. *Londerzeel, Brabant.* — Churn, on a new principle.

295 PEERS, BARON E. *Oostcamp, West Flanders.*—A plan for a farm.

297 ROMEDENNE, A. J. *Erpent, Namur.*—Agricultural implements.

298 TIXHON, J. *Fléron, Liége.*—Agricultural implements.

300 VAN MAELE, E. *Thielt.*—Ploughs, straw chopper, bread cutter, sowing machine.

CLASS 10.

301 ADEN, L. *Brussels.*—Patent door, opening and closing on four sides.

302 BEERNAERT, A. *Brussels.* — Marble chimney-pieces.

303 BOCH, BROS. *La Louviere, Hainault.*—Mosaic slabs for pavements.

304 BOUCHER, T. *Saint-Ghislain, Hainault.*—Refractory substances, bricks, crucible, retorts, stone for spreading melted glass.

305 BOUCNEAU, L. *Brussels.*—A marble chimney-piece, renaissance style.

306 BOUWENS, *Mechlin, Antwerp.*—Music-desk, door-lock.

307 CHAUDRON, J. *Brussels.*—A process of shaft-lining, for the formation of wells in humid soils.

308 DEFUISSEAUX, MRS. *Baudour, near Mons.*—Articles in fire-clay.

310 DELPERDANGE, V. *Brussels.* — New method of joining water pipes, gas pipes, &c.

311 DEWYNDT, J. & Co. *Antwerp.*—Cedar wood veneer.

312 GODEFROY, J. *Brussels.*—Double doors for rooms, with central handle, and masked lever-locks.

313 GUIBAL, T. *Mons.* — Ventilator for mines, capable of displacing more than 100 cubical yards of air per second.

314 JACOBS, *Mechlin.*—A chimney-piece in portor marble with interior and flooring.

315 JOSSON, N. & DE LANGLE, *Antwerp.*—Hydraulic cements, mastic, terra cotta, bricks, tiles, and flags.

316 KELLER, A. *Ghent.*—Refractory gas-retorts.

317 LAMBRETTE, J. *Brussels.*—Zinc roofing.

318 LECLERQ, A. J. *Brussels.*—Chimney-pieces in every style.

319 SIEGLITZ, J. *Brussels.*—Chimney-piece, with statuary-work.

320 VANDER ELST-BOURGOIS, *Brussels.*—A black chimney-piece.

321 VAN NEUSS, M. *Brussels.*—Inodorous water-closet.

322 WYNEN, G. *Schaerbeek, near Brussels.*—A specimen of flooring.

CLASS 11.

323 BAYET, BROS. *Liége.*—Ornamented fire-arms; Lefaucheux guns, and guns with ramrods; Swiss carbine and revolver.

324 BERNIMOLIN, BROS. *Liége.*—Lefaucheux-Bernimolin guns, pistols, carbines on Flobert's principle.

325 COOPPAL & Co. (Director: C. VAN CROMPHAUT), *Wetteren, East Flanders.*—Gunpowder of various kinds, refined saltpetre.

326 DANDOY, C. *Liége.*—Fire-arms of all kinds.

327 DE LEZAACK, A. *Liége.* — Fowling-pieces, &c.

328 DITS, A. J. *Saint-Gilles, near Brussels.*—Cartridges for Lefaucheux revolvers, patent balls.

329 DUHENT, L. *Brussels.*—Wheelbarrow convertible into a camp-bed, ambulance, tent, boat, or bridge.

330 DUMOULIN-LAMBINON, G. *Liege.*—Guns, revolvers, pistols, carbines, &c.

331 FAFCHAMPS, *Brussels.*—New kinds of fire-arms; works of art, and new system of defence.

332 FUSNOT & Co. *Brussels.*—Copper cartridges for pistols, bushes for Lefaucheux fire-arms.

333 HERMAN, J. *Liége.*—Designs for the manufacture of arms.

334 HUBAR, *Herstal, Liége.* — Various pieces used in gun-making.

335 JANSEN, A. *Brussels.*—Breech-loading gun.

336 JONGEN, BROS. *Liége.*—Fire-arms, &c.

337 LADRY, *Brussels.* — Apparatus for taking correct aim with portable fire-arms; instrument for measuring the bullet-marks in a target.

338 LARDINOIS, N. C. *Liége.* — Breech-oading carbine.

339 LEMAIRE, J. B. *Liége.* — Fowling-pieces and revolving pistol.

340 LEXIN, C. *Ghent.*—Cuirasses for infantry, cavalry, artillery, &c.

341 MALHERBE, P. J. & Co. *Liége.*—Guns, musketoons, pistols, gun-barrels.

342 MASU, BROS. *Liége.*—Fowling-pieces.

343 SIMONIS N. & Co. *Val-Benoit, near Liege.*—Laminated gun-barrels.

344 TINLOT, J. M. *Herstal, Liége.*—A carbine on Flobert's principle.

CLASS 12.

345 VAN BELLINGEN, A. J. *Antwerp.*—Proved chain cables, and rigging chains.

CLASS 13.

346 BULTINCK, E. *Ostend.* — Electro-galvanic apparatus with inodorous acid, giving any required current.

347 DUSAUCHOIT, E. *Ghent.* — Signal speaking-trumpet and whistles.

348 GÉRARD, A. *Liége.* — Electric clock, electric battery, an electro-magnet; plans of instruments and machinery.

349 GLOESENER, M. *Liége.* — Chronoscopes, registering multipliers, an electric clock.

350 JASPAR, *Liége.* — Chronoscope on Major Navez's principle; Doctor Stacquez's 'electro-medical;" a regulator of electrical light.

351 LIPPENS, P. *Brussels.*—Telegraph apparatus, and requisites for telegraphic lines.

352 SACRÉ, ED. *Brussels.*—Philosophical balance, eclemeter-compass, circle-level.

353 VANDEVELDE, N. *Ghent.*—A saccharometer.

CLASS 14.

355 DAVELUY, *Bruges.* — Photographic views of Bruges.

356 DUPONT, *Antwerp.*--Photographs: portraits selected from the collection named "The Antwerp School."

357 FIERLANTS, ED. *Brussels.* — Photographs, representing the master-pieces and monuments of Belgium: executed by order of the Government.

358 GHÉMAR, BROS. *Brussels.* — Photographs, natural size, and others; visiting cards.

359 MASCRÉ, J. *Brussels.*—Photographs from pictures, plaster casts, &c.

360 MICHIELS, J. J. *Brussels.* — Photographs.

361 NEYT, A. L. *Ghent.*—Photographic micrography (obtained through the agency partly of solar and partly of electric light).

362 NEYT, CH. *Brussels.*—Photographs.

CLASS 15.

363 GÉRARD, A. *Liége.*—Clocks and watches.

CLASS 16.

364 AERTS, F. G. *Antwerp.*—Oblique-trichord seven-octave pianos.

365 ALBERT, E. *Brussels.* — Clarinets, flutes, hautboys, bassoons.

366 BERDEN, F. & Co. *Brussels.*—Upright pianos, with oblique and vertical strings.

368 DARCHE, C. F. *Brussels.* — Tenor-violins, violins, violoncello, &c.

371 JASTRZEBSKI, F. *Brussels.* — Grand pianoforte; upright transpositional pianoforte.

372 MAHILLON, C. *Brussels.*—A complete collection of musical instruments.

373 STERNBERG, L. & Co. *Brussels.*—Four pianos of different shapes.

374 VUILLAUME, N. F. *Brussels.*—Violins, violoncello, counter-bass.

CLASS 17.

375 GLITSCHKA, H. *Ghent.*—Surgical instruments; artificial limb.

376 KAYSER, *Brussels.* — Escape box for the royal railway trains.

377 ODEURS, J. M. *Marlinne, Limburgh.* —Speculum uteri; a mouth-opener.

378 WAERSEGERS, J. *Antwerp.*—Herniary trusses; ventral and hypogastric belts; orthopœdic apparatus; artificial limbs.

CLASS 18.

BELGIAN GOVERNMENT. — Cotton goods produced in the Flemish Apprentice Schools:—

WEST FLANDERS.

379 APPRENTICE SCHOOL OF BECELAERE. —Cotton goods for summer.

380 —— OF MOORSEELE.—Cotton checks.

381 —— OF MOORSLEDE.—Summer goods.

382 —— OF POPERINGHE.—Cotton checks.

383 —— OF ROULERS.—Cotton checks.

384 —— OF RUDDERVOORDE. — Cotton woven fabrics.

385 —— OF YPRES.—Cotton checks.

386 LATE APPRENTICE SCHOOL OF BRUGES (owners: MM. DE RANTERE & CO. *Bruges*).—Dimity, &c.

387 —— OF COURTRAI (owner: M. SISENLUST, *Courtrai*).—Cotton velvet.

EAST FLANDERS.

388 APPRENTICE SCHOOL OF CALCKEN.—Cotton stuffs for dresses and window blinds.

389 —— OF OLSENE. — Dimity; cotton-satin.

390 —— OF OORDEGHEM.—Cotton velvet, stuffs for window blinds.

391 —— OF SINAY (owner: M. VERELLEN-RODRIGO, *Saint Nicolas*).—Cotton stuffs for dresses; cravats.

392 LATE APPRENTICE SCHOOL OF LEDE (owner: M. V. DERCHE, *Brussels*).—Dimity and frame embroidered muslins; "royaumont" dimity.

393 —— OF NAZARETH (owner: M. VAN DEN BOSSCHE-VERVIER, *Nazareth.*) — "Leather-dimity."

394 —— OF NEDERBRAKEL (owner: M. DE PROOST, *Opbrakel*).—Cotton-satins, &c.

395 —— OF SLEYDINGE (owners: MM. CEUTERICK & DE COCK, *Ghent.*—White Jacquard cotton stuffs.

396 —— OF WAESMUNSTER (owner: M. VAN HOOF, *Lokeren*).—Cravats; fine cotton checks.

---

397 DE BACKER, L. & N. *Braine-le-Château, Brabant.*—Short staple Georgia cotton yarn.

398 DE BAST, C. *Ghent.*—Woven goods from raw cotton.

399 DE BLOCK-DELSAUX, *Termonde, East Flanders.*—Cotton bed-covers.

401 DE SMET, BROS. *Ghent.*—Cotton warp dyed, and dressed; plain and printed fabrics.

402 DE SMET, E. & Co. *Ghent.*—Dressed and dyed raw cotton warp for mixed fabrics.

403 DIERMAN-SETH, F. *Ghent.*—Woven fabrics of Surat cotton.

404 DUCHAMPS, G. *Brussels.*—Cotton stuffs for trousers and other garments.

405 DUJARDIN, J. E. & L. *Bruges.*—Raw cotton spun, warp and weft.

406 DUPREZ & Co. *Dottignies, near Courtrai.*—Cotton stuffs for trousers and other garments.

407 HOOREMAN-CAMBIER & SON, *Ghent.*—Cotton fabrics for trousers and other garments.

409 LEMAIRE-DUPRET & SON, *Tournay.*—Cotton stuffs for trousers.

410 MOUSCRON, CITY OF, DISTRICT OF COURTRAI, WEST FLANDERS, COMMITTEE:—
Desprets, Bros. | Labis-Delecoeillerie.
Dujardin, L.
—Cotton stuffs for trousers.

411 PHILIPS-GLAZER, J. *Termonde.*—Cotton bed-covers, calicoes, pilous, Belgian leather-cloth, dimity half linen.

412 PIRON, J. *Tournay.*—Stuffs for trousers, all cotton.

413 ROELANDTS, F. *Courtrai.*—Cotton stuffs for trousers and other garments.

414 ROOS & VAN BELLE, *Termonde.*—Cotton bedcovers.

415 RYCX, A. & VERSPEYEN, *Ghent.*—Cotton spools; cotton fabrics.

416 SAEYS, BROS. *Termonde.*—Cotton bed-covers.

416A SCHMIDT & Co. *Courtrai.*—Cotton stuffs for trousers.

417 STAELENS, P. & Co. *Ghent.*—Surat-cotton yarns.

418 VAN HEE, BROS. *Mouscron, West Flanders.*—Cotton stuffs for trousers, &c.

419 VAN HEUVERSWYN, F. & Co. *Ghent.*—Counterpanes, petticoats, dimity, damasks (white and coloured), calicos.

420 VAN NESTE & VANDER MERSCH, *Rolleghem, near Courtrai.*—Cotton stuffs for trousers, dresses, and waistcoats.

CLASS 19.

BELGIAN GOVERNMENT.—Linen goods produced in the Flemish apprentice schools:—

WEST FLANDERS.

421 APPRENTICE SCHOOL OF AERSEELE.—Half-bleached linens.

422 —— OF AERTRYCKE.—Linen.

423 —— OF ANSEGHEM.—Linen.

424 —— OF ARDOYE.—Linens.

425 —— OF AVELGHEM.—Linens.

426 —— OF BECELAERE.—Linens and handkerchiefs.

427 —— OF CLERCKEN.—Linens.

428 —— OF CORTEMARCQ.—Linens.

429 —— OF COURTRAI.—Linens, and damasks.

430 APPRENTICE SCHOOL OF DEERLYK.—Linens.

431 —— OF DENTERGHEM.—Linens.

432 —— OF DESSELGHEM.—Linens and cambric handkerchiefs.

433 —— OF GHISTELLES.—Linens.

434 —— OF HEULE.—Linens.

435 —— OF HOOGHLEDE.—Linens.

436 —— OF HULSTE.—Linens.

437 —— OF INGOYGHEM.—Linens.

438 —— OF LANGHEMARCQ.—Linens, bleached and unbleached; linens for mattresses.

439 —— OF LENDELEDE.—Linens.

440 —— OF LICHTERVELDE.—Linens.

441 —— OF MENIN.—Linens, and handkerchiefs.

442 —— OF MEULEBEKE.—Linen.

443 —— OF MOORSEELE.—Linens.

444 —— OF MOORSLEDE.—Linens.

445 —— OF OOSTNIEUWKERKE.—Linens.

446 —— OF OOSTROOSEBEKE.—Linens.

447 —— OF OUCKENE.—Linen; cambric handkerchiefs.

448 —— OF OYGHEM.—Linens.

449 —— OF PASSCHENDAELE.—Linens.

450 —— OF PITTHEM.—Plain linen; linens for napkins and towels.

451 —— OF POPERINGHE.—Linen for mattresses, diaper handkerchiefs.

452 —— OF ROULERS.—Linens and damasks.

453 —— OF RUDDERVOORDE.—Linen fabrics.

454 —— OF RUYSSELEDE.—Linen.

455 —— OF STADEN.—Linens.

456 —— OF SWEVEGHEM.—Linens.

457 —— OF SWEVEZEELE.—Linens.

458 —— OF THIELT.—Linens.

459 —— OF THOUROUT.—Ticks and diapers.

460 —— OF WAEREGHEM.—Linens made of raw and half-bleached yarn, linen for towels.

461 —— OF WESTROOSEBEKE.—Bleached linens.

462 —— OF YPRES.—Linens, plain and damasked for mattresses.

463 LATE APPRENTICE SCHOOL OF BLANKENBERGHE (owner: M. L. DE LESCLUZE, *Bruges*).—Ticks, of flax only; and of flax and cotton, English mode of manufacture.

464 —— OF BRUGES (owner: M. MARLIER, *Bruges*).—Blue linens; diaper linens, blue and white.

465 —— OF BRUGES (owner: M. C. POPP, *Bruges*).—Linens and cambrics.

466 —— OF BRUGES (owner: M. ARDRIGHETTI, *Bruges*).—Fore-parts of shirts, with moveable breasts made of flax-yarn.

467 —— OF ISEGHEM (owner: M. MAES-VAN-CAMPENHANDT).—Linens, ticks, and handkerchiefs.

EAST FLANDERS.

468 APPRENTICE SCHOOL OF BAELEGEM (M. ROBYNS, *Baelegem*).—Linens, handkerchiefs, and linen ticks.

469 APPRENTICE SCHOOL OF CALCKEN.—Flax yarn fabrics.

470 —— OF EYNE (MM. L. & A. VAN DE PUTTE, *Ghent*).—Linens and handkerchiefs made of flax yarn, linens for mattresses.

471 —— OF NEDERBRAKEL (M. DE PROOST, *Opbrakel*).—Napkins.

472 —— OF OLSENE.—Linens.

473 —— OF OORDEGEM.—Damasked linens for mattresses, &c.

474 —— OF SYNGEM.—Unbleached linens.

475 —— OF URSEL.—Table-linens, &c.

476 LATE APPRENTICE SCHOOL OF ALOST (owners: MM. J. & P. NOËL, BROS.).—Damask and diaper table-linen, linen for mattresses.

477 —— OF BELLEM (owner: M. MOERMAN-VAN-LAERE, *Gand*).—Sail cloth, &c.

478 —— OF SLEYDINGE (owner: M. DOBBELAERE-HULIN, *Ghent*).—Sail cloth, plain linens, diapers for mattresses.

---

480 CAESENS, V. & SON, *Zele, East Flanders.*—Bolting cloth, cloth for stopping up casks, and other purposes.

482 DE BRANDT, J. *Alost, East Flanders.*—Damask and diaper table-linen, &c.

483 DE BROUCKERE, BROS. *Roulers.*—Tow-yarn.

483A DEVOS, F. & CO. *Courtrai.*—Plain unbleached hand-spun flax-yarn.

484 FRANCHOMME, L. *Brussels.*—Various sorts of ticks.

485 JELIE, J. B. *Alost.*—Flax sewing thread, hand and machine-spun.

487 LEFEBVRE V. *Alost.*—Plain linens.

488 MAES-VAN CAMPENHOUDT, *Iseghem, West Flanders.*—Linens, tickings, cambric handkerchiefs.

489 SAINT BERNARD HOUSE OF CORRECTION.—*Hemixem, Antwerp.*—Linens of various kinds.

491 SIREJACOB, E. & COUCKE, C. *Brussels.*—Diaper and damasked napkins with crests, towels, &c.

492 SOCIÉTÉ LINÈRE DE BRUXELLES, *St. Gilles, near Brussels.*—Machine-spun flax and tow yarns; flax and hemp fabrics, woven by hand and power looms.

493 SOCIÉTÉ LINIÈRE GANTOISE, *Ghent.*—Bleached and unbleached flax and tow-yarns.

494 SOCIÉTÉ LINIÈRE DE SAINT LÉONARD, *Liége.*—Flax and tow yarns.

495 TANT-VERLINDE, *Roulers.*—Flax, flax-yarns and linens, unbleached, &c.

496 THIENPONT, L. & SUNAERT, A. *Ghent.*—Damask and diaper table-linen, linen cloth for mattresses, towels.

498 VAN ACKERE, J. C. *Wevelghem and Courtrai, West Flanders.*—Unbleached and bleached linens; linen and cambric handkerchiefs.

499 VAN DAMME, BROS. *Roulers.*—Unbleached linen.

500 VAN DE WYNCKELE, BROS. & ALSBERGE, J. *Ghent.*—Flax-yarns, in every stage of bleaching.

501 VAN MELDERT, *Haeltert, near Alost.*—Unbleached linen, table-linen, &c.

502 VAN OOST, P. *Hooghledè, West Flanders.*—Linens made of machine and hand spun yarn.

503 VAN ROBAYS, A. J. *Waereghem, West Flanders.*—Sail cloths, russias, sackings made of jute.

504 VAN TIEGHEM & CO. *Courtrai.*—Linen made of machine and hand spun yarn.

505 VERRIEST, P. *Courtrai.*—Diaper and damask table-linen and cloths for mattresses.

506 VERTONGEN-GOENS, C. S. *Termonde.*—A piece of manille-hemp flat cable with eight strands.

CLASS 20.

BELGIAN GOVERNMENT.—Silk and velvet goods, produced in Flemish Apprentice Schools:—

WEST FLANDERS.

507 LATE APPRENTICE SCHOOL OF BRUGES (M. AVANZO), *Brussels.*—Ribbons for hats and caps.

EAST FLANDERS.

508 LATE APPRENTICE SCHOOL OF ALOST (owner: M. LEVIONNOIS-DEKENS, *Alost*).—Articles in plain black silk.

509 —— OF DEYNZE (owners: MM. LAGRANGE, BROS. *Deynze*).—Various articles in silk.

---

510 THYS, C. *Brussels.*—Thrown, unbleached, and dyed silks, for mercers' and lace-maker's goods.

CLASS 21.

BELGIAN GOVERNMENT.—Woollen and mixed fabrics produced in the apprentice schools of Flanders:—

WEST FLANDERS.

511 APPRENTICE SCHOOL OF BECELAERE.—Articles of wool and cotton.

512 —— OF BRUGES (MM. KAUWERZ & CO. *Brussels*).—Bournous half-wool, tartans, galaplaids, and goats hair cloth.

513 —— OF COURTRAI.—Stuffs for trousers.

514 —— OF DEERLYK.—Roubaix cloths, for trousers; materials for dresses; fancy stuffs of silk mixed with wool and cotton.

515 —— OF HULSTE.—Woollen fabrics.

516 —— OF LANGHEMARCQ.—Black paramata.

517 —— OF MENIN.—Stuff for trousers.

518 —— OF MOORSLEDE.—Siamese, an article of Roubaix.

519 —— OF MOUSCRON.—Woven fabrics, wool and cotton.

520 —— OF POPERINGHE.—Siamese, plain and twilled.

521 —— OF ROULERS.—Orléans.

522 —— OF THIELT (MM. SCHEPPERS, *Loth, near Brussels*).—Thibets, lastings, serges, &c.

523 —— OF YPRES.—Molletons.

EAST FLANDERS.

524 APPRENTICE SCHOOL OF CALOKEN.—Stuffs for dresses.

525 —— OF OLSENE.—Orléans, paramatas.

526 —— OF RUYEN.—Mixed fabrics for dresses and trousers; fabrics manufactured on Jacquart's principle.

527 —— OF SINAY (M. VERELLEN-RODRIGO, *St. Nicolas*).—Materials for dresses, &c. in wool cotton and silk.

528 —— OF URSEL.—Stuffs for trousers in wool and cotton.

529 —— OF WICHELEN (M. F. VAN BRABANDER, *Wichelen*).—Stuffs for mattresses.

530 LATE APPRENTICE SCHOOL OF NAZARETH (owner: M. VAN DEN BOSSCHE-VERVIER, *Nazareth*).—Tweed, corded stuff, satin, satin-reps (wool and cotton).

531 —— OF WAESMUNSTER (owner: M. VAN HOOFF, *Lokeren*).—Woven goods, in wool, and in wool and cotton.

---

532 ANDRIES & WAUTERS, *Mechlin, Antwerp*.—Woollen blankets.

533 BEGASSE, CH. *Liege*.—Woollen blankets; felts for paper factories; woollen stuffs.

COMMITTEE FOR THE EXHIBITION OF THE DISTRICT OF VERVIERS. — Woollen yarns and fabrics :—

535 BARAS-NAVAUX, *Hodimont, near Verviers*.—Light woollen stuffs for suits, and caps.

536 BERCK, CH. *Aerve*.—Spun goods for borders.

537 BIOLLEY, F. & SON, *Verviers*.—Cloths, satins, cashmeres, fancy woollen cloths.

538 BRULS-RIGAUX, *Goffontaine Cornesse*.—Thread of wool and cotton mixed.

539 CHANDELLE-HANNOTTE, *Dison*.—Beavers, and knitted articles.

540 CHAUDOIR & HOUSSAT, *Hodimont*.—Fancy stuffs for winter and summer.

541 COMMISSION VERVIÉTOISE.—Corded stuffs, billiard cloths, tweeds, satins, &c.

542 DEBEFVE-BLAISE, *Dison*.—Hangings, fancy cloths, military cloths.

543 DEHESELLE, *Thimister, near Verviers*.—Flannels, domets, gauzes, and swan-skins.

543A DELEVAL & SON, *Dison, Verviers*.—Fancy stuffs, &c.

544 DEL MARMOL, F. *Francomont, near Verviers*.—Domets, and flannels.

545 DORET, V. (LÉONARD DORET), *Verviers*.—Woollen cloth, dyed and undyed.

546 DUBOIS, GÉRARD, & Co. *Verviers*.—Stuffs of wool, and of silk and wool; wool-satin, and velvet cloths.

547 FLAGONTIER, J. J. *Verviers*.—Stuffs of wool, and of wool and silk for trousers, &c.

548 GAROT, J. *Hodimont, near Verviers*.—Stuffs of wool, and of wool and silk; fancy cloths for trousers, great coats, cloaks, &c.

549 GRANDJEAN, H. J. *Verviers*.—Stuffs of wool, and of wool and silk.

550 GRÉGOIRE & PELTZER, *Dison*.—Woollen stuffs for great coats and trousers.

551 HAUZEUR, P. & VIGAND, BROS. *Ensival, near Verviers*.—Stuffs of wool, and of wool and silk.

552 HAZEUR, GÉR. & SON, *Verviers*.—Thread of carded wool, for weaving.

552A HENROTTY, MARLOHAL, *Ensival, Verviers*.—Woollen stuffs.

553 HENRION, J. J. *Hodimont, Verviers*.—Cloths, and woollen stuffs for trousers, &c.

554 LAHAYE, M. & Co. *Verviers*.—Cloth and woollen stuffs for suits.

555 LAOUREUX, G. J. *Verviers*.—Cloth plain and twilled; woollen stuffs.

556 LECLERCQ, N. *Dison*.—Beavers, duffels, satins, moscows &c.

557 LEJEUNE-VINCENT, H. S. *Dison*.—Fancy stuffs, ladies cloaks, &c.

558 LEJEUNE-VINCENT, J. C. *Dison*.—Woollen stuffs, moscows, wool-satins.

559 LIEUTENANT & PELTZER, *Verviers*.—Thread, cashmeres, beavers, wool-satins, stuffs of fancy wool, and of wool and silk.

560 LINCÉ, WIDOW H. & SON, *Dison*.—Moscows, fancy stuffs, stuffs of wool and silk.

561 MARBAISE & SON, *Hodimont*.—Military and other cloth, woollen stuffs.

562 MASSON, L. *Verviers*.—Hangings, manufactured stuffs, black and coloured.

562A MATHIEU, J. F. *Dison, Verviers*.—Woollen stuffs.

563 MODION, A. & BERTRAND, M. *Verviers*.—Moscows, stuffs of wool and silk.

564 MULLENDORFF & Co. *Verviers*.—Thread made of carded wool for fancy cloths, stuffs, shawls, &c.

565 NAVAUX, R. & SON, *Hodimont*.—Reps and summer goods.

566 OLIVIER, J. J. & SON, *Verviers*.—Drapery and woollen stuffs.

567 PIRENNE & DUESBERG, *Verviers*.—Wool-satins, fancy and other woollen stuffs, military cloth.

568 PIRON-THIMISTER, *Francomont*.—Stuffs of wool and silk, double milled cloth used in garments for the Belgian army.

569 RAULENBEK & Co. *Verviers*.—Cloths, fancy stuffs, fabrics for gloves, cloth gloves.

570 SAGEHOMME-LUTASTER, S. *Dison*.—Moscows, cotelines, corded stuffs, &c.

571 SAUVAGE, A. J. *Francomont near Verviers*.—Woollen stuffs.

572 SERET & PIRARD, *Verviers*.—Thread made of white and other wool, washed and unwashed.

572A SIMAR, DRÈZE, *Dison, Verviers*.—Billiard cloths.

573 SIMON, J. & DIET, *Chaineux, near Verviers*.—Thread made of carded wool, unbleached, and mixed in different shades.

574 SIMONIS, I. *Verviers*.—Hangings, stuffs of wool, and of wool and silk.

575 SIRTAINE, F. *Verviers*.—Cloths and woollen stuffs, fancy goods.

577 SNOEK, E. *Charneux, near Verviers*.—Cloths, zephyrs, cashmeres, wool-satins, corded stuffs, moscows, and other woollen goods.

578 VAN DER MAESEN, L. C. *Verviers*.—Fancy stuffs for great coats, ladies' cloaks, &c.; stuffs in wool and silk.

578A SUHS, J. A.—Woollen stuffs.

579 Vervier & Gregoire, *Verviers.*—Fancy stuffs, velvets, stuffs in wool and silk.

580 Voos, J. J. *Verviers.*—Fancy stuffs, cloths, hangings, double milled cloth, &c.

580a Wimandy, Venster, *Dison.*—Woollen stuffs.

581 Xhibitte, *Charneux, near Verviers.*—Carded wool for fancy cloths, stuffs, &c.

581a Xhoffray, C. & Bruls, C. *Doltrain.*—Carded wool.

---

582 Duchamps, G. *Brussels.*—Stuffs of cotton and wool mixed, for trousers, &c.

583 Duprez & Co. *Dottignies, near Courtrai.*—Stuffs of wool and cotton for trousers, &c.

584 Gauchez, L. *Brussels.*—Blankets, felted threads, fancy and mixed fabrics woven from felted threads.

585 Kauwerz, P. & Co. *Brussels.*—Tartan shawls.

587 Lemaire-Dupret & Son, *Tournay.*—Stuffs for trousers, of wool cotton and thread.

588 Mouscron (City of), *District of Courtrai.*—Stuffs in cotton and wool, for trousers, &c.

589 Piron, J. *Tournay.*—Stuffs for trousers in wool and cotton; ticks in thread and cotton.

590 Roelandts, F. *Courtrai.*—Stuffs in wool, and in thread and cotton, for trousers, &c.

591 Rolin, H. Son, & Co. *Saint-Nicolas, East Flanders.*—Tartan shawls, fabrics all wool, or wool cotton and silk.

592 Schmidt & Co. *Courtrai.*—Fabrics for trousers, in cotton thread and wool.

593 Van Hee, Bros. *Mouscron, West Flanders.*—Stuffs, in wool and cotton, for trousers, &c.

594 Van Neste & Vander Mersch, *Rolleghem, West Flanders.*—Stuffs, in wool and cotton, for trousers, dresses, &c.

595 Wauters, A. & A. *Tamise, East Flanders.*—Wool-poplin, silk poplin, shawls.

CLASS 22.

596 Braquenie, Bros. & Co. *Ingelmunster, West Flanders.*—Carpeting, Flanders tapestry for furniture and hangings.

597 Moyersoen-Cammaerts, R. *Brussels.*—Pilous tapestry, carpetings, rugs, &c.

598 Schepens, L. *Ghent.*—Pattern, drawn for a high-warp carpeting.

599 Société de la Manufacture Royale de Tapis de Tournay, *Brussels.*—Carpets.

600 Timmermans, Miss M. *Ixelles, near Brussels.*—Design for a table-cover, &c.

CLASS 23.

601 Dewolf & De Mey, *Rouge-Cloitre-under-Auderghem, near Brussels.*—Cotton yarns in fast colours.

603 Idiers, E. *Auderghem, near Brussels*—Cotton yarns in Turkey red, and other fast colours.

604 Rave, N. Sen. *Curreghem, near Brussels.*—Dyed goods; wool, silk, cotton spun and raw.

CLASS 24.

Belgian Government.—Embroidered articles, manufactured in the Apprentice Schools of Flanders:—

WEST FLANDERS.

605 Apprentice School of Swevegheм (Girls).—Embroidered articles, style of St. Gall.

606 Late Apprentice School of Bruges (owner: M. Avanzo, *Brussels*).—Laces; galloons.

EAST FLANDERS.

607 Apprentice School of Calloo.—Embroidery on lace.

---

609 Boeteman, A. J. *Bruges.*—Handkerchief and collars in Valenciennes lace.

610 Bonnod, P. *Brussels.*—Designs for flounces and lace.

611 Bruyneel, Sen. *Grammont, East Flanders.*—Black silk lace.

612 Buchholtz, & Co. *Brussels & Valencienne.*—Lace embroidery, &c.

613 Christiaensen, G. H. J. *Antwerp.*—Embroidered lace.

614 Custodi-Besme, J. *Brussels.*—Handkerchief in gauze point, Chantilly veil.

615 Daimeries-Petitjean, *Brussels.*—White and black lace, antique style, &c.

616 De Clippèle, Mrs. C. *Brussels.*—Brussels lace.

617 Delaporte, Mrs. *Brussels.*—Galloons for carriages and livery-lace.

618 De Rantere & Co. *Bruges.*—Embroidered articles, in the style termed "Plumetis."

619 Everaert, J. & Sisters, *Brussels.*—Black and white lace.

620 Geffrier-Delisle, Bros. & Co.—Brussels lace, gauze point, &c.

621 Ghysels, V. & Co. *Brussels.*—Brussels lace transferred, gauze point, guipure.

622 Gregoir-Geloen, N. J. *Brussels.*—Brussels and Valenciennes lace.

623 Hanssens-Hap, B. *Vilvorde, near Brussels.*—Galloons for carriages.

624 Hoorickx & Steppe, *Brussels.*—Articles in lace.

625 Houtmans, A. J. *Brussels.*—Designs for lace, &c.

626 Houtmans, C. C. *Brussels.*—Designs for lace, &c.

627 Hutellier, *Brussels.*—Transferred lace, in point, cushion work, and gauze.

628 Keymeulen, H. *Brussels.*—Flemish black lace, and lace articles.

629 Lepage-Kina, J. G. *Grammont.*—Ladies' apparel in black lace, and other lace articles.

630 Melotte, E. *Brussels.*—Banner of the Brussels tennis club.

631 MINNIE-DANSAERT, C. *Brussels.* — Various descriptions of lace, and lace articles.

632 MULLIE-TRUYFFAUT, P. *Courtrai.*— Valenciennes lace, and articles made of it.

633 NAETEN, J. *Brussels.*—Designs for various articles in lace.

634 PHILIPPE, L. *Brussels.*—Gold embroidery; royal arms of Belgium.

635 REINHEIMER, C. (SOPHIE DEFRENNE), *Brussels.* — Brussels lace in cushion work, flounces, handkerchiefs.

636 ROOSEN, H. (SECLET-VANCUTSEM), *Brussels.*—Articles in lace.

637 SALIGO-VANDENBERGHE, *Grammont.* —Articles enriched with lace.

638 SASSE, MRS. P. F. *Brussels & London.* —Lace.

639 SCHUERMANS, G. *Brussels.*—Brussels transferred lace, gauze point, English point, embroidery, imitations.

640 STOCQUART BROS. *Grammont.*—Black lace, and lace articles.

641 STREHLER, J. *Brussels.*—Lace, transferred to gauze; Valenciennes lace, and embroidery.

642 VAN CAULAERT-STIÉNON, E. *Brussels.* — Articles in Brussels lace; head-dress in black lace.

643 VAN DER DUSSEN, B. J. *Brussels.*— Designs for various articles in Brussels lace.

644 VANDERHAEGEN & Co. *Brussels.*— Brussels transferred lace.

645 VAN DER PLANCKE, SISTERS, *Courtrai.*—Valenciennes lace.

646 VAN DER SMISSEN-VAN DEN BOSSCHE, *Alost.* — Specimens of Brussels and Valenciennes lace transferred.

647 VAN DER SMISSEN, V. *Brussels.*— Brussels lace transferred, and embroidery on net; various articles.

648 VAN ROSSUM, J. B. *Hal, Brabant.*— Gauze point lace, handkerchiefs, collars, sleeves, and lappets.

649 WASHER, V. *Brussels.*—Lace and imitations, and articles in lace.

650 WITTOCKX, H. *Saint-Josse-ten-Noode, near Brussels.*—Black silk lace, tunic, flounces, pelerine, &c.

CLASS 25.

651 BERTOU, BROS. J. J. & A. P. *Liége.* —Various tanned skins,

652 BULTER, CH. *Brussels.*—Furs.

653 DELMOTTE, H. *Ghent.*—Belgian hog's bristles.

654 DEVACHT, G. A. *Brussels.*—Articles in hair.

655 HANSSENS-HAP, B. *Vilvorde, near Brussels.*—Hair cloth; hog's bristles; painter's brushes.

656 HESNAULT, A. & BROTHER, *Ghent.*— Rabbit skins finished; hare and rabbit fur.

657 JONNIAUX, E. & Co. *Brussels.*—Tawed skins.

658 LONCKE-HAESE, *Roulers.* — Brushes, hog's bristles.

659 MOTTIE, *Brussels.*—Wigs on a new principle.

660 SCHMITZ, F. A. *Brussels.*—Morocco dressed sheep-skins; bands of cut leather for hat-making.

661 SOMZÉ, H. JUN. *Liége.* — Brushes, hog's bristles.

662 SOMZÉ-MAHY, H. *Liége.* — Brushes, hog's bristles.

663 VERRYCK-FLEETWOOD, *Brussels.* — Perukes, and hair fronts.

CLASS 26.

664 ARRETZ-WUYTS, G. *Aerschot, Brabant.*—Leather, vamp, &c.

665 BOONE, A. J. *Alost.*—Tanned, curried, and japanned skins.

666 BOONE, J. & Co. *Cureghem, near Brussels.*—Curried calf-skins.

667 BOUVY, A. *Liége.* — Calf-skins, and leather for various purposes.

668 COLLET, L. J. *Brussels.*—Leather and hides, japanned and plain, for saddlery, &c.

669 D'ANCRÉ, P. *Louvain, Brabant.*— Buenos-Ayres hides tanned but not beaten.

670 DAVID, P. *Stavelot, Liége.* — Strong sole leather.

671 DECLERQ-VANHAVERBEKE, L. *Iseghem, West Flanders.*—Tanned and curried skins, calf-skins, vamps, leather for soles.

672 DE CLIPPÈLE, CH. & Co. *Brussels.*— Engine-straps, joined on a new principle.

673 EVERAERTS, C. *Wavre, Brabant.*— Leather: calf-skin boot-legs and fronts.

675 FETU, J. G. J. & Co. *Brussels.*—Straps for machinery, leathern hose.

676 FONTEYNE, J. *Bruges.*—Foreign and native leather for soles, calf-skin, curried horse-hide, &c.

677 HEGH, F. & DUGNIOLLE, A. *Mechlin.* —Curried goods, morocco-leather, varnished leather and articles for hat making.

678 HOUDIN & LAMBERT, *Brussels.*—Sole-leather, calf-skins; French and Belgian military accoutrements.

679 JOREZ, L. & SON, *Brussels.*—Oil-cloths, American linen cloths, gummed taffetas, varnished leathers, American cloth panels.

680 LUYTEN, C. F. & J. *Cureghem, near Brussels.* — Leather, varnished linens and cottons.

681 MARÉCHAL, V. J. *Brussels.* — Harness.

682 MASSANGE, A. *Stavelot, Liege.* — American leather for soles and engine-straps; polished native cow-hides.

683 MOUTHUY, A. *Brussels.* — Engine-straps.

684 PERLEAU-TAZIAUX, MRS. *Saint-Hubert, Luxemburgh.*—Brazil tanned hides for soles.

685 PIRET-PAUCHET, E. *Namur.* — Sole-leather, tanned with oak bark.

687 ROUSSEL, E. *Tournay.*—Strong leather for cylinder-packings, sole-leather.

689 VAN MOLLE, L. L. *Lennick Saint Quentin, Brabant.*—Harness for a draught horse.

690 VAN SCHOONEN, E. *Ghent.*—Straps for machinery on an improved principle.

CLASS 27.

691 CANISIUS, G. *Huy, Liége.* — Caps, stuff-hats.

692 CÉSAR, A. & Co. *Brussels.*—Shoes and boots, ordinary and with wooden soles.

693 COLIN RENSON, H. *Brussels.*—Kid and leather gloves.

694 DE BLOCK, MRS. *Antwerp.* — Hygienic corsets.

695 DE COSTER, H. *Brussels.*—Shoes and boots.

697 FAGEL-VALLAEYS, B. *Ypres, West Flanders.*—Silk hats.

698 FRENAY, BROS. *Roclenge, Limburgh.*—Straw-plats, straw bonnets, and hats.

699 HANSEN, F. G. *Liége.* — Boots of morocco and varnished leather.

700 JONNIAUX, ED. & CO. *Brussels.*—Kid gloves.

701 LAINGLET, J. *Brussels.*—Silk corsets.

702 LECLERCQ, N. *Bruges.*—Boots, shoes, half-boots, fishing-boots, &c.

703 LIÉVAIN, L. *Mechlin.*—Silk and felt hats.

704 MASSON-FOUQUET, MRS. A. *Brussels.* —Horse-hair corsets.

705 SOITOUX, ET. *Saint Gilles, near Brussels.*—Galoches with wooden soles.

706 SOMZÉ-MAHY, H. *Liége.*—Shoes and boots.

707 TROOSTENBERGHE, D. *Bruges.* — Leather half-boots, waterproof shoes without seams, leather gaiters in a single piece.

708 VALENTYNS & VANDER PLAETSEN, *Saint-Josse-ten-Noode, near Brussels.* — Kid gloves, "gants duchesse" gloves.

709 VANDEN BOS-POELMAN, *Ghent.* — Waterproof sporting boots; other boots, fancy and plain.

710 VAN DE ROOST, M. *Brussels.*—Boot-trees and lasts; half-boots, &c.

711 VIMENET & SON, *Brussels.*—Hats of shorn-nap and velvet felt.

712 WATRIGANT, LATE ALLARD, *Brussels.* —Boots and shoes after a new pattern.

CLASS 28.

713 ASSELBERGHS-LEQUIME, *Brussels.* — Letter-paper.

714 BARBIER-HANSSENS, L. E. *Brussels.* —Packing and wrapping-paper.

715 BREPOLS, DIERCKX, & SON, *Turnhout, Antwerp.*—Playing cards, fancy paper; bound books.

716 BRIARD, J. H. *Brussels.*—Specimens of lithography.

717 BRUCK, P. A. *Arlon, Luxemburgh.*—Scientifical works on the manufacture of paper.

719 CALLEWAERT, BROS. *Brussels.*—Stationery.

720 DAVELUY, *Bruges.* — Playing cards chromo-lithographs.

721 DEMAEGT, J. *Saint-Josse-ten-Noode, near Brussels.*—Paper and pulp, made without rags or straw.

722 DESSAIN, H. *Mechlin.* — Liturgical, theological, and devotional works.

723 GLÉNISSON & SON, *Turnhout.*—Fancy paper, playing cards.

724 GOUWELOOS, A. *Brussels.* — Samples of registers, railway tickets, lithography, &c.

725 GREUSE, C. J. A. *Schaerbeek, near Brussels.*—Folio and quarto volumes.

726 HAYEZ, M. J. F. *Brussels.*—Books.

727 HENRY, P. *Dinant, Namur.*—Pressing boards, paste-board.

728 JERVIS, G. *Brussels.* — Diagram to illustrate a new method of printing chromo-lithographically with four impressions.

729 LELONG, C. *Brussels.*—Typographical works.

730 MORREN, ED. *Liege.* — An horticultural and botanical review, with chromo-lithographed illustrations of flowers.

731 MUQUARDT, C. *Brussels.*—Illustrated works.

732 OLIN & DEMEURS, G. *Brussels.* — Printing and packing paper.

733 PARENT, W. & SONS, *Brussels.* — Illustrated, and other works.

734 POISSONNIEZ, J. B. *Brussels.*—Paste-board and cards.

735 SCHAVYE, J. C. E. *Brussels.*—Ancient and modern bookbindings, designs, &c.

736 SEVEREYNS, G. M. C. *Saint-Josse-ten-Noode, near Brussels.*—Scientific and chromo-lithographic drawings, &c.

737 SOCIETY FOR PAPER MANUFACTURE OF BASSE-WAVRE AND GASTUCHE, *Basse-Wavre, Brabant.* — Writing and printing paper, paste-board.

739 TARDIF, BROS. *Brussels & Paris.*—Tracing and photographic paper.

740 TIRCHER, J. B. *Brussels.*—"History of Glass Staining."

741 VAN CAMPENHOUT, *Brussels.*—Specimens of registers.

742 VAN DOOSSELAERE, J. S. *Ghent.* — Typography; wood-cuts printed on vellum, silk, and enamelled paper.

743 VAN GENECHTEN, A. *Turnhout.* — Playing cards, fancy papers, enamelled pasteboard; typography, lithography, registers.

744 VAN VELSEN, E. F. *Mechlin.*—Illustrated, and other books.

746 WEISSENBRUCH, MISS, *Brussels.* — Books.

CLASS 29.

748 BELGIAN GOVERNMENT. — Collection of educational objects, formed under the superintendence of Professor Braun.

749 BRAUN, CH. *Rivelles, Brabant.*—Pedagogical and classical works.

750 CALLEWAERT, BROS. *Brussels.* — Method of writing adopted in the Belgian schools, &c.; atlases.

751 CAMPION, J. J. *Brussels.*—"Journal of Popular Education."

752 GÉRARD, JOSEPH, *Brussels.*—Tablets, for teaching history, &c.

CLASS 30.

753 DAEMS-SCHOY, J. B. *Brussels.*—Furniture in sculptured wood, framings, fancy articles.

754 DEBASIN-SCHMIDT, CH. *Namur.* — Painted imitations of woods and marbles.

755 DE GOBART, EM. *Ghent.* — Dining-room furniture.

756 DEKEYN BROS. *Saint-Josse-ten-Noode, near Brussels.*—Inlaid flooring, in wood of different colours.

757 DELEEUW-DEMARÉE, *Brussels.* — A gilt frame.

758 DERENNE, L. J. *Evelette, Namur.*—A sofa, table, chairs, &c.

759 DERUDDER, SON, & CO. *Brussels.*—Frames and console gilt, &c.

760 GODEFROY, J. *Brussels.*—Inlaid wood flooring.

761 GOYERS, BROS. J. & H. *Louvain.*—A pulpit in the Gothic style.

762 HODY, J. J. *Aubel, Liége.*—Piece of furniture serving four purposes; dressing-table, praying-desk, and writing-table.

763 LEARCH, A. *Brussels.*—Panels, in imitation of ancient leather.

765 LUPPENS, H. *Brussels.*—Clocks, vases, bronze model of a monument.

767 MARLIER, *Brussels.*—Buffet or cupboard.

768 OLIN & DEMEURS, G. *Brussels.*—Paper hangings.

769 PEETERS-VIERING, J. *Mechlin.*—Dining-room furniture in antique style.

770 POHLMANN, G. & DALK, A. *Brussels.* —Mouldings for panel and frame works, specimens of frames.

771 REISSE, CH. *Brussels.*—Gothic chimney-piece and clock, in oak.

772 RYCKERS, E. SON, *Brussels.*—Buffet in the style of Louis XIII.

773 VAN DEN BRANDE, BROS. *Mechlin.*—Carved and inlaid drawing-room furniture.

774 VAN DEN BROECK, D. *Brussels.*—A gilt wood toilet-console, in the style of Louis XV.

775 VAN HOOL, J. F. *Antwerp.*—A sculptured altar, and a crucifix.

776 WARIN, J. *Brussels.*—A book-case in black wood.

777 WATRISSE, L. *Dinant.*—Round claw-tables of Belgian marble.

778 VAN DE LAER, P. *Brussels.*—Panels, specimens of stained and gilt papers.

779 WAHLEN-FIERLANTS, MRS. *Brussels.* —Paper hangings.

---

863 GERMON-DIDIET, A. *Brussels.*—Artificial flowers in paper and muslin, wax fruits.

CLASS 31.

781 BAYARD, M. *Herstal, Liége.*—Bolts, screw-wrenches, compasses, squares, iron fittings for carriages, &c.

782 BECQUET, BROS. *Brussels.*—Samples of forged nails.

783 BOGAERTS, ALP. *Antwerp.*—Works of art in bronze.

784 BROERMANN, F. G. SEN. *Brussels.*—Iron bedsteads, an aviary flower stand, and garden chairs.

785 CANIVEZ, J. B. *Ath, Hainault.* — Zinc letters in relief, gilt, &c.

786 CARLIER, F. *Chenee, Liege.*—Wrought-iron anvil.

787 CHAUDOIR, CH. & H. *Liege.*—Unsoldered copper tubes for locomotive engine boilers, steam-boats, &c.

788 COMPANY FOR THE MANUFACTURE OF BRONZE AND ZINC, *Brussels.*—Statues, works of art, &c.

789 DARDENNE, T. & SON, *Chimay, Hainault.*—Screw-iron, for saddle and harness horses.

790 DAWANS, A. & ORBAN, H. *Liége.*—Nails.

791 DE BAVAY, P. & Co. *Brussels.*—Common iron, iron for wire, nails, &c.

793 DELLOYE-MASSON, E. & Co. *Laeken, near Brussels.*—Forged-iron tinned and galvanized, enamelled cast-iron.

794 DELLOYE-MATHIEU, C. *Huy, Liége.*—Sheet-iron, polished and unpolished.

795 FABRIQUE DE FER D'OUGRÉE, *Seraing, Liége.*—Sheet-iron, specimens of iron.

796 FAUCONIER-DELIRE, WIDOW, *Châtelet, Hainault.*—Hand-wrought iron nails.

797 FRAIGNEUX, BROS. *Liége.*—Fire and thief proof safes.

798 GAILLIARD, L. C. C. *Brussels.* — Models in chiseled and chased metal.

799 GÉRARD, H. & DIDIER, *Bouillon, Luxemburgh.*—Hooks, hinges; iron-work, for buildings and furniture.

800 GOFFIN, C. & J. *Brussels.*—Cast-iron tubes.

801 GROTHAUS, BROS. *Gosselies, Hainault.* —Wrought-iron nails.

802 HOORICKX, G. *Brussels.*—Iron safes.

803 LALMAND-LEFORT, F. J. *Bothey, Namur.*—Iron safe, with invisible key-hole.

804 LAMAL, P. & Co. *Brussels.*—Lead and tin pipes.

805 LAMBERT, W. G. J. *Charleroi.*—Rivets of all kinds.

807 LECHERF, IS. DE, *Brussels.*—Bronze articles.

808 LESAGE, V. *Saint-Josse-ten-Noode, near Brussels.* — Nails, rivets, springs, telegraph-wires.

809 MATHYS-DECLERCK, *Brussels.* — A lock, having 629 fixed and 414 moveable pieces, and a key with 84 different divisions.

812 NICAISE, P. & N. *Marcinelle, near Charleroi.*—Bolts, nuts, &c.

814 RAIKEM-VERDBOIS, H. J. *Liége.*—Sheet-iron, polished.

815 REMACLE, J. & PÉRARD, *Liége.*—Sheet-iron.

816 SIÉRON, L. *Brussels.*—Iron, copper and zinc nails.

817 SOCIÉTÉ ANONYME DE CORPHALIE, *Antheit, near Huy.*—Sheet zinc.

818 SOCIÉTÉ ANONYME DES FORGES DE LA PROVIDENCE, *Marchienne-au-Pont, Hainault.*—Rafters.

819 SOCIÉTÉ ANONYME DES HAUTS-FOURNEAUX ET LAMINOIRS DE MONTIGNY-SUR-

SAMBRE, *Montigny, near Charleroi.*—Drawn-iron, iron doo and shutter.

820 SOCIÉTÉ ANONYME DES HAUTS-FOURNEAUX, USINES ET CHARBONNAGES DE CHÂTELINEAU, *Châtelineau, near Charleroi.*—Drawn-iron.

821 SOCIÉTÉ ANONYME DES HAUTS-FOURNEAUX, USINES ET CHARBONNAGES DE MARCINELLE ET COUILLET, *Couillet, near Charleroi.*—Commercial iron, sheet-iron.

822 SOCIÉTÉ ANONYME DES HAUTS-FOURNEAUX, USINES ET CHARBONNAGES DE SCLESSIN, *Tilleur, Liege.*—Rolled and wrought-iron.

823 SOCIÉTÉ DES FORGES ET LAMINOIRS DE L'HEURE, *Marchienne-au-Pont, Hainault.*—Rolled iron, sheet-iron.

824 SOCIÉTÉ DES LAMINOIRS DE HAUT-PRÉ, *Ougrée, Liége.*—Sheet-iron.

825 SOCIETY FOR MANUFACTURING OF NAILS BY MACHINE, *Fontaine-l'Eveque, Hainault.*—Nails and tacks.

826 TREMOUROUX BROS. & DE BURLET, *Saint Gilles, near Brussels.*—Forged iron, household articles tinned and glazed.

827 VANDERMILEN, CH. *Brussels.*—Iron safes.

828 VAN NEUSS, M. *Brussels.*—A grilling oven.

CLASS 32.

829 BOMBOIR, G. *Houffalize, Luxemburgh.*—Sickles, scythes, axes, &c.

831 MONNOYER, P. J. *Namur.*—Knives, razors, scissors, &c.

832 NOTTE, F. *Gembloux, Namur.*—Cutlery.

833 OLIVIER, A. *Enghien, Hainault.*—Cast-steel hammers for dressing mill-stones.

834 ROBERT, J. & DE LAMBERT, *Liége.*—Files for watchmakers, jewellers, armourers, &c.: gravers.

835 SOCIÉTÉ ANONYME DES HAUTS-FOURNEAUX ET LAMINOIRS DE MONTIGNY-SUR-SAMBRE, *near Charleroi.*—Steel knife.

CLASS 33.

836 DEHIN, J. J. *Liége.*—A silver monstrance.

837 DUFOUR, J. & BROTHER, *Brussels.*—Various articles of jewellery, tea-service, &c.

838 GOUVERNEUR, C. & SON, *Brussels.*—Gold and silver wire.

839 HOKA, A. *Liége.*—Engraved bracelets, pins, brooches, &c.

840 PETERS, L. *Tongres, Limburgh.*—Silver-gilt pyxes and communion-cups, in the Gothic style: fire-gilt articles, &c.

CLASS 34.

841 ANDRIS-LAMBERT & CO. *Marchienne-au-Pont, Hainault.*—Window-glass.

842 BENNERT & BIVORT, *Jumet, Hainault.*—Bottles; window-glass.

844 BOURDON, J. & CO. *Liége.*—Bottles, butter-pots enclosed in wicker-work, milk-pans.

845 CAPPELLEMANS, J. B. SEN., BEBY, A. & CO. *Brussels and Saint-Vaast, Hainault.*—Bottles; window-glass, and glass tiles.

846 CAPRONNIER, *Brussels.*—Stained glass-window, for Howden Church, Yorkshire.

847 DAUBRESSE, BROS. *La Louviere, Hainault.*—Window-glass.

848 DE DORLODOT DE MORIAMÉ, L. & SON, *Lodelinsart, Hainault.*—Window-glass.

849 FLOREFFE CO. *Floreffe, near Namur.*—Silvered and unsilvered plate glass; window-glass and flint-glass.

851 JAMBERS, J. *Liege.*—Engravings on glass.

852 JONET, D. & CO. *Charleroi.*—Window-glass, coloured, polished, engraved, &c.

853 LEDOUX, J. B. & C. *Jumet, Hainault.*—Window-glass.

854 MONDRON, J. *Lodelinsart, Hainault.*—Window-glass.

855 SOCIÉTÉ ANONYME D'HERBATTE, *Herbatte, near Namur.*—Table service, goblets, drinking glasses, &c.

856 SOCIÉTÉ DES MANUFACTURES DE GLACES, VERRES A VITRES, CRISTAUX ET GOBELETERIES, *Brussels.*—Plate and window glass, bottles, drinking-glasses, silvered glass, &c.

857 VANDERPOORTEN, J. L. *Molenbeek-Saint-Jean, near Brussels.*—Painted church windows.

CLASS 35.

858 BARTH, D. *Andenne, Namur.*—Clay smoking-pipes.

859 BOCH, BROS. *Keramis, Hainault.*—Fine crockery-ware, plain and ornamental.

860 CAPPELLEMANS, J. B. SEN. *Hal.*—Table, coffee and tea services, and other articles.

861 DE FUISSEAUX, MRS. *Baudour, near Mons.*—Articles of crockery and porcelain.

862 DEMOL, *Brussels.*—Painted crockery-ware and porcelain.

BRAZIL.

N.E. Court, No. 7.

CLASS 1.

1 BURLAMAQUE, DR. F. L. C.—A collection of gold, diamonds, emeralds, topazes, and various Brazilian minerals.

2 TASSARA, A.—Slates from Minas Geraes.

3 MOULEVADE.—Asbestos and kaolin, from Minas Geraes.

4 COPPET.—Limestone, &c., from Rio Janeiro.

5 BARBACENA, VISCOUNT.—Coal from Laguna, Province of Santa Catharina.

6 LEÂO, J. A. F.—Malachite, iron ores, coal, and various other minerals.

7 LUZ, M. M. DA.—Diamond, in gravel, from Minas Geraes.

8 SOUZA, M. S. DA.—Diamond, in gravel, from Diamantina, Minas Geraes.

9 TEXEIR, J. I. JUN.—Sandstone from the banks of the Cahy, S. Pedro.

10 ROHAN, H. DE B.—Amethysts from S. Paulo.

11 BELLO, O.—Quartz crystals from San Pedro.

12 BOULIECH, G.—Jasper and kaolin from S. Pedro; minerals from the Jaquaro Mines, S. Pedro.

13 M., J. P.—Lignite from Ouro Preto, Minas Geraes.

14 CARVALHO, J. P. D. DE.—Gold, in quartz, from Ouro Preto, Minas Geraes.

15 ANCHIETA, J. DE.—Iron ore from Cocaes, Minas Geraes.

16 REIS, J. M. DOS.—Quartz from Goyaz.

CLASS 2.

17 PECKOLT, T. *Cantagallo, Rio Janeiro.*—Vegetable acids, and essential extracts from native plants.

18 SANTOS, M. E. C. DOS, & SON, *Rio Janeiro.*—Organic and inorganic chemicals.

19 MAGALHÂES, M. DA C.—Ipecacuanha.

20 GARY, M. M. ALEIXO, & CO. *Rio Janeiro.*—A collection of organic and inorganic chemicals.

21 CASTRO, M. M. & MENDES, *Nitherohy, Rio Janeiro.*—Chemical products.

22 BLANC, J. F. A. *Rio Janeiro.*—Chemical products.

CLASS 3.

23 PECKOLT, T. *Cantagallo, Rio Janeiro.*—Starch from various plants.

24 PIRAQUARA, BARON DE, *Rio Janeiro.*—Sugar cane rum; coffee.

25 HUET, D. D. H. *Rio Janeiro.*—Rum.

26 COATS, R. *Rio Janeiro.*—Rum, Hollands, and loaf sugar.

27 WENTEN, J. R.—Rum.

28 FARO, J. P. D. & J. D. DE, *Rio Janeiro.*—Orange rum, and loaf sugar; coffee in the husk and cleaned; maize, mandioc starch, arrowroot, and Jacatupè flour.

29 CALDERON, L. B.—Aniseed and Cajá rum.

30 RABELLO, J. H. DA SILVA.—Cashew wine, rum, and pine apple syrup.

31 PEREIRA, A. J. G.—Hollands.

32 RIBEIRO, M. R. J.—Rum.

33 HEWLER, S. *Campo, Rio Janeiro.*—Anhydrous alcohol, and sugar cane vinegar.

34 MARCHADO & REDONDO, *Rio Janeiro.*—Alcohol; white and brown vinegar.

35 COUTINHO, J. DE A.—Alcohol.

36 COSTA, F. G. DA, & SONS.—Paddy, rice, coffee, and tapioca.

37 GOMES, A. & CUNHA, A. DA.—Rice.

38 NITHEROHY SUGAR-REFINERY & DISTILLERY CO.—Refined sugar.

39 DOUS DE JULHO FACTORY, *Bahia.*—Crystallized sugar.

40 SOUZA, S. DE, & SILVA, *Pernambuco.*—Refined and loaf sugar.

41 MONTEIRO MANUFACTORY, *Pernambuco.*—White sugar.

42 GEREMOABO, T. P. *Bahia.*—White sugar.

43 LOURENÇO, BARON S. *Bahia.*—Refined sugar.

44 CARVALHO, J. P. D. DE, *Gavia, Rio de Janeiro.*—Specimens of coffee.

45 DIAS, H. J. *Rio Janeiro.*—Coffee.

46 CRUZ, J. B. DA, *Cantagallo, Rio Janeiro.*—Coffee.

47 FARO, A. P. DE, *Rio Janeiro.*—Coffee.

48 TAVARES, J. P. *Itaguahy, Rio Janeiro.*—Coffee.

49 ANDRADE, F. DE P.—Coffee.

50 ALMEIDE, DR. C. M. DE.—Coffee.

51 MUNIZ, H. F.—White and yellow carimam, prepared from the mandioc; mandioc starch.

52 TREASURER'S ESTATE, *Minas Geraes.*—Varieties of tea.

53 BOTANIC GARDENS, *Minas Geraes.*—Varieties of tea.

54 SILVA, C. I. DA.—Tea from Itú, S. Paulo.

55 ROSA, J. C. DA.—Tea from Constituçâo, S. Paulo.

56 BITTANCOURT, M. J. DA CUNHA.—Tea from Coritiba, Paraná.

57 FROUGETH, DR. J. F. *Rio Janeiro.*—Paquequer tea.

58 SILVA, J. J. DA, *S. Roque, S. Paulo.*—Green tea.

59 AMARAL, J. V. DE ARRUDA, *S. Paulo.*—Scented green tea.

60 BLANC, J. F. A. *Rio Janeiro.*—Chocolate.

61 BERRINI, G. *Rio Janeiro.*—Chocolate.

62 SRA. V. CASTAGNIER, *Rio Janeiro.*—Preserves.

63 DEROCHE & CO. *Rio Janeiro.*—Pin apple preserve.

64 VASCONCELLOS, F. P. DE, *Bahia.*—Preserved vegetables.

65 FREITAS, J. DA COSTA, *Rio Janeiro.*—Mandioc starch.

66 SOUZA, A. C. DE.—Tapioca.

67 FURTADO, J. C.—Tapioca.

68 OLIVEIRA, V. J. DE.—Wheaten flour.

69 AZEVEDO, J. F. DE.—Thirty varieties of Theresopolis beans, from the province of Rio Janeiro.

70 LAGOS, M. F.—Eleven varieties of Theresopolis beans, from the province of Ceará.

71 NATIONAL FACTORY, *Gamboa, Rio Janeiro.*—Various liqueurs.

72 FERREIRA, A. J. BRAGA, & ISMÃO, *Rio Janeiro.*—Liqueurs.

73 GOMEZ, A. J. *Rio Janeiro.*—Liqueurs and barley wine.

74 BASTOS, A. J. G. P. *Rio Janeiro.*—Liqueurs and syrups.

75 TAVEIRA, A. M.—Specimens of Mandioc starch.

76 LEÃO, J.C. DE M. JUN.—Mandioc starch.
77 GAMBÔA FACTORY, *Rio Janeiro.*—Rose-vinegar.
78 BRASIL, P. A. *Rio Janeiro.*—Coloured vinegar.
79 LOBO, J. F.—Brazilian wines.
80 RENDON, J. A. DE T.—Sweet grape wine.
81 AGUIAR, A. P. DE.—Orange wine.
82 BITTENCOURT, J. DE.—Orange wine.
83 PINHEIRO, J. H. — Mogy das Cruzes wine.
84 MARSE, C. *S. Leopoldo, S. Pedro.*—Grape wine.
85 BAUN & CASTANÊRA, *Rio Janeiro.*—Cigars.
86 PALOS, D. *Rio Janeiro.*—Cigars.
87 MONTES, J. J. & Co.—*Rio Janeiro.*—Cigars.
88 SOUZA FLORES, J. J. DE, *Rio Janeiro.*—Cigars.
89 PALOS, P. *Rio Janeiro.*—Cigars.
90 MACHADO, F. A.—Cigarettes.
91 GONÇALVES, J.—Cigarettes.
92 PERES, S.—Cigarettes.
93 SILVEIRA, P. *Rio Janeiro.*—Snuff.
94 CORDEIRO, J. P.—*Rio Janeiro.*—Snuff.
95 JAGUARARY, BARON DE.—Leaf tobacco.

CLASS 4.

96 PECKOLT, S. *Cantagallo.*—Fruits, seeds, roots, barks, vegetable fibres, gums, resins and dye stuffs, vegetable oils, &c.
97 PIMPARDE, H. *Rio Janeiro.* — Aloe-water and oil of aloes.
98 PINTO, J. DE A. — Indigo from Pernambuco.
99 M. C. O. *Rio Janiero.*—Tallow oil.
100 STRAUSS, H. A.—Manufactured Indian rubber.
101 HERBST & ROSSITER, *Rio Janeiro.*—Mexican variety of vanilla.
102 CASANOVA, —. —Charcoal and potash, from the coffee husk.
103 CARNEIRO, J. M. DOS S.—Wax.
104 SIQUEIROS, M. J. P. DE.—Wax.
105 LAGE, M. P. F.—Wax.
106 RAMOS, A. DA SILVA.—Wax.
107 ALBUQUERQUE, G. A. G. DE.—Black wax.
108 LAGOS, M. F.—Barks, and leaf tobacco: a collection of bees with their wax and honey.
109 GONÇALVES, J. A. *Rio Janeiro.*—Extract of Brazil-wood.
110 LEÃO, J. A. F.—Leaf tobacco.
111 SOARES, J. J.—Leaf tobacco.
112 STEARINE CANDLE COMPANY, *Rio Janeiro.*—Glycerine, soap, and candles.
113 BRELAZ, L. *Pará.*—Vegetable oils.
114 ARAUJO, J. A. DE, *Rio Janiero.*—Oils.
115 MARIA, S. *S. Pedro.*—Castor-oil.
116 ARAUJO, J. M. DE, & Co. *Penedo, Alagoas.*—Vegetable oils.
117 MAUA, BARON DE, CRUZ, M. D. DA, CRUZ, J. B. & OTHERS.—Woods of Brazil, comprising 410 varieties.
118 BRUSQUE, F. C. DE A.—Mosaic of the woods of the province of Pará.
119 MUNICIPAL CHAMBER OF DESTERRO.—Mosaic of the woods of the province of Santa Catharina.
120 CARVALHO, A. L. P. *Rio Janeiro.*—Soap.
121 MONTEIRO, J. F. C. *Aracaty, Ceará.*—Soap.
122 ARÊDE, J. B. DE, & Co. *Pará.*—Soap.
123 REGO, DR. P. DA S. *Bahia.*—Soap.
124 MARTELET, R. & Co. *Rio Janeiro.*—Soap.
125 BARCELLOS, A. P. S. *Pernambuco.*—Carnauba palm-oil candles.
126 ARAUJO & IRMÃO, *Rio Janeiro.*—Tallow candles.

CLASS 7.

127 SILVA, J. Y. DA.—A blacksmith's bellows.

CLASS 9.

128 SANTOS, M. C. DOS, *Rio Janeiro.*—Agricultural implements.

CLASS 11.

129 REAL, C. & PINTO, *Rio Janeiro.*—Embroidered scarf, and gold epaulettes.
130 MILITARY ARSENAL, *Rio Janeiro.*—Carbine, pistol, and Minié rifle.
131 MILITARY ARSENAL, *Pernambuco.*—A pistol.

CLASS 12.

132 NAVAL ARSENAL, *Rio Janeiro.*—Models of ships, &c.
133 MIERS, IRMÃO, & MAYLOR, *Rio Janeiro.*—Models of ships, &c.
134 PONTA DA ARÊA COMPANY.—Models of ships, &c.
135 NAVAL ARSENAL, *Pernambuco.*—Model of a ship.

CLASS 13.

136 MASCARENHAS, A. M. DE, *Rio Janeiro.*—A ship's compass.
137 REIS, J. M. DOS, *Rio Janeiro.*—Spectacles, reading glasses, &c.

CLASS 14.

138 PACHECO, J. I. *Rio Janeiro.*—Photographic portraits of the Imperial Family of Brazil, &c.
139 DAER, —, *Rio Janeiro.*—Photographic views of the Botanic Gardens, Rio Janeiro.

CLASS 15.

140 GONDOLO & Co. *Rio Janeiro.*—A gold watch.

CLASS 17.

141 BLANCHARD,—*Rio Janeiro.*—A set of surgical instruments.

CLASS 18.

142 PEREIRA, M. N. B.—White and yellow cotton.

143 RODRIGUES, C. J. A.—Raw cotton from Rio Janeiro.

144 MELLO, L. C. DE.—Raw cotton from Pernambuco.

145 REZENDE, L. R. DE S.—Cotton in the pod from Alagôas.

146 MASCARENHAS, D. L. DE A.—Cotton counterpanes.

147 ALBUQUERQUE, A. P. DE.—Cotton piece goods from Todos os Santos, Bahià.

148 USMAR, J. C. M. DE.—Cotton piece goods from Andarahy, Rio Janeiro.

149 FILGUEIRAS, J. A. DE A. & Co.—Cotton piece goods from Magé, Rio Janeiro.

150 ANDRADE, J. DAS C.—Cotton piece goods from Passa Tempo, Minas Geraes.

151 COSTA, M. DE A.—Cotton piece goods from Campo Grande, Rio Janeiro.

152 JUMBEBA, F. R. DA C.—Cotton piece goods from Brumado de Suassuhy, Rio Janeiro.

153 PADUA, F. N. N. DE.—Cotton piece goods from Queluz, Minas Geraes.

154 LAGOS, M. F.—Cotton piece goods from Crato, Ceará.

CLASS 19.

155 BARBACENA, VISCOUNT DE.—Guaxima from Pilar, Rio Janeiro.

156 BURLAMAQUE, GENERAL F. L. C.—Aloe fibre cloth.

157 MOTTA, F. L. DA, *Rio Janeiro.*—Aloe-fibre cloth, embroidered with gold.

CLASS 20.

158 UBATUBA, DR. M. P. DA S., ARAUJO, D. C. R. DE, and CAPBDEBILA, V. F.—Cocoons; raw and manufactured silks from the province I. Pedro.

CLASS 21.

159 SILVA, S. V. DA.—Woollen counterpane from Minas Geraes.

160 PADUA, F. N. N. DE.—Woollen counterpanes from the same province.

CLASS 25.

161 PINGARILHO, J. M. DA S.—Skin of the red socuryú snake, tanned.

162 THE IMPERIAL MORDOMIA.—Feather flowers.

CLASS 26.

163 GUIMARÃES, L. & SOUZA, *Rio Janeiro.*—Coloured morocco, and other leathers.

164 ROMANN, BRET, & KILIAN, *Rio Janeiro.*—Coloured morocco, and other leathers.

165 GUIMARAES, C. J. DE A, *Rio Janeiro.*—A saddle.

166 GUIMARAES, A DE A. *Rio Janeiro.*—A saddle.

167 JANSEN, G. *Rio Janeiro.*—A saddle.

168 PEIXE, G. DE S. *Pernambuco.*—A saddle.

169 SILVA, J. M. DA, & Co. *Rio Janeiro.*—A saddle.

170 DIAS, J. R. *Pernambuco.*—A saddle.

171 GUIMARAES, T. T. DE A. *Rio Janeiro*—A saddle.

CLASS 27.

172 MURIAMÉ, A. M. *Rio Janeiro.*—Boots.

173 CARREIRO, J. C. *Rio Janeiro.*—Boots.

174 PINGARILHO, J. M. DA S. *Pará.*—Boots.

175 QUEIRÓS, J. M. DE, *Rio Janeiro.*—Boots.

176 CAMPAS, J. & SON, *Rio Janeiro.*—Boots.

177 GUILHERME, P. A. & SON, *Rio Janeiro.*—Boots.

178 THER, P. *Porto Alegre, S. Pedro.*—Boots.

179 PINHEIRO, J. DE L. *Rio Janeiro.*—Felt hats.

180 COSTA, F. A. DA, & Co. *Rio Janeiro.*—Felt hats.

181 GOMES, V. J. & Co. *Rio Janeiro.*—Felt hats.

182 BARCELLOS & VIANNA, *Rio Janeiro.*—Felt hats, made from hares' fur.

183 ALMEIDA, R. A. DE, *Rio Janeiro.*—Felt hats.

184 CHASTEL & Co. *Rio Janeiro.*—Silk hats.

185 MELLO & ALMEIDA, *Rio Janeiro.*—Silk hats.

186 CASTRO, P. DE, & Co. *Rio Janeiro.*—Silk hats.

CLASS 28.

187 RENSBURG, E. *Rio Janeiro.*—An Atlas and Report on the S. Francisco river.

188 LEUZINGER, G. *Rio Janeiro.*—Merchants' ledgers, &c.

189 LAEMMERT, E. & H. *Rio Janeiro.*—Merchants' office books.

190 LOMBAERTS, —.—Merchants' office books.

191 OLIVEIRA, M. J. DE, JUN. *Rio Janeiro.*—Vegetable writing ink.

192 AZEVEDO, J. V. R. DE, *Rio Janeiro.*—Writing ink.

CLASS 30.

193 JOHN, A. *Santa Isabel, Espirito Santo.*—Work-box, of various Brazilian woods,

194 ZANCHI, —, *Rio Janeiro.*—Work-box of various Brazilian woods.

195 CAPÔTE, J. A. *Rio Janeiro.*—An inlaid work-box.

196 NASCENTES DE AZAMBUJA, B. A—An inlaid work-table.

197 VALLIM, M. DE A.—A work-table of Candêa wood.

198 QUINTANILHA, B.—*Rio Janeiro.*—Flowers made of insects' wings.

199 GARCIA, C. A. G. *Rio Janeiro.*—Paper-hangings.

200 PEREIRA, G. G. *Rio Janeiro.*—Printed paper-hangings.

201 HORN, F. DE.—Mosaic of Brazilian woods.

202 G. B. S.—A mosaic of flooring woods.

203 LANDOT, J. B. S. *Rio Janeiro.*—A mosaic of Vinhatico wood,

204 LEITE, J. A. JUN.—A vase of shell-flowers.

205 FERRAZ, A. M. DA S.—A vase of fish scale and shell flowers.

206 SILVA, E. F. DA.—A vase with artificial rose tree.

CLASS 31.

207 ANDRADE, A. R. DE.—A bar of wrought iron from Minas Geraes.

208 BARROS, L. A. M. DE.—A bar of wrought iron from Congonhas do Campo, Minas Geraes.

209 COTTA, M. P.—A bar of wrought-iron from Antonio Pereira, Minas Geraes.

210 ANDRADE, J. C. DA C.—A bar of wrought iron from Itabira de Mato Dentro, Minas Geraes.

211 MOULEVADE, J. A. DE.—A bar of iron from the same locality.

212 SANTOS, M. C. DOS. *Rio Janeiro*—A lock.

213 FERREIRA, J. V. *Rio Janeiro.*—A secret door-lock.

214 URBACH, A. *Rio Janeiro.*—Cast-iron medallion.

215 HARGREAVES, —, *Rio Janeiro.*—Cast-iron medallion.

216 SANTOS, M. C. DOS, *Rio Janeiro.*—Cast-iron ornaments and panel.

217 BEUCHON, —, *Rio Janeiro.*—Nails, screws, &c.

CLASS 32.

218 PRADÍNES, J. *Pernambuco.*—Knives and other cutlery.

219 BLANCHARD.—Specimens of razors.

CLASS 33.

220 DOMINGOS, FARINI, & IRMÃO, *Rio Janeiro.*—The Imperial arms of Brazil; a silver medallion.

221 LOPES, A. J.—Gold lace.

222 REIS, J. M. DOS.—Eye-glass, the property of His Imperial Majesty.

CLASS 34.

223 FONSECA, S. DA E, SÁ, *Rio Janeiro.*—Engraved flint-glass.

224 CASTRO, PAES, & CO. *Praia, Formosa, Rio Janeiro.*—Ornamental glass.

225 LOMBOS, M. & ROQUE, S. *Rio Janeiro.*—Ornamental glass.

CLASS 35.

226 ESBERARD, F. *Rio Janeiro.*—Earthenware.

227 SARVILLO, P. A. & CO. *Nitherohy, Rio Janeiro.*—Tiles, bricks, and pipes.

228 LAGE, M. P. F.—Tiles from the União e Industria Co.

229 FEREIRA, J. S. *Bahia.*—Tiles.

CLASS 36.

230 FORESTE, A. *Rio Janeiro.*—Jewellery-cases.

CHINA.

Nave, North Side, near Horticultural Society's Entrance.

1 MICHEL, GEN. SIR JOHN.—A carved screen, from behind the Emperor's throne in the Summer Palace; jars.

2 COPLAND, C.—Backgammon board.

3 DUNCANSON, E. J.—Chinese paper, and manufactured goods.

4 SWINHOE, R. *H.M. Vice-Consul, Taiwan-foo, Formosa.*—Various articles from Formosa.

5 FORREST, R. J. *H.M. Acting Vice-Consul at Kiukiang.*—Autograph of first rebel chief. Coins made by the rebel authorities at Nanking.

6 LEGGE, REV. J., D.D.—Specimens of Chinese types.

7 MERCER, W. T.—Two screens.

8 KANE, DR.—Porcelain vases and stands.

9 ROWLAND, J. C.—A table, vases, bronzes.

10 RENNIE, W. H.—Bath tub in porphyry.

11 JACOB, CAPT. 99*th Regt.*—Sundry articles.

12 WALKINSHAW, W.—A pagoda-stand, model of scaffolding, and carved ivory ball.

13 MURRAY, DR. *Chairman of Hong Kong Committee.*—Silver vase, and ivory articles.

14 MALCOLM, CAPT. C. D., R.E.—Carved ivory chessmen.

15 HOACHING, MR.—Carved ivory casket.

16 FLETCHER, ANG.—Jade ornaments and Chinese medicines.

17 MONTEIRO, MR.—Jewelled cups.

18 AN ARTILLERY OFFICER.—Ancient bronze incense burner and two candlesticks.

19 TAIT, CAPT. W.—A human skull richly set in gold; reported to be the skull of Confucius.

20 HEWITT, W. & Co. 18, *Fenchurch-st. E.C.*—Mandarin jars; tea, &c. services, enamels, &c.

21 HALL, CAPT. R.N.—Chinese pictures.

22 CAMPBELL, P.—A jade-stone sceptre bowl, vase, &c.

23 OLDING, J. A.—The Emperor's jade-stone seal, used to stamp documents certifying literary proficiency.

24 ROSARIO, R. A.—Pharmaceutical articles.

25 CAREY, H. W.—Chinese drugs, and miscellaneous articles.

26 RÉMI, SCHMIDT, & Co.—Raw silks, vases, bronzes, lacquer ware, cups of jade and agate, carpets from the Summer Palace, &c.

COSTA RICA.

N.E. Court, No. 1.

1 EXHIBITORS, THE GOVERNMENT OF COSTA RICA:—

CLASS 1.

2 Ores of gold, silver, copper, and lead, from various mines; gold, after separation of the mercury, and after having been melted; volcanic sulphur.

CLASS 2.

3 Sarsaparilla; balsams; medicinal, and other roots; gums; medical and chemical substances.

CLASS 3.

4 Fruits, beans, rice, coffee, sugar, tobacco, cacao, rum, &c.

CLASS 4.

5 Twine, &c. made of the Agave leaf, and other vegetable fibres; dye-stuffs, and matters used for tanning; nuts for the production of oils, &c.; numerous specimens of indigenous woods; tortoise and mother-of-pearl shells; cotton, caoutchouc, ocre, &c.

CLASS 22.

6 A tule mat.

CLASS 25.

7 Bird skins.

CLASS 26.

8 Tanned tapir skins; otter and jaguar skins: whips of deer skin and tapir skin.

CLASS 27.

9 Articles manufactured with English yarn; palm-leaf hats.

CLASS 33.

10 Gold, silver, and filligree work.

CLASS 36.

11 Cigar-cases and purses, made of pita and tule; halters of Cabuga; calabashes; cocoa-nut goblet.

DENMARK.

N.W. Court, No. 4, and N.W. Gallery, No. 3.

CLASS 1.

1 FORCHAMMER, G. *Copenhagen.*—Minerals from Denmark and her colonies.

2 SOUTH GREENLAND MINING CO.—Tin, copper, lead, cryolite, and other minerals.

3 WEBER, TH. & Co. *Copenhagen.* — Cryolite and its products.

CLASS 2.

Sub-Class A.

4 BENZON, A. *Copenhagen.* — Chemical, photographic, technical, and economical preparations and articles.

5 FREUDENREICH, A. G. *Flensborg.* — Chrome-colours.

6 HEYMANN & RÖNNING, *Sophiehaab, near Copenhagen.*—Chemical preparations and colours.

7 KEDENBERG & BLECKER, *Uetersen.*— Manures, superphosphate, and crushed bones.

8 MEIER, F. C. S. *Copenhagen.*—Linseed oil, varnishes, and drying extracts.

9 MÖLLER, H. C. *Kiel.*—Crushed bones, chemically clean and pulverized.

10 NISSEN & VOLKENS, *Heide.*—Bituminous sand (raw produce), asphalte tar (half manufactured), solar oil, asphalte oil, and mineral asphalte.

12 WEIL, M. & Co. *Copenhagen.*—Phosphate of lime.

Sub-Class B.

13 BENZON, A. *Copenhagen.* — Pharmaceutical preparations.

14 ERIKSEN, J. *Copenhagen.*—Artificial mineral-waters.

15 RIISE, A. H. *St. Thomas.*—Oil of lemon-grass (of *Andropogon Citratum*, D.C.).

16 ROSENBORG MANUFACTORY OF MINERAL WATERS, *Copenhagen.*—Artificial mineral waters.

17 STOLTZENBERG & UFFHAUSEN, *Altona.* — Hydro-chloride of ammonia, camphine, glycerine, citric chromate of potass, cariophyl oil, and nitric ether.

CLASS 3.

Sub-Class A.

18 ROYAL AGRICULTURAL HIGH SCHOOL, *Copenhagen.*—Grain and sheaves.

19 AGRICULTURAL SOCIETY, *Kiel.*—Holstein agricultural produce.

20 A SCHOOLMASTER, *Fyen.*—Hops.

21 HOLST, H. *Bredvad Mill, near Horsens.* —Flour.

22 JORDY, A. *Hómbgaard, near Ringsted* —Cheese.

23 KJORBOE, F. W. *Copenhagen.*—Grain from Jutland.

24 MARSTRAND, T. *Wodroff Mill, near Copenhagen.*—Wheat, flour, and groats.

25 NIELSEN, H. M. C. *Lystofte, near Lyngby.*—Pressed bran-cakes.

26 NIELSEN, C. G. *Flensborg.*—Grain from Slesvig.

27 PASCHKOWSKY, G. *Flensborg.*—Starch (amidon) and potato-flour.

28 PUGGAARD & HAGE, *Nakskov.*—Grain and seeds.

29 PUGGAARD, H. & Co. *Copenhagen.*—Chevalier barley.

30 RADBRUCH, H. *Kiel.*—Flour starch, common-glaze and blue-glaze starch and powder.

31 SCHLIEMANN, C. *Rastorff, near Kiel.*—Manufactures from flour.

32 SCHMIDT, H. & Co. *Copenhagen.*—Flour and bran.

33 SCHOENFELDT, A. *Heiligenhafen.*—Amidon (flour starch), common and glazed amidon.

34 SCHOU, H. H. *Slagelse.*—Grain from Zealand.

35 TESDORFF, E. *Ourupgaard, Falster.*—Grain from Falster.

36 VEIS, A. *Aarhuus-mill.*—Flour and bran.

37 WINNING & Co. STEAM-MILL, *Horsens.*—Flour and groats.

Sub-Class B.

38 BEAUVAIS, J. B. D. *Copenhagen.*—Hermetically sealed boxes, containing meat and fish.

39 HANSEN, A. N. & Co. *Copenhagen.*—India pork and mess pork.

40 HANSEN, J. J. & Co. *Copenhagen.*—Danish West India sugar.

41 HILL, MISS R. *St. John's.*—Arrow-root.

42 JÜRGENSEN, D. *Flensborg.*—India pork, mess pork, lard in bladders, sausages, and hams.

43 MEYER, J. C. F. & SON, *Altona.*—Cocoa paste, cocoa, and preparations of cocoa, vanille, and powder chocolate.

44 NEWTON, F. R. *St. Croix.*—Muscovado sugar.

45 PARTSCH, J. W. F. *Flensborg.*—India pork, mess pork, India beef, and lard in bladders.

46 PLASKETT, W. *St. Croix.*—Muscovado sugar.

47 RESTORFF, M. C. & Co. *Thorshavn, Faro Isles.*—Dried fish.

48 ROTHE, L. *St. Croix.*—Muscovado sugar.

49 ROTHE, MRS. C. *St. Croix.*—Guava jelly, limes preserved in pickled vinegar.

50 STEVENS, J. Y. *St. Croix.*—Muscovado sugar.

51 WENDT, MISS C. DE, *St. John's.*—Pickles

Sub-Class C.

53 HERRING, P. F. *Copenhagen.*—Cherry cordial.

54 HILL, MISS R. *St. John's.*—Guaverberry rum, old rum, and shrub.

55 PETERSEN, L. E. *Kolding.*—Danish corn spirits.

56 RIISE, A. H. *St. Thomas.*—Bay spirits.

57 ROTHE, MRS. C. *St. Croix.*—Old rum.

58 WILMS, H. B. *Flensborg.*—Vinegar.

---

300 MAACK, F. *Flensborg.*—Vinegar.

CLASS 4.

Sub-Class A.

59 ASMUS, G. E. A. *Kiel.*—Raw and refined rape oil.

60 BENZON, A. *Copenhagen.*—Stearine candles.

61 CLAUSEN, H. A. C. *Copenhagen.*—Fish oil.

62 DRIESHAUS, *Altona.*—Wax and composite candles.

63 HOLM, J. & SONS, *Copenhagen.*—Oils, oil-cake, and composite candles.

64 HOLMBLAD, L. F. *Copenhagen.*—Stearine candles and oil cakes.

65 KRACKE, C. W. *Flensborg.*—Linseed cakes and linseed oil.

66 NIELSEN, J. *Frederiksberg.*—Dzierson's beehives, improved.

67 RESTORFF, M. C. & Co. *Thorshavn, Faro Isles.*—Whale oil, and cod-liver oil.

68 UFFHAUSEN, J. F. *Mölln.*—Brilliant blacking, deep black varnish for preserving leather.

Sub-Class B.

69 THE GREENLAND TRADING CO. *Copenhagen.*—Raw products, skins, &c., from Greenland.

70 CLAUSEN, H. A. C. *Copenhagen.*—Iceland produce, wool, and eider-down.

72 A FARMER, *Iceland.*—Spoons of horn.

74 HOSKIER, F. *Copenhagen.*—Wools, and specimens of Greenland industry.

76 LUND, J. *Iceland.*—Spoons of horn.

77 MAGNUSSON, G. *Stokkholt, Iceland.*—Travelling knife and fork, the handles of whale tooth, mounted in brass.

79 THORSTEINSSON, J. *Vindás, Iceland.*—Travelling bottle of horn, mounted with brass.

Sub-Class C.

80 ANDERSEN, ORLOW, *Frederiksborg.*—Flax in different stages of dressing.

81 BEVENSEE, T. *Seegeberg.*—Turning in grey alabaster.

82 BORNHOFT, T. *Segeberg.*—Turning in grey alabaster.

83 CHRISTENSEN & KJELDSEN, *Copenhagen.*—Works in cork and cork-shavings.

84 HILL, A. C. *St. John's.*—Baskets.

85 JEBSENS, WIDOW P. H. *Seegeberg.*—Turning in grey alabaster.

86 LANGMAACK, E. *Plöen.*—Turning and carving in meerschaum.

87 LUND, J. *Iceland.*—Snuff-box (baukr) of mahogany, mounted with brass.

88 PETERS, J. F. C. *Windloch, near Flensborg.*—Improved trough.

89 ROSENÖRN LEHN, BARON, *Guldborgland.*—Samples of wood.

90 ROTHE, MRS. C. *St. Croix.*—Box containing wild cotton, in buds and blown; miniature fish-pots, the conic formed one for catching eels.

91 SCHWARTZ & SON, J. G. *Copenhagen.*—Specimens of turning, manufactures of whalebone, umbrellas, and combs.

92 SKIFBÖGGER, JÖRGEN, *Elstrup, near Sönderborg.*—Wood articles for ship, domestic, and dairy purposes.

93 TAYLOR, J. W. *Greenland.*—Baskets of Greenland grass.

94 VESSUP, *St. John's.*—Fish pot.

95 WEBER, TH. & Co. *Copenhagen.*—Paper pulp made from wood.

CLASS 6.

96 FIFE, HENRY, *Copenhagen.* — A carriage.

97 SCHRODER, H. A. *Flensborg.* — A phaeton.

CLASS 7.

Sub-Class A.

98 DITTMANN & BRIX, *Flensborg.* — Turned and polished case-hardened cast-iron roller, and piece of a smaller one.

99 MARSTRAND, TH. *Copenhagen.* — Weaving appliances, shuttles, &c.

100 NÖRHOLM, NIELS, *Copenhagen.*—An apparatus for measuring and cutting out clothes.

Sub-Class B.

101 DALHOFF, J. B. *Copenhagen.*—Machine for making files.

102 HAMMER & SORENSEN, V. *Copenhagen.*—Lasts and boot-trees.

103 RÜINNING & KROLL, *Preetz.*—Specimens of cooperage.

CLASS 8.

104 GAMEL & WINSTRUP, *Copenhagen.*—Fire-engines, constructed for large farms and small towns.

CLASS 9.

105 ALLERUP, M. P. *Odense.*—Agricultural implements.

106 MARSTRAND, TH. *Copenhagen.*—Agricultural, domestic, and gardening implements.

CLASS 10.

110 DITHMER, H. H. *Rønneberg.*—Tile-work, polished flower-vase of burnt clay.

111 NIELSEN, P. E. (SCHELLERS, Suc.) *Copenhagen.*—Sepulchral monuments.

---

301 MULZENBECHER, T. H. *Rensing.*—Moulded and pressed bricks.

302 VIDAL, C. *Fernsicht.*—Tiles, &c.

CLASS 11.

Sub-Class B.

112 COHEN, I. *Copenhagen.*—Camp-kettle and appurtenance.

Sub-Class C.

113 KRONBORG MANUFACTORY OF ARMS *Hellebek, near Elsinore.*—Rifles.

CLASS 12.

Sub-Class A.

114 THE NAVY-YARD, *Copenhagen.* — Models of ships.

115 WILDE, CAPT. A., R.N. *Copenhagen.*—Drawing and model of a line-of-battle ship.

Sub-Class B.

116 THE HOME DEPARTMENT, *Copenhagen.*—Model of a life-boat.

117 MÜLLER, H. C. *Thorshavn, Faro Isles.*—Whale and fishing boat, with weapons.

118 SOUTH GREENLAND MINING CO.—A cajak for seal-hunting with weapons.

Sub-Class C.

119 BRÜTZ, *Rendsburg.*—Ropes, &c.

120 HOLM, JACOB & SONS, *Copenhagen.*—Ropes and sail-cloth.

121 THÖL, W. *Rendsburg.*—Ornamental work for ship's stern; wheel, and specimen of clock turning.

122 WINGE, P. W. *Randers.*—Ropes.

CLASS 13.

123 DANISH STATE TELEGRAPH, *Copenhagen.*—Isolators and galvanic battery.

124 FAXÖ, *Stubbekjöbing.*—Machine worked by heated air.

125 KYHL, C. C. *Copenhagen.*—Relays and translators for telegraphic purposes.

126 ORNSTRUP, L. *Copenhagen.* — Gas boiling apparatus for chemical purposes.

CLASS 14.

127 HANSEN, G. E. *Copenhagen.*—Photographs.

129 KIRCHHOFF, A. W. *Copenhagen.* — Photographs.

130 KRIEGSMANN, M. *Flensborg.*—Photographs.

131 LANGE, E. *Copenhagen.*—Photographs.

132 MOST, P. H. C. *Copenhagen.*—Photographs.

133 STRIEGLER, R. *Copenhagen.*—Photographs.

CLASS 15.

134 FUNCH, A. *Copenhagen.*—Tower clock (improved construction), and case chronometer.

135 JÜRGENSENS, URBAN, & SONS, *Copenhagen.*—Sea and portable chronometers.

136 KRILLE (KESSEL'S successors), *Altona.*—Pendulum clock with quicksilver compensation; a chronograph (galvanic registering apparatus), and galvanic interrupter.

137 RANCH, CARL, *Copenhagen.*—Chronometers.

---

303 TENSEN, T. C. *Bornholm.*—Clocks.

CLASS 16.

139 ALPERS, O. F. *Copenhagen.*—Pianoforte.

140 CARLSEN, D. & Co. *Uetersen.*—Grand piano, brass tuning instrument, and iron sounding board.

141 HANSEN, O. *Flensborg.* — Upright piano.

142 HORNUNG & MÖLLER, *Copenhagen.*—Grand and upright pianos.

143 JACOBSEN, J. *Haderslev.*—Organ æolodicon, æolodicon with one stop, upright pianoforte.

144 KNUDSEN, CHR. *Copenhagen.*—Pianoforte.

145 LARSEN, CHR. *Odense.*—Stringed instruments.

146 MARSCHALLS, A. & SON, *Copenhagen.*—Demi-oblique upright piano.

147 PETERSEN, P. & SUNDAHL, *Copenhagen.*—Pianoforte.

148 SCHMIDT, P. E. *Copenhagen.*—Brass instruments.

149 SORENSEN, J. P. *Copenhagen.*—Pianoforte.

150 WULFF, L. *Copenhagen.*—Pianoforte.

CLASS 17.

151 NYROP, PROF. CAMILLUS, *Copenhagen.*—Surgical instruments, bandages, orthopœdical machines, and apparatus.

152 RASMUSSEN, A. *Copenhagen.*—Bandages, surgical, orthopœdical and electro-galvanic apparatus.

153 WULFF, CARL, *Copenhagen.*—Artificial leeches.

CLASS 18.

154 BIERFREUND, LOR. *Odense.*—Specimens of cotton manufactures.

CLASS 19.

156 OLSEN, O. F. *Wintersbölle, near Vordingborg.*—Damask and drill, all linen.

CLASS 21.

157 USSERÖD FACTORY, *near Hilleröd.*—Military cloth, blankets, and horse cloth for the army.

158 BECH, MARCUS, *Aarhuus.*—Shoddy, &c.

159 CHRISTIANSEN, H. *Thorshavn, Faro Isles.*—Woollen and worsted goods.

160 CLAUSEN, H. A. C. *Copenhagen.*—Woollen goods.

161 DAVIDSEN, I. & Co. *Thorshavn, Faro Isles.*—Woollen and worsted goods.

162 EHLEN, MARIE, *Lutterbeck.*—Hand-spun wool.

163 FISCHER, C. *Vestbirk, near Horsens.*—Woollen fabrics and worsted.

164 MODEVEG, J. C. & SON, *Brede, near Lyngby.*—Broad cloths.

165 RESTORFF, M. C. & Co. *Thorshavn, Faro Isles.*—Woollen and worsted goods.

166 SCHLIEMANN, CHR. *Rastorff, near Kiel.*—Shoddy of various kinds.

---

304 ALBECH, C. E. & SON, *Copenhagen.*—Shoddy.

CLASS 22.

167 GROTH & SONS, *Flensborg.*—Wax cloth table cover, oil-cloth, and lacquered calf-skins.

168 IÓNSDÓTTIR, T. *Iceland.*—Sewed carpet, old fashioned.

169 MAGNÚSDÓTTIR, MISS H. *Reykjavik, Iceland.*—Sewed carpet, old fashioned.

170 MEYER, J. E. *Copenhagen.*—Oil-cloth, and lacquered goods.

171 STEPHENSEN, MRS. *Videy, Iceland.*—Woven carpet, old fashioned.

CLASS 23.

172 SCHRIEVER, *Rendsburg.*—Dyed yarns.

CLASS 24.

173 BRIX, MISS, *Industrial Depôt, Copenhagen.*—Specimens of sewing by the country-women, Hedeboerne.

174 HANSEN, DETLEV, *Mögeltönder, near Tonder.*—Specimens of lace, trimmings, and collars.

175 KRAGH, MISS EMILIE, *Frederiksborg.*—Specimens of sewing by the country-women, Hedeboerne.

176 LEVISOHN, J. C. *Copenhagen.*—Embroidery in wool.

177 LOHSE, MISS HENRIETTE, *Copenhagen.*—Ladies' sets in point lace.

178 RICHTER, MRS. S. *Holstebro.*—White embroidery.

179 TOPP. MISS MATHILDE, *Copenhagen.*—White embroidery.

---

305 BOIESEN, MRS. M. *Copenhagen.*—Embroidery.

CLASS 25.

Sub-Class A.

180 BANG, J. C. *Copenhagen.*—Carpet and fur manufactures.

181 BRINCKMANN, FR. *Copenhagen.*—Fur manufactures.

182 SCHMID, *Kiel.*—Fox and cat furs.

183 TAYLOR, J. W. *Greenland.*—Female Greenlanders' costumes; Esquimaux hunting dress, with sundry Esquimaux articles; seal and dog-skin mat; footstool, and seal-skin gloves; white haired skin of seal fœtus; prepared bird skins for articles of dress; hand spun yarn, from hair of the white hare; sinews, from which thread is made.

184 TROLLE, C. A. *Copenhagen.*—Fur coat for travelling; carpets; trimming and lining for ladies' dress.

Sub-Class C.

185 LANGE, MISS HENRIETTE, *Altona.*—Articles worked in human hair.

CLASS 26.

Sub-Class A.

186 BORCH, BROS. *Copenhagen.*—Skins and leather.
188 ERIKSEN, S. *Horsens.*—Tanned lamb-skins.
191 MESSERSCHMIDT, E. *Copenhagen.*—Tanned hides and skins.
193 WIENGREEN & FIRJAHN, *Slesvig.*—Lacquered leather, skins, &c.

Sub-Class B.

194 BARTH, MAJOR S. C. cavalry, *Copenhagen.*—Riding equipage for cavalry.
195 DAHLMANN, F. & L. *Copenhagen.*—Set of double harness, saddles, &c.
196 HINTZ, C. O. *Kiel.*—Set of double harness.
197 SÖRENSEN, C. P. *Copenhagen.*—A stuffed horse.

Sub-Class C.

198 JENSEN, H. C. *Flensborg.*—Copper clinched fire-engine hose.

CLASS 27.

Sub-Class A.

199 BODECKER, A. F. *Copenhagen.*—Silk and felt hats.
200 BRET, A. H. *Copenhagen.*—Hats.

Sub-Class C.

201 CHRISTENSEN, PETER, *Copenhagen.*—Frock coats, vest, and trousers.
202 COHEN, H. *Copenhagen.*—Gentlemen's linen, underclothing, and neckties.
203 LANDER, P. JUN. *Copenhagen.*—Gloves.
204 LARSEN, H. C. *Copenhagen.*—Gloves.
205 LORENTZEN, P. J. *Flensborg.*—Gentlemen's linen.
207 RASMUSSEN, HANS, *Copenhagen.*—A uniform.
208 RUBEN, M. M. *Copenhagen.*—Gentlemen's linen.
209 SCHOTTLÆNDER & GOLDSCHMIDT, *Copenhagen.*—Shirts, shirt-fronts, surtout, and vest.

Sub-Class D.

210 BENDAHL, J. V. *Copenhagen.*—Spring shoes, with pasteboard bottoms, for flat-footed persons.
211 CORDWAINERS' GUILD, *Preetz.*—Boots and shoes.
212 DÜRING, N. P. *Copenhagen.*—Boots and shoes.
213 HJORTH, M. H. *Copenhagen.*—Boots and galoshes.
214 RUSCHE, *Altona.*—Boots.
215 SCHWARZ, *Altona.*—Specimens of boot and shoe making.
216 VOGES, J. C. *Altona.*—Specimens of bootmaking.

CLASS 28.

Sub-Class A.

217 DREWSEN & SONS, *Silkeborg.*—Colombier, chart, and writing, printing, and cartoon paper.
219 HOLMBLAD, L. F. *Copenhagen.*—Playing cards.
220 ROSENBERG, CAROLINE, *Hoffmansgave, Fyen.*—Writing paper decorated with moss and fern.

Sub-Class B.

222 GÖTTSCH, WILHELMINE, *Kiel.*—Specimens of cutting in leather and paper.

Sub-Class C.

223 BÆRENTZEN, EM. & Co. *Copenhagen.*—Chromo-lithograph, mezzotints, and lithographs.
224 HENNEBERG & ROSENSTAND, *Copenhagen.*—Frames containing woodcuts.
225 KLEIN, LOUIS, *Copenhagen.*—A book set by Sörensen's compositor.
226 LUNO, BIANCO, *Copenhagen.*—Printed books.

Sub-Class D.

228 CLEMENT, D. L. *Copenhagen.*—Bound books, typography, xylography, and copper and steel engravings.
229 JUNGE, CHR. *Copenhagen.*—Picture books.

CLASS 29.

Sub-Class A.

230 ROYAL ORDNANCE SURVEY, *Copenhagen.*—Maps.
231 DIRECTION OF PUBLIC SCHOOLS, *Copenhagen.*—Maps.
232 DIRECTOR OF PUBLIC SCHOOLS, *Slesvig.*—Maps.
233 NISSEN, J. V. *Ramten, near Grenaa.*—A Bible historical map.
234 STEEN, CHR. & SON, *Copenhagen.*—Maps.

Sub-Class B.

235 DIRECTOR OF PUBLIC SCHOOLS, *Slesvig.*—Books and apparatus for instruction from the Duchy of Slesvig.
236 DIRECTION OF PUBLIC SCHOOLS, *Copenhagen.*—Books and apparatus for instruction.
237 CONRADSEN, RUDOLF, *Copenhagen.*—Apparatus for educational purposes; stuffed animal.
238 HESTERMANN, *Altona.*—Model for educational purposes, specially adapted for natural philosophy in elementary schools.
239 SCHIÖTT, *Copenhagen.*—New writing apparatus for the blind.
240 STEEN, CHR. & SON, *Copenhagen*—Globe.
241 THORNAM, J. C. *Copenhagen.*—Zoological drawings for educational purposes.

---

306 GALDBERG, *Copenhagen.*—Writing apparatus for the blind.

CLASS 30.

Sub-Class A.

242 ART AND INDUSTRIAL UNION, *Copenhagen.*—Furniture and domestic utensils.

243 DAHL, EMANUEL, *Haderslev.*—Carved table inlaid with German silver.

244 FREESE, H. C. *Kiel.*—Furniture of wicker-work; chairs easily taken to pieces.

246 GRIMM, *Neustadt.*—Furniture in rosewood.

247 HANSEN, F. DUMONT, *Copenhagen.*—A commode.

248 HEINSEN, N. H. *Altona.*—Furniture.

249 HELLMANN, S. D. *Altona.*—Wicker-work furniture.

251 LARSEN, L. *Copenhagen.*—Couch, arm-chair, and chair.

252 LUND, J. G. *Copenhagen.*—China and plate cupboard, and chairs.

253 NIELSEN, O. *Odense.*—Sideboard, and model of a secretaire.

254 RAMCKE, H. H. *Altona.*—Furniture of rosewood and mahogany.

255 SCHIRMER, F. *Kiel.*—Veneering.

Sub-Class B.

257 CLAUDIUS, S. *Kiel.*—Wall-painting (new invention), Pompeian style.

258 DAHL, A. *Copenhagen.*—Printed blinds, Venetian blinds, and Persiennes.

259 FJELDSKOV, W. *Copenhagen.* — A figure (Christian IV.) tankard, with stand.

260 FREESE, F. *Kiel.* — Tapestry from wood-shavings.

261 FRÖLICH, L.—Decorative paintings, illustrative of Northern mythology; allegory of "Morning."

262 HARBOE, J. O. *Copenhagen.*—Blinds Venetian blinds, Persiennes, and floor paper.

263 HENRICHSEN, J. *Copenhagen.* — Frames.

264 HULBE, C. *Kiel.*—Beading.

265 MASSMANN, F. *Kiel.*—Rough and polished beading.

266 NIELSEN, O. *Odense.*—Mirror frames.

268 WARNHOLZ, H. D. *Neumünster.* — Paper-hangings, designs, roller and hand sten-cilling.

CLASS 31.

Sub-Class A.

269 BUHLMANN, C. & Co. *Heide.*—Gas-meters and water-cistern which do not require constant pressure.

270 HOLLER & Co. *Iron Foundry, Carlshütte, near Rendsburg.*—Enamelled milkpans, with appendages.

271 MARTIN, L. A. *Copenhagen.*—Pattern card of buttons.

272 RAMES, C. A. *Copenhagen.*—Nails.

273 UNION IRON WORKS, *Pinneberg.* — Tinned cooking-apparatus, curry-combs, lacquered iron sugar-loaf moulds.

---

307 SCHWEFFEL & KOWALD, *Kiel.*—Iron milk-dishes, for large dairies.

Sub-Class B.

274 CRUSAA COPPER WORKS, *Flensborg.*—Yellow metal, copper in plates, and brass pans.

275 HALLVARD, *Thvera, Iceland.*—Padlock of brass, with two appertaining keys.

276 UNION IRON WORKS, *Pinneberg.* — Brass goods.

Sub-Class C.

277 HÖY, HANS, *Copenhagen.* — Pewter utensils.

278 JRGENS, C. & SON, *Copenhagen.*—Tin plate goods.

279 MEYER, F. *Copenhagen.*—Brass, tin and japanned goods.

280 RASMUSSEN, L. *Copenhagen.*—Figures cast in zinc.

CLASS 33.

281 ART & INDUSTRIAL UNION.—*Copenhagen.*—Silver-plate.

282 CHRISTESEN, V. *Copenhagen.*—Silver-plate.

283 CLAUSEN, N. C. *Odense.*—Spoons and forks.

284 DAHL, E. F. *Copenhagen.*—Works in gold and silver.

285 DAHLHOFF, J. B. *Copenhagen.* — Chased bust.

286 DIDRICHSEN, JUL. *Copenhagen.* — Chased figures and animals in gold and silver.

287 DRAGSTED, *Copenhagen.* — Drinking horns in the northern-antique; silver plate.

288 DREWSEN, H. O. *Copenhagen.* — Electro plate.

289 FERSLÉV, O. & Co. *Copenhagen.*—Seals and arms.

290 HERTZ, P. *Copenhagen.* — Epergne with figures.

291 MAYENTZHUSEN H. C. V. *Copenhagen.*—Articles in gold and silver.

292 MÖLLER, CASPAR, *Copenhagen.*—Galvano plastic works, plated and bronzed.

293 THORNING, J. C. *Copenhagen.* — Jewellery.

294 VIGFUSSON, S. *Reykjavik, Iceland.*—Works in silver.

---

308 HOLM, C. *Copenhagen.* — Chased bronze.

CLASS 34.

Sub-Class B.

295 JENSEN, H. *Flensborg.*—Articles in glass.

CLASS 35.

296 ROYAL PORCELAIN MANUFACTORY, *Copenhagen.*—Porcelain, table services, bisquit figures, table ornaments, vases, &c.

297 BING & GRÖNDAHL'S PORCELAIN WORKS, *Copenhagen.*—Bisquit figures, bas-reliefs, domestic and apothecary utensils, and telegraph insulators.

298 MATZENBECHER, J. H. *Rensing.*—Modelled and pressed bricks.

299 VIDAL, C. *Fernsicht.*—Stoves, architectural ornaments, tiles, figures, and vases of baked clay.

309 SONNE, F. M. *Rönne, Bornholm.*—Stone ware.

ECUADOR.

N.W. Gallery, Court No. 1.

1 Gold dust from the mines of Cachabi.
2 Gold ornaments found in different parts of the country.
3 Silver ore; ditto roasted and crushed; ditto crushed.
4 Copper ore mixed with emeralds.
5 Set of emeralds mounted in gold by native workmen; exhibited by Mrs. Prichard.
6 Cotton.
7 Cacao.
8 Leaf tobacco and cigars.
9 Coffee.
10 Orchella weed.
11 Ivory nuts.
12 Cinbona bark, flat and round.
13 Caoutchouc.
14 Silk produced by Senor Chiriboia.
15 A collection of woods exhibited by the Ecuador Land Company.
16 Pita, or the fibré of aloe.
17 Panama hats, or palm-leaf hats; Panama straw.
18 Embroidery work by Indians, exhibited by his Excellency Senor Flores.
19 Antiquities of pottery, found six feet below the level of the sea, at the Pailon.
20 Paintings from churches at Quito, by native Indians, exhibited by Senor Sanquirico y Ajesa.
21 Paintings representing views of Ecuador, by native Indians, exhibited by Mr. Mocatta.
22 Head of an Inca, reduced to tenth part of its natural size by an unknown process; an idol, from the Temple of Jivaros.

FRANCE.

S.W. Court and S.W. Gallery.

CLASS 1.

1 DUPONT & DREYFUS, *Ars-sur-Moselle (Moselle).*—Iron in bars, and for special purposes.
2 SCHOOL OF THE MASTER-MINERS OF ALAIS (*Gard*).—Mineralogical collection of the Department of the Gard.
3 COLLECTIVE EXHIBITION OF THE DEPARTMENT OF CORSICA.—Ores and marbles (4 Exhibitors).
4 JAMES, JACKSON, SON, & Co. *St. Seurin-sur-l'Isle (Gironde).*—Steel, by Bessemer's process: bars, springs, &c.
5 DE DIETRICH & Co. *Niederbronn (Bas-Rhin).*—Ornamental castings, steel, forgings.
6 BONNOR, DEGROND, & Co. *Eurville (Haute-Marne).*—Castings, rolled iron, wire, and chains.
7 MARTIN, E. O. *Sireuil (Charente).*—Ores, castings, steel, rails, axles, &c.
8 LALOUËL DE SOURDEVAL & MARGUERITTE, *Paris.*—Converted and cast steel: tools.
9 BARON DE ROSTAING & BAUDOUIN, BROS. *Paris.*—Metals granulated by centrifugal force, and their products; steel, oxides, and salts.
10 BAUDRY, A. & COTTREAU, *Athis-Mons (Seine-and-Oise).*—Iron, and cast steel.
11 JOINT-STOCK MINING AND RAILWAY CO. OF CARMAUX, *Avalats (Tarn).*—Iron and steel: axles.
12 DURAND, JUN. & GUYONNET, P. *Perigueux (Dordogne).*—Cast iron, refined and welded iron, iron-wire, &c.
13 JACQUINOT, F. & Co. *Solenzara (Corsica).*—Iron ores, wood-charcoal, cast iron.
14 CHENOT, A. & E. BROS. *Clichy-la-Garenne (Seine).*—Models of furnaces; iron, steel.
15 GUILLEM & Co. *Marseille (Bouches-du-Rhône).*—Lead, silver; copper nails.
16 OESCHGER, MESDACH, & Co. *Biache-St. Vaast (Pas-de-Calais).*—Ores: lead, copper, zinc.
17 COMMITTEE OF COAL PROPRIETORS OF THE DEPARTMENT OF THE LOIRE, *St. Etienne (Loire).*—Pit-coal, coke, and agglomerates.
18 JOINT STOCK ARGENTIFEROUS LEAD MINE AND FOUNDRY CO. OF PONTGIBAUD (*Puy-de-Dôme*).—Lead-ores, lead-pigs.
19 CHAMBER OF COMMERCE OF CHAMBÉRY (*Savoy*).—Marbles, ores, cements.
20 DELMAS, E. *St. Capraix (Dordogne).*—Alluvial iron ore.
21 TAMISIER & Co. *St. Gervais-les-Bains (Haute-Savoie).*—Red jasper.
22 FOMMARTY & Co. *Perigueux (Dordogne).*—Hydraulic lime.
23 MAGNEUR, R. *Hautefort (Dordogne).*—Manganesiferous alluvial iron ore.
24 GUÉRIN, DOCTOR J. *Paris.*—Marls, lime, hydraulic cements, plan of a lime-kiln.
25 BICKFORD, DAVEY, CHANU, & Co. *Rouen (Seine-Inf.).*—Safety fusees.
26 DE PAGÈZE DE LAVERNÈDE, *Salles (Gard).*—Pit-coal, regulus of antimony.
27 DUBRULLE, A. N. *Lille (Nord).*—Safety lamps.
28 BEAU, D. *Alais (Gard).*—Regulus of antimony.
29 MICHEL, ARMAND, & Co. *Marseille (Bouches-du-Rhône).*—Lignite.
30 LEBRUN-VIRLOY, A. *Lanty (Haute-Marne).*—Plan of a portable apparatus for drying and carbonizing wood and peat.
31 SPIERS, *Paris.*—Coal agglomerated with, and without bituminous matter.
32 MATHIEU, BROS. *Anzin (Nord).*—Air and water counter-pressure apparatus, &c.
33 MARQUIS DE CHAMBRUN, *Marvéjols (Lozère).*—Argentiferous lead ore.
34 CROS, J. *Albi (Tarn).*—Manganese ore.
35 NICOLI, J. B. *Ajaccio (Corsica).*—Rocks and ores.

36 COUVRAT - DESVERGUES - GEOFFROY, *Excideuil* (*Dordogne*).—Manganesiferous alluvial iron ore.

37 MINING CO. OF BÉDOUÈS AND COCURÈS, *Meyrueis* (*Lozère*). — Ores of argentiferous lead, and of copper.

38 MINING CO. OF PALLIÈRES, *Alais* (*Gard*).—Sulphates of lead.

39 MINING CO. OF RICHALDON, *Collet-de-Dèze* (*Lozere*).—Argentiferous lead ore, and products obtained from it by mechanical means.

40 MULOT, SON, & DRU, *Paris.*—Sounding and boring tools.

41 JACQUET, N. J. SEN. *Arras* (*Pas-de-Calais*).—Parachute for coal mines.

42 DEPLAYE, JULLIEN, & CO. *Paris.*—French lithographic stone.

43 COLETTES KAOLIN CO. *Château de Veauce* (*Allier*). — China clay: vases and other articles.

44 MINING CO. OF MARSAC, *Coussac-Bonneval* (*Haute-Vienne*).—Kaolin: sulphate of alumina.

45 PLANTIÉ & SON, *Bayonne* (*Basses-Pyrénées*).—Kaolin, crude, ground, and elutriated.

46 LIÉNART, L. T. *Mortcerf* (*Seine-and-Marne*).—Lime, cements, pipes.

47 PARQUIN, L. P. *Chelles* (*Seine-and-Marne*).—Plaster, and model of a mill.

48 CHAPUIS, P. & A. BROS. *Paris.*—Ores of platina, and articles made of it.

49 MORIN P. & CO. *Nanterre* (*Seine*).—Aluminium, and aluminium bronze.

50 LÉTRANGE, L. & CO. *Paris.*—Copper, lead, and zinc.

51 HERNIO, E. *Clohars* (*Finistère*). — Kaolin.

52 MAIRE, E. *Plessis-en-Coësmes* (*Ille-and-Vilaine*).—Slates.

53 MAIRE, X. *Moisdon* (*Loire-Inf.*).—Roofing slate.

54 DEGOUSÉE & LAURENT, C. *Paris.*—Model of sounding apparatus; tools.

55 JOINT STOCK MINING CO. OF LA GRAND-COMBE (*Gard*).—Coal, ordinary and agglomerated coke.

56 GAILLARD, SEN., PETIT, & HALBOU, A. *La Ferté-sous-Jouarre* (*Seine-and-Marne*).—Millstone.

57 DANGREVILLE-CHERRON & VALOND, J. *La Ferté-sous-Jouarre* (*Seine-and-Marne*).—Millstone.

58 CHASSAING-PEYROT & CO. *Domme* (*Dordogne*).—Millstones, slabs.

59 ALLARD, SON, & CO. *Sarlat* (*Dordogne*).—Millstones, nut oil.

60 BAILLY & CO. *La Ferté-sous-Jouarre* (*Seine-and-Marne*). — Millstones.

THIERRION, *Épernay* (*Marne*).—Circular guide for dressing millstones.

61 GILQUIN, P. S. *La Ferté-sous-Jouarre* (*Seine-and-Marne*). — Millstones, slabs, and panels.

62 DESMOUTIS, CHAPUIS, & QUENNESSEN, *Paris.*—Platina, and apparatus made of it.

63 DUPETY, THEUREY-GUEUVIN, BOUCHON, & CO. *La Ferté-sous-Jouarre* (*Seine-and-Marne*).—Millstones, slabs, and panels.

64 GAILLARD, J. F. *La Ferté-sous-Jouarre* (*Seine-and-Marne*).—Millstones.

65 LEVEAU-BAUDRY, *Villaine-la-Gonais* (*Sarthe*).—Millstones.

66 CHAUVEAU, *Villaine-la-Gonais* (*Sarthe*).—Millstones.

67 DELÉPINE, C. & THOMAS, A. *La Ferté-sous-Jouarre* (*Seine-and-Marne*).—Millstones.

68 ROGER, SON, & CO. *La Ferté-sous-Jouarre* (*Seine-and-Marne*). — Millstones, slabs, and panels.

69 BARDEAU, E. *Fleury* (*Yonne*).—Millstones.

70 TIGER & JONQUET, *Cloyes* (*Eure-and-Loire*).—Millstones.

71 BESNARD, *Epernon* (*Eure-and-Loire*).—Millstones.

72 MATHER & SON, *Toulouse* (*Haute-Garonne*).—Ingot, sheet, and cupola copper.

73 VISCOUNT A. N. DESSERES, *Caylus* (*Tarn-and-Garonne*).—Lithographic stones.

74 MALBEC, A. A. *Paris.*—Artificial stones, for mills and other purposes.

75 MESNET, T. A. *Cinq-Mars-la-Pile* (*Indre-and-Loire*).—Millstones.

76 DESPAQUIS, P. A. *Harol* (*Vosges*).—Lithographic stones.

77 DELESSE, A. *Paris.*—Hydrological map of the Department of the Seine.

78 DORMOY, E. *Valenciennes* (*Nord*).—Subterranean map of the coal field of Valenciennes, &c.

79 SENS, E. *Arras* (*Pas-de-Calais*).—Typographical map of the coal field of the Department of the Pas-de-Calais.

80 POUGNET M. & CO. *Landroff* (*Moselle*).—Model of shaft-lining.

81 CABANY, A. *Valenciennes* (*Nord*).—Waggon, and plan of a mine.

82 LECOQ, H. *Clermont-Ferrand* (*Puy-de-Dôme*).—Geological map of the Department of the Puy-de-Dôme.

83 DEHAYNIN, F. *Paris.*—Agglomerated substances, and plan of the machine employed.

84 OHALAIN, E. *Riadan* (*Ille-and-Vilaine*).—Slates for flooring, roofing, and billiard tables.

85 BORDE, REYMOND, PALAZZI, *Corte* (*Corsica*).—Streaked copper ore.

86 BARON O. DE BARDIES, *Oust* (*Ariège*).—Argentiferous galena ores.

87 CHALLETON, J. F. F. *Montauger* (*Seine-and-Oise*). — Peat, purified, condensed, and carbonized; oils, ammoniacal compounds, manures, &c.

88 BRIÈRE, A. *Brassac-les-Mines* (*Puy-de-Dôme*).—Ore, regulus of antimony, arsenious acid (Schweinfurth green), &c.

89 CHAPERON, PERRIGAULT, & CO. *Libourne* (*Gironde*).—Millstones.

90 LAVALLÉE, E. *Fontenay* (*Seine*).—Moulding sand, for founders.

Addenda.

TERQUEM, *Metz* (*Moselle*).—Geological section of Mount St.-Quentin, near Metz.

BONHOMÉ, *Paris.*—Designs having reference to mines and metallurgical establishments.

CHUART M. *Paris*.—Safety lamp, gazoscope for preventing explosions in mines.

POUYAT, BROS. *Limoges* (*Haute-Vienne*).—Kaolin, felspath.

ROSIER, WIDOW, & BAROCHE, *Tain* (*Drôme*).—Kaolin.

VERDIÉ, F. F. & Co. *Firminy* (*Loire*).—Iron, puddled and cast steel.

VIEILLARD, J. & Co. *Bordeaux* (*Gironde*).—Kaolin, and felspath.

CLASS 2.

91 CALLOU, A. & VALLÉE, *Paris*.—Salts, &c. obtained from the waters of Vichy.

92 CHERBOUQUET-BADOIT & CHAMPAGNON, *St. Galmier* (*Loire*).—Mineral waters.

93 BOULOUMIÉ, L. *Vittel* (*Vosges*).—Mineral waters, and ferruginous products of the springs of Vittel; corks for preventing the decomposition of these waters.

94 THE CITY OF BAGNÈRES-DE-LUCHON (*Haute-Garonne*).—Plans of a Thermal establishment.

95 FRANÇOIS, J. *Paris*.—Collection of rocks accompanying certain mineral waters, &c.

96 THE PROPRIETORS OF THE MINERAL WATERS OF FRANCE.—Mineral waters, from 46 localities.

97 CROU, L. *Aubusson* (*Creuse*).—Telegraphic and writing ink.

98 CHARVIN, F *Lyon* (*Rhône*).—Green dye-stuff.

100 GERTOUX, J. *Bagnères-de-Bigorre* (*Hautes-Pyrenees*).—Labassère water, containing sulphuret of sodium.

101 BURGADE & SISTERS, *Garost* (*Hautes-Pyrénées*).—Iodurated sulphurous water of Garost.

102 MANINAT, JUN. *Ossun* (*Hautes-Pyrénées*).—Sulphurous mineral water.

103 BRUN, M. *Puteaux* (*Seine*).—Mordants for dyeing.

104 ROSELEUR, A. *Paris*.—Chemical products.

105 ARRAULT, H. *Paris*.—Chemical products, &c.

106 LALOUËL DE SOURDEVAL & MARGUERITTE, *Paris*.—Alkaline cyanides and ammoniacal salts.

107 PENNÈS, J. A. *Paris*.—Mineral salts for baths.

108 LEFRANC & Co. *Paris*.—Colours, varnishes, and ink.

109 LE PERDRIEL & MARINIER, *Paris*.—Pharmaceutical products.

110 FUMOUZE-ALBESPEYRES, *Paris*.—Dressings for blisters, &c.

111 DUROZIEZ, M. E. A. *Paris*.—Artists' materials, photographic chemicals.

112 GARZEND, A. *Paris*.—Prepared woods for dyeing.

113 USÈBE, C. J. *St. Ouen* (*Seine*).—Carmine of saffron in paste, &c.

114 ROCQUES & BOURGEOIS, *Ivry* (*Seine*).—Chemical products obtained in the carbonization of wood.

115 POIRRIER & CHAPPAT, JUN. *Paris*.—Chemical products for dyeing.

116 PETERSEN, F. & SICHLER, *Villeneuve-la-Garenne near St. Denis* (*Seine*).—Chemical products, dye-stuffs, and colours.

117 MALLET, A. A. P. *Paris*.—Various salts.

118 PARISIAN GAS-LIGHTING AND HEATING CO. *Paris*.—Chemical products obtained during the destructive distillation of coal.

119 CAMUS, C. & Co. *Paris*.—Chemical products.

120 PIVER & RONDEAU, A. *Paris*.—Colours, and varnishes.

121 DEROCHE, C. *Paris*.—Chemical products.

122 KUHLMANN & Co. *Lille* (*Nord*).—Chemical products, &c.

123 DRION-QUÉRITÉ, PATOUX, & DRION, A. *Aniche* (*Nord*).—Chemical products.

124 DESESPRINGALLE, A. *Lille* (*Nord*).—Chemical products obtained from alcohol and tar; salts of cadmium.

125 PÉRUS, J. & Co. *Lille* (*Nord*).—White lead.

126 RICHTER, B. & F. *Lille* (*Nord*).—Ultramarine blue.

127 DORNEMANN, G. W. *Lille* (*Nord*).—Ultramarine blue and green.

128 CHAPUS, A. *Lille* (*Nord*).—Ultramarine blue.

129 SERRET, HAMOIR, DUQUESNE, & Co. *Valenciennes* (*Nord*).—Chemical products: alcohol, sugar.

130 DEHAYNIN, M. G. *Valenciennes* (*Nord*). Chemical products from tar, for dyeing.

131 SERBAT, L. *St. Saulve* (*Nord*).—Mastic, oils, and fats, &c.

132 GAUTIER-BOUCHARD, L. J. *Paris*.—Chemical products, colours and varnish.

133 BONZEL, BROS. *Haübourdin* (*Nord*).—White lead, chicory.

134 JOINT-STOCK MINING CO. OF SAMBRE-AND-MEUSE, *Hautmont* (*Nord*).—Chemical products.

135 MINING CO. OF BOUXWILLER (*Bas-Rhin*).—Chemical products, and various salts.

136 KESTNER, C. *Thann* (*Haut-Rhin*).—Chemical products, and dye-stuffs.

137 SCHAAFF, & LAUTH, *Strasbourg* (*Bas-Rhin*).—Madder, and its extracts.

138 RIESS, M. *Dieuze* (*Meurthe*).—Gelatine, and phosphate of lime.

139 MERLE, H. & Co. *Alais* (*Gard*).—Various salts, chemicals, and dye stuffs.

140 PLANCHON, S. *St. Hippolyte* (*Gard*).—Glue and gelatine.

141 CHIRAUX, L. *Cambrai* (*Nord*).—Blacking.

142 CAZALIS, H. & Co. *Montpellier* (*Hérault*).—Sulphuric and other acids; salts.

143 LE BŒUF, F. *Bayonne* (*Basses-Pyrénées*).—Pharmaceutical products, &c.

144 TACHON, SON, & Co. *Roanne* (*Loire*).—White lead, obtained directly with carbonic acid evolved from mineral springs.

145 BERJOT, F. *Caen* (*Calvados*).—Pharmaceutical extracts, &c.

146 PARQUIN, LEGUEUX, ZAGOROWSKI, & SONNET, *Auxerre* (*Yonne*).—Ochres, raw and manufactured.

147 CAROF, A. & Co. *Portsal-Ploudal-*

mézeau (*Finistère*).—Chemical products, obtained from sea-wrack.

148 COURNERIE, SON, & CO. *Cherbourg* (*Manche*).—Chemical products, obtained from sea-wrack.

149 MAUMENÉ & ROGELET, *Reims* (*Marne*). —Potassa, and its salts, obtained from sheep-grease.

150 HUILLARD & GRISON, *Deville-lez-Rouen* (*Seine-Inf.*).—Products for dyeing.

151 DELACRETAZ & CLOUET, *Hâvre*.—Chrome oxide, and salts of chromic acid.

DELACRETAZ, *Paris*.—Chemical products, &c.

152 MULLER, P. *Rouen* (*Seine-Inf.*).—Gelatine.

153 LAURENTZ, P. C. *Rouen* (*Seine-Inf.*).— —Chlorides.

154 TISSIER & SON, *Conquet* (*Finistère*). —Chemical products, obtained from sea-wrack.

155 PICARD & CO. *Granville* (*Manche*).—Chemical products, obtained from sea-wrack.

156 JOINT-STOCK GLASS AND CHEMICAL CO. OF S. GOBAIN, CHAUNY, & CIREY, *Paris*. —Chemical products.

158 CHEVÈNEMENT, L. *Bordeaux* (*Gironde*).—Blacking, black and coloured inks.

159 FOURNIER-LAIGNY & CO. *Courville* (*Eure-and-Loire*).—Products obtained from pyroligneous acid.

160 DESCHAMPS, BROS. *Vieux-Jean-d'heures* (*Meuse*). — Ultramarine blue and green.

161 BARTHE, DURRSCHMIDT, PORLIER, & CO. *Pont-St. Ours* (*Nièvre*).—Acetic acid, and various salts.

162 BAZET, HAPPEY, & CO. *Paris*.—Gazogène apparatus.

163 LUTTON, A. LOLLIOT, & CO. *Neuvy-sur-Loire* (*Nièvre*).—Pyroligneous acid, and its products.

164 BRUZON, J. & CO. *Portillon* (*Indre-and-Loire*).—Chemical products, &c.

165 BERTRAND & CO. *Dijon* (*Côte-d'or*).—Ultramarine blue.

166 DANIEL H. *Paris*.—Blacking and inks.

167 GUINON, MARNAS, & BONNET, *Lyon* (*Rhône*).—Chemical products, obtained during the destructive distillation of pit-coal.

168 ALESMONIÈRES, A. *Lyon* (*Rhône*).—Chemical products, &c. from pit-coal.

169 GILLET & PIERRON, *Lyon* (*Rhône*).—Chemical products, &c.

170 MONNET & DURY, *Lyon* (*Rhône*).—Products for dyeing.

171 FAYOLLE & CO. *Lyon* (*Rhône*).—Chemical products.

172 PLATEL, L. J. & BONNARD, J. *Lyon* (*Rhône*).—Products for dyeing, &c.

173 BRUNIER, JUN. & CO. *Lyon* (*Rhône*). —Prussiates of potash.

175 RENARD. BROS. & FRANC, *Lyon* (*Rhône*).—Chemical products, and dye-stuffs.

176 GUIMET, J. B. *Lyon* (*Rhône*).—Ultramarine blue.

177 BLUM-GAY & CO. *Lyon* (*Rhône*).—Chemical products, &c.

178 COIGNET, SON, & CO. AND COIGNET, BROS. & CO. *Lyon* (*Rhône*).—Chemical products obtained from bone; &c.

179 MESSIER, *Paris*.—Lacs for paper-staining.

180 LANGE-DESMOULIN, J. B. C. *Paris*.—Colours.

181 BOYER & CO. *Paris*.—Chemical products, &c.

182 CHEVÉ, L. J. JUN. *Paris*.—Chemical products.

183 FOURCADE, A. & CO. *Paris*.—Chemical products.

184 JAVAL, J. *Paris*.—Products for dyeing.

185 MATHIEU-PLESSY, E. *Paris*.—Chemical products.

186 JACQUES-SAUCE, *Paris*. — Cochineal carmine, &c.

187 ROQUES E. & CO. *Paris*.—Chemical products.

188 POMMIER & CO. *Paris*.—Chemical products.

189 DALEMAGNE, L. *Paris*.—Silicate of potash, for preserving calcareous stone.

190 DURET, SEN. & BOURGEOIS, *Paris*.—Non-poisonous colours.

191 PERRA, B. *Petit-Vanvres* (*Seine*).—Pharmaceutical and dyeing products.

192 BOBŒUF, P. A. F. *Paris*.—Chemical products.

193 DEISS, E. *Paris*.—Fatty substances from refuse matter.

194 ADVIELLE, L. B. *Paris*.—Liquid for silvering.

195 ACCAULT, C. *Paris*.—Calcined magnesia.

196 ARMET DE LISLE, J. *Nogent-sur-Marne* (*Seine*).—Ultramarine blue, salts of quinine.

197 LAROCQUE, A. *Paris*.—Benzine, &c.

198 GÉLIS, A. *Paris*.—Various salts.

199 BLANCARD, H. *Paris*.—Iodide of iron pills.

200 BURDEL & CO. *Paris*.—Liquids for cleansing and reviving cloth.

201 DEFAY, J. B. & CO. *Paris*.—Dried blood, for manure, &c.; albumen, from blood.

202 JORET, E. M. F. & HOMOLLE, E. *Paris*. Apiol.

HOMOLLE & DEBREIUL, *Paris*.—Digitaline.

203 COLLAS, C. & CO. *Paris*.—Benzine, &c.

204 MENIER, E. J. *Paris*.—Pharmaceutical and chemical products.

205 COËZ, E. & CO. *St. Denis*, (*Seine*).—Extracts from dyewoods, &c.

206 LAURENT, F. & CASTHÉLAZ, *Paris*.—Chemical products, &c.

207 POULENC-WITTMANN, E. J. *Paris*.—Chemical and pharmaceutical products.

208 DUBOSC, F. & CO. *Paris*.—Chemical products.

209 FREZON, J. B. *Neuilly* (*Seine*) —Mordants.

210 LAURENT, C. & LABÉLONYE, C. *Paris*. —Pharmaceutical extracts, prepared in vacuo.

211 SCHOEN & REUTER, *Paris*.—Colouring substances.

212 BEZANÇON BROS. *Paris*.—White lead.

213 FERRAND, M. *Paris*.—Artist's colours.

214 STRAUSS-JAVAL & CO. *Paris*.—Dry extracts from dye-woods.

215 HARDY-MILORI, G. *Montreuil-sous-Bois* (*Seine*).—Colours in the dry and pasty state.
216 LATRY, A. & Co. *Paris*.—Zinc-white.
217 GELLÉ, J. B. A. SEN. & Co. *Paris*.—Perfumes, and toilet soap
218 MOLLARD, A. A. *Paris*.—Toilet soap.
220 SARDOU, L. *Cannes* (*Alpes-Maritimes*).—Perfumes, pomades.
221 HUGUES, SEN. *Grasse* (*Alpes-Maritimes*).—Perfumes; alcoholic extracts.
222 MÉRO, J. D. *Grasse* (*Alpes-Maritimes*).—Essences, perfumed oils, &c.
223 ISNARD-MAUBERT, *Grasse* (*Alpes-Maritimes*).—Essences, orange-flower water.
224 RANCÉ, F. & LAUTIER, JUN. *Grasse* (*Alpes-Maritimes*).—Essences, &c.
225 ARDISSON & VARALDI, *Cannes* (*Alpes-Maritimes*).—Essences, &c.
226 COUDRAY, P. E. *Paris*.—Perfumes, &c.
227 CLAYE, V. L. *Paris*.—Perfumes, &c.
228 PINAUD, E. & MEYER, E. *Paris*.—Perfumes, &c.
229 SICHEL, J. *Paris*.—Perfumes and toilet soap.
230 BLEUZE-HADANCOURT, *Paris*.—Perfumes, &c.
231 DELABRIERE-VINCENT, *Paris*.—Perfumes, &c.
232 MOUILLERON, A. *Paris*.—Vinegar, toilet soap, &c.
233 PORTE, F. X. L. *Paris*.—Eau de Cologne, &c.
234 LANDON-LEMERCIER, *Paris*.—Vinaigre de Bully.
236 GUERLAIN, P. F. P. *Paris*.—Essences, cosmetics, &c.
237 BOUTRON-FAGUER, *Paris*.—Essences, perfumes, &c.
238 MAILLY, F. *Paris*.—Perfumes and toilet soap.
239 PIVER, A. *Paris*.—Perfumes and toilet soap.
240 TITARD, J. L. *Paris*.—Vegetable rouge and blanc, for the toilet, &c.
241 JASSAU-RAIMOND, *Paris*.—Rouge for the toilet.
242 GIRAUD, BROS. *Paris*.—Perfumed oils, extracts, &c.

CLASS 3.

Sub-Class A.

251 DELAFONTAINE & DETTWILLER, *Paris*.—Cocoa, chocolate, &c.
252 MÉNIER, E. J. *Paris*.—Chocolate.
253 DEVINCK, F. J. *Paris*.—Chocolate.
254 GUÉRIN-BOUTRON, M. L. A. *Paris*.—Cocoa and chocolate.
255 CHOQUART, C. F. *Paris*.—Chocolate.
256 IBLED, BROS. & Co. *Paris*.—Chocolate.
257 LEGUERRIER, C. L. M. *Paris*.—Chocolate, and roasted coffee.
258 HERMANN, G. *Paris*.—Chocolate.
259 ALLAIS, E. *Paris*.—Ordinary, and ferruginous chocolate.
260 LABRIC, P. E. *Paris*.—Cocoa and chocolate.
261 PELLETIER, E. & Co. *Paris*.—Butter of cacao and chocolate.
262 POTIN, L. E. *Paris*.—Chocolate.
263 TRÉBUCIEN, BROS. *Paris*.—Coffee chocolate, tapioca.
264 FAGALDE, P. *Bayonne* (*Basses-Pyrénées*).—Chocolate.
265 PENIN, C. & Co. *Bayonne* (*Basses-Pyrénées*).—Chocolate.
266 RUBIÑO, A. *Nice* (*Alpes-Maritimes*).—Cocoa and chocolate.
267 LOUIT, BROS. & Co. *Bordeaux* (*Gironde*).—Chocolate, alimentary pastes, &c.
268 ASSOCIATION OF THE PASTE MANUFACTURERS, &c. OF AUVERGNE, *Clermont-Ferrand* (*Puy-de-Dôme*).—Alimentary pastes (13 exhibitors).
269 BOUDIER, F. *Paris*.—Alimentary pastes.
270 GROULT, JUN. *Paris*.—Pastes, flour, fecula.
271 FRELUT & Co. *Clermont-Ferrand* (*Puy-de-Dôme*).—Preserved fruits.
272 NOEL-MARTIN & Co. *Paris*.—Alimentary pastes, &c.
273 COLLECTIVE EXHIBITION OF THE CITY OF EPINAL (*Vosges*).—Potatoe-fecula, &c. (8 exhibitors).
274 COMBIER-DESTRE, *Saumur* (*Maine-and-Loire*).—Elixir of "Raspail."
275 LERVILLES, J. *Lille* (*Nord*).—Roasted chicory.
276 BÉRIOT, C. *Lille* (*Nord*).—Chicory and varnishes.
277 BOYER, A. *Paris*.—Eau de Melissc, eau des Carmes.
278 GIRAUD, BROS. *Paris*.—Olive and perfumed oils, &c.
279 ROUSSIN, ELIAS, *Rennes* (*Ille-and-Vilaine*).—Groats and pearled barley.
280 ARNAUD, SEN. & Co. *Voiron* (*Isère*).—Liqueurs.
281 OLIBET, *Bordeaux* (*Gironde*).—Sea-biscuits.
282 CAUSSEROUGE, BROS. *Paris*.—Liqueurs, syrups, &c.
283 DAVID, J. & Co. *Orléans* (*Loiret*).—"French bitter."
284 GOURRY & Co. *Cognac* (*Charente*).—Liqueurs.
285 ROBIN, L. P. JUN. *L'Ille d'Espagnac* (*Charente*).—Coffee and chicory.
286 ROCHER, BROS. *Côte-St. André* (*Isère*).—Liqueurs.
287 LEGIGAN & LEFÈVRE, *Paris*.—Liqueurs.
288 SAINTOIN, BROS.—*Orléans* (*Loiret*).—Curaçoa, chocolate, &c.
289 MARIE BRIZARD & ROGER, *Bordeaux* (*Gironde*).—Liqueurs.
290 JOURDAN-BRIVE, G. SEN. *Marseilles* (*Bouches-du-Rhône*).—Wines, liqueurs, &c.
291 COLLECTIVE EXHIBITION OF THE DEPARTMENT OF THE CÔTE D'OR (5 exhibitors).—Liqueurs.
292 HOFFMANN-FORTY, F. *Phlasbourg* (*Meurthe*).—Liqueurs.
293 PAULIN-FORT, DESPAX, & BACOT, *Toulouse* (*Haute-Garonne*).—Wines, liqueurs, and syrups.

294 ROUSSEAU & LAURENS, *Paris.*—Liqueurs, preserved fruits.

295 TESSON, A. *Pantin-lez-Paris (Seine).*—Wines and liqueurs; preserved fruits.

296 COLLECTIVE EXHIBITION OF THE CITY OF DUNKERQUE (*Nord*).—Juniper-berry liqueurs.

297 GALLIFET & Co. *Grenoble* (*Isere*).—Liqueurs.

298 FAIVRE, DOCTOR C. *Paris.*—"Mont Carmel" liqueur.

299 LASSIMONNE, C. *Paris.*—Liqueurs, syrups, preserved fruits, &c.

300 MAGNÈ, A. *Rouen* (*Seine-Inf.*).—Jellies, and apple-sugars.

301 POURCHIER, J. B. *Avignon* (*Vaucluse*).—Chocolate, cocoa nuggets.

302 NÈGRE, J. *Grasse* (*Alpes-Maritimes*).—Preserved fruits; orange-flower water.

303 BAUDOT-MABILLE, *Verdun* (*Meuse*).—Sugar-plums and liquorice.

304 CAIZERGUES, A. *Montpellier* (*Hérault*).—Preserved fruits, confectionery.

305 MALSALLEZ, C. *Paris.*—Liqueurs, syrups, chocolate, sweetmeats.

306 AUVRAY, JUN. *Orleans* (*Loiret*).—Bon-bons.

307 BRUNET, L. *Paris.*—Concentrated extracts, for the manufacture of liqueurs by mere mixture.

308 GELLER, G. *Marseille* (*Bouches-du-Rhône*).—Fruits preserved in sugar, and in brandy; sweetmeats.

309 MUSSO, *Nice* (*Alpes-Maritimes*).—Fruits, liqueurs, and syrups.

310 BONFILS, BROS. & Co. *Carpentras* (*Vancluse*).—Conserve of truffles.

311 JACQUIN, WIDOW, & SON, *Paris.*—Sugar-plums and crisped almonds.

312 CHOLLET & Co. *Paris.*—Preserved vegetables, chocolate, alimentary pastes.

313 REY, F. A. *Paris.*—Preserved fruits.

314 DEMEURAT, DOCTOR L. *Tournan* (*Seine-and-Marne*).—Preserved meats and vegetables; biscuits.

315 CORMIER, E. *Neuilly* (*Seine*).—Preserved eggs and vegetables; sardines in oil.

316 CARNET & SAUSSIER, *Paris.*—Preserved alimentary substances.

317 GALOPIN, P. JUN. *Paris.*—Preserved truffles.

318 CHEVET, C. J. *Paris.*—Preserved alimentary substances.

319 HENRY, L. *Strasbourg* (*Bas-Rhin*).—Goose-liver pasty.

320 GUILLOUT, E. *Paris.*—Gingerbread, biscuits, &c.

321 SIGAUT, J. J. *Paris.*—Gingerbread, &c.

322 DRIOTON, *St. Seine-l'Abbaye* (*Côte-d'-Or*).—Epines-vinettes preserve.

323 BLANC, *Perigueux* (*Dordogne*).—Preserved alimentary substances, containing truffles.

325 SAUSOT, J. *Bordeaux* (*Gironde*).—Preserved alimentary substances containing truffles; preserved vegetables, fish, &c.

327 REBOURS-GUIZELIN, DIONE, & Co. *Paris.*—Preserved alimentary substances, &c.

328 QUILLET, A. & SON, *Paris.*—Mustard, vinegar, pickles.

329 BORDIN-TASSART, A. *Paris.*—Mustard, vinegar, and pickles.

330 MAILLE ET SEGOND, *Paris.*—Mustard, vinegar, and pickles.

331 DUBOSC, *Paris.*—Mustard.

332 AMAND-GUENIER, *Auxerre* (*Yonne*).—Mustard.

333 COLLECTIVE EXHIBITION OF THE DEPARTMENT OF THE CÔTE D'OR.—Mustard. (7 Exhibitors).

334 DIETRICH, BROS. *Strasbourg* (*Bas-Rhin*).—Mustard.

335 ROUZÉ, H. *Paris.*—Preserved fruits.

336 PERRIER, J. P. F. *Crest* (*Drôme*).—Preserved truffles.

337 BATTENDIER, A. F. JUN. *Paris.*—Preserved truffles.

338 GALLOIS, H. *Paris.*—Ground pepper.

339 JOURDAIN, E. *Paris.*—Preserved fruits, and sweetmeats.

340 PHILIPPE, C. & CANAUD, WIDOW, *Nantes* (*Loire-Inf.*).—Preserved alimentary substances.

341 PELLIER, BROS. *Mans* (*Sarthe*).—Sardines in oil.

342 RODEL & SONS, *Bordeaux* (*Gironde*).—Preserved alimentary substances.

343 SAUCEROTTE & PARMENTIER, *Lunéville* (*Meurthe*).—Preserved fruits and vegetables.

344 HÉRON, *Paris.*—Fish, preserved without the bones; essence of coffee.

345 BALESTRIÉ, R. *Concarneau* (*Finistere*).—Sardines in oil.

347 LIREUX, S. *Le Havre* (*Seine-Inf.*).—Caramel.

348 CONNIÉ & MARTIN, *La Rochelle* (*Charente-Inf.*).—Sardines.

349 VOISIN, A. *Paris.*—Chestnut conserve.

350 SALLES A. & SON, *Paris.*—Preserved alimentary substances.

Addendum.

ALLARD, SON, & Co. *Sarlat* (*Dordogne*).—Nut oil.

Sub-Class B.

(*Nord*).

351 AGRICULTURAL ASSOCIATION OF LILLE.—Collection of cereals, forage, oils, alcohols, &c. (19 Exhibitors).

352 AGRICULTURAL SOCIETY OF BOURBOURG.—Collection of cereals, forage, colza, flax, &c. (21 Exhibitors).

353 AGRICULTURAL SOCIETY OF HAZEBROUCK.—Cereals, forage, flax, hops, tobacco, &c. (19 Exhibitors).

354 COMMUNE OF REXPOÈDE.—Wheat, oats, flax, beans (11 Exhibitors).

355 VANDERCOLME, A. *Rexpoede.*—Wheat, oats in the sheaf, forage, &c.

356 HAMOIR, G. *Saultain.*—Soils, cereals, forage, sugar.

357 GOUVION-DEROY, *Denain.*—Cereals, sugar, alcohol, potash.

358 FIEVET, *Masny.*—Wheat, oats in the sheaf, flax, sugar, &c.

359 CHEVAL, B.—Agricultural products.
360 VARDAELE, F. *Warhem.*—Corn, beet-root, flax, oil, oil-cake.
361 PORQUET-DOURIN, *Bourbourg.* — Wheat, oats, flax, pease.
362 RYCKELYNCK, *Beaudignies.*—Agricultural products.
363 SPIERS, J. A. *Valenciennes.*—Natural and artificial guano, made of the refuse of fish.
364 MESSERSCHMIDT, *St. Amand-les-Eaux.* —Strong vinegar, obtained by a new process.

(*Pas de Calais.*)

365 DECROMBECQUE, G. *Lens.*—Cereals, loaf-sugar, alcohol.
366 DELABY, A. & Co. *Courcelles-lez-Lens.* —Wheat, flax, beet-root sugar.
367 DELAUNE, A. *Courrières-lez-Lens.*—Sugar from beet-root molasses: alcohol from cane-sugar molasses.
368 MARQUIS D'HAVRINCOURT. — *Corbehem.*—Fleece; plan of a manure pit.
369 DE PLANCQUE, *St. André-lez-Gouy.*—Colza seed.
370 PROYART, *Hendecourt-lez-Gagnicourt.* —Cereals in sheaves, forage, flax.

(*Aisne.*)

371 COLLECTIVE EXHIBITION BY THE DEPARTMENT OF THE AISNE.—Agricultural products, &c. (52 Exhibitors).

(*Oise.*)

373 NORMAL AGRICULTURAL INSTITUTE OF BEAUVAIS.—Agricultural products.
374 FAULTE DE PUYPARLIER, A.—Compressed bread, a substitute for military and naval biscuit.
375 FLAMAND-SEZILLE, *Noyon.*—Shelled and husked peas.
376 COLLECTIVE EXHIBITION OF THE WHEAT AND WINE DISTRICTS.—Flour (14 Exhibitors).

(*Somme.*)

377 COLLECTIVE EXHIBITION OF THE DEPARTMENT OF SOMME.—Cereals, forage, oils, sugars, wood, leather, wool, clays, bricks, &c. (52 Exhibitors).

(*Seine-and-Marne.*)

378 COLLECTIVE EXHIBITION OF THE DEPARTMENT OF SEINE AND MARNE.—Wheat, forage, alcohols, vegetables, honey, wax, cheese, &c. (55 Exhibitors).

(*Seine-and-Oise.*)

379 COLLECTIVE EXHIBITION OF THE AGRICULTURAL ASSOCIATION OF THE DEPARTMENT OF SEINE-AND-OISE.—Fecula, oils, alcohol, cereals, &c. (23 Exhibitors).

(*Seine.*)

380 CHODZKO, *Neuilly, near Paris.*—Model of a drying apparatus for fecal matters.
382 GRIVEL, CHATEAU, & BAYLE, *Paris.* —Manure from sewerage, &c.
383 KRAFFT, *Le Paris.*—Manure made from offal of an abattoir.
384 ROHART & SON, *Paris.* — Animal matters for manures.
385 ROUILLIER, E. *Paris.*—Beer.
386 VOLLIER, J. B. A. *Paris.*—Malt, hops, beer.
387 LABADY, *Paris.*—Beer.
388 BOUCHEROT, *Puteaux.*—Beer.
389 L'HOMME-LEFORT, *Paris.*—Mastic for grafting and for curing unhealthy trees and shrubs.
390 DESCROIX, *Paris.*—Wine-vinegar.
391 VOIRIN, *Paris.*—Liqueurs.
392 FENAILLE & CHATILLON, *Paris.*—Resinous products.
394 THOURET, E. *Paris.*—Model of preservative granary.
395 VICAT, *Paris.*—Insect-killing powder.
396 BEAUSSIER, *Paris.*—Indigenous tea.
397 BIGNON, *Paris.*—Products obtained by improved cultivation, and notices of the method pursued. A collection of the chief wines of France.

(*Seine-Inf.*)

398 AGRICULTURAL SOCIETY OF THE ARRONDISSEMENT OF HÂVRE (7 Exhibitors). —Corn, flax, fleeces, cyder, &c.
399 DESMAREST, *Bully.*—Cereals.
400 MOISSON, *Luzy.*—Oats in the sheaf.
401 MULLOT, *St. Aubin-Cilloville.*—Oats in the sheaf.
402 RASSET, *Minterollier.* — Corn, oats, barley.
403 SÉMICHON, JUN. *Vieux-Rouen.*—Corn in the sheaf and in grain.
404 MAMBOUR DELAGRAVE, *Foucarmont.* —Hops.
405 DUVIVIER, *St. Martin.* — Cyder, cheese.
406 JOLY, *La Mobraye.*—Cyder.
407 LESUEUR, *Forges-les-Eaux.*—Refractory clays.

(*Manche.*)

408 SOCIETY OF THE POLDERS OF THE WEST.—Specimens of the soil, and products of the Polders.
409 MOSSELMANN & Co. *La Rocque-Genest.* —Lime-stone, lime, &c.
410 LAJOYE, *St. Lô.*—Animal manure.
411 LEMOIGNE-DULONGPRÉ.—Cyder, kaolin.

(*Calvados.*)

412 DELAUNEY, A. *St.-Désir.*—Liquid resin.
413 DE VILADE, L. C. *Surire.*—Cyder-brandy.

(*Allier.*)

414 AGRICULTURAL ASSOCIATION OF MONTLUÇON.—Rye in the sheaf; casket made of different kinds of wood, &c.
415 AGRICULTURAL ASSOCIATION OF EBREUIL.—Wines.
416 BARON DE VEAUCE, *Château de Veauce.*—Wines.
417 DE FINANCE, *Trévelles.* — Various agricultural products.

(*Corrèze.*)

418 AGRICULTURAL ASSOCIATION OF THE CANTON OF MEYSSAC.—Wines (8 Exhibitors).

419 COUNT J. DE COSNAC, *Château du Pui.*—Wheat, cereals, nuts, &c.
420 MAVIDAL, *Bronceilles.*—Wine.

(*Puy-de-Dôme.*)

422 MARQUIS DE LA SALLE, *St. Germain-Lembrou.*—Wines.
423 AUBERGIER, *Clermont.*—Indigenous opium, &c.
424 CHESNEAU, *Clermont.*—Vinegar.
425 DELMAS, *Besse-en-Chandèse.*—Liqueurs.
426 DUMAS-GIRAUD, *Courpiere.*—Artificial guano.

(*Seine-and-Oise.*)

427 IMPERIAL AGRICULTURAL SCHOOL AND AGRONOMIC SOCIETY OF GRIGNON.—Collection of cereals, honey, soils, &c.

(*Sarthe.*)

428 SOCIETY OF AGRICULTURE, SCIENCES, AND ARTS, OF THE DEPARTMENT OF SARTHE.—Cereals, and other agricultural products.

(*Ille-and-Vilaine.*)

429 DEPARTMENTAL AGRICULTURAL SOCIETY, AND GENERAL COMMITTEE OF THE ASSOCIATIONS OF THE DEPARTMENT OF ILLE AND VILAINE.—Flax, hemp, linen.
430 RITTER, *Fougères.*—Kirschwasser.

(*Côtes-du-Nord.*)

431 COLLECTIVE EXHIBITION OF THE DEPARTMENT OF THE CÔTES-DU-NORD.—Wheat, oats, flax hemp, &c. (10 Exhibitors).
432 LECOQ, *Dinan.*—Flax seed.

(*Finisterre.*)

433 COMMUNE OF ROSNOEN.—Cereals, &c.
434 BRIOT DE LA MALLERIE, *Kerlogotu.*—Corn, buck-wheat, wines.
435 COMTE DU COUEDIC, DIRECTOR OF THE IRRIGATION SCHOOL OF LÉZARDEAU.—Plans in relief, &c.
436 HERTEL, *Kerbourg.*—Agricultural products.
437 COLLECTIVE EXHIBITION OF THREE DISTRICTS.—Sugars (8 Exhibitors).
438 COLLECTIVE EXHIBITION OF THE WHEAT DISTRICT.—Alcohol (6 Exhibitors).
439 ——— Fecula, and starch (6 Exhibitors).
440 CHIRADE, P. P. *Paris.*—Eggs, butter.
441 BOURDOIS & SON, *Paris.*—Cheese.
442 COLLECTIVE EXHIBITION OF THE WHEAT DISTRICT.—Cheese (7 Exhibitors).
443 COLLECTIVE EXHIBITION.—Rape cakes, &c. (4 Exhibitors).
444 IMPERIAL AND CENTRAL HORTICULTURAL SOCIETY OF PARIS.—Fruits grown in the wheat region, and round Paris.

(*Aisne.*)

445 VICOMTE DE COURVAL, *Pinon.*—Specimens of wood cut by the new and old methods.
446 ROBERT, DR. E. *Bellevue* (*Seine-and-Oise*).—Treatment and cure of diseased elms illustrated, &c.

447 COLLECTIVE EXHIBITION OF THE WHEAT DISTRICT.—Wools (3 Exhibitors).
448 ——— Flax (2 Exhibitors).
449 BOURSIER-DELAPLACE, *Chevrières* (*Oise*).—Hemp.
450 LÉONI & COBLENZ, *Vaugenlieu* (*Oise*).—Hemp, mechanically prepared by a new process.
451 DEMOLON & COCHERY, *Paris.*—Phosphate of lime.
452 DURIVAU, *St. Jean-de-la-Motte* (*Sarthe*).—Wines.
453 PERS, A. *Paris.*—Artificial manures.
454 CAILLEAUX *Melun* (*Seine-and-Oise*).—Plans for drainage.
455 RICHARD DE JUVENCE, *Versailles* (*Seine-and-Oise*).—Plans, agricultural statistics, &c.
456 HEUZE, G. *Grignon* (*Seine-and-Oise*).—Plain and coloured engravings of cereals, &c.
457 ABOILARD, C. *Paris.*—Proposed methods of draining.
458 DUVILLERS, *Paris.*—Plans of parks and gardens.
459 BRUNIER, *Rouen* (*Seine-Inf.*).—Plan of a distillery.

(*Ardennes.*)

481 CHANAL, *Mézières.*—Hops.
482 GOSSIN, C. *Latour-Audry.*—Osiers.

(*Meuse.*)

483 JURY OF THE MEUSE.—Wines.
484 MAUPAS & SCHLAÏSSE, *Bar-le-Buc.*—Fossil phosphate of lime.
485 MÉRION, *Bar-le-Buc.*—Sparkling wine.

(*Moselle.*)

486 MANGIN, C. E. *Metz.*—Fecula.
487 ST.-JACQUES, *Metz.*—Starch.
488 CHAMPIGNEUL, *Metz.*—Soft corn.
489 HIRT, J. *Sarreguemines.*—Wines.
490 MACHETAY, JUN. *Metz.*—Wines.

(*Meurthe.*)

491 BLOCH & SON, *Tomblaine.*—Tapioca fecula, &c.
492 DERMIER, *Nancy.*—Prepared plants.
493 VOIRIN, JUN. *Nancy.*—Liqueurs.

(*Vosges.*)

494 COLLECTIVE EXHIBITION, BY THE AGRICULTURAL ASSOCIATION OF THE ARRONDISSEMENT OF EPINAL.—Cereals, oils, honey, tiles (10 Exhibitors).
495 AGRICULTURAL ASSOCIATION OF RAMBERVILLERS.—Hops, farina, groats, &c. (20 Exhibitors).
496 COLLECTIVE EXHIBITION BY THE ARRONDISSEMENT OF REMIREMONT.—Kirschwasser and gentian brandy.
497 COLLECTIVE EXHIBITION BY THE DEPARTMENT OF VOSGES.—Fecula (15 Exhibitors).
498 CUNY, GERARD, *St.-Dié.*—Corn.
499 FLEUZOT & THIERRY, *Val d'Ajol.*—Kirschwasser.
500 LEMASSON, *Val d'Ajol.*—Kirschwasser.
501 PARIS, *Remiremont.*—Kirschwasser.

(*Bas-Rhin.*)

502 AGRICULTURAL COLONY OF OSTWALD.—Agricultural products; tobacco.

503 SCHATTENMANN, *Rouxvillers.*—Geological collections, cereals, &c.

504 VOCLKER, *Strasbourg.*—Grains, &c.

505 ANDÉOUD, *Avolsheim.*—Wines.

506 DARTEIN, *Oltrott.*—Wine.

507 PASQUAY BROS. *Wasselonne.*—Wine.

508 PROST, *Strasbourg.*—Wine.

509 REISSER, *Oltrott.*—Kirchwasser.

510 REYSZ, *Traenheim.*—Wine.

511 SPIELMANN, *Werthoffen.*—Wine.

512 STOLZ, SEN. *Andlau.*—Wines and Kirschwasser.

513 ZEYSSOLFF, *Strasbourg.*—Wines.

514 ZIMMER, *Wangen.*—Wines.

515 COLLECTIVE EXHIBITION BY THE DEPARTMENT OF THE HAUT-RHIN, INCLUDING THE ARRONDISSEMENT OF COLMAR.—Wines, brandy, and liqueurs (18 Exhibitors).

(*Marne.*)

516 AGRICULTURAL SOCIETY, AND CENTRAL ASSOCIATION OF THE DEPARTMENT OF MARNE.—Agricultural products.

517 ROQUEPLAN, N. *Reims.*—Sparkling wine of Champagne.

518 RICBOUR MEUNIER, *Avenay.*—Semoule, farina of groats.

519 CHEMERY, *Noirmont.*—Agricultural products.

(*Haute-Marne.*)

521 PASSY, A. & CO. *Arc-en-Barrois.*—Wrought indigenous wood.

522 DELETTRE-COURTOIS, *Arc-en-Barrois.*—Preserved truffles and other eatables.

(*Haute-Saône.*)

523 LOCAL COMMITTEE OF GREY.—Wines.

524 JURY OF LURE.—Wines.

525 MARQUIS D'ANDELARRE, *Lure.*—Plans in relief.

526 LUZET, *Luxeuil.*—Wines.

(*Jura.*)

527 SOCIETY OF AGRICULTURE, SCIENCE, AND ARTS, OF POLIGNY.—Wines and brandy (6 Exhibitors).

528 BURY, *Lons-le-Saunier.*—Wines.

529 GAUDARD, *Courbouzon.*—Wines.

530 GENOT, BROS. *Lons-le-Saunier.*—Wines.

531 MANGIN & GIROD, *Lons-le-Saunier.*—Wines.

532 MONARD, *Lons-le-Saunier.*—Wines.

533 MOREAU, *Quintigny.*—Wines.

534 RENAUD, *Lons-le-Saunier.*—Wines.

(*Aube.*)

535 BEAU, SEN. *Riceys.*—Wine.

536 GRATTEPAIN, *Loches-sur-Ource.*—Wines.

(*Cote-d'Or.*)

538 STRONG WINES CO. OF BURGUNDY.—Wines.

539 BOUTON, E. *Montigny-sur-Aube.*—Preserved truffles, wine.

540 COUQUAUX-JOLY & CO. *Dijon.*—Liqueurs.

541 DEVILLEBICHOT, WIDOW J. *Dijon.*—Liqueurs.

542 MARQUIS DE LAGARDE.—Wines.

543 HUAN & FONTAGNY. *Dijon.*—Vinegar.

544 SAGLIER, *Dijon.*—Truffles.

545 VIEILHOMME, H. *Paris.*—Wines.

546 CHOLET-LHUILLIER, *Fixin.*—Wines.

547 CRÉTIN-CHOLET, *Fixin.*—Wines.

548 GRAY, M. *Dijon.*—Mustard.

(*Yonne.*)

549 AGRICULTURAL SOCIETY OF JOIGNY.—Corn and various products.

550 BARDEAU, E. *Fleury.*—Corn and oats in the sheaf.

551 ROY, *Tonnerre.*—Alcohol.

552 BONNEVILLE, A. A. *Villeneuve-sur-Yonne.*—Confection of grapes, wine.

553 LE PÈRE, C. *Auxerre.*—Wines.

(*Saône-and-Loire.*)

554 COLLECTIVE EXHIBITION OF THE MÂCONNAIS.—Wines (44 Exhibitors).

555 THE COMMUNE OF ROMANÉCHE.—Wines.

556 DESMARQUEST & CO. *Macon.*—Wines.

557 ANDELLE, G. *Epinac.*—Wines.

558 COMTE DE BÉTHUNE, *Macon.*—Wines.

559 COLLECTIVE EXHIBITION OF MÂCON.—Wines (8 Exhibitors).

560 DE MURARD, *Macon.*—Wines.

561 RUFFARD, *Mâcon.*—Vinegar.

562 BEAUPÈRE & CO. *Chalons-sur-Saône.*—Beet-root sugar.

(*Rhône.*)

563 ASSOCIATION OF BEAUJEN.—Wines (107 Exhibitors).

564 BLAIN, *Lyon.*—Wine.

565 TREVOUX, E. *Lyon.*—Manures.

(*Loire.*)

566 AGRICULTURAL ASSOCIATION OF PERREUX.—Wines.

567 MARQUIS DE VOUGY, *Roanne.*—Wines.

(*Haute-Loire.*)

568 COLLECTIVE EXHIBITION OF THE SOCIETY OF AGRICULTURE, SCIENCE, ART, AND COMMERCE, OF THE PUY.—Cereals, leguminous plants, forage, &c. (3 Exhibitors).

(*Ain.*)

569 GL. BAR. GIROD DE L'AIN, *Gex.*—Merino wool.

570 JACQUIN, *Seyssel.*—Wine.

(*Savoie.*)

571 ROUX-VOLLON, *St.-Jean-de-Belleville.*—Gruyère cheese.

572 TATOUT, J. *St. Bon.*—Gruyère cheese.

573 CHRISTIN, *St.-Pierre-de-Belleville.*—Wine.

(*Eure-and-Loir.*)

574 AGRICULTURAL AND HORTICULTURAL SOCIETY OF THE DEPARTMENT OF EURE-AND-LOIR.—Results of forest culture by a new method; wool.

575 RICOUR, *Chartres.*—Cereals in the sheaf and in the ear.

(*Loiret.*)

576 COLLECTIVE EXHIBITION BY THE DEPARTMENT OF LOIRET.—Wines (17 Exhibitors).

577 ——— Vinegar (6 Exhibitors).

578 ——— Honey and wax (5 Exhibitors).

579 ——— Saffron (6 Exhibitors).

580 DE BÉHAGUE, *Dampierre.*—Fecula.

581 ANSELMIER, DIRECTOR OF THE FARM SCHOOL OF MAUBERNEAUME.—Cereals and roots.

582 DAVID, *Orléans.*—Bitters.

583 HOARAU, *Orleans.*—Plums.

(*Loire-Inf.*)

584 IMPERIAL AGRICULTURAL SCHOOL OF GRAND-JOUAN.—Cereals, plants for forage, &c.

585 LIAZARD, A. *Treguel.*—Collection of cereals, oleaginous plants, forage, &c.

586 JOUBERT, *St. Herblon.*—Wines.

587 LEROUX & Co. *Nantes.*—Manure.

588 DERRIEN, E. *Chantenay-Nantes.*—Manure.

(*Maine-and-Loire.*)

589 INDUSTRIAL AND AGRICULTURAL SOCIETY OF ANGERS.—Wines.

590 HENNEQUIN, D. *Angers.*—Grains for soups and forage.

591 COMBIER-DESTRE, *Saumur.*—Brandy.

592 BOURDON & JAGOT, C. *Saumur.*—Wine.

593 BOLOGNÈSI, *Saumur.*—Elixir "Raspail."

(*Loir-and-Cher.*)

594 EXHIBITION BY THE DEPARTMENT OF LOIR-AND-CHER.—Wines, vinegars, and alcohols (11 Exhibitors).

595 SOYER, *Nouan.*—Wooden poles.

596 DESVAUX-SAVOURÉ, *Beauchéne.*—Cyder.

597 BRETHEAU-AUBRY, *Meusnes.*—Flints.

(*Cher.*)

598 AGRICULTURAL SOCIETY OF THE DEPARTMENT OF THE CHER.—Wines (7 Exhibitors).

599 AGRICULTURAL ASSOCIATION OF AUBIGNY.—Collection of grains; prunes.

600 LALOUEL DE SOURDEVAL, *Laverdines*—Soils, cereals, sugar, alcohol.

(*Indre.*)

601 COUSIN-MONOURY, *Issoudun.*—Vinegar.

602 GODEFROY, M. *Reuilly.*—Red wine.

603 AGRICULTURAL SOCIETY OF CHÂTEAUROUX.—Specimens of soils, and their products, &c. (13 Exhibitors).

(*Inare-and-Loire.*)

604 AGRICULTURAL ASSOCIATION OF THE ARRONDISSEMENT OF CHINON.—Agricultural products, wines, liqueurs &c. (17 Exhibitors).

605 ASSOCIATION OF THE PROPRIETORS VOUVRAY.—Wines.

606 HÉBERT, A. *Athée, near Tours.*—Starch, fecula, flour.

607 DELABROUSSE, *Civray-sur-Cher.*—Wine from the slopes of the Cher.

608 HARDY, *Joué-lès-Tours.*—Wine.

609 PETIT DE VAUZELLÈS, WIDOW.—Wines.

610 ROUILLÉ-COURBÉ, *Tours.*—Red and white wine.

611 VAUGONDY, *Rochecorbon.*—Wine.

612 DESBORDES & VOISIN, *Chinon.*—Vinegar.

(*Deux-Sèvres.*)

613 HORTICULTURAL SOCIETY OF NIORT.—Plants of fruit and forest trees (9 Exhibitors).

614 APERCÉ, *Giffont.*—Corn, maize, barley, nuts, trefoil, &c.; a fleece.

615 DE MESCHINET.—Specimens of soils, wheat, oats, barley.

616 MICHCAUD, *La Charrière.*—Trefoil seed.

617 PINARD, *St. Etienne.*—Wines and brandy.

618 DAVID, *Niort.*—Brandy.

619 FONTAINE, *Greffier.*—Brandy.

620 DESCOLLARD, *Epannes.*—Ray-grass.

621 PRIEUR, *Epannes.*—Hemp.

622 GRIFFIER-VERRASSON, *Niort.*—Osiers.

(*Vienne.*)

623 AGRICULTURAL ASSOCIATION OF THE ARRONDISSEMENT OF CHÂTELLERAULT.—Wheat, plants for forage, &c. (26 Exhibitors).

624 ASSOCIATION OF CIVRAY.—Collection of grain and sheaves, &c.

625 DE LARCLAUSE, DIRECTOR OF THE FARM SCHOOL OF MONTS.—Agricultural products.

(*Haute Vienne.*)

626 BRUCHARD, DIRECTOR OF THE FARM SCHOOL OF CAVAIGNAC.—Geological specimens.

(*Dordogne.*)

627 DE LENTILLAC, DIRECTOR OF THE FARM SCHOOL OF LAVALLADE.—Collection of cereals, leguminous plants, tobacco, silk-worms' eggs, silks, cocoons, &c.

628 LASALVÉTAT, H. *Périgueux.*—Alimentary preserves.

629 HOARAU, DE LA SOURCE, *Chateau de Ponthet.*—Prunes.

630 GOURSALLE, *Périgueux.*—Yellow wax.

631 BOURSON, E. *Farcies.*—Leaf tobacco prunes, and wine.

632 BLANC, *Perigueux.*—Alimentary preserves.

633 ALLARD, SON, & Co.—Nut oil, nut-bread.

634 COLLECTIVE EXHIBITION OF THE DEPARTMENT OF THE DORDOGNE.—Wines and liqueurs (34 Exhibitors).

(*Gironde.*)

635 AGRICULTURAL SOCIETY OF THE GIRONDE.—Cones, seeds, &c. of the maritime pine; resin, tar, oils, hops, &c. (10 Exhibitors).

636 CLAMARGERAN, *La Lambertie, near St. Foy.*—Agricultural products.

637 CONSTANTIN, *Bordeaux.*—Rich wines of Bordeaux.
638 ROUSSE, J. *Bordeaux.*—Alcohol, &c., obtained by a new method of distillation.

(*Lozere.*)

639 SOCIETY OF AGRICULTURE, INDUSTRY, SCIENCE, AND ARTS OF THE DEPARTMENT OF THE LOZÈRE.—Agricultural products.

(*Vendée.*)

640 JURY OF THE VENDÉE.—Agricultural products.

(*Charante-Inf.*)

641 CHAMBER OF COMMERCE OF ROCHEFORT.—Wood, wheat, and other agricultural products; sand, sulphuret of iron, refractory clays, &c.
642 BOUSCASSE, DIRECTOR OF THE FARM SCHOOL OF PUILBOREAU.—Beet-root seed, brandy, wine.
643 GUILLON-DESAMIS, *La Côte, near Nieul-sur-Mer.*—Oysters.
644 DR. KEMMÈRES, *La Côte de Rivadoux, Ile de Ré.*—Oysters, &c.
645 LEM, WIDOW, *St. Martin, Ile de Ré.*—Honey.
646 COLLECTIVE EXHIBITION BY THE DEPARTMENT OF THE CHARENTE INF.—Wines and brandy (10 Exhibitors).
647 DR. A. MENUDIER, *Pleaud-Chermignac.*—Wine.
648 CONTE & CO. *St. Pierre d'Oleron.*—White wine vinegar.
649 OLIVIER, *La Flotte, Ile de Ré.*—Vinegar.
650 ROBINEAU, P. & CO. *La Tremblade.*—Strong and clarified vinegar.

(*Charente.*)

651 COLLECTIVE EXHIBITION BY THE DEPARTMENT OF THE CHARENTE.—Brandy, and alcohol (9 Exhibitors).
652 E. THIAC, *Puyréaux.*—Specimens of the soil, potatoes, beet-root, plan of the farm, wine, &c.
653 BRUMAULD DES ALLÉES, *St. Claud.*—Cement, and hydraulic lime.
654 GALLAND, *Ruffec.*—Corn for poultry.
660 COLLECTIVE EXHIBITION.—Agricultural implements and produce (37 Exhibitors).
661 COLLECTIVE EXHIBITION.—The wines of Champagne (19 Exhibitors).
662 COLLECTIVE EXHIBITION.—The wools of the wine district (12 Exhibitors).
663 COLLECTIVE EXHIBITION.—The wines of Burgundy (248 Exhibitors).
664 GENERAL ADMINISTRATION OF FRENCH TOBACCOS.—Indigenous tobacco, &c.
665 COLLECTIVE EXHIBITION.—Indigenous tobacco (2 Exhibitors).
666 VOELCKER, *Benfeld* (*Bas-Rhin*).—Products obtained from chicory-root.
667 SENGENWALD, *Strasbourg* (*Bas-Rhin*).—Madder and its products.
668 COLLECTIVE EXHIBITION.—Hops of the wine district (4 Exhibitors).
669 COLLECTIVE EXHIBITION.—The wines of Bordeaux (289 Exhibitors).

670 NORMAL SCHOOL OF THE DEPARTMENT OF THE HAUT-RHIN.—Cereals and farinaceous grain (5 Exhibitors).
671 TOLLARD, P. *Paris.*—Collection of grains, and forage.
672 TAMISET, C. *Plombières-lez-Dijon* (*Côte-d'Or*).—Wheat, and bean flour.
673 PERTHUY-MARTINEAU, *Nantes* (*Loire-Inf.*).—Wines.
674 ROUCHIER, SEN. *Ruffec* (*Charente*).—Preserves, liqueurs, and biscuits.

(*Lot.*)

690 BOUTAREL-MEMBRY, *Luzech.*—Wine.
691 CAPMAS, *Prayssac.*—Wine.
692 IZARN, C. *Cahors.*—Wine.
693 LABICHE, C. *Cahors.*—Wine.
694 VIEULS, JUN. *Gaillac.*—Wine.
695 CABANÈS & MALGOUISARD, *Gourdon.*—Liqueurs, and nut oil.

(*Lot-and-Garonne.*)

696 DÉFFEZ, C. G. A. *Nerac.*—Corn, maize, wine, brandy.
697 DUCOS-BERNARD, *Beauziac.*—Ears of corn.
698 NADAU, *St. Livrade.*—Prunes.
699 CUZOL, SON, & CO. *Castelmoron-sur-Lot.*—Prunes.
700 TRUANT, E. *Domaine de Pader.*—Red and white wine.
701 MARGUES & DUVIGNEAU, *Nérac.*—Brandy.
702 DÉCHOO, LAROZE, & CO. *Mézin.*—Brandy.
703 SIGAUD, A. *Nérac.*—Liqueurs and fruits.

(*Tarn-and-Garonne.*)

704 HORTICULTURAL AND ACCLIMATIZATION SOCIETY OF MONTAUBAN.—Cocoons and silk.
705 AGRICULTURAL ASSOCIATION OF MONTAUBAN.—Corn, millet, maize, &c. (7 Exhibitors).
706 SOCIETY OF SCIENCE AND AGRICULTURE OF MONTAUBAN.—Cocoons, and silk.
707 COUDERC, & SOUCARET, JUN. *Montauban.*—Raw silk.
708 GASCOU, NEPH. & ALBRESPY, *Montauban.*—Cocoons, and silk.
709 AGRICULTURAL ASSOCIATION OF NÈGREPELISSE.—Montricoux marble.
710 AGRICULTURAL ASSOCIATION OF MONCLAR.—Corn, maize, large chestnuts, raw hemp.
711 SOL, *Verdun.*—Ears of corn and maize.
712 COLLECTIVE EXHIBITION OF MONTAUBAN.—Wine and brandy (13 Exhibitors).

(*Tarn.*)

713 ASSOCIATION OF THE WINE GROWERS OF GAILLAC (5 Exhibitors).—Wine.
714 THE MAYOR OF GAILLAC.—Wine.
715 THE MAYOR OF GRAULHET.—Trefoil seed.
716 RAYNAL & SON, *Gaillac.*—Trefoil seed, aniseed, prunes.
717 MARAVAL & CO. *Lavaur.*—Raw silk.

(*Landes.*)

718 COLLECTIVE EXHIBITION OF ST. SEVER.—Red and white wine (2 Exhibitors).

719 COLLECTIVE EXHIBITION OF PARLEBOSCQ, AND THE SURROUNDING COMMUNES.—Brandy (16 Exhibitors).

720 DUPRAT, *Hontaux.*—Brandy and wine.

721 LABADIE, P. *Arthez.*—Brandy.

722 DUPUY, *Mont-de-Marsan.*—Oil.

723 DIVES, H. *Mont-de-Marsan.*—Resinous products.

724 COLLECTIVE EXHIBITION OF ARMAGNAC.—Brandies (22 Exhibitors).

725 DARQUIER, *Lectoure.*—Red wine.

726 LAFFITTE, J. *Castres.*—White vinegar.

(*Haute-Garonne.*)

727 FORT DESPAX & BACOT, *Toulouse.*—Liqueurs and wine.

728 DELORME & Co. *Toulouse.*—Vegetable horse-hair, the produce of the dwarf palm.

(*Basses-Pyrénées.*)

729 PÉCAUT, *Salies.*—Wine, refined salt.

(*Hautes-Pyrénées.*)

733 FONTAN, *Bernadets-Debat.*—Wine.

734 NABONNE, *Madiran.*—Wine.

735 MANINAT, JUN. *Ossun.* — Mineral waters.

(*Pyrénées-Orientales.*)

736 BONET-DESMARRES, *St. Laurent-de-la-Salanque.*—Wines.

737 SALLENS, P.—Wine, manure.

(*Ardèche.*)

738 MALLET-FAURE & SON, *St. Péray.*—Wines.

739 ROY, *Privas.*—Model of a farm wagon.

740 RICHARD, H. *Tournon-sur-Rhône.*—Wine.

741 PRADIER, J. *Annonay.*—Agricultural products, raw and prepared silk.

742 CHANGEA, *Lamastre.*—Raw and prepared silk.

743 NICOD & SON, *Annonay.*—Silkworms' eggs, and cocoons.

744 BUISSON, C. *la Tronche.*—Raw and prepared silk.

(*Isère.*)

745 AGRICULTURAL AND HORTICULTURAL SOCIETY OF THE ARRONDISSEMENT OF GRENOBLE.—Corn, hemp, nuts, oils, manure, &c. (12 Exhibitors).

746 ARNAND, SEN. & Co. *Voiron.* — Liqueurs.

747 HEURARD D'ARMIEU, *Armieu-St.-Gervais.*—Nuts, and nut-oil.

748 PERBOST, SEN. *Cigalieres.*—Raw silk.

(*Drôme.*)

749 COMBIER, BROS. *Livron.*—Raw silk.

750 GAUTHIER, A. *Chabeuil.*—Raw and prepared silk.

751 HELME, A. *Loriol.*—Raw and prepared silk.

752 LACROIX, P.—Raw and prepared silk.

753 LASCOUR, *Crest.*—Raw and prepared silk.

754 LEYDIER, BROS. *Buis-lès-Barronies.*—Raw and prepared silk.

755 NOYER, BROS. *Dieulefit.* — Raw and prepared silk.

756 SAUVAGEON, *Valence.*—Cocoons, obtained under the influence of electricity.

757 COLLECTIVE EXHIBITION (26 Exhibitors).—Hermitage wines.

758 COLLECTIVE EXHIBITION BY THE DEPARTMENT OF THE DRÔME. — Wines (12 Exhibitors).

759 CHARRAS & SON, *Nyons.*—Liqueurs.

760 CHEVALLIER-ROBERT, & CUILLERIER, *Romans.*—Cherry liqueurs and ratafias.

748 ARLÈS DUFOUR.—Raw and prepared silk.

761 MARKERT, G. *Tain.* — Wine, crême de l'Hermitage.

762 BLANC-MONTBRUN, *Chateau dè la Rolière.*—White wine, raw silk.

763 THE CURÉ OF CHARVAT, PRESIDENT OF THE ASSOCIATION OF RÉAUVILLE.—Cereals, almonds, madder.

764 AGRICULTURAL ASSOCIATION OF RÉAUVILLE.—Cereals, wines.

765 BRUN, JUN. *Réauville.*—Maize, oats, French beans, madder.

766 BOUTAREL-MAUBRY, *Valence.* — Nuts, prunes, wines.

767 GIRARD.—Yellow wax.

768 GUERBY, V.—Agricultural products, Crozes wine.

769 DELHOMME, *Larnage.*—Wine, kaolin.

770 MARRON-STOUPANI, *Montelimart.* — Nugget, &c. of Provence.

771 PREMIER, & SON, *Romans.* — Preserved fruits, liqueurs.

772 ROBEUX, *Valence.*—Liqueurs.

773 GALOPIN.—Alimentary preserves.

774 PERRIER, J. *Crest.*—Alimentary preserves.

775 CHARBONNET & SON, *Montelimart.* —Preserved black truffles.

(*Gard.*)

776 COLLECTIVE EXHIBITION BY THE DEPARTMENT OF THE GARD. — Wines and liqueurs (15 Exhibitors).

777 LACOMBE, I. *Alais.*—Raw silk, &c.

778 VERNET, BROS. *Beaucaire.*—Raw and prepared silk.

779 DE FOURNÈS, *Remoulins.* — Long-stapled upland cotton.

780 CARENON,-BONIFAS, & Co. *Moussac.* —Liquorice juice.

781 DAVID, P.—Liquorice wood and juice.

(*Herault.*)

783 COLLECTIVE EXHIBITION BY THE DEPARTMENT OF HERAULT.—Wines and brandy (11 Exhibitors).

784 JURY OF THE ARRONDISSEMENT OF MONTPELLIER.—Wine, oil, wool.

785 NOURRIGAT, *Lunel.*—Raw silk, vegetable and animal substances, &c.

786 BOYER & HEIL, *Gignac.* — Preserved truffles, olives, essences.

(*Pyrénées-Orient.*)

787 JURY OF PRADES.—Honey.

(*Aude.*)

788 COLLECTIVE EXHIBITION BY THE DEPARTMENT OF AUDE.—Wines (5 Exhibitors).

789 DELCASSE, G. *Limoux.*—Fleeces, yarn, and wines.

790 DÉBOSQUE, *Esperaza.* — Ferruginous water.

791 DENILLE, DIRECTOR OF THE FARM SCHOOL OF BESPLAS.—Teasels, corn, maize, forage.

792 DE MARTIN, J. *Narbonne.*—Sea salt, wine.

(*Vaucluse.*)

794 CHABAUD, A. *Avignon.*—Raw and prepared silk.

795 BERTON BROS. *Avignon.*—Wines of different growths.

796 COMTE DE MALEYSSIE, *Chateau-neuf.* —Wine.

797 SAUTET, A. *Sorgues.* — Alcohol and sulphuric ether.

798 FAURE, P. *Avignon.*—Madder, and wine.

799 JULIAN, JUN. & HOQUER, *Sorgues.*—Alizarine, madder, and its products.

800 LEPLAY, H. & Co. *Avignon.*—Alcohol, derived from various sources.

801 REYNAUD, *Pertuis.*—Alimentary preserves.

802 BONNET, *Aps.*—Bark of the green oak.

(*Boüches du Rhône.*)

803 AGRICULTURAL SOCIETY OF THE BOUCHES-DU-RHÔNE.—Wheat, forage, madder, &c. ((12 Exhibitors).

804 AGRICULTURAL ASSOCIATION OF THE ARRONDISSEMENT OF AIX.—Wheat, beans, resinous products, tobacco, teasels, madder, &c. (3 Exhibitors).

805 AGRICULTURAL ASSOCIATION OF AUBAGNE.—Corn, maize, farina, oils.

806 BRUNET, *Marseilles.*—Wheat, flour, Semoule.

807 DE BEC, P. DIRECTOR OF THE FARM-SCHOOL OF MONTAURONNE—Collection of almonds.

808 AUBERT, F. *Aix.*—Wheat, farina, &c.

809 POMIROL, *Marseilles.*—Wines.

810 MONIER, *Aubagne.*—Wines.

811 REINAUD, CHAPPAZ, & Co. *Marseilles.* —Liqueurs.

812 JOURDAN, G. & BRIVE, SEN.—Wines, liqueurs, preserved fruits, &c.

813 OLIVE, NEPHEW, & MICHEL, *Marseilles.*—Liquorice juice.

(*Var.*)

814 AGRICULTURAL AND COMMERCIAL SOCIETY OF THE DEPARTMENT OF THE VAR. —Cereals, woods, cork, tobacco, wine.

815 AGRICULTURAL ASSOCIATION OF THE ARRONDISSEMENT OF TOULON.—Wines.

816 CORNEILLE & FABRE, *Trans.* — Cocoons, raw silk, &c.

(*Corsica.*)

817 JURY OF AJACCIO. — Specimens of rocks and minerals; animal and vegetable products (14 Exhibitors).

818 JURY OF CALVI.—Wine, oil, tobacco.

819 COLLECTIVE EXHIBITION OF CORSICA. —Corsican wine (10 Exhibitors).

820 BATTIONI, *Bastia.*—Liqueurs.

821 GASPARINI, *I. L'ille-Rousse.* — Alimentary pastes.

822 LINGÉNIEUR.—Cedrats, preserves.

823 GARINI & MARIOTTI, *Campele.*—Dried fruits, chestnuts, &c.

824 CAFFARELLI, J. *Bastia.* — Italian pastes.

825 BREGANTI, J.—Cigars.

826 LICCIA, *Monticello.*—Leaf tobacco.

827 JURY OF BASTIA. — Marbles and minerals.

(*Hautes-Alpes.*)

828 COLLECTIVE EXHIBITION BY THE DEPARTMENT OF THE UPPER ALPS.—Madder, teasels, honey, wines, &c. (10 Exhibitors).

(*Basses-Alpes.*)

829 RAYBAUD-L-ANGE, DIRECTOR OF THE FARM SCHOOL OF PAILLEROLS.—Collection of grains specially cultivated in the silk region; flour, teasels, olives, honey, &c.

EXHIBITION OF SILK, &c., FROM VARIOUS LOCALITIES.

830 DE BAILLET, *St. Germain-et-Mons (Dordogne).*—Silk.

831 BÉRARD & BRUNET, *Lyon (Rhône).*—Silks, raw and prepared.

832 CHABOD, JUN. *Lyon (Rhône).* — Cocoons.

833 COUNTESS C. DE CORNEILLAN, *Paris.* —Cocoons, raw silk, &c.

834 DUSEIGNEUR, P. *Lyon (Rhône).*—Cocoons.

835 FARA, JUN. *Bourg-Argental (Loire).*—Silks, for Caen lace.

836 FRIGARD, *Bourg-Argental (Loire).*—Raw silk, white and yellow.

837 GUERIN-MENEVILLE, DOCTOR F. *Paris.*—Silk-worms.

838 COLLECTIVE EXHIBITION.—Products of the olive tree (11 Exhibitors).

839 BLANCHON, L. *St. Julien-en-St. Alban (Ardèche).*—Yellow cocoons, raw and prepared silk.

840 BARRÈS, BROS. *St. Julien-en-St. Alban (Ardèche).*—Raw and prepared silk.

841 SÉRUSCLAT, L. *Etoile (Drôme).*—Prepared silk.

842 MONESTIER, SEN. *Avignon (Vaucluse).* —Raw and prepared silk.

843 BLANCHON, SEN. *Flaviac (Ardèche).*—Raw and prepared silk.

844 PALLUAT & Co. *Lyon (Rhône).* — Prepared silks.

845 BOISRAMEY, JUN. *Caen (Calvados).*—Raw and prepared silk, for laces.

847 TEISSIER DU CROS, *Valleraugue* (*Gard*).—Cocoons, raw and prepared silk.
848 CHAMBON, WIDOW, *St. Paul-Lacoste* (*Gard*).—Raw silk, and organzine.
849 BONNET & BOUNIOLS, *Vigan* (*Gard*).—Cocoons, raw silk.
850 BROUILHET & BAUMIER, *Vigan* (*Gard*).—Cocoons, white and yellow; raw and prepared silk.
851 MARTIN, L. & Co. *Lasalle* (*Gard*).—Raw silk.
852 BOUDET, F. *Uzès* (*Gard*).—Raw silk.
853 CHAMPANHET-SARGEAS BROS. *Vals* (*Ardèche*).—Cocoons, raw and prepared silk.
854 FOUGEIROL, A. *Ollières* (*Ardèche*).—Raw silk, organzine, cocoons.
855 REGARD, BRO. *Privas* (*Ardèche*).—Raw and prepared silk.
856 BISCARRAT, P. *Bouchet* (*Drôme*).—Raw and prepared silk.
857 CHARTRON & SON, *St. Vallier* (*Drôme*).—Silk.
858 FRANQUEBALME & SON, *Avignon* (*Vaucluse*).—Wrought Chinese and Japanese silk.
859 BANNETON, *St. Vallier* (*Drôme*).—Raw silk, organzine.
862 MAHISTRE, A. JUN. *Vigan* (*Gard*).—Raw silk.
863 COLLECTIVE EXHIBITION.—Rough and manufactured cork (8 Exhibitors).

SPECIAL EXHIBITIONS.

880 IMPERIAL SOCIETY OF ACCLIMATIZATION.—Results of its labours.
881 ROUYER, L. *Paris.*—Paintings, &c. of animals, and agricultural products.
882 VILMORIN-ANDRIEUX & Co. *Paris.*—Agricultural products.
883 CHAMBRELENT, *Bordeaux* (*Gironde*).—Forest products of the Landes.
884 JAVAL, *Arès* (*Gironde*).—Forest products of the Landes.
885 FLORENT-PRÉVOST, F. *Paris.*—Collection of preparations regarding the food of French birds, &c.
886 MUSEUM OF NATURAL HISTORY, *Paris.*—The principal types of mamifers and birds of France, both useful and mischievous.
888 ÉLOFFE & Co. *Paris.*—Geological collection: soils and subsoils of the three agricultural regions of France.

CLASS 4.

941 CUSINBERCHE JUN. *Paris.*—Stearic and oleic acids, wax candles, soap.
942 LEROY, C. & DURAND.—*Gentilly* (*Seine*).—Stearic and oleic acids, wax and other candles, soap.
943 DE MILLY, L. A. *Paris.*—Stearic and oleic acids, glycerine, wax candles, soda-soap.
944 PETIT, BROS. & Co. *Paris.*—Oleic acid, glycerine, and wax candles.
945 TREMEAU & MALEVAL, *Vienne* (*Isère*).—Stearic and oleic acids, wax candles, and soap.
946 GAILLARD, BROS. *Paris.*—Stearic and oleic acids, wax candles, wax, and tapers.
947 BUREAU, C. *Bordeaux* (*Gironde*).—Wax; wax, and other candles.
948 AUTRAN, L. *Paris.*—Candles.
949 AMENE, L. *Clermont-Ferrand* (*Puy-de-Dôme*).—Animal and olive oils, &c.
950 BLANCHARD, G. *Lyon* (*Rhône*).—Purified oils and soap.
951 FAULQUIER-CADET & Co. *Montpellier* (*Hérault*).—Stearic and oleic acids, tapers, wax and tallow candles.
952 LE TAROUILLY, A. & Co. *Rennes* (*Ille-and-Vilaine*).—Bleached wax tapers, &c.
953 GOHIN, SEN. *Vire* (*Calvados*).—Teazles.
954 ROBERT GALLAND & Co. *Paris.*—Bituminous schist and its products.
955 D'AMBLY, C. & Co. *Paris.*—Buffalo-horn, imitating whalebone.
956 D'ENFERT, BROS. *Paris.*—Gelatine and glue.
957 COGNIET C., MARÉCHAL, & Co. *Paris.*—Spermaceti, paraffine, and lubricating oils.
958 ROUSSEAU DE LAFARGE, L. & Co. *Persan-Beaumont* (*Seine-and-Oise*)—Vulcanised india-rubber.
959 AMBERT, A. & GÉRARD, *Paris.*—Articles in caoutchouc.
960 ARNAVON, H. *Marseille* (*Bouches-du-Rhône*).—White and marbled soaps.
961 DELATTRE & Co. *Dieppe* (*Seine-Inf.*).—Fish oil and manure.
962 MONTALAND, C. & Co. *Lyon* (*Rhône*).—Stearic and wax candles.
963 FOURNIER, F. *Marseille* (*Bouches-du-Rhône*).—Stearic candles and soap.
964 CAUSSEMILLE, JUN. & Co. *Marseilles* (*Bouches-du-Rhône*).—Lucifer matches of wood and wax.
965 MILLIAU, JUN. *Marseille* (*Bouches-du-Rhône*).—White soap.
966 ROUX, C. JUN, *Marseille* (*Bouches-du-Rhône*).—Marbled soap.
967 ROCCA, BROS. & NEPHEWS, *Marseille* (*Bouches-du-Rhône*).—Marbled soap, oil from seed, oil-cake.
968 GRESLAND. C. *Paris.*—Candle wicks.
969 ROULET, C. H. & CHAPONNIÈRE, *Marseille* (*Bouches-du-Rhône*).—Marbled soap, oil from seeds, oil-cake.
970 GOUNELLE, C. *Marseilles* (*Bouches-du-Rhône*).—Soap, and seed-oils.
971 SEMICHON, J. JUN. *Paris.*—Lamp black.
972 JACQUEMART & Co. *Paris.*—Varnishes.
973 NOIROT & Co. *Paris.*—India-rubber tubing without joint, and apparatus for making it.
974 VANSTEENKISTE, WIDOW (DORUS), *Valenciennes* (*Nord*).—Starch.
975 SOCHNÉE, BROS. *Paris.*—Varnishes.
976 MICHAUD, C. H. *Paris.*—Seed-oils, and purified animal oils.
977 MORISOT, C. T. *Vincennes* (*Seine*).—Lubricating oils.
978 SAUVAGE & Co. *Paris.*—Illuminating and lubricating oils, tar, paraffine, and lamp-black.

979 CAHOUET & MORANE, *Paris*.—Candle moulds and candles.

980 TESTON, J. *Nyons, (Drôme)*.—Oil for clocks and telegraph apparatus.

981 RIOT, L. M. T. *Paris*.—Soap made without heat.

982 GONTARD, A. & Co. *Paris*.—Marbled soap.

983 STEINBACH, J. J. *Petit-Quevilly, near Rouen (Seine Inf.)*.—Starch and gums.

984 PLICHART & CUVELIER, *Valenciennes (Nord)*.—Bone, and animal blacks.

985 LARTIGUE, J. *Bayonne (Basses-Pyrénées)*.—Animal black and manure.

986 BRIEZ, F. JUN. *Arras (Pas-de-Calais)*.—Fabrics in horse-hair.

987 FERMIER DE LA, PROVOTAIS & GAUMONT, *Paris*.—Pulp, fibrous substances, and paper, from the common broom.

988 CABANIS, F. *Paris*.—Sprigs and bark of the mulberry, for making paper-pulp, and for the manufacture of thread or woven fabrics.

989 DETHAN, A. *Paris*.—Lubricating oils.

990 FENAILLE & CHATILLON, *Paris*.—Fats, with resinous base for carriages; resin-oils.

991 PASQUIER, DE RIBAUCOURT, & Co. *Paris*.—Oils, and grease for lubrication, &c.

992 BARBIER & DAUBRÉE, *Clermont-Ferrand (Puy-de-Dôme)*.—Products resulting from the manufacture of caoutchouc.

993 DELACRETAZ, *Paris*.—Stearic and oleic acids, soap, and candles.

994 MULER, P. *Rouen (Seine-Inf.)*—Gelatine.

995 PLANCHON, S. *St. Hippolite (Gard)*.—Glue, gelatine.

996 SERBAT, L. *St. Saubie (Nord)*.—Putty for making the joints of steam-engines; oils, and grease for manufacturing purposes.

CLASS 5.

1011 CASTOR, A. *Paris*.—Collection of steam apparatus.

1012 ORLEANS RAILWAY CO. *Paris*.—Smoke-consuming locomotive; tender, and first-class carriage.

1013 SAGNIER, L. & Co. *Paris*.—Sextpule weigh-bridge for locomotives, and other apparatus for weighing.

1014 GENERAL RAILWAY PLANT CO. *Paris*.—Carriage, switches, &c.

1015 BARANOWSKI, J. J. *Paris*.—Automatic signals to prevent the collision of trains.

1016 LYONS RAILWAY CO. *La Croix Rousse, Paris*.—Self-acting break.

1017 JOINT-STOCK IRON-MASTERS' CO. OF MAUBEUGE (*Nord*).—Turn-table, cast-steel rails, hydraulic crane, &c.

1018 MAZILIER, *Paris*.—Designs and models of an iron road, without cast-iron or wood.

1019 EVRARD, *Douai (Nord)*.—Axles; specimen of a new method of lubrication.

1020 MATHIEU, *Anzin (Nord)*.—Miner's waggon.

1021 CABANY, A. *Anzin (Nord)*.—Miner's waggon of galvanized iron.

1022 CAIL, J. F. & Co. *Paris*.—Locomotive.

1023 ACHARD, A. *Paris*.—Application of electrical apparatus to securing a uniform water-level in steam-boilers.

1024 DE JOANNES, E. *Valenciennes (Nord)*.—Plan of an American sleeping carriage.

1025 MEYER & SON, J. J. & A. *Vienna (Austria)*.—Plans of an articulated locomotive, cut off apparatus, &c.

1026 DELANNOY, A. F. *Paris*.—Grease box, &c.

1027 COQUATRIX, J. B. *Paris*.—Self-acting lubricating apparatus.

1028 RASTOUIN, A. *Château-Renault (Indre-and-Loire)*.—Grease box, articulated waggon-axle, for turning in sharp curves, &c.

1029 FONTENAY, T. *Grenoble (Isère)*.—Plan of a smoke-consuming furnace for a locomotive.

1030 GUERIN, E. *Paris*.—Self-acting break.

1031 GARGAN & Co. *Paris*.—Cistern-waggon for transport of various liquids; feed apparatus, &c.

1032 NORTHERN RAILWAY CO. *Paris*.—Locomotive for steep inclines, &c.

1033 CATENOT-BERANGER & Co. *Lyon (Rhône)*.—Decuple weigh-bridge for heavy locomotives; and other apparatus for weighing.

1034 ARBEL, L., DEFLASSIEUX, BROS. and PEILLON, *Rive-de-Gier (Loire)*.—Wheels for locomotives.

1035 VERDIÉ, J. F. & Co. *Firminy (Loire)*.—Wheels, tires, springs, &c.

1036 VIGNIER, *Paris*.—A system of railway signals.

1037 ALEXANDRE-LESEIGNEUR & Co. *Paris*.—Tires and switches.

1038 DEYEUX, N. T. *Liancourt (Oise)*.—Cast-steel tires.

1039 DÉZELU & GUILLOT, *Paris*.—Railway carriage lamp.

1040 POMME DE MIRIMONDE, L. *Paris*.—Railway grease boxes.

Addenda.

POLONCEAU, WIDOW, *Paris*.—Model of locomotive with eight coupled wheels, for curves of small radius.

DIDIER, *Paris*.—Model of break.

CLASS 6.

1041 DESOUCHES-TOUCHARD & SON, *Paris*.—Chariot, coupé d'Orsay, model of a handle for a coach door.

1042 PARIS GENERAL OMNIBUS COMPANY.—An omnibus.

1043 BELVALLETTE BROS. *Paris*.—A landau, designs for carriages.

1044 PERRET, C. *Paris*.—A Victoria vis-à-vis, designs for carriages.

1046 BECQUET, J. F. *Paris*.—A calash.

1047 BENOIST, BROS. *Nogent-sur-Seine (Aube)*.—Axle-tree ends, which may be greased without taking off the wheels.

1049 POITRASSON, P. *Paris*.—A chariot.

1050 MOINGEARD, BROS. *Paris.*—High-springed Berline.

1051 MOUSSARD & Co. *Paris.* — Coupé d'Orsay.

1052 MUHLBACHER, BROS. *Paris.*—Four-wheel carriage.

1053 COLAS, DELONGUEIL, & COMMUNAY, *Courbevoie (Seine).*—Carriage wheels.

1054 DUFOUR, BROS. *Perigueux (Dordogne).* —Carriage.

1055 CLIQUENNOIS BROS. *Lille (Nord).*—Calash.

Addenda.

ALEXANDRE-LESEIGNEUR & Co. *Paris.*—Wheel tires of cast steel.

DEYEUX, N. T. *Leancourt (Oise).* Wheel tires of cast steel.

VERDIÉ, F. F. & Co. *Firminy (Loire).* —Tires formed of a combination of iron and cast steel; carriage-springs.

CLASS 7.

1061 ONFROY & Co. *Paris.*—Mechanism for printing woven fabrics.

1062 BRISSET, WIDOW, *Paris.* — Lithographic press.

1064 CALVET-ROGNIAT & Co. *Louviers (Eure).*—Ribbons and plates for cards.

1065 DECELLES, *Paris.*—Sewing machines, which fasten each stitch by a weaver's knot.

1066 CALLEBAUT, C. *Paris.*—Sewing machines.

1067 JOURNAUX-LEBLOND, J. F. *Paris.*—Sewing machines.

1068 TAILBOUIS, E. *St. Just-en-Chaussée (Oise).*—Machine for making net.

1069 BRUNEAUX, L. JUN. *Rethel (Ardennes).*—Power-loom for wool.

1070 BACOT, P. *Sedan (Ardennes).* — Power-loom for wool.

1071 TRIQUET, JUN. *Lyon (Rhône).* — Glazing machine.

1072 DELCAMBRE, *Lille (Nord).*—Composing machine.

1073 LANEUVILLE, J. B. V. *Paris.*—Machine for making watch-guards and purses.

1074 DESHAYS, A. *Paris.*—Machine for making various kinds of cords.

1075 LEMAIRE, E. F. *Paris.*—Automatic machine for making silk lace.

1076 SCRIVE, H. *Lille (Nord).*—Requisites for carding cotton, wool, &c.

1077 HARDING-COCKER, *Lille (Nord).*—Cards for spinning mills, &c.

1078 RONZE, R. *Lyon (Rhône).*—Economical jaquart-loom.
presses on a new principle.

1079 FILLIETTE & SON, *Paris.*—Copying

1080 BETHELOT, N. *Troyes (Aube).*—Circular, and rectilinear looms.

1081 PESIER, E. *Valenciennes (Nord).*—Model of apparatus for refining beet-root juice, by the application of alcohol.

1082 HERMANN-LACHAPELLE & GLOVER, *Paris.*—Apparatus for making gaseous drinks. Portable steam-engine.

1084 FRANÇOIS, E. S. *Paris.*—Apparatus for making seltzer water.

1085 BARIL, BROS. *Amiens (Somme).* — Machine for shearing velvet.

1086 ANDRÉ & GUILLOT, *Paris.*—Apparatus for making seltzer water.

1087 MONDOLLOT, BROS. *Paris.*—Apparatus for making seltzer water.

1088 FÈVRE, G. D. *Paris.*—Apparatus for making seltzer water.

1089 CAIL, J. F. & Co. *Paris.*—Apparatus for distilling in vacuo: sugar-cane mill and engine, &c.

1090 CHOUILLOU & JAEGER, *Paris.*—Currying machine.

1091 DE COSTER, WIDOW, *Paris.*—Punching machine; machine for making sugar by centrifugal force; portable expansive engine.

1092 BERNIER, SEN. & ARBEY, *Paris.*—Machines for working wood.

1093 DE DIETRICH & Co. *Reichshoffen (Bas-Rhin).*—Machine for bending metals.

1094 PERIN, J. L. *Paris.*—Endless saw, and fittings.

1095 TOUGARD, E. F. JUN. *Bapaume-lez-Rouen (Seine-Inf.).*—Machine for preparing engravers' plates.

1096 SCHMERBER, BROS. *Mulhouse (Haut-Rhin).*—Mechanical pestle.

1097 FREY, P. A. & SON, *Paris.*—Portable saw, for forest work.

1098 CORSEL, *Paris.*—Machine for reeling cocoons.

1099 MARESCHAL, J. *Paris.*—Machine for shaping wood.

1100 MERCIER, A. *Louviers (Eure).* — Spinning and weaving apparatus for wool.

1101 FRESNÉ, C. *Louviers (Eure).*—Plates and ribbons for cards.

1102 BAUDOUIN, BROS. & JOUANIN, *Paris.* —Machine for making fishing nets.

1104 DURAND, F. & PRADEL, *Paris.*—Jaquart-loom, in which paper is substituted for pasteboard.

1105 BERNARD, F. *Bourges (Cher).*—A jack.

1106 DAVID, BROS. & Co. *St. Quentin (Aisne).*—Dynamometer, for measuring the tension of threads for warp.

1107 VILLAIN, E. P. *Montmartre, Paris.*—Machine for making twisted fringes.

1108 CRESSIER, E. *Gras (Doubs).*—Horological tools.

1109 THOUROT, BROS. *Vandoncourt (Doubs).* —Horological tools.

1110 CAMBRAY, *Valenciennes (Nord).*—Apparatus for the combination of frames of different sizes.

Addenda.

ALAUZET, P. *Paris.*—Mechanical press.

BAZET, HAPPEY, & Co. *Paris.* — Apparatus for making gaseous waters.

BERJOT, F. *Caen (Calvados).* — Model of an apparatus for gaseous waters.

BOURGEOIS BOTZ, *Reims (Marne).* —Plates and ribbons for cards.

DUTARTRE, P. *Paris.* — Typographic press for two colours.

LEGAL, F. *Nantes* (*Loire-Inf.*). — Model of an apparatus for refining sugar in vacuo.

MATHIEU, BROS. *Anzin* (*Nord*). — Sugar apparatus.

SILBERMANN, J. J. *Paris.*—Universal presses.

CLASS 8.

1111 CHÉRET, M. J. *Paris.*—Model of a movement for a beam-engine.

1112 BOUILLON, MÜLLER & Co. & MÜLLER, E. *Paris.*—Washing and bath apparatus.

1113 EGROT, E. A. *Paris.*—Continuous distilling apparatus.

1114 MOUQUET, H. *Lille* (*Nord*).—Model of an apparatus for concentrating syrups in vacuo.

1115 DROUOT, E. *Paris.*—Steam bread making machine, moved by the waste heat of the oven.

1116 BOLAND, O. J. JUN. *Paris.*—Steam kneading machine.

1117 LESOBRE, C. *Paris.* — Rolland's revolving hot air oven, and mechanical kneading machine.

1118 MALBEC, A. A. *Paris.*—Apparatus for sharpening saws: grinding stones, &c.

1119 BEZIAT, J. C. M. *Paris.*—A jack for racking wines.

1120 GLEUZER, J. L. *Paris.*—Cork-cutting machine.

1121 LEMERCIER, E. *Paris.*—Machine for screwing shoes, &c.

1122 RATEL, *Saulieu* (*Cote d'Or*).—Machine for beating out scythes.

1123 TUSSAUD, F. *Paris.*—Various machines.

1124 VANÇON, J. A. *La Bresse* (*Vosges*).—Apparatus for the carriage of fish alive.

1125 VANDEVILLE, BROS. *Fèrin* (*Nord*).—Hammers for dressing mill-stones.

1126 GODIN-CHAMBRIAND, *Guise* (*Aisne*).—Hammers for dressing mill-stones.

1127 DUFOURNET & Co. *Clichy-la-Garenne* (*Seine*).—Pasteboard sugar moulds.

1128 LECOQ, E. F. *Paris.*—Machine for making and printing railway tickets.

1129 GERVAIS, A. *Paris.*—Apparatus for warming green-houses.

1130 FAUCONNIER, F. L. *Paris.*—Mill for pounding and sifting dry substances.

1131 HÉDIARD, A. *Paris.* — Inexplosive steam-boiler.

1132 NILLUS, E. *Havre* (*Seine-Inf.*).—Screw propeller apparatus.

1133 FAIVRE, BROS. *Nantes* (*Loire-Inf.*).—Immoveable screw propeller.

1134 SUCKFULL, L. *Déville - lez - Rouen* (*Seine-Inf.*).—Pistons.

1135 LEMIELLE, T. *Valenciennes* (*Nord*).—Ventilators for mines.

1136 TOUAILLON, *Paris.*—Model of a mill, &c.

1137 ZAMBEAUX, *St. Denis* (*Seine*). — Portable vertical steam-boiler.

1138 PARENT, SCHAKEN, CAILLET, & Co. *Oullins* (*Rhône*).—Crane.

1139 DESBORDES, L. & RONDAULT, E. A. *Paris.*—Metallic dial barometers, manometers, &c.

1140 LETHUILLIER-PINEL, *Rouen* (*Seine-Inf.*).—Magnetic indicators of the level of of water in boilers, and automatic feeders.

1141 RENAUD, P. *Nantes* (*Loire-Inf*).—Whistle-float for steam-boiler.

1142 PERREAUX, L. G. *Paris.*—Caoutchouc valves for pumps.

1143 POUGAULT, A. *Decize* (*Nièvre*). — Steam cleansing machine.

1144 CAIL, J. F. & Co. *Paris.*—Steam apparatus for washing pit-coal; crane, &c.

1145 HERMANN, G. *Paris.* — Chocolate grinding machine, &c.

1146 BERLIOZ, A. & Co. *Paris.*—Magneto-electric machine, for the production of light.

1147 BAUCHET-VERLINDE, & Co. *Lille* (*Nord*).—Ruling machines for account books.

1148 THIERS, A. *Paris.*—Electric lamp, and light regulator.

1149 DEVINCK, F. J. *Paris.* — Machines for weighing, grinding, and wrapping chocolate.

1150 VORUZ, SEN. *Nantes* (*Loire-Inf.*).—Machines for moulding bricks, and projectiles.

1151 LAURENS & THOMAS, *Paris.*—Steam-engine; boiler, &c.

1152 FARCOT & SONS, *Port St. Ouen* (*Seine*).—Condensing steam-engine, exhausting pump, pestle, &c.

1153 BRÉVAL, *Paris.*—Portable steam-engines.

1154 SAGEBIEN, A. *Amiens* (*Somme*). — Drawing of a water-wheel.

1155 CHENAILLER, P. C. *Paris.*—Evaporating apparatus.

1156 BOURDON, E. *Paris.*—Steam-engine, &c.

1157 LÉGAL, F. *Nantes* (*Loire-Inf.*). — Model of an apparatus for refining sugar in vacuo.

1158 CLERC, E. *Lyon* (*Rhône*).—Metallic manometers.

1159 DEDIEU, C. *Lyon* (*Rhône*).—Metallic manometers.

1160 DESBORDES, L. J. F. *Paris.*—Manometers, safety valves.

1161 FAUCONNIER, C. *Paris.* — Various cranes.

1162 SILBERMANN, J. J. *Paris.*—Universal presses acting by water pressure, and capable of printing any kind of surface, &c.

1163 DURENNE, J. F. *Courbevoie* (*Seine*).—"Hydratmopurificateur," for purifying the water of steam-boilers.

1164 FLAUD, H. *Paris.*—Steam fire-engine Giffard's injectors.

1165 BOLLÉE & SON, *Mans* (*Sarthe*). — Hydraulic ram.

1166 LECOINTE, J. *St. Quentin* (*Aisne*).—Apparatus for working a certain number of hydraulic presses under a continuous pressure.

1167 LETESTU, M. A. *Paris.* — Various pumps.

1168 DESPLAS, H. *Elbeuf* (*Seine Inf.*).—Fulling machine.

1169 BARRÉ, ROUGNON, & Co. *Paris.*—Machine for washing minerals.

1170 GAUTRON, B. J. *Paris.* — Hydro-extractor, for manufacturing fecula and starch.

1171 MOISON, F. T. *Mouy (Oise).*—Regulators and dynamometers, &c.

1172 BARRÉ & BESNARD, *Paris.*—Tubular boiler, on a new principle.

1173 FONTAINE & BRAULT, *Chartres (Eure-and-Loir).*—Turbines, ventilators.

1174 GIGNOUX, G. G. *Lige, near Andenge (Gironde).*—Drawing of a machine for the preservation of spawn.

1175 LECOUTEUX, H. *Paris.*—Double-cylinder beam-engine.

1176 MIGNOT, H. *Paris.*—Metallic manometers.

1177 ALAUZET, P. *Paris.* — Mechanical press.

1178 DUTARTRE, A B. *Paris.* — Typographic press for two colours.

1179 NORMAND, F. *Paris.*—Photograph of a press: model of unversal joint, &c.

1180 WORMS DE ROMILLY, M. *Paris.*—Lifting machine.

1181 ARMENGAND, SEN. *Paris.*—Drawings of steam-engines, &c.

1182 VERNAY, L. *Paris.*—Machine for lifting and weighing heavy articles.

1183 POIRIER DE ST. CHARLES, *Gentilly (Seine).*—Type-founding machine.

1184 JUTTEAU, A. *Orléans (Loiret).*—Specimens in relief and drawings of a system of stone veneering, for the restoration of buildings.

1185 DE CHODSKO, N. F. B. *Paris.*—Smoke-consuming apparatus, for steam-boilers.

1186 PALAZOT, *Bordeaux (Gironde).* — Smoke-consuming apparatus for steam-boilers.

1187 DUMERY, C. J. *Paris.*—Plan of a smoke-consuming apparatus, &c.

1188 LENOIR & Co. *Paris.*—Air-engine, acting by means of coal-gas inflamed by electricity.

1189 FORTIN-HERMANN, BROS. *Paris.*—Apparatus for the distribution of water for public use, &c.

1190 DARDONVILLE, V. *Paris.*—Filters of charcoal, &c.

1191 CARRÉ, F. MIGNON, & ROUART, *Paris.*—Machine for making ice by successive evaporations and liquefactions of sal-ammoniac.

1192 PROFESSIONAL SCHOOL OF DOUAI *(Nord).*—Small horizontal steam-engine.

1193 VARRAL, EHVELL, & POULOT, *Paris.* —Steam-engine, saws, &c.

1194 QUILLACQ, L. A. *Anzin (Nord).* — Horizontal double cylinder steam-engine for working coal mines.

1195 SOCIÉTÉ NOUVELLE DES FORGES ET CHANTIERS DE LA MEDITERRANÉE, *Paris.*—400-horse-power marine engines for screw propeller.

1196 PRINCE DE POLIGNAC, *Paris.*—Curved-cylinder steam-engine.

1197 MAUZAIZE, J. N. *Chartres (Eure-and-Loir).*—Apparatus for flour mills.

1198 HOËL-RENIER, F. *Lille (Nord).*—Manometer.

1199 MARQUIS DE MONTAGU, *Paris.* — Apparatus for preventing chimneys from smoking.

1200 ANTONY MASSON, *Paris.*— Smoke-box.

Addenda.

1200A HUBERT, H. *Paris.*—Feed-pump for fountains.

1200B FELDTRAPPE, BROS. *Paris.*—Design for a cylinder for printing stuffs and paper.

1200C COMTE DE EPRÉMESNIL, *Bernay (Eure).*—Design for a system of transmission to great distances.

DE COSTER, WIDOW. A. *Paris.* — Expansive portable engine.

DEHAYNIN, F. *Paris.* — Design for a machine to agglomerate small pit-coal.

HERMANN-LACHAPELLE & GLOVER, *Paris.*—Portable engine, with furnace within the boiler.

CLASS 9.

1201 MAZIER, DR. *L'Aigle (Orne).*—Two-horse reaping and mowing machine.

1203 BELLA F. *Grignon (Seine-and-Oise).* —Agricultural instruments.

1204 AGRICULTURAL ASSOCIATION OF SEINE-AND-OISE, *Versailles.* — Agricultural implements (10 Exhibitors).

1205 BARBIER & DAUBRÉE, *Clermont-ferrand (Puy-de-Dôme).*—Portable four-horse engine, reaping machine, &c.

1206 CUMMING J. *Orleans (Loiret).* — Portable engine, and thrashing machine.

1207 ALBARET & Co. *Liancourt, (Oise).*—Portables team-engine, thrashing-machine, &c.

1208 GANNERON, E. *Paris.*—Agricultural machines.

1209 PINET, J. JUN. *Abilly (Indre-and-Loire).*—Horse-gin, thrashing machine, winnowing machines.

1210 JACQUET-ROBILLARD, *Arras (Pas-de-Calais).*—Sowing machines.

1211 AGRICULTURAL COLONY OF METTRAY *(Indre-and-Loire).*—Ploughs.

1212 PERRIGAULT, J. *Rennes (Ille-and-Vilaine).*—Apparatus for the graduated aëration of mill stones while grinding, and the preservation of the mill dust.

1213 LANET & SON, *Cette (Hérault).* — Round and oval oak tuns.

1214 DOYÈRE & Co. *Paris.*—Model of subterranean cellars.

1215 PAVY, E. *Chemillé-sur-Dème (Indre-and-Loire).*—Model and drawings of a preservative granary.

1216 ROBIN, H. *Nantes (Loire-Inf.).* — Steam-reaping machine.

1217 RADIDIER & SIMONEL, JUN. *Paris.*—Root-cutters, &c.

1218 GRANDVOINET, J. A. *Paris.*—Plough.

1219 LEGACHEUX, A. *Pieux, (Manche).*—Plough, with instantaneous regulator.

1220 PARQUIN, L. V. *Villeparisis (Seine-and-Marne).*—Ploughs.

1221 CARDEILHAC & SON, *Toulouse (Haute-Garonne).*—Agricultural instruments.

1222 HUCK, J. M. *Paris.*—Apparatus for the manufacture of potatoe-fecula.

1223 DENET, E. *Paris.* — Moulds, &c. for alimentary pastes.

1224 Creuzé des Roches, R. *Chateau-de-Grandmaison (Indre)*.—Model of a moveable manége.

1225 Clauzel & Co. *Sauves (Gard)*.—Pitchforks of wood, &c.

1226 Dorléans, E. *Paris*.—Machine for making mats for vines.

1227 Redier, A. & Co. *Paris*.—Machines for sowing seed and spreading manure at once.

1229 Herbeaumont, P. F. *Paris*.—Designs for iron greenhouses.

1230 Vandercolme, A. *Rexpoëde (Nord)*.—Plans of the commune of Rexpoëde.

1231 Comte du Couëdic, *Quimperlé (Finistère)*.—Plan of his domain, &c.

1232 Bourgerie, E. *Remilly, Ardennes*.—Plough.

1233 Dumont, C. *St. Ouen Railway-station (Seine)*.—Machine for decorticating colza.

1234 Guiguet, *Paris*.—Agricultural drawings.

1235 Barbier, C. *Paris*.—Plans of experiments made in the application of liquid manures, &c.

1236 Lavoisy, *Paris*.—Mechanical churn.

1237 O'Reilly & Dormois, *Paris*.—Plans of iron greenhouses.

1239 Huart, H. *Cambrai (Nord)*.—Plan of a preservative granary.

1240 Lebœuf, *Paris*.—Venetian blinds for green-houses.

1241 Hamoir, G. *Saultain (Nord)*.—Horse rake, &c.

1242 Devaux, A. C. L. *London*.—Model of a ventilating and preservative granary.

1243 Buisson, *Tullins (Isère)*.—Flour-mill.

1244 Cail, J. F. & Co. *Paris*.—Straw and root cutters.

Addenda.

Roy, *Privas (Ardèche)*.—Model of an agricultural wheel-barrow.

Thouret, E. *Paris*.—Model of a preservative granary.

CLASS 10.

1251 The Minister of Agriculture, Commerce, and Public Works.—Collection of models and drawings relating to the public works of the French empire.

PASCAL, ANDRÉ, & DUSSAUD.—Model of a portion of the pier of the Napoleon Basin, Port of Marseilles: a plan of that port, with drawings of the means used in constructing it (1).

BOUNICEAU, LEMAITRE, ESCARRAGUEL, J., MONET, ESCARRAGUEL, A., ROULET, DUFFIEU, & PERRIN.—Sluice-gates of the citadel of Havre (2).

BOUNICEAU, BELLOT, COUCHE, E., ESCARRAGUEL, A., ROULET, DUFFIEU, & BATTAILLE.—Model of a graving-dock of the basin of the Eure, at Havre (3).

WATIER, CHATONEY.—Sluice-gate &c. of St Nazaire (4).

REYNAUD, L., FORESTIER, & MARIN.—Light-house of Barges, built on a rock, four miles from the Port of Sables d'Olonne (5).

LEPAUTE, H., REYNAUD, L., & ALLARD.—Lenticular light-house apparatus, white light varied alternately by red and green flashes, without obscuration (6).

SAUTTER, REYNAUD, & ALLARD.—Lenticular light-house apparatus, red light with obscurations (7).

CADIAT, OUDRY, MATHIEU, MAITROT DE VARENNES, AUMAITRE, & ROUSSEAU.—A turning-bridge of sheet-iron, constructed at Brest (8).

PERRIER, GENDARME, DE BAVOTTE, & CONTE.—Collection of drawings relating to the irrigation Canal at Carpentras, (Vaucluse) (9).

CHANOINE & DE LAGRENE.—Model of a moveable dam, constructed on the Upper Seine (10).

POIRÉE, CAMBUZAT, & MARINI.—Model of a moveable dam, constructed on the Youne, near Auxerre (11).

LOUICHE-DEFONTAINES, CARRO, & HOLLEAUX.—Model of a moveable dam constructed on the Marne (12).

ROZAT DE MANDRES.—Model of a portion of the dam of the reservoir of Settons, for supplying the navigation of the Youne (13).

VERON-DUVERGER & SCIAMA.—Model of a part of the dam of the reservoir of Montaubry, for feeding the Canal du Centre (14).

MARX & BRUNIQUEL.—Photographs of the Napoleon Bridge, at St. Sauveur, on the road from Paris to Spain (15).

MATHIEU, JOLY, A., & VIGOUROUX.—Model of the Bridge of St. Just, on the Ardèche (16).

TARBE RUINET, REGNIER, & DOURDET.—Chart of the Pools of the Dombes (Ain), in two colours (17).

BOURDALOUE.—Map of the levels taken throughout France (18).

FRANCOIS, J., & DUBRIEU.—The Baths of Ussat: subterranean works, searches for, and discovery of the sources of the mineral waters of that place (19).

FRANCOIS, J., & CHAMBERT.—Subterranean works, searches for, and discovery of the sources of the thermal waters of Bagnères-de-Luchon (20).

DAUBRÉE & JUTIER.—Plan and drawings of the search for, and conduction of, the waters of Plombières (21).

GRUNER.—Geological and mineralogical map of the Department of the Loire (22).

ELIE DE BEAUMONT & DE CHANCOURTOIS.—Geological map of the Department of the Haute Marne (23).

LEVALLOIS, M.—Geological map of the Department of the Meurthe (24).

DORMOY, M.—Map of the coal-basins of Valenciennes and Mons (25).

JAQUOT, M.—Agricultural map of the Arrondissement of Toul (Meurthe) (26).

REIBELE, G., VIRLA, MAHYER, & BONNIN.—Cherbourg breakwater (27).

NOEL & CALMAN.—Models of the docks of Castigneau, Nos. 1, 2, and 3 (28).

CHATONEY, LAROCHE, LE BOUËDEC, & KIEZELL.—Dam, with vertical shafts, used in laying the foundations of the new basins of the Port of Lorient (29).

DEHARNE & VERRIER, M.—Process of gradual removal of rock La Rose, at Brest (30).

GARNIER, A., ANGIBOUST, & CHARVIN.—New dry-dock at the military port of Rochfort (31).

REYNAUD, L., & ALLARD, E.—Metallic light-house for New Caledonia (32).

VUIGNIER, FLEUR-ST.-DENIS, & SAPPEL.—Lattice railway bridge over the Rhine at Kehl; caisson used in laying the foundation, fragment of a pile in process of driving, and of a pile in its place (33).

ZEILLER & DECOMBLE.—Viaduct of Chaumont, on the Paris and Mulhouse Railway (34).

VUIGNER, COLLET-MEYGRET, & PLUYETTE.—Viaduct of Nogent, on the Paris and Mulhouse Railway (35).

BOMMART, A., SURELE, DE LA ROCHE-TOLAY REGNAULT, PAUWELLS, NEPVEU, & EIFFEL.—Sheet-iron railway bridge, over the Garonne, at Bordeaux (36).

MAGUES & SIMONNEAU.—The Canal du Midi at the level of the torrent of Librons (37).

MICHAL, BELGRAND, DELAPERCHE, ROUSSELLE, & VALLÉE.—Map of the sewers of Paris; section of the sewer of Sebastopol, with its adjuncts; section of the great receptacle of the waters of Paris, with its adjuncts. Statistics of the water supply and sewerage of Paris, and of the works of the canal of St. Martin (38).

MICHAL, BELGRAND, & ROZAT DE MANDRES.—Map of the distribution of waters in the City of Paris; section of the reservoir of Passy, constructed for the supply of a part of Paris (39).

HERDEVIN, GUINIER, & FORTEIR HERMANN.—Apparatus used in connection with the Parisian fountains (40).

MICHAL, ALPHAND, DARCEL, & KIND.—The artesian wells of Passy, near Paris; geological section of the strata crossing them; boring tools and apparatus used with them; plan of the workshop (41).

LAUDET.—Machine for making the paving-stones used in Paris (42).

MICHAL, ALPHAND, DARCEL, GREGOIRE, & FOULARD.—A collection of drawings of the works executed by the City of Paris in the Bois de Boulogne, the wood of Vincennes, and various squares of the capital (43).

LEFEBURE DE FOURGY.—Maps of the subterranean quarries of the City of Paris (44).

DELESSE.—Subterranean geological map of the City of Paris (45).

LEFEBURE DE FOURGY.—Hydrological map of the City of Paris (46).

MATHIEU, JACQMIN, & DURBACH.—Sheet-iron railway bridge constructed at Fribourg (47).

CHAUBART.—Self-regulating sluice-gate (48).

BEAUDEMOULIN & BOUZIAT.—Improved apparatus for striking the centres of bridges, by means of sand, on the principle of Baudemoulin, used for the Bridge of Austerlitz (49).

1252 GARNAUD, E. F. JUN. *Paris.*—Objects in terra cotta.

1253 MIGNOT, L. *Paris.*—Objects in vegeto-mineral mastic.

1254 FACONNET, CHEVALLIER, & CO. *Paris.*—Tiles.

1255 GUICESTRE & CO. *Paris.*—Non-bitumenized pasteboard for roofs, &c.

1256 FERARY, C. A. *Grenoble (Isère).*—Cement, and articles made of it.

1257 LIPPMANN, SCHNECKENBURGER, & CO. *Paris.*—Artistic articles in factitious marble and stone.

1258 COIGNET, F. BROS. & CO. *Paris.*—Plain and decorated artificial stone, called agglomerated beton.

1259 MONDUIT, N. & BÉCHET, *Paris.*—Ornaments in lead and copper for building and decorative purposes.

1260 CRAPOIX, J. *Paris.*—Decorative objects in stucco.

1261 GÉRUZET, L. *Bagnères-de-Bigorre (Hautes-Pyrenées).*—Works in marble, for ornamentation and furniture.

1262 DE TINSEAU, P. *St. Ylie (Jura).*—Articles in Jura marble.

1263 SOYER, A. *Mareuil-lez-Meaux (Seine-and-Marne).*—Objects in factitious stone.

1264 ARNAUD, J. VENDRE, & CARRIERE, *La Porte de France (Isere).*—Cements, and fountain-pipes made of it.

1265 LINGÉE & CO. *Paris.*—Cement.

1266 GODIN-LEMAIRE, J. B. A. *Guise (Aisne).*—Cast-iron heating apparatus, &c.

1267 GASTELLIER, C. A. *Montauglaust (Seine-and-Marne).*—Tiles, bricks, and tubes.

1268 ZELLER & CO. *Ollwiller (Haute-Rhin).*—Glazed earthenware pipes for conveying water and gas.

1269 DAVID, L. *Uzès (Gard).*—Refractory clay and bricks.

1270 ALGOUD, BROS. DUPUY DE BORDES, & CO. *Grenoble (Isère).*—Cement, and cement tubes for fountains.

1271 VICAT, J. *Grenoble (Isère).*—Artificial cements.

1272 COUISSINIER, A. *Saint-Henry (Bouches-du-Rhône).*—Tiles for flooring.

1273 TROUILLIET, P. *Sens (Yonne).*—Refractory bricks.

1274 DUPONT, P. H. *Cherbourg (Manche)*—Metallic varnish for preserving wood and metals.

1275 LÉDIER, A. *Auffay (Seine-Inf.).*—Bricks.

1276 AGOMBART, A. *St. Quentin (Aisne).*—Hydraulic lime.

1277 BIGOT-DUVAL, WIDOW, & CO. *Mancelière (Eure-and-Loir).*—Natural hydraulic lime.

1278 DESFEUX, P. A. *Paris.*—Waterproof pasteboard for roofing.

1279 DUMONT, E. H. *Roanne (Loire).*—Tiles.

1280 MACHABÉE, L. *Paris.*—Hydraulic mastic.

1281 BAUDON. F. & SON, *Paris.*—Economic cooking apparatus.

1282 BREBAR. C. R. *Lille (Nord).*—Siliceous paintings.

1283 GRADOS, L. *Paris.*—Repoussé ornaments in zinc and lead.

1284 MICHELET, H. F. *Paris.*—Stamped articles in zinc and lead, for exterior decoration.

1285 PLANIER, E. & CO. *Paris.*—Elastic pads for stopping chinks in apartments.

1286 CANDELOT, L. F. SEN. *Paris.*—Various kinds of cement.

1287 VIEILLARD, G. (LACROIX), *Compiegne (Oise).*—Waterproof coating for wood, plaster, and stone.

1288 DERENUSSON, C. *Paris.*—Safety apparatus for scaffolding.

1289 LE BLANC, C. *Rennes (Ille-and-Vilaine).*—Drawings of the viaduct of Corbinières.

1290 MOULIN, A. *Bonsecours-lez-Rouen (Seine-Inf.).*—Imitation of Florentine mosaics.

1291 CAIL, J. F. & CO. *Paris.*—Model of a bridge, &c.

1292 BARTHÉLEMY, H. *Paris.*—Designs for theatres.

1293 FONTENELLE, C. C. *Paris.*—Cement for flagging.

1294 BOULANGÉ, *Auneuil (Oise).*—Enamelled mosaic tiles.

1295 JALOUREAU, A. *Paris.*—Pipes of bitumenized paper, for water or gas.

Addenda.

PARISIAN GAS-LIGHTING AND HEATING CO. *Paris.*—Refractory bricks.

DUPRAT, M. C. V. *Canéjan (Gironde).*—Refractory bricks.

JUTTEAU. A. *Orléans (Loiret).*—Specimen and drawings of veneering in stone.

ROSIER, WIDOW, & BAROCHE, *Tain (Drôme).*—Refractory bricks.

VIEILLARD, J. & CO. *Bordeaux (Gironde).*—Refractory bricks.

CLASS 11.

1301 BOCHE-TORDEUX, A. *Paris.*—Requisites for sporting.

1302 CARON, A. *Paris.*—Breach-loading cannon, carbines, &c.
1303 BRUN, A. *Paris.* — Breach-loading and other guns, pistols, &c.
1304 LESPIAUT, A. *Paris.*—Requisites for sporting.
1305 THOMAS, J. M. *Paris.*—Guns, pistols, and carbines.
1306 FLOBERT, L. N. A. *Paris.* — Guns, revolver pistols, cartridges, &c.
1307 GÉVELOT, *Paris.* — Percussion caps, and cartridges.
1308 JAVELLE-MAGAUD & SON, *St. Etienne* (*Loire*).—Double gun-barrels.
1309 RONCHARD-SIAUVE, *St. Etienne* (*Loire*).—Double gun-barrels, &c.
1310 DIDIER - DREVET, *St. Etienne* (*Loire*).—Double gun-barrels.
1311 BLACHON, J. *St. Etienne* (*Loire*).—Double gun-barrels.
1312 JAVELLE-MICHEL, *St. Etienne* (*Loire*).—Revolvers, cartridges.
1313 MURGUES, *St. Etienne* (*Loire*). — Fowling-pieces, pistols, &c.
1314 GABION - FOURNEL, *St. Etienne* (*Loire*).—Double-barrel fowling-pieces.
1315 BERGER, F. *St. Etienne* (*Loire*).—Fowling-pieces.
1316 VERNEY - CARRON, *St. Etienne* (*Loire*).—Fowling-pieces.
1317 ESCOFFIER, F. *St. Etienne* (*Loire*).—Military fire arms.
1318 PONDEVAUX & JUSSY, *St. Etienne* (*Loire*).—Double-barrel fowling-pieces.
1319 AURY, L. *St. Etienne* (*Loire*). — Double-barrel fowling-pieces.
1320 GEREST, A. *St. Etienne* (*Loire*).—Fancy guns.
1321 BOURGAUD & Co. *St. Etienne* (*Loire*).—Fowling-pieces, &c.
1322 GIRARD, L. *Chatellerault* (*Vienne*).—Seamless leather sword-sheaths.
1323 BLANC - MARTY & Co. *Paris.* — Military accoutrements.
1324 BERNARD, A. *Paris.*—Gun-barrels.
1325 GÉERINCKX, F. E. *Paris.*—Hunting guns and knives, &c.
1326 MARQUIS, F. *Paris.* — Weapons for the chase.
1327 PERRIN, L. *Paris.*—Guns, carabines, and pistols.
1328 DELVIGNE, H. G. *Paris.* — Duck-guns, rifled barrels, &c.
1329 TARDY & BLANCHET, *Paris.*—Percussion caps.
1330 GAUPILLAT, SON, & ILLIG, *Paris.*—Percussion caps and cartridges.
1331 SUDRE, F. *Paris.*—Military signals.
1332 GRANGER, E. *Paris.* — Copies of ancient arms and accoutrements.
1333 DELACOUR, L. F. *Paris.*—Burnished weapons.
1334 HOULLIER-BLANCHARD, C.H. *Paris.*—Guns and pistols.
1335 GASTINNE-RENETTE, L. J. *Paris.*—Guns, carbines, and pistols.
1336 CLAUDIN, F. *Paris.*—Fowling-pieces and pistols.
1337 BERNARD, L. *Paris.* — Barrels of fire-arms.
1338 DEVISME, L. F. *Paris.*—Gun-barrels, fancy guns; weapons for whaling and the chase.
1339 LEPAGE-MOUTIER, M. L. *Paris.* — Sporting guns, carbines, revolvers, &c.
1340 CORDIER & Co. *Paris.*—Revolvers and carbines.
1341 CHAUDUN, Á. SON, & DERIVIÈRE, N. *Paris.*—Cartridges.

Addenda.

GEIGER, Z. *Paris.*—Requisites for sporting.
WALCKER, W. *Paris.* — Requisites for sporting.

CLASS 12.

1361 LÉFEVRE, BROS. *Sotteville-lez-Rouen* (*Seine-Inf.*).—Rigging.
1362 LAFAYE, G. & Co. *Bordeaux* (*Gironde*).—Hemp and wire rigging.
1363 DELVIGNE, H. G. *Paris.*—Howitzers for carrying a line to a shipwrecked vessel.
1364 SALETTE, J. *Marseille* (*Bouches-du-Rhône*).—Model of apparatus for weighing anchors.
1365 PÉCOUL, A. *Marseille* (*Bouches-du-Rhône*).—Sounding log.
1366 CAVAYÉ, F. *Montpellier* (*Hérault*).—Safety and swimming belt.
1367 LABAT, T. *Bordeaux* (*Gironde*).—Apparatus for hauling vessels on land.
1368 TISSERANT, J. G. *Orléans* (*Loiret*).—Apparatus for saving from shipwreck.
1369 CABIROL, J. M. *Paris.*—Cork jacket; submarine lamp.
1370 GALLOIS-FOUCAULT, *St. Martin* (*Charante-Inf.*).—Floating fog-bells, &c.
1371 BOUQUIÉ, F. *Paris.*—Model of a system of towing by steam.
1373 BROQUANT, HOCHARD, & Co. *Dunkerque*, (*Nord*).—Machine-made fishing nets.
1374 DELAGE-MONTIGNAC, *Paris.* — Implements for fishing.
1375 DAVID & Co. *Hâvre* (*Seine-Inf.*).—Chain-cables, anchors, &c.
1376 OUARNIER - MATHIEU, *Compiêgne* (*Oise*).—Chain and hemp cables.
1377 BESNARD, F. RICHOU, & GENEST, *Angers* (*Maine-and-Loire*).—Hemp and wire-rope; fishing requisites.

CLASS 13.

1391 NAUDET & Co. *Paris.* — Metallic barometer.
1392 FASTRÉ, J. T. SEN. *Paris.*—Philosophical instruments.
1393 HARDY, E. *Paris.* — Electric, magnetic, and other philosophical apparatus.
1394 KŒNIG, R. *Paris.*—Acoustic instruments.
1395 COLOMBI, C. JUN. *Paris.*—Geodesic and other instruments.

S.W. Court and S.W. Gallery.

1396 MOLTENI, J. *Paris.*—Mathematical and other instruments.

1397 SILBERMANN, JUN. *Paris.*—Scientific diagrams, &c.

1398 GAVARD, A. *Paris.*—Improved pantographs, &c.

1399 BURDON, J. A. *Paris.*—Ruler for calculation.

1400 BONIS, P. F. *Paris.*—Insulated wire for electric and telegraphic apparatus.

1401 BILORET, A. *Paris.*—Insulated wire for electric apparatus, &c.

1402 PRUD'HOMME, P. D. *Paris.*—Electric bells and signals.

1403 POULLOT, V. N. *Paris.*—Eye-glasses, spectacles, &c.

1404 LEMAIRE, *Paris.*—Opera-glasses, &c.

1405 BRETON, BROS. *Paris.*—Philosophical instruments.

1406 MABRU, *Paris.*—Apparatus for maintaining a vacuum.

1407 THOMAS, C. X. *Paris.*—Calculating machines.

1408 BARDOU, P. G. *Paris.*—Opera glasses, and spectacles of various kinds.

1409 LEBRUN, A. *Paris.*—Optical instruments.

1410 DUTROU, E. P. *Paris.*—Philosophical instruments.

1411 SANTI, A. *Marseille* (*Bouches-du-Rhône*).—Mathematical and other instruments.

1412 BALBRECK, M. *Paris.*—Geodesic and other instruments.

1413 BREGUET, L. C. F. *Paris.*—Philosophical and horological instruments; electric telegraph apparatus.

1414 DIGNEY, BROS. & CO. *Paris.*—Telegraphic apparatus.

1415 BRUNNER & SON, *Paris.*—Optical instruments, &c.

1416 NACHET & SON, *Paris.*—Microscopes.

1417 HARTNACK, E. F. *Paris.*—Microscopes.

1418 MIRAND, A. SEN. *Paris.*—Microscopes.

1419 MOUCHET, A. *Rochefort.*—Microscopic apparatus.

1420 DUBOSCQ, L. J. *Paris*—Optical apparatus.

1421 DELEUIL, J. A. *Paris.*—Instruments for weighing, &c.

1422 BOURETTE, E. H. *Paris.*—Thermometers.

1423 DIGEON, R. H. *Paris.*—Scientific diagrams.

1424 TIFFEREAU, T. *Paris.*—Nautical hour-glasses; apparatus for receiving, measuring, and transferring gas.

1425 SAGEY, *Boulogne-sur-Mer* (*Pas-de-Calais*).—Gas-meter.

1426 RICOURT, C. *La Machine* (*Nièvre*).—Calculating cylinder.

1427 COLLOT, M. & A. BROS. *Paris.*—Balances of aluminium, &c.

1428 RATTIER & CO. *Paris.*—Submarine telegraph cables.

1429 PERREAUX, L. G. *Paris.*—Instruments for measurements of various kinds.

1430 TAURINES, J. M. H. A. *Paris.*—Balances, dynamometers.

1431 PRUDENT, L. *Paris.*—Opera glasses.

1432 CAM, J. C. *Paris.*—Opera glasses, spectacles, &c.

1433 COIFFIER, A. *Paris.*—Opera glasses, &c.

1434 LAFLEUR, A. *Paris.*—Opera glasses.

1435 CHUARD, M. *Paris.*—Safety-lamp; gazoscope for mines, &c.

1436 WARREN THOMSON, *Paris.*—Telegraph printing instruments.

1437 SERRIN, V. *Paris.*—Self-acting regulator for electric light.

1438 MOUILLERON & VINAY, *Paris.*—Electric telegraph apparatus.

1439 DUJARDIN, P. A. J. *Lille* (*Nord*).—Printing telegraph instrument.

1440 HOFFMANN, J. G. *Paris.*—Optical instruments and articles.

1441 LAMOTTE-LAFLEUR, C. G. *Paris.*—Boxes of mathematical instruments.

1442 BERTAUD, A. S. *Paris.*—Photographic object lenses, astronomical and nautical glasses, &c.

1443 MOREAU, A. *Paris.*—Electro-medical apparatus.

1444 GUYOT, D'ARLINCOURT, L. C. *Paris.*—Alphabetical printing telegraph instrument.

1445 LARGEFEUILLE, *Châlons-sur-Saône* (*Saône-and-Loire*).—Telegraphic requisites.

1446 ROBERT, H. *Paris.*—Astronomical clock.

1447 SORTAIS, T. *Lisieux* (*Calvados*).—Morse's telegraph apparatus.

1448 DU PUY DE PODIO, *Courbevoie* (*Seine*).—Stadiometer.

Addenda.

BERLIOZ, A. & CO. *Paris.*—Magneto-electric machine, for giving light.

COLLECTIVE EXHIBITION OF THE CITY OF CLUSES (*Haute-Savoie*).—Metallic thermometers.

DESHAYS, A. *Paris.*—Compass.

PATRY, E. L. *Paris.*—Spectacles, eye-glasses, &c.

CLASS 14.

1451 TITUS-ALBITÈS, *Paris.*—Photographic apparatus and photographs.

1452 SILBERMANN, J. JUN. *Paris.*—Table of the photogenic effects of the principal colours, on different substances; &c.

1453 ROBIN, A. *Paris.*—Photographic copies of plans, &c.

1454 DU MONT, H. *Paris.*—Photographic apparatus, representing the different phases of motion.

1455 ANTHONI, G. *Paris.*—Portable photographic laboratory.

1456 VILLETTE, E. *Paris.*—Large photographs, &c.

1457 DUBOSCQ, L. J. *Paris.*—Photographic apparatus.

1458 BERTAUD, *Paris.*—Photographic apparatus, and large object glass.

1459 BERTSCH, A. *Paris.*—Photographic apparatus.

1460 BILORDEAUX, A. *Paris.* — Photographs, &c.

1461 DISDÉRI, *Paris.*—Photographs.

1462 ALOPHE, M. *Paris.*—Photographs of the natural size, and obtained directly, &c.

1463 BISSON, BROS. *Paris.*—Photographs from nature.

1464 DELBARRE, P. J. & LELARGE, A. *Paris.*—Portraits of the natural size, obtained directly.

1465 CAMMAS, *Paris.* — Photographs: views in Egypt.

1466 YVON, *Paris.*—Specimens of photography obtained rapidly.

1467 DELESSERT, E. *Paris.* — Large photographs obtained directly, &c.

1468 DELTON, *Paris.*—Photographs taken quickly.

1469 BALDUS, E. *Paris.*—Large photographs obtained directly.

1470 POTTEAU, *Paris.* — Application of photography to the sciences.

1471 TOURNACHON, A. JUN. *Paris.*—Photographs.

1472 JAMIN, *Paris.*—Photographs.

1473 ROLLOY, JUN. *Paris.*—Photographic chemicals, &c.

1474 MARION, *Paris.* — Photographic paper.

1475 PUECH, L. *Paris.*—Photographic chemicals and apparatus.

1476 BRIOIS, C. A. *Paris.*—Photographic chemicals and apparatus.

1477 RICHARDIN, J. B. *Paris.*—Machine for polishing plates, &c.

1478 POIRIER, *Paris.*—Press for glazing photographs.

1479 LECU, F. N. *Paris.*—Photographic requisites.

1480 DE POILLY, SEN. *Boulogne (Pas-de-Calais).* — Photographic apparatus for the country.

1481 DUMONTEIL, *Paris.* — Photographic apparatus.

1482 KOCH, *Paris.*—Large photographic apparatus.

1483 RELANDIN, *Paris.* — Photographic apparatus.

1484 LEFÈVRE, *Paris.* — Requisites for photography.

1485 DERIVEAU, *Paris.* — Photographic alembic for travelling.

1486 GARIN & CO. *Paris.*—Photographic chemicals and paper.

1487 QUINET, A. M. *Paris.*—Photographic apparatus and requisites.

1488 DEROGY, *Paris.* — Photographic apparatus, &c.

1489 MILLET, A. *Paris.* — Photographic apparatus.

1490 HERMAGIS, *Paris.* — Photographic apparatus.

1491 DARLOT, *Paris.*—Photographic apparatus.

1492 LAVERDET, *Paris.*—Photographs.

1493 MATHIEU-PLESSY, E. *Paris.*—Photographic chemicals and paper.

1494 NUMA-BLANC & CO. *Paris.*—Photographs.

1495 D'ORZAGH, *Paris.*—Transferred collodion photographs.

1496 PLAISANT, *Paris.*—Oil-coloured photographic portraits.

1497 MAYER & PIERSON, *Paris.*—Photographs.

1498 NADAR, *Paris.*—Photographs, many of them taken by electric light.

1499 PESME, *Paris.*—Photographs.

1500 KEN, A. *Paris.*—Photographs.

1501 LEMERCIER, *Paris.* — Litho-photographs.

1502 COUVEZ, H. & COLOMBAT, *Paris.*—Photographs on wood, &c.

1503 NIEPCE DE ST. VICTOR, *Paris.*—Photographs on glass, &c.

1504 NÈGRE, C. *Paris.*—Heliographs obtained on steel, &c.

1505 DUFRESNE, *Paris.* — Photographic engraving.

1506 LAFOND DE CAMARSAC, *Paris.*—Unchangeable photographs on enamels and porcelain, &c.

1507 JOLY-GRANGEDOR, *Paris.*—Artistic photographs.

1508 POITEVIN, A. *Paris.*—Carbon photographs, &c.

1509 VIDAL, L. *Marseille (Bouches-du-Rhône).*—Carbon photographs.

1510 PETIT, P. *Paris.*—Photographs.

1511 CORBIN, H. *Paris.*—Photographs.

1512 TAUPENOT (dec.).—Photographs.

1513 GAUMÉ, *Mans (Sarthe).* — Photographs.

1514 FARGIER, *Lyon (Rhône).* — Photographs.

1515 CHARAVET, *Paris.* — Carbon photographs.

1516 GARNIER & SALMON, *Paris.*—Carbon photographs, heliographic engraving, &c.

1517 ROBERT, *Sèvres (Seine).* — Photographs, taken from Sèvres articles, &c.

1518 DAVANNE, A. *Paris.*—Photographs.

1519 GIRARD, A. *Paris.*—Photography of an eclipse.

1520 DAVANNE & GIRARD, *Paris.*—Specimens of photography.

1521 MAGNY, *Paris.*—Photographs.

1522 BRETON, MADAME, *Rouen (Seine-Inf.).*—Photographs.

1523 MARVILLE, *Paris.*—Photographs.

1524 BAYARD & BERTALL, *Paris.*—Photographic portraits and copies.

1525 RENARD, *Bourbonne-les-Bains (Haute-Marne).*—Photographs from nature, &c.

1526 JEANRENAUD, *Paris.* — Views obtained with dry collodion.

1527 BRAUNN, A. *Dornach (Haut-Rhin).* —Photographs.

1528 JOUET, E. *Paris.*—Photographs.

1529 MAILAND, E. *Paris.*—Photographs.

1530 DE BRÉBISSON, *Falaise (Calvados).* —Photographs.

1531 ADAM-SALOMON, *Paris.* — Photographs from nature.

1532 GAILLARD, P. *Paris.*—Moist collodion photographs.

1533 BINGHAM, R. *Paris.*—Photographs.

1534 MICHELEZ, C. *Paris.*—Photographs.
1535 CARJAT & Co. *Paris.*—Moist collodion photographs.
1536 LAFFON, J. C. *Paris.*—Photographs.
1537 MAXWEL-LYTE, *Bagnères-de-Bigorre (Hautes-Pyrénées).*—Views in the Pyrenees.
1538 ALEO, *Nice (Alpes-Maritimes).*—Photographs by various processes.
1539 MUZET, *Grenoble (Isère).*—Views of the Isere and Savoy.
1540 BERTHIER, P. *Paris.*—Photographs.
1541 COMTE O. AGUADO, *Paris.*—Enlarged photographs.
1542 VICOMTE O. AGUADO, *Paris.*—Enlarged photographs.
1543 SILVY, *Paris.*—Views in Algeria, &c.
1544 BRAQUEHAIS, *Paris.*—Stereoscopic photographs, &c.
1545 DAGRON, E. *Paris.*—Microscopic photographs.
1546 FERRIER & SON, *Paris.*—Large photographs on glass, &c.
1547 MARLÉ, C. A. *Paris.*—Moist collodion photographs.
1548 WARNOD, *Hâvre (Seine-Inf.).*—Photographs.
1549 MASSON, *Seville (Spain) and Paris.*—Photograph views in Spain, &c.
1550 BOUSSETON & APPERT, *Paris.*—Portraits.
1551 CREMIERE, *Paris.*—Instantaneous portraits, &c.
1552 MARQUIS DE BÉRENGER, *Paris.*—Photograph views of the department of the Isère.
1553 DELONDRE, P. *Paris.*—Photographs.
1554 DE CLERCQ, L. *Paris.*—Photographs.
1555 DE VILLECHOLLE, F. *Paris.*—Photographs.
1556 ANTHONY-THOURET, JUN. *Paris.*—Photographs from nature, and copies.
1557 CHARNAY, D. *Mâcon (Saône-and-Loire).*—Photographs.
1558 LACKERBAUER, *Paris.*—Natural history and anatomical photographs, of the natural size, &c.
1559 ROMAN, D. *Arles (Bouches-du-Rhône).*—Photographs.
1560 TILLARD, F. *Bayeux (Calvados).*—Photographs.
1561 COLLARD, *Paris*—Photographic views, &c.
1562 RICHEBOURG, *Paris.*—Photographs.
1563 DE LUCY, L. G. *Paris.*—Portraits and groups.
1564 DE CHAMPLOUIS, *Paris.*—Views in Syria.
1565 DUVETTE & ROMANET, *Amiens (Somme).*—Photographs of the cathedral of Amiens.
1566 MOULIN, F. *Paris.*—Photographs.
1567 AUTIN, *Caen (Calvados).*—Photographs.
1568 GUESNÉ, *Paris.*—Photographs.
1569 BACOT, *Caen (Calvados.)*—Portraits, and studies from nature, with collodion.

CLASS 15.

1581 DESFONTAINE, LEROY, & SON, *Paris.*—Watches, chronometers, time-pieces.
1582 ROBERT, H. *Paris.*—Time-pieces, chronometers, graver's tools, oils, &c.
1583 COLIN, A. *Paris.*—Clocks, regulators, travellers' time-pieces, bronzes, tools.
1584 ANQUETIN, M. *Paris.*—Watches, and universal time-pieces.
1585 DETOUCHE, C. L. *Paris.*—Time-pieces, watches, regulators, &c.
1586 MONTANDON, BROS. *Paris.*—Springs, for clocks, watches, &c.
1587 FARCOT, H. A. E. *Paris.*—Time-pieces, alarms.
1588 BROCOT, L. A. *Paris.*—Regulators, half-second time-pieces, and perpetual calenders.
1589 REDIER, A. *Paris.*—Time-pieces, alarms, &c.
1590 CHARPENTIER, P. A. *Paris.*—Watches, time-pieces, chronometers.
1591 GUYARD, F. V. *Dieppe (Seine-Inf.).*—Electric clocks.
1592 GONTARD, *Besançon (Doubs).*—Watches, &c.
1593 FERNIER, N. *Besançon (Doubs).*—Watches, &c.
1594 BERTHELOT, *Besançon (Doubs).*—Watches, &c.
1595 MONTANDON, *Besançon (Doubs).*—Watches, &c.
1596 JEANNOT-DROZ, *Besançon (Doubs).*—Watches, &c.
1597 RICHARDEY, *Besançon (Doubs).*—Watches, &c.
1599 SAVOYE, BROS. & CO. *Besançon (Doubs).*—Watches, &c.
1600 CRESSIER, E. *Besançon (Doubs).*—Watches, &c.
1601 GILET, E. *Besançon (Doubs).*—Watches, &c.
1602 BOSSY, *Besançon (Doubs).*—Watches, &c.
1603 BOUTEY & SON, *Besançon (Doubs).*—Watches, &c.
1604 ADLER, N. *Besançon (Doubs).*—Watches, &c.
1605 FAVRE-HEINRICK, *Besançon (Doubs).*—Watches, &c.
1606 PERROT, E. *Besançon (Doubs).*—Watches, &c.
1607 GROZ BROS. *Besançon (Doubs).*—Watches, &c.
1608 HUMBERT, A. *Besançon (Doubs).*—Watches, &c.
1609 CHALONS, V. *Besançon (Doubs).*—Watches, &c.
1610 LAMBERT, H. *Besançon (Doubs).*—Watches, &c.
1611 STERKY, A. *Besançon (Doubs).*—Watches, &c.
1612 BERTHET, *Besançon (Doubs).*—Watches, &c.
1613 PIGUET, *Besançon (Doubs).*—Watches, &c.
1614 FUMEY, *Besançon (Doubs).*—Watches, &c.

1615 GONDELFINGER & BICHET, *Besançon (Doubs)*.—Watches, &c.

1616 AMET, F. *Besançon (Doubs)*.—Watches, &c.

1617 PONSOT & HÉRIZÉ, *Besançon (Doubs)*.—Watches, &c.

1618 BRÉDIN, *Besançon (Doubs)*.—Watches, &c.

1619 CROUTTE A. & Co. *Rouen (Seine-Inf.)*.—Time-pieces, alarms.

1620 LÉGER, P. J. *Paris*.—Watches.

1621 GINDRAUX, A. & SONS, *Paris*.—Rubies and diamonds for wire drawers, and diamond points for engravers on steel; files.

1622 VISSIÈRE, S. *Hâvre (Seine-Inf.)*.—Chronometers, &c.

1623 LEROY, T. *Paris*.—Chronometers, &c.

1624 JACOT, H. L. *Paris*.—Traveller's time-pieces.

1625 FLÉCHET, P. *Paris*.—Sun dials.

1626 COUËT, L. C. *Paris*.—Watches, travellers' time-pieces, regulators.

1627 PATAY, P. *Paris*.—Traveller's time-pieces.

1628 DROCOURT, P. *Paris*.—Traveller's time-piece.

1629 PIERRET, V. *Paris*.—Watches, &c.

1630 COLLECTIVE EXHIBITION OF THE CITY OF CLUSES *(Haute-Savoie)*.—Parts of watches and clocks, tools, &c.

1631 SCHARF, B. *St. Nicolas-d'Aliermont (Seine-Inf.)*.—Astronomical clock, chronometers, &c.

1632 STRÉBET, L. *Paris*.—Steeple-clock.

1633 PERREAUX, L. G. *Paris*.—Public clocks.

1634 SCHIRRMANN, E. *Paris*.—Time-pieces, &c. in carved wood.

1635 BREQUET, L. C. F. *Paris*.—Time-pieces, regulators, chronometers, repeaters, &c.

CLASS 16.

1641 MUSTEL, C. V. *Paris*.—Organ-harmonium.

1642 LABBAYE, J. C. *Paris*.—Brass instruments.

1643 RODOLPHE, A. *Paris*.—Organ-harmonium.

1644 HENRY E. & MARTIN, J. *Paris*.—Military musical instruments, in brass, &c.

1645 BEAUCOURT, H. C. *Lyon (Rhône)*.—Organ-harmonium.

1646 DAVID, L. *Paris*.—Brass military instruments.

1647 MAYER-MARIX, *Paris*.—Harmoni-flutes.

1648 LECOMTE, A. & Co. *Paris*.—Wind instruments, of brass and wood.

1649 KASRIEL, L. M. *Paris*.—Harmoni-flutes, &c.

1650 ALEXANDRE & SON, *Paris*.—Organs.

1651 REMY & GROBERT, *Mirecourt (Vosges)*.—Non-metallic wind instruments; an organ.

1652 BUSSON, C. *Paris*.—Accordions, harmoni-flutes.

1653 DERAZEY, J. J. H. *Mirecourt (Vosges)*.—Stringed instruments, &c.

1654 MANGEOT, BROS. & Co. *Nancy (Meurthe)*.—Upright piano, with oblique strings.

1655 À CAUDÈRES, J. J. *Bordeaux (Gironde)*.—Upright piano, with a triple clavier.

1656 POIROT, D. *Mirecourt (Vosges)*.—Organ, and stringed instruments.

1657 CAVAILLÉ-COLL, A. *Paris*.—Organ, &c.

1658 KELSEN, E. *Paris*.—Self-acting organ.

1659 FARRENC, J. H. A. *Paris*.—Piano music.

1660 GRANDJON, J. *Mirecourt (Vosges)*.—Stringed instruments.

1661 GÉRARD, E. & Co. *Paris*.—Select music.

1662 VINCENT, *Paris*.—Organ, having a double clavier with quarter tones.

1663 DEBAIN, A. F. *Paris*.—Mechanical piano, harmonichords, &c.

1664 MARTIN, P. & SON, *Toulouse (Haute-Garonne)*.—Piano.

1665 KRIEGELSTEIN, J. G. *Paris*.—Pianos.

1666 HUSSON-BUTHOD, & THIBOUVILLE, *Paris*.—Wind and stringed instruments.

1667 COURTOIS, ANTONY, *Paris*.—Brass instruments.

1668 ELCKÉ, F. *Paris*.—Piano.

1669 LEMOINE, H. A. P. *Paris*.—Music.

1670 VUILLAUME, J. B. *Paris*.—Violins and bows.

1671 SAVARESSE, H. *Paris*.—Strings for instruments.

1672 KLEINJASPER, *Paris*.—Upright piano.

1673 BOISSELOT & SON, *Marseille (Bouches-du-Rhône)*.—Pianos.

1674 WIART, F. S. *Châteauroux (Indre)*.—Piano mechanism.

1675 GEHRLING, C. *Paris*.—Mechanism for pianos.

1676 GAUDONNET, P. *Paris*.—Upright piano.

1677 MIRMONT, C. A. *Paris*.—Stringed instruments.

1678 MONTAL, C. *Paris*.—Upright and grand pianos.

1679 DE ROHDEN, F. *Paris*.—Mechanism for pianos.

1680 BONNET, C. *Marseille (Bouches-du-Rhône)*.—Stringed instruments.

1681 FAVRE, J. *Lyon (Rhône)*.—Harmonichords.

1682 JAULIN, L. J. *Paris*.—Harmonichord.

1683 BAUBIER, V. *Paris*.—Requisites for the manufacture of pianos.

1684 BORD, A. *Paris*.—Pianos.

1685 AUCHER, L & J. BROS. *Paris*.—Pianos.

1686 PLEYEL, WOOLF, & Co. *Paris*.—Pianos.

1687 JACQUOT, C. *Nancy (Meurthe)*.—Violins, violoncellos, &c.

1688 PAPE, J. H. *Paris*.—Pianos.

1689 HERZ, H. *Paris*.—Pianos.

1690 BLANCHET, P. A. C. *Paris*.—Pianos.

1691 WÖLFEL, F. *Paris*.—Pianos.

1692 COTTIAU, P. F. J. *Paris*.—Reeds.

1693 LOT, L. *Paris*.—Flutes.

1694 GODEFROY, C. SEN. *Paris*.—Flutes.

1695 BUFFET, L. A. JUN. *Paris.*—Clarinets, hautboys, &c.
1696 TRIÉBERT, F. *Paris.*—Wind instruments.
1697 BUFFET-CRAMPON, & Co. *Paris.*—Wind instruments.
1698 BRETON, J. D. *Paris.*—Wind instruments.
1699 THIBOUVILLE, SEN. *Paris.*—Wind instruments.
1700 BARBU, J. P. *Paris.*—Reeds.
1701 SAX, A. J. *Paris.*—Wind instruments, &c.
1702 GAUTROT, P. L. SEN. *Paris.*—Wind and percussion instruments, of brass and wood.
1703 DE TILLANCOURT, E. *Paris.*—Silk acribelles-strings.
1704 FRELON, L. F. A. *Paris.*—Apparatus for teaching music.
1705 BAUDASSÉ-CAZOTTES, *Montpellier (Hérault).*—Harmonic strings.
1706 GICHENÉ, L'ABBÉ, *St. Medard (Landes).*—Symphonista.
1707 SAX, A. JUN. *Paris.*—Wind instruments.

CLASS 17.

1711 MERICANT, E. *Paris.*—Instruments for veterinary surgeons.
1712 CHARRIÈRE, J. J. *Paris.*—Surgical instruments.
1713 LUËR, G. G. A. *Paris.*—Surgical instruments.
1714 MATHIEU, L. J. *Paris.*—Surgical apparatus, artificial limbs, &c.
1715 SALES-GIRONS, DR. *Paris.*—Apparatus for *pulverizing* mineral waters.
1716 CHARLES, G. *Paris.*—Baths, with heating apparatus, &c.
1717 FOUQUET, A. *Paris.*—Bathing apparatus.
1718 MARTRÈS, DR. A. *St. Cyr (Seine-and-Oise).*—Tent answering for a litter.
1719 LÉCUYER, F. J. *Paris.*—Bath, with warming apparatus.
1720 PAQUET, DR. F. *Roubaix (Nord).*—Ferruginated gutta percha, for surgical uses.
1721 GALANTE, H. & Co. *Paris.*—Vulcanized India rubber, for surgical apparatus.
1722 REBOLD, E. *Paris.*—Electro-medical apparatus.
1723 JUNOD, V. T. *Paris.*—Cupping-glasses.
1724 GRANCOLLOT, L. P. *Paris.*—Orthopœdic and herniary apparatus.
1725 TOLLAY, MARTIN, & LEBLANC, *Paris.*—Shower-baths.
1726 PICHOT, J. A. & MALAPERT, *Poitiers (Vienne).*—Carboniferous disinfecting paper, &c.
1727 THIERS, L. P. T. *Paris.*—Hygienic, &c. apparatus.
1728 BURG, *Paris.*—Metallo-therapeutic apparatus.
1729 MORIN, F. J. *Paris.*—Electro-medical apparatus, &c.
1730 NACHET & SON, *Paris.*—Optical apparatus for anatomy.
1731 LEPLANQUAIS, P. F. *Paris.*—Trusses, &c.
1732 LE PERDIEL, C. & MARINIER, J. *Paris.*—Trusses, &c.
1733 WICKHAM, BROS. *Paris.*—Trusses.
1734 LORIEL, H. F. *Paris.*—Trusses.
1735 LE BELLEGUIC, P. J. *Paris.*—Orthopœdic apparatus, trusses, &c.
1736 BÉCHARD, R. L. *Paris.*—Trusses, artificial limbs, orthopœdic apparatus.
1737 VASSEUR, P. N. *Paris.*—Osteological and comparative anatomical preparations.
1738 BAILLIÈRE, J. B. & SON, *Paris.*—Works on natural history, anatomy, &c.
1739 DUCHENNE, DR. G. *Paris.*—Physiognomical photographs.
1740 LAKERBANER, P. *Paris.*—Drawings, lithographs, engravings, and photographs, for the natural and medical sciences.
1741 GION, D. J. *Paris.*—Artificial teeth, &c.
1742 LÉGER, DR. E. V. *Paris.*—Anatomical and pathological models in paper.
1743 SIMON, P. *Paris.*—Dental instruments.
1744 LAVEZZARI, E. *Montreuil-sur-Mer (Pas-de-Calais).*—Plans of an hospital.
1745 TALRICH, J. V. J. *Paris.*—Anatomical wax models, &c.
1746 GUÉRIN, J. J. B. *Paris.*—Comparative osteological preparations.
1747 LEFEVRE, A. A. *Paris.*—Taxidermic preparations of animals' heads.
1748 AUZOUX, DR. L. *Paris.*—Elastic anatomical preparations.
1749 BOISSONNEAU, A. SEN. *Paris.*—Moveable artificial eyes in enamel.
1750 COULOMB, J. *Paris.*—Moveable artificial eyes.
1751 LUCAS, L. P. *Paris.*—Preserved plants.
1752 JULIENNE, MADAME M. J. E. *Paris.*—Bathing-belt for infants.
1753 DARBO, F. *Paris.*—Hygienic and other instruments.
1754 PARZUDAKI, E. *Paris.*—Taxidermic preparations.
1755 BOURGOGNE, J. SEN. *Paris.*—Microscopic anatomical, &c. objects.
1756 PRÉTERRE, P. A. *Paris.*—Articles for the repair of injuries to the mouth.
1757 BOURGOGNE, BROS. *Paris.*—Microscopic anatomical preparations.
1758 LAMI, A. *Paris.*—Anatomical models.
1759 LÉTHO, F. H. *Paris.*—Artificial eyes for taxidermic preparations.
1760 DESJARDINS DE MORAINVILLE, DR. J. B. L. *Paris.*—Artificial eyes, &c.
1761 BOISSONNEAU, A. P. JUN. *Paris.*—Moveable artificial eyes, &c.
1762 MAREY, DR. J. *Paris.*—Surgical apparatus.
1763 DAMOISEAU, A. *Alençon (Orne).*—A substitute for leeches.
1764 FRANÇOIS, J. CONTE-GRANDCHAMP, & DESBUISSONS, *Paris.*—Plans of a proposed thermal establishment at Amelie-les-Bains.
1765 FRANÇOIS, J. & DURRIEU.—Plans of an establishment at the baths of Ussat.

1766 BILLET, FRANÇOIS, J. & PELLEGRINI, B.—Plans of the thermal establishment of Marlioz.

1767 FRANÇOIS, J. & NORMAND.—Plans of the baths of Barèges.

1768 CARON, A. *Paris*.—Medical spoon, for administering disagreeable medicine.

1769 MOUILLARD, P. F. V. *Paris*.—Process for purifying tobacco smoke, or combining it with matters intended to be inhaled.

CLASS 18.

1771 MALLET BROS. *Lille* (*Nord*).—Cotton thread, plain, twisted, &c.

1772 LOYER, H. *Lille* (*Nord*).—Twisted cotton thread.

1773 COLLECTIVE EXHIBITION OF THE CITY OF ROANNE (*Loire*).—Grain-dyed cotton fabrics, &c. (13 Exhibitors).

1774 COLLIN & Co. *Bar le Duc* (*Meuse*).—Cotton, and wool and cotton fabrics.

1775 HUMBERT & Co. *Gamaches* (*Somme*).—Cotton thread, plain and twisted.

1776 LEJEUNE, L. & VOITRIN, *Paris*.—Wadding.

1777 DUPONT-POULET, *Troyes* (*Aube*).—Spun and dyed cotton.

1778 DURET, *Brionne* (*Eure*).—Spun and dyed cotton.

1779 RITAUD, C. FLEURY, V. & Co. *Paris*.—Thread made of cotton dyed before being spun.

1780 CARTIER-BRESSON, *Paris*.—Twisted cotton thread.

1782 FAUQUET-LEMAITRE & PREVOST, *Pont-Audemer* (*Eure*).—Cotton thread.

1783 TABOUEL & LÉMERY, *Darnetal* (*Seine-Inf.*).—Dressed lustrings and fabrics.

1784 PIMONT, P. *Rouen* (*Seine-Inf.*).—Fabrics made of bleached and prepared cotton.

1785 FANQUET & LHEUREUX, *Rouen* (*Seine-Inf.*).—Cotton handkerchiefs.

1786 FANQUET, O. & Co. *Oissel-sur-Seine* (*Seine-Inf.*).—Spun cotton.

1788 DERLY, A. & CHABOY, *Bellemcombre* (*Seine-Inf.*).—Spun cotton.

1789 PEYNAUD & Co. *Remilly-sur-Andelle* (*Seine-Inf.*).—Spun cotton, and calico.

1792 ROUSSELIN, S. *Darnetal* (*Seine-Inf.*).—Spun cotton.

1793 VAUSSARD, *Nôtre-Dame-de-Boudeville* (*Seine-Inf.*).—Spun cotton, and cotton fabrics.

1794 BOUCLY-MARCHAND, *St. Quentin* (*Aisne*).—Plaited shirt-fronts, embroidered, &c. by machine.

1795 HUGUES-CAUVIN, *St. Quentin* (*Aisne*).—Figured muslins for furniture.

1796 LEDOUX-BEDU, & Co. *St. Quentin* (*Aisne*).—Cotton fabrics, handkerchiefs of silk and cotton.

1797 TROCMÉ, P. L. *Hervilly* (*Somme*).—White and figured quilting.

1798 LEROY-DAUPHIN, *St. Quentin* (*Aisne*).—Plain cotton fabrics, &c.

1799 HUET-JACQUEMIN, A. *St. Quentin* (*Aisne*).—Worked muslins, &c.

1800 DERCHE-GIRARDE, *St. Quentin* (*Aisne*).—White worked muslins, &c.

1801 ROLAND, C. *St. Quentin* (*Aisne*).—White and coloured quilting for petticoats.

1802 DELACOURT, C. *Epehy* (*Somme*).—Quilting.

1803 COLOMBIER BROS. *St. Quentin* (*Aisne*).—Fine quilting, for various purposes.

1804 CARCENAC & ROY, *Paris*.—Plain and other cotton fabrics.

1805 ODERIEU, C. & CHARDON, L. *Rouen* (*Seine-Inf.*).—White and figured quilting.

1807 MOTTE-BOSSUT, & Co. *Roubaix* (*Nord*).—Various articles in cotton.

1808 BOIGEOL-JAPY, *Giromagny* (*Haut-Rhin*).—Spun and woven articles in cotton.

1809 GROS, ODIER, ROMAN, & Co. *Wesserling* (*Haut-Rhin*).—White cotton fabrics.

1810 DOLFUS, MIEG, & Co. *Mulhouse* (*Haut-Rhin*).—Cotton yarn, and cotton fabrics.

1811 FÉROUELLE & ROLLAND, *St. Quentin* (*Aisne*).—Figured window blinds and curtains.

1812 MENNET-POSSOZ, DAVID, & TROULLIER, *Paris*.—Worked muslins and gauzes, for curtains.

1813 DUBOIS, V. *Paris*.—Window curtains and blinds, &c.

1814 CHATELUS-DUBOST, *Tarare* (*Rhône*).—Light cotton fabrics, tarlatans, &c.

1815 THIVEL-MICHON, *Tarare* (*Rhône*).—Tarlatans.

1816 COLLECTIVE EXHIBITION OF THE CITY OF TARARE (*Rhône*).—Figured muslins, &c., made with a Jacquart machine (8 Exhibitors).

1817 RUFFIER-LEUTNER, *Tarare* (*Rhône*).—Muslin, plain and figured.

1818 MAC-CULLOCH BROS. *Tarare* (*Rhône*).—Dyed cotton fabrics.

1819 PATUREAU, L. *Paris*.—Cases for sewing cotton.

CLASS 19.

1831 CASSE, J. & SON, *Lille* (*Nord*).—Damasked table-linen, mixed flax and cotton fabrics.

1832 LAUWICK, BROS. *Commines* (*Nord*).—Flax and cotton tapes.

1833 CORNILLEAU, L. SEN. & Co. *Mans* (*Sarthe*).—Machine-made hempen cloth.

1834 BARY, JUN. & Co. *Mans* (*Sarthe*).—Hemp-yarn, unbleached and bleached cloth.

1835 RÉVEILLÈRE, *Mans* (*Sarthe*).—Hemp; yarn and cloth of hemp.

1836 JOURNÉ, P. *Paris*.—Ticking of flax, of flax and cotton, &c. for trowsers.

1837 DICKSON & Co. *Dunkerque* (*Nord*).—Duck.

1838 FROMAGE, L. *Darnetal* (*Seine-Inf.*).—Ribbed sail-cloths.

1839 LEVEAU, A. *Evreux* (*Eure*).—Ticking of cotton, and of flax and cotton.

1840 DEQUOY, J. *Lille* (*Nord*).—Flax-yarn; tow-yarn, and cloth.

1841 DRUMMOND, BARTES, & Co. *Moulins-Lille* (*Nord*).—Jute-yarn.

1842 VRAU, P. *Lille* (*Nord*). — Sewing thread.

1843 MONCHAIN, Z. *Lille* (*Nord*).—Flax-yarn.

1844 POUCHAIN, V. *Armentières* (*Nord*). —Cloth of flax, table linen.

1845 GRASSOT & Co. *Lyon* (*Rhône*).—Damask table linen, fabrics of wool and cotton, for furniture.

1847 DENEUX, BROS. *Hallencourt* (*Somme*). —Table linen.

1848 BUCHHOLTZ & Co. *Valenciennes* (*Nord*).—Cambrics and lawns.

1849 DELAME-LELIÈVRE & SON, *Valenciennes* (*Nord*).—Cambrics, and lawns: handkerchiefs.

1850 GUYNET, L. H. *Paris & Cambrai* (*Nord*).—Linens of Cambrai, &c.

1851 LUSSIGNY, BROS. *Paris & Valenciennes* (*Nord*).—Cambrics, of hand and machine-spun yarn, &c.

1852 GODARD, A. & BONTEMPS, *Valenciennes* (*Nord*). — Cambrics, of hand and machine-spun yarn, &c.

1853 THOURY, BROS. & FLAIR, *Mans* (*Sarthe*).—Carded hemp, thread, and pack-thread.

1854 VERSTRAETE & Co. *Lille* (*Nord*).—Flax-yarn, and undressed tow-yarn; twisted sewing-thread.

1855 HEUZE, RADIGUET, HOMON, GOURY, & LEROUX, *Landerneau* (*Finistère*).—Flax-yarn and tow-yarn; duck, &c.

1856 DEFREY, HOUSSIER, & LEPRÊTRE, *Alençon* (*Orne*).—Waterproof fabrics, and cordage.

LEONI & COBLENZ, *Vaugenlieu* (*Oise*). —Hemp, peeled mechanically without steeping, by a new process.

CLASS 20.

1861 MATHEVON & BOUVARD, *Lyon* (*Rhône*).—Silk fabrics, for furniture.

1862 CARQUILLAT, M. M. *Lyon* (*Rhône*).—Silk fabrics.

1863 HECKEL, SEN. BROSSET, *Lyon* (*Rhône*).—Silk fabrics.

1864 BLACHE & Co. *Lyon* (*Rhône*).—Silk velvet.

1865 BRUNET-LECOMTE, DEVILLAINE, & Co. *Lyon* (*Rhône*).—Silk fabrics, plain or printed.

1866 PONCET, L., LENOIR, V. & Co. *Lyon* (*Rhône*).—Silk fabrics.

1867 TEILLARD, C. M. *Lyon* (*Rhône*).—Plain silk fabrics.

1868 BROSSET, SEN. & BOISSIEU, *Lyon* (*Rhône*).—Plain and figured silks.

1869 GIRARD, NEPHEW, QUINZON, & Co. *Lyon* (*Rhône*).—Plain velvet.

1870 BONNET, C. J. & Co. *Lyon* (*Rhône*).—Taffetas and satins.

1871 SCHULZ BROS. & BERAUD, *Lyon* (*Rhône*).—Silk fabrics for dresses, shawls.

1872 MARTIN, J. B. & P. *Tarare* (*Rhône*). —Dyed and milled silks, plush.

1873 GANTILLON, D. *Lyon* (*Rhône*). — Dressed silk handkerchiefs.

1874 VILLARD & JACKSON, *Lyon* (*Rhône*). —Black and coloured velvet.

1875 ROUGIER & Co. *Lyon* (*Rhône*).—Plain and worked silk fabrics.

1876 RIBOUD, J., PRAVAZ, H. & Co. *Lyon* (*Rhône*).—Silk crapes, and stuffs.

1877 GAUTHIER, J. & Co. *Lyon* (*Rhône*).—Plain velvet.

1878 SÉVÈNE, BARRAL, & Co. *Lyon* (*Rhône*).—Silk fabrics, plain and figured.

1879 DONAT, A. & Co. *Lyon* (*Rhône*).—Silk fabrics, velvet.

1880 FONT, CHAMBEYRON, & BENOIT, *Lyon* (*Rhône*).—Plain velvet.

1881 CAQUET-VAUZELLE, & COTE, *Lyon* (*Rhône*).—Plain and figured silks.

1882 BOCOUP, VILLARD, & Co. *Lyon* (*Rhône*).—Silk fabrics, Chambery gauze.

1883 BARBEQUOT, CHENAUD, & Co. *Lyon* (*Rhône*).—Silk shawls.

1884 BOYRIVEN, BROS. *Lyon* (*Rhône*).—Carriage lace.

1885 ALGOUD, BROS. *Lyon* (*Rhône*).—Black taffetas.

1886 ARAUD, BROS. *Lyon* (*Rhône*).—Silk fabrics for umbrellas.

1887 YÉMÉNIZ, *Lyon* (*Rhône*).—Silk for furniture.

1888 TAPISSIER, JUN. & HUTET, P. *Lyon* (*Rhône*).—Silk fabrics velvet.

1889 SÈVE & Co. *Lyon* (*Rhône*).—Plain black and coloured velvet.

1890 VANEL, L. & Co. *Lyon* (*Rhône*).—Worked silk fabrics.

1891 MILLION, J. P. & Co. *Lyon* (*Rhône*). —Silk fabrics, velvet.

1892 TAPISSIER, JUN, & DEBRY, *Lyon* (*Rhône*).—Black taffetas and stuffs of silk.

1893 SILO, COUSINS, *Lyon* (*Rhône*).—Silk fabrics.

1894 RIGOD, *Lyon* (*Rhône*).—Velvet.

1895 ROSSET, RENDU, & Co. *Lyon* (*Rhône*). —Silk fabrics, shawls.

1896 BRISSON, BROS. *Lyon* (*Rhône*). — Plain velvets, plushes, &c.

1897 BAYARD, BROS. *Lyon* (*Rhône*).—Plain and figured silk fabrics, galoons.

1898 BELLON, BROS. & CONTY, *Lyon* (*Rhône*).—Plain silk fabrics.

1899 COCHAUD, ADAM, & Co. *Lyon* (*Rhône*).—Plain and figured silks.

1900 CHARBIN & TROUBAT, *Lyon* (*Rhône*). —Plain velvet.

1901 HAMELIN, A. *Paris*.—Twisted silk for sewing and embroidery.

1902 FABRE, C. *Paris*.—Twisted silk, for sewing and embroidery, plain and dyed.

1903 CHILLIAT, E. *Paris*.—Silks, milled, undyed, and dyed.

1904 PLAILLY, N. P. *Paris*. — Sewing silk.

1905 HOCK, A. *Strasbourg* (*Bas-Rhin*).—Cottons and silks, yellow, and in imitation of straw and wood.

1906 BALLY, J. *Paris*.—Ribbons for orders of knighthood.

1907 DEBARY-MERIAN, *Guebwiller* (*Haut-Rhin*).—Plain and figured ribbons.

1908 WILLIAM, S. *Soultz* (*Haut-Rhin*).—Black silk ribbons, satined and figured taffetas.

1910 DENIS, A. *St. Etienne* (*Loire*).—Velvet, and articles in lace.

1911 GIRINON, JUN. *St. Etienne* (*Loire*).—Articles in lace.

1912 FAURE, E. *St. Etienne* (*Loire*).—Articles in lace.

1913 LARCHER, FAURE, & CO. *St. Etienne* (*Loire*).—Plain and figured silk ribbons.

1914 REBOURG, C. *St. Etienne* (*Loire*).—Figured silk ribbons.

1915 CALEMARD, J. *St. Etienne* (*Loire*).—Plain and figured silk ribbons.

1916 EPITALON, BROS. *St. Etienne* (*Loire*).—Plain satin ribbons.

1917 JOUCERAND, C. *St. Etienne* (*Loire*).—Plain and figured silk ribbons.

1918 COLLECTIVE EXHIBITION OF THE CITY OF ST. CHAMOND (*Loire*).—Hanks of silk, sewing silk, &c. (4 Exhibitors).

1919 RENARD, BROS. *Sarreguemines* (*Moselle*).—Silk plushes for hats.

1920 MASSING BROS. & CO. *Puttelange* (*Moselle*).—Silk plush for hats.

1921 LACOUR & WALTER, *Sarreguemines* (*Moselle*).—Silk, and cotton plush for hats.

1922 MASSING, P. & CO. *Sarreguemines* (*Moselle*).—Silk plush for hats.

1923 BARALLON & BROSSARD, J. *St. Etienne* (*Loire*).—Ribbons made of raw silk.

1924 BALAY, J. & CO. *St. Etienne* (*Loire*).—Ribbons made of raw silk, and dyed in the piece.

1925 NEYRET, J. *St. Etienne* (*Loire*).—Ribbons for special purposes.

1926 DUGNAT-GAUTHIER, *St. Etienne* (*Loire*).—Silk and velvet ribbons.

1927 GÉRENTET & COIGNET, *St. Etienne* (*Loire*).—Plain and figured silk ribbons.

1928 DONZEL, L. *St. Etienne* (*Loire*).—Silk and velvet ribbons.

1929 DESCOURS, A. *St. Etienne* (*Loire*).—Silk and velvet ribbons.

1930 AVRIL, A. & CO. *St. Etienne* (*Loire*).—Plain and figured velvet ribbon.

1931 GIRON BROS. *St. Etienne* (*Loire*).—Velvet fabrics, and ribbons.

1932 DAVID, J. B. *St. Etienne* (*Loire*).—Ribbons of silk, and velvet.

1933 PILLET-MEAUZÉ, & SON, *Tours* (*Indre-and-Loire*).—Silk fabrics for furniture, &c.

1934 FEY, MARTIN, EUDE, & VIEUGUÉ, *Tours* (*Indre-and-Loire*).—Silk fabrics for furniture, &c.

1935 LACHARD & BESSON, *Lyon* (*Rhône*).—Silk for dresses; shawls.

1936 BERNARD-JOLY, & CHAPPET, *Lyon* (*Rhône*).—Velvet.

1937 GALLAND, F. *Lyon* (*Rhône*).—Figured silks.

1938 BADOIL, G. & CO. *Lyon* (*Rhône*).—Plain and figured silks, shawls.

1939 BERLIE, A. SON, & CO. *Lyon* (*Rhône*).—Plain and figured silk fabrics.

1940 BARDON & RITTON, *Lyon* (*Rhône*).—Plain and figured silks.

1941 SERVANT DEVIENNE & CO. *Lyon* (*Rhône*).—Plain and figured silks for waistcoats and cravats.

1942 PONCET, JUN. & CO. *Lyon* (*Rhône*).—Silk fabrics for umbrellas.

1943 NICOLAS, F. & CO. *Lyon* (*Rhône*).—Fabrics of silk, of silk and wool, of wool and cotton, and of silk and cotton.

1944 GOURD, CROIZAT, SON, & DUBOST *Lyon* (*Rhône*).—Plain and figured silks.

1945 MAUVERNAY & DUBOST, *Lyon* (*Rhône*).—Black and coloured silks.

1946 GUISE & CO. *Lyon* (*Rhône*).—Plain silk and velvet fabrics.

1947 MERCIER, VUILLEMOT, & NEYRET, *Lyon* (*Rhône*).—Velvets, and fabrics for waistcoats.

1948 CHARBONNET & VILLATTE, *Lyon* (*Rhône*).—Silk fabrics.

1949 VERPILLAT, J. *Lyon* (*Rhône*).—Plain silk fabrics.

1950 BRÉBANT, SALOMON, & CO. *Lyon* (*Rhône*).—Silk fabrics.

1951 BELMONT-TERRET, & CO. *Lyon* (*Rhône*).—Plain silk fabrics.

1952 BÉRARD E. PONCET, & CO. *Lyon* (*Rhône*).—Silks for dresses.

1953 FAVROT, BROS. *Lyon* (*Rhône*).—Plain, printed, and figured silk handkerchiefs.

1954 LYON, A. & CO. *Lyon* (*Rhône*).—Black silk fabrics, silk handkerchiefs, velvet.

1955 KUPPENHEIM, *Lyon* (*Rhône*).—Printed handkerchiefs.

1956 EMERY, L. *Lyon* (*Rhône*).—Plain and figured silks.

1957 TRAPADOUX, A. & CO. *Lyon* (*Rhône*).—Silks for handkerchiefs.

1958 RÉROLLE, G. & CO. *Lyon* (*Rhône*).—Printed handkerchiefs.

1959 RONZE & VACHON, *Lyon* (*Rhône*).—Silk fabrics.

1960 MENET, J. H. & DURINGE, S. *Lyon* (*Rhône*).—Silk fabrics.

1961 LABORÉ, RODIER, & CO. *Lyon* (*Rhône*).—Figured silks.

1962 VALANSOT, M. *Lyon* (*Rhône*).—Plain silk fabrics.

1963 BOUVARD & SON, *Lyon* (*Rhône*).—Silk fabrics for furniture, and church ornaments.

1964 TABART, G. F. & CO. *Lyon* (*Rhône*).—Plain and coloured velvet.

1965 SAVOYE, RAVIER & CHANU, *Lyon* (*Rhône*).—Plain and figured silks.

1966 ROCHE & DIME, *Lyon* (*Rhône*).—Shawls; figured silks.

1967 PONSON, C. *Lyon* (*Rhône*).—Silk fabrics; velvets.

1968 MEYNIER, P. *Lyon* (*Rhône*).—Figured silks.

1969 JANDIN, C. & DUVAL, A. *Lyon* (*Rhône*).—Printed silk handkerchiefs.

1970 GINDRE & CO. *Lyon* (*Rhône*).—Plain silk fabrics.

1971 MONTESSUY, A. & CHOMER, A. *Lyon* (*Rhône*).—Silk fabrics; crape, muslin, &c.

1972 LAMY, A. *Lyon* (*Rhône*).—Figured silk for dresses.

1973 GONDRE & Co. *Lyon (Rhône).*—Plain silk fabrics; velvet.

1974 FRANC, SON, & MARTELIN, *Lyon (Rhône).*—Fancy silk thread.

1975 COLLET-LEFRANCQ, *Amiens (Somme).*—Raw silk waste, carded, and spun.

1976 RÉVIL, C. & Co. *Amilly (Loiret).*—Spun silk wadding.

1977 BINDSCHEDLER, LEGRAND, & FALLOT, *Thann (Haut-Rhin).*—Spun silk wadding.

1978 BLONDEAU-BILLIET, *Lille (Nord).*—Spun silk wadding.

1979 HENNECART, J. F. *Paris.*—Silk fabrics for bolting flour, &c.

1980 GASCOU, NEPHEW, & ALBRESPY, A. *Montauban (Tarn-and-Garonne).*—Fabrics for bolting flour; raw silk.

1981 COUDERC A. & SOUCARET, JUN. *Montauban (Tarn-and-Garonne).* — Silk abrics for bolting flour; raw silk.

Addenda.

1982 REYBAUD, J. *Lyon (Rhône).*—Portraits on silk, executed by a Jacquart loom.

DE BAILLET, *St.-Germains-es-Mons (Dordogne).*—Silk.

BANNETON, *St.-Vallier (Drôme).*—Raw silk, organzine.

BARRÈS, BROS. *St.-Julien-en-St.-Alban (Ardèche).*—Raw and prepared silk.

BERARD & BRUNET *(Rhône).*—Raw and prepared silk.

BISCARRAT, P. *Bouchet (Drôme).*—Raw and prepared silk.

BLANCHON, L. *St.-Julien-en-St.-Alban.*—Raw and prepared silk.

BLANCHON, JUN. *Flaviac (Ardèche).*—Raw and prepared silk.

BLANC-MONTBRUN, *La Rolière (Drôme).*—Raw and prepared silk.

BOISRAMEY, JUN. *Caen (Calvados).*—Raw and prepared silk, for blondes and net, &c.

BOUNET & BOUNIOLS, *Vigan (Gard).*—Raw silk.

RONDET, F. *Uzès (Gard).*—Raw silk.

BROUILHET & BAUMIER, *Vigan (Gard).*—Raw and prepared silk.

BUISSOU, C. *Tronche (Isère).*—Raw and prepared silk.

CHABAUD, A. *Avignon (Vaucluse).*—Raw and prepared silk.

CHAMBOU, WIDOW, *St.-Paul-Lacoste (Gard).*—Raw silk, and organzine.

CHAMPAUHET-SARGEAS, BROS. *Vals (Ardèche).*—Raw and prepared silk.

CHANGEA, *Lamastre (Ardèche).*—Raw and prepared silk.

CHARTRON & SON, *St.-Vallier (Drôme).*—Silk.

COMBIER, BROS. *Livron (Drôme).*—Raw silk.

COMTESSE C. DE CORNEILLAN, *Paris.*—Raw silk.

CORNEILLE & FABRE, *Trans (Var).*—Raw silk, &c.

FARA, JUN. *Bourg-Argental Loire).*—Silk for laces.

FOUGEIROL A. *Ouières Ardèche).*—Raw silk organzine.

FRANQUEBALME & SON, *Avignon (Vaucluse).*—Worked Chinese and Japanese silk, for woof and nap.

FRIGARD, *Bourg-Argental (Loire).*—White and yellow raw silk.

GAUTIER, F. *Chabeuil (Drôme).*—Raw and prepared silk.

HELME, A. *Loriol (Drôme).*—Raw and prepared silk.

LACOMBE, J. *Alais (Gard).* — Raw silk.

LACROIX, P. *(Drôme).*—Raw and prepared silk.

LASCOUR, *Crest (Drôme).*—Raw and prepared silk.

LEYDIER, BROS. *Buis-lès-Baronnies (Drôme).*—Raw and prepared silk.

MAHISTRE, A. JUN. *Vigan (Gard).*—Raw silk.

MARAVAL & Co. *Lavaur (Tarn).*—Raw silk.

MARTIN, L. & Co. *Lasalle (Gard).*—Raw silk.

MONESTIER, SEN. *Avignon (Vaucluse).*—Raw and prepared silk.

AGRICULTURAL ASSOCIATION OF MONTAUBAN *(Tarn-and-Garonne).*—Silk.

SOCIETY OF SCIENCE AND AGRICULTURE OF MONTAUBAN *(Tarn-and-Garonne.)*—Silk.

NOURRIGAT, *Lunel (Herault).*—Raw silk.

NOYER, BROS. *Dieulefit (Drôme).*—Raw and prepared silk.

PALLUAT & Co. *Lyon (Rhône).*—Prepared silk.

PASQUAY, BROS. & Co. *Wasselonne (Bas-Rhin).*—Spun silk wadding.

PRADIER, J. *Annonay (Ardèche).*—Raw and prepared silk.

RAIBAUD-L'ANGE, DIRECTOR OF THE FARM SCHOOL OF PAILLEROLS.—

REGARD, BROS. *Privas (Ardèche).*—Raw and prepared silk.

SÉRUSLAT, L. *Étoile (Drôme).*—Prepared silk.

TESSIER DU CROS, *Valleraugue (Gard).*—Raw and prepared silk.

VERNET, BROS. *Beaucaire (Gard).*—Raw and prepared silk.

CLASS 21.

1991 LELARGE. F. & AUGER, A. *Reims (Marne).*—Flannels.

1992 BOUFFARD, FERRIER, & Co. *Paris, and Reims.*—Flannel, various kinds of cloth.

1994 PRADINE & Co. *Reims (Marne).*—Merinos, wove by power-loom.

1995 LUCAS, BROS. *Rèims (Marne).*—Yarn from carded wool; merinos.

1996 PHILIPPOT, J. M. *Reims (Marne).*—Merinos, figured fabrics.

1998 VILLEMINOT, *Reims (Marne).*—Yarn, merinos.

1999 ROGELET, C., GAND, BROS. GRANDJEAN, IBRY, & Co. *Reims (Marne).*—Yarn, merinos.

2000 GILBERT & Co. *Reims* (*Marne*).—Yarn, merinos.

2001 CROUTELLE, ROGELET, GAND, & GRANDJEAN, *Reims* (*Marne*).—Yarns, flannel.

2002 SAUTRET, A. T. *Bétheniville* (*Marne*).—Merinos.

2003 ROBERT-GALLAND, *Pontfaverger* (*Marne*).—Merinos.

2004 OUDIN, BROS. *Reims* (*Marne*).—Merinos.

2005 MAUGRAS, H.—Merinos.

2006 CHATELAIN-FÉRON, *Reims* (*Marne*).—Cloth, flannel, plain and mixed.

2007 DESTEUQUE, BOUCHEZ & QUENOBLE, BROS. *Reims* (*Marne*).—Cloth, flannel, figured tissues.

2008 JOLTROIS, C. *Reims* (*Marne*).—Cloth, flannel.

2009 BENOIST-MALOT, & WALBAUM, *Reims* (*Marne*).—Cloth, shawls.

2011 BENOIST & GREVIN, *Reims* (*Marne*).—Flannel, fabrics of cotton, and of wool figured.

2012 APPERT-TARTAT, *Reims* (*Marne*).—Cloth.

2014 VIÉVILLE & Co. *Reims* (*Marne*).—Cloth.

2017 HARMEL, BROS. *Warmeriville* (*Marne*).—Woollen yarn.

2018 PIERRARD-PARPAITE, *Reims* (*Marne*).—Combed wool.

2019 HOLDEN, *Reims* (*Marne*).—Combed wool.

2020 LEGRAND, T. & SONS, *Fourmies* (*Nord*).—Combed wool, woollen yarn, and woollen fabrics.

2021 COLLECTIVE EXHIBITION BY THE DEPARTMENT OF THE SOMME.—Cotton, linen, and woollen yarn; velvet; various fabrics (20 Exhibitors).

2022 COLLET-DUBOIS & Co. *Amiens* (*Somme*).—Fabrics of wool, of wool and cotton, and of wool and silk.

2023 CRIGNON, SON, & HUE, *Amiens* (*Somme*).—Combed wool, yarn for embroidery, lace, &c.

2024 GAUTHIER, E. *Amiens* (*Somme*).—Shawls of wool, of wool and cotton, and of wool and silk; fabrics of wool, and of wool and silk, &c.

2025 VULFRAN-MOLLET, *Amiens* (*Somme*). *and Paris*.—Plain and figured fabrics of wool and silk.

2026 PAYEN & Co. *Amiens* (*Somme*).—Utrecht velvet.

2027 HORDÉ, E. & Co. *Amiens* (*Somme*).—Plain and worked fabrics, satins, fabrics of wool and silk.

2028 BARIL, SON, & Co. *Amiens* (*Somme*).—Utrecht velvet made with the power-loom, cotton velvet.

2029 GAMONNET-DEHOLLANDE, *Amiens* (*Somme*).—Satins.

2030 BOQUET, J. & Co. *Amiens* (*Somme*).—Utrecht velvet, satins, &c.

2031 FÉRY, J. JUN. *Metz* (*Moselle*).—Circular felts, for paper machines.

2032 CRÉTIEN, DEBOUCHAUD, MATTARD, VERIT, & Co. *Nersac* (*Charente*).—Circular felts.

2033 VOUILLON, F. *Louviers* (*Eure*).—Felted thread, and cloth made of it.

2038 DAVID-LABBEZ & Co. *Saint-Richaumont* (*Aisne*).—Twisted wool and cotton yarn; merinos.

2039 AUDRESSET, SON, & MENUET, *Paris*.—Cachemire wool, &c. combed, spun, and woven.

2040 NOIRET, CHOPPIN, & Co. *Paris*.—Combed wool, woollen yarn, &c.

2041 BELLOT, C. & COLLIÈRE, O. *Angecourt* (*Ardennes*).—Yarn of carded wool.

2043 COLBECK & GREVEN, *Douai* (*Nord*)—Unravelled wool.

2044 VERNET, BROS. *Nimes* (*Gard*).—Carded wool.

2045 PASQUAY, BROS. & Co. *Wasselonne* (*Bas-Rhin*).—Spun wool and wadding, &c.

2046 PALLIER, P. *Nimes* (*Gard*).—Laces and cords of cotton, silk and caoutchouc; silk, alpaca, &c.

2047 DE FOURMENT & SON, *Cercamp-lez-Frévent* (*Pas-de-Calais*).—Yarn of combed wool.

2048 ARRECKX-COLLETTE, WIDOW, *Tourcoing* (*Nord*).—Woollen yarn.

2050 BUIRETTE-THIAFFAIT, & FARAGUET, *Dijon* (*Côte-d'Or*).—Carded and spun wool.

2051 ROBIN, A. *Dijon*, (*Côte-d'Or*).—Woollen yarns.

2052 KOECHLIN-DOLLFUS, *Mulhouse* (*Haut-Rhin*).—Unbleached wool.

2053 MASSE, H. & CRESSIN, JUN. *Corbie* (*Somme*).—Twisted yarns.

2054 CRESSIN, C., DEWAILLY, E. & Co. *Fouilloy* (*Somme*).—Twisted yarn of cotton, of wool, of silk, &c.

2055 OUIN, A. & Co. *Paris*.—Dyed unravelled wool.

2056 HARTMANN, SCHMALZER, & Co. *Malmerspach* (*Haut-Rhin*.)—Woollen yarn, unbleached, plain, twisted, and milled.

2057 CHEGUILLAUME, P. & Co. *Cugand* (*Vendée*).—Cloth; fabrics of wool, and of cotton.

2058 HONNORAT, E. JUN. *St. André* (*Basses-Alpes*).—Cloth; fleeces, washed and unwashed.

2059 SIGNORET, P. *Vienne* (*Isère*).—Drapery, &c.

2060 ROUSSETT, SEN. & Co.—

2061 POIX-COSTE, WIDOW, BARBARIN, & Co. *Vienne* (*Isère*).—Drapery.

2062 GOUET, NEPHEW, & SON, *Vienne* (*Isère*).—Drapery.

2063 LAFONT & GAY, *Vienne* (*Isère*).—Drapery, &c.

2064 GONDALBERT & BURLE.—

2065 PERTUS & JULLIEN, A. *Vienne* (*Isère*).—Cloth; fancy articles for trousers.

2066 BONON, & ALEX BROS. *Vienne* (*Isère*).—Drapery.

2067 REYMOND & BAYARD-BARON, *Vienne* (*Isère*).—Drapery.

2068 SAVOYE, BLANC, & Co. *Vienne* (*Isère*).—Double milled cloth.

2069 PONCHON, SEN. *Vienne* (*Isère*).—Double-milled cloth, &c.

2070 GAUDCHAUX-PICARD (SONS), *Nancy, (Meurthe)*.—Plain and figured cloth.

2071 BLIN, SON, & BLOC, *Bischwiller (Bas-Rhin)*.—Black and coloured cloth.

2072 ROUSTIC, WIDOW A. *Carcasonne (Aude)*.—Plain and figured cloth.

2073 COLLECTIVE EXHIBITION OF THE CITY OF CASTRES *(Tarn)*.—Fabrics of wool, of wool and cotton, of wool and silk; drapery, &c.

2074 COLLECTIVE EXHIBITION OF THE CITY OF MAZAMET *(Tarn)*.—Flannel, figured woollen fabrics, &c.

2075 VERNAZOBRES, BROS. *Bédarieux (Hérault)*.—Plain and figured cloth.

2077 DAUTRESME, D. JUN. *Rouen (Seine-Inf.)*.—Figured cloth.

2078 MAUZIÈRES, SEN. & CO. *Castres (Tarn)*.—Flannel, figured cloth.

2079 BATUT-LAVAL, P. & CO. *Castres (Tarn)*.—Drapery.

2080 HOULÈS, SON, & CORMOULS, *Mazamet (Tarn)*.—Velvet, figured cloth.

2081 OLOMBEL, BROS. *Mazamet (Tarn)*.—Woollen fabrics, and velvets, &c.

2082 JUHEL-DESMARES, J. *Vire (Calvados)*.—Cloth, satins.

2083 LENORMAND, A. *Vire (Calvados)*.—Cloth, and woollen velvet.

2084 DEMAR, L. & CO. *Elbeuf (Seine Inf.)*.—Fancy cloth.

2085 THILLARD, J. & CO. *Elbeuf (Seine-Inf.)*.—Stuffs for garments.

2086 LEGRIX & MAUREL, *Elbeuf (Seine-Inf.)*.—Stuffs for garments.

2087 GERIN-RÔSE, H. *Elbeuf (Seine-Inf.)*.—Figured fabrics.

2088 CHARY F. & LAFENDEL, J. *Elbeuf (Seine-Inf.)*.—Cloth.

2089 VAUQUELIN, F. *Elbeuf (Seine-Inf.)*.—Fancy stuffs for trousers.

2090 COSSE, L. *Elbeuf (Seine-Inf.)*.—Stuffs for trousers; drapery.

2091 BELLEST, E., BENOIST, & CO. *Elbeuf (Seine-Inf.)*.—Cloth, satins.

2092 MIGNART & CO. *Elbeuf (Seine-Inf.)*.—Stuffs for garments.

2093 DUCLOS, L. & NEPHEW, *Elbeuf (Seine-Inf.)*.—Cloth for uniforms, &c.

2094 LEVIEUX, F. A. *Elbeuf (Seine-Inf.)*.—Fancy goods.

2095 OSMONT, A. & LHERMUZEAX, *Elbeuf (Seine-Inf.)*.—Fancy goods.

2096 POUSSIN, A. & SON, *Elbeuf (Seine-Inf.)*.—Cloth, black satins.

2097 CHENNEVIÈRE, WIDOW, & SON, *Elbeuf (Seine-Inf.)*.—Fancy goods.

2098 DESBOIS, P. & CO. *Elbeuf (Seine-Inf.)*.—Cloth, &c.

2099 BORDEREL, JUN. *Sedan (Ardennes)*.—Cloth, woollen velvet.

2100 DE MONTAGNAC, E. & SON, *Sédan (Ardennes)*.—Fabrics for garments, woollen velvet.

2101 LECOMTE, BROS. *Sédan (Ardennes)*.—Cloth, satins, velvet.

2102 LABROSSE, BROS. *Sedan (Ardennes)*.—Black and coloured cloths; plain and figured velvets.

2103 CUNIN-GRIDAINE, SON, & CO. *Sédan (Ardennes)*.—Cloth, satin, velvet, drapery.

2104 BERTÈCHE, BOUDOUX-CHESNON, & CO. *Sédan (Ardennes)*.—Cloth, satins, &c.

2105 RENARD, A. *Sédan (Ardennes)*.—Cloth. satins, &c.

2106 HULIN, E. G. & PAQUIN, H. JUN. *Sedan (Ardennes)*.—Cloth, satin, &c.

2107 BACOT, P. & SON, *Sedan (Ardennes)*.—Cloths, fancy goods, &c.

2108 BACOT, F. & SON, *Sédan (Ardennes)*.—Cloth, satins, &c.

2109 POITEVIN & SON, *Louviers (Eure)*.—Cloth and fancy goods.

2110 NOUFFLARD C. & CO. *Louviers (Eure)*.—Cloth, stuffs, &c.

2111 BRUGUIÈRE, H. & CO. *Louviers (Eure)*.—Cloth, stuffs, &c.

2112 JEUFFRAIN & SON, *Louviers (Eure)*.—Cloth, satins, stuffs, &c.

2113 PELLIER & TRUBERT, *Louviers (Eure)*.—Fancy goods.

2114 REMY, E. & PICARD, P. *Louviers (Eure)*.—Stuffs, fancy goods.

2115 GASTINNE H. *Louviers (Eure)*.—Fancy goods.

2116 POITEVIN, C. *Louviers (Eure)*.—Cloth, and stuffs.

2117 DANNET & CO. *Louviers (Eure)*.—Fancy goods, &c.

2118 RENAULT, R. & CO. *Louviers (Eure)*.—Cloth for various purposes.

2119 BRETON, L. & BARBE, L. *Louviers (Eure)*.—Cloth, and fancy goods.

2120 BERTIN, J. & PENNELLE, J. *Louviers (Eure)*.—Fancy goods.

2121 HURSTEL, BROS. *Ribemont (Aisne)*.—Combed and spun wool.

2122 LEFEBVRE-DUCATTEAU, BROS. *Roubaix (Nord)*.—Combed, carded, and spun wool, stuffs, &c.

2123 CATTEAU, P. *Roubaix (Nord)*.—Stuffs for dresses.

2124 DELATTRE, H. & SON. *Roubaix (Nord)*.—Combed and spun wool; plain and mixed fabrics.

2125 CORDONNIER, L. *Roubaix (Nord)*.—Mixed fabrics for dresses.

2126 HARINKOUCK & CUVILLIER, *Roubaix (Nord)*.—Fabrics for furniture.

2127 LAGACHE J. *Roubaix (Nord)*.—Fabrics for waistcoats.

2128 BULTEAU, BROS. *Roubaix (Nord)*.—Fabrics for dresses, and shawls.

2129 SCRÉPEL-LEFEBVRE, *Roubaix (Nord)*.—Fabrics of wool, of wool and cotton, of wool and silk.

2130 SCREPEL, L. & SON, *Roubaix (Nord)*.—Fabrics of wool, and of wool and silk.

2131 SCREPEL-ROUSSEL. *Roubaix (Nord)*.—Fabrics of wool, of wool and cotton, of wool and silk.

2132 DILLIES, BROS. *Roubaix (Nord)*.—Plain and figured fabrics mixed with cotton and silk.

2133 SADON & CO. *Roubaix (Nord)*.—Figured stuffs of wool and silk; shawls, &c.

2134 CAZIER, E. *Roubaix (Nord)*.—Woollen yarn, for poplins and shawls.

2135 PROUVOST A. & Co. *Roubaix* (*Nord*).—Combed wool.

2136 ERNOULT-BAYART & SON, *Roubaix* (*Nord*).—Yarns of carded wool.

2137 CATTEAU-LEPLAT, *Roubaix* (*Nord*).—Woollen stuffs for dresses, satins.

2138 SCRÉPEL, C. *Roubaix* (*Nord*).—Woollen fabrics.

2139 PIN-BAYART, *Roubaix* (*Nord*).—Woollen fabrics.

2140 VAN-DONGHEN, E. & V. *Roubaix* (*Nord*).—Fancy goods for dresses.

2141 POUILLIER-DELERUE, *Roubaix* (*Nord*).—Woollen fabrics, fancy goods.

2142 HEYNDRICKX-DORMEUIL, WIDOW, *Roubaix* (*Nord*).—Stuffs for waistcoats.

2143 MOURCEAU, C. H. *Paris.*—Tapestries for furniture.

2144 PIN, A. & Co. *Lyon* (*Rhône*).—Worked shawls.

2145 RIVOIRON, PERRAUD, & GUIGNARD, *Lyon* (*Rhône*).—Worked shawls of wool and cashmere.

2146 GELLIN, C. & Co. *Lyon* (*Rhône*).—Worked shawls of wool and cashmere.

2147 DAVIN, F. *Paris.*—Combed and spun wool.

2148 BOUTEILLE, COUSINS, *Paris.*—Woollen shawls.

2149 CHAPUSOT, PREVOST, & BOÏNG, *Paris.*—Worked woollen shawls.

2150 RIBES, ROUX, & DURAND, *Nîmes* (*Gard*).—Worked shawls.

2151 BIÉTRY, L. *Paris.*—Cashmere fabrics for dresses, &c.; shawls.

2152 SEYDOUX, A. SIEBER, & Co. *Le Cateau* (*Nord*).—Merino wool combed and spun, pure and mixed fabrics.

2153 LAIR, H. & LAIR, H. *Paris.*—Worked woollen shawls.

2154 HESS, G. *Paris.*—Fabrics of goose-down; fancy goods.

2155 DUCHÉ, A., DUCHÉ, JUN., BRIÈRE, & Co. *Paris.*—Worked shawls.

2156 FABART & Co. *Paris.*—Shawls.

2157 CHAMPION, L. *Paris.*—Worked shawls of cashmere and wool.

2158 AUBÉ, NOURTIER, & Co. *Paris.*—Shawls of cashmere and wool; shawls embroidered with gold and silver.

2159 BOUTARD & LASSALLE, *Paris.*—Figured shawls of wool and cashmere.

2160 BOAS, BROS. & Co. *Paris.*—Figured shawls of wool and cashmere.

2161 HÉBERT, E. F. & SON, *Paris.*—Shawls of wool and cashmere, &c.

2162 GÉRARD, C. & CANTIGNY, *Paris.*—Shawls of wool and cashmere.

2163 CATHERINE, V. & Co. *Paris.*—Shawls of wool and cashmere.

2164 DACHÈS & DUVERGER, *Paris.*—Worked shawls of wool and cashmere.

2165 BOURGEOIS, BROS. *Paris.*—Worked shawls of wool and cashmere.

2166 ROBERT & GOSSELIN, JUN. *Paris.*—Worked shawls of wool and cashmere.

2167 HOOPER, G., CARROZ, & TABOURIER, *Paris.*—Pure and mixed woollen fabrics; silk fabrics; net, laces.

2168 LEGRAND, A. *Paris.*—Worked shawls.

2169 DREYFOUS, F. *Paris.*—Mixed fabrics.

2170 VATIN, F. JUN. & Co. *Paris.*—Fabrics of silk, and of wool and silk; gauze, bareges.

2171 ZADIG, J. B. *Paris.*—Gauze, and bareges for shawls, &c.

2173 PLANCHE, L. & Co. *Paris.*—Fabrics of combed wool, muslins, &c.; plain, figured, and printed shawls; shawls and tissues of gauze and barege.

2174 SABRAN, V. & JESSÉ, G. *Paris.*—Woollen and mixed fabrics.

2175 DUNCAN & CHARPENTIER, *Paris.*—Plain and figured fabrics.

2176 WEISGERBER, BROS. & KIENER, *Paris.*—Mixed fabrics.

2177 COCHETEUX, SON, & Co. *Templeuve* (*Nord*).—Fabrics of wool and silk.

2178 JOURDAIN-HERBERT, JUN. *Amiens* (*Somme*).—Goats' hair velvet for furniture.

2179 MAZURE-MAZURE, *Roubaix* (*Nord*).—Fabrics for dresses and furniture.

2180 BERCHOUD, L. & GUERREAU, *Paris.*—Stuffs for furniture.

2181 DAGER, E., MENAGER, H., & WALMEZ H. *Paris.*—Reps for furniture; imitation tapestry.

2182 ARNAUD-GAIDAN, J. & Co. *Nîmes* (*Gard*).—Stuffs for furniture.

2183 DAMIRON & Co. *Lyon* (*Rhône*).—Shawls of wool, and cashmere.

2184 CHANEL, J. *Lyon* (*Rhône*).—Worked woollen shawls.

2185 MANTELIER, P. & Co. *Lyon* (*Rhône*).—Worked woollen shawls.

Addenda.

DESPRÉAUX, A. *Versailles* (*Seine-and-Oise*).—Stuffs for furniture.

GRASSOT & Co. *Lyon* (*Rhône*).—Fabrics of wool and cotton for furniture.

CLASS 22.

2191 IMPERIAL MANUFACTORIES OF THE GOBELINS, THE SAVONNERIE, & BEAUVAIS.—Tapestry.

2192 BRAQUENIÉ BROS. *Aubusson* (*Creuse*).—Tapestry, for hangings, &c.

2193 REQUILLART, ROUSSEL, & CHOCQUEEL, *Aubusson* (*Creuse*).—Aubusson and Moquette tapestry, tapestry panels, and tapestry for furniture.

2195 ARNAUD-GAIDAN, J. & Co. *Nîmes* (*Gard*).—High-napped tufted carpet.

2196 GRAVIER, C. & Co. *Nîmes* (*Gard*).—Tufted carpet.

2197 PLANCHON, F. & Co. *Neuilly* (*Seine*).—Neuilly tapestry.

2198 IMBS, *Brumath* (*Bas-Rhin*).—Carpets.

2199 GADRAT, P. *Meaux* (*Seine-and-Oise*).—High-napped, &c. carpets.

2200 SALLANDROUZE, J. J. & SON, *Aubusson* (*Creuse*).—Tufted carpets, moquettes.

2201 POLONCEAU, *Paris.*—Moquettes, with coating to make the nap adhere to the warp.

CLASS 23.

2211 Gros, Odier, Roman, & Co. *Wesserling* (*Haut-Rhin*).—Cotton yarn, cotton, and mixed fabrics.

2212 Paraf-Javal, Bros. & Co. *Thann* (*Haut-Rhin*).—Dyed and printed fabrics.

2213 Eck, D. *Cernay* (*Haut-Rhin*).—Cotton and mixed fabrics.

2214 Steinbach, Koechlin, & Co. *Mulhouse* (*Haut-Rhin*).—Printed fabrics.

2215 Thierry, Mieg, & Co. *Mulhouse* (*Haut-Rhin*).—Fabrics of cotton and of wool.

2216 Huguenin-Schwartz,& Conilleau, *Mulhouse* (*Haut-Rhin*).—Printed fabrics.

2217 Koechlin Bros. *Mulhouse* (*Haut-Rhin*).—Printed fabrics.

2218 Dollfus, Mieg, & Co. *Mulhouse* (*Haut-Rhin*).—Printed cotton fabrics.

2219 Lefebvre, C. *Paris*.—Printed cotton fabrics.

2220 Cantel & Co. *Rouen* (*Seine-Inf.*).—Leather-cloths, plain, printed, &c.

2221 Lamy-Godard, Bros. *Darnetal* (*Seine-Inf.*). — Printed cotton fabrics, for cravats and furniture.

2222 Gruel, *Rouen* (*Seine-Inf.*).—Printed cotton fabrics.

2223 Petel, *Darnetal* (*Seine-Inf.*).—Dyed cottons and chintzes.

2224 Bayle-Tabouret, *Darnetal* (*Seine-Inf.*).—Dyed cottons.

2225 Legras, F. *Rouen* (*Seine-Inf.*). — Dyed cottons.

2226 Gueroult, N. *Rouen* (*Seine-Inf.*).—Dyed yarns of wool, silk, and cotton.

2227 Drevon, E. Sen. *Lyon* (*Rhône*).—Black silk.

2228 Freppel, F. A. *Saint-Mandé* (*Seine*).—Dressing materials, &c.

2229 Lacuffer, Sen. *Annecy* (*Haute-Savoie*).—Cotton-yarn, fabrics, and prints.

2230 Henry & Son, *Savonnières-devant-Bar* (*Meuse*).—Dyed cotton yarn.

2231 Messier, A. *Paris*. — Dyed wools in powder, for the manufacture of velvet paper, &c.

2232 Jacques-Sauce, *Paris*.—Dyed wools and cottons in powder, for printing on stuffs, &c.

2233 Dessaint & Daliphar, *Radepont* (*Eure*).—Printed cotton fabrics.

2234 Parent, *Saint-André-lez-Lille* (*Nord*).—Dyed flax and cotton yarns.

2235 Feau-Béchard, A. *Paris*.—Dyed yarns of wool, and cashmere for dresses and shawls.

2236 Rouquès, A. *Clichy-la-Garenne* (*Seine*).—Fabrics of wool and cashmere, dyed and dressed.

2237 Pourchelle, F. *Amiens* (*Somme*).—Dyed cotton velvets.

2238 Guiliaumet, A. *Puteaux* (*Seine*).—Dyed and dressed fabrics of wool, and of wool and silk.

2239 Boulogne & Houpin, *Reims* (*Marne*).—Dyed fabrics.

2240 Veissière, *Puteaux* (*Seine*).—Dyed woollen fabrics.

2241 Seib, Hoffman, & Co. *Strasburg* (*Bas-Rhin*). — Plain and printed waxed cloths.

2242 Baudouin Bros. *Paris*.—Plain and printed waxed cloths.

2243 Le Crosnier, M. L. *Paris*.—Fabrics of cotton, of flax, and of silk gummed or waxed, plain and printed.

2244 Descat, Bros. *Flers* (*Nord*).—Dyed, printed, and dressed fabrics.

2245 Motte, A. & Co. *Roubaix* (*Nord*).—Dyed and dressed fabrics.

2246 Duparet, Madame A. *Paris*.—Printed fabrics.

2247 Petitdidier, *Paris*.—Dyed fabrics of wool, of cashmere, and silk.

2248 Guinon, Marnas, & Bonnet, *Lyon* (*Rhône*).—Dyed silks and cottons.

2249 Steiner, C. *Ribeauvillé* (*Haut-Rhin*).—Turkey-red dyed cotton fabrics, plain and printed, for furniture.

2250 Klotz, M. *Paris*.—Light fabrics for ball-dresses.

2251 Werner & Michniewicz, *Paris*.—Laces, printed fabrics imitating lace and fur.

2252 Hofer-Grosjean, E. *Mulhouse* (*Haut-Rhin*).—Printed fabrics for dresses.

2253 Onfroy & Co. *Paris*.—Fabrics of cotton, wool, and silk, printed by hand, &c.

2254 Troester, B. & Co. *Bourgoin* (*Isere*).—Printed silk fabrics.

2255 Renard Bros. *Lyon* (*Rhône*).—Silks and dyed silken fabrics.

2256 Brunet-Lecomte, H. *Bourgoin* (*Isère*).—Printed silks.

2257 Chocqueel, L. *Puteaux* (*Seine*).—Shawls, dresses, and printed fabrics for furniture.

2258 Guillaume & Son, *Saint Denis* (*Seine*).—Printed fabrics.

2259 Delamotte & Faille, *Reims* (*Marne*) Fabrics of washed merino-wool, dyed and dressed.

2260 Francillon & Co. *Puteaux* (*Seine*).—Fabrics of wool, of wool and cotton, and of wool and silk, dyed and dressed.

2261 Palloteau-Guyotin & Marquant, F. *Reims* (*Marne*).—Dressing material, used by the manufacturers of Rheims in the articles exhibited.

2262 Bernoville, Bros. Larsonnier Bros. & Chenest, *Paris*.—Combed and spun wool, woollen fabrics raw and dyed, woollen and mixed fabrics.

2263 Kœchlin-Dolphus & Co. *Mulhouse* (*Haut-Rhin*). — Wools, raw and dyed, for various purposes.

2264 Japuis, Kastner, & Carteron, *Claye* (*Seine-and-Marne*).—Printed fabrics for furniture.

Addenda.

Dopter, A. *Paris*.—Stuffs printed by lithography.

Weisgerber, Bros. & Kiener *Paris*.—Dyed wool, cotton, and silk yarn.

CLASS 24.

2271 Champailler, A. *Lyon* (*Rhône*).—Machine-made black lace.
2272 Burnier, A. *Lyon* (*Rhône*).—Fancy net.
2273 Bourry, Bros. *Paris*. — Muslins, cambrics, &c.
2274 Sajou, J. J. *Paris*.—Designs for tapestry and embroidery, &c.
2275 Dubus, T. *Paris*.—Embroidered church ornaments.
2276 Ferguson & Son, *Paris*.—Machine-made lace.
2277 Experton, R. *Le Puy* (*Haute-Loire*).—Spindle-made lace.
2278 Monard, F. T. *Paris*.—Machine-made imitation Chantilly lace.
2279 Lefort, Widow L. P. *Grand-Couronne* (*Seine-Inf.*).—Imitations of silk lace.
2280 Foulquié, Miss A. & Co. *Paris*.—Articles in net, &c.
2281 Lecomte, C. & Co. *Paris*.—Woven net, of silk and cotton.
2282 Galoppe, H. & Co. *Paris*.—Silk net, embroidered by hand, an imitation of Chantilly.
2284 Dognin & Co. *Paris*. — Various kinds of lace.
2285 Pannier, A., Ramibert, & Co. *Paris*.—Articles in lace, &c.
2286 Girard-Thibault, *Paris*. — Plain and worked articles in net, &c.
2287 Helbronner, Madame S. *Paris*.—Tapestry worked with the needle, for furniture.
2288 Garapon, L. *Paris*.—Lace articles in aluminium thread.
2289 Truchy & Vangeois, *Paris*—Embroideries and laces in gold and silver, for various purposes.
2290 Brouillet, A. *Paris*.—Laces.
2291 Chevalier, L. *Paris*.—Laces.
2292 Glénard, E. & Co. *Paris*.—Worked articles in net, &c.
2293 Pariot-Laurent, *Paris*.—Silk buttons and lace.
2294 Collective Exhibition of the City of St. Pierre-lèz-Calais (*Pas-de-Calais*).—Machine-made lace and net (10 Exhibitors).
2295 Brunot & Lefèvre, *Calais* (*Pas-de-Calais*).—Machine-made net.
2296 L'Heureux, Bros. *St. Pierre-lèz-Calais*).—Imitation Chantilly lace, silk net.
2297 Houette, L. *St. Pierre-lèz-Calais*.—Imitation of lace and silk guipure, in thread and cotton.
2298 Herbelot, Jun. & Genet-Dufay, *Calais* (*Pas-de-Calais*).—Silk net in imitation of Chantilly, and imitation silk blondes.
2299 Dubont & Sons, *Calais* (*Pas-de-Calais*).—Black silk imitation Chantilly lace.
2300 Mallet, Bros. *Calais* (*Pas-de-Calais*).—Machine-made silk blondes.
2301 Rebier, L. & Valois, F. *St.-Pierre-lèz-Calais*.—Machine-made net.
2302 Tropham, Bros. *St.-Pierre-lèz-Calais* (*Pas-de-Calais*).—Machine-made lace and blondes.
2303 Geffrier, Delisle, Bros. & Co. *Paris*.—Lace, made with spindles, and with the needle.
2304 Lecornu, *Caen* (*Calvados*).—Spindle-made lace.
2305 Lachez-Bleuze, *Paris*.—White embroidery on muslin and cambric.
2306 Lallemant, *Paris*. — White embroideries.
2307 Driou, Morest, & Co. *Paris*.—White embroideries.
2308 Moreau, Bros. *Paris*.—Worked shirt-fronts.
2309 Chapron, L. *Paris*.—Worked handkerchiefs.
2310 Husson-Hemmerlé, *Paris*.--Worked muslins and cambrics.
2311 Planche, L., Lafon, G., & Sival, D. *Paris*.—Machine-made spindle-lace.
2312 Bacouël-Troussel, A. *Arras* (*Pas-de-Calais*).—Spindle-made lace.
2313 Berr, D. & Son, *Nancy* (*Meurthe*).—Worked shirt-fronts, &c.
2314 Aubry-Febvrel, *Mirecourt* (*Vosges*).—Spindle-made guipures.
2315 Loiseau, J. B. *Paris*.—Spindle-made lace.
2316 Seguin, J. *Paris*.—Spindle-made lace.
2317 Robert-Faure, *Paris*.—Spindle-made lace.
2318 Poiret, Bros. & Nephew, *Paris*.—Articles of tapestry, &c.
2319 Roguier, C. *Paris*.—Articles of tapestry.
2320 Blazy, Bros. *Paris*.—Articles of tapestry.
2322 Raffard, E. & Gonnard, *Lyon* (*Rhône*).—Figured net.
2323 Idril, *Lyon* (*Rhône*).—Figured net, and machine-made lace.
2324 Dollfus-Moussy, *Lyon* (*Rhône*).—Figured net, and machine-made lace.
2325 Tarpin & Son, *Lyon* (*Rhône*).—Embroidery in gold.
2326 Cliff, Bros. *St. Quentin* (*Aisne*).—Machine-made blondes.
2327 Hinaut-Col. & Co. *St. Quentin* (*Aisne*).—White embroidery.
2328 Carpentier, A. *Caudry* (*Nord*).—Plain and fancy net.
2329 Maison de St. Marie, *Cherbourg* (*Manche*.)—Black spindle-made lace.
2330 Costallat-Laforque, *Bagneres-de-Bigorre* (*Hautes-Pyrenees*).—Articles in woollen net.
2331 Poignat, G. *Paris*.--Borders for hats.

Addenda.

Hooper, G., Carrez, & Tabourier, *Paris*.—Laces and nets.
Pallier, P. *Nimes* (*Gard*).—Articles in lace.
Werner & Michniewier, *Paris*.—Articles in lace.

CLASS 25.

2351 BENEDICK, *Paris.* — Buffalo-horn brushes.
2352 DUPONT & DESCHAMPS, *Beauvais (Oise).*—Brushes, buttons, bristles.
2353 PITET, SEN. & LIDY, *Paris.*—Hair-pencils, and brushes.
2354 HÉNOC, *Paris.*—Dusters.
2355 RENNES, A. J. M. *Paris.*—Brushes.
2356 ROMANCEY & RICARD, *Paris.*—Brushes.
2357 LAURENÇOT, J. E. *Paris.* — Toilet brushes.
2358 HASSE, F. *Lyon (Rhône).*—Skins and furs.
2359 BEAUFOUR-LEMONNIER, *Paris.* — Hair-work.
2360 LHUILLIER, *Paris.*—Skins and furs.
2361 PASQUIER, V. *Poitiers (Vienne).* — Dressed goose-skins.
2362 CAVY, A. *Nevers (Nièvre).* — Garments made with indigenous furs.

CLASS 26.

2371 CHANEY, J. A. *Nantes (Loire-Inf.).* —Black and white calf-skin, for boots and shoes.
2372 COURVOISIER, P. *Paris.*—Kid-skin, and kid-skin gloves.
2373 BUNEL, BROS. *Pont-Andemer (Eure).* —Leather.
2374 COLLECTIVE EXHIBITION OF THE CITY OF ANNONAY *(Ardèche).* — Tawed skins (6 Exhibitors).
2375 ROQUES, *Montpellier (Herault).* — Tanned sheep-skins.
2376 AMIC, L. & CO. *Toulon (Var).*— Polished, and morocco-dressed leather; bark.
2377 SCELLOS, E. *Paris.* — Leather, and straps.
2378 FORTIER-BEAULIEU, JUN. *Roanne (Loire).*—Legs of military boots.
2379 SUSER, H. *Nantes (Loire-Inf.).* — Leather of various kinds, for boots, shoes, &c.
2380 CAMUSAT-GUYON, *Auxerre (Yonne).* —Bridles.
2381 PINAULT-BRISOU E. *Rennes (Ille-and-Vilaine).*—Leather.
2382 ROBAUT, A. *Valenciennes (Nord).*— Leather for cards, driving belts, saddlery.
2383 LEGAL, *Chateaubriant (Loire-Inf.).* —Tanned and tawed leather.
2384 PRIN, A. *Nantes (Loire-Inf.).* — Calf skins.
2385 LES FILS COPPIN-LEJUNE, *Douai (Nord).* — Leather for cards, and spinning mills.
2386 PELLOQUIN, *Niort (Deux-Sèvres).*—
2387 BOUCHON, V. *Niort (Deux-Sèvres).* —Cow, horse, and calf-skins.
2388 HOCHEDÉ, *Montdidier (Somme).* — Strong leathers, for military shoes and boots.
2389 SUEUR G. & STOFFT, *Amiens (Somme).* —Tanned and curried leather.
2390 LES FILS D'HERRENSCHMIDT, G. F. *Strasbourg (Bas-Rhin).*—Strong leathers, for various purposes.
2391 POULLAIN-BEURRIER P. I. E. *Paris.* —Leather for spinning mills, cards, &c.
2392 CHEVILLOT, BROS. *Paris.* — Cow-skin, for uppers.
2393 MASSERANT & ROULLEAU, *Paris.*— Harness traces.
2394 JODOT, J. *Paris.* — Calf-skin and boot legs.
2395 VINCENT, J. *Nantes (Loire-Inf.).*— Boot legs.
2396 SUEUR, T. & LUTZ, *Paris.*—Varnished leather.
2397 OGEREAU, BROS. *Paris.*—Leather for boots and shoes, saddlery, &c.
2398 MARTEAU, A. & CO. *Paris.*—Black and white calf-skin; boot legs.
2399 WALTZ, B. BROS. *Paris.*—Saddlery, military boots and shoes, driving belts.
2400 GLATARD, L. *Roanne (Loire).*—Harness for unyoking instantaneously.
2401 LAMBIN & LEFÈVRE, *Paris.*—Saddles, bridles, and harness.
2402 PATUREL & BOYER, *Paris.*—Whips, and sticks.
2403 LEROUX, F. *Paris.*—Saddlery.
2404 DURAND, A. & L. BROS. *Paris.*— Tanned, curried, and polished leather, &c.
2405 PELTEREAU, A. *Château-Renault (Indre-and-Loire).* — Polished leather, for soles.
2406 PELTEREAU, P. *Château-Renault (Indre-and-Loire).* — Polished and other leather, for soles, &c.
2407 PAILLART. G. *Paris.*—Leather for spinning mills, cards, &c.
2408 LEROUX, E. & LE BASTARD, A. *Rennes (Ille-and-Vilaine).* — Strong, and polished leather.
2409 SENDRET, R. & SON, *Metz (Moselle).* —Leather for soles and uppers, for military accoutrements, &c.
2410 GALLIEN & CO. *Longjumeau (Seine-and-Oise).*—Strong leather, &c.
2411 FORTIER-BEAULIEU, C. A. *Paris.*— Saddlers' leather.
2412 STERLINGUE, *Bourges (Cher).* — Strong leather.
2414 TREMPÉ & SON, *Paris.* — Skins for shoes and boots.
2415 LOIGNON & CASSE, *Amiens (Somme).* —Muzzle for horses.
2416 COURTOIS, E. C. *Paris.*—Varnished calf-skin, for boots and shoes.
2417 LEVEN, M. & SON, *Paris.*—Curried, varnished, and black calf-skin.
2418 ALDEBERT, L. *Milhau (Aveyron).*— Black and white calf-skin.
2419 CARRIÈRE, P. BROS. DUPONT, *Milhau (Aveyron).*—Black and white calf-skin.
2420 CORNEILLAN, BROS. *Milhau (Aveyron).*—Tanned calf-skin.
2421 ALDEBERT, A., SON, & BROS. *Milhau (Aveyron).*—Calf, sheep, and lamb skin.
2422 HOUETTE, A. & CO. *Paris.*—Varnished calf-skin, for boots and shoes.
2423 JUMELLE, F. C. *Paris.*—Varnished leather, for boots and shoes, &c.

2424 NYS & Co. *Paris.*—Varnished leather, for boots and shoes.

2425 LANDRON, JUN. *Orléans* (*Loiret*).—Strong leather.

2426 LATOUCHE - ROGER, *Avranches* (*Manche*).—Skins and leather.

2427 BAYVET, BROS. *Paris.* — Morocco-dressed sheep-skin and calf-skin, for binding, &c.

2428 EHMANN, HERING, & GOERGER, *Strasbourg* (*Bas - Rhin*). — Morocco-dressed sheep-skin, tawed kid-skin.

2429 CERF - LANGENBERG, *Strasbourg* (*Bas-Rhin*).—Morocco-dressed sheep-skin, &c.

2430 VERGER, G. & C. BROS. *Pont-Audemer* (*Eure*).—Strong leather.

2430A ROULLIER, *Paris.*—Driving belts.

2430B DORÉ & Co. *Bordeaux* (*Gironde*).—Calf-skin, boot-legs.

2430C FANIEN & SONS, *Liliers* (*Pas-de-Calais*).—Tanned leather and curried skins.

CLASS 27.

2431 BEZON, J. *Lyon* (*Rhône*). — Gloves made by power-loom; Chinese net.

2432 DUVELLEROY, P. *Paris.*—Fans, and fire-screens.

2433 VANDEVOORDE, A. *Paris.*— Carved fan-mountings.

2434 ALEXANDRE, P. F. V. *Paris.*—Fans.

2435 MOLÉON, L. A. *Paris.*—Ready-made garments for men.

2436 MALLET BROS. *Lille* (*Nord*). — Ready-made garments for men.

2437 LELEUX, A. *Paris.* — Ready-made garments for men.

2438 GIBORY, *Paris.*—Court dresses, &c.

2439 WUY, A. *Paris.*—Ready-made garments for men.

2440 LEMANN, & Co. *Paris.*—Ready-made garments for men, &c.

2441 MASSEZ, M. *Paris.*—Boots and shoes.

2442 BROE, *Paris.*—Boots, and half boots.

2443 ROND, T. *Paris.*—Bottines and shoes.

2444 MÉLIÈS, L. S. *Paris.* — Boots and shoes.

2445 BESSARD, E. J. B. *Paris.*—Boots and shoes, for ladies and children.

2446 PRADEL-HUET, SEN. *Paris.*—Boots and shoes, for city wear, hunting, &c.

2447 PROUST, H. G. *Paris.*—Boots, half-boots, shoes, and slippers.

2448 DUPUIS, S. & Co. *Paris.* — Screw-boots and shoes.

2449 PELLETIER, L. *Paris.* — Boots and shoes.

2450 DELAIL, G. *Paris.*—Hunting boots.

2451 PICARD, BROS. *Paris.* — Children's boots and shoes.

2452 PINET, F. J. L. *Paris.*—Boots and shoes for ladies.

2453 CABOURG, T. *Paris.* — Screw-boots and shoes.

2454 TOUZET, J. C. *Paris.*—Boots and shoes, for men and women.

2455 LATOUR, P. *Paris.* — Riveted-boots and shoes.

2456 WALTER, F. *Paris.* — Breeches, gloves, &c.

2457 GEIGER, Z. *Paris.*—Articles for the chase, and travelling.

2458 NOËL, E. *Paris.* — Requisites for boots and shoes.

2459 CHAZELLE, E. *Tours* (*Indre-and-Loire*).—Boots and shoes, screwed and clouted by machine.

2460 MONTEUX & GILLY, *Paris.* — Boots and shoes, for men and women.

2461 ROUSSET & Co. *Blois* (*Loir-and-Cher*).—Boots and shoes, for men and women.

2462 POIRIER, P. *Chateaubriant* (*Loire-Inf.*).—Boots, half-boots, &c.

2463 HAULON, S. JUN. *Bayonne* (*Basses-Pyrénées*).—Boots and shoes.

2464 PETIT, J. A. *Paris.* — Boots and shoes, for various purposes.

2465 ETANCHAUD, F. *Paris.*—Fancy boots and shoes, for ladies.

2466 DORÉ & Co. *Bordeaux* (*Gironde*).—Riveted boots and shoes, for men, boot-tops, &c.

2467 FANIEN & SONS, *Lillers* (*Pas-de-Calais*).—Sewed, riveted, and screwed boots and shoes, for men, &c.

2468 SUSER, H. *Nantes* (*Loire-Inf.*). — Boots and shoes, &c.

2469 CLERCX, A. M. *Paris.*—Fancy boots and shoes, &c.

2470 MAYER, *Paris.* — Fancy boots and shoes.

2471 BARRÉ, WIDOW, *Paris.*—Fancy boots and shoes, for ladies.

2472 BERNARD, A. L. *Paris.*—Flowers for head-dresses.

2473 MEUREIN, A. *Paris.*—Artificial fruits, foliage, &c.

2474 PERROT, PETIT, & Co. *Paris.*—Artificial flowers, and feathers.

2475 BARDIN, J. L. F. *Paris.* — Flowers and fancy articles, made of materials obtained from quills,

2476 CHAGOT, D. A. *Paris.* — Artificial flowers, and feathers.

2477 HERPIN - LEROY, *Paris.*—Artificial flowers, and ornaments.

2478 KRAFFT, MADAME E. *Paris.*—Artificial flowers.

2479 TOURNIER, C. *Paris.* — Pistils, for artificial flowers.

2480 DEVRIÈS, & Co. *Paris.*—Feathers.

2481 MARIENVAL-FLAMET, L. *Paris.* — Artificial flowers, and feathers.

2482 JAVEY & Co. *Paris.* — Stuffs, and coloured papers for artificial flowers, &c.

2483 GÉROLD, L. A. *Paris.* — Artificial flowers.

2484 CHAGOT, A. *Paris.*—Feathers and artificial flowers.

2485 VANDEREYKEN, H. *Paris.*—Feathers.

2486 TRAVERSIER, A. *Paris.* — Bonnets and head-dresses, for ladies; mechanical hat, fitting into a box about three inches high.

2487 LANGEVIN-COULON, *Paris.* — Bonnets, caps, &c.

2488 DUFOUR, F. *Paris.*—Bonnets, head-dresses, &c.

2489 BRUN, MADAME M. *Paris.*—Head-dresses.

2490 BOUILLET, J. B. *Paris.*—Mantles, mantillas, scarfs, &c.

2491 GOSSEIN-JODON, MADAME, *Paris.*—Linen drapery, garments, head-dresses.

2492 BAPTEROSSES, *Briare* (*Loiret*).—Enamelled buttons, &c.

2493 LAVILLE, PETIT & CRESPIN, *Paris.*—Silk and felt hats.

2494 COUPIN, J. *Aix* (*Boûches-du-Rhône*).—Felt hats.

2495 BESSON, BROS. *Bordeaux* (*Gironde*).—Silk and felt hats.

2497 MOCH, F. *Paris.*—Head-dresses, for children.

2500 VIEL, WIDOW, & VALLAGNOSE, J. *Marseille* (*Boûches-du-Rhône*).—Pliable hats.

2501 HAAS, Y. *Paris.*—Hats, caps.

2502 MONROY, A. *Paris.*—Elastic hats.

2503 HISPA & BOQUET, *Toulouse* (*Haute-Garonne*).—Felt and woollen hats.

2504 VINCENDON, E. JUN. *Bordeaux* (*Gironde*).—Hats.

2505 MOSSANT, C. & SON, *Bourg-du-Péage* (*Drôme*).—Hats, felt mantles, &c.

2506 QUENOT & LEBARGY, *Paris.*—Hats of felt, silk, &c.

2507 DURST-WILD & Co. *Paris.*—Straw hats.

2508 BAMMES & Co. *Paris.*—Straw hats.

2509 LES FRERES AGNELLET, *Paris.*—Requisites for dresses, &c.

2510 GALLOT, A. SEN. *Paris.*—Straw and horse-hair hats.

2511 CHAUMONT & Co. *Paris.*—Straw hats.

2512 WILD, J. U. *Nancy* (*Meurthe*).—Straw hats.

2513 OPPENHEIM, WEILL, & DAVID, *Paris.*—Collars, shirt-fronts, &c.

2514 WERLY, R. *Bar-le-Duc* (*Meuse*).—Seamless corsets.

2515 CUNY, MADAME S. J. *Paris.*—Hygienic corsets, &c.

2516 GRINGOIRE, MADAME V. *Paris.*—Elastic corsets, &c.

2517 BOCQUET, J. T. *Paris.*—Corsets and belts.

2518 JOSSELIN, J. J. & Co. *Paris.*—Corsets.

2519 COSTALLAT-BOUCHARD, *Paris.*—Articles in crochet-work.

2520 HEMARDINGUER, *Paris.*—Shirts.

2521 DAUGARD, A. *Paris.*—Moveable busts, for hair-dressers and milliners.

2522 HAYEM, S. SEN. *Paris.*—Collars, cravats, shirts, &c.

2523 ALLAIN-MOULARD, L. A. F. *Paris.*—Articles in crochet-work, &c.

2524 RAVAUT, BOCKAIRY, BROS. & Co. *Paris.*—Ready-made garments; laces, &c.

2525 OPIGÈZ-GAGELIN & Co. *Paris.*—Court-dresses, &c.

2526 MATHIEU, F. & GARNOT, S. *Paris.*—Garments for ladies; worked shawls.

2527 CHARAGEAT, G. E. *Paris.*—Walking-sticks, umbrellas.

2528 CAZAL, R. M. *Paris.*—Walking-sticks, whips.

2529 ALLAMAGNY, P. *Paris.*—Brass tubing for umbrellas.

2530 ELLUIN, F. *Paris.*—Walking-sticks, whips, &c.

2531 LIPS, C. *Paris.*—Carved handles, for umbrellas.

2532 SALLES, JUN. & Co. *Paris.*—Gloves of woollen-web.

2533 TRÉFOUSSE, J. L. & D. *Chaumont* (*Haute-Marne*).—Kid-skin gloves.

2534 CALVAT, F. *Grenoble* (*Isère*).—Kid and lamb-skin gloves.

2535 VIAL, M. *Grenoble* (*Isère*).—Kid-skin gloves.

2536 CALVAT, E. & NAVIZET, H. *Grenoble* (*Isère*).—Kid-skin gloves.

2537 BAYOUD, A. *Grenoble* (*Isère*).—Dyed skins for gloves.

2538 GUIGNIÉ, *Grenoble* (*Isère*).—Kid-skin gloves.

2539 CURTIS, L. B. & Co. *Grenoble* (*Isère*).—Kid-skin gloves.

2540 BERTHOIN, JUN. *Grenoble* (*Isère*).—Kid-skin gloves.

2541 PIAUD, BROS. & Co. *Grenoble* (*Isère*).—Kid-skin gloves.

2542 MORIQUAND, SEN. *Grenoble* (*Isère*).—Kid-skin gloves.

2543 FRANCOZ, *Grenoble* (*Isère*).—Kid-skin gloves.

2544 ROUILLON, F. *Grenoble* (*Isère*).—Kid-skin gloves.

2545 CHEILLEY, JUN. *Paris.*—Kid-skin gloves.

2546 FONTAINE, A. & L. *Paris.*—Leather gloves.

2547 DESCHAMPS, P. J. T. *Paris.*—Kid-skin gloves.

2548 MARIOTTE, P. V. *Paris.*—Gloves, of tawed skins.

2549 GUÉRIMEAU-AUBRY, *Paris.*—Leather gloves.

2550 BONNEVAY, B. *Paris.*—Kid-skin gloves.

2551 BILLION, A. *Paris.*—Leather gloves.

2552 JOUVIN, WIDOW, X. & Co. *Paris.*—Leather gloves.

2553 ALEXANDRE, *Paris.*—Tawed and dyed skins; gloves.

2554 COURVOISIER, P. *Paris.*—Kid-skin gloves, &c.

2555 COMPÈRE, E. & DUFORT, *Paris.*—Kid-skin, and gloves.

2556 ROUQUETTE, MEUNIER, & Co. *Paris.*—Kid, beaver, and lamb-skin gloves.

2557 PRÉVILLE, A. L. *Paris.*—Kid-skin gloves.

2558 BOUDIER, MADAME M. A. *Paris.*—Long Victoria gloves.

2559 JOUVIN & Co. *Paris.*—Tawed skins; gloves.

2560 PARENT, A. & HAMET, T. & Co. *Paris.*—Silk and metal buttons.

2561 BAGRIOT, F. A. *Paris.*—Metal buttons.

2562 MARIE, E. & DUMONT, A. *Paris.*—Mother-of-pearl buttons.

2563 GRELLOU, H. *Paris.*—Silk buttons &c.

2564 LEMESLE, *Paris.*—Buttons.

2565 ROBINEAU, JUN. *Paris.*—Chaplets, medals.
2566 GOURDIN & Co. *Paris.*—Fancy buttons.
2567 RENÉE-BERNIER, *La Neuville-lèz-Corbie (Somme).*—Hosiery.
2568 BOULY-LEPAGE, *Moreuil (Somme).* —Cotton hosiery.
2569 BAIL, BROS. *Villers-Bretonneux (Somme).*—Stockings.
2570 DORÉ & Co. *Troyes (Aube).*—Stockings and under-stockings.
2571 VERJEOT-GOMIER, *Châtres (Aube).* —Cotton hosiery.
2572 MEURICE, A. *Arcis-sur-Aube (Aube).* —Hosiery.
2573 LAFAIST, V. *Aix-en-Othe (Aube).*—Wool and cotton hosiery.
2574 QUINQUARLET, A. *Aix-en-Othe (Aube).* —Stockings, with sides respectively of wool and cotton.
2575 QUINQUARLET, H. *Aix-en-Othe (Aube).* —Caps, with sides respectively of wool and cotton.
2576 MAROT, SEN. & JOLLY, *Troyes (Aube).* —Cotton hosiery.
2577 BAZIN-ARGENTIN, *Troyes (Aube)* — Cotton hosiery.
2578 FRÉROT, A. *Troyes (Aube).*—Cotton and wool under-stockings.
2579 DOUÉ & ROSENBURGER, *Troyes (Aube).*—Wool and cotton hosiery.
2580 EVRARD, E. *Troyes (Aube).*—Cotton hosiery.
2581 PORON, BROS. *Troyes (Aube).*—Cotton stockings, and under-stockings.
2582 BARADUC, J. *Paris.*—Steel buttons and buckles; steel articles for furniture.
2583 PHILIPPON-DEGOIS, *Romilly-sur-Seine (Aube).*—Stockings, and under-stockings.
2584 GORNET & BAZIN, *Romilly-sur-Seine (Aube).*—Cotton hosiery.
2585 LACOUR, *Romilly-sur-Seine (Aube).* —Cotton hosiery.
2586 COGNON, *Romilly-sur-Seine (Aube).* —Children's stockings, and under-stockings of cotton.
2587 BELLEMÈRE-VOISEMBERT, *Romilly-sur-Seine (Aube).*—Cotton stockings, and under-stockings.
2588 TAILBOUIS, E. & Co. *Paris.*—Hosiery of various materials.
2589 MEYRUEIS, BROS. *Paris.*—Hosiery, gloves.
2590 CONTOUR, A. F. *Paris.*—Hosiery, &c.
2591 MILON, SEN. *Paris.*—Hosiery, &c.
2592 BLANCHET, J. B. *Paris.*—Hosiery of various materials.
2593 PASQUAY, L. *Wasselonne (Bas-Rhin).* —Woollen socks and nets.
2596 DUCOURTIOUX, *Paris.*—Elastic stockings for varicose legs.
2597 LEDUC & CHARMENTIER, *Nantes (Loire-Inf.).*—Cotton and wool hosiery.
2598 COANET, *Nancy (Meurthe).*—Leather gloves.
2599 BRUNET, MADAME L. & Co. *Paris.*—Artificial flowers.
2600 GRUYER, *Paris.*—Umbrellas.
2601 GUYOT, C. V. *Paris.*—Braces and garters.
2602 DUPILLE, H. *Paris.*—Corsets.
2603 HUET & Co. *Rouen (Seine-Inf).*—Elastic caoutchouc fabrics, for braces, belts, garters, &c.

Addenda.
CHANDELIER, *Méru (Oise).*—Buttons.
COURVOISIER, P. *Paris.*—Gloves, and kid skins.

CLASS 28.

2631 IMPERIAL PRINTING OFFICE OF FRANCE, *Paris.*—Printed works of various kinds, maps, electrotype matrices, &c.
2632 GRUEL-ENGELMANN, *Paris.*—Fancy, and artistic bindings.
2633 LORTIC, P. M. *Paris.*—Artistic bindings.
2634 BLANCHET, BROS. & KLÉBER, *Paris.* —Various kinds of paper.
2635 ENGELMANN & GRAF, *Paris.*—Chromo-lithographs, &c.
2636 LEGRAND, *Paris.*—Letter-paper, envelopes.
2637 OUTHENIN-CHALANDRE, *Paris.*—Various kinds of paper.
2638 BOULARD, G. *Chamary (Nièvre).*—Paper for engraving and lithography.
2639 APPEL, F. A. *Paris.*—Fancy tickets, &c.
2640 OLIER, J. P. *Paris.*—Paper for bank notes and playing-cards.
2641 BOULARD, G. *Corvol-l'Orgueilleux (Nièvre).*—Various kinds of paper.
2642 BRETON, BROS. & Co. *Pont-de-Claix (Isère).*—Various kinds of paper.
2643 NISSOU, G. *Paris.*—Fancy cards; chromo-lithographs.
2644 JOINT STOCK PAPER MILLS CO. OF THE SOUCHE, *near St. Dié (Vosges).*—Various kinds of paper.
2645 ROMAIN & PALYART, *Paris.*—Fancy cards.
2646 BADOUREAU, M. P. L. *Paris.*—Lithographs, chromo-lithographs, &c.
2647 DUPUY, T. *Paris.*—Lithographs.
2648 LATUNC & Co. *Blacons (Drôme).*—Various kinds of paper.
2649 VORSTER, *Montfourat, near Angoulême.*—Various kinds of paper.
2650 DAMBRICOURT, BROS. *S. Omer (Pas-de-Calais).*—Various kinds of paper.
2651 OLLION, E. A. *Paris.* — Tracing paper.
2652 TALLE & Co. *Paris.*—Blacks, ink, varnishes, colours, paper.
2653 JOINT STOCK PAPER MILL CO. OF ESSONNE *(Seine-and-Oise).*—Various kinds of paper.
2654 BATAILLE, H. *Paris.*—Fancy paper for boxes, and fancy cards.
2655 BÉCOULET, C. & Co. *Angoulême (Charante).*—Paper for writing, lithography, and photography.

2656 PEULVEY, A. *Paris.*—Ink, carmine, &c.

2657 HÉRARD, L. *Paris.*— Copper-plate engravings.

2658 LANDA, *Chalons* (*Saone-and-Loire*).—Diagrams and cards for the Arts and Sciences; trademen's cards.

2659 NEWELL, T. F. *Paris.*—Letter-paper, &c.

2660 DESROZIER, *Moulins* (*Allier*). — Printed books.

2661 BARBAT, *Chalons* (*Marne*).—Lithographs and chromo-lithographs.

2662 BERNARD, J. & Co. *Prouzel* (*Somme*).—Various kinds of paper.

2663 JOINT STOCK PAPER MILL CO. OF THE MARAIS AND ST. MARIE, *Paris.*—Various kinds of paper.

2664 DOPTER, J. A. V. *Paris.*—Lithographic impressions on stuff, &c.

2665 COLLECTIVE EXHIBITION OF THE CITY OF ANNONAY (*Ardèche*).—Writing-paper (2 Exhibitors).

2666 COLLECTIVE EXHIBITION OF THE CITY OF ANGOULÊME (*Charente*).—Writing-paper (3 Exhibitors).

2667 LEBLOND, J. D. *Paris.*—Mannequin photographs.

2668 COSQUIN, J. *Paris.* — Topographic cards.

2669 TARDIF, M. J. A. *Paris.*—Stuffs and paper, crayons, and stamps for chalk drawing.

2670 AVRIL, BROS. *Paris.*—Plan of Paris; cards, and plans.

2671 WIESENER, P. F. *Paris.*—Specimens of typography, bank-notes, &c.

2672 JACOMME, C. & LELOUP, *Paris.*—Artistic lithographs.

2673 DE MOURGUES, C. BROS. *Paris.*—Albums and printed books.

2674 TAMBON, A. *Paris.*—Engraved copper-plates.

2675 COBLENCE, S. V. *Paris.* — Electrotypes in copper for the production of engraved plates: illustrated works.

2676 BOUDIN, L. A. E. *Paris.*—Black and coloured inks; sealing-wax.

2677 SALLE. E. *Levallois* (*Seine*).—Engravings in relief, on copper.

2678 DESJARDINS, J. L. *Paris.* — Imitations of painting, &c.

2679 CARLES, *Paris.*—Lithographs.

2680 PETITFRÈRE, T. P. *Paris.* — Composing sticks and moveable letters for gilders and book-binders.

2681 PORTHAUX, *Paris.*—Engravings and vignettes, &c.

2682 BERTAUTS, J. V. *Paris.* — Lithographs.

2683 SCHMAUTZ, C. *Paris.*—Rollers for lithographers and engravers.

2684 KNAPP, *Strasbourg* (*Bas-Rhin*).—Bronze powder.

2685 MATHIEU, C. *Paris.*—Chromo-lithographs.

2686 POUSSIN, *Paris.*—Cards.

2687 HARO, E. F. *Paris.*—Restored pictures: the paint having been separated from the wood or cloth.

2688 VINCENT & FOREST, *Paris.*—Sealing wax and wafers.

2689 BAULANT, SEN. *Paris.*—Fancy papers.

2690 FORESTIÉ, T. JUN. *Montauban* (*Tarn-and-Garonne*).—Printed books.

2691 BEAU, E. *Paris.*—Lithographs and chromo-lithographs.

2692 LEFRANC & Co. *Paris.*—Black and coloured inks.

2693 PÉQUÉGNOT, A. *Paris.*—Etchings.

2694 CREMNITZ, M. *Paris.*—Fancy cards, &c.

2695 DUNAND-NARAT, P. B. *Paris.*—Paintings on cloth, water-colour drawings, and engraving by topography.

2696 TROUILLET, A. *Paris.*—Mechanical counters.

2697 RIESTER, M. *Paris.* — Steel engravings.

2698 POUPART-DAVYL, *Paris.* — Books and engravings.

2699 LORILLEUX, C. & SON, *Paris.*—Black and coloured inks.

2700 LANÉE, *Paris.*—Maps.

2701 SAGANSAN, L. *Paris.*—Maps.

2702 SILBERMANN, G. *Strasbourg* (*Bas-Rhin*).—Specimens of typography.

2703 CHATELAIN, A. *Paris.*—Cards of the modes of communication by steam and telegraph throughout the world.

2704 DUNOD, P. C. *Paris.*—Architectural and other plates, &c.

2705 ARMENGAUD, J. E. SEN. *Paris.*—Printed works and diagrams.

2706 TURGIS, L. JUN. *Paris.*—Prints, &c. with lace borders.

2707 TANTENSTEIN, J. C. *Paris.*—Printed music books.

2708 CRÉTÉ, L. & SON, *Corbeil* (*Seine-and-Oise*).—Printed books.

2709 MOREL, A. & Co. *Paris.*—Treatises on the arts.

2710 BEST J. & Co. *Paris.* — Illustrated works, &c.

2711 RENOUARD, WIDOW J. *Paris.* — Books and engravings.

2712 BERGER-LEVRAULT, WIDOW, *Strasbourg* (*Bas-Rhin*).—Printed books.

2713 BARRE, A. *Paris.*—Coins, medals &c.

2714 DULOS, C. *Paris.*—Engravings, in intaglio and relief.

2715 DALY, C. *Paris.*—Architectural engravings.

2716 ARNOLD, WIDOW, & SON, *Lille* (*Nord*).—Bookbinding.

2717 BASSET, J. A. *Paris.* — Bordered prints, &c.

2718 BOUCHER-MOREAU, *Anzin* (*Nord*).—Account books, printed and ruled.

2719 COURT & Co. *Renage* (*Isère*). — Writing and drawing paper.

2720 GÉRAULT, H. *Paris.*—Account books.

2721 GONTHIER-DREYFUS, *Paris.* — Account books.

2722 CURMER, H. L. *Paris.* — Illustrated works.

2723 DUCROQUET V. *Paris.* — Account books, &c.

2724 DARRAS, A. D. *Paris.* — Account books.
2725 COUTTENIER-PRINGUET, *Lille (Nord).* —Specimens of book-binding.
2726 BRISSE, L. *Paris.* — Album of the French Exhibition of 1855.
2727 BRY, M. E. A. *Paris.*—Lithographs, black and coloured.
2728 FOSSEY, J. *Paris.*—Pasteboards.
2729 LATRY, A. *Paris.* — Cards, and enamelled paper.
2730 CHARPENTIER, H. D. *Nantes (Loire-Inf.).*—Illustrated works.
2731 HULOT, A. A. *Paris.*—Specimens of postage stamps, &c.
2732 DIDRON, V. *Paris.*— Archæological and other works.
2733 DUPONT, P. & Co. *Paris.*—Specimens of fancy printing, &c.
2734 BANCE, B. *Paris.*—Treatises on Architecture, and the arts.
2735 CLAYE, J. *Paris.* — Classical works, gift-books, &c.
2736 LENÈGRE, *Paris.*—Albums for drawings and photographs, &c.
2737 BARON I. S. J. TAYLOR, *Paris.* — Books of travels.
2738 MALLET-BACHELIER, A. L. J. *Paris.* —Scientific works.
2739 DÉRRIEY, J. C. *Paris.*—Specimens of type, &c.
2740 MAME, A. & Co. *Tours (Indre-and-Loire).*—Printed books; engravings.
2741 ANDRIVEAU-GOUGON, E. *Paris.* — Maps.
2742 ALMIN, A. S. *Paris.* — Pasteboard boxes.
2743 PAGNERRE, C. A. *Paris.*—Literary and popular publications, &c.
2744 CHARPENTIER, G. H. *Paris.*—Literary works.
2745 LEMERCIER, R. J. *Paris.* — Lithographs.
2746 GUILLEMOT, C. A. *Paris.*—Stamps.
2747 REPOS, E. *Paris.*—Liturgical works.
2748 PLON, H. *Paris.*—Matrixes, electrotype plates, &c.
2749 DANEL, L. *Lille (Nord).*—Specimens of typography.
2750 PERRIN, L. *Lyon (Rhône).*—Printed books.
2751 MARIÒN, A & Co. *Paris.* — Letter-paper, &c.
2752 MONCOURT, *Amiens (Somme).*—Lithographs.
2753 CABASSON, *Paris.*—Letter-paper with envelopes attached.
2754 REGNIER & DOURDET, *Paris.*—Maps on stone.
2755 HUMBLOT, CONTE & Co. *Paris.* — Crayons.
2756 BOUTON, N. V. *Paris.*—Imitation of old manuscripts, &c.
2757 LEFEVRE, T. *Paris.* — Educational works.
2758 MARICOT & VAQUEREL, *Paris.*— Coloured and fancy paper; pasteboard.
2759 ERHARD SCHIEBLE, *Paris.* — Maps on stone.
2760 FROMONT, *Paris.*—Wafers.
2761 CHARDON, F. C. SEN. *Paris.*—Copper-plate engravings.
2762 RAPINE, *Paris.*—Copper-plate relievo engravings.
2763 CORNILLAC E. & Co. *Chatillon-sur-Seine (Côte-d'Or).*—Specimens of fancy binding.
2764 JUNDT, J. & SON, *Strasbourg (Bas-Rhin).*—Pasteboard and enamelled paper.
2765 LEMAIRE-DAIMÉ, *Andrecy (Seine-and-Oise).*—Cigarette-paper.
2766 GARNIER, H. & SALMON, A. *Paris.*— Steeled copper-plates; plates engraved by autographic and photographic processes.
2767 DE CHOISEUL-BEAUPRÉ, *Chateau de Dury (Somme).*—Frame with specimens of paper cutting.
2768 WILLOUGHBY, J. *Paris.*—Pencil case with perpetual almanack.
2769 LEPRINCE, T. *Paris.*—Paper for the copying press.
2770 GODCHAUX, A. *Paris.*—Specimens of writing done in copper plate.

CLASS 29.

PUBLICATIONS EXHIBITED BY THE IMPERIAL COMMISSIONER—
2771 Works on the art of teaching.
2772 Journals of education.
2773 Works on law, rules, and discipline.
2774 Works on statistics, and reports.
2775 Works on school buildings.
2776 Information regarding infant schools and nurseries.
2777 Works on Roman Catholic religious instruction.
2778 Works on Protestant religious instruction.
2779 Works on Jewish religious instruction.
2780 Works on the art of reading.
2781 Reading books.
2782 Works on writing.
2783 Works on the French language.
2784 Works on arithmetic.
2785 Elementary works on geometry and algebra.
2786 Historical works.
2787 Geographical works.
2788 Works on the physical and natural sciences.
2789 Works on agriculture.
2790 Industrial treatises.
2791 Treatises on commerce, and the keeping of accounts.
2792 Works on living languages.
2793 Works on drawing.
2794 Works on vocal and instrumental music.
2795 Works on amusements and rewards.
2796 Works on school and communal libraries.
2797 Works on the special instruction of the blind.
2798 Works on the special instruction of deaf-mutes and idiots.

2799 Plans, &c. of school buildings and furniture.
2800 Apparatus for teaching reading and writing.
2801 Apparatus for teaching arithmetic and geometry.
2802 Apparatus for teaching geography and cosmography.
2803 Apparatus for teaching the physical and natural sciences.
2804 Apparatus for teaching drawing.
2805 Apparatus for teaching vocal and instrumental music.
2806 Gymnastic apparatus.
2807 Apparatus for the special instruction of the blind, and on deaf-mutes.
2808—60 Specimens of what has been done by the pupils of the primary schools, in various places.

CLASS 30.

2861 AHRENS, H. *Paris.*—Specimens of marquetry.
2862 GOUAULT, A. F. *Paris.* — Marble chimney-piece, style of Louis XIII.
2863 DELAPIERRE, P. & Co. *Paris.*—Frame with plate-glass, gilt and ornamented with Carton-pierre, article in carved oak.
2864 SAUVREZY, A. H. *Paris.*—Carved furniture in various styles.
2865 FOURNIER, A. M. E. *Paris.*—Sculptured and gilt furniture in various styles; hangings in the style of Louis XIV.; Venetian grass.
2866 CHARMOIS, C. *Paris.*—Ebony furniture, ornamented with bronze and mosaics.
2867 KNECHT, E. *Paris.* — Furniture in carved wood, &c.
2868 CHAIX, P. A. *Paris.*—Ebony bookcase.
2869 HUBER, BROS. *Paris.*—Carton-pierre decorations.
2870 LAURENT, E. & Co. *Paris.*—Inlaid floors and mouldings, &c.
2871 PALLU, A. & Co. *Paris*—Articles in Algerian onyx.
2872 LAROQUE, P. *Paris.*—Billiard-table &c.
2873 FOURDINOIS, H. *Paris.*—Furniture of carved ebony, decorated with marquetry, gilt, &c.
2874 GROHÉ, G. *Paris.*—Furniture of carved mahogany, ebony, &c. decorated with mosaics, and gilt bronze.
2875 DUVAL, BROS. *Paris.*—Transformable couch; fancy seats.
2876 GUÉRET BROS. *Paris.*—Carved furniture, &c.
2877 MAZAROZ-RIBAILLIER, P. *Paris.*—Furniture of carved walnut and oak, &c.
2878 GROS, J. L. B. *Paris.*—Ebony furniture, inlaid, decorated with mosaics, gilt bronze, &c.
2879 RIVART, J. N. *Paris.*—Artistic furniture, decorated with marquetry and painted porcelains.
2880 ROUX, F. *Paris.* — Inlaid furniture, decorated with gilt bronze.
2881 CRAMER BROS. *Paris.*—Bedsteads of wood, conveniently resolvable into many pieces.
2882 DEXHEIMER, P. *Paris.*—Buhl furniture, decorated with bronzes and Florentine mosaics.
2883 RAMONDENE, D. *Paris.*—Tables, capable of extension.
2884 BADÍN, MADAME R. *Paris.*—Articles in basket-work, &c.
2885 MUTET, E. *Paris.*—Cane furniture.
2886 JEANSELME, SON, & GODIN, *Paris.* —Cabinet furniture.
2887 MELLI, J. L. *Paris.* — Marbles and bronzes.
2888 MEYER, L. F. A. *Paris.*—Lacquered furniture.
2889 WIRTH, BROS. *Paris,*—Fancy carved furniture.
2890 QUIGNON, N. J. A. *Paris.*—Chairs, arm-chairs, book-case.
2891 DESPRÉAUX, A. *Versailles (Seine-and-Oise).*—Furniture stuffs.
2892 ARMÉ-MÉDARD, *Lyon (Rhône).* — Carved wood articles.
2893 ZIMMERMANN, P. *Paris.*—Gildings on wood, &c.
2894 FONTAINE, H. *Valenciennes (Nord).* —Carved oak furniture, &c.
2895 GANSER, L. G. *Paris.*—Imitation Bamboo chairs, &c.
2896 ZUBER, J. & Co. *Rixheim (Haut-Rhin).* —Paper hangings.
2897 LEGLAS-MAURICE, F. *Nantes (Loire-Inf.).*—Furniture.
2898 DULUD, J. M. *Paris.* — Leather for hangings, &c.
2899 GALLAIS, C. A. *Paris.*—Panels; lac furniture.
2900 BEAUFILS, *Bordeaux (Gironde).* — Fancy furniture.
2901 PECQUEREAU & SON, *Paris.*—Carved furniture.
2902 BALNY, J. P. *Paris.* — Side-boards, gun cases, &c.
2903 LÉROUX, C. H. F. *Paris.*—Sofa bed.
2904 LÉONARD, C. *Paris.*—Iron bedsteads, &c.
2905 DE LATERRIÈRE, J. & Co. *Paris.*—Iron-bed, &c.
2906 RIBAL, F. *Paris.* — Dining tables, capable of extension.
2907 VANLOO, P. *Paris.*—Bed, chest of drawers, work-table, &c.
2908 EVETTE, H. & YON, C. L. *Paris.*—Elastic hair quilts, mattresses, and pillows.
2909 DUBREUIL, SEN. *Paris.* — Paper hangings.
2910 REBEYROTTE, F. *Paris.*—Chairs, and arm chairs, of lacquered wood, gilt.
2911 FOSSEY, JUN. *Paris.* — Carved and decorated furniture.
2912 LEMOINE, H. *Paris.* — Carved and decorated rose-wood furniture.
2913 CREMER, J. *Paris.*—Furniture, enriched with marquetry, mosaics, &c.
2914 LOREMY & GRISEY, *Paris.* — Gilt frames and cabinets.

2915 JOSSE, C. L. *Paris*.—Paper hangings, gilt.

2916 GRANTIL, JUN. & DIDION. C. *Montigny-lez-Metz (Moselle)*.—Paper hangings.

2917 POLGE C. & BEZAULT, *Paris*.—Paper hangings.

2918 TURQUETIL & MALZARD, *Paris*.—Paper hangings in panels.

2919 DESFOSSÉ, J. *Paris*.—Paper hangings.

2920 BARTHÉLEMY & DUBREUIL, *Paris*.—Paper hangings.

2921 SEEGERS, A. *Paris*.—Paper hangings.

2922 GENOUX & Co. *Paris*.—Paper hangings.

2923 BINANT, L. A. *Paris*.—Stuffs for the decoration of ceilings, &c.

2924 LEROY, J. *Paris*.—Paper hangings made by machine.

2925 RIOTTOT, J., J. CHARDON & PACON, *Paris*.—Paper hangings.

2926 LEROY, C. V. *Paris*.—Painted window blinds.

2927 BACH-PÈRÈS, A. *Paris*.—Painted window blinds.

2928 GERFAUX, J. JUN. *Paris*.—Picture frames.

2929 THIAULT & CORNET, *Paris*.—Paper imitating oak.

2930 WAASER, C. A. *Paris*.—Kiosque, and specimens of trellis work decoration.

Addenda.

2930 (1) CORNU, T. JUN. *Paris*.—Furniture of ebony, enriched with bronze.

2930 (2) PILHOUÉE, A. *Paris*.—Chairs and arm-chairs.

2930 (3) MURE, C. *Paris*.—Steel nails and ornaments for furniture, &c.

BARBEDIENNE, F. *Paris*.—Artistic articles in wood, and sculptured marble.

BLUMER, C. *Strasbourg (Bas-Rhin)*.—Inlaid floors, artistic furniture.

DELACOUR, L. F. *Paris*.—Screen-curtains.

PLANIER, E. & Co. *Paris*.—Elastic pads for keeping apartments warm.

TRONCHON, N. J. *Paris*.—Iron seats and furniture.

CLASS 31.

2931 BRICARD & GAUTHIER, *Paris*.—Locks and hardware.

2932 HERDEVIN, J. M. *Paris*.—Cocks, and steam apparatus, &c.

2933 DOUCHAIN, C. *Paris*.—Apparatus for the distribution of water, &c.

2934 LEBOUC, J. L. J. *Paris*.—Apparatus connected with artificial light.

2935 CRIBIER, E. JUN. *Paris*.—Steel springs for petticoats.

2936 DUCROS, L. J. *Paris*.—Wrought iron balustrade, &c.

2937 ROUY, P. A. *Paris*.—Fecula for the moulding of metals.

2938 FRAISSINET, E. & Co. *Alais (Gard)*.—Metallic articles and fabrics.

2939 BARBOU & SON, *Paris*.—Iron bottle-cases.

2940 LEPAN, T. *Lille (Nord)*.—Foil and tubes of lead and tin.

2941 PASSAGER, J. D. D. *Paris*.—Lamp for congealed fats.

2942 AROUX, *Melun (Seine-and-Marne)*.—Hardware.

2943 LETURC, A. M. *Paris*.—Venetian blinds, of moveable sheet-iron plates, &c.

2944 BAUDRIT, A. *Paris*.—Iron work.

2945 CHAUVEL, E. J. B. *Evreux (Eure)*.—Buckles, thimbles, &c.

2946 JUBERT, BROS. *Charleville (Ardennes)*.—Iron fittings for carriages.

2947 MIETTE BROS. *Braux près Charleville (Ardennes)*.—Ironwork for wagons, &c.

2948 DEMANS & Co. *Chambon-Feugerolles (Loire)*.—Bolts and files.

2949 DEMANDRE, C. *Aillevillers (Haute-Saône)*.—Iron wire for weaving-cards and combs.

2950 DE POILLY, JUN. *Boulogne-sur-Mer (Pas-de-Calais)*.—Locks.

2951 DE LA HAYE DE BARBEZIÈRES, S. N. *Paris*.—Iron sashes.

2952 VALÉRY & SON, & SEROUX, *Beauchamps (Somme)*.—Keys.

2953 GOLDENBERG, G. *Zornhoff (Bas-Rhin)*.—Iron-work, cutting instruments, &c.

2954 BOURGEOIS-BOTZ, *Reims (Marne)*.—Plates and ribbons for cards.

2955 TRUCHE, J. M. J. *Lyon (Rhône)*.—Mirrors for catching larks.

2956 GAILLARD, C. C. *Paris*.—Metallic cloths.

2957 WEBER, *Nancy (Meurthe)*.—Metallic cloths.

2959 CRIBIER, *Viroflay (Seine-and-Oise)*.—Pins.

2960 LAPERCHE, A. JUN. *Paris*.—Military accoutrements.

2961 BOHIN, B. F. *L'Aigle (Orne)*.—Hardware, &c.

2962 DELAGE, J. P. *Paris*.—Articles in malleable cast-iron.

2963 ANDRÈ, J. L. *Paris*.—Lever-buckles for driving belts, &c.

2964 TRONCHON, N. J. *Paris*.—Garden-locks, &c.

2965 FICHET, A. *Paris*.—Iron safes.

2966 SAUVE L., & MAGAUD, *Marseille (Bouches-du-Rhône)*.—Iron safes, and locks for iron safes.

2967 DELACOUR, L. F. *Paris*.—Screens, &c.

2968 LHERMITTE, B. *Paris*.—Steel safes, &c.

2969 HAFFNER BROS. *Paris*.—Iron safes.

2970 LANG, L. *Schelestadt (Bas-Rhin)*.—Metallic cloths.

2971 FRANCK & Co. *Schelestadt (Bas-Rhin)*.—Metallic cloths.

2972 TRONCHON, A. *Paris*.—Models of portable iron buildings.

2973 GANDILLOT, *Paris*.—Iron tubing.

2974 ESTIVANT BROS. *Givet (Ardennes)*.—Unsoldered copper tubes.

2975 BOUTTEVILLAIN, L. F. *Paris*.—Iron tubes, for boilers, &c.

2970 VICAIRE, L. F. *Paris*.—Copper tubing.

2977 MAQUEMNEHEU, E., & IMBERT, *Escarbotin (Somme).*—Locksmiths' work.

2978 MAGE, SEN. *Lyon (Rhône).*—Metallic cloths.

2979 ROSWAG, V. *Schelesladt (Bas-Rhin).* —Metallic thread and cloth.

2980 TAILFER, A. & CO. *L'Aigle (Orne).* —Galvanized-iron pins, &c.

2981 SIROT-WAGRET, *Trith-St.-Leger (Nord).*—Pegs for boots and shoes.

2982 CANTAGREL, F., & CO. *Paris.*—Instrument for finding leaks in gas apparatus, &c.

2983 FOURNIER, C. A. *Paris.*—Safety apparatus for finding leaks in gas pipes, &c.

2984 REBOUR, C. J. N. *Paris.*—Locks.

2985 BRODIN, C. J. *Soissons (Aisne).*—Apparatus for cleaning lamps.

2986 LAMBERT, S. *Paris.*—Tin foil &c.

2987 HUBERT-REGNAULT & CO. *Charleville (Ardennes)*—Machine-made nails.

2988 GAILLY, SEN. *Charleville (Ardennes).* —Machine-made nails.

2989 FERGUSSON, E. *Paris.*—Regulators of gas-pressure.

2990 BOUBILLA, J. R. *Paris.*—Apparatus for securing bags for despatches.

2991 GAUSSERAN & CO. *Paris.*—Smoothing irons.

2992 MONIER, H. & CO. *Paris.*—Crystal gas jets.

2993 PRADEL, P. *Paris.*—Mode of fastening without a key.

2994 NOS-D'ARGENCE, *Rouen (Seine-Inf.).* —Metallic teazles.

2995 DUPUCH, G. *Paris.* — Copper and bronze articles.

2996 BOUCHER, E. *Fumay (Ardennes).*—Kitchen articles, &c.

2997 PAJOT & CO. *Randonnay (Orne).*—Cast-iron household requisites.

2998 SICARD, J. A. *Paris.*—Copper kitchen requisites.

2999 FONDET, J. B *Paris.*—Heating apparatus.

3000 POTTECHER, B. *Bussang (Vosges).*—Curry-combs, &c.

3001 DALIFOL & DALIFOL, A. *Paris.*—Articles in malleable cast-iron.

3002 BRUN, P. *Lyon (Rhône).*—Portable forge.

3003 MASSON, J. B. F. *Paris.*—Tin foil.

3004 MENAND, F. R. *Paris.*—Lanterns.

3005 CHÊNE, J., & SON, *Paris.*—Heating apparatus.

3006 PAROD, E. *Paris.* — Cooking utensils.

3007 AULON, C. L. E. *Darney (Vosges).*—Sheet-iron covers.

3008 PARET, N. E. A. *Paris.*—Portable kitchen-furnaces.

3009 PIAT, J. J. D. *Paris.*—Mechanism for numbering cotton threads.

3010 JAVAL, E. *Paris.*—Hardware.

3011 COLETTE, *Paris.*—Hardware.

3012 PENANT, *Paris.*—Coffee pots.

3013 MORISOT, N. J. *Paris.*--Bronze fittings for grates.

3014 GRANDRY-GRANDRY & SON, *Nouzon (Ardennes).*—Fittings for grates.

3015 BROQUIN & LAINE, *Paris.*—Ornamental and other bronzes.

3016 BION, V. *Paris.*—Bronze fittings for grates.

3017 CLAVIER, E. *Paris.*—Bronze and iron andirons.

3018 BESNARD, H. *Paris.* — Ornamental articles in zinc and bronze.

3019 BLOT & DROUARD, *Paris.* — Ornamental articles in zinc, imitating bronze.

3020 ROBIN BROS. *Paris.*—Candelabras; lustres.

3021 LEFEVRE, E. J. *Paris.*—Time-pieces, &c. of bronzed or gilt zinc.

3022 GAUTROT & LATOUR, *Paris.*—Ornamental zinc articles.

3023 BRUNEL, L. *Paris.*—Ornamental zinc articles.

3024 LAMBIN, SAGUET, & FOUCHET, *Paris.* —Ornamental zinc articles.

3025 RIGOLET, R. M. *Paris.* — Imitation bronzes.

3026 GRILLOT, C. *Paris.* — Imitation-bronze groups, statuettes, &c.

3027 FOUBERT, *Paris.* — Time-pieces, groups in bronzed or gilt zinc.

3028 HOTTOT, L. *Paris.*—Time-pieces, &c. in zinc.

3029 LECLERCQ, F. P. *Paris.*—Time-pieces, and ornamental articles in zinc.

3030 BOY, J. *Paris.*—Ornamental articles in zinc.

3031 MOURCY, P. *Paris.*—Articles in gilt aluminium, and in gilt zinc.

3032 MIROY FRÈRES, *Paris.*—Articles in zinc and bronze.

3033 MARCHAND, L. *Paris.*—Bronzes.

3034 DUCEL, J. J *Paris.* — Ornamental articles in cast iron.

3035 DURENNE, A. *Paris.* — Ornamental articles in cast iron.

3036 GRAUX-MARLEY, L. J. B. *Paris.*—Bronzes.

3037 BARBEZAT & CO. *Paris.* — Ornamental articles in cast iron.

3038 THIÉBAUT, V. *Paris.*—Ornamental articles in bronze.

3039 DELAFONTAINE, A. M. *Paris.* — Bronzes.

3040 CAIN, A. *Paris.*—Bronzes.

3041 DUPLAN & SALLES, *Paris.*—Bronzes.

3042 MOIGNIEZ, J., JUN. *Paris.* — Bronzes.

3043 PEYROL, H. *Paris.*—Bronzes.

3044 MÈNE, P. J. *Paris.*—Bronzes.

3045 GALOPIN, A. *Paris.* — Bronze and porcelain lamps.

3046 GONON, E. *Nantes (Loire-Inf.).* — Bronzes.

3047 POILLEUX, J. B. *Paris.*—Bronzes.

3048 ROLLIN, F. G. *Paris.*—Bronze time-pieces, &c.

3049 MAGE, F. *Paris.*—Plaster groups, for models of bronzes.

3050 FÉTU, J. L. *Paris.*—Bronzes.

3051 CHAUMONT & LANGUEREAU, *Paris.* —Bronzes.

3052 POPON, N. *Paris.*—Bronzes.

3053 MARERCHAL & BERNAND, *Paris.*—Porcelain lamps mounted in bronze.

3054 DAUBREÈ, A. *Nancy* (*Meurthe*). — Bronzes.
3055 BARBEDIENNE, F. *Paris.*—Bronzes.
3056 GAGNEAU, E. *Paris.*—Bronzes.
3057 SCHLOSSMACHER, J. & Co. *Paris.*—Bronze and other lamps.
3058 HADROT, L. JUN., BONNET, C. & BORDIER, *Paris.*—Bronze lamps.
3059 LACARRIÈRE, A. SON, & Co. *Paris.*—Gas-fittings, in cast-iron, zinc, and bronze.
3060 CARLHIAN & CORBIÈRE, *Paris.* — Lamps.
3061 HADROT, L. N. *Paris.* — Bronze lamps, &c.
3062 LEROLLE, L. *Paris.* — Articles in bronze.
3063 LEVY, J. E. *Paris.*—Artistic bronzes.
3064 CHARPENTIER & Co. *Paris.*—Groups, statues, &c. in bronze.
3065 DENIÈRE, G. JUN. *Paris.* — Bronzes.
3066 GAUTIER, F. *Paris.*—Bronzes.
3067 RAINGO, BROS. *Paris.*—Bronzes.
3068 PAILLARD, A. V. *Paris.*—Bronzes.
3069 HOUDEBINE, C. H. A. *Paris.*—Bronzes.
3070 BOYER, V. & SON, *Paris.* — Bronzes.
3071 LIONNET BROS. *Paris.*—Objects of art, &c., obtained by the electrotype process.
3072 VITTOZ, E., JUN. *Paris.*—Bronzes.
3073 MERCIER, D. *Paris.*—Bronzes.
3074 PERROT, H. *Paris.*—Fancy bronzes.
3075 SUSSE, BROS. *Paris.*—Bronzes.
3076 SERVANT, JUN. & DEVAY, *Paris.*—Bronzes.
3077 ZIER, A. J. *Paris.*—Bronzes.
3078 LEMAIRE, A. *Paris.*—Bronzes.
3079 PICKARD, C. & PUNANT, A. *Paris.*—Bronze chimney ornaments, &c.
3080 VANVRAY, BROS. *Paris.*—Bronzes.
3081 HERMANN, G. *Paris.* — Artistic objects mounted in bronze.
3082 FENQUIERÈS, J. *Paris.*—Bronzes.
3083 LEVY, BROS, *Paris.*—Bronzes.
3084 OUDRY, L. *Paris.*—Objects in cast-iron, coated with copper, &c. by the electrotype.
3085 MATIFAT, C. S. *Paris.* — Bronzes.
3086 NADAULT DE BUFFON, *Paris.*—Tubular filters.
3087 BURG, DR. V. *Paris.*—Apparatus for purifying and cooling water.
3088 DESCOLE, P. *Paris.*—Apparatus for giving light.
3089 SERVAU, J. J. *Paris.*—Articles in chiselled and repousséd copper.
3090 GONSSE, C. L. *Paris.*—Bronzes.
3091 LARGAUT, WIDOW, *Paris.*—Polishing powder.
3092 FARJAT, *Rouen* (*Sen. Inf.*). — Scrapers.
3093 CUBAIN, R. & Co. *Verneuil* (*Eure*).—Laminated copper, &c.
3094 BRUNT & Co. *Paris.*—Gas-meters.

CLASS 32.

3131 VITRY, BROS. *Nogent* (*Haute-Marne*).—Cutlery, surgical instruments.
3132 THUILLIÉR-LEFRANC, *Nogent* (*Haute-Marne*).—Cutlery.
3133 PERRAY, F. C. *Nogent* (*Haute-Marne*).—Cutlery.
3134 GIRARD, C. *Nogent* (*Haute-Marne.*)—Knives, cutters, &c.
3135 MARESCHAL-GIRARD, *Nogent* (*Haute-Marne*).—Cutlery.
3137 PLUMEREL-DAGUIN, *Nogent* (*Haute-Marne*).—Common cutlery.
3138 GUILLEMIN-RENAUT & Co. *Nogent* (*Haute-Marne*).—Cutlery.
3140 LÉCOLLIER, V. *Nogent* (*Haute-Marne*).—Razors.
3142 ROZE, C. *St. Dizier* (*Haute-Marne*).—Cutlery.
3143 COLLECTIVE EXHIBITION BY THE CITY OF THIERS (*Puy-de-Dôme*).—Cutlery (9 Exhibitors).
3144 ROMMETIN, C. M. *Paris.*—Dentists' instruments.
3145 GUERRE, C. JUN. *Langres* (*Haute-Marne*).—Common and fancy cutlery.
3146 PIAULT, J. *Nogent* (*Haute-Marne*).—Cutlery.
3147 TOURON-PARISOT, E. *Paris.*—Fancy cutlery.
3148 PAGÉ, BROS. *Châtellerault* (*Vienne*).—Cutlery, razors.
3149 MERMILLIOD BROS. *Prieuré* (*Vienne*).—Cutlery.
3150 CARDEILHAC, A. E. *Paris.*—Fancy cutlery, &c.
3151 PICAULT, G. F. *Paris.*—Common and fancy cutlery.
3152 CHARRIERE, J. J. *Paris.*—Common and fancy cutlery.
3154 PROUTAT, MICHOT, & THOMERET, *Arnay-le-Duc* (*Cote-d'Or*).—Steel files and tools.
3158 CLICQUOT, R. M. *Courbevoie* (*Seine*). Gravers' tools, &c.
3159 PETREMENT, F. *Paris.*—Gauges.
3160 MAILLET, C. P. A. *Compiègne* (*Oise*).—Shoemakers' tools.
3161 DELAHAYE, E. *Paris.* — Screws, broaches, &c.
3162 PERRIN, A. *Nogent* (*Haute-Marne*).—Cutlery.
3163 LEPAGE, J. H. *Paris.* — Cast-steel files.
3164 MANGIN, SEN. & Co. *Paris.*—Files, rasps, and gravers' tools.
3165 LANGUEDOCQ, C. F. *Paris.*—Cutlery.
3166 BLANZY & Co. *Boulogne* (*Pas-de-Calais.*—Steel pens and holders, &c.
3167 MALLAT, J. B. *Paris.*—Steel pens of various kinds, &c.
3168 LIBERT & Co. *Boulogne* (*Pas-de-Calais*).—Pens, springs, &c..
3169 MONCHICOURT, V. *Paris.*—Steel pens and holders.
3170 PARYS, A. *Paris.*—Plates for wire-drawers.
3171 RENARD, J. A. *Paris.*—Steel tools.
3172 LIMET, LAPAREILLÉ, & Co.—*Paris.*—Steel files and cutlery.
3173 HUËT, J. *Paris.*—Steel beads.
3174 ESSIQUE, L. *Paris.*—Steel thimbles, &c.

3175 JACQUEMIN & SORDOILLET, *Paris.*—Jewellery and ornaments of steel.

3176 BOURGAIN, J. B. *Paris.*—Caskets, jewels, and ornaments of steel.

3177 GOLDENBERG, G. & Co. *Lornhoff* (*Bas-Rhin*).—Steel tools, &c.

3178 CRIBIER, E. JUN *Paris.*—Steel springs for petticoats.

CLASS 33.

3201 ODIOT, *Pàris.*—Goldsmith's work; toilet requisites, table services, &c.

3202 DEBAIN, A. *Paris.*—Silver candelabras, tea services, &c.

3203 CHERTIER, A. *Paris.*—Church plate.

3204 CHRISTOFLE, C. *Paris.*—Table ornaments, &c. in gold and silver.

3205 MARRET & BEAUGRAND, *Paris.*—Jewellery in gold and silver.

3206 PETITEAU, E. *Paris.*—Jewellery.

3207 ROUVENAT, L. *Paris.*—Jewellery.

3208 LHOMME, L. *Paris.*—Bracelets.

3209 BELLEAU, F. A. *Paris.*—Work-requisites.

3210 PHILIPPI, A. *Paris.*—Jewellery.

3211 PAYEN & SON, *Paris.*—Jewels and filigree work.

3212 LEMOINE, V. *Paris.*—Gold chains and bracelets.

3213 OGEZ, WIDOW, & CADET-PICARD, *Paris.*—Gold jewellery, &c.

3214 JARRY, L. G. SEN. *Paris.*—Agate cup; jewellery.

3215 CONSTANT-VALÈS, & Co. *Paris.*—Imitation pearls; necklaces and bracelets.

3216 JARDIN-BLANCOUD, *Paris.*—Metals, inlaid stones, &c.

3217 PURPER, L. *Paris.*—Onyx cameos.

3218 LANGEVIN, H. *Paris.*—Copper and silver jewellery.

3219 TOPART BROS. *Paris.*—Imitation pearls and coral.

3220 GUEYTON, A. *Paris.*—Gilt jewellery

3221 BON, L. A. *Paris.*—Imitation lapis-lazuli, &c.

3222 CHEVALIER, A. E. *Paris.*—Seals, coins, &c.

3223 STUBLER, A. *Paris.*—Diamonds, for graving tools, &c.

3224 VILLEMONT, C. H. *Paris.*—Gilt jewellery.

3225 FAASSE, *Paris.*—Imitation jewellery.

3226 BENDER, L. A. E. *Paris.*—Imitation jewellery.

3227 TRUCHY, C. E. *Paris.*—Imitation pearls.

3228 GRANGER, E. *Paris.*—Historic jewellery, for the theatre, &c.

3229 BACHELET, L. C. *Paris.*—Church plate and bronzes.

3230 CHANUEL, WIDOW. *Paris.*—Ornamental repoussé articles in silver.

3231 HUBERT, L. J. L. *Paris.*—Repoussé and enamelled silver articles.

3232 RUDOLPHI, F. J. *Paris.*—Ornamental plate.

3233 BALAINE, C. *Paris.*—Plated articles.

3234 GOMBAULT, A. & Co. *Paris.*—Maillechort table service.

3235 POUSSIELGUE-RUSAND, *Paris.*—Church plate and bronzes.

3236 DOTIN, A. C. *Paris.*—Enamelled articles.

3237 FLORANGE, E. H. *Paris.*—Plated articles.

3238 ARMAND-CALLIAT, *Lyon* (*Rhône*).—Devotional articles, in gold and silver.

3239 TRIOULLIER, E. C. *Paris.*—Church plate and bronzes.

3240 COMTE, H. C. C. DE RUOLZ & FONTENAY, *Paris.*—Various articles made of a new alloy.

3241 SAVARD, A. F. *Paris.*—Double-gold jewellery.

3242 DOBBÉ, V. & HÉMON, *Paris.*—Double-gold jewellery.

3243 MURAT, C. *Paris.*—Double-gold jewellery.

3244 CORBEELS, S. *Paris.*—Inlaid tortoise-shell jewellery.

3245 SAVARY, A. *Paris.*—Imitation jewellery.

3246 CASALTA, L. & ISLER, L. *Paris.*—Carved corals, cameos.

3247 DECAUX, E. *Paris.*—Jewellery.

3248 BONNET-GROLLIER, *Bourg* (*Ain*).—Enamelled-silver jewellery.

3249 CELLIER, L. *Marseille* (*Bouches-du-Rhône*).—Gold and silver jewellery, &c.

3250 DE LAURENCEL, *Paris.*—Auriferous agate, of San Francisco.

3251 JOUANIN, C. V. *Paris.*—Cameos.

3252 BOURET & FERRÉ, T. *Paris.*—Jewellery obtained by stamping.

3253 BARBARY, E. *Paris.*—Ivory work-boxes, &c.

3254 THÉNARD, F. *Paris.*—Plate; jewellery.

3255 FRIBOURG, G. A. *Paris.*—Gold jewellery.

3256 LOBJOIS, A. *Paris.*—Jewellery.

3257 BRUNEAU & Co. *Paris.*—Jewellery, plate.

3258 MAGNIADAS, MADAME, SEN. *Paris.*—Fancy jewellery.

3259 COFFIGNON, BROS. *Paris.*—Silver and plated jewellery, &c.

3260 GENTILHOMME, J. H. *Paris.*—Gold and silver jewellery.

3261 CAILLOT, JUN., JECK & Co. *Paris.*—Jewellery.

3262 JACTA, E. & Co. *Paris.*—Jewellery.

3263 MELLERIO (MELLER), BROS. *Paris.*—Jewellery, &c.

3264 WIESE, J. *Paris.*—Jeweller's and artistic goldsmith's work.

3265 DURON, C. *Paris.*—Artistic jewellery.

3266 DESURY, A. *St. Briene* (*Côtes-du-Nord*).—Plate; silver plateau.

3267 CARMANT, T. A. NORMAND, P. *Paris.*—Gilt jewellery.

3268 BRISSON, T. A. *Paris.*—Designs for bronze and goldsmith's work.

3269 FANNIERE, *Paris.*—Silver plate.

3270 GUEYTON, A. *Paris.*—Artistic goldsmith's work.

3271 DUPONCHEL, H. *Paris*.—Candelabra, plate.
3272 ROUCOU, *Paris*.—Articles damascined and inlaid with gold and silver.
3273 AUCOE, L. SEN. *Paris*.—Table ornaments.
3274 FEUQUIÈRES, J. *Paris*.—Gold, silver, and enamelled jewellery.

CLASS 34.

3281 JOINT STOCK CO. OF THE MANUFACTURES OF GLASS AND CHEMICAL PRODUCTS OF SAINT-GOBAIN, CHAUNY, AND CIREY.—*Paris*.—Flint glass.
3282 ROUGET DE LISLE, T. A. *Paris*.—Bottles, and alimentary cases of glass.
3283 WALTER, BERGER, & CO. *Paris*.—Watch glasses, &c.
3284 DE POILLY, DE FITZJAMES, & LABARBE, *Follembray* (*Aisne*).—Bottles, glasses, decanters, garden bell-glasses.
3285 PASQUES, A. & CO. *Blanzy* (*Saône-and-Loire*).—Bottles, garden bell-glasses, &c.
3286 VAN LEEMPOEL, *Quiquengronne* (*Aisne*).—Champagne bottles.
3287 GOBBE, O., FOGT, A. & CO. *Aniche* (*Nord*).—Window glass.
3288 DELHAY, H. *Aniche* (*Nord*).—Glass for windows and photography.
3289 RAABE, C. & CO. *Rive-de-Gier* (*Loire*).—Bottles, window glass, &c.
3290 CHAPPUY, L. *Frais-Marais-lez-Douai* (*Nord*).—Bottles, of large and small sizes.
3291 CHARTIER, P. *Douai* (*Nord*).—Large bottles.
3292 HOUTARD, F. & CO. *Lourches* (*Nord*).—Bottles.
3293 RENARD & SON, *Fresnes* (*Nord*).—Window glass, and glass for optical purposes.
3294 DEVIOLANE, BROS. *Vauxrot* (*Aisne*).—Champagne bottles.
3295 JOINT STOCK CO. OF THE GLASS WORKS OF EPINAC (*Saône and Loire*).—Bottles.
3296 DEROCHE, C. *Paris*.—Glass, porcelain, and chemical ware.
3297 MAËS, L. J. *Clichy-la-Garenne* (*Seine*).—Flint-glass.
3298 OUDINOT, E. *Paris*.—Glass windows, window glass.
3299 LAFAYE, P. *Paris*.—Glass windows.
3300 NICOD, *Paris*.—Glass cases, church windows.
3301 HONER, *Nancy* (*Meurthe*).—Stained glass: a window.
3302 BOURGEOIS, G. *Reims* (*Marne*).—Glass cases, windows.
3303 DRION-QUÉRITÉ, PATOUX, & DRION, A. *Aniche* (*Nord*).—Glass windows, glass for roofs, &c.
3304 MARÉCHAL, C. L. *Metz* (*Moselle*).—Windows.
3305 LUSSON, A. *Paris*.—Windows.
3306 DIDRON, A. N. SEN. *Paris*.—Painted windows.
3307 LAURENT & GSELL, *Paris*.—Church window.
3308 MONOT, E. S. *Pantin* (*Seine*).—Flint-glass.
3309 THOMAS, KUHLIGER, BROS. & CO. *Paris*.—Plated glass.
3310 ALEXANDRE, JUN. *Paris*.—Flint-glass.
3311 SIMEON & SON, *Paris*.—Ornamented glass.
3312 COFFETIER, N. *Paris*.—Stained glass.
3313 MARÉCHAL, C.R. JUN. *Metz* (*Moselle*).—Windows.
3314 VIEILLARD, J. & CO. *Bordeaux* (*Gironde*).—Bottles.
3315 PICARD BROS. *Sarrebourg* (*Meurthe*).—Watch glasses.
3316 CATTAERT, *Paris*.—Decanters, closing hermetically.
3317 DORSY, A. *Paris*.—Flasks and cases.
3318 LAHOCHE & PANNIER, *Paris*.—Cut flint glass for the table, &c.

CLASS 35.

3321 IMPERIAL MANUFACTORY OF SÈVRES.—Articles in porcelain, crockery, and terra cotta.
3322 ROZIER, WIDOW, & BAROCHE, *Tain* (*Drôme*).—Kaolin, refractory bricks.
3323 DUMERIL, C. & LEURS, H. *St. Omer* (*Pas-de-Calais*).—Earthenware pipes.
3324 LAUDET (BEAUFAY), SOUCHART, & CO. *Paris*.—Laboratory articles in refractory clay.
3325 DEYEUX, N. T. *Liancourt* (*Oise*).—Refractory crucibles.
3326 DUPRAT, M. C. V. *Canéjan* (*Gironde*).—Refractory crucibles and bricks.
3327 FIOLET, L. *St. Omer* (*Pas-de-Calais*).—Earthen pipes.
3328 GISCLON, *Lille* (*Nord*).—Earthen pipes.
3329 GOSSE, F. A. *Bayeux* (*Calvados*).—Hard porcelain articles, made to stand the fire.
3330 HAVILAND, C. F. & CO. *Limoges* (*Haute-Vienne*).—Decorated porcelain.
3331 POUYAT, BROS. *Limoges* (*Haute-Vienne*).—Articles in hard porcelain.
3332 ARDANT, H. & CO. *Limoges* (*Haute-Vienne*).—Fancy porcelain articles.
3333 JULLIEN, *Saint-Leonard* (*Haute-Vienne*).—Porcelain vases, and table services.
3334 BATIER, H. *Limoges* (*Haute-Vienne*).—Ornamented porcelain.
3335 VIEILLARD, J. & CO. *Bordeaux* (*Gironde*).—Crockery ware and porcelain.
3336 MÉNARD, C. *Paris*.—Artistic porcelain articles, &c.
3337 ROUSSEAU, E. *Paris*.—Porcelain, crockery ware, and enamels.
3338 LAHOCHE & PANNIER, *Paris*.—Porcelain and flint-glass.
3339 LAURIN, F. *Bourg-la-Reine* (*Seine*).—Crockery ware.
3340 DECK, T. *Paris*.—Fancy crockery ware, panels, medallions, &c.
3341 PINART, H. A. *Paris*.—Ornamented plates and dishes.
3342 AVISSEAU, E. *Tours* (*Indre-and-Loire*).—Artistic articles in pottery.

3344 CHABLIN, N. L. *Paris.*—Porcelain, flint-glass, enamels encrusted with engraved gold and silver.

3346 LYONS, G. *Nevers (Nièvre).*—Common, and artistic crockery.

3347 DANIEL, S. *Paris.*—Porcelain imitations of precious stones.

3348 DE ST. ALBIN, MADAME, C. H. *Paris.*—Fruits and flowers painted on porcelain.

3349 BARBIZET, V. *Paris.*—Glazed pottery, style of Palissy.

3350 MACÉ, L. A. C. *Paris.*—Ornamented porcelain.

3351 LAVALLE, J. *Premières (Cote-d'Or).*—Artistic pottery painted on coarse enamel, imitation of Etruscan.

3352 DEVERS, *Paris.*—Enamelled earthenware.

3353 JEAN, A. *Paris.*—Artistic decorations in porcelain.

3354 GILLET & BRIANCHON, *Paris.*—Enamelled ceramic pastes, imitating mother-of-pearl, ivory, and emerald.

3355 GILLE, JUN. *Paris.*—White and ornamented porcelain statues, groups, &c.

3356 LÉTU & MAUGER, *Isle-Adam (Seine-and-Oise).*—Statue of the Virgin, in biscuit.

3357 HACHE, A. & PEPIN-LEHALLEUR, *Paris.*—White and ornamented porcelains.

3358 PILLIVUYT, C. & Co. *Paris.*—White and ornamented porcelain.

3359 PRÉVOST, J. R. *Paris.*—Ornamented porcelain.

3360 DE BETTIGNIES, M. *St. Amand-les-Eaux (Nord).*—Soft porcelain, imitating old Sèvres.

CLASS 36.

3371 POISSON, P. L. M. *Paris.*—Fancy articles, in ivory, &c.

3372 HUMERY, J. J. L. *Paris.*—Porcelain, &c. mounted in bronze; malachite jewellery, &c.

3374 GIRAUDON, S. A. *Paris.*—Cases, &c.

3375 COLLECTIVE EXHIBITION OF THE CITY OF DIEPPE (*Seine-Inf.*).—Articles in ivory (11 Exhibitors).

3376 JEANTET-DAVID, *St. Claude (Jura).*—Snuff-boxes, pipes, combs, &c.

3377 SCHOTTLANDER, H. *Paris.*—Photograph-albums.

3378 LEFORT, V. M. *Paris.*—Articles of carved ivory.

3379 DROUARD, BROS. *Paris.*—Photograph-albums.

3380 MARX, W. *Paris.*—Albums, portfolios, &c.

3381 LATRY, SEN. & SON. *Paris.*—Fancy articles made of the saw-dust of exotic wood, hardened and compressed.

3382 GERSON & WEBER, *Paris.*—Requisites and fancy articles in carved wood.

3383 LERUTH, F. L. *Paris.*—Small cabinet-work articles in morocco, &c.

3384 MERCIER, C. V. *Paris.*—Snuff boxes in exotic wood, &c.

3385 TRIEFFUS & ETTLINGER, *Paris.*—Morocco articles, tortoise-shell and ivory jewellery.

3386 HORCHOLLE, A. *Paris.*—Articles in ivory.

3387 HOCHAPFEL, BROS. *Strasbourg (Bas-Rhin).*—Pipes.

3388 NŒTINGER, C. *Mutzeg (Bas-Rhin).*—Stone balls.

3389 CHANDELIER, *Méru (Oise).*—Buttons.

3390 BEUGNOT, C. H. *Paris.*—Requisites in plain, carved, and encrusted ivory; fancy articles.

3391 BONHOMÉ, *Paris.*—Designs for metallurgical establishments.

3392 BONDIER-DONNINGER, & ULBRICH, *Paris.*—Pipes, cigar-cases.

3393 PATRY, E. L. *Paris.*—Spectacles, &c. of aluminium, tortoise-shell, gold, &c.

3394 BEAUDOIRE-LEROUX, & Co. *Paris.*—Albums, note books.

3395 GRUMEL, F. R. *Paris.*—Photograph-albums, &c.

3396 BLUMER, C. *Strasbourg (Bas-Rhin).*—Inlaid floors; artistic furniture.

3397 WALCKER, *Paris.*—Travellers', sportsmen's, and military requisites.

3398 MOREAU, J. L. *Paris.*—Carved ivory articles.

3399 BONTEMS, B. *Paris.*—Mechanical toys and automata.

3400 DUCLOS & RUBALTO, *Paris.*—A fume-cigarre.

3401 BÉREUX, MISS J. *Paris.*—Dresses, &c. for dolls.

3402 LEMAIRE-DAIMÉ, *Andrèsy (Seine-and-Oise).*—Atmospheric playthings.

3403 THÉROUDE, A. N. *Paris.*—Automatic toys; self-acting organ.

3404 PEZET, C. *Paris.*—Articles in morocco, and toilet requisites.

3405 MIDOCQ, N. E. & GAILLARD, E. A. *Paris.*—Morocco articles; requisites for travelling, &c.

3406 SORMANI, P. *Paris.*—Travelling requisites; small and fancy cabinet work.

3407 GELLÉE, BROS. *Paris.*—Sheaths, caskets, &c.

3408 SCHLOSS, S. & NEPHEW, *Paris.*—Morocco articles, cigar-cases, &c.

3409 AUCOC, L. SEN. *Paris.*—Morocco articles requisite for the toilet, tea and coffee services, &c.

3410 ULMANN, P. *Paris.*—Designs for fabrics, shawls, &c.

3411 DELAYE, P. V. *Paris.*—Designs for shawls, &c.

3412 NAZE, SON, & Co. *Paris.*—Designs for shawls, dresses, &c.

3413 BERRUS, A. BROS. *Paris.*—Designs for long shawls.

3414 AUBRY, C. H. *Paris.*—Designs for lace, &c.

3415 GONELLE, BROS. *Paris.*—Designs for long shawls.

3416 HENRY, H. F. *Paris.*—Designs for paper hangings, and stuffs for furniture.

3417 CAPTIER, E. V. *Fontainebleau (Seine-and-Marne).*—Designs for silks, &c.

3418 MATHIEU, E. *Paris.* — Designs for shawls.
3419 VAILLANT, BROS. *Paris.*—Designs for shawls, &c.
3420 FAURE, E. *Paris.*—Designs for lace.
3421 MADELEINE, BROS. *Paris.*—Designs for lace.
3422 GATTIKER, G. *Paris.*—Designs for dresses.
3423 GOURDET & ADAN, *Paris.*—Designs for carpets, &c.
3424 GUICHARD, E. A. D. *Paris.*—Industrial designs.
3425 BOISSIER, *Paris.*—Designs for worked shawls.
3426 FAUVELLE-DELEBARRE, *Paris.* — Combs.
3427 DESRUELLES, A. *Paris.*—Combs.
3428 JEUNIAUX, D. *Paris.*—Combs.
3429 MASSUE, E. *Paris.*—Combs.
3430 LEGAVRE, G. B. *Paris.*—Combs.
3431 PICARD, F. A. *Paris.*—Combs.
3432 CASELLA, E. *Paris.*—Combs.
3433 HEUDE, *Paris.*—Combs.
3434 PINSON, BROS. *Paris.*—Panel imitating tortoise-shell, mother-of-pearl and ivory.
3435 RAMBERT, C. D. *Paris.*—Designs for the goldsmith and cabinet-maker.
3436 LIENARD, *Paris.*—Industrial designs.

Addenda.
ALLAIN-MOULARD, L. A. F. *Paris.* —Morocco articles.
DUPONT & DESCHAMPS, *Beauvais (Oise).*—Fancy articles.

FRENCH COLONIES.

GUIANA.

CLASS 2.

3451 HERARD.—Citrate of lime.

CLASS 3.

3452 Flour and fecula, (2 Exhibitors).
3453 Sugar (3 Exhibitors).
3454 Alcohol, liqueurs, syrups, and preserves (5 Exhibitors).
3455 Cocoa (1 Exhibitor).
3456 Spices (2 Exhibitors).
3457 Tobacco (2 Exhibitors).
3458 Wood (3 Exhibitors).
3459 Organic industrial products—cocoons cotton, dye-stuffs (5 Exhibitors).
3460 Animal products—salt fish, fish-glue, tortoise-shell (3 Exhibitors).
3461 Collection of colonial products (1 Exhibitor).

CLASS 18.

3462 Thread and woven fabrics made of Guiana cotton (5 Exhibitors).

CLASS 25.

3463 Articles made of feathers (2 Exhibitors).

CLASS 27.

3464 Costumes, arms, and garments (1 Exhibitor.

ST. PIERRE AND MIQUELON.

CLASS 3.

3465 Products of the fisheries—salt cod, salt herrings, cod-liver oil (7 Exhibitors).
3466 Cereals, and medicinal plants (1 Exhibitor).

COCHIN-CHINA.

3467 Collection of colonial products (1 Exhibitor).

NEW CALEDONIA.

CLASS 1.

3468 Pit-coal (1 Exhibitor).

CLASS 2.

3469 Medicinal substances (2 Exhibitors).

CLASS 3.

3470 Tuberous roots and alimentary grain (3 Exhibitors).
3471 Sugar and coffee (2 Exhibitors).
3472 Wool (2 Exhibitors).
3473 Products of the fisheries—tortoise-shell, pearl-oysters, &c. (4 Exhibitors).
3474 Specimens of wood (5 Exhibitors).
3475 Industrial organic products—cotton and textile fibres, gums, resins, oleaginous plants and oils, dye-stuffs (9 Exhibitors).

CLASS 27.

3476 Arms and garments (2 Exhibitors).

TAHITI AND ITS DEPENDENCIES.

CLASS 3.

3477 Alimentary substances—fecula, sugar alcohol, coffee, vanilla, &c. (7 Exhibitors).
3478 Organic industrial products—wood, cotton, textile fibres, oleaginous substances, tobacco, pearl-oysters, and pearls (7 Exhibitors).

MAYOTTE AND NOSSI-BE.

CLASS 3.

3479 Alimentary substances—rice, sugar, coffee, rum (4 Exhibitors).

3480 Organic industrial products—wax, tortoise-shell (1 Exhibitor).

ST. MARY OF MADAGASCAR.

CLASS 3.

3481 Collection of colonial products (1 Exhibitor).

EAST INDIES.

CLASS 2.

3482 Medicinal substances (3 Exhibitors).

CLASS 3.

3483 Alimentary substances—cereals, fecula, sugar, spices, models of fruit from nature, preserved fish (4 Exhibitors).

3484 Organic industrial products—silk, cotton, wood, oleaginous substances, indigo and other dye-stuff, tanning matters (9 Exhibitors).

CLASS 4.

3485 Articles in straw and cane (3 Exhibitors).

CLASS 20.

3486 Mixed cotton and silk fabrics (2 Exhibitors).

CLASS 27.

3487 Boots, shoes, &c. (1 Exhibitor).

WEST COAST OF AFRICA.

CLASS 3.

3488 Alimentary substances—Rio Nuñez coffee (1 Exhibitor).

3489 Organic industrial products—cotton and textile fibres, medicinal plants, oleaginous substances, colouring matters, peltry (4 Exhibitors).

3490 Collection of the products of the country (1 Exhibitor).

CLASS 27.

3491 Arms and costumes (1 Exhibitor).

GUADALOUPE.

CLASS 1.

3492 Mineral products (1 Exhibitor).

CLASS 2.

3493 Medicinal substances (1 Exhibitor).

CLASS 3.

3494 Alimentary substances—flour and fecula, sugar, coffee, alcohol, preserves and liqueurs, spices, sea salt (10 Exhibitors).

3495 Organic industrial products—silk, cotton, oleaginous substances, dye-stuffs, tobacco (19 Exhibitors).

3496 Collection of colonial products (1 Exhibitor).

3497 Woven fabrics made with colonial cotton (4 Exhibitors).

ISLE OF REUNION.

CLASS 1.

3498 Mineral products (4 Exhibitors).

CLASS 2.

3499 Medicinal substances (3 Exhibitors).

CLASS 3.

3500 Alimentary substances—alimentary grain, flour and fecula, sugar alcohol, preserves, coffee, cocoa, spices, wax, honey, &c. (59 Exhibitors)

3501 Organic industrial products—wood, cotton and textile fibres; gums, balsams and resins; oleaginous substances, dye-stuffs, and tanning matters; tobacco (25 Exhibitors).

CLASS 30.

3502 Furniture and ornamental articles (13 Exhibitors).

MARTINIQUE.

CLASS 1.

3503 Mineral products (3 Exhibitors).

CLASS 3.

3504 Alimentary substances—flour and fecula, sugar, alcohol, liqueurs, preserves, coffee, cocoa, salt meats (33 Exhibitors).

3505 Organic industrial products—wood, cotton, gums, oleaginous substances, manures (8 Exhibitors).

3506 Collection of colonial products (1 Exhibitor).

3507 Birds—group of Martinique birds (1 Exhibitor).

ALGERIA.

3601 COLLECTIVE EXHIBITION.—Metalliferous minerals.

Province of Algiers: ores of iron, copper, and lead (3 Exhibitors).

—— of Oran: ores of iron, copper, lead, zinc, and nickel (5 Exhibitors).

—— of Constantine: ores of iron, copper, lead, zinc, antimony, manganese, and mercury (13 Exhibitors).

3602 COLLECTIVE EXHIBITION.—Mineral non-metalliferous products.

Province of Algiers: marbles, calcareous substances, onyx marbles, gypsum, phosphyroidal diorite (7 Exhibitors).

—— of Oran: marbles, calcareous substances, lime and mortar, alabaster, plaster, saltpetre, sea salt (9 Exhibitors).

—— of Constantine: marbles, calcareous substances, lime, plaster, alabaster, porphyry, granite, grit stone, sal gemma (11 Exhibitors).

3603 Metallurgical products—

Province of Constantine: cast iron from the ores of Allelick (1 Exhibitor).

CLASS 2.

3604 COLLECTIVE EXHIBITION.—Mineral and thermal waters—

Province of Constantine (2 Exhibitors).

3605 COLLECTIVE EXHIBITION.—Medicinal substances.

Province of Algiers: lichens, poppies, umbelliferous aromatic plants, mallows, borage, liquorice, mint, indigenous pepper and senna, oak-galls, opium (6 Exhibitors).

—— of Oran: absynthe, hops, hachish, pimenta, liquorice, cumin, anise, pastes, syrups (10 Exhibitors).

—— of Constantine: kef, a kind of hemp used instead of tobacco; indigenous tea, fenugreek, pimenta, liquorice, coriander seed, poppies, opium, &c. (12 Exhibitors).

3606 COLLECTIVE EXHIBITION.—Essences and essential oils for perfumery.

Province of Algiers: essential oil of geranium, peppermint water (2 Exhibitors).

—— of Constantine: orange-flower water; the leaves and flowers of odoriferous plants (2 Exhibitors).

CLASS 3.

3607 COLLECTIVE EXHIBITION.—Cereals.

Province of Algiers: wheat, barley, rice, maize, oats, &c. (9 Exhibitors).

—— of Oran: wheat, barley, oats, maize, sergho (23 Exhibitors).

—— of Constantine: wheat, barley, rice, maize, oats, millet, sergho (65 Exhibitors).

3608 COLLECTIVE EXHIBITION.—Farinaceous vegetables, and plants for forage.

Province of Algiers: French beans, peas, lentils, beans, &c. (2 Exhibitors).

—— of Oran: French beans, peas, chick-peas, lentils, beans (9 Exhibitors).

—— of Constantine: French beans, peas, chick-peas, lentils, beans (12 Exhibitors).

3609 COLLECTIVE EXHIBITION.—Tuberous roots.

Province of Algiers: Tubers, colocases, &c. (1 Exhibitor).

—— of Oran: Tubers, beet-root, potatoes, &c. (5 Exhibitors).

—— of Constantine: Bulbs of the taronda, potatoes, &c. (4 Exhibitors).

3610 COLLECTIVE EXHIBITION.—Fruits.

Province of Algiers: almonds (1 Exhibitor).

——of Oran: various fruits, raisins, capers, &c. (10 Exhibitors).

——of Constantine: various fruits, jujubes, raisins, &c. (6 Exhibitors)

3611 COLLECTIVE EXHIBITION.—Flour and alimentary pastes.

Province of Algiers: wheat-flour, semoule macaroni, small pastes, biscuit for the army (4 Exhibitors).

——of Oran: wheat and maize flour, semoule (3 Exhibitors).

——of Constantine: wheat-flours, fecula of the various tubers, semoules, and semouleka, pastes (15 Exhibitors).

Metropolitan France: flour, alimentary pastes, and other articles, prepared with the wheat and maize of Algeria (3 Exhibitors).

3612 COLLECTIVE EXHIBITION.—Honey and wax.

Province of Algiers: honey, wax, bee-hive of a new construction (1 Exhibitor).

——of Oran: honey, candles made of indigenous wax (4 Exhibitors).

——of Constantine: honey, wax, candles made of indigenous wax (7 Exhibitors).

3613 COLLECTIVE EXHIBITION.—White and red wine.

Province of Algiers (24 Exhibitors).

——of Oran (37 Exhibitors).

——of Constantine (17 Exhibitors).

3614 COLLECTIVE EXHIBITION.—Alcohol, and alcoholic drinks, liqueurs, preserves, and confectionery.
Province of Algiers (3 Exhibitors).
——of Oran (9 Exhibitors).
——of Constantine (2 Exhibitors).

3615 COLLECTIVE EXHIBITION.—Leaf and manufactured tobacco.
Province of Algiers (5 Exhibitors).
——of Oran (10 Exhibitors).
——of Constantine (21 Exhibitors).

3616 COLLECTIVE EXHIBITION.—Plants and oleaginous substances.
Province of Algiers: flax, &c.; olives, oil of olives, &c. (7 Exhibitors).
——of Oran: olives, and oil of olives, flax, colza, &c. (21 Exhibitors).
——of Constantine: olives, and oil of oil of olives, flax, white mustard, oil of sweet almonds, &c. (15 Exhibitors).

3617 COLLECTIVE EXHIBITION.—Wools.
Province of Algiers: Sheeps' wool and camel's hair (7 Exhibitors).
——of Oran: sheep's wool (13 Exhibitors).
——of Constantine: sheep's wool, camel's hair (20 Exhibitors).

3618 COLLECTIVE EXHIBITION.—Raw silk and cocoons.
Province of Algiers (10 Exhibitors).
——of Oran (12 Exhibitors).
——of Constantine (17 Exhibitors).

3619 COLLECTIVE EXHIBITION.—Woods and forest products.
Province of Algiers: 47 specimens of woods, &c. (4 Exhibitors.)
——of Oran: 18 specimens of woods (6 Exhibitors).
——of Constantine: 104 specimens of woods (3 Exhibitors).

3620 COLLECTIVE EXHIBITION.—Raw and prepared corks.
Province of Algiers (1 Exhibitor).
——of Oran (2 Exhibitors).
——of Constantine (5 Exhibitors).

3621 COLLECTIVE EXHIBITION.—Vegetable textile fibres.
Province of Algiers: fibres of the Nain palm, termed vegetable horse hair, hemp, flax, agave, &c. (12 Exhibitors).
——of Oran: fibres of the Nain palm, agave, flax, &c. (18 Exhibitors).
——of Constantine: flax, hemp, and various other textile fibres (8 Exhibitors).

3622 COLLECTIVE EXHIBITION.—Cotton.
Province of Algiers (24 Exhibitors).
——of Oran (19 Exhibitors).
——of Constantine (13 Exhibitors).

3623 COLLECTIVE EXHIBITION.—Dye stuffs, and other industrial organic substances.
Province of Algiers: cochineal, indigo, kermes, madder, carthamum, glunes of sorgho, and teasels (5 Exhibitors).
——of Oran: hemp, carthamum, kermes, sumach, &c. (15 Exhibitors).
——of Constantine: madder, indigo, nut-galls, tanning bark, pine-resin, sugar (7 Exhibitors).

CLASS 9.

3624 Agricultural instruments (3 Exhibitors).

CLASS 12.

3625 Instrument ascertaining the rate at sea (1 Exhibitor).

CLASS 18.

3626 Yarns, and woven fabrics made with Algerian cotton (10 Exhibitors).

CLASS 19.

3627 Thread, and woven fabrics made of Algerian flax (1 Exhibitor).

CLASS 20.

3628 Woven fabrics, made of Algerian silk (1 Exhibitor).

CLASS 21.

3629 COLLECTIVE EXHIBITION.—Woven fabrics, made of Algerian wool (3 Exhibitors).

CLASS 22.

3630 COLLECTIVE EXHIBITION.—Indigenous carpets.
Province of Algeria (3 Exhibitors).
——of Oran (4 Exhibitors).
——of Constantine (2 Exhibitors).

CLASS 26.

3631 COLLECTIVE EXHIBITION.—Leather, peltry, and other matters derived from animals.
Province of Oran: goose down, coral (3 Exhibitors).
——of Constantine: calf, sheep, and goat-skins; skins of the lion, jackall, hyæna; feathers and down of birds; coral (3 Exhibitors).

3632 COLLECTIVE EXHIBITION.—Saddlery and harness.
Province of Constantine (5 Exhibitors).

CLASS 27.

3633 COLLECTIVE EXHIBITION.—Costumes, arms, garments.
Province of Algiers (8 Exhibitors).
—— of Oran (15 Exhibitors).
—— of Constantine (21 Exhibitors).

3634 Flowers and head-dresses made of Algerian products (1 Exhibitor).

CLASS 28.

3635 COLLECTIVE EXHIBITION.—Books and manuscripts.
Province of Algiers (3 Exhibitors).
—— of Oran (3 Exhibitors).
—— of Constantine (1 Exhibitor).

CLASS 30.

3636 COLLECTIVE EXHIBITION.—Articles of furniture and decoration.
Province of Algiers (3 Exhibitors).
—— of Oran (3 Exhibitors).
—— of Constantine (4 Exhibitors).
3636 COLLECTIVE EXHIBITION.—Furniture and articles made of indigenous wood (8 Exhibitors).

GERMANY.

AUSTRIA.

N.W. Transept and N.W. Transept Gallery.

CLASS 1.

1 BAADER'S SON, DR. *Vienna*, 797, *Wieden.*—Minerals.
2 BOCHDANOVICS, A. *Zimbró, Arad Com. Hungary.*—Mining products.
3 BRUNICKI, BARON M. *Pisarzowa, Neu-Sandez, Galicia.*—Petroleum.
4 BRUNN-TAUBITZ, MINERS' COMPANY AT, *Brunn am Walde, near Gföhl, Lower Austria.*—Graphite.
5 CURTI, DR. A. *Muthmannsdorf, near Wr.-Neustadt.*—Portland cement in powder and cakes.
6 D'ELIA, J. *Alt-Orsova, Milit. Border.*—Chromium ores.
7 DOMOKOS, I. R. MINES, COPPER, SMELTING AND HAMMER WORKS AT, *Balánbánya, Transylvania.*—Ores and slags.
8 DOPPLER, J. *Salzburg.*—Various kinds of Salzburg marble.
9 EGGER, S. *Pesth, Dorotheengasse* 11.—Minerals.
10 EGGERT, A. & CO. *Mugrau, District Krumau, Bohemia.*—Graphite.
11 EIBISWALD, I. R. ADMINISTRATOR OF THE STEEL AND IRON WORKS AT, *Styria.*—Cast-steel; puddled and cemented steel; iron.
12 ESCHER, H. *S. Andrea di Rovigno, Trieste.*—Hydraulic cement.
13 FALKENHAIN, COUNT T. *Kyowitz, near Troppau, Austrian Silesia.*—Roofing slate.
14 FINANCES, I. R. MINISTRY OF (for the administration of the salt works and forests at Gmunden, Hallein, Wieliczka, Klausenburg, Maros-Ujvár, and Marmaros-Sziget), *Vienna.*—Rock salt and culinary salt from the Government works; marble, sandstone, chalk, and gypsum.
15 GEOLOGICAL INSTITUTION, I. R. *Vienna.*—Collection of pit coal, brown coal, and peat, dug in Austria.
16 GESSNER E., POHL W., & ULBRICH, O., *Müglitz, Moravia.*—Graphite.
17 GOLDSCHMIDT, S. *Vienna.*—Emerald, specimens and rock crystal from Salzburg.
18 GROHMANN, A. *St. Wolfgang, near Ischl, Upper Austria.*—Pit coal and peat.
19 HOFFMANN, E. *Alt-Orsowa, Milit. Border.*—Chromic iron ore and other minerals.
20 IDRIA, I. R. MINING OFFICE AT, *Carniola.*—Cinnabar.
21 JACOMINI-HOLZAPFEL-WAASEN, F. DE, *Bleiberg, Carinthia.*—Block of lead; grey sulphuret of zinc.
22 JENBACH, IRON-WORK ADMINISTRATION AT, *Tirol.*—Refined steel.
23 JOACHIMSTHAL, I. R. OFFICE OF MINES AT, *Joachimsthal, Bohemia.*—Mining products.
24 KAISERSTEIN, BARON, *Raabs, Lower Austria.*—Graphite.
25 KERTSCHKA, F. *Brunn am Walde, District of Gföhl.*—Graphite.
26 KÖTTIG, A. H. *Teplitz, Bohemia.*—Brown coal; coke.
27 KRATZER, L. *Rosenberg, Com. of Lipto Hungary.*—Regulus of antimony.
28 KRIEG, C. *St. Marein, near Brunn-on-the-Wald, Lower Austria.*—Graphite, raw and washed.
29 KRONSTADT MINING COMPANY, *Zillthal, Transylvania.*—Pit coal and iron-stone.
30 LACZAY, SZABÓ C. *Sárospátak, Com. of Zemplin, Hungary.*—Quartz mill-stones.
31 LENGYEL, A. I. R. ASSAYER, *Nagy-Bánya, Com. of Száthmar, Hungary.*—Crystallized quartz, and antimony.
32 LEOBEN, THE CHAMBER OF COMMERCE, *Stiria.*—Collective Exhibition of that District, from the following manufacturers:—

Drasche, H.	Pfeifer, J.
Eisenerz, Company at.	Radmeister Association.
Fürst, I. & M.	Schwarzenberg, Prince.
Henckel, Count.	Sessler, V. F.
Neuberg, I. R. Administration at.	

—Pit coal, ores, pig iron, and cast steel.
33 LIECHTENSTEIN, J. PRINCE OF, *Aloisthal, Moravia.*—Plate cast from the furnace.
34 MANGER, R. *Prague* 852/2.—Iron pyrites.
35 MAYR, F. *Ironworks, Leoben, Stiria.*—Steel and iron.
36 MAYR, M. *Vienna, Landstrasse* 703.—Steel, soft and hardened.
37 MERAN, COUNT, *Krems, Stiria.*—Iron and coal.
38 MITTROWSKY, COUNT WLADIMIR, *Pernstein, Galicia.*—A cube of lepidolite.
39 MOSLAVIAN ROCK-OIL MINING CO. *Borik, Croatia.*—Rock-oil, mineral tar, and raw asphalt.
40 NAGY, CH. VON, *Vienna* 276.—Manganese, pyrolusite.
41 NICKEL, G. *Vienna* 1088.—Mining and smelting products.
42 NOWICKY, C. & HAUSOTTER, F. *Graslitz, Bohemia.*—Copper ores, and map of locality.

43 PILLERSEE, I. R. MINING, SMELTING, AND HAMMER WORKS AT, *Tyrol.*—Refined steel.

44 PRAGUE, COMPANY OF IRON INDUSTRY AT.—Mining products and iron wares.

45 PRASCHNIKER, A. *Stein, Carniola.*—Hydraulic lime.

46 PRIBRAM, I. R. MINING OFFICE AT (from the I. R. silver and lead works near Birkenberg), *Pribram, Bohemia.*—Mining products.

47 PUTZER, P. VON, *Gonze, near Tüffer, Stiria.*—Brown coal.

48 PUTZER, P. VON, *near Cilli, Stiria.*—Brown coal.

49 PUTZER, P. VON, *Store, near Cilli, Stiria.*—Cast-steel, wrought and pig-iron.

50 QUAGLIO, G. 140, *Weissgärber, Vienna.*—Products of hauyofels (newly discovered minerals), from Transylvania.

51 RAUSCHER & CO. *Heft and Mosing, Carinthia.*—Pig-iron and ores; sulphate of barytes.

52 RIEDMAYER, E. *Giesenbach, near Seefeld, Tyrol.*—Asphalt-stone and mineral oil.

53 RIEGEL, A. *Mohacs and Fünfkirchen, Hungary.*—Agglomerated coal, compressed, grit-coal, and other coal.

54 ROTT, V. J. *Prague.*—Polishing stones.

55 RUDOLFSTADT, SILVER, LEAD, AND COAL MINING CO. AT, *Bohemia.*—Auriferous and argentiferous ores; anthracite.

56 SALGÓ-TARJANY, ST. STEPHAN COAL MINING CO. AT, *Pesth, Obere Donauzeile.*—Pit coal and mining products.

57 SALT WORKS, OWNERS OF, AT PIRANO, *Pirano, Istria.*—Sea salt.

58 SALT WORKS, DIRECTION OF, *Venice.*—Sea salt of the year 1861; plans of the salt works and buildings.

59 SAULLICH, A. & KRAFT, *Kufstein, Tyrol.*—Hydraulic lime, and cement.

60 SCHEMNITZ, I. R. ADMINISTRATION OF MINES, FORESTS, AND ESTATES AT (works at Windschacht and Rhonitz), *Schemnitz, Hungary.*—Mining products.

61 SCHWARZENBERG, J. A. PRINCE OF, *Prague.*—Mining and mineral products.

62 SEYBEL, E. *Vienna*, 26, *Wieden.*—Chromic ironstone and magnesite from Kraubath, in Stiria, iron pyrites from Bösing, in Hungary.

63 SIEGL, A. *Lemberg.*—Raw bitumen.

64 SILBERNAGL, BARON G. *Waidisch, Carinthia.*—Pig and refined iron; scoria.

65 STATES RAILWAY CO. *Vienna.*—Products of iron and metal works.

66 STIRIAN PATENT STEEL WORKS CO. *Vienna* 1047.—Steel.

67 SWOSOWICE, I. R. MINING AND SMELTING OFFICE AT, *Swosowice, District of Cracaw, Galicia.*—Raw and refined sulphur.

68 SZUMRÁK, J. *Libethen, Com. of Sohl, Hungary.*—Ores of cobalt and nickel.

69 THROR & CO. *Eperies, Com. of Sáros, Hungary.*—Mill-stones.

70 THURN-VALLE-SASSINA, COUNT G. *Streieben, Carinthia.*—Puddled steel.

71 TÖPPER, A. *Scheibbs, Lower Austria.*—Rolled iron, steel.

72 TREMBOWL, QUARRY AT, *Galicia.*—Hones, slabs.

73 UPPER HUNGARIAN FOREST AND MINE CO. *Iglo, Com. of Zips, Hung.*—Ores, smelting products, and metals.

74 VOLDERAUER, F. *Rothgulden, near Lungau, Salzburg.*—Arsenic ores.

75 WAIDHOFEN, IRON AND STEEL WORKS AT, *Klein-Hollenstein, Lower Austria.*—Rod iron, pig-iron, and refined steel.

76 WALDSTEIN, COUNT E. *Boros-Shebas, Com. of Arad, Hung.*—Mining and smelting products.

77 WALLAND, I. *Vienna* 303.—Lead, and lead ores.

78 WANG, J. *near Schönberg, Bohemia.*—Antimony.

79 WINTERNITZ, BROS. *Prague and Wiesenthal, Bohemia.*—Mining products.

80 WISSIAK, CH. *Vienna*, 134, *Burgglacis.*—Quartz, refractory argillaceous earth, gold ochre, limonite, benzine earth.

81 ZEH, J. *Lemberg.*—Mineral tar, montanine, raw naptha.

CLASS 2.

82 ACHLEITNER, L. *Salzburg.*—Lucifer matches.

83 AUSTRIAN COMPANY FOR CHEMICAL DESTRUCTIVE METALLURGIC PRODUCTS, *Aussig, Bohemia.*—Chemicals.

84 BÁNYAI, C. *Klausenburg, Transylvania.*—Chemicals.

85 BATKA, W. *Prague* 3571.—Chemicals.

86 BAUER, J. *Brünn.*—Tooth-paste.

87 BLUMBERG & RINDSKOPF, *Zuckmantel, Aust. Silesia.*—Chemicals.

88 BRAZZANOVICH, N. *Zara.* — Insect destroying powder.

89 BREITENLOHNER, DR. *Klumetz, near Schwarzbach, Bohemia.*—Paraffine and peat oil.

90 BRUNICKI, BARON M. *Village Pisarzowa, District of New Sandez, Galicia.*—Petroleum preparations.

91 BURKHART, H. *Aussig, Bohemia.*—Ultramarine, and green free from metal.

92 COLLECTIVE EXHIBITION OF MINERAL WATERS, BY DR. W. VON WELL, *Vienna* 1133.—Austrian mineral waters.

93 DIEZ, E. *St. Johann, near Villach, in Carinthia.*—White lead.

94 ENGELMANN, S. *Karolinenthal* 168, *near Prague.* — Albumen, dextrine, Leiogomme-starch and dextrue.

95 FICHTNER, J. & SONS, *Atzgersdorf near Vienna.*—Manure, and glue.

96 FINANCIAL DIRECTION OF THE DISTRICT OF BUKOWINA, I. R. (works at Gurahumora), *Czernowitz, Bukowina.*—Potash.

97 FISCHER, F. *Vienna*, 280, *Landstrasse.*—Saltpetre and soda.

98 FRAUNDORFER, F. *Gaudenzdorf, near Vienna.*—Chemicals.

99 FÜRTH, B. *Schüttenhofen, Bohemia.*—Lucifer matches.

100 GESSNER E., POHL W., ULBRICH O. *Müglitz, Moravia.*—Starch.

101 GÖDL, C. & Co. *Baern, Moravia.*—Chemical lights.

102 GOSLETH, F. VON, *Trieste-Hrastnigg.*—Chemicals.

103 GRAF, M. *Vienna, Schaumburger-grund,* 62, *Liniengasse.*—Blacking.

104 GUST, A. *Kronstadt, Transylvania.*—Economic wicks.

105 HEINDL, J. B. *Ottakring, near Vienna.*—Photogen, and solar oil.

106 HERBERT, BARON F. P. *Klagenfurt, Carinthia.*—White lead.

107 HERRMANN & GABRIEL, *Vienna* 426.—Lucifer matches.

108 HIRSCH, F. *Vienna* 689.—Washing and scouring preparations.

109 JÄCKLE, G. *Gratz, Stiria.*—Tartar.

110 KAISER, J. E. *Pesth, Herminenplatz.*—Chemical colours and starch.

111 KEIL, A. *Vienna,* 2, *Wieden.*—Spirit of wine lac, and polish.

112 KOPPERL, S. *Jungbunzlau, Bohemia.*—Albumine.

113 KÜHN, E. & C. *Sechshaus, near Vienna.*—Prussian blue, ammoniacal compounds.

114 KUNERLE, F. *Prague.* — Mineral waters.

115 KURZWEIL, F. *Freudenthal, Aust. Silesia.*—Chemical colours.

116 KUTJEWO, ESTATE AT, *Slavonia.*—Potash.

117 KUTZER, J. *Prague.*—Ultramarine.

118 KWIZDA, F. *Korneuburg, Lower-Austria.*—Veterinary drugs.

119 LAMATSCH, DR. J. *Vienna* 8/9, *Wieden.*—Chemicals.

120 LARISCH-MÖNNICH, COUNT, *Petrowitz, Aust. Silesia.*—Calcined soda and Glauber's salt.

121 LEHNER, E. *Vienna* 71, *Gumpendorf.*—Aniline, and substances manufactured from it.

122 LEHRER, A. *Fürnitz, Bohemia.*—Ultramarine.

123 LEWINSKY BROS. *Dobrisch, District of Prague, Bohemia.*—Acetic acid and acetates.

124 MADÁCS, E. & Co. *Ployeschti,* by C. STIEHLER, *Kronstadt, Transylvania.* — Photogen gas.

125 MANUFACTORY OF CHEMICALS, *Fiume.*—Chemical products.

126 MARASPIN BROS. *Trieste.*—Chemical products.

127 MILLER & HOCHSTAETTER, *Hruschau, Aust. Silesia.*—Chemicals.

128 MOLL, A. *Vienna* 562.—Chemical and pharmaceutical products.

129 MÜLLER, A. J. *Schwechat, near Vienna.*—Inks for typography, copper-plate printing, and lithography.

130 NACKH, J. & SON, *Vienna, Wieden* 118, *Heugasse.*—Chemicals.

131 NEUBURG & ECKSTEIN, *Pilsen, Bohemia.*—Chemicals.

132 NICOLAUS, C. *Kronstadt, Transylvania.*—Photogen gas.

133 NOWACK, J. *Karolinenthal, near Prague* 138. — Dextrine, and extracts of colouring matter.

134 OTTENREITER, L. *Pesth, Bruckgasse.*—Gout-salve; pharmaceutical preparations.

135 PANESCH, A. *Vienna* 1100.—Blacking, and lac.

136 PARGER, J. *Vienna* 851. — Lac, indigo lac, and blacking.

137 PEUSENS, F. C., WERSCHOWSKY, W. *Vienna, Alservorstadt* 73.—Benzine.

138 PFEFFERMANN, DR. *Vienna.*—Perfumery.

139 PIERING, C. F. *Karolinenthal, near Prague* 157.—Acetate of lead.

140 POLLAK, A. M. *Vienna* 768.—Matches and chemical lights.

141 POLLAK, B. JUN. *Vienna, Praterstern* 396.—Illyrian potash.

142 POLLEY, C. *Simmering, near Vienna.*—Coal and mineral tar; petroleum, raw matters for manufactures, plans of apparatus.

143 POPP, J. G. *Vienna* 557.—Perfumery.

144 POSELT, L. *Vienna, Landstrasse* 472.—White bleached shel-lac.

145 PRASCHNIKER, A. *Stein in Carniola*—Metallic polishing powder.

146 PRIMAVESI, P. *Gross-Wisternitz and Bedehosch, Moldavia.*—Potash.

147 PUNSCHART F. & RAUSCHER, *St Veit, Carinthia.*—White lead.

148 QUAPILL, R. *Znaim, Moravia.*—Liquorice-juice.

149 RADAUZ, I. R. MILITARY STUD ESTABLISHMENT AT, *Bukowina.*—Calcined potash.

150 RICHTER & CLAR BROS. *Hernskretschen, near Tetschen, Bohemia.*—Archil, weeds and extract; indigo, carmine.

151 RIEDMAYER, E. *Giesenbach, Tyrol.*—Photogen and asphalt tar.

152 SAPIEHA, PRINCE A. *Krasiczyn, District of Przemisl, Galicia.*—Turpentine.

153 SCHAUMBURG-LIPPE, PRINCE OF, *Veröcze, Croatia.*—Potash.

154 SETZER, J. *Weitenegg on the Danube, Lower Austria.*—Ultramarine.

155 SEYBEL, E. *Bösing, in Hungary, and Vienna* 26, *Wieden.*—Iron pyrites.

156 SIEGL, A. *Lemberg.*—Rectified naptha.

157 STATES RAILWAY COMPANY I. R. *Vienna.*—Paraffine in blocks and candles; blue vitriol; sulphuric acid, &c.

158 STELZL, F. *Baern, Moravia.*—Matches and chemical lights.

159 STOC, A. *Delnice, Croatia.*—Potash.

160 STROBENTZ, BROS. *Pesth, Zrinyi Gasse.*—Chemical products.

161 TOBISCH, A. JUN. *Königgrätz, Bohemia.*—Pharmaceutical preparations.

162 TSCHELIGI, R. *Villach, Carinthia.*—Protoxide of lead.

163 UNTERWEGER, G. B. *Trient, Tyrol.*—Varnish.

164 VOLDERAUER, F. *Rothgulden and Langau, Salzburg.*—Metallic and white arsenic.

165 WAGEMANN, G. *Vienna* 1047.—Products of petroleum, resin oil, coal tar oil.

166 WAGENMANN, SEYBEL & Co. *Liesing, near Vienna.*—Chemical products.

167 WAGNER, DR. D. *Pesth.*—Chemicals, and pharmaceutical products.

168 WILHELM, F. & Co. *Vienna* 1100.—Drugs and chemicals.

169 ZARZETSKY, J. *Pesth, zwei Mohrengasse.*—Chemical lights.

170 ZACHERL, J. *from Tiflis, Vienna* 624.—Pyrethrum roseum Caucasicum, in flowers, seeds, and as powder.

171 ZEH, J. *Lemberg.*—Benzine, leather and cart grease, purified Montanine.

CLASS 3.

172 ALTHAN, COUNT M. C. *Swoischitz, District of Czaslau, Bohemia.*—Agricultural products.

173 ANDRASSY, COUNT G. *Hosszurét, Com. of Gömor, Hungary.*—Wines.

174 AUGUSZ, BARON A. *Szegszard, Com. of Tolna, Hungary.*—Wines.

175 BÄCK, W. *Gross-Meseritsch, Moravia.*—Liqueurs.

176 BALÁZS, G. *Alsónyék, Com. of Hont. Hung.*—Wines, plum-whiskey.

177 BALTZ, P. *Pesth, Donauzeile.*—Wines.

178 BÁNYAI, B. C. *Klausenburg, Transylvania.*—Wines.

179 BARBER'S SONS, *Buda.*—Various kinds of flour.

180 BARTHA, L. *Fehér-Gyármath, Com. of Száthmár, Hungary.*—Tobacco.

181 BARZDORF, SUGAR MANUFACTORY AT, *Aust. Silesia.*—Raw-sugar and loaf-sugar.

182 BATTHYANYI, COUNT G. *Reichnitz, Com. of Eisenburg, Hungary.*—Agricultural products.

183 BAUER, K. 316, *Leopoldstadt, Vienna.*—Liqueurs.

184 BAUER, T. *Königsfeld and Tischnowitz, Moravia.*—Sugar.

185 BELGIAN TURBINE MILL, *Carlstadt, Croatia.*—Flour.

186 BERCHTOLD, COUNT S. JUN. *Buchlau, Moravia.*—Wine.

187 BIZEK, F. *Alt-Brese, Com. of Bacs-Bodrogh, Hungary.*—Products.

188 BLAESS, G. *Waitzen, Com. Pesth, Hungary.*—Vinegar.

189 BLASCHÜTZ BROS. *Bogsan, Com. of Krassó, Hungary.*—Spirits.

190 BLÜCHER VON WAHLSTADT, COUNTESS M. *Radun, near Troppau. Aust. Silesia.*—Sugar.

191 BLUM, J. *Buda, Hauptgasse.*—Mill products.

192 BOHEMIAN STEAM MILLS CO. *Smichow, near Prague.*—Flour.

193 BOGESCHDORF LOCAL BAILIWICK, *Bogeschdorf, near Mediasch, Transylvania.*—Wines.

194 BOGYAI, L. *Haláp Village, Com. of Zala, Hungary.*—Wines.

195 BRAUN, BROS. *Pesth.*—Wines, liqueurs, &c.

196 BREUNER, COUNT A. *Zselisz, Com. of Barse, Hungary.*—Wines.

197 BUDACKER, T. *Bistritz, Transylvania.*—Wines.

198 BURCHARD, S. & Co. *Tokay, Com. of Zemplin, Hungary.*—Wines.

199 CALIGARICH, C. *Zara.*—Maraschino.

200 CATTICH, BROS. *Zara.*—Wine and olive-oil.

201 CHWALIBOG, J. *Lipowce, District of Zloczow, Galicia.*—Brandy and liqueurs.

202 COMER, F. *Trieste.*—Maraschino and rosoglio.

203 CSERNOVICS, P. *Szatárd, Com. of Bihar, Hungary.*—Wheat.

204 CZALOGOVIC, B. *Esseg, Slavonia.*—Maize and wheat.

205 DEGENFELD, COUNT E. *Nyir-Bakta, Com. of Szabolcz, Hungary.*—Tokay of the first press.

206 DÉSENSY, M. *Banat-Komlós, Com. of Torontal, Hungary.*—Spirits.

207 DEVRIENT, A. *Pesth.*—Wines.

208 DITZL & PONINGER, *Karolinenthal* 145, *near Prague.*—Wines.

209 DREHER, A. *Klein-Schwechat, near Vienna.*—Beer.

210 DRESSLER, F. *Obora, Bohemia.*—Loaf-sugar.

211 DUDA, J. F. *Prague* 707/2.—Chocolate.

212 EDER, F. M. *Pesth, Waitznergasse.*—Hungarian champagne.

213 EGAN, E. *Bernstein, Com. of Eisenberg.*—Agricultural products.

214 ELTZ, COUNT, C. *Vucovar, Slavonia.*—Wine; maize; distillery products.

215 ENGELMANN, S. *Karolinenthal* 168, *near Prague.*—Starch made from wheat and potatoes.

216 ERDÖDY, COUNT F. *Somlovar, Com. of Veszprim, Hungary.*—Wine.

217 FABER, C. M. *Vienna* 947.—Stirian wine of Frauenheim.

218 FARKAS, L. *Grosswardein, Com. of Bihar, Hungary.*—Wines.

219 FEHER, BROS. *Törok-Beese, Com. of Torontal, Hungary.*—Agricultural products and wines.

220 FEKETE & Co. *Erlau, Com. of Heves Hungary.*—Wines.

221 FESTETITS, COUNT G. *Oedenburg, Hungary.*—Products.

222 FESTETITS, COUNT T. *Keszthely, Com. of Zala, Hungary.*—Wines.

223 FISCHER, C. F. *Pesth, Servitenplatz.* Preserves and marmalade.

224 FISCHER VON RÖSLERSTAMM, E. *Vienna* 407.—Vermicelli.

225 FIUME, MANUFACTORY OF VERMICELLI AT.—Pastes.

226 FIUME, ESTABLISHMENT OF FLOUR COMMERCE, *Fiume.*—Groats and biscuits.

227 FLANDORFER, I. *Oedenburg, Hungary.*—Wines.

228 FORGACH, COUNT K. *Nagy-Szalantz, Com. of Abauj, Hungary.*—Wheat, and specimens of soils.

229 FRANK, J. *Kronstadt. Transylvania.*—Mill products.

230 FRÖLICH, W. *Klausenberg, Transylvania.*—Honey.

231 FUCHS, G. & Co. *Pesth.*—Wines.

232 FÜNK, E. *Gratz, Stiria.*—Stirian liqueurs.

233 FÜNK, H. *Gratz, Stiria.*—Spirits.

234 FÜNK, M. F. *Gratz, Stiria.*—Liqueurs.

235 GALLINY, F. *Lugos, Com. of Krásso, Hungary.*—Wines.
236 GANZ, J. & Co. *Pesth.*—Wines.
237 GAVELLA, N. *Agram, Croatia.*—Wine.
238 GEMPERLE, J. *Vienna, Leopoldstadt* 311.—Chicory, and other substitutes for coffee.
239 GÖGL, Z. *Krems, Lower Austria.*—Mustard.
240 GOLDNER & ZINNER, *Pesth.*—Paprika.
241 GÖRZ, AGRICULTURAL SOCIETY AT, *Görz.*—Agricultural products.
242 GREIS, J. *Vienna, Jesuitenhof Laimgrube.*—Fine pollard, middlings, and groats.
243 GYÖNGYÖS, AGRICULTURAL SOCIETY OF THE HEVES COMITATE AT. *Com. of Heves, Hungary.*—Collection of agricultural products.
244 GYULAI, COUNT D. *Czombárd, Com. of Somogy, Hungary.*—Wheat and wines.
245 HAACK, A. *Gratz.*—Liqueurs.
246 HAAS, G. *Grosswardein, Com. of Bihar, Hungary.*—Wines.
247 HAASE, A. & SONS, *Znaim Moravia.*—Moravian products and wines.
248 HAASE'S SON, *Znaim, Moravia.*—Moravian wines.
249 HANNAK, G. *Brandeis (an der Adler), Bohemia.*—Flour.
250 HAVÁS, J. VON. *Buda, Hungary.*—Wines.
251 HEGYALJA, WINE GROWERS' CO. AT. *Mad, Com. of Zemplin, Hungary.*—Tokay wines.
252 HIZGERN, J. G. *Krems, Lower Austria.*—Wine and mustard.
253 HOFGRÄFF, J. *Bistritz, Transylvania.*—Wines.
254 HOLZER, I, *Vienna, Altlerchenfeld* 52.—Viennese wine and tea-cakes.
255 HOP GROWERS AT AUSCHA, *Auscha, Bohemia.*—Hops.
256 HUMMER, J. *Brünn.*—Austrian and Moravian wines.
257 JALICS, F. & Co. *Pesth.*—Wines.
258 JANKÓ, V. VON. *Pesth.*—Collective exhibition of Hungarian raw products; illustrations of costumes, wine districts, &c.
259 JANKOVITZ, V. *Galábocs, Com. of Neograd, Hungary.*—Wines.
260 JASZAY, D. *Abauj-Szánthó, Com. of Abauj, Hungary.*—Tobacco of Szt. Andras.
261 JEKEL, J. P. *Husszufalu, Transylvania.*—Mill products, pearl barley, and groats.
262 JOÓ, J. *Erlau, Com. of Heves, Hungary.*—Wines.
263 JORDAN & SONS, *Tetschen, Bohemia.*—Flour.
264 JORDAN & TIMÄUS, *Bodenbach, Bohemia.*—Chocolate and chicory.
265 JURENAK, A. *Pressburg, Hungary.*—Wheat.
266 KAISER, J. E. *Pesth, Herminenplatz.*—Starch.
267 KAMMERGRUBER, J. *Lugos, Com. of Krassó, Hungary.*—Wines.
268 KAMPLMILLNER, H. *Gratz, Stiria.*—Wine.
269 KAPPEL, F. *Kis-Tur, Com. of Hont, Hungary.*—Products.
270 KASSOWITZ, J. H. *Pesth* 5, *Felbergasse.*—Wine.
271 KATTUS, J. *Vienna* 330.—Wines.
272 KISS DE NEMESKED, P. *Misslán, Com. of Tolna, Hungary.*—Wines.
273 KLEIN, D. W. *Vukovar, Slavonia.*—Maize, &c.
274 KLEINOSCHEG BROS. *Gösting, near Gratz, Stiria.*—Champagne and dessert wines.
275 KLOSTERNEUBURG CHAPTER-HOUSE *Klosterneuburg, Lower Austria.*—Wines.
276 KONKOLY, F. *Békés, Com. of Békés, Hungary.*—Wines.
277 KORNIS, J. *Gyorok, Com. of Arad, Hungary.*—Wines.
278 KOSZTHA, J. *Pesth.*—Wines of Ermelék.
279 KRAETSCHMAR, C. A. *Rima Szombat, Hungary.*—Wines.
280 KRATOCHWILL, W. *Lounky, Bohemia.*—Hops.
281 KRAUS, BROS. *Karolinenthal* 9, *near Prague.*—Mill products.
282 KREMS, CITY OF, *Lower Austria.*—Collective exhibition of wines.
283 LAIBACH, I. R. AGRICULTURAL SOCIETY, *Laibach, Carniola.*—Different kinds of wheat.
284 LARISCH-MÖNNICH, COUNT, *Ober-Suchau, Aust. Silesia.*—Beet-root, clarified, and loaf sugar.
285 LAY'S, MICHAEL, HEIRS, *Essegg, Slavonia.*—Wheat and maize.
286 LEIBENFROST, F. & Co. *Vienna* 1107.—Wines.
287 LEITNER, J. *Gratz.*—Chicory, and other substitutes for coffee.
288 LENK, S. *Oedenburg, Hungary.*—Wines.
289 LENKAY, A. VON, & Co. *Vienna.*—Wines.
290 LIEBL, V. & SON, *Retz, Lower Austria.*—Wine.
291 LITTKE, J. & J. *Fünfkirchen, Com. of Baranya, Hungary.*—Wines.
292 LÖFFLER, J. P. *Langhalsen, near Neufelden, Upper Austria.*—Hops.
293 LONYAI, M. VON, *Pesth,* 2 *Adlergasse* 3.—Wines.
294 LOVASSY, A. *Nagy-Szalonta, Com. of Bihar, Hungary.*—Wheat.
295 LÖWENFELD & HOFMANN, BROS. *Kleinmünchen, Upper Austria.*—Mill products of wheat and rye.
296 LUKACS, S. *Palosa, Com. of Stuhlweissenburg, Hungary.*—Various kinds of flour.
297 LUXARDO, G. *Zara.*—Maraschino.
298 MAGYAR, E. *Maria-Theresiopel, Com. of Bacs-Bodrogh, Hungary.*—Wheat.
299 MAKAY, A. *Lugos, Com. of Krassó, Hungary.*—Slivovitz.
300 MÁLNAY, DR. I. *Pesth, Museumgasse* 1.—Wines.
301 MARBURG-STIRIAN CO. OF WINE GROWERS AT, *Marburg, Stiria.*—Wines.
302 MATKOWSKA, S. *Jezierzany, District of Stanislau, Galicia.*—Barley.
303 MAUTHNER, A. J. & SON, *Vienna* 338.—Beer, spirit, solidified yeast.
304 MERAN, COUNT, *Marburg, Stiria.*—Wine.
305 MOLNAR, G. *Debreczin, Com. of Bihar, Hungary.*—Wheat, rye, tobacco, wine.

306 MOLNÁR & TORÖK, *Pesth.*—Tokay of the first press.
307 MUNCH-BELLINGHAUSEN, COUNT J. *Vienna* 543.—Wines.
308 NAGY, M. *Ráab, Hungary.*—Pastes.
309 NAWRATIL, P. *Romanovka, District of Zloczow, Galicia.*—Hops.
310 NITSCH, J. *Strassnitz, Moravia.*—Wine.
311 OROSZY, N. VON, *Szentes, Com. of Csongrad, Hungary.*—Wheat.
312 PANETH. J. S. *Vienna* 154.—Wines.
313 PARRAGH, G. *Pesth.*—Wines.
314 PASTNER, J. *Gratz.*—Liqueurs.
315 PERGHAMMER, F. VON, *Eppan, Tirol.*—Champagne.
316 PESTH ROLLING MILL CO. *Pesth.*—Flour.
317 PIERING, C. F. *Karolinenthal* 157, *near Prague.*—Spirit, and wine vinegar; acetic acid.
318 PIERING & DONAT, *Karolinenthal* 157, *near Prague.*—Mustard.
319 PODMANITZKY, BARON A. *Aszód, Com. of Pesth, Hungary.*—Wines.
320 POHRLITZ ARCHDUCAL MILL ADMINISTRATION, *Pohrlitz, Moravia.* — Mill products.
321 POTOCKI, COUNT A. *Tenczynek, Galicia.*—Flour.
322 PRASCH, L. *Mitterrötzbach, Lower Austria.*—Wines.
323 PRIMAVESI, P. *Gross-Wisternitz and Bedihoscht, Moravia.*—Loaf-sugar and spirits.
324 RAMASETTER, V. *Sümeg, Com. of Somogy, Hungary.*—Wines.
325 RANOLDER, J. DIOC. BISHOP, *Veszprim, Hungary.*—Wines.
326 RAUSSNITZ, E. *Vienna, Hernals* 5.—Liqueurs.
327 REHBERG, MILL AT, *near Krems, Lower Austria.*—Mill products.
328 REISENLEITNER, J. *Vienna* 599 *and* 600.—Wine.
329 REISSENBERGER, W. *Hermannstadt, Transylvania.*—Wine.
330 RICHTER, A. *Königsaal, Bohemia.*—Sugar.
331 RIEMERSCHMIED, A. *Vienna, Wieden* 237.—Potato spirit, completely free from fusel-oil.
332 RIESE - STALLBURG, BARON F. *Vienna.*—Photographs of domestic animals.
333 RIGYITSKY, P. *Skrebesztye, Com. of Krassó, Hungary.*—Slivovitz and products.
334 RIMANOCZY, P. *Töhöl, Com. of Bars, Hungary.*—Agricultural products.
335 RINGLERS, J. SONS, *Botzen, Tyrol.*—Candied fruits.
336 RIPKA, J. M. & CO. *Brünn.*—Mill products.
337 RITTER, H. VON ZAHONY, *Strazig, Görz.*—Flour.
333 ROBERT & CO. *Selowitz, Moravia.*—Sugar and spirits.
339 ROHONCZY, I. VON, *Nagy-Bogdany, Com. of Veszprim, Hungary.*—Spirits and liqueurs.
340 ROSCONI, N. *Buda.*—Wines.
341 RULIKOWSKY, C. *Switazoco, District of Zolkiew, Galicia.*—Wheat.
342 RUNGG, A. *Trient, Tirol.*—Wine.
343 RUPP, F. *Neufelden, Upper Austria.*—Hops.
344 SAAZ AGRICULTURAL SOCIETY, *Saaz, Bohemia.* — Agricultural products from the district of Saaz.
345 SAUER, DR. I. *Pesth.*—Wines.
346 SCHARY, J. M. *Prague.*—Beer and barley-malt.
347 SCHÄSSBURG, COMMUNITY AT, *Schässburg, Transylvania.*—Wines.
348 SCHAUMBURG - LIPPE, PRINCE OF, *Veröce, Slavonia.*—Wine, flour, and groats.
349 SCHLUMBERGER, R. *Vöslau, near Vienna.*—Wine.
350 SCHMITT, A. *Deutschbrod. Bohemia.*—Clover-seed.
351 SCHNEIDER, A. *Vienna* 726.—Wine.
352 SCHÖFFEL, J. *Saaz, Bohemia.*—Hops.
353 SCHÖLLER, A. *Ebenfurth, Lower Austria.*—Flour in all stages of manufacture.
354 SCHORKOPF, C. H. *Also-Rakos, Transylvania.*—Wheat.
355 SCHOTTEN, CHAPTER-HOUSE, *Vienna.*—Wines.
356 SCHÜRER, F. P. *Stein on the Danube, Lower Austria.*—Wine.
357 SCHWARZENBERG, J. A. PRINCE OF, *Vienna.*—Hops.
358 SCHWARTZERS, A. SUCCESSOR, *Vienna* 1102.—Wine.
359 SEEGER, F. *Pesth,* "*Marokkaner-house.*"—Wine.
360 SELBSTHERR BROS. *Maad, Com. of Zemplin, Hungary.*—Wine of Hegyalja.
361 SIMON, A. *Vienna, Neubau,* 145.—Biscuit.
362 SINA, BARON J. *St. Miklos. Hungary, and Rossitz, Moravia.*—Loaf-sugar and spirit.
363 SMETANA, J. *Mileschau, Moravia.*—Corn, pulse, &c. the products of the Mittelgebirg, in Bohemia.
364 SPRINGER, M. *Reindorf, near Vienna* 32.—Solidified yeast.
364A STAMM MARTIN, *Wartmann, Brenndorf, near Kronstadt, Transylvania.*—Wheat.
365 STEAM MILLS CO. *Vienna.* — Mill products.
366 STEPHEN, STEAM MILL CO. *Debreczin, Com. of Bihar, Hungary.*—Flour.
367 STEPNICZKA, F. *Prague, Neustadt* 1269.—Rocks-drops.
368 STIASNY, H. *Vienna, Jägerzeile* 503.—Liqueurs.
369 STIRIAN AGRICULTURAL SOCIETY, I. R. *Gratz.*—Wine and corn.
370 SUCHANEK, A. *Brünn.* — Moravian products.
371 SZALAY, P. *Tissa Varkoni, Com. of Szolnok, Hungary.*—Theiss wheat.
372 SZEGZARD, WINE TRADE CO. AT, *Com. of Tolna, Hungary.*—Wines.
373 SZIRÁNYI, A. & CO. *Gyöngyös, Com. of Heves, Hungary.*—Comfits.
374 TAUBER, M. & BETTELHEIM C. *Vienna, Jägerzeile* 522.—Plum whisky, and cherry-spirit.
375 TAUSSIG, J. A. & BROS. *Vienna, Sechshaus* 16.—Spirits.
376 TELEKY, COUNT D. *Klausenburg, Transylvania.*—Wines.

377 THUN HOHENSTEIN, COUNT F. *Tetschen on the Elbe, Bohemia.*—Mill products.

378 THUN HOHENSTEIN, COUNT F. *Peruc, Bohemia.*—Sugar.

379 TOBACCO MANUFACTORY CENTRAL ADMINISTRATION OF THE, *Vienna.*—Tobacco, raw and manufactured.

380 TOMASI, A. *Gaya, Moravia.*—Vinegar, and spirts.

381 TÖRÖK, G. VON, *Arad, Hungary.*—Wines.

382 TRAITLER, L. A. *Arad, Hungary.*—Flour.

383 TROPPAU, SUGAR REFINING CO. AT, *Aust. Silesia.*—Loaf-sugar.

384 TSCHINKEL'S, A. SONS, *Schönfeld and Lobositz, Bohemia.*—Chicory, and other substitutes for coffee.

385 URBANEK, BROS. *Prerau, Moravia.*—Corn and flour.

386 VALERIO, A. *Triest.*—Chocolate.

387 VARGA, L. *Bath Füred, Com. of Veszprim, Hungary.*—Wines.

388 VASICZ, S. E. *Klausenburg, Königsgasse* 51, *Transylvdnia.*—Rosoglio.

389 VERSETZ, CITY CORPORATION OF, *Banat.*—Wines.

390 WALLIS, F. COUNT, *Koleschowitz, Bohemia.*—Hops.

391 WANKA, F. *Prague* 796/2.—Flour, beer.

392 WARHANEK, C. *Vienna, Leopoldstadt* 4.—Preserves in vinegar, and oil.

393 WAWRA, V. JUN. *Prague* 207/2. — Tlacenka mill products, in the solid state, and fit for being kept many years.

394 WEINMAYER, L. *Grinzing, near Vienna.*—Wine.

395 WEISS, J. *Bistritz Transylvania.*—Wines.

396 WERTHER, F. *Buda.* — Agricultural and manufacturing products.

397 WIDTER, J. *Pesth.*—Flour.

398 WINNICKI, S. *Boryszkowee, District of Csortkow, Galicia.*—Leaf tobacco, and maize.

399 WÖHRL, J. G. *Vienna* 511.—Wine.

400 WOLF, S. & CO. *Weisskirchen, Moravia.*—Spirits.

401 ZALLINGER, C. VON, *Botzen Tyrol.*—Wines.

402 ZICHY-FERRARIS, E. COUNT, *Nagy-Szöllös, Com. of Veszprin, Hungary.*—Wines.

403 ZWICKL, J. *Esterház, Hungary.* — Essence of wormwood.

CLASS 4.

404 AFH, F. *Vienna, Neubau* 19.—Bird-cage, flower-stands.

405 ALTHAN, M. C. COUNT, *Swoischitz, District of Czaslau, Bohemia.*—Fleece.

406 BARATTA, CHEVALIER DE C. *Budischau, Moravia.*—Fleeces.

407 BARTENSTEIN, J. BARON, *Hennersdorf, Aust. Silesia.*—Fleeces.

408 BAUER, M. *Warasdin, Croatia.* — Cross-cuts of oak-trunks.

409 BAUMGARTEN, F. M. *Unterweissenbach, Upper Austria.*—Flax.

410 BEISIEGEL, P. *Vienna, Wieden* 925.—Articles in meerschaum.

411 BELLEGARDE, R. COUNTESS, *Gross Herrlitz, Aust. Silesia.*—Fleeces.

412 BEZEREDY, N. *Cseb. Com. of Bacs Bodrogh, Hungary.*—Flax.

413 BIACH, E. & CO. *Theresienfeld, near Wr.-Neustadt, Lower Austria.*—Resin and products.

414 BIENERT, D. & SON, *Maderhäuser Bohemia.*—Wood for musical instruments.

415 BIONDEK, M. *Baden, Lower Austria*—Agriot-sticks.

416 BIRNBAUM, J. *Pesth.*—Flax.

417 BISTRITZ, VENEER CUTTING CO. AT *Transylvania.*—Veneers.

418 BIZEK, F. *Alt-Becse, Com. of Bacs-Bodrogh, Hungary.*—Wool.

419 BLÜCHER, VON WAHLSTADT M. COUNTESS, *Radun near Troppau, Aust. Silesia.*—Fleeces.

420 BÖHM, J. & GENSCHEL, J. *Vienna, Neubau* 59.—Fancy works in brass, mother-of-pearl, and ivory.

421 BOSCHAN, J. & SONS, *Angern, Lowe Austria.*—Oils.

422 BRICHTA, A. *Prag.* 36/11, *Breitegasse*—New toilet soap.

423 BUKOWINA, I. R. PROVINCIAL FINANCE DIRECTION AT KIMPOLUNG, *Czernovitz, Bukowina.*—Timber.

424 BURKHARDT, C. *Osterberg, Carniola.*—Oil.

425 BURSCHIK, A. *Vienna, Wieden* 852.—Pipe-sticks.

426 BÜTTNER, C. F. *Steieregg, Upper Austria.*—Teasels.

427 CHIOZZA, C. L. & SON, *Trieste.*—Soap

428 CZILCHERT, R. DR. *Gutor, Com. of Pressburg, Hungary.*—Merino-fleeces.

429 DAPSY, W. *Rima-Szombath, Com. of Gömör, Hungary.*—Gall-nuts.

430 DAUN, H. COUNT, *Vöttau, Moravia.*—Fleeces.

431 DIEDEK'S, A. C. SON, *Vienna, Altlerchenfeld* 117.—Washing and toilet soap.

432 DOBLINGER, F. *Vienna, Matzleinsdorf*—Wax.

433 ENGL, H. *Traiskirchen, Upper Austria.*—Specimens of wood.

434 FALKENHAIN, T. COUNT, *Kyowitz, near Troppau, Aust. Silesia.*—Fleeces.

435 FERENCZY, J. *Pesth, Herrengasse.*—Billiard cues.

436 FINANCES, MINISTRY OF (I. R. direction of forests at Vienna), *Vienna.*—Forest products.

437 FINANCES, MINISTRY OF (provincial direction at Gratz), *Vienna.* — Forest products.

438 FINANCES, MINISTRY OF (direction of mines and forests at Gratz), *Vienna.*—Forest products.

439 FINANCES, MINISTRY OF (direction of forests at Montona in Istria), *Vienna.*—Forest products.

440 FINANCES, MINISTRY OF (direction of salters and forests at Gmunden), *Vienna.*—Forest products.

441 FINANCES, MINISTRY OF (direction of

salters and forests at Salzburg and Hallein), *Vienna*.—Forest products.

442 FINANCES, MINISTRY OF (provincial direction at Lemberg), *Vienna*.—Forest products.

443 FINANCES, MINISTRY OF (provincial direction at Pressburg), *Vienna*.—Forest products.

444 FINANCES, MINISTRY OF (direction at Venice), *Vienna*.—Forest products.

445 FISCHA, P. *Bistritz, Moravia*.—Bleached wax.

446 FISCHER, F. *Vienna, Landstrasse* 280.—Tallow and soap.

447 FISCHER, F. *Smichow, near Prag*.—Specimens of wood.

448 FRIEDRICH, J. *Vienna* 1037.—Meerschaum manufactures.

449 FRÖHLICH, W. *Klausenburg, Transylvania*.—Wax.

450 GÄRTNER, J. F. JUN. *Ranersdorf, near Vienna*.—Samples of dextrine.

451 GERMER, W. & Co. *Baden, near Vienna*.—Agriot-sticks.

452 GOLDMANN, M. *Vienna, Gumpendorf* 621.—Meerschaum pipes.

453 GÖRZ, AGRICULTURAL SOCIETY AT, *Görz*.—Agricultural products.

454 HANNIG, M. L. *Debreczin, Com. of Bihar, Hungary*.—Rush and cane mats.

455 HARTL, G. & SON, *Vienna, Rossau* 98.—Soaps.

456 HARTMANN, L. *Vienna, Laimgrube* 325.—Turnery.

457 HERMANNSTADT, STEARINE MANUFACTORY AT, *Hermannstadt, Transylvania*.—Candles and soap.

458 HERTRUM, J. *Trieste*.—Fats and volatile oils.

459 HILLINGER, J. *Vienna, Alservorstadt* 113.—Chess boards and turnery.

460 HIMMELBAUER, A. & Co. *Stockerau, Lower Austria*.—Stearine candles, and soap.

461 HOFMANN, J. *Gratz*.—Stearine candles.

462 HOFRICHTER, F. *Reichenau, Lower Austria*.—Boxes.

463 HOSCHEK, A. *Vienna, Neubau* 263.—Meerschaum pipes, cigar pipes, and pipe-sticks.

464 HOSSNER, J. J. *Schluckenau, Bohemia*.—Spanning plates.

465 HOYOS-SPRINZENSTEIN, COUNT E. *Stixenstein and Gutenstein, Lower Austria*.—Cross cuts of trees.

466 JABUREK, F. *Vienna, Mariahilf* 92.—Meerschaum articles.

467 JOSEFSTHAL, MANUFACTORY AT, *Carniola*.—Dye-wood and oil.

468 JUNG, J. G. *Vienna, Braunhirschen* 161.—Combs.

469 KELL, G. & E. VON RUDZINSKY-RUDNO, *Endersdorf, Aust. Silesia*.—Fleeces.

470 KLEIN, C. *Wedtzierz, Galicia*.—Indestructible incrustations on wood, brass, and stone.

471 KORIZMICS, L. *Pesth, Ulloërstrasse* 12.—Fleeces.

472 KOUF, F. *Hinterbrühl, Mödling, near Vienna*.—Colophany, pitch, resin, turpentine, tools, and sections of trees.

473 KRÄTSCHMAR, C. A. *Rima-Szombath, Hungary*.—Gall-nuts, Hungarian red cloverseed, yellow wax, sheep's wool.

474 KREUTER, F. *Vienna, Landstrasse* 744.—Telegraph poles impregnated by Boucherie's method.

475 KUMPF. P. *Schluckenau, Bohemia*.—Articles made of Spanish broom.

476 KUTJEWO, ESTATE OF, *Kutjewo, Slavonia*.—Specimens of wood, and gall-nuts.

477 KUZEL, J. *Vienna, Schottenfeld* 23.—Turnery, mounted with bronze.

478 LANGE, W. *Vienna, Schottenfeld* 3.—Mother-of-pearl buttons.

479 LARISCH-MONNICH, COUNT J. *Freistadt, Aust. Silesia*.—Fleeces.

480 LAY, M. HEIRS OF, *Essegg, Slavonia*.—Rape, linseed, madder, and various oils.

481 LEOPOLD, J. *Pesth, Hochstrasse* 7.—Gall-nuts.

482 LIECHTENSTEIN, J. PRINCE OF, *Vienna*.—Fleeces.

483 LÖFFLER, J. P. *Langhalsen, near Neufelden, Upper Austria*.—Flax.

484 LUKASCH, J. *Vienna, Spittelberg* 143.—Mother-of-pearl manufactures.

485 MACHACEK, J. *Horovitz, Bohemia*.—Fleeces.

486 MAYER, J. A. & Co. *Niedersdorf, Tirol*.—Brewers' pitch.

487 MEINERT, H. HEIRS OF, *Partschendorf and Erb-Sedlitz, Moravia*.—Fleeces.

488 MERGENTHALER, C. *Essegg, Slavonia*.—Wax.

489 MITTAK, J. *Skalitz, Com. of Pressburg*.—Hungarian woad (isatis tinctoria).

490 MÜNCH-BELLINGHAUSEN, COUNT J. *Vienna* 543.—Wood-pitch, and pitch-wood.

491 MUNDY, BARON J. *Ratschitz, Moravia*.—Fleeces.

492 MURALT, D. *Vienna* 1075.—Fancy articles made of cockle shells.

493 NITSCH, J. *Strassnitz, Moravia*.—Dried herbs, and fleeces.

494 NOCKER, P. *Botzen, Tirol*.—Wood carved in high relief.

495 PAGET, F. *Vienna, City* 487.—Waterproofed stuffs for awnings.

496 PARTEL & ZESCHKO, *Laibach, Carniola*.—Soaps.

497 PETRI, C. A. *Thereseinfeld, Lower Austria*.—Raw fleeces.

498 PETRICIOLI-SALGHETTI, BROTHER, *Wax Manufactory, Zara*.—Wax.

499 PFEIFFER, H. *Trieste*.—Combs.

500 PICHELMAYER, T. *Leoben, Stiria*.—Oils.

501 PITTNER, J. *Algersdorf, near Gratz, Stiria*.—Tlower-resin, a drug which prevents the sterility of domestic animals.

502 POLLAK, B. JUN. *Vienna*.—Wood for gun-carriages, &c.; coopers' wood.

503 PRANDAU, G. BARON OF HILLEPRANDT, *Valpo and Miholac, Slavonia*.—Wood products, and gall-nuts.

504 RADAUZ, J. R. *Military Stud. Establishment, Bukowina*.—Timber.

505 RÖMER, C. *Pressburg, Hungary*.—Joiners' glue, gelatine.

506 ROSCHÉ, M. & PRAUSE, A. *Vienna*.—Cod-oil.

507 RÖSNER, H. *Olbersdorf, Aust. Silesia.*—Wax-drawers' products.
508 ROTSCH, F. *Gratz, Stiria.*—Teasels.
509 ROTT, V. J. *Prague.*—Turnery.
510 SAAZ, AGRICULTURAL SOCIETY AT, *Bohemia.*—Clover, woods, and commercial seeds.
511 SAPIEHA, PRINCE A. *Krasiczin, Galicia.*—Fleeces.
512 SARG, F. A. *Liesing, near Vienna.*—Candles, soaps, glycerine.
513 SARTYNI, A. *Smorce, District of Stry, Galicia.*—Wood prepared for lucifer matches.
514 SCHÄDLBAUER, V. *Vienna, Ottakring* 105.—Mother-of-pearl buttons.
515 SCHAUMBURG-LIPPE, PRINCE, *Veröce, Slavonia.*—Wood products, staves.
516 SCHINDLER, E. F. VON *Kunewald, Moravia.*—Fleeces.
517 SCHMID, E. *Vienna, Jagerzeile* 484.—Walking-sticks and pipe-tubes.
518 SCHREIBER, E. *Vienna, Margarethen* 201.—Meerschaum pipes and cigar-pipes.
519 SCHÜPLER, J. *Vienna, Schottenfeld* 17.—Turnery, fancy goods.
520 SCHWARZENBERG, J. A. PRINCE, *Vienna.*—Sheep's wool.
521 SEILLER, A. & Co. *Görz.*—Soap.
522 SOAP MANUFACTURERS' CO. FIRST AUSTRIAN. *Vienna and Penzing.*—Candles and soap.
523 SPRINGER & BECHER, *Vienna, Mariahilf* 71.—Perfumery.
524 STATES RAILWAY SOCIETY I.R. *Vienna.*—Forest products, impregnated woods.
525 STABILE, A. *Görz.*—Sumach.
526 THUN-HOHENSTEIN, F. COUNT, *Prague.*—Fleeces, samples of wool.
527 THURN & TAXIS, H. PRINCE, *Laucin, Bohemia.*—Fleeces.
528 TREBITSCH, A. *Vienna, Neu-Wieden* 723.—Meerschaum pipes and cigar-pipes.
529 TRENNER, J., & SON, *Baden, near Vienna.*—Agriot-sticks.
530 TYROL & VORARLBERG, CENTRAL SOCIETY OF AGRICULTURISTS AT, *Innsbrück, Tyrol.*—Flax from Oetzthal and Axam.
531 VITORELLI, P. *Borgo di Valsugana, Tyrol.*—Nutwood.
532 WALDSTEIN-WARTENBERG, E. COUNT, *Münchengrätz, Bohemia.*—Unwashed fleeces.
533 WALLIS, F. COUNT, *Koleschowitz, Bohemia.*—Fleeces.
534 WEISS, J. *Bistritz, Transylvania.*—Specimens of wood.
535 WERTHEIM, E. *Vienna* 1009.—Ivory goblet, with carvings, and fancy turnery.
536 WIDMANN, L. BARONESS, *Lodnitz and Stremplowitz in Aust. Silesia.*—Unwashed fleeces.
537 WIROWATZ, BROS. *Apatin, Com. of Bacs-Bodrogh, Hungary.*—Hemp.
538 WOLF, J. *Vienna, Mariahilf* 64.—Fancy turnery.
539 WOLLNER, A. *Vienna, Leopoldstadt* 564. Very finely combed flax.
540 WSETIN, ADMINISTRATION OF ESTATE AT, THROUGH RAIKEM EMILIUS, *Wsetin, Moravia.*—Sections of silver leaved fir-tree.
541 WÜNSCHE, ANTON, & SON, *Ehrenberg, Bohemia.*—Wood plaits, wood hats and caps.
542 ZICHY-FERRARIS, F. COUNT, *Orosyvár Com. of Wieselberg, Hungary.*—Fleeces.

CLASS 5.

543 GANZ, A. *Buda.*—Cast iron articles for railways.
544 HERZ, DR. *Vienna, Haidenschuss.*—Plans and drawings.
545 KÖSTLIN & BATTIG, A. *Vienna.*—Full-size model, &c. of a new railway, not requiring wooden sleepers.
546 ROTHSCHILD BROS. *Witkowitz, Moravia.*—Rails and wheels.
547 SIEGEL, C. G. *Vienna, Währingerlinie.*—Locomotives and various machines.
548 STATES RAILWAY SOCIETY I.R. *Vienna* 12.—Locomotive for sharp curves, and gradients.
549 STATES RAILWAY SOCIETY I.R. *Vienna* 12.—4-cylinder full-speed locomotive; model of improved break and slider, &c.

CLASS 6.

550 LONNER, J. *Vienna, Rossau* 237.—Coach.
551 WAENZEL, F. & SON, *Vienna.*—Chaise and waggon axle-trees.

CLASS 7.

552 BEARZI, J. B. *Vienna, Wieden* 114.—Sleys.
553 BLOBL, A. *Vienna, Schottenfeld* 238.—Rods for making velvet.
554 BRAND, C. & F. L'HUILIER, *Brünn.*—Sugar moulds.
555 FICHTNER, A. *Vienna, Leopoldstadt* 551.—Lithographic and other presses.
556 GASTEIGER, G. VON, *Wilten, near Innsbruck, Tyrol.* — Pantographs, spinning wheels, &c.
557 KRONIG, C. *Vienna.*—Sugar moulds.
558 PORTHEIM, J. VON KÜNDIG H. & BERTSCHY, G. *Smichow, near Prague.*—Electro-coppered roller, for calico printing.
559 REZNICEK, A. M. *Smichow, near Prague.*—Sugar moulds of tin.
560 RIEFFEL, C. *Pesth, Toleranzgasse.*—Self-adjusting spanner.
561 SCHALLER, J. *Vienna, Leopoldstadt* 426.—Bellows, army forges.
562 SCHRAMM, W. *Vienna, Gumpendorf* 486.—Jacquard loom.
563 THUMB, V. *Unterhimmel, near Steyr, Upper Austria.*—Weaving apparatus.
564 WALLERSTEIN, S. *Vienna, Landstrasse* 128.—Machine for cutting out linen. 600 pieces an hour.

CLASS 8.

565 BEER, L. *Vienna, Wieden* 51.—Models of fire engines.

566 DINGLER, H. *Vienna, Wieden* 120.—Pump for hydraulic press; wine press, &c.

567 DOBBS, W. *Vienna, Landstrasse* 286.—Fire-box of steam boiler.

568 HORAK, A. *Vienna, Gumpendorf* 324.—Machines.

569 HUBAZY, G. *Vienna, Erdberg* 152.—Portable 4-horse engine, to be heated with straw.

570 JENBACH, IRON WORKS AT,—*Jenbach, Tyrol.*—Turbine, stamp for medals.

571 KLOTZ, J. *Gratz, Stiria.*—Safety-valves.

572 KOHN, H. & SONS, *Vienna, Weissgärber* 18.—Portable spirit and wine pump.

573 MICHALEK, F. *Vienna.*—Boring machine, shears.

574 NEMELKA, L. *Fischamend, Lower Austria.*—Model of mill.

575 NETREBSKY, CHEVALIER DE J. *Cracow, Galicia.*—Model of a steam-engine; the governor in the piston.

576 PLUMER, M. *Borgo di Valsugana, Tyrol.*—Parts of an hydraulic pump.

577 POSZDECK, J. *Pesth, Gezogasse.*—Model of a bell with belfry.

578 REACH, F. *Prague,* 1049.—Decimal weights.

579 REZNICEK, A. M. *Smichow, near Prague.*—Candy-sugar vats, and fire engines.

580 SASSE, F. *Gratz* 1167.—Jacks.

581 SCHEMBER, C. *Vienna, Leopoldstadt* 799.—Centesimal and decimal weights.

582 SCHEMBER, L. *Vienna, Leopoldstadt* 799.—Autographic multiplying press and glazing machine.

583 SCHMID, H. D. *Simmering, near Vienna.*—Horizontal steam-engine, caloric engine, screw and nut-cutting engines, balances, &c.

584 STATES RAILWAY SOCIETY I. R.—Drawing of a new steam forging-press.

585 SZABÓ, F. *Pesth, Schlangengasse* 5.—Drawings of a mill.

586 TIBÉLY, F. *Bries, Hungary.*—Excavators' apparatus, without screws.

587 VITTORELLI, P. *Borgo di Valsugana, Tyrol.*—Sausage machine.

588 WARCHALOVSKY, J. *Vienna, Josephstadt* 143.—New sewing-machine.

589 WINIWATER, G. CHEVALIER DE, *Gumpoldskirchen, near Vienna.*—Machine for punching 6 holes at once.

CLASS 9.

590 BOKOR, F. *Gr. Zinkendorf, Com. of Oedenburg, Hungary.*—Hoeing-plough, turnip-cutter.

591 BORROSCH & EICHMANN, *Prague.*—Agricultural implements and machines.

592 BUKOWINA, BOARD OF COMMERCE IN THE, *Czernowitz, Bukowina.*—Hand mill.

593 CRISTALLNIGG, COUNT C. *St. Johann am Brückl, Carinthia.*—Machine for husking maize.

594 FARKAS, STEPHEN, *Pesth.*—Ploughs.

595 FICHTNER, J. & SONS. *Atzgersdorf, near Vienna.*—Drill-plough, rotatory-harrow, &c.

596 GOLARZEWSKI, L. VON, *Targowiska, Galicia.*—Thrashing-machine.

597 GUBITZ, A. *Pesth.*—Socs of ploughs.

598 HOFMANN, F. W. *Vienna, Josephstadt.*—Illustration of hop-cultivation.

599 KOLB, F. J. *Maria Enzersdorf, Lower Austria.*—Honey and wax, implements, machines.

600 LEICHT, M. *Essegg, Slavonia.*—Plough.

601 MELICHER, DR. L. *Vienna, Alservorstadt* 166.—Bee-hives and their products.

602 MÉZÁROS, J. *Gross-Zinkendorf, Com. of Oedenburg, Hungary.*—Ploughs with socs.

603 NACHTMANN, J. *Hermagor, Carinthia.*—Carinthian bee-hives.

604 OETTL, J. N. *Puschwitz, Bohemia.*—Mechanical bee-hive of straw, wax-filtering pot of tin.

605 QUAGLIO, G. *Vienna, Weissgärber* 140.—Corn-measuring-machine.

606 REACH, F. *Prague* 1049.—Cattle-weighing-machine.

607 REISS, A. *Vienna.*—*Laimgrube* 87.—Garden implements.

608 SZABÓ, F. *Pesth, Schlangengasse* 5.—Plan of a corn-cleaner.

609 VIDATS, BROS. *Pesth*, 2 *Hasengasse* 9.—Field implements.

610 VITTORELLI, P. *Borgo di Valsugana* 80, *Tyrol.*—Butter-van, model of a thrashing-machine.

CLASS 10.

611 BENCZUR, J. *Mogyoroska, Com. of Zemplin, Hungary.*—Articles in cement.

612 CURTI, DR. A. *Muthmannsdorf, near Wr.-Neustadt, Lower Austria.*—Ornaments in cement.

613 DOPPLER, J. *Salzburg.*—Marble fountain.

614 EGGSPÜLER, A. *Vienna, Wieden* 920.—Filtering machine and requisites.

615 FILLUNGER, J. *Vienna, Landstrasse.*—Arched bridge with cast-iron ribs.

616 HALL, SALTHOUSE I. R. ADMINISTRATION AT, *Tyrol.*—Model for furnace.

617 HALLER, C. *Gratz* 217.—Roof-felt, and roof-slate.

618 LANGER, J. *Vienna, Landstrasse* 472.—Models of iron bridges.

619 LÖSSL, CHEVALIER DE F. *Linz, Upper Austria.*—Architectural isopedimetrical ground plans.

620 NAPHOLTZ, M. *Nagy Becskerek, Com. of Torontal, Hungary.*—Bricks.

621 PALESE, *Trieste.*—Stone-masonry.

622 PLEISCHL, A. M. *Vienna, Alservorstadt* 109.—Drawings, showing a new kind of suspension bridges.

623 SAULICH, A. & KRAFT, *Kufstein, Tyrol.*—Architectural ornaments.

624 RAMSAUER, I. *Ischl, Upper Austria.*—Model of hydraulic pile-driving machine.

625 SCHAUMBURG, C. *Gratz, Stiria.*—Architectural drawing.

626 SCHNIRCH, F. & FILLUNGER, J. *Vienna, Landstrasse.*—Suspension bridge, photographs.

627 STATES MINISTRY, I. R. BUILDING SECTION, *Vienna.*—Charts of the Theiss and Danube, with explanation.

628 STATES RAILWAY SOCIETY, I. R. *Vienna.*—Plans of bridges (drawings and photographs), grate-bars.

629 STREFFLEUR, V. *Vienna.*—Hysometric-plastic maps and charts, &c.

630 VIENNA COMMISSION, I. R. OF VENTILATION-HEATING OF MILITARY HOSPITALS, *Vienna.*—Ventilating and heating apparatus, from the Garrison Hospital.

631 WINIWARTER, G. VON & PFANKUCHE, G. *Vienna.*—Account-office fitted up, fire and thief-proof.

CLASS 11.

632 DIEZ, E. *St. Johann, near Villach, Carinthia.*—Small shot.

633 FENDT, A. *Steyr, Upper Austria.*—Rifled gun and cross bow.

634 GASTEIGER, A. VON, *Wilten, near Innsbruck, Tyrol.*—Model cannon, and revolver pistol.

635 JIRKU, B. *Reichenberg, Bohemia.*—Guns.

636 JUNG, F. *Vienna* 1049.—Arms, engraved, and inlaid with gold and silver.

637 JURMANN, C. *Neunkirchen, Lower Austria.*—Arms, and parts of arms.

638 KIRNER, J. *Pesth, Servitenplatz.*—Guns.

639 LENGYEL, S. *Pesth, Spitalgasse.*—Newly-invented guns and arms.

640 MASCHECK, W. *Vienna, St. Ulrich* 75.—Guns and pistols.

641 MAURER, BROS. *Vienna* 1146.—Arms; sportsman's outfit.

642 OHLIGS, B. W. *Vienna.*—Fire-arms, and fancy swords in gold, silver, and steel.

643 STRIBERNY, A. *Vienna* 1134.—Arms.

644 UMFAHRER, F. *Klagenfurt, Carinthia.*—Guns and pistols.

645 WAENZEL, F. & SON, *Vienna Margarethen.*—Guns.

CLASS 12.

646 DANUBE STEAM NAVIGATION SOCIETY, I. R. *Vienna.*—Models of ships.

647 JÄGER, F. J. *Prague* 849—1.—Rope work.

648 REICH, S. *Zaschau, Moravia.*—Rope work.

CLASS 13.

649 APPOLONIO, A.—*Ampezzo, Tyrol.*—Mathematical instruments.

650 HAAS, C. & Co. *Gratz.*—Specimens of tinning, zincing, coppering, and electrotyping.

651 JERÁK, F. *Prague, Bohemia.*—Philosophical instruments.

652 KAPELLER, L. S. *Vienna.*—Meteorological instruments.

653 KAVALIER, J. *Sazawa, Bohemia.*—Glass, and apparatus for chemical purposes.

654 KRAFT, E. & SON, *Vienna, Wieden* 447.—Mathematical and philosophical instruments.

655 LEITER, J. *Vienna, Alservorstadt* 150.—Philosophical instruments.

656 LENOIR, G. A. *Vienna* 1019.—Philosophical instruments.

657 LEOPOLDER, J. *Vienna Wieden* 132.—Philosophical apparatus.

658 MARCUS, S. *Vienna, Neubau* 313.—Telegraph instruments, and other philosophical apparatus.

659 NAVY OFFICE, I. R. CENTRAL, *Trieste.*—Nautical instruments, inventions of Profs. Zesevich & Zamara.

660 NEUHOFER & FRIEDBERGER, *Vienna* 1149.—Optical instruments.

661 NUSS, A. & GALLITZER, T. *Pesth, blaue Hahngasse* 12, and *Elisabethplatz* 4.—Improved morse-telegraph apparatus.

662 PERFLER, J. *Vienna, Landstrasse* 311.—Mathematical instruments.

663 SCHABLASS, J. *Vienna, St. Ulrich* 136.—Mathematical instruments.

664 SEYSS, L. *Atzgersdorf, near Vienna.*—Philosophical instruments, manometer, indicators, spring balances, and bell-signal apparatus.

665 SIEMENS & HALSKE, *Vienna, Erdberg* 46.—Mine-match apparatus, on Ebner's principle; telegraph apparatus.

666 STATES RAILWAY SOCIETY, I. R. *Vienna,*—System of telegraph without batteries at intermediate stations: drawing and description.

667 TELEGRAPH DIRECTION, I. R. *Vienna.*—Morse-apparatus, with electro-motor: telegraph maps of Austria.

668 VOIGTLÄNDER & SON, *Vienna* 949.—Optical instruments.

669 WALDSTEIN, J. *Vienna, Michaelerplatz.*—Optical instruments.

CLASS 14.

670 ANGERER, L. *Vienna, Wieden* 1061.—Photographs.

671 Dietzler, Ch. *Vienna, Wieden* 102.—Photographic apparatus; astronomical double object camera; photographs; photographic copies of paintings, engravings, &c.

672 Giessendorf, Ch. von, *Vienna, Wieden* 508.—Panicographs and photographs.

673 Lemann, Ch. *Vienna, Gumpendorf* 24.—Photographs of archæological and art objects.

674 Mayer, G. *Pesth.*—Photographs.

675 Melingo, A. *Vienna, Praterstrasse* 512.—Photographs.

676 Oberhausen, E. *Vienna* 613.—Photographic art-productions.

677 Perini, A. *Venice.*—Fac-simile of the Breviario Grimain, from St. Mark's at Venice.

678 Rupp, W. *Prague.*—Photographs.

679 Voigtländer & Son, *Vienna* 949.—Photographic apparatus and photographs.

680 Widter, A. *Vienna, Landstrasse* 136.—Photographs.

681 Wünsch, G. *Vienna* 648.—Photographs.

CLASS 15.

682 Effenberger, F. *Vienna* 1146.—Clocks and watches.

683 Fröhlich, Fr. C. *Hartberg, Stivia.*—Oil for clocks.

684 Kralik, S. *Pesth.*—Clocks, watches.

685 Krespach, A. *Vienna, Schottenfeld* 293.—Clock cases, clock works, pendulums.

686 Luz, I. *Vienna, Alservorstadt* 276.—Tell-tale clocks for factories.

687 Marenzeller, I. *Vienna* 641.—Pendulum clocks.

688 Mayer, A. W. *Vienna* 647.—Table clocks.

689 Mösslinger, F. *Vienna, Laimgrube* 195.—Clock and watch dial-plates; enamelled labels.

690 Oberleitner, J. *Hall Tyrol.*—Clocks, watches.

691 Schönberger, W. *Vienna* 648.—Clocks.

CLASS 16.

692 Ast, J. *Vienna, Wieden* 412.—Pianofortes.

693 Bauer, E. J. *Prague* 780/2.—Musical instruments.

694 Beregszaszy, L. *Pesth.*—Pianoforte.

695 Bittner, D. *Vienna* 1038.—Violins.

696 Blümel, F. *Vienna, Wieden* 300.—Pianoforte.

697 Bock, F. *Vienna, Neulerchenfeld* 171.—Metal wind instruments.

698 Bohland, G. *Graslitz* 522, *Bohemia.*—Brass wind instruments.

699 Bösendorfer, L. *Vienna, Neu-Wien* 377.—Pianofortes.

700 Brandl, C. *Pesth.*—Violins, modelled after Straduarius and Quarnerius.

701 Cerveny, V. J. *Königgrätz, Bohemia.*—Metal instruments.

702 Cramer, G. *Vienna, Landstrasse* 37.—Pianoforte.

703 Eberle, C. *Verona.*—Musical instrument strings.

704 Ehrbar, F. *Vienna, Wieden* 753.—Pianoforte.

705 Holzschuch, M. *Vienna, Laimgrube* 100.—Improved pianoforte pedal-treddles.

706 Hörbiger, A. *Atzgersdorf, near Vienna.*—Automaton "vox humana."

707 Kandler, D. *Vienna Strozzengrund* 20.—Metal wind instruments.

708 Kiendl, A. *Vienna, Josephstadt* 18.—Citherns.

709 Kluibenschädl, J. *Innsbruck* 114, *Tyrol.*—Violins.

710 Kohlert, V. *Graslitz* 308, *Bohemia.*—Wind instruments in wood.

711 Lausmann, J. W. *Linz, Upper Austria.*—Wind instruments.

712 Lemböck, G. *Vienna* 838.—Violins.

713 Martin, J. *Graslitz* 198, *Bohemia.*—Brass wind instrument.

714 Meinl, D. *Vienna, Landstrasse* 94.—Metal wind instruments.

715 Miller, M. *Vienna, Gumpendorf* 351.—Pianoforte strings.

716 Patzelt, J. F. *Vienna* 31.—Violins, viola, violoncello, cithern.

717 Pick, Widow R. *Vienna, Alservorstadt* 27.—Musical instruments.

718 Pohl, E. *Kreibeitz, Bohemia.*—Glass-harmonica.

719 Pottje, J. *Vienna, Wieden* 88.—Pianofortes.

720 Riedel, J. *Pressburg, Hungary.*—Musical wind instruments.

721 Röder, Ch. *Vienna, Wieden* 1.—Bassoon tubes.

722 Rott, A. H. *Prague* 799/11.—Brass wind instruments.

723 Schamal, W. *Prague* 1025/11.—Brass wind instruments.

724 Schneider, J. *Vienna, Wieden* 447.—Pianofortes.

725 Srpek, J. *Vienna, Mariahilf* 2.—Wind instruments.

726 Stehle, J. *Vienna, Landstrasse* 379.—Wind instruments, with photographs of them.

727 Stöhr, J. *Prague* 391/1.—Brass wind instruments.

728 Stowasser, I. *Vienna, Josephstadt* 66.—Musical instruments.

729 Stratzinger, R. *Linz* 129.—Cithern, with table.

730 Streicher, J. B. & Son, *Vienna, Landstrasse* 875.—Pianofortes, and a string-tester.

731 Thie, W. *Vienna, Mariahilf* 95.—Wind-harmonica, harmonica.

732 Uhlmann, L. *Vienna, Mariahilf* 25.—Pianoforte.

733 Ziegler, J. *Vienna, Leopoldstadt* 693—Wind instruments.

CLASS 17.

734 CZERMAK, DR. *Prague.*—Autolaryngoscopical apparatus case: laryngoscopical and chinoscopical photographs, &c.

735 DREHER, I. *Pesth.*—Surgical instruments.

736 FISCHER, P. *Pesth, Franziskanerplatz.*—Surgical instruments.

737 HEBRA, DR. F. *Vienna.*—Continuous baths for medical purposes.

738 KRANZLBAUER, J. *Pesth, Morgengasse* 3.—Surgical band.

739 LEITER, J. *Vienna, Alservorstadt* 150.—Surgical instruments.

740 TEICHMANN, DR. L. *Cracow, Galicia.*—Anatomical preparations.

741 TÜRCK, DR. L. *Vienna.*—Laryngoscopical and pharingoscopical apparatus, &c.

CLASS 18.

742 BREVILLIER & CO. *Schwadorf, Lower Austria.*—Cotton yarn.

743 BURCKHART, C. *Vienna, Schottenfeld* 219.—Articles in cotton chenille.

744 CORDELLA, BROS. *Weissbach, Bohemia.*—Cotton yarn.

745 EHINGER, A. *Oberlangenau, near Hohenelbe, Bohemia.*—Cotton fabrics and yarns.

746 FECHNER, P. *Vienna* 541.—Cotton fabrics.

747 FIAL, J. *Vienna, Gumpendorf.*—Articles in chenille.

748 FÖRSTER, J. JUN. *Rumburg, Bohemia.*—Cotton fabrics.

749 GARBER, J. & SON, *Vienna, Gumpendorf* 206.—Cotton fabrics.

750 GOLDBERG, C. R. *Warnsdorf* 64, *Böhmen.*—Cotton trouserings.

751 GOLDBERGER & SONS, S. F. *Old Buda.*—Cotton fabrics.

752 HELLMAN, N. *Prague, Alstadt* 710.—Raw materials.

753 KLAUS, I. E. *Niedergrund, Bohemia.*—Cotton coatings and trouserings.

754 LANG, J. N. *Vienna, Mariahilf* 45.—Cotton llama, and fustian: cotton stuffs for dresses.

755 MICHEL, P. *Gärten, near Schönlinde, Bohemia,*—Cotton articles.

756 MILLER, F. *Georgswald, Bohemia.*—Muslins and cambrics.

757 MITSCHERLIK, A. & R. *Teplitz, Bohemia.*—Assortment of webs and cotton fabrics.

758 MÜLLER, J.'s SON, *Rumburg, Bohemia.*—Cotton fabrics.

759 POTTENDORF, COTTON SPINNING AND WEAVING MILL AT, *Lower Austria.*—Cotton, yarn, and fabrics.

760 RICHTER, F. & CO. *Smichow* 204, *near Prague.*—Cotton yarn.

761 RICHTER, I. & SONS, *Niedergrund, Bohemia.*—Cotton velvets, and trouserings.

762 SPINNING AND DYEING MILL, *Aidussina, Görz.*—Cotton yarns.

763 SPITZER, M. A. *Vienna Mariahilf* 327.—Chenille articles.

764 WITSCHEL & REINISCH, *Warnsdorf, Bohemia.*—Cotton fabrics.

765 WOLF. C. *Bielitz, Aust. Silesia.*—Machine carding.

CLASS 19.

766 EHINGER, A. *Oberlangenau, near Hohenelbe, Bohemia.*—Bleached linen fabrics, and yarns.

767 GROHMANN, J. & SON, *Sternberg, Moravia.*—Linen.

768 HEIDENPILTSCH MILL, *Heidenpiltsch, near Freudenthal, Aust. Silesia.*—Machine-made yarn.

769 HEINZ, F. & A. *Freudenthal, Aust. Silesia.*—Linen.

770 JERIE, W. *Hohenelbe, Bohemia.*—Flax-yarns.

771 JERIE, W. *Marklow, Bohemia.*—Linen.

772 KLAUS, J. E. *Niedergrund, Bohemia.*—Linen stuffs for coats and trousers.

773 KRATKY, J. *Freudenthal, Aust. Silesia*—Table-cloths.

774 KÜFFERLE, A. & CO. *Vienna.*—Linen and damask table-cloths, napkins.

775 KÜHNEL, J. *Engelsberg, Aust. Silesia.*—Linen and thread.

776 MAY, J. & SON, *Schönlinde, Aust. Silesia.*—Flax yarn dyed and bleached.

777 MICHEL, P. *Gärten, Bohemia.*—Linen webs.

778 MÜLLER, J. *Schönlinde, Bohemia.*—Flax yarn of all colours and sorts. Flax yarn socks and stockings.

779 MÜLLER, J. & SON, *Rumburg, Bohemia,*—Linen.

780 PICK, I. D. *Nachod, Bohemia.*—Linen.

781 PLISCHKE, J. *Freudenthal, Aust. Silesia.*—Table-cloths.

782 RAYMANN & REGENHART, *Vienna.*—Rumburg inland linen, ticken, damask linen cloths.

783 SCHÖNBAUMSFELD, D. *Vienna, Altlerchenfeld.*—Thread-buttons.

784 TÄUBER, F. *Untermeidling, near Vienna.*—Waterproof woven hemp-yarn hose, 200 feet long.

785 VONWILLER & CO. *Vienna.*—Linen, and linen and cotton trouserings and coatings.

786 WITSCHEL & REINISCH, *Vienna* 350.—Linen.

787 WURST, J. N. & SONS, *Freudenthal, Aust. Silesia.*—Table-cloths.

CLASS 20.

788 BACKHAUSEN, J. *Vienna, Schottenfeld* 206.—Grenadin stuffs.

789 BALOGH, M. *Klausenburg, Transylvania.*—Raw silk.

790 BRAUN, A. *Vienna, Schottenfeld* 280. —Silk band.

791 BURGSTALLER, J. *Hruskovatz, Com. of Agram, Croatia.*—Silk.

792 CATTICH, BROS. *Zara.*—Raw silk.

793 COLLECTIVE EXHIBITION OF THE VIENNESE SILK STUFF MANUFACTURERS, *Vienna:*—

Bujatti, F. *Schottenfeld* 476;	Lemann, J. & Son, *Gumpendorf* 24;
Flemmich's Widow, *Gumpendorf* 563;	Mayer & Son, *Schottenfeld* 437;
Frischling, Arbesser, & Co. *City* 1105;	Siebert, F. & Son, *Wieden* 656;
Giessauf, J. & Sons, *Margarethen* 38;	Woitech, F. *Gumpendorf* 44;
Hornbostel, C. G. & Co. *Laimgrube* 1;	Wottitz, W. *Schottenfeld* 154;

—Silk goods.

794 DUSSINI, J. *Lana, Tyrol.*—Fine spun-silk.

795 FIGAROLLI, L. *Riva, Tyrol.*—Silk.

796 FRANZOT, A. *Farra, near Gradisca.*—Raw silk.

797 GANAHL, C. *Feldkirch, Vorarlberg.*—Raw silk produced in Feldkirch.

798 GARBER, J. & SON, *Vienna, Gumpendorf* 206.—Silk stuffs and dyed silk.

799 GENTILLI, J. J. *Görz.*—Raw silk.

800 GIAVONI, F. *Verona, Venetia.*—Raw silk.

801 GROHMANN, BROS. *Schönlinde, Bohemia.*—Silk-petinet articles.

802 HAAS, P. & SONS, *Vienna.*—Silk furniture stuffs.

803 HETZER, C. & SONS, *Vienna, Gumpendorf* 534.—Silk ribbons.

804 HOFFMANNSTHAL, I. VON, DR. *Vienna* 733.—Raw silk.

805 INSTITUTE OF DEAF AND DUMB, GEN. AUSTR. ISRAELITE, *Vienna.*—Raw silk.

806 LENASSI, B. *Görz.*—Raw silk.

807 MATUSIK, J. N. *Klausenburg, Transylvania.*—Silk, reeled off.

808 RITTER-ZAHONY, W. VON, *Görz.*—Floret silk.

809 SCHEY, F. *Vienna* 1157.—Raw silk.

810 SCHWARZ, J. *Vienna, Hundsthurm* 90. —Silk hat-bands.

811 SILK MILL, *Gross-Zinkendorf, Hungary.*—Raw silk.

812 SILKWORM REARING CO. AT STIRIA, *Gratz.*—Silk.

813 SILKWORM REARING CO. IN AUSTRIAN SILESIA, *Troppau, Aust. Silesia.*—Cocoons, raw silk, and silk damask.

814 STOFFELLA, D. A. *Roveredo, Tyrol.*—Raw silk.

815 WIEDNER, F. *Raab, Hungary.*—Silk counterpane.

816 WINKLER, T. *Klausenburg, Transylvania.*—Raw silk.

CLASS 21.

817 AUSPITZ, L. *Brünn.*—Woollen cloths.

818 BAUER, T. *Brünn.*—Woollen cloths.

819 BAUM, G. *Bielitz, Aust. Silesia.*—Woollen cloths.

820 BEUER, B. *Reichenberg, Bohemia.*—Cloths.

821 BLASCHKA & CO. *Liebenau, Bohemia.* —Plain woollen cloths.

822 BRÜNN CHAMBER OF COMMERCE, *Brünn.*—Woollen cloths.

823 BUM, M. *Brünn.*—Woollen cloths.

824 CAFFIER & POHLENZ, *Vienna, Neubau* 194.—Shawls.

825 DÖPPER, A. F. & RAINER HOSCH, *Neutitschein, Moravia.*—Woollen cloths.

826 EKSTEIN, F. *Brünn.*—Woven coverlet, a new invention.

827 ELGER, J. *Reichenberg* 55/3, *Bohemia.* —Cloth fabrics.

828 ELGER, J. *Reichenberg* 43/3, *Bohemia.* —Cloth fabrics.

829 FELS, *Gratz.* — Machine paper-felt, without list.

830 FÖRSTER, C. *Bielitz, Aust. Silesia.*—Woollen cloths.

831 FÖRSTER, J. JUN. *Rumburg, Bohemia.* —Mixed woollen goods.

832 GACS, WOOLLEN CLOTH MANUFACTORY AT, *Gacs, Hungary.*—Cloth-buckskin, shawls, coverlets.

833 GINSKEY, I. *Maffersdorf, Bohemia.*—Woollen coverlets.

834 GINZEL, *Reichenberg* 14/1, *Bohemia.*—Cloth.

835 GOEBEL, C. *Jägerndorf, Aust. Silesia.* —Woollen cloths.

836 GOEBL, J. *Vienna, Schottenfeld* 257.—Mixed silks for dresses. &c.

837 GOECKEL, S. C. *Hermannstadt, Transylvania.*—Cloths.

838 GRUNER, F. *Reichenberg* 145/3, *Bohemia.*—Cloths.

839 HAAS, P. & SONS, *Vienna.*—Woollen goods for furniture.

840 HÄRTLT, H. *Reichenberg* 183/3, *Bohemia.*—Cloths.

841 HÄRTLT, S. *Reichenberg* 74/4, *Bohemia.* —Cloths.

842 HAYDTER, S. *Vienna, Gumpendorf* 328. —Woollen shawls.

843 HEIDE, J. JUN. *Jägerndorf, Aust. Silesia.*—Woollen cloths.

844 HEIDE, R. *Jägerndorf, Aust. Silesia.* —Woollen stuffs, for coats and trousers.

845 HENNINGER, F. *Reichenberg* 157/1, *Bohemia.*—Cloths.

846 HENNINGER, J. *Reichenberg* 336/1, *Bohemia.*—Cloths.

847 HERRSCHMANN, H. *Brünn.*—Woollen fabrics.

848 HIEBEL, J. *Reichenberg* 137/4, *Bohemia.* —Cloths.

849 HIEBEL, W. *Reichenberg* 349/1, *Bohemia.*—Cloths.

850 HIRT, C. & SONS, *Wagstadt, Aust. Silesia.*—Fancy woollen fabrics.

851 HLAWATSCH & ISBARY, *Vienna, Gumpendorf* 224.—Shawls.

852 HOFFMANN, A. *Reichenberg* 88/2, *Bohemia.*—Cloths.

853 HOFFMANN, J. *Reichenberg* 61/3, *Bohemia.*—Cloths.

854 HOFFMANN, W. *Reichenberg* 24/3, *Bohemia.*—Cloths.
855 HOFFMANN, GÖNNER, & Co. *Klein-Beranau, Moravia.*—Woollen fabrics.
856 HOFMANN, A. & Co. *Tischnowitz, Moravia.*—Satin-cloth.
857 HORN, A. *Reichenberg* 97/4, *Bohemia.*—Cloths.
858 HORN, C. *Reichenberg* 97/4, *Bohemia.*—Cloths.
859 HÜBNER, F. *Reichenberg* 128/1, *Bohemia.*—Cloths.
860 HÜBNER, F. *Reichenberg* 114/1, *Bohemia.*—Cloths.
861 JAKOB, F. *Reichenberg* 126/3, *Bohemia.*—Cloths.
862 JAKOWITZ, W. *Reichenberg* 68/3, *Bohemia.*—Cloths.
863 ILLEK, F. *Brünn.*—Woollen fabrics.
864 KAFKA, H. *Brünn.*—Woollen fabrics.
865 KAHL, F. *Reichenberg* 46/3, *Bohemia.*—Cloths.
866 KAHL, BROS. *Reichenberg, Bohemia.*—Cloths.
867 KALLAB, F. & SONS, *Gross-Meseritsch, Moravia.*—Woollen fabrics and cloths.
868 KASPER, J. *Reichenberg* 333/4, *Bohemia.*—Cloths.
869 KASPER, F. *Reichenberg* 138/3, *Bohemia.*—Cloths.
870 KEIL, A. *Reichenberg* 167/4, *Bohemia.*—Cloths.
871 KELLER, J. *Brünn.*—Worsted yarns.
872 KERN, H. *Altenberg and Vienna* 625.—Cloths.
873 KIRSCH, F. J. SONS, *Brünn.*—Woollen fabrics.
874 KLAUS, J. E. *Niedergrund, Bohemia.*—Mixed woollen fabrics, goods for coats and trousers.
875 KLEIBER, A. *Vienna, Schottenfeld* 336.—Shawls.
876 KLINGER, F. *Reichenberg* 335/1, *Bohemia.*—Cloths.
877 KOHN, M. *Brünn.*—Woollen fabrics.
878 KÖNIG, J. I. *Reichenberg* 160/3, *Bohemia.*—Cloths.
879 KÖNIG, J. II. *Reichenberg* 274/3, *Bohemia.*—Cloths.
880 KÖNIG, W. A. *Reichenberg* 167/3, *Bohemia.*—Cloths.
881 KRAMER, B. S. *Vienna, Gumpendorf* 570.—Shawls.
882 KRECZY, BROS. *Brünn.* — Woollen fabrics.
883 KRZYSZTOFOWICZ, F. *Trybuchowce, District of Czortkow, Galicia.* — Coarse cloth.
884 LANG, J. N. *Vienna, Mariahilf* 45.—Woollen scarves and travelling shawls.
885 LARISCH, A. *Jägerndorf, Aust. Silesia.*—Cloths.
886 LEIDENFROST, E. & SONS, *Brünn.*—Worsted yarns.
887 LEUBNER, F. *Reichenberg* 276/3, *Bohemia.*—Cloths.
888 LIEBIEG, F. *Reichenberg, Bohemia.*—Printed woollen shawls, printed and dyed woollen goods.
889 LIEBIEG, J. & Co. *Reichenberg, Bohemia.*—Mixed woollen fabrics, printed, dyed, and plain.
890 LÖW, A. & SCHMAL, *Brünn.*—Woollen and Vigogne fabrics.
891 MAY, C. *Vienna, Gumpendorf* 502.—Long shawls.
892 MAYER, C. *Brünn.*—Woollen fabrics.
893 MAYER, E. & SCHILLER, *Brünn.*—Woollen fabrics.
894 MICHEL, P. *Gärten, Bohemia.*—Woollen textures.
895 MORO, BROS. *Klagenfurt, Carinthia.*—Woollen stuffs.
896 MÜLLER, A. E. *Reichenberg* 323/3, *Bohemia.*—Cloths.
897 MÜLLER, A. J. *Reichenberg* 123/4, *Bohemia.*—Cloths.
898 MÜLLER, A. L. *Reichenberg* 145/4, *Bohemia*—Cloths
899 MÜLLER, W. *Reichenberg* 338/4, *Bohemia.*—Cloths.
900 NAMIEST, CLOTH MANUFACTORY OF C. BARON PUTHON, & Co. AT, *Namiest, Moravia.*—Woollen fabrics.
901 OFFERMANN, J. H. *Brünn.*—Woollen fabrics.
902 PFEIFER, G. *Rumburg, Bohemia.*—Woollen fabrics.
903 PINTNER, W. *Brünn.*—Fancy woollen cloths.
904 PISKO, H. *Brünn.*—Woollen fabrics.
905 POPPER, BROS. *Brünn.*—Woollen fabrics.
906 POSSELT, F. A. *Reichenberg* 342/3 *Bohemia.*—Cloths.
907 QUITTNER, J. *Troppau, Aust. Silesia.*—Cloths.
908 RASCHKA, J. JUN. *Freiberg, Moravia.*—Cloths.
909 RATHLEITNER, J. & SON, *Gratz.*—Stirian unfulled cloth.
910 REDLICH, F. *Brünn.*—Woollen fabrics.
911 REDLICH, M. *Brünn.* — Woollen fabrics.
912 REICHENBERG, WEAVING SCHOOL AT, *Bohemia.*—Model textures.
913 ROHN, J. *Reichenberg* 136/4, *Bohemia.*—Cloths.
914 ROHN, C. *Reichenberg* 156/3, *Bohemia.*—Cloths.
915 SALOMON, C. *Reichenberg, Bohemia.*—Cloths.
916 SALOMON, F. *Reichenberg* 255/3, *Bohemia.*—Cloths.
917 SALOMON, I. *Reichenberg* 342/3, *Bohemia.*—Cloths.
918 SALOMON, J. J. *Reichenberg* 78/4, *Bohemia.*—Cloths.
919 SALOMON, J. *Reichenberg* 21/2, *Bohemia.*—Cloths.
920 SALOMON, W. *Reichenberg* 98/4, *Bohemia.*—Cloths.
921 SCHMIEGER, A. & Co. *Neudek, Bohemia.*—Worsted, and carded yarn.
922 SCHMITT, F. *Böhmisch-Aicha, Bohemia.* — Unprinted and printed woollen and mixed woollen stuffs.
923 SCHÖLL, A. *Brünn.*—Woollen fabrics.
924 SCHÖLLER, A. *Brünn.*—Felts.

925 Schöller, Bros. *Brünn.*—Woollen fabrics.

926 Siebeneicher, L. Jun. *Reichenberg* 151/3, *Bohemia.*—Cloths.

927 Siebeneicher, W. *Reichenberg* 284/3, *Bohemia.*—Cloths.

928 Siegmund, W. *Reichenberg* 84/4, *Bohemia.*—Cloths.

929 Siegmund, W. *Reichenberg, Bohemia.* —Cloths.

930 Spitz, S. *Brünn.*—Woollen fabrics.

931 Stossimmel, A. *Reichenberg* 198/3, *Bohemia.*—Cloths.

932 Strakosch, Bros. *Brünn.*—Woollen fabrics.

933 Strakosch, R. *Butschowitz, Moravia.* —Woollen fabrics.

934 Strakosch, S. Sons, *Brünn.*—Woollen fabrics.

935 Strnischtie, C. & Co. *Brünn.*—Woollen fabrics.

936 Teuber, J. & Sons, *Brünn.*—Worsted yarns.

937 Trenkler, A. *Reichenberg* 164/2, *Bohemia.*—Cloths.

938 Trenkler, A. & Sons, *Reichenberg* 31/5, *Bohemia.*—Cloths.

939 Trisztinovitz, M. *Pesth.*—Galloon, and double fabrics.

940 Tschörner, J. *Reichenberg* 96/4, *Bohemia.*—Cloths.

941 Tugemann, G. *Reichenberg* 22/2, *Bohemia.*—Cloths.

942 Ullrich, A. *Reichenberg* 164/3, *Bohemia.*—Cloths.

943 Ullrich, A. S. *Reichenberg* 97/2, *Bohemia.*—Cloths.

944 Ullrich, A. Jun. *Reichenberg* 163/4, *Bohemia.*—Cloths.

945 Ullrich, F. *Reichenberg* 48/3, *Bohemia.*—Cloths.

946 Vonwiller & Co. *Vienna* 727.—Fabrics for coats and trousers; woollen shawl.

947 Vöslau, Worsted Yarn Co. at, *Vöslau, Lower Austria.*—Worsted yarn for various purposes; a specimen of dyeing, containing 1,600 tints.

948 Weinlich, F. *Vienna, Gumpendorf* 538.—Mixed woollen shawls.

949 Westhauser, J. *Vienna, Gumpendorf* 345.—Fancy fabrics.

950 Wilsch, J. *Jägerndorf, Aust. Silesia.*—Woollen goods for coats and trousers.

951 Wolff, A. *Gr. Sieghart, and Vienna* 500—Fabrics for furniture.

CLASS 22.

952 Ginskey, I. *Maffersdorf, Bohemia.*—Carpets.

953 Haas, P. & Sons, *Vienna.*—Carpets.

954 Harbander, Ch. & Son, *Freudenthal, Aust. Silesia.*—Carpets.

955 Zuschik, J. *Vienna, Gumpendorf* 124. —Printed carpet.

CLASS 23.

956 Bindeles, J. *Prague* 118/5.—Patterns for printers.

957 Blaschka & Co. *Liebenau, Bohemia.* —Printed woollen fabrics.

958 Bossi, J. *Vienna* 648.—Printed fabrics; shawls.

959 Dormitzer, L. & Sons, *Holeschowitz, Bohemia.*—Printed calicoes and muslin-delaines.

960 Fröhlich's, G. A. Son, *Warnsdorf, Bohemia.*—Printed woollen fabrics.

961 Ganahl, C. & Co. *Feldkirch, Tyrol.*—Turkey-red cotton yarn.

962 Getzner & Co. *Feldkirch, Tyrol.*—Turkey-red cotton yarn.

963 Goldberger, J. F. & Sons, *Pesth, Hungary.*—Printed calicoes.

964 Heintschel, E. & Co. *Heinersdorf, Bohemia.*—Printed woollen fabrics.

965 Hiller, F. *Jungbunzlau, Bohemia.*—Printed cashmeres; merino and woollen shawls.

966 Hottinger, A. *Vienna.*—Dyed wool.

967 Leitenberger, F. *Cosmanos, Bohemia.*—Printed calicoes.

968 Liebieg, F. *Reichenberg, Bohemia.*—Printed woollen shawls and gowns.

969 Liebieg, J. & Co. *Reichenberg, Bohemia.*—Printed, dyed, plain, and mixed woollen fabrics.

970 Mayer, V. & Sons. *Vienna* 553.—Printed mixed woollen stuffs.

971 Porges, Bros. *Prague, Bohemia.*—Printed calicoes.

972 Rödel, G. *Brünn.*—Drawings for weaving and other industrial branches.

973 Schmitt, F. *Böhmisch-Aicha, Bohemia.* —Printed woollen fabrics.

974 Seebach, Turkish-Red Dyeing at, *Seebach, Lower Austria.*—Red yarns.

975 Spinning and Dyeing Mill, *Aidussina, Görz.*—Turkey-red cotton yarns.

976 Thomas, L. *Vienna.*—Printed woollens.

CLASS 24.

977 Bellmann, A. *Prague, Bohemia.*—Embroidered curtain.

978 Benkowitz, M. Jun. *Vienna* 1100.—Embroidery.

979 Drakulic, S. *Korenica, Military Confines.*—Embroidered robe.

980 Farkas, Caroline, *Pesth.*—Embroidery.

981 Fuchs, J. Sons, *Vienna,* 441.—Laces and embroidery.

982 Gossengrün, Town of, *Bohemia.*—Laces.

983 Hoffmann, L. *Freudenthal, Aust. Silesia.*—Embroidery.

984 Hüber, J. *Hirschenstand, Bohemia.*—Tamboured lace shawls.

985 Mariotti, C. *Trentino, Tyrol.*—Laces and embroidery.

986 NOWOTNY, F. *Vienna*, 1144.—Indian embroidered cashmere shawls.

987 PIFFER, F. *Trient, Tyrol.* — Embroidery.

988 PRAGUE, COMMITTEE FOR THE PROMOTION OF INDUSTRY AT.—Bone lace, embroidery, and crotchet work.

989 SARTORELI, Z. *Trentino, Tyrol.* — Altar cloths.

990 SCHMID, A. *Vienna, Margarethen* 123. —Embroidered long shawl.

991 SOMMER, E. *Vienna, Alservorstadt* 14. —Embroidery.

992 SOUPPER, T. *Pesth, Obere Donauzeile* 9. —Embroidered pictures.

993 Sováthy, S. *Miskolcz.* — Hungarian embroidered cloaks.

994 ULLMANN, A. *Hirschenstand, Bohemia.* —Tamboured lace manteau.

995 ULLMANN, J. F. *Neudegg, Bohemia.*— Black silk lace.

996 ZELGER, T. *Vienna, Schottenfeld* 469. —Embroidery.

CLASS 25.

997 BANYAY, L. C. *Klausenburg, Transylvania.*--Lamb skins.

998 BERGER, T. *Vienna, Mariahilf* 64.— Fancy feathers.

999 CHLADEK, J. *Pesth, Palatingasse.* — Brushes.

1000 FLEISCHL, S. D. *Pesth, Felbergasse.*— Bed feathers and eider down.

1001 GOLDSTEIN, A. & SON, *Pesth.* — Various furs.

1002 HUDOVERNIG, PRIMUS, *Strassis, near Laibach, Carniola.*—Horse-hair sieves.

1003 JONAS, M. (for the Furriers' Guild at Bistritz), *Bistritz, Transylvania.*—Furs.

1004 KAPICKA, C. JUN. *Prague.*—Wigs.

1005 KIBITZ, G. *Pilsen.*—Furs.

1006 PAIDLY, F. *Lemberg* 137/2/4, *Galicia.* —Bed feathers.

1007 PATTAK, G. *Hermannstadt, Transylvania.*—Brushes.

1008 PERELES, M. *Prague* 918.—Bohemian felt of hare's fur.

1009 PERELIS & POLLAK, *Prague* 1196/2. —Bed feathers and down.

1010 SCHINDLER, R. JUN. *Prague* 953/1.— Wigs.

1011 SCHWER, J. *Vienna, Neubau* 256.— Fancy feathers.

1012 STETTNER, J. *Pesth*, 3 *Kronengasse.* —Brushes.

1013 TEXTORIS, A. *Bistritz, Transylvania.* —Goat's hair oat-sacks.

CLASS 26.

1014 BRÜNNER & SONS *Liebenau, near Gratz.*—Leather.

1015 CZERWENKA, F. *Chrudim, Bohemia.* —Leather.

1016 EINHAUSER, J. F. *Uderns, Zillerthal, Tyrol.*—Saddlery and harness.

1017 FOGES, J. *Prague* 1171. — Dressed leather.

1018 GALLASCH, F. *Vienna, Hundsthurm* 98. —Pianoforte-hammer leather.

1019 GIRARDET, C. *Vienna* 1100.—Fancy goods of leather.

1020 GOLDSCHMIDT, J. S. *Prague.* — Leather and skins.

1021 HABENICHT, A. *Vienna, Neubau* 100. —Leather hangings.

1022 KOTTBAUER, L. *Vienna, Rossau* 111. —Leather pipe and bucket.

1023 KRAETSCHMAR. K. A. *Rima-Szombath, Com. of Gömör, Hungary.*—Saddlery.

1024 MANSCHÖN, M. F. *Pesth, Badgasse.*— Leather goods.

1025 NORILLER, B. *Roveredo* 655, *Tyrol.*— Skins and hides.

1026 OTTENREITER, L. *Pesth, kleine Bruckgasse.*—Fancy purses.

1027 REIKH, F. *Gratz* 712.—Calf-skins.

1028 SCHMITT, BROS, *Krems* 40, *Lower Austria.*—Calf-skins.

1029 SCHÖPPEL, J. *Komotau, Bohemia.*— Dyed leather.

1030 SEYKORA, J. *Kosteletz* 233, *Bohemia.* —Brown leather, black elastic leather.

1031 STAUDINGER, I. *Marburg, Stiria.*— Horse-hide for roofing.

1032 SUESS, A. H. & SONS, *Vienna, Sechshaus* 110.—Leather.

1033 SZEPESSY, A. *Debreczin, Com. of Bihar, Hungary.*—Whips.

1034 TOPERCZER, J. *Vienna, Praterstrasse* 513. — Saddle-girths, stirrup-leathers, and straps.

1035 TOSI, C. *Scariano, near Görz.* — Hides.

1036 WOLLF, F. *Hermannstadt, Transylvania.*—Leather.

CLASS 27.

1037 ADLER, V. *Vienna*, 1105. — Ladies' shoes.

1038 BACH, T. *Vienna, Landstrasse* 376.— Boots and shoes.

1039 BACHRICH, J. *Vienna, Neubau,* 211.— Shirts.

1040 BAUMANN, M. *Pesth.*—Ladies' shoes.

1041 BIRNBAUM, A. M. *Teplitz, Bohemia.*— Waterproof and elastic articles of clothing.

1042 BODNAR, J. *Pesth, Universitätsgasse* 6. —Gentlemen's clothing; ladies' riding habits.

1043 BRUNNER, F. F. *Pesth, Weiznergasse.* —Straw bonnets.

1044 BUDAN, J. *Prague* 861/2.—Leather gloves.

1045 BUSCH, J. *Prague* 209/5. — Gentlemen's boots and shoes.

1046 CHRISTI, J. *Vienna* 941.—Boots and shoes.

1047 DEPOLD, J. *Pesth.*—Boots.

1048 DRAKULIC, S. *Korenica, District of Ottowan, Military Borders.*—Clothing.

1049 ELSINGER, M. S. & SON, *Vienna, Mariahilf, Rittergasse.*—Waterproof clothing.

1050 FALLER, TRITSCHELLER, & Co. *Vallonara, near Vicenza.*—Straw bonnets and plaitings.

1051 FEDRIGOTTI, J. *Val di Ledro, Tyrol.*—Boots and shoes.

1052 FRESE, A. *Prague* 592/1.—Gloves, &c.

1053 FÜRTH, WOLF, & Co. *Strakonitz, Bohemia.*—Oriental tesse.

1054 GASTEIGER, A. VON, *Wilten, near Innsbruck, Tyrol.*—Artificial flowers.

1055 GINTER BROS. *Pesth, Weiznergasse* 17.—Gentlemen's clothing.

1056 GULCHER'S, T. & SON, *Vienna* 720.—Turkish caps.

1057 HAAN, BROS. *Vienna.*—Boots and shoes.

1058 HAHN, L. *Vienna* 737.—Boots and shoes.

1059 HALLER, C. *Gratz.* — Waterproof articles of clothing.

1060 HIRSCH, J. P. *Vienna, Wieden* 923.—Hats and caps.

1061 HÖNIG, I. *Vienna, Mariahilf* 45.—Stocks and cravats.

1062 HUBER, A. *Pesth.*—Ladies' dresses.

1063 HÜBSCH, J. *Prague* 583.—Hats.

1064 JAMBOR, A. *Pesth.*—Clothes.

1065 JANOWITZ, S. *Brünn.*—Hats.

1066 JAQUEMAR, G. *Vienna* 251.—Gloves.

1067 KLEIN, C. *Pesth.* — Ready-made clothes.

1068 KRACH, BROS. *Prague.*—Clothes and millinery.

1069 KRAETSCHMAR, K. A. *Rima-Szombáth, Com. of Gömör, Hungary.*—Clothes.

1070 KRAUZBERGER BROS. *Marienbad, Bohemia.*—Hats and caps.

1071 KRISCHONIG, A. *Vienna, Wieden* 307.—Mignon-bouquet.

1072 KRISTIAN, I. *Vienna, Laimgrube* 1.—Hats.

1073 MADLBAUER, A. *Vienna, Fünfhaus* 225, 226.—Linen shirts.

1074 MALATINSZKY, E. *Pesth, Landstrasse* 17.—Halina, for gentlemen and ladies.

1075 MAUTHNER, BROS. *Leippa, Bohemia.*—Clothes.

1076 MAYERHAUSER, E. *Vienna, St. Ulrich* 38.—Straw flowers.

1077 MICSEI & Co. *Pesth, Weiznergasse* 8.—Straw hats.

1078 MORTÁN, P. *Pesth, Theaterplatz.*—Clothes.

1079 MOTTL, M. & SONS, *Prague.*—Clothes.

1080 NITSCH, J. *Pesth.*—Clothes.

1081 PAGET, F. *Vienna* 487.—Waterproof clothes, sailor's hat.

1082 PANESCH, A. *Vienna* 1100.—Boots and shoes.

1083 PERNER, A. *Budweis, Bohemia.*—Wooden pegs.

1084 PETRY, M. *Vienna, Strozzengrund* 19.—Straw bouquets.

1085 POLLAK, D. H. *Vienna, Windmühle* 109.—Boots and shoes.

1086 PORFY, F. *Pesth, Weiznergasse* 19.—Hats.

1087 PRAGUE, COMMITTEE FOR THE PROMOTION OF INDUSTRY AT.—Gloves.

1088 RACZKOVICS, C. *Nagy-Kikinda, Com. of Torontal, Hungary.*—National costumes.

1089 RATTICH, J. & SON, *Krumau, Bohemia.*—Wooden pegs for shoemakers.

1090 REGENSBURGER, M. A. *Predazzo, District of Cavalese, Tyrol.*—Spatterdashes.

1091 REICH, M. *Vienna, Gumpendorf* 342.—Boots and shoes.

1092 RICHTER, J. *Kaaden* 323, *Bohemia.*—Gloves.

1093 RÖMISCH, C. *Prague* 590/1.—Clothes.

1094 ROTH, M. *Hermannstadt, Transylvania.*—Boots.

1095 ROTHBERGER, J. *Vienna* 627.—Clothes.

1096 SCHILLING, F. *Vienna* 638.—Boots and shoes.

1097 SCHMID, B. *Vienna, Mariahilf* 74.—Shirts.

1098 SCHMIDT, A. *Vienna, Kärnthnerstrasse.*—Clothes.

1099 SERVATIUS, A. *Kronstadt, Transylvania.*—Woollen felt-hats.

1100 SINGER, J. *Vienna* 1120.—Dress coat.

1101 SPITZMÜLLER, F. *Vienna* 441.—Gloves.

1102 STEIDL, A. *Vienna* 819.—Boots.

1103 STRASCHITZ, B. *Prague* 147/1.—Clothes.

1104 SZÁBO, F. *Pesth, Schlangengasse.*—Clothes.

1105 SZÓVÁTHY, S. *Miskolcz, Com. of Borsod, Hungary.*—Cloaks.

1106 SZURGENT, A. *Pesth, Gr. Bruckgasse* 7.—Lady's fur cloak.

1107 VADASZ, F. *Pesth, Herrngasse.*—Boots and shoes.

1108 ZEIDLER & MENZEL, *Schönau, near Schluckenau, Bohemia.*—Wooden pegs for shoemakers.

CLASS 28.

1109 BELLMANN, C. *Prague.*—Wood engravings.

1110 BOLDOG, L. *Pesth.*—Specimens of bookbinding.

1111 DIAMANT, M. *Petersdorf, near Mühlbach, Transylvania.*—Paper for various purposes.

1112 GEROLD, M. *Vienna, Dominicanerplace.*—Specimens of typography.

1113 GIESSENDORF, C. *Ober-Meidling, near Vienna.*—Lithographic prints, and anastatic plates, &c.

1114 GUCKER, J. *Pesth.*—Bookbinding.

1115 HABENICHT, A. *Vienna, Neubau*, 100.—Specimens of bookbinding.

1116 HARDTMUTH, L. & C. *Budweis, Bohemia.*—Pencils and elastic writing-boards.

1117 HARTINGER, A. & SON, *Vienna., Mariahilf* 71.—Illustrated works, and prints in oil colours.

1118 HASLINGER, T. *Vienna.*—Engraved and printed music.

1119 JOSEFSTHAL, PAPER MANUFACTORY AT, *Carniola.*—Paper.

1120 KNEPPER, W. & CO. *Vienna, Wieden* 348.—Fancy paper.

1121 LEIDESDORF, F. & CO. *Ebenfurth, Wiener - Neustadt, Ober - Eggendorf, and Vienna* 427.—Machine-made paper.

1122 LORENZ & SONS, *Arnau, Vienna* 896. —Paper.

1123 LUIGI, JACOB & CO. *Roveredo, Tyrol.*—Paper.

1124 LUKSICH, A. *Karlstadt, Croatia.*—Specimens of printing.

1125 NEUMANN, L. T. *Vienna.*—Lithography.

1126 PATERNO, F. *Vienna* 1064.—Lithography.

1127 POSNER, C. L. *Pesth.*—Specimens of printing.

1128 REIFFENSTEIN & ROESCH, *Vienna, Leopoldstadt*—Lithographs.

1129 REISSIG, J. *Vienna, Windmühle* 79. —Cartoons.

1130 ROLLINGER, BROS. *Vienna, Haarmarkt.*—Account books, and bookbinding.

1131 SIEGER, E. *Vienna, Leopoldstadt* 642. —Lithographs.

1132 SMITH & MEYNIER, *Fiume.*—Machine paper.

1133 SOMMER, L. *Vienna.*—Specimens of printing.

1134 SPINA, C. A. *Vienna, City* 1133.—Engraved and printed music.

1135 TRENTSENSKY, M. *Vienna, Leopoldstadt* 642.—Machine-ruled paper.

1136 WALDHEIM, R. VON *Vienna, City* 324.—Woodcuts.

1137 WINKLER, A. *Vienna, Gumpendorf* 142.—Samples of plate printing.

1138 ZAMARSKI, L. & C. DITTMARSCH, *Vienna, City* 24. — Illustrated books, and chromo-typography.

CLASS 29.

1139 AGRAM, CHAMBER OF COMMERCE AT.—Printed books.

1140 ARTARIA & CO. *Vienna, Kohlmarkt.* —Atlases, printed books, maps, &c.

1141 BOTZEN, CHAMBER OF COMMERCE AT, *Botzen, Tyrol.*—Statistical publications.

1142 BRAUMÜLLER, W. *Vienna.*—Educational works.

1143 BRODY, CHAMBER OF COMMERCE AT, *Brody, Galicia.*—Printed books.

1144 BRÜNN, CHAMBER OF COMMERCE AT, *Brünn.*—Works of the Moravian Upper Weaving School, drawings, models and modellings, printed books, &c.

1145 BRUNNER, F. F. *Pesth, Weiznergrasse.*—Toys.

1146 BUDWEIS, CHAMBER OF COMMERCE AT, *Budweis, Bohemia.*—Statistical publications.

1147 CRACOW, CHAMBER OF COMMERCE AT, *Cracow, Galicia.*—Printed books.

1148 CZERNOWITZ, CHAMBER OF COMMERCE AT, *Czernowitz, Bukowina.*—Industrial map of the Bukowina.

1149 EGER, CHAMBER OF COMMERCE AT, *Eger, Bohemia.*—Printed books.

1150 ESSEGG, CHAMBER OF COMMERCE AT, *Essegg, Slavonia.*—Printed books, general reports.

1151 FINANCES, I.R. MINISTRY OF, *Vienna.* —Statistical publications.

1152 FIUME, CHAMBER OF COMMERCE AT, *Fiume.*—Printed books, reports, &c.

1153 FRIESE, F. M. *Vienna.*—Statistic-geographical Review of the Austrian Mining Industry, with its progress for the last 36 years.

1154 GEOGRAPHICAL SOCIETY, I.R. *Vienna.* —Printed books.

1155 GEOLOGICAL INSTITUTE OF THE EMPIRE, I.R. *Vienna.*—Maps and printed books.

1156 GÖRZ, CHAMBER OF COMMERCE AT, *Görz.*—Printed books.

1157 GRATZ, CHAMBER OF COMMERCE AT, *Gratz.*—Printed books, statistical report.

1158 HOFMANN F. W. *Vienna.*—Books for popular instruction in agriculture.

1159 INNSBRUCK, CHAMBER OF COMMERCE AT, *Innsbruck, Tyrol.*—Printed books, annual reports, industrial map of Tyrol, and tables.

1160 KRONSTADT, CHAMBER OF COMMERCE AT, *Kronstadt, Transylvania.*—Printed books.

1161 LAIBACH, CHAMBER OF COMMERCE AT, *Laibach.*—Printed books.

1162 LAUFFER & STOLP, *Pesth, Waiznergasse.*—Travelling album.

1163 LECHNER, R. *Vienna.*—Books and games for children.

1164 LEMBERG, CHAMBER OF COMMERCE AT, *Lemberg.*—Printed books.

1165 LEOBEN, CHAMBER OF COMMERCE AT, *Leoben, Stiria.*—Printed books.

1166 LIEBICH, C. *Prague.*—Tableau representing mulberry leaves and wood products.

1167 LIEBSCHER, L. & J. Kunz, *Vienna* 258.—Mechanical toys and automatons.

1168 LIHARZIK, DR. F. *Vienna* 1142.—Plastic representation of the law of human growth.

1169 LINZ, CHAMBER OF COMMERCE AT, *Linz.*—Printed books, industrial map of Upper Austria.

1170 LOWER AUSTRIAN CHAMBER OF TRADE, *Vienna.*—Transactions of the Board since 1851.

1171 MECHITARIST'S COLLEGE, *Venice.*—Printed books.

1172 MILITARY GEOGRAPHICAL INSTITUTE, I.R. *Vienna.*—Maps, lithographic and photographic works.

1173 MINISTRY OF TRADE, I.R. *Vienna.* —Works on the state of mining in Austria, geographical representation of the Austrian railroads, &c.

1174 NAVY, I.R. CHIEF COMMAND OF THE, *Vienna.*—Objects relating to the Novara expedition.

1175 OLMÜTZ, CHAMBER OF COMMERCE AT, *Olmütz, Moravia.*—Printed books.

1176 PATRIOTICAL ŒCONOMICAL SOCIETY, I.R. *Prague, Bohemia.*—History of the Society and its Proceedings during the last 10 years.

1177 PERINI, PROFESSOR A. *Trient, Tyrol.*—Work on the disease of silkworms.

1178 PILSEN, CHAMBER OF COMMERCE AT, *Pilsen, Bohemia.*—Books and maps.

1179 PRAGUE, COMMITTEE FOR THE ADVANCEMENT OF INDUSTRY AT, *Prague.*—Printed books.

1180 PRAGUE, CHAMBER OF COMMERCE AT, *Prague.*—Printed books.

1181 PRAGUE, SOCIETY FOR THE ENCOURAGEMENT OF BOHEMIAN TRADE AT, *Prague.*—History and annual reports of the Society.

1182 PRÜFER, C. *Vienna, Landstrasse* 413.—Crystal models.

1183 PURGER, J. B. *Gröden, Tyrol.*—Toys, wood-carvings.

1184 REICHENBERG, CHAMBER OF COMMERCE AT, *Reichenberg, Bohemia.*—Printed books.

1185 REISS, H. *Vienna,* 1167.—Missale Romanum.

1186 ROVEREDO, CHAMBER OF COMMERCE AT, *Roveredo, Tyrol.*—Industrial map of Tyrol.

1187 SALZBURG, CHAMBER OF COMMERCE AT, *Salzburg.*—Printed books.

1188 STATES MINISTRY, I. R. *Vienna.*—Illustrative exhibition of the course, progress, and present state of public instruction.

1189 STIRIAN SOCIETY OF INDUSTRY, *Gratz.*—Printed books.

1190 STUMMER, J. *Vienna, Wieden* 309.—Pictorial Tabular History of the Emperor Ferdinand Northern Railway.

1191 TRENTSENSKY, M. *Vienna, Leopoldstadt* 642.—Pictures, and other objects for the instructive occupation of youth.

1192 TRIESTE, CHAMBER OF COMMERCE AT, *Trieste.*—Statistic reports.

1193 TROPPAU, CHAMBER OF COMMERCE AT, *Troppau, Aust. Silesia.*—Printed books.

1194 UDINE, CHAMBER OF COMMERCE AT, *Udine.*—Printed books.

1195 VIENNA, ARCHÆOLOGICAL SOCIETY AT, *Vienna.*—Books, maps, photographs.

1196 VIENNA, I. R. OFFICE OF ADMINISTRATIVE STATISTICS AT, *Vienna.*—Statistic and cartographic representation of the Austrian Empire.

1197 VIENNA, CHAMBER OF COMMERCE AT, *Vienna.*—Annual statistic reports, books for industrial schools.

1198 WEISS, J. P. —*Vienna, N. Wieden* 667.—Samples of tools for wood-workers, to be used for instruction in technology.

1199 WINTERNITZ, C. & LECHNER, R. *Vienna.*—Educational aids for children.

1200 WURZBACH, C. VON. DR. *Vienna.*—Literary works.

1201 ZOOLOGICAL-BOTANICAL SOCIETY, I. R. *Vienna.*—Books.

CLASS 30.

1202 APPOLONIO, A. *Ampezzo, Tyrol.*—Case.

1203 BAUMANN, L. *Ghymes-Kosztolan, Com. of Neutra, Hungary.*—Inlaid floors.

1204 FROYDA, F. *Vienna, Schottenfeld* 138—Joiner's work.

1205 GRÜNER, J. *Vienna, Altlerchenfeld* 27.—Fancy cabinet work.

1206 KEITEL, H. *Vienna, Schottenfeld* 83.—Furniture and fancy works of stone-paste and antlers.

1207 KINSKY, PRINCE, *Vienna.*—Carved furniture.

1208 KLEYHONZ, J. *Vienna Wieden* 704.—Desks.

1209 KLOBASSER, J. *Vienna, Alservorstadt* 334.—Paper-hangings.

1210 KNEPPER & SCHMIDT, *Vienna, Wieden* 420.—Paper-hangings.

1211 KRONIG, C. *Vienna, Landstrasse* 94.—Papier-mâché articles.

1212 KUHN, C. *Olmutz* 4.—Gothic pontifical of wood and gilt.

1213 MAURER, BROS. *Vienna* 1146.—Stone-paste.

1214 PODANY, F. & M. *Vienna, Schottenfeld* 298.—Veneers.

1215 RÖSNER, PROF. *Vienna* 435.—Oratory of Her Imperial Highness the Archduchess Sophia.

1216 SCHECK, F. *Linz* 895.—Carvings, with joiner's and upholsterer's work.

1217 SCHMIDT, P. *Vienna, Gumpendorf* 235.—Sideboard.

1218 SCHÖNTHALER, F. *Vienna, A. Wieden* 213.—Furniture and carver's work.

1219 SCHUBERT, J. *Vienna, Ottakring* 413.—Furniture, cases, purses, boxes.

1220 SPOERLIN & ZIMMERMANN, *Vienna, Gumpendorf* 368.—Paper-hangings.

1221 THONET, BROS. *Vienna, Leopoldstadt* 586.—Chairs of bent wood, carriage wheels, inlaid floors.

1222 VITTORELLI, P. *Borgo di Valsugana, Tyrol.*—Presses.

1223 WILDE, J. *Zircz, Com. of Vesprin, Hungary.*—Chiffoniers.

1224 WOLF, J. *Vienna, Margarethen* 22.—Shop-front.

CLASS 31.

1225 BARTELMUS, A. JUSA, J. W. & Co. *Brünn.*—Enamelled iron-plate wares.

1226 BARTELMUS, E. *Neu-Joachimsthal Bohemia.*—Enamelled and non-enamelled cast-iron wares.

1227 BAUER, J. J. *Vienna, Goldschmiedgasse.*—Samples of screws and rivets.

1228 BAUMGARTEN-FÜRST, M. *Unterweissenbach, Upper Austria.*—Iron wares.

1229 BODE, F. M. *Vienna, Wieden* 704.—Match boxes.

1230 BREVILLIER & Co. *Neunkirchen Lower Austria.*—Screws, rivets, and nails.

1231 CHADT, J. *Vienna, St. Ulrich* 157.—Engraved and enamelled table bronzes.

1232 CHRISTALLNIG, COUNT, *St. Johann am Brückl, Carinthia.*—Cast wares.

1233 COLLECTIVE EXHIBITION OF IRON WARES OF THE FOLLOWING STEEL AND IRON MANUFACTURERS AT WAIDHOFEN AND YBSITZ, *Waidhofen and Ybsitz, Lower Austria*:—

Grossmann, J. V. & Sons.	Schweinecker, Bros.
Leik, F.	Wagner, A.
Plankh, L.	Wertich, J.
	Windischbauer, F.

—Iron wares.

1234 COLLECTIVE EXHIBITION OF THE IRON WARES OF THE DISTRICT OF THE STADT STEYER, *Upper Austria*, by—

Almeroth, J.	Philipp, A.
Amort, J.	Pöpperl, J.
Brandegsky, F.	Preitler, M.
Degenfellner, L.	Reichl, J.
Dunger, S.	Reinweger, J.
Eisgruber, M.	Reschauer, M.
Engel, M. A.	Riess, F.
Ernst, P.	Riess, J.
Heindl, A.	Riess, Th.
Koller, A. von	Schartner, G.
Kollers von, widow and heirs	Schönthann F. von
	Schütz, B.
Lechner, M.	Sergl, M.
Mayerhofer, M.	Sonnleitner, A.'s widow
Michel, E.	
Mitter, J.	Unzeitig, F.
Mitter, J.	Voith, J. J.
Moser, A.	Werndl, J.
Peuker, J.	

—Iron wares.

1235 DIENER, C. *Vienna, Landstrasse* 34.—Zinc ornaments.

1236 DITMAR, R. *Vienna, Erdberg* 108.—Moderator lamps for the table and for hanging up.

1237 DOMOKOS, I. R. COPPER MINING AND HAMMER WORKS AT, *Balánbánya, Transylvania.*—Hammered copper-plate.

1238 EDER, F. *Wöllersdorf and Vienna.*—Sheet-iron, tinned iron-plate, and confectionery moulds.

1239 EGGER, COUNT F. *Lippitschbach and Freudenberg, Carinthia.*—Wire, nails, &c.

1240 FEIWEL, L. *Pesth.* — Locksmith's work.

1241 FERNKORN, CHEVALIER DE A. *Vienna, Wieden* 318. — Fountain, monument (Archduke Charles).

1242 FLORENZ, J. *Vienna, Leopoldstadt.*—Scales.

1243 GROSSAUER, F. *Steyer, Upper Austria.*—Mathematical instruments.

1244 HARRACH, COUNT F. E. *Janowitz, near Römerstadt, Moravia.* — Tin-plate and iron-plate, wires.

1245 HELLRIEGEL, P. *Insbruck* 249.—Iron kitchen-utensils of all kinds, and tools.

1246 HOLLENBACH, D. *Vienna, Josefstadt* 167.—Bronzes.

1247 KERL'S, F. A. HEIRS, *Platten, Bohemia.*—Tinned iron spoons.

1248 KOLBENHEYER, E. *Vienna, Wieden* 856.—Metal wares.

1249 KRONIG, C. *Vienna, Landstrasse* 94—Lacquered plate.

1250 KÜHN, E. & C. *Vienna, Sechshaus.*—Tin and tin articles.

1251 LANDERL. L. *Steyer, Upper Austria.*—Machine-made nails.

1252 LEOBEN CHAMBER OF COMMERCE AT, *Leoben*, (Collective Exhibition of that District):—

Fürst, I.	Sessler, V. F.

—Mining products, articles in iron and steel.

1253 LUCKNER, A. & FABRICIUS, C. *Vienna, Wieden* 48. — Fire and rust-proof safe.

1254 MÄRKY & GECMEN, *Komorau, near Horovic, Bohemia.* — Enamelled cast-iron hollow-ware.

1255 MONNIER, J. *Vienna, Wieden* 1098.—Pianoforte pegs.

1256 MORGENTHALER, C. *Pesth.* — Brass and tin-plate work.

1257 OETL, J. *Pesth.*—Fire-proof safe.

1258 OSTWALD & RITTIG, *Vienna, Wieden* 752.—Bronze fancy goods.

1259 PANLEHNER, F. *Waidhofen, Lower Austria.*—Iron wares.

1260 PFANNKUCHE, G. & SCHEIDLER, *Vienna, Althan* 1 *and* 2.—Fire-proof safes.

1261 PICHLER, J. *Steinbach, Lower Austria.*—Chopping-blade, shovels, hoes, and hatchets.

1262 PLEISCHL, A. M. *Vienna, Alservorstadt* 109.—Sanitary cooking vessels of iron-plate.

1263 PRAGUE, SOCIETY OF IRON INDUSTRY AT, *Prague.*—Iron ware.

1264 REACH, F. *Prague* 1049.—Decimal weights, cattle scales, and metal ware.

1265 REISS, A. *Vienna, Laimgrube* 87.—Shower and bath apparatus; articles in tin-plate.

1266 REWOLT, F. *Vienna*, 604.—Articles in tin-plate.

1267 REZNICEK, A. M. *Smichow, near Prague.*—Sugar moulds and steam pipes.

1268 ROTHSCHILD'S, BARON, IRON WORKS, *Witkowitz, Moravia.*—Flat iron and plates.

1269 RUBENZUCKER, M. *Ramingsteg, near Steyer, Upper Austria.*—Machine-made nails.

1270 SCHOLTZ, CH. *Matzdorf, Com. of Zips, Hungary.*—Articles in tin.

1271 SEIDAN, W. *Vienna, Gumpendorf* 407.—Medals and superscription boards in frames.

1272 SIKINGER, CH. *Steyer, Upper Austria.*—Hammered and machine-made nails.

1273 THIERING, CH. *Neu-Gaudenzdorf, near Vienna* 199.—Tin, separated from tin-plate.

1274 WACHTLER, P. *Pesth, Waiznergasse* 16.—Engravings.

1275 WAND, S. *Vienna, Michelbeuern* 56.—Bronze wares in connection with glass.

1276 WERTHEIM, F. & WIESE, *Vienna* 436.—Safes.

1277 WIESE, F. *Vienna, Alservorstadt* 188.—Tin-plate, and cooking vessels.

1278 WINIWARTER, J. & G. PLATE AND LEAD WARE MANUFACTORY, *Gumpoldskirchen, Lower Austria.*—Plates, tubes, vessels, wire, lead tubes, &c.

1279 WINKLER, M. *Vienna, Gumpendorf* 583, 589.—Combination safety locks and cast metal signboards.

1280 ZIEGLER, A. W. *Wilhelmshof, post Klentsch, Bohemia.*—Tinfoil and tin leaf.

CLASS 32.

1281 BACHNER, F. *Steyer, Upper Austria.* —Shoemakers' tools.

1282 BAUMGARTEN-FÜRST, M. *Unter-Weissenbach, Upper Austria.*—Steel wares.

1283 BOSCH, G. *Neuzeug, Upper Austria.*—Table knives.

1284 BRUNNER, A. *Vienna, Mariahilf* 116. —Piercing (buhl); samples of saws.

1285 COLLECTIVE EXHIBITION OF STEEL WARES FROM THE STEEL AND IRON MANUFACTURERS AT WAIDHOFEN AND YBSITZ, *Waidhofen and Ybsitz, Lower Austria*, by—

Grossmann, J. V. & Sons.	Schweinecker, Bros.
Leik, F.	Wagner, A.
Plankh, L.	Wertich, J.
	Windischbauer, F.

—Steel and iron wares.

1286 COLLECTIVE EXHIBITION OF STEEL WARES FROM THE DISTRICT OF STADT STEYER, IN UPPER AUSTRIA, *Stadt Steyer, Upper Austria*, by—

Degenfellner, L.	Reiweger, J.
Dunger, S.	Riess, F.
Eigruber, M.	Riess, Th.
Mayerhofer, M.	Schütz, B.
Michl, E.	Sergl, M.
Mitter, J.	Sonnleitner, A.
Philipp, A.	Unzeitig, F.
Pöpperl, J.	

—Steel wares.

1287 COLLECTIVE EXHIBITION OF STEEL WARES FROM STADT STEYER, *Stadt Steyer, Upper Austria*, by—

Almeroth, J.	Moser, A.
Amort, J.	Peuker, J. F.
Brandegsky, F.	Preter, M.
Engel, M. A.	Reichl, J.
Ernst, P.	Reschauer, M.
Heindl, A.	Riess, T
Koller, A. v.	Schartner, G.
Koller, v.	Schönthan, F. v.
Lechner, M.	Voith, J.
Mitter, J.	Werndl, J.

—Steel wares.

1288 EGGER, M. *Sterzing, Tyrol.*—Scythes and sickles.

1289 EIGRUBER, M. *Steyer, Upper Austria.*—Cutlery.

1290 ERNST, J. *Steyer, Upper Austria.*—Tacks.

1291 FISCHER, G. *Hainfield, Lower Austria.*—Steel castings.

1292 GROSSAUER, J. *Steyer, Upper Austria.*—Table-knives.

1293 HACK, F. *Steyer, Upper Austria.*—Table-knives.

1294 HAIDER, E. *Waidhofen-on-the-Ybbs, Lower Austria.*—Steel wares, chopping-blades, scythes.

1295 HASLINGER, J. *Garsten, near Steyer, Upper Austria.*—Tacks.

1296 HAUSER, J. *Steyer, Upper Austria.*—Gimlets.

1297 HEIDER, L. *Königswiesen, Upper Austria.*—Piercing-saws.

1298 HOLZINGER, J. *Neuzeug, Upper Austria.*—Clasp and table-knives.

1299 HUBER, J. *Sierning, Upper Austria* —Table-knives.

1300 KERBLER, J. *Grünberg, Upper Austria.*—Table-knives.

1301 KERBLER, F. *Sierninghofen, Upper Austria.*—Table-knives.

1302 KERNREUTER, F. *Vienna, Hernals* 205.—Tools for metal workers.

1303 KRAUPA, A. *Sierning, Upper Austria* —Knives.

1304 KRAUS, E. *Ossegg, near Teplitz, Bohemia.* — Steel ribs, for parasols and umbrellas.

1305 LECHNER, M. *Steyer, Upper Austria.* —Files.

1306 LEOBEN, CHAMBER OF COMMERCE AT, *Leoben, Stiria* (Collective Exhibition of that District), by—.

Weinmeister, C.	Zeilinger J.

—Mining and smelting products; iron and steel articles.

1307 LÖSCHENKOHL, J. *Trattenbach, Upper Austria.*—Pocket-knives.

1308 LOVREK, A. *Vienna* 213.—Tools of tungsten-steel.

1309 MILLER, M. & SON, *Vienna, Gumpendorf* 351.—Steel wares.

1310 MITTER, J. *Steyer, Upper Austria.*—Knives.

1311 MOSDORFER, B. *Weitz, Stiria.* —Sickles.

1312 MOSER, J. *Sierning, Upper Austria.* —Pocket-knives.

1313 MOSER, J. *Grünburg, Upper Austria.* —Knives.

1314 OFFNER, J. M. *Wolfsberg, Carinthia.* —Scythes.

1315 PACHERNEGG, J. *Uibelbach, Stiria* Scythes.

1316 PANLECHNER, C. *Vienna, Fünfhaus* 142.—Piercing-saws.

1317 PANLEHNER, F. *Waidhofen, Lower Austria.*—Steel-wares.

1318 PESSL, A. *Sierninghofen, Upper Austria.*—Knives.

1319 PESSL, T. *Sierninghofen, Upper Austria.*—Knives.

1320 PHILIPP, A. *Steyer, Upper Austria* —Files.

1321 PIKL, M. J. *Himmelberg, Carinthia* —Scythes.

1322 PILS, C. *Neuzeug, Upper Austria.*—Table-knives.

1323 REICHENAU, M.—*Waidhofen-on-the-Ybbs, Lower Austria.*—Chopping-blades and scythes.

1324 ROHRAUER, M. *Molle, Upper Austria.* —Several kinds of Jews' harps.

1325 SCHAFFENBERGER, F. *Steyer, Upper Austria.*—Awls.

1326 SCHWARZ, I. *Steyer, Upper Austria.* —Jews' harps.

1327 SCHWINGHAMMER, T. *Steinbach, Upper Austria.*—Knives.
1328 STARKE, C. *Steinbach, Upper Austria.*—Knives.
1329 STIRIAN STEEL WORKS COMPANY, *Vienna.*—Files.
1330 STORNIGG, P. *Steyer, Upper Austria.*—Tools, in metal.
1331 VOLBERT, F. *Steyer, Upper Austria.*—Tools.
1332 WEINMEISTER, J. *Leonstein, Upper Austria.*—Scythes.
1333 WEISS, J. & SON, *Vienna, Neue Wieden* 667.—Tools for wood-workers.
1334 WERTHEIM, F. *Vienna* 436.—Tools.
1335 WINKLER, BROS. *Waidhofen-on-the-Ybbs, Lower Austria.*—Scythes.
1336 WINTER, W. *Neuzeug, Upper Austria.*—Knives.
1337 ZEILINGER, F. *Uibelbach, Stiria.*—Scythes.
1338 ZEITLINGER, C. *Strub, near Molle, Upper Austria.*—Scythes.
1339 ZEITLINGER, M. & SON, *Blumau, near Kirchdorf, Upper Austria.*—Scythes.

CLASS 33.

1340 APPOLONIO, A. *Ampezzo, Tyrol.*—Filigree-work, chalice.
1341 BOLZANI & CO. *Vienna, Laimgrube* 132.—Articles in gold.
1342 BUBENICEK, W. *Prague* 549/1.—Articles in gold.
1343 EGGER, D. *Pesth, Gr. Brückgasse.*—Articles in gold.
1344 EGGER, S. *Pesth, Dorotheergasse* 11.—Antique jewellery.
1345 FABER, C. M. *Vienna* 282.—Plastic crystal-gold, for stopping hollow teeth.
1346 GLANZ, J. *Vienna.*—Articles in cast gold and silver.
1347 GROHMANN, H. *Prague.*—Articles in gold and silver.
1348 KOBEK, F. *Vienna* 768.—Jewellery.
1349 LEONIC MANUFACTORY, J. OBERLEITNER, *Stans, near Schwaz, Tyrol.*—Leonic wires.
1350 LERL, G. *Vienna, Strozzengrund*, 36.—Bronze fancy goods.
1351 LOBKOWITZ, PRINCE F. *Raudnitz, near Bilin, Bohemia.*—Bohemian garnets, in the rough state and cut.
1352 NETZ, B. *Vienna, St. Ulrich* 152.—Diamonds, rings, and jewellery.
1353 NEUSTADTL, M. H. *Prague* 703/1.—Articles in gold.
1354 PAUL, C. *Vienna, Schottenfeld* 364.—Bronze fancy goods.
1355 PICHLER, L. *Prague* 762/2.—Articles in gold.
1356 PODEBRAD, H. *Prague*—Works in gold.
1357 POPPE, I. G. *Vienna, Schottenfeld.*—Golden bracelet, of new and safe construction.
1358 SCHEIDL, T. *Vienna, Neubau* 258.—Articles in silver.
1359 SCHMITT, A. *Deutschbrod, Bohemia.*—Garnets.
1360 SCHÖNBORN, E. COUNT, *Prague.*—Bohemian garnets.
1361 ZLOCH, A. R. DR. *Prague.*—Eighteen large rock-crystals set in silver, for a candelabrum, 5 feet high.

CLASS 34.

1362 ADAM, J. H. & CO. (firm), JANKE BROS. *Blottendorf, Bohemia.*—Flint-glass table articles.
1363 BATTISTI, J. *Innsbruck, Tyrol.*—Glass-pieces for stropping razors.
1364 CZECH, A. L. *Haida, Bohemia.*—Flint-glass table articles.
1365 FRIEDRICH, A. *Köflach, Stiria.*—Glass.
1366 HARRACH, COUNT, *Neuwelt, Bohemia.*—Flint-glass table and ornamental articles.—
1367 HEGENBARTH, A. *Haida, Bohemia.*—Glass.
1368 HOFMANN, W. *Prague* 864/11.—Flint-glass table and ornamental articles.
1369 KOSSUCH, J. *Pesth, Palatingasse.*—Glass.
1370 LOBMAYR, J. & L. *Vienna* 940.—Flint-glass table articles, chandeliers, and candelabra.
1371 MAYER'S NEPHEWS, *Adolph and Leonorenhain, Bohemia.*—Flint-glass table articles and candelabra.
1372 MOSER, L. *Karlsbad, Bohemia.*—Glass goblet.
1373 PALME, I. KÖNIG, & CO. *Steinschönau, Bohemia.*—Glass articles.
1374 PALME, R. *Hayda, Bohemia.*—Flint-glass plateaux.
1375 PALME-KÖNIG, F. *Steinschönau, Bohemia.*—Flint-glass and fancy glass articles.
1376 PELIKAN'S, J. NEPHEWS, *Hayda, Bohemia.*—Glass.
1377 REICH, S. & CO. *Krasna, Moravia.*—Glass articles.
1378 SCHREIBER, J. *St. Stephen, St. Sidonia, Gr. Ullersdorf, Moravia.*—Glass articles.
1379 STÖLZLE, C. *Nagelberg, near Krems, Lower Austria.*—Glass articles.
1380 UNGER, F. & CO. *Tiefenbach. Marchendorf, Hayda, Bohemia.*—Glass articles.
1381 VOGEL, I. *Jägerndorf, Bohemia.*—Glass articles.
1382 WEIDLICH, S. *Steinschönau, Bohemia.*—Glass articles.
1383 WINTERNITZ, BROS. *Prague* 1187/11.—Glass articles.
1384 ZAHN, J. & CO. *Steinschönau, Bohemia.*—Glass articles.

CLASS 35.

1385 ASCAN, C. *Charlottenhütte, near Salzburg.*—Crockery and china.
1386 BEHR, C. *Karlsbad, Bohemia.*—Stoneware.

1387 Erndt, F. Jun. *Vienna, Alservorstadt* 237.—Articles in terra-cotta.

1388 Fischer, M. *Herend, Com. of Veszprim, Hungary.*—Porcelain services, vases, jugs, &c.

1389 Huffsky, V. *Hohenstein, near Teplitz, Bohemia.*—Terralith and earthenware.

1390 Miller & Hochstetter, *Hruschau, Aust. Silesia.*—Earthenware.

1391 Prague Porcelain and Earthenware Manufactory, *Smichow, near Prague.*—Porcelain.

1392 Ress, A. *Wr.-Neustadt, Lower Austria.*—Clay pipes.

1393 Richter, A. *Königsaal, Bohemia.*—Earthenware.

1394 Vienna, I. R. Porcelain Manufactory, *Vienna, Rossau* 137.—Porcelain.

CLASS 36.

1395 Breul & Rosenberg, *Vienna, Graben.*—Fancy goods, in leather and wood.

1396 Enders, F. *Vienna, Gumpendorf* 405.—Fancy goods in leather.

1397 Griensteidl, F. *Vienna* 746.—Fancy goods.

1398 Hirner & Singer, *Vienna, Laimgrube* 201.—Turnery and toys, fancy goods.

1399 Jaff, J. *Vienna, Leopoldstadt.*—Fancy goods.

1400 Klein, A. *Vienna, Mariahilf.*—Fancy goods, in leather, wood, and bronze.

1401 Koch, A. & C. and Neiber & Breiter, Nawratil, J. and Laschitzka, L. Schreiber, E. Fischer, J. Haan Bros. Bauer, M. Heimann, J. Mannstein, J. von, *Vienna, Laimgrube* 184.—Vienna fancy goods.

1402 Krammer, H. Jun. *Vienna, Weissgärber* 70.—Travelling trunk.

1403 Krebs, A. *Vienna* 771.—Fancy goods.

1404 Loysch, A. *Vienna* 628.—Vienna fancy goods.

1405 Müllner, G. & Co. *Vienna, Leopoldstadt* 34.—Leather fancy goods.

1406 Reif, J. *Kuschwarda, Bohemia.*—Articles in wood.

1407 Rodek, Bros. *Vienna* 1150.—Fancy goods, in leather, wood, and bronze.

1408 Schmöle, F. R. *Vienna, Mariahilf* 74.—Fancy goods.

1409 Stenzel, C. *Vienna, Wieden.*—Fancy goods.

1410 Theyer, F. *Vienna* 905.—Fancy goods, in bronze and wood.

ZOLLVEREIN.

S.W. Transept and S.W. Transept Gallery

DUCHY OF ANHALT-BERNBURG.

CLASS 3.

1 Brumme, A. F. & Co. *Dröbel.*—Raw beet-root sugar.

2 Brumme, A. F. & Co. *Waldau.*—Beet-root sugar in loaves; lump, and moist sugar.

3 Cuny & Co. *Bernburg.*—Beet-root sugar in loaves and refined.

DUCHY OF ANHALT-DESSAU, CÖTHEN.

CLASS 2.

6 Anhalt Co. *Roslau.*—Chemical products, mineral tar, and pit coal.

CLASS 3.

8 Heiden, R. *Cöthen.*—Sanitary chocolate, medicated pectoral caramels.

CLASS 4.

8a Kämmerer, C. G. *Dessau.*—Perfumes.

CLASS 8.

9 Achilles, W. & Heckert, H. *Cöthen and Gröbzig.*—Beer-conservator and beer-cooler.

CLASS 11.

11 Berger, R. *Cöthen.*—Arms.

CLASS 20.

13 Vollschwitz, W. F. *Zerbst.*—Silk hat-plushes.

CLASS 21.

14 Dessau Wool-spinning Mill.—Samples of yarn, from carded wool.

15 Meinert, S. *Dessau.*—Fancy fabrics.

16 Peuckert & Körner, *Jessnitz.*—Carpets, woollen cloth, and winter articles.

CLASS 23.

17 Languth, H. (F. Robitzsch), *Dessau.*—Samples of dyed goods.

18 PLAUT & SCHREIBER, *Jessnitz.*—Printed woollen table-covers.

CLASS 26.

19 SPIELER, L. *Dessau.*—Halter for glandered horses.

CLASS 27.

20 ROSENTHAL, C. *Cöthen.* — Japanned boots.

CLASS 28.

21 KATZ, BROS. *Dessau.* — Books, and specimens of printing.

CLASS 29.

21A ACHILLES, W. & HOBUSCH, H. *Cöthen.* —Apparatus for writing and reading.

CLASS 31.

22 POLYSIUS, G. *Dessau.*—Combination-ocks.

GROSSHERZOGTHUM BADEN.

CLASS 1.

26 KILIAN, v. *Waldshut.*—Mill-stones.
27 ZIMMERMANN, *Oppenau.* — Minerals from the Renchthal.

CLASS 2.

28 BENKISER, *Pforzheim.* — Crystallized and ground tartaric acid.
29 CLEMM, LENNIG, *Mannheim.*—Chemically-prepared manure; barytic mercury, sulphate of copper.
30 GOEHRINGER, J. *Rippoldsau.*—Mineral waters.
32 RÖTHER, H. *Mannheim.* — "Diamond paint," for protecting wood, iron, and masonry, &c.
33. ULTRAMARINE MANUFACTORY, *Heidelberg.*—Ultramarine.

CLASS 3.

34 BASSERMANN, HERRSCHEL, & DIEFFENBACHER, *Mannheim.*—Maccaroni, sago, starch, and other pastes.
38 HETZEL, *Willstedt.*—Maize.
39 GRAND DUCAL AGRICULTURAL SCHOOL, *Carlsruhe.*—Seeds.
40 SCHOLLENBERGER, *Carlsruhe.*—Maize.
41 BADEN ASSOCIATION FOR THE CULTIVATION AND COMMERCE OF TOBACCO, *Carlsruhe.*—Baden tobaccos.
42 BÖCKLIN, R. v. *Orschweier.* — Baden tobacco.
43 KÜBLER, H. *Ilgen.* — Tobacco grown in the Palatinate.
45 SEITZ, *Oftersheim.*—Palatinate tobacco.
46 BADER, A. T. *Lahr.*—Cigars of Palatinate tobacco.
47 LANDFRIED, T. P. *Rauenberg.*—Cigars and Palatinate tobacco.
48 MAYER, BROS. *Mannheim.*—Cigars.
49 PFEIFFER, T. M. *Ziegelhausen, near Heidelberg.*—Cigars of Palatinate tobacco.
50 SEELIG, DAVID, *Mannheim.*—Cigars.
51 MAYER, F. *Heidelberg.* — Palatinate hops.
52 KÜBLER, *St. Ilgen.*—Palatinate hops.
53 ALBRECHT, JOS. *Freiburg.* — Cherry-brandy.
54 BABO, L. v. *Weinheim.*—Wines grown near Bergstrasse.
55 BLANKENHORN, BROS. *Müllheim.* — Wines, called "Markgräfler" and "Kaiserstuhler," and cherry-brandy.
56 BÖRSIG, T. *Oberkirch.* — Wines from the Renchthal, and spirits.
57 DILGER, O. *Triberg.* — Cherry-brandy of the Black Forest.
58 SEXAUER, C. F. *Sulzburg.*—Wines.
60 FISCHER, F. H. *Offenburg.*—Wines and cherry-brandy.
61 HANOVER, ABR. *Schmieheim.*—Wine and spirits.
62 KUENZER & CO. *Freiburg.* — Baden sparkling wines.
63 LANDFRIED, P. J. KÜNZLE, & CO. *Heidelberg.*—Baden wines.
64 AGRICULTURAL OFFICE, *Briesach.*—Wines called "Kaiserstuhler."
66 LOTHER, TH. E. *Eppingen.* — Wines and cherry-brandy.
67 MERK, C. & SONS, *Bühl.*—Wine and cherry-brandy.
68 SCHÜTT, F. *Affenthal.*—Affenthal wines and cherry-brandy.
69 SPITZMÜLLER, C. A. *Biberach.*—Cherry-brandy of the Black Forest.

CLASS 8.

74 MÜRRLE, G. JAC. *Pforzheim.* — A steam apparatus for chemical, and two for pharmaceutical laboratories.
74A KÜHFUSS, *Carlsruhe.*—Portions of fireman's equipment.

CLASS 13.

75 SIKLER, C. *Carlsruhe.* — Theodolites, levels, and scales for chemical analyses.

CLASS 14.

76 LORENT, DR. *Mannheim.* — Photographs.

CLASS 15.

77 JOINT-STOCK CO. FOR THE MANUFACTURE OF CLOCKS, *Lenzkirch*.—Parts of clocks, and finished clocks.

78 BEHA, J. A. *Eisenbach*. — Cuckoo clocks.

79 BOB, L. *Furtwangen*.—Regulators.

79A BOB, M. *Triberg*. — Regulators and clocks.

80 BOB, V. *Furtwangen*.—Regulators.

80A BÜHLER, C. H. *Triberg*.—Clocks.

81 DILGER, O. *Triberg*.—Black Forest clocks.

81A DOLD & HETTICH, *Furtwangen*.—Parts of clocks.

82 FURTZWÄNGLER, L. *Gütenbach*.—Black Forest clocks.

83 HEINE & DILGER, *Neustadt*.—Dials painted in oil.

84 HERZER, V. & STOCKER, *Villingen*.—Regulators and Black Forest clocks.

85 KALTENBACH, L. *Furtwangen*.—Regulators and Black Forest clocks.

86 KAMMERER, S. *Furtwangen*.—Spring clocks.

86A KETTERER, A. *Vöhrenbach*.—Parts of clocks.

87 KÖRNER & HEILBOCK, *Villingen*.—Wooden clock-cases.

87A KREUZER, R. & A. *Furtwangen*.—Halyalophames.

87B LAULE, T. *Furtwangen*.—Clocks with dials painted in oil.

88 MARTENS, J. H. & Co. *Furtwangen*.—Pocket chronometer, and watches with anchor-escapement, and single parts of them.

88A MAURER, R. *Eisenbach*.—Regulators.

89 METZGER, J. *Carlsruhe*.—Carved clock-cases and frames.

91 THOMANN, P. *Furtwangen*.—Springs for clock-works, and lamellæ for suspending pendulums.

91A TRITSCHLER, T. *Vöhrenbach*.—Spring clocks.

92 WEHRLE, C. *Neustadt*. — Clock-case with clock-works.

93 WEHRLE, E. *Furtwangen*.—Trumpet clock.

93A WELTE, M. *Vöhrenbach*.—Grand musical clock-work—"orchestrion."

93B WINTERHALTER, C. *Vöhrenbach*.—Automatical clocks.

CLASS 16.

94 PADEWET, J. *Carlsruhe*. — Repaired violono, bass-viol, two new violins made of old wood.

CLASS 18.

95 HEUSS, FR. *Hassmersheim, near Mosbach*.—Cotton sewing-yarn.

98 WOLF, J. & SONS, *Mannheim*.—Cotton ropes for cotton spinning and weaving.

99 ZÜRCHER, BROS. *Lahr*.—Summer and winter waistcoatings.

CLASS 19.

100 FINGADO, C. *Mannheim*.—Ship ropes made of Baden split hemp.

102 FREI, J. *Ettenheim*.—Hemp for spinning.

103 HANOVER, ABR. *Schmieheim*.—Hemp for spinning.

104 HAUS, *Altfreistett*.—Samples of hemp.

104A HUTH & Co. *Neufreistädt*.—Samples of hemp.

105 SCHOCH, A. & F. *Lichtenau*. — Raw hemp, hemp for spinning.

106 SCHOLLENBERGER, J. *Carlsruhe*.—Hemp-seed.

107 WAGNER, C. *Emmendingen*.—Hemp-seed and hemp.

CLASS 20.

108 METZ, BROS. *Freiburg*. — Raw silk, double woof, dyed sewing silk, sewing-machine silk.

CLASS 25.

110 KAHN, M. & SONS, *Mannheim*.—Cleaned bed-feathers and eider-down.

111 PLATT, FR. *Eberbach*.—Spun horse-hair.

CLASS 26.

112 HEINTZE & FREUDENBERG, *Weinheim*.—Varnished and blackened calf skins.

113 KONSTANZER, A. *Villingen*. — Calf skins for coach-roofs, aprons, and harness.

117 SCHWEIKHARDT & KURZ, *Lahr*.—Calf skin.

CLASS 28.

122 BOHNENBERGER & Co. *Pforzheim*.—Printing and copper-plate printing paper.

CLASS 30.

124 HASSLINGER & Co. *Carlsruhe*.—Household furniture.

CLASS 31.

126 HELMEICH, MOLL, & Co. *Mannheim*.—Nails, tags, rivets of iron, brass, and copper.

127 SCHULTHEISS, BROS. *St. Georgen*.—Objects of enamelled copper and sheet-iron.

CLASS 32.

128 LACHMAN, A. *Rastatt*.—Cutlery.

CLASS 33.

129 PFORZHEIM JEWELLERY EXHIBITORS: C. Becker, Benkiser, & Co. Schlesinger & Weber, C. Gülich, A. Noesgen, J. Kiehnle, Gschwindt & Co. W. F. Grumbach, G. Mayer, Widow Kämpff & Co. L. Fischer & Co. G. Müller, C. Ehrardt & Co. Dillenius & Bohnen-

berger, T. Tschopp, J. Hiller, C. Becker, H. Keller, C. E. Rohreck, A. Roller, C. Steinbrenner, A. Eisenmenger, G. Saacke & Co. —Jewellery of 14-carat gold.

CLASS 34.

133 MANNHEIM LOOKING-GLASS MANUFACTORY (branch of the joint-stock Co. of the manufactories of St. Gobain, Chauny, and Cirey).—Looking-glass plates.

BAVARIA.

CLASS 1.

141 FRÄNKEL, L. H. *Massbach, Lower Franconia.*—Brown coal and clay.

142 KISSINGEN, ROYAL ESTABLISHMENT FOR MINERAL WATERS, *Kissingen, Lower Franconia.*—Mineral waters. Rakoczy, Pandur, Maxbrunnen, bitter-water.

143 KLINGENBERG, MUNICIPALITY OF. —Refractory clay.

144 KROHER, A. *Staudach, Upper Bavaria.*—Cement roofing plates, cement paving stones.

145 MARTIUS, Dr. TH. W. C.—Phosphorite.

146 SOLENHOFEN, JOINT-STOCK CO. OF, *Solenhofen, Middle Franconia.*—Lithographic lime-stones.

CLASS 2.

148 ADAM, F. M. *Rennweg, near Nuremberg.*— Ultramarine, Prussian blue, prussiate of potash.

149 GRAF & CO. *Nuremberg.* — Aniline, benzine, photogen, camphine oil, creosote, asphalt, varnishes, aniline-colours, &c. &c.

149A GROSSBERGER & KURZ, *Nuremberg.* —Dried yeast.

150 HENFELD *(Upper Bavaria)* BAVARIAN JOINT-STOCK CO.—Chemicals, artificial manure.

151 HOFFMAN, G. *Schweinfurt, Lower Franconia.*—Mineral and chemical colours.

152 KAISERSLAUTERN, ULTRAMARINE MANUFACTORY OF, *Kaiserslautern, Palatinate.*—Ultramarine.

153 LICHTENBERGER, C. *Hambach, Palatinate.* — Œnantic-ether, œnantic-acid, tartar, tartrate of lime, spirit, wine-dregs, essential oil of French brandy.

154 MEYER, H. (MITTLER, F.) *Augsburg.* — Chromic-acid-green, called "Mittler's giftfreies grün" (green-colour, free from poison). Green-coloured window-blinds.

155 MARTIUS, DR. W. E. *Erlangen.* — Zanzibar, Mogadore, Angola, West Indian, and Manilla copals dissolved in alcohol (0·803), or spirit (0·839).

156 RÖSCH, F. *Nuremberg.* — Deckweiss, white colour.

157 SATTLER, W. *Schweinfurt, Lower Franconia.* — Samples of colours, colour-grinding mill.

158 STOLLREITHER, C. P. *Munich.*—Blue colour (substitute for indigo).

159 TOUSSAINT, G. F. *Fürth.* — Dry chemical substances.

CLASS 3.

160 BARTH, S. & CO. *Würzburg.*—Fine Franconian wines.

161 BIFFAR, A. A. (RHEINISCHE FRÜCHTEHANDLUNG), *Deidesheim, Palatinate.*—Preserved fruits.

162 DÖRING, F. *Würzburg.* — Sparkling wines.

163 EICHHORN & CO. *Speyer, Palatinate.* —Cigars.

164 HÄUTLE, T. *Munich.*—Chocolate.

165 LEUCHS, T. C. (C. LEUCHS & CO.) *Nuremberg.*—Wine made of water and various substances.

166 NEUBAUER, J. *Winzingen, Palatinate.* —Starch.

167 OPPMANN, W. C. *Würzburg.* — Wines from Bavarian vineyards.

167A OPPMANN, M. *Würzburg.*—Sparkling Franconian wine.

CLASS 4.

168 HECKEL, G. *Munich.*—Bone-glue, assortment of buttons.

169 ROTH, MUNICIPALITY OF, *District Spalt, Middle Franconia.*—Hops, not cured.

170 UHLMANN, S. *Fürth.*—Cured Bavarian hops.

170A SECKENDORF, LE VINO & CO. *Nuremberg.*—Hops.

CLASS 5.

171 KOLB, L. *Bayreuth.*—Tilt for railway waggons.

CLASS 7.

172 DINGLER, T. & WOLFF, T. B. (DINGLER'S MASCHINEN-FABRIK AT) *Zweibrücken.*— Printing-presses.

173 HEINTZ, DR. C. FRED. *Munich.*— Printing machine.

CLASS 8.

175 KLINGENFELD, PROFESSOR C. A. *Nuremberg.*—Scales, invented by the exhibitor.

176 PFANZEDER, G. *Munich.*—Scale beam on four edges.

CLASS 9.

177 DANZER, B. *Munich.*—Apparatus for watering delicate plants and for destroying pucerons.

CLASS 10.

179 ECKHARDT, P. *Grosshesselohe, near Munich.*—Sideway paving-stones.

180 KLETT & CO. *Nuremberg.*—Model of the railway-bridge across the Rhine, at Mayence.

181 TÖLZER, J. *Tegernsee, Upper Bavaria.* —Designs of wooden farm-houses, barns, stables, &c. in the Bavarian highlands.

CLASS 11.

182 UTENDÖRFFER, H. *Nuremberg.*—Percussion-caps.

CLASS 13.

184 HAFF, BROS. *Pfronten, Suabia.*—Cases of mathematical instruments.

186 RIEFLER, CL. *Maria Rhein, Suabia.* —Cases of mathematical instruments.

186A RODLER, A. *Nuremberg.*—Carbon-cylinders for electro-galvanic apparatuses.

187 ULLRICH, BROS. *Maykammer, Palatinate.*—Measuring rules.

CLASS 14.

188 ALBERT, J. *Munich.*—Photographic portraits in life size, reproductions of objects of art.

189 GYPEN & FRISCH, *Munich.*—Photographs of modern religious works of art.

CLASS 15.

190 MANNHARDT, J. *Munich.*—Steeple clock; two sets of wheel-work, peculiarly constructed.

CLASS 16.

191 BÖHM, T. *Munich.*—Newly-invented wind instruments.

192 HASELWANDER, J. *Munich.*—Citherns and violins.

193 HENTSCH, JAC. *Lindberg, Lower Bavaria.*—Wood for sounding-boards.

195 PFAFF, F. *Kaiserslautern, Palatinate.* —Musical instruments.

196 STEGMAIER, F. *Ingolstadt, Upper Bavaria.*—Musical instruments.

197 GSCHWENDTUER, *Oberstdorf.*—Sounding boards.

CLASS 17.

198 WOLFFMÜLLER, A. *Munich.*—Pharmaceutical steam apparatus and utensils.

CLASS 18.

201 SALCHER, TH. *Passau, Lower Bavaria.* —Canvas for embroidery.

CLASS 19.

202 TEGELER, E. *Otterberg, Palatinate.*—Linen-thread, yarn, and linen-cloth. Samples of bleaching.

CLASS 20.

203 ESCALES, BROS. *Zweibrücken, Palatinate.*—Hat-plushes.

204 RITTER & THIEL, *Kaiserslautern, Palatinate.*—Silks.

CLASS 21.

206 GEORG, JAC. *St. Lambrecht, Palatinate.* —Broadcloth and buckskins.

CLASS 23.

208 LEUCHS, G. (LEUCHS & Co.) *Nuremberg.*—Turkey-red cotton.

CLASS 24.

209 WÜNSCH, J. B. *Nuremberg.*—Gold and silver embroideries for ecclesiastical and other purposes.

CLASS 26.

210 EICHTHAL, J. BARON OF (IGNAZ MAYER'S LEATHER MANUFACTORY), *Munich.* —Leather.

211 SCHWARZMANN, F. X. *Munich.*—Leather.

CLASS 27.

212 GREIDER, G. *Tegernsee, Upper Bavaria.*—Shoes for the mountains, shoes and buskins.

213 HERBIG, C. *Kaiserlautern, Palatinate.* —Boot-trees and lasts.

CLASS 28.

214 BECKER, A. *Munich.*—Lithographic prints in oil colours.

215 BEISBARTH, I. C. & SON, *Nuremberg.* —Painters' brushes.

216 BEROLZHEIMER & ILLFELDER, *Fürth.* —Lead pencils of all sorts.

217 BRAUN & SCHNEIDER, *Munich.*—Literary works, illustrated with woodcuts.

219 DESSAUER, A. *Aschaffenburg, Lower Franconia.*—Stained paper.

220 FABER, A. W. *Stein, near Nuremberg.* —Blacklead pencils, crayons, pastils, &c.

221 GROSSBERGER & KURZ, *Nuremberg.*—Pencils, crayons, oil-chalk.

222 GYPEN & FRISCH, *Munich.*—Copper and steel engravings, lithographic prints in oil-colours.

223 HÄNLE, LEO, *Munich.*—Gold and silver paper, edgings, frames, corner ornaments.

224 KATHAN, P. *Augsburg.*—Gold and silver paper of all sorts.

225 KIMMEL, T. G. (C. G. RÖSER), *Nuremberg.*—Stained paper, gold and silver paper edgings, medalions, &c. &c.

226 KITZINGER, T. G. *Munich.*—Oil colour prints.
228 MOZET, T. *Munich.*—Pictures printed in oil colours.
230 STÄDTLER, T. G. *Nuremberg.*—Lead, slate pencils, pastils, chalk, &c.
231 STERN, W. *Fürth.*—Stained paper.
232 SUSSNER, G. W. *Nuremberg.*—Creta Polycolour, or coloured oil chalk pencils.

CLASS 29.

233 TOPOGRAPHIC BUREAU OF THE GENERAL QUARTERMASTER-STAFF OF THE R.B. ARMY, *Munich.*—40 leaves of the topographic map of Bavaria.
235 MUNICH ASSOCIATION FOR INDUSTRIAL IMPROVEMENTS, *Munich.*—Publications of the Association for 1851-1862. Drawings showing the method employed in the drawing-school of the Association.
237 MÜLLER, T. *Nuremberg.* — Stencil-painting-plays.
238 NEUSSNER, L. *Nuremberg.* — Magic lantern.

CLASS 30.

240 CNOPF, P. *Nuremberg.* — Articles of papier mâché.
240A GREMSER, T. *Augsburg.*—Model of an altar, cut in ivory.
240B ROYAL INDUSTRIAL ART SCHOOL AT NUREMBERG.—The show-cases of Nos. 182, 220, 221, 230, and 232.
241 KÜBLER, BROS. *Munich.*—Floor inlaid with 12 kinds of wood.
242 LÖHNER, T. *Nuremberg.*—Boxes and fancy articles.
242A SHEIDIG, L. *Fürth.* — Picture and mirror frames in the show-cases Nos 230 and 254.
243 TRIMBORN, CHR. *Munich.*—Figures, &c. of papier mâché.
244 ZINN, S. & THURNAUER, M. (SAM. ZINN & Co.), *Redwitz, Upper Franconia.*—Willow wares.

CLASS 31.

245 BAUER, AD. (T. P. AMMON), *Nuremberg.*—Plated wire.
246 BIBERBACH, J. C. *Nuremberg.*—Brass plates, brass wire.
247 BRANDEIS, J. *Fürth.*—Bronze powder, leaf metal, and orsedew.
248 BRUNNBAUER, H. *Munich.*—Gold and silver leaf.
250 CONRADTY, C. *Nuremberg.*—Bronze powder.
250A EBERLE, T. N. *Augsburg.*—Piercing-saws.
251 EYERMANN, M. & LÖWI, C. *Fürth.*—Leaf metal and bronze powder.
253 FUCHS, G. L. & SONS, *Fürth.*—Leaf metal, bronze powder and orsedew.
254 FÜRTH SOCIETY FOR PROMOTING INDUSTRIAL PROGRESS, *Fürth.*—Tableau representing the leading articles made at this city.
255 HÄNLE, LEO, *Munich.* — Bronze powder.
255A HECKEL, T. G. HEIRS, *Allersberg, near Nuremberg.*—Gilt and silvered wire.
256 HEININGER, A. *Passau, Lower Bavaria.* — Monstrance of gilt bronze, after a design of the sixteenth century.
257 HUTTULA, W. *Nuremberg.*—Gold and silver leaf.
258 ISSMAIER, J. M. *Nuremberg.*—Tin-plate-ware.
259 LINZ, J. L. *Fürth.*—Leaf metal.
260 MEYER, J. C. *Fürth.*—Leaf metal and bronze powder.
261 PAULI, G. C. *Nuremberg.*—Gold and silver leaf.
263 SCHÄTZLER, G. E. *Nuremberg.*—Gold and silver leaf, gold sponge.
264 SCHMIDTMER, E. (E. KUHN), *Nuremberg.*—Gilt and silvered wire, &c.
264A SCHEIBLEIN, FR. *Roth.*—Gilt and silvered wire.
265 SCHWEIZER, A. *Fürth.*—Steel frames for spectacles and eye-glasses.

CLASS 33.

269 BIRNBÖCK, TH. *Munich.* — Impressions of seals, medals.

CLASS 34.

270 ARNDT, BROS. *Pirmasens, Palatinate.*—Watch glasses.
271 BACH, J. *Fürth.* — Looking-glass plates, silvered and unsilvered.
272 HILDEBRAND, C. *Munich.* — Muslin-glass.
273 KOCH, C. W. *Einbuch, near Ratisbonne.*—Glass plates for photographers.
274 STEIGERWALD, FR. *Munich and London.*—Cut crystal and ornamental glass.

CLASS 35.

275 GRUBER & RAUM, *Nuremberg.* — Pressed graphite crucibles.
276 SCHMIDT, C. *Bamberg.*—Paintings on porcelain.
277 WIMMERS, H. *Munich.*—Paintings on porcelain.

CLASS 36.

279 SCHWARZ, J. v. *Nuremberg.*—Steatite wares, gas-burners.
280 STÄDTLER, J. G. SEN. *Nuremberg.*—Buttons and gas-burners.
280A SPERL, H. *Nuremberg.* — Xylochromic and xyloplastic objects. Dyed wood embossed and pressed objects of dyed wood.

BRUNSWICK.

CLASS 3.

282 WITTEKOP, & Co. — Chocolate and mill products, maccaroni, &c.

CLASS 8.

283 SEELE, FR. & Co.—Centrifugal machine, for extracting the juice of sweet turnips.

CLASS 11.

284 SEYDLITZ, W. A. E. — Fast-loading percussion rifle, &c.

CLASS 13.

285 BORNHARDT, A.—Balances for chemical purposes.

CLASS 16.

286 PAULUS, A.—Wind instruments (cornet-à-piston).

CLASS 21.

289 DEGENER & DABELSTEIN.—Woollen and mixed cloaking.

CLASS 28.

290 VIEWEG, F. & SON.—Paper, types, and engravings.
291 WESTERMANN, G.—Stereotypes and engravings.

CLASS 30.

292 WALTER, E. & SON.—Varnished basket work.

CLASS 31.

292A GOLDBERG & JACOBS, late WRIED, before STOBWASSER.—Pictures on wood and plate-iron, varnished.
295 WILKE, A.—Apparatus for beating eggs.

FRANKFORT-ON-MAIN.

CLASS 2.

302 FRANKFORT JOINT STOCK CO. FOR CHEMICO-AGRICULTURAL PRODUCTS.—Soda, chloride and superphosphate of lime, artificial guano, crushed bones, manure.
303 ZIMMER, C.—Chemical preparations of quinine.

CLASS 3.

304 ECKERT, W. & Co.—Cigars.
305 GIORGI, DE, BROS.—Cocoa.

CLASS 4.

308 MOUSON, J. G. & Co.—Transparent and perfumed soaps and perfumery.
309 RIEGER, W.—Perfumed soaps and perfumery.

CLASS 13.

311 KNEWITZ, P. J. F.—Scales for chemical analysis, set of weights, orthograph, circle-dividing machine.

CLASS 14.

312 HAMACHER, G. — Literary publications on art and science, and photographs.

CLASS 16.

314 ANDRÉ, C. A.—Grand Mozart pianoforte in polisander wood.

CLASS 22.

315 VACONIUS, J. J.—Velvet carpets.

CLASS 24.

316 DANN, L. & Co. — Lace, braids, tassels, trimmings, &c.

CLASS 26.

317 HAUSMANN, BROS.—Dyed calf skins and leather.
318 LANDAUER-DONNER, G. F.—Leather and dyed calf skins.
319 REGES, J. A. B.—Miniature photograph albums.
320 ROTH, J. A. SONS.—Danish and tanned leather.
321 ROTH, C. W.—Leather.
322 SCHEIDEL, GOUDA, & Co.—Portfolios, workboxes, cases, &c.

CLASS 27.

323 QUILLING, J. F.—Shawl, capouches, children's jackets, &c.
324 SCHNAPPER, J. JUN.—Drawers, jackets, covers, &c.

CLASS 28.

324A DRESSLER. — Specimens of type founding.
325 KNATZ, C.—Books and lithographs.

CLASS 29.

326 KÖBIG, J.—Lithographic prints.
327 KRUTHOFFER, C.—Literary publications on art and science, lithographic works.

CLASS 30.

328 BÖHLER, F.—Household furniture of hart's horn, objects of ivory.
329 JACQUET, H. SON.—Fancy furniture of carved wood.

CLASS 31.

330 LAUSBERG, C.—Metal caps for corks.

CLASS 33.

331 GOLDSCHMIDT, M. SON.—Jewellery.
332 FRIEDMANN, J.—Gold and silversmith's ware.

HANOVER.

CLASS 1.

341 GEORGE MARIA SMELTING WORKS, *near Osnabrück*.—Iron ores and pig-iron.
342 EGESTORFF, G. *Linden, near Hanover*.—Salt.
343 IN DER STROTH, H. *Bentheim*.—Jet coal.
344 MUNICIPALITY, *Osnabrück*. — Coal (Anthracite) from the Piesberg.
345 MOSQUA, *Hildesheim*.—Mill stones.

CLASS 2.

346 CHEMICAL WORKS, *Nienburg*.—Chemical products.
347 EGESTORFF, A. *Linden, near Hanover*.—Ultramarine.
348 EGESTORFF, G. *Linden, near Hanover*.—Chemical products, ultramarine.
349 HEINS, E. *Harburg*.—Crystals of saltpetre, tin-salt.
350 HEUER, A. & Co. *Lichtenstein-on-the-Harz*.—White lead, litharge, &c.
351 HILKENKAMP & Co. *Osnabrück*.—Bremen green.
352 KOCH, R. *Goslar*.—Matches, quick-matches for miners, gelatine, bone-glue.
353 RÖHRIG, GEBRÜDER, *Linden, near Hanover*.—Ultramarine.
354 SCHACHTRUPP & Co. *Osterode*. — White lead, sugar of lead.
355 STACKMANN & RETSCHY, *Lehrte*.—Prepared bones for manuring, and materials belonging to them.

CLASS 3.

356 BREDE, C. L. *Hanover*.—Maccaroni, vermicelli, &c.
357 BREUL & HABENICHT, *Hanover*.—Tobacco, snuff, and chewing tobacco.
358 GRÜTTER & Co. *Nienburg*.—Sparkling wines.

CLASS 4.

359 COHEN, A. VAILLANT, & CO. *Harburg*. Caoutchouc manufactures.
360 GUMMI-KAMM-CO. *Harburg*.—Combs of vulcanized caoutchouc.
361 HENNECKE, JAC. *Goslar*.—Oil and oil-cakes.
362 HURTZIG, BROS. *Linden, near Hanover*.—Manufactures of bleached and hardened caoutchouc.
363 TIEDGE, C. *Hanover*. — Pitch for chasing.
364 WILHARM & MÜLLER, *Melle*.—Machine-made corks.

CLASS 7.

365 KNÖVENAGEL, A. *Linden, near Hanover*.—Wood-sawing machine.
366 VOIGTLÄNDER, F. *Schladen*. — Tobacco spinning machine.

CLASS 8.

367 EGESTORFF, G. *Linden, near Hanover*.—Loco mobil steam-engine, iron casts.

CLASS 10.

368 EGESTORFF, A. *Linden, near Hanover*.—Asphalt.
369 HENNING, D. H. *Limmer, near Hanover*.—Asphalt and mineral tar.
370 HEYN, C. F. *Lüneburg*.—Statue of cement. Blocks united by cement.
371 MOSQUA, *Hildesheim*.—Cement.

CLASS 11.

372 EGESTORFF, G. *Linden, near Hanover*.—Percussion caps.
373 HUISMAN, J. G. *Leer*.—Shot.
374 KLAWITTER, TH. *Herzberg*.—Guns.
375 SCHACHTRUPP & Co. *Osterode*.—Shot.
376 STÖRMER, G. FR. *Herzberg*.—Rifle.
377 WESTPHAL, L. *Peine*.—Guns.

CLASS 13.

378 LANDSBERG & PARISIUS, *Hanover*.—Mathematical and philosophical instruments.
379 LOEHDEFINK, D. A. *Hanover*.—Manometers.

CLASS 16.

380 MEYER & Co. *Hanover*.—Musical instruments.
381 RIECHERS, A. *Hanover*.—Two violins, viol and violoncello.

CLASS 18.

382 MECHANISCHE WEBEREI, *Linden, near Hanover*.—Cotton velvets.

CLASS 19.

383 MEYERHOF, J. *Hildesheim*. — Linen fabrics.
384 SCHULZE, D. *Bodenteich*. — Linen cloth, yarn, and flax.
385 SIEMSEN, J. *Hanover*. — Bell-ropes, bags, &c. of Manilla hemp.

CLASS 21.

386 HENNECKE, JAC. *Goslar.*—Flannel, coating, &c.
387 LEPPIEN, G. *Lüneburg.*—Plush.
388 LEVIN & WESTERMANN, *Herzberg.*—Velours (woollen cloth).

CLASS 22.

389 ROSKAMP, *Springe.*—Carpets.

CLASS 25.

390 LEPPIEN, G. *Lüneburg.*—Hair-cloth.

CLASS 26.

391 GOLDSCHMIDT, *Wölpe.*—Horse-leather.
392 SCHRITZMEYER, T. H. *Harburg.*—Leather.
393 WOLFF & ROHTE, *Walsrode.*—Leather.

CLASS 27.

394 FROELING, C. & S. *Münden.*—Hats. Felted goods.
395 HANSEN, H. *Peine.*—Shoemakers' work.
396 HIRSCH, I. *Hanover.*—Clothes.
397 HUGO BROS. *Celle.*—Umbrellas and parasols.
398 KEUNECKE, W. *Peine.*—Shoemakers' work.
399 LIPPE, H. *Peine.*—Shoemakers' work.
400 POLLMANN, C. A. *Hanover.*—Umbrellas and parasols.
401 VOGES, C. *Peine.*—Silk buttons and similar articles.

CLASS 28.

402 EDLER & KRISCHE, *Hanover.*—Ledgers and other commercial books.
403 EPPEN, J. H. *Winsen a. L.*—Printing paper.
404 HOSTMANN, C. *Celle.*—Printing ink.
405 JÄNECKE BROS. *Hanover.*—Products of typography, type-founding, and lithography.
406 JÄNECKE BROS. & SCHNEEMANN, FR. *Hanover.*—Printing ink, black and coloured, for typography and lithography.
407 KLINDWORTH, F. *Hanover.*—Typographic and lithographic prints.
408 KOCH, J. H. A. *Freyburg a. E.*—Writing ink for steel pens.
409 KÖNIG & EBHARDT, *Hanover.*—Ledgers and other books.
410 LACHMUND & CO. *Göttingen.*—Frames for photographs.
411 SCHNELL, P. *Münden.*—Samples of stained and glazed paper.

CLASS 30.

412 BEHRENS & CO. *Alfeld.*—Shoe-lasts, table covers, window blinds.
413 HERTING, C. *Einbeck.*—Paper-hangings, borders, and decorations.
414 SCHWIKKARD, FR. *Goslar.*—Vases, &c. of alabaster.
415 TAEGER, L. *Hanover.*—Heads of animals, &c. papier mâché.

CLASS 31.

416 BERNSTORFF & EICHWEDE, *Hanover.*—Colossal lions of bronze.
417 BREITENBACH, J. *Nörten.*—Sheet-iron ware.
418 DU BOIS, J. E. *Hanover.*—Pewter toys.
419 HÄNDLER & NATERMANN, *Münden.*—Lead, tin, wire, shot and bullets, &c.

CLASS 32.

420 GRABOH, A. *Hanover.*—Cutlery
421 LAGE-SCHULTE, *Hanover.*—Files.
422 PFUHL, H. C. *Linden, near Hanover.*—Files.

CLASS 34.

423 PEZOLT & SCHELE, *Osterwald.*—Hollow glassware.

HESSE-CASSEL.

CLASS 2.

431 HABICH, G. C. SONS, *Cassel.*—Painters and washing colours.
432 SCHNELL, FR. *Cassel.*—Matches.

CLASS 4.

433 RUPPERT, C. & CO. *Cassel*—Toilet soaps and perfumes.

CLASS 8.

434 HENSCHEL & SON, *Cassel.*—Boiler-tubes and circumgyral valve-motion for steam-engines.

CLASS 13.

436 BREITHAUPT, F. W. & SON, *Cassel.*—Mathematical instruments.
437 FENNEL, O. *Cassel.*—Mathematical instruments.

CLASS 21.

439 BRAUN, BROS. *Hersfeld.*—Woollen cloth.
440 MÜLLER, B. *Fulda.*—Woollen yarn, plush, canvas, lastings, &c.

CLASS 22.

441 GEIBEL, FR. *Kesselstadt, near Hanau.*—Carpet patterns.

CLASS 25.

442 ALLENDÖRFFER, J. C. *Cassel.*—Furs.
443 RIEBE, C. *Cassel.*—Furs.

CLASS 29.

445 FISCHER, TH. *Cassel.*—Relief-maps of Jerusalem and Palestine.

446 SCHELLER, WITTICH, & SCHERB, *Cassel.* —Toys.

447 WEBER, A. & ALLMEROTH, *Marburg.* —Toys.

CLASS 31.

449 ZIMMERMANN, E. G. *Hanau.*—Bronze, iron, and zinc casts, and marble wares.

CLASS 33.

450 WEISHAUPT, C. M. SONS, *Hanau.*—Jewellery.

CLASS 35.

451 GÖBEL, J. P. & SONS, *Grossallmerode.* —Graphite crucibles.

452 PFEIFFER, C. H. *Cassel.*—Cassel china.

—

GRAND DUCHY OF HESSE.

CLASS 1.

461 GOVERNMENT MINE WORKS, *Thalitter.*—Rose-copper.

462 TASCHÉ, *Salzhausen.*—Minerals from the Wetterau and the Vogelsberg.

CLASS 2.

468 ALBRECHT, J. *Mentz.*—Varnish, sealing-wax, bleached and coloured gum-lac.

463 BLAUFARBWERK, *Marienberg, near Bensheim.*—Ultramarine.

464 BUECHNER, W. *Pfungstadt, near Darmstadt.*—Ultramarine.

582 HEUMANN, O. *Oberramstadt.*—Drying oil, baryta green.

469 MAYER, F. H. *Mentz.*—Varnish, sealing-wax, bleached and coloured gum-lac.

470 MEHL & MOSKOPP, *Mentz.*—Varnish, sealing-wax, bleached and coloured gum-lac.

471 MELLINGER, C. *Mentz.*—Varnish, sealing-wax, bleached and coloured gum-lac.

465 MERCK, E. *Darmstadt.*—Chemical products.

466 OEHLER, K. & R. *Offenbach.*—Products from the distillation of coal-tar, &c. Products for dyeing, printing, &c.

467 PETERSEN & CO. *Offenbach.* — Products from the distillation of coal-tar.

472 SCHRAMM, C. *Offenbach.*—Varnish, and ivory-black for printing.

473 VOLTZ, J. A. *Mentz.*—Metal-putty.

CLASS 3.

585 ADMINISTRATION OF THE "LUDWIGSBRUNNEN," *near Gross, Karben.*—Samples of mineral-water.

477 CANTOR & SON, *Mentz.*—Sparkling wines.

478 DAEL, G. A. *Mentz and Geisenheim.*—Still and sparkling Rhine and Moselle wines.

583 DOLLES, J. *Bodenheim.*—Rhine wines.

475 ENGELHARDT, F. *Rüsselsheim, near Mentz.*—Samples of chicory and chicory-flour.

479 FRIDBERG & GUTTMANN, *Mentz.* — Sparkling wines.

481 HENKELL, A. & CO.—Sparkling Rhine wines.

491 HEYL, L. & CO. *Worms.*—Tobacco, cigars, and cigarettes.

482 KUPFERBERG, C. A. *Mentz.*—Sparkling wines.

483 LAUTEREN, C. & SON, *Mentz.*—Still and sparkling Rhine and Moselle wines.

484 MÜLLER, J. JUN. *Bensheim.*—Samples of wines.

485 PABSTMANN, G. M. SON, *Mentz and Hochheim.*—Rhine wines.

486 PROBST, F. A. *Mentz.*—Rhine wines.

487 RAISER, C. *Oppenheim.* — Still and sparkling Rhine wines.

488 RODENSTEIN, V. A. *Bensheim.*—Samples of wine.

490 SICHEL, M. *Mentz.*—Wine-oil, cognac tartar, oil of tartar, tartrate of lime.

480 GEORGE, E. *Büdesheim, near Bingen.* —Wines from "Büdesheimer Scharlachberg."

476 JUNGBLUTH, J. V. *Worms.*—Samples of chicory.

489 VALCKENBERG, P. J. *Worms.*—Samples of wine.

474 WAGNER, J. P. & CO. *Mentz.*—Samples of chocolate.

492 WENCK, F. A. *Darmstadt.*—Samples of cigars.

CLASS 4.

496 BÖHM, G. *Offenbach.*—Toilet soaps.

497 GLÖCKNER, G. *Darmstadt.*—Soaps and bleached tallow.

500 HAMM, F. *Offenbach.* — Samples of glue.

507 HEIDEN, HEIMER, & CO. *Mentz.* — Hops.

510 HEIDENREICH, *Affolterbach.* — Wood for matches.

504 HEINRICH & HEYL, *Offenbach.*—Phosphates and bone powder.

505 HESSE, GRAND DUCHY, FOREST ADMINISTRATION.—Oak barks for tanning.

501 HEYL, C. *Worms.*—Samples of glue.

506 KELLER, H. SON, *Darmstadt.*—Forest grass and clover seeds.

493 KIEFER & PIRAZZI, *Offenbach.*—Turkish tobacco, opium, rose oil.

502 LYON, L. *Michelstadt.* — Samples of glue.

494 MARTENSTEIN, J. D. SON, *Worms.*—Lubricating oil from the bones of animals, and lamp oil.

499 MICHEL, J. *Erbach.*—Articles in ivory.

495 MIELCKE, K. *Worms.*—Amber, amber beads.

498 SCHMITT, F. *Darmstadt.*—Soap.

508 SCHRÖDER, SANDFORT, *Mentz.*—Hops.

509 WAGNER, H. *Aschbach, near Wald-Michelbach.*—Matches.

503 ZERBAN, C. *Worms.* — Samples of glue.

CLASS 6.

511 DICK & KIRSCHTEN, *Offenbach.* — Waggon, and models of axletrees, springs, &c.

CLASS 8.

513 DE BARY, J *Offenbach.*—Machines for making cigars.

514 HEIM, BROS. *Offenbach.*—Presses for lithographic and other printing bookbinding, and gilding

515 JORDAN J. & SON, *Darmstadt.*—Machines for making drain-pipes, tiles, bricks, &c.

512 MASCHINENFABRIK & EISENGIESSEREI, *Darmstadt.*—Locomotive engine, and machine for cutting keys.

584 MEIER, L. *Darmstadt.*—Machine for making sausages.

517 SCHRÖDER, J. *Darmstadt.*—Sewing machines.

516 SCHULTZ, BROS. *Mentz.*—Machine for cutting the teeth of wheels.

CLASS 11.

519 DELP, K. *Darmstadt.*—Double-barrelled gun.

518 KAUFFMANN, E. H. *Mentz.*—Steel cannon with new mode of closing the chamber.

CLASS 13.

581 HESSE, GRAND DUCHY OF, BOARD OF MEASURES AND WEIGHTS.—Hessian system of weights and measures.

520 SIENER, G. *Darmstadt.*—Philosophical instruments.

CLASS 14.

521 HERZ, S. *Darmstadt.*—Photographs.

CLASS 15.

522 SCHÖPPLER, P. J. *Mentz.*—Steeple clock.

CLASS 16.

523 GEBAUER, G. D. *Alsfeld.*—Piano.

524 GLÜCK, C. L. *Friedberg.*—Pianos.

525 KÜHNST, G. *Darmstadt.*—Pianos.

526 SEIDEL, JOS. *Mentz.*—Musical wind-instruments.

CLASS 21.

527 HESSE, GRAND DUCHY OF, BOARD OF AGRICULTURE, *Darmstadt.*—English long-stapled (Cotswold) wool, and wool from the German sheep.

530 LANG & HESS, *Marbach, near Erbach, in the Odenwald.*—Woollen cloth.

529 LIST, C. & T. *Lauterbach.*—Carpets, and damask table-cloths.

528 VERHÖVEN, F. *Worms.*—Woollen yarns and shoddy.

CLASS 26.

531 DÖRR & REINHART, *Worms.*—Japanned leather.

532 HERTZ, BROS. *Oppenheim.*—Brown and waxed calf-skin and leather.

533 HEYL, C. *Worms.*—Japanned, waxed, and brown calf-skins.

534 MAYER, MISHEL, & DENINGER, *Mentz.*—Skins; saddle-leather; hides for bridles, stirrups, and saddles; moroccos, &c.

535 MAYER, P. *Mentz.*—Brown and waxed calf-skin; boot trees and fore-shoes; hides for engine-straps, &c.

536 MAYER, J. & FEISTMANN, *Offenbach.*—Leather in skins.

537 MELAS, L. & Co. *Worms.*—Calf-skin, japanned and for japanning.

538 MÜLLER, G. *Bensheim.*—Leather for soles, and calf-leather.

539 PFEIFFER, C. *Eberstadt, near Darmstadt.*—Brown calf-skins, leather.

541 PREETORIUS, W. & Co. *Alzey.*—Calf-skins, leather, boot-trees, &c.

540 SPICHARZ, J. *Offenbach.*—Japanned and grained hides and skins.

542 WORMATIA, SOCIETY OF LEATHER MANUF. *Worms.*—Japanned leather.

CLASS 27.

548 HÄNLEIN & LESCHEDITZKY, *Offenbach.*—Cap-peaks, hat-bands, and other japanned cloth.

543 HERZ, O. & NASS, *Mentz.*—Boots.

547 KOLBE, L. & Co. *Bessungen, near Darmstadt.*—Buttons.

546 SCHUCHARD, H. *Darmstadt.*—Silk and felt-hats; military and other caps.

544 SCHUHMACHER, J. & SON, *Mentz.*—Boots.

545 WOLF, S. *Mentz.*—Boots.

CLASS 28.

556 DEUFEL, R. *Darmstadt.*—Cover for a diploma.

550 FREUND, E. A. *Offenbach.*—Coloured and embossed paper; bookbinding, &c.; enamelled cards, labels, &c.

554 FROMMANN, M. *Darmstadt.*—Playing cards.

549 HEMMERDE, C. & Co. *Darmstadt.*—Straw-paper and paste-board.

555 REUTER, W. *Darmstadt.*—Playing cards.

552 SCHOTT, B. SONS, *Mentz.*—Music-printings.

553 WIRTZ, F. *Darmstadt.*—Geological and geographic cards.

551 WUEST, BROS. *Darmstadt.*—Coloured and marble paper.

CLASS 29.

558 BEYERLE, W. *Darmstadt.*—Drawing models for schools.

557 DIEHL, J. P. *Darmstadt.*—Geographical charts for schools.

559 LÖSSER, F. *Darmstadt.*—Drawing instruments.

560 SCHRÖDER, J. *Darmstadt.*—Technological school apparatus.

CLASS 30.

566 BÜTTNER, P. *Darmstadt.*—Looking-glasses in bronze frames, and composition ornaments.

567 HOCHSTÄTTER, F. *Darmstadt*.—Paper-hangings.
561 KNUSSMANN, W. *Mentz*.—Furniture.
562 NILLIUS, F. C. SON, *Mentz*.—Furniture.
564 REINHARDT, T. M. *Mentz*.—Straw-chairs.
563 REITMAYER, A. *Mentz*.—Furniture.
565 THÜRING, P. *Darmstadt*.—Carved wood picture-frames.

CLASS 31.

568 BUSCH, L. *Mentz*.—Statue in bronze.
569 FISCHER, J. & Co. *Offenbach*.—Cast-iron, enamelled, and bronzed inkstands, clocks, lamps, candelabra, and other articles.
570 GASAPPARAT & GUSSWERK, *Mentz*.—Bronze, and bronzed zinc lustres.
571 SEEBASS, A. R. & Co. *Offenbach*.—Cast-iron bronzed lamps, clocks, candelabra, inkstands, boxes, and other articles.

CLASS 32.

572 BATZ, G. W. *Offenbach*.—Bows for provision-bags, wallets, &c.

CLASS 36.

573 DOLLMANN, A. *Offenbach, and* 51, *Aldermanbury, London*.—Leather purses, cigar-cases, dressing-cases, &c.
580 KALTENHÆUSER, D. & Co. *Schotten*.—Articles in carved wood.
574 KNIPP, J. F. & Co. *Offenbach*.—Cigar-cases, leather purses, pocket-books, albums, dressing-cases, and other articles.
575 POSEN, E. & Co. *Offenbach*.—Ornaments, cigar-cases, leather purses, boxes, &c.
576 SPIER, P. *Offenbach*.—Cigar-cases, leather purses, bags, &c.
577 STAUTZ & Co. *Offenbach*.—Pocket-books, cigar-cases, purses, bracelets in glass, pearls.
578 STEINHART & GÜNZBURG, *Offenbach*.—Wallets, game-bags, &c.
579 WEINTRAUD, C. JUN. *Offenbach*.—Cigar-cases, leather purses, pocket-books, dressing-cases, work-boxes, &c.

PRINCIPALITY OF LIPPE.

CLASS 2.

591 SCHIERENBERG, G. A. B. *Meinberg*.—White lead.

CLASS 3.

592 WIPPERMANN, F. W. *Lemgo*.—Vinegar.

CLASS 4.

593 HOFFMANN, F. *Salzuflen*.—Wheat starch.

CLASS 19.

594 KOTZENBERG, F. L. *Salzuflen*.—Linen thread. bleached and dyed.

CLASS 36.

595 REHTMEYER, G. *Lemgo*.—Meerschaum bowls, with silver mounting.

GRAND DUCHY OF LUXEMBURG.

CLASS 1.

601 GODDIN, D. (Director of the Stolzenburg Mining Co.)—Copper ore and copper.
602 GÖSDORF CO. FOR THE WORKING OF THE ANTIMONY PITS.—Ore and metallic antimony.
603 DE LA FONTAINE, E. *Remich*.—Alabaster, lime-stone, paving-stones.

CLASS 18.

606 REMICH WEAVING MILL, *Luxemburg*.—Cotton manufactures.

CLASS 26.

607 CHARLES, A. & Co. *Luxemburg*.—Leather. Gloves.

CLASS 31.

611 NEUEN, G. (NEUEN-THERER), *Luxemburg*.—Iron bedstead.

NASSAU.

CLASS 1.

617 DÖPPENSCHMIDT, FR. *Caup*.—Roofing slate.
619 HAAS, L. & Co. *Dillenburg*.—Nickel ore, and products of the smelting process.
620 HEUSLER, C. L. *Dillenburg*.—Nickel ore, and products of the smelting process.
621 LOSSEN, A. SONS, *Michelbacher Hütte*.—Wrought-iron, showing the resistance to torsion and other forces.
622 SCHÄFER, *Diez*.—Roofing slate.
623 WIESBADEN, THE DUKE'S MINING OFFICERS.—Products of the Nassau mines and smelting works; samples of marble and marble work.

CLASS 2.

624 DIETZE, H. & Co. *Mentz*.—Verdigris, creosote obtained from beech-wood.

CLASS 3.

624A ASCHROTT, H. S. *Hochheim*.—Hock.
625 DIETRICH & EWALD, *Rüdesheim*.—Sparkling wines.
626 DILTHEY, SAHL, & Co. *Rüdesheim*.—Hock.
627 MÜLLER, M. *Eltville*.—Sparkling hock and Moselle wines.
628 SIEGFRIED, G. W. *Rauenthal*.—Hock.

629 WIESBADEN DIRECTORS OF THE DUKE'S DOMAINS, *Wiesbaden.*—Hock; mineral waters.

630 WIESBADEN DIRECTORS OF THE ASSOCIATION OF THE NASSAU FARMERS AND FOREST CULTIVATORS, *Wiesbaden.* — Red wheat, flour, hasty-barley, and oats for sowing.

CLASS 16.

631 SCHELLENBERG, A. *Wiesbaden.*— Citherns.

CLASS 19.

632 WIESBADEN DIRECTORS OF THE ASSOCIATION OF THE NASSAU FARMERS AND FOREST CULTIVATORS, *Wiesbaden.* — Flax from the "Westerwald."

CLASS 21.

633 STIRN, M. SONS, *Oberurstel.*—Artificial wool.

CLASS 28.

634 KREIDEL, C. W. *Wiesbaden.*—Literary works, some illustrated.

GRAND DUCHY OF OLDENBURG.

CLASS 1.

636 LÜERSSEN, B. A. *Delmenhorst.*—Gunflints.

CLASS 3.

637 HOYER & SON, *Oldenburg.*—Stout, beer, and pale ale; chocolate.

CLASS 4.

638 LÜERSSEN, B. A. *Delmenhorst.*—Cork.

639 HOYER & SON, *Oldenburg.*—Stearic acid, stearine candles, and soaps.

CLASS 7.

640 BRADER & Co. *Zwischenahn.*—Bobbins.

CLASS 11.

641 WIECHMANN, *Oldenburg.* — Musket, with empty cartridge.

CLASS 17.

642 KIRCHMEYER, PH. *Ovelgönne.* — Trusses.

CLASS 26.

643 KIRCHMEYER, PH. *Ovelgönne.*—Saddle and saddle-tree.

CLASS 36.

645 CASSEBOHM, H. S. *Oldenburg.*—Phelloplastic work.

PRUSSIA, KINGDOM OF.

CLASS 1.

651 ACTIEN-GESELLSCHAFT FOR MINING; LEAD AND ZINC FABRICATION, *Stolberg, and in Westphalia.*—Ores and metallurgical products.

652 ACTIEN-GESELLSCHAFT FOR MINING AND SMELTING, *Stolberg and Nordhausen.*—Raw and dressed copper ores.

652A ALUM-WORKS (General-director Rieloff), *Muskau, Silesia.*—Products of alum and vitriol.

653 ALTENBEKER HÜTTENWERK.—Ironstone.

654 AMALIA COLLIERY, *near Minden.*—Coal and ironstone.

655 AM SCHWABEN COLLIERY, *near Dortmund.*—Fire-clay.

656 ANANIAS IRON MINE, *near Prussian Oldendorf.*—Ironstone.

657 ÄTZROTT & Co. *Cologne.*—Table-top of roofing-slate.

658 BACHEM & Co. and SPINDLER & Co. *Königswinter, near Bonn.* — Trachyt and basaltic-lava.

659 BAUTENBERGER EINIGKEIT IRON MINE, *near Unterwilden.*—Ironstone.

660 BELGISCH - RHEINISCHE GESELLSCHAFT FOR MINING, UPON RUHR, *Düsseldorf.*—Coal.

661 BERGER & Co. *Witten.*—Coke.

662 BERG - INSPECTION, ROYAL, *Ibbenbüren.*—Coal and specimens of intermediate rocks.

663 BERG-INSPECTION, ROYAL, *Königshütte.*—Coal and clay ironstone.

664 BERG-INSPECTION, ROYAL, *Tarnowitz.*—Raw and dressed lead ores.

665 BERG-INSPECTION, ROYAL, *Wettin.*—Coal and coke.

666 BERG-INSPECTION, ROYAL, *Zabrze.*—Coal.

667 BERG UND SALINEN - INSPECTION, ROYAL, *Stassfurt, near Magdeburg.*—Rock-salt, compounds of potassium, and minerals.

668 BERGISCHER GRUBEN UND HÜTTEN-VEREIN, *Hochdahl.*—Common pig-iron; pig-iron for steel making; lead ore.

669 BESSELICH, N. *Secretary to the Chamber of Commerce, Treves.*—Building-stones from the district of Treves.

670 BLEIBTREU, DR. H. *Director of Mines and Works at Obercassel, near Bonn.*—Lime-coke.

671 BÖCKING, BROS. *Asbacher Hütte, near Kirn.*—Ironstone.

672 BÖRNER, M. *Siegen.*—Ores and products of iron-works.

673 BOLLNBACH IRON & MANGANESE MINE, *near Neunkirchen.*—Iron and manganese ores.

674 BONNER BERGWERKS UND HÜTTEN-VEREIN, *Bonn.*—Iron ores; brown coal, and its products.

675 BOLTZE, *Salzmünde, near Halle.*—Fire and china clay.

676 BONZEL, J. *Mecklinghausen, near Olpe.*—Slabs of marble.

677 BORUSSIA COLLIERY, *near Dortmund.*—Coal and coke.

678 BRAUT COLLIERY, *near Werden.*—Coal.

679 BROICHER EISENSTEIN BERGWERK, *Mühlheim upon Ruhr.*—Ironstone.

680 BREWER, XAVER VON, *Nieder-Mendig, Coblenz.*—Volcanic-tuff, roofing-slate.

682 BUDERUS, G. W. SONS, *Germania, near Neuwied.*—Tin-plates.

683 BUDERUS, I. W. *Wetzlar.*—Ironstone.

684 BURGHOLDINGHAUSER HÜTTENWERK, *near Siegen.*—Ironstone and products of blast furnaces.

685 CAMP & Co. *Wetzlar.*—Ironstone.

686 CAPITAIN, TH. *Obercassel, near Siegen.*—Fine and coarse-grained basalt in flags, from Obercassel.

687 CAPPELN.—Jura sandstone.

689 CAROLINE ERBSTOLLEN COLLIERY, *near Unna.*—Coal.

690 CHARBON, P. E. *Friedrichsglück, near Lintorf.*—Iron pyrites and lead ore.

691 CHARLOTTE COLLIERY, *near Hattingen.* — Coal and specimens of intermediate rocks.

692 CHRISTIAN GOTTFRIED COLLIERY, *near Tannhausen.*—Anthracite.

693 CHURFÜRST ERNST LEAD MINE, *near Bönkhausen.*—Lead ores.

694 CÖLN MÜSENER BERGWERKS VEREIN, *Müsen, near Siegen.*—Iron, lead, copper, and silver ores, and smelting products.

695 CONCORDIA CO. OF ESCHWEILER, FOR MINING AND SMELTING, *Ichenberg, near Eschweiler.*—Iron ores and blast furnace products.

696 CONRADINE IRONSTONE MINE, *near Porta, Minden.*—Ironstone.

697 CONSTANTIN DER GROSSE COLLIERY, *near Bochum.*—Coal and coke.

698 DAHLBUSCH COLLIERY, *near Gelsenkirchen.*—Coal.

699 DEUTSCH-HOLLÄNDISHER ACTIEN-VEREIN FOR SMELTING AND MINING, *Johannishütte, near Duisburg.* — Coal, ironstone, blast furnace and rolling-mill products.

700 DIRECTION ROYAL OF MINES, *Saarbrücken.* — Collection of coal, coke, maps, and models of coal veins.

701 DOINET, *Zülpich.* — Iron and manganese ores, brown coal and products.

701A DOBSCHUETZ, V., *at Friedersdorf, near Greiffenberg, Silesia.* — Brown coal of colliery Heinrich, near Langenöls.

702 DORSTFELD COLLIERY, *near Dortmund.*—Coal.

703 DRECKBANK COLLIERY, *near Werden.*—Ironstone.

704 DRESLER, I. H. SEN. *Siegen.* — Iron, lead, zinc, copper, nickel ore, blast furnace, and rolling-mill products.

705 ECKARDT, GILES, *Gefell, District of Ziegenrück.*—Washed ochre.

707 EINIGKEIT COLLIERY, *near Steele.*—Fire clay, ironstone.

708 EISENZECHE IRON MINE, *near Siegen.*—Ironstone and manganese ore.

709 EMONDS, H. J. *Bergisch, Gladbach.*—Brown coal.

710 ENDEMANN, WILHELM, & Co., *Bochum*—Coke.

711 ESCHWEILER BERGWERKS VEREIN *Eschweiler.*—Coal, coke, and maps.

712 ESCHWEILER GESELLSCHAFT FOR MINING AND SMELTING, *Stolberg, near Aix-la-Chapelle.*—Lead, zinc, and silver ores; smelting products.

713 (EUNICKE, L.) LEOPOLD KAYSER, & Co. *Naumburg-upon-Bober.* — Cubes of nickel.

714 FANNY COLLIERY, *near Kattowitz.*—Calcined rocks of the coal measures.

715 FELTHAUSS, H. *Wetzlar.*—Ironstone and quicksilver ores.

716 FINA COLLIERY, *near Mettmann.*—Fire clay, ironstone.

717 FLEITMANN, DR. *Iserlohn.*—Metallic cobalt and nickel.

718 FRAUENBERGER EINIGKEIT IRON MINE, *near Neunkirchen.*—Ironstone.

719 FRIEDERIKE COLLIERY, *near Bochum*—Ironstone.

720 FRIEDRICH WILHELM IRON MINE, *near Siegen.*—Ironstone.

721 FRIEDRICH WILHELM HÜTTE (Joint Stock Co. for mining at Mühlheim-upon-Ruhr.)—Pig iron and foundry objects.

722 FRIEDRICH WILHELM HÜTTE, *near Siegburg.*—Iron ores, products of blast furnace, and rolling-mill; waggons.

723 FUCHS COLLIERY, *near Waldenburg.*—Coal and specimens of intermediate rocks.

724 GENERAL No. 1. TIEFBAU COLLIERY, *near Dahlhausen.*—Coal.

725 GERLACH, GABRIEL & BERGENTHAL, *Germaniahütte, near Grevenbrück - upon - Lenne.* — Ironstone, and products of blast furnaces.

726 GIBRALTAR ERBSTOLLEN COLLIERY, *near Dahlhausen.*—Coal.

727 GLÜCKAUF-TIEFBAU COLLIERY, *near Brüninghausen.*—Coal.

728 GLÜCKHILF COLLIERY, *near Waldenburg.*—Coal.

729 GOTTESSEGEN CONCESSION ON BLEIBERG, *near Commern.*—Knobby copper ore.

730 GRAF BEUST COLLIERY, *near Essen.*—Coal.

731 GREVEL IRON MINE, *near Unna.*—Ironstone.

732 GRÜNEBERGER CONSOLIDIRTE BRAUNKOHLENGRUBEN, *Grüneberg.*—Brown coal of different sizes.

733 GÜTTLER, *Reichenstein.*—Products of arsenic, and gold works.

734 GUTER TRAUGOTT COLLIERY, *near Myslowitz.*—Coal.

735 HÄUSLINGS-TIEFE IRON MINE, *near Siegen.*—Ironstone.

736 HARPEN, MINING SOCIETY OF, *Dortmund.*—Coal.

737 HASENWINKEL COLLIERY, *near Hilgenstock.*—Specimens of intermediate rocks.

738 HECTOR IRONSTONE MINE, *near Ibbenbüren.*—Ironstone.

739 HEINRICH GUSTAV COLLIERY, *near Bochum.*—Coke.

740 HEINRICHSSEGEN LEAD MINE, *near Müsen*.—Grey copper ore.

741 HELENE AND AMALIE COLLIERY, *near Essen*.—Coal.

742 HELLE IRON MINE, *near Sundwig*.—Ironstone.

743 HENKEL, HUGO, EARL OF DONNERSMARK, *Siemanowitz*.—Coal, ironstone, products of zinc and iron works.

744 HENRICHSHÜTTE, ADMINISTRATION OF THE, *near Hattingen*.—Pig-iron, rolling-mill products, and lead ore.

745 HIEPE, S. *Wetzlar*.—Specimens of marble.

746 HÖRDER BERGWERKS UND HÜTTEN VEREIN, *Hörde*.—Coal, specimens of intermediate rocks.

747 HOLLAND COLLIERY, *near Bochum*.—Coal.

748 HOLLERTERZUG IRON MINE, *near Dermbach*.—Manganese ores.

749 HORHAUSEN MINE, *near Hamm-upon-Sieg*.—Ironstone.

750 HOYER, GUSTAV, & CO. *Schönebeck*.—Salt for cattle, and mangers for it.

751 HÜBNER, DR. BERNARD (Manufactory of mineral oil and paraffin), *Rehmsdorf, near Zeitz*.—Brown coal and products.

752 HÜTTENAMT, ROYAL, *Creutzburgerhütte, Silesia*.—Ironstone, and products of iron works.

754 HÜTTENAMT, ROYAL, *Königshütte, Silesia*.—Calamine, iron ores, and products of zinc and iron works.

755 HÜTTENAMT, ROYAL, *Malapane*.—Iron ores, and products of iron works.

756 HÜTTENAMT, ROYAL, *Messingwerk, near Neustadt-Eberswalde*.—Products of yellow and common copper.

757 HÜTTENAMT, ROYAL, *Rybnik*.—Rolled and hammered iron.

757A HÜTTEN INSPECTION, ROYAL, *Friedrichshütte, near Tarnowitz, Silesia*.—Lead ores, lead and silver products.

758 HUYSSEN, DR. *Royal Director, Breslau*.—Maps of the coal measures of Lower Silesia.

759 JACOBY, HANIEL, HUYSSEN, & CO. *Oberhausen*.—Coal, coke, ores, products of blast furnace, and rolling-mill.

760 IBBENBÜREN.—Coal-sandstone.

761 JUNG, C. *Dahl, near Hagen*.—Steel bars.

762 JUNGE SINTERZECHE IRON MINE, *near Siegen*.—Ironstone.

763 KÄRGER, C. H. L. *Willmannsdorf, near Jauer*.—Red iron ore.

764 KARL GEORG VICTOR COLLIERY, *near Alt-Lässig*.—Coal.

765 KELLER, KÜPPERS, RAABE, & CO. *Aix-la-Chapelle*.—Sandstone and crystal-sand.

766 KLOSE, *Kupferberg*.—Copper ores, products of copper-smelting works.

767 KOLLMANN, CLOTTEN, *Coblenz*.—Roofing-slate from Müllenbach.

767A KRAMSTA, *Freyburg, Silesia*.—Coal, iron-stone, zinc plates.

768 KRAUSE, *Berlin*.—Iron ore and products of Silesian iron works.

769 KRAUSS, *Bensberg*.—Lead, silver, and zinc ores.

770 KREUTZ, JAMES, *Siegen*.—Ironstone and products of blast furnaces.

770A KRUPP, FR. *Essen, Düsseldorf*.—Samples of steel manufacture.

771 KULMITZ, *Saarau, Liegnitz*.—Granite from the "Streitberg," near Striegau, Silesia. Raw and dressed coal, coke, and brown coal.

772 LAMARCHE CO. *Velbert*.—Alum-shale, iron-stone.

773 LANDAU, S. *Coblenz and Andernach-upon-Rhine*.—Lava millstones and trass.

774 LANDESKRONE LEAD MINE, *near Siegen*.—Lead ores.

775 LANDSBERG-VELEN, EARL, *Velen, near Münster*.—Iron pyrites and iron ores.

777 LAURA COLLIERY, *near Minden*.—Coal.

778 LENNE-RUHR JOINT STOCK CO. FOR MINING AND SMELTING, *near Altenhundem*.—Iron ores, iron pyrites, iron and steel products.

779 LIMBURGER FABRIK & HÜTTEN-VEREIN, *Limburg-upon-Lenne*.—Steel products.

780 LIPOWITZ, A. *Berlin*.—Bricks of coal and coke.

782 LOHMANNSFELD LEAD MINE, *near Alten-Seelbach*.—Lead ores.

783 LOMMERSDORF IRON MINE, *near Adenau*.—Ironstones.

783A LUESCHWITZ, *Administration of Mines and Smelting Works, Breslau*.—Ores and products of arsenic of Bergmamstrost, near Altenberg.

784 MÄRKISCH WESTPHÄLISCHER BERGWERKS VEREIN, *Iserlohn*.—Zinc and lead ores, and metallurgical products.

785 MAIKAMMER IRON MINE, *near Wülfrath*.—Ironstone.

786 MANSFELDSCHE GEWERKSCHAFT, CO. FOR COPPER-SLATE MINING, *Eisleben*.—Brown coal, copper-slate, and products of smelting-works. Maps.

787 MARIA COLLIERY, *near Höngen, in the district of Worm*.—Coal, coke. Maps.

788 MARIE LOUISE COLLIERY, *near Dahlhausen*.—Ironstone.

789 MECHERNICHER BERGWERKS ACTIEN-VEREIN, *Mechernich in Rhenish Prussia*.—Raw and separated lead ores and products.

790 MECKLINGBÄNKER ERBSTOLLN COLLIERY, *near Steele*.—Ironstone.

791 MERKELL, *Halle-on-Saale*.—Mottled sandstone from Nebra, near Querfurt.

792 MEURER, W. *Cologne*.—Ores and smelting products.

793 MINERVA JOINT-STOCK CO. *Breslau*.—Products of the iron and steel works of Minerva.

794 MORGENSTERN VITRIOL WORKS, *near Landeshut*.—Iron pyrites, products of sulphur, vitriol, and alum works.

795 MÜHLMANN, ALBERT V., *Plato, near Siegburg*.—Brown coal, clay, ironstone.

796 MÜLLENSIEFEN, G. *Crengeldanz, near Witten*.—Coal.

797 NEU-ALTSTÄDDE III. IRON MINE, *near Ibbenbüren*.—Ironstone.

798 NEU-DÜSSELDORF COLLIERY, *near Dortmund*.—Coal and stones from the coal measures.

799 NEUMANN & ZIMMERMANN, *Lindlar, near Wipperfürth, Rhenish Prussia.* — Sandstone.

800 NEU-SCHOTTLAND JOINT-STOCK CO. FOR MINING AND SMELTING, *Steele.* — Pig-iron.

801 NIEDERFISCHBACHER GESELLSCHAFT FOR MINING AND SMELTING, *Niederfischbach, district Siegen.*—Ores, products of separating and smelting works.

802 H.R.H. PRINCESS MARIANNE OF THE NETHERLANDS.—Marble from Seitenberg, near Habelschwerdt, Silesia.

803 NIEMANN, *Beverungen, Münster.* — Sandstone.

804 NORDSTERN IRONSTONE MINE, *near Siegen.*—Ironstone.

805 OBERBERGAMT, ROYAL, *Bonn.* — Geological map of Rhenish Prussia and West phalia.

806 OBERBERGAMT, ROYAL, *Breslau.*— Geological maps, and maps of the coal seams. Specimens of rock and ore.

807 OBERBERGAMT, ROYAL, *Dortmund.*— Maps of the Westphalian coal districts.

808 OBERBERGAMT, ROYAL, *Halle.* — Geological maps and sections.

810 OSTERMANN, A. & CO. *Bochum.*— Coke.

811 PHILIPPSHOFFNUNG, COBALT MINE, *near Siegen.*—Cobalt ores.

812 PHŒNIX JOINT-STOCK CO. FOR MINING AND SMELTING, *Laar, near Ruhrort.*— Coals, iron-ores, blast furnace, rolling-mill products.

813 PÖRTINGSSIEPEN COLLIERY, *near Werden.*—Specimens of intermediate rocks.

814 PORTA WESTPHALICA HÜTTENGESELLSCHAFT, *near Porta.*—Ironstone.

815 PRANG, J. B. *Münster, Alme, near Brilon.*—Slabs of marble.

816 PRIMAVESI & CO. *Gravenhorst.*—Ironstone.

817 PRINZ WILHELM MINE, *near Langenberg.*—Raw and dressed lead and zinc ores.

818 RAAB, LEWIS, *Wetzlar.*—Ironstone.

819 REDENSHÜTTE, *near Zabrze.* — Products of iron-works.

820 REHM, H. *Wetzlar.*—Pieces of marble.

821 REICHER TROST MINE, *near Reichenstein.*—Arsenic ores, and products of arsenic manufacture.

822 REMY, BROS. *Wendener Hütte, district Olpe.* — Ironstones and products of blast-furnaces.

823 RHONARD MINE, *near Siegen.*—Copper pyrites.

824 RITTER, F. *Bochum.*—Coke.

825 RÖMEL IRON MINE, *near Neunkirchen.*—Ironstone.

826 ROLAND COLLIERY, *near Mühlheim-upon-Ruhr.* — Coal, and specimens of intermediate rocks.

827 RUFFER, *Breslau.*—Products of iron and zinc rolling-mills.

828 RUGE, *Wildschütz, near Hohenmölsen.*—Brown coal and its products.

829 RUMPE, J. C. & W. *Altena.*—Products of antimony smelting works.

830 RUNGE, *Breslau.*—Geological sections through the giant-mountains and the basin of Waldenburg-Nachod.

831 SÄCHSISCH THÜRINGISCHE ACTIEN GESELLSCHAFT, *Halle.* — Products of brown coal.

832 SACK, *Sprockhövel.*—Coal, ironstone, and coke.

833 SÄLZER & NEUACK COLLIERY, *near Essen.*—Coal.

834 SALZAMT, ROYAL, *Schönebeck.*—Common salt.

835 SATURN RHENISH JOINT STOCK MINING CO. *Cologne.*—Lead, copper, zinc ores of the mines Blücher and Katzbach.

836 SCHEUR IRON MINE, *near Siegen.*— Copper ores.

837 SCHMIEDEBERG IRON MINE, *near Siegen.*—Ironstone.

838 SCHNEIDER, H. D. F. *Neunkirchen, near Betzdorf.* — Ironstone and products of blast-furnaces.

839 SCHRAN, FRANCIS, *Gleidorf, near Oberkirchen.*—Ironstone and products of roofing-slate of the Slate Co. at Fredeburg.

840 SCHRECKENDORFER - HÜTTE, *near Landeck.* — Iron ores and products of iron works.

841 SCHRUFF, HERBST, & EISLEB, *Call, near Commern.* — Separated ores and lead-smelting products.

842 SCHULTEN, F. *Metal Foundry at Duisburg.*—Metal blast-pipes.

843 SCHULTZ & WEHRENBOLD, *Justushütte, near Gladenbach.*—Ironstone.

843A SCHWIDTHAL, mining director, *Bromberg.*—Pressed brown coal.

844 SHAMROCK COLLIERY, *near Bochum.* —Coal and coke.

845 SIEG-BERGWERKSGESELLSCHAFT, *Cologne.*—Crystallized galena.

846 SILBERART LEAD AND ZINC MINE, *near Altenseelbach.*—Blende.

847 SOCIÉTÉ DES MINES DE L'EIFEL, *Heistert, near Call.*—Knobby lead ores.

848 SOLMS MINING ADMINISTRATION OF THE PRINCE SOLMS-BRAUNFELS, *Braunfels, near Wetzlar.*—Iron ores.

849 STADTBERGER GEWERKSCHAFT, *Altena.*—Copper ores, products of the copper-works; vitriols.

850 STEINHAUSER HÜTTE, *Witten.*—Products of rolling-mill, and stamped waggon-mountings, &c.

851 STEINHEUER, C. H. *Lindlar, near Wipperfürth.*—Sandstone.

852 STINNES, G. *Mühlheim-upon-Ruhr.*— Screened coal.

853 ST. JOSEPHSBERG COPPER MINE, *near Linz.*—Copper ores.

854 STÖCKLEIN, *Sangerhausen.* — Brown coal-sand-stone from quarries near Sangerhausen.

855 STOLBERG-WERNIGERODE, COUNT OF, *Ilsenburg.*—Rolled wire, card wire.

856 STORCH & SCHÖNEBERG IRON MINE, *near Siegen.*—Ironstone.

857 STROTHE IRON MINE, *near Tecklenburg.*—Ironstone.

858 STUMM, BROS. *Neunkirchen.*—Iron and manganese ores

859 TEUTONIA HÜTTENGEWERKSCHAFT, *near Willebadessen.*—Ironstone.

860 THIELE-WINKLER, v., *Kattowitz.* — Iron ores, coals, and specimens of intermediate rocks; products of iron works.

861 THISQUEN, PAUL VAL. *Neu-Hattlich, near Montjoie.*—Peat-charcoal.

862 TREMONIA COLLIERY, *near Dortmund.* —Coal.

863 UJEST, DUKE OF, *Hohenlohehütte.*— Clay ironstone.

864 ULRICH, T. *Bredelar.*—Ironstone and products of blast furnace.

865 USLAR, R. Manager of the Co. W. GESSNER & Co. *Nuttlar.*—Roofing slate.

866 VEREINIGTE PÖRTINGSSIEPEN COLLIERY, *near Werden.*—Coal.

867 VEREINIGTE HANNIBAL COLLIERY, *near Bochum.*—Coal.

868 VEREINIGTE PRÄSIDENT COLLIERY, *near Bochum.*—Coal and coke.

869 VEREINIGTE SELLERBECK COLLIERY, *near Mühlheim-upon-Ruhr.*—Coal.

870 VEREINIGTE WIESCHE COLLIERY, *near Mühlheim-upon-Ruhr.*—Coal bricks.

871 VEREINIGUNGS - GESELLSCHAFT FOR COAL-MINING IN THE DISTRICT OF WORM, *Kohlscheid, near Aix-la-Chapelle.*—Coal, coke. Maps.

872 VICTORIA MATHIAS COLLIERY, *near Essen.*—Coal and coke.

873 VICTORSFELD LEAD MINE, *near Lippe.* —Galena.

874 VOLLMOND COLLIERY, *near Bochum.* —Coal and coke.

875 VORWÄRTS-HÜTTE, *near Waldenburg.* —Magnetic-iron ore of Schmiedeberg.

877 WERSCHEN - WEISSENFELS ACTIEN GESELLSCHAFT, *Weissenfels.* — Brown coal, and its products.

878 WERTHER COLLIERY, *near Halle, Westphalia.* — Specimens of intermediate rocks.

879 WETZLAR STOCK-CO. *Wetzlar.*—Ironstone.

880 WIESEHAHN, GILES, *Dortmund.*— Coke.

881 WIESMANN, A. & Co. *Bonn.*—Brown coal, tar oil; candles.

882 WILDBERG LEAD AND IRON MINE, *near Siegen.*—Ores of lead and copper.

883 WILDE WIESE MINE, *near Arnsberg.* —Ironstone.

884 WILHELMINE VICTORIA COLLIERY, *near Gelsenkirchen.*—Coal.

885 WINKLER, E. Representative of the JOINT-STOCK CO. OF ALTONRATH, *Overath.* —Copper ores.

886 WINTER. G. & Co. Proprietor of the WEIDENSTAMM MINE, *near Braunfels, Frankfort-upon-Main.*—Black oxide of manganese.

887 WOLFF, *Gollenfels, Coblenz.*—Marble from Stromberg.

888 ZOLLVEREIN COLLIERY, *near Essen.*— Coal and specimens of intermediate rocks.

CLASS 2.

951 AMENDE, R. 146-7, *Müller Berlin.* —Dung-powder, crushed bones, crushed horn, phosphate of lime, blood powder.

952 ANDRÆ & GRÜNEBERG, *Stettin.* — Chemical products.

953 BEHREND, G. *Hirschberg.* — Vermilion, sesqui-oxide of manganese, gum copal, varnish-blacking, sealing-wax.

954 BERINGER, A. *Charlottenburg, near Berlin.* — Samples of colours, stained paper, and paper-hangings.

955 BODEN, TH. *Cologne.*— Spirit of vinegar, sugar of lead.

957 BREDT, O. *Barmen.*—Colours.

958 BÜHRING, C. 100, *Wilhelm-str. Berlin.* —Crystal-water.

960 COHN, DR. W. *Moabit, near Berlin.*— Ground bones and bone-cake.

962 CUNTZE, E. *Cologne.*—Eau de Cologne.

963 CURTIUS, I. *Duisburg, Düsseldorf.*— Ultramarine.

966 ENGEL & v. SCHAPER, 17, *Mohren-str. Berlin.*—Chemical products.

967 FARINA, J. M. *Jülicher-place, Cologne.* —Eau de Cologne.

968 FARINA, J. M. *oppos. Jülicher-place, Cologne.*—Eau de Cologne.

969 FARINA, J. M. 23, *Rhine-str. Cologne.* —Eau de Cologne; lavender water.

970 FARINA, J. M. 54, *oppos. the Altenmarkt, Cologne.*—Eau de Cologne.

971 FARINA, J. M. *oppos. the Neumarkt, Cologne.*—Eau de Cologne.

972 FARINA, F. M. 47, *Glockengasse, Cologne.*—Eau de Cologne.

973 FARINA, J. A. 129, *Hochstrasse, Cologne.*—Eau de Cologne.

974 FRANK, A. *Stassfurt, Magdeburg.*— Rock-salt; rock-salt used as manure; kitchen-salt, and alkalies.

975 GASSEL, RECKMANN, & Co. *Bielefeld.* —Drops.

976 GEISS, F. G. *Aken on Elbe, Magdeburg.* —Oils.

977 GEORG-HÜTTE, *Aschersleben.*—Paraffin and solar oil.

979 GRAHN, WIDOW A. *Halle on Saale.*— Polishing composition; clay.

980 GRASS, M. 2, *Domhof, Cologne.*—Eau de Cologne; eau de melisse.

982 HARTMANN, W. *Cologne.*—Tartar.

983 HERMANN, O. *Schönebeck, Magdeburg.* —Chemical products.

984 HERSTADT, CHR. & Co. 17, *Fettenhennen, Cologne.*—Eau de Cologne.

985 HEYL, R. & Co. *Charlottenburg, near Berlin.*—Colours; chemical products.

986 HIRSCH, G. *Königsberg.*—Animal charcoal, ground bones, gypsum used as manure, ground magnesite, blacking.

988 HOFFMANN & Co. *Müngersdorf, Cöln.* —Ground bones.

990 HUGUENEL, C. *Breslau.*—Vegetable dyes.

991 HUTTER, L. & Co. 51, *Alte Jacob-str. Berlin.* — Liquid perfumes; metallic solutions.

992 JÄGER, C. *Barmen.* — Extract of safflower.

993 KAYSER, JUN. & Co. *Naumburg-on-Queiss.*—Sulphate of copper.

994 KOBER, TH. *Sömmerda, Erfurt.* — Chemical products.

995 KOCH, A. *Zell-on-Mosel.* — Lactucarium.

996 KÜDERLING, H. F. *Duisburg, Düsseldorf.*—Chemical products.

997 KUHNHEIM & Co. 26, *Linden-str. Berlin.*—Chemical products.

998 KULMITZ, C. *Saarau, Breslau.*—Chemical products.

999 LEHMANN, J. R. *Labagienen, Königsberg.*—Ground bones and guano.

1000 LEVERKUS, DR. C. *Wermelskirchen, Düsseldorf.*—Ultramarine.

1002 LINDGENS & SONS, *Mühlheim, Cologne.*—White lead, sugar of lead.

1004 LUCAS, M. *Cunersdorf, Liegnitz.*—Vermilion, oxide of manganese, siccative powder.

1006 MARQUARDT, Dr. L. *Bonn.*—Chemical products.

1007 MARTIN, M. Kl. *Cologne, South portal of the Cathedral.*—Eau de Cologne, eau de Melisse, eau des Carmes.

1008 MATHES & WEBER, *Duisburg, Düsseldorf.*—Chemical products.

1009 NACHTWEYH, C. *Dziewentline, Breslau.*—Grape sugar.

1010 NEUDORF, W. *Königsberg.*—Sulphur soap with an admixture of bromine and iodine.

1013 OSTER, *Königsberg.*—Succinic acid.

1015 POMMERENSDORF, *Stettin.*—Chemical products.

1016 PRAGER, L. *Erfurt.*—Varnish, blacking, pulses, and agricultural seeds.

1017 RATHKE, G. 41, *Alexandrinen-str. Berlin.*—Medicines, pastilles, and gelatinous capsules.

1018 REIFFEN, G. *Saarbrücken.* — Perfumes.

1019 RUFFER & Co. *Breslau.* — White zinc.

1020 SCHEER, SONS, *Treves.* — Artificial mineral waters.

1021 SCHINDLER & MÜTZEL, *Stettin.*—Perfumes.

1022 SCHÖNFELD, DR. FR. *Düsseldorf.*—Painter's colours.

1023 SCHÜR, DR. O. *Stettin.*—1, Mineral waters; 2, Representative of the Stettin "Kraftdünger Fabrik." — Specimen of powerful manure.

1024 SCHUSTER & KÄHLER, *Danzig.*—Artificial mineral waters.

1027 SPENDECK, J. P. & Co. 18, *Gr. Neugasse, Cologne.*—Eau de Cologne.

1028 SPIRO, P. 8, *König-str. Berlin.*—Bronze colour in powder.

1030 SZITTNICK, O. *Königsberg.*—Ground bones.

1031 THURN, WIDOW, F. P. 17, *Fettenhennen, Cologne.*—Eau de Cologne.

1032 VILTER, F. W. 4, *Kl. Waldemar-str. Berlin*—Sulphate of ammonia, phosphate of potash.

1033 VOIGT & HAVELAND, *Breslau.* — Chemical products.

1034 VORSTER & GRÜNEBERG, *Kalk, Cologne.*—Chemical products.

1035 WEISS, J. H. & Co. *Mühlhausen, Erfurt.*—Samples of madder-lac.

1036 WEITZE, DR. C. G. *Stettin.*—Perfumes, mineral waters.

CLASS 3.

1040 APPEL, H. *Schwedt-on-Oder, Uckermark.*—Cigars and tobacco.

1041 AXMANN, R. *Erfurt.*—Wheat vermicelli.

1042 BÄVENROTH, C. FR. *Stettin.*—Spirits.

1043 BARRE, E. *Lübbecke, Minden.* — Wheat-starch.

1044 BAUTE & Co. *Camen, Arnsberg.*—Extract of bitters; smoked ham.

1045 BECKHARD & SONS, *Kreuznach.*—Hock and Haardt wine, sparkling hock and Moselle wine.

1046 BEISERT, A. *Sprottau, Liegnitz.*—Flour, groats.

1047 BIESCHKY, A. *Danzig.*—Bitter cordial.

1048 BLUME, C. 51, *Königs-str. Berlin.*—Honey wine.

1049 BONNE, W. *Rheda, Minden.* — Preserved sausage.

1051 BROMBERG, ROYAL ADMINISTRATION OF THE MILLS AT.—Wheat and rye-flour, groats and bran.

1052 BUDDE, E. *Herford, Minden.*—Mustard.

1053 BÜHL, A. & Co. *Coblenz.*—Sparkling hock and Moselle wine.

1054 CARSTANJEN, A. F. SONS, *Duisburg.* —Tobacco, snuff, and cigars.

1055 CARSTANJEN, C. & W. *Duisburg.*—Tobacco, snuff, and cigars.

1056 CRESPEL & SONS, *Ibbenbüren.* — Wheat-starch.

1057 DELIUS, R. & Co. *Bielefeld.*—Flour.

1058 DITTRICH, H. *Seitendorf, Breslau.*—White wheat.

1059 DOMMERICH & Co. *Magdeburg.*—Chicory.

1060 DREBS, *Ottomin, Danzig.*—Flour and groats (semoule).

1061 DRUCKER, J. *Coblenz.* — Rhenish brandy.

1062 EISENMANN, R. 23, *Alexander-str. Berlin.*—Potato spirit.

1063 ENGELBRECHT & VEERHOF, *Herford, Minden.*—Wheat-starch.

1064 ERMELER, W. & Co. 11, *Breite-str. Berlin.*—Cigars, tobacco, snuff, and tobacco for chewing.

1065 FASQUEL, D. 119, *Linden-str. Berlin.* —Arrack, Cognac, rum, essence of absinthe, and cumin oil.

1067 FIER, JOS. *Zell-on-Moselle.*—Moselle wine, 1857, 1858, and 1859.

1069 FLATAU, J. J. *Berlin and Neutomysl, Posen.*—Hops.

1070 FÖRSTER, FR. JUN. *Grünberg, Liegnitz.*—Dried and preserved fruits.

1071 FÖRSTER & GREMPLER, *Grünberg, Liegnitz.*—Wine.

1073 FRANKE, J. H. *Magdeburg.*—Liqueurs.

1074 FRENZEL, E. *Baugsch - Korallen, Königsberg.*—Grey pease, beans.

1076 GERTEN, J. H. alias BRÜGGEMANN, *Obrighoven, Wesel.*—Corn, fruits, and jellies.

1077 GEYGER, AD. & CO. *Kreuznach.*—Sparkling hock and Moselle wine.

1078 GILKA, J. A. 9, *Schützen-str. Berlin.*—Spirit, rum, French brandy (Cognac), and liqueurs.

1079 GRAFF, H. *Janischken, Königsberg.*—Wheat, pease.

1080 GRASHOFF, M. *Quedlinburg, Magdeburg.*—Seeds, vegetables, and flowers.

1081 GREVE-STERNBERG, PH. *Cologne.*—Bitter cordials.

1082 GRONOW, M. E. v., *Kalinowitz, Oppeln.*—Rye, barley, esparcet seed.

1083 GRÜNEBERG, J. H. 56, *Oranienburger-str. Berlin.*—Preserved meat, broth, fruits, vegetables, craw-fish tails.

1084 GURADZE, A. *Tost Peiskretscham, Oppeln.*—Vermicelli, &c.

1085 HÄUSLER, C. S. *Hirschberg.*—Cider and champagne.

1086 HALLE-ON-SAALE ASSOCIATION OF THE BEET-ROOT SUGAR MANUFACTURERS OF THE ZOLLVEREIN.—Samples of raw and refined sugar.

1087 HANDEL, FR. CHR. v. & CO. *Treves.*—Wines.

1088 HARTMANN, F. A. *Münster.*—Vinegar.

1089 HEERS, F. W. *Telgte, Münster.*—Bitter cordial.

1090 HEES, W. VAN, *Cleve.*—Brandy made of corn.

1092 HÜRTER, H. & SONS, *Coblenz.*—Sparkling wines.

1093 JTZENPLITZ, COUNT, *Kunersdorf, near Wrietzen-on-Oder.*—Beet-root, raw sugar, lump and moist sugar.

1094 JODOCIUS, BROS. & CO. *Coblenz.*—Sparkling hock and Moselle wine.

1095 JOSSMANN, J. 1, *Alexander-str. Berlin.*—Seeds.

1096 KANTOROWICZ, E. *Posen.*—Rye-brandy.

1097 KEILER, J. S. *Danzig.*—Liqueurs.

1099 KÖBKE & BERGENER, *Madgeburg.*—Liqueurs.

1100 KRAFORST, P. J. *Leichlingen, Düsseldorf.*—Grass and clover seed.

1101 KRAUSE, C. H. *Langensalza, Erfurt.*—Saveloy.

1102 KREUZBERG, M. J. *Ahrweiler.*—Ahr wine.

1103 KREUZBERG, P. JOS. & CO. *Ahrweiler.*—Ahr wine.

1104 KULMITZ, C. 39, *Adalbert-str. Berlin.*—Loaves of sugar.

1105 KULMITZ, C. *Saarau, Breslau.*—Refined sugar and sugar candy.

1106 KRUSE, A. T. *Stralsund.*—Wheat-starch and hair powder.

1108 LANDER & KRUGMANN, *Beuel, Bonn.*—Wheat and potato-starch.

1109 LANGGUTH & KAYSER, *Trarbach.*—Moselle and sparkling wines.

1110 LAUE, H. *Wehlau, Königsberg.*—Wheat flour, rape, and linseed cakes.

1111 LEHMANN, R. *Nitsche, Posen.*—Samples of rye and barley.

1112 LEHMANN, J. C. *Potsdam.*—Preserved fruits.

1114 LUDORFF, FR. *Geist, Münster.*—Roasted acorns, as substitute for coffee, chicory.

1116 MEYER, H. G. *Herford, Minden.*—Vinegar, mustard.

1117 MINDEN-RAVENSBERG STEAM-MILL, *Bielefeld.*—Flour.

1118 MOLLARD, E. O. PH. *Gora.*—Preserved green pease; partridges and snipe.

1119 NERNST, H. *Taplacken, Königsberg.*—Tobacco and cigars.

1121 NIESSEN, A. VON, *Danzig.*—Liqueurs.

1121A OHLSEN-BAGGE, *Frankfort-on-Oder.*—Purified potato spirit.

1122 POPPELSDORF, *Royal Agricultural Academy, Bonn.*—Wheat, hops.

1122A PRITZKOW, H. 28, *König-str. Berlin.*—Liqueurs and extracts.

1123 PROCHNOW, J. G. *Rügenwalde, Cöslin.*—Smoke-dried geese.

1123A RACZYNSKI, COUNT VON, *Woÿnowitz, Posen.*—Hops.

1124 RADICKE, A. & CO. *Grünberg, Liegnitz.*—French brandy (Cognac).

1125 RATH, H. VOM, *Lauerfort, Crefeld.*—Samples of flour.

1126 RAUSSENDORF, H. 28, *Neue Jacob-str. Berlin.*—Rum, arrack, and essence of French brandy (cognac).

1126A REIFFEN, G. *Saarbrüken.*—Liqueurs.

1127 RÖDER, J. A. *Cologne.*—Punch syrup; liqueurs.

1128 RUNGE, DR. F. F. *Oranienburg, near Berlin.*—Dried vegetables.

1129 SACHS & HOCHHEIMER, *Coblenz.*—Sparkling hock and Moselle wine.

1130 SCHERR, J. SONS, *Treves.*—Liqueurs.

1131 SCHIFFER, C. *Düsseldorf.*—Essence of punch.

1134 SCHMELLITSCHEK, W. *Wesel.*—Chocolate.

1135 SCHMIDT, A. *Labes, Stettin.*—Compressed vegetables.

1136 SCHMITZ, J. A. *Hübsch, near Wesel.*—Fruit and beet-root jelly.

1137 SCHÖLLER, L. *Schwieben, Oppeln.*—Wheat, barley, esparcet seed.

1138 SCHULTZE, J. C. 82, *Charlotten-str. Berlin.*—Home-made French brandy (cognac).

1140 STEIFF, G. F. A. *Danzig.*—Stout, or double beer; liqueurs, and brandy.

1142 STETTIN, POMMERANIAN PROVINCIAL SUGAR REFINERY.—Loaf-sugar.

1143 STETTIN, STEAM-MILL JOINT STOCK CO.—Products of the corn-mill.

1145 UHLENDORF, L. *Hamm, Arnsberg.*—Vermicelli, and other pastes.

1146 UNDERBERG-ALBRECHT, H. *Rheinberg, Düsseldorf.*—Bitter cordial, vinegar, essence of punch, fruit-jellies.

1148 VISSER, BROS. *Cologne.*—Wheat starch.

1149 VÖLCKER & RICKMANN, *Schwedt-on-Oder, Uckermark.*—Cigars and tobacco.

1150 WAGNER, *Proskau, Oppeln.*—Hops.

1151 WED-LING, WIDOW J. *Danzig.* — Liqueurs and brandy.

1152 WELLMANN, A. *Dittersbach, near Stettin.*—Flour and starch.

1156 WETTENDORF, J. W. *Treves.*—Hops.

1157 WIGANCKOW, O. 93, *Bad-str. Berlin.* —Cigars.

1158 WILKE, F. 16, *Mohren-str. Berlin.*— Spirits and liqueurs.

1160 WITTEN - ON - RUHR STEAM - MILL JOINT STOCK CO. — Vermicelli, and other pastes.

CLASS 4.

1161 ADENAU-ON-MOSEL AGRICULTURAL ASSOCIATION.—Oak bark.

1163 BENNECKE & HEROLD, 43, *König-str. Berlin.*—Specimens of lac and varnish.

1164 BLANKE, C. W. J. *Magdeburg.*— Caoutchouc and gutta percha articles.

1167 BRÜNNECK, S. v., *Bellschwitz, Marienwerder.*—Fleeces.

1168 BÜRGERS, W. J. *Cologne.*—Glue.

1169 BURRMEISTER, H. *Stettin.*—Rape-seed oil.

1170 DEHMEL, H. *Quaritz, Liegnitz.*— Soap.

1171 DOUGLAS, *Amalienau, Königsberg.*— Fleeces.

1172 DYHRENFURTH, L. *Jacobsdorf, Breslau.*—Fleeces.

1173 ELDENA CENTRAL AGRICULTURAL ASSOCIATION, *Stralsund.*—Specimens of wool and cereals.

1174 ELS, C. VAN & CO. *Düsseldorf.*—Oil for greasing machinery.

1175 ENGEL & SCHAPER, VON, 17, *Mohren-str. Berlin.*—Scented soaps.

1177 FÖRSTER, SEN. *Halbmeilmühle, near Grünberg, Liegnitz.*—Oil and oil-cake.

1178 FONROBERT & REIMANN, 17, *Tempelhofer Ufer, Berlin.* — Caoutchouc and gutta percha articles.

1180 FRIEDENTHAL, C. PH. *Domslau, Breslau.*—Teazles.

1181 GAILLARD, C. F. 51, *Friedrich-str. Berlin.*—Samples of perfumes and soaps.

1183 GIESLER, N. H. *Tröchtelborn, Erfurt.* —Woad, in balls.

1184 GÖBEL, J. *Siegen, Arnsberg.*—Glue.

1185 GOLDFUS, A. VON, *Niclasdorf, Breslau.*—Fleeces.

1186 GRONOW, M. E. VON, *Kalinowitz, Oppeln.*—Fleeces of the Escurial breed.

1187 GROSSMANN, C. E. 38A, *Alexander-str. Berlin.*—Scented soap and perfumes.

1191 GURADZE, S. *Kottlischwitz, Oppeln.*— Fleeces.

1192 GURADZE, A. *Tost, Peiskretscham, Oppeln.*—Fleeces.

1194 HALLER, J. *Halle-on-Saale.*—Starch.

1196 HERZ, S. 1, *Dorotheen-str. Berlin.* —Rape, linseed, dottercake ("camelina sativa"), and oil of the same.

1197 HEYL, J. F. & CO. 75, *Leipziger-str. Berlin.*—Rape, hemp, linseed, castor, radish, and cumin oil; scoured wool; grease obtained by scouring.

1198 HOFFMANN, J. G. *Breslau.*—Lamp oil and oil for greasing.

1199 HOLTZ, H. *Wogenthin, Cöslin.* — Oak bark.

1200 HOMEYER, FR. *Ranzin, Greifswalde.* —Fleeces.

1201 JANSSEN MICHELS, & NEVEN, *Cologne.* —Stearine candles.

1202 KAMPER, TH. *Cologne.* — Oil for greasing machinery.

1203 KIRSTEIN, C. *Hirschberg, Liegnitz.* —Officinal plants from the neighbourhood of Hirschberg.

1204 KISTENMACHER & GÜRCKE, *Sprottau Liegnitz.*—Soaps.

1205 KLEIST, F. W. 54, *Oranienburger-str. Berlin.*—Oil for watchmaking.

1207 LEDER, BROS. 113, *Alte Jacob-str. Berlin.*—Perfumes and scented soaps.

1208 LEHMANN, R. *Nitsche, Posen.* — Fleeces.

1209 LICHNOWSKY, PRINCE C. OF, *Kuchelna, Oppeln.*—Fleeces.

1211 LÜBBERT, E. *Zweibrodt, Breslau.*— Fleeces.

1214 MAHR, BROS, *Naumburg-on-Saale.*— Ivory dandruff-comb.

1215 MOTARD, DR. A. 11, *Brüder-str. Berlin.*—Elaine, samples of soap, stearine and paraffin candles.

1216 NERNST, H. *Taplacken, Königsberg.*— Turnsol oil.

1217 OPPERSDORF, COUNT E. OF, *Ober-Glogau, Oppeln.*—Fleeces.

1218 OTTO, J. F. *Frankfurt-on-Oder.*— Paraffin candles, palm oil, rape-seed oil, Sidney cocoa-nut oil, soaps.

1220 POHL, G. *Canth, Breslau.*—Teazles.

1222 POPPELSDORF AGRICULTURAL ACADEMY, *Bonn.*—Black mallows.

1223 PRINGSHEIM, E. *Oberschönau, Breslau.*—Fleeces.

1225 RATH, H. VON, *Lauersfort, Crefeld.*— Oak-tan.

1226 RATIBOR, DUKE OF, *Niedane, Oppeln.* —Fleeces.

1226A REIBNITZ, J. VON, *Hochkirch, Liegnitz.*—Fleeces.

1227 RISTOW, C. *Repkow, Cöslin.*—Fleeces.

1228 RÖMER & HACKENBERG, *Cologne.*— Oil for greasing machinery.

1229 RUDZINSKI, C. VON, *Liptin, Oppeln.*— Fleeces.

1230 RUNGE, DR. F. F. *Oranienburg, near Berlin.*—Wool, yarns, leather.

1231 SARRE, H. & CO. 16, *Stralauer-str. Berlin.*—Tallow and motled soap.

1232 SARRE, H. JUN.—*Moabit, near Berlin.* —Residue of grease for obtaining asphalt of stear-elaine.

1234 SAURMA, E. v. *Zülzendorf, Breslau.* —Fleeces.

1235 SCHEIDT, V. *Cologne.*—Hairdressers' combs.

1237 SCHINDLER & MÜTZEL, *Stettin.* — Soap, grease.

1238 SCHÖLLER, L. *Schwieben, Oppeln.*— Fleeces.

1239 SCHRAM, P. & JOS. BROS. *Neuss, Düsseldorf.*—Samples of starch.
1240 SCHULTZE, J. C. 82, *Charlotten-str. Berlin.*—Lac varnish.
1241 SIEGEN AGRICULTURE AND TRADES' ASSOCIATION, *Arnsberg.*—Oak bark.
1243 THAER, DR. A. P. *Möglin, near Wrietzen-on-Oder.*—Fleeces.
1244 VILTER, F. W. 4, *Kleine Waldemar-str. Berlin.*—Specimens of oil.
1245 VOGT, A. G. *Mühlhausen, Erfurt.*—Apothecary's utensils of horn, scales, spoons, pill machines.
1246 WÄCHTER, JOH. *Tilsit.*—Linseed, rapeseed cakes, oil, varnish, soap.
1247 WEHOWSKY, *Graase, Oppeln.*—Fleeces.
1248 WEITZE, DR. C. G. *Stettin.*—Oil, soap.
1248A WELLMANN, *Stettin*, representative of the mill at Bärwalde, and the manufactory at Dittersbach. Flour and starch.
1249 WIEDEBACH, F. VON, *Beitzsch, Frankfurt-on-Oder.*—Fleeces.
1250 WUNDER, L. *Liegnitz.*—Soap.
1251 ZECHER, A. *Mühlhausen, Erfurt.*—Apothecary's utensils of horn.

CLASS 5.

1252 BERLIN JOINT-STOCK CO. FOR THE MANUFACTURE OF RAILWAY REQUISITES.—Railway carriages, railway mail coach.
1253 BOCHUM MINING AND CAST-STEEL MANUFACTURING CO. *Arnsberg.*—Cast-steel axletrees and wheels, cast-steel tires, cast-steel bell.
1254 BORSIG, A. *Berlin.*—Locomotive and tender.
1255 COLOGNE JOINT-STOCK CO. FOR THE MANUFACTURE OF MACHINES.—Iron work for railway switches.
1258 HÖRDE MINING AND FORGING CO. *Arnsberg.*—Wrought-iron wheels, locomotive tires of puddled steel, wrought-iron telegraph poles, steam-boiler bottom.
1258A KRUPP, FR. *Essen, Düsseldorf.*—Locomotive, tender, carriage tyres, crank-axles, crank-pins, wheels, springs, &c.
1259 LEHRKIND, FALKENROTH, & CO. *Haspe, Arnsberg.*—Wrought railway waggon wheel, steel tires, puddled steel showing the fracture.
1261 RUFFER, G. H. *Breslau.*—Cast-iron railway waggon wheels.
1261A SCHWARZKOPF, L. 20, *Chaussee-str. Berlin.*—Cast-iron railway waggon wheels.
1262 SIEGBURG, F. *Wilhelmshütte, Cologne.*—Cast-iron railway waggon wheel and two dray-carts.

CLASS 6.

1264 KARWIESE, A. G. *Graudenz.*—State coach.
1265 NEUSS, J. 225, *Friedrich-str. Berlin.*—State coach.
1267 SCHRAN, FR. *Gleidorf, Arnsberg.*—Axle-trees for coaches, waggons, and carts.

CLASS 7.

1269 BRAUN, J. M. *Düren.*—Machine cards.
1271 HAMANN, A. 23 & 24, *Schilling-str. Berlin.*—Turning lathe, planing machine.
1274 KAMP, J. 56, *Kraut-str. Berlin.*—Reeds for silk and woollen weaving.
1276 KÜHNEN, F. *Wesel.*—Card mountings.
1278 MÜLLER, J. 59 & 60, *Mühlen-str. Berlin.*—Paper machine.
1279 OFFENHAMMER, F. 20, *Neue Schönhauser-str. Berlin.*—Tools for watchmaking and engraving.
1280 SAUER, FR. *Lennep.*—Steel reeds, wire combs.
1281 SCHULZE, W. VELLINGHAUSEN & CO. *Stockum, Arnsberg.*—Hard roller.
1282 SCHWARZKOPF, L. 20, *Chaussee-str. Berlin.*—Steam-hammer.
1283 SIGL, G. 29, *Chaussee-str. Berlin.*—Lithographic steam-press.

CLASS 8.

1286 BIALON (C. HUMMEL), 1, *Johannis-str. Berlin.*—Steam cylinder drying machine, printing machine, double pantograph, calandre.
1287 BLEYENHEUFT, J. H. *Aix-la-Chapelle.*—Machine straps, strings, hose for fire-engine, fire-bucket.
1287A BLEYENHEUFT, M. F. MILLIARD, *Eupen.*—Machine straps, fire-buckets, and hoses.
1288 BÖCKE, F. 167, *Garten-str. Berlin.*—Sewing machine.
1290 BROSOWSKY, W. A. *Jasenitz, Stettin.*—Peat-cutting machine.
1291 BÜCHNER, C. W. *Pyritz, Stettin.*—Distilling apparatus.
1292 CADURA, H. *Breslau.*—Machine straps.
1293 CAHEN-LEUDESDORF, A. & CO. *Mühlheim-on-Rhine.*—Machine straps.
1294 COLOGNE JOINT-STOCK CO. FOR THE MANUFACTURE OF MACHINES.—Locomobil-engine, steam-pump, transmission-rod, water-slide, stopping-valves, heater.
1295 DREWITZ, E. *Thorn.*—Alcoholometer.
1297 EGELLS, F. A. 2, *Chaussee-str. Berlin.*—Wolf's steam-engine of 30 horse power.
1298 FRAUDE, W. O. & CO. 68, *August-str. Berlin.*—Steam apparatus for making mineral waters and infusions.
1299 GÄRTNER, THEYSON, & EDE, *Borgholzhausen, Minden.*—Cork cutting-machine.
1301 GRESSLER, J. & CO. 34, *König-str. Berlin.*—Apparatus for making mineral waters.
1303 HECKMANN, C. 18 & 19, *Schlesische-str. Berlin.*—Vacuum pan.
1307 KROPF, C. *Nordhausen.*—Model of a

mash-cooling apparatus, bung-manometer, model of heating-apparatus of a forge.

1311 LORENZ, H. & VETTE, TH. 15, *Engelufer, Berlin.*—Filtering machine.

1312 MAGDEBURG - HAMBURG, STEAM-NAVIGATION CO. *Magdeburg.*—High-pressure expansion steam-engine with standard lever, centrifugal sugar drying apparatus.

1314 PHILIPPSSON, F. C. 1, *Münz-str. Berlin.*—Fastenings for steam-boilers, hydraulic windlass, meat mincing-machine.

1315 RULAND, W. *Bonn.*—Leather for machine-straps.

1316 SCHÄFFER & BUDENBERG, *Magdeburg and Manchester.*—Steam-engine and steam-boiler braces, &c.

1317 SCHLICKEYSEN, C. 11, *Köpnicker-str. Berlin.*—Brick-machine.

1318 SCHULZ, KNAUDT, & CO. *Essen, Düsseldorf.*—Steam-boiler cupola, steam-boiler bottom, boiler plates.

1319 SCHWARZKOPF, L. 20, *Chaussee-str. Berlin.*—Caloric-engine.

1320 SIEGBURG, FRIEDRICH, *Wilhelmshütte, Cologne.*—Smoke consuming grate for steam.

1321 SPROTTAU IRON FOUNDRY, *Wilhelmshütte, Liegnitz.*—Horizontally constructed steam-engine, high pressure, with self-acting distributing regulator.

1322 THOMAS, H. 31, *Graben-str. Berlin.*—Longitudinal shearing-machine, friezing-machine.

1323 UHLHORN, H. *Grevenbroich, Düsseldorf.*—Coining machines.

1324 WAGENER, F. G. 72, *Mauer-str. Berlin.*—Machine for copying in relief, pantograph, number-printing machine, frame with specimen impressions.

1325 WATREMEZ & KLOTH, *Aix-la-Chapelle.*—Black's safety apparatus to prevent the bursting of steam boilers.

CLASS 9.

1329 CEGIELSKY, H. *Posen.*—Ploughs, thrashing machine, straw cutter, sowing machine, roller.

1330 ECKERT, H. F. 1, *Kleine Frankfurter-str. Berlin.*—Ploughs, sowing, thrashing, cleaning machine, straw cutter, miners' scooping iron, valve pump, machine for washing potatoes.

1332 PINTUS & CO. *Brandenburg, near Berlin.*—Ploughs, barrows, sowing, mowing, thrashing machine. Drawings of agricultural machines.

1335 TONNAR, A. *Eupen.*—Drying and winnowing machine.

CLASS 10.

1336 ADLER, M. 47, *Neue Friedrich-str. Berlin.*—Smoke consumers, of iron and clay.

1337 BARHEINE, R. R. 61, *Friedrich-str. Berlin.*—Marble chimney, baptismal font, table and monuments.

1338 BERLIN MINISTER OF COMMERCE AND PUBLIC WORKS.—Model of the bridge across the Vistula, near Dirschau, and that of the plain of the Oberländische Canal, Elbing.

1339 BONN MINING AND FORGING CO.—Bricks, cement, and cement-work.

1341 BÜSSCHER & HOFFMANN, *Neustadt-Eberswalde, near Berlin.*—Carton-pierre, asphalt slabs, pipes coated with asphalt, model of a circular brick-kiln.

1342 CARSTANJEN, J. *Duisburg.*—Roofing pasteboard.

1343 FÖRSTER, F. S. *Krampe, Liegnitz.*—Paper and roofing pasteboard.

1343A GÄNICKE, L. *Wittenberge-on-Elbe.*—Asphalte roofing pasteboard.

1344 GALOPIN, J. *Aix-la-Chapelle.*—Marble chimney, table ornaments of marble.

1345 GASSEL, RECKMANN, & CO. *Bielefeld.*—Asphalt felt.

1347 HÄUSLER, C. S. *Hirschberg, Liegnitz.*—Model of an asphalt paper roof.

1348 HAURWITZ, L. & CO. *Stettin.*—Asphalt floor slabs, asphalt paper, roofing pasteboard.

1352 KRZYZANOWSKI, A. *Posen.*—Madonna di S. Sisto, bust of Lelewel, cast-iron.

1353 KULMITZ, C. 39, *Adalbert-str. Berlin.*—Capital, and granite monument.

1355 MEISSNER, W. *Stargard, Stettin.*—Asphalt roofing pasteboard, model of a roof covered with roofing pasteboard.

1356 MEWS, H. *Stettin.*—Objects made of Stettin cement.

1357 MICHELI, BROS. 52, *Jäger-str. Berlin.*—Plaster of Paris and marble work.

1358 MÖHRING, L. 24, *Karl-str. Berlin.*—Asphalt roofing pasteboard, asphalt varnish for roofing.

1359 POHL, H. 21, *Alte Jacob-str.*—Architectural ornaments of zinc.

1360 QUISTORP, J. *Stettin.*—Samples of cement, cement floor-slabs.

1362 SCHLESING, F. 3, *Werdersche Markt, Berlin.*—Asphalt tar, asphalt cement used in the construction of telegraphic lines.

1363 SCHOLZ, J. & SON, *Klitschdorf, Liegnitz.*—Roofing pasteboard.

1364 SCHOTTLER & CO. *Lappin, Danzig.*—Roofing pasteboard.

1365 SCHRÖDER & SCHMERBAUCH, *Stettin.*—Conduit pipes of asphalted paper.

1367 SCHÜTTLER, R. J. *Moabit, near Berlin.*—Cement manger and cement tiles.

1368 SCHULZE & WILHELM, *Nordhausen.*—Chimney of artificial marble.

1369 SIEGEN AGRICULTURE & TRADES' ASSOCIATION, *Arnsberg.*—Relief map of the environs of Siegen, and model of an artificial meadow.

1370 STETTIN PORTLAND-CEMENT MANUFACTORY.—Specimens of cement, cement table-top.

1372 STRACHWITZ, COUNT H. VON, *Gross-Stein, Oppeln.*—Refractory bricks.

1373 TARNOWITZ CEMENT WORKS.—Cement and cement-work.

1374 WEIMAR, J. P. 118, *Oranien-str. Berlin.*—Chimney of artificial marble, bust of King William of Prussia, and medals.

1375 WOLFFHEIM, W. *Stettin.*—Asphalt paper, roofing pasteboard.

1376 ZEYLAND, J. *Posen.* — Four-sash window to be closed hermetically.

CLASS 11.

1377 ALBRECHT, F. *Bromberg.*—Double-barrelled gun.

1378 BEERMANN, B. *Münster.* — Rifled sporting gun, double-barrelled gun, and pistols.

1379 BERGER & Co. *Witten, Arnsberg.*—Cast-steel cannon and gun barrels.

1383 HÄNEL, C. G. *Suhl.*—Military arms.

1384 HÖSTEREY, J. P. *Barmen.*—Percussion caps.

1386 KIRSCHBAUM, C. R. *Solingen.* — Sabres, daggers, cutlasses.

1386A KRUPP, FR. *Essen, Düsseldorf.*—Steel as applied to ordnance.

1387 LAUTE, G. 32, *Brüder-str. Berlin.*—Swords, sabres, daggers, and cutlasses.

1389 LEONHARDT, J. E. 46, *Wilhelm-str. Berlin.* — Machine for casting bullets for Minié-rifle guns.

1390 LÜNESCHLOSS, P. D. *Solingen.* — Sword presented to King William I. of Prussia.

1391 PRETZEL, A. *Glogau, Liegnitz.* — Pistols.

1392 RÖDDER, M. *Cologne.*—Arms.

1392A SCHALLER, A. *Suhl.*—Revolver.

1392B SCHILLING, V. *Suhl.* — Military arms, pistols, revolver, sporting-guns, rifles.

CLASS 13.

1393 BELLÉ, R. *Aix-la-Chapelle.*—Brass scale-beams, inductor apparatus, magnifying apparatus.

1394 BREDEMEYER, J. *Frankfurt-on-O.*—Mariners' compasses, levels, hydraulic press, Dubois Reymond's inductor.

1395 ELSTER, S. 67, *Neue König-str. Berlin.*—Apparatus for testing carburetted hydrogen, photometer, gasometer, regulator, pressure-gauge.

1397 FESSEL, FR. *Cologne.*—Myographion, to represent the course of muscular convulsion.

1400 GREINER, J. C. & SON, 15, *Kur-str. Berlin.*—Barometer, thermometer, and alcoholometer.

1401 GREINER, F. F. *Stützerbach, Erfurt.*—Philosophical and chemical apparatus of glass.

1402 GRESSLER, J. & Co. 34, *König-str. Berlin.*—Galvanic apparatus.

1403 HAGER, F. *Stettin.*—Brass octant, mariner's compass, amplitude compass.

1404 HEMPEL, O. M. 98, *Zimmer-str. Berlin.*—Manometer for the pressure of steam.

1406 LÜTTIG, C. 11, *Post-str. Berlin.*—Drawing instruments, theodolite, mariner's compass, level.

1407 LUHME, J. T. & Co. 51, *Kur-str. Berlin.*—Chemical, pharmaceutical, philosophical, meteorological apparatus and utensils.

1408 MEISSNER, A. 71, *Friedrich-str. Berlin.*—Theodolite, level, mariner's compass, sundial, cases of mathematical instruments.

1410 NOBERT, F. A. *Barth, Stralsund.*—Microscope; optical apparatus.

1411 OERTLING, AUG. 57, *Oranienburger-str. Berlin.*—Optical instrument.

1411A PINTUS & Co. *Brandenburg, near Berlin.*—Weighing scales.

1412 REICHEL, C. 65, *Alte Jacob-str. Berlin.*—Tubular levels.

1413 SIEMENS & HALSKE, 94, *Markgrafen-str. Berlin.* — Telegraphic apparatus, voltaic inductors, and other scientific instruments.

1415 SCHMIDT, F. 74, *Alexandrinen-str. Berlin.*—Philosophical instruments.

1417 WARMBRUNN, QUILITZ, & Co. 40, *Rosenthaler-str. Berlin.* — Apparatus, glass and porcelain vessels, and instruments for pharmaceutical and chemical laboratories.

CLASS 14.

1419 BEYRICH, F. 101, *Friedrich-str. Berlin.*—Chemical products and paper for photography.

1420 BUSCH, E. *Rathenow, Potsdam.*—Photographic apparatus; photograph by Albert, of Munich.

1421 DUNKER, A. 21, *Französische-str Berlin.*—Photographic tableau.

1422 FESSLER & STEINTHAL, 48, *Französische-str. Berlin.*—Photographs.

1424 HAMMERSCHMIDT, W. A. *Neu Schöneberg, Berlin.*—Photographs.

1425 KLITZING, VON, *Glass-works, Bernsdorf, Liegnitz.*—Caps *(cuvettes)* for photography.

1427 KUNZMANN, H. 218, *Friedrich-str. Berlin.* — Chemical products and paper for photography.

1428 MINUTOLI, VON, *Liegnitz.* — Models and patterns.

1429 MOSER & SENFTNER, 44, *Linden, Berlin.*—Stereoscopic views.

1430 NICOLAI, 13, *Brüder-str. Berlin.*—Kaulbach's Shakespeare Album.

1431 OEHME, G. & JAMRATH, J. 19, *Jäger-str. Berlin.*—Photographs.

1432 SCHAUER, G. 188, *Friedrich-str. Berlin.*—Specimens of photography.

1433 SCHERING, E. 21, *Chaussee-str. Berlin.*—Chemical products; photograph.

1435 WOTHLY, J. *Aix-la-Chapelle.*—Photographs.

CLASS 15.

1437 BECKER, G. *Freiburg, Breslau.*—Pendulum clock.

1438 EPPNER & Co. 31, *Behren-str. Berlin.*—Watches and chronometers.

1439 FELSING, C. 1, *Schlossfreiheit, Berlin.*—Regulator.

1442 TIEDE, FR. 20, *Jäger-str. Berlin.*—Box chronometer, pendulum clock.

1443 WEISS, C. *Glogau, Liegnitz.*—Eight-day clock for a steeple.

1444 WIESE, R. *Landsberg-on-Warthe.*—Time-piece.

CLASS 16.

1445 ADAM, G. *Wesel.*—Pianofortes.

1446 BECHSTEIN, C. 56, *Behren-str. Berlin.* Pianofortes.

1450 ENGEL, F. A. *Danzig.* — Tuning machine.

1451 ESSLINGER, C. W. 53, *Jerusalemer-str. Berlin.*—Guitar.

1453 GRIMM, C. 15, *Kur-str. Berlin.*—Violins, violoncellos, violas.

1455 HARTMANN, W. 94, *Alte Jacob-str. Berlin.*—Concert pianino, low pianino.

1456 JBACH, J. SONS, *Barmen.* — Grand pianoforte with double touch, oblique pianoforte.

1457 KLOSS, E. *Bernstadt, Breslau.* — Violin.

1458 KNACKE, BROS. *Münster.* — Grand piano; pianino.

1459 KÜNTZEL, L. 75, *Kronen-str. Berlin.* —Quintetto of string and bow instruments; violin, viola, violoncello.

1463 MAHLITZ, E. 82, *Kloster-str. Berlin.* —Grand pianoforte, pianino.

1464 MANN, TH. *Bielefeld.*—Pianino.

1465 OBERKRÜGER, FR. *Cologne.*—Oblique pianino.

1466 OECHSLE, A. SONS, 86, *Wall-str. Berlin.* — Kettledrums, rolling drum, large parchment for drums.

1467 OTTO, L. *Cologne.*—Quintetto (bass, violoncello, counter-alto, violins, and citherns).

1468 SCHMIDT, F. A. *Cologne.* — Wind instruments of brass and German silver euphonium, C bass.

1469 SCHWECHTEN, G. 40, *Linden-str. Berlin.*—Grand polysander pianoforte, polysander pianino.

1471 SPANGENBERG, W. 60, *Charlotten-str. Berlin.*—Grand polysander pianoforte polysander pianino.

1472 WESTERMANN & Co. 97, *Leipziger-str. Berlin.*—Grand polysander pianoforte.

CLASS 17.

1473 GOLDSCHMIDT, S. 28, *Dorotheen-str Berlin.*—Surgical bandages and trusses.

1474 IMME, J. & Co. 60, *Oranienburger-str. Berlin.*—Volta-electric metal-brush.

1475 KÄSTNER, FR. *Cologne.* — Dentists instruments and utensils.

1476 LANGGAARD, DR. O. 104, *Leipziger-str. Berlin.*—Orthopœdic instruments.

1478 LUTTER, A. 53, *Französische-str. Berlin.*—Surgical instruments and apparatus.

1479 PISCHEL, E. *Breslau.* — Case with four of Groove's elements, case of galvano-caustic instruments.

1480 RÖTTGEN, FR. *Bonn.*—Ear-trumpets, respirators.

1481 TRESCHINSKY, F. W. G. 62, *Krausen-str. Berlin.* — Orthopœdic instruments and apparatus.

1482 WINDLER, H. 64, *Mittel-str. Berlin.* —Surgical instruments and apparatus.

CLASS 18.

1484 BORNEFELD, W. *Gladbach, Düsseldorf.*—Cotton and worsted canvas.

1485 COLSMANN, J. F. & Co. *Barmen.*—Thread and cotton yarn.

1486 ERMEN & ENGELS, *Barmen.*—Twist, knitting and sewing yarn.

1487 FLEISCHER, L. *Mühlhausen, Erfurt.*—Silk canvas.

1488 GRUNSFELD, J. & SONS, *Heiligenstadt, Erfurt.*—Cotton fabrics.

1490 KLEMME & GRUBE, *Crefeld.*—Cotton ribbons for the manufacture of buttons.

1492 MITSCHERLICH, F. A. *Eilenburg, Merseburg.*—White quiltings, undress fabrics.

1493 PFERDMENGES & SCHMÖLDER, *Rheydt, Düsseldorf.*—Cotton-cloth and yarns.

1494 ROLFFS & Co. *Cologne.* — Printed pocket handkerchiefs.

1495 SENNE, FR. *Erfurt.*—Lamp-wicks.

1496 STERNENBERG, F. H. & SONS, *Schwelm, Arnsberg.* — Cotton window-blinds, ticking.

1497 WOLFF, SCHLAFFHORST, BRUEL, & Co. *M. Gladbach, Düsseldorf.*—Cotton manufactures.

1498 WOLFF & SCHLAFHORST, *M. Gladbach, Düsseldorf.*—Cotton manufactures.

CLASS 19.

1499 ADLER, BROS. *Neustadt, Oppeln.*—Damasks.

1500 BENDER, W. SEN. *Bleicherode, Erfurt.*—Flax and tow.

1501 BIELEFELD. — COMMITTEE OF THE BIELEFELD LINEN TRADE, acting for A. W. Kisker, E. A. Delius & Sons, Krönig & Jung, C. H. Potthoff, Fr. W. Krönig & Sons, Rabe & Consbruch, H. M. Wittgenstein, Bertelsmann & Son, Goldbeck & Vieler, Ferd. Lüder, C. F. Gante & Sons, L. Meyer & Co. F. Piderit, C. Colbrunn, L. Heidsick.—Flax, tow, linen cloth, handkerchiefs, ready-made linen, table-cloth.

1502 BRAUNSBERG AGRICULTURAL ASSOCIATION, *Königsberg.*—Samples of flax.

1503 CADURA, H. *Breslau.*—Hemp hose.

1504 DELIUS, C. W. & Co. *Versmold, Minden.*—Sail cloth.

1505 EICHELBAUM, S. *Insterburg.*—Fishing nets of hemp.

1506 ENGEL, F. E. *Görlitz, Liegnitz.*—Pack-thread, travelling bags, and hunting bags.

1507 ERDMANNSDORF ROYAL SPINNING AND WEAVING MILL, *Liegnitz.*—Linen cloth.

1508 EXNER & STOCKMANN, *Schweidnitz, Breslau.* — Indigo-dyed and printed linen cloth.

1509 Fränkel, S. *Neustadt, Oppeln.* — Plain linen, table cloth, napkins, towels.

1509a Goszlau, C. *Nitsche, Posen.* — Damask towels.

1511 Huhn, N. von, *Obergerlachsheim, Liegnitz.*—Flax, linseed.

1512 Helling, Fr. *Borgholzhausen, Minden.*—Sail cloth, linen cloth.

1513 Herford Association for the Improvement of Hand-spinning, *Minden.*—Linen, yarn.

1514 Hohlstein, B. *Bollstedt, Erfurt.*—Flax.

1516 Kirstein, C. *Hirschberg. Liegnitz.*—Linen cloth.

1516a Lehmann, R. *Nitsche, Posen.*—Flax.

1517 Lüttwitz, R. von, *Simmenau, Oppeln.*—Flax.

1518 Mevissen, G. *Dülken, Düsseldorf.*—Flax, linen-yarn, thread.

1519 Müller, M. H. *Münster.*—Cart tilts without seam.

1521 Pannwitz, W. von, *Burgsdorf, Oppeln.*—Flax.

1522 Pohl, A. *Stralsund.*—Damask tablecloth and napkins, towels.

1523 Ravensberg Spinning Mill, *Bielefeld.*—Linen yarn.

1524 Reisswitz, B. von, *Wendrin, Oppeln.*—Flax.

1525 Rösner, C. *Wüste-Waltersdorf, Breslau.*—Linen cloth.

1526 Schöller, Mevissen & Bücklers, *Düren.*—Machine-spun linen yarn.

1526a Schönfeld, F. L. *Herford.*—Flax and yarns of flax and tow.

1529 Stoltenburg, E. *Stralsund.* — Damask table cloth and napkins, towels.

1529a Vorwärts Spinning Mill, *Bielefeld.*—Linen yarn.

1530 Wagener, for the Agricultural Academy, *Proskau, Oppeln.*—Flax.

1531 Willmann, A. & Sons, *Patschkey, Breslau.*—Specimens of flax.

1532 Zöllner, A. *Stralsund.* — Linen damask, towels.

CLASS 20.

1534 Andreä, Chr. *Mühlheim-on-Rhine.*—Velvet, velvet-ribbons, worsted and silk drapery.

1535 Baur, F. H. vom & Son, *Ronsdorf, Düsseldorf.*—Silk ribbons.

1536 Bendorf Central Silk-growing Establishment of Rhenish Prussia, *near Coblenz.*—Silk and cocoons.

1537 Bökemann & Wessel, Bertelsmann & Son, Delius, C. A. & Son, Krönig, C. & Th., Wahner Bros., Wertheimer, M., Wittgenstein, C. H. & Son, *Bielefeld*, and Bartels Bros. *Gütersloh.*—Velvets, silk textures, and silk ribbons.

1538 Bruck, H. vom & Sons, *Crefeld.*—Velvet and velvet ribbons.

1540 Dappen, Böckner, & Schnütgen, *Crefeld.*—Silk fabrics (slips and shawls).

1541 Diergardt, Fr. von, *Viersen, Düsseldorf.*—Velvet and velvet ribbons.

1542 Draemann & Dellmann, *Crefeld.*—Silk fabrics.

1543 Engelmann & Bohnen, *Crefeld.*—Silk fabrics.

1544 Fränkel, S. *Neustadt, Oppeln.*—Silk table-covers.

1547 Gressard & Co. *Hilden, Düsseldorf.*—Printed silk textures.

1549 Heimendahl, G. *Crefeld.*—Twisted silk.

1550 Hipp, H. G. & Better, *Crefeld.*—Velvet and silk fabrics.

1551 Jacobs, J. M. & Co. *Crefeld.*—Silk fabrics.

1553 Küppers, L. & Co. *Crefeld.*—Silk fabrics.

1554 Küppers & Kniffler, *Crefeld.*—Silk and mixed fabrics.

1556 Mähler & Tappen, *Crefeld.*—Silk fabrics.

1557 Oehme, C. W. 74, *Spandauer-str. Berlin.*—Black hat plushes.

1558 Ristow, C. *Repkow, Cöslin.*—Raw silk.

1559 Scheibler & Co. *Crefeld.* — Silk velvet, velvet ribbons and shawls.

1561 Schröder, W. & Co. *Crefeld.*—Taffetas and satin.

1562 Schrörs, G. & H. *Crefeld.*—Silk fabrics.

1563 Schumacher & Schmidt, *Wermelskirchen, Düsseldorf.*—Silk and mixed ribbons.

1564 Seyffardt & te Neues, *Crefeld.*—Silk and mixed fabrics.

1565 Töpfer, G. A. *Stettin.* — Cocoons, silk, breeding apparatus.

1566 Westen, von den & Co. *Crefeld.*—Velvet and velvet ribbons.

CLASS 21.

1570 Arendt, E. *Zielenzig, Frankfurt-on-O.*—Woollen yarn.

1571 Aron, J. 9, *Dorotheen-str. Berlin.*—Woollen and mixed shawls.

1572 Ax, H. *Rheydt, Düsseldorf.*—Woollen mixed coatings.

1573 Becker & Auerbach, 26, *König-str. Berlin.*—Woollen shawls.

1574 Bellingrath, C. H. *Barmen.* — Worsted, mixed and silk ribbons.

1575 Berger, M. *Peitz, Frankfurt-on-O.*—Summer buckskins.

1575a Bernstein & Lichtenstein, *Königsberg.*—Sheddy and Mungo wool.

1576 Bertelsmann & Niemann, *Bielefeld.*—Plushes and damask.

1577 Bieger, G. M. *Finsterwalde, Frankfurt-on-O.*—Black cloth.

1578 Blecher & Clarenbach, *Hückeswagen, Düsseldorf.*—Woollen coatings.

1579 Bleissner, J. G. *Neudamm, Frankfurt-on-O.*—Woollen cloth.

1580 Bluhm, S. J. *Haynau, Liegnitz.*—Woollen cloth.

1581 Bockhacker, C. Successor, *Hückeswagen, Düsseldorf.*—Woollen cloth, paletot stuffs.

1582 Bockhacker, Fr. & Sons, *Hückeswagen, Düsseldorf.*—Woollen yarn.

1583 BOCKMÜHL, FR. SONS, *Düsseldorf.*—Worsted.

1584 BÖHMER & ERKLENTZ, *M. Gladbach, Düsseldorf.*—Mixed woollen.

1585 BORNEFELD & KNÖTGES, *M. Gladbach, Düsseldorf.*—Mixed manufactures.

1586 BRACH & CO. 76, *Spandauer-str. Berlin.*—Long and square shawls, cloakings, mixed fabrics for dresses, plushes.

1587 BRUCK, BROS. 10, *Papen-str. Berlin.*—Long and square shawls.

1588 BRÜGMANN, W. & CO. *Burtscheid.*—Samples of yarn.

1589 BUDDE & MÜNTER, *Herford, Minden.*—Toilet fabrics.

1591 CAMPHAUSEN & KÜPPERS, *M. Gladbach, Düsseldorf.*—Cotton and mixed fabrics.

1592 CARO & ROSENHAIN, 27, *Post-str. Berlin.*—Woollen long and square shawls.

1593 CLEFF, BROS. *Barmen.*—Silk ribbons and slips.

1596 DAVID & CO. 27 *Post-str. Berlin.*—Plushes and woollen fabrics.

1598 DELLMANN, W. & CO. *Elberfeld.*—Shawls, mufflers, cravats, slips.

1599 DEUSSEN, J. *Sagan, Liegnitz.*—Woollen cloth, twilled cloth.

1600 DEUTZ, JUN. & STROM, *Aix-la-Chapelle.*—Woollen cloth and satins.

1601 EICKHOFF, H. *Meschede, Arnsberg.*—Woollen cloth and buckskins.

1602 ERKENS, J. SONS, *Burtscheid.*—Dyed woollen cloth.

1603 ESCHENHAGEN, F. A. *Cottbus, Frankfurt-on-Oder.*—Worsted summer stuffs.

1603A FELLER, J. G. & SON, *Guben.*—Woollen cloth and tricots.

1604 FEULGEN, BROS. *Werden, Düsseldorf.*—Woollen cloth.

1605 FÖRSTER, F. S. *Grünberg, Liegnitz.*—Woollen cloth, twilled cloth, electorals, satins, tricots.

1606 FÖRSTER, F. S. *Suckau, Liegnitz.*—Wicuna-yarn.

1607 FRÄNKEL, J. 17, *Grüner Weg, Berlin.*—Long and square shawls.

1608 FREMEREY, J. P. *Eupen.*—Woollen cloth.

1609 FRIEDHEIM, S. M. & SONS, 18, *Spandauer-str. Berlin.*—Orleans.

1610 GEBHARDT & WIRTH, *Frauenmühle, Frankfurt-on-Oder.*—Woollen cloth.

1611 GEISSLER, E. *Görlitz, Liegnitz.*—Woollen cloth.

1612 GEVERS & SCHMIDT, *Görlitz, Liegnitz.*—Wollen cloth.

1613 GRÄSER, BROS. & CO. *Langensalza, Erfurt.*—Buckskins, paletot-cloth.

1614 GREIFF & CO. *Barmen.*—Ribbons for coat-trimmings.

1615 GRÖSCHKE, C. A. *Forst, Frankfurt-on-Oder.*—Buckskins.

1617 GRÜNDER, TR. *Peitz, Frankfurt-on-Oder.*—Coatings and cloakings.

1619 HABERLAND, G. A. *Finsterwalde, Frankfurt-on-Oder.*—Woollen cloth and buckskins.

1021 HAHN, A. & CO. 20, *Neue Friedrich-str. Berlin.*—Shoddy and Mungo-wool.

1622 HALBACH, W. & WOLFERTS, L. *Barmen.*—Silk and mixed ribbons.

1623 HEEGMANN & MESTHALER, *Barmen.*—Worsted lasting for the manufacture of buttons.

1625 HENDRICHS, FRZ. *Eupen.*—Woollen cloth.

1626 HERRMANN, PH. *Bromberg.*—Wool and woollen yarn of the district of Posen.

1629 HILGER, BROS. *Lennep.*—Woollen cloth.

1630 HIRNSTEIN, FR. *Meschede, Arnsberg.*—Worsted knitting yarns.

1631 HOFFMANN, E. *Sorau, Frankfurt-on-Oder.*—Worsted fabrics.

1632 HOFFMANN, GÖNNER, & CO. *Görlitz.*—Woollen and tweeled cloth.

1634 HÜFFER & MORKRAMER, *Eupen.*—Woollen cloth.

1635 JTZIGSOHN, M. *Neudamm, Frankfurt-on-Oder.*—Woollen cloth and twills.

1636 JAHN, C. G. *Neudamm, Frankfurt-on-Oder.*—Woollen cloth and woollen yarns.

1637 JANSEN, J. W. *Montjoie.*—Trouserings and coatings.

1638 KAUFFMANN, M. *Tannhausen, Breslau.*—Worsted and mixed fabrics.

1639 KAYSER, ALF. *Aix-la-Chapelle.*—Woollen cloth, beavers, and satins.

1640 KEBEN, S. 7, *Burg-str. Berlin.*—Plushes and woollen fabrics.

1641 KELLER & STRÖTER, *Barmen.*—Trimming lace and ribbons.

1644 KLEMM, G. *Forst, Frankfurt-on-Oder.*—Woollen cloth.

1645 KRAGE, P. H. *Quedlinburg.*—Woollens.

1646 KRAUSE, C. F. & SONS, *Görlitz, Liegnitz.*—Woollen cloth.

1647 KRUGMANN & HAARHAUS, *Elberfeld.*—Furniture and coach stuffs, velours d'Utrecht, table covers.

1648 LANGENBECK & WEX, *Barmen.*—Buttons covered with worsted.

1649 LEHMANN, D. J. 64, *Spandauer-str. Berlin.*—Plushes, double stuffs, long and square shawls, travelling blankets.

1650 LENDER, C. & CO. *Rheydt.*—Coatings and trouserings.

1651 LEVI & ARON, 17, *Magazin-str. Berlin.*—Long and square shawls.

1654 LÖWEN, P. & HILDESHEIMER, *Brandenburg.*—Woollen and silk mixed buckskins.

1655 LÜTGENAU & WIEHAGER, *Hückeswagen, Düsseldorf.*—Woollen fabrics.

1656 MARGGRAFF, BROS. *Schwiebus, Frankfurt-on-Oder.*—Woollen cloth.

1657 MARX, N. & SONS, *Aix-la-Chapelle.*—Woollen cloth.

1658 MATTHESIUS, F. C. & SON, *Cottbus.*—Woollen cloth.

1659 MAYER, FERD. *Cologne.*—Serge de Berry, woollen and cotton shoes, and cravats.

1660 MAYER, J. F. *Eupen.*—Woollen cloth, Spanish stripes, broadcloth.

1661 MEBUS & RÜBEL, *Barmen.*—Trimming lace and ribbons.

1662 MEYER, M. & CO. *Aix-la-Chapelle.*—Coatings and trouserings.

1663 NAUSESTER, W. *Lötmaringhausen,*

Arnsberg.—Worsted knitting yarns, jackets, and caps.

1664 NETTMANN, H. D. & SON, *Limburg-on-Lenne, Arnsberg.*—Woollen cloth, twills, and tricots.

1665 NIEDERHEITMANN & BUCHHOLZ, *Aix-la-Chapelle.*—Buckskins, paletot cloth, and cloakings.

1666 NIEMANN & GUNDERT, *Barmen.*—Ribbons, lace, and cords.

1667 NOSS, CH. *Cologne.* — Furniture plushes.

1668 OFFERMANN, FR. W. *Imgenbroich, Aix-la-Chapelle.*—Trouserings and coatings.

1669 OSTERROTH, W. & SON, *Barmen.* —Worsted and mixed ribbons, lace, and cords.

1670 PASTOR, ED. FR. *Burtscheid.*—Woollen yarns.

1671 PASTOR, G. *Aix-la-Chapelle.*—Woollen yarns.

1672 PETERS, L. *Eupen.*—Woollen cloth.

1673 PFERDMENGES, J. H. & SON, *Rheydt, Düsseldorf.* — Worsted, cotton and linen trouserings.

1675 POLZIN CLOTH WEAVERS' GUILD, *Cöslin.*—White and blue flannels.

1677 RICHTER, A. *Forst and Muskau, Frankfurt-on-Oder.*—Buckskins.

1679 RITTINGHAUS & BRAUNS, *Kettwig, Düsseldorf.*—Woollen trouserings.

1680 RITZ & VOGEL, *Aix-la-Chapelle.*—Black and coloured woollen cloth.

1681 RUFFER, S. B. & SON, *Liegnitz.*—Woollen cloth.

1685 SCHEIFCHEN & SON, *Güntersberg, Frankfurt-on-Oder.*—Black woollen cloth.

1686 SCHLIEF, E. P. *Guben, Frankfurt-on-Oder.*—Woollen fabrics.

1686A SCHLIEF, S. *Guben.*—Woollen cloth, and figured woollen stuffs.

1687 SCHMIDT, FR. & CO. *Sommerfeld, Frankfurt-on-Oder.*—Woollen cloth.

1688 SCHNABEL, BROS. *Hückeswagen, Düsseldorf.*—Woollen cloth and paletot stuffs.

1689 SCHNEIDER, AUG. 30, *Kaiser-str. Berlin.*—Long and square woollen shawls.

1690 SCHÖLLER, J. P. *Düren.*—Woollen cloth.

1691 SCHÖLLER, L. & SONS, *Düren.*—Woollen cloth and stuffs.

1692 SCHÜRMANN, P. & SCHRÖDER, H. *Lennep.*—Woollen cloth.

1693 SCHWAMBORN, NEUHAUS & KRABB, *Aix-la-Chapelle.*—Paletot cloth; buckskins.

1695 SOUTER & ALT, *Eupen.* — Woollen yarns.

1696 STERKEN, H. *Aix-la-Chapelle.*—Coatings, trouserings, and cloakings.

1697 STERNICKEL & GÜLCHER, *Eupen.*—Woollen cloth.

1699 TESCHEMACHER & KATTENBUSCH, *Werden, Düsseldorf.* — Plain and twilled woollen cloth.

1703 ULENBERG & SCHNITZLER, *Opladen.* Woollen and knitting yarn, worsted.

1705 VOSS, C. & J. *Kettwig, Düsseldorf.*—Woollen cloth and buckskins.

1706 WALDTHAUSEN, C. *Aix-la-Chapelle.* —Woollen cloth and fancy cloth.

1708 WEIGERT, BROS. 60, *Oranienburger-str. Berlin.*—Mohair plushes, cotton plushes, woollen and mixed fabrics, cotton fabrics.

1710 WERNER, H. *Torst, Frankfort-on-Oder.*—Twilled woollen cloth.

1711 WIPPRECHT, E. 76, *Mauer-str. Berlin.* —Woollen and linen horse rugs.

CLASS 22.

1714 BURCHARDT, B. & SONS, 19, *Brüder-str. Berlin.*—Oil-cloth, window-blinds.

1715 DINGLINGER, A. F. 18, *Spittelbrücke, Berlin.*—Carpets.

1716 GEVERS & SCHMIDT, *Schmiedeberg, Liegnitz.*—Imitation Turkey carpets.

1717 HERRMANN, PH. *Bromberg.*—Carpets made of cow-hair yarn.

1718 KÜHN, TH. & CO. *Cottbus, Frankfurt-on-Oder.*—Drawing-room carpets.

1719 LEHMANN, M. 16, *Brüder-str. Berlin.* —Oil-cloth, oil floor-cloth.

1720 PRÄTORIUS & PROTZEN, 4, *Kölnischer Fischmarkt, Berlin.*—Carpets.

1721 STEIDEL & SOMMER, 5, *Weber-str. Berlin.*—Carpets, travelling bags.

1722 TÖPFER, G. A. & BLEUDORN, *Stettin.* —Coir rugs.

CLASS 23.

1723 BERGMANN & CO. 39, *Krausen-str. Berlin.*—Dyed Berlin wool.

1724 HAMERS, A. *Crefeld.* — Dyed silk fabrics.

1725 LAUEZZARI, C. *Barmen.*—Turkey-red marking thread.

1728 NEUHAUS, C. A. *Crefeld.*—Samples of dyed silk.

1729 RITTERSHAUS, J. P. *Düsseldorf.*—Turkey-red yarn.

1730 SPINDLER, W. 12, *Wall-str. Berlin.*—Dyed woollen, cotton, and silk fabrics. Samples of dyes.

1731 WOLFF, I. F. *Elberfeld.*—Turkey-red and grey twist.

CLASS 24.

1732 BESSERT-NETTELBECK, PAULINE, 52, *Kronen-str. Berlin.*—Gold and silver embroidery.

1733 FRIEDBERG, L. 22, *Niederwall-str. Berlin.* —Chenille, tassels, lace, braids, buttons.

1736 HANCKE, MISS CL. *Düsseldorf.*—Model embroidery.

1738 KRISTELLER, H. 32, *Heiligegeist-str. Berlin.*—Undress caps.

1739 PAREY, C. F. W. 39, *Leipziger-str. Berlin.*—Embroidery.

1740 PLASMAN, F. J. *Barmen.* — Cotton lace.

1741 SCHÄRFF, R. *Brieg, Breslau.*—Lace, tassels, &c. for coachmakers and saddlers.

1742 STEINER, J. *Breslau.*—Lace, tassels, &c. for coachmakers and saddlers.

1744 WECHSELMANN, J. 42, *Französische-str. Berlin.*—Lace, flounces, pocket-handkerchiefs.

CLASS 25.

1745 BECKE, C. G. 8, *Barnim-str. Berlin.*—Coarse brushes.

1747 ENGELER, H. M. & SON, 36, *Behren-str. Berlin.*—Assortment of brushes.

1748 FRIESECKE, W. *Wittenberg, Merseburg.*—Hair and cloth brushes.

1749 HEGEWALD, H. *Bromberg.*—Wigs.

1750 HERMANN, PH. *Bromberg.*—Cow-hair yarn.

1751 HORNEMANN, A. *Goch, Düsseldorf.*—Painters' brushes.

1752 KELLER, J. C. & SON, *Weissenfels, Merseburg.* — Minever-backs and squirrel-lining.

1753 KOCH, C. F. *Zeitz, Merseburg.*—Brushes, skin-rubber, table-covers, and chess-board.

1754 KRAFFT, G. *Wetzlar, Treves.*—Toupee, wig.

1755 LANGE, J. T. 60, *Alte Schönhauser-str. Berlin.*—Feather duster.

1758 NANNY, A. *Königsberg.*—Bristles and brushes.

1760 RÖGNER, C. H. *Striegau, Breslau.*—Brushes.

1761 SAMTER, L. *Lissa.*—Bristles.

1763 SCHULZE, R. *Glogau, Liegnitz.* — Fancy articles in hair.

1764 STANGE, F. D. *Aschersleben, Magdeburg.*—Horse and pig's hair.

CLASS 26.

1765 BARTSCH, FR. & SONS, *Striegau, Breslau.*—Leather, machine-straps.

1766 BENJAMIN, M. A. *Cologne.* — Leather.

1770 EHRHARDT, C. T. *Nordhausen.*—Trunk, game-bag.

1771 GAMMERSBACH, F. W. *Roisdorf, Bonn.*—Calf-skin, goat's-skin.

1772 HARFF, P. J. *Cologne.*—Sole-leather.

1773 HARTMANN, F. 166, *Friedrich-str. Berlin.*—Saddle, brougham-harness.

1776 HÜTTENHEIN, H. *Hilchenbach, Arnsberg.*—Sole-leather.

1777 JACOBI, B. JUN. *Weissenfels, Merseburg.* — Three tanned skins, for equestrian uses.

1779 KÄRNBACH, C. 46, *Louisen-str. Berlin.* — Harness for a saddle-horse.

1780 KLEINSCHMIDT, C. C. *Mühlhausen, Erfurt.*—Glove-leather.

1781 KLEIST, T. W. 54, *Oranienburger-str. Berlin.*—Sole-leather, boots.

1782 KORN, R. A. *Saarbrücken.*—Morocco leather.

1783 KORNFELDT, L. & CO. 62, *Spandauer-str. Berlin.*—Tanned calf and horse skins.

1784 KÜHLING, A. *Düsseldorf.* — Elastic saddle.

1785 LAMM, I. H. 29, *Ross-str. Berlin.*—Boot and shoe trees, calf skins.

1786 MEYER, A. *Gumbinnen.*—Harness.

1789 PASCHEN, W. *Königsberg.*—Saddles, with elastic bows.

1791 ROLKE, F. *Breslau.*—Portmanteaus and bags.

1792 ROSENBAUM, F. W. *Breslau.*—Harness.

1796 STERNEFELD, BROS. *Goch, Düsseldorf.*—Calf leather.

1797 STRATHMANN, H. 44, *Hirschel-str. Berlin.*—Roof leather without varnish.

1799 WIEHR, F. & SONS, 22, *Fischer-str. Berlin.*—Whips.

CLASS 27.

1803 BERLO, J. & A. v. *Aix-la-Chapelle.*—Worsted and kid gloves.

1806 BRESLAU SHOEMAKERS' ASSOCIATION, (*Weintraube*).—Boots and shoes.

1807 CADURA, H. *Breslau.*—Waterproof coats.

1808 CIESIELSKI, A. *Bromberg.*—Boots and shoes.

1809 CLASSEN-KAPPELMANN, *Cologne.*—Hosiery.

1811 DOMBROWSKY, ST. *Posen.*—Gentlemen's boots.

1812 DODECK, R. *Burg, Magdeburg.*—Gloves and leather.

1813 EIGEL, F. *Cologne.* — Gentlemen's boots.

1814 ELSTER, A. 124, *Frankfurter-str. Berlin.*—Straw hats and bonnets, straw tresses and straw.

1818 HACKENBERG, FR. *Elberfeld.*—Buttons.

1818 A HELFEIER, S. 47, *Alexander-str. Berlin.*—Gloves.

1819 HERRMANN, E. & CO. 7, *Breite-str. Berlin.*—Umbrellas.

1820 HESSTHAL, W. J. *Aix-la-Chapelle.*—Worsted gloves.

1821 HOLLÄNDER, B. *Leobschütz, Oppeln.*—Woollen children's jackets, tippets, and ladies' caps.

1823 KIPPMEYER, *Crefeld.*—Boots and shoes.

1825 KÜHN, I. W. *Crefeld.*—Boots and shoes.

1826 LANGETHAL, G. *Erfurt.* — Boots, buskins, and shoes.

1827 LANGMEYER, C. A. 105 & 106, *Friedrich-str. Berlin.*—Hats and bonnets of woven fabrics.

1828 LAUFFER, E. & CO. 4, *Nicolaï-kirch-gasse, Berlin.*—Straw hats and bonnets.

1829 LENZ, C. F. 15, *Tauben-str. Berlin.*—Caps, shawls, sleeves, mantillas.

1834 MÜLLER, J. L. 76, *Schützen-str. Berlin.*—Boots and buskins.

1837 NOACK, E. 3, *Spittelmarkt, Berlin.*—Felt and felt-cloth shoes.

1838 OPPERMANN, E. F. 60, *Linden, Berlin.*—Boots and shoes.

1840 SALKOWSKI, J. *Posen.*—National dress.

1842 SCHMIDT, C. R. *Breslau.*—Boots and spatterdashes.

1843 SCHRÖDER, CHR. *Laasphe, Arnsberg.*—Boots.

1844 SEELIG, S. 53, *Alexander-str. Berlin.*—Clothes made of different kinds of hair.

1846 SOMMERFELD, L. 30, *König-str. Berlin.*—Great coat, waistcoat.

1847 SONDERMANN, FR. W. *Gummersbach, Cologne.*—Worsted hosiery.

1848 STEGMANN, C. A. F. 8, *Scharren-str. Berlin.*—Umbrellas.

1850 TEICHMANN, M. *Leobschütz, Oppeln.*—Woollen shawls, jackets, caps, and sleeves.

1851 TESCHEMACHER, R. & KATTENBUSCH, A. *Werden, Düsseldorf.*—Felt and leather boots and shoes.

1853 WECHSELMANN & Co. 42, *Leipziger-str. Berlin.*—Caps, tippets, and millinery.

1854 WEIDLING, A. *Erfurt.*—Boots and buskins.

1855 WEISSLER, S. *Leobschütz, Oppeln.*—Woollen caps, sleeves, and jackets.

1856 WLOSCIBORSKI, L. *Posen.*—Shoes and spatterdashes.

1859 ZARRAD, F. A. *Crefeld.*—Boots and shoes.

1860 ZARRAD, J. A. *Crefeld.*—Boots and shoes.

CLASS 28.

1861 BEHREND, B. *Cöslin.*—Writing and printing paper, straw paper, asphalt paper.

1862 BERLIN PATENT-PAPER MILL, 75, *Mühlen-str.*—Samples of paper.

1863 BORNEFELD, H. W. *Barmen.*—Wafers.

1864 BRASELMANN & VORSTER, *Stennert, Arnsberg.*—Letter, common writing, printing, and packing paper.

1866 CADURA, H. *Breslau.*—Ink.

1867 CRAMER, J. *Cologne.*—Aleppo ink.

1867A DECKER, R. *Berlin.*—Specimens of typography.

1868 DUNKER, A. 21, *Französische-str. Berlin.*—Printed books and albums.

1869 EBART, C. E. 13 & 14, *Mohren-str. Berlin.*—Paper and paste-board; press-board.

1870 EBBINGHAUS, FR. W. *Letmathe, Arnsberg.*—Writing and drawing paper.

1871 EHLERT, H. & Co. 104, *Oranien-str. Berlin.*—Specimens of type founding.

1871A FÖRSTER, F. S. *Crampe, Liegnitz.*—Writing and packing paper.

1872 FRIEDLÄNDER, J. 217, *Friedrich-str. Berlin.*—Plates for music printing.

1873 FRIEDLÄNDER, R. & SON, 9, *Kur-str. Berlin.*—Fac-simile of ancient prints.

1874 FRIEDERICH, TH. *Camen, Arnsberg.*—Straw paper.

1875 GÄDICKE, J. 34, *Linden-str. Berlin.*—Straw paper.

1877 GLÜER, L. 52, *Friedrichspracht, Berlin.*—Embroidery patterns.

1878 GOGLER, L. 65, *Friedrich-str. Berlin.*—Embroidery patterns.

1879 GUNDLACH, W. *Breslau.*—Fancy articles of pasteboard and leather.

1882 HASSEL, W. *Cologne.*—Books.

1883 HAYN, A. W. 29, *Zimmer-str. Berlin.*—Specimens of typography.

1884 HENDLER, FR. *Alt Friedland, Breslau.*—Writing and printing paper.

1885 HEYMANN, C. 29, *Bellevue-str. Berlin.*—Specimens of typography.

1887 HÖSCH, L. & E. *Düren.*—Paper.

1888 HÖSCH, BROS. *Düren.*—Paper.

1889 HOFERDS, J. & Co. *Breslau.*—Account books.

1890 HOHENOFEN PATENT-PAPER MILL, *near Neustadt-on-Dosse.*—Paper.

1890A HÜTTENMÜLLER, TH. *Lorenzdorf, Liegnitz.*—Press-boards.

1891 KATZSCHKE, R. *Weissenfels, Merseburg.*—Lanterns.

1892 KOCH, C. A. *Kippemühle, B. Gladbach, Cologne.*—Paper.

1895 KÜHN, C. & SONS, 25, *Breite-str. Berlin.*—Commercial account books; waterproof money-paper.

1897 KUHLHOFF, W. *Neheim, Arnsberg.*—Straw paper.

1898 KULLRICH, F. F. 34, *Leipziger-str. Berlin.*—Bookbinders' work.

1899 LAMBERTS, W. *M. Gladbach, Düsseldorf.*—Ledger, journal, copying-book.

1900 LEHMANN & MOHR, 114, *Linien-str. Berlin.*—Bill forms, types for the blind, games, &c.

1901 LESIMPLE, A. *Cologne.*—Specimens of typography.

1905 LUNGE, DR. G. *Breslau.*—Wheat straw paper.

1906 MAY, C. F. 5, *Stralauer-platz, Berlin.*—Cards and parchment.

1907 MEISSNER, C. F. & SON, *Raths-Damnitz, Cöslin.*—Writing and printing paper, paper-hangings, &c.

1908 MEISTER, TH. W. 16, *Breite-str. Berlin.*—Patterns for Berlin fancy work.

1909 MÖSER, W. 34, *Stallschreiber-str. Berlin.*—Illustrated typographic tableau.

1910 MOOLEN, VAN DER H. *Geldern, Düsseldorf.*—Sealing-wax, ink.

1911 MÜLLER, W. *Kettwig, Düsseldorf.*—Paper.

1912 MÜNCH, C. H. *Königsberg.*—Lady's writing portfolio.

1913 NICOLAI, 13, *Brüder-str. Berlin.*—Various books.

1917 RHEINEN, H. J. *Elberfeld.*—Transparent drawing paper.

1918 ROSENTHAL, J. 45, *Neue Friedrich-str. Berlin.*—Commercial account books.

1920 SCHMIDT, G. A. *Halle-on-Saale.*—Photographic albums.

1921 SCHMITZ, BROS. *Düren.*—Paper dyed in the stuff.

1922 SCHNEIDER, F. 9, *Links-str. Berlin.*—Stamped pasteboards and frames for photographs.

1923 SCHÖLLER, H. A. & SONS, *Düren.*—Paper.

1924 SCHULZE, FR. R. H. 49, *Ritter-str. Berlin.*—Illustrated chronicle of the orders of knighthood.

1926 STANGE, E. 5, *Neue Schönhauser-str. Berlin.*—Paper.

1928 STERN, A. 21, *Fischerbrücke, Berlin.*—Ruled paper.

1929 TENGE'S PAPER MILL, *Dalbke, Bielefeld.*—Hand-made paper and press-boards.

1930 TROWITZSCH & SON, 112, *Leipziger-str. Berlin.*—Specimens of printing and type-founding.

1933 WEISS, E. & Co. *M. Gladbach, Düsseldorf.*—Ledger, journal, sample book.

1935 ZECHENDORF & BERTHOLD, 1, *Wilhelm-str. Berlin.* — Galvanoplastic copper-plates, galvanoplastic type plates, specimens of typography.

CLASS 29.

1936 BRENNECKE, DR. *Posen.* — English grammar, compendium of arithmetic, work on mathematics.

1937 BURO, L. 6, *Gyps-str. Berlin.* — Map of Europe in relief.

1938 CADURA, H. *Breslau.*—Caoutchouc balls and toys.

1939 DÜMMLER, F. 36, *Mohren-str. Berlin.*—Engravings and books, educational works, illustrated according to Fröbel's designs.

1940 FLEMMING, C. *Glogau, Liegnitz.* — Topographic map of Europe. Books, &c.

1941 FRANZ, C. F. *Oranienburg, near Berlin.*—Book with geometrical figures.

1942 GRUNERT, C. 20, *Wollank-str. Berlin.*—System of Arend's stenography.

1943 GUTTENTAG, J. 8, *Unterwasser-str. Berlin.*—The joiner's guide-book, with map.

1945 HENNIG, AUG. *Halle-on-Saale.* — Printed figures to be applied to caoutchouc balls, &c.

1946 HERMES, W. 26, *König-str. Berlin.*—Copy-books, albums, &c.

1947 HILDEBRANDT, *Burg.*—Plan of education, specimens of pupils' work.

1948 HIRSCH, M. DR. 65, *Schützen-str. Berlin.*—Educational and other works.

1950 HUPP & WÜLFING, *Düsseldorf.* — Copy-tables to teach children and the blind to write.

1952 KÖHLER, A. 28, *Schöneberger-str. Berlin.*—Globe and map of Europe in relief. Photographic map in relief.

1953 KRANTZ, DR. A. *Bonn.* — Wood models of crystals, minerals, rocks, petrifactions.

1954 KÜHN, G. *Neu Ruppin, near Berlin.*—Prints.

1955 LANGENSCHEID, J. L. G. 43, *Hirschel-str. Berlin.*—Letters for studying the French and English languages.

1956 LENZ, G. F. 24, *Gertranden-str. Berlin.*—Lithographic prints and works on gymnastics.

1957 MOHR, C. W. & Co. 114, *Linien-str. Berlin.* — Method for teaching German, English, and Russian. Juvenile library.

1960 REIMER, D. 11, *Anhalt-str. Berlin.*—Globes and maps.

1961 RUNGE, DR. F. F. *Oranienburg, near Berlin.*—Works on chemistry.

1963 SCHNABEL, DR. C. *Siegen, Arnsberg.*—Glass crystals.

1965 STOLLE, A. *Erfurt.* — Toys (horse, sheep, goat).

1966 THÄRMANN, H. *Königerode, Merseburg.*—Explanatory table of the method of tuition.

1967 WARNSDORF, L. VON, 33, *Köthener-str. Berlin.*—Jacobi's letters for the study of languages.

CLASS 30.

1969 ARON & JACOBI, 29, *Brüder-str. Berlin.*—Window blinds.

1970 BASCH, W. & Co. 9 & 10, *Neue Friedrich-str. Berlin.*—Safety lock.

1972 RISSING, FR. *Lissa, Posen.* — Chess-board table and stool.

1973 BOMBÉ, A. *Ehrenfeld, Cologne.* — Floor of inlaid work.

1974 CARL, J. F. C. *Spandau, Berlin.*—Gilt fillets for picture frames, &c.

1975 DREYKLUFT, A. *Merseburg.*—Jewellery table of walnut-tree.

1976 FERRENHOLZ, G. J. *Wesseling, Bonn.*—Wood fillets, gilt and polished.

1977 FRANZ, O. 88, *Wilhelm-str. Berlin.*—Clockcases, inkstands, picture frames.

1978 FRILING & Co. *Cologne.* — Wood fillets, gilt and polished.

1979 FUHRBERG, F. 190, *Friedrich-str. Berlin.*—Household furniture of wicker work, &c.

1980 GERICKE, H. & PFITZNER, 21, *Prinzessinnen-str. Berlin.*—Lady's bureau, picture-frames.

1983 HEINRICH, C. *Nordhausen.*—Eider-down quilt.

1984 HEINRICH, J. *Bonn.*—Wood fillets, gilt and polished.

1985 HERBST, A. *Bonn.*—Household furniture of wicker work.

1987 KELTERBORN, R. 57, *Behren-str. Berlin.*—Tables, chairs, cases.

1988 KEMPEN, W. & Co. *Emmerich, Düsseldorf.*—Gold and polished fillets.

1989 KILIAN, GAB. *Bonn.*—Brooms and brushes of straw.

1990 KLEIN, G. W. J. 48A, *Zimmer-str. Berlin.*—Gilt fillets for picture frames, &c.

1993 KOKSTEIN, A. 1A, *Wollank-str. Berlin.*—Looking-glass frames.

1997 LÜDECKE, A. *Brandenburg, Berlin.*—Gilt fillets for picture frames, &c.

1998 LÖVINSON, L. & S. 8, *Linden, Berlin.*—Carved household furniture.

2000 MELLER, F. & Co. *Bonn.* — Wood fillets, gilt and polished.

2001 METHLOW, E. & Co. 109A, *Köpnicker-str. Berlin.*—Frames for photographs.

2002 NEESS, A. F. *Cologne.*—Wood fillets, gilt and polished; frames; flowers of carton-pierre.

2003 PAASCHEN, E. *Stendal, Magdeburg.*—Household furniture.

2004 REICHARDT, J. F. *Erfurt.*—Household furniture of wicker-work.

2005 SCHIEVELBEIN, J. F. E. 46, *Spandauer-str. Berlin.*—Arm chairs.

2006 SCHIROW, C. A. & Co. 68A, *Leipziger-str. Berlin.*—Wicker-work.

2007 SCHMIDT, L. *Culm, Marienwerder*—Polysander bureau.

2008 SCHULZE, F. A. 6, *Neue Grün-str. Berlin.*—Key-hole lids of bone.

2010 SONNENBURG, BOARD OF DIRECTORS OF THE PENITENTIARY, *Frankfurt-on-Oder.*—Ornamental chair.

2014 UNGER, J. *Erfurt.*—Pillow for railway passengers; elastic under-bed.

2015 VALLENTIN & SCHÄFER, *Schweidnitz.*—Mountain-pine ware.

2016 WERKMEISTER, A. 11, *Michaelkirch-str. Berlin.*—Gilt fillets for picture frames, &c.

2017 WIEDEMANN, D. P. 54, *Jerusalemer-str. Berlin.*—Household furniture of wicker-work.

2019 WINKLER, H. 29, *Kur-str. Berlin.*—Varnished baskets.

CLASS 31.

2020 ARNHEIM, S. J. 36, *Rosenthaler-str. Berlin.*—Iron safe, complicated locks.

2022 BERG, W. *Lüdenscheid, Arnsberg.*—Metal buttons, clasps, &c.

2023 BERLIN ROYAL IRON FOUNDRY.—Monument of Frederick William III., busts of the Prince and Princess Royal, chandelier, candlestick, vase, pedestal to baptismal font.

2025 BOGDANSKI, J. VON, *Posen.*—Locks.

2026 BRIX, J. 25, *Karl-str. Berlin.*—Zinc groups; architectural ornaments.

2029 COLOGNE JOINT STOCK CO. FOR THE MANUFACTURE OF MACHINES.—Iron gas and water pipes.

2030 COHN, E. J. & Co. 55, *Alexander-str. Berlin.*—Doors for stoves.

2031 COSACK & Co. *Hamm, Arnsberg.*—Telegraph wire, springs, iron axle-trees.

2032 DAHM, KNÖDGEN, & KIRCHNER, *Fraulautern, Saarlouis.*—Kitchen utensils of sheet iron.

2033 DIEBITSCH, C. VON, 4, *Hafenplatz, Berlin.*—Zinc vases.

2035 EPSTEIN, L. *Lublinitz, Oppeln.*—Wrought iron and tin spoons.

2036 FABIAN, M. 75, *Spandauer-str. Berlin.*—Iron safe.

2037 GECK, A. TH. *Iserlohn.*—Stamped brass ware; mouldings, picture frames, and photograph albums.

2038 GEISS, M. 31, *Behren-str. Berlin.*—Four statues, zinc cast and bronzed by the electro-galvanic process.

2039 GLADENBECK, H. 10, *Münz-str. Berlin.*—Bronze group, by Wolf, bust of Schiller, figures of Russian soldiers.

2041 HASEMANN, M. A. 154, *Friedrich-str. Berlin.*—Scales with columns.

2042 HAUSCHILD, C. 3, *Neander-str. Berlin.*—Safe.

2043 HECKEL, G. *St. Johann, Treves.*—Wire ropes.

2044 HEINTZE & BLANKERTZ, 4, *Flieder-tr. Berlin.*—Steel pens.

2045 HENRIETTENHÜTTE IRON FOUNDRY, *near Primkenau, Liegnitz.*—Cast-iron enamelled pottery.

2046 HERRMANN, C. *Danzig.*—Bronze chandelier.

2048 HITSCHLER, J. *Crefeld.*—Iron safe, decimal weighing machine, copying press.

2049 HOBRECKER, WITTE & HERBERS, *Hamm, Arnsberg.*—Iron wire, wire tags.

2051 HUPP, C. *Düsseldorf.*—Mounting of a rifled sporting gun.

2052 HÜSTEN MINING Co. *Arnsberg.*—Tin plate, iron plate.

2053 ILSENBURG, COUNT OF STOLBERG WERNIGERODE'S FACTORY.—Ornamental cast-iron work.

2056 KAHN, J. & Co. *Cologne.*—Iron chains, hinges.

2057 KISSING & MÖLLMANN, *Iserlohn.*—Lustres, chandeliers, bronze ornaments.

2058 KNOLL, L. 113, *Linien-str. Berlin.*—Bronze statue inlaid with silver.

2059 KOCH & BEIN, 49, *Neue Friedrich-str. Berlin.*—Metal and glass letters.

2060 KÖPPEN, J. W. 32, *Alte Jacob-str. Berlin.*—Chandeliers, lustres, lamps.

2061 KÖRNER, H. 91, *Zimmer-str. Berlin.*—Assortment of lamps.

2062 KOLESCH, H. *Stettin.*—Iron safe.

2064 KRAUSE, F. W. *Neusalz, Liegnitz.*—Cast-iron enamelled pottery.

2065 KRIEG & TIGLER, *Wesel.*—Iron-wire, wire ropes.

2067 LAUCHHAMMER, COUNT EINSIEDEL'S IRONWORKS, *near Mückenberg, Merseburg.*—Chimneys, stoves, statuettes, figures of animals, household furniture.

2068 LENZMANN, C. W. *Hagen, Arnsberg.*—Locks for caskets and doors, padlocks, &c.

2069 LIEBL, A. 19, *Gollnow-str. Berlin.*—Copper kettle.

2071 LUDWIG, BROS. *Breslau.*—Curry-combs.

2072 MACIEJEWSKI, ST. *Posen.*—Large door-lock.

2074 MERTINS, C. P. 90, *Linden-str. Berlin.*—Arms of the Queen of England, two lions. Corinthian capital.

2075 MEVES, A. 86, *Chaussee-str. Berlin.*—Group cast in zinc and chased, chased silver chess-board and figures.

2076 MULACK, H. 21, *Kur-str. Berlin.*—Gothic church window, Gothic cross-flower, vase of zinc, &c.

2077 NELCKE, O. 41, *Krausen-str. Berlin.*—Engraved stamps.

2078 NEUMANN, L. H. *Königsberg.*—Iron safe.

2079 NÖLLE, BROS. *Lüdenscheid, Arnsberg.*—Hardware, articles of Britannia metal.

2080 OHLÉS, E. F. HEIRS, *Breslau.*—Lead-ware.

2080A PASCHMANN, F. *Hörstgen, Düsseldorf.*—Door latches, key-hole lids.

2081 PETERS, FR. 22, *Köthener-str. Berlin.*—Gothic church-window.

2083 POHL, H. 21, *Alte Jacob-str. Berlin.*—Cast-zinc table, gas-chandelier, vase, four figures, full-sized.

2084 POKORNY, J. A. 17, *Oberwall-str. Berlin.*—Gold-weights and scale-beams.

2085 PUTH, H. *Blankenstein, Arnsberg.*—Rope of cast-steel wire.

2086 REIMANN, L. 104, *Oranien-str. Berlin.*—Scales and weights.

2090 ROTHENBURG, *Merseburg-on-Saale, Kupferhammer.*—Vacuum-pan.

2091 SCHÄFFER & WALKER, 19, *Linden-str. Berlin.*—Glass chandeliers, lustres, and lamps.

2095 SCHMALZ & SIMSON, *Magdeburg.*—Steel and iron hardware.

2097 SCHWARTZ, R. *Posen.*—Lock, with safety-contrivance.

2100 SOMMERMEYER & Co. *Magdeburg.*—Iron safes.

2101 SPINN, J. C. & SON, 13, *Wasserthor-str. Berlin.*—Bronze chandelier.

2102 STOBWASSER, C. H. & Co. 98, *Wilhelm-str. Berlin.*—Fire-screen, lamps, chandelier.

2103 STROBEL, G. *Frankenstein, Breslau.*—Crucifix of electro-plated metal.

2105 TURK, WIDOW R. C. *Lüdenscheid, Arnsberg.*—Buttons, brooches, clasps, buckles, tinder-boxes.

2106 UHLHORN, D. *Grevenbroich, Düsseldorf.*—Samples of card.

2107 VARENKAMP, FR. *Düsseldorf.*—Book on the treatment of the horse's hoof; assortment of horse-shoes.

2108 WILD & WESSEL, 26, *Prinzessinnen-str. Berlin.*—Select specimens of paraffin-oil lamps.

CLASS 32.

2109 BEISSEL, WIDOW S. & SON, *Aix-la-Chapelle.*—Needles.

2110 BÖLLING & VON DER CRONE, *Haspe, Arnsberg.*—Scythe, straw-knife.

2111 CORTS, G. *Remscheid, Düsseldorf.*—Files and edge-tools.

2112 DICKERTMANN, BROS. *Bielefeld.*—Files.

2113 EDELHOFF, J. W. & Co. *Remscheid, Düsseldorf.*—Saws and files.

2115 JUNG, C. *Dahl, Arnsberg.*—Anvils, scythes, straw-knives.

2116 KLEB, W. *Allenbach, Arnsberg.*—Hatchet, hoe, axe.

2116A KRUPP, FR. *Essen, Düsseldorf.*—The manufacture of steel in large masses and the production of heavy forgings, as demonstrated by an ingot of 21 tons' weight broken in half; a double crank shaft, weighing 9 tons, partly finished; a forging suitable for an intermediate shaft, weighing 16 tons; several pieces of heavy ordnance; tyres, locomotive crank-axles; an anchor, a screw, &c.

2117 LAMMERTZ, L. *Aix-la-Chapelle.*—Needles.

2120 MANNESMANN, A. *Remscheid, Düsseldorf.*—Cast steel, refined steel, files.

2122 PRINTZ, G. & Co. *Aix-la-Chapelle.*—Needles.

2124 REINSHAGEN, G. *Remscheid, Düsseldorf.*—Files and rasps.

2125 SCHLEICHER, C. *Schönthal, Aix-la-Chapelle.*—Needles.

2126 SCHUMACHER, F. *Aix-la-Chapelle.*—Glass breast-pins and buttons, steel and brass pins, needles.

2129 WERMINGHAUS, J. C. *Hagen, Arnsberg.*—Knives, called "Holländermesser."

2130 WITTE, ST. & Co. *Iserlohn, Arnsberg.*—Needles and fishing-hooks.

CLASS 33.

2132 BRANDT, E. F. & H. (firm, Friedrich & Brandt) *Stettin.*—Silver drinking-cup, and various silver articles.

2134 FRIEDEBERG, S. & SONS, 42, *Linden, Berlin.*—Set of plate, silver escutcheon, jewels, and trinkets.

2135 GOSCHE, H. 9, *Stallschreiber-str. Berlin.*—Chased silverwork.

2136 GRÄGER, KLUG, & HARTUNG, *Mühlhausen, Erfurt.*—Silver-plated and cast-iron image, silver-plated iron nails.

2137 KÖHLER, A. *Liegnitz.*—Stamped silver jewellery.

2138 LÖWENTHAL, A. M. & Co. *Cologne.*—Plated table-service.

2139 MOSGAU, F. 49, *Markus-str. Berlin.*—Silversmith's ware.

2140 SCHÜTZ & HOFFMEISTER, 112, *Linden-str. Berlin.*—Silver-plated coach lanterns.

2141 SCHWARZ, C. 26, *Mohren-str. Berlin.*—Gold and silver work.

2142 SY & WAGENER, 28, *Kronen-str. Berlin.*—Silversmith's ware.

2144 VOLLGOLD, D. & SON, 14, *Kommandanten-str. Berlin.*—Silversmith's ware.

2145 WINKELMANN, H. F. 11, *Mauer-str. Berlin.*—Plated cork stoppers.

CLASS 34.

2147 BIVER, *St. Gobain, Chauny, and Cirey, Aix-la-Chapelle.*—Looking-glass plates.

2148 HAARMANN, SCHOTT, & HAHNE, *Witten, Arnsberg.*—Plate-glass.

2149 HECKERT, C. 33, *Kronen-str. Berlin.*—Crystal sconces and chandeliers, lamps, glass flowers, looking-glasses, glass letters, cut glass.

2150 HECKERT, E. *Halle-on-Saale.*—Stained window glass.

2151 HERB, A. G. *Gersweiler, Saarbrücken.*—Bottles, jars, glasses, latches, buttons.

2155 SCHAFFGOTSCH, COUNT, *Josephinenhütte, Silesia.*—Cut crystal and ornamental glass.

2152 KLITZING, VON, *Bernsdorf, Liegnitz.*—Lamp shade, chimney, bottles, glasses.

2154 MÜLLENSIEFEN, BROS. *Crengeldanz, Arnsberg.*—Plate-glass.

2155 OIDTMANN, DR. & Co. *Linnich, Aix-la-Chapelle.*—Stained church window.

2156 SCHMIDTBORN, R. *Friedrichsthal, Saarbrücken.*—Window glass, wine bottles, large bottles for oil of vitriol.

2157 SCHULZE, TH. O. *Rauscha, Liegnitz.*—Lamp shade, chimney, glass plates.

2159 SOLMS-BARUTH, COUNT F. H. L. *Baruth, Berlin.*—Stained plate-glass, cylinders, bell glasses and covers, globes, &c.
2160 STRILACK, A. *Waitze, Posen.*—Glass.

CLASS 35.

2164 AUGUSTIN, A. *Lauban, Liegnitz.*—Conduit pipes, vase of baked clay.
2166 BERLIN ROYAL PORCELAIN MANUFACTORY.—Assortment of porcelain.
2167 BERLIN ROYAL GESUNDHEITSGESCHIRR MANUFACTUR.—Assortment of porcelain called "Gesundheitsgeschirr."
2168 DRYANDER & Co. *Saarbrücken.*—Objects of clay.
2171 KULMITZ, C. *Saarau, Breslau.*—Crucibles, refractory bricks.
2172 LINDEN, G. *Ratingen, Düsseldorf.*—Glazed tiles, drain pipes.
2173 MARCH, E. *Charlottenburg, Berlin.*—Figures, chandeliers, vases, font, architectural ornaments, conduit pipes of baked clay.
2174 MÜLLER, J. H. 61, *Mohren-str. Berlin.*—Breakfast service, vase, cup, &c.
2174A MÜLMANN, v. A. *Plato, Siegburg.*—Terra-cotta ware.
2175 ROTHENBACH, W. & Co. *Breslau.*—Painted porcelain.
2177 SCHOMBURG, H. *Moabit, Berlin.*—Porcelain.
2178 STRAHL, O. *Frankfurt-on-Oder.*—Clay and pottery.
2179 TIELSCH, C. & Co. *Altwasser, Breslau.*—Painted porcelain.
2180 VYGEN, H. J. & Co. *Duisburg.*—Coal tar retorts, refractory bricks.

CLASS 36.

2181 ALVES, H. 49, *Spandauer-str. Berlin.*—Leather work, whips, gold borders.
2182 BRAUNE, B. *Danzig.*—Amber work.
2183 BÜLOW, C. *Görlitz, Liegnitz.*—Masks; wire-gauze.
2186 GALOPIN, J. *Aix-la-Chapelle.*—Stair-step, banisters, column.
2187 GESELL, J. FR. *Görlitz, Liegnitz.*—Looking-glass frames.
2188 GOLDSCHMIDT, J. P. 35, *Mittel-str. Berlin.*—Razor strops.
2188A HAANEN, G. v. *Cologne.*—Painting imitating marble and tortoise-shell.
2189 HERZIG, G. *Hermsdorf. Liegnitz.*—Children's toys made of wood, polished wooden ware for household use.
2190 JANTZEN, A. F. *Danzig.*—Amber work.
2191 JANTZEN, G. E. *Stolp, Cöslin.*—Amber jewellery.
2192 KAHDE, E. 8, *Gertrauden-str. Berlin.*—Necessaries.
2193 KÖHLER, W. G. *Zeitz, Merseburg.*—Wood-carvers' work.
2194 LAUE, C. F. *Naumburg-on-Saale, Merseburg.*—Chess-board.
2196 NIESE, F. *Danzig.*—Amber work.
2197 PERLBACH, H. L. *Danzig.*—Amber work.
2198 RÖMPLER & TÖLLE, *Barmen.*—Articles of caoutchouc.
2200 VITÉ, F. 7, *Kommandanten-str. Berlin.*—"Ladies' Companion;" bags; albums.
2201 WELLHÄUSER, E. L. *Elberfeld.*—Articles of caoutchouc.
2202 WESTPHAL, C. A. *Stolp, Cöslin.*—Amber-beads, and other amber work.

KINGDOM OF SAXONY AND PRINCIPALITY OF REUSS (J. L.).

The Exhibitors of Reuss are marked with an *.

CLASS 1.

2301 SOCIETY OF THE MARBLE QUARRIES OF FÜRSTENBERG, *Grünhain.*—Chimney-pieces and other articles of marble.
2302 THE MILLSTONE MANUFACTORY, *Johnsdorf, near Zittau.*—Millstone.
2303 ZWITTERSTOCKS FACTORY, *Altenberg.*—Pure metallic tin in blocks, balls, and bars.

CLASS 2.

2304 DUVERNAY, PETERS & Co. *Leipzig.*—Preparations of archil, indigo and aniline.
2305 HEINE & Co. *Leipzig.*—Ethereal oils and mineral products.
2306 KRAUSE, H. H. *Freiberg.*—Collections for the study of chemistry.
2307 POMMIER & Co. *Neuschönefeld, near Leipzig.*—Preparations of archil, cochineal, picric acid, salts of alumina, &c.
2308 SACHSSE, E. & Co. *Leipzig.*—Ethereal oils and chemical products.
2309 SCHIMMEL & Co. *Leipzig.*—Ethereal oils and essences.
3310 SCHUETZ, A. *Wurzen.*—Dyed wool flock for paper hangings.
2311 THEUNERT & SON, *Chemnitz.*—Artificial ultramarine.
2312 WÜRTZ, T. *Leipzig.*—Chemical products and dye-stuffs.

CLASS 3.

2313 JORDAN & TIMÆUS, *Dresden.*—Chocolate, cocoa, confectionery.
2314 STENGEL, W. *Leipzig.*—Spirits of wine.

CLASS 4.

2315 BURCHARDI, F. v., *Hermsdorf, near Königstein.*—Honey, bees' wax, and similar products.
2316 KIND, J. C. H. *Gleina, near Bautzen.*—Fleeces.
2317 SCHÖNBERG, A. v., *Rothschönberg.*—Fleeces.
2318 STEIGER, E. *Löthain, near Meissen.*—Fleeces.

CLASS 5.

2319 HARTMANN, R. *Chemnitz.*—Locomotive; steam-engine; machines for working in metal; wool-carding machines.

CLASS 7.

2320 PURSCH, E. T. *Dresden.*—Grinding tools.

2321 SAUER, J. *Plauen.*—Flat ropes and driving-straps for machines; harness-cords for weaving.

2322 SONDERMANN & STIER, *Chemnitz.*—Cutting engine for boiler tubes.

2323 ZIMMERMANN, J. *Chemnitz.*—Machines for working metal and wood.

2324 LINDNER, W. F. *Chemnitz.*—Tools for planting trees.

2325 Vacat.

CLASS 8.

2326 PFITZER, E. *Oschatz.*—Decimal weighing machines, table balances.

2327 RUF, L. *Leipzig.*—Weighing machine and table balance.

CLASS 11.

2328 BÖSENBERG, L. H. *Leipzig.*—Percussion rifle-gun, rifle-gun and revolver.

CLASS 13.

2329 HUGERSHOFF, F. *Leipzig.*—Chemical apparatuses; balances, pneumatic pumps, apparatuses for testing gas, for spectrum analysis, &c.

2330 KOOSEN, J. H. *Burkhardswalde, near Pirna.*—Galvanic chronometers.

2331 RUETE, T. *Leipzig.*—Ophthalmotrope built by Stoehrer, Dresden. Model of the human eye, &c. by Tauber, Leipzig.

2332 SCHICKERT, H. *Dresden.*—Balance for chemical purposes.

2333 SCHNEIDER, K. H. E. *Leipzig.*—Balances and cases of mathematical instruments.

2334 Vacat.

CLASS 14.

2335 MANECKE, F. *Leipzig.*—Photographic portrait, whole size.

2336 BROCKMANN, F. & O. *Dresden.*—Photographs of pictures of the Dresden gallery, after drawings by Prof. Schurig.

CLASS 15.

2337 ASSMANN, J. *Glashütte.*—Watches, anchor escapement.

2338 GROSSMANN, M. *Glashütte.*—Watches and watchmakers' tools.

2339 LANGE, A. & Co. *Glashütte.*—Watches, anchor escapement.

2340 SCHNEIDER, A. *Glashütte.*—Watches, anchor escapement.

CLASS 16.

2341 BREITKOPF & HÆRTEL. *Leipzig.*—Pianofortes, music prints.

2342 GLÄSEL, C. W. *Markneukirchen.*—Guitar (viola-mela).

2343 HAUPT, C. A. F. *Leipzig.*—Pianino.

2344 IRMLER, E. *Leipzig.*—Grand piano, pianino.

2345 KAPS, E. *Dresden.*—Grand piano.

2346 OBERREIT, R. *Klingenthal.*—Accordions. Boxes of inlaid work.

2347 OTTO, F. A. *Markneukirchen.*—Violin bows.

2348 SCHUSTER, BROS. *Markneukirchen.*—String and wind instruments, &c.

2349 SCHUSTER, M. JUN. *Markneukirchen.*—Musical instruments.

2350 SEYFERT, F. W. *Chemnitz.*—Pianino.

2351 *WAGNER & Co. *Gera.*—Accordions and harmoniums.

2352-54 Vacat.

CLASS 18.

2355 BAUMGÄRTEL, C. E. & SON, *Lengenfeld.*—Cotton curtain-stuffs, spotted cambrics.

2356 CHALYBÄUS & MÜHLMANN, *Chemnitz.*—Cotton yarns and threads.

2357 Vacat.

2358 FÖRSTER, O. *Chemnitz.*—Knitting yarns, threads, spindle-strings, wick-yarns.

2359 Vacat.

2360 GRUNER, H. *Ebersbach, near Löbau.*—Cotton goods.

2361 HÄBLER, BROS. *Gross-Schönau.*—Cotton and half linen tickings.

2362 HERZOG, H. W. *Neugersdorf, near Löbau.*—Cotton and half-woollen stuffs.

2363 HEYDENREICH, R. *Witzschdorf, near Zschopau.*—Cotton yarns and threads.

2364 HOFFMANN, C. G. *Neugersdorf, near Löbau.*—Cotton and mixed stuffs.

2365 HÜTTIG, B. & Co. *Leutersdorf, near Löbau.*—Cotton and mixed stuffs.

2366 KELLER & GRUBER, *Chemnitz.*—Cotton yarns for knitting and crocheting.

2367 KLEMM, R. *Plauen.*—Cotton gauze curtains, mixed stuffs.

2368 MARX, H. R. *Seifhennersdorf, near Löbau.*—Cotton and mixed stuffs.

2369 Vacat.

2370 SCHÖNE, J. G. *Grossröhrsdorf.*—Ribbons and small articles of cotton.

2371 SEYFERT, G. *Auerbach.*—Cotton gauze curtains.

2372 TETZNER, C. A. & SON, *Burgstädt.*—Twists, yarns, merino weaving yarns.

2373 Vacat.

2374 WÆNTIG & Co. *Zittau.*—Cotton and mixed-linen tickings.

2375 ZSCHIMMER & GRIMM, *Plauen.*—Cotton gauze curtains.

2376-80 Vacat.

CLASS 19.

2381 BURSCHE, J. G. *Pulsnitz.*—Sail-cloth, twillings, covers for railway lorries.

2382 LANGE, J. *Waltersdorf, near Zittau.*—Linen twillings and figured articles.

2383 MEYER, J. *Au petit bazar, Dresden (Grosschönau).*—Linen damasks and diapers.

2381 NEUMANN, C. F. *Eybau, near Loebau.*—Creas, listados, &c.

CLASS 21.

2385 ALBRECHT, R. *Chemnitz.*—Stuffs for furniture, curtains, carriages, table-covers, &c.
2386 BAUCH, C. A. *Rosswein.*—Duffel cloth.
2387 BAUCH, T. *Rosswein.*—Duffel cloth.
2388 BECKERT, R. *Chemnitz.* — Mixed woollen furniture stuffs, mixed-woollen and silk Ponchos.
2389 BLEYL. F. *Camenz.*—Woollen cloth.
2390 BÖTTIGER, F. W. *Crimmitzschau.*—Vigogne yarns.
2391 BÖTTGER, BROS. *Leisnig.*—Woollen stuffs.
2392 BRODENGEYER, F. & CO. *Annaberg.* —Cravats of woollen and cotton chenille.
2393 CASPARI, J. F. *Grossenhain.*—Woollen cloth.
2394 CLAUS R. & CO. *Schedewitz, near Zwickau.*—Mixed-silk stuffs.
2395 DIETERICH, H. *Meerane.*—Half-woollen and mixed silk stuffs.
2396 ECKHARDT, BROS. *Grossenhain.* — Buckskins.
2397 FACILIDES & WIEDE, *Plauen.* — Vigogne yarns.
2398 Vacat.
2399 FROHBERG, C. G. *Rosswein.*—Cylinder-cloth.
2400 FROHBERG, BROS. *Rosswein.*—Woollen cloth.
2401 GLASS & SON, *Reichenbach.*—Woollen and mixed-woollen stuffs. Printed table-covers.
2402 HERRMANN, F. G. & SON, *Bischofswerda.*—Woollen cloth.
2403 HERTEL & BÜCHELEN, *Meerane.*—Mixed-woollen and mixed silk goods.
2404 HÜFFER, H. *Crimmitzschau.* — Vigogne yarns.
2405 Vacat.
2406 KRAH, C. *Camenz.*—Woollen cloth.
2407 KRÄMER & MARKENDORF, *Glauchau.* —Woollen and mixed stuffs.
2408 KRATZ & BURK, *Glauchau.*—Woollen and mixed stuffs.
2409 KÜRZEL, H. F. *Crimmitzschau.* — Vigogne yarns.
2410 LANGE, A. *Camenz.*—Woollen cloth.
2411 LEHMANN, F. G. *Böhrigen, near Rosswein.*—Flannels and other woollen stuffs.
2412 LEONHARDT, F. & SON, *Haynichen.*—Flannels.
2413 LEONHARDT, G. F. *Am Markt.—Haynichen.*—Flannels.
2414 LIPPERT, C. F. *Camenz.* — Woollen cloth.
2415 LOHSE, E. *Chemnitz.*—Stuffs for furniture, curtains, &c.
2416 MEHLHORN & SON, *Glauchau.*—Woollen and mixed stuffs.
2417 MEISSNER, F. T. *Grossenhain.*—Woollen cloth.
2418 METZLER, G. *Rosswein.* — Woollen cloth and duffel cloth.
2419 METZLER, W. *Rosswein.* — Woollen cloth.
2420 MINOKWITZ, A. *Camenz.*—Woollen cloth.
2421 *MORAND & CO. *Gera.* — Coloured stuffs of soft worsted yarn, mixed with silk.
2422 MÖRBITZ, C. G. E. *Bautzen.* — Woollen cloth and woollen stuffs.
2423 MÜLLER & CO. *Crimmitzschau.*—Buckskins.
2424 NOSSKE, E. *Camenz.*—Woollen cloth
2425 NOSSKE, F. *Camenz.*—Woollen cloth and satin.
2426 NOSSKE, W. *Camenz.*—Woollen cloth.
2427-28 Vacat.
2429 RAUE, F. *Rosswein.*—Cloth and cylinder-cloth.
2430 REICHEL, C. F. *Rosswein.*—Woollen cloth.
2431 REISSMANN & TRÄGER, *Reichenbach.* —Soft worsted yarns.
2432 RESCH & CO. *Meerane.*—Mixed woollen stuffs.
2433 SIEVERS & ENGELL, *Meerane.* — Mixed woollen and silk goods.
2434 SOLBRIG. C. F. *Harthau, near Chemnitz.*—Soft worsted yarns.
2435 SOELLHEIM, G. F. *Chemnitz.*—Furniture stuffs.
2436 SPENGLER, C. *Crimmitzschau.* — Buckskins.
2437 STEIN, J. G. *Haynichen.*—Flannels.
2438 STRAFF & SON, *Meerane.* — Stuffs for ladies' dress.
2439 STRÜBELL & MÜLLER, *Meerane.* — Woollen and mixed silk stuffs.
2440 THIEME, L. & CO. *Meerane.* — Woollen and mixed stuffs.
2441 UNGER, C. G. *Kirchberg.*—Woollen cloth.
2442 VORWERG, E. *Camenz.* — Woollen cloth.
2443 WAGNER, J. & CO. *Grünroda, near Döbeln.*—Buckskins and other stuffs.
2444 * WEISSFLOG, E. F. *Gera.* — Soft worsted goods.
2445 WIPPERN, C. & WIEHE, *Crimmitzschau.*—Mixed angola yarns.
2446 * WITTMER & SCHÖNHERR, *Gera.*—Soft worsted goods.
2447 WOLF, J. G. SEN. *Kirchberg.* — Woollen cloth.
2448 WOLFF, J. H. *Burgstädt.*—Woollen and Vigogne yarns.
2449 * ZETZSCHE & MÜNCH, *Gera.* — Thibets, cashmeres, shawls.
2450 ZSCHILLE, F. & CO. *Grossenhain.*—Woollen stuffs.
2451 ZSCHILLE, BROS. *Grossenhain.* — Woollen cloth.
2452-62 Vacat.

CLASS 22.

2463 QUAST, F. *Leipzig.* — Oilcloth-plafonds for lining the walls of railway waggons.
2464 RÖLLER & HUSTE, *Leipzig.* — Oilcloth. Geographical maps printed on oilcloth.
2465 SCHÄFER, J. H. *Chemnitz.* — Oilcloth.
2466 SCHUMANN, A. *Leipzig.* — Floorcloth.
2467 WÄNTIG, E. F. *Leipzig.* — Oilcloth imitation of leather.

CLASS 23.

2468 CHEVALIER, L. & SON, *Leipzig.*—Printed cashmere shawls.

2469-70 Vacat.

2471 RÖMER, BROS. *Hainsberg.* — Cotton yarns, dyed in Turkish red and rose.

2472 * SCHLOTT, BROS. *Gera.* — Dyed woollen stuffs.

2473 Vacat.

2474 UNGER & CO. *Schönhaide.* — Printed, plain, and embroidered shawls and table-covers.

2475 WINTER, W. *Chemnitz.* — Printed shawls and table-covers.

CLASS 24.

2476-77 Vacat.

2478 HIETEL, J. A. *Leipzig.* — Embroideries.

2479 KLEMM, H. T. *Lössnitz and Schneeberg.*—Embroideries.

2480 REIM, C. G. *Buchholz.*—Laces, belts, galloons, &c.

2481 SCHUBERT, E. *Annaberg.*—Crotchet articles.

2482 UNGER, M. *Johanngeorgenstadt.*—Laces.

2483 Vacat.

CLASS 26.

2484 BECK, D. *Döbeln.*—Leather, boot-trees, rings, &c.

2485 Vacat.

2486 LANGE, F. *Oschatz.*—Coloured sheep's and morocco leather.

2487 NEUBERT, A. *Leipzig*—Baskets and fancy articles of leather.

2488 *SCHLESSIGER & LUMMER, *Gera.*—Leather for piano-forte makers.

CLASS 27.

2489 BRAUNSDORF, W. *Leipzig.*—Boots and boot-trees.

2490 HAUGK, H. *Leipzig.*—Hats and hat-makers' articles.

2491 HAUSDING, L. *Chemnitz.*—Umbrellas, parasols, en-tous-cas.

2492 HECKER, G. & SONS, *Chemnitz.*—Hosiery.

2493 HERRMANN, F. G. *Oberlungwitz, near Chemnitz.*—Cotton, woollen, and silk hose, gloves, drawers, and jackets.

2494 HILLER, C. & SON, *Chemnitz.*—Cotton hose and socks.

2495 MÜHLE, A. *Pirna.*—Bootmakers' articles of felt.

2496-7 Vacat

2498 REICHEL, H. H. *Dippoldiswalde.*—Straw plaitings.

2499 RUDLOFF, H. *Leipzig.*—Boots and shoes.

2500 SCHMIDT & HARZDORF, *Hartmannsdorf, near Chemnitz.*—Cotton hose and socks.

2501 UHLE & CO. *Neustadt, near Chemnitz.*—Hosiery.

2502 VÖCKLER, T. & CO. *Coelln, near Meissen.*—Umbrella rods of artificial whale-bone (wallosine).

2503 WEX & SONS, *Chemnitz.*—Hosiery.

2504 WOLLER, F. E. *Stollberg, near Chemnitz.*—Cotton hose, socks, and gloves.

CLASS 28.

2505 BACH, J. G. *Leipzig.*—Lithographic print with oil colours.

2506 BRANDSTETTER, F. *Leipzig.*—Printed books.

2507 BROCKHAUS, F. A. *Leipzig.*—Printed books and tableaux.

2508 FISCHER, C. F. A. *Bautzen.*—Papers for printing with copper and steel plates, and for lithography.

2509 FLINSCH, F. *Leipzig.*—White and coloured papers.

2510 GIESECKE & DEVRIENT, *Leipzig.*—Typographic, lithographic, copper and steel-printing, engraving in copper and steel, lithography, galvanoplastic, &c.

2511 GRUMBACH, C. (formerly E. KRETZSCHMAR), *Leipzig.*—Patterns of artistic typographical printing.

2512 HINRICHS, J. C. *Leipzig.*—Books, geographical maps, and globes.

2513 HOFMEISTER, F. *Leipzig.*—Music printing.

2514 KRÄTZSCHMER, F. *Leipzig.*—Lithographic title-pages for musical works.

2515 LORCK, C. B. *Leipzig.*—Books, mostly in oriental characters.

2516 MEINHOLD, C. C. & SONS, *Dresden.*—Printed books.

2517 MEISSNER & BUCH, *Leipzig.*—Fancy papers, stationery.

2518 RÖDER, C. G. *Leipzig.*—Patterns of musical engraving and type-cutting.

2519 SIEGEL, C. F. W. *Leipzig.*—Music printing.

2520 THE ROYAL STENOGRAPHIC INSTITUTION, *Dresden.*—Books, mostly in stenographic characters.

CLASS 29.

2530 HAWSKY, A. *Leipzig.*—Toys.

2531 HÜLSE, E. R. A. *Dresden.*—Toys.

2532 Vacat.

CLASS 30.

2533 BIBO & EALES, *Coelln, near Meissen.*—Picture and mirror frames, fancy borders, &c.

2534 GEYER, W. *Plauen.*—Frames and borders, fancy boxes of wood.

2535 GÜNTHER, J. *Waldheim.*—Furniture in walnut and rosewood.

2536 MADACK, R. JUN. *Leipzig.*—Furniture of wicker-work.

2537 MERZ, O. *Dresden.*—Arm-chair in cane work.

2538 SCHMIDT, T. F. C. *Leipzig.*—Paper-hangings, printed as imitation of wood, marble, granite, &c.

2539 TÜRPE, A. *Dresden.*—Furniture of inlaid work.

2540 WÖLFFERT, C. *Dresden.*—Tooth-picks.

CLASS 31.

2541 HERRMANN, L. *Dresden.*—Pattern of wire-plaiting for malt kilns.

2542 LENK, C. *Dresden.*—Steel pens, supplying themselves with ink.

2543 MÜNNICH, A. & Co. *Chemnitz.*—Articles of plaited wire work.

2544 *WEISKER, A. & Co. *Schleiz.*—Lamps, candlesticks, &c. Spun.

2545 WINCKELMANN, J. *Leipzig.*—Galvanoplastic sculptures.

CLASS 32.

2546 SAXON MANUFACTORY OF CAST STEEL, *Döhlen, near Dresden.*—Rollers, tools, springs, &c.

CLASS 33.

2547 WIMMER & DIETRICH, *Annaberg.*—Leonic gold and silver ware, coloured tinsel and tinsel work.

2548 Vacat.

CLASS 35.

2549 BUCKER, H. *Dresden.*—Paintings on china.

2550 FIKENTSCHER, F. C. *Zwickau.*—Vessel of earthenware for chemical purposes.

2551 FISCHER, C. *Zwickau.*—China plates, bowls, cups, pots, &c.

2552 KRAMER, H. E. *Leipzig.*—Painted china plates.

2553 ROYAL SAXON CHINA MANUFACTORY, *Meissen.*—Chimney-piece, mirrors, lustres, candelabra, pendule cases, vases, figures, china services, &c.

2554 THORSCHMIDT, C. L. & Co. *Pirna.*—Vases, figures, drinking vessels of siderolithe.

2555 WENTZEL, M. *Dresden.*—Vases of serpentine.

2556 HARKORT, C. & G. *Leipzig.*—Earthenware water-coolers and butter-coolers.

GRAND DUCHY OF SAXONY.

CLASS 27.

2602 ZIMMERMANN, C. & SON, *Apolda.*—Worsted hosiery.

CLASS 29.

2602A SCHREINER, O. *Weimar.*—Butterfly caterpillars, artificially preserved for natural-philosophical purposes.

CLASS 31.

2603 BARDENHEUER, C. *Ruhla.*—Brass, steel, and German silver ware, &c.

DUCHY OF SAXE-ALTENBURG.

CLASS 2.

2606 KÜHNEMANN, B. G. *Kahla.*—Alkaline metals, glacial phosphoric acid, dextrine, starch sugar.

CLASS 11.

2607 HEU, A. *Altenburg.*—Bows, arrows, and springs.

CLASS 16.

2609 VOLLRATH, K. G. *Altenburg.*—Violoncello, violin.

CLASS 21.

2611 MÜNZER, C. H. & SON, *Ronneburg.*—Flannels.

CLASS 25.

2612 MEUSCHKE, J. C. & SON, *Altenburg.*—Various brushes.

CLASS 26.

2614 RANNIGER, J. L. & SONS, *Altenburg.*—Gloves and sheep skins.

CLASS 30.

2615 SPRINGER, E. J. *Altenburg.*—Articles of wicker-work.

DUCHY OF SAXE-COBURG GOTHA.

CLASS 2.

2621 HOLZAPFEL, C. F. & S. F. *Grub, near Coburg.*—Paris and steel-blue, prussiate of potash, manganese.

CLASS 11.

2624 KLEY & BARTHELMES, *Zella St. Blasii.*—Sporting guns.

CLASS 13.

2625 AUSFELD, H. *Gotha.*—Astronomical and electro-magnetical apparatus.

CLASS 19.

2626 BURBACH & Co. *Gotha.*—Hemp hose for fire-engines, &c.

CLASS 23.

2627 FISCHER, F. G. *Coburg.*—Plush shoes.

CLASS 26.

2628 ARNOLDT, W. *Gotha.*—Harness and other leather.

CLASS 27.

2629 RABUS, C. *Gotha.*—Boots and shoes.

CLASS 29.

2630 BENDA, G. & Co. *Coburg.*—Fancy articles.

2631 TRADING CO. AT NEUSTADT, o H.—Toys.

2632 HELM & WELLHAUSEN, *Friedrichrode.*—Toys.

2633 KRAUSE, T. *Gotha.*—Toys.

2634 PERTHES, J. *Gotha.*—Geographical maps, copper-plates, &c.

CLASS 30.

2635 HOFFMEISTER, T. & GRASSER, *Coburg.*—Household furniture.

CLASS 31.

2636 GRÜNEWALD, M. *Coburg.*—Articles of pewter.

CLASS 35.

2637 DORNHEIM, H. *Gräfenrode, Gotha.*—Articles of clay.

2638 HENNEBERG, F. E. & Co. *Gotha.*—Chinaware.

CLASS 36.

2639 ARNOLDI, H. *Gotha.*—Fruit imitations of clay.

2641 WENIGE, E. *Ohrdruf, Gotha.*—Shirt buttons.

DUCHY OF SAXE-MEININGEN.

CLASS 2.

2646 MERLET & Co. *Sophienau.*—Ultramarine.

2647 ORTLOFF, DR. F. *Eisfeld.*—Porcelain, enamel, and lustre colours.

CLASS 4.

2648 SCHMIDT, C. *Pösneck.*—Toilet soaps.

CLASS 21.

2649 DIETRICH, J. F. & SON, *Pösneck.*—Woollens and flannels.

CLASS 26.

2650 DIESEL & WEISE, *Pösneck.*—Dressed calf skins.

CLASS 29.

2651 HUTSCHENREUTHER & Co. *Wallendorf.*—Papier mâché dolls.

CLASS 31.

2652 DEHLER, BROS. & Co. *Saalfeld.*—Wire-cloth, horse-hair sieve-bottoms, &c.

CLASS 35.

2655 CONTA & BÖHME, *Pösneck.*—White, painted, and gilt porcelain.

2656 EBERLEIN, J. C. *Pösneck.*—China ornaments.

2657 HEUBACH KÄMPF, & SONNTAG, *Wallendorf.*—Porcelain.

2658 MÜLLER & STRASBURGER, *Sonneberg.*—Chinaware and toys.

2660 UNGER, SCHNEIDER, & Co. *Gräfenthal.*—Toys and ornaments.

PRINCIPALITY OF SCHWARZBURG-RUDOLSTADT.

CLASS 29.

2667 SPECHT, B. & Co. *Rudolstadt.*—Water-colours.

PRINCIPALITY OF SCHWARZBURG-SONDERSHAUSEN.

CLASS 1.

2670 LUTHERSTEUFE (ILMENAU) MINING Co.—Hyperoxide of manganese.

CLASS 31.

2671 BRÖMEL, A. *Arnstadt.*—Decimal weighing-machines.

CLASS 35.

2672 SCHIERHOLZ, C. G. & SONS, *Plaue.*—China figures, diaphan plates, &c.

PRINCIPALITY OF WALDECK.

CLASS 4.

2676 BACKHAUS, BROS. *Affoldern.*—Brewers' pitch for barrels.

CLASS 8.

2677 DUNCKER, E. *Friedensthal.*—Bread cutter, with adjusting screw.

WÜRTEMBERG.

CLASS 2

2681 BÖHRINGER, C. F. & SONS, *Stuttgart.*—Chemicals; preparations of quinine and santonine.

2682 BÜRCKLE, J. F. *Grossheppach.*—Sulphured paper, free from arsenic, for wine purposes.

2683 FRANKEN, J. H. *Stuttgart.*—Eau de Cologne, called "Stuttgarter Wasser."

2684 FÜRST & SIEGLE, *Stuttgart.*—Carmine and other lacs.

2685 HÄCKER, C. *Stuttgart.* — Bronze colours.

2686 JOBST, FR. *Stuttgart.*—Sulphate of quinine.

2687 KNOSP, R. *Stuttgart.*—Indigo, carmine, cudbear, and aniline colours.

2688 REUSS & Co. *Heilbronn.*—Alum, free from iron; artificial manure; animal charcoal, produced by a new process.

2689 RUND, G. FR. *Heilbronn.*—Refined sugar of lead.

2690 SCHWEICKHARDT, DR. *Tübingen.*—Artificial manure.

2691 SIEGLE, H. *Stuttgart.*—Colours of all kinds, flower paper, &c.

2692 VEIT-WEIL, *Oberdorf, near Bopfingen.*—Artificial manure.

2693 WÜRTEMBERG GLUE AND MANURE MANUFACTURING Co. *Reutlingen.*—Artificial manure.

2694 ZIEGLER, E. *Heilbronn.*—Substitute for bone black.

CLASS 3.

2695 ROYAL BOARD OF AGRICULTURE (in the name of different wine producers of Würtemberg), *Stuttgart.*—Wines produced in Würtemberg.

2696 DAUR, H. *Ulm.*—Ulm pearl-barley.

2697 ENGELMANN & Co. *Stuttgart.*—Sparkling wines and liqueurs.

2698 ROYAL HOFCAMERALAMT (Office of Domains of the King of Würtemberg), *Stuttgart.*—Wines.

2699 KESSLER, G. C. & Co.'s SUCCESSORS, *Esslingen.*—Sparkling wines.

2700 KIRSNER, W. *Rottweil.*—Essential oil of cumin.

2701 LAIBLIN, ED. & Co. *Stuttgart.*—Sparkling Neckar wine, Hock, and Moselle.

2702 LUDWIG, J. F. & Co. *Stuttgart.*—Candied and dried fruits; preserved vegetables.

2703 MITTLER & ECKHARDT, *Stuttgart, and 6, Grocers'-hall-ct. London.*—Sparkling and other wines.

2704 RENNER, J. A. *Hall.*—Starch crystallized and in lumps; farina.

2705 SCHÖLLKOPF, JOH. *Ulm.*—Starch.

2706 SELIG, E. *Heilbronn.*—Chicory-coffee.

CLASS 4.

2707 ROYAL BOARD OF AGRICULTURE, *Stuttgart.*- Samples of Würtemberg sheep's wool.

2708 GRUNER, FR. *Esslingen.*—Soaps for technical and other purposes.

2709 HEDINGER, C. *Stuttgart.*—Walking and umbrella sticks.

2710 ROYAL HOFCAMERALAMT (Office of Domains of the King of Würtemberg), *Stuttgart.*—Sheep's wool, Angora goats' hair and down, raw and prepared; yak's hair.

2711 KAUZMANN, BROS. *Geislingen.*—Articles of ivory, stag horn, and bone.

2712 KÖLLE, TH. *Ulm.*—Glue and spunk.

2713 LINSE, F. *Bopfingen.*—Glue.

2714 METZ & Co. *Heilbronn.*—Glue.

2715 REUSS & Co. *Heilbronn.*—Glue and leather cuttings.

2716 SCHWARZ, FR. *Göppingen.*—Glue.

2717 VEIT-WEIL, *Oberdorf, near Bopfingen.*—Glue, gelatine, ground bone, bone grease.

2718 WÜRTEMBERG GLUE AND MANURE MANUFACTURING Co. *Reutlingen.*—Glue.

CLASS 7.

2719 BALDAUF, G. (formerly BRÖSTERLI & Co.) *Stuttgart.*—Carpenters' and joiners' tools.

2720 DÖRTENBACH & SCHAUBER, *Calw.*—Cards for wool and cotton spinning.

2721 FREY, CHR. *Pfedelbach.*—Drawing plates for goldsmiths and silversmiths.

2722 LANCASTER, KLEEMANN & Co. *Obertürkheim.*—Power-loom.

2723 STEINER, JOS. *Laupheim.*—Carpenters' and joiners' tools.

2724 VOGEL, M. *Ulm.*—Joiners' tools.

CLASS 8.

2725 FOUQUET & FRAUZ (formerly NOPPER & FOUQUET), *Stuttgart.*—Circular knitting machines.

2726 KURTZ, H. *Stuttgart.*—Portable fire-engine, with sucking and forcing-pump.

CLASS 9.

2727 DITTMAR, BROS. *Heilbronn.*—Garden tools.

2728 RAU, DR. PROF. *Hohenheim.*—150 models of ploughs.

2729 WEITZEL, E. (Society for Bees' Breeding), *Sonnenberg.*—Bee-hives, model of a bee-house, various implements.

CLASS 10.

2730 CHAILLY, J. PROF. *Kirchheim-u-T.*—Portland and Roman cement, bricks showing the strength of the cement, articles, &c.

2731 LEUBE, BROS. *Ulm.*—Samples of cement work.

2732 WÜRTEMBERG GLUE AND MANURE MANUFACTURING Co. *Reutlingen.*—Roman cement.

2733 ZIEGLER, CHR. *Heilbronn.*—Floating bricks.

CLASS 11.

2734 BEUTTER, BROS. *Reutlingen.* — Percussion rifle gun, with needle adapted at once for breech and muzzle-loading.

CLASS 13.

2735 ENGLER, R. *Ellwangen.* — Horoscopes.

2736 SAUTER, A. *Ebingen.*—Balances for chemical purposes.

2737 WOLFF, F. A. & SONS, *Heilbronn*—Chemical and pharmaceutical distilling apparatus.

CLASS 14.

2738 SPRÖSSER, C. *Stuttgart.* — Photographs.

CLASS 15.

2739 BENZING, RAPP, & CO. *Stuttgart.*—Dutch clocks from the Black Forest.

2740 SAUTER, A. *Ravensburg*—Watches.

2741 SCHUCHMANN, W. *Ravensburg.* — Watches.

2742 WÜRTEMBERG CLOCK MANUFACTORY (J. BÜRK, *Schwenningen*).—Various Dutch clocks and component parts of clocks. Controlling watches.

CLASS 16.

2743 BLÄDEL, C. *Stuttgart.*—Pianoforte.

2744 BONZELIUS, W. *Esslingen.*—Æolian harps and accordions.

2745 EBERMAYER, *Ellwangen.* — Novel musical instrument called "Tastenblas Jupwüment."

2746 HARDT & PRESSEL, *Stuttgart.* — Pianino.

2747 HUNDT & SON, *Stuttgart.*—Cottage and square pianoforte.

2748 MISSENHARTER, C. JUN. *Ulm.*—Collection of musical instruments of brass and German silver.

2749 OEHLER, CHR. *Stuttgart.* — Pianoforte.

2750 PROSS, GESCHWIND, & CO. *Stuttgart.* —Harmoniums.

2751 SCHIEDMAYER, J. & P. *Stuttgart.*—Harmonicorde and harmoniums.

2752 SCHIEDMAYER & SONS, *Stuttgart.*—Piano and pianino.

2753 TRAYSER & CO. *Stuttgart.*—Harmoniums.

2754 WALKER, E. F. & CO. *Ludwigsburg.* —Parts of a large organ.

CLASS 18.

2755 BAUMANN, CARL, *Leonberg.*—Cotton fabrics.

2756 GUTMANN, BROS. *Göppingen.*—Cotton, and cotton mixed fabrics.

2757 KAUFMANN & SONS, *Göppingen.*—Cotton and cotton mixed fabrics.

2758 KOLB & SCHÜLE, *Kirchheim-u-T.*—Various cotton fabrics.

2759 KRUMBHOLTZ, L. *Böblingen.*—Fancy tickings.

2760 LANG & SEIZ, *Stuttgart.* — Cotton blankets.

2761 LEVINGER, LEOP. *Ulm.* — Cotton fabrics.

2762 OTTENHEIMER & DETTELBACH, *Jebenhausen.*—Cotton fabrics.

2763 VAIHINGER & CO. *Göppingen.*—Cotton, and cotton mixed fabrics.

2764 WINDRATH, C. A. *Heidenheim.* — Cotton fabrics, white and coloured.

CLASS 19.

2765 FABER, C. *Stuttgart.*—Damask-linen and quiltings.

2766 GUTMANN, BROS. *Göppingen.*—Linen tickings.

2767 KAUFMANN & SONS, *Göppingen.*—Linen tickings.

2768 LANG, A. F. *Blaubeuren.*—Bleached linen and linen handkerchiefs.

2769 LANG & SEIZ, *Stuttgart.* — Linen-damask and linen tickings.

2770 MÜNSTER, W. F. *Freudenstadt.* — Flax and linen yarn.

2771 OTTENHEIMER & DETTELBACH, *Jebenhausen.*—Linen fabrics.

CLASS 20.

2772 BAUMANN, C. *Leonberg.* — Silk fabrics.

2773 ROYAL HOFCAMERALAMT (Office of Domains of the King of Würtemberg), *Stuttgart.*—Raw silk.

CLASS 21.

2774 BAUMANN & BÜRGER, *Göppingen.*—Woollen cloths.

2775 BAUMANN, C. *Leonberg.*—Woollen fabrics.

2776 HARTMANN, BROS. *Esslingen.* — Woollen cloth and buckskins.

2777 KAUFFMANN, S. *Stuttgart.*—Mixed woollen and cotton stuffs.

2778 KISSEL, A. *Böblingen.*—Mixed woollen stuffs.

2779 LAMPARTER, BROS. *Reutlingen.*—Buckskins.

2780 MÜLLER, J. G. JUN. *Metzingen.*—Woollen cloth.

2781 RAIFSTÄNGER, M. *Metzingen.*—Woollen cloths.

2782 SCHILL & WAGNER, *Calw.*—Woollen cloth.

2783 WIEDENMANN, G. F. *Heidenheim.*—Manchons and felts for paper manufacturing.

2784 ZÖPPRITZ, BROS. *Heidenheim.* — Woollen fabrics.

CLASS 22.

2785 ERLENBUSCH, *Stuttgart.*—Carpet of a peculiar tissue.

CLASS 24.

2786 BECK & SALZMANN, *Ulm.*—Window blinds, curtains, stuffs for furniture and clothing, &c.

2787 DEFFNER, O. *Ravensburg.*—Stuffs for curtains.

2788 HUMMEL, MISS S. *Stuttgart.* — An embroidered portrait of silk, representing H.R.H. Prince Albert.

2789 NEUBURGER & SONS, *Stuttgart and Ulm.*—White cotton stuffs, embroideries, and shirts.

2790 WEISS, W. A. *Ravensburg.* — Embroidered and damask stuffs for curtains.

2791 ZWERGER, FR. VON, *Ravensburg.*—Jaconets and mousselins, power-loom manufacture.

CLASS 26.

2792 BERINGER, C. *Stuttgart.* — Leather for machines.

2793 DITTMANN, BROS. *Stuttgart.*—Cowhide.

2794 ECKART, F. M. & Co. *Ulm.*—Varnished leather and oil-cloth.

2795 KIDERLEN & MARIUS, *Ulm.*—Various kinds of leather.

2796 KLEMM, TH. *Pfullingen.* — Harness-leather, straps, peculiar preparation of leather, &c.

2797 LINSE & Co. *Crailsheim.*—Varnished leather.

2798 METZ & Co. *Heilbronn.*—Sole leather.

2799 MÖLLEN & Co. *Bopfingen.*—Numerous specimens of leather.

2800 ROSER, C. *Stuttgart.*—Various kinds of leather.

2801 SCHMID, CHR. *Stuttgart.* — Cow leather.

CLASS 27.

2802 BAUMANN, C. *Leonberg.* — Corsets without seams.

2803 BINDER, FR. W. *Ebingen.*—Various articles of hosiery, wool, and cotton.

2804 FALKENSTEIN, C. G. *Balingen.* — Boots and shoes.

2805 FAUTH & Co. *Lorch.*—Hosiery.

2806 GRÜBER, FR. *Riedlingen.*—Hosiery.

2807 HAAS, F. P. & Co. *Schramberg.*—Straw bonnets and hats, Panamas, and others made of whalebone, &c.

2808 HAAS, CL. *Schramberg.*—Hosiery.

2809 KIENZLE, B. *Balingen.*—Hosiery.

2810 KIESEL, FR. *Ludwigsburg.*—Shoes.

2811 KISPERT & STICHLING, *Ulm.* — Hosiery.

2812 KNAPP, B. *Reutlingen.*—Corsets.

2813 LIEB, FR. *Ulm.*—Hosiery.

2814 MUNDORFF & MÜLLER, *Calw.* — Hosiery.

2815 OCHS, J. F. *Reutlingen.*—Hosiery in wool, cotton, and silk.

2816 PFLÜGER, CHR. FR. *Rottweil.* — Hosiery.

2817 OTTENHEIMER & SONS, *Stuttgart.*—Corsets without seams.

2818 ROSENTHAL & Co. *Göppingen.* — Corsets.

2819 SCHUMM, FR. *Calw.*—Hosiery.

2820 STEINHART, HERZ, & Co. *Göppingen.*—Corsets without seams.

2821 VOTTELER, BROS. *Reutlingen.* — Hosiery, in cotton, wool, and silk.

2822 WÜRTEMBERG TRADING CO. *Stuttgart.*—Hosiery in wool; corsets, clothes, boots and shoes.

2823 ZEILE, J. P. *Reutlingen.*—Hosiery in wool.

CLASS 28.

2824 ENSLIN & CLOSTERMAYER, *Kirchheim-u-T.*—Fancy leather articles.

2825 FABER, G. F. *Crailsheim.* — Lead pencils and creta polycolour.

2826 ADE, E. *Stuttgart.*—Wood-cuts.

2827 MÜLLER & RICHTER, *Stuttgart.*—Albums for photographs.

2828 ROMETSCH, C. *Stuttgart.* — Metal slates for the use of schools, offices, &c.

2829 SCHÄUFFELEN, G. *Heilbronn.* — Paper.

2830 SCHWENK, PROF. *Ludwigsburg.* — Artificial parchment.

2831 VÖLTER, H. & SONS, *Heidenheim.*—Wood-pulp paper (H. Voelter's patent invention), for writing and other purposes.

CLASS 29.

2832 BENZ, *Ellwangen.*—Drawing-book.

2833 BLUMHARD & Co. *Stuttgart.*—Japanned tin toys.

2834 ROYAL COMMISSION FOR EDUCATION, *Stuttgart.*—Models in plaster of Paris, and copies for drawing schools.

2835 DIETERICH, C. F. (A. KATZ), *Ludwigsburg.*—Toys.

2836 ENGLER & LUTZ, *Ellwangen.* — Japanned tin toys.

2837 GROSS, CARL, *Stuttgart.*—Wood toys.

2838 GUTEKUNST, FR. *Ulm.* — Petrefactions.

2839 MALTÉ, *Stuttgart.*—Maps and globes.

2840 NITZSCHKE, W. *Stuttgart.*—Weiser's Illustrated Map of Universal History. Braun's Transparent Map of Constellations. Wende's Map of Natural History.

2841 NÖRDLINGER, PROF. *Hohenheim.*—Different sorts of wood in form of a library.

2842 ROCK & GRANER, *Biberach.* — Japanned tin toys.

CLASS 30.

2843 GROSS, C. *Stuttgart.*—Fancy articles of wood.

2844 KIENLE & Co. *Stuttgart.*—Polished furniture.

2845 VETTER, C. *Stuttgart.*—Cornices and frames, gilt and varnished.

2846 WAIDELICH, K. *Ulm.*—Gilt cornices.

2847 WEBER & Co. *Esslingen.*—Boxes for gloves, &c., and other articles of wood.

2848 WOLBACH, W. *Ulm.*—Cornices prepared for gilding.

2849 WÜRTEMBERG TRADING CO. *Stuttgart.*—Furniture and articles of wood.

CLASS 31.

2850 BAHNMAYER, J. L. *Esslingen.* — Pastry moulds of copper, rules, gas-fittings.

2851 BÜHRER & KALLENBERG, *Ludwigsburg* —Copper and iron hardware.

2852 DEFFNER, C. *Esslingen.*—Articles of brass and bronzed copper.

2853 MANUFACTORY OF IRON AND IRON WIRE, *Erlau, near Aalen.*—Wire, wire nails, and chains.

2854 ERHARD & SONS, *Gmünd.*—Articles of bronze, cast, stamped, and electrotyped.

2855 MARTIN, N. *Tübingen.*—Smoothing irons.

2856 REXER, C. *Stuttgart.*—Garden-furniture of wrought-iron, and other articles of wire-work.

2857 RUEFF, DR. PROF. *Hohenheim.* — Nose-rings for cattle.

2858 STOTZ, A. *Stuttgart.* — Castings of malleable iron.

2859 STRAUB & SCHWEIZER, *Geislingen.*— Bronzed copper-ware.

2860 VETTER, FR. *Ludwigsburg.* — Bird cages, sugar boxes, fruit baskets, tea-trays.

2861 WAGNER, G. *Esslingen.* — Copper pastry moulds.

2862 WÜRTEMBERG TRADING CO.—*Stuttgart.*—Hardware.

CLASS 32.

2863 DITTMAR, BROS. *Heilbronn.*—Knives.

2864 HAUEISEN & SON, *Stuttgart.*—Scythes, sickles, and chopping blades.

CLASS 33.

2865 DEFFNER, C. *Esslingen.*—Silver-plated wares.

2866 DEYHLE & BÖHM, *Gmünd.*—Gold and silver-ware.

2867 ERHARD & SONS, *Gmünd.* — Silver-ware, cast, stamped, electrotyped.

2868 FORSTER, D. *Gmünd.*—Silver-ware.

2869 GABLER, BROS. *Schorndorf.*—Thimbles of silver and German silver.

2870 KOTT, D. *Gmünd.*—Silver-ware.

2871 OTT & CO. *Gmünd.*—Jewellery.

2872 STRAUB & SCHWEIZER, *Geislingen.*— Silver-plated ware.

2873 WÖHLER & CO. *Gmünd.*—Jewellery.

CLASS 35.

2874 STAIB-WASSEROTT, *Ravensburg.* — Gothic window, architectural ornaments, and other articles in terra cotta.

2875 WEYSER, CARL, *Liebenzell.* — Artificial whetstones.

MECKLENBURG-SCHWERIN.

CLASS 2.

Sub-Class A.

1 WINKLER, *Niegleve, near Gustrow.*— Bleached stuffs.

CLASS 3.

Sub-Class A.

3 HILLMANN, *Scharstorf, near Rostock.*— Oats.

4 KLOCKMANN, *Harmshagen, near Wismar.*—Wheat and other grain.

5 PFANNENSTIEL, *Brusenbeck, near Kleinen.*—Groats and flour.

6 SCHLIEFFEN, COUNT, *Schlieffenberg, near Gustrow.*—Wheat, rye, and pease.

6A WIRKEDE, C. VON, *Klein, Leukow, near Peuzlin.*—Wheat.

7 SCHUBART, *Gallentin, near Wismar.*— Roots of cereal plants.

Sub-Class B.

8 CLEVE, V. *Carow, near Plau.*—Cheese.

9 MICHAEL, V. *Gantzkow, near Neubrandenburg.*—Smoked goose-breast.

10 OERTZEN, V. *Woltow, near Tessin.*— Cheese.

11 SCHLIEFFEN, COUNT, *Schlieffenberg, near Gustrow.*—Cheese.

Sub-Class C.

12 SANITER & WEBER, *Rostock.*—Liqueurs.

12B ENGELL & CO. *Wismar.*—Spirits and beer.

CLASS 4.

Sub-Class A.

13 BRUNNENGRÄBER, H. *Schwerin.* — Medical and other soaps.

Sub-Class B.

14 BEHR, V. *Rentzow, near Gadebusch.*— Wool.

15 SCHLIEFFEN, COUNT, *Schlieffenberg, near Gustrow.*—Wool.

16 CÖLLE, *Meetzen, near Gadebusch.*— Fleeces.

17 SCHACK, VON, *Retchendorf, near Schwerin.*—Fleeces.

18 SCHALBURG, *Herzberg, near Parchim.* —Fleeces.

19 HOFFSCHLAEGER, J. F. *Weisin, near Lübz, Estate.*—Fleeces and samples of wool.

CLASS 6.

24 FLORKOWSKY, *Schwerin.* — Carriage with swan's neck, and without axle-beam or box-seat.

CLASS 8.

26 MEMMERT, *Schwerin.* — Machines for cutting angles for cabinet-makers, and for cutting almonds.

CLASS 9.

28 MEYER, *Schwaan.* — A Mecklenburg hook-plough.

CLASS 10

30 DAHSE, B. *Rostock.*—Roofing and isolating material of stone paste-board with asphalte.

CLASS 11.

33 SCHMIDT, *Schwerin.* — Lefaucheux changing double-barrelled rifle, Rifle for deer-stalking, with cast-steel barrel. Percussion needle-gun, after Tervy.

CLASS 12.

Sub-Class A.

36 DAHSE, B. *Rostock.*—Tightening material for paste-board with asphalte.

CLASS 13.

37 BÖCKENHAGEN, *Gustrow.* — Spirit-gauge for distilleries.

38 DOLBERG, *Rostock.* — Reflecting-circle, after Borda, with repetition.

CLASS 14.

39 DETHLEFF, *Rostock.*—Ambrotypes.

CLASS 15.

40 DREYER, *Kehna.* — Pendulum; second clock, with regulator.

CLASS 16.

43 JENTZEN, W. *Grabow.*—Pianino.

CLASS 17.

44 AHRENS, R. *Neubukow.*—Trusses.

45 MÖSSINGER, G. *Rostock.* —Models of artificial foot and hand.

CLASS 19.

46 KRASEMANN, G. *Rostock.*—Covers for upholstery, woven from unspun Manilla hemp fibres.

CLASS 24.

Sub-Class A.

47 SIEGERT, C. L. *Rostock.*—Embroidery.

CLASS 26.

Sub-Class B.

48 BLIEFFERT, C. *Schwerin*—Horse-collars, curved at the top, changeable.

Sub-Class C.

49 AHRENS, R. *Neubukow.* — Leather game-bag.

50 LEHMANN, C. *Schwerin.*—Gloves and dyed glove leather.

CLASS 27.

Sub-Class A.

51 SIEGERT, C. L. *Rostock.*—Straw hats.

Sub-Class C.

52 CRULL, E. *Rostock.*—Waterproof oil-cloth dress for mariners.

Sub-Class D.

54 MÜLLER, H. *Rostock.*—Boots.

CLASS 28.

Sub-Class D.

56 DAHSE, B. *Rostock.*—Ledgers in leather and moleskin, with spring.

57 GARBE, *Rostock.*—Bookbinding articles.

57A LAN, E. *Neu-Brandenburgh.*—Bookbinding articles.

CLASS 29.

Sub-Class B.

58 JENTZEN, F. *Schwerin.*—Drawing examples for the industrial schools of Mecklenburg, executed for the Government.

CLASS 30.

Sub-Class A.

60 BEHR, H. *Rostock.*—Jewel-press.

63 HERMES, F. — Walnut drawing-room table.

65 PETERS, C. *Schwerin.*—Parquet floors. Cloth press of oak, with carvings.

67 PETERS, C. — Portions of the inlaid floors in the saloons of the Grand Ducal Castle at Schwerin, and a folding-door.

CLASS 34.

Sub-Class A.

71 GILLMEISTER, E. *Schwerin.* — Glass paintings.

CLASS 36.

74 CRULL, E. *Rostock.*—Life-buoy.

75 JÜRSS & CROTOGINO, *Rostock.*—Garden table and seat of Portland cement. Cement.

77 BERLING, L. *Rostock.*—Matches.

78 OPPERMANN, *Niegleve, near Gustrow.* —Knitted travelling stockings and woollen gloves.

79 BURMEISTER, F. *Schwerin.* — Lithographic stone with drawing, electrotype copy taken from this, second electrotype taken from the first, and print taken from the latter; electrotype copy, taken from a gutta percha copy of the stone, and print obtained from it; lithographic print from the stone.

HANSE-TOWNS.

S.W. Transept, No. 2.

BREMEN.

CLASS 2.

1 SCHMIDT & ROHLAND, *Hemelingen, Bremen.*—White lead (dry, and as paint).

CLASS 3.

2 BOLLMANN, E. & M.—French, German, and malt vinegar.
3 SCHOMBURG & Co.—Caraway-cordial.
4 WALTJEN, E.—Flour.

CLASS 7.

5 BRUNS, G. H.—Hand-blowing machine for forges.

CLASS 8.

6 WALTJEN, C. & Co.—Friction-balance for ascertaining the laws of friction, and testing lubricating substances.

CLASS 12.

7 GLEISTEIN, G. & SON, *Vegesack, nr. Bremen.*—Machine-manufactured cordage for ship's use.

CLASS 19.

8 WALTJEN, F. W.—Kyanised linen.

CLASS 24.

9 EBEL, H.—Picture in embroidery.
10 ZIEGENHIRT, F.—Picture in embroidery.

CLASS 26.

11 MEYER, A.—Saddle and bridle.

CLASS 27.

12 ALBRECHT, J. H.—Dress-boots.
13 BÖDEKER, C. F. & KRÜGER, BROS.—Dress-coat.
14 BORTFELDT, C.—Silk and felt hats.
15 CORSSEN, J. F.—Boots.
16 DONOP, H. & Co —Trousers and waistcoat.
17 HENNING, H.—Varnished leather riding-gaiters.
18 HOFFMEYER, C. H.—Jacket.
19 LEONHARDT, F. W.—Coat à deux mains.
20 SCHMIDT, J. H.—Coat.
21 SICK, C.—Silk hats.
22 WEBER & EEG.—Boots and shoes.

CLASS 28.

23 GEFFKEN, D.—Ledger, cash-book, photograph album, portfolio.

CLASS 30.

24 ARMERDING, J. H.—Secretaire.
25 BRUNS, G. H.—Iron work-table and embroidering tables.
26 FORQUIGNON, J.—Boudoir furniture.
27 LAMPE, C.—Sofa of basket work.
28 SCHLEEF, J. H.—Sideboard and sofa table.

CLASS 31.

29 ASENDORPF, C.—Iron money-safe, with safety-locks and letter-combination.

CLASS 33.

30 KOCH & BERGFELD.—Silver tea and coffee service, German antique style; silver tea tray.
31 WILKENS, M. H. & SONS.—Silver table service, &c.

CLASS 36.

32 BIERMANN & FREVERT.—Glass pyramid with articles of cigar-manufacture.
33 BÖDEKER, H. & SONS.—Brematin and alabaster candles, with the raw material and composition in blocks.
34 DÖDING, F. H.—Cocks of tin composition, injured neither by corrosion nor friction.
35 OETTING, D.—Tallow candles with prepared wicks.
36 WALTJEN, F. W.—Kyanised wood. Fire-lighters.

HAMBURG.

CLASS 1.

1 KLEUDGEN & Co.—Hardened Portland cement.
2 STAUB, J.—Grinding and sand stone, from the Deichbrüche Quarries, on the Bohemian frontiers.

CLASS 2.

3 BEHRMANN & V. SPRECKELSEN.—Mineral waters.

4 DOUGLAS. SONS, J. J. — Soap, fancy shapes and slabs.

5 FLÜGGER, J. D.—Lac varnish and dryers.

6 GILLMEISTER, CROP, & Co.—Chemical preparations.

7 HASPERG & SCHÄFER.—Concentrated non-ferruginous alum, from argillaceous cryolite.

8 OBERDÖRFFER, A. & Co.—Artificial mineral waters.

9 VÖLKLEIN, STEINERT, & Co.—Soap pomatum, tooth-powder, and perfumery.

CLASS 3.

10 DAMPFZUCKERSIEDEREI VON 1848.—Refined and raw sugar, and treacle.

11 ECKHARD, A. F. — Sweetmeats and comfits. Plastic fancy picture, of sugar.

12 FETT, J. M. & Co.—Smoked beef.

13 GREEN, H. F.—Cherry cordial.

14 MULSON, L. & Co.—Preserves.

15 PETERS, J. J. W.—Liqueurs, brandy, and spirits.

16 REESE & WICHMANN. — Chocolate, cocoa, sweetmeats, and tea.

17 REESING, W. C.—White crystallized candy.

18 SCHRÖTER, G.—Fruit-drops and rocks.

19 TREDE, G. H. G.—Anti-cholera and stomachic bitters.

CLASS 4.

20 GOLDSCHMIDT & Co.—Merino-sheep wool.

21 KAEMMERER, A.—A glass containing colour.

22 PORTEN, I. v. D.—Cigars.

23 WERNER, P. O. E.—Splinter for upholstery work.

24 ZIPPERLING, KESSLER, & Co.—Pulverized and ground dye-wood.

CLASS 6.

25 SACHS, F. B. C.—A barouche.

CLASS 7.

26 BECKER, J. C.—Sewing machine.

27 CATHOR, M. & GRABAU, A. C.—oiner's tool-case.

28 HEBERLING, H. P.—Gilder's press.

29 KOLTZAU, H. C.—Shoemaker's tools.

30 KÖNIGSLÖW, H. v.—Sewing machine.

31 WINTER, E.—Diamond pencils for lithographers, engravers, &c. Rubies for ruling.

CLASS 8.

32 GÄBLER & VEITSHAUS, *Hamburgh*, makers; DEWIT, W. C. & H. *Amsterdam*, patentees.—Manometer.

CLASS 10.

33 PLATH, C. C.—Mathematical instruments.

CLASS 12.

34 STEINHAUS, C. F.—Model of a merchant frigate.

35 EISERN BOOT BAU Co.—Corrugated and galvanized metal articles.

CLASS 13.

36 KRÜSS, A.—Dissolving-view apparatus. Photographical objective barometer.

37 SCHRÖDER, H.—Microscope, with minor apparatus. Optical articles.

CLASS 15.

38 BRÖCKING, W.—Double-regulator.

39 KNOBLICH, T.—Astronomical constant pendulum clock.

40 NIEBERG, J. L.—Astronomical constant clock, with free escapement.

CLASS 16.

41 ALBRECHT, F. E. J.—Upright pianoforte.

42 BAUMGARDTEN & HEINS.—Grand pianoforte.

43 DOLL, J. & KAMPRATH.—Upright pianoforte.

44 MELHOP, W.—Æolian harps.

45 MÜLLER, L. W.—Upright pianoforte.

46 MEJER, A. W. A. & Co.—Mechanism for pianos.

47 PIERING, T. — Bass-tube, bassoon, German silver cornet-à-piston.

48 PLASS, C. H. L.—Pianoforte.

49 RACHALS, M. F. & Co.—Upright pianoforte.

50 ROTT, J. A. W.—Upright pianoforte.

51 RUPPACH, R.—Upright pianoforte.

52 SCHLÜTER, C. E. L.—Vertical grand pianoforte. Organ-box for children.

53 SCHRÖDER, C. H.—Piano, half oblique.

CLASS 17.

54 BRÜNING, G. H.—Paralytic and valetudinarian's self-acting mahogany arm-chair with wheels.

55 MATTHIAS, B. & Co.—Artificial teeth.

56 NEBEL, J. C. R.—Surgical wire leg and arm ligatures.

CLASS 22.

57 WERNER, P. O. E.—Floor-carpets of splint.

CLASS 23.

58 MYLIUS & HASENOHR.—Printed woollen shawls, table-cloths, and furniture-cloth.

CLASS 24.

59 FRANCK, GESCHW.—Gold and silver-embroidered table-cloth.

CLASS 26.

60 EISSFELDT, J. E. & Co.—Calf and goat-skins.
61 FEIDEL & Co.—Calf-skins, varnished knee and shade-leather and horse-skins.
62 LAHRMANN, A. W. & Co.—Calf-skins and skivers.
63 MÖNCKEBERG, G. A. C.—Horse-collar, travelling-trunk and box.
64 WAMOSY, D. — Calf-skins, varnished knee and shade-leather. Brown skins for saddlery, and black oil-cloth.

CLASS 27.

65 DOSSE, F. A.—Boots.
66 LADENDORF, H. W. F. — Boots and shoes.
67 PAPE, J. C. W.—Boots and slippers.
68 PREHN, G. C. W.—Samples of boots.
69 SANDER, A.—Various kinds of boots.

CLASS 28.

70 ADLER.—Frame, containing samples of steel-engravings and lithographs.
71 ASPERN, W. M. v. — Counting-house books.
72 BADE, W.—Counting-house books.
73 BAHRS, T —Chemical manifold copying-books.
74 BERENDSOHN.—Three albums.
75 KITTLER, R.—Dictionary, English and Portuguese.
76 MÖLLER, C. H. A. — Counting-house books.
77 SCHMIDT, C. O.—Watch-stand, cigar-salver, almanack, and stationery.

CLASS 29.

78 KÖHLER, A.—Writing-books, containing prototypes of alphabets.

CLASS 30.

79 AHRENS, H. — Chairs, looking-glass, flower-stand, foot-stool, book-stand, sofas, &c.
80 BERNHARDT, J. H. A.—Upholstery.
81 BEYER, F.—Stuffed basket-work arm-chair, reed-sofa, &c.
82 BOCK, L. & Co. Looking-glass and table with marble slab. Gilt pedestal.
83 BOCK, F. D.—Furniture.
84 DIÈBER, J. H. — Mahogany writing-desk.
85 EHRENSMANN, R. — Osier arm-chairs, flower-stand, and basket-work.
86 FREESE, H.—Window, with wood-chip-weft for blinds. Wood-chip hangings.
87 GRUPE, L.—Pattern for printed woollen table-cloth, curtains, furniture-cloth, and printed paper-hangings.
88 HANDWERKER-VEREINIGUNG v. 1853.—Furniture of walnut and mahogany.
89 HARDEN, C. H.—Tables with round and oval leaf. Tobacco-box.
90 HARDER, F. JUN.—Flower-vase and stand, chairs, work-baskets, &c.
91 HAUER, H. W.—Oak sideboard.
92 KLÜVER, D. H. J.—Walnut sideboard.
93 KÖPCKE, C. C. J.—Oak dining-table.
„ NIESS, A.—Walnut sideboard,
„ SCHINDLER, H.—Oak chairs.
„ STIEHR, J. W.—Oak sideboard.
„ WEHRSPOHN, H. G.—Walnut centre-table.
94 KOSTE, H. N. & Co.—Chairs, sofa, table, and piano-chair, in curved wood.
95 LAGERFELDT, J. G. A.—Gilt sofa-looking-glass.
96 LOOSE, J. R.—Jacquaranda centre-table.
97 MAACK, P.—Basket-work furniture, and furniture for children.
98 MÜLLER, J. H. & KRÖGER. — Oak chandelier, Gothic style.
99 NIESS, A.—Walnut centre-table.
100 PATEIN, J. F. M.—Black varnished tables.
101 PLAMBECK, C. F. M.—Church door, Gothic style: inlaid work.
102 PLÖNSKI, FRAU J.—Leaf of a table, painted.
103 RAMPENDAHL, H. F. C.—Fancy furniture of antlers, and articles of amber, &c.
104 SCHMIDT, E.—Wicker-work for door-plates; round centre-table, with leaf in wicker-work.
105 SCHULTZE, C. F.—Cot, trimmed with wicker-work.
106 SCHULTZ, J. E. F.—Window-blinds.
107 TENGER, J.—Centre-table, with leaf inlaid in Mosaic work and pedestal.
108 WEHRSPOHN, H. G.—Ladies' work-table.
109 WERNER, C. F. & PIGLHEIN.—Book-shelf and chimney-piece; ebony etagere library table; arm and other chairs.
110 WULF, E. & KELTER.—Field-beds fo railway travelling; railroad cushions.
111 ZIMMERMANN, F.—Rocking and oth chairs, and perambulator.

CLASS 31.

112 BLECHER, I. H.—An iron safe.
113 CLASSEN, F.—Bird-cage.
114 KLEIN, T.—Spelter hip-bath, hexagonal lantern, hearth-blower in zinc.
115 KRAHNSTOVER, F. B.—Shower-bath portable water-closets.

116 KRAMER, J. G. W. JUN. — Brass parrot-cage.
117 RÄCKE, F. W.—Spelter hearth-crown, brass bird-cage.
118 SCHULTZ, F. J. — Fire-screen, bird-cages.
119 TIMCKE, H. F. — Brass parrot-cage, coal-scuttle, German silver pendant kettle.
120 WEBER, J. F.—Varnished bird-cages.
121 WOLTERS, J.—Fancy figure, in embossed iron-plate: iron-plate vase.

CLASS 32.

122 WEBER, W.—Pocket-knives, razors, and carving-knives.

CLASS 33.

123 PETERS, J. H. & Co.—Crystal spirit-gas lamps.

CLASS 34.

124 EIMBCKE, E. — A skylight for ship's use, in gutta percha, framed.
125 MEYER, BROS.—Demijohns, covered with wicker-work.
126 SCHULZ, A.—Transparencies, and various articles of glass.

CLASS 35.

127 HAUTHAL, J. G. & Co.—Painted tea-cups in porcelain, flower-vase.
128 ROTH, G. A. — Plaster-cast figure, chandeliers.

CLASS 36.

129 FEDERWISCH, G. L.—Kites.
130 MEYER, H. C. JUN.—Hamburg Arms, composed of walking-sticks, figure, &c.
131 NATHANSEN, W.—Coloured, stamped, and other paper.
132 SCHNACK, H. F. — Sample card of mother-of-pearl escutcheons.
133 WÖBCKE, H.—Meerschaum and cigar-tubes.
134 ZUBER, J.—Brooches of pink and green shells, ivory flower-basket, &c.

LÜBECK.

S.W. Transept, No. 2.

CLASS 2.

1 FRITZ & BÖNING.—Eau de Cologne.

CLASS 3.

2 CARSTENS, D. H. — Hermetically preserved victuals.
3 GRELL, A. F.—Hand-made marchpane.
4 HAHN, G. C. & Co.—Preserved alimentary substances in boxes.
5 JÜRGENSEN, F. — Aromatic stomachic liqueur.
6 PETERSEN, J. C. C.—Hand-made marchpane.

CLASS 11.

7 FISCHER, C. A. & SON. — Rifles and fowling pieces.

CLASS 24.

8 AMANN, C.—Embroidered handkerchief.

CLASS 27.

9 FRITZ, J. H. & BÖNING.—Shirts.

CLASS 28.

10 HEINRITZ, C. C. J.—Specimen of bookbinding.

CLASS 29.

11 SÖHLBRAND, J. M. — Caligraphical copy.

CLASS 30.

12 LEDERHAUSEN, J. F. C.—Bed-screen of wicker-work.

CLASS 36.

13 HERMBERG, H. M. C. — Cigar boxes portfolios, and portemonnaies.

GREECE.

N. Court, No. 2.

CLASS 1.

1 AGRICULTURAL SCHOOL OF TIRYNTH. —Lime-stone.
2 ALEXANDROPOULOS, C. *Tripolis.*—Black and variegated stone.
3 CAPARIAS, J. G. *Panormas.* — Black and white marble of Tinos.
4 CENTRAL COMMITTEE, ATHENS. — Pair of hand mills; hewn mill-stones; specimens of chromium ore, and of Naxos emery stone.
4A CLEANTHES, *Athens.* — Specimens of verde antique marble from Tinos.
5 DELENDAS, P. G. *Theræ.*—Theraic porcelain.
6 DEMOS OF CERONELEON, *Mantoudion.*—Magnesite or white stone.
7 DEMOS OF CHALCHIS, *Chalchis.*—Magnesite or white stone.
8 DEMOS OF DORION, *Soulima.* — Soft soap-clay.
9 DEMOS OF PANORMOS, *Panormos.* — Green and white marble paper presses.
10 HOME DEPARTMENT, ATHENS.—Seventeen specimens of marbles from Peloponnesus.
11 LANDERER, H. *Athens.* — Specimens of minerals, metals, and vegetable substances.

11A MALAKATES, J. *Athens.*—Specimens of marbles.

12 MELAS, B. *Athens.*—Sulphur from Milo.

13 MICHALACHACOS, C. *Panitza.*—Red or pophyry marble of Laconia.

14 PETRINOS, N. *Tripolis.*—Marble paper-press.

14A PHYTALÆ, L. & G. *Athens.*—Specimens of marbles.

15 SPIROPOULOS, C. *Gargaliani.*—Specimen of ceramic clay.

CLASS 2.

16 CONGOS, G. *Patras.*—Liquorice.

17 DEMOS OF CYTHNOS, *Cythnos.* — Mineral waters.

18 DEMOS OF PHARIS, *Xerocampi.* — Siderites or wild tea.

19 DEMOS OF PROSCHION, *Platanos.*— Wild tea.

20 DEMOS OF PSOPHIS, *Livartzi.*—Wild tea.

21 DEMOS OF SPARTA, *Sparta.*—Orange flower water.

21A LANDERER, H. *Athens.*—Collection of plants.

CLASS 3.

22 AGRICULTURAL SCHOOL OF TIRYNTH. —Wheat, barley, maize, rye, honeycomb, wine of 1851 and 1861, aniseed, colza, tobacco, &c.

23 ALEXANDRIS, P. *Mataranga.*—Maize, barley, oats, wheat.

24 BACHLOS, J. J. *Cyme.*—Wheat, maize.

25 BASILIADES, B. N. *Nauplia.*—Wine, spirits of wine, almonds.

26 BOUTOUNAS, B. N. *Karichi.* — Haricot beans.

26A CAIRIS, L, M. *Andro.*—Soft wheat and figs.

27 CARTEROULIS, C. K. *Kalamæ.*—Wine.

28 CONDYLIS, G. *Andros.*—White wheat.

29 CONGOS, G. *Patras.*—Currants, &c.

30 CONSTANTINIDES, D. *Villia.*—Wheat, chick-beans, garlick.

31 CONVENT OF MEGASPILION, *Megaspilion.*—Haricot beans.

32 DEMOS OF ÆGÆON, *Limne.*—Wheat, chick-beans, &c.

33 DEMOS OF ÆGINA, *Ægina.*—Almonds, barley.

34 DEMOS OF AMBRACHIA, *Carvassaras.* —Wheat, corn meal, &c.

35 DEMOS OF APODOTIAS, *Great Lombotina.*—Wheat, lentils.

36 DEMOS OF ASOPON, *Molae.*—Wheat, barley, beans, figs, tobacco, &c.

37 DEMOS OF ATALANTE, *Atalante.* — Wheat.

38 DEMOS OF CALAVRETA, *Calavreta.* — Broad haricot beans, barley.

39 DEMOS OF CALTHEZON, *Vlackokerassia.* —Wheat, barley.

40 DEMOS OF CASTORION, *Castanea.*— Dari.

41 DEMOS OF CERONELEON, *Mantoudion.* Wheat, maize, haricot beans, broad beans, vetches, oats, dari, lentils, rye.

42 DEMOS OF CERPENI, *Cerpeni.* — White wheat.

43 DEMOS OF CHALCHIDEON, *Chalchis.*— Wheat, chick-beans, lentils, haricot beans, &c.

44 DEMOS OF CLETORIA, *Mazeika.* — Maize.

45 DEMOS OF COLOCYNTHION, *Flomochori.*—Lupins, vetches, and peas.

46 DEMOS OF CORYTHION, *Stenon.* — Wheat, barley, colyander, peas, haricot and French beans, maize, dari, vetches, linseed.

47 DEMOS OF DAPHNESEON, *Livanates.*— Wheat, barley, oats, vetches.

48 DEMOS OF DORION, *Soulima.*—Wheat, lentils, barley.

49 DEMOS OF DRIMIAS, *Dadi.* — Wheat, maize, barley.

50 DEMOS OF ELLATIA, *Vroulia.*—Barley.

51 DEMOS OF ELLATIA, *Baischine.* — Wheat and almonds.

52 DEMOS OF ELOUS, *Apedea.* — Wheat, barley, haricot beans, tobacco, and maize.

53 DEMOS OF ERANIS, *Philiatra.* — Currants, wheat, wine.

54 DEMOS OF GERONTHRON, *Geraki.* — Tobacco.

55 DEMOS OF HERACLEOTON, *Moscochorion.* —Wheat, barley, sesame, chick-beans, maize.

56 DEMOS OF HYPATE, *Hypate.*—Tobacco.

57 DEMOS OF IDOMENIS, *Xerachia.*— Wheat.

58 DEMOS OF IPERCHIAS, *Aga.* — Tobacco.

59 DEMOS OF ISTIAEON, *Xerochorion.* — Wheat, beans, lentils, tobacco.

60 DEMOS OF LAMIA, *Lamia.* — Wheat, barley.

61 DEMOS OF LARYMNA, *Martinon.*— Wheat.

62 DEMOS OF LELENTION, *Steny.* — Walnuts.

63 DEMOS OF LETRINON, *Pyrgos.* — Wheat, maize, wine, and currants.

64 DEMOS OF LIVADIA, *Livadia.* — Rice, tobacco, wheat, maize, aniseed, sesame, and barley.

65 DEMOS OF LYCOUSOURAS, *Isaris.* — White maize, wheat, dari, haricot beans.

66 DEMOS OF MACRACOMIS, *Varybombe.* —Wheat.

67 DEMOS OF MALEVRION, *Panitza.* — Haricot beans.

68 DEMOS OF MEDEA, *Merbacca.* — Tobacco.

69 DEMOS OF MEGALOUPOLIS, *Synanon.* —Indian corn, &c.

70 DEMOS OF MESSOLONGHI.—Tobacco.

71 DEMOS OF MONEMBASIA, *Monembasia.* —Wheat, barley, beans, honey, &c.

72 DEMOS OF MYCENÆ, *Couzopodi.*—Tobacco.

72A DEMOS OF MYRTOUDION, *Lechena.*— Linseed.

73 DEMOS OF NASSON, *Nasson.* — Indian corn meal, and Dari.

74 DEMOS OF NAUPLIA, *Nauplia.*—Agegartus raisins, and currants.

75 DEMOS OF ŒNOUNTOS, *Vrestena.* — Wheat.

76 DEMOS OF ORCHOMENION, *Scripou.*— Wheat, barley, beans, and maize.

77 DEMOS OF ORCHOMENN, *Levidion.*— Wheat, barley, maize, lentils, dari, beans.

78 DEMOS OF PHALARON, *Stylis.*—Maize.

79 DEMOS OF PELANIS, *Agorgena.* — Wheat from American seed.

80 DEMOS OF PLATAMODOUS, *Gargagliani.*—Lentils, haricot beans, vetches.

81 DEMOS OF PROSCHION, *Platanos.* — Wheat, maize.

82 DEMOS OF PTELEOS, *Pteleos.*—Tobacco.

83 DEMOS OF SALAMIS, *Salamis.*—Honey, oil.

84 DEMOS OF SICHYONOS, *Chiatoa.* — Haricot beans.

85 DEMOS OF SPARTA, *Sparta.*—Wheat.

86 DEMOS OF TEGEA, *Achouria.*—Beans, lentils, barley, maize, wheat, oats, vetches.

87 DEMOS OF THERMOPYLÆ, *Molos.*—Red wheat and cocoons.

88 DEMOS OF THIAMON, *Valmada.* — Wheat, maize, wine, and haricot beans.

89 DEMOS OF TRINASSON, *Scala.*—Wheat.

90 DEMOS OF TRIPYLES, *Sarachinada.*— Wheat.

91 DEMOS OF TYMPHRISTOU, *Mavrillon.*— Haricot beans.

92 DEMOS OF VION, *Neapolis.*—Wheat, barley, figs, peas, vetches.

93 DEMOS OF ZARACOS, *Rigea.* — Wheat, barley, honey.

94 DIALETTES, G. *Missolonghi.*—Raisins, bottargo, maize, oats, barley, wheat.

95 DOUROUTES, A. *Athens.* — Flour.

96 ECONOMIDES, G. *Oropos.* — Wheat, barley, and tobacco.

97 EMMANUEL, A. *Ægina.*—Barley.

98 EPARCH OF DORIS, *Lidorikion.*—Haricot beans, maize, wheat.

99 ERIOTIS, N. G. *Ægina.*—Barley.

100 EVANGELIS, P. & NICHOLAOU, P. *Calamos.*—Honey.

101 GAMELIARIS, N. *Argos.*—Tobacco.

102 GEORGACOPULOS, G. A. *Cyparissia.*— Sesame, honey-seed.

103 GEORGANDAS, G. *Athens.*—Wine.

104 GHICAS, D. N. *Ægina.*—Honeycombs.

105 GHICAS, G. *Livanatae.*—Wheat.

106 IOANNOU, S. *New Corinth.*—Currant wine, currants, and almonds.

107 IRIOTTIS, N. G. *Ægina.*—Barley.

108 KEZEAS, V. *Dolæ.*—Honey.

109 LAMBRINIDES, L. G. *Argos.*—Currants.

109A LANDERER, H. *Athens.*—Wheat, &c.

110 LANGADAS, A. N. *Naxos.*—Wine.

111 MANOUSOS, P. *Cyparissia.*—Barley.

112 MARCOPULOS, TH. *Calamae.*—Figs.

113 MASTAKAS, D. A. *New Pelli.*—Honey.

114 MILAKIS, A. *Patras.*—Wine.

115 MOSTRAS, M. N. *Xerochorion.*—Muscat wine.

115A NICOLOPOULOS, G. D. *Gargagliani.*— Wine.

116 ŒTYLON, BISHOP OF, *Areoupolis.*— Almonds.

117 PANAGIATOPOULOS, C. N. *Gargagliani.* —Sultana raisins and currants.

118 PETRIDES, D. *Styles.*—Green olives.

119 PETROPULOS, G. *Tripolis.*—Sparkling and common wine.

120 PETROU, E. *Atalante.* — Tobacco, wheat.

121 PROVINCIAL COMMITTEE OF LOCRIS, *Atalante.*—Indian-corn.

121A PRINTESIS, X. *Syra.*—Naked barley.

122 SIMANDIRAS, I. *Argos.*—Wine.

123 SIROCHILOS, A. *Livanatae.*—Wheat, oats.

124 SOTIROPULOS, P. *Ætolicon.*—Wheat, barley, maize, oats, raisins.

124A SPIROPOULOS, C. *Garghagani.* — Broad beans.

125 STEPHANOPULOS, C. G. *Tripolis.*— Wheat.

126 STERIANOPULOS, L. *Sarmousakli.* — Wheat.

127 THEODOSIOS, A. *Megara.*—Honey.

128 THOMARAS, C. *Cyparissia.*—Almonds.

128A VARTHALITES, G. T. *Syra.*—Barley.

129 VOUCHÈRER, G. G. *Piræus.*—Spirits of wine.

130 VOULPIOTIS, N. *Baischine.*—Tobacco, wheat.

131 VOUTOUNAS, B. N. *Caryke.*—Fruit.

132 WINE COMPANY, PATRAS.—Champagne and other wines.

132A WINE COMPANY, *Thera, Santorin.* —Wines.

133 ZACHAROPULO, M. *Cyparissia.* — Currants.

134 ZIMBOURACHI, P. A. *Stamna.*—Wheat maize, oats, barley.

CLASS 4.

135 AGRICULTURAL SCHOOL, TIRYNTH.— Acacia wood, cotton, wool, silk-worms, wax.

136 ALEXANDRIS, P. *Mataranga.*—Fustic wood and oil.

137 BOUDOUNAS, B. *Lycosura.* — Cotton wool, and fustic.

138 BOURNAKIS, J. *Leonidion.*—Cochineal.

139 CALAMARAS, N. C. *Ægina.*—Sponges.

140 CARAMOUSSAS, P. *Steni.*—Fustic wood.

141 CENTRAL COMMITTEE, ATHENS.—Fustic, sumach, tragacanth, and specimens of the forest timber of Greece.

142 COMMITTEE OF TROEZINIA, *Poros.*— Green olive oil, common oil.

143 CONSTANTINIDES, D. *Villia.*—Oil, resin.

144 DEMARCH OF MYCENAE, *Coutzopodi.*— Valonea.

145 DEMOS OF ÆGITION, *Lidorichion.*— Tallow.

146 DEMOS OF ÆGEON, *Limne.*—Resin, naptha, fustic wood.

147 DEMOS OF ÆGINA, *Ægina.*—Madder-roots, sponges, oil.

148 DEMOS OF APODOTIA, *Great Lobotina.* —Cocoons.

149 DEMOS OF AROANIA, *Sopoton.*—Silk.

150 DEMOS OF CASTORION, *Castanea.*— Galls.

151 DEMOS OF CERONELEON, *Mantudion.* —Ash, maple, pine, elm, oak, platane woods, &c. Resin, naptha, Colophany resin-tar.

152 DEMOS OF CLEITORIA, *Mazeika.* — Yellow berries.
153 DEMOS OF CROKEON, *Levetzova.* — Yellow sumach, fustic, &c.
154 DEMOS OF DELONDEON, *Eubea.* — Fustic.
155 DEMOS OF DRYMIAS, *Dadi.*—Cotton.
156 DEMOS OF ERANIS, *Philiatra.* — Oil, soap.
157 DEMOS OF GYTHEON, *Gytheon.*—Wax, valonea.
158 DEMOS OF IRACLEOTON, *Moschochorion.*—Cotton wool.
159 DEMOS OF LIVADIA, *Livadia.*—Cotton wool.
160 DEMOS OF MALEVREON, *Panizza.*—Wax and valonea.
161 DEMOS OF MELITENES, *San Nicholas.* —Wax, galls.
162 DEMOS OF MONEMBASIA, *Monembasia.* —Wax, oil.
163 DEMOS OF MYRTOUDION, *Lechena.*—Linseed.
164 DEMOS OF NAUPLIA, *Nauplia.* — Sponges.
165 DEMOS OF PHELLIAS, *Torani.*—Wax, galls.
166 DEMOS OF PLATOMODUS, *Gargagliani.* —Cocoons and linseed.
167 DEMOS OF PROSCHION, *Platanos.* — Cocoons.
168 DEMOS OF TRIPYLES, *Sarakinada.*—Valonea.
169 DEMOS OF VOEON, *Neapolis.*—Sponges.
170 DEMOS OF VOUPRASION, *Vouprasion.* —Galls.
171 DEMOS OF VRYSEON, *Anavryte.* — Yellow berries, sumach, fustic.
172 DEMOS OF ZARACOS, *Richea.* — Wax, fustic.
173 DOUROUTES, A. *Athens.*—Silk, oil.
174 DURAND, F. *Andros.*—Silk.
175 ECONOMOU, C. *Leonidion.*—Cocoons.
176 ELIADES, P. *Gytheon.*—Oil.
177 EMMANUEL, A. *Ægina.* — Madder-roots.
178 EVANGELIS, C. *Athens.*—White and yellow wax, and candles.
179 FELLS & Co. *Calamae.*—Silk.
180 GEORGANDAS, G. *Athens.*—Oil.
181 KEZEAS, V. *Doloe.*—Oil, silk.
182 LAPIÈRE, J. *Salessi.*—Valonea.
183 LONDOS, A. C. *Patras.*—Cotton wool.
184 MARCOPOULOS, TH. *Calamae.* — Cocoons.
185 MERLIN, M. *Athens.* — Cotton from American seed.
186 ŒTYLON, BISHOP OF, *Areoupolis.*—Oil.
187 PETRITES, A. N. *Ægina.*—Madder-roots, oil.
188 POTIROPOULOS, B. *Calamae.*—Oil.
189 REMBOUTZIKAS, A. *Lower Achaïa.*—Valonea.
190 RHALLIS, L. *Piraeus.*—Silk.
191 SABBAS, G. *Mantoudion.*—Madder.
192 SOTIROPULOS, T. *Etolicon.*—Linseed.
193 THEOLOGOS, N. *Piræus.*—Soap.
194 TOMARAS, C. *Cyparissia.*—Cotton.
195 VRYSAKIS, C. & TH. *Athens.*—Vegeble red dye, kermes.
196 ZAFIRACOS, N. *Gytheon.*—Valonea.
197 ZOUCHLOS, *Leonidion.*—Fustic and oil.

209 PALEOLOGOS, D. *Steni.*—Box-wood.

CLASS 7.

198 DEMIDES, C. *Athens.*—Specimens of type casts.

CLASS 8.

199 FETICHANES, B. *Athens.* — A calico printing machine.
200 GREEK STEAM NAVIGATION Co. *Syra.* —A steam-engine, 4¼ h.p., working shafts in the Western Annex, near column 32.

CLASS 10.

204 CLEANTHES, *Athens.*—Two verd antique pillars, two round tables, and one chimney-piece of the same marble from Tinos.
205 COSSOS, I. *Athens.*—Marble busts, &c.
207 MALAKATES, F. *Athens.*—Marble busts and a marble paper press.
208 MALAKATES, I. *Athens.*— A funeral monument, a fountain, and two chimney-pieces, all of Pentelic marble.
210 PHYTALAE, L. & G. *Athens.*—Table of Pentelic marble, marble busts, &c.

CLASS 11.

214 CONSTANDOULAKIS, D. M. *Hydra.*—Military and naval officer's belts.
215 DEMOS OF NAUPLIA, *Nauplia.*—Gold lace and silver cords.

CLASS 14.

216 CONSTANTIN, D. *Athens.*—Specimens of photography.
217 MARGARITIS, F. *Athens.* — Photographs.

CLASS 18.

218 CARAMERZANES, D. *Dadi.*—Cotton towels and a bed cover.
219 DIALETIS, J. *Missolonghi.* — Table cloth, bed sheet, towels, pillow cases, and handkerchiefs.
220 GEORGIOU P. *Chalcis.*—Cotton towels.
221 KOUTZOUKOS, A. D. *Chalcis.*—Cotton fabrics.
222 PHOTINOS, P. *Patras.*—Cotton twist.
223 PISPIRINGOS, G. *Patras.*—Cotton cords.

CLASS 20.

224 CONSTANDOULAKIS, D. M. *Hydra.*—Barége dresses and silk fabrics.
225 DEMOS OF ERANES, *Philiatra.*—Silk fabrics.
226 DEMOS OF SPARTA, *Sparta.*—Silk fabrics.
227 FRANGOULIS, N. *Avlonarion.*—Barége dresses, handkerchiefs, scarves, silk fabrics, &c.

228 MELANE MONACHE, *Calamae.*—Barége handkerchiefs, mosquito curtains, &c.

CLASS 21.

229 CONSTANDOULAKIS, D. M. *Hydra.*—Scarf, and various mixed fabrics.

CLASS 22.

230 ALEXANDRIS, P. *Mallaranga.* — A carpet.
231 DEMOS OF CALAVRYTA, *Calavryta.*—A carpet.
232 DEMOS OF DISTOMON, *Distomon.*—A carpet.
233 DEMOS OF LAMIA, *Lamia.*—A carpet in two breadths.
234 DEMOS OF RACHOVA, *Rachova.*—A carpet.
235 DEMOS OF STRATON, *Adamas.*—Bed coverlet.
236 DEMOS OF TRIPOLIS, *Tripolis.* — A carpet.
237 DEMETRIOU, M. G. *New Peli.*—A carpet.
238 DEVA, ELLEN, *New Peli.* — Three carpets.
239 STYLOUDY, S. *New Peli.* — Three carpets.
240 ZERVAS, M. & E. *Krieza.*—A carpet.

CLASS 24.

241 DEMARCH OF HERMOUPOLIS.—Lace.
242 DEMOS OF SPARTA, *Sparta.*—Lace collars and sleeves.
243 PISPIRINGOS, G. *Patras.*—Lace, &c.
243A ISANTYLAS, C. *Athens.* — An embroidered table-cover.

---

256 CINGLIS, *Athens.*—Gold tassels, legging ties, &c.
257 CHALMOUCOPOULOS, Z. *Lamia.* — A gold and silk knit cap.
259 DEMOS OF LAMIA, *Lamia.*—A goat's hair capot, a mantlet, a worsted tunic, and a pair of leggings.
260 DEMOS OF LELANTION, *Steni.* — A female peasant's dress.
275 ANDREOU, A. & ZENOU, T. *Athens.*—A richly embroidered gold table cover.
281 EDIPIDES, I. *Athens.*—Lace bag.

CLASS 25.

244 YALISTRAS, S. *Athens.*—Furs of various kinds, and tobacco pouches.

---

247 GIATSIS, CH. *Messolonghi.*—Goatskins.
251 PLATANIOTIS, N. *Chalcis.*—Coloured skins.

CLASS 26.

245 ANASTASIOU, G. *Chalchis.* — Bull hides.
246 FACHIRIS, E. G. *Hermoupolis.*—Bull hides.
249 LAGOURAS, G. *Hermoupolis.*—Bull hide.
252 ROUMBAKES, E. *Hermoupolis.*—Bull hide.

CLASS 27.

255 ALEXANDRIS, P. *Mataranga.*—A pair of Greek sandals.
258 CRIANOPOULA, Q. *Syra.* — Printed head covering.
261 DEMOS OF NAUPLIA, *Nauplia.*—Red silk sashes, and silk fabric.
262 GALINOS, G. *Athens.* — Black dress coat.
263 GIALELI, A. *Syra.*—A handkerchief.
264 IOANNOU, N. P. *Athens.*—Red fez caps.
265 MELANE, MONACHE, *Calamae.*—Jackets, burnous, collars, sleeves, and sashes.
266 NICOLAIDIS, A. *Andros.*—A handkerchief.
267 NICOLAOU, D. *Athens.*—Gentleman's and lady's velvet embroidered costumes.
268 PAPACOSTAS, C. N. *Athens.*—Richly embroidered Greek costumes for a gentleman, lady, and a boy.
269 PIZOURAS, G. *Chalcis.*—A silk belt.
270 THEODOROU, G. *Athens.*—Silk embroidered gentleman's dress.
271 TSANTYLAS, C. *Athens.* — A lady's jacket, mantlets, and other embroidered goods.
272 VARIA, G. *Hermoupolis.*—A head covering.

---

248 GREGORIADIS, H. *Athens.*—Boot lasts.
250 LEVANDIS, G. *Athens.*—Boot lasts.
253 TATOS, AL. *Andros.*—A pair of man's boots.
254 ZIVARAKIS, BROS. *Hermoupolis.* — Boots of various kinds.

CLASS 28.

273 PENDEFRES, P. *Athens.* — Account books
213 SKIADOPOULOS, P. *Athens.*—Specimens of wood carving, and of the electrotype process.
274 VLASTOS, HEIRS OF S. G. *Athens.*—Printed books.

CLASS 29.

274A CONDES, H. *Athens.*—A manuscript book of Greek caligraphy.
282 MASTORES, M. *Gargagliani.*—Specimen of writing.

CLASS 30.

202 ALBANOPOULOS, A. *Chalcis.*—Wooden vase.
276 ATHANASIOU, I. *Poros.* — A brass lamp.
203 BARALES, M. *Chalchis.* — Cubical barrel.
206 DEMOS OF THYAMON, *Valmada.*—A tobacco pipe.
277 SARGENTES, M. *Chalcis.* — A small wooden box.
278 TSANGLIS, EM. *Hermoupolis.*—A brass tap.

---

201 AGATHANGELOS, *Athens.*—"The coming of our Lord," in carved wood.
211 PAPAGEORGIOU, C. *Athens.*—Wood carving.
212 PRAOUDAKES, M. *Athens.*—Wood carving.

CLASS 31.

279 COCOREMBAS, P. *Tripolis.*—Butcher's knife.

CLASS 33.

280 ANTONIADES, TH. *Chalcis.*—A gold ring.

HAWAIIAN OR SANDWICH ISLANDS.

Northern Courts, near the Horticultural Gardens entrance.

Samples of cocoa-nut oil, from Fanning's Island, manufactured by H. English & Co. Annual product, 120 to 150 tons.

FRANKLIN, LADY.—Valuable feather tippet and feather collar, made from the single yellow feather obtained from under the wing of a native bird, worn by persons of high rank; collar of human hair; sulphur lava, &c.; Tapa cloth; hats of native manufacture; kava root; photographs of natives; newspapers, books, and other specimens of native printing.

HAYTI.

North Courts, British Side.

THE HAYTIAN GOVERNMENT.—Cotton, cocoa, arrow-root, coffee, sugar, plantain meal, pitre, tea. Tanning beans, holly gum, pulse, sesame, or "hoholo." A large block of mahogany. Two mahogany wardrobes, one mahogany wash-stand, water-bottles and drinking vessels in lignum-vitæ, two mahogany pestles, eight earthenware syphon-pitchers. Gut-whips, wooden locks, a polished stone for a table-top, saddles, plated bit, specimens of minerals, iron ore. A *toque avec macoite*, or a *sac-paille.*

IONIAN ISLANDS.

North Courts, British Side.

A.—CEPHALONIA.

1 CEPHALONIAN COMMITTEE.—A collection of natural productions and manufactured articles, viz.:—

Lime and limestone, chalk, sand, red earths, and other cements; fossils, &c.; mineral waters.

Various specimens of timber, and miscellaneous vegetable productions.

Cereals—Wheat, barley, Indian corn, oats, and rye.

Pulse—Peas, beans, French beans, vetches, and avrios.

Textile materials—Flax, wool, silk, and cotton.

Oil, honey, almonds, walnuts, stone pine seeds, carob pods, capers, onions, garlic.

Jaws of sheep, with gold tint from Pilaros.

Woollen cloth, basket of rye-straw, embroidered handkerchiefs, slippers, woollen socks, sashes.

Model of a net.

2 LUSI, G.—Quince and gourd preserves.
3 MANZAVINO, D.—Boots.
4 CANGILLARI, C.—Dress tail-coat.
5 MORAITI, G.—Chest of drawers, chair, and small boxes.
6 ANGELLATO, P.—A table.
7 GASI, D. M.—A carpet.
8 BERDOHÉ, M. D. M.—A carpet.
9 BENEDETTO, A.—Banisters for balustrades.
10 CORAFAN, G.—Gold head-pins and ring.
11 INGLESSI, T.—Gown of aloe lace, doyleys, and collars.
12 BAGGALI, C.—Counterpane, burnoose, polkas, and sleeves.
13 CURCUMELI, V.—An apron and gloves.
14 METAXA, E.—An apron.
15 CARITATO, S.—A tobacco bag.
16 INGLESSI, E.—Polkas.
17 VULISMA, A.—Two toilette tables.
18 MONTALDO, A.—Coffee-pot and tray.
19 DALAPORTA, G.—Wax candles.

B.—CERIGO.

20 CERIGO COMMITTEE.—Mineral waters, everlasting flowers, sponges, marbles, slate, stalactites, clays, petrified bones, toys.
21 STAI, DR. N.—Oil and olives.
22 VARIPATI, G.—Oil, honey, salted quails, four kinds of grain, beans, two sorts of skins.
23 ALEXANDRACHI, A.—Wine.
24 CALIGERO, A.—Wine, honey.
25 CALIGERO, SOFIA.—Embroidered collar.
26 PISANO, Z.—Hams.
27 COSACHI, P.—Hams.
28 COSACHI, M.—Cloth stuff.
29 AVIERINO, S.—Hams.
30 CALUCI, G.—Hams; almonds.
31 MAVROMATI, G.—Capers.
32 MAVUDI, P.—Cheese.
33 FAZEA, A.—Cheese (ricotta secca.)
34 MORMORI, DR. E.—Barley meal; sweet herbs; lime.
35 MORMO, F.—Baskets.
36 CARIDI.—Almonds and figs; medicinal and other herbs.
37 CARIDI, M.—Dyed wool.
38 MELITA, E.—A silk sash; a carpet; waterproof bags; silk and cotton towels.
39 FARDULI, C.—Baskets.
40 CONDOLENI, M.—Baskets.
41 CASTRISSO, P.—Waterproof bags.
42 NEOPHOTISTO, G.—Dress sacket and kerch efs.

43 PLUSACHI, S.—Embroidered collar.
44 AVEPERINA, P.—Embroidered collars and handkerchiefs.
45 DIACOPULO, S.—Embroidered handkerchiefs.
46 VESE, G.—Boxes of olive wood.

C.—CORFU.

47 ALEXIS.—Montenegrin and Greek caps and costumes.
48 TAYLOR, N. worked by COSTI, S.—Albanian bags, cases, cuffs, caps and costume.
49 COSTI, S.—Embroidery.
50 PSORULA.—Albanian shoes, jacket and waistcoat, and other embroidery.
51 RUBAN, M. — Embroidered cushions, pouches, and costumes.
52 PAVIA.—Silversmith's work and jewellery, rings, &c.
53 MOSEO, PAPA. — Gold necklaces and other jewellery.
54 FLORIAS, A.—Filigree-work.
55 COSTA, PAPA.—Silver lamp, sacramental cup, &c.
56 PAUDIN, P. BROS.—Silversmith's work and jewellery.
57 BIASI, G. DE.—Knitted collars and cuffs.
58 SARACHINO, SA.—Gilt and silk sashes, &c.
59 MULATO, A.—Silk sashes.
60 PAPASTERI, E.—An embroidered dress.
61 MANDUCHIO SCHOOL. — Embroidered collars, cuffs, &c.
62 CURCULO, A.—Embroidered handkerchief.
63 CALOGEROPULO, SA.—Worked pocket-handkerchief.
64 GALDIES, G.—Specimens of small cabinet-work in jujube, acacia and olive wood, &c.
65 PETROVICH, S.—Inlaid works in native wood.
66 ALTAR, G. — Inlaid cabinet-work in native wood.
67 DIMOCASTO, N.—Inlaid paper-cutter.
68 SCHURLURLING, ROSA. — Vinewood walking sticks.
69 GALLO, G.—Whip handles.
70 MONTANARI.—Carved wooden frames.
71 ZERVO, A.—Flax.
72 ZERVO, N.—Straw hat and blue cloth.
73 SERVO, C.—Cotton, raw and carded; wool, raw and carded.
74 MUTHOLLAND.—Boots and shoes.
75 DUSMANI, CT. SIR A. L.—Marble, stalactites. &c.
76 SARACHINO, SA.—77 DIODACHI, G.—Blue cloths.
78 POLLITA.—Cotton cloths, yellow veils, and machine for extracting cotton.
79 PALURIA, A.—80 COSTANDI, C.—81 CALICHIA, N.—82 BALBI, BARON.—83 CORATA, S.—Tobacco.
84 CURCUMELLI, SIR D.—Tobacco, papyrus cloth, woods, grain, oil, wine, and vinegar.
85 SAULI.—Tobacco, olive oil and rice.
86 VENTURA.—Pack-saddles, wine and tobacco.
87 ZANONI.—Stuffed birds, &c.
88 MAMAS, C.—Model of an oil mill and a lamp.
89 WORSLEY.—Collection of insects.
90 VASSILACHI, E.—Wine casks, baskets, olive oil, wine and preparations of pork.
91 DELVINIOTTI.—Dyes.
92 DESSILA.—Medicinal plants.
93 COLLAS, BROS.—Almond and castor oil, medicines, and articles of toilette.
94 ALEXACHI, P.—Wax.
95 ALEXACHI, DR.—Rose-water, &c. and opium.
96 ZANINI, SA.—Flowers.
27 GIRONCI.—A lock.
98 BUFFA.—Model of staircase.
99 BARBIROLI.—Chronometer.
100 MIGLIARESSI.—101 CAMBISO.—102 PROVATÀ, C.—103 CAPO D'ISTRIA, CT.—Olive oil.
104 CARMENI.—Liqueurs.
105 THE PACKET ESTABLISHMENT.—Flags of the Islands.
106 CATHOLIC SISTERS OF CHARITY.—Bead embroidery.

D.—ITHACA.

107 DENDRINO, T.—Rug.
108 DOVA, G.—White marble.
109 GIANNIOTTI, G.—Sponges.
110 XANTHOPULO, P.—Anchor of wood.
111 CUZZUVELI, P.—Silk and cotton stockings.
112 PROCOPI, G.—Silk and cotton stockings.
113 ALIMERIATI, P. — Silk and cotton stockings.
114 PROCOPI, G.—Silk stockings and raw silk.
115 ZANNETTI, P.—Wine and oil.
116 PROCOPI, G.—Red and white Rosolio, and Rachi of Mastica.
117 FERENDINO, S.—Currants.
118 CENTRAL COMMISSION.—Coral.

E.—PAXO.

119 THE PAXO COMMITTEE.—Pitch and mineral water.
120 VEGLIANITI, ALOISIO.—Olive oil.
121 VEGLIANITI, ANASTASIO.—Olive oil.
122 VEGLIANITI, ANDREA.—Olive oil.
123 VEGLIANITI, ATT.—Stone trough.
124 BOGDANO, A.—Olive oil.
125 CARUSO, F.—Olive oil.
126 MORICHI, C.—Olive oil.
127 ARGIRO, V.—Olive oil.
128 MIZZIALI, DR.—Olive oil and preserved olives.
129 LECCA, S.—Olive oil.
130 PETRO, A. P. M.—Olive oil.
131 MACRÌ, N.—Red and white wine.
132 VLACOPULO, M.—Willow fibre cloth.
133 VLACOPULO, G.—Lintel.
134 MACI, S.—A female dress.

F.—SANTA MAURA.

135 SANTA MAURA COMMITTEE.—Conglomerate, limestone, marble, sandstone, timber, cochineal, flax (dressed and undressed), seeds, salt, cotton, and a carpet from Cania.
136 ARCHELLE.—Silk, raw and manufactured.
137 PEZZALI, THE MISSES.—Embroidery.
138 VRIONI, P.—Dye stuffs.
139 VALAORITI, CAV. A.—Cotton from Maduri.
140 WOLFF, H. D.—A carpet.

G.—ZANTÈ.

141 ZANTE COMMITTEE.—Mineral waters. A collection of minerals, raw silk, corn, pulse, linseed, sesamum, and coriander seed.
142 STRAVAPODI, DR. — Currants and raisins.
143 ARVANATACHI, G. — 144 DOMENEGHINI, DR. G.—145 PLANTERO, G.—146 SOLOMOS, COUNT SIR D.—147 COMUTO, G.—148 FLAMBURIARI, CT. SIR D.—149 BARFF, T.—150 MERCATI, CT. D.—151 CARAMALICHI, N.—Currants.
152 WODEHOUSE, THE HON. COL. B.—Red wine of 1851 and 1857. Sponge.
153 CHIEFALINO, D.—White wine of 1859.
154 CATEVATI, P.—Red wine of 1859.
155 CONOFAO, D.—White wine of 1860.
156 ROSSOLIMO, C.—White wine of 1857.
157 ROSSI, G.—Currant wine, 1857 and 1861.
158 RISURCA, A.—Porous jars and bottles.
159 PAPADATO, N.—Manufactured silk, hemp-cord, flax, and cotton.
160 ARGASSARI, D.—Manufactured articles and raw silk.
161 POLITI, N.—Silk handkerchiefs.
162 DRAGENA, I.—Manufactured articles, silk fringe and braid.
163 PAPANDRICOPULO. — Cotton dresses, table-cloths, bed-ticking, and towels.
164 GROSSU, M.—Towels.
165 ANDRAMIOTI, A.—Cotton socks, embroidered handkerchiefs.
166 PASSARIA. — Spirit made from currants.
167 CALAVATI, G.—168 RAFTANI, C.—169 ZAZICHI, D.—170 ANDULICO, C.—171 PILICAS, F.—Soap.
172 COSTANDACOPULOS, D.—Soap and bees' wax.
173 MOPURGO, R.—Soap and purified oil.

---

174 VEJA, DIONISIO.—A festa at Corfu, showing the Romaika or ancient Pyrrhic dance; a picture of Greek costume; "Madonna."
175 ASPIOTI.—A picture as used in Greek churches.
176 MANZAVINO. — Interior of a Greek church.
177 PIERI.—Picture of Greek costume.

ITALY.

S.C. Court, No. 2, S.C. Gallery, No. 2, and Western Annex.

[The names of provinces, according to the new administrative division of the kingdom, are placed within parentheses, thus (*Florence*)].

CLASS 1.

1 ADRAGNA, BARON G. *Trapani.*—Salt.
2 AGOGNA AND BROVELLO MINING CO. *Palanza (Novara).*—Argentiferous lead ore.
3 ALDISEO, G. & CO. *Turin (Iron-works at Bard, Val d'Aosta).*—Iron rod: gun barrels, wrought while cold.
4 ANGHIRELLI, G. *Montalcino (Sienna.)*—Floating bricks, made of "mountain meal" (siliceous skeletons), from Monte Amiata.
5 ARRIGONI, A. *Varese (Milan).*—Peat.
6 BARBA TROYSE, G. *Spezia (Genoa).*—Ores of manganese. Decomposed metamorphic jasper, for hydraulic cement (pozzolana).
7 BELTRANI, G. *Trani (Terra di Bari).*—Limestone and tufa.
8 BELTRAMI, COUNT P. *Cagliari.*—Lead slags of ancient smelting works, and pig of lead obtained from them. Ores of lead, manganese, iron, and copper. Refractory clay, hydraulic lime, lignite.
9 BENTIVOGLIO, CAV. C. *Modena.*—Lime.
10 BIRAGHI, G. & CO. *Milan.*—Lignite.
11 BOLOGNA MINERAL CO. *Bologna.*—Copper pyrites and purple copper ore from the mine of Bisano, with gangue, &c. Serpentine, steatite, &c.
12 BOTTINO CO. *Leghorn.*—Argentiferous and other ores of lead, from the mine of Bottino. Specimens illustrating the smelting process. Lead, refined copper, and refined silver. Model of the apparatus employed for transporting the ores.
13 BOUGLEUX, F. *Leghorn.*—Millstones.
14 BOUQUET & SERPIERI, *Cagliari (Smelting-works at Domusnovas and Flumini, Cagliari).* — Ancient slags. Pigs of lead from Domusnovas and Flumini Maggiore.
15 BRESCIA ATHENÆUM.—Specimens of the rocks and ornamental stones of the province.
16 BUCCI, G. *Campobasso (Molise).*—Campobasso marble.
17 BURGARELLA, A. *Trapani.*—Sulphur.
18 CAGLIARI SUB-COMMITTEE FOR THE EXHIBITION.—Specimens of building stone.
19 CALZA CRAMER, G. *Grugliasco (Turin).*—Peat pressed into moulds.
20 CHIAVARI ECONOMIC SOCIETY, *Chiavari (Genoa).*—Slates for various purposes, from Lavagna quarries, Chiavari.
21 CHIAVENNI SUB-COMMITTEE FOR THE EXHIBITION.—Gypsum, and limestone, from Madesimo, near Chiavenna.
22 CHIOSTRI, L. *Pomarance (Pisa).*—Specimens of rocks found in the neighbourhood of Pomarance and Libbiano: geological map, &c.
23 COCCHI CAV. BROTHERS, *Florence.*—

Jasper from Giarreto, Val di Magra; limestone, hydraulic lime, gypsum, plaster of Paris.

24 COJARI AVV. V. *Fivizzano (Massa-Carrara).*—Copper ore from S. Giorgio mine, Ajola, Fivizzano; statuary, and other marble, from Equi, Fivizzano.

25 CORBI ZOCCHI, C. *Sienna.* — Sienna earth.

26 CORNELIANI, L. *Milan (Iron-works at Premadio, near Bormio, Sondrio).*—Iron ore, pig iron, wrought iron.

27 COSTANZO, C. *Catania.*—Sulphur, crude and manufactured, from the solfatara of Cugno, &c.

28 COSTANZO, L. *Catania.*—Sulphuriferous marl, from the solfatara of Estricello, &c.

29 CURIONI, PROF. G. *Milan.*—Casting sand for iron furnaces.

30 DAMIOLO, S. *Pisogna (Brescia).*—Iron ore, pig iron, &c.; specimens illustrating metallurgical processes.

31 DELPRINO, M. *Vesima, near Acqui (Alexandria).*—Limestone, and gypsum found around Acqui.

32 DE MORTILLET, G. *Milan.*—Magnesiferous and hydraulic lime, from different parts of Italy; sand, cements, gypsum; plans of works.

33 DINI, P. *Camaiore (Lucca).*—Manganese ore from Fontanaccio.

34 DODERLEIN, PROF. P. *Modena.*—Collection of minerals, of the Provinces of Modena and Reggio, with catalogue, &c.

35 DOL, BALDASSARRE, LESSEE OF THE COMACCHIO ROYAL SALT WORKS, *Turin.*—Crystallized salt, table salt, curing salt.

36 D'URSO, F. P. *Salerno (Principato Citeriore).*—Building stones from Paterno, &c.

37 FERRATA & VITALE, *Brescia.*—Millstones, grindstones, and scythe stones.

38 FERRO, F. *Cagliari.*—Antimony, from Su Suergiu mine, Villasalto; lead ore, from Ringraxius mine.

39 FLORENCE ROYAL NATURAL HISTORY MUSEUM, GEOLOGICAL DEPARTMENT (Director, PROF. I. COCCHI).—Collection of ornamental stones, marbles, breccias, alabaster, serpentine, chalcedony, jasper, building materials, clays, &c.; iron ore from Val d'Aspra, lead ore from Castellaccia, &c.

40 FOICO, G. *Chiavenna, Sondrio.*—Collection of minerals.

41 FORNOVO JUNTA FOR THE EXHIBITION, *Parma.*—Specimens of petroleum from Neviano de' Rossi, &c.

42 FRANEL, E. & Co. *Turin.*—Copper, and copper pyrites from Ollomont mine, Aosta; specimens illustrating mode of dressing the ores, &c.

43 GABRIELE, A. *S. Bartolomeo in Galdo (Benevento).*—Spathose iron ore (carbonate of iron), from S. Bartolomeo; peat.

44 GANNA, S. *Turin.*—Flagstone, from the quarries of Guessi, &c. (Pinerolo.)

45 GARUCCIU, CAV. G. M. *Iglesias (Cagliari).*—Lead ore, from Is Cortis de Pubusinu mines, near Iglesias.

46 GAVIANO, A. *Lanusci (Cagliari).*—Lignite, from S. Sebastiano a Secis (Cagliari).

47 GENNAMARI & INGURTUSU MINING CO. *Arbus.*—Argentiferous lead ores, &c., from Jugurtusu and Gennamari (*Cagliari*); quartz gangue, with copper pyrites, blende, &c.

48 GIOVANNINI, BROTHERS, *Carmignano (Florence).*—Tables in serpentine, from Monteferrato, near Prato.

49 GLISENTI & RAGAZZONI, *Brescia.*—Iron ore, cast iron, wrought iron, and steel of Val Trompia and Val Sabbia.

50 GOUIN, LÉON, & Co. *Iglesias (Cagliari).*—Argentiferous lead ore (galena), mine of S. Giorgio, near Iglesias; rock accompanying the vein.

51 GRASSI, BROTHERS, *Schilpario (Bergamo).*—Iron ore and pig iron, from Valle di Scalve (Bergamo).

52 GREGORINI, A. *Lovere (Bergamo).*—Iron ore, pig iron, steel; specimens illustrating the manufacture of steel by the Bergamese process.

53 GUERRA, BROTHERS, *Massa (Massa-Carrara).*—Specimens of marble, from the Val del Palazzuolo, &c. (*Massa*); various articles in marble.

54 GUIDOTTI, F. *Lucca.*—Umber.

55 GUPPY & PATISSON, *Naples.* — Cold-bent iron bars.

56 HAUPT, T. *Florence.*—Plans of the mines of Tuscany: synoptical tables.

57 HENFREY & FRANEL, *Iglesias (Cagliari).*—Argentiferous lead from Monte Cour, &c.

58 ITALIAN MARBLE CO., HÄHNER & CO. *Leghorn.*—Specimens of marble from the hills above Massa.

59 JACOBELLI, A. *S. Lupo (Benevento).*—Specimens of marble, from Vitulano and Pietraroia (Benevento).

60 JERVIS, W. P. *of the Royal Italian Central Committee.* — Specimens of magnetic iron ore from Forno (*Massa-Carrara*), discovered by exhibitor.

61 LIPARI JUNTA FOR THE EXHIBITION (*Messina*).—Minerals.

62 MACERATA SUB-COMMITTEE FOR THE EXHIBITION.—Lignite.

63 MAFFEI, CAV. N. *Volterra (Pisa).*—Collection of minerals; articles in breccia; calcedony, from Monte Rufoli.

64 MAGGI, SANTI, & BECCHINI, *Montalcino (Sienna).* — Sienna earth, raw and burned; "mountain meal," with floating bricks, and bricks for polishing metals made of it.

65 MAGRI, D. *Bologna.*—Gypsum, plaster of Paris, scagliola.

66 MALMUSI, CAV. C. *Modena.*—Clay for pottery, marl from Biamana.

67 MANNA, E. *Iglesias (Cagliari).*—Lead ore, from the mine of S. Benedetto, near Iglesias.

68 MARCHESE, E. *Cagliari.*—Collection of minerals from the island of Sardinia.

69 MASSA, C. *Casale (Novara).*—Limestone, and gypsum.

70 MASSERANO, G. *Biella, Novara.*—Syenite from La Balma in Quittanga, near Biella.

71 MASSOLENI, M. *Genoa.* — Millstones, made at Genoa of fragments of stone imported from France.

72 MASSONE & MUSANTI, *Genoa.*—Argentiferous lead ore; galena with gangue, spa-

those iron, copper pyrites, &c.; lead desilverized by Pattinson's process.

73 MASSONE, CAV. M. *Cagliari.*—Minerals from a lead mine, commune of Lula, near Nuoro.

74 MELIS, S. *Cagliari.*—Slags from smelting works existing last century near Villacidro; pig of lead made from them, &c.

75 MESSINA SUB-COMMITTEE FOR THE EXHIBITION.—Lignite, marbles, metalliferous and other minerals from the province of Messina.

76 MILESI, A. *Bergamo.*—Iron ore, cast iron; steel capable of being welded like iron, and of scratching glass.

77 MINERALOGICAL MUSEUM, *Naples* (Director, PROF. SENATOR SCACCHI).—Collection of minerals from the southern provinces of Italy.

78 MODENA AGRICULTURAL INSTITUTION. —Marls, earths, &c.

79 MONTE ALTISSIMO MARBLE CO. *Florence.*—Statuary, and other marble, from the quarries of Monte Altissimo. They were worked by Michel Angelo.

80 MONTEPONI MINING CO. *Genoa.*—Lead ore, and other minerals from Monteponi mine (Sardinia).

81 NAPLES ROYAL FOUNDRY.—Ores of iron and lead, pig iron, coal.

82 NICOLAI, P. A. (President of Monteponi Mining Co. Sardinia), *Genoa.*—Carbonate of lead, sulphuret of lead, and galena from S. Giovanni mine.

83 NOCERA, MUNICIPALITY OF (*Umbria*). —Travertine, and various kinds of limestone partaking more or less of the character of marble.

84 OLLOMONT MINING CO. *Aosta* (*Turin*). —Copper ore, copper.

85 ORSINI, COUNT, *Fuligno* (*Umbria*).—Tables of veined Alberese (cretaceous limestone).

86 PEDEVILLA, F. *Tortona* (*Alexandria*). —Limestone, and lime.

87 PELLICCIA, L. *Naples.*—Felspar and refractory clay, from Parghelia (Calabria Ulteriore II.)

88 PÉTIN GAUDET & CO. *Cagliari.*—Magnetic iron ore, &c., from Perda Niedda mine, near Domusnovas, and Perda Sterria mine, Domus de Maria; gangue, &c.

89 PISTILLI, F. *Campobasso* (*Molise*).—Iron pyrites from Salcito (Molise).

90 PLATANIA, P. & CO. *Catania.*—Sulphur in cakes, &c.; sulphur melted with the heat naturally evolved from the solfatara.

91 PONTICELLI, G. *Grosseto.*—Stalactite from Poggetto, near Grosseto.

92 QUARTAPELLE, R. *Teramo* (*Abruzzo Citeriore I.*).—Lignite, clay, pozzolana, gypsum, limestone.

93 RAVENNA SUB-COMMITTEE FOR THE EXHIBITION.—Salt from Cervia, gypsum.

94 ROMAGNA SULPHUR MINING CO. *Bologna.*—Sulphur in rolls, &c.

95 ROMAN IRON MINING CO. *Rome.*—Ores of iron, and refractory bricks from Tolfa; specimens of cast and wrought iron from Terni; various kinds of iron wire from Tivoli, &c.

96 ROSSI, F. & N. *Lucca.*—Hydraulic lime.

97 SADDI, S. & CO. *Cagliari.*—Lead ore from Arcilloni mine, commune of Burcei; lead ore from Su Bacci di S. Arrideli mine, commune of S. Vito.

98 SADUN & ROSSELLI *Sienna.*—Ore of mercury, metallic mercury, &c., from S. Fiora mine, Sienna; plan of mine.

99 SANTI, C. *Montalcino* (*Sienna*).—"Mountain meal," with floating bricks, and other articles, made of it.

100 SANTINI, G. *Seravezza* (*Lucca*).—Statuary, and other marble, &c., from Campanice, Vaglisotto.

101 SARDINIAN SALT WORKS, *Genoa.*—Bay salt from Cagliari salt works.

102 SCACCHI, PROF. SENATOR A. *Naples.* —Artificial crystals.

103 SCLOPIS, BROTHERS, *Turin.*—Pyrites, Giobertite, &c.; plans of the mine of Brosso (Turin).

104 SERPIERI, E. *Cagliari.*—Ancient slags from Domusnovas and Villamassargia, and Flumini Maggiore; pigs of lead obtained from them, &c.

105 SERRA, L. *Iglesias* (*Cagliari*).—Granite, from Arbus and Guspini; trachite, from Carloforte.

106 SISTO, BARON, *Catania.*—Cake of sulphur, from Muglia.

107 SPANO, L. *Oristano* (*Cagliari*).—Ores of iron.

108 STREIFF, G. & CO. *Bergamo.*—Ores of copper and lead from the Valsassina (Bergamo), 1300 feet above the plain.

109 TALACCHINI, A. *Milan.*—Copper pyrites from the mine of Nibbio, near Palanza, Lago Maggiore.

110 TIMON, CAV. A. *Cagliari.*—Lignite from Terras de Collu, commune of Gonnesa, near Iglesias.

111 THOVAZZI, C. *Fornovo* (*Parma.*)—Petroleum.

112 ROYAL TUSCAN IRON MINES AND FOUNDRIES, *Leghorn.*—Ores of iron from the mines of Rio, Capo Calamita, Terra Nera; cast and wrought iron from Follonica.

113 VELLANO, S. *Turin.*—Glass and emery paper, emery cloth, emery leather: machines for pounding and sifting glass.

114 VICTOR EMMANUEL MINING CO. *Miggiandone* (*Novara*).—Copper ore.

115 VILLA, A. & G. *Milan.*—Classified collection of minerals, from the cretaceous system of Brienza.

116 VINCENTINI, COUNT P. O. *Rieti* (*Umbria*).—Calcareous sandstone, coralline, and yellow breccia, from Monte Alviano, near Rieti.

117 VOLTERRA SALT WORKS, *Volterra.*—Specimens of salt.

118 ZICCARDI, N. *Campobasso* (*Molise*).—Gypsum, from Ripalomosi, near Campobasso.

---

119 ACERBI, G. *Torpiana* (*Genoa*).—Copper ore, serpentine, decomposed jasper for making hydraulic cement.

120 BARBAGALLI, S. *Catania.*—Sublimed sulphur.

121 BOURLON & CO. *Pisa.*—Iron ore, from Monte Valerio.

122 COJOLI, E. *Leghorn.*—Copper ore, from Caggio; lignite, from Monterufoli.

123 COSSU, P. *Domusnovas (Cagliari).*—Lead ore, from Acquabona.

124 CRIVELLI, C. *Tortona (Alexandria).*—Flagstone, from Sorli.

125 DE BOISSY, MARCHIONESS T. *Setimello (Florence).*—White marble, serpentine, granite, ochres from Elba.

126 DEL GRECO, F. *Arezzo.*—Marl for manure.

127 D'ERCHIA, A. *Monopoli (Terra di Bari).*—Limestone, tufaceous stone.

128 FEDERICI, DR. M. *Arcola, Sarzana (Genoa).*—Manganese ore, from Arcola.

129 FOGGIA ROYAL ECONOMIC SOCIETY (*Capitanata*).—Collection of marble, from the Gargano; alabaster, travertine.

130 FORESI, L. *Portoferrajo, Elba (Leghorn).*—Copper and antimony ores, from Elba; plans of the mineral deposits.

131 FORLÌ SUB-COMMITTEE FOR THE EXHIBITION.—Sulphur, gypsum, lignite.

132 GELICHI, T. *Florence.*—Calcareous serpentine (oficalce), and calcareous exphotide, from Colle Salvetto.

133 GIACOMELLI, P. *Lucca.*—Steel.

134 GIUDICE, G. *Molo (Girgenti).*—Sulphur, raw and refined; gypsum, and selenite.

135 LA FONTANA MINING CO. *Domusnovas (Cagliari).*—Lead ore, from Monte Cervus.

136 LICCIARDELLO, S. *Catania.*—Powdered sulphur for vines.

137 LICATA, MUNICIPALITY OF (*Girgenti*).—Dolomite.

138 LUCCA SUB-COMMITTEE FOR THE EXHIBITION.—Collection of marbles of the province.

139 MAZZIOTTI, BARON, & CO. *Turin.*—Oxide of manganese, from Framura mine.

140 NAPLES SUB-COMMITTEE FOR THE EXHIBITION.—Sand from the Bay of Naples, rich in iron ore.

141 NOCITO, DR. G. *Girgenti.*—Agates; bituminous Tertiary schists, from the neighbourhood of Girgenti.

142 NURCHIS, R. *Domusnovas (Cagliari).*—Lead ore, from Buoncammino.

143 PAPARELLA, G. *Tocco, near Chieti (Abruzzo Citeriore).*—Petroleum.

144 PERELLI, G. *Laurino (Principato Citeriore).*—Marble from Laurino and Laurito.

145 PIRAZZI, MAFFIOLA, & CO. *Piedimulera (Novara).*—Native gold, from Val Toppa.

146 PIROLI, PROF. A. *Parma.*—Iron and copper ores, lignite, &c.

147 PODESTÀ, B. *Sarzana (Genoa).*—Copper ore, from Bracco mine.

148 PODESTÀ, D. *Sarzana (Genoa).*—Bardiglio and black marble, from the neighbourhood of Sarzana.

149 ORREGONI, A. *Varese (Como).*—Peat, from near the lake of Varese.

2078 ALBIANI TOMEI, CAV. F. *Seravezza. (Lucca).*—Slab of "bardiglio" marble, from La Cappella quarry, near Seravezza; marble squares for pavements, from Solais quarry, near Seravezza.

2079 FALLICA, A, *Catania.*—Crystallized sulphur and celestine.

2080 PARMA UNIVERSITY, NATURAL HISTORY MUSEUM OF.—Minerals and fossils, from the provinces of Parma and Placenza.

2081 PATE, BROS. *Leghorn.*—Regulus of antimony, from S. Stefano (*Grosseto*).

2082 RACALMUTO, MAYOR OF (*Girgenti*).—Rock salt, from Racalmuto.

2083 RACCHI, DR. G. *Casalduni (Benevento).*—Lignite, from Casalduni and Pagliara.

2084 REGGIO (CALABRIA) SUB-COMMITTEE FOR THE EXHIBITION.—Magnetic iron ore, from Aspremonte; antimonial nickel; ores of argentiferous lead and manganese; amianthus; marble.

2085 REGGIO (EMILIA) AGRICULTURAL ASSOCIATION.—Marl, plaster of Paris, lime, gypsum; earth rich in nitrogenous matter.

2086 RICCARDI DI NETRO, CAV. E. *Turin.* Iron and copper ores, from Traversella; electric apparatus, for separating magnetic oxide of iron from copper pyrites; dressed ore, some of it concentrated by the electric apparatus.

2087 REMEDI, MARQUIS A. *Sarzana (Genoa).*—Copper ore, from Marciano.

2088 ROCCA, GUERRIERO, & CO. *Levanto (Genoa).*—Copper pyrites, from La Francesca mine, Levanto.

2089 ROCCHETTA, MUNICIPALITY OF (*Massa-Carrara*).—Umber, serpentinous breccia, jasper for making hydraulic cement, manganese ore.

2090 RUSCHI, AVV. P. *Sarzana (Genoa).*—Red marble, from Monte Caprione.

2091 SARAGONI & TURCHI, *Cesena (Forlì).*—Sulphur.

2092 SCACCHI, D. *Gravina (Terra di Bari).*—Building stone.

2093 SCOVAZZO-CAMMERATA, BARON ROCCO, *Catania.*—Sulphur.

2094 SLOANE, HALL, BROS. & COPPI, *Florence.*—Copper ores, from Monte Catini, near Volterra (*Pisa*); copper.

2095 SPEZIA, BROS. *Pestarena (Novara).*—Gold, from Pestarena, in Val d'Anza.

2096 TANCREDI, P. *Trebiano, Sarzana (Genoa).*—Manganese ore, from Graziola and Guarcedo.

2097 TURIN ENGINEERING SCHOOL (*Scuola d'applicazione per gl'Ingegneri*).—Galena, from Monteponi mine (*Cagliari*); copper pyrites, from St. Marcel mine (*Turin*); copper pyrites, from Ollomont, Aosta (*Turin*); nickel pyrites, from Varallo; nickeliferous matt; lignite, from Cadibona, with plans of the mine.

2098 COCCHI, PROF. I. *Florence.*—Geological collection, from the mountains of Spezia, with sections.

2099 CURRO, *Catania.*—Flowers of sulphur.

2100 FLORI, A. *Forlì.*—Sulphur.

2101 GALLIGANI, DR. G. *Seravezza (Lucca).*—Marble table-tops and vase.

2102 GINORI-LISCI, MARQUIS, *Florence.*—Alabaster, lignite and copper ores, from near Volterra (Pisa).

2103 HÄHNER, CAV. *Leghorn.*—Marble, from Massa Carrara; copper and lead ores, from Val di Castello (*Lucca*); cinnabar, from Ripa (*Lucca*).

2104 PESARO AGRICULTURAL ACADEMY.—Iron ore, from Monte Nerone.

2105 RICHARD, G. *Milan.*—Turf—natural, and compressed without machinery, for distillation.

2106 ROYAL ENGINEER CORPS, *Turin.*—Topographical map of the former kingdom of Sardinia.

2107 ROYAL ENGINEER CORPS, *Naples.*—Topographical map of the Neapolitan provinces of the kingdom of Italy.

2108 RODRIGUEZ, *Lipari.*—Sulphur.

2109 SCALIA, L. *Palermo.*—Sulphur ore from Rabbioni mine, Serradifalco (*Caltanissetta*); sulphur.

CLASS 2.

Sub-Class A.

150 ALBERTI, F. *Naples.* — Collection of chemical products.

151 ARROSTO, G. *Messina.* — Citric acid, prepared by a new process; acid of the bergamotte citron; citrate of lime, &c.

152 ASQUER, CAV. A. *Cagliari.*—Soda produced from ashes of plants growing near Cagliari.

153 BOTTONI, DR. C. *Ferrara.* — Crystallized cream of tartar.

154 CAGLIARI SUB-COMMITTEE FOR THE EXHIBITION. — Sulphate of magnesia, crude soda from vegetable ashes, cream of tartar, potassa in cakes and powder.

155 CONVENT OF THE PADRI SERVITI, *Sienna.* — Bicarbonate of potassa, prepared with carbonic acid spontaneously evolved from mineral waters.

156 CORSINI, L. *Florence.*—Blacking, varnish for leather.

157 CURLETTI, A. *Milan.*—Crude potassa from vegetable ashes, raw and partly purified carbonate of soda, caustic soda, saltpetre.

158 DE LARDAREL, HEIRS OF COUNT, *Leghorn.* — Boracic acid, from the Lagoons, near Volterra; natural productions of the Lagoons.

159 FANNI, F. *Cagliari.*—Crude soda.

160 FERRONI, G. *Florence.*—Blacking.

161 GASPARE, M. *Teramo* (*Abruzzo Ulteriore I.*).—Cream of tartar.

162 GHIBELLINI, D. & V. *Persiceto* (*Bologna*).—Japan for ironwork.

163 LODINI, BROTHERS, *Persiceto* (*Bologna*).—Japan for ironwork.

164 MAFFEI, G. *Reggio, Emilia.*—Heliolene, a colourless and inodorous spirit for lamps.

165 MAJORANA, BROTHERS, BARONS DI NICORRA, *Catania.*—Crude soda.

166 MARINI, G. *Arezzo.*—Potassa.

167 MARRA, E. *Salerno* (*Principato Citeriore*).—Crude carbonate of potassa, from incinerated grape skins; purified potassa.

168 MASSEI, C. *Giulia* (*Abruzzo Ulteriore I.*).—Cream of tartar.

169 MELIS, B. *Quartu, near Cagliari.*—Crude soda, obtained from vegetable ashes.

170 MIRALTA, BROTHERS, *Savona* (*Genoa*).—Tartaric acid, cream of tartar.

171 ORSINI, ORSINO, & NEPHEW, *Leghorn.*—Specimens of saltpetre.

172 PARODI, P. *Savona* (*Genoa*).—Cream of tartar.

173 PETRI, G. *Pisa.*—Reduced iron, and protoiodide of iron, preserved from oxidation by a new process. Solidified cod-liver oil.

174 RIATTI, V. *Reggio, Emilia.*—Cyanide of aluminium and iron; aluminate of soda, and chloro-sulphuric acid.

174A SOERNO, E. *Genoa.*—White lead.

175 SCLOPIS, BROTHERS, *Turin.* — Sulphuric and other acids; green vitriol, blue vitriol, double sulphate of iron and copper: Epsom salts.

176 SINISCALCO, M. *Salerno* (*Principato Citeriore*).—Crude cream of tartar, from wine; purified cream of tartar.

177 SUPPA & CASOLINO, *Trani* (*Terra di Bari*).—Nitre; cream of tartar, crude, and purified.

178 TOVO, F. *Turin.*—Coral reduced to the state of mucilage, and capable of being coloured and moulded.

179 VERCIANI, A. *Lucca.*—Illustration of a process for staining ivory indelibly.

---

180 BELTRANI, G. *Trani* (*Terra di Bari*).—Crude cream of tartar.

181 CASASCO, G. *S. Antonino, Susa* (*Turin*).—Essence of peppermint.

182 CIUTI, N. & SON, *Florence.*—Photographic chemicals.

183 DE BELLIS, G. *Castellana* (*Terra di Bari*).—Cream of tartar.

184 DI CEVA, MARQUIS, COLONEL G. B. *Nocetto* (*Genoa*).—Saltpetre; refined sulphur.

185 DE VITA, N. *Giffone Velle Piane* (*Principato Citeriore*).—Crude and purified soda.

186 DURVAL, E. *Monterotondo* (*Grosseto*).—Common and purified boracic acid, borax, sulphate of ammonia.

187 FORLÌ SUB-COMMITTEE FOR THE EXHIBITION.—Sulphuric acid.

188 GULLI, G. *Reggio, Calabria* (*Ulteriore I.*).—Essence of bergamotte.

189 LEONI, A. *Leghorn.*—White lead.

2110 LOFARO, B. *Reggio* (*Calabria Ulteriore I.*). — Essence of bergamotte citrons, lemons, citrons, bitter oranges; concentrated citric acid.

2111 MELISSARI, F. S. *Reggio* (*Calabria Ulteriore I.*).—Essence of bergamotte citrons, bitter lemons, Portugal lemons, bitter oranges, and mandarin oranges.

2112 MINERVINOM, *Benevento.* — Crude Benevento potassa, prepared from the skins of white grapes.

2113 REGGIO (CALABRIA) SUB-COMMITTEE FOR THE EXHIBITION.—Crude and refined cream of tartar.

2114 SPANO, L. *Oristano* (*Cagliari*). — Hard and soft charcoal.

2115 TARTARONE, G. *Giffone Velle Piane* (*Principato Citeriore*).—Refined potassa.

2116 TORRISI, M. *Trecastagni* (*Catania*).—Cream of tartar.

2117 AMANTINI, *Urbino.*—Specimens of lacquered iron.

2118 CAMPISI, A. *Melitello* (*Catania*).—Citric acid.

2119 MESSINA SUB-COMMITTEE FOR THE EXHIBITION.—Potash alum; citric acid.

Sub-Class B.

190 ABBAMONDI, PROF. N. *Solopaca* (*Benevento*). — Sulphurous water from Jaliss, acidulous waters from Villa, sulphurous aluminous water from S. Antonio: all rising from an extinct volcanic crater.

191 ALDROVANDI, M. *Bologna.* — Anti-scorbutic water.

192 ARROSTO, G. *Messina.* — Mineral waters from the neighbourhood of Messina, with memoir.

193 BARRACCO, BARON, *Cotrone* (*Calabria Ulteriore II.*).—Liquorice root, liquorice.

194 BORTOLOTTI, P. *Bologna.* — Felsina water, a perfumed cosmetic.

195 BOLOGNA PROVINCIAL DEPUTATION. —Thermo-mineral waters of La Poretta (*Bologna*).

196 CAGLIARI SUB-COMMITTEE FOR THE EXHIBITION.—Saline, chalybeate, and alkaline mineral waters of the province of Cagliari.

197 CARINA, PROF. A. *Lucca.* — Mineral waters from the Baths of Lucca, with their natural deposit.

198 COJARI, AVV. V. *Soliera, near Fivizzano* (*Massa-Carrara*). — Sulphurous water from Equi, Fivizzano.

199 COMI, R. *Giulia* (*Abruzzo Ulteriore I.*). —Liquorice juice.

200 CUGUSI, E. *Cagliari.*—Mineral waters from Domusnovas.—Thermo-mineral water from Siliqua, Sardaru, Fordongianus, and Villacidro; chalybeate waters from Capoterro.

201 DE ROSA, R. *Atri* (*Abruzzo Ulteriore I.*).—Liquorice juice.

202 DUFOUR, BROTHERS, *Genoa.*—Mannite, quinine, cinchonine, cinidine, cinchonidine, and chinoline.

203 FAVILLI, G. *Pontesserchio* (*Pisa*).—Unoxidizable protoiodide of iron, reduced iron.

204 FOTI, S. *Acireale* (*Catania*).—Thermo-mineral water, called S. Venere del Pozzo, near Acireale.

205 GALLIANI & MAZZA, *Milan.* — Mannite, castor oil.

206 GARELLI, Dr. G. *Turin.*—Samples of the thermo-sulphurous and alkaline waters of Valdieri (Cuneo); medicinal deposit and scum of those waters.

207 GENNARI, PROF. P. *Cagliari.*—Collection of Italian medicinal plants.

208 GIORGINI, PROF. G. *Parma.*—Saline mineral water containing iodine, from Sassuolo (Modena): pamphlet on its analysis, and medical properties.

209 GIORGINI, Dr. G. *Radicofani* (*Sienna*). —Mineral waters, from S. Casciano dei Bagni.

210 GRASSI, P. *Acireale* (*Catania*). — Mineral waters from S. Tecla, near Acireale.

211 LIPARI JUNTA FOR THE EXHIBITION, (*Messina*).—Thermo-mineral water from the Lipari Islands.

212 MACERATA SUB-COMMITTEE FOR THE EXHIBITION.—Collection of mineral waters.

213 MADESIMO MINERAL WATER Co. *Madesimo* (*Sondrio*).—Mineral water, from Madesimo.

214 MONTINI, P. *Fabriano* (*Ancona*).—Seltzer water, seidlitz water, Vichy water, magnesian water, chalybeate water, soda water.

215 MORELLI, G. *Rogliano* (*Calabria Citeriore*). — Manna of the ash (*Fraxinus Ornus*), several kinds.

216 NAPLES SUB-COMMITTEE FOR THE EXHIBITION.—Liquorice.

217 NOCERA, MUNICIPALITY OF (*Umbria*). —Acidulous mineral water from Nocera; Samian or Nocera earth, used for polishing metals.

218 ORSI, A. *Montalcino* (*Sienna*).—Sulphurous alkaline mineral water, from Collalli, near Montalcino.

219 PATUZZI, L. *Limone* (*Brescia*).—Citron water.

220 PELLAS, C. F. *Genoa.*—Cod-liver oil: calcined, and fluid magnesia; acid to render carbonate of magnesia effervescent.

221 PERI, G. *Milan.*—Anti-rheumatic oil.

222 PIGHETTI, A. *Salò* (*Brescia*).—Citron water.

223 POLI, G. B. *Brescia.*—Pills made from the deposit of the principal mineral waters of France; anti-mercurial sarsaparilla pills.

224 RICCI, G. *Turin.*—Digestive pastiles, tincture of mint.

225 RIOLO, MUNICIPALITY OF (*Ravenna*). —Mineral waters.

226 REGGIO (EMILIA) SUB-COMMITTEE FOR THE EXHIBITION. — Saline, chalybeate, and sulphurous mineral waters.

227 RUSPINI, G. *Bergamo.*—Sulphurous saline, and saline ferruginous waters containing iodine; other mineral waters; and crystallized mannite, obtained by a new process.

228 SCERNO, E. *Genoa.*—Sulphate and citrate of quinine, sarsaparilla.

229 SCOLA, B. *Turin.*—Gelatine capsules, containing medicines.

230 SPANO, L. *Oristano* (*Cagliari*). — Mineral waters, from the thermal, tepid, and cold springs of Fordongianus; deposit of these springs.

231 TURIN ROYAL ACADEMY OF MEDICINE.—Mineral waters of the ancient Sardinian provinces; medicinal deposit formed by several of them; catalogue.

232 VERATTI, C. *Bologna.*—Cashew-nut pastiles.

233 VERGA, DR. A. *Milan.*—Saline mineral water containing iodine, lately discovered near Miradolo (Pavia); pamphlet regarding it.

---

234 AMICARELLI, V. *Montesantangelo* (*Capitanata*).—Manna.

235 BELLIA, S. *Caltagirone* (*Catania*).—Sulphurous mineral waters.

236 BRASINI, BROS. *Forlì.* — Purgative mineral waters.

237 CASTAGNACCI, A. *Florence.* — Compound lichen pastiles, sulphur pastiles, vermifuge sweetmeats.

238 CONTESSINI, F. & Co. *Leghorn.*—Salts

of quinine; caffeine; mannite; santonine; morphine.

239 CORRIDI, G. *Leghorn.*—Mannite; sulphate and citrate of quinine; santonine, castor-oil.

240 CROPPI, C. *Forli.*—Mineral waters.

241 GIORDANO, D. *Cetara (Principato Citeriore).*—Manna.

242 MAJORANA, G. & TORNABENE, F. *Catania.*—Mineral waters.

243 PIGNATELLI, V. *Cosenza (Calabria, Citeriore).*—Liquorice.

244 POLENGHI, C. *S. Fiorano (Milan).*—Mallow yarn for surgical use.

245 PONDI, G. *Palagonia (Catania).*—Chalybeate and acidulous mineral waters.

246 REGGIO (CALABRIA) SUB-COMMITTEE FOR THE EXHIBITION. — Liquid and solid nitrate of magnesia.

247 TORRI, DR. F. *Pisa.*—Mineral waters from the baths of Pisa and Asciano.

248 VITI, MARQUIS A. *Orvieto (Umbria).* Acidulous and chalybeate waters.

249 VALERI, & Co. *Lagnano (Brescia).*—Castor-oil and castor-oil seeds; residuum of castor oil manufacture for manure.

CLASS 3.

Sub-Class A.

260 ADRAGŃA, G., BARONE DI ALTAVILLA, *Trapani.*—Canary seed, and linseed.

261 ALEXANDRIA SUB-COMMITTEE FOR THE EXHIBITION.—Collection of various kinds of wheat cultivated in the province.

262 ALGOZINO, S. *Leonforte, Catania.*—Spelta wheat.

263 ANZALONE, F. *Catania.* — Kidney beans.

264 ARESU, S. *Selargius, Cagliari.*—Wheat in the ear.

265 BELLESINI, BROTHERS, *Imola, Bologna.*—Specimens of rice, to illustrate its cultivation; Chinese, Novara, and American rice.

266 BELTRAMI, COUNT P. *Cagliari.* — Wheat in the ear, rice, and beans.

267 BELTRANI, G. *Trani, Terra di Bari.*—Dried figs, raisins, sweet and bitter almonds, apricot kernels, and olives.

268 BERGAMI, P. *Ferrara.*—Wheat and Indian corn.

269 BOLOGNA AGRICULTURAL SOCIETY.—Collection of corn, and other seeds of the province of Bologna.

270 BURESTI, F. *Arezzo.*—Wheat, rye, and Indian corn.

271 CAGLIARI SUB-COMMITTEE FOR THE EXHIBITION.—Wheat, semolina, flour, bran, starch, barley, Indian corn, sorgho, beans, pease, chick-pease, mustard seed, canary seed, almonds, nuts of various kinds, fennel, raisins, and dried figs.

272 CAMPOBASSO SUB-COMMITTEE FOR THE EXHIBITION.—Various kinds of wheat and leguminous seeds, Indian corn, lentils, beans, and kidney beans.

273 CAO DI S. MARCO, COUNT, *Cagliari.*—Wheat, barley, chick-pease, beans, and lentils.

274 CASALI, A. *Calci, Pisa.*—Corn and flour.

275 CASAZZA, CAV. A. *Ferrara.*—Wheat, Indian corn, oats, barley, chick-pease, and grass seeds.

276 CASERTA PROVINCIAL AGRICULTURAL GARDEN, *Terra di Lavoro.*—Various kinds of wheat, barley, oats, beans, and pease; ground nuts.

277 CASSOLA, AVV. C. *Vercelli, Novara.*—American rice, from Anitre, near Vercelli.

278 CATANIA, HERMITS OF S. ANNA.—Chick-pease.

279 CELI, PROF. E. *Modena.*—Collection of arious grass seeds grown in the province.

280 CHELLI, F. *Leghorn.*—Starch, made by a new process.

281 CHERICI, N. *S. Sepolcro, Arezzo.*—Various kinds of wheat; Indian corn, flour, and semolina; Indian corn bread and polenta; chestnuts, chestnut flour, and polenta.

282 CHERICI, CLELIA, *S. Sepolcro, Arezzo.*—Dried edible fruit of the *carlina acaulis.*

283 CHIETI SUB-COMMITTEE FOR THE EXHIBITION, AND OTHER EXHIBITORS, *Abruzzo Citeriore.*—Various kinds of wheat (some exhibited by G. LAUCIANO and D. D'ONOFRIO); risciola, winter rye, barley, oats, Indian corn, beans, kidney beans (some exhibited by D. BUTOLO and DR. G. BONETTI); chick-pease (some exhibited by V. JULIANI VILLEMAGNA); lentils, lupins, pease, vetches, flax seed, hemp seed, and potatoes.

284 CONSIGLIO, M. *Lentini, Noto.*—Almonds.

285 CONVENT OF S. FRANCESCO D'ASSISI, *Catania.*—Barley.

286 DE' GIUDICI, A. *Arezzo.*—Wheat, oats, Indian corn, beans, chick-pease, millet, and lupins.

287 DI NISSA, MARQUIS G. *Cagliari.*—Sweet and bitter almonds, carob and other beans, pine cones.

288 DRAMMIS, BARON S. *Scandale, Calabria Ulteriore II.*—Various kinds of wheat; Peruvian barley.

289 DROUIN, G. *Naples.*—Flour, bran, and bread.

290 FERRARA ROYAL CHAMBER OF COMMERCE.—Indian corn and Indian corn flour: wheat of various kinds and wheat flour: oats, rice, chick-pease, and beans.

291 FIORENTINI, G. *Castrocaro, Florence.*—Sainfoin.

292 GABRIELE, DR. A. *S. Bartolomeo in Galdo, Benevento.*—Cereals and leguminous seeds.

293 GIORDANO, E. *Salerno, Principato Citeriore.*—Red and white yams, introduced into Italy by the exhibitor.

294 GRASSI, A. *Giarre, Catania.* — Almonds cultivated on the flanks of Mount Etna.

295 GREFFI, A. *Monterchi, Arezzo.*—Wheat.

296 GUACCI, F. *Campobasso, Molise.*—Varieties of wheat.

297 GUIDA, G. & G. *Gargarengo, Vicolonga (Novara).*—Various kinds of wheat,

barley, oats, rye, Indian corn, millet, sorgho, pease, chick-pease, beans, and lupins.

298 LAI, L. *Lanusei, Cagliari.*—Ground nuts, beans, dried figs, and dried prunes.

299 LEGA, M. *Brisighella, Ravenna.*—Aniseed.

300 LIGAS, A. *Selargius, Cagliari.*—Wheat in the ear.

301 LIPARI JUNTA FOR THE EXHIBITION, *Messina.*—Raisins and capers.

302 LUCCA SUB-COMMITTEE FOR THE EXHIBITION.—Collection of cereals and other agricultural produce of the province.

303 MACERATA SUB-COMMITTEE FOR THE EXHIBITION.—Various kinds of wheat and Indian corn.

304 MAJORANA BROTHERS, BARONS DI NICORRA, *Catania.*—Indian corn, oats, canary seed, beans, pease and chick-pease, lentils, sesame seed, mustard seed, hemp and flax seed, and acorns.

305 MAMELI, F. *Selargius (Cagliari).*—Wheat in the ear, and beans.

306 MANCUSO, M. *Catania.*—Wheat, canary seed, oats, beans, chick-pease, clover, flax, and mustard seed.

307 MARINI DEMURO AVV. T. *Cagliari.*—Wheat, barley, beans, pease, chick-pease, and figs.

308 MAROZZI, E. *Pavia.* — Cleaned and uncleaned rice; clover seed.

309 MASSONE, CAV. M. *Cagliari.*—Pease, and millet seed.

310 MAZZURANA, F. *Trent.* — Carraway seed.

311 MELIS, B. *Quartu (Cagliari).*—Wheat, barley, and beans.

312 MELONI, A. *Quartu (Cagliari).* — Dried figs, and raisins.

313 MINUTOLI TEGRIMI, COUNT E. *Lucca.*—Rice obtained without rotation of crops; Chinese and American rice, cleaned and uncleaned.

314 MODENA ROYAL BOTANICAL GARDENS.—200 varieties of Indian corn, collected by the late Professor G. Brignoli.

315 MODENA SUB-COMMITTEE FOR THE EXHIBITION.—Cereals, and vegetable seeds.

316 MONARI, C. & C. *Bologna.*—Several kinds of cleaned rice.

317 MONTERISI, G. *Bisceglie (Terra di Bari).*—Pease and lentils.

318 MONTORI, G. *Colonnella (Abruzzo Ulteriore I.).*—Agricultural products.

319 MOSCERO, G. *Cosenza (Calabria Citeriore).*—Dried figs.

320 MURRU-MURRU, A. *Sanluri Cagliari).*—Buck wheat.

321 NATOLI, A. D. *Patti (Messina).*—Clover seed.

322 NIEDDA DI S. MARGHERITA COUNT P. & BROTHER, *Cagliari.*—Stalks of sorgo for forage.

323 PADRI BENEDETTINI MONKS, *Monte Cassino (Catania).*—Various kinds of wheat; barley and beans.

324 PAGARELLI, DR. L. *Castrocaro (Florence).*—Aniseed and coriander seed.

325 PALUMBO, O. *Trani (Terra di Bari).*—Lentils and beans.

326 PALUMBO, P. *Cava (Principato Citeriore).*—Potatoe and other kinds of starch; dextrine obtained from potatoe starch.

327 PANNA, DR. G. *Quartu, near Cagliari.*—Dried figs.

328 PANTANO, F. P. *Asaro (Catania).*—Chick-pease.

330 PASI, G. *Ferrara.*—Various kinds of beans.

331 PAVANELLI, G. *Ferrara.* — Wheat, maize, and rape seed.

332 PICCALUGA, G. *Cagliari.*—Pistachio nuts.

333 PISTILLI, F. *Campobasso (Molise).*—Wheat and chick-pease.

334 PITTAU, M. *Sanluri, Cagliari.*—Chick-pease.

335 QUERCIOLI BROTHERS, *Modigliana (Florence).*—Aniseed.

336 RAMO, S. *Laconi (Cagliari).*—Wheat.

337 RAVENNA SUB-COMMITTEE FOR THE EXHIBITION.—Wheat, oats, rice, and beans.

338 REVEDIN, COUNT G. *Ferrara.*—Wheat and Indian corn.

339 RUNDEDDU, R. *Selargius (Cagliari).*—Beans, chick-pease, and pease.

340 SANNA, V. *Selargius, Cagliari.* — Beans, figs.

342 SAVONA JUNTA FOR THE EXHIBITION, *Genoa.*—Collection of leguminous seeds.

343 SCHLAEPFER, WENNER, & Co. *Salerno (Principato Citeriore).*—Potatoe starch: dextrine, and gum from potatoe starch.

344 SERRA, DR. L. *Iglesias (Cagliari).*—Wheat, barley, Indian corn, linseed, and raisins.

345 SPANO, L. *Oristano (Cagliari).*—Barley, wheat, chick-pease, beans, lucerne seeds, and pine cones.

346 TARDITI & TRAVERSA, *Brà, Turin.*—Various kinds of semolina, and flour.

347 TELLINI, V. *Calci (Pisa).*—Wheat flour.

348 TURIN AGRICULTURAL ACADEMY.—Collection of agricultural produce; beans, chick-pease and pease, lupins, and fenugreek. Artificial fruit.

349 VACCARO, L. *Cosenza (Calabria Citeriore).*—Seeds of the ground pistachio.

350 ASSOM VILLASTEL, *Turin.* — Hemp-seed.

---

351 AREZZO SUB-COMMITTEE FOR THE EXHIBITION.—Collection of the agricultural products of the province.

352 ARRANGA, G. *Serracapriola (Capitanata).*—Wheat.

353 ASCOLI SUB-COMMITTEE FOR THE EXHIBITION. — Collection of cereals, leguminous seeds, &c.

354 BARONE, BROS. *Foggia (Capitanata).*—Wheat.

355 BARACCO, BROS. *Catanzaro (Calabria Ulteriore I.).*—Agricultural products.

356 BOCCARDO, BROS. *Candela (Capitanata).*—Saragolla, Carlantino, and Carosella wheat.

357 BOFONDI, COUNT P. *Forlì.*—Cereals.

358 CAMMARATA SCOVAZZO, BARON ROCCO, *Palermo.*—Wheat.

359 CAPELLI, CAV. *Foggia (Capitanata).* —Wheat.

360 CARBONE, F. *Catania.*—Kidney beans.

361 CASERTA SUB-COMMITTEE FOR THE EXHIBITION. — Collection of cereals, leguminous seeds, potatoes, yams, beetroot, &c.

362 CASSANO, F. *Gioja (Terra di Bari.)*—Cereals, mustard and fennel seeds.

363 CHIARINI, P. *Faenza (Ravenna).*—Cleaned rice of different kinds.

364 CICCHESE, P. *Campobasso (Molise).*—Wheat.

365 CICCHESE, R. *Campobasso (Molise).*—Flour.

366 COSTANTINO, G. *S. Marco dei Cavoti (Benevento).*—Leguminous seeds.

367 DANZA, D. *S. Agata (Capitanata).*—Wheat.

368 DANZA, G. *S. Agata (Capitanata).*—Wheat.

369 DE FIDIO, G. *Casaltrinità (Capitanata).*—Wheat.

370 DE LUCA, P. *Catania.* — Pistachio nuts.

371 DEL BUONO, E. *S. Agata (Capitanata).*—Wheat.

372 DE LEO, A. *Casaltrinità (Capitanata).* —Wheat.

373 DELL'ERMA, V. *Castellano (Terra di Bari).*—Figs.

374 DEMURTAS, E. *Lanusei (Cagliari).*—Wheat, leguminous seeds.

375 FANTINI, *Bertinori (Forlì).*—Aniseed.

376 FIAMINGO, S. *Giano (Catania).* — Indian corn, rye, French beans, almonds, &c.

377 FOGGIA SUB-COMMITTEE FOR THE EXHIBITION.—Oats, barley, beans.

378 FORLÌ SUB-COMMITTEE FOR THE EXHIBITION.—Almonds, aniseed.

379 GARAU CARTA, L. *Sanluri (Cagliari).* —Almonds.

380 GIULIANI, L. *S. Marco in Lamis (Capitanata).*—Carosella wheat.

381 GIUDICE, G. *Favara (Girgentiz).*—Wheat, almonds, linseed.

382 GULINELLI, COUNT G. *Ferrara.*—Cereals.

383 LECCE SUB-COMMITTEE FOR THE EXHIBITION — Cereals, leguminous seeds, almonds, raisins, walnuts, &c.

384 LOFARO, B. *Reggio (Calabria).* — French beans.

385 LUCERA, MAYOR OF (*Capitanata*).—Wheat.

386 MELE, N. G. *S. Agata (Capitanata).*—Wheat.

387 MERCATILE, COUNT M. *Ascoli.* — Cereals.

388 MILAN CHAMBER OF COMMERCE.—Collection of the agricultural produce of the province; semolina, bran, flour.

389 NERI, A. *Bologna.*—Cleaned rice.

390 ORTONA, MUNICIPALITY OF (*Abruzzo Citeriore*).—Figs.

391 PACCA, MARQUIS G. *Benevento.* — Wheat and leguminous seeds.

392 PATERNÒ CASTELLO, PRINCESS M. *Catania.*—Wheat, leguminous seeds, &c.

393 PATERNÒ CASTELLO, MARQUIS DI S. GIULIANO, *Catania.*—Peas, &c.

394 PARMA SUB-COMMITTEE FOR THE EXHIBITION.—Cereals, grapes.

395 PASCAZIO, V. *Mola (Terra di Bari).* —Carob-beans, figs.

396 PELLEGRINO, D. *Casaltrinità (Capitanata).*—Wheat.

397 PESARO AGRICULTURAL SOCIETY.—Collection of cereals, leguminous seeds, castor oil seeds, &c.

398 REGGIO (CALABRIA) SUB-COMMITTEE FOR THE EXHIBITION.—Prickly pears, figs.

399 REGGIO (EMILIA) AGRICULTURAL ASSOCIATION.—Collection of the agricultural produce of the province.

2120 REGGIO (EMILIA) AGRICULTURAL SOCIETY.—Wheat, millet, rice, beans, peas, Chestnut and bean flour.

2121 ROMEO, L. *Acquaviva* (*Terra di Bari*).—Aniseed.

2122 RUBINO, M. *Foggia* (*Capitanata*).—Wheat.

2123 SANTORO, G. *S.Agata* (*Capitanata*). —Wheat.

2124 SARAGATU, AVV. P. *Sanluri* (*Cagliari*).—Wheat, starch.

2125 SARCINA, N. R. *Casaltrinità* (*Capitanata*).—Wheat.

2126 SCOCCHERA, *Canosa* (*Terra di Bari*). —Castor oil seeds.

2127 SINISCALCO, BROS. *Foggia* (*Capitanata*).—Wheat.

2128 TORRI, L. *Bondeno* (*Ferrara*). — Wheat.

2129 TREJAVILLA, A. *Cerignola* (*Capitanata*).—Wheat.

2130 TROIA, MAYOR OF, *Capitanata.* — Wheat.

Sub-Class B.

400 BELLENTANI, G. *Modena.*—Shoulder ham, sausages, lard, Italian paste, and tomato sauce.

401 BENEDETTI, P. BROTHERS, *Faenza* (*Ravenna*).—Italian paste.

402 BIANCHI, G. & C. *Lucca.*—Italian paste, &c.

403 BOSCARELLI, A. *Cosenza* (*Calabria Citeriore*).—Preserved larks.

404 BOTTAMINI, B. *Bormio* (*Sondrio*).—Honey.

405 CALDERAI, A. *Florence.*—Sausages, &c.

406 CESARI, L. *Torre Annunziata* (*Naples*) —Paste for soup.

407 DOZZIO, G. *Belgioioso* (*Pavia*). — Cheese.

408 DRAGHI, D. *Placenza.*—Salt meat, known as *coppe.*

409 FANNI, F. *Cagliari.*—Tunny eggs and mullet eggs (*Bottarghe*).

410 FERRARA ROYAL CHAMBER OF COMMERCE.—Cheese and sausages.

411 FORNI, A. *Bologna.*—Bologna sausages (*mortadella*), &c.

412 FRANZINI, B. *Pavia.*—Cheese.

413 GABRIELE, DR. A. *S.Bartolomeo in Galdo* (*Benevento*).—Cheese and honey.

414 GATTI, A. *Cosenza* (*Calabria Citeriore*). —Mushrooms.

415 GIULIANI, V. *Turin.*—Chocolate.

416 GUELFI, G. *Navacchio* (*Pisa*).—Biscuits.

417 JACCHINI, G. A. *Alexandria.*—Sausages, &c.

418 LAMBERTINI, G. *Bologna.*—Sausages, &c.

419 LANCIA, BROTHERS, *Turin.*—Preserved lard, salt, and preserved food.

420 LAVAGGI, G. *Augusta* (*Noto*). — Hyblean honey.

421 LIUZZI, C. *Reggio, Emilia.* — Sheep's milk cheese, from Loriano.

422 LUPINACCI, BARON L., BROTHERS, *Cosenza* (*Calabria Citeriore*). — Butter and cheese.

423 MAJORANA, BROTHERS, BARONS DI NICORRA, *Mostarda* (*Catania*).—Honeycomb, and honey; cheese, butter, preserved olives, artichokes.

424 MALMUSI, CAV. C. *Modena.*—Honey.

425 MARINI DEMURO AVV. T. *Cagliari.*—Honey, and saffron.

426 MILAZZO JUNTA FOR THE EXHIBITION.—Salt-fish.

427 MODENA SUB-COMMITTEE FOR THE EXHIBITION.—Sheep's-milk cheese.

428 NUNS OF S. LUCIA, *Cagliari.*—Bunch of flowers in sugar.

429 ORRÙ, S. *Burcei* (*Cagliari*).—Bitter honey.

430 ORSI, R. & Co. *Bologna.* — Bologna pork sausages.

431 PAOLETTI, F. *Pontedera* (*Pisa*).—Italian paste and biscuits.

432 PAOLETTI, G. *Pontedera* (*Pisa*).—Italian paste, biscuits, and flour.

433 PAOLETTI, O. *Florence.*—Biscuits.

434 PARMA SUB-COMMITTEE FOR THE EXHIBITION. — Cheese, Italian paste, and salame.

435 PETRUCCELLI, C. *Castelfranco* (*Benevento*).—Sheep's milk and other cheese.

436 RAINOLDI, G. *Milan.*—Salt and smoked pork.

437 REVEDIN, COUNT G. *Ferrara.*—Cheese.

438 SAGLIOCCA, G. *Pietro Elcina* (*Benevento*).—Cheese.

439 SALTARELLI, A. *Pisa.*—Candied fruit.

440 SAMOGGIO, G. *Bologna.*—Hog's lard.

441 SONA, C. *Alexandria.* — Mostarda, a kind of preserve.

442 SPANO, L. *Oristano* (*Cagliari*). — Pickled olives, smoked mullet, salted eels, honey, &c.

443 TORRICELLI, A. *Florence.*—Chocolate.

444 VALAZZA, G. *Turin.*—Sardines, and tunny.

445 VALERI, A. *Ferrara.*—Peach preserve.

446 ZANETTI, G. *Bologna.* — Bologna sausages, and *capocollo.*

---

447 AMICARELLI, D. V. *Montesantangelo* (*Capitanàta*).—Honey.

448 ASTENGO C. *Savona* (*Genoa*). — Vermicelli and other kinds of Italian paste for soup.

449 BARBETTI, S. *Fuligno* (*Umbria*). — Chocolate and sweetmeats.

450 BARACCO, BARON A. *Naples.*—Cheese, olives, chestnuts.

451 BARRACCO, BROS. *Cotrone* (*Calabria Ulteriore II.*).—Calabrian cheese.

452 BELTRAMI, COUNT P. *Cagliari.* — Cow's milk cheese.

453 BERGAMI, P. *Ferrara.*—Cheese.

454 BERNARDI, BROS. *Borgo a Buggiano, Val di Nievole* (*Lucca*). — Italian paste for soup.

455 BIBIANO (MUNICIPALITY OF) *Reggio, Emilia.*—Old cheese.

456 BIFFI, P. *Milan.*—Confectionery.

457 BODINO, L. *Genoa.*—Chocolate.

458 BOLLINI, G. *Alexandria.*—Dried salt pork, bondiola.

459 BOSIO, D. *Alexandria.*—Pork sausages.

460 BOTTAMINI, B. *Bormio* (*Sondrio*).—Honey.

461 BRASINI, BROS. *Forlì.*—Chocolate.

462 CAGLIARI SUB-COMMITTEE FOR THE EXHIBITION. — Maccaroni, vermicelli, and other kinds of paste for soup; tunny and mullet eggs; sweet and bitter honey.

463 CAPASSO, F. *Benevento.* — Benevento torrone, &c.

464 CASERTA SUB-COMMITTEE FOR THE EXHIBITION (*Terra di Lavoro*).—Maccaroni, vermicelli, &c.

465 CASSANO, F. *Gioja* (*Terra di Bari*).—Cheese.

466 CARPANETO & GHILINO, *Genoa.* — Preserved food and fruit.

467 CICCHESE, P. *Campobasso* (*Molise*).—Maccaroni, semolina made of Saragolla wheat.

468 CICCHESE, R. *Campobasso* (*Molise*).—Italian paste, semolina.

469 CIOPPI, L. & S. *Pontedera, Pisa.* — Italian paste for soup, salame.

470 COSTANTINO, G. *S. Marco de'Cavoti* (*Benevento*).—Cheese.

471 DE GAETANO, F. *Gallico* (*Calabria Ulteriore I.*).—Maccaroni and vermicelli.

472 DE GORI, COUNT A. *Sienna.*—Goat's milk cheese.

473 DEMURTAS, E. *Lanusei Cagliari.* — White Italian paste for soup, cheese.

474 FARINA, BROS. *Baronissi* (*Principato Citeriore*). — Provoloni and Caciocavallo cheese.

475 FOGGIA SUB-COMMITTEE FOR THE EXHIBITION. — Hand and machine-made Italian paste, for soup; pickled capers.

476 FORLÌ SUB-COMMITTEE FOR THE EXHIBITION.—Sausages.

477 FORNO MATTEI, A. *Prato* (*Florence*).—Biscuits.

478 GALASSO, G. *Benevento.*—Benevento torrone.

479 GIORDANO, D. *Ceara* (*Principato Citeriore*).—Salt anchovies.

480 LANZARINI, BROS. *Bologna.* — Pork sausages, &c.

481 LEMBO, P. *Minori* (*Principato Citeriore*).—Collection of Italian paste.

482 LOFARO, B. *Reggio* (*Calabria Ulteriore I.*).—Pickled olives.

483 MELISSARI, F. S. *Reggio* (*Calabria Ulteriore I.*).—Dried and pickled olives.

484 MOSCATO, BROS. *Salerno (Principato Citeriore)*.—Caciocavallo cheese.

485 PACCA, MARQUIS G. *Benevento.* — Cheese, honey.

486 PALUMBO, O. *Trani (Terra di Bari)*. —Olives, tomato sauce.

487 PASCAZIO, V. *Mola (Terra di Bari)*. —Tomato sauce.

488 RAMIREZ, G. *Reggio (Calabria Ulteriore I.)*.—Preserved tomatoes.

489 REGGIO (CALABRIA) SUB-COMMITTEE FOR THE EXHIBITION.—Sheep's-milk cheese, honey, eleozaccaro.

490 REGGIO (EMILIA) AGRICULTURAL ASSOCIATION.—Ham, shoulder ham, coppa, sausages.

491 SALERNO SUB-COMMITTEE FOR THE EXHIBITION *(Principato Citeriore)*. — Dried figs, pears, chesnuts, &c.; Cilento sausages.

492 SPEZI, D. *Fuligno (Umbria)*. — Confectionery and chocolate.

493 TRUCILLI, V. *Salerno (Principato Citeriore)*.—Provole, a buffalo milk cheese; butter preserved in an envelope of cheese.

494 VIVARELLI, C. *Pistoja (Florence)* — Sheep's milk cheese.

---

495 CAMPOBASSO SUB-COMMITTEE FOR THE EXHIBITION.—Biscuits.

496 DAMIANI, C. *Portoferajo, Elba (Leghorn)*.—Biscuits.

497 IANNICELLI, M. *Salerno (Principato Citeriore)*.—Italian paste.

498 MARINELLI, E. *Parma.* — Collection of Italian paste.

499 MATTEI, A. *Prato (Florence)*.—Rusks.

500 SGARIGLIA, *Dalmonte.* — Pickled olives.

Sub-Class C.

505 AGNINI, T. *Modena.*—Rosolio, rinfresco.

506 AGNELLO, BARON, *Siculiana (Girgenti)*.—Wine.

507 AGOZZOTTI, AVV. *Modena.* — Lambrusco wine.

508 ALONZO, A. *Catania.*—White wine.

509 ANSELMI, B. *Verona.*—Wine.

510 ALBINO P. *Campobasso (Molise)*.—Wine.

511 ALESSI, G. *Messina.*—Tobacco.

512 ALLEMANO, BROTHERS, *Asti (Alexandria)*.—Wines.

513 ALMERICI, MARQUIS G. *Cesena (Ferrara)*.—Wine.

514 ANGHIRELLI, G. *Montalcino (Sienna)*. —Wines.

515 ASQUER, VISCOUNT DI, *Flumini (Cagliari)*.—Red wine.

516 BALLOR, G. *Turin.*—Liqueurs, &c.

517 BARACCO, N. & CO. *Turin.*—Liqueurs.

518 BARTHOLINI, C. *Cosenza (Calabria Citeriore)*.—Wines.

519 BELLENTANI, G. *Modena.*—Vinegar.

520 BELTRANI, G. *Trani (Terra di Bari)*. —Wines.

521 BERNARDI, F. *Sienna.*—Red wine.

522 BOLOGNA ROYAL TOBACCO MANUFACTORY.—Tobacco, snuff, and cigars.

523 BONNET, G. *Comacchio (Ferrara)*.—Wine.

524 BONOLIS, F. *Teramo (Abruzzo Ulteriore)*.—Effervescing wines.

525 BORATTO, D. *Alexandria.*—Vermouth.

526 BORLASCA, C. *Govi (Alexandria)*.—Red wine.

527 BOTTI, A. *Chiavari (Genoa)*.—Wine.

528 BOZZO, M. *Benevento.*—Tobacco.

529 BRAGGIO, COUNT F. *Strevi (Alexandria)*.—Wines.

530 BRESCIA SUB-COMMITTEE FOR THE EXHIBITION.—Wine.

531 CADONI, A. *Quartu (Cagliari)*.—Wine.

532 CAGLIARI ROYAL EXCISE OFFICE.—Tobacco, snuff, and cigars.

533 CAGLIARI SUB-COMMITTEE FOR THE EXHIBITION.—Wines, spirits of wine, vinegar, and tobacco.

534 CAIMI, F *Sondrio.*—Wine.

535 CARA, CAV. G. *Cagliari.*—Wines.

536 CARAMORA, P. *Asti (Alexandria)*.—Liqueurs, and wine.

537 CASAZZA, CAV. A. *Ferrara.*—Wines.

538 CASTAGNINO, I. *Imola (Bologna)*.—Wine.

539 CASTIGLIONE BENDINELLI, *Novi (Alexandria)*.—Wines.

540 CASTIGLIONE DELLE STIVIERE JUNTA FOR THE EXHIBITION *(Brescia)*.—Tobacco and snuff.

541 CERRONE, G. *Teramo (Abruzzo Ulteriore I.)*—Effervescing wines.

542 CHERICI, N. *S. Sepolcro (Arezzo)*.—Wines.

543 CLARKSON, S. V. *Mazzara (Trapani)*. —Wine.

544 COBIANCHI & ARDIZZOLI, *Boca (Novara)*.—Wines.

545 COCCHI, F. *Reggio, Emilia.*—Rosolio.

546 COCOZZA, C. *Benevento.*—Wine.

547 CODIGORO, MUNICIPALITY OF, *Ferrara.*—Red wines.

548 COJARI, AVV. V. *Fivizzano (Massa and Carrara)*.—Wines.

549 CONTI, B. *Pontedera, Pisa.*—Wines, &c.

550 CUCCHI, T. *Parma.*—Effervescing wine.

551 D'ANTONIO, S. *Ornano (Abruzzo Ulteriore I.)*.—Wines.

552 DE ANGELIS, M. *Isola (Abruzzo Ulteriore I.)*.—Wine.

553 DE GORI, COUNT A. *Sienna.*—Wines.

554 DEL PRINO, DR. M. *Vesime, near Acqui (Alexandria)*.—Wines.

555 DEMICHELI, G. *Novi (Alexandria)*.—Wine.

556 DEMURTAS, E. *Lanusei (Cagliari)*.—Wines.

557 DENEGRI, G. *Novi (Alexandria)*.—Wine, some of it made of dried grapes.

558 DE RUBERTIS, L. *Lucito (Molise)*.—Wine, made of dried grapes.

559 DI BLASIO, F. *Bagnoli (Molise)*.—Wine.

560 FANTINI, G. *Comacchio (Ferrara)*.—Wine.

561 FAVARE VERDIRAME, V. *Mazzara (Trapani)*.—Wine.

562 FERRARINI, DR. A. *Reggio, Emilia.* —Wines.

563 FLORENCE ROYAL TOBACCO MANUFACTORY.—Tobacco, snuff, and cigars.

564 FLORIS COIANA, P. *Cagliari.*—Wines.

565 GARAU CARTA, L. *Sanluri* (*Cagliari*). —Wine.

566 GAVIANO, A. *Lanusei* (*Cagliari*).—Wines.

567 GENTA, AVV. P. *Caluso* (*Turin*).—Wines.

568 GINNASI, COUNT D. *Imola* (*Bologna*). —Wine.

569 GRISALDI DEL TAIA, DR. C. *Sienna.* —Wine.

570 GROSSO, E. *Turin.*—Liqueurs.

571 JACCHINI, B. *Modena.*—Wine made without grapes.

572 LAI, L. *Lanusei* (*Cagliari*).—Wines.

573 LIPARI JUNTA FOR THE EXHIBITION (*Messina*).—Wines.

574 LORU, CAV. A. *Cagliari.*—Wine vinegar.

575 LUCCA ROYAL TOBACCO MANUFACTORY.—Tobacco, snuff, and cigars.

576 LUPINACCI, BARON L. & BROS. *Cosenza* (*Calabria Citeriore*).—Wine, &c.

577 MADONNA, G. *Isola Abruzzo* (*Ulteriore I.*).—Wine.

578 MALMUSI, CAV. C. *Modena.*—Aromatic vinegar, 200 years old.

579 MARCHI, L. ROYAL ESTATE OF S. LORENZO, *Volterra* (*Pisa*).—Gin, and spirit from the arbutus.

580 MARINI DEMURO AVV. T. *Cagliari.*—Wine, and wine vinegar.

581 MARINI, P. *Cagliari.*—Wines.

582 MARYRETTI, G. *Savona* (*Genoa*).—Liqueurs.

583 MASSA AVV. C. *Casale* (*Alexandria*).—Wines.

584 MASSONE, M. *Cagliari.*—Red wine, and wine vinegar.

585 MAZZAROSA, MARQUIS G. B. *Lucca.* —Wine.

586 MELIS, B. *Quartu* (*Cagliari*).—Wine.

587 MELLUSI, G. *Torrecusi* (*Benevento*).—Wine.

588 MERLO, G. B. *Castelnuovo, Bormida* (*Alexandria*).—Wines.

589 MILAN ROYAL TOBACCO MANUFACTORY.—Tobacco, snuffs, and cigars.

590 MILIANI, F. *Peccioli* (*Pisa*).—Wines.

591 MILAZZO JUNTA FOR THE EXHIBITION.—Wine.

592 MODENA ROYAL TOBACCO MANUFACTORY.—Tobacco, snuff, and cigars.

593 MODENA SUB-COMMITTEE FOR THE EXHIBITION. — Wines, liqueurs, and vinegar.

594 MONCALVO, D. *Bisio, near Novi* (*Alexandria*).—Wines.

595 MONTEMERLO, E. *Novi* (*Alexandria*). —Red wine.

596 MONTERISI, G. *Bisceglie* (*Terra di Bari*).—Wines.

597 MONTINI, P. *Fabriani* (*Ancona*) —Wine and liqueurs.

598 MORANDO, I. & SONS, *Sampierdarena* (*Genoa*).—Liqueurs.

599 MORIANI, CAV. N. *Florence.*—Wines.

600 MORTINI, L. *S. Bartolomeo in Galdo* (*Benevento*).—Wine.

601 MURGIA, G. *Sanluri* (*Cagliari*).—Wines.

602 MURRU MURRU, A. *Sanluri* (*Cagliari*). —Wine.

603 NAPLES SUB-COMMITTEE FOR THE EXHIBITION.—Wine.

604 OREGGIA, C. *Savona* (*Genoa*).—Wine.

605 ORLANDO, G. *Pescolamazza* (*Benevento*).—Wine.

606 OUDART, L. *Genoa.*—Wines.

607 OVADA, MUNICIPALITY OF (*Alexandria*).—Red wines.

608 PACIFICO, G. *Salerno* (*Principato, Citeriore*).—Wines.

609 PALUMBO, O. *Trani* (*Terra di Bari*). —Wines, brandy, and rum.

610 PARENTE, C. *Monterocchetto* (*Benevento*).—Red wine.

611 PARENTE, G. *Ceppaloni* (*Benevento*).—Red wine.

612 PARMA SUB-COMMITTEE FOR THE EXHIBITION.—Wine and liqueurs.

613 PASOLINI, G. *Imola* (*Bologna*). —Wine.

614 PATRICO, DR. V. *Trapani*—Wine.

615 PAVANELLI, G. *Ferrara.*—Red wine.

616 PENNACCHI, F. *Orvieto* (*Umbria*).—Wine.

617 PERINI, P. *Desenzano* (*Brescia*).—Liqueurs.

618 PERRA, A. *Cagliari.*—Wines.

619 PERUSINO, V. *Sandamiano d'Asti* (*Alexandria*).—Wines and vinegar.

620 PICCHIO, COUNT P. *Alexandria.*—Wines.

621 PIZZI, L. *Petrella* (*Molise*).—Wines.

622 PLACENZA SUB-COMMITTEE FOR THE EXHIBITION.—Collection of wines of the province.

623 POTENZIANI, HEIRS OF, *Rieti* (*Umbria*).—Wine.

624 PRAMPOLINI, A. *Reggio, Emilia.*—Vinegar made with mother a century old.

625 PRUNAS, CAV. R. *Bosa* (*Cagliari*).—Wines and vinegar.

626 RAPPIS, P. *Andorno Cacciorna* (*Novara*).—Ratafia.

627 RAVIZZA, G. BROTHERS, *Orvieto* (*Umbria*).—Wine.

628 REGGIO (EMILIA) SUB-COMMITTEE FOR THE EXHIBITION.—Collection of wines of the province.

629 RICASOLI, BARON B. *Florence.*—Wines.

630 RICCARDI STROZZI, C. *Florence.*—Wines.

631 RICCI, G. B. *Asti* (*Alexandria*). —Wines.

632 RICCI, L. *Bruno, near Aqui* (*Alexandria*).—Red wine.

633 RIDOLFI, MARQUIS C. *Florence.*—Wine.

634 RONCHI, P. *Florence.*—Vinegar.

635 ROTA & CO. *Alexandria.*—Beer.

636 SAGLIOCCA, G. *Pietra Elcina* (*Benevento*).—Wine.

637 SALIS, F. *Lanusei* (*Cagliari*).—Wines.

638 SANNA, V. *Selargius* (*Cagliari*).—Wines.

639 SANTI, DR. C. *Montalcino* (*Sienna*).—Wines and liqueurs.

640 SANTOSPACO, N. *Castiglione alla Pescara* (*Abruzzo Ulteriore I.*).—Muscat wine.

641 SATTA FLORIS, R. *Cagliari*.—Wines.

642 SAVORINI, F. *Persiceto* (*Bologna*).—Liqueurs.

643 SCAZZOLA, G. D. *Cascine, near Alexandria*.—Wines.

644 SERRA, DR. L. *Iglesias* (*Cagliari*).—Wine.

645 SESIMA, V. *Alexandria*.—Liqueurs.

646 SIRIGU, G. *Cagliari*.—Vermouth.

647 SPANO, L. *Oristano* (*Cagliari*).—Wine, alcohol, and tobacco.

648 SPENSIERI, G. *Ferrazzano* (*Molise*).—Wine.

649 SUPPA & CASOLINO, *Trani* (*Terra di Bari*).—Spirit of aniseed, &c.

650 TARTAGLIOZZI, G. *Isola* (*Abruzzo Ulteriore I.*).—Wine.

651 TORO, B. F. & E. *Chieti* (*Abruzzzo Citeriore*).—Centerba.

652 TORRI, L. *Bondino* (*Ferrara*).—Red wine.

653 TOTORO, N. *Archi* (*Abruzzo Citeriore*).—Boiled wine.

654 TURIN ROYAL TOBACCO MANUFACTORY.—Tobacco, snuff, and cigars.

655 ULRICH, D. *Turin*.—Vermouth, &c.

656 VALLINO BROTHERS, *Brà* (*Cuneo*).—Wines.

657 VARVALLO, F. *Asti* (*Alexandria*).—Wines.

658 VENTURA, V. *Castiglione alla Pescara* (*Abruzzo Ulteriore I.*).—Muscat wine.

659 VICENTINI, P. O. *Rieti* (*Umbria*).—Imitation champagne.

660 VIETRI, D. A. *Salerno* (*Principato Citeriore*).—Wines.

661 VITTONE, F. *Milan*.—Wines and liqueurs.

662 ZICCARDI, V. *Foiano* (*Benevento.*)—Wine.

---

663 ARRANGA, G. *Serracapriola* (*Capitanata*).—Wine.

664 BALSAMO, G. N. *Catania*.—Tobacco leaves.

665 BARBAGALLO, S. *Catania*.—Wine.

666 BARI SUB-COMMITTEE FOR THE EXHIBITION (*Terra di Bari*).—Liqueurs.

667 BAZZIGER, L. & C. *Sassuolo*.—Liqueurs.

668 BERGAMI, P. *Ferrara*.—Common red wine.

669 BERTI, F. & G. *Rubbiera* (*Reggio, Emilia*).—Aniccione and alkermes.

670 BIFFI, P. *Milan*.—Liqueurs.

671 BOCCARDO, BROS. *Candela* (*Capitanata*).—Wine.

672 BONI, E. *Modena*.—Vinegar 150 years old.

673 BUELLI, E. *Bobbio* (*Pavia*).—Collection of wines.

674 CAMPOLONGHI, G. B. *Parma*.—Rosolio.

675 CANTON, G. *Turin*.—Liqueurs.

676 CASALTRINITÀ, MAYOR OF (*Capitanata*).—Wine.

677 CASERTA SUB-COMMITTEE FOR THE EXHIBITION (*Terra di Lavoro*).—Wine.

678 CASSANO, F. *Gioja* (*Terra di Bari*).—Wine.

679 CASSINESI BENEDICTINE MONKS, *Catania*.—Wine.

680 CESENA, C. *Bari* (*Terra di Bari.*)—Red wine.

681 CIANI, G. *Bisceglia* (*Terra di Bari*).—Wine.

682 COLLENZA, E. *Valenzano* (*Terra di Bari*).—Malvasia wine.

683 COPPOLI, MARQUIS R. *Perugia* (*Umbria*).—Wine.

684 CORRIDI, G. *Leghorn*.—Alcohol.

685 COSTARELLI, M. *Catania*.—Wine.

686 D'AMBROGIO, L. *Deliceto* (*Capitanata*).—Wine.

687 D'AMBROSIO, V. *Sansevero* (*Capitanata*).—Wine.

688 DELL' ERMA, N. *Castellana* (*Terra di Bari*).—Wine.

689 DELL' ERMA, V. *Castellana* (*Terra di Bari*).—Wine.

690 DELLA BELLA, D. *Vico* (*Capitanata*).—Muscat wine.

691 DEL TOSCANO, MARQUIS, *Catania*.—Wine.

692 DE MARTINO, G. *Salerno* (*Principato Citeriore*).—Wine.

693 DI GROSSI, G. *Riposto* (*Catania*).—Aniseed liqueur.

694 EBOLI, N. *Bari* (*Terra di Bari*).—Liqueurs.

695 FERRARA ROYAL CHAMBER OF COMMERCE.—Wine.

696 FASCIA, S. *Marco la Catola* (*Capitanata*).—Wine.

697 FERRAROTTO, G. *Catania*.—Muscat wine.

698 FERRI VITO, N. *Canneto* (*Terra di Bari*).—Musaglica wine.

699 FIAMINGO, G. B. *Riposto* (*Catania*).—Wine, alcohol.

700 FLORIO, BROS. *Asti* (*Alexandria*).—Wine, vermouth.

701 FORLÌ SUB-COMMITTEE FOR THE EXHIBITION.—Wine.

702 FREJAVILLE, A. *Cerignola* (*Capitanata*).—Wine.

703 GASPARRI, A. *Bicoari* (*Capitanata*).—Wine.

704 GERVASIO, G. *Canneto* (*Terra di Bari*).—Zagarese wine.

705 GIORDANO, G. *Salerno* (*Principato Citeriore*).—Alcohol from the Arbutus unedo and Helianthus tuberosus.

706 GIOVINE, G. B. *Canelli* (*Alexandria*).—Wine.

707 GIULIANI, L. *S. Marco in Lamis* (*Capitanata*).—Wine.

708 GIUSTI, G. *Modena*.—Balsamic vinegar, 150 years old.

709 GIVONI, V. *Catania*.—Wine, vinegar.

2140 GENOESE ZERBI, D. *Reggio* (*Calabria Ulteriore I.*).—Wine.

2141 GUARNASCHELLI, CAV. G. *Broni (Pavia).*—Wine, vinegar.

2142 GUIDI, C. *Volterra (Pisa).*—Wine.

2143 LADERCHI, A. *Faenza (Ravenna).*—White wine.

2144 LELLA, G. *Messina.*—Wine.

2145 LOFARO, B. *Reggio (Calabria Ulteriore I.).*—Wine.

2146 LOMBARDI, *Sansevero (Capitanata).*—Wine.

2147 LUCERA, MAYOR OF *(Capitanata).*—Wine.

2148 MAJORANA, BROS. BARONS OF NICORRA, *Catania.*—Wine, vinegar, rum, collection of tobacco.

2149 MANCUSO, M. *Catania.*—Wine.

2150 MANGINI, F. *Modena.*—Vinegar, a century old.

2151 MARCHI, P. *Florence.*—Alkermes.

2152 MARTINI, L. *S. Bartolomeo in Galdo (Benevento).*—Wine.

2153 MASSA-CARRARA ROYAL TOBACCO MANUFACTORY.—Snuff, tobacco, cigars.

2154 MASSELLI, A. *Sansevero (Capitanata).*—Wine, and vinegar.

2155 MASSETTI, COUNT P. *Florence.*—Wine, and vinegar.

2156 MELISSARI, F. S. *Reggio (Calabria Ulteriore I.).*—Wine.

2157 MENGAZZI, F. *Cesena (Forlì).*—Absinthe.

2158 MERLONI, BROS. *Bertinoro (Forlì).*—Wine.

2159 MESSINA SUB-COMMITTEE FOR THE EXHIBITION.—Wine.

2160 MONCADA, A. *Catania.*—Wine.

2161 MUNELLI GALILEI, L. *Pontedera (Pisa).*—Wine, vermouth.

2162 MURATO-SOLI, P.—Balsamic vinegar, 132 years old.

2163 NAPLES AND CAVA ROYAL TOBACCO MANUFACTORY.—Tobacco, cigars.

2164 NESII, A. *Reggio (Calabria Ulteriore I.).*—Wine.

2165 NOVA, D. A. *S. Agata (Capitanata).*—Wine.

2166 ORTANO, MUNICIPALITY OF *(Abruzzo Citeriore).*—Rosolio.

2167 PACCA, MARQUIS G. *Benevento.*—White and red wine.

2168 PAGANO, M. A. *Pisciotta (Principato Citeriore).*—Wine.

2169 PAGLIANO, F. *Asti (Alexandria).*—Wine, and vinegar.

2170 PALIZZI, BARON C. *Reggio (Calabria Ulteriore I.).*—Wine.

2171 PARLATORE, E. *Florence.*—Spirits and fruit of arbutus (Arbutus unedo).

2172 PARMA ROYAL TOBACCO MANUFACTORY.—Snuff, tobacco, cigars.

2173 PASCAZIO, V. *Mola (Terra di Bari).*—Wine.

2174 PETROSEMILO, A. *Ortona.*—Liqueurs.

2175 PICCARDI, G. *S. Casciano.*—Wine.

2176 PRATI, G. *Alexandria.*—St. Bernard's elixir.

2177 QUADRAT, L. *Genoa.*—White and red wine.

2178 RAVENNA SUB-COMMITTEE FOR THE EXHIBITION.—Wines.

2179 REGGIO (CALABRIA) SUB-COMMITTEE FOR THE EXHIBITION.—Wine, tobacco.

2180 SALIMBENI, L. *Modena.*—Wine; aromatic vinegar 100 years old; liqueurs.

2181 SALVAGNOLI MARCHETTI, CAV. A. *Corniola, Empoli (Florence).*—Wine.

2182 SANT' AGOSTINO MONASTERY, *Catania.*—Wine.

2183 STA. ANNA, HERMITS OF *(Catania).*—White wine.

2184 S. FRANCESCO MONASTERY, *Catania.*—Wine.

2185 S. PLACIDO MONASTERY, *Catania.*—White and red wine.

2186 S. SCOLASTICA MONASTERY, *Bari (Terra di Bari).*—Liqueurs.

2187 SANTORO, G. *S. Agata (Capitanata).*—White wine.

2188 SARACENO, V. *Catania.*—Vinegar.

2189 SAVORELLI, MARQUIS A. *Forlì.*—Wine.

2190 SCOCCHERA, S. *Canosa (Terra di Bari).*—White and red wine.

2191 SCUDERI, F. M. *Catania.*—Zambria liqueur, alcohol.

2192 SESTRI PONENTE ROYAL TOBACCO MANUFACTORY, *near Genoa.*—Tobacco, cigars.

2193 SINISCALCO, M. *Salerno (Principato Citeriore).*—Rectified spirits of Arbutus unedo, Asphodelus ramosus, and Pancratium maritimum; rum.

2194 SISTO, BARON A. *Catania.*—Wine.

2195 SPANO, CAV. P. *Oristano (Cagliari).*—Vernaccia wine.

2196 SYLOS LABINI, V. *Bitonto (Terra di Bari).*—Wine.

2197 TARANTINI, N. *Corato (Terra di Bari).*—Wine.

2198 TARELLO, M. *Viverone (Novara).*—Wine made of dried grapes.

2199 TESI, L. *Pistoja.*—Wine, vermouth.

2200 TRAPANI, G. *Gallico (Calabria Ulteriore I.).*—Wine.

2201 VAGLIASINDI, F. *Catania.*—Vinegar.

2202 ZERBINI, P. *Modena.*—Lambrusco wine.

2203 LI-GRESTI, *Catania.*—Zambu liqueur.

CLASS 4.

Sub-Class A.

710 ALFANI, C. *Nocera.*—Olive oil.

711 ASTENGO, BROTHERS, *Savona (Genoa).*—Manufactured wax.

712 BAFFONI, V. *Fermo (Ascoli).*—Grape-stone oil.

713 BANCALARI, L. *Chiavari (Genoa).*—Olive oil.

714 BARTOLINI, *Cosenza (Calabria Citeriore).*—Olive oil.

715 BARACCO, BROTHERS, *Cotrone (Calabria Ulteriore II.).*—Olive oil.

716 BELTRANI, G. *Trani (Terra di Bari).*—Olive oil.

717 BELLELLA, CAV. E. *Capaccio (Principato Citeriore).*—Olive oil.

718 BOTTI, A. *Chiavari (Genoa).*—Olive oil.

719 BOTTEGHI, A. *Chiavari* (*Genoa.*)—Olive oil.
720 CAROBBI, G. *Florence.*—Wax cakes, candles, tapers, and torches: spermaceti cakes and candles.
721 CATTANEO, G. B. *Chiavari* (*Genoa*).—Olive oil.
722 CONTI, B. *Villa Saletta, Pontedera* (*Pisa*).—Olive oil.
723 CONTI, E. BROTHERS, *Leghorn.*—Soap.
724 DANIELLI & FILIPPI, *Buti* (*Pisa*).—Olives, olive oil, olive kernels, &c.
725 DANZETTA, BARON, & BROTHERS, *Perugia* (*Umbria*).—Olive oil.
726 DE CESARE, A. *Penne* (*Abruzzo Ulteriore I.*).—Olive oil.
727 DE GORI, A. *Sienna.*—Olive oil; wax.
728 DEMURTAS, E. *Lanusei* (*Cagliari*).—Olive oil and wax.
729 DE RUBERTIS, L. *Lucito* (*Molise*).—Olive oil.
730 DUNANT, G. M. *Milan.*—Soap and perfumery.
731 FRANCIOSI, P. *Terriciola* (*Pisa*).—Olive oil.
732 FURLANI, G. *Florence.*—Soap.
733 GABRIELE, DR. A. *S. Bartolomeo in Galdo* (*Benevento*).—Collection of oils from the neighbourhood.
734 GAVIANO, A. *Lanusei* (*Cagliari*).—Olive oil.
735 GHIGO, C. *Saluzzo* (*Cuneo*).—Wax candles.
736 GIORDANO, G. *Naples.*—Collection of oils.
737 GIUSTI, N. *Pisa.*—Olive oil, wax.
738 GRISALDA DEL TAIA, *Sienna.*—Olive oil.
739 LUPINACCI, BARON L. & BROTHERS, *Cosenza* (*Calabria Citeriore*).—Olive oil.
740 MAJORANA, BROTHERS, BARONS DI NICORRA, *Catania.*—Unbleached wax; olive oil; hard soap.
741 MANGANONI, L. *Milan.*—Stearine candles, large tapers, and soap.
742 MARCHI, L. *Volterra* (*Pisa*).—Pistachio nut oil.
743 MASTIANI SCIAMANNA, MARQUIS C. *Pisa.*—Olive oil.
744 MAZZULLO, G. *Mandanici* (*Messina*).—Olive oil.
745 MAZZAROSA, MARQUIS G. B. *Lucca.*—Olive oil.
746 MESSINA SUB-COMMITTEE FOR THE EXHIBITION.—Collection of the oils of the province of Messina.
747 MILAZZO JUNTA FOR THE EXHIBITION.—Olive oil.
748 MILIANI, F. *Peccioli* (*Pisa*).—Olive oil.
749 MILAN CHAMBER OF COMMERCE.—Collection of the oils manufactured in the province of Milan.
750 MINUTOLI TEGRIMI, COUNT E. *Lucca.*—Olive oil.
751 MODENA SUB-COMMITTEE FOR THE EXHIBITION.—Grape-stone and other oils.
752 MOSCERO, G. *Cosenza* (*Calabria Citeriore*).—Oil.
753 NOBERASCO & AQUARONI, *Savona* (*Genoa*).—Soap.
754 OREGGIA, C. *Savona* (*Genoa*).—Olive oil.
755 ORLANDO, G. *Pescolamazza* (*Benevento*).—Olive oil.
756 OTTOLINI BALBANI, COUNTESS C. *Lucca.*—Olive oil.
757 PANCANI, BROTHERS, *Florence.*—Olive oil and other soaps.
758 PENSA, F. *Teramo* (*Abruzzo Ulteriore I.*).—Wax candles.
759 PIERI PECCI G. *Sienna.*—Olive oil.
760 PISTIS, G. *Elini* (*Cagliari*).—Olive oil.
761 PRUNAS, CAV. R. *Bosa* (*Cagliari*).—Olive oil.
762 REGGIO (EMILIA) AGRICULTURAL SOCIETY.—Linseed and other oils.
763 RICCARDI STROZZI, C. *Florence.*—Olive oil.
764 SARDINI, G. *Lucca.*—Olive oil.
765 SCUDERY, A. *Messina.*—Olive oil.
766 SPANO, L. *Oristano* (*Cagliari*).—Olive and other oils; wax.
767 SQUARCI, E. *Leghorn.*—Stearine candles.
768 TACCHI, G. *Bergamo.*—Wax candles, torches, and cakes.
769 TALENTI, COUNT L. *Lucca.*—Olive oil.
770 TURCHI, L. & Co. *Ferrara.*—Olive and other oils; toilet soap.
771 VACCARO, L. *Cosenza* (*Calabria Citeriore*).—Oil obtained from the ground nut.

---

772 AGAZZOTTI, AVV. F. *Modena.*—Walnut oil.
773 ALBIANI, CAV. F. *Pietrasanta* (*Lucca*).—Olive oil.
774 ARPINI, CAV. E. *Ascoli.*—Olive oil.
775 ARRANGA, G. *Serracapriola* (*Capitanata*).—Maccaroni, capers.
776 BARBAGALLO, S. *Catania.*—Hard and soft soap, tallow, linseed oil.
777 BARBATO, N. *S. Agata* (*Capitanata*).—Olive oil.
778 BASTONI, V. *Turin.*—Grape-stone oil.
779 BAZZIGER, L. & Co. *Sassiuolo* (*Modena*).—Liqueurs.
780 BERGAMO SUB-COMMITTEE FOR THE EXHIBITION,—Wax candles.
781 BERNARDI, F. *Sienna.*—Olive oil.
782 BOCCARDO, BROS. *Candela Foggia* (*Capitanata*).—Olive oil.
783 CAGLIARI SUB-COMMITTEE FOR THE EXHIBITION.—Olive, almond, and lentisc oil; wax.
784 CANOSA, MUNICIPALITY OF (*Terra di Bari*).—Olive oil.
785 CANTALLAMESSA, I. *Ascoli.*—Olive oil.
786 CARDUCCI, A. *Taranto* (*Terra di Otranto*).—Olive oil.
787 CASALTRINITÀ, MAYOR OF (*Capitanata*).—Olive oil.
788 CASERTA SUB-COMMITTEE FOR THE EXHIBITION (*Terra di Lavoro*).—Olive and ground nut-oil; oil of the Holcus cernuus and Sorghum saccharatum.
789 CASTORINA, *Catania.*—Soft soap.
790 COSENTINO, S. *Catania.*—Soft soap.
791 COSTANTINO, G. *S. Marco de' Cavoti* (*Benevento*).—Olive oil.

792 Corsini, *S. Casciano (Pisa).*—Olive oil.

793 D'Ambrosio, L. *Deliceto (Capitanata).* Olive oil.

794 De Biase, G. *S. Marco la Catola (Capitanata).*—Olive oil.

795 De Cataldis, O. *Giffoni (Principato Citeriore).*—Olive oil.

796 D'Erchia, A. *Monopoli (Terra di Bari).*—Olive oil, soap.

797 Della Bella, D. *Vico (Capitanata).* —Olive oil.

798 Dell'Erma, V. *Castellana (Terra di Bari).*—Common olive oil.

799 Delli Santi, F. *Manfredonia (Capitanata).*—Olive oil.

800 Di Rignano, Marquis, *Foggia (Capitinata).*—Olive oil.

801 Frejaville, A. *Cerignola (Capitanata).*—Olive oil.

802 Gasparri, A. *Biccari (Capitanata).* —Olive oil.

803 Gioia, A. *Corato (Terra di Bari).*—Olive oil.

804 Giovandonato, O. *Benevento.*—Olive oil.

805 Girardi, M. *Turin.*—Olive, hazelnut, walnut, colza, linseed, and castor oil.

806 Idone, G. *Lecce (Terra di Otranto).* —Olive oil.

807 Lamenaco, L. & G. *Corato (Terra di Bari).*—Olive oil.

808 Lanza, Bros. *Turin.*—Stearine, and stearine candles.

809 Lofaro, F. *Catania.*—Soft soap.

810 Macerata Sub-Committee for the Exhibition.—Olive and olive-kernel oil.

811 Manni, D. *Tocco (Abruzzo Citeriore).* Fine olive oil.

812 Masetti, Count P. *Florence.*—Olive oil.

813 Masselli, A. *S. Severo (Capitanata).* —Olive oil.

814 Melissari, F. S. *Reggio (Calabria Ulteriore I.).*—Olive oil.

815 Mercatile, M. *Ascoli.*—Olive oil.

816 Milella, G. *Bari (Terra di Bari).*—Fine olive oil.

817 Munnelli Galilei, L. *Pontedera (Pisa).*—Olive oil.

818 Mazzuchetti, E. *Turin.*—Castor oil.

819 Naldini, B. *Florence.*—Olive oil.

2205 Niedda di Sta. Margherita, Count P. *Cagliari.*—Olive oil.

2206 Novi, D. *S. Agata (Capitanata).*—Olive oil.

2207 Ortona, Municipality of *(Abruzzo Citeriore).*—Olive oil.

2208 Pacca, Marquis G. *Benevento.*—Olive oil.

2209 Palizzi, Baron C. *Reggio (Calabria Ulteriore I.).*—Olive oil.

2210 Palumbo, O. *Trani (Terra di Bari).* —Gum of the olive tree; tamarisk.

2211 Paolella, G. *Castelluccio, Val Maggiore (Capitanata).*—Olive oil.

2212 Pascazio, V. *Mola (Terra di Bari).* Olive oil.

2213 Paulucci, Marquis G. B. *Forlì.*—Oil from the seeds of the *Kolreuteria paniculata.*

2214 Pesaro Agricultural Society.—Olive, and olive-kernel oil.

2215 Pesci, G. *Fuligno (Umbria).*—Olive oil.

2216 Piccardi, Giuseppe, *S. Casciano.*—Olive oil.

2217 Porto Maurizio Sub-Committee for the Exhibition.—Olive oil.

2218 Reggio (Calabria) Sub-Committee for the Exhibition.—Olive oil.

2219 Ricasoli, Baron B. *Brolio (Sienna).* —Olive oil.

2220 Rignano, Marquis of, *Foggia (Capitanata).*—Olive oil.

2221 Rospigliosi, Prince, *Pistoja (Florence).*—Olive oil.

2222 Sansone, P. *Cagnano (Capitanata).* —Olive oil.

2223 Santoro, G. *S. Agata (Capitanata).* —Olive oil.

2224 Saulli, L. *Pisciotta (Principato Citeriore).*—Olive oil.

2225 Savorelli, Marquis A. *Forlì.*—Stearine, and stearine manufactures.

2226 Scocchera, S. *Canosa (Terra di Bari).*—Olive oil.

2227 Serracapriola, Municipality of, *(Capitanata).*—Olive oil.

2228 Serventi (Heirs of), *Parma.*—Wax, and wax candles.

2229 Sgariglia, M. *Ascoli.*—Olive oil.

2230 Sylos Labini, V. *Bitonto (Terra di Bari).*—Olive oil, and olive-kernel oil.

2231 Tesi, L. *Pistoja (Florence).*—Olive, and other oils.

2232 Valeri & Co. *Legnago (Verona).*—Cold-drawn castor oil, &c.

2233 Aicardi, F. & Co. *Bari (Terra di Terra).*—Fine, filtered, and common olive oil.

2234 Lauri, Count J.—Olive oil.

2235 Pisa Sub-Committee for the Exhibition.—Collection of common, best, and washed olive oil of the province of Pisa.

Sub-Class B.

820 Barracco, Brothers, *Cotrone (Calabria Ulteriore II.).*—Sheep's fleeces.

821 Bentivoglio, Cav. C. *Modena.*—Sheep's fleeces.

822 Bertone, *Turin.*—Sheep's fleeces.

823 Binda Grugnola & Co. *Milan.*—Combs.

824 Bussolati, Brothers, *Parma.*—Silkworm cocoons.

825 Cagliari Sub-Committee for the Exhibition.—Coral, fished on the coast of Sardinia; guano; burned dung.

826 Campi, Count G. *Dovadola (Florence).* —Silkworm cocoons.

827 Carro, Marianna, *Cagliari.*—Bunch of flowers, made of shells.

828 Chisoli, A. *Brignone (Bergamo).*—Cocoons and thrown silk.

829 Cucchi, T. *Parma.*—Cocoons.

830 Donini, S. *Bologna.*—Animal manure.

831 FINO, L. *Turin.*—Albumen, hematosine, &c.

832 GABRIELE, DR. A. *S. Bartolomeo in Galdo (Benevento).*—Wool.

833 GARAU, CARTA L. *Sanluri (Cagliari).*—Opened cocoons.

834 MACERATA SUB-COMMITTEE FOR THE EXHIBITION.—Sheep's fleeces.

835 MILAN.—ROYAL LOMBARD SCIENTIFIC INSTITUTION.—Collection illustrating the metamorphoses of the silkworm.

836 MONTALTI, E. *Bologna.*—Gelatine and glue.

837 ORLANDO, G. *Pescolamazza (Benevento).*—Sheep's fleeces.

838 PANICHI, *Perugia (Umbria).*—Carded wool.

839 PIZZETTI, F. *Parma.*—Silkworm cocoons, silkworm moths, and raw silk.

840 PONTICELLI, G. *Grosseto.*—Long merino wool.

841 PUPILLI, G. *Pontedera (Pisa).*—Glue.

842 SICCARDI, BROS. *Mondovi (Cuneo).*—Silkworm cocoons.

843 SOMMARIVA, B. *Palermo.*—Glue.

844 SPANO, L. *Oristano (Cagliari).*—Wool, silkworm cocoons, &c.

845 TONI, F. *Perugia (Umbria).*—Cocoons and raw silk.

846 VEGNI, L. *Città di Castelio (Umbria).*—Glue.

847 VETERE, G. *Gerchiara (Calabria Citeriore).*—Wool.

---

848 CASSANO, F. *Gioia (Terra di Bari).*—Wool.

849 COSTANTINO, G. *S. Marco de' Cavoti (Benevento).*—Wool.

850 GALANTI, PROF. A. *Perugia (Umbria).*—White Chinese cocoons.

851 GIOVANETTI, G. BROS. *Pisa.*—Bone buttons.

852 PACCA, MARQUIS G. *Benevento.*—Wool.

853 VACCARO, L. *Cosenza (Calabria Citeriore).*—Calabrian cantharides (*Cantharis vesicatoria*).

Sub-Class C.

N.B.—All the specimens of cotton are placed together in the collection formed by the Royal Italian Commission.

880 ANZI, DON M. *Bormio (Sondrio).*—Collection of lichens.

881 AUGIAS, S. *Tempio (Sassari).*—Three kinds of archil.

882 AVENTI, COUNT F. M. *Ferrara.*—Hemp.

883 AYMERICH, I. *Cagliari.*—Bark, and acorns of the cork oak.

884 BAFFONI, V. *Fermo (Ascoli).*—Rape-seed cakes, &c.

885 BARATELLI, BARON, *Ferrara.*—Green hemp.

886 BARTOLINI, C. *Cosenza (Calabria Citeriore).*—Indigenous cotton.

887 BARTOLINI, F. *Corigliano (Calabria Citeriore).*—Prepared flax.

888 BELLELLA, G. *Salerno (Principato Citeriore).*—Madder.

889 BELLELLA, CAV. E. *Capaccio (Principato Citeriore).*—Madder.

890 BELTRAMI, CAV. P. *Cagliari.*—Woods, charcoal, and cork.

891 BELTRANI, G. *Trani (Terra di Bari).*—Mustard seed, castor oil seed, and cotton.

892 BENZI, T. *Carpi (Modena.)*—Straw plait.

893 BERNARDUSI, M. *Ferrara.*—Green hemp.

894 BIAVATI, P. *Crevalcore (Bologna).*—Green hemp.

895 BOLOGNA AGRICULTURAL SOCIETY.—Preparation of the *Botys silicealis*, an insect that blights the hemp plant: by Prof. Bertoloni.

896 BONORA, *Ferrara.*—Raw hemp.

897 BOTTER, PROF. F. *Bologna.*—Collection illustrating the cultivation of hemp in the Emilia: hemp plants blighted and diseased: hemp seed oil, &c.

898 BURGARELLA, A. *Trapani.*—Sumac.

899 CAGLIARI SUB-COMMITTEE FOR THE EXHIBITION.—Flax, hemp, lentisc, madder, sumac, saffron, gum, cork, castor oil fruit, white and tawny yellow Siamese cotton (*Gossypium Siamense*); straw baskets.

900 CALANDRINI, PROF. F. *Florence.*—Collection of 185 species of wood, indigenous or acclimatized in Tuscany.

901 CAMPOBASSO SUB-COMMITTEE FOR THE EXHIBITION.—Hemp-seed.

902 CAVALIERI, P. *Ferrara.*—Hemp.

903 CERTANI, A. *Bologna.*—Hemp-seed and hemp.

904 CHERICI, N. *S. Sepolcro (Arezzo).*—Specimens of timber woods; woad (*Isatis tinctoria*), woad plant and seed.

905 CREMONA SUB-COMMITTEE FOR THE EXHIBITION.—Collection of building and other woods, grown in the province.

906 CRIPPA, IDA, *Florence.*—Pine cones, pine seeds, and pine seed oil.

907 D'ALESSIO, G. *Capaccio (Principato Citeriore).*—Madder.

908 DE LUCA, P. *S.Giovanni in Fiore (Calabria Citeriore).*—Pitch, pine-resin, turpentine, and spirits of turpentine.

909 FACCHINI, BROTHERS, *Bologna.*—Dressed hemp.

910 FAVARA, T. *Trapani.*—Sumac leaves; white and tawny Siamese cotton (*Gossypium Siamense*), and seeds of the same.

911 FERRARA ROYAL CHAMBER OF COMMERCE.—Hemp-seed and hemp.

912 FERRARA AGRICULTURAL SCHOOL.—Green hemp.

913 FIORELLI, G. *Salò (Brescia).*—Sumac leaves.

914 FRASSELLI, BROTHERS, *Castrocaro, Florence.*—Flax.

915 FRÖLICH & Co. *Castellamare (Naples).*—Garancine and madder.

916 GAROVAGLIO, S. Director of the Botanic Garden, *Pavia.*—Collection of dried lichens.

917 GATTI, A. *Cosenza* (*Calabria Citeriore*). —Thread and cloth, made of the fibres of the broom.

918 GENNARI, P. *Cagliari.*—Herbarium of 200 dried rare Sardinian plants, &c.

919 GRANOZIO, D. *Salerno* (*Principato Citeriore*).—Madder, and raw cotton.

920 GUIDA, BROTHERS, *Gargarengo* (*Novara*).—Colza, rape-seed, flax, and lupins.

921 GULINELLI, COUNT G. *Ferrara.*—Hemp stalks.

922 LORU, PROF. CAV. A. *Cagliari.*—Saffron, tamarisk leaves, and bulbs of the saffron-crocus.

923 LUPINACCI, BARON L. BROTHERS, *Cosenza* (*Calabria Citeriore*).—Flax stalks, and partly dressed flax.

924 MACCAFERRI & CO. *Bologna.*—Specimens of hemp, softened by machinery invented by the exhibitor.

925 MACERATA SUB-COMMITTEE FOR THE EXHIBITION. — Sumac leaves; merino and *vissana* wool; collection of woods.

926 MAFFEI, CAV. N. *Volterra* (*Pisa*).—Juniper berries, myrtle and lentisc leaves, collection of woods.

927 MAGGIORANA, F. *Milan.*—Sumac.

928 MAJORANA, BROTHERS, BARONS DI NICORRA, *Catania.* — Sumac, mustard, and other seeds; collection of Sicilian grown cotton.

929 MARATTI, V *Benevento.*—Woods.

930 MERCATILI, COUNT M. *Ascoli.* — Hemp.

931 MODENA SUB-COMMITTEE FOR THE EXHIBITION.—Collection of the various kinds of timber of the province, bark, &c.

932 MOZZANO, MUNICIPALITY OF (*Ascoli*). —Cotton seed.

933 MUSICÒ, D. *Messina.*—Fibre of the American aloe.

934 NAPLES SUB-COMMITTEE FOR THE EXHIBITION.—Madder; white Siamese and common cotton; dried specimen of the cotton plant.

935 NAPLES ROYAL FOUNDRY. — Specimens of the woods of the south of Italy.

936 NIEDDA DI STA. MARGHERITA, COUNT P. & BROTHER, *Cagliari.*—Sumac.

937 PACIFICI, COUNT D. *Ascoli.*—Hemp.

938 PACIFICO, G. *Salerno* (*Principato Citeriore*).—Madder.

939 PAGANELLI, DR. E. *Castrocaro* (*Florence*).—Saffron, and saffron-crocus seed.

940 PALLOTTA, C. *S. Giuliano di Sepino* (*Molise*).—Hemp-seed and hemp.

941 PALLOTTA, S. *Orvieta* (*Umbria*). — Dressed hemp.

942 PALUMBO, O. *Trani* (*Terra di Bari*). —Asclepias, or vegetable silk.

943 PASI, G. *Ferrara.*—Hemp-seed and dressed hemp.

944 PASOLINI, G. *Imola* (*Bologna*). — Hemp, raw and dressed.

945 PAVANELLI, G. *Ferrara.*—Hemp.

946 PICCALUGA, G. *Cagliari.*—Collection of woods of exotic trees acclimatized and grown around Cagliari.

947 PIÙ, F. *Lanusei* (*Cagliari*).—Madder.

948 RAVENNA SUB-COMMITTEE FOR THE EXHIBITION.—Wood from the pine forests around Ravenna.

949 RAMO, S. *Laconi* (*Cagliari*).—Holly, and cork bark.

950 RIETI COMMITTEE FOR THE FLORENCE EXHIBITION (*Umbria*).—Castor-oil seeds, gall-nuts, mustard, and various gums.

951 REGGIO (EMILIA) AGRICULTURAL SOCIETY.—Plait of willow bark, and plait made of the glumes or husks of Indian corn.

952 RENUCCI, V. *Parma.*—Straw for plaiting.

953 REVEDIN, COUNT G. *Ferrara.*—Hemp.

954 RIZZOLI, R. *Bologna.*—Dressed hemp.

955 SAGLIOCCA, G. *Pietralcina* (*Benevento*).—Wild madder.

956 SALADINI, COUNT M. *Ascoli.*—Hemp.

957 SERRA, L. *Iglesias* (*Cagliari*).—Linseed.

958 SONDRIO, ADMINISTRATION OF THE FORESTS OF.—Indigenous woods of the Valtellina, &c.

959 SPANO, L. *Oristano* (*Cagliari*). — Collection of woods, gums, archil, bark, seeds, berries, manufactured articles, and their materials, &c.

960 SACCONI, COUNT E. *Ascoli,*—Hemp.

961 TEDESCHI, L. I. *Reggio, Emilia.* — Brooms and their materials.

962 TIMON, CAV. A. *Cagliari.* — Various kinds of building wood.

963 TORRI, L. B. *Ferrara.*—Hemp-stalks.

964 VARSI, G. *Oristano* (*Cagliari*).—Cork.

965 VINCI, M. *Carpi* (*Modena*).—Straw plait.

966 VINCENZI, P. *Carpi* (*Modena*).—Straw plait.

967 VONWILLER & CO. *Naples.*—Cork.

---

968 AMICARELLI, V. *Montesantangelo* (*Capitanata*).—Pine resin.

969 ARNAUDON, PROF. G. *Turin.*—Collection of woods; dying and tanning substances.

970 ASCOLI SUB-COMMITTEE FOR THE EXHIBITION.—Weld (*Reseda luteola*).

971 AVELLINO SUB-COMMITTEE FOR THE EXHIBITION.—Collection of woods.

973 BARACCO, BROS. *Cotrone* (*Calabria Ulteriore II.*)—Flax from Siconia.

974 BERTERO, A. & GALLA, G. B. *Carmignola* (*Turin*).—Hemp for ropes.

975 BERTONI, MARQUIS, *Turin.*—Flax.

976 BIANCAVILLA, MUNICIPALITY OF (*Catania*).—White Siamese cotton, cultivated at Biancavilla.

977 BRINDISI SUB-COMMITTEE FOR THE EXHIBITION (*Terra di Otranto*).—Best white and yellow Siamese cotton.

978 BISCARI, PRINCE OF, *Catania.*—White Siamese cotton, with the seeds, cultivated at Paterno.

979 CANOSA, MUNICIPALITY OF (*Terra di Bari*).—White cotton (*G. herb*).

980 CASERTA SUB-COMMITTEE FOR THE EXHIBITION (*Terra di Lavoro*).—Raw and dressed hemp, tow; safflower, madder, &c.

981 CATANZARO SUB-COMMITTEE FOR THE EXHIBITION (*Calabria Ulteriore II.*)—White and tawny yellow Siamese cotton (*Gos-*

sypium Siamense), cultivated in the province, and seeds of the same.

982 CAMPOBASSO SUB-COMMITTEE FOR THE EXHIBITION.—Oak wood, &c.

983 FINZI, M. *Carpi (Modena).*—Willow bark for making chip bonnets.

984 FOGGIA SUB-COMMITTEE FOR THE EXHIBITION *(Capitanata).*—White and tawny Siamese cotton *(G. Siamense)*; common white cotton (*G. herbaceum*).

985 FOGGIA ROYAL ECONOMIC SOCIETY *(Capitanata).*—Riga and Calabrian flax.

986 FORLÌ SUB-COMMITTEE FOR THE EXHIBITION.—Hemp.

987 GIRGENTI SUB-COMMITTEE FOR THE EXHIBITION.—White Siamese cotton; white cotton.

988 GIUDICE, GASPARE, *Girgenti.* — Powdered shumac.

989 GUIDI, C. *Volterra (Pisa).*—Leaves of the *Martino*, for tanning; bird-lime.

990 HENKEL, L. *Florence.*—Waterproof cloth, flannel and cambric; waterproof shawl.

991 LIBRA, F. *Catania.*—White cotton.

992 LICATA, MUNICIPALITY OF (*Girgenti*). —White cotton (*G. herbaceum*).

993 MAZZARA, MUNICIPALITY OF (*Trapani*).—Cotton seed.

994 MENFI, MUNICIPALITY OF (*Girgenti*). —White and tawny yellow cotton.

995 MONTALLEGNO, MUNICIPALITY OF (*Girgenti*).—White cotton *(G. herb).*

996 MUGGIONI, A. *Placenza.*—Roots of the *Andropogon Ischœmum*, for making brooms, brushes, &c.

997 ORLANDO, G. *Pescolamazza (Benevento).*—Wool.

998 ORTONA, MUNICIPALITY OF *(Abruzzo Citeriore).*—Oak wood.

999 PACE, V. *Castrovillari (Calabria Citeriore).*—White and tawny yellow Siamese cotton *(G. Siam.).*

2240 PASQUI, G. *Forlì.*—Hops.

2241 REGGIO AGRICULTURAL ASSOCIATION (*Reggio, Emilia*).—Collection of woods; flax, rape, and colza seed.

2242 REGGIO (CALABRIA) SUB-COMMITTEE FOR THE EXHIBITION.—Shumac, broom tow, and thread; broom thread.

2243 ROYAL ITALIAN COMMISSION.—Collection of Italian grown cotton, made by the Royal Italian Commission, and described under the head of the different producers. See Nos. 294, 886, 891, 899, 910, 919, 928, 934, 967, 976, 977, 978, 979, 981, 984, 987, 991, 992, 993, 994, 995, 999, 2244, 2245, 2247, 2248, 2249, 2251, 2253, 2266, 2267, 2268.

2244 SCIACCA, MUNICIPALITY OF (*Girgenti*).—White cotton.

2245 SALERNO SUB-COMMITTEE FOR THE EXHIBITION (*Principato Citeriore*).—White Siamese and common white cotton.

2246 SAVONA JUNTA FOR THE EXHIBITION, *Savona* (*Genoa*).—Wooden hoops of various kinds.

2247 SCOCCHERA, S. *Canosa* (*Terra di Bari*).—White cotton *(G. herb.)* and cotton seed; madder.

2248 SICULIANA, MUNICIPALITY OF (*Girgenti*).—White cotton (*G. herb.*).

2249 SINASTRA, C. *Noto.*—White Siamese cotton, with the seeds; tawny yellow Siamese cotton, with the seeds.

2250 SINISCALCO, M. *Salerno* (*Principato Citeriore*). — Pancratium maritimum, preserved in spirits, fecula of the same; teazles, &c.

2251 TAORMINA, MUNICIPALITY OF (*Messina*).—White Siamese cotton.

2252 TRUFFELI DI TREVIGLIO.—Flax.

2253 UGO, G. MARQUIS DELLE FAVARE, *Catania.*—White Siamese and common white cotton; common cotton, with the seeds.

2254 BALDI, G. *Florence.*—Lasts for boots and shoes.

2255 BUGGIANO, *A. Placenza.*—Roots of *Andropogon Ischœmum*, for making brooms.

2256 CATANIA SUB-COMMITTEE FOR THE EXHIBITION.—Brooms and rope.

2257 COSTANTINO, T. *Ascoli.*—Flax.

2258 FORCALLI, G. *Salò* (*Brescia*).—Shumac.

2259 GIORDANO, E. *Salerno* (*Principato Citeriore*).—Flax.

2260 ITALIAN CRYPTOGAMICAL SOCIETY, *Genoa.*—Herbarium of Italian Cryptogamous plants.

2261 MUNAFÒ, G. *Sicily.*—Shumac.

2262 PESARO AGRICULTURAL ACADEMY. —Woods, fungi for making tinder, &c.

2263 PICCHI, P. *Leghorn.*—Cork and corks.

2264 SEMMOLA, CAV. F. *Naples.*—Woods.

2265 SACCONI, COUNT E. *Ascoli.*—Hemp.

2266 PATERNÒ, MUNICIPALITY OF, *Catania.* —White Siamese and common cotton.

2267 CATANIA, MUNICIPALITY OF.—White Siamese cotton.

2268 DILGH, E. & Co. *Catania.*—White Siamese and common white cotton, and seeds of the same.

Sub-Class D.

2273 FONSIO, P. *Palermo.*—Essential oil of lemons and oranges.

2274 GARDNER, ROSE, & Co. *Palermo.*—Essential oils of lemons and oranges.

2275 FRANZINI, L. *Florence.*—Perfumery.

CLASS 5.

1000 FUSINI, V. *Pavia.* — Atmospheric railway, with valveless tube.

1001 GRIMALDI, F. *London.* — Rotatory steam-boiler.

1002 PIETRARSA ROYAL WORKS, *Portici* (*Naples*).—Locomotive with 6 coupled wheels.

1003 TURIN ENGINEERING SCHOOL.—Model of a locomotive in section.

1004 VANOSSI, G. *Chiavenna* (*Sondrio*).—Steam engines, for railways and steamers, on a new principle.

1005 VELINI & Co. *Verzaro* (*Milan*).—Model of locomotive tender, applicable to steep inclines.

1006 VINCENZI, E. *Modena.*—Model of an

electric signal to prevent the collision of railway trains.

1007 AGUDIO, T. *Turin.*—Model of machinery for railway trains on inclined planes, set in motion by two stationary engines.

1008 LUÉ, A. *Milan.*—Wooden model of a horse tram-road; model of a new system of rails, capable of being employed successively on four sides.

1009 SIPRIOT, C. *Milan.*—Tarpaulin for railway waggons.

CLASS 6.

1014 BERTI, P. *Milan.* — State carriage, with silver mountings; harness, with silver and chased steel mountings for the same.

CLASS 7.

Sub-Class A.

1020 BONELLI, CAV. G. *Turin.*—Electric loom, for weaving any kind of material.

1021 BOSSI, L. *Milan.*—Beam for warping silk, &c. with screw movement.

1022 DELPRINO, M. *Sesime* (*Alexandria*). —Model of new method of winding cocoons.

1023 DELAPIERRE, F. *Naples.*—Steel card for silk manufacture.

1024 FORNARA, G. *Turin.*—Combs for weaving.

1025 FRIGERIO, G. *Molteno* (*Como*).—New apparatus for preparing cocoons for winding.

1026 GUPPY & PATISSON, *Naples.*—Spinning machine.

1027 SANROME, M. BROTHERS, *Como.*—Cards for weaving.

1028 SILVATICI, G. *Vico Pisano* (*Pisa*).—Cards for wool and cotton.

1029 VINCENZI, E. *Modena.*—Loom.

1030 CAMPI, COUNT G. *Bellosguardo, Florence.*—New system of alternate motion for spinning machines.

1031 MANGANO, A. & SON, *Messina.*—Frame for winding and throwing silk.

1032 ROSSI, P. & Co. *Bibbiena* (*Arezzo*).—Spinning frames and spindles.

Sub-Class B.

1036 CIANFERONI, A. *Florence.*—Blocks for printing oil-cloth.

1037 DEI, F. *Florence.*—Brass and wooden blocks for printing cotton stuffs and handkerchiefs.

1038 PAVASI, G. *Pavia.*—Universal tap for manufacturing screws.

1039 PISA SUB-COMMITTEE FOR THE EXHIBITION.—Weaving machine and implements employed in the manufacture of cotton.

1040 RICCI, R. *Leghorn.*—Chest of carpenter's tools.

1041 VEROLE, P. *Turin.*—Universal fixed tool-holder, for planing machines.

1042 ROSSI, P. & Co. *Bibbiena* (*Arezzo*). —Cheap cooperage and other wood-work.

1043 TREVES, M. *Florence.*—Universal tap for manufacturing screws of any size.

1044 SOMMEILLER, GRANDIS, & GRATTONE, *Turin.*—Machinery employed in making the tunnel through Mont Cenis (drawing); topographical map and section of Mont Cenis, indicating the line of railway.

CLASS 8.

1050 BERNARD, A. *Naples.*—Lighthouse reflectors.

1052 CORTI, D. *Milan.* — New portable pump.

1053 FUSINA, V. *Pavia.* — Machines for filling up cart-ruts, and clearing away snow from streets; gear applicable to rotatory motion under various circumstances; model of pontoon for railway bridges; jets for fire-engines, &c.

1054 LANCIA, G. *Turin.*—Meat chopping machine, machine for filling sausages.

1055 MACRY, H. & Co. *Naples.*—Steam-engine cylinder.

1056 MURATTI, COL. A. *Naples.*—Model of a crane.

1057 PEREZ, V. C. *Lanciano* (*Abruzzo Citeriore*).—Drawing of water mill.

1058 PIETRARSA ROYAL WORKS, *Portici* (*Naples*).—Toothed wheels: steam case and admission valve for a large steam engine large cast-iron shaft for screw propeller, &c.

1060 ACQUADIO, B. *Biella* (*Novara*). —Firework apparatus.

1061 GAUTHIER, A. *Turin.*—New system of corking machine.

1062 LEVINSTEIN & Co. *Milan.*—Lustring machine for giving a gloss to dyed silk.

1063 ROYAL POST-OFFICE.—Two mechanical letter-boxes as used in Piedmont, in which the postman cannot change or see the letters.

1064 TEODORANI, S. *Forli.*—Lever of the first order.

1065 TOVO, *Vinadio* (*Turin*).—Apparatus for giving alarm in case of fires.

1066 TURCHINI, *Florence.* — Mechanical letter-box, in which the postman cannot change the letters.

CLASS 9.

1078 BATTAGLIA, G. *Cermignana* (*Como*). —Apparatus for winding cocoons with a single fire.

1079 BOLGÉ, T. *Brescia.*—Flax scutching machine; apparatus for compressing hay, and cleaning corn, &c.

1080 BACCIOLANI, C. L. *Modena.*—Harrows.

1081 BALDANTONI, G. & BROTHERS, *Ancona.*—Machine for thrashing Indian corn, wine press, straw cutter, and corking machine.

1082 BARGIONI, G. *Florence.*—Hempen, rush, and wicker work bags used in expressing olive oil, &c.

1083 BERTELLI, G. *Bologna.*—Reversible plough.

1084 BERTONE DI SAMBUY, GENERAL, MARQUIS E. *Turin.*—Ploughs.

1085 BOLOGNA AGRICULTURAL SOCIETY. —Model of Bolognese hemp farm, sowing machines, tanks for macerating hemp, &c.

1086 BORELLO, S. & BOANO, A. *Asti, Alexandria.*—Wine press, with screw workable in both directions.

1087 BOTTER, PROF. F. *Bologna.*—Collection illustrating instruments used in the culture of hemp.

1088 CAMBINI, E. *Florence.*—Apparatus for sulphurating vines.

1089 CASUCCINI, P. *Sienna.*—Agriculturists' levels.

1090 CERTANI, A. *Bologna.*—Plough for deepening the furrow made by an ordinary plough.

1091 CIAPETTI, B. *Castelfiorentino (Florence).*—Tuscan cast-iron plough, machine for thrashing Indian corn, cart, harrow, &c.

1092 CROSETTI, P. *Asti (Alexandria).*—Measures for wine.

1093 DE CAMBRAY DIGNY, COUNT L. G. *S. Piero a Sieve (Florence).*—Digny's ploughs, with helicoidal wing.

1094 DE FASSI, *Milan.*—Drawing of a machine for cleaning rice.

1095 DELLA BEFFA, G. *Genoa.*—Thrashing machine.

1096 DELPRINO, M. *Vesime (Alexandria).* —Silkworm nursery.

1096A DUINA, A. *Brescia.*—Plough-shares, &c.

1097 FÀA DI BRUNO, A. *Alexandria.*—Farmers' walking-stick, serving as a level, plumb-line, square, &c.

1098 FACCHINI, BROTHERS, *Bologna.* — Model of a machine for softening hemp; combs used in dressing it.

1099 FISSORE, G. B. *Tortona (Alexandria).* —Dombasle's ploughs, with spare shares.

1100 GELLI & DELLE PIANE, *Pistoia (Florence).*—Bresciana.

1101 GIUNTINI, O. *Piccioli (Pisa).*—Short handled cast-iron Tuscan, and other ploughs.

1102 GUPPY & PATISSON, *Naples.* — Hydraulic and screw presses, for making olive oil.

1103 JACUZZI, G. B. *Pistoia (Florence).*—Bresciana.

1104 KRAMER, E. *Milan.*—Models illustrating the Lombard system of farming and irrigation.

1105 LEOLI, N. *Brescia.*—Spade and shovel.

1106 MACCAFERRI, D. *Bologna.*—Model of apparatus for softening raw hemp.

1107 MAFFEI, CAV. N. *Volterra (Pisa).*—Models of bee-hives.

1108 MARCHI, L. *Volterra (Pisa).*—Machine for compressing faggots.

1109 MILAN ROYAL LOMBARD SCIENTIFIC, LITERARY, AND ARTISTICAL INSTITUTION.—Model of exit gate, for irrigatory canals.

1110 MORI, G. *Greve (Florence).*—Two-pronged fork.

1111 PAGNONI, A. *Ferrara.*—Model of an apparatus for breaking hemp-stalks by hand.

1112 PASQUI, G. *Forli.*—Agricultural implements.

1113 PIZZARDI, MARQUIS, & BROTHERS, *Bologna.*—Machine for chopping up horns and hoofs for manure.

1114 ROSSI, A. *Bologna.*—Clod breaker.

1115 SAJNO, F. *Milan.* — Apparatus for hatching silkworms' eggs.

1116 SANTINI, L. *Fuccechio (Florence).*—Spades and hoe.

1117 SPANO, L. *Oristano (Cagliari).* — Agricultural cart, and plough.

1118 SPINA SANTALA, F. *Acireale (Catania).*—Plough-shares and scythes.

1119 STAFFUTI, O. *Pesaro.*—Corking machine.

1119A SUPERCHI, P. *Parma.*—Parmisan plough on wheels.

1120 TORELLI, D. *Luco (Abruzzo Ulteriore II.).* — Spade, hoe, and shears for sheep-shearing.

1121 VAIRO, G. *Messina.*—Reaping machine.

1122 VIDO, F. *Codogno (Milan).*—Cart for soft ground.

1123 CONROTTO, C. *Turin.*—Machine for packing silk; apparatus for killing the silkworm chrysalis; silkworm cocoons.

1124 FUSINA, V. *Pavia.*—Thrashing machine for Indian corn.

1125 LUCHINI, G. *Florence.* — Copper churn.

1126 MUSSIARI, DR. G. *Parma.*— Parmesan sub-soil plough.

CLASS 10.

Sub-Class A.

1140 ALTOVITI AVILA, CAV. F. *Florence.* —Bricks and brick-pavements.

1141 CALZA, A. *Spezia (Genoa).*—Metamorphic-manganesiferous jasper from Beverone, near Spezia, with hydraulic cement made of it.

1142 CARAFA DI NOIA, P. *S. Giovanni a Teduccio (Naples).*—Tubes, tiles, cornices, of terra cotta; refractory bricks.

1143 COLONNESE, F. & G. *Naples.*—Enamelled bricks and tiles, pipes for water-closets.

1144 GALLIGANI, DR. G. *Seravezza, Lucca.* Oficalce and marble vase.

1145 GUALA, G. *Turin.*— Architectural model in 24 pieces.

1146 GUERRA, COUNT P. *Massa Carrar* —Slabs of polished marble.

1147 MOLINARI & DESCALZI, *Genoa.*—Drawing of apparatus for submarine constructions, and design for a port.

1148 PELAIS, G. *Pistoia (Florence).* — Argillaceous limestone, and hydraulic cement prepared from it.

1149 PETIT-BON, G. *Parma.* — Hollow bricks.

1150 PIANA G. *Bologna.*—Model of an improved tiled roof.

1151 RONDANI, T. *Parma.*—Tiles, hollow bricks, &c.

1152 SEMMOLA, CAV. F. *Naples.* — Collection of the building materials of the Neapolitan provinces.

1153 SPANO, L. *Oristano* (*Cagliari*).—Tiles and bricks.

1154 TREVES, M. *Florence*.—Artificial marbles for pavements, &c.

1155 ZECCHINI, I. *Milazzo* (*Messina*).—Artificial marbles.

1156 ARMAO, G. *S. Stefano di Camastra* (*Messina*).—Bricks.

1157 GAI, F. *S. Mato, Pistoia* (*Florence*).—Terra-cotta tiles of a new form; terra-cotta for paving.

1158 LEE, G. *Sarzana* (*Genoa*).—Bricks, tiles, &c. of different kinds; water pipes.

1159 LEONCINI, BROS. *Rotta* (*Pisa*).—Collection of bricks and tiles.

1160 PARADOSSI, O. *Leghorn*.—Drawings of a drawbridge.

1161 PULITI, C. *Pelago, Florence*.—Double glazed gas pipes; water pipes.

1162 SAVONA JUNTA FOR THE EXHIBITION, *Savona* (*Genoa*).—Refractory clay; lime, bricks.

1163 TAIANI, G. *Vietri* (*Principato Citeriore*).—Bricks for ornamental flooring.

1164 VALERIO, C. *Turin*.—Drawing of a new kind of graving dock for tideless seas, with manuscript report.

Sub-Class B.

1165 ATENOLFI, PRINCE OF, *Castelnuovo* (*Naples*).—Draining pipes.

Sub-Class C.

1166 BORELLA & BOIANO, *Asti* (*Alexandria*).—Model of a shop front, to form a private entrance at night.

1167 BACCI, F. *Impruneta* (*Florence*).—Ornamented vase, flower pots, cornices, roses, capitals, &c.

1168 BRUNETTI, G. *Florence*.—Model of a mechanical self-supporting staircase.

1169 BRUSA, G. B. *Milan*.—Model of a new kind of stove.

2282 CAMPANA, MARQUIS G.—Collection of artificial stone and marble work-tables, vases, statues, pedestals, sphynx, Egyptian figure, &c.

2283 CAPRONI, G. *Perugia* (*Umbria*).—Architectural design.

2284 DELLA VALLE, P. *Leghorn*.—Scagliola and inlaid work on terra cotta.

2285 GALIZIOLI, B. *Brescia*.—Two frescoes on linen removed from a wall.

2286 LEGA, M. *Forlì*.—Artificial marbles.

2287 MATTARELLI, G. *Lecco* (*Como*).—Model of Milan cathedral, in inlaid wood.

2288 PIEGAJA, R. *Lucca*.—Terra cotta model of mediæval cornices, &c.

2289 RABBINI, CAV. A. *Turin*.—Collection of statistical documents.

CLASS 11.

Sub-Class A.

1170 BINDA, A. *Milan*.—Silk, military, and other cravats; scarves, military buttons.

1171 JAMOLI, G. *Turin*.—Plumes used by the *Bersaglieri*, and naval officers.

Sub-Class B.

1180 EXCOFFIER, G. *Asti* (*Alexandria*).—Model of a military observatory.

1181 GALLI, G. *Milan*.—Oiled cloth.

1182 SIPRIOT, C. *Milan*.—Oiled cloth for artillery.

1183 HENKEL, LUIGI, *Florence*.—Waterproof military tent.

Sub-Class C.

1190 BERNARDI, P. *Rimini*.—Muskets.

1191 COLOMBO, C. M. *Milan*.—Fowling-piece, revolvers, and side arms.

1192 COMINAZZI, M. *Gardone* (*Brescia*).—Pistol and gun barrels.

1193 DE STEFANO, *Campobasso* (*Molise*).—Sword, sabre, and foil.

1194 FUSEO, F. *Vitulano* (*Benevento*).—Double-barrelled pistol, both barrels in one piece, &c.

1196 IZZO, A. *Naples*.—Double-barrelled guns: and revolver rifle, the barrel made with iron wire.

1197 LANCIA, G. *Turin*.—Breech-loading cannon, &c.

1198 LABRUNA, G. *Naples*.—Sword.

1199 MARELLI, A. *Milan*.—Gun barrel, long gun, and pistol.

1200 MAZZA, S. *Naples*.—Rifled revolver, fowling-pieces, rifles, &c.

1201 MEROLLA, S. *Naples*.—Double-barrelled guns.

1202 MINOTTINI, G. & LANCETTI, F. *Perugia* (*Umbria*).—Engraved sword, and gun in inlaid case.

1203 MURATTI, COL. A. *Naples*.—Machine for compressing fulminating powder into percussion caps; model of a carriage for a mortar.

1204 PARIS, M. *Brescia*.—Musket, and fowling-piece in several parts.

1205 PILLA, G. *Benevento*.—Six-barrelled pistol, each barrel going off separately by once pulling the trigger.

1206 PRIORA, G. *Milan*.—Revolvers.

1207 RISSONE, L. *Parma*.—Shot of various kinds.

1208 SICHLING, A. *Turin*.—Sabres, swords, hunting knives, and steel for making damasked blades.

1209 SQUINZO, L. *Cagliari*.—Revolvers.

1210 TORRE ANNUNZIATA ROYAL MANUFACTORY OF ARMS.—Locks, damasked barrel, and sabre blade.

1211 TOSCHI, A. *Lugo* (*Ravenna*).—Two fire-arms, one to fire six, and the other sixty times.

1212 TRAVAGLINI, C. *Pisa*.—Eight-gun naval battery, workable by six men.

1213 BEVILACQUA, P. *Campobasso* (*Molise*).—Gun barrel.

1214 DI TORO, C. *Campobasso* (*Molise*).—Walking sticks containing arms.

1215 FABBRICA SOCIALE, *Brescia*.—Rifles.

1216 LANDI, G. *Salerno* (*Principato Citeriore*).—Model of a revolver cannon.

1217 MONGIANA ROYAL METALLURGICAL WORKS, *Mongiana* (*Calabria Ulteriore II.*).—Arms.

1218 ROYAL ARSENAL, *Turin*.—Model of Cavalli's steel-protected battery as employed at the siege of Gaeta; Cavalli's breech-loading cannon; field and mountain pieces, with Cavalli's carriage.

1219 ROYAL MANUFACTORY OF ARMS, *Turin*.—Arms.

CLASS 12.

Sub-Class B.

1225 TAGLIACOZZO, P. *Rome*.—Combination of river barges, capable of being put together so as to form a single sea-going steam-boat.

CLASS 13.

1226 BIFEZZI, G. *Naples*.—Telegometer, for surveying and mensuration.

1227 BONELLI, CAV. G. *Turin*.—Typo-electric telegraph, capable of transmitting 500 messages hourly; four compositors' tables for the above.

1228 CASSANI, E. *Milan*.—Spectacles.

1229 CASUCCINI, P. *Sienna*.—Level.

1230 FÀA DI BRUNO, A. *Alexandria*.—Ellipsograph.

1231 GONNELLA, T. *Florence*.—Calculating machines, for whole numbers and fractions, with descriptive pamphlet.

1232 JEST, C. *Turin*.—Collection of philosophical instruments for schools.

1233 LUCIFERO, T. *Messina*.—Constant chloride of sodium battery.

1234 MARCHI, U. *Florence*.—Self-registering maximum and minimum thermometer.

1235 AMICI, PROF. G. B. *Florence*.—Achromatic refractor, diameter 17 7-10th inches; parabolic speculum for a large telescope; ocular micrometer, with double image, for measuring the diameter of planets, and similar small angular distances; reflecting sextant, and repeating circle; telescopes; surveying cross, without parallax; levels; several kinds of camera lucida; polarizing, and other microscopes; microscope camera lucida for drawing small objects, &c.

1236 BANDIERI, G. *Naples*.—Chemical balance, electro-dynamic apparatus, &c.

1237 MANUELLI, G. *Reggio, Emilia*.—Economic pile, with charcoal diaphragm, for obtaining a constant current.

1238 PAVIA UNIVERSITY, PHILOSOPHICAL MUSEUM OF.—Electrophorus, condenser, electrometer, Voltaic piles: interesting as having belonged to Volta. Prof. Belli's apparent hygrometer, double-action psicrometer, and air pump; Cantoni's calorimeter and thermographs.

1239 ROBERTO, P. *Naples*.—Genometer, for making any kind of metrical scales.

1240 MARONI, M. *Milan*.—Modification of Morse's telegraph.

1241 MILESI, A. *Bergamo*.—Electrical apparatus for taking votes in large public meetings.

1242 MINOTTO, G. *Turin*.—New constant pile.

1243 MURE, BROS. *Turin*.—Standard for taking the height of recruits; steel standard meter.

1244 SELLA, COMMENDATORE Q. *Turin*.—Tripsometer, for measuring the co-efficient of friction.

CLASS 14.

1245 DURONI, A. *Milan*.—Photographs.

1246 MAZA, E. *Milan*.—Photographs.

1247 MODENA SUB-COMMITTEE FOR THE EXHIBITION.—Photographs.

1248 RANCINI, C. *Pisa*.—Photographic miniature of a fresco in the Composanto, Pisa.

1249 RONCALLI, A. *Bergamo*.—Photographs of microscopic objects executed directly.

1250 VAN LINT, E. *Fine Art Studio, Pisa*.—Photographs.

1251 ALINARI, BROS. *Florence*.—Views of Florence; portfolio of photographs of paintings in the galleries of Florence, Venice, and Vienna.

1252 CHIAPELLA, F. M. *Turin*.—Photographs on silk.

1253 FRATACCI, C. *Naples*.—Views.

CLASS 15.

1255 BERNARD, A. *Naples*.—Church clock.

1256 DECANINI, C. *Florence*.—Escapement for watches.

1257 MANUELLI, G. *Reggio, Emilia*.—Escapement for watches.

1258 OLETTI, P. *Turin*.—Small astronomical chronometer.

CLASS 16.

1265 AIELLO, S. *Naples*.—Strings for musical instruments.

1266 BOCCACCINI, A. *Pistoia* (*Florence*).—Improved and common drums.

1267 DE MEGLIO, L. *Naples*.—Grand piano.

1268 FORNI, E. *Milan*.—Flutes and clarinets.

1269 FUMMO, A. *Naples*.—Piano-melodium with two rows of keys, vertical piano-melodium, new kind of flute.

1270 MARZOLO, G. *Padua.*—Organ with melographic apparatus, for repeating and printing any music played.

1271 PANNUNZIO, D. *Agnone* (*Molise*).—Model of a brass bridge for pianofortes.

1272 RUGGIERO, C. *Naples.* — Straight horn, in B flat and A flat.

1273 SIEVERS, F. *Naples.* — Pianos and piano-stools.

1274 VINATIERI, F. & SONS, *Turin.*—Flutes and clarinets.

1275 BOLGÈ, T. *Brescia.*—Drum.

1276 PELITTI, G. *Milan.*—Wind instruments.

1277 PIETRASANTA, L. & SONS, *Lucca.*—Wind instruments.

CLASS 17.

1285 ARIANO, G. *Turin.*—Veterinary surgical instruments.

1286 BARBERIS, A. *Turin.*—Surgical, veterinary, and dentists' instruments.

1287 BELTRAMI, G. *Placenza.*—Surgical instruments.

1288 BERTINARI, G. *Turin.* — Surgical instruments.

1289 COMERIO, BROTHERS, *Brescia.*—Artificial legs, orthopœdic apparatus, apparatus for fractured limbs, ruptures, &c.

1290 FERRERO, G. *Turin.*—Bandages for curing inguinal and crural hernia.

1291 GADDI, PROF. CAV. M. *Modena.*—Injections of the auditory organs of man, quadrupeds, and birds.

1292 GIORDANO, S. *Turin.*—Surgical instruments.

1293 LOLLINI, P & P. *Bologna.*—Collection of surgical instruments.

1294 MONTI, ELVIRA & CO. *Florence.*—Herniary bandages, and surgical apparatus of various kinds.

1295 OBIGLIO, L. *Turin.*—Artificial teeth, variously mounted.

1296 PAVIA UNIVERSITY.—Surgical instruments.

1297 PAVIA UNIVERSITY, ANATOMICAL MUSEUM OF.—Preparation of the facial nerves; preparation of an abnormal human trunk, remarkable for the transposition of both thoracic and abdominal viscera; various injections and preparations.

1299 SERGI, P. *Messina.*—Surgical instruments.

1300 TUBI, G. *Milan.*—Orthopœdic shoes for horses.

1301 BRIZIANO, A. *Milan.*—Sticking and corn plasters.

1302 OLMETA, A. *Cagliari.*—Dentists' instruments.

1303 PARMA VETERINARY INSTITUTION (ANATOMICAL MUSEUM OF THE).—Preparation of the muscular system of the dog.

1304 PARMA VETERINARY INSTITUTION (PATHOLOGICAL MUSEUM OF THE).—Monstrous fœtus of a cow.

CLASS 18.

1307 ALEXANDRIA PENITENTIARY.—Cotton goods.

1308 CALAMINI, M. & CO. *Pisa.*—Coloured cotton goods.

1309 CREMONCINI, A. *S. Vivaldo* (*Florence*).—Cotton counterpanes.

1310 CANTONI, C. *Milan.*—Cotton yarn, cotton stuffs, fustians, calicoes, dimity, damask, &c.

1311 COBIANCHI, P. & SON, *Intra* (*Novara.*—Cotton yarn.

1313 LUALDI, E. *Brescia.*—Cotton yarn.

1314 MILAN CHAMBER OF COMMERCE.—Samples of the cotton manufactures of the province of Milan.

1315 MODENA SUB-COMMITTEE FOR THE EXHIBITION.—Cotton cloth with coloured threads.

1316 MORELLI, F. *Florence.*— Coloured cotton stuffs for trousers and dresses.

1317 OSCULANI PIROVANO & CO. *Monza* (*Milan*).—Cotton trousering, damask, and fustians, &c.

1318 PERSICHETTI, S. *Ancona.* — Cotton sail cloth.

1319 PIATTI & CO. *Placenza.* — Cotton stuffs.

1320 PISA SUB-COMMITTEE FOR THE EXHIBITION.—Cotton stuffs manufactured in the province of Pisa.

1321 SCHLAEPFER, WENNER, & CO. *Naples.*—Unbleached calico, and printed cotton goods.

1322 STEINAUER, J. A. *Chiavenna* (*Sondrio*).—White and coloured wadding.

1323 THOMAS, A. *Milan.*—Machine-made cotton stuffs.

1324 VONWILLER & CO. *Naples.*—Cotton yarn.

1325 ZEPPINI, F. *Pontedera* (*Pisa*).—Cotton counterpanes.

1326 CAMPANA, J. & F. *Gandino* (*Milan*).—Cotton counterpanes.

1327 HOZ & FONZOLI, *Terni* (*Umbria*)—Cotton goods.

1328 LAZZARI, R. *Lucca.*—White cotton mosquito gauze.

CLASS 19.

1329 BERTERO, A. & GALLO, G. B. *Carmagnola* (*Turin*).—Dressed hemp for cordage.

1330 BORZONE, G. *Chiavari,* (*Genoa*).—Fringed towelling.

1331 CAGLIARI SUB-COMMITTEE FOR THE EXHIBITION.—Linen fabrics made by peasants.

1332 CAMPOBASSO SUB-COMMITTEE FOR THE EXHIBITION (*Molise*).—Linen for household purposes.

1333 COSTA, GIULIA, *Chiavari* (*Genoa*).—Linen.

1334 DE-ANGELIS, BROTHERS, *Naples.*—Sailcloth.

1335 DEVOTO L. *Chiavari* (*Genoa*). — Linen.

1336 FERRARA ROYAL CHAMBER OF COMMERCE.—Hand-made ship's cable, marlines, ropes, and sail-cloth.

1337 LUPINACCI, BARON L. & BROTHERS.—*Cosenza* (*Calabria Citeriore*). — Hand-made linen.

1338 MEZZANO, P. *Celle* (*Genoa*).—Fishing nets.

1339 MILAN CHAMBER OF COMMERCE.—Collection illustrating the linen manufacture of the province.

1340 MORELLI, G. *Cosenza* (*Calabria Citeriore*). — Hand-made napkin, linen for shirting.

1341 NOBERASCO, L. *Savona* (*Genoa*).—Sail-cloth.

1342 OSCULATI, PIROVANO, & Co. *Monza* (*Milan*).—Flaxen trousering.

1343 PADOA, P. *Cento* (*Ferrara*).—Sail-cloth, sacking, and bed ticking.

1344 PARTHENOPE MANUFACTURING Co. *Naples*.—Flax and hemp, with linen and yarn made of them.

1345 PELLEGRINETTI, F. *Florence*.—Linen, damask, &c.

1346 PERSICHETTI, S. *Ancona*. — Yarn, twine, ropes, cables, &c.

1347 POLENGHI, C. *S. Fiorano* (*Milan*).—Flax, hand-made linen, &c.

1348 QUADRI, E. *Naples*. — Tarred rope made by machinery; hemp, scutched without steeping or preparation, &c.

1349 REGGIO AGRICULTURAL SOCIETY, *Reggio, Emilia*.—Hempen and flaxen cloth, and yarn.

1350 SANGUINETTI, F. *Chiavari* (*Genoa*).—Linen.

1351 ZILIANI, G. B. *Brescia*.—Nets.

1352 FERRIGNI, G. *Leghorn*.—Ropes and cordage.

1353 REGGIO (EMILIA) AGRICULTURAL ASSOCIATION.—Bruised, scutched, and carded flax and hemp.

1354 DE ANGELIS, BROTHERS, *Naples*.—Ships' cables.

CLASS 20.

1365 ABBATE, P. *Parma*.—Raw silk, spun by a steam-engine.

1366 ACQAVIVA, COUNT C. *Giulia* (*Abruzzo Ulteriore I.*).—Raw silk.

1367 ALEXANDRIA PENITENTIARY. — Brocade.

1368 ANDREIS, V. *Trincotto a Racconigi*.—Organzine.

1369 ARCANGIOLI, A. *Pistoia* (*Florence*).—Hanks of raw silk.

1370 ASCOLI, A. *Terni* (*Umbria*).—Hanks of good raw silk obtained from diseased cocoons.

1371 ASSOM, BROTHERS, *Villa Stellone* (*Turin*).—Cocoons, &c.

1372 BALDINI, L. *Perugia* (*Umbria*).—Raw silk.

1373 BANOALARI, G. *Chiavari* (*Genoa*).—Raw silk.

1374 BAROZZI, ANTOINETTA, *Milan*.—Waste of carded silk.

1375 BAVASSANO, G. B. *Alexandria*.—Raw silk.

1376 BELLETTI, G. *Bologna*.—Spun silk, and silk veils.

1377 BELLINI, G. *Osimo* (*Ancona*).—Raw silk.

1378 BELLINO, BROTHERS, *Turin*.—Raw silk, with cocoons.

1379 BERETTA, CAV. D. *Ancona*.—Raw silk and silk waste.

1380 BERETTA, BROTHERS, *Parlenghe* (*Brescia*).—Silk.

1381 BERIZZI, S. *Bergamo*.—Raw silk and organzine.

1382 BERTARELLI, C. *Cremona*.—Raw silk, brocade, galloon.

1383 BEVILACQUA, M. & SON, *Lucca*.—Raw silk, brocade, galloon.

1384 BINDA, A. *Milan*. — Waistcoating.

1385 BOLMIDA, BROTHERS, *Turin*.—Raw and carded silk, organzine.

1386 BOLOGNINI RIMEDIOTTI, *Pistoia* (*Florence*).—Raw silk.

1387 BOZZOTTI, C. *Milan*.—Raw silk, tram, and sewing silk.

1388 BRACCO, M. & SONS, *Turin*.—Raw silk and organzine.

1389 BRUNI, F. *Milan*. — Organzine and tram.

1390 CARRADORI, COUNT G. *Osimo, near Ancona*.—Raw silk.

1391 CASISSA, SONS, *Novi* (*Alexandria*).—Raw silk.

1392 CECCONI & SANTINI, *Lucca*.—Floss silk for embroidering.

1393 CERIANA, BROTHERS, & NOÉ, *Turin*.—Organzine, &c.

1394 CHABANON, A. *Portici* (*Naples*).—Galloon.

1395 CHICHIZOLA, G. & Co. *Turin*.—Silk velvets.

1396 COLLER, D. *Portici* (*Naples*).—Silk ribbons.

1397 COLLER, L. *Portici* (*Naples*).—Silk ribbons.

1398 COLOMBO, F. *Ceva* (*Cuneo*).—Raw silk.

1399 COMBONI, BROTHERS, *Limone* (*Brescia*).—Raw silk.

1400 COMPAGNO, P. *Cosenza* (*Calabria Citeriore*).—Organzine.

1401 CONTI, A. & Co. *Fossombrone* (*Pesaro and Urbino*).—Raw silk.

1402 CONTI, F. *Milan*.—Raw silk, tram, and organzine.

1403 CORNA, G. *Pisogne* (*Brescia*).—Raw silk.

1404 CORTI, BROTHERS, *Milan*.—Raw silk and silk yarn.

1405 COZZA, COUNT G. *Orvieto* (*Umbria*).—Raw silk.

1406 CRESTINI, D. *Asinalunga* (*Sienna*).—Silk yarn.

1407 DE ANTONI, *Milan*. — Carded silk waste.

1408 DE FERRARI, BROTHERS, *Genoa*.—Silk velvet; silver and gold brocades.

1409 De Filippi, Merzagora, & Co. *Arona* (*Novara*).—Silk waste for spinning.

1410 De Gori, Count A. *Sienna*.--Raw silk.

1411 Delprino, M. *Vesime* (*Alexandria*). —Raw silk, wound by a particular method.

1412 Demeo, F. *Messina*.—Silk ribbons and stuffs.

1413 Denegri, G. *Novi* (*Alexandria*). —Wansey silk, and organzine waste.

1414 Devincenzi, G. *Notaresco* (*Abruzzo Ulteriore I.*).—Raw silk.

1415 Dittaiuti, Count G. *Osimo* (*Ancona*).—Raw silk.

1416 Faraglia, M. *Terni* (*Umbria*).—Raw silk.

1417 Ferrari, F. *Codogno* (*Milan*).—Raw silk, and tram.

1418 Ferri, Brothers, *Grosseto*.—Raw silk.

1419 Fontana, B. & Co. *Turin*.—Organzine.

1420 Fossi & Bruscoli, *Florence*.—Raw silk.

1421 Franchi, Brothers, *Brescia*.—Raw silk and organzine.

1422 Gaddum, E. F. *Torre Pellice* (*Turin*). —Raw silk, and organzine.

1423 Galatti, G. *Messina*.—Raw silk. and cocoons.

1424 Gavazzi, P. *Milan*.—Raw silk, organzine, and tram.

1425 Giambarini, A. *Bergamo*.—Various kinds of tram.

1426 Giardinieri, Brothers, *Osimo* (*Ancona*).—Raw silk.

1427 Granozio, D. *Salerno* (*Principato Citeriore*).—Silk, spun by machinery.

1428 Grassi, F. *Vicofaro* (*Florence*).—Raw silk.

1429 Jaeger, G. *Messina*.—Raw silk.

1430 Keller, A. *Turin*.—Silk, raw and manufactured.

1431 Lanzani, L. & Brother, *Milan*.—Silk, from the inner coarse part of the cocoon, manufactured by hand and machine.

1432 Lardinelli, B. *Osimo* (*Ancona*).—Raw silk.

1433 Lazzari, R. *Lucca*.—Silk mosquito gauze.

1434 Levinstein & Co. *Milan*.—6000 tints of sewing silk; organzine, and tram.

1435 Lazzaroni, P. *Milan*.—Raw silk, and tram.

1436 Macerata Sub-Committee for the Exhibition.—Raw silk.

1437 Maffio, Brothers, *Sondrio*.—Raw silk.

1438 Magnani, E. *Florence*.—Raw silk.

1439 Massina, L. *Calvenzano* (*Brescia*).—Raw silk, and cocoons.

1440 Mazzeri, P. *Milan*.—Sewing silk, organzine, and tram.

1441 Mirabelli, F. *Cosenza* (*Calabria Citeriore*).—Raw silk.

1442 Modena, A. *Reggio, Emilia*.—Silk yarn.

1443 Moresco & Molinari, *Genoa*.—Velvets.

1444 Nefetti, A. *Sta Sofia* (*Florence*).—Raw silk.

1445 Nieri & Lenci, *Lucca*.—Raw silk.

1446 Novellis, C. G. *Savigliano* (*Cuneo*). —Organzine.

1447 Ottaviani, Brothers, *Messina*.—Raw silk.

1448 Padoa, P. *Cento* (*Ferrara*).—Silk yarn.

1449 Padovani, Brothers, *Codogno* (*Milan*).—Raw silk.

1450 Palazzeschi, G. *Città di Castello* (*Umbria*).—Raw silk and silk yarn.

1451 Pasqui, Cav. Z. *Florence*.—Raw silk.

1452 Pastacaldi, F. *Pistoia* (*Florence*).—Raw silk.

1453 Peripetti, C. *Placenza*.—Raw silk.

1454 Piatti & Co. *Placenza*.—Raw silk.

1455 Piazzoni, G. B. *Bergamo*.—Raw silk.

1456 Piccaluga, E. F. *Gavi* (*Alexandria*). —Raw silk.

1457 Piri Pecci, Count G. *Sienna*.—Hanks of raw silk.

1458 Pizzni, A. M. *Rossiglione*.—Raw silk and organzine.

1459 Porro, P. *Milan*.—Raw silk, tram, and organzine.

1460 Preiswerk, G. & Son, *Milan*.—Organzine and tram.

1461 Rampoldi, D. *Como*.—Woven picture, in colours.

1462 Ravenna Sub-Committee for the Exhibition.—Raw and spun silk.

1463 Ronchetti, Brothers, *Milan*.—Raw silk, organzine, and tram.

1464 Rossi, M. *Sondrio*.—Raw silk and silk yarn.

1465 Rossini, G. *Terni* (*Umbria*).—Raw silk.

1466 Rota, A. *Chiari* (*Brescia*).—Raw silk.

1467 Rubinacci, S. *Naples*.—Raw silk and sewing silk.

1468 Salari, D. *Fuligno* (*Umbria*).—Raw silk.

1469 Sari, B. *Lucca*.—Raw silk.

1470 Scola, G. *Villa d'Adda* (*Bergamo*). —Raw silk.

1471 Segre, S. *Vercelli* (*Novara*).—Raw silk.

1472 Sennochi, G. *Placenza*.—Raw silk.

1473 Sinigaglia, S. *Lugo* (*Ravenna*).—Raw silk.

1474 Solari, M. *Chiavari* (*Genoa*).—Raw silk.

1475 Sorlini, A. *Ospitaletto* (*Brescia*).—Raw silk.

1476 Spedaliere, P. *Portici* (*Naples*).—Galloons.

1477 Steiner & Sons, *Bergamo*.—Raw silk, tram, and organzine.

1478 Surtera Sopransi, M. *Codogno* (*Milan*).—Raw silk.

1479 Tallacchini, Brothers, *Milan*.—Raw silk, organzine, tram, and grenadine.

1480 Tesi, L. *Pistoia* (*Florence*).—Raw silk.

1481 Todi Vecchi, *Reggio, Emilia*.—Raw silk, sewing silk, and tram.

1482 Tomassoni, G. *Jesi* (*Ancona*).—Raw silk.

1483 VALVO, P. *Portici* (*Naples*).—Silk ribbons and stuffs.

1484 VANNUCCI, G. *Pistoia* (*Florence*).—Raw silk.

1485 VIOLA, G. *Cairo, near Savona* (*Genoa*).—Raw silk.

1486 ZAMERA, HEIRS OF, *Brescia*.—Raw silk.

1487 ZUPI, BROTHERS, *Cerisano* (*Calabria Citeriore*).—Raw silk.

1488 ZUPPINGER, SIBER, & Co. *Bergamo*.—Raw silk, tram, and organzine.

1489 BACCHINI ROSSI L. *Perugia* (*Umbria*).—Silk shawls made by a hand frame without any loom.

1490 CAMPANI, I. & F. *Gaudino* (*Milan*).—Counterpanes made of waste silk.

1491 COPPOLO, A. *Reggio, Calabria*.—Raw silk.

1492 CRISTOFANI & SON, *Florence*.—Figured stuffs and armoisine.

1493 DE CIANI, D. *Trent, Tyrol*.—Raw and thrown silk.

1494 DE FERRARI, G. (late F.) *Genoa*.—Velvet.

1495 DIENA, MARQUIS G. *Modena*.—Raw silk, cocoons.

1496 GIOVANELLI, A. & D. *Pesaro* (*Pesaro and Urbino*).—Raw silk.

1497 HALLAM, T. *Villa S. Giovanni* (*Calabria Ulteriore I.*).—Raw silk.

1498 HUTH, P. *Como*.—Dyed black silk.

1499 IMPERATORE, G. (late B. & SONS), *Intra* (*Novara*).—Thrown silk.

1500 LOFARO, A. *Reggio, Calabria*.—Raw silk.

1501 LOFARO, G. *Reggio, Calabria*.—Raw silk.

1502 MOSCHETTI, G. A. *Boves* (*Cuneo*).—Raw silk, organzine, cocoons.

1503 NAPLES SUB-COMMITTEE FOR THE EXHIBITION.—Raw silk; dyed silk yarn.

1504 RIZZI, BROS. *Pisogne* (*Brescia*).—Raw silk.

1505 SCIARRONI, M. *Reggio, Calabria*.—Raw silk.

1506 SOLEI. B. *Turin*.—Silk stuffs for furniture and decoration.

1507 VALAZZI, L. *Pesaro* (*Pesaro and Urbino*).—Raw silk.

1508 VERZA, BROS. (late C.), *Milan*.—Raw and thrown silk.

1509 VIALI & MASSETTI, *Fano* (*Pesaro and Urbino*).—Raw silk.

1510 VIGANOTTI, G. *Milan*.—Galloon.

1511 ZANARDINI, P. *Pisogne* (*Brescia*).—Raw silk.

1512 ZAMOLI, L. *Cesena* (*Forli*).—Raw silk.

CLASS 21.

1515 ANTONGINI BROTHERS, *Milan*.—Raw and coloured woollen yarn.

1516 CAGLIARI SUB-COMMITTEE FOR THE EXHIBITION.—Woollen cloth, &c. made by peasants of Aritzo.

1517 CASTELLI, C. *Milan*.—Woollen counterpane.

1518 HOZ & FONZOLI, *Terni* (*Umbria*).—Goods of mixed cotton and wool.

1519 LUPINACCI, BARON L. & BROTHERS, *Cosenza* (*Calabria Citeriore*).—Hand-made woollen cloth.

1520 MORELLI, F. *Florence*.—Textures of cotton and wool.

1521 MORELLI, G. *Rogliano* (*Calabria Citeriore*).—Knitted wool.

1522 ORLANDO, G. *Pescolamazza* (*Benevento*).—Woollen cloth.

1523 OSCULATI, PIROVANO, & Co. *Monza* (*Milan*).—Textures of cotton and wool, of cotton wool and silk, and of cotton and flax; cotton and wool damask.

1524 PIRAS, MARIA, *Samassi* (*Cagliari*).—Sardinian wallet.

1525 ROSSI, F. *Milan*.—Woollen stuffs.

1526 SELLA, BROTHERS, *Turin*.—Cloth, of various kinds and colours.

1527 SELLA, M. *Biella* (*Novara*).—Cloth, velvet, flannel, &c.

1528 SPANO, L. *Oristano* (*Cagliari*).—Woollen wallets; counterpanes of wool, cotton, and silk, and of linen and cotton.

1529 THOMAS A. *Milan*.—Machine-made stuffs, of mixed cotton and wool; cotton and flax thread, of several colours.

1530 CALAMINI, M. & Co. *Pisa*.—Plaid woollen shawls.

1531 COSTANTINO, G. *S. Marco de' Cavoti* (*Benevento*).—Woollen cloth.

1532 CROCCO, C. & L. *Genoa*.—Woollen hosiery.

1533 FLORENCE WORKHOUSE.—Woollen counterpanes, flannel.

1534 GIANNATASIO, G. *S. Cipriano* (*Principato Citeriore*).—Blankets.

2291 ALEXANDRIA PENITENTIARY.—Mixed fabrics.

CLASS 22.

1535 CAMPRA, C. *Graglia, near Biella* (*Novara*).—Woollen carpets.

1536 CASTELLI, C. *Milan*.—Carpets.

1537 MILAN BLIND ASYLUM.—Carpet made by the blind.

1538 PIRAS, V. *Samossi* (*Cagliari*).—Carpet.

1539 CIANFERONI, A. *Florence*.—Oil cloth.

1540 GALLI, G. *Milan*.—Oil cloth.

CLASS 23.

1545 BOSIO, F. & Co. *Castello di Lucento* (*Turin*).—Cotton yarn, dyed different colours.

1546 CECCONI & SANTINI, *Lucca*.—Wool dyed in 10 colours, for embroidery.

1547 FOLETTI, WEISS, & Co. *Milan*.—Turkey-red cotton yarn.

1548 SANTILLI, B. *Isernia* (*Molise*).—Specimens of cloth simultaneously dyed a different colour on either side.

1549 WISER, S. *Modena.*—Fleece, the wool coloured in various tints.

1550 HUBER & KELLER, *Pisa.*—Turkey red cotton yarn.

CLASS 24.

1555 BAFICO, ANGELA, *Chiavari* (*Genoa*). —Several kinds of point lace.

1556 BASSETTI, ANTOINETTA, *Sienna.*—Embroidered sleeves.

1557 BINDA, A. *Milan.*—Trimmings of silk, and of mixed silk and cotton.

1558 CALANDRA, CAMILLA, *Savigliano* (*Cuneo*).—Embroidered white counterpane.

1559 CUCCHIETTI, C. *Busca* (*Cuneo*).—Embroidery in worsted and silk, representing Mary Queen of Scots at Langsyde.

1560 FIESCHI (CONSERVATORY), *Genoa.* —Embroidered cambric handkerchiefs, collar, &c.

1561 FUMMO, MARIA, *Naples.*—Embroidered pocket-handkerchief.

1562 GARBESI, ERSELIA & ANGELA, *Vorno* (*Lucca*).—Silk shawl, embroidered to imitate ancient lace.

1563 MARTINI, L. *Milan.*—Embroidery, silk and brocade stuffs.

1564 MARTINI, ERSELIA, *Milan.*—Embroidery.

1565 NAPLES (ROYAL INSTITUTION OF CARMINELLO).—Embroidered handkerchiefs.

1566 PARLANTI, E. *Florence.*—Embroideries, representing a satyr and monkey, after Annibal Caracci, &c.

1567 TACCHINI, TERESA, *Modena.* — Chiaro-scuro embroidery, representing a halt at an inn; embroidery on cloth.

1568 BUONINI, M. *Lucca.*—Lace.

1569 BROCCI, D. & A. *Cantù* (*Como*).—Lace veil, mantle, &c.

1570 GENOA, POOR ASYLUM (ALBERGO DEI POVERI).—Embroidered cambric handkerchief; shirts; towels with lace border.

1571 LANDUZZI, F. *Bologna.*—Embroidery.

1572 LEPORATTI, E. *Pistoia* (*Florence*).—Embroidered silk.

1573 MARINO, P. *Turin.*—Trimmings.

1574 PANIZZI, M. *Parma.*—Embroidered cambric handkerchief.

2293 CAMPODONICO, E. *Genoa.*—Lace.

2294 PETTI, E. *Campobasso* (*Molise*).—Embroidery on wool.

2295 PETRUCCI, A. *Lucca.*—Embroidered handkerchief.

2296 SERVI, E. *Florence.*—Black Thibet scarf, embroidered with silk.

2297 TECCHI, A. *Pisa.*—Invisible mending and darning on various kinds of stuffs.

2298 TRAFIERI, A. *Lucca.*—Embroidery on cambric.

2299 TESSADA, F. *Genoa.*—Lace shawl.

CLASS 25

Sub-Class A.

1574A ORRÙ, S. & G. *Cagliari.*—Carpet made of stag, deer, and mufflone skins. Prepared skins.

1575 PRATTICO, F. *Naples.*—Bird's skin cap, and muff of Siberian fox fur.

1576 PILLONI, ARNETTA, *Cagliari.*—Prepared skins.

1577 SEVERI, A. *Reggio, Emilia.*—Furs.

1578 SPANO, L. *Oristano* (*Cagliari*).—Skins of Gangorra.

1579 WISER, S. *Modena.*—Sheepskin.

Sub-Class C.

1580 PICCINI, A. *Florence.*—Various kinds of brushes.

CLASS 26.

Sub-Class A.

1585 BERSELLI, CIRO, & Co. *Reggio, Emilia.*—Various kinds of leather.

1586 BOSSI, E. *Naples.*—Coloured tawed skins for gloves.

1587 CARLETTI, L. *Chiavenna* (*Sondrio*).—Collection of tawed skins.

1588 CONSIGLI, G. *Leghorn.*—Leather and buffalo hides.

1589 DEIDDA, A. *Cagliari.*—Sole leather, and shagreen kid skin.

1590 DEL SERE, G. *Florence.*—Leather for various purposes.

1591 DONATI & Co. *Sienna.*—Calf-skin and sole leather.

1592 DURIO, BROTHERS, *Turin.*—Leather prepared without lime.

1593 JAMMY BONNET, M. *Castellamare* (*Naples*).—Leather for various purposes.

1594 MANCINI, A. *Arezzo* (*Umbria*).—Sheep-skin.

1595 ORRU', S. & G. *Cagliari.*—Lamb-skin for gloves; sole, and other leather.

1596 PARMA SUB-COMMITTEE FOR THE EXHIBITION.—Hides and leather.

1597 PELLERANO, G. B. *Naples.*—Kid-skins.

1598 PIELLA, G. *Pavia.*—Calf-skin and sole leather.

1599 SANTONI, F. *Calci* (*Pisa*). — Sole leather, made with gelatine; calf and kid-skin, &c.

1600 TANNING Co. *Modena.* — Calf-skin for shoes.

1601 ARNAUDON, L. *Turin.*—Goat and sheep leather; morocco and varnished leather.

1602 AVELLANO SUB-COMMITTEE FOR THE EXHIBITION.—Sheep skins; morocco leather; and parchment.

1603 BALDINI, A. & Co.—Sole leather; white and black calf upper leather.

1604 BOLGÈ, T. *Brescia.*—Parchment for drums, &c.
1605 CAPON, G. *Venice.*—Sole and calf shoe leather.
1606 CAPRETTI, P. *Brescia.*—Hides.
1607 CERESOLE, BROS. *Turin.*—Black and coloured leather; portmanteau and hog leather.
1608 CIONI, L. *Florence.*—Black and coloured varnished leather.
1609 DE FABRITIIS, BROS. *Teramo (Abruzzo Ulteriore I.).*—Leather.
1610 DEROSA, P. *Benevento.*—Sole leather.
1611 FIORINI, G. *Darfo (Brescia).*—Calf hides.
1612 FORNARI, BROS. *Fabriano (Ancona).*—Saddlery, waxed, and morocco leather.
1613 GAMBAZZI, P. *Brescia.*—Hides.
1614 IMPACCIATORE, T. *Elice (Abruzzo Ulteriore I.).*—Leather.
1615 LANZA, BROS. *Turin.*—Saddlery and calf leather.
1616 MESSINA SUB-COMMITTEE FOR THE EXHIBITION.—Collection of skins for gloves; gloves.
1617 PIACENTINI, CECCHI, & CO. *Pescia (Lucca).*—Leather.
1618 PONCI, S. *Sarteano (Florence).*—Parchment.
1619 PRACCHI, A. *Lucca.*—Varnished leather.
1620 ROMANO, F. *Turin.*—Calf leggings; calf leather.
1621 SORBI, L. *Leghorn.*—Sole leather; waxed and chamois leather; shagreen kid leather.
1622 STICKLING, A. *Leghorn.*—Sole leather.
1623 VIGNOLI, *Forlì.*—Sheep skins, morocco leather.

Sub-Class B.

1625 BARBARO, L. *Naples.*—Bridle and harness.
1626 CORA, D. & SONS, *Turin.*—Harness, saddle and bridle, and saddle without bows.
1627 LICHTENBERGER, BROTHERS, *Turin.*—Saddle.

1628 ASTORRI, M. *Forlì.*—Harness.
1629 DEL PERO, G. B. *Brescia.*—Whip handles.
1630 MARINO, P. *Turin.*—Carriage trimmings.
1631 SANTI, TALAMUCCI, & SON, *Florence.*—Saddles.

Sub-Class C.

1635 MARZOCCHINI, C. *Pisa.*—Leather cigar-cases, match-boxes, and powder-flask.

1636 SANTI, TALAMUCCI, & SON, *Florence.*—Cigar cases, match boxes, tobacco pouches, &c.

CLASS 27.

Sub-Class A.

1640 AZZI, BROTHERS, *Lucca.*—Stiff and pliable hare-skin hats.
1641 BELTRAMI, P. *Milan.*—Military folding hats; hats in process of manufacture.
1642 BORELLO, P. & BROTHERS, *Biella (Novara).*—Felt hats made of mixed hair.
1643 CAMPOBASSO SUB-COMMITTEE FOR THE EXHIBITION.—Woollen hats for peasants.
1644 CAVIGLIONE, R. *Turin.*—Silk and gibus hats.
1645 FOSSATI, A. *Monza (Milan).*—Woollen, goat-skin, and hare-skin hats.
1646 GALISE, V. *Naples.*—Silk hats.
1647 PONZONE, A. *Milan.*—Military, flexible, and other hats.
1648 PUGLIESE, A. *Cagliari.*—Woollen caps.
1649 RAVENNA SUB-COMMITTEE FOR THE EXHIBITION.—Felt hats.

1650 LA FARINA, *Palermo.*—Collection of silk hats.
1651 MANTELLERO, S. *Sagliano d'Adorno (Novara).*—Otter, hare, and rabbit-skin hats lambs'-wool hats.
1652 PEONE, BROS. *Leghorn.*—Felt hats.
1653 PIEROTTI, U. & A. *Florence.*—Felt hats.

Sub-Class B.

1655 CALZAROSSA, M. *Parma.*—Bonnets and head-dresses.
1656 CLEMENTE, B. *Teramo (Abruzzo Ulteriore I.).*—Straw bonnets.

1657 BRAZZINI, D. *Florence.*—Fiesole straw plait, horse-hair, and chenille bonnet-trimmings.
1658 CONTI, C. *S. Giacomo (Florence).*—Collection of straw plait and trimming; Tuscan straw hats and bonnets; straw plait cigar cases.
1659 KUBLI, J. J. *Florence.*—Collection of Tuscan straw plait and trimmings; straw hats and bonnets; straw plait cigar cases.

2314 MASINI, A. *Florence.*—Collection of straw plait; Tuscan straw hats, &c.
2315 NANNUCCI, A. *Florence.*—Tuscan straw hats, slippers, straw work.
2316 VYSE & SONS, *Prato (Florence).*—Tuscan plait, Tuscan straw hats, trimmings, &c.

Sub-Class C.

1660 ALEPPI, L. *Parma.*—Seamless cotton drawers, made with a common frame.
1661 BINDA, A. *Milan.*—Buttons of various materials.
1662 BOSSI, E. *Naples.*—Gloves.
1663 BRACHETTI, G. *S. Giovanni (Arezzo).*—Reversible trousers and convertible clothes.
1664 DE ANGELIS, A. *Messina.*—Artificial flowers.
1665 DE MARTINO, G. *Naples.*—Ornamented parasols, and umbrellas.
1666 DESSI MAGNETTI AVV. V. *Cagliari.*—Byssus of the Pinna, with thread, gloves, &c. made of it.
1667 GILARDINI, G. *Turin.*—Umbrellas.

1668 MESSINA SUB-COMMITTEE FOR THE EXHIBITION.—Gloves.

1669 MONTECCHI, E. A. *Parma.*—Artificial leaves and fruit.

1670 PELLERANO, G. B. *Naples.*—Gloves.

1671 PRATTICO, F. *Naples.*—Gloves.

1672 RANDACCIU, M. *Cagliari.*—Shawl made with the byssus of the Pinna.

1673 SALA, F. *Milan.*—Gloves.

1674 TACCHINI, LERTORA, & Co. *Milan.*—Scarves, gloves, and cravats of silk; scarves of silk and cotton, and of silk and wool. Collection of buttons of every description.

1675 TESSADA, F. *Genoa.*—Cambric handkerchiefs, burnous, mantilla, shawls, &c.

1676 BERTI, A. *Florence.*—Gloves cleaned by a new method.

1677 CERNUSCHI, BROTHERS, *Milan.*—Silk, woollen, cotton, and other braid cord, and tape; buttons.

1678 DE BENEDETTI, BROTHERS, *Asti (Alexandria).*—Cotton shirts with coloured fronts.

1679 FESTA, G. *of Turin* [41, *Somerset-st. Portman-sq. London*].—Stays.

2317 GIOVANETTI, G. & SONS, *Pisa.*—Bone buttons.

Sub-Class D.

1680 BRUNO, G. *Turin.*—Boots and slippers.

1681 DELIA, P. *Leghorn.*—Boots and shoes.

1682 GALLI, N. *Pisa.*—Top boots, and cavalry boots.

1683 PERRATA, S. *Savona (Genoa).*—Shoes.

1684 ROLANDO, A. *Turin.*—Ladies' silk boots; shoes.

1685 FLORENCE WORKHOUSE (*Pia casa di Lavoro*).—Boots and shoes.

1686 GNESI, G. *Florence.*—Boots and shoes.

1687 PASQUERO, D. *Castiglione (Turin).*—Boots and shoes.

1688 SALANI, A. *Leghorn.*—Boots.

CLASS 28.

Sub-Class A.

1695 GHILIOTTI, B. *Pegli (Genoa).*—Hand-made paper.

1696 JACOB, L. & Co. *Milan.*—Specimens of paper, &c.

1697 MAGLIA, PIGNA, & Co. *Milan.*—Paper of various kinds.

1698 MARTELLI, D. *Florence.*—Fancy paper.

1699 MOLINO, P. A. *Milan.*—Pulp for paper manufacture, and paper.

1700 PICCARDO, A. *Genoa.*—Paper.

1701 PLONCHERI, G. *Chiavenna (Sondrio).*—Incombustible paper, made of amianthus.

1702 POLI, A. *Villa Basilica (Lucca).*—Straw paper and pasteboard.

1703 AVONDO, BROS. *Borgo Sesia (Turin).*—Drawing, writing, and office paper.

1704 MAGNANI, E. *Pescia (Lucca).*—Hand-made paper.

1705 MAGNANI, G. *Pescia (Lucca).*—Hand-made paper.

1706 MAFFIZOLI, A. *Toscolano (Brescia).*—Collection of drawing and writing paper.

1707 POLLERA, A. M. *Lucca.*—Hand-made paper.

1708 SORVILLO, N. *Naples.*—Drawing, writing, printing, lithographic, and other kinds of paper.

1709 VOLPINI, C. *Florence.*—Printing, drawing, letter, and other paper.

Sub-Class C.

1710 BENTIVOGLIO, CAV. C. *Modena.*—Nature printing, obtained by simple pressure.

1711 BERNARDONI, G. *Milan.*—Specimens of printing.

1712 BORZINO, U. *Milan.*—Chromo-lithography.

1713 CORNIENTI, G. *Milan.*—Lithographic portraits, drawn directly on stone.

1714 GAMBERINI, D. *Ravenna.*—Papyrus writing.

1715 GIOZZA, G. *Turin.*—New process for stereotyping, employing very thin moulds which dry instantaneously.

1716 MECHITARISTI MONKS, *Venice.*—Prayer of S. Narsete, translated into 24 languages; Milton's "Paradise Lost," and other works, in Armenian.

1717 PARIS, A. *Florence.*—Chromo-lithography, &c.

1718 PROSPERINI, P. *Padua.*—Lithography, and chromo-lithography; lithographic stones.

1719 RICCÒ, F. *Modena.*—Nature printing, image produced by simple pressure.

1720 SALARI, R. *Florence.*—Interesting fac-similes, written with the pen.

1721 VALLABREGA, G. *Bologna.*—Improved compositor's table.

1722 APPIANI & DUCCI, *Florence.*—Self-feeding cushion for stamps.

1723 BOLLINI, P. *Milan.*—Powder for making ink by simple addition of cold water.

1724 CANTI, G. *Milan.*—Music; catalogue of publications.

1725 GRAVINA, D. B. *Monreale (Palermo).* Chromo-lithographic views of the Cathedral of Monreale.

1726 LIVIZZANI, E. *Bologna.*—Silhouettes.

1727 TREVES, M. *Padua.*—Lithography and chromo-lithography.

1728 NOBILI, G. *Naples.*—A volume of chromo-lithographical views of the monuments of Pompeii.

Sub-Class D.

1730 BIANCONCINI, L. *Naples.*—Books bound in morocco.

1731 ELISEO, D. *Campobasso (Molise).*—Ornamental binding.

1732 FAGIUOLI, G. *Florence.*—Album for photographs, ornamented with Florentine mosaics; *Guerino il Meschino*, a code of the 15th century, bound in the style of the times.

CLASS 29.

Sub-Class A.

1739 BORSARI.—Educational works for the deaf and dumb.
1740 BARBÈRA, G. *Florence.*—Books.
1741 CELLINI, M. *Florence.*—Books.
1742 DELLA BEFFA, G. *Genoa.*—Books.
1743 FERRARIS, DR. C. *Alexandria.* — Books.
1744 GICCA, A. *Turin.*—Books.
1745 GUIDI, G. G. *Florence.*—Music.
1746 JERVIS, W. P. *Royal Italian Central Committee.*—Collection of the Italian newspaper and periodical press.
1747 LE MONNIER CAV. F. *Florence.* — Forty-three volumes of the *Biblioteca Nazionale.*
1748 LUCCA, F. *Milan.*—Musical publications.
1749 MARIETTI, G. *Turin.*—The Scriptures, stereotyped Latin translation.
1750 NAPLES SUB-COMMITTEE FOR THE EXHIBITION.—Books on various subjects.
1751 PAGANUCCI, PROF. L. *Florence.* — Two plates of the anatomy of the horse.
1752 PENDOLA, PROF. T. *Sienna.*—Books for the use of the deaf and dumb.
1753 PINELLI, L. *Florence.*—Nuovo Testamento, Diodati's translation; various Evangelical works.
1754 PUCCINELLI, M. *Lucca.*—Books.
1755 RABBINI, CAV. A. *Turin.*—Trigonometrical surveys, &c.
1756 RICCORDI, T. *Milan.*—Music.
1757 ROYAL COMMISSION OF THE ITALIAN EXHIBITION OF 1861.—Works descriptive of, and connected with the Italian Exhibition at Florence in 1861; photographs of the building, &c.
1758 SANSEVERINO, COUNT F. *Turin.*—Books.
1759 TIMON, CAV. A. *Cagliari.*—Books, printed at Cagliari.
1760 TRON, G. *Turin.*—The Scriptures in Italian; collection of Evangelical works.
1761 VILLA, A. & G. B. *Milan.*—Geological works.

1762 ABBATE, G. *Messina.*—Specimen of calligraphy.
1763 ARCOZZI MASIMO, AVV. L. *Turin.*—Treatise on the manner of rearing silkworms, with plates.
1764 BOLOGNA SUB-COMMITTEE FOR THE EXHIBITION.—Scientific periodicals.
1765 BORGO-CAVATTI, G. *Cuneo.*—Treatise on ornamentation.
1766 BOTARELLI, P. *Valiano (Sienna).*—New system of calligraphy.
1767 CASTIGLIONI, P. *Milan.*—Books.
1768 CELLINI, M. *Florence.*—Books.
1769 CIVELLI INSTITUTION, *Milan.*—Geographical school atlas.
1770 FLORENCE, MAGLIABECCHIAN LIBRARY.—Collection of books.
1771 LOFARO PIETRASANTA, D. DUKE OF SERRADIFALCO, *Florence.*—The antiquities of Sicily illustrated; notices on the Cathedral of Monreale; album of picturesque views of the antiquities of Sicily.
1772 MAZZEI, CAV. F. *Florence.*—Books.
1773 MIGLIACCIO, R. *Salerno (Principato Citeriore).*—Books.
1774 MUZZI, L. *Florence.*—Phonic system of teaching to read and write Italian.
1775 RAMO, S. *Naples.*—Books.
1776 RIZZETTI, DR. G. *Cagliari.*—Medical and chemical works of the Exhibitor.
1777 SANTERINI, BROS. *Cesena (Forlì).*—Calligraphy.
1778 TENERELLI, F. *Teramo (Abruzzo Ulteriore I.).*—A book—Easy method of learning to read Italian.
1779 VIGANO, F. *Milan.*—The Exhibitor's works on political economy, &c.

2318 LAMBRUSCHINI, CAV. PROF. R.—Various works on public instruction.
2319 MARZULLO, *Palermo.*—Grammar for the deaf and dumb.
2320 UNIONE TYPOGRAFICA TORINESE.—*Encyclopedia Populare Italiana; Nuova Biblioteca Italiana; Dizionario della Lingua Italiana, &c.*

Sub-Class B.

1780 CAIMI, E. *Sondrio.*—Topographical model of the Pass of the Stelvio.
1781 CAPURRO, REV. G. F. *Novi, Alexandria.*—Telegraphic alphabet, for teaching a large number of children without books.
1782 FÀA DI BRUNO, CAV. F. *Turin.*—Educational apparatus.
1783 FLORENCE ROYAL NATURAL HISTORY MUSEUM.—Botanical specimens. Collection of graminaceous plants. Specimens from the Royal Museum: Various kinds of rice, with semolina, and flour made of it. Indian corn, with semolina, flour, bread, and alcohol made from it. Various kinds of canary seed. Sorgho, with sugar, rum, alcohol, and bread made from it. Varieties of millet. Clothes, ropes, and matting, with the materials of which they were made. Brooms, baskets, plait, brushes, and wicker-work, with the materials of which they were made. Oats. Cane; and bamboo cane, with various objects made of it; bamboo preserve. Wheat, with flour, bran, starch, wafers, bread, biscuits, maccaroni, alcohol, &c. made from it, and articles manufactured with its straw. Barley, with flour, ale, and porter made from it, and articles made from its straw. Sugar-cane, sugar, sugar candy, and rum. Other cereals.
1784 GENNARI, PROF. P, *Cagliari.* — Minerals, and natural history specimens.
1785 MILAZZO JUNTA FOR THE EXHIBITION (*Messina.*) — Collection of non-metalliferous minerals, shells, seaweeds, and fossils.
1786 PAVIA UNIVERSITY, MACHINERY MUSEUM.—Drawings of machinery.

1787 PISA, NATURAL HISTORY MUSEUM, ROYAL UNIVERSITY OF.—Plaster and wax models of fossils.

1788 TURIN ENGINEERING SCHOOL.—Models of fossils and crystals.

1789 ZAPPALA, G. *Messina.* — Moses, a plaster cast.

1790 CALENZOLI, C. & S. *Florence.* — Anatomical wax figures and preparations; lymphatic system; the eye, ear, brain, &c.

1791 CASELLA, DR. G. *Laglio* (*Como*).—Fossils.

1792 CONVENT OF THE SIGNORE DELLE QUIETE, *Florence.*—Plan and elevation of the school attached to the convent.

1793 FLORENCE INFANT ASYLUMS. — Photographic drawings, plans, regulations, &c.

1794 FLORENCE, LAURENTIAN LIBRARY. —Views of the library, &c.

1795 FLORENCE MUNICIPAL SCHOOLS.—Regulations and statistics, &c.

1796 FLORENCE, RICCARDIAN LIBRARY. —Drawings.

1797 FLORENCE ROYAL ACADEMY OF FINE ARTS (LIBRARY OF).—Collection of illustrated works.

1798 FLORENCE, ROYAL ARCHÆOLOGICAL DEPARTMENT FOR THE TUSCAN PROVINCES. —Photographic views of the archive office.

2321 FLORENCE ARTISTIC SOCIETY.—Collection of engravings.

2322 FLORENCE, LIBRARY OF THE HOSPITAL OF STA. MARIA NOVELLA.—Books.

2323 FLORENCE, ROYAL NATURAL HISTORY MUSEUM.—Views of the museum.

2324 FLORENCE, ROYAL GALLERY OF MOSAICS IN PIETRE DURE.—Views of the gallery; historical notices.

2325 FLORENCE ROYAL LYCEUM AND GYMNASIUM.—Plan and photographs of the building, &c.

2326 FLORENCE ROYAL MARUCELLIAN LIBRARY.—Photographic views.

2327 FLORENCE ROYAL NATURAL HISTORY MUSEUM.—Wax preparation of the grape disease; anatomical wax preparations of the rabbit.

2328 FLORENCE ROYAL NORMAL GIRLS' SCHOOLS FOR THE PEOPLE.—Regulations and statistical notices; photographic view.

2329 FLORENCE ROYAL NORMAL BOYS' SCHOOL.—Plans, regulations, and statistics.

2330 FLORENCE ROYAL NORMAL GIRLS' SCHOOL. — Photographic views, regulations, and statistics of the school.

2331 FLORENCE S. MARCO DOMENICAN LIBRARY.—Photographic view of the library.

2332 FLORENCE WORKHOUSE. — Photographs; statistics; notices on the system of education adopted.

2333 MAZZEI, CAV. F. *Florence.*—Photographs of the restorations at the Palazzo del Podestà.

2334 MALATESTIAN LIBRARY, *Cesena* (*Forli*).—Photographs of the library.

2335 MILAN SOCIETY FOR THE PROMOTION OF ARTS AND MANUFACTURES.—Diagrams of agricultural implements.

2336 PAVIA UNIVERSITY, ZOOLOGICAL MUSEUM OF.—Drawings illustrating the development and diseases of the silkworm, by Dr. Maestri.

2337 PIEROTTI, P. *Milan.*—Plaster casts, &c.

2338 PISA ROYAL UNIVERSITY.—Photographs and plans.

2339 PRATO ORPHAN ASYLUM, *Prato* (*Florence*).—Photographic views of the establishment.

2340 PRATO ROYAL LYCEUM, *Prato* (*Florence*).—Plans of the Lyceum.

2341 RANDACCIU, G. *Sassari.* — Anatomical wax figures.

2342 RAVENNA, CLASSENSE PUBLIC LIBRARY.—Plans of the library.

2343 RAVENNA ROYAL ACADEMY OF FINE ARTS.—Plans of the academy.

2344 RIPOLI, CONSERVATORIO DI, *Florence.*—Photographs.

2345 RONDANI, PROF. C. *Parma.*—Collection of Italian dipterous insects, with account of the same, &c.

2346 S. ANDREA WORKHOUSE, *Leghorn.* —Photographs of the establishment, statistics, &c.

2348 S. S. ANNUNZIATA ROYAL INSTITUTE FOR GIRLS, *Florence.*—Photographs of the establishment, &c.

2349 VIAREGGIO, PROPOSED HOSPITAL (*Lucca*).—A photograph, &c.

2350 VILLA, I. *Florence.*—New terrestrial planisphere, indicating the time for every longitude; new celestial planisphere, indicating the passage of stars for every terrestrial meridian; cosmographical diagrams, &c.; collection of photographs of the Exhibitor's artistic works.

Sub-Class D.

1799 BOLGI, T. *Brescia.*—Toys.

CLASS 30.

Sub-Class A.

1800 BERTOLOTTI, G. *Savona* (*Genoa*).—Collection of marquetry tables.

1801 BETTI, F. *Florence.* — Table in Florentine mosaics.

1802 BIANCHINI, PROF. G. *Florence.*—Table in Florentine mosaics.

1803 BINAZZI, G. *Florence.*—Real, and imitation mosiac tables.

1804 BOCCHIA, E. *Parma.* — Inlaid-wood watch-stand.

1805 BOSI, E. *Florence.*—Tables and ebony case, inlaid with mosiac and bronze work, &c.

1806 CANTIERI, G. *Lucca.* — Work-table inlaid with tortoise-shell, wood, and ivory.

1807 CANEPA, G. B. *Chiavari* (*Genoa*).—Chiavari chairs.

1808 CENA, G. *Turin.*—Inlaid table, &c.

1809 CIACCHI, J. *Florence.*—Inlaid pavement in Byzanto-Gothic style.

1810 COEN, M. *Leghorn.*—Walnut-wood sideboard.

1811 COSTA, A. *Lavagna (Genoa)*.—Slate tables inlaid with various marbles.

1812 DELLA VALLE, P. *Leghorn*.—Inlaid scagliola tables.

1813 DELLEPIANE, L. *Savona (Genoa)*.—Chairs.

1814 DE MARTINO, G. *Naples*.—Mahogany toilet table.

1815 DESCALZI, E. *Chiavari (Genoa.)*—Chiavari chairs.

1816 DESCALZI, G. *Chiavari (Genoa.)*—Chiavari chairs; inlaid table.

1817 ESCOUBAS, M. A. & SCOTTI, I. *Genoa*.—Oval inlaid tables.

1818 FLORENCE WORKHOUSE, *Florence*.—Veneered sideboard.

1819 FLORENCE ROYAL MANUFACTORY OF MOSAICS IN PIETRE DURE.—Florentine mosaic table, mosaics, vases of Egyptian porphry, carved and inlaid chest: collection of 121 siliceous stones employed in the Royal Manufacture of Pietre-Dure mosaics.

1820 FRANCESCHI, E. *Florence*.—Looking-glass.

1821 FRULLINI, L. *Florence*.—Walnut-wood escritoire, style of the 14th century.

1822 GARGIULO, L. *Sorrento (Naples)*.—Marquetry furniture, made with Italian woods.

1823 GIUSTI, PROF. P. *Sienna*.—Sculptured chests.

1824 GUALA, G. *Turin*.—Walnut-wood fire-screen, and bedstead.

1825 HOLMAN, R. *Florence*.—Wardrobe, ancient style.

1826 IANNICELLI, M. *Salerno (Principato Citeriore)*.—Mosaic toilet table.

1827 LANCETTI, F. *Perugia (Umbria)*.—Ebony table and casket, inlaid with woods, ivory, &c.

1828 LEVERA, BROTHERS, & Co. *Turin*.—Various articles of furniture, inlaid, &c.

1829 LURASCHI, A. *Milan*.—Billiard tables.

1830 LUCCHESI, BROTHERS, *Lucca*.—Small writing table inlaid with metal, &c.

1831 MAINARDI, B. *Milan*.—Inlaid table, with mosaic painting.

1832 MARTINOTTI, G. & SONS, *Turin*.—Cornices, escritoires, tables, and sideboards, inlaid, gilt, &c.: folding travelling furniture.

1833 MERLINI, C. *Florence*.—Ebony chest, with compartments in Florentine mosaic.

1834 MONTELATICI, A. BROTHERS, *Florence*.—Round table in Florentine mosaic.

1835 MONTENERI, A. *Perugia (Umbria)*.—Marquetry views of Rome, Venice, Florence, Naples, &c.

1836 MOROZZI, F. *Florence*.—Machine-cut veneer.

1837 MUSICO, D. *Messina*.—Chairs.

1838 ODIFREDI, G. *Leghorn*.—Toilet and writing-table, inlaid.

1839 PASQUINI, G. *Florence*.—Roll of nut-wood veneer, cut by circular saw.

1840 RIGHINI, C. *Milan*.—Furniture.

1841 ROVELLI, C. *Milan*.—Window blinds.

1842 SCALETTI, A. *Florence*.—Ebony coffer, inlaid, &c.; inkstand.

1843 SCOTTI, I. *Genoa*.—Inlaid tables.

1844 SGUEZZO, V. *Savona (Genoa)*.—Chairs.

1845 TORRINI, G. & VIECCHI, C. *Florence*.—Tables in Florentine mosaics.

1846 VITI, CAV. A. *Volterra (Pisa)*.—Tables in coloured and indurated alabaster, in imitation of Florentine mosaics.

1847 ZORA, G. *Turin*.—Inlaid wood-flooring.

1848 BARBETTI, A. & SONS, *Florence*.—Carved walnut wood bookcase, carved bench, &c.

1849 BIGAGLIA, CAV. P. *Venice*.—Inlaid tables.

1850 BUCCI, R. *Ravenna*.—Inlaid table.

1851 CHALON & ESTIENNE, *Florence*.—Inlaid wood flooring, models of inlaid flooring.

1852 CORRIDI, P. *Leghorn*.—Inlaid table-top.

1853 FONTANA, D. *Milan*.—Monumental inlaid escritoire.

1854 GATTI, G. *Rome*.—Inlaid cabinet and looking-glass frame.

1855 GHIRARDI, G. *Brescia*.—Furniture.

1856 GRANDVILLE, M. *Sorrento (Naples)*.—Marqueterie work.

1857 INGEGNERI, P. *Scilla (Calabria Ulteriore I.)*.—Terra-cotta figures in the Abruzzan and Sicilian costume.

1858 MORESCHI, G. A. *Brescia*.—Table.

1859 NOVI, C. *Brescia*.—Inlaid table.

1860 SALVIATI, DR. A. *Venice*.—Roman and Venetian mosaic tables.

1861 TANGASSI, CAV. BROS. *Volterra (Pisa)*.—Inlaid alabaster table.

1862 ZAMPINI, L. *Florence*.—Imitation Chinese lacquered folding screen.

Sub-Class B.

1870 BACCI, F. *Florence*.—Cornices, brackets, Corinthian capital, &c.

1871 BARBENSI, G. *Florence*.—Napoleon III. on horseback, executed in Florentine mosaics; table in Florentine mosaics.

1872 BENSI, C. *Volterra (Pisa)*.—Candelabra in veined alabaster; alabaster vases.

1873 CHERICI, G. & BROTHERS, *Volterra (Pisa)*.—Alabaster vases, and candelabrum.

1874 FRANCESCHI, E. *Florence*.—Ornamental cornices, brackets, and frames.

1875 FRULLINI, L. *Florence*.—Figures in nut-wood; bracket, with grotesque animals' heads; ebony jewel-case, with relievo figures.

1876 GIUSTI, PROF. P. *Sienna*.—Sculptured and carved frames.

1877 LOMBARDI, A. *Sienna*.—Carved walnut-wood frame.

1878 PARENTI, G. *Volterra (Pisa)*.—Paper weights, and alabaster vase.

1879 PAPI, L. *Florence*.—Carved frame, purchased by H.M. the King of Italy.

1880 PICCHI, A. *Florence*.—Cornices in ebony, &c.

1881 RENZONI, A. *Pisa*.—Model of the church of S. Maria della Spina Pisa, and of the leaning tower of Pisa, in alabaster.

1882 TANGASSI, CAV. C. & BROTHERS, *Volterra, Pisa*.—Articles in alabaster.

1883 VITI, CAV. A. *Volterra (Pisa)*.—Alabaster statuettes.

1884 Van Lint, E. *Pisa.* — Alabaster ornament.

1885 Zambelli, G. B. *Milan.* — Wood carving.

1886 Ambrogio, G. *S. Alessandro* (*Brescia*).—View of Solferino, carved in cork.

1887 Billotti, Dr. P. *Turin.*—Copies of paintings executed on marble.

1888 Brilla, A. *Savona* (*Genoa*).—Crucifix.

1889 Colletti, M. *Florence.* — Carved frames.

1890 Fieschi (Conservatory), *Genoa.*—Bouquet, artificial and wild flowers.

1891 Garassino, V. *Savona* (*Genoa*).—Ivory statuette; painting in inlaid woods.

1892 Garnier Valletti, F. *Turin.*—Artificial fruit.

1893 Lippi, A. *Pietrasanta* (*Lucca*).—Statuary marble platter.

1894 Micali, G. & Sons, *Leghorn.* — Alabaster work.

1895 Negroni, G. *Bologna.* — Epergne, with chased work.

1896 Norchi, E. *Volterra* (*Pisa*).—Alabaster vases and statuettes.

1897 Pacinotti, F. *Florence.*—Engraved marbles.

1898 Sartori, G. *Venice.*—Carved picture frames, brackets, &c.

1899 Traballesi, P. *Florence.*—Bacchus sleeping in a barrel; plaster of Paris model, in one piece.

CLASS 31.

Sub-Class A.

1900 Alfano, A. & G. *Naples.*—Bedstead.

1901 Angiolillo, G. A. *Campobasso* (*Molise*).—Lock.

1902 Azzerboni, C. *Pontassieve* (*Florence*).—Locks.

1903 Baldantoni, G. B. *Ancona.*—Iron bedstead.

1904 Bargiani, F. *Pisa.*—Shoes for race-horses.

1905 Beccalossi, F. *Brescia.*—Hand and machine-made nails, screws, hammers, &c.

1906 Bolzani, S. *Milan.*—Wire gauze.

1907 Calegari, V. *Leghorn.* — Cast-iron work, &c.

1908 Campobasso Sub-Committee for the Exhibition.—Door locks, bells for cattle.

1909 Cecchetta, P. *Pisa.*—Horse-shoes.

1910 Cesari, G. *Cremona.* — Strong-box in hammered iron-work, with cast-iron and bronze ornaments.

1911 Ciani, G. *Florence.* — Lock, with countless combinations; other locks.

1912 Cima, G. B. *Lecco* (*Como*).—Hardware.

1913 Cobianchi, V. *Omegna* (*Novara*).—Iron wire.

1914 Florence Workhouse.—Bedsteads, table, chairs, and other articles of iron.

1915 Fornara, G. *Turin.*—Various kinds of wire gauze and fire guards.

1916 Franci, P. *Sienna.*—Iron gate, made with the hammer.

1917 Ghibellini, Brothers, *Persiceto* (*Bologna*).—Japanned iron table and sofa.

1918 Guppy & Pattison, *Naples.*—Collection of iron and brass nails.

1919 Londini, Brothers, *Bologna.*—Iron bedstead.

1920 Macry, Henry, & Co. *Naples.*—Cast-iron candelabrum; ornamental castings.

1921 Mombelli, G. *Milan.* — Pins and nails.

1922 Mossone, G. B. *Andorno, Cacciorna* (*Novara*).—Locks for strong-boxes and shops.

1924 Pietrarsa Royal Works, *Portici* (*Naples*).—Ornamental iron castings.

1925 Rusconi, A. *Breno* (*Brescia*). — Plough-shares, frying-pans, &c.

1926 Simion, G. *Pescia* (*Lucca*).—Frame for paper manufacture.

1927 Spano, L. *Oristano* (*Cagliari*).—Gate-lock used in Sardinia.

1928 Bolgè, T. *Brescia.*—Iron wire-work.

1929 Bevilacqua, P. *Campobasso* (*Molise*).—New system of locks.

1930 De la Morte, F. *Naples.*—Ornamental iron castings.

1931 Ignesti, F. *Florence.*—Helmet hammered from a single piece of iron.

1932 Theodorani, S. *Forlì.*—Steelyards.

1933 Rovelli, C. *Milan.*—Iron sofa, stool, and chairs.

Sub-Class B.

1935 Beccalossi, F. *Brescia.* — Candlesticks and other articles in brass.

1936 Cammilletti, A. *Perugia* (*Umbria*).—Candelabrum and other articles in bronze.

1937 Manuelli, G. *Prato* (*Florence*). — Domestic utensils in copper.

1938 Marinelli, T. *Agnone* (*Molise*). — Bronze bell, for a belfry, producing an entire semitonic octave.

1939 Penza, F. *Naples.*—Ornamental castings.

1940 Guppy & Patisson, *Naples.*—Brass work.

Sub-Class B.

1941 Giani, V. *Como.*—Bronze statue of Balilla, cast at the Royal Arsenal, Turin.

Sub-Class C.

1945 Colombo, N. *Milan.* — Leaden and tinned articles.

1946 Kramer & Co. *Milan.*—Lead pipes.

1947 Simmola, Cav. F. *Naples.*—Metal-work.

1948 Decoppet, L. *Turin.*—Patent lead pipes.

1949 Lau, A. *Naples.*—Metal tea service, &c.

CLASS 32.

Sub-Class A.

1960 Cerie, C. *Lucca.*—Steel bit and spurs.

Sub-Class B.

1965 BARBERI, A. *Turin.*—Various objects in cutlery.

1966 BECCALOSSI, F. *Brescia.*—Knives, forks, and files.

1967 DUINA, A. *Brescia.*—Cutlery.

1968 GRAVINA, M. *Campobasso* (*Molise*).—Scissors, knives, razors, &c.

1969 OLMETTA, A. *Cagliari.*—Large knife.

1970 SANTANGELO, S. *Campobasso* (*Molise*).—Clasp-knives, scissors, razors, &c.

1971 SELLA, L. *Massareno* (*Novara*).—Knives, penknives, razors, &c.

1972 SPETRINI, L. *Campobasso* (*Molise*).—Chased dagger.

1973 VENDITTI & TERZANO, *Campobasso* (*Molise*).—Knives, scissors, penknives, &c.

1974 VILLANI, R. *Campobasso* (*Molise*).—Damasked and chased table and dessert knives.

1975 VINEIS, G. B. *Biella* (*Novara.*)—Hunting knife, scythes.

1976 VINEIS-BARON, & BROTHERS, *Mongrando* (*Novara*).—Scythes.

1977 VINEIS, C. & BROTHERS, *Mongrando* (*Novara*).—Scythes.

1978 VINEIS, M. G. *Mongrando* (*Novara*).—Scythes.

1979 VINEIS, S. & BROTHERS, *Mongrando* (*Novara*).—Scythes.

1980 VINEIS, T. & BROTHERS, *Bologna.*—Scythes.

1981 VINEIS, T. & NEPHEWS, *Mongrando* (*Novara*).—Scythes.

1982 BUFFI, G. *Scarperia.*—Collection of cutlery.

1983 DE STEFANO, BROTHERS, *Campobasso* (*Molise*).—Cutlery.

1984 SPINA SANTALA, F. *Acireale* (*Catania*).—Scythes.

1985 TORO, P. A. *Campobasso* (*Molise*).—Cutlery.

CLASS 33.

1990 AMBROSINI, G. *Naples.*—Coral neck-lace, ear-rings, brooches, &c.

1991 CALVI, G. *Ripateatina* (*Abruzzo Citeriore*).—Jewel case with 5 stones from Vesuvius.

1992 DELLA VALLE, P. *Leghorn.*—Brooches in *scagliola*, in imitation of the Florentine mosaics.

1993 FORTE, E. *Genoa.*—Filligree work.

1994 FUSCO, G. *Naples.*—Red coral, and lava work.

1995 GHEZZI, A. & SONS, *Milan.*—Engraved silver challice, candelabra, &c.

1996 GRISETTI, E. *Milan.*—Gold brooch set with stones.

1997 GUIDA, C. *Trapani.*—Red coral work.

1998 LABRIOLA, F. *Naples.*—Various articles in tortoise-shell, inlaid with gold.

1999 MASINI, G. *Naples.*—Lava ornaments, electrotype articles, bronze and silver articles.

2000 MINOTTINI, G. *Perugia* (*Umbria*).—Reliquary with engraved work.

2001 PARAZZOLI, L. *Milan.*—Ring serving as perpetual calendar.

2002 PELUFFO, V. *Cagliari.*—Peasants' jewellery.

2003 PIERONI, A. *Lucca.*—Figure in silver.

2004 PIEROTTI, P. *Milan*—Galvanoplastic imitations of ancient shields, helmets, &c.

2005 ROCCA, R. & A. *Modena.*—Table ornament.

2006 SCALETTI, A. *Florence.*—Church ornament.

2007 AVOLIO & SONS, *Naples.*—Collection of coral and lava work.

2008 BOLOGNA ROYAL MINT.—Medals and coins.

2009 BORANI, CAV. *Turin.*—Swords presented by the Tuscans to General Alfonso della Marmora; another sword presented by the Legations and Marches; another sword presented as a remembrance of the Piedmontese troops in the Crimea. Laurel crown, with precious stones, presented to General Cialdini, after the taking of Gaeta.

2010 CAPURRO, N. *Pisa.*—Books.

2011 CASTELLANI, *of Rome and London.*—Large collection of archæological jewels from existing originals.

2012 ERCOLANI, E. *Florence.*—Copy of St. John by Donatello, beaten on a plate of metal.

2013 FLORENCE ROYAL MINT.—Medals.

2014 GERMANI, DR. G. *Cremona.*—Engraved quartz and hyacinth.

2360 NANNEI, G. *Florence.*—Turned silver cup.

2361 NAPLES ROYAL MINT.—Coins and medals.

2362 PANE, M. *Naples.*—Wrought silver work.

2363 ROGAI, L. *Florence.*—Steel dies.

2364 ROYAL ITALIAN COMMITTEE FOR THE EXHIBITION.—Swords, presented to H.M. the King of Italy, by the citizens of Rome; sword presented by several cities of Central Italy; sword presented by the armourers of Mongiana.

2365 SALVIATI, AVV. A. *Venice.*—Silver filagree model of the Church of St. Mark's, &c.

2366 SANTARELLI, PROF. E. *Florence.*—Collection of medals.

2367 SICHLING, A. *Turin.*—Sword, the property of H.M. the King of Italy.

2368 TORRONI, G. & VECCHI, C. *Florence.*—Florentine mosaic paper weights, brooches, &c.

2369 TURIN ROYAL MINT.—Collection of medals.

2370 PENNA, S. & C. *Leghorn.*—Cameos.

2371 BASSI, B. *Macerata.*—Cameos.

2372 FINIZIO, G.—Cameos, shell and lava.

2373 LAODICINI, G.—Shell cameos.

2375 CORTELLAZZO, A. *Vicenza.*—Sword belonging to the King of Italy; repoussé iron-plate, inlaid with silver gilt.

CLASS 34.

Sub-Class A.

2015 BIGAGLIA, CAV. P. *Venice.*—Collection of artificial aventurine, and other glass work.

2016 FRANCINI, G. *Florence.* — Painted glass window, style of the 14th century.

2017 PACINOTTI, F. *Florence.*—Engraving on coloured glass.

2018 SALVIATI, DR. A. *Venice.*—Collection of artificial chalcedony, mosaics, enamelled, and other glass work.

2019 TRARI, M. *Bologna.* — The three Graces of Canova, engraved on gilt glass.

2020 FRANCISCI, FATTORINI, & MORETTI, *Tod. (Umbria).*—Painted glass windows.

2021 BERTINI, *Milan.* — Painted glass window.

Sub-Class B.

2030 BRUNO, G. *Naples.*—Glass shades.

2031 CARAFA DI NOIA, P. *S. Giovanni a Teduccio, Naples.*—Bonbons for acids.

2032 NARDI, R. & SON, *Montelupo (Florence).* — Wine and oil flasks, protected by straw.

2033 VENICE UNITED MANUFACTORIES, —Large collection of Venice beads.

2034 MORGANTINI & BERNARDINI, *Ravenna.*—Blown glass.

2035 MENCACCI, M. & Co. *Lucca.*—Flasks protected by straw; coloured glass.

2036 SEVOULLE, B. & Co. *Vietri (Principato Citeriore).* — Window glass; glass shades.

CLASS 35.

2040 ARMAO, G. *S. Stefano di Camastra (Messina).*—Imitation Pompeian and Egyptian vases.

2041 BELTRAMI, COUNT P. *Cagliari.* — Terra-cotta stove, &c.

2042 CALVETTI, AVV. G. *Pianezza (Turin).* —Terra-cotta flower vase.

2043 COLONNESE, F. & G. *Naples.*—Terra-cotta flower vases, imitations of Etruscan and Greco-Siculean vases.

2044 FERNIANI, COUNT A. *Faenza (Ravenna).*—Two majolica vases, painted in the ancient style.

2045 GALEAZZO, G. A. *Castellamonte (Turin).*—Economical earthenware fire-grates.

2046 GINORI LISCI, MARQUIS L. *Florence.* —Collection of porcelain; collection of majolica, in imitation of that of Urbino and Pesaro, of the 14th and 15th century; imitation Lucca della Robbia ware; majolica for common use; earthenware; stoves.

2047 MARRAS, F. *Assemini (Cagliari).* — Collection of pottery.

2048 OLIVIERI & FERRO, *Savona (Genoa).* —Tobacco pipes.

2049 RICHARD & Co. *Milan.*—Porcelain services: white and coloured earthenware; crucibles, &c.; garden vases; fire bricks.

2050 RONDANI, T. *Parma.*—Earthenware diaphragms and cells for galvanic batteries; earthenware furnaces for jewellers.

2051 SAVONA JUNTA FOR THE EXHIBITION *(Genoa).*—Collection of common stoves.

2052 SPANO, L. *Oristano (Cagliari).* Collection of terra-cotta stoves.

2053 BERTE & STROBEL, *Parma.*—Fire pan; clay and quartz employed in the manufacture of the pans.

2054 CAROCCI, FABBRI, & Co. *Gubbio (Umbria).*—Vases, plates, &c. with historical figures.

2055 FURLANI, G. & Co. *Florence.*—Terra cotta stove.

2056 MOSSA, BROS. *Pianezza (Turin).* — Water pipes, flower pots, &c.

CLASS 36.

Sub-Class B.

2070 CORA, D. & SONS, *Turin.*—Portmanteau.

2071 GHEZZI, *Milan.* — Portmanteaux trunks, &c.

JAPAN.

North Court, British Side.

1 ALCOCK, R. Esq. H.M. Envoy Extraordinary and Minister Plenipotentiary at the Court of the Taikoon.

191 specimens of lacquer ware, lacquering on wood, and inlaid wood and lacquer mixed, consisting of lacquered and inlaid cabinets and stands; lacquered trays for various purposes, and waiters; toilet and luncheon boxes, tables, bowls, nests of drawers, &c.

Lacquer and enamelled objects on ivory, tortoise-shell, mother-of-pearl, &c.: boxes, pedestals, saucers, ivory tortoises, buttons, &c.

Inlaid and carved wood work: trays, boxes, writing-cases, Hakoni ware, clogs, &c.

Straw and basket-work and lacquer combined in articles of use and ornament: despatch and other boxes, bamboo and rattan baskets, drawers, trays, ladies and gentlemen's sandals, cigar-cases, &c.

China and porcelain: enamelled, lacquered, and plain; pottery, and quaint forms of earthenware: vases, jars, flower-baskets, &c.; cups, saucers, and bowls, in egg-shell china; cups, saucers, bowls, trays, tea-pots, dishes, and grotesque figures, in lacquered china; cups, dishes, tea-pots, bowls, &c. in pottery, from Osaca; an inkstand, jars, cups, bowls, &c. in porcelain, from Ocasaki.

A collection of bronzes: bronze jars, vases with dragons and other figures in relievo, stands, baskets, mirrors, candlesticks, grotesque figures, models of gardens, &c. and ingenious designs in bronze work for a variety of useful purposes.

Sets of metal buttons, medallions and intaglios in pure and mixed metals, brooches, scent-bottles, &c.

Arms and armour: Japanese shirt and cap of mail, fire-brigade helmet and leather surcoat, mask-armour and quarter-staff for fencing, swords, &c.

Specimens of lacquered Japanese bows and arrows, a quiver, archery gloves, a Daimio's ladies' bows and arrows, &c.

Specimens of minerals: copper in bar; lead ore, and stone for sharpening tools, from Hakodadi; lava from Fusiyama; coal from Fezin.

A collection of Japanese gold, silver, and copper coins.

A large collection of specimens of Japanese paper used for a great variety of purposes, and manufactured in imitation of leather, &c.; also paper money.

A collection of manufactured articles in silk, crape, cotton, tapestry, and Kowoo bark, and a cable of human hair.

Specimens of carving in ivory and wood; illustrated books and maps; lithochrome printing, illustrating Japanese manners, costumes, &c.; figures by a native artist; story-books in the Hirakana character, &c.

A Japanese Encyclopædia, illustrated works on natural history, &c., on chemistry, and a list of public officers; a quadrant and sundial, a compass, a pedometer, a clock, a thermometer, and a telescope.

A collection of toys, and models illustrating costumes, &c.

A box of very precious tea in powder, tooth-powder, and a collection of shells.

2 HAY, LORD J., H.M.S. Odin.—A gold-lacquered luncheon-box.

3 MACDONALD, J. Esq., H.M. Legation.—A matchlock.

4 VYSE, CAPT. F. H., H.M. Consul at *Yokohama.*—Screens and vases, ivory curios and charms, a porcelain bottle, and lacquer china cups, saucers, and bowls.

5 MYBURGH, DR. F. G.—198 Japanese medicines, and a collection of surgical instruments.

6 CRAWFORD, MRS.—A Japanese table.

7 COPLAND, C.—A cabinet.

8 NEAVE, MR.—Lacquered boxes.

9 REMI SCHMIDT & CO.—Raw silks and cocoons.

Inlaid cabinet work and lacquer ware.

Books, arms, bronzes, porcelain, &c.

10 BARTON, DR. A.—Old and choice lacquer and bronzes, swords, &c. from Japan.

11 BARING, MESSRS. — Two polished spheres of rock crystal.

12 HAY, COMMODORE LORD JOHN, C.B. —A collection of raw silk.

LIBERIA.

Northern Courts, under the Stairs, near Horticultural Gardens entrance.

1 LIBERIAN COMMISSION.—Coffee; cocoa; sugar; molasses; spices; preserves; meal; rice; starch; bacon; textile materials and fabrics; native weapons, implements, &c., in wood and iron; skin pouches; basket-work; earthenware; oil; timber; mineral ink, &c.

2 BARNARD, J. L. *Cornhill.*—Samples of medicinal oil.

MADAGASCAR.

North Court, British Side.

1 MAURITIUS GOVERNMENT.—Packets of gums, specimens of unwrought iron, iron ore, blocks of wood.

2 MIDDLETON, LT.-COL.—An iron chair, presented by H.M. King Radama II. A spear, a collection of spoons, and two musical instruments.

3 MELLISH, E. — Specimens of spades, knives, and tools; horn spoons, shoes, stockings, and silk lambas.

4 E. N.—Specimens of spades, axes, choppers, and chisels; silk lambas, and one of cotton.

5 MARINDIN, F. A.—A spade.

6 BEDINGFELD, F.—A spear, and a cotton lamba.

8 CALDWELL, J.—A collection of tools, a mat, Rafia cloth and cord, and a cotton lamba.

9 ELLIS, THE REV.—Lambas for different purposes; matting, boxes, cloth, &c. made from the rafia palm; an iron lamp.

10 MORRIS, J.—A native straw box, containing five others.

11 LONDON MISSIONARY SOCIETY.—Brass scales for weighing money, iron and silver weights (the latter made from dollars); silver ore, native ornaments, &c. of silver and beads; baskets made from a reed, grass for basket work; a pair of shoes, two wooden dishes, a sword and scabbard, knives, a walking-stick, an amulet of beads.

12 —— A branch of the rafia palm, with a cone.

THE NETHERLANDS.

N.W. Court, No. 2.

CLASS 1.

1 BOUVY, J. J. *Amsterdam.*—Refined salt.

2 BRANDHOFF ISSELMAN, J. J. *Leyden.*—Dutch table salt.

CLASS 2.

3 BOSSON, K. G. W. DE, *Dordrecht.*—Preparations of iron, and other chemicals.

4 DIEDERICHS, P. A. *Amsterdam.*—Box of water colours.

5 ELST AND MATTHES, VAN DER, *Amsterdam.*—Sublimed sal-ammoniac, and sulphate of ammonia.

6 NETHERLAND CARBONISATION FACTORY, *Hillegersberg, near Rotterdam.*—Carbonized peat.

8 FOCK VAN COPPENAAL, G., M.D. *Amsterdam.*—Chemicals for technical and pharmaceutical use. Pure acids, ethers, weights made of aluminium.

10 GARANCINE AND MADDER MANUFACTORY, *Tiel.*—Garancine, and other products of madder.

11 GROOTES, BROS. D. & M. *Westzaan.*—Powder-blue, cocoa, and chocolate.

12 GROOTE & ROMENY, *Amsterdam.*—Chemicals for photography.

13 GRINTEN, L. VAN DER, *Venlo.*—Tartrate of potash and iron, in crystals.

14 JACOBS, AZ. S. & A. *Zwolle.*—Samples of lac, varnish, and standoil.

15 JONG, H. DE, *Almelo.*—Varnish, with specimens of application.

16 LENSING COLLARD, H. *Leeuwarden.*—A map of 1739, and engravings, restored. Ink and varnish.

17 MENDEL BOUR & CO. *Amsterdam.*—Samples of garancine.

18 MEIJER, J. W. *Workum, Friesland.*—Butter and cheese preserver. Eau de Cologne. Cheese-dye; antifebrile herbs.

19 MULLER, M. & CO. *Utrecht.*—Chrysammic acid (polychromic acid).

20 NOORTVEEN & CO. *Leyden.*—Chromic and Dutch green, yellow chromate. Paris blue, lac, varnish, and standoil.

21 OCHTMAN, VAN DER VLIET, & CO. *Zierikzee.*—Madder, and its products.

22 RENTERGHEM, C. A. VAN, & CO. *Goes.*—Madder, garancine, fleur de Garance, alizarine, and colorine.

23 SPRUYT & CO. *Rotterdam.*—Purified cod-liver oil. Salts of ammonia. Samples of ink.

24 TACONIS, P. *Joure, Friesland.*—Dye-stuff (Friesland green), and oil.

26 VERHAGEN & CO. *Goes.*—Madder and garancine products.

28 VRIESENDORP, C. A. & SONS, *Dordrecht.*—Lac and varnish, &c.

29 FRANKEN & CO. *Amsterdam.*—Enamel coating for iron.

30 SOETENS, C. *the Hague.*—Charcoal manufactured by a new process from Dutch peat.

CLASS 3.

31 AGRICULTURAL SOCIETY, *Culemborg.*—Preserved provisions.

32 ANDEL, T. VAN, *Gorinchem.*—Wheat, raw and in the various stages of manufacture.

33 APKEN & SON, *Purmerend.*—Preserved sweet-meats.

34 BOGAARD, J. VAN DEN, & CO. *Gennep, Limburg.*—Wheat flour.

35 BOLS, ERVEN LUCAS LOOTSJE, *Amsterdam.*—Liqueurs.

36 BONT & LEYTEN, DE, *Amsterdam.*—Sugar, chocolate, and liqueurs.

37 BOOTZ, H. *Amsterdam.*—Liqueurs and elixirs.

38 CATZ & SON, *Pekela.*—Stomachic elixir.

39 CHARRÓ, F. DE, *Amersfoort.*—Dutch tobacco.

40 DYEL, L. VAN DER, & SON, *Weesp.*—Chocolate powder.

41 DIEVELAAR H. & DE BREUK, *'s Hertogenbosch.*—Chicory and pea-coffee.

42 DOYER & VAN DEVENTER, *Zwolle.*—Liqueurs.

44 DUYVIS, J. *Koog-on-Zaan.*—Urling's starch and glue from the refuse.

46 DUTCH SUGAR-REFINING SOCIETY, *Amsterdam.*—Sugar in loaves and crushed.

47 EBERSON, H. P. *Arnhem.*—Liqueurs.

48 EGBERTS, B. H. *Dalfsen.*—Chicory.

49 ELLEKOM & VISSER, VAN, *Amsterdam.*—Preserved provisions. Fruits in vinegar.

50 FOCKINK, W. *Amsterdam.*—Liqueurs.

51 GORTER, H. S. *Dokkum.*—Friesland clover-seed.

52 GRIENDT & LUYTEN, VAN DER, *Rotterdam.*—Preserved provisions, in tin boxes with crystal bottoms of the exhibitor's invention.

53 HAAGEN, BROS. VAN, *Utrecht.*—Cigars.

54 HENKES, J. H. *Delftshaven.*—Dutch gin.

55 HEYLIGERS, T. & SON, *Schiedam.*—Dutch gin.

56 HOPPE, P. *Amsterdam.*—Liqueurs, rectified and unrectified spirits.

58 HUNCK, H. P. *Amsterdam.*—Chocolate powder, flavoured with spices.

61 KAKEBEEKE, GZ. J. H. C. *Goes.*—Samples of Zealand wheat, wheat flour, bran, wheaten groats, &c.

64 KOPPEN, H. T. *Leerdam.*—Cigars, manufactured from Java tobacco. Dutch tobacco.

65 KORFF, F. *Amsterdam.*—Chocolate powder, cocao-butter or vegetable-fat.

66 LANS, H. & SON, *Haarlem.*—Samples of beer.

67 LEVERT & CO. *Amsterdam.*—Liqueurs.

69 MACKENSTEIN, A. F. & SON, *Amsterdam.*—Cavendish tobacco.

70 NICOLA KOECHLIN, & CO. *The Hague.*—Wheat-meal, wheat, and rye-flour.

71 OBENHUYSEN, D. P. A. *Amsterdam.*—Samples of vinegar.

74 OOLGAARD, D. & SON, *Harlingen.*—Liqueurs.

75 OOLGAARDT, J. *Amsterdam.*—Preserved fruits and sweetmeats.

76 PATERS, P. L. *Leyden.*—Dutch preparations of buckwheat.

78 RALAND, G. A. *Deventer.*—Preserved fruits and vegetables, pickles, &c.

79 REEKERS, L. & CO. *Haarlem.*—Raisin-vinegar.

80 REYNVAAN, A. J. *Amsterdam.*—Tobacco, snuff, and cigars.

82 RÖNTGEN, J. E. *Deventer.*—Liqueurs.

83 ROYAL LIQUEUR DISTILLERY, STIBBE, BROS. *Kampen.*—Liqueurs and bitters.

84 SCHONEVELD & WESTERBAAN, *Gouda.*—Potatoe-flour: white, yellow, and brown syrups, and sago.

87 ULRICH, J. S. & C. *Rotterdam.*—Ship-bread, table and dessert biscuits.

89 Zuijlekom Levert, van, & Co. *Amsterdam.*—Purified and rectified alcohol, amyl-alcohol, gin, brandy, liqueurs.

90 Aa, T. J. van der, *Rotterdam.* — Liqueurs.

91 Gevers Deijnoot, D. R. *near The Hague.*—Cheese.

92 Nieuwenhuis & Co. *Rotterdam.* — Preserved provisions.

CLASS 4.

93 Alberdingk, F. & Sons, *Amsterdam.* —Oil.

94 Benoist, J. L. & Mooi, J. *Heerenveen.* —Sealing or bottle-wax of various colours.

96 Bousquet, J. & Co. *Delft.*—Samples of soap.

97 Clercq, H. de, *Haarlem.* — Gutta-percha : collection, illustrative of its applications.

98 Crommelin, R. *Renkum.*—Rape-seed-cakes and oil.

99 Diemont, J. J. & Son, *Amersfoort.*— Raw and ground oak-bark.

100 Dobbelman, Bros. *Nijmegen.*—Samples of soap.

101 Dorssen, Gz. G. van, *IJsselstein.*— White and grey hoops.

102 Dutch Stearine Manufactory, *Amsterdam.*—Materials for the manufacture of stearine; stearine; stearine candles; oleine; spermaceti candles, &c.

106 Janssens, E. B. F. *Weert (Limburg).* —Hard and soft soap.

107 Kamer, W. P. van de, *Middelburg.* —Eau de Zélande.

108 Krol, G. J. & Co. *Zwolle.*—Bone-black.

109 Lobry & Porton, *Utrecht.*—Cod-liver oil.

111 Robette & Draijer, *Gouda.*—Grease, glue, gelatine for weaving purposes, phosphate of lime, animal manure.

112 Royal Wax Candle Manufactory, *Amsterdam.* — Products of stearine, stearine candles, oleine.

113 Röntgen, C. A. *Deventer.* — Oils, seeds and kernels of indigenous plants.

114 Smits, Widow P. & Son, *Utrecht.*— Bone-black; sulphuric, and nitric acid; sulphate of iron.

115 Society for the Manufacture of Varnish, Colours, &c. Molijn & Co. *Rotterdam.*—Lac, lacquered panels. Tools used for the imitation of wood.

116 Stearine Candle Manufactory, *Gouda.* — Materials, fat-acids, stearic acid, candles, &c.

118 Verkade, E. G. *Zaandam.* — Rape and linseed oil, raw and purified. Train oil, raw and purified.

119 Viruly, J. P. & Co. *Gouda.* — Soft green soap.

120 Visser, E. E. *Amersfoort.*—Yellow wax.

122 Wolff, M. *Amersfoort.*—Yellow wax.

123 Hilst, L. v. d. *The Hague.*—Corks.

CLASS 6.

124 Beynes, J. J. *Haarlem.* — Landau carriage.

125 Deventer, J. S. van, *Zwolle.*—Carriage (Victoria), with a new kind of springs.

126 Hermans, M. L. & Co. *The Hague.*— Carriages.

127 Preyer, J. Hz. B. *Amsterdam.*—A Victoria carriage.

128 Roeckx, H. W. *Maastricht.*—Chiselled springs for a gala carriage.

CLASS 7.

131 Hunck, Widow J. T. & Son, *Amsterdam.*—Tools for diamond cutters.

132 Watson, G. & Son, *Rotterdam.* — Rasps and files.

CLASS 8.

135 Pasteur, W. C. & Co. *Rotterdam.*— Fire-engine, model of others, and gasometer.

136 Peck, Bros. *Middelburg.*—Miniature fire-engine, with model of new metallic pistons, &c.

137 Arends, Alcmenum E. *Amsterdam.* —Fire engine.

138 Ham, T.—Fire escape.

139 Yserman, J. M. *The Hague.*—A centralisator for fire-engines.

CLASS 9.

140 Brakell van den Eng, Baronet. F. L. W. van, *Lienden.*—Model of sheep-house.

142 Geluk, J. Az. A. *Tholen.*—Clod-crusher, Cheddam wheat, red chaffed wheat, madder.

143 Jenken, W. *Utrecht.*—Foot plough, with dredges, coulter, and disk-coulter. Cylindric mangle.

144 Riphagen & Co. *Hattem.*—Cultivator, with levers; turning harrow.

145 Sarphati, Dr. S. *Amsterdam.* — Models of agricultural implements. Peat ash.

146 Six, Sir P. H. *'sGraveland.*—Meadow-sledge (model).

147 Starre, G. van der, *Sloten, North-Holland.*—Model of a Dutch hay or corn-rick.

148 Staring, Dr. W. C. H. *Haarlem.*— Three-wheeled tipping cart and whippletrees.

149 Stout, G. *Tiel.*—Paddy thrashing machine, with sieves.

150 Vogelvanger, E. *Hulst.*—Self-acting harrow, for one or two horses.

CLASS 10.

151 Andriessens, J. *Roermond.*—Samples of pipes manufactured from pewter, &c. Sheet of pressed lead.

152 ESTA, F. F. VAN, *Harlingen.*—Blue enamelled tiles; hard, red waterproof bricks.

153 HAART, C. H. DE, *Utrecht.*—Agate marble vases; alabaster; black marble chimney-piece.

154 HEUKELOM, N. VAN, *Erlecom, near Nijmegen.*—Artificial paving-stones; samples of bricks; specimens of Dutch masonry.

155 HULST, J. VAN, *Harlingen.*—White tiles.

158 MIRANDOLLE, C. *Fijenoord, near Rotterdam.*—Painted wood, creosoted, and not creosoted.

159 MULLER & Co. *Valkenburg.*—Samples of bricks.

160 RHEE, S. J. *Haarlem.*—Model of a joined beam. Staircase.

161 TRUFFINO, W. F. K. A. *Leyden.*—Bricks; Rhine stones.

162 VERWEYDE, C. C. *Amsterdam.*—Model of a moveable vertical window-blind.

164 CUIJPERS & STOLZENBERG, *Roermond.*—Sculpture in wood.

165 VENEMA, L. *Hertogenbosch.*—Freda or baldequin, in Portland stone.

CLASS 11.

166 DEPARTMENT OF WAR, *The Hague.*—Breech of a 6-pounder, filled with brass, by Maritz, to be bored as a 4-pounder.

167 STEVENS, P. *Maastricht.*—Fire-arms.

CLASS 12.

170 HOOGEN, J. VAN DEN, *Dordrecht.*—Hemp and iron rigging.

171 MAAS, A. E. *Scheveningen.*—Model and description of a life-boat.

172 VERVLOET, W. A. *Rotterdam.*—Apparatus for hauling ships upon a slip.

CLASS 13.

176 BECKER & BUDDINGH, *Arnhem.*—Brass balances and beams.

177 EDER, S. J. *Rotterdam.*—Scales for letters.

178 EIJK, DR. J. A. VAN, *Amsterdam.*—Thermometer adapted for public lectures.

179 EMDEN, A. VAN, *Amsterdam.*—Compass, the rose floating in spirit.

180 GEISSLER, W. *Amsterdam.*—Hygrometer, Prof. von Baumhauer's principle; &c.

181 HOLLEMAN, F. A. *Oisterwijk.*—Hygrometer for domestic use.

185 OLLAND, H. *Utrecht.*—Balance of aluminium for chemical use; sea barometer.

186 SPANJE, T. VAN, *Tiel.*—Electro-magnetic clock, with commutator.

187 SOCIETY "DE ATLAS," *Amsterdam.*—Water-meters for high and ordinary pressure, Prof. von Baumhauer's principle.

188 WELLINGHUYSEN, WIDOW, R. J. & SON, *Amsterdam.*—Gas lamps for chemical use. Prof. V. Baumhauwer's principle.

CLASS 14.

190 EIJK, DR. J. A. VAN, *Amsterdam.*—Photographic copies of etchings by Rembrandt, &c.

191 SANDERS, VAN LOO, *Amsterdam.*—Photographs on dry collodion.

CLASS 15.

191A HOWHU, A. *Amsterdam.*—Chronometer with secondary compensation, of the exhibitor's invention; chronometer with ordinary compensation.

CLASS 16.

192 BERGEN, A. H. VAN, *Midwolde, Groningen.*—Musical bells; Turkish cymbals.

193 BEVERSLUIS, P. *Dordrecht.*—Semi-melodiums.

194 CUYPERS, J. F. *The Hague.*—Inlaid pianino for hot climates

198 OSCH, E. P. VAN, *Maastricht.*—Brass wind-instruments.

CLASS 17.

203 BERGHUIS, J. *Groningen.*—Skeletons of various animals.

204 KOENAART, A. J. C. *The Hague.*—Moveable set of teeth.

205 LINDEN, J. & SON, *Rotterdam.*—Surgical, obstætrical, orthopœdical, and anatomical instruments.

206 SCHMEINK, BROS. A. & B. *Arnhem.*—Collection of surgical, anatomical, orthopœdical instruments and knives.

CLASS 18.

209 CATE, HZ. H. TEN & CO. *Almelo.*—Raw and bleached cotton manufactures, calicoes, shirtings, cambrics, drills, table-cloth, striped dimities.

210 SCHAAP, L. & A. *Amersfoort.*—Plain and striped dimities.

211 VELTMAN, BROS. *Amsterdam.*—Cotton blankets, quilts.

212 VISSER & CO. *Amersfoort.*—Linen and cotton Amersfoort, Marseilles.

CLASS 19.

216 CATZ, J. B. VAN & SON, *Gouda.*—Horse reins and yarns.

217 DUINTJER, J. J. *Veendam.*—Rope rigging, &c. made by machinery. Iron joining-screws.

218 ELIAS, J. *Stryp, near Eindhoven.*—White linens.

219 GALEN, B. VAN, *Gouda.*—Specimens of yarns for netting, packing, sail-cloth, &c. Strings and reins.

220 GORTER, H. S. *Dokkum.*—Bundle of flax.

221 KAMERLING, Z. & SON, *Almelo.*—Damask collation-table-cloth. Napkins of flax-yarn, woven in one piece.

222 KLEYN, A. VAN DER & SON, *Gouda.*—Hand-spun reins.

223 KORTENOEVER, J. *Gouda.*—Yarns for sailmakers, shoemakers, &c. Hemp and flax.

224 LEEUW, M. DE & SON, *Boxtel.*—Table-cloth of damask with heraldic designs.

225 LEYDEN, D. VAN & SON, *Krommenie.*—Sail-cloth of Dutch hemp and handspun yarn.

226 LEYDEN & DERKER, VAN, *Krommenie.*—Sail-cloth.

229 VERSLUYS, L. J. *Amsterdam.*—Twilled fire-engine hose, of tanned and untanned hemp.

230 VISSER, S. E. & SON, *Amersfoort.*—Mixed yarn and cotton twillings and Marseilles.

CLASS 20.

232 SARTINGEN, S. *Horst, near Venlo.*—Silk taffetas, satin, satin de Chine, velvet.

CLASS 21.

235 KERSTENS, J. A. A. *Tilburg.*—Wool-dyed pilot-cloth, Amadou. Beaverfrieze. Baize.

236 KRANTZ, J. J. & SON, *Leyden.*—Woollen cloth, &c.

237 LEDEBOER, L. V. & SONS, *Tilburg.*—Wool-dyed and mixed pilot-cloth. Baize and castor, piece dyed.

238 MEER, P. VAN DER & SONS, *Leyden.*—Leyden worsteds.

239 POSTHUMUS & Co. *Maastricht.*—Felts for paper-making.

240 SCHELTEMA, Jz. J. *Leyden.*—Woollen blankets. Wrappers. Horse blankets.

241 WILLINK, J., *Winterswyk.*—Mixed woollen and cotton cloth.

242 WYK, BROS. VAN & Co. *Leyden.*—Woollen blankets.

243 IJSSELSTEYN, J. F. J. *Leyden.*—Woollen blankets. Serge and pilot-cloth.

244 ZAALBERG, J. C. & SON, *Leyden.*—Woollen blankets.

245 ZUURDEEG, J. & SON, *Leyden.*—Woollen blankets.

CLASS 22.

248 HEUKENSFELDT, J. *Delft.*—Imitation Smyrna carpet, knitted and woven from wool.

249 KRONENBERG, W. F. *Deventer.*—Deventer Smyrna carpet, woven in one piece.

250 PRINS, WIDOW L. J. *Arnhem.*—Scotch carpet. Carpets of cow-hair and wool.

CLASS 23.

252 JANSSENS, BROS. & Co. *Roermond.*—Turkey-red dyed yarns.

254 ROOIJEN, H. VAN, *Utrecht.*—Silk dyed with chrysammic acid. Silk dyed with an extract of soot.

CLASS 24.

258 BLOOK, A. DE, *Amsterdam.*—Gold and silver embroidery, military ornaments, galloon, &c. Raw materials.

259 OLIE, C. H. *Amsterdam.*—Embroidered fire-screen, and table.

260 OVEN, BROS. VAN, *The Hague.*—Mantle richly embroidered with gold.

261 RENIER VAN WILLES, MAD. A. S *Sluis (Zeeland).*—Lace-work.

262 TEULINGS & Co. *'sHertogenbosch.*—Linen lace (Bosscher band), &c.

263 KRAFT & Co. *Amsterdam.*—Coach livery lace, &c.

CLASS 25.

264 CATZ, P. S. & Co. *Amsterdam.* Rough, drawn, and curled horse-hair.

265 DEVENTER, J. S. VAN, *Zwolle.*—Fur carpet made of indigenous skins.

266 DIRKS, H. J. *Dordrecht.*—Brooms, brushes, &c.

267 JONKER, BROS. *Amsterdam.*—Brushes.

268 DR. VAN LITH DE TEUDE, *Utrecht.*—Fur carpet.

269 NYMAN, H. *'sHertogenbosch.*—Brushes and prepared swine-hair.

270 REUS, Pz. N. *Dordrecht.*—Swine-hair and brushes.

271 SIEBERG, J. *Amsterdam.*—Artificial hair-work.

272 THIJSSEN, W. *Tiel.*—Foot, bath, and horse-brushes, &c.

273 GREEVE & SON, *Amsterdam.*—Fur carpet.

CLASS 26.

275 HÄGER, J. G. *The Hague.*—Saddle with pockets and boxes for military surgeons.

276 HAVEKOST, T. H. *Amsterdam.*—Head-stalls.

CLASS 27.

279 BOASSON, A. *Middelburg.*—Hygienic corsets.

280 CIERENBERG, A. *Zwolle.*—Stockings, knitted in blue and black worsted.

281 DONCKERWOLCKE, J. B. F. *Amsterdam.*—Shirts, flannel waistcoats, and trousers.

282 DONKER, H. *Ijsselmonde.*—Wooden shoes.

286 HOLSBOER, P. & SON, *Arnhem.*—Waterproof Russian-leather shooting-boots and bottines, &c.

287 LEENARTS, J. *Utrecht.*—Silk hats, indigenous genet skins.

288 VELDE, H. VAN DER, *Zwolle.*—Boots and bottines for deformed feet.

289 VIGELIUS, H. & Co. *Rotterdam.*—Kid gloves and skins.

CLASS 28.

290 ABRAHAMS, BROS. *Middelburg.* — Ledger, journal, diaries.

292 DEPARTMENT OF WAR (topographical section), *The Hague.*—Copies of topographical, geological, and other maps.

293 EMRIK & BINGER, *Haarlem.*—Lithography in crayon tints and chromo-typography.

294 ES, GÉRARD D. VAN, *Amsterdam.*—Nature printing. Specimens of wood-cut printing.

295 ENSCHEDÉ J. & SONS, *Haarlem.* — Types, specimens of printing, stereotype plates, &c.

297 LAMPERT, P. *Middelburg.* — Map of the Isle of Walcheren, lithographed for the Department of War.

298 LHOEST, LAMMERS, & Co. *Maastricht.* —Various samples of paper.

299 METZLER & BASTING, *Amsterdam.*—Printings in colours, with gold and silver.

301 PANNEKOEK, NEUY, *Heelsum.*—Specimens of paper.

302 PANNEKOEK, T. *Heelsum.*—Specimens of paper.

303 RINCK, F. W. *The Hague.*—Album for photographs, &c.

305 SMULDERS, J. & Co. *The Hague.*—School maps of the Netherlands. Map of Palestine, in chromo-lithography. Stamped Works.

306 SYTHOFF, A. W. *Leyden.* — Books printed in the Japanese, Chinese, and other languages.

307 TETTERODE, N. *Amsterdam.* — Printings in various oriental types.

309 WYT, M. & SONS. *Rotterdam.*—Specimens of printing.

311 SPANIER, E. *The Hague.* — Lithographs.

312 MULLER, F. *Amsterdam.*—Dictionaries in Oriental languages; anastatic printing, &c.

CLASS 29.

313 BACKER, J. *Arnhem.*—Slabs for entomologists.

314 BRUINSMA, J. J. *Leeuwarden.*—Specimens illustrating the anatomy and development of the silkworm.

315 BURG, J. L. VAN DEN, *Amsterdam.*—Plaster busts of Paul Potter and Van der Helst.

CLASS 30.

317 ANSLYN, H. J. *Haarlem.* — Ladies' work-table, teaboard, &c.

318 BORZO, WIDOW, J. & SONS, *s'Hertogenbosch.*—Looking-glasses and models of frames.

319 DRILLING, A. *Amsterdam.* — Inlaid etagère-tables.

321 HEYMANS, W. G. F. *Amsterdam.* — Tea-table painted in imitation of fine wood.

322 HORRIX, BROS. *The Hague.*—Saloon furniture.

323 KERREBYN, J. *Haarlem.* — Mosaic-painted table.

324 KEMMAN, J. H. W. *Amsterdam.* — Bookcase convertible into a desk; lady's bureau.

325 KLOESMEYER, W. F. *Amsterdam.*—Lacquered table with carved foot.

326 MICHIELS, J. & WEUSTENRAED, J. *Maastricht.*—Oil-cloth.

327 MOEN, P. *Baambrugge.*—Painted mosaic inlaid table.

328 PENNOCK, D. J. *Middelburg.*—Horse-hair mattress and beddings.

329 PIETERSE, P. & Co. *Gouda.*—Envelopes of straw and rush for packing bottles.

330 RUTTEN, J. H. *Maastricht.*—Paper-hangings.

331 SALA D. & SONS, *Leyden.* — Frames manufactured by machinery.

333 VIERLING, M. *The Hague.* — Mahogany buffet—etagère.

334 WILHELM, A. J. *Deventer.*—Fire-proof writing-table with secret lock.

335 WIJDOOGEN, JR. J. *Amsterdam.* — Paper-hangings.

336 ZIRKZEE, J. E. *Leyden.*—Transparent painted window-blinds.

337 BOSCH, A.—Inlaid tea-table and trays.

CLASS 31.

339 BEKKERS, WIDOW J. & SON, *Dordrecht.* —Japanned goods.

340 ENTHOVEN, M. J. & SON, *Zalt-Boemel.* —Nails and culinary utensils of tinned iron.

342 HESSELS, WIDOW J. P. *Dordrecht.*—Tinplate ewer and basin, painted in imitation of porcelain.

343 HEUS, H. DE, & SON, *Utrecht.*—Specimens of rolled and hammered copper.

344 HUBERS, F. W. *Deventer.*—Cocoa-nut fibre matting and cloth, woven wire.

346 LEEFERS, J. L. *The Hague.*—Circular fire grate (model.)

347 NOOYEN, L. J. *Rotterdam.*—Japanned goods.

349 PRESBURG, M. J. & Co. *Nijmegen.*—German silver and copper tobacco boxes, &c.

351 RUYVEN, A. H. VAN, *Amersfoort.* — Nails of 45 sorts.

356 VOLKERS, JR. J. *Zwolle.*—Domestic and culinary utensils.

357 WALL BAKE, VAN DEN, *Utrecht.*—The arms of the Netherlands in cast iron.

359 ZWART, D. H. *Oudewater.* — Collection of horse-shoes for preventing and healing diseases.

CLASS 33.

361 BONEBAKKER, A. & SON, *Amsterdam.* —Silver and gold articles.

362 COSTER, M. E. *Amsterdam.*—Rough, cut, and polished diamonds.

363 GREVINK, G. *Schoonhoven.*—Silversmiths' work.

365 LITTEL, W. & Co. *Schoonhoven.*—Silver and goldsmiths' work.

366 LOON, H. W. VAN, *Rotterdam.*—Netherlanders' gold and silver head-dresses; silver articles.

CLASS 34

370 BOUVY, J. H. B. J. *Dordrecht.*—Bent glass.
371 CASTRO, D. H. DE, *Amsterdam.*—Glass cups engraved with diamond.

CLASS 35.

374 BOSCH, N. A. *Maastricht.*—Earthenware.
375 DRAAISMA, D. *Deventer.*—Water-closets, kilnstones, &c.
376 LAMBERT, G. & Co. *Maastricht.*—Earthenware, stoneware, and terra-cotta.
378 MAAS, B. VAN DER, *Gouda.*—Tobacco and cigar-pipes.
379 PRINCE, J. & Co. *Gouda.*—Tobacco and cigar-pipes.
380 REGOUT, F. *Maastricht.*—Articles of glass, crystal, and fine earthenware.
381 SPARNAAY & SONS, *Gouda.*—Tobacco and cigar-pipes.
383 WANT, AZ. P. J. VAN DER, *Gouda.*—Tobacco-pipes.

CLASS 36.

385 PILGER, L. *Amsterdam.*—Trunks.

DUTCH COLONIES.

387 HOWARD, *Java.* — Specimens of Cinchona barks, grown in Java, and products obtained from them.
388 NETHERLAND COMMISSION.—Collection of produce from the Netherland Colonies, for the European market.
389 MINING DEPARTMENT AT JAVA.—Minerals and metals from India.
390 CARIMON TIN MINE COMPANY.—Tin ore and tin from Carimon.
391 PREANGER GOVERNMENT.—Iron ore, coals, and flint from the Preanger, West Java.
392 NETHERLANDS COMMISSION. — Gypsum from Cheribon, Java.
393 PREANGER GOVERNMENT.—Sulphur from West Java.
394 ————Naphtha from West Java.
395 PADANG GOVERNMENT.—Gold dust from Corintje.
396 TEYSMAN, T. E. *Buitenzorg.*—Pakoe-kidang.
397 PADANG GOVERNMENT. — Camphor baros, camphor wood, dragon's blood.
398 TEYSMAN, T. E. *Buitenzorg.*—Substances used for food, prepared at Java.
399 AMBON GOVERNMENT. — Sago in several forms.
400 PADANG GOVERNMENT. — Cassia lignea, from the west coast of Sumatra.
401 TEYSMAN, T. E. *Buitenzorg.*—Oils.
402 MAKASSER GOVERNMENT. — Macassar oil.
403 PADANG GOVERNMENT. — Dammar oil.
404 PREANGER GOVERNMENT.—Oils.
405 MOLUCCO GOVERNMENT.—Sereh and clover oil.
406 NETHERLAND GOVERNMENT IN INDIA.—Leaves and flowers from plants cultivated at Java.
407 PADANG GOVERNMENT.—Ivory, horn of the rhinoceros.
408 PREANGER GOVERNMENT.—Malemsala, wax and gummi.
409 TEYSMAN, T. E. *Buitenzorg.*—Wax and gums.
410 MAKASSAR GOVERNMENT. — Getah soesoh.
411 PADANG GOVERNMENT.—Samples of getah pertja from the interior of Sumatra.
412 WEBER, L. *Buitenzorg.* — Gambir (Nauslia Gambier), from Java.
413 PADANG GOVERNMENT. — Gambir from the West Coast of Sumatra, and amboina moeda (dye stuff.)
414 PREANGER GOVERNMENT.—Specimen of wood.
415 MOLUCCO GOVERNMENT.—Wood.
416 WEBER, L.—Prepared material for paper.
417 TEYSMAN, A. E.—Flax from Koffo. and yuta, cultivated at Java; gemoeti.
418 ROYAL NETHERLAND COMMISSION.—Flax and thread of Koffo.
419 WEBER, L.—Flax.
420 TEYSMAN, T. E.—Fruits and leaves, dried, &c.
421 PREANGER GOVERNMENT.—Benzoïn and gum malabar from Java.
422 PADANG GOVERNMENT. — Benzoïn from Sumatra W. Coast.
423 TEYSMAN, T. E. — Stangee from Makassar and Amboina.
424 DEPARTMENT OF PUBLIC WORKS AT JAVA.—Model of the irrigating sluices in the Kedirie river, near Soerabaya.
425 MAKASSAR GOVERNMENT.—Stangee and doepa from Makassar.
426 ROCHUSSEN, J. J.—A rifle from Borneo.
427 PREANGER GOVERNMENT.—A knife or sword, used by the Javanese.
428 PADANG GOVERNMENT.—Swords and krisses from the Malays of Sumatra, manufactured by Datoe Panghoelve, Radja from Soengipoeor and Agam.
429 TEYSMAN, A. E.—Cleaned cotton from Buitenzorg.
430 PREANGER GOVERNMENT. — Cotton cloth and yarn from the west part of Java.
431 SAMARANG GOVERNMENT.—Shirtings in the various stages of manufacture for dyeing, and tools used.
432 SAMARANG GOVERNMENT. — Sarong and kainpanjangs.
433 CHERIBON GOVERNMENT.—Kainpanjangs, slindangs, and cloth, batikked and woven.
434 ROCHUSSEN, J. J. — Three sarongs

from Makassar, and cloth, batikked at Buitenzorg.

435 TEYSMAN, A. E.—Three sarongs from Buitenzorg, cloth of bamboe and bark.

436 PADANG GOVERNMENT.—A sarong of silk from Sumatra.

437 SAMARANG GOVERNMENT.—Sarongs and kainpanangs of silk.

438 ROCHUSSEN, J. J.—Silk cloth batikked, and a crape shawl embroidered at Batavia.

439 SAMARING GOVERNMENT.—Carriage harness and shoes.

440 ———Samples of embroidery.

441 PREANGER GOVERNMENT. — Two bedayas, two capala Java, and fruits carved in stone by Javanese.

442 SAMARANG GOVERNMENT. — Two bedayas, carved in stone by Javanese.

443 ROCHUSSEN, J. J.—Two Hindoo gods, carved in ivory at Bali.

444 PREANGER GOVERNMENT.—Baskets of rattan from the west part of Java.

445 ROCHUSSEN, Y. J.— Baskets and spoons for rice, box for pickles.

446 PALEMBANG GOVERNMENT. — Lacquered wood.

447 CHERIBON GOVERNMENT.—Precious stones, used by the Javanese.

448 SAMARANG GOVERNMENT.—A pair of gold bracelets with jewels.

449 BATAVIA COMMISSION.—A silver box for sirh (tampat sirh.)

450 PADANG GOVERNMENT. — Filigree and other gold and silver works.

451 ROCHUSSEN, J. J.—A cigar box from Soerakarta.

452 POOLMAN, W.—Filigree and other gold and silver works.

NORWAY.

N.W. Court, No. 5, and N.W. Gallery, No. 4.

CLASS 1.

1 AAS, A. *Throndhjem.*—Copper-ores.

2 BRÜNECH, J. *Bergen.*—Copper-ores.

3 DAHLL, J. *Kragero.*—Magnetic pyrites, about one-fourth nickel; and apasite, chiefly phosphate of lime.

4 DAHLL, T. *Kragerö.*—A collection of Norwegian minerals.

5 EGERSUND MINING CO. *Egersund.*—Ilmenite.

6 ESMARK, M. *Tönsberg.* — Norwegian minerals, discovered by Esmark.

7 GOVERNMENT SILVER MINE OF KONGSBERG.—Silver-ores, pugworks, and smelting products.

8 GRÖNNING, B. *Bergen.* — Limestone, burnt lime, and bricks.

9 HOLE, H. *Langöén.*—Model of a lime-kiln.

10 IRON WIRE MANUFACTORY OF KJELSAAS, *near Christiania.*—Wire of different metals.

11 JOHANNESEN, *Bergen.*—Ores, and other minerals.

12 KJEER, N. *Næs Iron Works at Tvedestrand.*—Feldspath and quartz.

13 PETERSEN, F. *Stavanger.*—Copper-ore and magetic pyrites, containing nickel.

14 REINERTSEN, H. *Christiansand.*—Limestone and lime from Södal.

15 ROSCHER, *the nickel-mine of Ringerige.*—Magnetic pyrites, one-fiftieth nickel: and nickel-metal, one-half nickel.

16 THE CEMENT MANUFACTORY OF LANGÖ, *Christiania.*—Cement and lime.

17 THE COPPER MINE OF SELBO, *Throndhjem.*—Copper ores and smeltings.

18 THE IRON WORKS OF ÖIENSJÖFOS, *Christiania.*—Limonite, and the iron produced from it.

19 THE LEAD AND SILVER MINE OF RANEN, *in Helgeland, Throndhjem.*—Ores of lead and silver.

20 THE ROLLING WORKS OF LEEREN, *Throndhjem.*—Rolled copper from the mines of Röraas.

21 WEDEL-JARLSBERG, BARON H. *The Iron Works in Bærum.*—Iron ores and smeltings.

22 WEDEL-JARLSBERG, BARON H. *The Iron Works at Moss.*—Iron smeltings.

AALL & SON, *Næs Iron Works, near Tvedestrand.*—Magnetic iron ore.

CLASS 2.

23 BUCHNER, *Kaupanger.*—Oil of turpentine.

24 CHROME MANUFACTORY OF LEEREN, *Throndhjem.*—Salts of chromium.

25 FINCKENHAGEN, *Hammerfeet.* — Medical cod-liver oil.

26 IBENFELDT, S. *Aalesund.* — Medical cod-liver oil.

27 KNUTZON, A. *Christiansund.*—Medical cod-liver oil.

28 KRAMER, J. *Bergen.*—Medical cod-liver oil.

29 MANUFACTORY OF LYSAKER, *Christiania.*—Pyrites, and superphosphate of lime.

30 MÖLLER, P. *Christiania.*—Medical cod-liver oil.

31 STEEN, D. *Christiania.* — Soaps, and cod liver oil.

32 SVENDSEN, L. & S. *Bergen.*—Medical cod-liver oil.

HAUSEN, F. *Aalesund.*—Cod-liver oil.

CLASS 3.

33 BÖ, P. *Gausdal, near Lillehammer.*—Cheese from the sweet milk of goats.

34 BORNHOLDT, G. I. *Christiania.*—Prepared mustard.

35 LIND, MISS, *Christiania.*—Anchovies.

36 MÖLLER, P. *Christiania.*—Norwegian wild medical plants.

37 NORDBYE, J. *Christiania.* — Cheese made of the sweet milk of goats.

38 NORMAN, *Tromsöe.*—Specimen of grass grown in Finmark.

39 SCHÜBELER, DR. F. C. *Christiania.*—Cereals and other vegetable products of Norway.

40 SMITH, MRS. G. *Christiania.* — Anchovies.

41 THE AGRICULTURAL SOCIETY OF TROMSÖ.—Cereals from Finmarken (latitude 70 deg. north) :—

- ANCUNDSEN, T. *Tromsöe.*—Seeds of wild vetches.
- CHRISTENSEN, C. P. *Trondenæs.*—Rye.
- KROGSENG.—*Maalselven.*—Barley.
- LUDVIGSEN, J. H. *Gïsund.*—Barley.
- MAURSUND, *Tromsöe.*—Barley.
- OXAAS, *Lyngen.*—Barley.
- POULSEN, P. *Dyrö.*—Potatoes.
- SCHJÖLBERG, *Sengen.*—Rye, barley, oats, and pease.
- STENSOHN, D. C. *Tromsöe.*—Barley.
- STENSOHN, D. C. *Tromsöe.*—Rye.
- STENSOHN, D. C. *Tromsöe.*—Turnip-seeds.
- STRÖM, LARS, *Dyrö.*—Rye.
- STRÖM, LARS, *Dyrö.*—Oats.

42 THORNE, CHR. AUG. *Drammen.*—Hermetically preserved articles of food: anchovies.

CLASS 4.

43 FROST, *Tranöe.*—Eider-down.

44 IHLEN, J. S. *Christiania.*—Deals and boards of fir and Norway spruce-fir

45 KLINGENBERG, H. *Throndhjem.*—Furs of Norwegian wild animals.

46 NISSEN, A. *Kirkesnæs.*—Wild reindeer horns.

47 THE AGRICULTURAL SOCIETY OF TROMSÖ.—Samples of forest products, from Finmarken (latitude 70 deg. north).

48 WEDEL-JARLSBERG, BARON H. *Christiania.*—Deals and boards of fir and Norway spruce-fir.

CLASS 6.

49 GRÖNNEBERG, I. H. *Drammen.* — A cariole and sledge.

50 HEFFERMEHL, *Drammen.* — A sledge and apron.

51 NORMAN, *Tromsöe.*—A Fin sledge.

52 STEEN, A. *Drammen.*—Sledges, and a cariole.

CLASS 7.

53 JACOBSON, O. *Christiania.*—Model of a mangle.

54 NILSEN, I. *Christiania.* — Sliding-rest, with double cutting apparatus for metallic cogwheels.

55 THE MECHANICAL WORKS OF AKER, *Christiania.*—Lund's ball-press.

56 THE MECHANICAL WORKS OF THE ROYAL NAVY, *Horten.*—A set screw taps.

CLASS 8.

57 THE MECHANICAL WORKS OF AKER, *Christiania.*—A high pressure expansive steam-engine, 6 h. p.

58 THE MECHANICAL WORKS OF NYLAND, *Christiania.* — A steam-engine: cocks and valves for water and steam: a turning lathe.

59 THE MECHANICAL WORKS OF THE ROYAL NAVY, *Horten.* — A donkey-engine steam-pump.

60 THE WORKS OF LAXEVAAG, *Bergen.*—Machinery for ropemaking.

CLASS 9.

61 BISETH, *Christiania.* — Models of agricultural implements: a cultivator for root-crops: Biseth's double mould-board plough, &c.

62 JACOBSON, O. *Christiania.* — A chaff-cutter, and a Norwegian revolving harrow.

63 ROLFSEN, W. *Bergen.*—An iron plough.

64 ROSING, C. W. *Frederikstad.*—An iron plough, with a steel mould-board.

65 SCHÜBELER, Dr. F. C. *Christiania.*—A hand-cultivator, invented by the exhibitor, and other horticultural implements.

66 SKJETNE, G. *Tiller by Trondhjem.* — A double plough, constructed by exhibitor.

67 WINGAARD & BOUILLY, *Bergen.* — A chaff-cutter.

CLASS 10.

68 THE ROYAL OFFICE FOR THE PUBLIC ROADS.—Models of a wheel-barrow and ballast-waggon.

69 ——— System of the Norwegian highways.

CLASS 11.

70 BUE, E. *Lillehammer.*—A gun.

71 HJELMELAND, O. *Bergen.*—A revolver-rifle, a gun, and a pistol with four barrels.

72 LARSEN, H. *Drammen.* — A breech-loading rifle and a revolver rifle.

73 TENDEN, R. *Bergen.*—A rifle.

74 THE BOARD OF ORDNANCE, *Christiania.*—Collection of arms for the Norwegian army, a hunting-rifle, and a field carriage.

CLASS 12.

75 BALCHEN, H. *Bergen.*—Specimens of rope-makers' work.

76 BENDIXEN, C. *Stavanger.* — Ship's augers, constructed by exhibitor.

77 BRUNCHORST, C. H. *Bergen.*—Model of a ship.

78 CHRISTOFFERSEN, *Bergen.*—A grapnel.

79 DEKKE, A. *Bergen.*—Models of ships.

80 ELLERTSEN, *Bergen.*—Model of a sloop,

81 GRAN, J. *Bergen.*—Models of ships.

82 HOEL, *Christiania.* — Model of a ship, with several new inventions.

83 HOLMBOE, O. *Wefsen.* — Model of a Nordland boat.

84 JORDAN, H. *Bergen.* — Specimens of rope-makers' work.
85 THE MECHANICAL WORKS OF THE ROYAL NAVY, *Horten.*—Anchors and chains: models of Norwegian boats.

CLASS 13.

86 ENGER, CHR. *Christiania.* — Instruments for surveying.
87 LUNDGREN, J. *Bergen.* — Optical instruments.
88 QVAMME, L. *Bergen.*—Instruments for surveying, invented by exhibitor.

CLASS 14.

89 SELMER, M. *Bergen.*—Photographs of Norwegian national dresses, and of Norwegian scenery.

CLASS 15.

90 CHRISTOPHERSEN, *Christiania.* — An astronomical clock.
91 IVERSEN, J. *Bergen.*—A chronometer.
92 PAULSEN, M. *Christiania.*—An astronomical clock.

CLASS 16.

93 BRANTZEG, P. *Christiania.*—A monocord, and an upright piano.
94 HALS, BROTHERS, *Christiania.* — A grand, and an upright piano.
95 HELDAL, A. *Bergen.* — A Hardanger fiddle, used by the Norwegian peasantry.

CLASS 17.

96 GALLUS, *Christiania.* — Surgical instruments.
97 STEGER, A. *Christiania.* — A medical syringe apparatus.
98 SUNDBY, H. *Christiania.*—An artificial leg.

CLASS 18.

99 HALVOR SCHOU, *Christiania.*—Striped cotton stuffs and checks.
100 HOLST, MRS. *Bergen.*—Quiltings.
101 JEBSEN, P. *Bergen.*—Dyed shirtings and twills, bleached and grey calico.
102 LILLEHAMMER COTTON SPINNING MILL, *Lillehammer.*—Grey cotton yarn.
103 RICHTER, MISS I. *Inderöen.*—Bed-quilts.
104 RICHTER, MISS G. *Inderöen.*—Quilting.
105 THE SPINNING MILL OF NYDALEN, *Christiania.*—Cotton yarn.
106 THE SPINNING MILL OF FOSS, *Christiania.*—Cotton yarn.

CLASS 19.

107 CHRISTIANIA SAILCLOTH FACTORY *Christiania.*—Linen-yarn and sailcloth.
108 ECHE & SON, *Bergen.* — Window-blinds.

109 KNOPH & BLICHFELDT, *Christiania.*—Hosiery
110 RICHTER, MISS I. *Inderöen.*—Damask.
111 RICHTER, MISS G. *Inderöen.*—Damask.
112 THE HOUSE OF CORRECTION AT BERGEN.—Fishing-nets for herrings.

CLASS 21.

113 DONS, N. *Senjen.*—Home-made stuffs, from Finmark; woollen gloves, and specimen of wool.
114 HANSEN, H. & Co. *Bergen.*—Woollen stuffs.
115 SCHOU, HALVOR, *Christiania.*—Half-woollen stuffs.

CLASS 22.

116 BRUN, MISS A. B. *Fosnæss.*—Quiltings.
117 KJELVEN, MISS G. *Sogn.*—Quiltings.
118 KNAGENHJELM, *Kaupanger.* — Quiltings.
119 WESTREM, *Kaupanger.*—Quiltings.

CLASS 24.

120 NEUMANN, MISS, *Bergen.*—Embroidery.
121 WIENCKE, L. *Bergen.*—Trimmings.

CLASS 25.

122 ECKHARDT. I. M. *Christiania.*—Hair-work.
123 PETTERSEN, *Bergen.*—Hair-work.

CLASS 26.

124 AMUNDSEN, T. *Frederikshald.*—Skins and leather.
125 BÉRGMANN, G. *Drammen.*—Harness.
126 BERNER, I. *Christiania.*—Skins and leather.
127 BRANDT, C. F. *Bergen.*—Furs.
128 CHRISTOFFERSEN, O. *Værdalsoren.*—Skins.
129 ELLERHUSEN, J. *Bergen*—Leather.
130 ENDSTRUP, C. *Christiania.*—Harness.
131 ERICHSEN, E. *Bergen.* — Saddler's work.
132 GRÖNNEBERG, I. H. *Drammen.*—Harness.
133 HALLÉN, C. *Christiania.*—Leather.
134 HEYERDAHL, C. *Christiania.*—Skins.
135 ROHDE, E. *Bergen.*—Saddler's work.
136 VALEUR, P. *Nordland.*—Harness for a reindeer.

CLASS 27.

137 CHRISTIANIA INDUSTRIAL UNION FOR INDIGENT WOMEN, *Christiania.* — Needle-work and embroidery.
138 ELIASSEN, A. *Bergen.*—Boots.
139 FOUGNER, J. *Bergen.*—Hats.
140 FREY, M. *Christiania.*—Hats, and a seamless dress made of felt.

141 HANSEN, C. *Christiania.* — Gentlemen's boots and shoes.
142 HANSEN, H. F. *Christiania.*—Ladies' boots and shoes.
143 LOTZ, P. *Bergen.*—Gloves.
144 SALVESEN, J. *Bergen.*—Hats.
145 SOLBERG, N. *Christiania.* — Gentlemen's boots and shoes.
146 SVENSEN, F. *Bergen.*—Boots.
147 THISTED, A. *Bergen.*—Boots.
148 TRESSING, E. *Christiania.*—Boots.
149 TYSLAND, G. *Bergen.*—Boots.

CLASS 28.

150 BERG, G. *Christiania.*—Printed books.
151 BEYER, F. *Bergen.*—Printed books, and bookbinding.
152 BROGGER & CHRISTIE, *Christiania.*—Printed books.
153 DAHL, J. *Christiania.*—Printed books.
154 GJERTSEN, E. *Bergen.*—Bookbinding.
155 HELDAL, J. *Bergen.*—Pocket-books.
156 HENRICHSEN, C. *Christiania.*—Playing cards.
157 HERMANN, L. *Bergen.*—Paste-work.
158 HJORTH, G. *Christiania.*—Paste-work.
159 JENSEN, H. *Christiania.* — Printed books.
160 THE WORKS OF GAUSA, *Lillehammer* —Pasteboard.

CLASS 29.

161 THE GOVERNMENT DEPARTMENT FOR THE CHURCH, *Christiania.*—Drawings and models of buildings and furniture for schools.
162 ——— Books and instruments for teaching generally.
163 ——— Appliances for physical education.
164 ——— Specimens of school-work.
165 ——— Museums.
166 ——— Drawings and objects for the Philanthropic Congress.

CLASS 30.

167 HENNEMOE, M. *Throndhjem.*—Furniture made of veined birch-wood.
168 LOSTING, I. *Bergen.*—Looking-glass with console.

CLASS 31.

169 BERGENDAHL, H. *Bergen.*—Lacquered work.
170 BONGE, R. *Bergen.*—Articles in iron and zinc.
171 FLOOD, G. *Bergen.*—Smith's-work.
172 JOHANNESEN, C. *Christiania.*—Lacquered work.
173 NILSEN, J. *Christiania.*—Door handles.
174 OPSAHL, P. *Christiania.*—Locks.
175 ROKNE, K. *Sogn.*—Augers.
176 SCHMIDT, E. *Bergen.*—Horse-shoes.
177 THE FACTORY OF LAXEVAAG, *Bergen.* —Various castings.
178 WINGAARD & BOUILLY, *Bergen.* — Various castings.
179 WALLENDAHL, B. *Bergen.*—A stove.

CLASS 32.

180 EIMSTAD, ASLAK, *Bergen.*—Cutlery.

AALL & SON, *Næs Iron Works, near Tvedestrand.*—Polished blistered steel.

CLASS 33.

181 HALVORSEN, N. *Bergen.*—Norwegian peasants' bridal ornaments, of silver.
182 HAMMER, L. *Bergen.* — Norwegian peasants' ornaments, of silver.
183 TOSTRUP, I. *Christiania.* — Flower-vase with flowers of silver, an epergne, a beer-can, a coffee-pot, and other articles.

CLASS 35.

184 SCHWARZENHORN & BEYER, *Christiania.*—Paintings on china.

CLASS 36.

185 BLYTT, H. *Bergen.* — Artificial flies, for fishing.
186 BROSTRÖM, G. W. *Bergen.*—Carvings in wood.
187 BUCHER, H. *Bergen.*—Lithographs.
188 BUCK, *Oexfjord.* — Finmark dresses and a cradle.
189 CHRISTIE, W. *Bergen.*—Objects illustrating the life and industry of the Norwegian peasants.
190 CHRISTIANIA PENITENTIARY, *Christiania.*—Carvings in wood.
191 EGE & CO. *Bergen.*—Matches.
192 FANDREM, *Karasjok.*—Fin-shoes.
193 FLADMOE, T. *Christiania.*—Carvings in wood and stone.
194 HANNO, V. *Christiania.*—A baptismal font of sand-stone.
195 HANSEN, O. *Christiania.*—Persiennes.
196 HAUGSE, O. *Bergen.*—Turners' work.
197 IVERSEN, I. *Sandefjord.*—A knife with sheath; carvings.
198 JANSEN, N. *Bergen.*—Coopers' work.
199 JÖRGENSEN, H. *Skjærvöe.*—Model of a mussel-dredge from Finmark.
200 KLUTE, I. *Christiania.* — Basket-makers' work.
201 KNUDSEN, A. *Bergen.*—Articles made of huldrabromms, a sickly development of the birch-tree.
202 KÜHNE, V. *Christiania.* — Basket-makers' work.
203 LARSEN, G. *Lillehammer.*—Carvings in meerschaum.
204 LARSEN, MISS L. *Christiania.*—Dried and pressed flower-work.
205 LÖBERG, O. *Bergen.*—Fishing-lines.
206 LOSTING, *Bergen.* — Illustrations of skin diseases (leprosy).
207 MÜLLER, E. *Bergen.*—Basket-makers' work.

208 NORMAN, *Tromsöe.*—Collection of national costumes from Finmark.

209 OLSEN, J. *Bergen.*— Novel window-frame.

210 PRAHL, G. *Bergen.*—School atlas.

211 SCHWENGEN.—Maps of coast survey.

212 SÖRUM, O. *Hadeland.*—A knife, with carved-wood sheath, and belt.

213 STEENSTRUP, CHR. *Horton.*—A moveable jet-piece for fire-engines: apparatus for fishing, sounding, and preventing rust in boilers, &c.

214 STOCHFLEDT, FR. *Bergen.*—Drawings of architectural ornaments, applicable to furniture.

215 THE COMMITTEE AT BERGEN, *Bergen.* —A collection of fishing implements, and models of fishing-boats.

216 THE HOUSE OF CORRECTION OF AGERSHUUS, *Christiania.*—An economical, adjustable oil can; an apparatus for communicating with the diver, by speaking; articles in polished granite and porphyry: and objects illustrating the life and customs of the Laplanders.

217 THE TOPOGRAPHICAL SURVEY OFFICE. —Maps of some Norwegian provinces.

218 VALEUR, P. *Nordland.* — Finmark dresses, and a cradle.

219 VEDELER, E. *Bergen.*—National costumes, after a picture of A. Fedemand.

PERU.

N.C. Court, No. 2.

KENDALL, H. Consul for Peru, 11, *New Broad-st. E.C.*—

1 Silver ores, from the mines at Cerro de Pasco.

2 Quicksilver ore, from the mines of Huancavelica.

3 Nitrate of soda, or cubic nitre, from Iquique.

4 Borax from the same district.

5 Guano from the Chincha Islands.

6 Remarkable specimen of guano, containing a pure white crystallized ammoniacal salt.

7 Cascarilla, or Peruvian bark, from Huanuco.

8 Coca-leaf from Huanuco; it supports Indian miners and couriers under great fatigue with very little food.

9 Matico-leaves, a very powerful styptic and astringent.

10 Sarsaparilla from the department of Piura.

11 Large maize from Cuzco, in grain and in the ear.

12 Rice from Lambayeque and San Pedro.

13 Coffee from the mountains near Huanuco.

14 Samples of sheep's wool from Junin and Puno.

15 Alpaca wool from Arequipa and Puno.

16 White alpaca skin.

17 Llama and Vicuna, or Vigonia wools.

NAYLOR, J. E. Consul for Peru, *Liverpool.*—

18 Silver vessel, from the ruins of the ancient Indian city of Grand Chimu.

19 Silver hammer, seals, buttons, and medals, from the same ruins.

20 Drinking cup mounted in silver.

21 Poncho of Peruvian cotton, made by Indians.

22 Small piece of poncho cloth, taken from an Indian grave some centuries old.

23 Jaquima, or halter of plaited hide, mounted in silver.

24 Wooden stirrups mounted in silver.

25 Ancient stirrups of carved wood mounted in silver.

26 Crupper of stamped leather.

27 Four huaqueros, or earthen jars.

28 A face carved in wood, from the ruins of the Temple of the Sun at Pachacamac.

ELIAS, DON D. *Estate at Nasca.*—

29 Cotton from *Nasca*, producing 1,200,000 lbs. yearly: it might be 3,000,000 lbs.

30 Cochineal from Nasca; annual produce, 50,000 lbs.

31 Wine from Hoyos, near Pisco, similar to Madeira wine; annual produce, 100,000 gallons.

32 Wine from Urrutia estate, near Pisco; annual produce, 200,000 gallons.

33 Aguardiente Italia, produced at Yca near Pisco; a liqueur made from Muscatel grape.

HOYLE, DON. J. *Truxillo.*—

34 Blankets or coverlets made from cotton, of great antiquity, and recently found in the ruins of an Indian city.

KENDALL, MRS. *The Limes, Mortlake.*

35 A silver ornament, known in Lima as "Briscado" work, made by the nuns.

36 Specimens of silver filigree work, made in Huamanga.

37 A piece of plata-pina, or crude porous silver, moulded in the shape of a dog.

EASTTED, W. & Co. 27, *Regent-st.*—

38 Wines from Moquegua resembling sherry and Muscat.

Addenda.

EGUSQUIZA, DA. PAULA, of *Lima.*—

39 Two straw hats, plaited by the Indians of Moyobamba.

40 Six cigar cases, made in the Department of Yca.

41 Silk and cotton poncho, as made by the Indians of Cuzco before the Conquest.

42 Cotton coverlet, made by the Indians of Eten.

43 Silk braces and garters, made by the Indians of Huamanga.

44 Two filigree turkeys, made at Ayacucho.
45 Four bottles of Moscatel wine, from the Falconi estates at Yca.
46 Four bottles of wine, sherry class, from the same.
47 Two pounds of chocolate of Cuzco cocoa.
48 Coffee from Carabaya, Department of Arequipa.
49 Coffee from Huanuco, Department of Junin.
50 Quinua, a grain indigenous to all the Sierra of Peru.
51 Pallares, a large kind of bean, from Yca.

DE LA JARA, DA. M. A., of *Lima*.—

52 Two filigree baskets, with dishes, made at Ayacucho.
53 Cotton and silk table-cover, made in the Province of Huaylas.
54 Cotton from the valley of Viso, Province of Jauja.

GALVEZ, HIS EXCELLENCY DN. PEDRO, *Paris*.—

55 Silver ores from different mines.
56 Felt of Vicuña wool.

MAIZ, DN. THOMAS MORENO.—

57 Coca leaf (vide No. 8).

COTES, SRA. ALTHAUS DE.—

58 Cloth Poncho.
59 Counterpane from Lambayeque.
60 Towelling from Loja, Caxamarca, Arequipa, and Yca.
61 Embroidery work from Moquegua.
62 Specimens of lace from Camana, Arequipa, and Carameli.
63 Purse from Cuzco, in figure of an Indian.
64 Relics of the nuns at Cuzco.
65 Garters made at Cuzco.
66 Collar used by the Indian women of San Jose
67 Fine lace cap.
68 Vicuna gloves from Cuzco.
69 Cloth made in Arequipa.

HAYNE, MRS. *Gloucester-square*.—

70 Painting, by a native artist at Cuzco, representing portraits of the Incas, from Manco Capac to Atahualpa, with historical notes.
71 Jewellery, forming a nun's rosario, made in Lima.
72 Collection of metallic ores.

WENT, MRS. *Cleveland-square*.—

73 Silver filigree baskets and flowers.
74 Specimens of lace and work from Arequipa.

PORTUGAL.

S.C. Court, No. 1, and S.C. Gallery No. 1.

The Portuguese Commissioners having forwarded their Catalogue alphabetically arranged according to Christian instead of Surnames, it has not been considered expedient to alter the arrangement, owing to the necessity of preserving the consecutive order of the numbers.

CLASS 1.

1 A. J. BOTELHO, *Evora*.—White marble.
2 A. A. DA SILVEIRA PINTO, *Porto*.—Minerals.
3. A. A. RIBEIRO, *Bragança, Macedo de Cavalleiros*.—Amianthus.
4 A. F. LARCHER, *Portalegre*.—Ochres, and moulding sand.
5 A. J. FERREIRA, *Santarem, Rio Muior*.—Salt.
6 A. J. M. RELVAS, *Portalegre*.—Hydraulic lime.
7 A. P. DA COSTA, *Bragança, Miranda do Douro*.—Copper pyrites, oxide of tin, and metallurgic products.
8 COMMISSION (DISTRICT) OF AVEIRO.—Clays used for earthenwares, salt and lime, anthracite coal.
9 COMMISSION (DISTRICT) OF COIMBRA.—Limestone, marbles, clays, &c.
10 COMMISSION (DISTRICT) OF EVORA—Copper ore.
11 COMMISSION (DISTRICT) OF FARO.—Sulphuret of antimony, copper pyrites.
12 COMMISSION (FILIAL) OF ALANDROAL, *Alemtejo, Evora*.—Copper ore.
13 COMMISSION (FILIAL) OF BORBA, *Alemtejo, Evora*.—Marbles.
14 COMMISSION (DISTRICT) OF LEIRIA.—Marble and ochres.
15 COMMISSION (DISTRICT) OF PORTALEGRE.—Marbles and granites.
16 COMPANY (PERSEVERANÇA MINING), *Porto*.—Antimony and tin ores.
17 COMPANY OF LEZIRIAS, *Lisboa*.—Peat.
18 C. DE ALMEIDA E SOUSA, *Coimbra, Penacova*.—Models of mill-stones.
19 COUNT DE FARROBO, *Lisboa*.—Coal.
20 ——— Lignites.
21 DEJEANTE, L. B. *Lisboa*.—Marbles.
22 DIRECTION OF THE PUBLIC WORKS OF ANGRA, *Açores*.—Volcanic grit.
23 DIRECTION OF THE PUBLIC WORKS OF COIMBRA.—Hydraulic lime.
24 DIRECTION OF THE PUBLIC WORKS OF LEIRIA.—Capital of a column.
25 E. DELIGNY, *Paris*.—Cupriferous iron-pyrites.
26 E. I. PARREIRA, *Açores, Angra*.—Ochres.
27 FEUERHEERD, D. M. *Porto*.—Ores of lead, zinc, and copper, metallurgic products, &c.
28 FIDIÉ, A. M. A. G. *Faro*.—Sulphuret of antimony, lime-stone, &c.

29 F. P. M. Furtado, *Beja, Moura.*—Copper pyrites with calcareous spath.

30 Garcia, C. A. C. *Bragança.*—Bars of tin.

31 G. Croft, *Lisboa.*—Lignites and magnetic iron.

32 G. J. de Salles, *Lisboa.*—Lime-stone, marbles, &c.

33 Inspector of Public Works at Açores, *Lisboa.*—Pozzolana.

34 Inspector of the Second Mining District, *Lisboa.*—Magnetic iron ore, oxide of iron, marbles, lignites, &c.

35 J. Pring, Jun. L. F. Defferari e A. C. de Almeida, *Lisboa.*—Lead ore and blende.

36 J. F. Braga, *Lisboa.*—Marbles.

37 J. G. Roldan, *Lisboa.*—Tin ore.

38 J. J. Ramos, *Evora, Redondo.*—Marl.

39 J. M. S. Gouveia, *Coimbra, Figueira da Foz.*—Salt.

40 J. Pereira, *Leiria.*—Common limestone and marble.

41 J. A. dos Santos, *Lisboa.*—Marble.

42 J. U. Peres, *Coimbra, Penella.*—Marble.

43 J. F. P. Basto, *Lisboa.*—Minerals.

44 J. J. de Lemos Sousa e C. e A. L. Batalha, *Lisboa.*—Cupriferous iron-pyrites.

45 J. J. da Silva Pereira Caldas, *Braga.*—Granite, magnetic iron ore.

46 J. M. de Figueiredo Antas, *Bragança, Vimioso.*—Marble.

47 J. R. Tocha, *Evora, Extremoz.*—Copper ores, &c.

48 Lacerda, R. V. de Sousa, *Leiria, Alcobaça.*—Gypsum, coal.

49 L. de Abreu Magalhaes Figueiredo, *Guarda. Ceia.*—Slate.

50 L. F. Defferari, *Lisboa.*—Lead ore.

51 M. J. V. Novaes, *Lisboa, Setubal.*—Salt.

52 M. Monteiro, *Santarem, Rio Maior.*—Flints.

53 Marquis da Bemposta, L. G. C. de Salle, *Lisboa.*—Lime-stone, grit, and pitch.

54 O. Machado & Irmaos, *Lisboa.*—Pozzolana.

55 S. A. da Cunha, *Santarem.*—Lime.

56 V. B. Pires, *Faro.*—Salt.

57 V. B. Pires, Jun. *Faro.*—Ores of copper.

58 Viscount de Bruges, *Açores.*—Freestone, pumice-stone, &c.

59 Viscount de Villa Maior, *Lisboa.*—Sulphate of barites.

CLASS 2.

60 A. A. Andrade, *Porto.*—Pharmaceutical products.

61 A. Simoes, *Coimbra.*—Ricinus-seed.

62 A. J. Cardoso, *Portalegre.*—Seeds, roots, herbs, &c.

63 A. X. da Silva, *Lisboa, Villa Franca de Xira.*—Black mustard.

64 B. J. P. da Mota, *Coimbra.*—Calcined bones

65 B. J. de Sousa, *Aveiro.*—Lesser centaury.

66 C. J. Pinto, *Lisboa.*—Pharmaceutical products.

67 C. A. C. Garcia, *Bragança.*—Charcoal.

68 Commission (Central), *Aveiro, Feira.*—Charcoal, common ergot, and squalus-liver oil.

69 Company of Tejo and Sado Lesirias, *Lisboa, Villa Franca de Xira.*—Mustard.

70 F. J. da Costa, *Vizeu, Lamego.*—Eau de Cologne.

71 General Portuguese Wood Administration, *Leiria.*—Turpentine, acids, essences, resin, charcoal, &c.

72 General Society of Chemical Products, *Lisboa.*—Chemical products.

72a G. T. de Magalhaes Collaço, *Coimbra, Soure.*—Marsh-mallow root.

73 I. de Ciria, *Coimbra, Figueira da Foz*—Fish oil.

74 J. A. D. Grande, *Portalegre.*—Syrup of currants; charcoal.

75 J. F. Norberto, *Lisboa.*—Pharmaceutical products.

76 J. P. Duarte, *Beja.*—Bitter almond oil.

77 J. C. Pucci, *Lisboa.*—Orange flower water.

78 J. C. da Costa, *Coimbra, Louza.*—Cantharides, poppy and mustard seed.

79 J. F. da Silva, *Lisboa.*—Oil, nitrate of potash, jelly, &c.

80 J. G. da Cruz Viva, *Faro.*—Orange oil essence.

81 J. M. dos Santos, *Evora, Montemór o Novo.*—Syrups.

82 M. M. da Terra Brum, *Horta.*—Fish oils.

83 M. da Cunha de Abreu, *Braga.*—Tallow candles.

84 M. J. de Sousa Ferreira, *Porto.*—Pharmaceutical products.

85 Viscount de Bruges, *Angra do Heroismo.*—Tortoise oil, camomile, mustard seed, &c.

CLASS 3.

86 A. F. Garcia de Andrade, *Horta.*—Brandy.

87 A. M. Pereira, *Portalegre, Alter de Chão.*—Wheat.

88 A. B. Ferreira, *Aveiro, Mealhada.*—Wine.

89 Agricultural Institute of Lisboa.—Maize, wines, wheat.

90 A. J. de Moraes, *Bragança, Mogadouro.*—Wine, olive oil.

91 A. Simoes, *Coimbra.*—Maize.

92 A. J. Pereira, *Faro.*—Wine.

93 A. P. Soares, *Vizeu, Carregal.*—Cheese, brandy.

94 A. Allen, *Porto.*—Port wine.

95 A. da Fonseca Coutinho, *Portalegre.*—Acorn coffee.

96 A. de Brito Montoso, *Portalegre, Monforte.*—Wheat and rye.

97 A. da Fonseca Corsino, *Castello Branco, Covilha.*—Wine brandy.

98 A. G. Ponges, *Beja.*—Wheat.

99 A. R. COUTINHO, *Santarem, Almeirim.* —Rice.

100 A. DE ALMEIDA PEREIRA, *Beja, Alustrel.*—Wheat.

101 A. A. DE MORAES CAMPILHO, *Bragança, Vinhaes.*—Wheat, potatoes.

102 A. A. DA COSTA SIMOES, *Aveiro, Mealhada.*—Wine.

103 A. A. RIBEIRO, *Bragança, Macedo de Cavalleiros.*—Rye, wheat, and honey.

104 A. B. DE MORAES, *Bragança, Villa Flor.*—Wine, brandy.

105 A. C. DE FIGUEIREDO, *Lisboa, Alcacer do Sal.*—Wheat.

106 A. DE CARVALHO, *Portalegre, Aviz.* —Red wine, French beans, and maize.

107 A. CEIA, *Portalegre.*—Cherry brandy.

108 A. C. PERDIGAO, *Beja, Vidigueira.*—Cheese.

109 A. DA COSTA LIMA, *Santarem, Almeirim.*—Maize, French beans.

110 A. D. MIRANDA, *Santarem.*—Wheat.

111 A. E. B. FREIRE, *Beja.*—White wine.

112 A. DO ESPIRITO SANTO FERREIRA, *Coimbra.*—Rye, rye flour, brandy.

113 A. F. LARCHER, *Portalegre.*—Maize, rice, oil, &c.

114 A. F. DA MOTTA, *Santarem, Thomar.* —Wine.

115 A. DA FONSECA ESGUELHA, *Lisboa, Villa Franca de Xira.*—Chick-beans, beans.

116 A. FORTIO, *Portalegre, Fronteira.*—Chick-beans.

117 A. F. R. M. TAVARES, *Bragança, Freixo de Espada á Cinta.*—Chestnuts.

118 A. FRANCO, *Beja, Aljustrel.*—Wheat.

119 A. F. CARVAO, *Santarem, Chamusca.* —White wine, olive oil.

120 A. G. C. COTRIM, *Lisboa, Azambuja.* —Wheat, maize, French beans.

121 A. G. P. SALAZAR, *Coimbra, Penella.* —Maize, plums, walnuts, plum brandy.

122 A. H. DOS SANTOS, *Santarem.*—Wheat and maize.

123 A. I. PEREIRA, *Evora, Redondo.*—Brandy of sugar cane.

124 A. J. DA COSTA, *Portalegre, Gaviao.* —Olive oil, honey.

125 A. JOAO, *Beja, Odemira.*—Rice.

126 A. J. DE CASTRO, *Evora, Montemór o Novo.*—Rice.

127 A. J. DA FONTE, JUN. *Santarem, Torres Novas.*—Rye, French beans.

128 A. J. G. DE BARAHONA, *Beja, Aljustrel.*—Brandy.

129 A. J. LOPES, *Beja, Moura.*—Goats' milk cheese, wheat.

130 A. J. MOREIRA, *Beja, Ferreira.*—Chick-pease.

131 A. J. P. DE MATOS, *Faro.*—Wine, dried figs.

132 A. J. P. DE CAMPOS, *Evora.*—Wheat, sausages, cheese.

133 A. J. DE SOUSA, *Beja, Ferreira.*—Red wine.

134 A. J. G. BRANCO, *Lisboa, Alcacer do Sal.*—Rice.

135 A. J. M. RELVAS, *Portalegre, Crato.*—Honey.

136 A. J. NOGUEIRA, *Faro.*—Wheat, rye, maize.

137 A. J. DOS SANTOS, *Coimbra, Oliveira do Hospital.*—Arbutus-berry brandy.

138 A. DE LEMOS TEIXEIRA DE AGUILAR, *Porto, Pesqueira.*—Port wine.

139 A. L. FERREIRA, *Coimbra, Tábua.*—Raisins.

140 A. L. DE GUSMAO, *Portalegre, Alter de Chão.*—Olive oil.

141 A. L. DUARTE, *Beja.*—Wheat.

142 A. L. DA SILVEIRA, *Santarem.* — Wheat, maize, chick-pease, French beans.

143 A. M. P. RODRIGUES, *Beja, Serpa.*—Wheat, olive oil, barley, sheep's milk cheese.

144 A. M. DA CUNHA E SÁ, *Portalegre.*—Red wine.

145 A. M. S. C. FAJARDO, *Castello Branco, Belmonte.*—Cereals, olives, &c.

146 A. M. C. BELLO, *Portalegre, Castello de Vide.*—Vinegar, honey, wine, cheese, olive oil.

147 A. M. MURTEIRA, *Portalegre, Campo Maior.*—Barley.

148 A. M. PAIVA, *Lisboa, Villa Franca de Xira.*—Wheat.

149 A. DE MATOS, *Portalegre, Campo Maior.*—White wine.

150 A. M. P. CABRAL, *Bragança, Macedo dos Cavalleiros.*—Potatoes.

151 A. M. ALMEIDA, *Lisboa, Alcacer do Sal.*—Wheat flour.

152 A. M. GODINHO, JUN. *Beja, Alvito.*—Potatoes.

153 A. N. DOS REIS, *Lisboa.* — Wheat, wine, cyder.

154 A. P. SALGADO, *Lisboa.*—Wheat.

155 A. P. C. M. CERTA, *Lisboa, Alcacer do Sal.*—Wines and liqueurs.

156 A. P. N. DE VELLEZ, JUN. *Portalegre.* —Potatoes.

157 A. PEREIRA, *Lisboa, Barreiro.* — Wine.

158 A. P. DE CARVALHO, *Faro.*—Wine.

159 A. P. DA SILVA, *Lisboa, Villa Franca de Xira.*—Potatoes.

160 A. PINTO, *Coimbra, Montemór o Velho.* —Wheat.

161 A. PIRES, *Portalegre, Marvão.*—Hazelnuts, wheat.

162 A. R. PEREIRA, *Lisboa, Azambuja.* —Barley, chick-beans, and chick-pease.

163 A. DE SAMPAIO COELHO E SOUSA, *Guarda.*—Olive oil, white wine.

164 A. S. FRANCO, *Portalegre, Fronteira.* —Cheese.

165 A. DE SOUSA ZUZARTE MALDONADO, *Portalegre, Fronteira.*—Corn, wine, cheese, &c.

166 A. T. DE SOUSA, *Vizeu, Lamego.*—Wines.

167 A. T. DE SOUSA FRANCO, *Evora, Portel.* —Olive oil.

168 A. V. DE ALMEIDA FERNANDES, *Santarem, Benavente.*—Chick-beans, wine.

169 A. X. DA SILVA, *Lisboa, Villa Franca de Xira.*—Wheat.

170 A. C. BRANCO, *Lisboa, Alcacer do Sal.* —Wine.

171 A. C. DA COSTA BARBOSA, *Santarem, Azambuja.*—Chick-pease.

172 A. F. BAIRRAO RUIVO, *Santarem, Abrantes.*—Wheat, French beans, maize.

173 A. M. DE ALMEIDA GARCIA FIDIÉ, *Faro.*—French beans, maize, barley, carob.

174 A. PEREIRA, *Santarem, Torres Novas.*—Maize, French beans, barley, chick-beans, &c.

175 A. P. BRETES, *Santarem, Torres Novas.*—White wine, brandy, wheat, almond.

176 A. V. FALCAO, *Fundao.*—Red wine.

177 B. J. C. DAS NEVES E COSTA, *Coimbra, Pampilhosa*—Hams.

178 B. J. DE ASSIS E BRITO, *Beja.*—Wheat.

179 B. J. DE ARAUJO, *Beja, Vidigueira.*—Honey.

180 B. M. DA CUNHA OSORIO, *Portalegre, Elvas.*—Preserved olives, olive oil.

181 BARON DE PRIME, *Vizeu.*—Walnuts.

182 BARON DE SEIXO, *Porto.*—Port wine.

183 BARON DE VIAMONTE DA BOA VISTA.—*Villa Real, Sabrosa.*—Port wine.

184 B. F. JORGE, *Aveiro, Mealhada.*—Wine.

185 B. I. D. DE ALMEIDA, *Coimbra, Tábua.*—Maize, chick-pease, French beans, &c.

186 B. J. DE MELLO PINTO, *Braga.*—Rye, maize, maize flour, French beans, &c.

187 B. COELHO, *Beja, Ourique.*—Honey.

188 B. J. C. DAS NEVES, *Coimbra.*—French beans, chestnuts, honey, oil, wine, &c.

189 B. P. F. T. COELHO, *Braga, Celorico de Basto.*—Wheat, maize, French beans, walnuts.

190 B. P. R. PARENTE, *Portalegre, Fronteira.*—Barley, wheat.

191 B. DO CANTO MEDEIROS, *Ponta Delgada, Villa Franca.*—Wine.

192 B. J. DE SOUSA, *Aveiro, Cambra.*—French beans, potatoes, honey, &c.

193 B. P. DE ARAGAO, *Vizeu, Lamego.*—Maize, arbutus-berry brandy.

194 B. DA P. FIGUEIRA, *Beja, Cuba.*—Wheat, raisins.

195 B. DA SILVA CONSOLADO, *Santarem, Abrantes.*—French beans, walnuts, dried figs.

196 C. J. FERREIRA, *Bragança, Vinhaes.*—Walnuts.

197 C. VELLEZ, *Portalegre, Aviz.*—French beans.

198 C. PECEGUEIRO, *Coimbra, Montemór o Velho.*—Lupines, beans, French beans, chick-beans.

199 C. DA SILVEIRA BARBOSA, *Santarem, Almeirim.*—French beans.

200 C. FERREIRA, *Coimbra.*—Pease.

201 C. M. DA SILVEIRA ALMENDRO, *Santarem, Almeirim.*—Rice.

202 C. PEREIRA, *Santarem, Thomar.*—Dried plums.

203 C. A. CARNEIRO, *Bragança, Freixo de Espada á Cinta.*—Almonds, walnuts, honey, &c.

204 C. A. C. GARCIA, *Bragança.*—Corn, flour, potatoes, &c.

205 C. J. GOMES, *Castello Branco.*—French beans.

206 COLLARES & IRMAO, *Lisboa.*—Preserved food.

207 COMMISSION (CENTRAL), *Aveiro, Ilhavo, Ovar, Feira.*—Cereals, fruits, oils, fish, honey, and other edibles.

208 COMMISSION OF ALANDROAL, *Evora, Alandroal.*—Corn, oil, and cheese.

209 COMMISSION OF THE DISTRICT OF ANGRA DO HEROISMO.—Corn, fruit, potatoes, brandies, &c.

210 COMMISSION (FILIAL) OF BORBA, *Evora.*—Cheese, olive oil, wine.

211 COMMISSION OF THE DISTRICT OF COIMBRA.—Maize, maize flour, vegetables, and fruit.

212 COMMISSION OF THE DISTRICT OF EVORA.—Bacon.

213 COMMISSION OF THE DISTRICT OF FARO.—French beans, and various fruits.

214 COMMISSION OF PONTA DELGADA.—Maize, millet, sweet potatoes, French beans, nuts.

215 COMPANY OF ALTO-DOURO, *Porto.*—Port wines.

216 COMPANY OF TEJO AND SADO LESIRIAS, *Santarem, Benavente, Lisboa.*—Wheat, rye, French beans, lentils, chick-pease, canary-grass, rice, maize.

217 C. DE ALMEIDA A. E SOUSA, *Coimbra.*—French beans, maize, maize flour.

218 C. F. DE MENEZES, *Beja.*—Vinegar, filtered olive oil.

219 C. G. DE CAMPOS, *Portalegre, Aviz.*—Rye, barley, maize, chick-beans.

220 COUNT D'ATALAIA, *Santarem, Almeirim.*—Rye.

221 COUNT D'AZAMBUJA, *Lisboa.*—Port wine.

222 COUNT DA GRACIOSA, *Aveiro, Anadia.*—Orange brandy, orange wine, geropiga, white wine.

223 COUNT DE SAMODAES, *Porto, Lamego. Varzea. Armamar.*—Cereals, fruits, wine, &c.

224 COUNTESS DE SAMODAES, *Porto, Marco de Canavezes.*—Wine and preserved fruits.

225 COUNT DO SOBRAL, *Santarem, Almeirim.*—Honey, wine, wheat, chick-pease.

226 DABNEY, *Horta.*—Wine.

227 D. A. M. DE ALMEIDA, *Portalegre Alter de Chao.*—Maize.

228 D. J. DA COSTA LEAO, *Bragança, Vinhaes.*—Red wine.

229 DELGADO & PEREIRA, *Evora, Montemór o Novo.*—Wine.

230 D. A. TALLÉ RAMALHO, *Evora.*—Cereals, seeds, cheese, wine, dried fruit, &c.

231 D. A. FIUZA, *Evora.*—Wheat, rye, barley.

232 D. A. DE FREITAS, *Coimbra.*—Wheat, flour, biscuits, vermicelli, maccaroni, &c.

233 D. J. FIALHO, *Beja, Alvito.*—Wheat.

234 D. J. GONÇALVES, *Portalegre.*—Pork sausages.

235 D. P. DA SILVA, *Beja, Serpa.*—Wheat.

236 E. A. MORA, *Santarem, Sardoal.*—Wine, olives, fruits, wheat, maize, &c.

237 E. A. DE SOUSA, *Bragança, Vinhaes.*—Brandy, wine, honey, chick-pease, wheat.

238 E. LARCHER, *Portalegre.*—Olive oil, vinegar.

239 E. DO CANTO, *Ponta Delgada.*—Wine.

240 D. E. M. M. SMITH, *Lisboa, Olivaes.*—White wine.

241 E. C. A. C. DE VASCONCELLOS, *Portalegre, Elvas.*—Wine, olives, wheat, &c.

242 F. A. P. DE ARAUJO, *Beja, Serpa.*—Sheep's milk cheese.

243 F. J. F. S. DE CARVALHO, *Beja.*—Wheat flour.

243A F. M. DE AQUINO FIALHO, *Beja, Vidigueira.*—Wheat.

244 F. T. DA SILVA, *Santarem, Thomar.*—Raisins.

245 F. M. S. PINÇAO, *Beja, Aljustrel.*—Olive oil.

246 F. DE SOUSA FERREIRA, *Coimbra.*—Araucaria Brasilienses kernels.

247 F. M. B. PINTO, *Vizeu, Armamar.*—Wine.

248 F. A. GOMES, *Beja, Alvito.*—Honey.

249 F. A. SOBRINHO, *Beja, Alvito.*—French beans, maize, panicum, almonds.

250 F. DA SILVA ROBALLO, *Castello Branco, Idanha a Nova.*—Wheat.

251 F. T. FERRAO, *Vizeu, Carregal.*—Geropiga, wine.

252 F. M. L. GONCALVES, *Beja, Alvito.*—Olive oil, brandy.

253 F. DE C. THEMEZ, *Vizeu, Lamego.*—Honey, wine, olive oil.

254 F. RITA (D.) *Beja.*—Rye.

255 F. SANCHES GUERRA (D.) *Bragança, Freixo de Espada á Cinta.*—Sheep's milk cheese.

256 F. DE ABREU CALADO, *Portalegre, Aviz.*—Wheat, maize, goat and sheep's milk cheese.

257 F. A. DA S. PRADO, *Beja, Odemira.*—Wheat.

258 F. A. BERNARDES, *Bragança, Mogadouro.*—Wheat, barley.

259 F. A. CARNEIRO DE MAGALHAES, *Bragança, Moncorvo.*—French beans.

260 F. A. DINIZ, *Coimbra.*—Almonds.

261 F. A. DA GAMA, *Santarem, Chamusca.*—French beans.

262 F. A. MALATO, *Portalegre.*—Side of bacon.

263 F. A. MARQUES, *Coimbra.*—Walnuts.

264 F. A. MENDES, *Coimbra, Montemor o Velho.*—Wheat.

265 F. A. DA SILVA GRENHO, *Evora, Montemór o Novo.*—Olive oil.

266 F. A. MAGRO, *Castello Branco, Idanha a Nova.*—Honey.

267 F. DE ASSIS CALEJO, *Bragança, Mogadouro.*—Chick-pease.

268 F. DE ASSIS LEDESMA E CASTRO, *Bragança.*—Olives, pease, potatoes, French beans.

269 F. DE ASSIS SOBRINHO *Beja.*—Rice flour.

270 F. BARRETO CASTELLO BRANCO, *Portalegre, Alter do Chao.*—Wheat.

271 F. BERNARDES DE SARAIVA, *Coimbra, Condeixa.*—Wine, pear brandy.

272 F. DE BRITO, *Portalegre.*—Pork sausages.

273 F. CABRAL PACHECO METELLO DE NAPOLES, *Guarda, Celorico.*—Wheat, wine.

274 F. C. GIRALDES, *Castello Branco, Idanha a Nova.*—Rye.

275 F C. DE MORAES CARVALHO E MACHADO, *Bragança, Mogadouro.*—Wine.

276 F. C. N. DE CARVALHO, *Portalegre, Fronteira.*—Wheat.

277 F. CORREIA DA COSTA, *Coimbra.*—Acorns.

278 F. DA COSTA FIALHO, *Beja, Alvito.*—Wheat.

279 F. E. DA SILVA BARROS, *Bragança.*—Wine.

280 F. F. CARNEIRO, *Beja, Vidigueira.*—Wine.

281 F. F. DE OLIVEIRA, *Beja, Serpa.*—Wine.

282 F. DE FREITAS MACEDO, *Santarem.*—Fruit, pickles, wine, oil, and condiments.

283 F. H. RIPADO, *Portalegre, Elvas.*—Olives.

284 F. J. DE SOUSA, *Portalegre, Fronteira.*—Red wine, olive oil.

285 F. J. DA COSTA, *Coimbra.*—French beans, and various liqueurs.

286 F. LEITE BASTO, *Braga, Cabeceiras de Basto.*—Brandy.

287 F. DE LEMOS RAMALHO A. COUTINHO, *Coimbra.*—Sorgho.

288 F. MAGALHAES MASCARENHAS, *Coimbra, Louza.*—Millet.

289 F. M. DE CAMPOS CARVALHO PACHECO, *Lisboa, Alcacer do Sal.*—Wheat.

290 F. M. DA COSTA, *Braga.*—French beans, potatoes.

291 F. M. LOUREIRO, *Beja, Serpa.*—Wild honey.

292 F. M. FELGUEIRAS LEITE, *Bragança, Mogadouro.*—Wheat, barley, potatoes.

293 F. MARQUES DE FIGUEIREDO, *Coimbra.*—Olives, liqueurs, dried fruits, wine, &c.

294 F. M. FERREIRA SERRANO, *Santarem.*—Honey, maize.

295 F. M. PALMA, *Beja.*—Wheat, maize.

296 F. NETO PRATAS, *Santarem, Chamusca.*—Sheep's milk cheese.

297 F. DE PAULA DA FONSECA ESGUELHA, *Lisboa.*—Barley, lentils, canary grass.

298 F. DE PAULA PARREIRA, *Beja, Serpa.*—Wheat, aniseed.

299 F. DE PAULA RISQUES, *Portalegre, Alter do Chao.*—Wheat.

300 F. P. DA VEIGA, *Vizeu, Lamego.*—Wheat, vinegar, olive oil.

301 F. PEREIRA DA COSTA, *Vizeu, Tondella.*—Maize.

302 F. PESSANHA DE MENDONÇA, *Beja.*—Wheat, pease.

303 F. R. DE CARVALHO, *Coimbra.*—Rice.

304 F. R. DA SILVA, *Coimbra, Penella.*—Sheep's milk cheese.

305 F. DA SILVA, *Portalegre, Aviz.*—Chick-beans, oats, maize.

306 F. DA SILVA LOBAO, *Portalegre, Arronches.*—Wheat, chick-pease.

307 F. DA SILVA PAES, *Portalegre, Aviz.*—Wheat, maize.

308 F. DE SOUSA, *Vizeu, Carregal.*—Honey.

309 F. TAVARES DE ALMEIDA PROENÇA, *Castello Branco.*—Wheat, rye, French beans, olives, olive oil.

310 F. T. SEQUEIRA DE SÁ, *Beja, Vidigueira.*—Potatoes.

311 F. V. DA COSTA CARDOSO, *Portalegre, Aviz.*—Cheese.

312 F. X. DE MORAES PINTO, *Bragança, Mirandella.*—Olive oil.

313 F. X. DE MORAES SOARES, *Villa Real, Chaves.*—Cereals, fruit, wine, vegetables, &c.

314 F. X. DA MOTTA PORTOCARRERO, *Santarem, Thomar.*—Olive oil.

315 F. X. NEVES, *Bragança, Mogadouro.* —Wheat.

316 F. DA SILVA OLIVEIRA, *Aveiro, Mealhada.*—Wine.

317 G. ROXO LARCHER, *Portalegre.* —Lard.

318 G. T. DE MAGALHAES COLLAÇO, *Coimbra, Soure.*—Cereals, dried fruits, pickles, &c.

319 G. C. GARCIA, *Beja, Almodovar.*—Barley and cheese.

320 G. A. TEIXEIRA, *Bragança, Vinhaes.* —Wheat, rye, maize, &c.

321 G. F. PEREIRA NUNES, *Coimbra.*—Cereals, maize, seeds, &c.

322 G. C. HERLITZ & Co. *Lisboa.*—Food preserved in olive oil.

323 H. C. DE MACEDO, *Braga, Guimaraes.* —Maize, French beans, wine.

324 H. J. F. DE LIMA, *Bragança.*—Wheat, rye, wine, &c.

325 H. L. DE AGUIAR, *Beja, Cuba.* —Wheat.

326 I.A. DA GAMA, *Santarem, Chamusca.* —Maize.

327 I. F. G. RAMALHO, *Evora.*—Wheat.

328 I. X. DE ORIOL PENA, *Santarem, Torres Novas.*—Olive oil, walnuts.

329 I. L. B. GODIM, *Beja.*—Sweet wine.

330 I. COLLACO, (D.) *Beja.*—Chick-pease.

331 I. M. R. VALENTE, & T. ARCHER, *Porto.*—Port wine.

332 J. R. M. COELHO, *Coimbra, Figueira da Foz.*—Wine.

333 J. BARATA, *Evora, Mora.*—Wine.

334 J. C. DE TORRES, *Evora.*—Wheat.

335 J. G. CURADO E SILVA, *Lisboa, Villa Franca de Xira.*—Chick-pease.

336 J. J. DA MOTTA, *Beja.* —Olive oil, raisins.

337 J. ALVES DE SÁ BRANCO, *Lisboa, Alcacer do Sal.*—Wheat, olive oil.

338 J. A. D. GRANDE, *Portalegre.*—Wheat and other cereals; wine, fruit, brandy, vinegar, olive oil, coffee, honey, &c.

339 J. A. DE CAMPOS, *Bragança, Moncorvo.*—Wheat, maize, almonds, olives, wine.

340 J. A. CARDOSO, *Evora, Extremoz.*—Wheat.

341 J. A. MARQUES ROSADO, *Evora, Redondo.*—Wine.

342 J. A. DE OLIVEIRA E SILVA, *Evora, Montemór o Novo.*—Chick-pease, French beans.

343 J. A. ROCO, *Bragança, Mogadouro.*—Wheat.

344 J. A. RODRIGUES, *Coimbra.*—French beans.

345 J. B. DE CARVALHO, *Bragança.*—Wine.

346 J. B. CASIMIRO, *Bragança, Mirandella.* —Cereals, honey, olive oil, &c.

347 J. B. DUCTEL, *Bragança, Vinhaes.*—Wine.

348 J. BARRETO DA COSTA REBELLO, *Portalegre, Aviz.*—Olive oil.

349 J. BATALHA, *Portalegre, Aviz.*—Rye, cheese.

350 J. B. PACHECO TEIXEIRA, *Braga Celorico de Basto.*—Rye, chick-beans, French beans, millet.

351 J. DE BRITO PIMENTA DE ALMEIDA, *Beja, Moura.*—Wheat, barley, olive oil.

352 J. CARDOSO DE SOUSA, *Portalegre, Alter da Chao.*—Maize, French beans, cheese. Indian-fig brandy.

353 J. CARLOS, *Portalegre, Niza.*—Rye.

354 J. C. DE ARAUJO BASTO, *Braga, Cabeceiras de Basto.*—Wine.

355 J. DO CARMO RAPOSO, *Beja, Moura.*—Olives, olive oil.

356 J. DE CASTRO SAMPAIO, *Braga, Guimaraes.*—French beans.

357 J. DO CARAÇAO DE JESUS FIGUEIREDO, *Vizeu, Carregal.*—French beans, chick-pease.

358 J. DA COSTA CALLADO, *Portalegre, Alter do Chao.*—Wheat.

359 J. DE DEUS, *Portalegre, Aviz.*—Chick-pease, chick-beans.

360 J. FERREIRA DE CARVALHO, *Coimbra.* —Maize flour and maize.

361 J. FERREIRA DE LIMA, *Coimbra.*—Maize and maize flour.

362 J. GONÇALVES VIEIRA, *Braga, Cabeceiras de Basto.*—Wine.

363 J. H. PINHEIRO, *Castello Branco, Idanha a Nova.*—Wheat.

364 J. I. PEREIRA, *Bragança, Vinhaes.*—Chestnuts.

365 J. J. RAMOS, *Evora, Redondo.*—Ears of wheat, French beans, dried plums.

366 J. J. LE COCQ, *Portalegre, Castello de Vide.*—Sparkling and other wines.

367 J. J. DE SOUSA, *Beja, Serpa.*—Maize.

368 J. M. DA COSTA, *Santarem, Almeirim.* —French beans.

369 J. M. DA COSTA BARBOSA, *Santarem, Cartaxo.*—Chick-beans, French beans.

370 J. M. HENRIQUES, *Coimbra.* —Maize flour.

371 J. M. P. VASQUES, *Beja, Cuba.*—Wine.

372 J. M. DA SILVA, *Lisboa.*— Olive oil, and dried plums.

373 J. DE MATOS DE FARIA BARBOSA, *Braga, Barcellos.*—Millet, seeds, confections, wine, &c.

374 J. N. P. GIRAO, *Faro.*—Wheat, dried fruit, wine, olive oil, &c.

375 J. N. R. VALENTE, *Aveiro, Oliveira de Azemeis.*—Wheat.

376 J. N. R. VALLADA, *Aveiro, Oliveira de Azemeis.*—Maize.

377 J. N. DA CONCEIÇAO, *Portalegre, Elvas.*—Preserved fruits.

378 R. DE OLIVEIRA CABECAS, *Braga, Guimaraes.*—Flour.

379 J. P. DE FARIA LACERDA, *Beja, Vidigueira.*—Wheat, olive oil, and olives.

380 J. P. CORDEIRO, *Portalegre, Elvas.* —Brandy, orange wine.

381 J. P. MARTINS, *Lisboa, Setubal.*—Wine.

382 J. P. ROXO, *Portalegre, Portalegre.*—Potatoes, and pickles.

383 J. P. DA SILVA, *Portalegre, Aviz.*—Wheat.

384 J. P. GUIMARAES, *Vizeu, Carregal.*—Common wine.

385 J. P. Frausto, *Portalegre, Marvao.* —Rye, walnuts.

386 J. P. T. Clero, *Evora.*—Vinegar.

387 J. R. de Paiva Lobato, *Portalegre, Fronteira.*—Wheat, olive oil, and cheese.

388 J. R. de Azevedo, *Santarem.*—Wheat.

389 J. de Sacadeira Robe Corte Real, *Vizeu, Nellas.*—Wine.

390 J. S. X. Leitao, *Portalegre, Aviz.*—Ears of maize.

391 J. de Sousa Falçao, *Santarem, Almeirim.*—Olives.

392 J Tavares de Azevedo Lemos, *Braga, Cabeceiras de Basto.*—Dried chestnuts.

393 J. T. P. da Maia, *Evora.*—Rye, wine, brandy.

394 J. V. de Almeida, *Santarem, Benavente.*—Wheat, maize, honey, and wine.

395 J. de Almeida Campos, *Vizeu.* — Hazel-nuts, dried fruits.

396 J. A. de Sá Branco, *Lisboa, Alcacer do Sal.*—Olive oil.

397 J. A. Monteiro, *Portalegre, Monforte.*—Wheat.

398 J. A. Murteira, *Portalegre, Campo Maior.*—Wheat.

399 J. A. Pereira de Matos, *Faro.*—Dried figs and fig brandy.

400 J. A. Ribeiro, *Lisboa, Alcacer do Sal,* —Red wine and rice.

401 J. A. Simoes, *Coimbra, Figueira da Foz.*—Olive oil, wine, brandy.

402 J. A. da Silva Cordeiro, *Santarem.* —Red wine.

403 J. da Costa, *Beja, Cuba.*—Maize.

404 J. da Cruz Freire, *Coimbra, Cantanhede.*—Red wine.

405 J. F. da Cunha Osorio, *Portalegre, Fronteira, Arronches.*—Wheat, olive oil.

406 J. F. Fernandes, *Beja.*—Wheat and other cereals, cheese, wine, and brandy.

407 J. F. Levita, *Portalegre.*—Potatoes.

408 J. Gavino de Vasconcellos, *Santarem, Gollega.*—Wheat, maize, lupines, and beans.

409 J. I. Cabrito, *Beja, Cuba.*—Red wine.

410 J. I. de Saldanha Machado, *Santarem, Benavente.*—Wheat, barley, maize, and French beans.

411 J. José, *Portalegre, Elvas.*—Wheat.

412 J. J. de Araujo, *Lisboa, Barreiro.*—Wine.

413 J. J. de Castro, *Portalegre, Aviz.*—Wheat and chick-beans.

414 J. J. da Costa, *Portalegre, Elvas.*—Wheat.

415 J. J. Maria, *Santarem.*—Wheat.

416 J. M. Clemente, *Portalegre, Monforte.*—Wheat.

417 J. M. Penteado, *Portalegre, Aviz.*—Cheese.

418 J. M. Telles, *Portalegre, Aviz.* — Mixed wheat, olive oil.

419 J. M. Camacho, *Beja, Alvito.*—Honey, olive oil.

420 J. M. Ferreira Pestana, *Coimbra, Figueira da Foz.*—Wheat, maize, white wine, brandy.

421 J. M. Rodrigues, *Coimbra.* — Pineapple kernels.

422 J. M. R. de Brito, *Coimbra.*—Pease French beans, and maize.

423 J. M. da Silva, *Evora, Extremoz.*—Olive oil.

424 J. M. L. de Carvalho, *Lisboa, Alemquer.*—Wine.

425 J. P. Duarte, *Beja.*—Chick-pease.

426 J. Pereira Claro, *Portalegre, Arronches.*—Cheese.

427 J. Pereira da Costa, *Portalegre Fronteira.*—Wheat.

428 J. Ribeiro do Amaral, *Coimbra Oliveira do Hospital.*—Dried pears.

429 J. R. do Nascimento, *Coimbra.*—Dried chestnuts.

430 J. Sotero Soares Couceiro, *Coimbra, Montemór o Velho.*—Orange vinegar.

431 J. de Sousa Guimaraes, *Porto.*—Port wine.

432 J. U. Peres, *Coimbra, Penella.* — Olive oil, arbutus-berry brandy.

433 J. Ferrao Castello Branco. — Bucellas wine.

434 J. Maxima, *Coimbra, Soure.*—Dried cherries.

435 J. de Aguilar, *Porto, Pesqueira.*—Wine.

436 J. A. de Lemos Trigueiros, *Vizeu S. Joao das Areias.*—Olive oil.

437 J. de Andrade, *Portalegre, Arronches.*—Wheat and chick-pease.

438 J. A. da Cruz Camoes, *Evora.*—French beans, cheese, pork sausages.

439 J. A. de Castro, *Portalegre, Fronteira.*—Wheat and chick-pease.

440 J. A. Garcia Blanco, *Faro, Silves.* —Wheat and other cereals.

441 J. A. Junqueiro, Jun. *Bragança.*—Wine.

442 J. A. Lopes Maia, *Braga.* — Brandy.

443 J. A. da Silva, *Santarem, Abrantes.* —Wine.

444 J. A. de Oliveira, *Coimbra.* — Brandy.

445 J. A. Mendes Pereira, *Guarda, Pinhel.*—Wine.

446 J. A. S. de Aboim, *Beja.*—Wheat.

447 J. de Beires, *Vizeu, Lamego.* — Wheat and maize.

448 J. do Canto, *Ponta Delgada.*—Arrowroot starch.

449 J. C. Pucci, *Lisboa.*—Wine, liqueurs, potatoes, confections, olive oil, &c.

450 J. Carrilho Garcia, *Beja, Almodovar.*—Wheat, lupines, oats, and honey.

451 J. Carvajal Vasconcellos Gama, *Portalegre, Campo Maior.*—Cheese.

452 J. da Conceiçao Guerra, *Portalegre, Elvas.*—Preserved fruit.

453 J. da Costa, *Portalegre, Fronteira.*—Wheat.

454 J. da Costa Malhau, *Santarem, Almeirim.*—White wine and chick-beans.

455 J. Correia Monteiro Gorjao, *Santarem, Torres Novas.*—Wine.

456 J. Cupertino da Fonseca e Brito, *Coimbra, Arganil.*—Dried pears and plums.

457 J. D. de Carvalho Monteiro. *Coimbra.*—Potatoe flour.

458 J. Diogo, *Portalegre, Alter do Chao.*—Wheat.

459 J. D. Fazenda, *Beja, Vidigueira.*—Wine and brandy.

460 J. Eschrich, *Lisboa.*—Wine.

461 J. Farinha Relvas de Campos, *Santarem, Torres Novas.*—White wine.

462 J. Ferreira da Silva, *Lisboa.*—Wine, liqueurs, and potatoe-starch.

463 J. Fialho Coelho, *Beja, Moura.*—Olive oil and oats.

464 J. F. Levita, *Portalegre.*—Red wine.

465 J. F. da Cruz, *Coimbra.*—Wheat, flour, and biscuits.

466 J. F. da Gama Freixo, *Evora.*—Red wine.

467 J. F. da Silva, *Beja.*—Quince brandy, lentils, and beans.

468 J. da Gama Caldeira Jun. *Portalegre, Crato.*—Chick-pease.

469 J. Gonçalves, *Coimbra.*—Maize ears.

470 J. Gonçalves da Cruz Viva, *Faro.*—Maize, chick-beans, and brandies.

471 J. Gonçalves de S. Thiago, *Santarem, Almeirim.*—Rice.

472 J. Guedes Coutinho Garrido, *Coimbra, Penella.*—Maize flour.

473 J. Guerreiro da Lança Sobrinho, *Beja, Ferreira.*—Wheat and red wine.

474 J. I. Pinto Guerra, *Bragança, Miranda do Douro.*—Rye, chick-pease, and almonds.

475 J. Jeronymo de Faria, *Portalegre, Aviz.*—Wheat.

476 J. J. Cardoso, *Portalegre.*—Side of bacon, and sausages.

477 J. J. de Carvalho, *Evora, Extremoz.*—Red wine.

478 J. J. Correia, *Ponta Delgada, Ribeira Grande.*—Wine.

479 J. J. da Costa, *Lisboa, Alemquer.*—Olive oil.

480 J. J. da Costa, *Braga, Guimaraes.*—Vinegar and pease.

481 J. J. Fiusa Guiao, *Evora, Montemor o Novo.*—French beans.

482 J. J. G. A. de Vilhena, *Beja, Ferreira.*—Wheat.

483 J. J. Lampreia, *Beja, Vidigueira.*—Brandy.

484 J. J. Pinto Guerra, *Bragança, Miranda do Douro.*—Wheat.

485 J. J. Ramos, *Evora.*—Wine.

486 J. J. Rodrigues, *Bragança, Vinhaes.*—Potatoes.

487 J. J. da Silveira, *Santarem.*—Wheat, barley, and French beans.

488 J. J. do Valle, *Vizeu, Santa Comba Dao.*—Wheat, rye, and barley.

489 J. J. Marçal, *Coimbra, Montemor o Velho.*—Rice.

490 J. Leal de Gouveia Pinto, *Coimbra, Miranda do Corvo.*—Maize, maize flour, and olive oil.

491 J. Leite Gonçalves Basto, *Braga, Cabeceiras de Basto.*—Wine.

492 J. de Lima Guimaraes, *Santarem, Gollega.*—Wheat, maize, and French beans.

493 J. L. Coelho, *Portalegre, Aviz.*—Rye.

494 J. L. Guimaraes, *Coimbra.*—Brandy and wine.

495 J. Lopes Serra, *Coimbra, Montemór o Velho.*—Cereals, potatoes, and wine.

496 J. L. da Costa e Silva, *Beja, Vidigueira.*—White wine and brandy.

497 J. M. Ferreira, *Bragança, Vinhaes.*—Rye.

498 D. J. M. de Menezes de Alarcao, *Santarem, Coruche.*—Wheat, rice, acorns, marmalade, and confections.

499 J. M. do Monte, *Evora, Redondo.*—Wheat and other cereals; honey and wine.

500 J. M. A. Cardoso, *Portalegre.*—Beans, fruits, honey, and olive oil.

501 J. M. Ayres de Seixas, *Portalegre, Gaviao.*—Red wine.

502 J. M. de Brito, *Beja, Odemira.*—Maize and French beans.

503 J. M. do Couto Gançoso, *Evora.*—Honey.

504 J. M. de Figueiredo Antas, *Bragança, Vimioso.*—Rye.

505 J. M. de Figueiredo, *Guarda, Gouveia.*—Cheese.

506 J. M. da Fonseca, *Lisboa.*—Wine.

507 J. M. Henriques, *Coimbra, Poiares.*—Maize, honey, and olive oil.

508 J. M. L. Falcao, *Beja, Odemira.*—Maize.

509 J. M. de Matos, *Portalegre, Campo Maior.*—Chick-pease.

510 J. M. da Motta Cerveira, *Santarem, Almeirim.*—French beans.

511 J. M. Raposo, *Ponta Delgada.*—Wine.

512 J. M. Roxo Larcher, *Portalegre.*—Gin and pickled olives.

513 J. M. de Sá Pereira e Moura, *Santarem, Benavente.*—Wheat.

514 J. M. da Silva Reguinga, *Santarem, Almeirim.*—Beans.

515 J. M. Correia Bello, *Faro.*—Common wine.

516 J. M. Godinho, *Beja, Alvito.*—Aniseed, fruit, honey, olive oil, wine, and brandy.

517 J. M. Leitao, Sen. *Beja, Vidigueira.*—Wheat and wine.

518 J. M. Leitao Sobrinho, *Beja, Vidigueira.*—Wheat and wine.

519 J. M. Lopes, *Faro, Olhao.*—Dried figs.

520 J. de Matos Machado, *Portalegre, Ponte de Sor.*—Maize and rice.

521 J. M. da Silva Azevedo, *Belmonte, Castello Branco.*—French beans, olive oil.

522 J. de Mello Pitta, *Vizeu, Tarouca.*—Cereals, raisins, and wine.

523 J. Militao de C. e Sousa, *Beja.*—Canary-grass.

524 J. de Moraes Pinto de Almeida, *Coimbra.*—Carolina rice.

525 J. N. da Motta, *Coimbra, Penella.*—Cheese.

526 J. do Nascimento Noronha, *Vizeu Lamego.*—Wine.

527 J. P. Cortez de Lobao, *Beja, Serpa.*—Wheat, olive oil, and vinegar.

528 J. P. Leal, *Beja.*—Wheat.

529 J. P. de Mira, *Evora, Redondo.*—Wheat, wheat-ears, and oats.

530 J. P. Burguete de Magalhaes e

OLIVEIRA, *Santarem, Abrantes*.—Vinegar and wine.

531 J. P. DA SILVA, *Portalegre, Aviz*.—Wheat, oats, and chick-beans.

532 J. PERFEITO PEREIRA PINTO, *Vizeu, Lamego*.—Rye, French beans, and wine.

533 J. PIRES DE CARVALHO, *Portalegre, Aviz*.—Rice.

534 J. QUIRINO THADEO DE ALMEIDA, *Beja, Ourique*.—Wheat and olive oil.

535 J. R. CORTEZ DE LOBAO, *Beja, Serpa*.—Wheat.

536 J. R. L. DE CARVALHO, *Santarem, Torres Novas*.—Raisins.

537 J. RILVAS, *Portalegre, Fronteira*.—Wheat.

538 J. RIBEIRO MACHADO GUIMARAES, *Coimbra*.—Maize and pearl barley.

539 J. ROBALLO, *Castello Branco, Fundao*.—Olive oil.

540 J. R. P. BASTO, *Aveiro, Oliviera de Azemeiz*.—Wheat, wheat-ears, maize, and pine-apples.

541 J. R. DE OLIVEIRA, *Santarem, Torres Novas*.—Dried figs.

542 J. S. DE BRITO, *Coimbra, Tábua*.—Dried plums and figs.

543 J. SERRAO DO VALLE, *Beja, Odemira*.—French beans.

544 J. DA SILVA, *Lisboa, Alcacer do Sal*.—French beans, and rice.

545 J. SOARES MASCARENHAS, *Faro*.—Wine, dried plums, geropiga, and vinegar.

546 J. SOARES TEIXEIRA DE SOUSA, *Angra do Heroismo, Vélas*.—Maize and wine.

547 J. DE SOUSA FALCAO, *Santarem, Constancia*.—Olive oil and wine.

548 J. DE VASCONCELLOS NORONHA, *Vizeu, Lamego*.—Wheat.

549 J. V. DE ALMEIDA, *Portalegre, Arronches*.—Cheese.

550 J. G. DE ALMENDRO, *Bragança, Vinhaes*.—Red wine.

551 J. A. BORGES DA SILVA, *Bragança, Vinhaes*.—Wheat, French beans, and ham.

552 J. COELHO PALHINHA, *Evora, Montemór o Novo*.—Honey.

553 J. M. BAIAO MATOSO, *Beja, Vidigueira*.—Honey and cheese.

554 KEMPES & Co. *Lisboa*.—Olive oil.

555 L. RODRIGUES, *Bragança, Vinhaes*.—Red wine.

556 L. V. PEREIRA PINTO GUEDES, *Bragança, Vinhaes*.—French beans and walnuts.

557 L. X. DE FIGUEIREDO, *Coimbra, Cantanhede*.—Brandy and wine.

558 L. PEREIRA DE CASTRO, *Braga, Cabeceiras de Basto*.—Wine.

559 L. DE ABREU MAGALHAES E FIGUEIREDO, *Guarda, Ceia*.—Dried fruits.

560 L. A. DIAS, *Coimbra, Miranda do Corvo*.—Barley.

561 L. A. DE MAGALHAES, *Castello Branco, Fundao*.—Wine.

562 L. C. FERREIRA, *Porto, Figueira de Castello Rodrigo*.—Preserved olives, olive oil, almonds, and wine.

563 L. CARDOSO DE ALARCAO, *Coimbra, Penella*.—Maize and maize flour.

564 L. J. RODRIGUES, *Santarem, Benavente*.—Wine.

565 L. J. DA ROSA LIMPO, *Portalegre*.—Salep-root flour.

566 L. M. DO SIDRAL, *Portalegre, Alter do Chao*.—Wheat.

567 L. DE PINA CARVALHO FREIRE FALCAO, *Castello Branco*.—Cheese.

568 L. PINTO TAVARES, *Castello Branco, Fundao*.—Olive oil.

569 L. X. DE BARROS CASTELLO BRANCO, *Portalegre*.—Olive oil.

570 M. J. TAVARES MENDES VAZ, *Coimbra, Montemór o Velho*.—Wine, geropiga.

571 M. M. DA TERRA BRUM, *Horta*.—Wine

572 M. S. MACEDO, *Angra do Heroismo, Vélas*.—Butter.

573 M. A. PEREIRA, *Faro*.—Wheat.

574 M. A. DO RIO, *Lisboa, Belem*.—Olive oil.

575 M. A. GIRALDES, *Bragança, Vimioso*.—Chick-pease and wheat.

576 M. A. L. NAVARRO, *Bragança, Freixo de Espada á Cinta*.—Sausages and raisins.

577 M. A. DE ALMEIDA VALLEJO, *Santarem, Abrantes*.—Pine-apple kernels.

578 M. B. PESTANA GONTAO, *Portalegre, Niza*.—Wheat and other cereals, wine, and cheese.

579 M. BRANDAO, *Coimbra, Cantanhede*.—Wheat, and red wine.

580 M. C. CABRAL VELHO DE LEMOS CALHEIROS, *Santarem, Benavente*.—Wheat and lentils.

581 M. DE CARVALHO, *Coimbra*.—Maize and maize flour.

582 M. C. G. COELHO, *Lisboa*.—Acorn coffee.

583 M. CORDEIRO, *Coimbra, Montemór o Velho*.—Husked barley.

584 M. DA CRUZ AMANTE, *Coimbra*.—Wheat and wheat flour.

585 M. D. BAPTISTA, *Portalegre, Ponte de Sor*.—Maize and husked rice.

586 M. DAS DORES NUNES, *Portalegre, Elvas*.—Wine.

587 M. FERREIRA DE AZEVEDO, *Aveiro, Mealhada*.—Wine.

588 M. FERREIRA BRETES, *Santarem, Torres Novas*.—Wine.

589 M. FREIRE DE ANDRADE, *Coimbra*.—Vinegar.

590 M. GAIFAO BELLO, *Santarem, Macao*.—Wheat, maize, honey, olive oil, and wine.

591 M. GOMES, *Santarem, Thomar*.—Wine.

592 M. G. GAMEIRO CARDOSO, *Santarem, Torres Novas*.—Dried figs and raisins.

593 M. GUERRA *Bragança*.—Dried plums and brandy.

594 M. DE GUIMARAES DE ARAUJO PIMENTEL, *Braga, Celorico de Basto*.—Wine.

595 M. H. DE M. FEIO, *Beja*.—Aniseed.

596 M. J. MOCINHA, *Portalegre, Campo Maior*.—Rye.

597 M. J. FIALHO TOJA, *Evora, Portel*.—Cheese.

598 M. J. GUERRA, *Bragança Frexio de Espada á Cinta*.—Almonds.

599 M. J. DE OLIVEIRA, *Bragança, Miranda do Douro*.—Alcohol, sausages.

600 M. J. DE OLIVEIRA ESTACAL, *Lisboa, Barreiro.*—Wine.

601 M. J. DE BIVAR GOMES DA COSTA, *Faro.*—Wheat and other cereals, fruits, wine, olive oil, &c.

602 M. J. COUTINHO, *Santarem, Almeirim.*—Wheat.

603 M. J. FERREIRA DA SILVA GUIMARAES, *Braga, Guimaraes.*—Lupines, French beans, and maize.

604 M. J. DE PINHO SOARES DE ALBERGARIA, *Leiria.*—Honey, olive oil.

605 M. J. VIEIRA DE NOVAES, *Lisboa, Setubal.*—French beans, lupines, and pease.

606 M. L. VARELLA, *Evora, Arraiollos.*—Cheese.

607 M. L. AÇO, *Beja.*—Barley.

608 M. LUIZ, *Coimbra.*—Chick-pease and chick-beans.

609 M. DE MAGALHAES COUTINHO, *Coimbra, Cantanhede.*—Wine.

610 M. M. CALISTO, *Coimbra.*—French beans.

611 M. MARQUES, *Santarem, Thomar.*—Wine.

612 M. MARQUIS DE FIGUEIREDO, *Coimbra.*—Beans, maize, panicle, seed, and vinegar.

613 M. MARQUES & IRMA, *Bragança, Freixo de Espada á Cinta.*—Almonds in confection.

614 M. M. DA SILVA, *Lisboa, Villa Franca de Xira.*—Wheat.

615 M. NUNES MOUZACO, *Castello Branco, Belmonte.*—Wheat and French beans.

616 M. NUNES SERRAO, *Beja, Alvito.*—Rye, maize, and French beans.

617 M. P. DE OLIVEIRA, *Bragança.*—Wine.

618 M. DE OLIVEIRA, *Santarem, Chamusca.*—Rice.

619 M. PINTO DE ALBUQUERQUE, *Coimbra, Arganil.*—Olive oil.

620 M. DOS PRAZERES E SILVA, *Braga, Guimaraes.*—French beans and chestnuts.

621 M. A. CAIEIRO, *Beja, Serpa.*—Red wine.

622 M. ANTUNES, *Coimbra, Miranda do Corvo.*—Honey.

623 M. C. DE ASSIS ANDRADE (D.), *Beja.*—Walnuts.

624 M. CANDIDA DA FONSECA (D.), *Vizeu.*—Sweet-meats.

625 M. DA CONCEIÇAO, *Braga.*—Confections.

626 M. G. FRANCO (D.), *Portalegre, Fronteira.*—Wheat.

627 M. J. FERRAO CASTELLO BRANCO (D.), *Coimbra, Louza.*—Walnuts and dried plums.

628 M. DA NAZARETH, *Coimbra.*—Cereals and brandy.

629 M. R. FERREIRA DE CASTRO, *Braga, Guimaraes.*—Wine.

630 M. J. DE SOUSA FEIO, *Beja.*—Wheat, chick-beans, chick-pease, maize, cheese, and olive oil.

631 M. J. DE MEDEIROS, *Ponta Delgada, Villa Franca.*—Dragon-wort starch.

632 MARQUIS OF ALVITO, *Beja, Alvito.*—Olives.

633 M. J. RAPOSO, *Beja, Moura.*—Wheat, oats, and cheese.

634 M. MARQUIS AYRES DE SEIXAS, *Portalegre, Gaviao.*—Maize and rice.

635 M. A. MALHEIROS, *Porto.*—Wine.

636 M. A. MARRECO, *Coimbra, Miranda do Corvo.*—Maize and maize flour.

637 M. C. MARTINS, *Portalegre, Aviz.*—Brandy.

638 M. J. DA FONSECA ESGUELHA, *Lisboa, Villa Franca de Xira.*—Maize.

639 M. L. DA SILVA ATHAIDE, *Leiria.*—Maize and wine.

640 M. OSORIO CABRAL DE CASTRO, *Coimbra.*—Cereals, fruits, wine, and olive oil.

641 M. R. DE CARVALHO, *Portalegre.*—Wine.

642 N. J. PEDROSO, *Santarem, Chamusca.*—Maize, French beans, and wheat.

643 N. OF BELLAS, *Coimbra.*—Preserved fruits.

644 N. OF CELLAS, *Coimbra.*—Preserved apricots.

645 N. OF FERREIRA, *Vizeu.*—Preserved plums.

646 N. OF S. DOMINGOS, *Braga, Guimaraes.*—Preserved fruits.

647 N. DE SANT' ANNA, *Coimbra.*—Preserved fruits.

648 N. DE SANTA ROSA DE GUIMARAES, *Braga, Guimaraes.*—Preserved fruits.

649 N. DE SEMIDE, *Coimbra, Miranda do Corvo.*—Confection of turnip.

650 N. C. DE MATOS FERRAO CASTELLO BRANCO, *Coimbra, Louza.*—Maize, maize flour, and olive oil.

651 O. S. LEITE, *Lisboa.*—Dried fruit.

652 P. FERREIRA, *Bragança.*—Confection.

653 P. J. DE MESQUITA, *Coimbra, Tábua.*—Olive oil.

654 P. L. DE OLIVEIRA VELHO, *Santarem, Abrantes.*—French beans.

655 P. M. COELHO MACHADO, *Portalegre.*—Cheese.

656 P. VIEIRI GORJAO, *Santarem.*—Olive oil.

657 P. A. DA SILVA REBELLO COELHO VASCONCELLOS MAIA, *Braga, Povoa de Lanhoso.*—Wheat, barley, chestnuts, honey, olive oil, and wine.

658 R. LARCHER, *Portalegre.*—French beans.

659 R. B. SOBRINHO, *Beja, Alvito.*—Wheat and chick-beans.

660 R. GONÇALVES MONIZ, *Beja, Alvito.*—Wheat, honey, and brandy.

661 R. J. SOARES MENDES, *Santarem, Abrantes.*—Olive oil and wine.

662 R. WIGHAM & Co. *Porto.*—Wine.

663 R. PEREIRA MENDES, *Santarem, Thomar.*—French beans and maize.

664 R. SOARES CASTELLO BRANCO, *Faro, Lagoa.*—Dried figs.

665 R. DE VITERBO (D.), *Vizeu, Lamego.*—Pears.

666 S. PIRES DE OLIVEIRA, *Lisboa.*—Red wine.

667 S. GIL TOJO, *Beja, Vidigueira.*—White wine.

668 S. GIL TOJO BORJA DE MACEDO, *Evora, Portel.*—Wine.

669 S. J. DA GUERRA, *Bragança, Freixo de Espada á Cinta.*—Wheat and olives.

670 S. DE MELLO FALCAO TRIGOSO, *Lisboa, Torres Vedras.*—Wheat, olive oil, maize, and wine.

671 S. PINHEIRO, *Portalegre, Campo Maior.*—Red wine.

672 S. REI, *Coimbra.*—Rice.

673 S. A. DA CUNHA, *Santarem.*—French beans.

674 S. FARIA GARCIA, *Lisboa, Alcacer do Sal.*—Honey.

675 T. A. DE CARVALHO E ALMEIDA, *Braga, Cabeceiras de Basto.*—Wine.

676 T. DA COSTA, *Coimbra, Tábua.*—Dried pears.

677 T. GUERREIRO, *Beja.*—Wheat.

678 T. J. DUARTE, *Coimbra, Figueira da Foz.*—Butter.

679 T. J. DA SILVA & Co. *Angra do Heroismo.*—Maize.

679A TOBACCO Co. *Lisboa.* — Snuff and cigars.

680 V. B. PIRES, *Faro.*—Cereals, fruits, honey, wine, &c.

681 V. B. PIRES, JUN. *Faro.*—Walnuts and fig brandy.

682 V. J. DE ALCANTARA, *Coimbra, Oliveira do Hospital.*—Cheese.

683 V. J. DIAS, *Coimbra, Miranda do Corvo.*—Honey-comb.

684 V. A. MONTEIRO, *Santarem, Chamusca.*—Chick-pease.

685 VISCOUNT DE BRUGES, *Angra do Heroismo.*—Cereals, butter, confection, brandy, &c.

686 VISCOUNT DA ESPERANÇA, *Beja, Cuba.*—Cereals, fruits, cheese, honey, wine, brandy, &c.

687 VISCOUNT DA FOZ, *Evora, Extremoz.*—Red wine.

688 VISCOUNT DE GUIAES, *Vizeu, Lamego.*—Honey, olive oil, and wine.

689 VISCOUNT DE OLEIROS, *Castello Branco.*—Cheese.

690 VISCOUNT DE SÁ, *Santarem, Almeirim.*—Wheat, rye, maize, French beans, and barley.

691 VISCOUNT DE TAVEIRO, *Coimbra.*—Cereals, fruits, seeds, &c.

692 VISCOUNT DO TORRAO, *Beja, Alvito.*—Wheat and rice.

693 VISCOUNTESS DE ALPENDURADA, *Porto.*—Red wine.

694 VISCOUNTESS DE FONTE BOA, *Santarem.*—Wheat and olive oil.

695 VISCOUNTESS DA VARZEA, *Vizeu. Lamego.*—Dried plums and figs.

696 V. J. PEREIRA FRANCO, *Guarda.*—Wheat, rye, barley, beans, cherries, vinegar, and olive oil.

697 W. M. DE CARVALHO, *Coimbra.*—Walnuts.

CLASS 4.

698 ABBOT DE CRESPOS, *Braga.*—Flax.

698A A. CAETANO, *D'Oliveira, Bragança.*—Soap.

699 AGRICULTURAL INSTITUTE OF LISBOA.—Portuguese wool and silk.

700 A. FERNANDES, *Bragança, Vinhaes.*—Flax.

700A A. MOREIRA DOS SANTOS, *Porto.*—Soap.

701 A. P. PEREIRA (D.) *Coimbra, Louza.*—Potatoe starch.

702 A. A. RIBEIRO, *Bragança, Macedo de Cavalleiros.*—Cocoons.

703 A. C. BANHA, *Beja.*—Cocoons.

704 A. I. PEREIRA, *Evora, Redondo.*—Virgin and white wax.

705 A. J. DURAES CASTANHEIRA, *Vizeu, Carregal.*—Black wool.

706 A. J. GONÇALVES BRANCO, *Alcacer do Sal, Lisboa.*—Cork.

707 A. J. DA SILVA FRANCO, *Leiria, Peniche.*—Archil.

708 A. M. PACHECO RODRIGUES, *Beja, Serpa.*—Wool.

709 B. PERES, *Beja, Ferreira.*—Virgin wax.

710 B. J. PINTO DA MOTTA, *Coimbra.*—Pitch.

711 B. J. DE SOUSA, *Aveiro, Cambra.*—Wax, goat's hair, raisins, and berries.

712 C. PECEGUEIRO, *Coimbra.*—Linseed.

713 C. J. DE MATOS VEIGA, *Beja.*—Macereed flower.

714 CARRILHO, J. G. *Beja, Almodovar.*—Wax.

715 C. J. MORA E VAZ, *Bragança, Mogadouro.*—Wood.

716 C. A. C. GARCIA, *Bragança.*—Sumach and cocoons.

717 COMMISSION OF THE ALANDROAL, *Evora, Alandroal.*—Mace-reed.

718 COMMISSION OF THE DISTRICT OF ANGRA DO HEROISMO.—Mat-weeds.

719 COMMISSION (CENTRAL) PORTUGUESE, *Aveiro, Cambra Vagos.*— Beehive, linseed, Armenian bole, wool, and seeds; mace-reed, thistle, and clay.

720 COMMISSION OF THE DISTRICT OF COIMBRA.—Wood, lentisk, and linseed.

721 COMMISSION OF THE DISTRICT OF EVORA.—Wood.

722 COMMISSION OF THE DISTRICT OF FARO.—Saffron and cork.

723 COMMISSION OF THE DISTRICT OF GUARDA, *Subugal.*—Flax.

724 COMMISSION OF PONTA DELGADA.—Wood and bark.

725 COUNT DE ARROCHELLA, *Braga Guimaraes.*—Wood and cork.

726 COUNT DE SAMODAES, *Porto, Lamego*—Silk and cocoons, vine ashes, elder-berries.

727 COUNT DO SOBRAL, *Santarem, Almeirim.*—Wax and olive husks.

728 D. J. DA CUNHA GUIMARAES, *Coimbra*—Ox-horn.

729 D. J. DA COSTA LEAO, *Bragança Vinhaes.*—Cocoons.

730 DELGADO & PEREIRA, *Evora, Monte mór o Novo.*—Woollen fleece.

731 D. A. TALLÉ RAMALHO, *Evora, Redondo.* — Woollen fleeces, and thistle to curdle milk.

732 D. GARCIA, *Faro, Silves.*—Cork,

733 D. M. DE SÁ MORAES, *Bragança Vinhaes.*—Wool.
734 E. A. DE SOUSA, *Bragança, Vinhaes.*—Hops and wax.
735 E. LARCHER, *Portalegre.*—Sumach.
736 F. M. S. PINÇAO, *Beja, Aljustrel.*—Linseed.
737 F. DE SOUSA ROMEIRAS, *Evora, Montemór o Novo.*—Cork tree bark.
738 F. A. GOMES, *Beja Alvito.*—Virgin-wax.
739 F. CABRAL PAES PINTO, *Vizeu, Sernancelhe.*—Raw silk and cocoons.
740 F. DE FREITAS MACEDO, *Santarem.*—Linseed, lavender, and elder berry.
741 F. J. WENCESLAU, *Portalegre, Gaviao.*—Juniper berries.
742 F. MAIA, *Coimbra, Montemór o Velho.*—Linseed.
743 F. DE MAGALHAES, *Coimbra.*—Cork.
744 F. M. MAGALHAES MASCARENHAS, *Coimbra, Louza.*—Cork and flax.
745 F. M. FELGUEIRAS LEITE, *Bragança, Mogadouro.*—White wool, wood, and cork.
746 F. MARQUIS DE FIGUEIREDO, *Coimbra.*—Flax.
747 F. T. DE ALMEIDA PROENÇA, *Castello Branco.*—Black and white wool.
748 GENERAL PORTUGUESE WOOD ADMINISTRATION, *Leiria.*—Carbonized and resinous billet, pitch, tar, colophany, and wood.
749 ROBINSON, G. *Portalegre.*—Cork.
750 G. T. DE MAGALHAES COLLAÇO, *Coimbra, Soure, Louza.*—Juniper, aloe, mastich-tree leaves, and hops.
751 G. F. PEREIRA NUNES, *Coimbra, Oliveira do Hospital.*—Ray-grass.
752 J. R. M. COOK, *Coimbra.*—Cork.
753 J. J. DA MOTTA, *Beja.*—Cocoons.
754 J. A. DIAS GRANDE, *Portalegre.*—Cork-tree bark, virgin-wax, wood, linseed, and juniper berries.
755 J. A. DE CAMPOS, *Bragança, Moncorvo.* Flax, hemp, and wood.
756 J. A. GARCIA, *Coimbra.*—Madder.
757 J. B. CASIMIRO, *Bragança, Mirandella.*—Wood.
758 J. B. DE MIRA, *Beja.*—Cocoons.
759 J. M. AFFONSO, *Bragança, Mogadouro.*—Linseed.
760 J. DE MATOS DE FARIA BARBOSA, *Braga, Barcellos.*—Woollen fleeces.
761 J. N. PESTANA GIRAO, *Faro.*—Guinea aloe, cotton seeds.
762 J. N. REBELLO VALENTE, *Aveiro, Oliveira de Azemeis.*—Wood.
763 J. A. PEREIRA DE MATOS, *Faro.*—Virgin and white wax.
764 J. FERREIRA, *Coimbra, Louza.*—Willow bark.
765 J. GONÇALVES FINO, *Coimbra.*—Wood.
766 J. P. DA SILVA PINÇAO, *Beja, Aljustrel.*—Flax.
767 J. ALVES, *Portalegre, Gaviao.*—Cork.
768 J. A. C. DE CARVALHO, *Beja, Aljustrel.*—Flax.
769 J. A. SOUSA, *Portalegre.*—Pastil.
770 J. A. DA CRUZ CAMOES, *Evora.*—Woollen fleece.
771 J. A. DIAS ROMANO, *Bragança, Vimioso.*—Flax.
772 J. B. TEIXEIRA DE ALMEIDA, *Bragança, Mogadouro.*—Wood.
773 J. DO CANTO, *Ponta Delgada.*—Archil.
774 J. CARVAJAL VASCONCELLOS GAMA, *Portalegre, Campo Maior.*—Woollen fleece.
775 J. D. DE CARVALHO MONTENEGRO, *Coimbra, Louza.*—Wood.
776 J. J. DA COSTA, *Braga, Guimaraes.*—Linseed.
777 J. LEAL GOUVEIA PINTO, *Coimbra, Miranda do Corvo.*—Cork, lees of olive oil.
778 J. M. LOPES FALCAO, *Beja, Odemira.*—Cork.
779 J. M. DE ALARCAO E MENEZES (D.), *Santarem, Coruche.*—Bark, wood, and cork.
780 J. M. FERREIRA, *Bragança, Vinhaes.*—Cocoons.
781 J. M. HENRIQUES, *Coimbra, Poiares.*—Wax.
782 J. M. CORREIA BELLES, *Faro.*—Nankin cotton.
783 J. PARREIRA CORTEZ DE LOBAO, *Beja, Serpa.*—Woollen fleece.
784 J. R. P. BASTO, *Aveiro, Oliveira de Azemeis.*—Woollen fleece.
785 J. SERRAO DO VALLE, *Beja, Odemira.*—Cork.
786 J. DA SILVA, *Lisboa, Alcacer do Sal.*—Cork.
787 J. SOARES TEIXEIRA DE SOUSA, *Angra do Heroismo, Vélas.*—Wool.
788 J. T. RABITO, *Portalegre, Niza.*—Woollen fleece.
789 J. COELHO PALHINHA, *Evora, Montemór o Novo.*—Wax.
790 L. A. DE CARVALHO, *Coimbra, Louza.*—Linseed and flax.
791 L. DA CUNHA MARTINS, *Guarda, Manteigas.*—Fleeces.
792 M. A. GERALDES, *Bragança, Vimioso.*—Wool.
793 M. GUERRA, *Bragança.*—Cocoons.
794 M. J. FERREIRA DA SILVA GUIMARAES, *Braga, Guimaraes.*—Flax.
794A M. G. D'OLIVEIRA, *Lisboa.*—Soap.
795 M. M. HOLBECHE CORREIA, *Santarem.*—Seed, olive-husks, aloe-threads, goat's hair, thistle-flower and seed.
796 M. N. FURTADO, *Bragança.*—Potatoe-starch.
797 M. J. FERRAO CASTELLO BRANCO (D.), *Coimbra, Louza.*—Cocoons.
798 M. DA NAZARETH, *Coimbra.*—Flax.
799 M. OSORIO CABRAL DE CASTRO *Coimbra.*—Flax, tow, and linseed.
800 N. C. MATOS FERRAO CASTELLO BRANCO, *Coimbra, Louza.*—Cherry-tree wood and oak-bark.
801 P. A. DA SILVA REBELLO COELHO VASCONCELLOS MAIA, *Braga, Povoa de Lanhoso.*—Cocoons and flax.
802 P. L. GUIMARAES, *Braga, Guimaraes.*—Silk.
803 R. B. SOBRINHO, *Beja, Alvito.*—Raisins.
804 R. L. DE MESQUITA PIMENTEL, *Angra do Heroismo.*—Raw silk, blue silk, and bored cocoons.
804A J. DU GUERRA, *Bragança.*—Soap.

805 S. DE MELLO FALCAO TRIGOSO, *Lisboa, Torres Vedras.*—Sheep's wool.

806 T. DE MELLO SERRAO, *Beja, Ourique.*—Fleeces.

807 V. B. PIRES, *Faro.*—Linseed.

808 V. B. PIRES, JUN. *Faro.*—Madder and oak-bark.

809 VISCOUNT DE BRUGES, *Angra do Heroismo.*—Hemp, moss, herbs, seeds, roots, &c.

810 VISCOUNT DA ESPERANÇA, *Beja, Cuba.*—Wool, cork, and wood.

811 VISCOUNT DE GUIAES, *Vizeu, Lamego.*—Silk.

812 VISCOUNT DE TAVEIRO, *Coimbra.*—Cocoons, flax-seed, and hops.

813 VISCOUNT DE TORRAO, *Beja, Alvito.*—Elder-berries.

814 V. J. PEREIRA FRANCO, *Guarda.*—Linseed, and elder-berries.

815 W. M. DE CARVALHO, *Coimbra.*—Bark and root of walnut tree.

CLASS 7.

816 A. DA COSTA PEREIRA, *Braga.*—Wooden hoops for sieves.

817 J. M. F. THOMÁS, *Coimbra.*—Pulleys, and blocks.

818 NATIONAL PRINTING OFFICE, *Lisboa.*—Printing apparatus.

819 V. J. DE CASTRO, *Lisboa.*—Typographic plate.

CLASS 8.

820 B. POTIER, *Lisboa.*—Thrashing machine.

821 PERSEVERANCE CO. *Lisboa.*—Steam-engine, for the manufacture of olive-oil.

CLASS 9.

822 A. POLYCARPO BAPTISTA, *Lisboa.*—Garden-tools.

823 A. SERRA, *Portalegre.*—Agricultural implements.

824 COMMISSION (CENTRAL), *Aveiro, Oliveira de Azemeis, Ovar.*—Shovel to turn grain, rake, sieve, and hatchet.

825 COUNT DE SAMODAES, *Porto, Lamego.*—Hatchet.

826 F. DE FREITAS MACEDO, *Santarem.*—Model of agricultural implement.

827 J. FELICIANO, *Coimbra.*—Sieve.

828 J. G. FERREIRA, *Coimbra.*—Casks.

829 J. R. P. BASTO, *Aveiro, Oliveira de Azemeis.*—Shovel and rake for maize.

830 O. DA COSTA, *Santarem, Anciao.*—Hammer and hoe.

831 PERSEVERANCE CO. *Lisboa.*—Distilling apparatus.

CLASS 10.

832 COMMISSION (CENTRAL), *Aveiro, Oliveira de Azemeis.*—Model of syphon.

833 L. LOUGE, *Lisboa, Setubal.*—Refractory bricks.

CLASS 13.

834 INDUSTRIAL INSTITUTE OF LISBOA, *Alcantara.*—Philosophical instruments.

CLASS 14.

835 M. N. GODINHO, *Lisboa.*—Photograph of a picture drawn with a pen by Exhibitor.

CLASS 16.

836 F. A. TEIXEIRA DE CARVALHO, *Braga.*—Viol.

837 P. J. TEIXEIRA, *Braga.*—Violin.

CLASS 17.

838 A. POLYCARPO BAPTISTA, *Lisboa.*—Surgical instruments.

CLASS 18.

839 A. G. PORTO, *Portalegre, Niza.*—Crochet work.

840 A. J. VIEIRA RODRIGUES FARTURA, *Angra do Heroismo.*—Cotton quilt.

841 A. P. DE OLIVEIRA, *Braga.*—Cotton drills.

842 B. DAUPIAS & CO. *Lisboa.*—Cotton under-waistcoats.

843 CO. OF COTTON FABRICS, *Lisboa.*—Cotton fabrics.

844 J. J. ANTUNES, *Braga.*—Cotton drills.

845 J. P. DABNEY, *Horta.*—Cotton stockings.

846 J. D. DA COSTA, *Braga.*—Cotton drills.

847 LISBON SPINNING AND WEAVING CO. *Lisboa, Belem.*—Cotton fabrics, ropes and strings.

848 L. BERAUD, *Lisboa.*—Stockings, shirts, and caps.

849 M. A. RODRIGUES, *Braga, Villa Verde.*—Striped stuff and aprons.

850 M. DA GRAÇA ALFAIA, *Portalegre, Niza.*—Crochet work.

851 SPINNING CO. OF CRESTUMA, *Aveiro Villa da Feira.*—Woof and warp.

CLASS 19.

852 COMMISSION (CENTRAL), *Aveiro.*—Linen.

853 COMMISSION OF THE DISTRICT OF GUARDA, *Guarda, Sabugal.*—Skein of thread.

854 COMMISSION OF PONTA DELGADA.—Cords made of New Zealand flax.

855 F. J. DE OLIVEIRA, *Braga, Guimaraes.*—Sewing thread.

856 G. F. PEREIRA NUNES, *Coimbra, Oliveira do Hospital.*—Cloth and linen.

857 J. DO PILAR, *Braga.*—Linen.

858 LISBON SPINNING & WEAVING CO *Lisboa, Belem.*—Thread.

859 M. DOS DESAMPARADOS SOARES, *Braga.*—Knitted flax stockings.

860 R. VIEIRA, *Braga.*—Linen.

861 TORRES NOVAS NATIONAL SPINNING & WEAVING CO. *Lisboa.*—Flax and hemp fabrics.

862 VISCOUNT DE BRUGES, *Angra do Heroismo.*—Flax fabrics, and hemp cord.

CLASS 20.

863 A. J. BARBOSA ARAUJO, *Braga.*—Shaded satin and damasks.

864 A. M. G. DA SILVA RAMOS, *Braga.*—Damasks, satins, and silk ribbons.

865 B. J. C. DAS NEVES, *Coimbra, Pampilhosa.*—Silk.

866 C. A. C. GARCIA, *Bragança.*—Silk.

867 COUNT DO FARROBO, *Lisboa, Villa Franca de Xira.*—Silk, woof, &c.

868 COUNT DE SAMODAES, *Porto, Lamego.*—Silk.

869 CORDEIRO & IRMAO, *Lisboa.*—Silks and velvets.

870 E. M. RAMIRES, *Lisboa.*—Silks and satins.

871 J. L. DE ALMEIDA ARAUJO, *Vizeu.*—Silk thread.

872 J. J. FERREIRA DE MELLO E ANDRADE, *Braga, Povoa de Lanhoso.*—Skein of silk.

873 M. J. GUERRA, *Bragança, Freixo de Espada á Cinta.*—Silk for sieves.

874 J. DA SILVA PEREIRA DE VASCONCELLOS, *Braga.*—Velvets.

875 M. J. RODRIGUES, *Bragança, Mirandella.*—Silk.

876 R. L. DE MESQUITA PIMENTEL, *Angra do Heroismo.*—Sewing-silk and silk caps.

CLASS 21.

877 A. J. DE LIMA, *Braga.*—White wool.

878 A. PINTO DE OLIVEIRA, *Braga.*—Kerseymere.

879 B. DAUPIAS & CO. *Lisboa, Belem.*—Woollen and worsted fabrics, &c.

880 CENTRAL COMMISSION, *Aveiro.*—Fids.

881 COMMISSION (DISTRICT) OF FARO.—Cloth.

882 CO. OF WOOLLEN MANUFACTURES OF CAMPO GRANDE, *Lisboa, Olivaes.*—Cloth.

883 CORSINO, IRMAO, & CO. *Guarda.*—Blankets.

884 E. I. PARREIRA, *Angra do Heroismo.*—Quilt, and wool.

885 F. J. DE ALMEIDA, *Vizeu.*—Blankets.

886 F. M. MASCARENHAS, *Coimbra, Louza.*—Cloth.

887 IGREJA, ROLDAN, & CO. *Lisboa, Seixal.*—Cloth.

888 J. D. DA COSTA, *Braga.*—Coatings.

889 L. COMETUDO, *Braga.*—Flannel shirt.

890 LARCHER & CUNHADOS, *Portalegre.*—Cloth and kerseymeres.

891 LARCHER & SOBRINHOS, *Portalegre.*—Cloths and kerseymeres.

892 M. ANTONIA, *Guarda, Pinhel.*—Stockings.

893 VISCOUNT DE BRUGES, *Angra do Heroismo.*—Woollen apron.

CLASS 23.

894 A. C. MIRANDA & CO. *Lisboa.*—Shawls, handkerchiefs, and printed calicoes.

895 F. J. DA LUZ, *Lisboa; Cintra.*—Shawls.

896 PINTO & CO. *Lisboa, Belem.*—Bilbao shawls.

897 P. J. L. DOS ANJOS, *Lisboa, Belem.*—Handkerchiefs and printed calicoes.

CLASS 24.

898 A. E. DE SALLES (D.), *Angra do Heroismo.*—Satin towel embroidered in embossed gold.

899 BARONESS DE PRIME, *Vizeu.*—Embroidered pincushion.

900 COMMISSION, CENTRAL, *Aveiro, Ovar.*—Lace.

901 F. P. DOS SANTOS, *Lisboa.*—Embroideries in gold.

902 G. CONDERT, *Lisboa, Setubal.*—Lace.

903 G. R. VIEIRA MACHADO (D.), *Braga.*—Embroidered towel.

904 J. P. DABNEY, *Horta.*—Embroidered handkerchief.

905 M. C. PRATA, *Lisboa, Setubal.*—Lace.

906 M. J. TEIXEIRA DE CARVALHO E SAMPAIO (D.), *Vizeu.*—Silk embroidery, representing Conway Castle, N. Wales.

907 M. L. DO AMARAL (D.), *Lisboa.*—Embroidery, representing the monument of Joseph I.

908 F. WILKINSON & CO. *Madeira, Funchal.*—Embroideries.

CLASS 25.

909 F. X. DE MORAES SOARES, *Villa Real, Chaves.*—Kid skin.

910 J. G. PEREIRA CALLADO, *Lisboa, Alfama.*—Morocco.

911 VISCOUNT DE BRUGES, *Angra do Heroismo.*—Skin and furs.

CLASS 26.

912 ABBOT DE CRESPOS, *Braga.*—Heifer skins.

913 A. J. DE PASSOS, *Braga.*—Goat and sheep skins.

914 A. DA FONSECA CARVAO PAIM, *Angra do Heroismo.*—Skins and hides.

915 COMMISSION, CENTRAL, *Aveiro, Oliveira de Azemeis.*—Saddle and tanned leather.

916 FONSECA & FERREIRA, *Porto.*—Tanned hide.

917 M. B. MONTEIRO, JUN. *Guarda, Pinhel.*—Tanned skins.

918 M. GUEIFÃO BELLO, *Santarem, Maçao.*—Tanned skins.

919 TANNING FABRIC ASSOCIATION OF EXTREMOZ, *Evora.*—Leather.

CLASS 27.

920 A ROXO, *Lisboa.*—Hats.

921 A. MOREIRA E SILVA, *Aveira, Oliveira de Azemeis.*—Hats.

922 COMMISSION OF THE DISTRICT OF COIMBRA, *Figueira*.—Shoes.
923 F. DA COSTA, *Vizeu*.—Shoes.
924 F. J. MAIA, *Braga*.—Caps.
925 F. LAURENCE, *Coimbra*.—Gloves.
926 J. P. DABNEY, *Horta*.—Shawls and ornament made from the threads of the Guinea aloe.
927 J. J. ROBALLO DA FONSECA, *Lisboa*.—Coat.
928 J. DA CUNHA ALVES DE SOUSA, *Braga*.—Boots.
929 J. CURRY DA CAMARA CABRAL, *Horta*.—Silk mantle with lace made from the threads of the Guinea aloe.
930 M. J. DE SOUSA, *Angra do Heroismo*.—Hat.
931 M. J. DE CARVALHO, *Aveiro, Oliveira de Azemeis*.—Hats.
932 M. L. DA SILVA, *Aveiro, Oliveira de Azemeis*.—Hats.
933 M. DIAS DE AFFONSECA, *Braga*.—Woollen hats.
934 R. L. PESSOA, *Coimbra, Figueira da Foz*.—Wooden shoes.

CLASS 28.

935 COMMISSION, CENTRAL, *Aveiro, Feira*.—Paper and pasteboard.
936 E. M. M. SMITH (D.), *Lisboa, Olivaes*.—Paper.
937 FERIN, *Lisboa*.—Bookbinding.
938 GYMNASIUM GODINHO COLLEGE, *Lisboa*.—Caligraphy.
939 J. L. DE CARVALHO, *Lisboa*.—Caligraphy.
940 J. A. CABRAL DE MELLO, *Angra do Heroismo*.—Ode dedicated to H.R.H. Prince Alfred.
941 M. DIAS CESARIO, JUN. *Lisboa*.—Caligraphic frame.
942 M. N. GODINHO, *Lisboa*.—Caligraphy.
943 NATIONAL PRINTING OFFICE, *Lisboa*.—Printing and lithography.
944 RUAES PAPER MANUFACTORY, *Braga*.—Paper.
945 VICOUNTESS DE VILLA NOVA DA RAINHA, *Santarem, Thomar*.—Paper.

CLASS 29.

946 WEIGHTS AND MEASURES BOARD, *Lisboa*.—Models of educational instruments; coins, &c.

CLASS 30.

947 DISTRICT COMMISSION OF ANGRA DO HEROISMO.—Bay-tree-wood flower work.
948 I. CAETANO, *Lisboa*.—Escutcheon of the Portuguese arms.
948A DECOMBES & CO. SOCIÂTY, *Lisboa*.—Articles made with the endless saw, and boring machine.

CLASS 31.

949 A. GOMES, *Braga*.—Horse bits and stirrups.
950 H. SCHALCK, *Lisboa*.—Buttons and nails.
951 J. CORREIA, *Braga*.—Nails.

CLASS 32.

952 A. P. BAPTISTA, *Lisboa*.—Scissors, knives, and razors.

CLASS 34.

953 A. MICHON C. PIERRE, *Porto, Villa Nova de Gaia*.—Glass covers and plates.

CLASS 35.

954 A. A. DA LAPA, *Portalegre, Elvas*.—Pots.
955 COMMISSION, CENTRAL, *Aveiro*.—Earthenware.
956 COMMISSION OF THE DISTRICT OF COIMBRA.—Earthenware.
957 GENERAL SOCIETY OF CHEMICAL PRODUCTS, *Porto, Villa Nova de Gaia*.—Earthenware (grit stone) for acids.
958 G. J. HOWARTH, *Lisboa*.—Stone china service.
959 J. CORREIA DA COSTA, *Coimbra*.—Earthenware.
960 J. A. BRAAMCAMP, *Lisboa*.—Drainage tubes.
961 J. FRANCISCO, *Vizeu, Tondella*.—Earthenware.
962 J. J. CESAR, *Coimbra*.—Vases and jars.
963 J. LUIZ, *Portalegre, Niza*.—Earthenware.
964 M. LOBO, *Portalegre, Crato*.—Earthenware.
965 VISCOUNT DE VILLA MAIOR, *Porto, Villa Nova de Gaia*.—Earthenware.

CLASS 36.

966 A. CORREIA DE LEMOS, DR. *Vizeu*.—Broom made of maize straw. Ivory carving.
967 A. P. CARDOSO CRUZ, *Braga*.—Cards.
968 A. DE BETTENCOURT, *Lisboa*.—Map of Portugal.
969 BIESTER, FALCAO, & CO. *Lisboa*.—Cork.
970 C. DA PURIFICAÇAO REIS GUEDES (D.), *Lisboa*.—Flowers made of threads of the Guinea aloe.
971 COMMISSION, CENTRAL, *Aveiro, Feira, Ovar, Estarreja*.—Tobacco, halters, shackles, bed, and mattress.
972 COMMISSION OF THE DISTRICT OF COIMBRA, *Soure*.—Razor strop, made from the Guinea aloe.
973 COMMISSION OF THE DISTRICT OF EVORA.—Wax and cork.
974 C. DE ALMEIDA A. E SOUSA, *Coimbra, Penacova*.—Tooth-picks.
975 COUNT OF SOBRAL, *Santarem, Almeirim*.—Horse-hair halter.

976 D. G. BLANCO, *Faro, Silves.*—Cork.

977 F. A. PEREIRA, *Braga.*—Objects in horn, &c.

978 F. A. DE VASCONCELOS, *Lisboa.*—Tooth-picks.

979 F. DOMINGOS, *Santarem.*—Sofa and chair of mace-reed.

980 F. J. GOMES, *Portalegre.*—Girth-cloth, crupper, and halter.

981 F. LUDGERO MARQUES, *Lisboa.*—Spectacles.

982 F. M. B. PINTO DE CARVALHO, *Lamego.*—Bellows.

983 F. DA SILVA, MARQUES & Co. *Porto.*—Cork.

984 F. MARQUIS DE FIGUEIREDO, *Coimbra.*—Mat.

985 I. M. PROFIRIA DE CASTRO, *Lisboa.*—Wax flowers and fruits.

986 T. DABNEY, *Angra.*—Baskets.

987 J. R. M. COOKE, *Coimbra, Figueira da Foz.*—Cork.

988 J. A. DE OLIVEIRA, *Braga.*—Inkstands.

989 J. M. DA COSTA BARBOSA, *Santarem, Cartaxo.*—Mace-reed cord, and horse-hair halter.

990 J. MARQUES DIAS, *Braga.*—Horse-hair cloth for sieves.

991 J. J. DOS REIS, *Lisboa.*—Umbrellas, parasols, and sticks.

992 J. A. DE SOLEDADE, *Leiria, Peniche.*—Artificial flowers, made of shells and sea productions.

993 J. F. DA C. TERRA BERQUÓ, *Horta.*—Straw flowers.

994 J. HENRIQUES & FILHOS, *Coimbra, Poiares.*—Manufactured wax.

995 J. J. RIBEIRO, *Lisboa.*—Spectacles.

996 J. M. DE AZEVEDO GIRAO, *Santarem, Alpiaça.*—Straw cloak.

997 J. R. DA SILVA, *Braga.*—Strings for viols and violins.

998 M. DA COSTA, *Braga.*—Straw cloaks.

999 M. DA S. SOUSA, *Coimbra.*—Models of casks.

1000 M. M. HOLBECHE CORREIA, *Santarem.*—Secret lock; gourd-bark fruit-stand and sugar-pot; beehive.

1001 M. J. DA SILVA DOMINGUES (D.), *Lisboa.*—Glazed frame, containing a nosegay, and ornamented with threads of the Guinea aloe.

1002 MARIA DO O. *Coimbra, Penacova.*—Tooth picks.

1003 M. DA PIEDADE, *Santarem.*—Goat-skin dressed to carry wine.

1004 M. OSORIO CABRAL DE CASTRO, *Coimbra.*—Mats and straw cloaks.

1005 PERSEVERANCE CO. *Lisboa.*—Lead pipes and copper chocolate pots.

1006 P. A. BRANDAO, *Coimbra.*—Statue of Pedro V.

1007 R. J. DE ALMEIDA, *Lisboa.*—Mat.

1008 T. J. FERREIRA, *Lisboa.*—Mats.

1009 T. C. DE OLIVEIRA & FILHOS, *Lisboa.*—Billiard balls, &c.

1010 VISCOUNT DE BRUGES, *Angra do Heroismo.*—Cords and ropes made with different kinds of vegetable fibre.

1011 VISCOUNT OF TAREIRO. *Lisboa.*—Goat-skin bottles.

1012 Z. J. PINTO, *Braga.*—Box-tree pedestal.

1013 B. DA SILVA, *Lisboa.*—Mats.

1014 J. J. DE MACEDO, *Madeira, Funchal.*—Thirty-one samples of plaited straw.

COLONIES.

CLASS 1.

1015 A. H. DA COSTA MATOS, *Ilhas de S. Thomé e Principe.*—Red ochres and tabatinga.

1016 DEJEANTE L. B. *Lisboa.*—Marbles from the Cape Verde Islands.

1017 COMMISSION OF CABO VERDE, *Ilhas de Cabo Verde.*—Lime, sand, and salt.

1018 F. R. BATALHA, *Lisboa.*—Salt from Angola, and Timor.

1019 J. N. DE SALLES, *Ilhas de Cabo Verde.*—Volcanic products.

1020 MARTINS & LIMA, *Ilhas de Cabo Verde.*—Salt.

1021 P. M. TITO & Co. *Ilhas de Cabo Verde*—Salt.

1022 B. J. BROCHADO, *Angola, Mossamedes.*—Ochre in powder; iron and copper rings, iron and copper poniard.

1023 F. A. PONCE DE LEAO, *Angola, Mossamedes.*—Magnetic iron, copper ore, plastic stone.

1024 F. DA COSTA LEAL, *Angola, Huilla.*—Copper bracelet.

1025 F. WELWITCH, *Lisboa.*—Micaceous iron, from Cacula; iron pyrites; pemba stone, from Angola, Cazengo.

1026 GOVERNOR OF DOMBE GRANDE, *Angola, Dombe Grande.*—Sulphur, copper ore, ochre.

1027 J. D. D'ALMEIDA, *Angola, Mossamedes.*—Salt.

1028 J. J. D'ALMEIDA, *Angola, Golungo Alto.*—Salt.

1029 J. J. DE PAIVA, *Angola, Mossamedes.*—Gypsum stone, copper ore, limestone.

1030 J. TEIXEIRA XAVIER, *Angola, Benguella.*—Gypsum, rough and calcined, in masses and in powder; limestone.

1031 T. M. BESSONE, *Timor.*—Asphalt.

1032 ULTRAMARINE BOARD, *Lisboa.*—Copper ore from Angola, Cuio.

CLASS 2.

1033 COMMISSION OF CABO VERDE ISLANDS, *Ilhas de Cabo Verde.*—Ashes.

1034 F. R. BATALHA, *Lisboa.*—Bindweed from Angola, and purgative cassia from Timor.

1035 G. DA CRUZ LIMA, *Ilhas de Cabo Verde.*—Palma-Christi seeds.

1036 J. F. ANTONIO SPENCER, *Ilhas de Cabo Verde.*—Senna plant.

1037 J. J. BOMTEMPO, *Ilhas de Cabo Verde.*—Seed.

1038 J. M. DE SOUSA E ALMEIDA, *Ilhas de S. Thomé e Principe.*—Palm oil and cocoa oil.

1039 J. DO PINO, *Ilhas de S. Thome e Principe.*—Palm oil.

1040 M. J. DA COSTA PEDREIRA & J. V. DE CARVALHO, *Ilhas de S. Thomé e Principe.*—Cassia and cocoa oil.

1041 R. M. X. DE RAMOS, *Ilhas de S. Thomé e Principe.*—Palm oil and cocoa oil.

1042 F. DA C. LEAL, *Angola, Huilla.*—Flax-seed.

1043 F. A. P. BAYAO, *Angola, Duque de Bragança.*—Sawdust fecula, roots, hemp.

1044 F. WELWITSCH, *Angola.*—Chemical substances and products.

1045 GOVERNOR OF ENCOGE, *Angola, Encoge.*—Ginger, bastard saffron, and N-cassa bark.

1046 GOVERNOR OF MASSANGANO, *Angola, Massangano.*—Palm oil.

1047 GOVERNOR OF ZEUZA OF COLUNGO, *Angola, Zeuza do Colungo.*—Butua root, and palm oil.

1048 J. J. D'ALMEIDA, *Angola, Golungo Alto.*—Ginguba oil, dong a luto (a root).

1049 J. T. XAVIER, *Angola, Benguella.*—Alcohol, bunze (a plant).

1050 M. P. S. VENDUNEM, *Angola, Barra do Bengo.*—Palm oil, gimbunze, &c.

CLASS 3.

1051 F. R. BATALHA, *Lisboa.*—Canary-almond, cinnamon, coffee, sago, &c. from Timor; cocoa from Prince's Island, cinnamon from Goa.

1052 J. J. DE CARVALHO, *Ilhas de S. Thomé e Principe.*—Cocoa.

1053 J. M. DE SOUSA E ALMEIDA *Ilhas de S. Thomé e Principe.*—Coffee, cocoa, manihot, balsam of S. Thomé.

1054 J. F. A. SPENCER, *Ilhas de Cabo Verde.*—Tamarinds.

1055 J. J. DE MELLO, *Ilhas de S. Thomé e Principe.*—Cocoa.

1056 J. M. DE FREITAS, *Ilhas de S. Thomé e Principe.*—Coffee, arrow-root, cocoa, and brandy.

1057 J. DO PINO, *Ilhas de S. Thomé e Principe.*—Tapioca, coffee, and cocoa.

1058 J. RIBEIRO DA CUNHA AZURAR, *Ilhas de S. Thomé e Principe.*—Safú.

1059 M. J. DA COSTA PEDREIRA & J. VELLOSO DE CARVALHO, *Ilhas de S. Thomé e Principe.*—Tapioca, manihot flour, pulp of tamarinds, fruits, coffee, &c.

1060 M. DOS REIS BORGES, *Ilhas de Cabo Verde.*—Rice, coffee, Yuca manihot flour.

1061 P. R. TAVARES, *Ilhas de Cabo Verde.*—Coffee.

1062 ULTRAMARINE BOARD, *Lisboa.*—Cinnamon, rice, and coffee, from Timor.

1063 A. A. SEQUEIRA THEDIM, *Cabo Verde.*—Rum, and coffee.

1064 F. A. P. BAYAO, *Angola, Duque de Bragança.*—Giéfu seeds, cola nuts.

1065 F. WELWITSCH, *Angola.*—Maize, different seeds, maboca (a fruit).

1066 GOVERNOR OF BENGUELLA, *Angola, Benguella.*—Different seeds, and manihot starch.

1067 GOVERNOR OF CACONDA, *Angola, Caconda.*—Uindo (bark).

1068 GOVERNOR OF ENCOGE, *Angola, Encoge.*—Fruits, seeds, and coffee.

1069 GOVERNOR OF ZEUZA OF GOLUNGO, *Angola, Zeuza of Golungo.*—Rice, French beans, tapioca, and honey.

1070 J. A. G. PEREIRA, *Angola, Cazengo.*—Coffee.

1071 J. D. D'ALMEIDA, *Angola, Mossamedes.*—Honey.

1072 J. J. D'ALMEIDA, *Angola, Golungo Alto.*—Rum, and brandy extracted from maize.

1073 J. L. D'ALBUQUERQUE, *Angola, Bumbo.*—Sugar.

1074 J. T. XAVIER, *Angola, Benguella.*—Brandy.

1075 M. P. DOS SANTOS VENDUNENI, *Angola, Barra do Bengo.*—French beans.

1076 ULTRAMARINE BOARD, *Angola, Encoge.*—Coffee.

CLASS 4.

1077 E. A. DE SOUSA, *Ilhas de Cabo Verde.*—Corals, lichen for dyeing, archil, and seed.

1078 F. DE ALVA BRANDAO, *Ilhas de S. Thomé e Principe.*—Bastard-saffron root, and cotton.

1079 F. R. BATALHA, *Lisboa.*—Archil, lichen, cotton, &c. from various Portuguese colonies; filaments of caroco, dye-wood, &c. from Timor and Mozambique.

1080 G. DA CRUZ LIMA, *Ilhas de Cabo Verde.*—Seeds.

1081 HORTET RAYMUNDO, *Ilhas de Cabo Verde.*—Archil.

1082 J. M. DE SOUSA E ALMEIDA, *Ilhas de S. Thomé e Principe.*—Wood, archil, and white cotton.

1083 J. B. DE OLIVEIRA, *Ilhas de S. Thomé e Principe.*—Palm-tree wool, and cotton.

1084 J. R. DA CUNHA AZURAR, *Ilhas de S. Thomé e Principe.*—Wood.

1085 L. J. MONIZ, *Ilhas de Cabo Verde.*—Cotton.

1086 M. J. DA COSTA PEDREIRA & J. VELLOSO DE CARVALHO, *Ilhas de S. Thomé e Principe.*—Cotton, wood, and gum.

1087 M. DOS REIS BORGES, *Ilhas de Cabo Verde.*—Cotton.

1088 P. M. TITO & Co. *Ilhas de Cabo Verde.*—Archil.

1089 P. A. DE OLIVEIRA, *Ilhas de Cabo Verde.*—Cotton and indigo.

1090 R. DE SÁ NOGUEIRA, *Ilhas de Cabo Verde.*—Cotton.

1091 T. DA SILVA BASTOS VARELLA *Ilhas de S. Thomé e Principe.*—Wood.

1092 ULTRAMARINE BOARD, *Lisboa.*—Tobacco, archil, and wood, from Timor.

1093 Z. PEREIRA MAFRA, *Ilhas de S. Thomé e Principe.*—Wood.

1094 A. C. DE SOUSA E CUNHA, *Angola, Encoge.*—Resin.

1095 A. J. DE SEIXAS, *Angola.*—Wax from Loanda and Benguella.

1096 B. J. Brochado, *Angola, Mossamedes.* —Wax.

1097 F. da Costa Leal, *Angola, Bumbo.* —Samples of woods.

1098 F. A. Pinheiro Baijao, *Angola, Duque de Bragança.*—Tobacco.

1099 F. Welwitsch, *Angola.* — Gum Tragacanth from Loanda; copal from Benguella; dragons'-blood from Huilla; muance gum and cabella from Golungo; copal from Zenza do Golungo; mumbango gum from Ambaca; resin from Cazengo; fifty-two samples of wood.

1100 Governor of Benguella, *Angola, Benguella.*—Copal, archil, and dragons'-blood.

1101 J. D. d'Almeida, *Angola, Huilla.*—Tobacco and archil.

1102 T. M. Bessone, *Moçambique.*—Gum arabic and caoutchouc.

CLASS 9.

1103 M. P. dos Santos Venduneni, *Angola, Barra do Bengo.*—Press for extracting palm oil.

CLASS 18.

1104 F. R. Batalha, *Lisboa.* — Holes worked with cotton thread, spun by the savages of Agra.

1105 J. R. de Carvalho (D.), *Ilhas de Cabo Verde.*—Cotton cloth.

1106 Ultramarine Board, *Lisboa.* — Cotton from Timor, raw and manufactured.

1107 B. F. de Figueiredo e Castro, *Angola, Mossamedes.*—Cotton in the pod, and ginned.

1108 J. D. d'Almeida, *Angola, Mossamedes.*—Cotton in the pod.

1109 M. J. Corrêa, *Angola, Moçambique.* —Yellow cotton in the pod.

CLASS 23.

1110 M. T. Moutel (D.), *Ilhas de Cabo Verde.*—Fabrics of cotton and silk.

1111 Ultramarine Board, *Lisboa.* — Cloth of wool and silk, from Timor.

CLASS 25.

1112 B. J. Brochado, *Angola, Mossamedes.*—Hart-hide.

1113 F. A. P. Bayao, *Angola, Duque de Bragança.*—Fox, monkey, and deer-skins.

1114 G. dos R. C. e Barros, *Angola, Muxima.*—Deer-skins.

1115 Governor of Zeuza of Golungo, *Angola, Zeuza of Golungo.*—Stag-hide.

CLASS 26.

1116 F. da C. Leal, *Angola, Huilla.*—Tanned leather, hides.

CLASS 36.

1117 F. R. Batalha, *Lisboa.* — Cauris from Timor, carved walking-stick of sandal-wood.

1118 J. M. de Sousa e Almeida, *Ilhas de S. Thomé e Principe.*—Tobacco and cigars.

1119 P. A. Ferreira, *Ilhas de S. Thomé e Principe.*—Tobacco.

1120 Ultramarine Board, *Lisboa.* — Ramé and cords made of it, caroco, and wild banana-tree, from Timor.

1121 F. A. Pinheiro Baiao, *Angola, Duque de Bragança.*—Pipes.

1122 F. da Costa Leal, *Angola, Huilla.* —Bricks and tiles.

1123 F. Welwitsch, *Angola,*—Filaments, and articles made of them; feathers; elephant's mane, and articles made of it; subi sieve.

1124 G. dos R. Claro e Barros, *Angola, Muxima.*—Teeth of the hippopotamus.

1125 Governor of Catumbella, *Angola. Catumbella.*—Palm wool.

1126 Governor of Encoge, *Angola, Encoge.*—Gimbusu straw, and articles made of it.

1127 Governor of Zenza do Golungo, *Angola, Zenza do Golungo.*—Empacassa horns.

1128 J. J. d'Almeida, *Angola, Golungo Alto.*—A bag made of the filaments of embondeira; porcupine bristles.

1129 M. P. S. Vendunem, *Angola, Barra do Bengo.*—Empalanca and empacassa horns.

1130 T. M. Bessone, *Moçambique.*—Filaments of Guinea aloe, and hemp.

APPENDIX
TO PORTUGUESE CATALOGUE.

CLASS 1.

1131 A. P. da Fonseca Vaz, *Santarem.* Lime, and phosphate of lime.

1132 J. N. Rebello Valente, *Aveiro.*—Slate from serra da Gualva; quartz from which glass is made; clay.

1133 M. M. da Silva, *Porto, Vallongo.*—Slate.

CLASS 2.

1134 Commission (Central Portuguese), *Aveiro, Feira.*—Charcoal.

1135 Count of Samodaes, *Porto, Lamego.* —Charcoal.

1136 Domingos Sant'Agatha, *Lisboa.*—Artificial guano.

1137 J. Ferreira, *Coimbra, Louza.*—Charcoal.

1138 J. R. Correia Belem & Bro. *Lisboa.*—Pharmaceutical products.

1139 T. E. Aylores, Jun. *Lisboa.*—Verdigris.

1140 Y. V. d'Almeida, *Santarem.*—Mustard.

1141 M. F. da Silva, *Porto.*—Fish oil.

1142 Viscount of Taveiro, *Coimbra.*—Mustard.

CLASS 3.

1143 A. M. DOS SANTOS, *Porto*.—Brandy.

1144 A. L. REBELLO DA GAMA, *Porto*.—Wine; honey.

1145 A. ALLEN, *Porto*.—French beans, pease, maize, wheat.

1146 ANGELINA ROSA CARNEIRO, *Porto, Santo Thyrso*.—Wine.

1147 A. B. DE ALMEIDA SOARES LENCASTRE, *Porto*.—French beans, maize, honey.

1148 A. B. FERREIRA, *Porto*.—Port wine.

1149 A. E. R. DE SOUSA PINTO, *Porto*.—Rye, maize, hazel-nuts, vinegar, &c.

1150 A. F. BAPTISTA, *Porto*.—Oil of olives.

1151 A. F. MENEZES, *Porto*.—Port wine.

1152 A. G. DA COSTA, *Porto*.—Dried fruits, honey.

1153 A. J. CARREIRA, *Lisboa*.—Dried fruits, pease and beans, olives.

1154 A. J. AYRES DE MENDONÇA, *Faro, Olhao*.—Wine.

1155 A. P. SOARES DA COSTA, *Porto*.—Wine.

1156 A. DE SOUSA CARNEIRO, *Porto*.—Agricultural products, flour, wine.

1157 A. T. DE QUEIROZ, *Porto*.—Dried fruits.

1158 A. V. DE TOVAR MAGALHAES E ALBUQUERQUE, *Guarda*.—Dried pears, cheese.

1159 BARON OF VAZZEA, *Porto*.—Acorns.

1160 B. DO C. DE M. PEREIRA, *Porto*.—Maize, walnuts, hazel-nuts.

1161 B. J. JACOME, *Vianna do Castello, Espozende*.—Wheat.

1162 C. DE A. GUIMARAES, *Lisboa*.—Collares wine.

1163 COMMISSION (CENTRAL PORTUGUESE), *Aveiro, Anadia*.—Wine from Bairrada; vinegar.

1164 COUNT OF ARROCHELLA, *Braga, Guimaraes*.—French beans, wheat, barley, walnuts.

1165 COUNT OF SAMODAES, *Porto, Lamego*.—Wheat, flour, chick-beans, dried figs and raisins, honey, potatoes, French beans.

1166 COUNT OF VILLA REAL, *Lisboa*.—Wine.

1167 E. A. DE SOUSA, *Bragança, Vinhaes*.—French beans.

1168 E. DE SEQUEIRA, *Lisboa*.—Collares wine.

1169 F. A. DA ROCHA, *Lisboa, Setubal*.—Alimentary preserves.

1170 F. M. B. P. DE CARVALHO, *Lamego, Peso da Regua*.—Wine.

1171 F. DE O. CALHEIROS, *Lisboa*.—Oil of olives.

1172 F. R. BATALHA, *Lisboa*.—Coffee from Madeira.

1173 F. X. DE MORAES SOARES, *Villa Real, Chaves*.—Cheese.

1174 FREIRAS BENEDICTINAS, *Porto*.—Preserved fruits.

1175 G. P. M. AGUIAR, *Porto*.—Wine, and vinegar.

1176 G. G. DE CARVALHO, *Porto*.—Oil of olives.

1177 H. THOMÁS, *Lisboa*.—Collares wine.

1178 I. F. DE CARVALHO. *Lisboa*.—Collares wine.

1179 J. MARTINS & SON, *Lisboa*.—Muscadine wine.

1180 J. E. DE BRITO E CUNHA, *Porto*.—French beans.

1181 J. N. REBELLO VALENTE, DR. *Aveiro, Oliveira de Azemeis*.—Walnuts.

1182 J. P. ARAUJO DE SEQUEIRA, *Lisboa*.—Collares wine.

1183 J. DE V. CARNEIRO MENEZES, *Porto*.—French beans, walnuts, oil of olives, and wine.

1184 J. A. FERREIRA, *Porto*.—Wheat, French beans.

1185 J. A. RODRIGUES COIMBRA, *Porto*.—Wheat, maize, French beans, wine.

1186 J. F. PEREIRA, *Porto*.—Walnuts, pine kernels.

1187 J. L. MARTINS, *Porto*.—Wheat, rye, and maize.

1188 J. A. DA CUNHA MACEDO, *Porto*.—Wine.

1189 J. A. DA SILVA, *Porto*.—French beans.

1190 J. B. MASCARENHAS, *Faro, Villa Nova de Portimao*.—Dried figs.

1191 J. B. VAZ, *Porto*.—Hazel-nuts.

1192 J. F. DA SILVA, *Porto*.—Dried whiting, and dried roach.

1193 J. F. PINTO, *Porto*.—French beans.

1194 J. G. SOBRINHO, *Porto*.—Rye.

1195 J. J. T. DA COSTA GUIMARAES, *Porto*.—Agricultural products, honey, oil of olives, and wine.

1196 J. L. FERNANDES, *Porto*.—Maize.

1197 J. M. DE SOUSA RODRIGUES, *Porto, Santo Thyrso*.—Rye, French beans, chestnuts, and acorns.

1198 J. M. DA V. CABRAL SAMPAIO, *Villa Real*.—Port wine.

1199 J. DA R. RIBEIRO, *Porto*.—Walnuts and French beans.

1200 J. R. DE AZEVEDO, *Santarem, Benavente*.—Maize.

1201 J. R. C. BELEM & BRO. *Lisboa*.—Preserves and liqueurs.

1202 J. DE S. MAGALHAES CABRAL, *Porto*.—Oil of olives.

1203 L. P. DE CASTRO, *Braga, Guimaraes*.—Rye, millet, oil of olives, &c.

1204 L. A. M. DA SILVA ARAUJO, *Porto, Santo Thyrso*.—French beans and wheat.

1205 M. A. PEREIRA, *Faro*.—Mendobi.

1206 M. A. P. DE SAMPAIO & BRO. *Villa Real*.—Port wine of different years.

1207 M. A. FERNANDES, *Porto*.—Wheat, barley, maize, and French beans.

1208 M. D. P. DE ARAGAO (D), *Faro*.—Almonds.

1209 M. J. F. DA SILVA GUIMARÃES, *Braga, Guimaraes*.—Acorns.

1210 M. F. F. DA COSTA ARAUJO, *Porto, Santo Thyrso*.—Vinegar.

1211 M. G. BELLO, *Santarem, Maçao*.—Rye.

1212 M. MARQUES DA SILVA, *Porto, Vallongo*.—Agricultural products.

1213 M. M. ALVES, *Porto.*—Oil of olives and honey.

1214 M. P. DA SILVA, *Porto.*—Maize.

1215 M. P. P. DE S. VILLAS BOAS, *Porto.* —Brandy.

1216 M. DOS P. E SILVA, *Braga, Guimaraes.* —Chestnuts.

1217 M. DA R. G. CAMOES, *Porto.*—Agricultural products, dried fruits, wine and vinegar, oil of olives, and honey.

1218 M. V. GUEDES DE ATAIDE (D.), *Porto.* —French beans, walnuts, barley, dried figs, and maize.

1219 M. J. DE SOUSA FEIO, *Beja.*—Sorgho.

1220 R. WANZELLER, *Porto.*—Wheat and French beans.

1221 S. DE M. FALCAO TRIGOSO, *Lisboa, Torres Vedras.*—Virgin oil.

1222 S. P. DE MESQUITA, *Porto.*—Maize, wheat, wine, oil of olives, chestnuts, and French beans.

1223 S. R. FERREIRA, *Porto.*—French beans.

1224 V. B. PIRES JUNIOR, *Faro.*—Wine.

1225 V. N. CORADO, *Lisboa.*—Collares wine.

CLASS 4.

1226 ABBOT OF CRESPOS, *Braga.*—Cocoons.

1227 A. C. DE SOUSA E SA, *Porto.*—Linseed, tow, and flax.

1228 A. P. MONTEIRO, *Porto.*—Wood.

1229 A. GRANT, *Porto.*—Wood.

1230 A. ALLEN, *Porto.*—Wool.

1231 A. B. DE A. S. LENCASTRE, *Porto.*—Wood, and virgin wax.

1232 A. E. R. DE S. PINTO, *Porto.*—Wood.

1233 A. M. ARRISCADO, *Braga, Espozende.* —White and black wool.

1234 A. P. SOARES DA COSTA, *Porto.*—Wood.

1235 A. DE S. CARNEIRO, *Porto.*—Wood.

1236 ARSENAL OF MARINE, *Lisboa.*—Wood.

1237 B. DO C. DE MARIA PEREIRO, *Porto.* —Flax.

1238 BIESTER FALCÃO & Co. *Lisboa.*—Corks.

1239 CASTRO SILVA & SONS, *Porto.*—Soap.

1240 COMMISSION (CENTRAL PORTUGUESE), *Aveiro.*—Vegetable ashes, and flax.

1241 COMMISSION OF EVORA, *Evora.*—Wax.

1242 COUNT OF ARROCHELLA, *Braga, Guimaraes.*—Cork.

1243 D. A. F. RAMALHO, *Evora, Redondo.* —Wool.

1244 E. I. PARREIRA, *Angra.*—Wool.

1245 F. M. DA SILVA PINÇAO, *Beja* —Flax.

1246 F. B. CASTELLO BRANCO, *Portalegre.* —White wool.

1247 F. R. BATALHA, *Lisboa.*—Archil.

1248 J. P. VALVERDE, *Porto, Miranda do Corro.*—Cocoons.

1249 J. N. R. VALENTE, DR. *Aveiro, Oliveira de Azemeis.*—Flax.

1250 J. DE V. C. DE MENEZES, *Porto.*—Wood.

1251 J. V. DE ALMEIDA, *Santarem, Benavente.*—Mustard.

1252 J. A. RODRIGUES COIMBRA, *Porto.*—Wood and cork.

1253 J. C. F. CORREIA FALCAO, *Castello Branco.*—White wool.

1254 J. J. TEIXEIRA DA COSTA GUIMARAES, *Porto.*—Flax and wax; wood.

1255 J. M. DE SOUSA RODRIGUES, *Porto, Santo Thyrso.*—Linseed and flax.

1256 J. MARIANNI, *Porto.*—Cocoons.

1257 J. DE S. MAGALHAES CABRAL, *Porto.* —Linseed and flax.

1258 KEMPES & Co. *Lisboa.*—Soap.

1259 L. HUET BACELLAR, *Porto.*—Cocoons.

1260 M. FRANCISCO, *Castello Branco.*—Wool.

1261 M. J. DA SILVA, *Porto.*—Flax.

1262 M. L. DA C. VASCONCELLOS, *Braga.* —Wool.

1263 M. M. DA SILVA, *Porto, Vallongo.*—Flax, wax.

1264 M. P. DE CARVALHO, *Porto.*—Black wool.

1265 M. P. P. DE SOUSA VILLAS BOAS, *Porto.*—Wood.

1266 M. DA C. DO AMARAL, *Braga, Guimaraes.*—Wood.

1267 M. DA R. GONÇALVES DE CAMOES, *Porto.*—Wood, wool, raw silk, linseed and flax, cocoons and silk, virgin wax, husks of maize, &c.

1268 P. M. DE MACEDO, *Porto.*—Wax.

1269 R. WANZELLER, *Porto.*—Flax.

1270 VISCOUNT OF TAVEIRO, *Coimbra.*—Wood.

1271 WIDOW OF J. B. BURNAY, *Lisboa.*—Oils.

CLASS 7.

1272 D. J. DE A. BUBONE, *Lisboa.*—Block and tackle.

CLASS 8.

1273 L. F. DE S. CRUZ, *Porto, Cedofeita.*—Wheel for rudder and binnacle, on a new principle: rolling machine, hydraulic machine.

CLASS 9.

1274 J. M. DA SILVA, *Guimaraes.*—Garden tools.

CLASS 18.

1275 A. M. DE AGUIAR ALVARO, *Porto Bomjardim.*—Yellow nankeen.

1276 VISCOUNT OF BRUGES, *Angra do Heroismo.*—Cotton thread prepared for fishing-nets.

CLASS 19.

1277 A. C. DE SOUSA E SÁ, *Porto.*—Sacking and linen.

1278 B. DO C. DE MARIA PEREIRA, *Porto.* —Linen.

1279 J. N. REBELLO VALENTE, DR. *Aveiro.* —Linen.

CLASS 20.

1280 A. DE A. PERES, *Porto.*—Spun silk.
1281 J. P. VALVERDE, *Miranda do Corvo, Porto.*—Spun silk.
1282 J. M. BRANDAO, *Porto.*—Spun and twisted silk.
1283 J. MARIANNI, *Porto.*—Organzine, woof, spun and twisted silk.
1284 L. HUET BACELLAR, *Porto.*—Spun silk.

CLASS 24.

1285 FERREIRAS MADRUGAS (SENHORAS), *Horta.*—A representation of the English arms, made with the pulp of the fig tree.
1286 M. L. DA SILVA MAFRA, *Lisboa.*—Palace of Mafra in embroidery.

CLASS 27.

1287 J. JORGE, *Porto.*—Hessian boots of satin, silk, &c.; boots.
1288 K. KEIL, *Lisboa.*—Articles of clothing.

CLASS 28.

1289 J. DE SÁ COUTO, *Aveiro, Feira.*—Paper.

CLASS 32.

1290 J. M. DA S. GUIMARAES, *Porto.*—Scissors and a clasp-knife.

CLASS 36.

1291 J. DE SÁ COUTO, *Aveiro, Feira.*—Corks.
1292 J. DA COSTA, *Portalegre.*—Basket.
1293 J. F. DA PIEDADE, *Porto.*—Umbrellas and parasols.
1294 J. M. DA C. BARBOSA, *Santarem, Cartaxo.*—Agricultural implements.
1295 J. M. DE C. N. L. E VASCONCELLOS, *Lisboa.*—Cocks for gas.
1296 J. P. CARDOSO & SON, *Porto.*—Shot, tin-foil, silver, and gold-leaf.
1297 L. B. DA SILVA, *Porto.*—Silver purses.
1298 L. CARNEIRO, *Arganil, Coimbra.*—Winnowing sieves.
1299 M. M. DA S. RAMOS, *Castello Branco, Covilha.*—Medals; image of the Virgin in box-wood.
1300 PRISONERS OF THE GAOL OF PORTALEGRE, *Portalegre.*—A mat.
1301 VALENTE, DR. J. N. R., *Aveiro, Oliveira de Azemeis.*—Ornamented vase of box-wood.
1302 VISCOUNT OF ESPERANÇA, *Beja.*—Model of a triangular harrow.

COLONIES.

CLASS 1.

1303 ADMINISTRATION OF THE FIRST DIVISION OF INDIA, *Embarbacem, India.*—Iron from India.
1304 F. R. FLORES.—Copper ore from Angola.
1305 J. DA GAMA, *Moçambique.*—Salt.
1306 J. Z. X. ALVES, *Moçambique.*—Gold from Moçambique.
1307 R. TAYLOR & CO.—Malachite from Angola.

CLASS 2.

1308 ULTRAMARINE BOARD, *Lisboa.* — Rock oil from Timor.
1309 ADMINISTRATION OF BARDEZ, *Bardez, India.* — Ginger, bastard saffron, and oicond (a root) from India.
1310 C. DA C. SOARES, *Moçambique.*—Dried apples from Moçambique.
1311 GENERAL GOVERNOR OF MOÇAMBIQUE, *Moçambique.*—Cocoa-oil and Calumba from Moçambique.
1312 GOVERNOR OF DAMAO, *Damao, India*—Alcohol from India.
1313 GOVERNOR OF DIU, *Diu, India.*—A liqueur from India.
1314 J. B. DE OLIVEIRA, *S. Thomé e Principe.*—Palm oil from S. Thomé.
1315 J. R. DA C. AZURAR, *S. Thomé e Principe.*—Balsam of S. Thomé.

CLASS 3.

1316 ADMINISTRATOR OF BARDEZ, *Bardez. India.*—Pepper, teflam, vinegar and liqueur, husked rice (calloqui), and Muscavado sugar.
1317 ADMINISTRATOR OF THE SECOND DIVISION OF INDIA, *India, Satary.*—Rice, coffee, and different seeds.
1318 ADMINISTRATOR OF THE FOURTH DIVISION OF INDIA, *India, Bardez.*—Cinnamon.
1319 A. C. H. DE CADILHOS, *S. Thomé e Principe.*—Coffee.
1320 A. P. DE BORJA, *Cabo Verde.*—Sugar.
1321 B. F. DA COSTA & CO. *Goa, India.*—Almond liqueur.
1322 C. DA COSTA SOARES, *Moçambique.*—Brandy, coffee, and dry manihot.
1323 CAPTAIN CARLOS DUARTE, *India, Bardez.*—Wild pepper.
1324 E. A. DE SOUSA, *Cabo Verde.*—Rum.
1325 F. Q. DA SILVA, *S. Thomé e Principe.*—Tapioca.
1326 GENERAL GOVERNOR OF MOÇAMBIQUE, *Moçambique.*—Oils of sesame and mendobi, jugo, vinegar, rum, French beans, and tobacco.
1327 GOVERNOR OF DAMAO, *Damao, India.*—Maurá, husked rice (dangue).
1328 J. M. DE SOUSA E ALMEIDA, *S. Thomé.*—Dendem (a fruit).
1329 J. DA SILVA CARRAO, *Moçambique.*—Manihot flour and rice.
1330 J. V. DA GAMA, *Moçambique.*—Olanga and tapioca extracted from manihot.
1331 J. M. SANTA ANNA MASCARENHAS, *Salsete, India.*—Wine, brandy, and vinegar.
1332 J. DA GAMA, *Moçambique.* — Caju brandy.
1333 J. Z. X. ALVES, *Moçambique.*—Coffee.

CLASS 4.

1334 ADMINISTRATOR OF BARDEZ, *Bardez. India.*—Gums.

1335 ADMINISTRATOR OF THE SECOND DIVISION OF INDIA, *India.*—Vegetable tallow, and sesame.

1336 ADMINISTRATOR OF THE FOURTH DIVISION OF INDIA, *India.*—Catto de quer (for dyeing), and different sorts of seeds.

1337 A. L. MOREIRA, *India.*—Oil.

1338 A. M. GUEDES, *Moçambique.*—Mulala, irety, and camogy in filaments and manufactured.

1339 C. DA C. SOARES, *Moçambique.*—Filaments of cairo.

1340 CAPTAIN CARLOS DUARTE, *Goa, India.*—Wood. *India.*—Seeds of kert, and oil.

1341 C. DO R. MIRANDA, *India.*—Cotton.

1342 F. R. BATALHA, *Timor.*—Leaves of loba, for dyeing.

1343 GENERAL GOVERNOR OF MOÇAMBIQUE, *Moçambique.*—Sesame, cotton, bastard saffron, and ginger.

1344 J. DA SILVA CARRAO, *Moçcambique.*—Pearls, and mendobi.

1345 M. J. DA C. PEDREIRA, *S. Thomé e Principe.*—Jobo (seeds).

1346 T. M. BESSONE, *Moçambique.*—Caoutchouc.

CLASS 18.

1347 ADMINISTRATOR OF GOA ISLANDS, *Goa.*—Caps and stockings.

1348 GOVERNOR OF DAMAO, *Damao, India.*—Cotton fabrics of different patterns.

1349 ULTRAMARINE BOARD, *Lisboa.* — Cotton fabrics of different patterns.

CLASS 24.

1350 ADMINISTRATOR OF THE GOA ISLANDS, *Goa.*—Lace.

CLASS 27.

1351 C. DO ROSARIO MIRANDA, *Bardez, India.*—Cap, made of the bark of poddono.

CLASS 36.

1352 ADMINISTRATOR OF SALSETE, *Salsete, ndia.*—Wax candles.

1353 A. B. NEVES, *Moçambique.*—Tobacco.

1354 A. M. GUEDES, *Moçambique.*—Purses and caps, made from irrety; camogy.

1355 C. DO ROSARIO MIRANDA, *Bardez, India.* — Yarns made of different fibrous matters.

1356 J. Z. XAVIER ALVES, *Moçambique.*—Cords made of cairo.

1357 NARANA XELK, *India, Conculim.*—Boxes, made of sandal-wood.

ADDENDA.

CLASS 14.

1358 J. N. SILVEIRA, *Lisboa.* — Photography.

CLASS 27.

1359 A. M. DA SILVA, *Lisboa.*—Boots.

CLASS 28.

1360 NOGUEIRA DA SILVA.—Clichets in wood.

CLASS 29.

1361 ROYAL GEODESICAL COMMISSION, *Lisboa.*—Geodesical maps.

CLASS 33.

1362 MOUZADO & BRO. *Porto.*—Filigrane, and other works in silver.

CLASS 35.

1363 FERREIRA PINTO, *Lisboa.*—Articles in hard porcelain.

ROME.

S.C. Court, No. 3.

1 Breviary; a present from H.H. the Pope to H.E. Card. Wiseman.

2 Ebony case for the above, which also forms a reading desk.

CLASS 1.

3 BALDINI, HIS EXCELLENCY THE BARON COMMENDATORE, P. D. Minister of Commerce and Public Works.—Calcareous stones of Monticelli, St. Angelo in Capoccia, and Tivoli; pozzolane tuf; argillaceous earths, and plasters; with vases, artificial marbles, &c. made of them.

4 ——— Travertini, peperini, lave macchi, with other building stones; marbles for decorations; millstones, refractory materials, asphalt, &c.

5 BONDI, G. & Co.—Argillaceous earth, and bricks made of it.

6 BONIZI, A.—Roman cement from Tolfa.

7 BONIZI, G. & Co.—Minerals.

8 SOCIETÀ ROMANA.—Ores of iron and metallurgical products; bricks.

CLASS 2.

9 DE PAOLIS, A. B.—Potash.

10 GOVERNMENT ESTABLISHMENT OF THE ALUMS OF TOLFA.—Specimens of alum.

11 GOVERNMENT SALT WORKS (THE COM. BALDASSARE, Director).—Specimens of marine salt.

12 MOSAIC MANUFACTORY OF THE VATICAN.—Smalts.

BARBERI, M.—Mosaics.

MOGLIA, L.—Copy of the Madonna della Seggiola.

BARBERI, L.—Mosaic table.

TADDEI, L.—Mosaic table.

BARZOTTI, B.—Mosaic representing St. Peter's at Rome.

SIBBIO.—Panorama of Rome, in mosaic.

ROSSI, A.—Mosaic table.

DESTRADA, D.—Specimens of Etruscan, Roman, Greek, and Byzantine goldsmith's work.

DIES, G.—Specimens of Etruscan, &c., goldsmith's work.

ODELLI, A.—Cameos.

SAULINI, T.—Cameos.
LISTRUCCI, E.—Cameos.
LISTRUCCI, B.—Cameos.
PICKLER.—Engraved sardonyx.
LUPI, F.—Engraved sardonyx.
BIANCHI, G.—Bronze medal.
CELLI, V.—Bronze medal, &c.
VESPIGNANI, R.—Ebony frame.
MARCHETTI, L., & BAVADIN.—Door of the Vatican.
ERCOLI, P.—Bas-relief in ivory.

CLASS 3.

JACOBINI, BROS.—Wine, vinegar, and oil.

CLASS 4.

13 ANTONELLI, CONTE F.—Indian corn from the Pontine marshes.
14 ARVOTTI, G.—Raw and spun silk.
15 ERBA, B.—Asphalt, crude and prepared, for various purposes.
16 CASTRATI, G. B.—Wax candles.
17 MUTI PAPAZZURRI, MARQUIS S.—Stearine candles.
18 ORTO AGRARIO OF THE ROMAN UNIVERSITY.—Cereals, and textile plants.

CLASS 7.

19 SOCIETÀ ROMANA.—Crucibles, and furnace for laboratory.

CLASS 8.

20 GRAIZIOS, N.—Machine for shewing new movement.
21 ROSSI, P.—Oil mill.

CLASS 9.

22 PFEIFFER, F.—Arched saws for gardeners, pruning tools.
23 ROSSI, M. S. DE.—Stenographic and orthographic machine.

CLASS 10.

24 ROSSI, P.—Anti-concussion apparatus, applicable to various purposes.

CLASS 11.

25 BRAND, R.—Double-barrel gun, with appendages.
26 TONI, T.—Revolver gun.

CLASS 13.

27 TESSIERI, PROF. P.—Medal-holder, for examining medals and gems.

CLASS 14.

28 ANDERSON, G.—Photographic views of Rome, and of ancient and modern sculpture.
29 CUCCIONI, T.—Photographs of paintings by A. Caracci, and of the Roman Forum, Colliseum, Piazza of St. Peter's, &c.
30 DOVIZIELLI, P.—Photographs of paintings in the Farnesina, and of the Colliseum, Roman Forum, &c.
31 MACPHERSON, R.—Photographs.
32 ROCCHI, D.—Photographs.

CLASS 20.

33 ARVOTTI, G.—Silk fabrics, corded, coloured, embroidered with gold, &c.
34 PASQUALE, S.—Articles in silk; and silk fabrics, plain, coloured, enriched with gold, &c.
STEFANI, P.—Silk.
BIANCHI, A.—Silk.

CLASS 24.

35 ADMINISTRATION OF PRISONS.—Articles in lace of various kinds, made by the prisoners.
36 HOSPITAL OF S. MICHAEL.—Tapestry copied from an ancient mosaic, &c.

CLASS 28.

37 ANGELINI, CAV. A.—Treatise on perspective.
38 BERTINELLI, G.—Richly bound Psalter.
39 OLIVIERI, L.—Monuments of the Lateran Museum, richly bound.

CLASS 30.

40 MANZI, L. M.—Table madde of a rare stone, found in the ruins of Rome; tables made of breccia found in Adrian's villa.
41 MARTINORI, P.—Articles in Oriental alabaster.
42 MUTI PAPAZZURRI, MARQUIS S.—Marble table, inlaid work; table of petrified lumachello.
43 SOCIETÀ ANONIMA DEI MARMI ARTIFICIALI.—Tables, &c. in imitation malachite, lapis lazuli, porphyry, &c.; inlaid tables of imitation Oriental alabaster.

CLASS 32.

44 PFEIFFER, F.—Scissors and razors.

CLASS 33.

45 ARVOTTI, G.—Roman pearls, and articles made of them.
PAZZI, V.—Roman pearls.

CLASS 35.

46 SOCIETÀ ANONIMA DEI MARMI ARTIFICIALI.—Pavements in imitation marbles, porphyry, granite, &c.

CLASS 36.

47 DIES, G.—Tazza of Giallo, and rosso antiquo.
48 CHIALLI, B.—Bronzed lamp, Pompeian style; other lamps.
49 RAINALDI, G.—Déjeunés and tazze of of Egyptian alabaster; Pompeian lamp of rosso antiquo.
50 Beaitier, jewelled.
51 Box, with miniature of H.H. Pius VII.
52 Reliquary.
53 Two statuettes, copies of statues in front of St. Peter's.

RAINALDI, G.—Tazza in Oriental alabaster.
LUCATELLI, G.—Tragan's column, and obelisks in rosso antico.
MONACHESI, A.—Gothic table, and alabaster vases.
PRINCE ALDOBRANDINI.—Etruscan and Chinese vases.
RICCIANI, SISTERS.—Flowers embroidered on cloth.

RUSSIA.

N.W. Court, No. 6, and N.W. Gallery, No. 5.

CLASS 1.

1 ADMINISTRATION OF KOZAKS SETTLEMENT, *Orenburg.*—Collection of minerals.

2 ALIBERT, N. P.—Specimens and ornaments of black lead; blocks of nephrite; collection of other Siberian minerals.

3 BELOSSELSKI-BELOZERSKI, PRINCE K. *Katava Iron Works, Oofim Circ. Orenburg Gov.* —Steel; wrought and sheet iron.

4 BELOSSELSKI-BELOZERSKI, PRINCE, HEIRS OF, *Katava-Ivanofski Iron Works, Orenburg Gov.*—Collection of minerals; iron and steel.

5 BOGOSLOVSKI CROWN COPPER WORKS, *Perm Gov.*—Specimens of copper ore, and refined copper, &c. &c.

6 CABINET OF HIS IMPERIAL MAJESTY, *St. Petersburg.* — Collection of samples of polished hard stones, from different Russian quarries.

7 CAUCASIAN AGRICULTURAL SOCIETY, *Tiflis.*—Collection of ores, coal, and other Transcaucasian minerals.

8 DEMIDOF, P. P. *Nijne-Tagil Works, Perm Gov.* — Cast steel, sheet iron, and copper.

9 GOLITZYN, PRINCE S. *Nytvinsk Foundry. Okhansk Circ. Perm Gov.*—Iron ore, and cut iron bars.

10 GOOBIN, HEIRS OF, *Nijne-Sergin, and Michaïlof Mining Works, Perm Gov.*—Bar and sheet iron.

11 GRECHISHCHEF, *Michael, near Riazan.* —Lime and plaster.

12 KIRGHIZ DISTRICT ADMINISTRATION, *Orenburg.*—Chalk and common salt, from Lake Inder.

13 KNAUFS MINING CO. *Ossinsk and Perm Circuits, Perm Gov.*—Bar iron.

14 LATKIN, M. *Ustsissolsk, Vologda Gov.* —A grindstone, from the quarry of Mount Broossianaïa.

15 ADMINISTRATION OF CROWN DOMAINS, *Tchernigof Gov.*—China clay from Poloski village, near Glookhof.

16 MINING DEPARTMENT OF POLAND, *Warsaw.*—Coal, iron ore, calamine, fire-clay, cast-iron, zinc sheets, &c. &c.

17 OORAL KOZAKS, *Orenburg.*—Samples of copper ore, chalk, and gypsum.

18 PASHKOF, A. *Bogoyavlensky Copper Mines, Orenburg Gov.*—Copper ores, lingots, sheets and wire.

19 PASHKOF, M. V. *Preobrajensk Mining Works, Orenburg Gov.*—Geological specimens; plate and foil copper.

20 PASHKOF, HEIRS OF, *Voskressensk Iron Works, Orenburg Gov.*—Collection of ores, and other minerals.

21 PERMIAN MINING DISTRICT ADMINISTRATION. — Collection of ores and other minerals, and of metallurgical products.

22 POPOF, A. & N. BROS. *Kirghiz District, Siberia.*—Native copper, different ores, coal, &c.

23 RACHETTE, V. *Nijnetagilsk, Perm Gov.* —Models of universal high furnaces and cupola ovens.

24 ROCHEFORT, COUNTESS, *Olghinski Copper Works, Perm Gov.*—Copper ore, and metallic copper.

25 SAMSONOF, S. & MAMONTOF, *Sernopol Semipalatinsk District, Siberia.*—Specimens of black lead.

26 SIDOROF, M. *Eastern Siberia, Tooroohansk Circ. Yenisseisk Gov.*—Black lead, and other minerals.

27 TCHERKASSOF, N. *Moscow.*—Samples of peat and fire-clay.

28 VÖLKERSAM, BARON G. VON, *Papenhof, Coorland Gov.* — Samples of yellow amber, containing insects.

29 VOTKINSK CROWN WORKS, *Perm Gov.* —Assortment of iron and steel.

30 YAKOVLEF, P. HEIRS OF, *Navialof Iron Works, Yekaterinburg Circuit, Perm Gov.* —Bar iron.

31 YAKOVLEF, S. HEIRS OF, *Alapaef Iron Works, Verkhotoorsk Circ. Perm Gov.*—Sheet iron, bluish and bright.

32 ZEITLER, M. *Michalof Works, near Slawkof, Radom Gov.*—Fire-clay, coal, and iron ore.

CLASS 2.

Sub-Class A.

33 CHETVERTAKOF, N. M. *Moscow.*—Red paint and white lead.

34 COLLEGE OF FORRESTERS, *Lissino, Tsarskoe-Selo Circ. St. Petersburg Gov.* — Birch bark, oil, and tar.

35 EGGERS, *Reval.*—Acetate of lead, and vinegar.

36 EPSTEIN, A. & LEVY, *Warsaw.*—Copperas, Roman vitriol, white lead, Glauber's salt, and saltpetre.

37 HIRSCHENFELD, R. *Warsaw.*—Matches without phosphorus.

38 IRTEL, I. VON, *Tiflis.*—Raw and refined soda from Erivan.

39 KRAUSE, J. *Warsaw.*—Oil and spirit varnishes, oil colours, floor rubbing wax, and marking ink.

40 KRUSE, G. *Reval.*—Varnish for furniture.

41 LEPESHKIN, BROS. *Moscow.*—Garancin.

42 OOSSACHEF, B. *St. Petersburg.* — Colours.

43 REICHEL, A. *Somin Chemical Works, Borovitch Circ. Novgorod Gov.*—Birch bark, oil, and turpentine.

44 SANIN, V. J. *near Borovsk, Kalooga Gov.*—Chemical products.

45 SCHMIDT, K. E. & Co. *Svatoi Island, Caspian Sea.*—Paraffin.

46 SHIPOF, A. *Kineshma Circ. Kostroma Gov.*—Chemicals.

47 SPIES, LEWIS, *Tarkhomina, near Warsaw.*—Bone dust for manure, and artificial guano, &c.

48 TORNAU, BARON, & Co. *near Bakoo.*—Raw naphtha and napthadchil, and products of their distillation.

49 VOLOSKOF, J. *Rjef, Tver Gov.*—Carmine.

Sub-Class B.

50 ANOKHIN, A. J. *St. Petersburg.*—Scents, pomatum, hair powder, and other cosmetics.

51 GLAZER, F. *Tiflis.*—Blossom of pyretrum, carneum, and Persian powder.

52 IRTEL, I. VON, *Tiflis.*—Blossom and seed of pyretrum, carneum, and Persian powder.

53 JDANOF, BROS. *St. Petersburg.*—Aromatic waters, perfumed and deodorising fluids.

54 NATANSON, J. & SHEEMAN, *Warsaw.*—Perfumery, toilet powders, cold cream, &c.

55 OORAL KOZAKS, *Orenburg.*—Roots of rhubarb, glycyrrhiza glabra, inula helenium, and dried salvia leaves.

CLASS 3.

Sub-Class A.

56 AHMET, BEKIR-OGLOO, *Derekoi Vil. Alooshta District, Crimea.*—Walnuts.

57 ADMINISTRATION OF CROWN DOMAINS, *Vilno.*—Cereals.

58 ADMINISTRATION OF THE KOZAK SETTLEMENTS, *Orenburg.*—Collection of cereals, peas, poppy, lin and hemp seed.

59 AGRICULTURAL DEPARTMENT, *St. Petersburg.*—Collection of cereals, flour, groats, pulse, oil and grass seeds, nuts, chicory, and malt.

60 ALEXANOF, M. *Alexandropol Circ. Erivan Gov.*—Lentils.

61 ARUTINOF, A. *Alexandropol, Erivan Gov.*—Carmeline and Lucerne seeds.

62 ARUTINOF, A. *Elizabethopol Circ. Tiflis Gov.*—Hemp seed.

63 ARUTINOF, A. *Guzander Vil. Alexandropol Circ. Erivan Gov.*—French beans and lentils.

64 ARUTINOF, VANO, *Boluis Vil. near Tiflis.*—Peas.

65 BOBIN, B. *Slavianka Vil. Telav Circ. Tiflis Gov.*—Millet.

66 BOBYSHEF, I. *Elisabethopol, Tiflis Gov.*—Linseed.

67 BOGOLUBSKI, S. *Protopresbyter, Nerchinsk Circ. Irkootsk Gov.*—Cereals, oil seeds, flour, and pine nuts.

68 BORISSOF, J. *Rojestvensk Village, Korotoiak Circ. Voronesh Gov.*—Canary seed.

69 BRUKHOVETSKY, N. *Markovka Village, Bogoochar Circ. Voronesh Gov.*—Buck wheat.

70 CAUCASIAN AGRICULTURAL SOCIETY, *Tiflis.*—Mountain rice (chaltik) and saffron.

71 CLAYHILL & SONS, *Reval.*—Samples of Estonian cereals.

72 COLONISTS OF ANNENFELD SETTLEMENT, *Tiflis Gov.*—Barley, common and spelt wheat, millet, peas, lentils, linseed.

73 COLONISTS OF EKATERINFELD SETTLEMENT, *near Tiflis.*—Barley, oats, samples of wheat, and French beans.

74 CORNIES, J. *Taschchinak Freehold, Melitopol Circ. Tauride Gov.*—Wheat, rye, millet, and lucerne seed.

75 DENGINK, A. *Kishenef, Bessarabia.*—Cereals, oil seeds, madder seed, &c.

76 DOROSHENKOF, P. *Beerooch, Voronesh Gov.*—Sunflower seeds.

77 EKATERINSTADT COLONISTS, *Saratof Gov.*—Common Russian and Turkish wheat.

78 ELIOZOF, H. *Kvarlee Vil. Telav Circ. Tiflis Gov.*—Millet.

79 ERISTOF, E. *Goree, Tiflis Gov.*—Millet, common and French beans, peas, and linseed.

80 GENT, G. E. & Co. *Pskof.*—Linseed.

81 HARTMANN, *Riga.*—Assortment of cereals.

82 HASSAN-OGLOO, M. *Mashadee Vil. Bakoo Circ.*—Rice.

83 HENNER, T. *Helendorf Settlement, Alexandropol Circ. Erivan Gov.*—Beans.

84 IVANOF, G. *Mikhaïlovka Vil. Elisabethopol Circ. Tiflis Gov.*—Hemp-seeds.

85 KELBLER, J. *Tonkoroonofka Settlement, Saratof Gov*—Wheat and rye.

86 KERBELAÏ-SADECH-MEKHTI, O. *Kschil-Agatch Vil. Lenkoran Circ. Bakoo Gov.*—Barley.

87 KHANAGOF, J. *Djelal-Ogloo Vil. Alexandropol Circ. Erivan Gov.*—Common and French beans.

88 KOOSHELEF-BEZBORODKO, COUNT NICHOLAS, *Illinsko, Anchekrak Estate, near Odessa.*—Samples of wheat.

89 KOVESHNIKOF, J. *Markovka Village, Bagoochar Circ. Voronesh Gov.*—Spring wheat.

90 LANDAU, G. *Warsaw.*—Rye meal.

91 LEVSHIN, T. A. *Pasheelino Village, Yefremof Circ. Toola Gov.*—Soft white peas.

92 LUH, J. *Paninsk Settlement, Saratof Gov.*—Sunflower seed.

93 LYSACK, B. P. *Kistero Village, Starodoob Circ. Tchernigof Gov.*—Hemp seed.

94 MANHOLD, C. *St. Petersburg.*—Flour of beans and peas, prepared without grinding.

95 MANOOKIANTZ, *Martyros, Elisabethopol Circ. Tiflis Gov.*—Spelt wheat.

96 MARIINSKAÏA MODEL FARM, *near Saratof.*—Millet.

97 MOROZOF, P. *Pantzerevka Village, Gorodischi Circ. Penza Gov.*—Green rye and buck wheat groats.

98 MUSTIALA AGRICULTURAL INSTITUTION, *Tavasthus Gov. Finland.*—Samples of wheat, oats, barley, buck-wheat, and fir seeds.

99 NASHROOLEE-OGLOO, *Sopkooli, Lenkoran Circ. Bakoo Gov.*—Wheat.

100 NESTEROF, J. *Yusenkof Village, Nijnedevitzk Circ. Voronesh Gov.*—Hemp seed.

101 VOLOGDA CROWN DOMAINS, *Velikii-Oostug Circ. Vologda Gov.*—Rye and barley.

102 OOMANSKI, *Imperial Appanage Estates in the Gov. of Tver.*—Pensylvania rye.

103 OORAL KOZAKS, *Orenburg Gov.*—Wheat, rye, oats, millet, and sunflower seed.

104 PETROOSSOF, O. *Alexandropol, Erivan Gov.*—Rape seed.

105 PETSCHKE, *Coorland.*—Assortment of cereals.

106 PLEININGER, G. *Marienfeld Settlement, near Tiflis.*—Barley.

107 PLOTNIKOF, S. *Pessok Village, Novokhopersk Circ. Voronesh Gov.*—Millet and winter rye.

108 PNIOWER, J. & I. BROS. *Piotrkow, Warsaw Gov.*—Wheat flour and groats.

109 POLEJAEF, BROS. *Belozersk Circ. Novgorod Gov.*—Wheat flour, 1st, 2nd, and 3rd qualities.

110 POOSANOF, M. A. *Schigrof Circ. Koorsk Gov.*—Oats, buck-wheat, and millet.

111 RIGA COMMITTEE FOR EXHIBITION.—Linseed, pease, &c.

112 ROTHHAR, A. *Tonkoroonofka Settlement, Saratof Gov.*—Barley.

113 SAMARIN, S. *Mikhaïlovka Vil. Elisabethopol Circ.*—Hemp-seed.

114 SARKISSOF, A. *Zeiva Vil. Etchmiadzin Circ. Erivan Gov.*—Wheat.

115 SCHALE, A. *Marienfeld Settlement, near Tiflis.*—Oats.

116 SEESSOEF, P. *Novo Saratof Settlement, Elisabethopol Circ.*—Hemp seed.

117 SHVEELEE, N. M. *Kvarelec Vil. Telav Circ. Tiflis Gov.*—French beans.

118 SHVEELEE, S. *Ooriatuban Vil. Telav Circ. Tiflis Gov.*—Wheat.

119 SHESTOF, J. *Mshaga Village, Nijni-Novgorod Circ.*—Linseed.

120 SOOLKHANOF, E. *Goree, Tiflis Gov.*—Oats and wheat.

121 SOUTH-EASTERN MODEL FARM, *Kazan Gov.*—Seeds of spring rye, buck-wheat, millet, &c. &c.

122 STEIGERWALD, H. *Krasny-Yar Settlement, Saratof Gov.*—Turkish wheat.

123 STOROSHEF, D. *Trostianka Village, Ostrogoshsk Circ. Voronesch Gov.*—Winter wheat.

124 TOMILIN, J. *Pelagiada Vil. near Stavropol.*—Oats.

125 TSHOOKMALDIN, N. *Tiumen Circ. Tobolsk Gov.*—Cereals.

126 TURINE, A. *Konstantinovka Vil. Novobaïazet Circ. Erivan Gov.*—Oats.

127 VÖLKERSAM, BARON G. VON, *Papenhof, Coorland Gov.*—Samples of sunflower seeds and heads, and winter linseed.

128 YAKOVLEF, G. *Proossinich Village, Mohileff Gov.*—Linseed.

129 YAZDOONOF, G. *Dogkuz Vil. near Erivan.*—Ricinus and sesamum seeds.

130 SADYRIN, PH. *Kotelnich Circ. Viatka Gov.*—Cereals, peas, and grass seeds.

Sub-Class B.

131 BOCHAREF, M. I. *Yamskaïa Sloboda, near Koorsk.*—Buck-wheat, flour, and groats.

132 BOGOLUBSKI, S. *Protopresbyter, Nerchinsk Circ. Irkootsk Gov.*—Groats.

133 BORISSOVSKY, M. *Moscow.*—Refined beet-root sugar.

134 CAUCASIAN AGRICULTURAL SOCIETY, *Tiflis.*—Dried apricots, plums, figs, raisins, and Tchoorkhel (dainties of the natives).

135 EKATERINHOF SUGAR-REFINING CO. *St. Petersburg.*—Refined sugar.

136 EPSTEIN, H. *Hermanovo and Lyshkowicy, Lowicz Circ. Warsaw Gov.*—Refined beetroot sugar.

137 FOONDOOKLEY, J. J. *Ossota Factory, Chighirin Circ. Kief Gov.*—Specimens of raw beet-root sugar.

139 HAUF, BARON, *St. Petersburg.*—Refined sugar.

140 JACKOWSKI & CO. *Przasnysz Circ. Plotsk Gov.*—Sugar.

141 KIRGHIZ DISTRICT ADMINISTRATION, *Orenburg.*—Dried mutton and smoked beef.

142 KLIKOVSKY, PROF. *Kazan.*—Crystallized honey.

143 KOOSHELEF-BEZBORODKO, COUNT N. *Novochigly Refinery, Bobrof Circ. Voronesh Gov.*—Refined beet-root sugar.

144 KRICH, K. *Reval.*—Estonian anchovy.

145 LANDAU, G. *Warsaw.*—Groats and biscuits.

146 MONAKHOF, A. *Klin, Moscow Gov.*—Potato syrup.

147 NATANSON, BROS. I. & J. *Goozof, Lovicz Circ. Warsaw Gov.*—Refined beetroot sugar.

148 NATANSON, S. & I. *Sanniki, Gastyn Circ. Warsaw Gov.*—Raw and refined beetroot sugar.

149 GERKE, BROS. L. *Gliadkovo Vil. Elatom Circ. Tambof Gov.*—Raw beetroot sugar.

150 KLAASSEN, F. *Ladekop Settlement, Berdiansk Circ. Tauride Gov.*—Dried Mirabelle plums.

151 OORAL KOZAKS, *Orenburg Gov.*—Caviar, isinglass, viasiga, balyk (dried fish).

152 PNIOWER, BROS. J. & I. *Piotrkow, Warsaw Gov.*—Groats.

153 RAVICZ, A. & CO. *Elsbetowo, Sedletz Circ. Lublin Gov. Poland.*—Refined beet-root sugar.

154 ROCHEFORT, COUNTESS OLGA, *Ossa Circ. Perm Gov.*—Lime-tree blossom honey, obtained by cold pressure.

155 ROTHERMUND, A. *Bobrik Refinery, Soomy Circ. Kharkof Gov.*—Raw beet-root sugar.

156 SOOMAKOF, T. *St. Petersburg.*—Tablets of portable veal soup.

157 SPIES, L. *Tarkhomino, near Warsaw.*—Pulverized bones for food.

158 TCHERKASSOF, N. *Moscow.*—Refined beet-root sugar.

159 TERENTIEF, M. *Novotorjsk Circ. Tver Gov.*—Wheat and rye starch.

160 WISNOWSKY, R. *Warsaw.*—Biscuits.

Sub-Class C.

161 ABHAZOF, PRINCE D. *Kakhetia.*—Red and white wine.

162 AGRICULTURAL DEPARTMENT, *St. Petersburg.*—Assortment of tobacco leaves from American, Turkish, and Russian seeds.

163 BEKIR, I. OGLOO, *Derekoi Vil. Alooshta District, Crimea.*—Samson tobacco leaves.

164 CHARENTON, B. I. *Akkerman and Chabog Vineyards, Bessarabia.*—Red and white wines.

165 AUTORHUFFEN, T. *St. Petersburg.*—Cigarettes.

166 BOSTANJOGLO, M. & SONS, *Moscow.*—Tobacco, cigars, and cigarettes.

167 GABAÏ & MIGRI, *St. Petersburg.*—Samples of tobacco from Turkish seeds.

168 GROOTE, VON, *Distillery, Livonia.*—Cumin liqueur.

169 HEINRICHS, F. *St. Petersburg.*—Tobacco, cigars, and cigarettes.

170 HELLER, T. H. *St. Petersburg.*—Cigars, made by machinery.

171 INGLESY, A. *Tatareshty Vil. Orgey Vil. Bessarabia.*—Tobacco leaves.

172 JAPHA, B. *Moscow.*—Spirituous liqueurs.

173 ADMINISTRATION OF TCHERNIGOF CROWN DOMAINS, *Tchernigof.*—Bakoon tobacco and seeds.

174 MÜLLER, A. TH. *St. Petersburg.*—Cigars, cigarettes, and tobacco.

175 MUSTAPHA, HALIL-OGLOO, *Goorsoof Vil. Alooshta District, Crimea.*—Samson tobacco leaves.

176 ONANOF, *near Kootaïs.*—Turkish tobacco leaves.

177 RIGA COMMITTEE FOR EXHIBITION.—Assortment of tobacco.

178 SCHWABE, H. *Riga.*—Rectified spirits of wine and liqueurs.

179 SCHWEINFURTH & SEECK, *Riga.*—Wine.

180 STADLER, J. *Podstepnoi Settlement, Saratof Gov.*—Tobacco leaves from American seeds.

181 STRIEDTER, *Distillery, St. Petersburg.*—Raw, purified, and rectified spirits; rum, cognac, and liqueurs.

182 TÖPFER, A. *St. Petersburg.*—Cigars and cigarettes.

183 UNGERN-STERNBERG, BARON VON, *Distillery, Estonia.*—Cumin liqueur.

184 WIGAND, C. *Krasny-yar-Settlement, Saratof Gov.*—Tobacco leaves.

185 WINKELSTERN, A. *Ernestinendorf Settlement, Saratof Gov.*—Tobacco leaves.

186 WICKEL, *Riga.*—Currant wines.

187 WISNOWSKY, R. *Warsaw.*—Liqueurs.

188 WORONZOF, PRINCE, *Crimea.*—Wines of his estate.

CLASS 4.

Sub-Class A.

189 BORODOOLIN, N. wax chandler, *St. Petersburg.*—Wax candles.

190 COMPANY OF THE ST. PETERSBURG STEARINE CANDLES, SOAP, AND OLEIN FACTORY.—Assortment of stearine candles.

191 COMPANY FOR PREPARING SZAR SOAP AND RUSSIAN COSMETICS.—Egg-yolk oil, soap cosmetics, &c.

192 EPSTEIN, A. & LEVI, M. *Warsaw.*—Block of stearine, and stearine candles.

193 KRESTOVNIKOF, BROS. *Kazan.*—Stearine candles.

194 METEOR OIL MILL, *Gorodishche Circ. Penza Gov.*—Hemp seed oil-cake.

195 NATANSON, J. & SIMON, *Warsaw.*—Common and scented soap.

196 NENNINGER, A. *Sevsk Circ. Orel Gov.*—Hemp and linseed oil, raw and refined.

197 ALFTHAN & CO. *Finland.*—Stearine candles.

198 STEAM OIL MILL CO. *Riga.*—Hemp and linseed oil refined, and linseed cake.

199 KUEMMEL & CO. *Odessa.*—Rape and linseed oil, and oil cake.

200 OIL STEAM MILL CO. *St. Petersburg.*—Linseed cake.

201 OORAL KOZAKS, *Orenburg.*—Photogen and solar oil, train oil, tallow, and soap.

202 PROKHOROF, A. *Belef-Toola Gov.*—Tallow.

203 SAPELKIN, V. A. *Vladimerovka Vil. near Moscow.*—Wax candles.

204 STEINER, *St. Petersburg.*—Samples of common soap.

205 TIMOFEEF, BROS. *Stary Oskol, Koorsk Gov.*—Common soap.

Sub-Class B.

206 ADMINISTRATION OF THE KOZAKS' SETTLEMENTS.—Samples of sheep's wool.

207 AGRICULTURAL DEPARTMENT, *St. Petersburg.*—Collection of fleeces, raw and washed, wether, ewe and goat, and goat's hair.

208 BABARYKIN, J. *Kholm, Pskof-Gov.*—Samples of bristles.

209 BOGOLUBSKI, S. *Protopresbyter, Nerchinsk Circ. Ircootsk Gov.*—Wool, goats' and camels' hair.

210 SHER, N. *St. Petersburg.*—Medallion carved in ivory.

211 DOLGANOF, J. *St. Petersburg.*—Ivory and tortoiseshell combs, &c.

212 DÖRING, *Livonia.*—Fleeces.

213 DORONIN, J. *Archangel.*—Basso-relievo, folding knives, and other articles carved in ivory.

214 ERISTOF, M. *Ossetia District, Caucasus.*—Wether, ewe, and lamb wool.

215 FOONDOOCLEY, J. J. *Reshbairaki Estates, Robrinetz Circ. Kherson Gov.*—Fleeces in raw state.

216 GRAND DUCHESS HELEN PAVLOVNA.—Merino fleeces and wool, from Her Imperial Highness's estate, Karlovka, Gov. of Poltava.

217 GROODININ, P. *Velikii-Looki, Pskof Gov.*—Assortment of bristles.

218 KABYZEF, M. K. *St. Petersburg.*—Bone black and bone dust.

219 KAZAKOF, L. *Velikii-Oostug, Vologda Gov.*—Assortment of bristles.

220 KELBI-OGLOO, I. *Nakhitchevan.*—Samples of wool.

221 MAMONTOF, A. *Moscow.*—Samples of bristles: okatka, 1st and 2nd quality.

222 VOLOGDA CROWN DOMAINS, *Velikii-Oostug Circ. Vologda Gov.*—Bristles, various.

223 PHILIBERT, A. *Atmanay Farm, near Ghenitchesk, Melitopol Circ. Tauride Gov.*—Merino fleeces.

224 OORAL KOZAKS, *Orenburg.*—Wool, goats' and camels' hair, and glue.

225 RUSSIAN AMERICAN CO. *St. Petersburg.*—Walrus teeth.

226 SHVEELEE, G. B. *Kvarelee Vil. Telav Circ.*—Sheep's wool.

227 SHVEELEE, S. M. *Oorsatossany Vil. Telav Circ. Tiflis Gov.*—Wool.

228 SOOLTANOF, K. K. *Reshish-Rend Vil. Nakhichevan Circ. Erivan Gov.*—Combed wool.

229 STEPANOF, N. *Eysk.*—Washed wool.

230 VOKOOYEF, TH. *Mezen Circ. Archangel Gov.*—Walrus tooth.

231 VONSOWSKY, *Zeiva Vil. Echmiadzin Circ. Erivan Gov.*—Wool of a Kurtinsk wether.

Sub-Class C.

232 ABHAZOF, PRINCE D. *Kardansky Vil. Signah Circ.*—Safflower.

233 ABKHAZOF, PRINCE D. *Kakhetia.*—Safflower-seeds and blossom.

234 ADMINISTRATION OF THE KOZAK SETTLEMENTS, *Orenburg.*—Samples of flax, hemp, and wild madder root.

235 AGRICULTURAL DEPARTMENT, *St. Petersburg.*—Samples of flax from Poodosh and Vladimir, madder and statice roots.

236 BABARYKIN, J. *Kholm, Pskof Gov.*—Samples of flax.

237 BEK-HADJINOF, J. *Shemakha.*—Madder-root.

238 CAUCASIAN AGRICULTURAL SOCIETY, *Tiflis.*—Dendrological collection of eighty-two Transcaucasian trees and shrubs.

239 CLAYHILL & SONS, *Reval.*—Samples of Estonian flax.

240 LEESSIN, S. *Illinskoe Vil. Lookaïanov Circ. Nijni-Novgorod Gov.*—Mats of lime-tree bast.

241 FEDOROF, G. *Koshelevo, Rogachef Circ. Mohilef Gov.*—Lime-tree bast mats.

242 GENT, G. E. & Co. *Pskof.*—Samples of flax.

243 GRZYMALA, V. *Obrowec, Grubeszow Circ. Lublin Gov. Poland.*—Log of oak.

244 HADJI-DJAVAT-BEEK-ALI-OGLOO, *Koola Gov. of Bakoo.*—Madder-root.

245 IVANOF, D. *Boody, Rogachef Circ. Mohilef Gov.*—Lime-tree bast mats.

246 PERCY, JACOBS, *Riga.*—Corks.

247 KARAPET-MIKIRTOOMOF, *Tiflis.*—Cotton, cleaned and hackled by machinery.

248 KARAPET-SHAGIANOF, *Nigri Vil. Ordoobat Circ.*—Samples of cotton, raw, clean, and hackled.

249 KARDAKOF, M. *Kotelnich Circ. Viatka Gov.*—Flax.

250 KAZAN-KHAN-MUSTAFA-OGLOO, *Kooba, Bakoo Gov.*—Madder-root.

251 KIRGHIZ DISTRICT ADMINISTRATION, *Orenburg.*—Wild madder root.

252 KOOZMIN, J. *Boody Village, Rogachef Circ. Mohilef Gov.*—Lime-tree bast mats.

253 KOPILOF, M. *Potchep Borough, Mglin Circ. Tchernigof Gov.*—Clean hemp.

254 KORNILOF, J. *Doorashkovo Village, Lookaïanof Circ. Nijni-Novgorod Gov.*—An oaken barrel.

255 KRIEGSMANN, A. *Riga.*—Samples of corks.

256 MASHADÉE-HADJI-ALI-OGLOO, *Kooba, Bakoo Gov.*—Madder-root.

257 MEDVEDEF, T. *Medvedef Village, Mohilef Gov.*—Flax.

258 MOROZOF, P. *Pantzerevka Village, Gorodishche Circ. Penza Gov.*—A bag in lime tree bast.

259 MUSTIALA AGRICULTURAL INSTITUTION, *Tavasthus, Finland.*—Tanning bark used in Finland.

260 NEMILOF, A. M. *Orel.*—Samples of half-clean and clean hemp.

261 MALOKROSHECHNOY, I. *Pudosh Olonetz Gov.*—Flax.

262 VOLOGDA CROWN DOMAINS, *Sol Vychegodsk Circ. Vologda Gov.*—Vychegodsk flax.

263 OBRASTZOF, B. *Rshef, Tver Gov.*—Clean flax.

264 ONANOF, *near Kootaïs.*—Clean cotton.

265 OORAL KOZAKS, *Orenburg.*—Wild madder root.

266 PETCHORA TIMBER TRADE CO. *St. Petersburg.*—Larch logs, from Petchora River.

267 PHILEMONOF, T. *Semakof Settlement, Mohilef Gov.*—Flax

268 POOZANOF, M. A. *Schigrof Circ. Koorsk Gov.*—Specimens of fine hemp.

269 PROKHOROF, A. *Belef-Toola Gov.*—Clean hemp.

270 ROTCHEF, I. *Mezen Circ. Archangel Gov.*—Larch-tree sponges.

271 SAVITCH, I.—Walnut wood, exported from Caucasus by the Russian Steam Navigation and Trading Company.

272 SCHEGLOF, N. *Schershof Village, Nijni-Novgorod Gov.*—Fishing net.

273 SEIDLITZ, N. *Nookha, Tiflis Gov.*—Safflower.

274 SHESTOF, J. *Mshaga Village, Nijni-Novgorod Circ.*—Samples of flax.

275 SOROKIN, P. *Belef, Toola Gov.*—Samples of hemp.

276 SOROKIN, R. & S. BROS. *Belef, Toola Gov.*—Clean hemp.

277 SOROKIN, B. *Belef, Toola Gov.*—Hemp.

278 TCHOOKMALDIN, N. *Tiumen Circ. Tobolsk Gov.*—Hemp.

279 RUSSIAN AMERICAN INDIA RUBBER CO. *St. Petersburg.*—Elastic bands, tubes, interlayers, &c.

280 VASSILTCHIKOF, PRINCE A. *Vybit Village, Starorooss Circ. Novgorod Gov.*—Flax.

281 VOLKERSAM, BARON G. VON, *Papenhof, Coorland Gov.*—Samples of flax.

282 SADYRIN, PH. *Kotelnitch Circ. Viatka Gov.*—Flax.

CLASS 5.

283 DEMIDOF, P. P. *Nijni-Tagilsk Iron Works, Perm Gov.*—Rails, with bolts and fittings.

CLASS 6.

285 FROEBELIUS, T. *Carriage Factory, St. Petersburg.*—A town carriage.

286 JAKOVLEF, P. *St. Petersburg.*—Droshky and sledge for racing.

287 KOOPIDONOF, T. & A. BROS. *Moscow.*—Carriage-springs.

288 LIEDTKE, A. *Warsaw.*—Town carriage for four persons.

289 LUBLINSKI, L. *Warsaw.*—A pony-chaise.

290 MOKHOF, J. *Pagost Vil. Pereïaslof Circ. Vladimir Gov.*—Set of coach springs.

291 NELLIS, C. SEN. *Carriage Factory, St. Petersburg.*—A coach.

292 NELLIS, CH. JUN. *St. Petersburg.*—Drosky (egoistka).

293 POLIAKOF, J. *Mooravikha Village, Nijni-Novgorod Gov.*—Oaken wheels for a town carriage.

294 RENTEL, J. *Warsaw.*—Two-seated town carriage.

295 SCHWARTZE, H. *St. Petersburg.*—Two-seated town calash.

296 WAGNER, T. *St. Petersburg.*—Four-seated calash-landau.

CLASS 7.

297 BOSTANJOGLO, B. *Moscow.*—Crucibles.

298 TECHNICAL SCHOOL, ORDNANCE DEPARTMENT, *St. Petersburg.*—Models of different machinery.

CLASS 8.

299 HECKER, H. *Engine and Agricultural Implement Manufactory, Riga.*—Linseed sorter, colour mills, decimal balances.

300 LIKHATCHEF, COL. *Yaroslaff.*—Machinery for making staves.

CLASS 9.

301 ADMINISTRATION OF KOZAK SETTLEMENT, *Orenburg.*—Share of a sokha (Russian plough).

302 CIEHOWSKI, R. *Linow, Sandomir Circ. Radom Gov.*—Ploughs, grubber, and triangular iron harrow.

303 KLIKOVSKY, PROF. *Kazan.*—Models of beehives.

304 KOSZARSKI, C. *Warsaw.*—Plough, invented by the exhibitor.

305 MUSTIALA AGRICULTURAL INSTITUTION, *Tavasthus Gov. Finland.*—Model of a common drying-kiln for corn.

306 IANITZKY, *Boleslas, Kaligorka Estate, Cherson Gov.*—Implement for beetroot harvesting.

CLASS 10.

307 BOSTANJOGLO, M. *Moscow.*—Bricks and artificial stones, made of Gshelsk clay.

308 CABINET OF HIS IMPERIAL MAJESTY.—Candelabra, vase, and column in jasper.

309 CIECHANOWSKI, J. *Cement Works, Gorodetz & Slawkow, Olkush Circ. Radom Gov. Poland.*—Cement, cement castings, sandstone.

310 EKATERINBURG STONE-POLISHING FACTORY.—Ornamental cups, inkstands, and letter weights, in jasper, porphyry, &c.

311 MUSTIALA AGRICULTURAL INSTITUTION, *Tavasthus Gov. Finland.*—Specimen of shingle-roofing, and shingles.

312 PETERHOF STONE-POLISHING FACTORY, *St. Petersburg Gov.*—Paper weights and vase in nephrite.

313 SCHMIDT, *St. Petersburg.*—Specimens of asphalte.

314 STURM, H. *Dorpat.*—Dutch tile mantel-piece.

315 TSEPENNIKOF, J. *St. Petersburgh.*—Models of pneumatic and bath ovens.

316 ZIMARA, R. *St. Petersburg.*—Model of a pneumatic heating-oven, patented 1860.

CLASS 11.

Sub-Class A.

317 JIGOONOF, N. *St. Petersburg.*—Epaulets and other military accoutrements.

318 SOULKHANOFF, *Goree, Tiflis Gov.*—Cartridge box.

Sub-Class B.

319 NISSEN, W. *St. Petersburg.*—Camp bedstead, mattress, and cushion.

Sub-Class C.

320 AGADJANOF, A. *Alexandropol, Erivan Gov.*—Child's dagger.

321 BOORUNSOOZOF, M. *Akhalzih, Kootaïs Gov.*—Asiatic sabre and dagger.

322 THE COMMANDER OF BASHKIR MILITIA, *Orenburg.*—A wooden cross-bow, with sheath, quiver, and 25 arrows.

323 CROWN STEAM-SHIP FACTORY OF THE PORT OF CRONSTADT.—Gun-carriage, designed by Colonel Pestitch.

324 HADJI-SEID-AGHI-SEID-OGLOO, *Nookha, Bakoo Gov.*—Dagger.

325 KOBATEE-IBRAKHIM-MAHMED-OGLOO, *Daghestan District, Caucasus.*—Dagger.

326 OBOOKHOF, COL. *Orenburg Gov.*—Twelve-pounder cast-steel gun, and a steel ring sawn from the gun before polishing.

327 POPOF, J. *Tiflis.*—Fowling-piece, pistol, powder flask, sabre, dagger, and girdle.

328 TCHIFTALAROF, M. *Akhaltzyk.*—Rifle, fowling-piece, and pistol.

329 VISHNEVSKY, F. *St. Petersburg.*—A revolver, invented by the exhibitor.

330 YOOST-SAMAN-OGLOO, *Nookha, Bakoo Gov.*—Gun barrel.

331 ZLATOÜST CROWN ARMOUR FACTORY, *Southern Oooral, Orenburg Gov.*—Sword-blades, swords and scythes, polished cast-steel breastplate.

CLASS 12.

Sub-Class A.

332 CROWN STEAM-SHIP FACTORY OF THE PORT OF CRONSTADT.—Ship fittings.

333 HAAKER, A. *St. Petersburg.*—Model of the man-of-war Victory.

334 IJORA ADMIRALTY IRON WORKS, *near St. Petersburg.*—Specimens of chain-cable and cat-block.

335 MANUFACTORY & MODEL ROOM OF NAVAL ARCHITECTURE, *St. Petersburg.*—Models of ships for the Imperial Russian navy.

336 YEGOROF, LIEUT. J. *St. Petersburg.*—Model of the 111 gun-ship Emperor Nicholas I.

CLASS 13.

337 KADINSKY, PROF. K. *St. Petersburg.*—Logarithmical or calculating sliding rule.

338 NAUTICAL INSTRUMENT MANUFACTORY, *St. Petersburg.*—Sea barometer, compass, with illuminator, and patent log.

339 PIK, J. *Warsaw.*—Model of hydraulic press, electro-galvanic appâratus, magnifying glass, spectacles in filagree setting, crystal thermometer, and apparatus for assaying.

340 REISSNER, PROF. *Dorpat.*—Microscopic objects.

341 STEINBERG, T. *Osseenovoy-Koost, near Saratof.*—Controlling apparatus for distilleries.

CLASS 14.

342 DENIER, *St. Petersburg.*—Portraits.

343 FAJANS, M. *Warsaw.*—Photographs.

344 LEVITZKY, S. 22, *Rue de Choiseul, Paris.*—Portraits.

345 LORENS, A. *St. Petersburg.*—Photographic portraits and stereoscopic prints.

346 MIECZKOWSKI, J. *Warsaw.*—Photographic prints and visiting cards, on albumenized paper.

347 PETROFSKI, *St. Petersburg.*—Photographic copies: Bruni's picture "The Brazen Serpent," and Ch. Brulof's "Last Day of Pompei."

348 ROSENBERG, *Riga.*—Coloured photographs, without after-touch (elaiögraphy).

349 RUMINE, G. 5, *Lower Gore, Kensington, London.*—Life-size portraits, photographed with carbon, on canvas, oil painted, &c.

350 SHPAKOFSKI, A. *St. Petersburg.*—Portraits.

CLASS 15.

351 SON, H. *Moghilef.*—A clock, with the inventor's new mechanism.

CLASS 16.

352 BECK, P. *St. Petersburg.*—A pianoforte.

353 RUDERT, H. *Warsaw.*—Musical instruments.

CLASS 17.

354 CROWN FACTORY FOR SURGICAL INSTRUMENTS, *St. Petersburg.*—Various sets of surgical instruments.

355 VARIPAEF, TH. *Pavlovo, Gorbatof Circ. Nijni-Novgorod Gov.*—Surgical instruments.

CLASS 18.

356 THE ADMINISTRATION OF THE ORENBURG KOZAK SETTLEMENTS.—A knitted cover, from English cotton yarn.

357 AREKOF, G. *Novobayazet, Erivan Gov.*—Cotton stuffs.

358 BOORNAYEF, A. *Kazan.*—Long cloth.

359 BORISSOVSKY, M. *Pereslavl-Zalessky, Vladimir Gov.*—Cotton yarn.

360 CAUCASIAN AGRICULTURAL SOCIETY, *Tiflis.*—Cotton stuff, called "Noshoree."

361 DEINESS, J. *Norki Settlement, Saratof Gov.*—Cotton cloth (sarpinka).

362 FINLAYSON & CO. *Tammerfors, Finland.*—Cotton cloth and yarn.

363 KINDSVATER, O. *Splavnookha Settlement, Saratof Gov.*—Cotton cloth (sarpinka).

364 MZIREOOLOF, SOPHIA, *Tarsky District, Caucasus.*—Cotton yarn.

365 FORSSA, *Cotton Mill Company.*—Finland shirting.

366 SPADI, W. *Norki Settlement, Saratof Gov.*—Cotton-cloth (sarpinka).

367 PYCHLAU, *Riga.*—Cotton twist.

368 REINEKE, A. *Popovka Settlement, Saratof Gov.*—Cotton cloth (sarpinka), and a checked head kerchief.

369 SCHEFER, J. *Goly Karaslish Settlement, Saratof Gov.*—Red checked cotton cloth; shirting.

370 SMIDT BROS. *Oost-Zolikha Settlement, Saratof Gov.*—Six pieces of cotton cloth (sarpinka).

CLASS 19.

371 THE ADMINISTRATION OF THE ORENBURG KOZAK SETTLEMENTS.—Hand-spun thread; linen, plain and twilled.

373 ALEXANDROF, EUDOXIA, *Sopelki Village, Yaroslaf Circ.*—A piece of linen.

374 CAZALET, A. & SONS, *St. Petersburg.*—Rope yarn, bolt rope, cordage, white rope log lines, &c.

375 DOMBROWICZ, C. *Dobrovolia, Mariampol Circ. Augustowo Gov.*—Table-cloth, napkins, and towels.

376 HIELLE, C. & DITTRICH, CH. *Girardovo, Lovicz Circ. Warsaw Gov.*—Linen, tablecloths, napkins, and towels.

378 NEMILOF, A. *Orel.*—Hemp yarn.

379 LUCKS, C. *Halberstadt Settlement, Berdiansk Circ. Tauride Gov.*—Ropes.

380 RIGA LOCAL COMMITTEE.—Hemp yarn.

381 NOVIKOF, A. & J. *Briansk, Orel Gov.*—Hemp rope-yarn, white and tarred, and clean hemp (mas plant).

382 STIEGLITZ, BARON, *near Narva, St. Petersburg Gov.*—Assortment of hemp and flax, sail cloth.

383 TSHOOKMALDIN, N. *Tiumen Circ. Tobolsk Gov.*—A towel and linen.

CLASS 20.

384 ABDOOL-BEK-HADJI-MEERAM-BECK-OGLOO, *Shemakha, Bakoo Gov.*—Red and yellow móv (silk stuff).

385 AGA-KISHI-BEK-MAHMED-HASSEIN-BECK-OGLOO, *Lenbaran Vil. Shoosha Circ, Bakoo Gov.*—Silk stuff (Djeedjim).

386 ANDRONIKOF, PRINCE S. *Bakoortzyk Vil. Signah Circ. Tiflis Gov.*—Raw silk and cocoons.

387 ATAKEESHEE-OGLOO, *Khachmaz Vil. Nookha Circ. Bakoo Gov.*—Raw silk.

388 BEGLIAROF, *Akoolissy Vil. Ordoobat Circ. Erivan Gov.*—Samples of raw silk.

389 CAUCASIAN AGRICULTURAL SOCIETY, *Tiflis.*—Silk stuff (móv) and blankets.

390 CLEMENTZ, LÖH, & Co. *Quellenstein, Livonia.*—Samples of silk.

391 DAVYDOFF, *Varagirt Vil. Ordoobat Circ. Erivan Gov.*—Raw silk and cocoons.

392 DEUTSCHMAN, A. *Tiflis.*—Raw silk.

393 FET-ALI-OGLOO, *Rostadir Vil. Shemacha Circ. Bakoo Gov.*—Raw silk.

394 KOMAROVSKY, COUNTESS MARIA, *Prilooki Circ. Poltava Gov.*—Raw silk.

395 KRIPNER, P. G.—Raw silk, cocoons.

396 MAHMUD-ALI-YUSSOOF-OGLOO, *Buck Degniz Vil. Nookha Circ. Bakoo Gov.*—Silk stuff, called "Djajeem."

397 MARTHI-BOODOOGHIA-SHVEELEE, *Kvareli Vil. Telav. Circ. Tiflis Gov.*—Raw silk.

398 MASHADI - ALI - ZARAB - OGLOO, *Shoosha, Bakoo Gov.*—Silk stuff, called "Kassabec."

399 MESHADI-HUSSEIN-OGLOO, *Shoosha, Bakoo Gov.*—Silk stuff, called "Aleeshee."

400 MORTIEROSSOF, M. *Nookha, Bakoo Gov.*—Raw silk and cocoons.

401 MOSCOW COMMERCIAL SCHOOL.—Cocoons and raw silk from common silkworms hatched, fed, and reared in the school-garden at Moscow, on white mulberry leaves.

402 NASSIR-ABDOOLARTZIZ-OGLOO, *Bakoo Gov.*—Red and black móv (silk stuff).

403 NISSEN, A. *near St. Petersburg.*—Silk stuffs, raw silk, and organzine spun from cocoons produced in the south governments of Russia.

404 PETAÏL-OGLOO-MULLAH, *Zuzzid Vil. Nookha Circ. Bakoo Gov.*—Raw silk.

405 STROOKOF, P. *Ekaterinslov Gov.*—Raw silk.

406 REDA-OGLOO, *Zengishali Vil. Koobinsk Circ. Bakoo Gov.*—Raw silk.

407 SAPOJNIKOF, BROS. *Astrakhan.*—Raw silk and cocoons.

408 SEMENOF, *Vartaly Vil. Nookha Circ. Bakoo Gov.*—Raw silk and cocoons.

409 SHOROEF, *Avak-Shemakha, Bakoo Gov.*—Blue móv (silk stuff).

410 VORONIN, BROS. & ALEXEYEF, B. *Nookha, Bakoo Gov.*—Raw silk and cocoons.

411 YOORÏEVA, MISS L. *Wladimirovka Village, Stavropol Gov.*—Raw silk.

CLASS 21.

412 THE ADMINISTRATION OF THE ORENBURG KOZAK SETTLEMENT.—Woollen yarn, cloths from camels' hair.

413 ALI-PANAH-OGLOO, *Dash-Sala-Ogloo Vil. Elisabethopol Circ. Tiflis Gov.*—Mafrash and Palas.

414 ARMAND, E. *Moscow.*—Plain and figured Orleans, Paramata satin, cotton warp, and Cashmere, pure worsted.

415 BABKIN, BROS. HEIRESSES OF, *Koopavna, Bogorodsk Circ. Moscow Gov.*—Broadcloth called "Mezeritski."

416 BAKHROOSHINA, N. S. & SONS, *Moscow.*—Broad-cloth, black and blue.

417 CAUCASIAN AGRICULTURAL SOCIETY, *Tiflis.*—Ropes of woollen yarn.

418 CLEMENTZ, LÖH, & Co. *Quellenstein, Livonia.*—Woollen yarns and cloth.

419 COMMANDER OF BASHKIR MILITIA, *Orenburg Government.*—Bashkir cloth.

420 FIEDLER, G. A. *Opatovka, Kalish Circ. Warsaw Gov.*—Broad cloths.

421 GOOCHKOF, E. & SONS, *Moscow.*—Plain worsted fabric.

422 HADJI-YOOSOF-SHABAN-OGLOO, *Zakatala Circ.*—Common Lesghin cloths, and shawl.

423 BAERG, H. *Halbstadt Settlement, Berdiansk Circ. Tauride Gov.*—Cloth and flannel.

424 IOKISH, B. *Michalkof Manufactory, Moscow Circ.*—Cloth.

425 KOOBAREF, M. *Klintzy Borough, Soorash Circ. Tchernigof Gov.*—Broad cloth, black and grey.

426 KOPALIANTZ, N. *Alexandropol, Erivan Gov.*—Samples of woollen yarn (dyed).

427 MAHOM - AKBAR - OGLOO, *Zakatala Circ.*—Common Lesghin cloth, and shawl.

428 NAVRUS, G. *Gorskee Circ.*—Lesghin cloth shawl.

429 OORALIAN KOZAKS, *Orenburg Gov.*—Common cloth.

430 SCHEFER, J. *Goly Karaslish Settlement, Saratof Gov.*—Half silk kerchief.

431 SCHEPELER, T. *Riga.*—Mixed fabrics.

432 SELIVERSTOF, COLONEL N. *Roomianstof Factory, Korsun Circ. Simbirsk Gov.*—Samples of cloth.

433 SOOLHANOF, *Goree, Tiflis Gov.*—Common cloths and tiftik; pouch and palas, with felted lining.

434 THILO, A. *Riga.*—Cloth.

435 TURPEN, E. *Nikolaevsky Worsted Manufactory, St. Petersburg.*—Wool and worsted.

436 UNGERN, BARON STENBERG, *Isle of Esel.*—Cloth.

437 VADYM-AHVERDI-OGLOO, *Arab Redine Vil. Shemakha Circ. Bakoo Gov.*—Palas.

438 VONSOVSKY, *Zeiva Vil. Etchmiadzin Circ. Erivan Gov.*—Cloth of Koortinsk wool.

439 WERGAU, J. *Lodz, Lenczic Circ. Warsaw Gov.*—Woollen table covers.

440 WOEHRMANN & SON, *Zintenkhoff Factory, near Pernau, Livonia.*—Cloths.

CLASS 22.

441 FLANDIN & Co. *Klin Circ. Moscow Gov.*—Carpets.

442 HASSAN - ALI - DJEBRAEL - OGLOO *Sharbachee Vil. Shemakha Circ. Bakoo Gov.*—Carpet.

443 MUSTAPHA - HADJI - OGLOO, *Oodoola Vil. Shemakha Circ. Bakoo Gov.*—Carpet.

444 MUSTAPHA - KARA - MIRZA - OGLOO, *Ymam-Koolee Vil. Kooba Circ. Bakoo Gov.*—Carpet.

445 NOOR-ALI-FET-ALI-OGLOO, *Sharbachee Vil. Shemakha Circ. Bakoo Gov.*—Carpet.

446 TSHOOKMALDIN, N. *Tiumen Circ. Tobolsk Gov.*—A carpet,—specimen of peasant-women's work.

CLASS 23.

447 ABKHAZOF, PRINCE D. *Kakhetia.*—Silk, dyed with safflower.

448 ADAM, CH. *Th. Bittepage Manufactory, Shlisselburg Circ. St. Petersburg Gov.*—Chintzes.

449 STROGANOF SCHOOL OF TECHNICAL DRAWING, *Moscow.*—Napkins, horsecloth, and a carpet.

450 BARANOVA, ALEXANDRA, *Troitzko-Alexandrov Manufactory, Alexandrovsk Circ. Vladimir Gov.*—Dyed and printed cotton goods.

451 GOOTCHKOF, E. SONS OF, *Moscow.*—Woollen, cotton, and mixed stuffs.

452 GOOTCHKOF, J. *Moscow.*—Dyed and printed woollen, silk, and cotton mixed stuffs.

453 HUBNER, A. *Moscow.*—Chintzes.

454 PROKHOROF, BROS. *Moscow.*—Cotton print.

455 REZANOF, T. *Moscow.*—Printed cotton kerchiefs.

456 TZINDEL, E. *Moscow.*—Chintz.

457 ZOOBKOF, P. *Vosnecensk Borough, Shooya Circ. Vladimir Gov.*—Chintzes.

CLASS 24.

458 THE ADMINISTRATION OF THE ORENBURG KOZAK SETTLEMENTS.—Embroidered linen towels, towel trimmings, and laces.

459 ALMAZOF, S. *Torjok. Tver Gov.*—Articles of embroidered Morocco leather.

460 BAHCHINOF, G. *Alexandropol, Erivan Gov.*—Embroidered slippers.

461 CAUCASIAN AGRICULTURAL SOCIETY, *Tiflis.*—Embroidered table-cloth and arm-chair coverings.

462 DIATCHKOVA, P. *Torjok, Tver Gov.*—Girdles, sashes, caps, and aprons.

463 DUTACQ, MRS. M. L. *St. Petersburg.*—Crotchet-work carpet, both sides alike.

464 KARELIENA, *Torjok, Tver Gov.*—Embroidered boot and shoe fronts. Caps and cushions in Morocco leather, velvet, satin, and leather. Patchwork.

465 KISSELEVSKY, CATHERINE.—Embroidered cushion.

466 KRZYWICKA, MARY, *Warsaw.*—Bed coverings and small table-covers in worsted work.

467 PROHASKA & STENZEL, *St. Petersburg.*—Table cover of cloth patchwork.

468 SAPOJNIKOF, HEIRS OF, *Moscow.*—Gold and silver brocades.

469 TER-POGOSSOF, *Nookha, Bakoo Gov.*—Embroidered caparison.

470 WITKOWSKA, NATALIE, *Warsaw.*—A carpet in worsted work.

471 WORONCOW-WELIAMINOV, *Nepomucena. Warsaw.*—A carpet in worsted work.

CLASS 25.

Sub-Class A.

472 AGRICULTURAL DEPARTMENT, *St. Petersburg.*—Collection of sheep, lamb, and goat skins, raw and tanned.

473 ALIBERT, N. P. *Irkootsk Gov. Siberia.*—Stuffed sables.

474 ARCHANGEL CHAMBER OF CROWN DOMAINS.—Dressed skins of wolf, fox, and swan.

475 BOOKIN, J. *Bolshoe - Moorashkino, Kniaghinin Circ. Nijni-Novgorod Gov.*—Black lamb and sheep skins.

476 DITZEL, N. *St. Petersburg.*—Carpet made from seal skins.

477 GOLOF, T. *Mezen Circ. Archangel Gov.*—Dressed bear skin.

478 OORALIAN KOZAKS, *Orenburg Gov.*—Furs and skins.

479 PHILIPOF, H. *Mezen Circ. Archangel Gov.*—Undressed blue fox skins.

480 POPOF, M. *Mezen Circ. Archangel Gov.*—Dressed skin of reindeer.

481 PRESNIAKOF, P. *Bolshoe-Moorashkino, Kniaghinin Circ. Nijni-Novgorod Gov.*—White and black lamb skins.

482 ROOSHNIKOF, J. *Mezen Circ. Archangel Gov.*—Blue fox skins, undressed.

483 ROTCHEF, I. *Mezen Circ. Archangel Gov.*—Young reindeer skins, dressed.

484 ROTCHEF, J. *Mezen Circ. Archangel Gov.*—Dressed skins of reindeer, and black and blue foxes.

485 RUSSIAN-AMERICAN CO. *St. Petersburg.*—Skins of sea-otter, fox, and seal.

486 SHRAPLAU, H. *Moscow.*—A carpet, and a morning gown made of fur.

487 TYRKASOF, N. *Mezen Circ. Archangel Gov.*—Dressed otter skins.

488 YERMILOF, J. *Ostasheva Village, Romano-Borisoglebsk Circ. Jaroslaf Gov.*—Tanned sheep skin coats, called "Romanoffsky."

Sub-Class B.

489 ARTEYEF, S. *Mezen Circ. Archangelsk Gov.*—Eider down.

490 MIRONOF, G. *Troobino Village, Maloiaroslavez Circ. Kalooga Gov.*—Goose down, feathers, and wings.

Sub-Class C.

491 BEZROOKAVNIKOF-SOKOLOF, A. S. *St. Petersburg.*—Samples of horse-hair, raw, cleaned, and curled.

492 BRÄUTIGAM, MARY, *Hair-cloth and Curled Hair Manufactory, St. Petersburg.*—Samples of horse-hair.

493 KONDRATENKO, P. *St. Petersburg.*—Stage wigs.

CLASS 26.

Sub-Class A.

494 ARCHANGELSK CHAMBER OF CROWN DOMAINS.—White chamois dressed reindeer hides.

495 ARTEYEF, S. *Merzen Circ. Archangel Gov.*—Yellow chamois-dressed reindeer hides.

496 BAHROOSHINA, N. & SONS, *Moscow.*—Boot, enamelled, and morocco leather.

497 BAUERFEIND, T. F. *Warsaw.*—Leather.

498 BROOSNITZIN, N. *St. Petersburg.*—Assortment of leather.

499 GORIACHKIN, G. *Klintzy-Borough, Sooraj Circ. Tchjernigof Gov.*—Calf and morocco leather.

500 HEINRICH, *Riga.*—Leather.

501 HUBNER, N. *St. Petersburg.*—Leather and tanned skins.

502 KELBI-KHAN-OGLOO, *Nakhichevan, Erivan Gov.*—Deer-skin.

503 KHECHATOOROF, S. *Signah, Tiflis Gov.*—Sheep-leather.

504 LIEDTKE, J. H. *Warsaw.*—Samples of leather.

505 MILLER, ERDMAN, *St. Petersburg.*—Boot-legs and vamps.

506 MESNIKOF, *Stavropol.*—Oxen and buffalo leather.

507 MILLER, C. *St. Petersburg.*—Boot-legs and vamps.

508 MILLER, J. T. *St. Petersburg.*—Boot-legs and vamps.

509 POPOF, M. *Mezen Circ. Archangelsk Gov.*—Chamois dressed reindeer hides.

510 SHEVNIN, A. *St. Petersburg.*—Samples of leather.

511 SHOOVALOF, P. & T. BROS. manufacturers, *Moscow.*—Boot leather.

512 TEMLER, C. & A. & SZWEDE, L. *Warsaw.*—Sole, saddle, and enamelled leather, calf and morocco leather, strap leather.

513 VASSILIEF, A. *Moscow.*—Samples of parchment.

514 SKOOBEEF TANNING CO. *Alexandria Circ. Kherson Gov.*—Leather.

515 YEREMEYEF, T. *Perm.*—Russian leather (yooft), calf and morocco leather.

516 ZAREENOF, M. *Akhalzik, Kootaïs Gov.*—Dyed kid-leather.

Sub-Class B.

517 CAUCASIAN AGRICULTURAL SOCIETY, *Tiflis,*—Kabardian saddle, bridle, and whip.

518 TSHOOKMALDIN, N. *Tiumen Circ. Tobolsk Gov.*—A common bridle.

519 KIRGHIZ DISTRICT ADMINISTRATION *Orenberg Gov.*—Kirghiz saddlery.

CLASS 27.

Sub-Class A.

520 ALEXANDRO, A. *Kootaïs.*—Imeritian caps (papaneeki).

521 CAUCASIAN AGRICULTURAL SOCIETY, *Tiflis.*—National fur-cap (papakha).

522 DADASH-MOLLA-HOOSSEIN-OGLOO, *Shoosha.*—Persian cap.

523 KHECHATOOROF, A. *Signah, Tiflis Gov.*—Tooshinsk felted caps.

524 KIRGHIZ DISTRICT ADMINISTRATION, *Orenburg Gov.*—Kirghiz caps.

525 RODIONOF, T. *Novostarinska Village, Kniaghinin Circ. Nijni-Novgorod Gov.*—Lambskin caps.

526 ROTCHEF, I. *Mezen Circ. Archangelsk Gov.*—Samoyed woman's cap of dressed skins.

527 SOODAKOF, S. *Moscow.*—Hats.

528 SYROF, N. *Zaproodna Village, Kniaghinin Circ. Nijni-Novgorod Gov.*—Bookharian white lambskin cap.

529 ZIMMERMAN, F. *St. Petersburg.*—Hats, caskets, and caps.

Sub-Class B.

530 FLEROVSKY, D. *Tomsk. Siberia.*—A woman's head-dress (Kokoshnik), made of birch-bark.

Sub-Class C.

531 ADMINISTRATION OF THE KOZAK SETTLEMENTS, *Orenburg.*—Goats'-hair gloves and worsted mittens.

532 AGRICULTURAL DEPARTMENT, *St. Petersburgh.*—Fur clothes (Malitzas) of the Zyrian and Samoyed.

533 ARTEYEF, J. *Mezen Circ. Archangel Gov.*—Samoyed coat of dressed reindeer skin.

534 AVASTERKOF, J. P. *Lookoyanof Circ. Nijni-Novgorod Gov.*—Woollen mittens.

535 CAUCASIAN AGRICULTURAL SOCIETY.—Caucasian dress and felt cloak.

536 CHANCERY OF THE GOVERNMENT GENERAL OF ORENBURG & SAMARA.—Hand-spun and knit goats' hair scarfs.

537 CHILINCHAROF, M. *Akhalzik, Kootaïs Gov.*—Black wrapper, with gold and silver embroidery.

538 THE COMMANDER OF THE BASHKIR MILITIA, *Orenburg Gov.*—A belt with pouch.

539 KASSUM-YVOZBASH-ALI-OGLOO, *Leubaran Vil. Shoosha Circ. Bakoo Gov.*—Black shawl.

540 KIRGHIZ DISTRICT ADMINISTRATION, *Orenburg.*—Kirghiz belts.

541 MAHMET-KEEZI, WIDOW, *Sala-Ogly Vil. Elisabethopol Circ. Tiflis Gov.*—A pouch.

542 MALAVAL, O. *St. Petersburg.*—Specimens of gloves.

544 OORALIAN KOZAKS, *Orenburg Gov.*—Sashes.

545 PAJER, M. *Warsaw.*—Plaited corset, invented by the exhibitor.

546 ROTCHEF, I. *Mezen Circ. Archangel Gov.*—Samoyed coats of dressed skins.

547 STOLZMAN, A. *Warsaw.*—Ladies' sashes.

548 TSHOOKMALDIN, N. *Tiumen Circ. Tobolsk Gov.*—Mittens of sheepskin.

549 YOOSBASHEF, A. *Khachmaz Vil. Nookha Circ. Bakoo Gov.*—Silk trousers.

Sub-Class D.

550 HUBNER, N. *St. Petersburg.*—Boots, shoes, and over-shoes; boots made of plaited leather.

551 LERCH, W. *St. Petersburg.*—Boots and goloshes.

552 LOOJIN, A. *Korchef Circ. Tver Gov.*—Boots.

553 MASHADI-KALI-OGLOO, *Shoosha, Bakoo Gov.*—Shoes (Koshi).

554 MOKRIAKOF, A. *Korchef Circ. Tver Gov.*—Boots.

556 ROTCHEF, I. *Mezen Circ. Archangel Gov.*—Samoyed boots and half boots.

557 RUSSIAN-AMERICAN INDIA-RUBBER Co. *St. Petersburg.*—Caoutchouc goloshes and half-boots.

558 LAUBE, W. *St. Petersburg.*—Pair of boots.

559 SELIVERSTOF, P. *Pochinok, Nijni-Novgorod Gov.*—Fell boots.

560 SHIRMER, E. *Moscow.*—Boots.

561 SITNOF, P. *Ankino Village, Korchef Circ. Tver Gov.*—Boots and goloshes.

562 SOOLKHANOF, *Goree, Tiflis Gov.*—Worsted socks.

563 STOLAREF, A. *Korchef Circ. Tver Gov.*—Boots.

564 STOLAREF, S. *Korchef Circ. Tver Gov.*—Boots.

565 SVEDONTSEF, A. *Korchef Circ. Tver Gov.*—Boots.

566 TER-POGOSSOF, *Nookha, Bakoo Gov.*—Pachichi (boot fronts).

567 TONENKOF, J. *Baykovo Village, Lookoyanof Circ. Nijni-Novgorod Gov.*— Plaited bark shoes.

568 TSELIBEYEF, T. *St. Petersburg.*—Boots and shoes.

569 TSHOOKMALDIN, N. *Tiumen Circ. Tobolsk Gov.*—Peasants' shoes of sheepskin.

570 VANEZOF, M. *Tiflis.*—Common and travelling boots.

571 VOKOOYEF, T. *Mezen Circ. Archangel Gov.*—Boots of reindeer skin.

572 YOOSBASHIANETZ, G. *Damboolak Vil. Nookha Circ. Bakoo Gov.*—Worsted socks.

CLASS 28.

Sub-Class A.

573 EPSTEIN, J. *Sorewka, Gostyn Circ. Warsaw Gov.*—Samples of paper.

574 TROITZKO-KONDROVSKI PAPER FACTORY Co. *Medyn Circ. Kalooga Gov.*— Assortment of paper.

575 VARGOONIN, A & P. BROS. *Nevski Paper Factory, St. Petersburg.*—Samples of paper.

Sub-Class B.

576 GERKE, A. *St. Petersburg.*—Red marking and printing ink.

577 KRAUSE, J. *Warsaw.*—Samples of sealing-wax.

Sub-Class C.

578 FAJANS, M. *Warsaw.*—Lithographs and chromolithographs.

579 KANTOR, A. *Warsaw.*—Samples of bookbinding.

580 LEMAN, T. *St. Petersburg.*—Samples of types and puncheons.

Sub-Class D.

581 HAAG, C. *St. Petersburg.*—Album.

CLASS 29.

Sub-Class A.

582 NAMANSKI, A. *St. Petersburg.*—Table for facilitating the demonstration of elementary arithmetical rules.

583 NOWOLECKY, A. *Warsaw.*—Illustrated historical alphabet, printed on glazed calico.

584 ROGOJSKI, J. *Kelce, Radom Gov.*—Two synoptic tables, facilitating the study of chemistry.

585 ZOLOTOF, B. *St. Petersburg.*—Set of his publications for elementary instruction.

Sub-Class B.

586 GOTLUND, *Helsingflors.*—Collection of mushrooms, dried by the exhibitor's peculiar process.

587 HAN, C. *Dorpat.*—Figures in national dresses.

588 HEISER, H. *St. Petersburg.*—Collection of ethnological and zoological models.

589 IMPERIAL MINING DEPARTMENT *St. Petersburg.*—A standard set of Russian monies, weights and measures.

590 OOSPENSKY, P. *St. Petersburg.*—Collection of stuffed birds.

591 UNIVERSITY OF DORPAT.—Wax models of fruits, and parts of the human body.

CLASS 30.

Sub-Class A.

592 FREIBERG, A. *Billiard Factory, St. Petersburg.*—Billiard, improved structure, patented.

593 PETERHOF STONE-POLISHING FACTORY, *near St. Petersburg.*—Inlaid tables, cupboard, chair, jardiniere, &c.

Sub-Class B.

594 CAMUSET & Co. *St. Petersburg.*—Paper-hangings.

595 COMMANDER OF THE BASHKIR MILITIA, *Orenburg.*—Scoop and cup made from maple root.

596 FLEROVSKY, D. *Irkootsk, Siberia.*—Dish carved in birchwood (root and bark).

597 SALZMAN, J. *Warsaw.*—A wood carving, after a French bronze, by MENE.

598 TOLSTINSKI, J. *St. Petersburg.*—Paper-hangings.

599 VETTER, A. & Co. *Warsaw.*—Paper-hangings.

CLASS 31.

Sub-Class A.

600 BOORAKOF, A. *Tver.*—Nails, hand-made.

601 KOROLEF, G. *Tarki Village, Gorbatof Circ. Nijni-Novgorod Gov.*—Padlocks.

602 NASONOF, N. *Oostug, Vologda Gov.*—A chest with secret locks.

603 RAÏVOLOVO WORKS, *near St. Petersburg, Finland.*—Stonecutter's and locksmith's tools, and locks.

604 SHEBAROF, J. *Tarki Village, Gorbatof Circ. Nijni-Novgorod Gov.*—Padlocks.

605 SEID IBRAHIM KOORTAY-OGLOO, *Shoomy Vil. Alooshta District, Crimea.*—Spades.

606 TER-KREEKOROF, D. *Shoosha.*—Iron bridle bit.

607 VARYPAEF, T. *Pavlovo, Gorbatof Circ. Nijni-Novgorod Gov.*—Locks.

608 VORONOF, A. *Ooliagii Vil. near Olonetz.*—Spades and drainage implements.

609 WOJNICKI, *Warsaw.*—Padlocks.

Sub-Class B.

610 CHOPIN, F. *St. Petersburg,*—Monument representing the Empress Catherine II.

611 HAYDOOKOF, A. *St. Petersburg.*—Specimens of tin and lead foil.

612 JONOVA, *Toola.*—Tea-urns.

613 KUMBERG, J. *St. Petersburg.*—Lamps and lanterns used in the Imperial navy.

614 LOMOF, E. J. *Toola.*—Tea-urns, in tombac (yellow metal).

615 MORAND & CO. *St. Petersburg.*—A group in bronze—The overthrow of idols in Russia, in the tenth century; and the model of a monument to Admiral Lazaref; both after designs and moulds of Professor Pimenof.

616 ROCHEFORT, COUNTESS OLGA, *Perm Gov.*—Medallion of Peter the Great, bronze cast, chased, and gilt.

617 ROODAKOF, S. *Toola.*—Tea-urns and coffee-pot.

618 SAMGHIN, D. *Moscow.*—Church bells.

619 TCHERNIKOF, N. *Toola.*—Tea-urns and church bells.

CLASS 32.

Sub-Class B.

620 BAKANOF, M. *Reebino Vil. Gorbatof Circ. Nijni-Novgorod Gov.*—Penknives.

621 DOORACHKIN, M. *Vorsma Village, Gorbatof Circ. Nijni-Novgorod Gov.*—Edge tools.

622 IVANOF, J. *Tiumen Circ. Tobolsk Gov.*—A mortice chisel.

623 JARKOF, P. *Reebino Vil. Gorbatof Circ. Nijni-Novgorod Gov.*—Penknives.

624 KIREELOF, B. *Reebino Vil. Gorbatof Circ. Nijni-Novgorod Gov.*—Pocket and penknives.

625 KIREELOF, M. *Reebino Vil. Gorbatof Circ. Nijni-Novgorod Gov.*—Penknives.

626 KIRGHIZ DISTRICT ADMINISTRATION, *Gov. of Orenburg.*—Kirghiz knives.

627 LEPESHKIN, J. *Martova Vil. Gorbatof Circ. Nijni-Novgorod Gov.*—Penknives.

628 LEVIN, T. *Dolotkovo Village, Gorbotof Circ. Nijni-Novgorod Gov.*—Cutlery.

629 OSMAN, OMER-OGLOO, *Goorsoof Vil. Alooshta District, Crimea.*—Axes.

630 MYSHIN, T. *Pavlovo Village, Gorbatof Circ. Nijni-Novgorod Gov.*—Scissors.

631 REESEF, J. *Boolatnikova Vil. Gorbatof Circ. Nijni-Novgorod Gov.*—Penknives and files.

632 VARIPAEF, T. *Pavlovo, Gorbatof Circ. Nijni-Novgorod Gov.*—Cutlery and edge tools.

633 VOROTILOF, A. *Pavlovo, Nijni-Novgorod Gov.*—Table and penknives.

634 ZAVIALOF, BROS. *Vorsma Village, Gorbatof Circ. Nimi-Novgorod Gov.*—Cutlery, files, &c.

CLASS 33.

635 ALEXEEF, V. *Moscow.*—Gold and silver thread, wire, and spangles.

636 ARAPET-AGADJANOF, *Shoosha.*—Silver buckle for a belt.

637 BEK-TARKHANOF, J. *Shoosha.*—Silver buttons worn by Tartar women.

638 BELIBEKOF, M. *Tiflis.*—Enamelled silver cup, tray, and koolas.

639 BOIANOFSKY, C. *St. Petersburg.*—Silver vases, basket and ostensorium.

640 GALAMKAROF, K. *Akhaltzyk.*—Box in filigree.

641 GOOBKIN, S. *Moscow.*—Works in silver for ecclesiastical and for household use.

642 MORAND & CO. *St. Petersburg Foundry and Galvano-Plastic Establishment.*—A group, representing St. George, by Professor Pimenof.

643 THE REV. E. POPOV, 32, *Welbeck-st.*—The New Testament in chaste silver-gilt binding; a crucifix in chaste silver-gilt, made by Sazikoff, Moscow.

644 SAZIKOF, V. *St. Petersburg & Moscow.*—Ornamental articles in silver, cups, goblets, &c.

645 VERKHOVZEF, T.—Silver plates for a cabinet, with portraits of Tzars.

CLASS 34.

Sub-Class A.

646 IMPERIAL GLASS WORKS, *St. Petersburg.*—Mosaic pictures representing St. Nicholas and two angels.

647 TCHETCHER GLASS WORKS (belonging to the COUNT TCHERNISHOF—KROOGLIKOF), *Rogatchef Circ. Mohilef Gov.*—Window glass.

Sub-Class B.

648 HORDLICZKA, W. & E. BROS. *Czechy, Gov. of Lublin.*—Ornamental and other glass.

649 IMPERIAL GLASS FACTORY, *St. Petersburg.*—Ornamental vases and glass.

CLASS 35.

650 GOLDOBIN, A. S. *Bov Village, near Nijni-Novgorod.*—Common earthenware used by peasants.

651 GRECHISHCHEF, M. *near Riazan.*—Glazed tiles.

652 IMPERIAL CHINA MANUFACTORY, *St. Petersburg.*—China figures, ornaments, and tea services.

653 KOOZNETSOF, *Riga.*—Chinaware.

CLASS 36.

Sub-Class B.

654 KADALOF, T. *Agrapino Village, Lookaianof Circ. Nijni-Novgorod Gov.*—Plaited bark pouch.

655 KEPHAR-MASHADI-DJAFAR-OGLOO, *Oodoola Vil. Shomakha Gov.*—A travelling pouch.

656 NISSEN, W. *St. Petersburg.*—Portmanteaus, travelling bags, and pouches.

657 STOLZMAN, A. *Warsaw.*—Leather cases and trunks; basket, portfolio.

ADDENDA.

CLASS 1.

658 DMITRIEF, N. *near Odessa.*—South Russian prairie soil.

CLASS 2.

Sub-Class A.

659 HESEN, A. *Moscow.* — Specimens of phosphoric matches.

660 PITANCIER, G. & Co. *Odessa.*—Sulphuric and hydrochloric acids, soda, copperas, and other chemical products.

CLASS 3.

Sub-Class A.

661 ARGOOTINSKY-DOLGOROOKOF, PRINCE N. *Tiraspol Circ. Cherson Gov.* — Winter wheat and maize.

661A ABBEY, W. *near St. Petersburg.*—A truss of Timothy grass.

662 BENEDSKI, A. *Katerinovka Vil. near Odessa.*—Winter wheat.

663 BONELLIS, W. *Schoensee Settlement, Berdiansk Circ. Tauride Gov.*—Samples of wheat flour.

664 IVASHCHENKO, N. *Odessa,*—Wheat, rye, and French beans.

665 KOORT, SEID HUSSEIN-OGLOO, *Derekoi Vil. Alooshta District, Crimea.*—Nuts.

666 LOBODA, P. *Pavlovsk Vil. Alexandrovsk Circ. Ekaterinoslaf Gov.* — Wheat (Arnaootka).

667 ROMANDIN, *Kishenef, Bessarabia.*—French beans, maize.

668 SEID OMER ABDOOL OGLOO, *Derekoi Vil. Alooshta District, Crimea.*—Nuts.

669 SEMENIUTA, ARTEMIUS, *Goolaïpole, Alexandrovsk Circ. Ekaterinoslaf Gov.*—Linseed.

670 SHABELSKI, C. *Rostof Circ. Ekaterinoslaf Gov.*—Wheat and wild rape seed.

671 STROONNIKOF, T. & V. BROS. *Gov. of Tver and St. Petersburg.* — Wheat-flour and semoulia.

672 TRITHEN, O. *Odessa.* — Samples of wheat, maize, rye, and wild rape seed.

673 VINEIEF, *Nikolaevka Vil. Bobrynetz Circ. Cherson Gov.*—Winter wheat, wheat-flour, and seed of Triticum repens.

674 VOLKOF, T. *Jerebetz Vil. Alexandrovsk Circ. Ekaterinoslaf Gov.*—Ghirka wheat.

675 YOORIN, A. *Konstantinofka Vil. Novobaiazet Circ. Erivan Gov.*—Oats.

Sub-Class B.

676 DENGINK, A. MANAGER OF THE KISHENEF HORTICULTURAL SCHOOL, *Bessarabia.*—Dried plums, pears, and apples.

677 SHABELSKI, C. *Rastof Circ. Ekaterinoslaf Gov.*—Isinglass.

678 WIEBE, P. *Youshanly Freehold, Berdiansk Circ. Tauride Gov.*— Ham, cheese, dried apples, pears, and plums.

Sub-Class C.

679 CORAY, TN. & TB. *Ackerman Circ. Bessarabia.*—Red and white wine.

680 DECHESKOOL, V. *Galegit Vil. Orgey Circ. Bessarabia.*—Samples of tobacco.

681 DENGINK, A. MANAGER OF THE KISHENEF HORTICULTURAL SCHOOL, *Bessarabia.*—Tobacco.

682 FOONDOOCKLEY, J. *Goorsoof Estate, Southern Crimea.*—Wine.

683 MAGARATCH MODEL VINEYARDS, *near Ialta, Southern Crimea.*—Wines.

684 MALMBERG, C. *Moscow.*—Cigars and cigarettes.

685 MATVEÏEF, M. *Moscow.*—Snuff.

686 PHILIBERT, A. *Limena Vil. near Aloopka, Southern Crimea.*—Wines.

687 ROMANDIN, *Kishenef, Bessarabia.*—Red and white wine, and tobacco leaves.

688 ROTHE, R. *near Odessa.*—Wine.

689 SEID OMER ZMAIL OGLOO, *Aootky Vil. Alooskta District, Crimea.* — Samson tobacco leaves.

690 TRITHEN, O. *Odessa.*—Red and white wine.

691 WOLFSCHMIDT, A. *Riga.* — Liqueurs and fine brandies.

CLASS 4.

Sub-Class A.

692 PITANCIER, G. & Co. *Odessa.*—Stearic and oleic acids, stearine candles and soap.

Sub-Class B.

693 SHABELSKI, C. *Rastof Circ. Ekaterinoslaf Gov.*—Goats' hair.

694 TRITHEN, O. *Odessa.* — Samples of washed wool from merinos, half-bred and common Crimea sheep.

695 WIEBE, P. *Youshanby Freehold, Berdiansk Circ. Tauride Gov.*—Washed merino fleeces.

696 YEAMES, W. & Co. *Rostof-on-the-Don, and Eysk, near Azof. Wool-washing Establishment.*—Donskoy, short and combing wools.

Sub-Class C.

697 ALABIN, K. *Peasant, Kotchkoorof Vil. Lookyanof Circ. Gov. of Nijninovgorod.* — Mats of lime-tree bast.

698 CORNIES, D. *Orlof Settlement, Berdiansk Circ. Tauride Gov.*—Baskets.

699 RIAZAN CROWN DOMAIN ADMINISTRATION.—Hemp and hempseed.

700 RIGA LOCAL COMMITTEE FOR THE INTERNATIONAL EXHIBITION.—Collection of woods; samples of flax and hemp.

701 SHABELSKY, *Port Caton Vil. Rastov Circ. Cherson Gov.*—Flax.

CLASS 10.

702 RAÏVOLOVO WORKS, *near St. Petersburg, Finland.*—Military engineering tools.

CLASS 11.

Sub-Class A.

703 SYTOF, J. *St. Petersburg.*—Epaulets, shoulder-knots, &c.

704 TOOMASSOF, K. *Stavropol.*—Silver dagger and girdle.

CLASS 12.

Sub-Class C.

705 RAÏVALOVO WORKS, *near St. Petersburg, Finland.*—Naval implements and tools.

CLASS 17.

706 SHIMANOFSKY, Professor of the University of Kief.—Surgical instruments of the exhibitor's invention.

CLASS 20.

707 BEPHANI, P. *Yeremovka Vil. near Novogeorgiefsk, Cherson Gov.*—Raw silk.

708 DENGINK, A. Manager of the Kishenef Horticultural School, *Bessarabia.*—Raw silk and cacoons.

709 HAMM, M. *Orlof Settlement, Berdiansk Circ. Tauris Gov.*—Raw silk.

710 TCHORBA, *Cherson Gov.*—Raw silk.

CLASS 21.

711 PELTZER, G. *Moscow.*—Woollen stuffs for paletots, &c.

712 SOKOLOF, N. & M. BROS. *Moscow.*—Cloths.

713 STEPOONIN, A. *Klintsee Borough, Tchernigof Gov.*—Cloths.

714 VINOGRADOF, A. *Nijni-Novgorod.*—Paletot and kerchiefs, made of goats' hair.

CLASS 24.

715 SYTOF, J. *Moscow.*—Gold and silver cloth, thread, &c.

CLASS 25.

Sub-Class A.

716 SIDOROF, M. *Eastern Siberia, Tooroohansk Circ. Yenisseisk Gov.*—Furs and skins.

CLASS 26.

Sub-Class A.

717 KOOSSOF, J. SONS OF, *St. Petersburg.*—Leather and morocco.

718 SIDOROF, M. *Eastern Siberia, Tooroohansk Circ. Yenisseisk Gov.*—Leather made of deer-skin, plain and embroidered.

Sub-Class B

719 KERBELAY ZOOLAL OGLOO, *Shoosha, Caucasus.*—Harness for Georgian saddle and bridle.

720 LUCKS, C. *Halberstadt Settlement, Berdiansk Circ. Tauride Gov.*—Horse trappings, whips, &c.

CLASS 27.

Sub-Class A.

721 CHOORKIN, *Moscow.*—Hats.

722 KALLO, U. *Odessa.*—Cap adapted to different seasons.

723 TOOMASSOF, *Karapet, Stavropol.* —Fur cap (papakha).

CLASS 28.

Sub-Class C.

724 PAULY, VON. *St. Petersburg.*—Ethnological essays "Les peuples de la Russie."

CLASS 30.

Sub-Class A.

725 MERCKLING, *Odessa.*—Walnut cabinet. modern style.

Sub-Class B.

726 GOETSCHY, T. *St. Petersburgh.* —Paper-hangings.

727 ABROSIMOF, *Great Okhta, St. Petersburg.*—A shrine carved in cedar wood.

CLASS 31.

Sub-Class B.

728 CLODT, BARON, *St. Petersburg.* —Bronze groups.

729 LIBERIH, *St. Petersburg.* —Hunting groups.

SIAM.

North Courts, British Side.

1 SCHOMBURGK, SIR R. H. *H.M. Consul at Bangkok.*—Silk petticoats, worn by the Lao and Burmese females, do. of a Karen, ornamented with Job's tears (seeds of Coix); cotton coverlets of Lao and Karens, travelling bags, cotton thread and spindle, fauteuils of carved teak-wood, collection of walking-sticks, samples of paper, native-made Lao cutlasses, and a collection of trade; products medicines.

2 MARKWALD, A. & Co. *Bangkok.* —A collection of Siam products, consisting of silk, cotton, and other fibres, rice and pulse, resins, dye-stuffs, and tanning substances, sugars, coffee, woods, and animal products.

3 SIMMONDS, P. L. 8, *Winchester-st. London.*—Specimens of horns, feathers, edible birds'-nests, shark-fins, beche-de-mer, stick-lac.

4 HAMMOND, W. P.—A collection of ornamental articles.

5 BOWRING, SIR J.—Various articles illustrating Siamese art.

SPAIN.

S.W. Court, No. 2, and S.W. Gallery, No. 2.

CLASS 1.

1 ABAD, V. *Nijar, Almeria.*—Per-oxide of manganese.

2 ADMINISTRACION DE HACIENDA PUBLICA DE ZARAGOZA.—Rock salt from Remolinos.

3 ALCALDE DE CABRA.—Marbles.

4 ALCALDE DE OVIEDO.—Marbles, refractory clays, gypsum, hyacinth of Compostela.

5 ALFONSO, M. *San Miguel en Tenerife.*—Blue-stone delfts for flooring.

6 ALVAREZ, S. *Riotinto, Huelva.*—Sulphate of copper.

7 ASTIZ, M.R. *Aldar, Navarra.*—Calamine.

8 ASTORGA, M. DE, *Madrid.*—Marine salt of Huelva.

9 AURRE, G. *Siero, Asturias.*—Hydraulic cements.

10 AYUNTAMIENTO DE BERNARDOS, *Segovia.*—Slates.

11 AYUNTAMIENTO DE CARTAGENA. — Stone, gypsum.

12 AYUNTAMIENTO DE COLUNGA, *Asturias.* —Breccia marbles.

13 AYUNTAMIENTO DE MORON.—Marbles.

14 AYUNTAMIENTO DE SIERO, *Asturias.*—Hydraulic cements, refractory clays.

15 AYUNTAMIENTO DE VELILLA, *Soria.*—Red and yellow ochre.

16 BALLESTEROS, S, *Parres, Asturias.*—Bituminous slate, and oil obtained from it.

17 BARRON, T. *Almeria.*—Sulphate and carbonate of copper.

18 BATIER, L. *Santander.*—Calamine, sulphate of iron, oxide of iron.

19 BERINERE & Co. *Santander.*—Calamine.

20 BLANCO, J. *Sevilla.*—Marbles.

21 BLAZQUEZ, J. 1, *Calle du Luchana, Madrid.*—Magnesite.

22 BOIVIN & Co. *Alava.* — Asphalt minerals of Maestu.

23 BOULAY, L. *Cabrales, Asturias.* — Copper ores.

24 BRAVO, J. M. *Santa Eulalia de Oscos, Asturias.*—Malleable iron.

25 BURGOS, J. *Córdova.*—Copper ore, pit-coal.

26 BURGOS, J. *Almeria.*—Baryta.

27 CARRIAS, BLANCO, & Co. *Almeria.*—Carbonate and silicate of zinc.

28 CASALS, J. *Balaguer, Lérida.*—Iron.

29 CASTILLO, M. *Sevilla.*—Marbles.

30 CHIEF OF THE MINERAL DISTRICT OF ALMERIA.—Earths, phosphates, porphyry, marbles.

31 ——— OF BADAJOZ. — Copper, sulphate of lead.

32 ——— OF BARCELONA. — Lignites, pit coal, galena, nickel pyrites, hydraulic cement, coke, rock salt of Cordova.

33 ——— OF BURGOS.—Pit coal, coke, galenas, grey and other copper, manganese, iron, sulphate of soda, tin.

34 ——— OF CÁCERES.—Lime-stone.

35 ——— OF CÓRDOVA. — Calamine, grey copper, marble; native and distilled mercury; cinnabar, lead, argentine iron.

36 ——— OF GALICIA. — Refractory clay, semi-pegmatite, quartz steatite, serpentines, white and other marbles, lignites, cupreous iron pyrites, hydrated oxide of iron, oxide of manganese, oxide of tin, metallic tin.

37 ——— OF GRANADA. — Iron ore, copper, nitre, zinc, lead, marbles, salt.

38 ——— OF GUADALAJARA. — Silver ore, hydraulic cements, coke, iron ore, wrought iron, alabaster, salt.

39 CHIEF OF THE MINERAL DISTRICT OF MADRID.—Kaolin, grey hematites, marbles, lime-stone, magnesite, cupreous and blende pyrites.

40 ——— OF MURCIA.—Iron ore, lead ore, sulphur, alum, building-stone, marbles.

41 ——— OF VALENCIA. — Lignites, marbles, gypsum, alabaster, vermilion, lead, copper, sulphur, salt.

42 ——— OF VIZCAYA.—Lignite, soft mena, magnetic iron ore, siderose, calamine, blende, galena, copper pyrites, sulphate of soda, asphalt mineral, marl, iron, lead.

43 ——— OF ZARAGOZA. — Grey copper, sulphate of soda, pit-coal, asphalt mineral, marbles, coal, manganese, earths for colouring.

44 CILLERUELO, M. DE, *Santander.*—Iron.

45 COLLANTES, A. *Palencia.*—Coal and iron.

46 COMPANIA DEL GUADALQUIVIR, *Sevilla.* —Huelva iron.

47 COMPANIA GENERAL DE MINAS DE ESPANA.—Sulphate of lead.

48 COMPANIA MINERA DE PEDROSO, *Sevilla.*—Iron.

49 COMPANIA SAN MIGUEL ARCANGEL, *Madrid.* — Calcined ore, black copper, fine copper, copper obtained from the province of Huelva.

50 COMUNIDAD DE SALINAS DE ANANA, *Alava.*—Common salt.

51 DAGUERRE DOSPITAL, *Sevilla.*—Copper pyrites

52 DUCLER, E. *Langres, Asturias.*—Coal.

53 DURO & Co. *Langres, Asturias.*—Iron, ferruginous sands.

54 SHAW, D. *Córdova.*—Lead.

55 ESTABLECIMIENTO NACIONAL DE RIO-TINTO, *Huelva.* — Cupreous iron pyrites, calcined ore, native copper, fine copper.

56 FABRICA DE SAL DE FORREVIEJA.—Common salt.

57 FABRICA DE SAN JUAN DE ALCARAZ, *Albacete.*—Metallurgic products.

58 FABRICA NACIONAL DE FRUVIA. — Pit-coal, coke, iron ore.

59 FALCONI, T. *Almeria.* — Lead ore (carbonate).

60 FERNANDEZ, V. *Mieres, Asturias.* — Coal, coke.

61 FRASINELLI, R. *Cangas de Onis, Asturias.*—Red ochre, coal, amber, peat.

62 FORCADA, A. *Lérida.*—Iron.

63 FORCADA, V. *Lérida.* — Pit-coal of Almatres.

64 FOSSEY & Co. *San Sebastian.*—Castings.

65 FRANCO, M. — Asphalt from the mine of Maceda (*Soria*), asphaltic tar.

66 GALLEGO, M. *Palencia.*—Mineral coal from the mine Joven Ildefonso.

67 GIL & Co. *Langreo, Asturias.*—Iron.

68 GOBERNADOR DE LA PROVINCIA DE JAEN.—Salt, argentine galenas, lead, marbles.

69 GOMEZ DE SALAZAR, Y. *Almeria.*—Mill-stones, agate, lignites.

70 GOROSABEL, A. *Vittoria.* — Blende, copper.

71 GRAN DUQUESA DE LENGTEMBERG, *Asturias.*—Coals of Siero and Langres.

72 Guillen, G. *Sevilla.*—Native sulphur.

73 Gutierrez & Quevedo, *Santander.*—Peat.

74 Hartley, Zafra, & Co. *Huelva.* — Native peroxide of manganese, cupreous iron pyrites, sulphate of lead, amianthus, red marble, sulphur.

75 Henares, Corredor, & Co. *Montoro, Córdova.*—Galena, lead.

76 Heredia, hijos de M.A. *Málaga.*—Lead, white lead, red lead, litharge, silver in nugget.

77 Heredia, T. *Málaga.*—Iron.

78 Hernandez, L. *Sevilla.* — Cupreous iron pyrites.

79 Herrero, T. *Espinar, Segovia.*—Crystallizations of quartz.

80 Huellin, M. *Garrucha, Almeria.*—Sulphate of antimony.

81 Iceta, M. *San Sebastian.*—Cement.

82 Ingenieros de Minas de la Provincia de Leon.—Hydrated oxide of iron, coal, fire-bricks, metallurgic products, &c. Iron, slates, marbles, grey copper, coke.

83 Ingenieros de Minas de la Provincia de Santander. — Salt, bituminous schist.

84 Iris Amarillo, Sociedad de Minas, *Madrid.*—Cava alta, ochre, sulphur.

85 Junta de Agricultura, Industria and Comercio de Granada.—Coal.

86 Lafiguera, E. *Langreo, Asturias.*—Coal, coke.

87 Llana, J. *Castro Urdiales.*—Refractory earths, cement.

88 Llanos, R. *Vitoria.*—Coal.

89 Monterola, Cortazar, & Co. *S. Sebastian.*—Cement.

90 Marron, V. M. *Valladolid.*—Tin ore.

91 Martinez, J. *Lorca.*—Sulphur.

92 Massia, E. *Tortosa.*—Jasper.

93 Mercier, V. *Huelva.*—Sulphur, copper from the mines of Tharsis.

94 Miguel, L. *Soria.*—Espejon marbles.

95 Mora, F. *Alicante.*—Marbles.

96 Morentin, F. M. *Allo, Navarra.*—Sulphate of soda.

97 Noriega, F. *Onis, Asturias.* — Grey copper, carbonate of copper, malachite.

98 Orozco, R. *Almeria.* — Argentiferous lead ore, &c.

99 Orozco, G. *Almeria.*—Silicated carbonate of zinc, spar.

100 Orta, J. J. *Alosno, Huelva.* — Red and yellow ochre.

101 Ortigosa, M. *Valverde del Camino.*—Cupreous iron pyrites.

102 Pelayo, M. *Buicenes, Asturias.* — Coal.

103 Peña, M. *Espeja, Soria.*—Iron ore.

104 Perez, B. *Soria.*—Argentine lead.

105 Perez, E. *Almeria.*—Argentiferous lead ore.

106 Perez, Cardenal A. *Zamora.* — Foundry earths.

107 Picazo, A. *Zaragoza.*—Salt obtained by evaporation.

108 Porta, F. *Alicante.*—Mercury, copper.

109 Real Compania Asturiana, *Castrillon, Asturias.*—Zinc, iron, coal.

110 Redondo, J. *Pereruela, Zamora.*—Refractory earths.

111 Restoy, A. *Fondon Almeria.*—Lead ore; sulphuratere.

112 Reyes, F. *Huelva.*—Pyrolusites.

113 Picken, J. *Huelva.*—Pyrolusites.

114 Rodriguez, V. *Valencia.*—Marbles.

115 Rubin, F. *Llanera, Asturias.* — Refractory pudinga.

116 Ruiz Reyes, M. *Almeira.* — Ore of nickel, cobalt, cinnabar.

117 Sanz, D. *Riazas, Segovias.*—Earths, flat paving-stones of slate, grind-stones.

118 Serrano, C. *Coruna.*—Chromium.

119 Sociedad Belga de Samuno, *Langreo, Asturias.*—Coal.

120 Sitchá, J. *Valladolid.*—Gypsum.

121 Sociedad Candido Conde & Co. *Zaragoza.*—Rock-salt.

122 Sociedad Carbonera de Sta. Ana, *San Martin del Rey, Aurelio, Asturias.*—Coal.

123 Sociedad de las Minas Unidas del Castillo de las Guardas, *Sevilla.* — Subsulphate of iron, copper, sulphur.

124 Sociedad del Herrerito, *Sevilla.*—Cupreous iron pyrites, fine copper.

125 Sociedad Especial Minera, *El Porvenir en Asturias, Madrid.* — Cinnabar, mercury, pit-coal.

126 Sociedad Fusion Carbonifera y Metalifera de Belmez y Espiel, *Cordova.*—Coal, coke.

127 Sociedad Hullera y Metalúrgica de Asturias, *Mieres.*—Iron.

128 Sociedad Justa, *Langreo, Asturias.*—Coal, coke.

129 Sociedad "La Concepcion," *Sevilla.*—Cupreous iron pyrites, fine copper, copper of Huelva.

130 Sociedad "La Poderosa," *Sevilla.*—Cupreous iron pyrites, fine copper, copper of Huelva.

131 Sociedad "La Valiente," *Madrid.*—Sulphate of lead.

132 Sociedad "Los Santos," *Belmez, Cordova.*—Coal, coke, iron.

133 Sociedad Minera Argentífera de Estremadura Cáceres. — Silver ore, zinc, lead, and copper.

134 Sociedad Minera "Amistad," *Alicante.*—Mercury.

135 Sociedad Minera "Chispa," *Abulense, Avila.*—Copper ore.

136 Sociedad Minera "El Consuelo," *Chinchon, Madrid.*—Glauberite, sulphate of soda, barilla, soda, chalk.

137 Sociedad Minera "Feliz Hallazgo," *Alicante.*—Iron.

138 Sociedad Minera "Fraternidad," *Zaragoza.*—Common salt.

139 Sociedad Minera "La Campurriana," *Santander.*—Copper ore.

140 Sociedad Minera "La Lealtad," *Santander.*—Calamine.

141 Sociedad Minera "Providencia," *Santander.*—Calamine.

142 Sociedad Nuestra Senora de la Salud, *Sevilla.*—Cupreous iron pyrites, fine copper from Huelva.

143 Sociedad Nuestra Senora de Los

REYES, *Sevilla.*—Cupreous iron pyrites; calcined ore, from Huelva.

144 SOCIEDAD PALACIOS DE GOLONDRINAS, *Cáceres.*—Silver ores, zinc, lead, and and copper.

145 SOCIEDAD PROTECTORA, *Ciempozuelos, Madrid.*—Sulphate of soda, calcined, and as a hydrate; crystallized gypsum.

146 SOCIEDAD SAN FELMO, *Sevilla.* — Cupreous iron pyrites, bark for cementation, copper from Huelva.

147 SOCIEDAD SEVILLANA. — Silver, zinc, lead, and copper ores, from the mine of Giraldo (*Cáceres*).

148 SOCIEDAD UNION ASTURIANA, *Mierés, Asturias.*—Mercury.

149 SOCIEDAD UNION DEL COMERCIO, *Sevilla.*—Cupreous iron pyrites.

150 SOTO, M. *Orihuela.*—Pure copper for fusion, marbles, alabaster, gypsum.

151 TEUREIRO, N. *Madrid.*—Gypsum.

152 TORRES, MUNOZ Y LIMA, R. *Madrid.* —Apatita de Jumilla (*Murcia*).

153 VILLAFRANCA, M. DE, *Mazarron, Murcia.*—Alum.

154 VIÑAS, T. *Lérida.*—Lead.

155 VRIE, R. *Cangas de Tineo, Asturias.* —White marble.

156 UNZUETA, T. *Villarrubia de Santiago, Toledo.*—Sulphate of soda, calcined, and as a hydrate; gypsum.

CLASS 2.

157 AYUNTAMIENTO DE COCA, *Segovia.*— Spirits of turpentine, varnish, resin.

158 BERRENS, H. *Gracia, Barcelona.*— Mercurial products, lac, verdigris, minium.

159 BOTT, E. *Oviedo.* — Pharmaceutical products.

160 CALLEJA, S. *Villaviciosa, Asturias.*— Pharmaceutical products.

161 CANALES, HEREDEROS DE, *Malaga.*— Essence of lemon, citric acid.

162 CARRASCOSA, F. *Ariza, Zaragoza.*— Opium.

163 CASTELLO, N. *Avenys, Barcelona.*— Verdigris.

164 COURT É HIJO, *San Juan de Azualfarache, Sevilla.*—Perfumery.

165 CROS, J. T. *Barcelona.* — Acids, sulphates, and other chemical productions.

166 GARCIA DE VINUESA, J. *Sevilla.*— Artificial guano.

167 GRAU, J. *Sevilla.*—Bitumen.

168 MANJARRES, R. *Sevilla.*—Manure and other substances, prepared with phosphorite of Logrosanz, extracted from algas.

169 MANZANO, L. B. *Villardeciervo, Salamanca.*—Chemical substances.

170 MARQUÉS & MATAS, R. *Barcelona.*— Pharmaceutical products.

171 MAYER & BARTRINA, 1, *Prado Madrid.* —Typographic and lithographic inks; varnish for lithography.

172 ROYO, M. *Valencia.*—White and red lead.

173 RUVILLART, F. *Valencia.*—Essences.

174 SALVÁ, M. *Llurmayor, Baleares.* — Verdigris.

175 SIMON, T. *Madrid, 3, Caballero de Garcia.*—Zeiodetita, a newly discovered substitute for lead.

176 TOLOSA, R. *Valdemoro, Madrid.* — Artificial barilla.

CLASS 3.

177 ADALID, J. *Sevilla.* — Agricultural products.

178 ADANERO, CONDE DE, *Cáceres.* — Cheese.

179 AGUADOS MUNOZ, F. *Madridanos, Zamora.*—Chick pease.

180 AGUDO, F. CEHEGIN, *Murcia.* — Brandy.

181 AGUIRRE, T. *Coruna.*—Chocolate.

182 AGUIRRE, S. *Soria.*—Honey.

183 AICART, V. *Valencia.*—Wines, vinegar, olive oil, smooth podded tares, maize, peanuts.

184 AIGE, A. *Serós, Lérida.*—Figs.

185 ALBERREAH, J. *Reus.*—Oil.

186 ALBERT, J. *Valencia.*—Maize, wine.

187 ALCALÁ, B. DE, *Huesca.*—Wine.

188 ALCALDE DE VILLASABARIEGO, *Leon.* —Short wheat.

189 ALEMANY, A. *Tortosa.*—Oils.

190 ALMECH, E. *Zaragoza.*—Oil.

191 ALONSO, G. *Moraleja del Vino, Zamora.*—Common wine.

192 ALONSO DE PRADO, M. *Leon.*—Flour.

193 ALONSO DE PRIDA & Co. *Leon.* — Chocolate.

194 ALÓS, J. A. *Balaguer, Lérida.* — Wheat.

195 ALVAREZ, G. *Ricote, Murcia.*—Common wine.

196 ALVAREZ, N. *Vinetres, Tarragona.*— Wines, almonds.

197 ALVARGONZALEZ, R. *Gijon.*—Alimentary preserves.

198 ALZUETA, J. *Peralta, Navarra.* — Wines.

199 ALZUGARAY, VIUDA DE & Co. — Flour.

200 AMORES, M. *Sevilla.* — Agricultural products.

201 ANDRES, B. *Rioseco, Soria.*—Garlic.

202 ANEZCAR & Co. *Pamplona.*—Wines.

203 ANGUERA, C. *Reus.* — Wines, dried fruits.

204 ANGUERA, J. *Reus.*—Dried fruits.

205 ARANA, L. *Villaneuva de Puente, Navarra.*—Wheat.

206 AREVALO, J. *Matapozuelos, Valladolid.* —White wine.

207 ARIAS, M. *Zaragoza.*—Polish flour.

208 ARIAS, S. *Zamora.*—Common wine.

209 ARMERO, J. *Sevilla.* — Agricultural products.

210 ARROU, L. *Llubi, Balcares.*—Capers.

211 ARROYO, P. *Navalmoral, Toledo.* — Oils.

212 AUNON, J. *Sevilla.* — Agricultural products.

213 AURORA, L. *Fabrica de Harinos, Rioseeo.*—Flour.

214 AVEDILLO, Y. *Moraleja del Vino, Zamora.*—Common wine.

215 AVILA, C. *Lepe, Huelvo.*—Figs.

216 AYUNTAMIENTO DE ALCALA DE GUADAIZA.—Agricultural products.

217 AYUNTAMIENTO DE COCA, *Segovia.*—Pine-nuts, with and without the shell.

218 AYUNTAMIENTO DE GARROBILLAS, *Cáceres.*—Sausages.

219 AYAMANS, C. DE, *Manacor, Baleares.*—Wheat.

220 AYUNTAMIENTO DE MEDINA DEL CAMPO.—Wines, wheat, muela.

221 AYUNTAMIENTO DE MONTANCHEZ, *Cáceres.*—Ham.

222 AYUNTAMIENTO DE MULA, *Murcia.*—Wheat.

223 AYUNTAMIENTO DE RIOSECO, *Soria.*—Wheat.

224 AYUNTAMIENTO DE SAN MAUCIO, *Valladolid.*—Red and white wheat, barley.

225 AYUNTAMIENTO DE SANTO EULALIA, *Baleares.*—Almonds.

226 AYUNTAMIENTO DE SON SERVERA, *Baleares.*—Beans.

227 AYUNTAMIENTO DE SOTO DE SAN ESTEBAN, *Soria.*—Siliginose wheat, barley.

228 AYUNTAMIENTO DE TORDEHERMOSOS, *Valladolid.*—White wheat.

229 BAETA, M. *Zarragoza.*—Garden beans.

230 BALLESTER, L. *Barcelona.*—Wines.

231 BALLESTOROS, P. *Zaragoza.*—Oil.

232 BANCO, M. *Tafalla.*—Brandy.

233 BARRERA, J. *Tudela.*—Oil.

234 BARRERA, L. *Almonte, Huelva.*—Wines.

235 BEIGADA, J. *Verin, Orense.*—Garlic, chestnuts.

236 BELMONTE, A. *Corrales, Zamora.*—Chick pease.

237 BELDA, A. *Valencia.*—Wheat, barley, lupines, sorgho, maize, almonds, smooth podded tares, wine, oil.

238 BELDA, RAÑO, & CO. *Bocaivente, Valencia.*—Brandy.

239 BENITO, C. *Avila.*—Wheat, barley, oats, chick pease.

240 BENJUMEA, P. *Sevilla.*—Agricultural products.

241 BERASTEGUI, L. *Lacunza, Navarra.*—French beans, maize.

242 BERENGUER, J. *Valencia.*—Wheat, maize, beans.

243 BERNALDEZ, VIUDADE, *Villanueva, Sevilla.*—Wines.

244 BERNER, J. *Elche, Alicante.*—Almonds, dry figs.

245 BERRO, E. A. *Jaen.*—Oil.

246 BESSO, F. *Tarragona.*—Brandy.

247 BLANQUER, M. *Zaragoza.*—Red wine.

248 BLAT, J. *Valencia.*—Rice.

249 BOIX, M. *Lérida.*—Garden beans, maize.

250 BOFIL, R. *Orihuela, Alicante.*—Olive oil, sweet pepper, bird pepper.

251 BORRÁS, J. *Portella, Lérida.*—Beans.

252 BOST, J. *Carabaca, Murcia.*—Wheat.

253 BROSCA, T. *Valencia.*—Honey, wheat, oil.

254 BRUCART, J. *Manresa.*—Beans, wine.

255 BRUCE, HAMILTON, & CO.—Teneriffe wine.

256 BRUNENGO, J. *Rota, Cadiz.*—Sweet wine, Tintilla, Muscat, and Pajarete.

257 BUCHACA, A. *Valencia.*—Wine.

258 BULL & WEALE, *London.*—Sherry wine.

259 CABALLERO, T. *Barajas de Melo Cuencas.*—White wheat.

260 CABALLERO, J. *Chinchon Madrid.*—Red wine, brandy.

261 CABALLERO, J. 17, *Corredera vaja de S. Pablo, Madrid.*—Wine.

262 CABALLERO, MARQUÉS DE, *Valladolid.*—Common white wine, red wine.

263 CEBALLOS, B. *Nava del Rey, Valladolid.*—Dry wine.

264 CALCANO, J. *Sevilla.*—Pastes.

265 CALVO, P. *Fregeneda, Salamanca.*—Almonds, oil.

266 CALVO, R. *Casaseca de las Chanas Zamora.*—Pease, French beans.

267 CALVO, T. *Moraleja del Vino, Zamora.*—Common wine.

268 CALZADA, G. *Alcald de Honores.*—Red wine.

269 CALZADA, J. *Astorga.*—Chocolate.

270 CALZADA, T. *Sevilla.*—Agricultural products.

271 CALZADILLA, MA. *Sevilla.*—Olives.

272 CAMACHO, A. *La Palma, Huelva.*—Wines.

273 CÁMARA, M. *Sevilla.*—Agricultural products.

274 CAMERO DE ALDEA, T. *Sanzoles Zamora.*—Common wine.

275 CAMPOFRANCO, M. DE, *Palmo.*—Wine.

276 CAMPS, J. *Tortosa.*—Oil.

277 CANDALIJA, A. *Andojar.*—Wheat, barley, garden beans, pease.

278 CÁNOVAS, J. *Caravaca, Murcia.*—Potatoes, honey, olive oil.

279 CAPELLA, VIUDADE, *Barcelona.*—Chocolate, Llobregat wine.

280 CARRABIAS, J. *Salamanca.*—Wheat.

281 CARDENAS, A. *Valencia.*—Wines.

282 CARMONA, J. *Avilo.*—Wheat.

283 COREY HERMANOS, *Zaragoza.*—Wines.

284 CARO, R. *Constantina, Sevilla.*—Agricultural products.

285 CARRASCOSA, Y. *Bunol, Valencia.*—Olive oil.

286 CARREÑO, S. *San Cristobal de la Vego, Segovia.*—Wheat.

287 CARRERAS, A. *Mahon.*—Cheese.

288 CARRETERO, P. *Cordova.*—Montilla wines.

289 CARRETERO, R. *Mercadal, Baleares.*—Wheat.

290 CARVAJAL, H. *Zamora.*—Almonds.

291 CASADO, S. *Olmedo, Valladolid.*—Wheat.

292 CASAJUANA, J. *Castellgali, Barcelona.*—Oil.

293 CASARES, R. *Toro.*—Sweet red-wine.

294 CASO, M. *Ynfiesto, Asturias.*—Salted lard.

295 Castell, A. *Pous, Barcelona.*—Wine.

296 Castell de Pons, *Esparraguera, Barcelona.*—Wine, oil, olives, vinegar.

297 Castellet, B. *Tarrasa.*—Wine.

298 Castillo, P. 6, *Calle del Bano, Madrid.*—Wine.

299 Catala, A. *Tavea, Alicante.*—Almonds.

300 Cemeli, F. *Zaragoza.*—Oil, wine.

301 Centeno, G. *Acebo, Cáceres.*—Oil.

302 Cervera, C. *Valencia.*—Raisins.

303 Cinos, A. *Villamayor, Salamanca.*—Wine.

304 Calariana, V. *Reus.*—Hazel-nuts, smooth podded tares.

305 Clemens, — *Málaga.*—Almonds and raisins.

306 Colás, L. *Velamazan, Soria.*—Honey.

307 Collantes, A. *Madrid.* — Wheat, pease, maize, barley, garden beans, honey, olives.

308 Colldeforms, J. *Belliras, Barcelona.*—Wine.

309 Comesaña, T. *Sevilla.*—Agricultural products.

310 Corbacho, A. *Sevilla.*—Agricultural products.

311 Córdoba, M. *Tortosa.*—Nuts.

312 Corro de Bresca, L. *Málaga.*—Raisins; Muscatel wine.

313 Cotoner, F. *Baleares.*—Wines.

314 Creibach, V. *Valldeuró, Castellon.*—Oil.

315 Crespo, F. *Estepa.*—Oil.

316 Crespo, M. *Nava del Rey, Valladolid.*—Barley.

317 Dalmau, J. *Reus.*—Maize.

318 Dameto, A. *Palma.*—Millet.

319 Davidson & Co. *Sta. Cruz de Teneriffe.*—Teneriffe wine.

320 Delgado, V. *Zalamea la Real, Huelva.*—Honey, honey-combs.

321 Diaz, B. *Tudela.*—Wines, brandy.

322 Diaz, E. *Huelva.*—Oats, barley, almonds.

323 Diaz, J. *La Palma, Huelva.*—Chick pease.

324 Diaz, P. *Tudela.*—Wines.

325 Diaz Obregon, M. *Sevilla.*—Liqueurs.

326 Die, V. *Alicante.* — Fondellol and Aloque wines.

327 Domeneck, M. *Esparraguera, Barcelona.*—Wine, oil.

328 Dominguez, C. *Lillo, Leon.*—Mantua, imitation of Flanders.

329 Domingo, C. *Valencia.*—Earth-nuts, rice, almonds.

330 Duff, Gordon, *Jerez.*—Sherry wine, sherry Amontillado, Pedrojimenez.

331 Elvira, E. *Moraleja del Vino, Zamora.*—Common wine.

332 Enrique, J. A. *Toro.* — Aniseed brandy, dry wine, and spirits of wine.

333 Escofet, S. *Zaragoza.*—Brandy.

334 Escudero, M. *Orihuela, Alicante.*—White wine.

335 Escuela Practica de Agricultura de Alava.—Agricultural products.

336 Escuer Ordaz & Co. *Huesca.*—Wheat flour.

337 Eslaba, G. *Tudela.*—Wine, oil.

338 Estarico, R. *Valencia.*—Wines.

339 Esteban, P. *Valencia.* — Garuacha wine.

340 Estellés, G. *Valencia.*—Wheat, wine, oil.

341 Estevez, H. *Zamora.* — Albillo's natural wine, two years old; Siliginose wheat, almonds.

342 Esteller, G. *Benicarló, Castellon.*—Wine.

343 Estepa, T. *Urrea de Jalon, Zaragoza.*—Wheat.

344 Eugenio, R. *Zaragoza.* — Wheat, maize, oil.

345 Fabrica de Ntra, *Señora de los Remedios, Málaga.*—Native hard wheat flour.

346 Fajardo, J. L. *Constantina, Sevilla.*—Agricultural products.

347 Febrer, J. *Benicarlo, Castellon.*—Wine.

348 Fernandez, A. *Bullas, Murcia.*—Saffron, brandy, wine.

349 Fernandez, J. *Masroig, Tarragona.*—Almonds.

350 Fernandez, M. *Tafalla, Navarra.*—Wheat, barley.

351 Fernandez de Córdoba, M. *Constantina Sevilla.*—Agricultural products.

352 Fernandez, Elvira L. *Sanzoles, Zamora.*—Natural wine, three years old.

353 Fernandez, V. *Cáceres.*—Oil.

354 Fernandez & Ventosa, *Coruña.*—Wheat flour.

355 Fernandiz, J. *Valencia.*—Fondellol's wines and vinegar.

356 Ferrer, B. *Sevilla.* — Agricultural products.

357 Fierro, T. *Santa Cruz de la Palma.*—Sweetmeats, Malmsey wine, arrow-root.

358 Fiballer, J. *Barcelona.*—Oil, beans, almonds, hazel nuts.

359 Flores, A. *Moquer Huelva.*—Alcohol.

360 Fonollar, C. de, *Barcelona.*—Wines.

361 Fontes, J. *Murcia.*—Olives.

362 Fontellas, M. de, *Fontellas, Navarra.*—Oil.

363 Forces, M. *Zaragoza.* — Red wine, Muscat wine.

364 Fornes, M. *Zaragosa.*—Wine.

365 Fosalva, M. *Pierola, Barcelona.*—Wine.

366 Freixas, T. *Castellgali, Barcelona.*—Wine.

367 Fuentes del Sauce, C. de, *Constantina, Sevilla.*—Agricultural products.

368 Fuemnayor, P. *Sevilla.*—Wines.

369 Fuemnayor, V. *Caltojar, Soria.*—Filtered honey.

370 Fuente el Salce, C. de, *Córdoba.*—Figs, oil.

371 Fuentes, A. *Córdoba.*—Montilla wine.

372 Gago, Roperuelos, M. *Zamora.*—Barley, cheese.

373 Gaitan, F. *Tudela.*—Oil.

374 Galés, B. *Esparraguera, Barcelona.*—Wine, vinegar.

375 Galí, A. *Tarrasa.*—Wine.

376 Galí, J. *Fals, Barcelona.* — Blue vetches, wines.

377 GALNIDO, A. *Sevilla.*—Agricultural products.
378 GAMERO, M. *Sevilla.*—Agricultural products.
379 GARCÉS, R. *Huesca.*—Maize, barley.
380 GARCIA, A. *Córdoba.*—Wheat, barley. pease, acorns.
381 GARCIA, D. *Guadalajara.*—Torrontes wine, Siliginose wheat.
382 GARCIA, F. *Avila.*—Wheat.
383 GARCIA, J. *Barco de Avila.*—Beans.
384 GARCIA, J. R. *Valencia.*—Kitchen garden seeds.
385 GARCIA, R. *Onil Alicante.*—Spirits of wine.
386 GARCIA, R. 79, *Calle-Mayor, Madrid.*—Chocolate.
387 GARCIA ALFONSO, A. *Moral de Orbigo, Leon.*—Wheat.
388 GARCIA CALATRAVA, F. *Aleobendas, Madrid.*—Wheat, barley, pease, wine.
389 GARCIA, MORENO, J. *Valencia.*—Figs, wine.
390 GARCIA, VALENCIA, M. *Pedroso, Sevilla.*—Agricultural products.
391 GARRIGA, S. *Manresa.*—Wine.
392 GAYEN F. *Málaga.*—Preserved fruits, olives.
393 GIL & BORRAS, F. *Reus.*—Wines, oil, almonds.
394 GINER, T. *Valencia.*—French beans.
395 GIOL, T. *Reus.*—Wines, oils.
396 GISBERT, T. A. *Onil, Alicante.*—Olive oil.
397 GOBERNADOR DE CUENCA.—Honey.
398 GOBERNADOR DE GUADALAJARA.—Honey from the Alcarria.
399 GOMBAN, T. *Reus.*—Wines, almonds.
400 GOMEZ, J. *Jérica, Castellon.*—Wine.
401 GOMEZ, J. *Orense.*—Beans.
402 GOMEZ, M. *Jérica, Castellon.*—Brandy, pears.
403 GOMEZ, V. *Zaragoza.*—Red wine.
404 GOMEZ ALONSO, Y. *Serradilla Cáceres.*—Wine.
405 GOMEZ DE ALIA, J. *Navalmoral, Toledo.*—Oil.
406 GOMEZ DE BARREDA, F. *Sevilla.*—Agricultural products.
407 GOMIS, T. *Manresa.*—Wines.
408 GONZALEZ, J. *Moron.*—Agricultural products.
409 GONZALEZ ARCAINA, J. *Cartagena.*—Wines.
410 GONZALEZ, DUBOST, *Jerez.*—Wines, and plan of wine vaults.
411 GONZALES DE MESA, A. *Laguna, Canarias.*—Mexican black beans.
412 GONZALVO, G. *Huesca.*—Common wine.
413 GORMAN & CO. *Puerto de Sta. Maria.*—Sherry wines and Montilla.
414 GOYENETA, M. *Sevilla.*—Agricultural products.
415 GRANJA PROVINCIAL, *Leon.*—Polish oats.
416 GRAU, J. *Reus.*—Wines, oils.
417 GREMIO DE CHORIZEROS DE CANDELARIO, *Salamanca.*—Large and small sausages.
418 GÜELL, J. *Almacella, Lérida.*—Wheat.
419 GUENDULAIN, C. DE, *Tafalla, Navarra.*—Wines.
420 GUILLE, HERMANOS, *Zaragoza.*—Wines, brandy.
421 GUILLEN, J. *Zaragoza.*—Honey.
422 GUILLEN, L. *San Esteban de Litera Huesca.*—Wines.
423 GUILLEN, M. *Huesca.*—Vinegar.
424 GUTIERREZ, E. *Corrales, Zamora.*—Imitation of port wine; pease.
425 HAMILTON & GRIEVE, *London.*—Sherry wine.
426 HARTLEY, ZAFRA, & CO. *Huelva.*—Mazagan garden beans.
427 HELBANT, S. *Jerez.*—Very old sherry Amontillado, white Manzanilla.
428 HEREDIA, M. *Malaga.*—Sugar.
429 HERNANDEZ, A. *Murcia.*—Olive oil.
430 HERNANDEZ, Y. *San Frontis, Zamora.*—Common wine.
431 HERNANDEZ, F. *Avila.*—Wheat, pease.
432 HERNANDEZ, V. *Miranda del Castanar, Salamanca.*—Oil.
433 HERNAN SANZ, E. *Iscar, Valladolid.*—Pine-nut seed.
434 HERRERA, J. A. *Sevilla.*—Agricultural products.
435 HERP, J. *Manrresa.*—Wine, oil.
436 HIDALGO, E. *Jerez.*—Pedrojimenez, Muscat, Manzanilla.
437 HIGUERA BARBAGERO, R. DE LA, *Toro.*—Sweet and dry red wine, common red wine.
438 YBARRA, M. *Valencia.*—Wine, oil, vinegar.
439 YBIRICA, H. *Tudela.*—Oil.
440 YLLAN, J. *Zamora.*—Common wine.
441 YNGUANZO, F. *Rivadesellas, Asturias.*—Chestnuts.
442 YÑIGO, Z. *Zaragoza.*—Beans, fresh vegetables, red wines.
443 JIMENEZ, A. *Sevilla.*—Crackers.
444 JIMENEZ, A. *Cascante, Navarra.*—Wine.
445 JIMENEZ, G. *Avila.*—Pease.
446 JIMENEZ DE PEDRO, J. *Madrid.*—Arganda wine.
447 JIMENEZ DE TEJADA HERMANOS, *Moguer, Huelva.*—Wines, spirits of wine.
448 JIMENO, J. *Zaragoza.*—Oil.
449 JIMENO, J. C. *Valladolid.*—Garden pease.
450 JIMENO & AZPEITIA, *Ateca, Zaragoza.*—Wines, brandy, liqueurs.
451 JORDANA, J. *Albalate, Teruel.*—Wines, maize flour.
452 JUNCO, P. *Rivadesella, Asturias.*—Apples.
453 JUNTA DE AGRICULTURA, INDUSTRIA & COMERCIO DE CASTELLON.—Brandy, wines, oil, figs, raisins, vegetables.
454 JUNTA DE AGRICULTURA, INDUSTRIA & COMERCIO, *Granada.*—Olive oil, wheat, barley, wines, vinegar.
455 JUNTA DE AGRICULTURA DE OVIEDO.—Avile's jams, garden-beans, vetches, hazelnuts, apples, chestnuts.
456 JUVÉS, F. *Manresa.*—Wheat, garlics.
457 LACAMBRA, J. *Zaragoza.*—Pastes.

458 LACASA J. A. *Jaen.*—White and yellow maize.

459 LACAVE, J. P. *Sevilla.*—Olives.

460 LACORTE, J. A. *Cabra.*—Honey, wine, vinegar, figs, prunes.

461 LAGORIO, F. DE P. *Cartagena.*—Wines.

462 LAGUARTA, V. *Savayes, Huesca.* — Garnocha wine.

463 LAHIGUERA, A. *Sevilla.* — Agricultural products.

464 LAHOZ, V. *Zaragoza.*—Flour.

465 LA INDUSTRIAL HARINERA, *Barcelona.*—Wheat, flour.

466 LAMAS, M. *Cebreros, Avila.*—White and red wine.

467 LA NAVARRA, *Tudela.* — Vinegar, brandy.

468 LARA, J. *Pedroso, Sevilla.*—Agricultural products.

469 LASALA, Y. *Valencia.* — Wine, oil, syrup.

470 LASERNA, J. J. *San Juan del Puerto, Huelva.*—Wheat.

471 LASERNA, J. P. *Moguer, Huelva.*—Orange wine.

472 LASERNA, M. *Sevilla.*—Agricultural products.

473 LASIERRA, J. *Quinzano, Huesca.*—Wheat, barley.

474 LASTRA, J. *Sevilla.* — Agricultural oducts.

475 LEON BENDICHO J. *Almeria.*—Maize.

476 LEZCANO, M. *Zaragoza.*—Oil, wine.

477 LINARES, F. J. *Sevilla.*—Agricultural products.

478 LOPE, J. A. *Alagon, Zaragoza.* — Oil.

479 LOPEZ, M. *Carabaca, Murcia.*—Millet.

480 LOPEZ, M. 32, *Calle de Tudescos, Madrid.* —Chocolates.

481 LOPEZ, S. *Alhama, Murcia.*—Figs.

482 LOPEZ, ARRUEGO M. *Velilla de Ebro, Zaragoza.*—Dried fruit.

483 LOPEZ, CABALLERO S. *Murcia.*—Figs, acorns, nuts, almonds, rice, wheat, honey, pepper.

484 LORENZANA, R. *Leon.*—Beans.

485 LORENZO, T. *Reus.*—Wines, spirits.

486 LORETO, M. DE, *Sevilla.*—Agricultural products.

487 LOSADA, D. R. *Illescas, Toledo.*—Wines, oils, wheat, pease, &c.

488 LOZANO, A. *Fermoselle, Zamora.* — Olive oil.

489 LUELMO, A. *Moraleja del Vino, Zamora.*—Natural wine.

490 LUJAN, L. *Arenas, Avila.*—Oil, olives, chestnuts.

491 LUNA, CONDE DE, *Elche, Alicante.*—Wheat, barley, black figs, almonds, olive oil, Muscat wine.

492 LLADÓ, M. *Campos, Baleares.*—Teja wheat.

493 MAESTRE, G. *Petrel, Alicante.*—Fondellol wine.

494 MALEGUÉ, T. *Reus.*—Wine, brandy, nuts.

495 MARCH, M. *Castellvell, Barcelona.*—Wines.

496 MARCO, M. *Esplus, Huesca.*—Wheat.

497 MAROTO, E. *Carmona, Toledo.*—Siliginose wheat, pease, wine.

498 MARQUET, F. *Zaragoza.* — Garden beans, vetches.

499 MARRACO, P. *Zaragoza.* — Wheat, flour.

500 MARRACO, J. *Zaragoza.*—Flour, and bran.

501 MARRON, B. *Sevilla.*—Oils.

502 MARSILLA, B. *Bullas, Murcias.* — Brandy, wine.

503 MARTI, J. *Reus.*—Plums, raisins.

504 MARTI, R. *Reus.*—Spirits.

505 MARTIN, M. *Olmedo, Valladolid.*—White wine.

506 MARTIN, P. *Poboleda, Tarragona.*—Wine, vinegar.

507 MARTINEZ, A. *Jaen.*—White wine.

508 MARTINEZ, F. *Tarrazona.*—Beans.

509 MARTINEZ, F. *Tarrazona.*—Honey.

510 MARTINEZ, VIUDA DE R. *Murchante, Navarra.*—Wines.

511 MARTINEZ DE EUGENIO J. *Navalmoral, Toledo.*—Oils.

512 MARTINEZ GUTIERREZ, J. *Jerez.*—Manzanilla, Pedrogimenez, Pajarete.

513 MARTINEZ MADRID, J. *Cartagena.*—Wines.

514 MARTIN SUAREZ, A. *Sevilla.*—Agricultural products.

515 MARTORELL, P. *Ciudadela, Baleares.* —Wine.

516 MASANET, A. *Muro, Baleares.*—Beans.

517 MASPONS, T. *Vina Roz, Castellon.*—Alcohol.

518 MASSIA, E. *Tortosa.*—Oil.

519 MATEOS, C. *Cebreros, Avila.*—White and Muscat wine.

520 MATEOS, Y. *Cebreros Avila.*—Brandy.

521 MATEU, J. *Bacarisas, Barcelona.* — Wine.

522 MATTIEPHEN, FURTON, & Co. *London and Jerez.*—Sherry wines.

523 MAYOL, M. *Baleares.*—Dried figs, tares.

524 MAZARRON, M. 3, *Bordadores, Madrid.* —Valdepenas wine.

525 MEDRANO, J. *Almonacid de Zorita, Guadalajara.*—Olive oil.

526 MELA, J. *Madrigal, Avila.*—Wines.

527 MENSA, J. *Lérida.*—Oil.

528 MERCADAL, M. *Malson.*—Wine, garden beans.

529 MERIC & Co. *Co. Colonial Gran Establecimiento al Vapor.*—Chocolates, and other alimentary substances.

530 MERINO, M. *Berlanga, Soria.*—Potatoes.

531 MICHANS, A. *Lecumberri, Navarra.*—Garden beans.

532 MIGUEL, J. *Tarraga, Lérida.*—Blue vetches.

533 MIRAFLORES, C. DE, *Sevilla.* — Agricultural products.

534 MIRALLES, ARAGON, & Co. *Barcelona.*—Wine.

535 MIRET & TERSA, *Barcelona.*—Beer.

536 MIURA, A. *Sevilla.*—Agricultural products.

537 MOLPECERES, V. *Oknedo, Valladolid* —Wheat.

538 MONFORT, A. *Barcelona.*—Olive oil.
539 MONFORT, F. *Torrente de Cima, Huesca.*—Wheat, beans, dry sliced fruits, dried figs, almonds, oil.
540 MONJAS DE SAN PELAYO, *Oviedo.*—Preserved fruits.
541 MOMPEAU, J. C. *Callosa de Segara, Alicante.*—Wines.
542 MOMPÓ, J. *Valencia.*—Wine.
543 MOMPUEY, P. *Valencia.* — Raisins, tares, pea-nuts, pea-nut oil, &c.
544 MONSERRAT, F. A. *Maella.*—Figs.
545 MONSERRAT, R. *Puigpelat, Tarragona.*—Wheat, maize, almonds, hazel-nuts, and wine.
546 MONTANER, J. *Reus.* — Wine, oil, vinegar.
547 MONTERO, S. *Sevilla.*—Agricultural products.
548 MONTÓLIN, P. M. *Tarragona.* — Beans, barley, tares, wines, oils, &c.
549 MONTIEL, M. *Trigueros, Huelva.* — Wine.
550 MORELL, P. *Baleares.*—Oil.
551 MORENO, A. *Guadalcanal, Sevilla.*—Agricultural products.
552 MORRAS, M. *Tafalla, Navarra.* — Wheat, wine.
553 MOTILLA, M. DE LA, *Sevilla.*—Agricultural products.
554 MOYANO SANCHER, P. *Nava del Rey, Valladolid.*—Wheat.
555 MUNOZ, P. *Valladolid.*—Sausages.
556 MURUVE, M. *Los Palacios, Sevilla.*—Agricultural products.
557 NARANJO, L. *Caralla, Sevillâ.*—Agricultural products.
558 NASARRE, A. *Lupinen, Huesca.*—Oil, wheat, beans, &c.
559 NEBOT, S. *Son Servera, Baleares.*—Dried figs.
560 NOCHDO, F. *Orense.*—Maize.
561 NUNEZ, J. *Coruna.*—Flour.
562 OJEITO, J. *Fregenada, Salamanca.*—Wine.
563 OLANIER, A. *Játiva, Valencia.*—Rice.
564 OLIVA, D. *Reus.*—Wine.
565 OLIVA, T. *Salamanca.* — Wheat, French beans, pease.
566 OLIVER, B. *Ciudadela, Baleares.* — Honey.
567 OLMEDO, L. *Borlanga, Soria.*—Beans.
568 ORDUNA, C. *Cascante, Navarra.* — Liqueurs.
569 ORTEGA, L. *Reus.*—Wine, fruits.
570 ORTIZ, V. *Brozas, Cáceres.*—Wines.
571 ORNS, M. *Manresa.*—Wine.
572 ORNS & Co. *Huesca.* — Spirits of wine.
573 OSACAR, G. *Santesteban, Navarra.*—French beans, maize, &c.
574 OSORNO, VIUDA DE, *Villanueva, Sevilla.*—Wines.
575 OTERIM, J. *Gùizo de Luinia, Orense.*—Turnips.
576 OTERO, E. *Verin, Orense.*—Onions.
577 PALACIO, J. M. *Espeluy, Jaen.*—Olives, oil, honey, wine, wheat, &c.
578 PALAZUELOS, VISCONDE DE, *Toledo.*—Siliginose wheat, Polish oats, oil, olives.
579 PANES, D. *Manzanilla, Huelva.* — Wine.
580 PARADELL, J. *Sampedor, Barcelona.* —Wine.
581 PARDO, J. *Valencia.*—Maize, tares.
582 PARÉS, J. *Collbato, Barcelona.*—Wine.
583 PASCUAL, B. *Reus.*—Wine, oil, dried fruits.
584 PASCUAL, P. *Bruck, Barcelona.* — Wine.
585 PASCUAL, T. *Cosuenda, Zaragoza.*—Wine.
586 PATERNA, M. DE, *Cavalla, Sevilla.*—Agricultural products.
587 PAULES, J. *Zaragoza.*—Oil.
588 PAYO, M. *Bermillo de Sdyago, Zamora.*—Rye.
589 PEDROSA, J. *Esparraguera, Barcelona.* —Wine, oil.
590 PENA, F. *Zaragoza.*—Wine.
591 PENILLAS, J. *Bollullos del Condado, Huelva.*—White wine.
592 PENAFLOR, CONDE DE, *Sevilla.*—Agricultural products.
593 PEREIRA, J. *Sevilla.* — Agricultural products.
594 PEREZ, J. *Tudela.*—Oil.
595 PEREZ, L. *Cebreros, Avila.*—Wines.
596 PEREZ, F. *Orense.*—Hazel-nuts.
597 PEREZ, F. DE P. *Ybi, Alicante.*—Olive oil.
598 PEREZ, J. *Orense.*—Maize.
599 PEREZ BAERLA, M. *Magallon, Zaragoza.*—Wheat, oil, wine.
600 PEREZ DE LOS COBOS, C. *Jumilla, Murcia.*—Wines.
601 PEREZ DE LOS COBOS, P. *Jumilla, Murcia.*—Olive oil.
602 PEREZ MARCO, F. *Rellen, Alicante.*—Raisins, figs, almonds, olive oil.
603 PEREZ PAULINO, D. *Fregeneda, Salamanca.*—Almond oil.
604 PEREZ VILLORIA, R. *Fregeneda, Salamanca.*—Almonds, olives, wine, oil.
605 PEREZ ZAMORA, A. *Puerto de la Orotava, Canarias.*—Arrow-root.
606 PERIBANEZ, L. *Zaragoza.*—Oil.
607 PIMENTEL & Co. *Valladolid.*—Wines.
608 PINOS, M. *Zaragoza.*—Honey.
609 PLÁ, J. *Tafalla.*—Spirits of wine.
610 PLÁ, F. *Reus.*—Wines, spirits.
611 PONS, L. *Mahon.*—Honey.
612 PORTA, M. *Zaragoza.*—Liqueurs.
613 PORTILLA, D. DE LA, *Sevilla.*—Agricultural products.
614 PRAT, J. *Fals, Barcelona.*—Pease, wine.
615 PRESAS, S. *Coruña.*—Salt, sardines.
616 PRIEGO, J. DONA, *Mercia, Córdoba.*—Wine, brandy, wheat.
617 PUGA, WIUDA DE É HIJOS, *Zamora.*—Mulberry wine, cherry wine, liqueurs.
618 PUIG, D. *Sevilla.*—Agricultural products.
619 PUIGGENER, J. *Oleza, Barcelona.*—Very old wine.
620 PUJADAS, P. *Tudela.*—Wheat, wine.
621 QUER, M. *Aytoria, Lerida.*—Wine.
622 QUINTANA, M. *Baleares.*—Olives.
623 RABASSÓ, J. *Vendrell, Tarragona.*—Wine, brandy.

624 Ramirez, F. *Zaragoza.*—Oil.
625 Ramirez de Arellano, J. *Bunuel, Navarra.*—Oil.
626 Requesso, M. *Zamora.*—Barley.
627 Reyero, B. *Leon.*—Honey.
628 Ribot, J. *Petra, Baleares.*—Wheat.
629 Rico, D. 18, *Humilladaros, Madrid.*—Morcat wine.
630 Riera, J. *Huelva.*—Pease.
631 Riera, J. *Payá, Barcelona.*—Wine.
632 Riesco, J. P. *Alicante.*—Wine, olive oil.
633 Rincon, A. *Tembleque, Toledo.*—Siliginose wheat, cheese.
634 Rincon, M. C. *Sevilla.*—Agricultural products.
635 Ripalda, M. *Pamplona.*—Ezcaba wine.
636 Ripoll, M. *Zaragoza.*—Barley, wine.
637 Ripoll, N. *Palma.*—Almonds.
638 Rives, A. *Manreza.*—Beans.
639 Roca, B. *Palma.*—Alimentary substances.
640 Roca de Togores, B. *Orihuela, Alicante.*—Olive oil, wheat, maize.
641 Roca, Hermanos, *Murcia.*—Ground pepper.
642 Roca Mora, J. *Reus.*—Beans, and fruits.
643 Rodriguez, B.—Manzanilla, Amontillado.
644 Rodriguez, C. *Orense.*—Beans.
645 Rodriguez, Lorenzo, J. *Toro.*—Honey.
646 Rodriguez Modenes, T. *Córdoba.*—Montilla wine.
647 Rodriguez, Tejedor A. *Toro.*—Wine of 1858.
648 Roman, Vinda é hijos de, *Leon.*—Chocolate.
649 Romero, M. *Sevilla.*—Agricultural products.
650 Romero de la Bandera, M. *Málaga.*—White wine.
651 Romo, T. M. *Hinojosa, Salamanca.*—Almonds.
652 Rotova, C. de, *Valencia.*—Dates.
653 Royo, M. *Valencia.*—Wines, olive oil.
654 Rubert, J. *Baleares.*—Oil.
655 Rubio Velazquez, M. *Málaga.*—Wines.
656 Ruiz, J. *Leganés, Madrid.*—Wine, brandy.
657 Ruiz, M. *Burgo de Osma, Soria.*—Beans.
658 Sadaba, F. *Palencia.*—Liqueurs.
659 Sagasti, J. *Tudela.*—Oil.
660 Sala, E. *Zaragoza.*—Flour.
661 Salva, M. *Palma.*—Oil.
662 Salvador, C. *Villaralvo, Zamora.*—Common wine, Siliginose wheat.
663 Salvador, G. *Tafalla.*—Brandy.
664 Salvador, J. *Tortosa.*—Oil, almonds.
665 Sallés, V. *Bellinds, Barcelona.*—Wine.
666 Sampol, P. J. *Baleares.*—Oil.
667 San Adrian, M. de, *Monteagudo, Navarra.*—Oil.
668 San Andrés, M. Vindo de, *Puerto de Cruz, Canarias.*—Arrow-root.
669 Sanchez, A. *Murcia.*—Olive oil.
670 Sanchez Chicarro, A. *Leon.*—Claret wine, honey, &c.
671 Sanchez Pascual, J. *Elche, Alicante.*—Brandy, Malmsey wine.
672 Sanchez Vida, T. *Guadalcanal, Sevilla.*—Agricultural products.
673 San Roman, A. *Fermoselle, Zamora.*—Spirits of wine, brandy, imitation of port wine, olive oil, &c.
674 Santa Maria, F. *San Juan del Puerto, Huelva.*—Wines.
675 Santana, C. *Salamanca.*—Wheat, barley, rye, &c.
676 Santiago, Y. *Zamora.*—Spirits of wine, brandy.
677 Santorras, A. *Tarragona.*—Wine, oil, almonds, barley, wheat.
678 Santonja, B. *Valencia.*—Wines.
679 Santillan, P. *Riara, Segovia.*—Smoked flesh of a mountain goat.
680 Santos, A. *Leon.*—Flour.
681 Sanz, E. *Olmodo, Valladolid.*—Wheat.
682 Sanz, M. *Zamora.*—Common wine.
683 Sarder, J. *Valencia.*—Wine.
684 Semitier, C. de, *Tortosa.*—Oil, tares.
685 Simo, P. *Porrera, Tarragona.*—Wine.
686 Sobradiel, C. de, *Zaragoza.*—Oil.
687 Sociedad Economica de Murcia.—Pepper, white maize.
688 Solar de Espinosa, B. del, *Jumilla, Murcia.*—Wheat, oil.
689 Soler, J. *Mahon.*—Wheat.
690 Soler, R. *Manresa.*—Wine.
691 Soler, Hermanos, *Tarragona.*—Wine, brandy.
692 Sorá, J. *Yviza.*—Wheat.
693 Soriano, *Madrid.*—Sweet acorns from Murcia.
694 Soto, M. *Orihuela, Alicante.*—White maize.
695 Suarez Centi, J. *Valladolid.*—Flour.
696 Subdelegacion Agricola de San Ysidro de Reus.—Wines, brandy, vinegar, oil, grains, dried fruits.
697 Subirat, J. *Pierola, Barcelona.*—Wine.
698 Suelves, J. *Tortosa.*—Oil, tares, maize.
699 Sulla, J. *Tremp, Lérida.*—Wine, brandy.
700 Tamarit, M. de, *Tortosa.*—Rice, wine, oil.
701 Taraza, P. *Piera, Barcelona.*—Wine, spirits of wine.
702 Tena, C. *Guadalcanal, Sevilla.*—Agricultural products.
703 Teruero, Y. *Guadalajara.*—Flour.
704 Thorices, F. R. *Moguer, Huelva.*—Wine, vinegar.
705 Teruero, J. *Marchena, Sevilla.*—Agricultural products.
706 Tio, M. *Valencia.*—Wines.
707 Tipping, W. R. *London.*—Sherry wine.
708 Tomé, T. *Corrales, Zamora.*—Spirits of wine.
709 Tomar, Y. *Torres de Segre, Lérida.*—Gamacha wine.
710 Torrella, J. *Castellgali, Barcelona.*—Wine.

711 Torres, C. *Arahall, Sevilla.*—Agricultural products.
712 Torres, J. *Corbin, Lérida.*—Millet.
713 Torres, L. *Guadalcanal, Sevilla.*—Agricultural products.
714 Torres, Marques de las, *Sevilla.*—Agricultural products.
715 Tous, M. *Palma.*—Almonds.
716 Tous, M. de, *Sevilla.*—Agricultural products.
717 Trapero, F. *Rota, Cadiz.*—Tintilla wine, Muscat, Pajarete.
718 Travez, G. *Toro.*—Sweet almonds.
719 Trell, P. *Adra, Almeria.*—Pedrogimenez wine.
720 Trias, P. *Esporlas, Baleares.*—Wheat.
721 Uceda, M. *Berlanga, Soria.*—Onions.
722 Urrutia, M. *Tudela.*—Oil.
723 Urzaiz, *Lucena del Puerto, Huelva.*—Orange wine, Muscat, Pedrogimenez.
724 Vado, M. *Guadalajara.*—Siliginose wheat, oil.
725 Valero, J. *Elche, Alicante.*—Muscat wine.
726 Valle, L. *Ocana.*—White wheat.
727 Valle, V. *Rioadesella, Asturias.*—Cider.
728 Vallés, A. *Castilsabas, Huesca.*—Wine.
729 Valls, J. A. *Sevilla.*—Agricultural products.
730 Varios Propietarios de Villena, *Alicante.*—Wines, spirits of wine.
731 Vargas, F. *Caralla, Sevilla.*—Agricultural products.
732 Vargas, R. *Córdoba.*—Honey, wine, brandy.
733 Vazquez, J. *Sevilla.*—Agricultural products.
734 Vazquez, L. R. *Moraleja del Vino, Zamora.*—Common wine.
735 Vela, J. *Rota, Cádiz.*—Tintilla wines.
736 Vera, G. *Baleares.*—Red wine, dried figs.
737 Verdeja, J. A. *Sevilla.*—Agricultural products.
738 Verdugo, J. *Toledo.*—Wheat.
739 Veruet, J. *Bardellos, Tarragoná.*—Brandy.
740 Vicente, P. *Zaragoza.*—Beans, maize, wine.
741 Victor, J. & Co. *Jerez.*—Sherry wines.
742 Vidal, A. *Cartagena.*—Sorgho.
743 Vidal, G. *Cartagena.*—Figs, bread of figs.
744 Vidal, R. *Zaragoza.*—Wheat, flour.
745 Vides, F. *Sevilla.*—Crackers.
746 Vila, F. *Almatret, Lérida.*—Almonds.
747 Vila, J. *Valencia.*—Wines.
748 Vilanova, P. *Alcalá, Castellon.*—Wine, oil.
749 Vilaplana, P. *Onil, Alicante.*—Pickles.
750 Vilches, F. *Almeria.*—Wine.
751 Villa Alcazar, M. de, *Salamanca*—Flour.
752 Villalta, L. *Jaen.*—White wheat.
753 Villalonga, M. *Palma.*—Oil, honey
754 Villalonga, M. *Lurmayor, Baleares.*—Wheat.
755 Villalonga, M. *Porreras, Baleares.*—Brandy.
756 Villanueva, M. *Tudela.*—Wine, oil.
757 Villanueva, M. *Tafalla.*—Wine.
758 Villapineda, C. de, *Sevilla.*—Agricultural products.
759 Villerí, J. M. *Zamora.*—Wine.
760 Villores, M. de, *Alcalá, Castellon.*—Wine, oil, raisins.
761 Vinagre, J. *Toro.*—Sweet almonds.
762 Vinez, J. *Reus.*—Wines.
763 Yegros, S.—Olive oil, wines.
764 Yuste, A. *Valladolid.*—Sausages.
765 Zaforteza, J. *Palma.*—Almonds.
766 Zayas, J. J. *Sevilla.*—Agricultural products.
767 Zerpa, J. *Villanueva, Sevilla.*—Wine.
768 Zorrilla & Co. *Burgo de Osusa, Soria.*—Flour.
769 Zozaya, M. *Errazu, Navarra.*—French beans.
770 Zubiri, T. *Tafalla, Navarra.*—Wheat.
771 Zuloaga, F. *Coruña.*—Preserves.

Agreda, J. A. de, *Jerez.*—Wines, natural and prepared.

CLASS 4.

772 Arazo, J. *Valencia.*—Cork.
773 Alonso, Y. *Mojados, Valladolid.*—Madder.
774 Arbones, J. *Almatret, Lerida.*—Aniseed.
775 Arguindey, Viuda de, *Puerto de Bejar, Salamanca.*—Glue.
776 Arnao, J. *Belloch, Lerida.*—Barilla.
777 Aynes & Co. *Valladolid.*—Madder, garancine.
778 Ayuntamiento de Cartagena.—Spanish hemp-grass and palm-hemp; palm-leaf brooms.
779 Ayuntamiento de Sarreaus, *Orence.*—Leeches.
780 Baller, J. *Córdoba.*—Sumach.
781 Belda, A. *Valencia.*—Bastard saffron.
782 Berner, F. *Elche, Alicante.*—Green almond bark, almond gum, &c.
783 Beynon, Stoken, é Hijos, *London.*—Havana cigars.
784 Borderias, V. *Montmesa, Huesca.*—Camomile.
785 Brizuela, J. M. *Salamanca.*—Oil of aniseed, crystallized.
786 Casado, M. *Málaga.*—Sugar-cane.
787 Catalan, J. *Monreal del Campo, Teruel.*—Brown saffron.
788 Cifra, P. *Santa Cruz de Tenerife.*—Aloes.
789 Cobaleda, J. A. *Porteros, Salamanca.*—Hay.
790 Comision de la Exposicion, *Almeria.*—Articles made of Spanish hemp-grass.
791 Compañia Gurri, *Barcelona.*—Oak, &c.
792 Cowen & Co. *London.*—Havana cigars

793 Credito Moviliario Barcelones, *Barcelona.*—Woods, from Muniello, Austurias.

794 Cuerpo de Ingenieros de Montes.—Woods.

795 Cunningham, J. *Sevilla.*—Liquorice.

796 Delgado, V. *Zalamea del Real, Huelva.*—Wax.

797 Duran, A. *Almeria.*—Coloquintida.

798 Fernandez, M. *Almeria.*—Spanish hemp-grass.

799 Fierro, J. *Santa Cruz de la Palma, Canarias.*—Cochineal, woods.

800 Flujá, F. *Elche, Alicant.*—Palm wood.

801 Gago Estéban, M. *Sta. Clara de Avedillo, Zamora.*—Sumach, sage.

802 Gago, Roperuelos, M. *Zamora.*—Camomile.

803 Gallardo, L. *Barcelona.*—Starch from the yarrow.

804 Garcia, J. R. *Valencia.*—Cochineal.

805 Garcia, M. L. *Cortegana, Huelva.*—Corks.

806 Garcia Acena, P. *Sevilla.*—Starch.

807 Gaston, V. *Sevilla.*—Corks.

808 Gobernador de Cuenca.—Bastard saffron.

809 Guerra, A. *Cuellar, Segovia.*—Madder.

810 Guerra, A. *Valladolid.*—Madder.

811 Heras, M. *Zamora.*—Oil of aniseed.

812 Yngeniero de Montes de la Provincia de Zaragoza.—Specimens of woods, liquorice, roots.

813 Yngeniero de Montes del Distrito de Gerona.—Articles in cork.

814 Junta de Agricultura Industria & Comercio de Granada.—Sugar cane.

815 Junta de Agricultura de Palencia.—Indigenous woods.

816 Lacave, J. P. *Sevilla.*—Corks.

817 Lopez, A. *Corcos, Valladolid.*—Aniseed.

818 Miranda, A. *Oviedo.*—Wood from the forests of Quiros.

819 Monfort, F. *Torrente de Cinca Huesca.*—Aniseed.

820 Morá, B. *Porrera, Balearcs.*—Saffron.

821 Mora, F. *Villanueva de los Castillejos, Huelva.*—Wax.

822 Oliver, F. *Sevilla.*—Corks.

823 Oroz, F. *Pamplona.*—Linseed oil.

824 Padilla, F. *Almeria.*—Spanish hemp-grass, in paste.

825 Pelaez, M. *Salamanca.*—Starch.

826 Perez, Cardenal A. *Zamora.*—Dyer's weed, lavender.

827 Perez Zamora, A. *Puerto de la Orotava, Canarias.*—Havana cigars.

828 Priego, J. *Doña Mencia, Cordova.*—Sumach.

829 Quemada, J. *Cuellar, Sogavia.*—Pulverized madder.

830 Ramirez, J. F. *Zaragoza.*—Liquorice.

831 Rivas, F. *Zaragoza.*—Sabine wood.

832 Royero, J. *Salorico, Cáceres.*—Corks.

833 San Andres, M. Conde de, *Puerto de la Cruz, Canarias.*—Cochineal.

834 Sancho, A. *Gandia, Valencia.*—Sugar-cane.

835 San Roman, A. *Fermocelle, Zamora.*—Water hemlock, hemlock-seed, digitalis leaves, &c.

836 Serrano, L. *Zalamea la Real, Huelva.*—Wax.

837 Sol, J. *Liñola-Lérida.*—Camomile.

838 Spencer & Roda, *Almeria.*—Barilla.

839 S. M. La Reina de Espana.—Woods.

840 Tamarit, M. de, *Tortosa.*—Soda barilla.

841 Traber, G. *Toro.*—Gum, from fruit trees.

842 Trias, P. *Esporlas, Baleares.*—Lignin of ash tree.

843 Valdes, T. *Pedraja de Portillo Valladolid.*—Pulverized and natural madder.

844 Valero, J. *Elche, Alicante.*—Soda, barilla.

845 Vidal, A. *Cartagena.*—Barilla.

846 Vidés, F. *Sevilla.*—Corks.

847 Vilches, F. *Almeria.*—Palmetto.

848 Zozaya, M. *Errazu, Navarra.*—Wax.

CLASS 5.

849 Astrua, D. *Córdova.*—A waggon wheel, constructed so as to move without friction at the axle.

850 Gallardo, L. *Barcelona.*—Model of a locomotive moved by hydrogen gas.

851 Sorguie, A. *Avilés, Asturias.*—Model of rails.

CLASS 7.

852 Cuchillo Hermanos, *Barcelona.*—Cards.

CLASS 8.

853 Bergue, M. *Barcelona.*—Pressing machine.

854 Berner, F. *Elche, Alicante.*—A press for various purposes.

855 Bridgman, E. *Tarragona.*—A gas-meter.

856 Ciervo & Co. *Barcelona.*—Gas apparatus.

857 Gallegos.—Carriage.

858 Tosser & Co. *San Sebastian.*—Machinery.

CLASS 9.

859 Aspe, J. *Sevilla.*—Agricultural implements.

860 Escuela de Veterinaria de Córdoba.—The "hiponcetro," for measuring horses.

CLASS 10.

861 Sédo, J. *Valencia.*—Model of a wooden bridge.

862 Munoz, A. *Madrid.*—Model for masonry.

CLASS 11.

863 Carbonell, A. *Alcoy, Alicante.*—Military uniform.

864 Cuerpo de Artilleria.—Rifled artillery, projectiles, fire arms, &c.

865 DIRECCION GENERAL DE INFANTERIA. —Military uniform.

866 DIRECCION DE LA GUARDIA CIVIL.—Equipage for a foot and a mounted guard.

867 DIRECCION GENERAL DE ESTA MAYOR. —Typographical plans, &c.

868 FABRICA NACIONAL DE TRUVIA.—Rifled cannon, fire-arms, projectiles.

869 FABRICA DE ARMAS DE OVIEDO.—Rifled gun, rifled carabine, &c.

870 LORENZALE, M. 60, *Calle de Alcalá, Madrid.*—Military articles.

CLASS 12.

871 MASDEN, M. *Málaga.*—Diving apparatus, of cast-iron.

CLASS 15.

872 JUSTE VILLANUEVA, Y.—Pocket chronometer.

CLASS 16.

873 MONTANO, V. 3, *Calle de In. Bernardino, Madrid.*—A grand piano, and piccolo.

874 ALBAT, J. *Valencia.* — Strings for musical instruments.

CLASS 17.

875 BOUSQUET, A. *Valencia.*—A set of teeth.

876 DIRECCION DE SANIDAS MILITAR.—Sanitarium.

877 GALLEGOS, J.—Artificial arm.

878 POUS, Y. *Barcelona.*—Lints.

879 RILÁS, R. *Valencia.*—Set of teeth.

880 RONAULT, HERMANOS, *Madrid.*—Orthopœdic apparatus.

881 TORRES MUÑOS Y LUNA, R. *Madrid.* —Medical and surgical case.

CLASS 18.

882 ACHON, J. *Barcelona.*—Chintz.

883 CIFRA, P. *Santa Cruz de Tenerife, Canarias.*—Cottons.

884 GONUS, T. *Manreza.*—Cottons.

885 FERRER & Co. *Villaneuva y Gettru.* —Handkerchiefs, &c.

886 GÜELL & Co. *Barcelona.*—Fine cord, coloured beaver.

887 JUNTA DE AGRICULTURA INDUSTRIA Y COMMERCIO DE GRANADA.—Cotton.

888 LARA, M. *Valladolid.*—Coloured muletones.

889 LARA, VILLARDELL É HIJOS, *Valladolid.*—Cotton goods, twilled.

890 LUNA, CONDE DE, *Elche, Alicante.*—Cotton in the pod.

891 RICART É HIJOS, *Barcelona.*—Chintz.

892 RUIZ DE LA PARRA, G. *La Cavada, Santander.*—Unbleached cotton goods.

893 SADÓ, J. *Barcelona.*—Table-linen, towels.

CLASS 19.

894 ALCALDE DE AGREDA, *Soria.*—Raw hemp and flax.

895 ALFARO, T. *Valladolid.*—Damasked linen.

896 ALONSO, T. *Orense.*—Flax.

897 ALÓS, T. A. *Balaguer, Léridæ.* —Hemp.

898 ANTON GUTIEREZ, B. *Zamora.*—Flax.

899 ARSENAL DE CARTEJANA.—Black and white rigging, sail cloth.

900 BRAÑA, B. *Coruña.*—Sail cloth.

901 CHICO, A. *Cehegin, Murcia.*—Hemp.

902 ESCUDERO, M. *Orihuela, Alicante.*—Hemp.

903 ESTEPA, T. *Vrrea de Jalon, Zaragoza.* —Flax.

904 ESTEVEZ, J. C. *Orense.*—Table linen.

905 FERREIRO, J. *Orense.*—Flax.

906 FRUTOS, A. *Sevilla.*—Linen goods.

907 GAGO ROPERUELOS, M. *Zamora.*—Flax.

908 GARCIA ALFONSO, A. *Moral de Orbigo, Leon.*—Flax.

909 GUIER, J. *Valencia.*—Hemp.

910 GOBERNADOR DE LA PROVINCIA DE GRANADA.—Hemp.

911 JIMENO, J. *La Milla del Rio, Leon.*—Flax.

912 JUNTA DE AGRICULTURA, INDUSTRIA, Y COMERCIO, DE CASTELLON.—Cordage, and articles in hemp.

913 JUNTA DE AGRICULTURA, INDUSTRIA, Y COMERCIO, DE GRANADA.—Hemp, flax.

914 LAFUENTE, R. M. *Valladolid.*—Linen.

915 LOZANO VERA, E. *Orihuela, Alicante.* —Hemp, sail cloth.

916 MALDONADO, M. *Orense.*—Gallician linen.

917 MARTINEZ, F. *Tarazona.* — Flax, hemp.

918 MAT & Co. *Barcelona.*—Flax, damask; a hammock.

919 MASERES, J. *Orihuela, Alicante.*—Hemp.

920 MUNOZ, F. *Leon.*—Flax.

921 MORA, J. *Callosa de Segura, Alicante.* —Hemp.

922 PALOZUELOS, VISCONDE DE, *Toledo.* —Cordage, hemp.

923 PARRA, J. R. *Santibañez de Toro, Zamora.*—Flax.

924 PERCIAS, *Palmas.*—Cordage and canvas.

925 PUIG, F. *Barcelona.*—Linen thread.

926 RIPALDA, CONDE DE, *Valencia.*—Hemp.

927 ROCA DE TOGORES, B. *Orihuela, Alicante.*—Flax.

928 RODRIGUEZ, J. *Elche, Alicante.*—Cordage.

929 RUBIO & RODRIGUEZ, *Sevilla.*—Linen goods.

930 SANZ, B. *Burgo de Osma, Soria.*—Raw flax.

931 SOCIEDAD ECONÓMICA DE MURCIA.—Sail cloth.

932 VACA, M. *Hospital de Orbigo, Leon.*—Flax.

933 VEGAS, A. *Hospital de Orbigo, Leon.* —Flax.

934 VIDAL, SEMPRUM, & Co. *Valladolid.*—Unbleached linen.

CLASS 20.

935 Almansa, D. *Murcia.*—Sewing silks.
936 Bonell, T. R. *Valencia.*—Goods in silk and gold.
937 Calonge, M. R. *Sevilla.*—Galloons.
938 Carrere, E. *Graus, Huesca.*—Silk, silk gauze.
939 Castillo, M. *Sevilla.*—Silk goods.
940 Escuder, Viuda é hijos de, *Barcelona.*—Goods in silk, and in silk and gold.
941 Fierro, T. *Santa Cruz de la Palma, Canarias.*—Raw, spun, and woven silk.
942 Garin M. *Valencia.*—Goods in silk and gold.
943 Garriga, L. *Manreza.*—Silk.
944 Horns, T. *Barcelona.*—Glazed silk.
945 Hortal, J. *Salamanca.*—Cocoons.
946 Junta de Agricultura Industria y Comercio de Castellon.—Silk.
947 Lopez, Ubeda T. *Valencia.*—Spun silk.
948 Lozano-Vera, E. *Orichuela, Alicante.*—Raw and spun silk.
949 Marin, B. *Murcia.*—Cocoons.
950 Martinez & Co. M. *Sevilla.*—Silks.
951 Molina-Nuñez, F. *Orihuela, Alicante.*—Spun silk.
952 Moreno Miguel, V. *Madrid.*—Valencia silk.
953 Peñafiel, E. *Murcia.*—Silk, spun and twisted by a single process.
954 Perez Villoria, R. *Fregeneda, Salamanca.*—Spun silk.
955 Portales & Co. *Zalavera de la Rèina.*—Spun silk.
956 Pujal, Vinda de, & Co. *Valencia.*—Spun and twisted silk.
957 Real, J. *Sevilla.*—Galloons.
958 Rodriguez Lara, M. *Orihuela, Alicante.*—Cloth for sieves.
959 Rubio Hermanos, *Valencia.*—Woven silk.
960 Satorras, A. *Tarragona.*—Raw silk.
961 Sever y Tena, M. *Valencia.*—Goods in silk and gold.
962 Sociedad Economica de Murcia.—Cocoons; twisted and dyed silk.
963 Sorní, Vinda de é Hijos, *Valencia.*—Ribbons and mohair.
964 Torder, M. *Valencia.*—Spun and twisted silk.
965 Trell, J. *Berja, Almeria.*—Cocoons.
966 Vilumara, Hermanos & Co. *Barcelona.*—Glazed silk, satin, &c.

CLASS 21.

967 Andrés, J. B. *Alcoy, Alicante.*—Woollen goods.
968 Arroyo, G. *Palencia.*—Shawls.
969 Bernardos, M. *Bernardos, Segovia.*—Coarse cloth.
970 Bonaplata, N. *Sevilla.*—Worsted of various colours.
971 Buzaren & Masoliver, *Barcelona.*—Merino wool, and webs.
972 Cantó, Viuda de é Hijos, *Alcoy, Alicante.*—Woollen cloth, samples of silk.
973 Capusany, J. *Barcelona.*—Satin.
974 Casanovas é Hijos, *Sabadell.*—Eider-down, woollen goods.
975 Casanovas & Bosch, *Sabadell.*—Woollen goods.
976 Coma, T. *Barcelona.*—Spun worsted.
977 Conquista, Marques de la, *Trujillo, Cáceres.*—Merino and Saxon wool.
978 Galí & Co. *Tarraia.*—Broad cloth, beavers, satins, cashmeres, fine woollen cloth.
979 Garcia, A. *Córdova.*—Woollen cloth, &c.
980 Gomez Rodulfo, G. *Bejar, Salamanca.*—Broadcloth.
981 Gonzalez, M. *Valdeavellano de Tera Seria.*—Articles in fine and mixed Saxon wool.
982 Gorina, T. *Sabadell.*—Broadcloth, beavers, &c.
983 Herrero, D. *Almeira, Zamora.*—Worsted.
984 Jordá, F. *Alcoy, Alicante.*—Woollen goods.
985 Losada, D. R. *Illescas, Toledo.*—Churra wool.
986 Maicas, T. *Valencia.*—Woollen goods.
987 Maña é Yseru, *Barcelona.*—Ratteens, kerseymeres, drills.
988 Maro, M. *Villalpando, Zamora.*—Wool of native sheep.
989 Miarens & Doria, *Barcelona.*—Broadcloth, eider-down, fine woollen cloth, &c.
990 Palazuelos, Visconde de, *Toledo.*—Worsted.
991 Perez Marco, *Rellen, Alicante.*—Wool.
992 Pinto da Costa, R. *Lumbrales, Salamanca.*—Shawls, sackcloth.
993 Pujol, Buxeda, & Co. *Barcelona.*—Woollen velvets, double-milled woollen-cloth, beavers, eider-down, wool-satins.
994 Sacristan, R. *Segovia.*—Wool.
995 Sallares, J. *Sabadell.*—Beavers, broadcloth, fine woollen cloth.
996 Samuel, E. *Sevilla.*—Worsted of different colours.
997 Santos, N. *San Sebastian.*—Broadcloth.
998 Sociedad Económica de Murcia.—Rugs.
999 Solá & Co. *Barcelona.*—Shawls, mantles, handkerchiefs, vests.
1000*Spencer & Roda, *Almeira.*—Raw wool.
1001 Tastel Hermanos, *Sevilla.*—Spun wool, of different colours.
1002 Tello, I. V. *Valencia.*—Woollen cloth.
1003 Terol & Gisbert, *Alcoy, Alicante.*—Fine woollen cloth.
1004 Turrull, P. *Sabadell.*—Fine woollen cloth, beavers.
1005 Vera, R. *Soria.*—Fine and Merino wool, and worsted.
1006 Vieta & Co. *Tarrasa.*—Broadcloth, wool-satins, fine woollen cloth.

CLASS 24.

1007 Capmany y Volart, *Barcelona.*—Blonds and laces.
1008 Fiter, J. *Barcelona.*—Blonds and laces.
1009 Garcia & Atienza, *Valencia.*—Embroidery.
1010 Hidalgo, R. *Palma del Rio, Cordova.*—Embroidery.
1011 Margarit, J. *Barcelona.*—Blonds and laces.
1012 Guiblier, A. & Co. *Barcelona.*—Tapestry.
1013 Serrano, B. *Limares, Jaen.*—Embroidery.
1014 Suarez, T. *Coruña.*—Embroidery.

CLASS 25.

1015 Diaz, J. *Sevilla.*—Gazelle furs.

CLASS 26.

1016 Arguindey, Vinda de, *Puerto de Bejar, Salamanca.*—Hides.
1017 Cadena, J. *Málaga.*—Andalusian horse harness.
1018 Carabias, J. *Salamanca.*—Hides.
1019 Fernandez, Vinda de, *Sevilla.*—Hides.
1020 Garcia Dorado, G. *Valladolid.*—Saddle and harness.
1021 Manzanos, A. *Sevilla.*—Spanish leather.
1022 Marticorena, R. *San Sebastian.*—Hides.
1023 Muños, M. *Plasencia, Cáceres.*—Sole leather.
1024 Ramos, J. *Navalmoral, Cáceres.*—Sole leather.
1025 Rodriguez, A. *Coruña.*—Hides.
1026 Romero, P. *Fuentepelayo, Segovia.*—Hides.
1027 Teja, J. *Coruña.*—Sole leather.
1028 Yglesias, M. *Sevilla.*—Shagreen, hides.

CLASS 27.

1029 Abat, P. *Valencia.*—Materials for fans.
1030 Acebedo, A. *Valencia.*—Hats.
1031 Anquidad, F. *Coruña.*—Gloves, kid skin.
1032 Baladó, F. *Barcelona.*—Muslin; flowers.
1033 Bosch, J. *Vich.*—Leather, and gloves.
1034 Castilla, M. *Málaga.*—Gloves.
1035 Colomina, J. *Valencia.*—Fans.
1036 Creixach, V. *Valencia.*—Sandals.
1037 Cubedo, J. *Valéncia.*—Gloves.
1038 Enseñat, M. *Barcelona.*—Artificial flowers.
1039 Espeso, D. *Valladolid.*—Buttons.
1040 Esquerdo, T. *Villajoyosa, Alicante.*—Palm-leaf hats.
1041 Fortuñ, T. *Zaragosa.*—Hats, and illustration of their manufacture.
1042 Gelí, F. & P. *Sevilla.*—Gloves.
1043 Gand, E. *Valladolid.*—Gloves.
1044 Gomez, J. *Madrid.*—Hats.
1045 Guerrero, J. B. *Ecija.*—Hats.
1046 Horna, M. *Zamora.*—Waterproof hats.
1047 Martin, F. *Valencia.*—Abamios.
1048 Martin, M. J. *Salamanca.*—Shoes.
1049 Masfaruer, J. *Valencia.*—Gloves.
1050 Mendieta & Co. *Riara, Segovia.*—Pins.
1051 Molina, S. *Iaen.*—Shoes.
1052 Perrier, I. *Sevilla.*—Gloves, and leather.
1053 Ray, F. *Coruña.*—Hats.
1054 Reinaldo, I. 28, *Carrera de Sn. Geron. Madrid.*—Shoes.
1055 Rojan, Viuda de, e Hijo, *Sevilla.*—Brass buttons.
1056 Tejedor, R. *Bermillo de Sayago, Tramora.*—A sayaguesa mantelet

CLASS 28.

1057 Carbó, L. *Riba, Zaragoza.*—Pasteboard.
1058 Carderera, V. *Madrid.*—Spanish ichnography.
1059 Comision para la Publicacion de la Obra "Monumentos Arquitectonicos de España, *Madrid.*—Proofs of Engravings of Spanish national edifices.
1060 Fabricantes de Alcoy, *Alicante.*—Smoking paper.
1061 Fort, E. *Alcoy, Alicante.*—Smoking paper.
1062 Junta de Agricultura Industria y Comercio, *Granada.*—Continuous paper.
1063 Lizarbe & Co. *Cascante, Navarra.*—Smoking paper.
1064 Lopez, H. 16, *Calle de Santa Ana, Madrid.*—Smoking paper, prepared by a new process.
1065 Masustegui, M. *Castellano.*—Paper.
1066 Peña, J. M. *Candelario, Salamanca.*—Continuous paper.
1067 Peña, J. B. *Candelario, Salamanca*—Fine paper.
1068 Romaní & Miró, *Barcelona.*—Paper.
1069 Romaní & Olivella, *Barcelona.*—Paper.
1070 Romaní & Tarres, *Barcelona.*—Smoking paper, &c.

CLASS 30.

1071 Administracion de la Alhambra, *Granada.*—Arabesques.
1072 Aleman, V. *Alicant.*—Bedstead, changeable into a column.
1073 Arraez, A. *Madrid.*—Arabesques.
1074 Ballesteros, S. 34, *Carrera de Sn. Geronimo, Madrid.*—Stained paper.
1075 Botana, J. *Sta. Eulalia de Dena Pontevedra.*—Group of animals, in mother-of-pearl.
1076 Castels & Serra, *Barcelona.*—Carved work in wood.
1077 Dubuison & Co. *Sevilla.*—Iron bedsteads.
1079 Perez, F. *Valencia.*—Mosaic in wood.

1080 ROMERO, J. *Sevilla.*—Brushes.

1081 YSAURA, *Barcelona.*—A metal chandelier, medals.

1082 ZULOAGA, P. *Madrid.*—Desk, mirror, album, vases, &c. in bas-relief, and damascened with iron, gold, and silver.

CLASS 31.

1083 CANALES, D. *Barcelona.* — Tin articles.

1084 COTARELO, J. *Oviedo.*—Unoxidizable nails.

1085 FABRICA NACIONAL DE TRUVIA.—Iron, steel, files.

1086 MALABUCH, F. *Valencia.* — Scales, steelyards, &c.

1087 MATA, R. *Valencia.*—Nails.

1088 PEREZ, J. B. *Taen.*—Iron decomposed without losing its elasticity.

1089 SOCIEDAD HULLERA Y METALURGICA DE ASTURIAS, *Mieres.*—Casting in unoxidizable iron.

CLASS 32.

1090 ELENA, M. *Salamanca.*—Jewellery and filigree work.

1091 GOMEZ, L. *Salamanca.* — Filigree work.

1092 SOLER, P. *Barcelona.*—Jewellery.

1093 TELLEZ, F. *Salamanca.* — Filigree work in silver.

CLASS 33.

1094 CIFUENTES POLA & Co. *Gijon.* — Glass, hollow and flat.

1095 COLLANTES, A. *Santader.*—Glass.

CLASS 34.

1096 ALCADE DE MUELAS, *Zamora.*—Refractory clay.

1097 BACAS, P. *Andugjar.*—White clay.

1098 BAIGNOL HERMANOS & Co. *Sñ. Sebastian.*—Porcelain.

1099 CUBERO, J. *Málaga.*—Málaga figures.

1100 GARCIA, B. *Tamamer, Salamanca.*—Common earthenware.

1101 GOBERNADOR DE LA PROVINCIA DE TAEN.—Articles in white clay.

1102 GONZALEZ VALLS, R. *Valencia.* — Ornamental glazed tiles.

1103 GUTIERREZ DE LEON, A. *Málaga.*—Málaga figures.

1104 JUAN & SEVA, M. *Madrid.*—A large earthen jar.

1105 MIGUEZ, J. *Sevilla.* — Ornamental glazed tiles.

1106 NOLLA Y SAGRERA, *Valencia.* — Ornamental glazed tiles.

1107 OJEDA & Co. *Sevilla.* — Ornamental glazed tiles.

1108 PEIKMAN & Co. *Sevilla.* — Fine earthenware.

1109 REDONDO OLEA, J. *Porezeuela, Zamora.*—Refractory clay.

1110 RODRIGUEZ, R. *San Juan de Azualfarache, Sevilla.*—Fine earthenware.

1111 SANCHEZ CABALLERO, L. *Málaga.* —Specimens of pottery.

CLASS 35.

1112 ALVAREZ Y PINILLA, *Coruna.* — Soap.

1113 BAUL & Co. *Cartagena.* — Safety match-rope.

1114 CARREÑO, J. *Sevilla.*—Sperm candles.

1115 CONRADI, J. *Sevilla.*—Soaps.

1116 ESQUERDO, J. *Villàjoyosa, Alicante.*—Palm-leaf case.

1117 ESTER, F. *Sevilla.*—Soaps.

1118 FABRA, C. *Barcelona.*—Fishing nets.

1119 GALLIANO, M. *Sevilla.*—Barrels.

1120 GARRET, SAENZ, & Co. *Málaga.* — Stearine candles.

1121 GIMENEZ, H. *Mora, Toledo.*—Hard soap.

1122 GOTZENS, DELOUSTAL, & Co. *Barcelona.*—Soaps.

1123 GRACIAN & Co. *Málaga.*—Soap.

1124 GUERRERO É HIJOS, VINDA DE, *Mora Toledo.*—Hard soap.

1125 GUEVENZÚ, E HIJOS, *Cascante, Navarra.*—Stearine candles.

1126 GIMENO, J. *Zaragoza.*—Soap.

1127 LIZARBE, P. *Bertanga, Soria.* — Stearine candles.

1128 LUZ, B. *Burgos.* — Stearine candles, Oleine candles.

1129 MARTICORENA, R. *San Sebastian.*—Stearine soap.

1130 MARTINEZ, J. M. *Sevilla.*—Cases.

1131 PERLA, F. 24 *&* 26, *Gobernador Madrid.*—Candles, soaps.

1132 SANTA CRUZ, J. *Sevilla.*—Soaps.

1133 TOSA, A. *Zaragoza.* — Artificial jaspers.

SWEDEN.

N.W. Court, No. 5, and N.W. Gallery, No. 4.

The Swedish Alphabet has three letters more than the English Alphabet, viz., Å, Ä, and Ö. These letters were originally the diphthongs Ao, Ae, and Oe, and follow in order after the letter Z, at the end of the alphabet.

CLASS 1.

1 ADELSWÄRD, S. E. BARON, *Åtvidaberg, Ostgothland.*—Specimens illustrative of the processes used at Åtvidaberg copper works.

2 ADLERS, A. *Yxhult, Nericia.* — Manger of lime-stone.

3 ANDERSSON, N. *Södvik, Isle of Öland.* —Öland stone for flooring.

4 ARBORELIUS, E. G. *Elfdal, Dalecarlia.* —Manufactures of porphyry.

5 ASCHAN, N. N. *Ohs, Smaland.*—Alloy of nickel and iron.

6 BALDERSNÄS IRON WORKS, *Dalsland.*—Steel-iron in bars.

7 BESKOW, J. W. *Wrethammar, Nericia.* —Iron ores from Stora Blanka mine.

8 BISPBERG MINING CO. *Dalecarlia.* — Iron ore, weighing 16 cwt.

9 DANNEMORA MINING CO. *Upland.* — Iron ore, and the strata in which it is found.

10 EDLUND, J. *Hammarbacken, Nericia.* —Iron ore from the Moss mine.

11 EKMAN, C. *Finspong, Ostgothland.* — Iron ore, pig-iron, and bar-iron.

12 EKMAN, G. *Lesjöforss, Wermland.*— Pig-iron, with analysis attached.

13 ELFSTORP IRON WORKS, *Nericia.*— Iron ore, pig-iron, bar-iron, scoria.

14 FINNÅKER IRON WORKS, *Nericia.*— Bar-iron.

15 FREDRIKSBERG COPPER CO. *Smaland.* —Copper ore and copper.

16 GAMMALKROPPA AND SÄLBODA IRON WORKS, *Wermland.*—Series of iron ores from Persberg mines, pig-iron, and bar-iron.

17 GARPENBERG IRON WORKS, *Dalecarlia.*—Iron ore from Långvik and Bispberg mines, pig-iron, and bar-iron.

18 GULDSMEDSHYTTE SILVER WORKS, *Nericia.*—Ores of lead (sulphurets), lead.

19 GUSTAF AND CARLBERG COPPER WORKS, *Jemtland.*—Copper ores, refined copper, copper-sheets and wire.

20 GYSINGE IRON WORKS, *Gefle.* — Pig-iron, steel-iron in bars.

21 HAMILTON, H. BARON, *Boo, Nericia.* —Bar-iron, steel, and peat.

22 HARMENS, AF, H. O. HEIRS OF, *Berga, Calmar.* — Bog-iron ore, pig-iron, specimens of porphyry.

23 HEIJKENSKÖLD, S. *Granhult, Nericia.* —Pig-iron.

24 HELLSTRÖM, V. *Dalkarlshyttan, Nericia.*—Iron ores from Stripa mine.

25 HERMANSSON, C. COUNT, *Ferna, Westmanland.*—Specimens of bar-iron and steel.

26 HORNDAL IRON WORKS, *Dalecarlia.*— Iron ores from Bispberg, Holm, and Westersjö mines, pig-iron, and bar-iron.

27 JANSSON, *Westgothland.* — Tombstone from quarries at Kinnekulle.

28 JOHANNESSEN, W. *Sörbytorp, Scania.* —Marl.

29 KILLANDER, F. *Hook, Smaland.*—Iron ores from Taberg mine, pig-iron, scoria, bar-iron, iron-wire.

30 KJELLSON, J. F. *Hofmanstorp, Smaland.*—Iron ore from Westanå mine: pig-iron, bar-iron, produced from the ore.

31 KLOSTER IRON WORKS, *Dalecarlia.*— Iron ore from Rällingsberg Mine, pig-iron: Bessemer steel in ingots, rolled, and in sheets.

32 KOCKUM, F. H. *Malmö.*—Copper ores from Virum mines.

33 KULLANDER, C. G. *Kjellsviken, Dalsland.*—Slate.

34 LAGERHJELM, P. *Bofors, Nericia.*— Steel.

35 LAMM, A. & HOFFMAN, G. L. *Stockholm.*—Monuments in granite.

36 LANDSTRÖM, C. *Österplana, Westgothland.*—Articles from the quarries at Kinnekulle.

37 LESSEBO NICKEL CO. *Smaland.* — Nickel ore from Klefva mines, mat and refined nickel.

38 LIDÉN, J. *Töksmark, Wermland.* — Cooking utensils of Talc stone.

39 LILLIECREUTZ, J. BARON, *Åminne, Smaland.*—Bog-iron ore and pig-iron.

40 LJUSNARSBERG COPPER WORKS, *Nericia.*—Copper ores, mat, black, and refined copper.

41 LUNDBOM, P. J. *Löfnäs, Nericia.* — Iron ores from Nya Kopparberg, pig-iron, bar-iron.

42 LUNDHQVIST, G. A. *Nyköping.*—Sheet-iron and cold-bent iron bars; iron pyrites; roll-sulphur.

43 MALMSTEN, J. A. *Jönköping.* — Manganese, in ore and powdered.

44 MARÉ, DE, G. *Tofverum, Westerwik.*— Bog-iron ore, pig-iron.

45 MOSSGRUFVE MINING CO. *Westmanland.*—Iron ore from Norberg mines.

46 MÖLNBACKA IRON WORKS, *Wermland.* —Bar-iron.

47 NORDENFELDT, O. *Björneborg, Wermland.*—Pig-iron, bar-iron, steel.

48 BREFVEN IRON WORKS, *Nericia.*—Pig-iron, bar-iron, steel-iron.

49 PETRÉ, T. HEIRS OF, *Hofors Iron Works, Gefle.*—Iron.

50 RETTIG, C. A. *Kihlaforss, Gefle.*—Bar-iron for gun-barrels. Steel made from pig-iron and bar-iron. Razors.

51 ROCZYCKI, C. BARON, *Kolmården, Ostgothland.*—Manufactures of marble.

52 RÄMEN IRON WORKS, *Wermland.*— Series of ores from mines in Wermland, pig-iron, bar-iron.

53 RÖNNÖFORSS IRON WORKS, *Jemtland.* —Bog-iron ore, pig-iron, bar-iron.

54 SANDERSON, P. B. *Spexeryd, Smaland.*—Manganese ore.

55 SEDERHOLM, J. *Näfveqvarn, Sudermannia.*—Cobalt ore from Tunaberg mines.

56 SILJANSFORSS IRON WORKS, *Dalecarlia.*—Iron ores from Sörskog mines, pig-iron, Bessemer iron and steel.

57 SJÖGREEN, C. W. *Bruzaholm, Smaland.*—Aedelforsit: a mineral consisting of trisilicate of lime.

58 STABERG & BJÖRLING, *Stockholm.* —Cross of granite.

59 STOCKENSTRÖM, VON A. *Åker Cannon Foundry, Sudermannia.*—Iron ores from Skotvång and Förola mines. Cannon-iron.

60 STORA KOPPARBERG MINING CO. *Dalecarlia.*—Iron ores from Windtjern, Skinnaräng, Tuna Hästberg. Copper ores from Fahlu mine. Iron and copper.

61 SUBER & SJÖGREEN, *Bruzaholm, Smaland.*—Series of bog-iron ore and its products.

62 SVANÅ IRON WORKS, *Westmanland.*— Pig-iron and cold-bent bar-iron.

63 SÖDERFORSS IRON WORKS, *Upland.*— Pig-iron, bar-iron.

64 UDDEHOLM IRON WORKS, *Wermland.* —Pig-iron, bar-iron, steel, and iron-wire.

65 ULFF, C. R. *Wikmanshyttan, Dalecarlia.*—Pig-iron, bar-iron. Steel produced by the method of Uchatius.

66 VIEILLE MONTAGNE ZINC MINING Co. *Åmmeberg, Nericia.*—Ores of zinc (sulphurets), and minerals from Åmmeberg mines.

67 WAHLSTEDT, A. *Källered Steel Works, Gothenburg.*—Steel.

68 WERMLAND LOCAL COMMITTEE, *Carlstad.*—Iron ores from Persberg and other mines. Scoria.

69 ZETHELIUS, W. *Surahammar, Westmanland.*—Specimens of iron and steel. Iron plates and sheets.

70 ÖREBRO LOCAL COMMITTEE, *Nericia.*—Collection of iron ores from Nora.

71 ÖSTBERG, C. *Elfkarleö, Upland.*—Pig-iron, bar-iron, blistered and hammered steel.

72 ÖSTERBY IRON WORKS, *Upland.*—Pig-iron, bar-iron, blistered and hammered steel.

---

73 CARLSDAL IRON WORKS, *Nericia.*—Pig-iron; steel.

74 FREDRIKSBERG IRON WORKS, *Dalecarlia.*—Bar-iron; pig-iron; blooms; steel.

75 GÖRANSSON, F. *Högbo, Gefle.*—Iron ores; pig-iron; Bessemer iron, steel, and manufactures.

76 HAMMARBY IRON WORKS, *Nericia.*—Pig-iron.

77 HELLEFORSS IRON WORKS, *Nericia.*—Pig-iron; bar-iron.

78 SCHRAM, J. R. *Singö, Grisslehamn.*—Marble table.

79 SMEDJEBACKEN ROLLING MILL, *Dalecarlia.* — Bar-iron.

80 SÅGMYRE NICKEL WORKS, *Dalecarlia.*—Granulated nickel copper.

80A OSTERHOLM, O. *Stockholm.*—Monument of granite.

80B CARLÉN, MISS, *Stockholm.*—Letterpress of marble, from the Isle of Gottland; petrifactions.

80C SALA BERGSLAG, *Sala.*—Silver ore.

80D STRIBERG MINING CO. *Nericia.*—Iron ore from the Åsboberg mine.

CLASS 2.

Sub-Class A.

81 ASSOCIATED COPPER-SMELTERS, FAHLUN, *Dalecarlia.*—Granulated copper, oxide of copper, copperas, sulphur.

82 DJURÖ TECHNICAL MANUFACTORY, *Norrköping.*—Sulphate of copper.

83 DUFVA, C. F. *Stockholm.*—Blacking.

84 FRIESTEDT, A. W. *Stockholm.*—Steamed bone-dust for manure. Animal charcoal. Blacking for horses' hoofs.

85 GOTHENBURG LUCIFER-MATCH CO. *Gothenburg.*—Matches.

86 HAMILTON, COUNT H. D. *Hönsäter, Westgothland.*—Alum, sulphate of iron, red ochre.

87 HASSELGREN, O. S. *Rådaneforss, Dalsland.*—Peat, carbonized peat. Steamed bone-dust manure.

88 HEDENDERG, V. L. A. *Umeå.*—Saltpetre made by peasants.

89 HJERTA, L. J. & MICHAELSSON, J. *Stockholm.*—Sulphuric acid.

90 JÖNKÖPING LUCIFER-MATCH CO. *Jönköping.*—Safety matches, &c.

91 KYLBERG, H. *Ryssbylund, Calmar.*—Poudrette.

92 KYLBERG, I. *Såtenäs, Westgothland.*—Poudrette.

93 LEWENHAUPT, COUNT C. M. *Claestorp, Sudermannia.*—Turpentine (raw and rectified), pinol, resin, lamp-black, &c.

94 LOFVERS ALUM WORKS, *Isle of Oland.*—Alum slate, alum.

95 MAJERAN, J. M. *Jönköping.*—Fibrous matter prepared from pine needles. Extracts therefrom.

96 PIPER, COUNT C. E. *Andrarun, Scania.*—Alum, sulphate of iron, red ochre, alum wash.

97 STORA KOPPARBERG MINING CO. *Dalecarlia.*—Iron pyrites, red ochre.

98 WAERN, C. F. & Co. *Gothenburg.*—Superphosphate of lime.

99 WALLBERG, F. *Stockholm.*—Specimens of steamed bone-dust.

100 WESTERWIK LUCIFER-MATCH CO. *Westerwik.*—Various kinds of matches.

101 WIKLAND, P. *Swartsång, Wermland.*—Carbonized wood.

---

102 DYLTA SULPHUR WORKS, *Nericia.*—Sulphur; green vitriol; red ochre; jewellers rouge.

103 HAZELIUS, A. K. *Stockholm.*—Sulphate of ammonia.

Sub-Class B.

106 CAVALLI, J. G. *Gothenburg.*—Pharmaceutical preparations. Compressed medicinal vegetables.

107 DUFVA, C. F. *Stockholm.*—Eau de Cologne from Swedish spirits.

108 HYLIN & CO. *Stockholm.*—Perfumery.

109 JACOBSSON, L. A. *Stockholm.*—Perumery. Tooth-paste, tincture for the teeth.

110 LINDMAN, B. *Stockholm.*—Gelatine capsules containing various medicines.

111 PAULI, F. *Jönköping.*—Perfumery.

112 WIKSTRÖM, Z. *Stockholm.*—Various kinds of pulverized drugs.

---

113 ALEXIS, T. *Stockholm.*—Eau de Cologne.

114 WILLMAN, L. *Ystad.*—Tooth powder.

CLASS 3.

Sub-Class A.

116 AGRICULTURAL SOCIETY, OSTGOTHLAND, *Linköping.*—Grain produced in the province.

117 AGRICULTURAL SOCIETY, NORRA MÖRE, *Calmar.*—Grain. Seeds of forage and other plants.

118 AMEEN, H. *Hogsta, Stockholm.*—Rye.

119 ARRHENIUS, J. *Ultuna, Upland.*—Grain and agricultural seeds from the middle part of Sweden.

120 BÖÖS, J. *Gräfsnäs, Westgothlund.*—Oats.

121 CELSING, V. L. G. *Lindholmen, Sudermannia.*—Wheat, rye.

122 DUGGE, H. *Latorp Nericia.*—Autumn wheat.

123 ENHÖRNING H. *Knista, Nericia.*—Alsike clover-seed.

124 ERIKSSON, E. J. *Umeå.*—Wheat.

125 EVERS, O. *Ström, Gothenburg.* — Wheat, oats, clover-seed.

126 FORSELL, C. A. *Mellösa, Nericia.*—Pease.

127 FRIESTEDT, A. W. *Stockholm.* — Steamed bone-dust, used as food for cattle.

128 GIERTSSEN, G. *Malmö.*—Flour.

129 GYLLENCREUTZ, Y. *Gyljen, Luleå.*—Wheat, grown ten miles south of the Arctic circle.

130 HEDIN, J. G. *Snaflunda, Nericia.*—Rye.

131 HJORTH, C. W.—*Sundswall.*—Wheat, rye, barley.

132 HOFSTÉN, S. E. v. *Sund, Nericia.*—Wheat and wheat-flour, oats.

133 KEY, E. *Sundsholm, Calmar.*—Wheat, rye, oats, pease.

134 KLOCKHOFF, O. D. *Ytterstforss, Umeå.* —Autumn wheat, grown at 65° lat.

135 KYLBERG, H. *Ryssbylund, Calmar.*—Hops.

136 KYLBERG, J. *Såtenäs, Westgothland.*—Rye.

137 LIEPE, C. F. *Gothenburg.*—Grain.

138 LIGNELL, C. A. *Carlslund, Jemtland.* —Wheat, rye, barley.

139 LUND AGRICULTURAL SCHOOL.—*Nericia.*—Barley, red clover-seed.

140 MALMSTEN, J. A. *Jönköping.*—Potatoe-flour.

141 MALMÖ LOCAL COMMITTEE, *Malmö.*—Grain.

142 NENSEN, J. A. *Umeå.*—Ears of wheat.

143 NYBERG, J. *Brefven, Nericia.*—Autumn and spring wheat, rye.

144 ROSSANDER, F. J. *Isle of Öland.*—Wheat, rye, barley, oats.

145 SAHLMARK, C. P. *Ljungsnäs, Westgothland.*—Barley.

146 SPRENGTPORTEN, BARON J. W. *Sparreholm, Sudermannia.*—Wheat, oats, grass-seeds.

147 SÖDERSTRÖM, A. G. *Hernösand.*—Barley, rye.

148 UNANDER, F. *Yttertafle, Umeå.*—Timothy-seeds.

149 WEDBERG, A. *Yxe, Nericia.*—Wheat, black tares.

150 WULFF, H. A. *Applerum, Calmar.*—Wheat, rye, barley, oats, pease, tares.

151 ÅKERBLOM, C. M. *Norrköping.* — Starch prepared from wheat.

152 ÅNGSTRÖM, J. *Berglunda, Umeå.*—Four-rowed barley.

---

153 KIHLMAN, O. S. *Tofta, Gothenburg.*—Yellow beans.

Sub-Class B.

156 DUFVA, C. F. *Stockholm.*—Chocolate, mustard.

157 FROMELL, C. J. *Gothenburg.*—Confectionery.

158 GRAFSTRÖM, C. J. *Stockholm.*—Confectionery, chocolate, syrup.

159 HERNÖSAND LOCAL COMMITTEE, *Hernösand.*—Berries of rubus arcticus, preserved in sugar.

160 JUHLIN-DANNFELT, C. *Stockholm.*—Cranberries and blackberries, preserved in sugar.

161 KYLBERG, H. *Ryssbylund, Calmar.*—Cheese.

162 KYLBERG, J. *Såtenäs, Westgothland.*—Cheese.

163 LANDSKRONA SUGAR-MANUFACTURING CO. *Landskrona.*—Beet-root sugar.

164 LINDBLOM, A. E. HEIRS OF, *Ronneby, Carlskrona.*—Articles prepared from potatoes.

165 LUNDGREN, P. W. *Stockholm.*—Starch, treacle, gum, &c. prepared from potatoes.

166 OLOFSSON, A. *Åse, Jemtland.*—Goats'-milk cheese.

167 PETTERSSON, C. & SON, *Stockholm.*—Sugar, refined by hand.

168 WAHLBOM, G. B. *Köhlby, Calmar.*—Cheese.

169 WEDBERG, MRS. L. *Nora.*—Cranberries, cloudberries, and raspberries, preserved in sugar.

Sub-Class C.

171 BODACH, C. *Stockholm.*—Cigars.

172 BOMAN, J. A. & CO. *Gothenburg.* — Manufactured tobacco.

173 BRINCK, HAFSTRÖM, & CO. *Stockholm.* —Cigars, snuff, roll tobacco.

174 GRAFSTRÖM, C. J. *Stockholm.*—Cordials.

175 KOCKUM, F. H. *Malmö.* — Manufactured tobacco.

176 KYLBERG, H. *Ryssbylund, Calmar.*—Tobacco grown by the exhibitor.

177 LJUNGLÖF, J. F. *Stockholm.* — Cigars, snuff.

178 LITTSTRÖM, E. G. *Fahlun.*—Refined corn brandy.

179 LUNDGREN, P. W. *Stockholm.*—Wines prepared from fruits.

180 PRYTZ & WIENCKEN, *Gothenburg.*—Manufactured tobacco.

181 ROSÉN, E. A. & STRÖMBERG, *Stockholm.*—Cigars, tobacco, snuff.

182 STERNHAGEN & CO. *Gothenburg.* — Snuff.

183 SUNDIN & CO. *Westerås.* — Manufactured tobacco.

184 SUNDSTEDT & CO. *Stockholm.*—Roll tobacco.

185 WÖLNER, K. P. *Stockholm.*—Brandy, chemically purified.

---

186 HELLGREN, W. & CO. *Stockholm.*—Cigars; snuff; tobacco.

187 HÖGSTEDT & CO. *Stockholm.*—Punch of various kinds.

188 LILJEHOLMEN WINE MANUFACTORY, *Stockholm.*—Wines made of various kinds of berries.

189 ODELBERG, A. *Enskede, Stockholm.*—Rectified corn brandy.

190 SCHRÖDER, C. *Stockholm.*—Punch.

CLASS 4.

Sub-Class A.

191 HANSON & SEWAN, *Malmö.*—Various kinds of soap.

192 HIERTA, L. J. & MICHAELSSON, J. *Stockholm.*—Stearine candles and soap.

193 HYLIN & CO. *Stockholm.* — Soap. Flower vase made of soap.

194 JACOBSSON, L. A. *Stockholm.*—Soap, pomatum.

195 MONTÉN, L. *Stockholm.* — Stearine candles.

196 PAULI, F. *Jönköping.*—Various kinds of soap, &c.

Sub-Class B.

201 CELSING, L. G. v. *Helleforss, Sudermannia.*—Fleeces of merino sheep.

202 ROSSING, J. A. *Gothenburg.* — Silk, produced by worms fed on Scorzonera humilis and hispanica.

203 SAHLSTRÖM, C. G. *Jönköping.* — Albumen extracted from fish roe.

204 SWEDISH SILKWORM BREEDING SOCIETY, *Stockholm.*—Swedish silk and cocoons.

205 TROLLE-WACHTMEISTER, COUNT H. G. *Årup, Scania.*—Merino, wether, and ewe wool.

206 UNANDER, F. *Yttertafle, Umeå.*—Reindeer horns.

---

207 RAMSTEDT, MISS F. *Stockholm.*—Wax flower.

208 ÅNGSTRÖM, J. *Lycksele, Umeå.*—Reins and shoes made from gut and skin of reindeer; rope of pine-roots.

Sub-Class C.

211 BALDERSNÄS IRON WORKS, *Dalsland.* —Red deals.

212 CARLSFORSS SAW MILLS, *Dalecarlia.* —Red deals.

213 DICANDER, A. J. *Uppbo, Westgothland.* —Speckled birch.

214 HAMILTON, COUNT H. *Mariedal, Westgothland.*—Red and white deals. Sections of a pine-tree.

215 HARMENS, AF, H. O. HEIRS OF, *Berga, Calmar.*—Pine, fir, oak, beech, and juniper.

216 HASSELGREN, O. S. *Rådaneforss, Dalsland.*—Pine and fir seeds.

217 KOSTA GLASS WORKS, *Smaland.*—Pine and fir; pine seeds.

218 KRAUSE, H. E. *Gothenburg.*—Cork, cut by hand.

219 LEMAN, H. M. *Gothenburg.* — Rush mats.

220 LINDQVIST, A. P. *Bergsgården, Dalsland.*—Pine and fir seeds.

221 MAGNUSSON, N. P. *Grankulla, Isle of Öland.*—Ropes made of pine-bast.

222 MÖLNBACKA IRON WORKS, *Wermland.*—Pine.

223 NYBERG, J. *Brefven, Nericia.*—Pine.

224 NYGREN, C. *Klamstorp, Calmar.* — Milk-pan, and butter-tub of wood, lined with glass.

225 RITTER, F. *Glimminge, Scania.*—Pine and fir; pine seeds.

226 SPARRE, COUNT E. T. *Torpa, Westgothland.*—Pine.

227 SPRENGTPORTEN, BARON J. W. *Sparreholm, Sudermannia.*—Fir, pine, birch, and aspen.

228 STORA KOPPARBERG MINING CO. *Dalecarlia.*—Red and white deal. Aspen deal.

229 TROLLE-WACHTMEISTER, COUNT H. G. *Årup, Scania.*—Oak and beech.

230 UDDEHOLM IRON WORKS, *Wermland.* —Pine and fir.

231 WESTERWIK LOCAL COMMITTEE, *Westerwik.*—Oak, ash, birch, pine, and fir.

232 WREDE, BARON E. *Fjellskäfte, Sudermannia.*—Pine and fir.

---

233 GYLLENKROK, BARON T. *Oby. Wexiö.* —Basket-work.

CLASS 5.

241 LUNDHQVIST, G. A. *Nyköping.* — Wrought-iron railway wheels, axle and tires of puddled steel. Axles and tires.

242 MORGÅRDSHAMMAR IRON WORKS, *Dalecarlia.*—Railspikes.

243 BREFVEN IRON WORKS, *Nericia.*—Rail-spikes.

244 ZETHELIUS, W. *Surahammar, Westmanland.*—Railway wheel tires.

---

245 SKYLLBERG IRON WORKS, *Nericia.*—Rail-spikes.

CLASS 6.

251 NORMAN, J. E. *Stockholm.*—Carriage.

252 OLSEN, N. *Gothenburg.*—Carriage.

253 SJÖSTEEN, B. F. & TILLBERG, L. *Stockholm.*—Sledge with fur apron.

254 WEGELIN, J. F. & ÖSTBERG, J. W. *Stockholm.*—Carriages.

CLASS 7.

Sub-Class B.

258 BOLINDER, J. & C. G. *Stockholm.*—Iron planing machine. Copying press.

259 FROMELL, C. J. *Gothenburg.*—Chocolate mill.

260 GUSTAFSSON, C. *Hellstorp, Jönköping.* —Machine for making rolled nails.

261 KRAUSE, H. E. *Gothenburg.* — Cork-counting machine.

262 LIBERG, A. G. *Stockholm.*—Watchmakers' tools.

263 SLÖÖR, J. *Stockholm.* — Potatoe-peeling machine.

CLASS 8.

267 BOLINDER, J. & C. G. *Stockholm.*—Steam engine.

268 HALLSTRÖM, O. G. *Köping.*—Rotatory steam engine.

269 HOLMGREN, J. *Sala.* — Model of a blowing machine.

270 KNORRING, BARON V. J. V. *Stockholm.* —Rotatory steam engine.

271 LINDAHL & RUNER, *Gefle.* — High-pressure steam engine.

272 MUNKTELL, T. *Eskilstuna.* — Groat-grinding mill; forcing pump.

273 SCHEUTZ, E. *Stockholm.* — Rotatory steam engine.

---

274 FRESTADIUS, A. W. *Bergsund, Stockholm.*—Marine steam engine.

275 WREDE, BARON F. *Stockholm.*—Steam hammers.

CLASS 9.

276 CENTRAL COMMITTEE FOR SWEDEN, *Stockholm.*—Plans and perspectives of farms and farm buildings, agricultural maps, photographs.

277 CELSING, v. L. G. *Helleforss, Sudermannia.*—Plough, clover thrashing machine.

278 DAHLGREN, V. *Stockholm.*—Maps, indicating the division of the land belonging to a village.

279 GEFLE PROVINCIAL LOCAL COMMITTEE, *Gefle.*—Model of a one-horse cart.

280 HARMENS, AF, HEIRS OF, H. O. *Berga, Calmar.*—Iron halter for bulls.

281 HOLMGREN, J. *Sala.*—Churn.

282 KYLBERG, H. *Ryssbylund, Calmar.*—Dressing machine.

283 LYCKEBY IRON FOUNDRY, *Bleking.*—Rotatory harrow, plough.

284 ROSSING, J. A. *Gothenburg.*—Forestry tools and implements.

285 ROYAL AGRICULTURAL ACADEMY, *Stockholm.*—Model of a kiln.

286 SEDERHOLM, J. *Näfvequarn, Sudermannia.*—Plough.

287 ULTUNA AGRICULTURAL SCHOOL, *Upland.*—Iron plough.

288 ÖFVERUM IRON WORKS, *Calmar.* — Dressing machine, chaff cutter, ploughs, carrot drills, clover sowing machine.

---

289 BERGELIN, J. TH. *Hammarby, Stockholm.*—Plough.

290 EKMAN, P. J. *Stockholm.*—Models of a kiln and of a new method of stacking hay.

291 GUSSANDER, P. U. *Gammelstilla, Gefle.* —Set of dairy utensils.

292 HAGLUND, P. *Näfvequarn, Nyköping.* —Grain and corn-crushing mill.

292A LAGERBERG, COUNT F. *Stockholm.*—Milk can, with warming apparatus.

292B LINDQVIST, C. M. *Stockholm.*—Churn; sowing and planting machines.

292C ODELBERG, A. *Enskede, Stockholm.*—Model of a kiln.

292D PALMAER, C. W. *Forsvik, Westgothland.*—Drain-tile machine—Schlosser's principle.

292E STJERNSVÄRD, G. M. *Stockholm.*—Churn, with self-skimming milk-pail, strainer, and butter gatherer.

CLASS 10.

293 BARK & WARBURG, *Gothenburg.* — Doors, window frame with wainscotting, inlaid floor, joinery work.

294 CELSING, v. L. G. *Helleforss, Sudermannia.*—Gas welding furnaces, for gas from wood and coals.

295 GEFLE PROVINCIAL LOCAL COMMITTEE, *Gefle.* — Model of a roof formed of shingles.

296 KLEMMING, C. *Stockholm.* — Water-closet.

297 KYLBERG, J. *Sätenäs, Westgothland.* —Draining tiles.

298 LAGERHEIM, BARON E. *Nyqvarn, Stockholm.*—Draining tiles.

299 LARSSON & LUNDGREN, *Örebro.* — Glazed porcelain stove.

300 MALMBERG, G. A. & Co. *Stockholm.* —Watercloset.

301 MARINO, J. *Stockholm.*—Watercloset. Model of a binn for sweepings, with conveniences.

302 NABSTEDT, H. *Gothenburg.*—Water-closets.

303 RINGNÉR, A. *Gothenburg.* — Glazed porcelain stoves.

304 ROSSING, J. A. *Gothenburg.*—Ventilating apparatus.

305 WALLIN, O. F. W. *Stockholm.*—Model of an oven.

306 ÅKERLIND, O. H. *Stockholm.*—Glazed porcelain stoves.

---

307 EKMAN, P. J. *Stockholm.*—Specimens of joinery work.

308 ERICSON, BARON N. *Stockholm.*—Cement.

309 ROSÉN, C. G. *Waxholm.* — Water-warming apparatus.

310 LINDSKOG, S. & FRYCKHOLM, P. B. *Gothenburg.*—Water-closet.

CLASS 11.

Sub-Class C.

311 ANKARCRONA, J. *Husqvarna, Smaland.* —Rifled gun, adopted 1860 for the Swedish army.

312 CELSING, L. G. v. *Helleforss, Sudermannia.* — Five-pounder rifled cannon, invented by C. Engström, R.S.N. Grenade.

313 EKMAN, C. *Finspong, Ostgothland.*—Breech-loading cannon, invented by C. Engström, R.S.N. Six-pounder field cannon.

314 HULT, A. *Grafås, Wermland.*—Gun.

315 MARÉ, A. DE, *Aukarsrum, Calmar.*—Bombs.

316 SASSE, E. *Carlskrona.* — Models for cannon foundries of breech-loading and other cannon, projectiles for rifled bore.

317 STOCKENSTRÖM, A. v. *Åker, Sudermannia.*—Rifled breech-loading field cannon, with suitable conical ball.

318 SVENGREN, J. *Eskilstuna.* — Sword-blades.

319 WAY, J. *Upsal.*—Gun invented by the exhibitor.

---

320 SCHARIN, O. M. T. *Stockholm.*—Model of a breech-loading rifled cannon.

CLASS 12.

Sub-Class A.

323 GRÖNDAHL, C. E. M. *Stockholm.*—Drawing and model of a ship.

Sub-Class C.

326 FURUDAL IRON WORKS, *Dalecarlia.*—Chains and chain-cables of iron.
327 SÖDERFORSS IRON WORKS, *Upland.*—Anchor.

CLASS 13.

331 BYSTRÖM, O. F. *Stockholm.*—Hydro-pyrometer, for measuring high temperatures in furnaces.
332 HULTGREN, F. A. *Stockholm.*—Hydrometer.
333 KINDBOM, J. P. *Calmar.*—Photometer.
334 KLINTIN, J. F. *Stockholm.* — Logs, compass, vacuometer, manometer, spring balance.
335 LEFFMAN, C. R. *Stockholm.*—Laboratory table, containing apparatus, retorts, &c. for chemical analysis and research.
336 MEURTHIN, J. *Gefle.* — Miner's compass.
337 WIBERG, M. *Malmö.* — Calculating machines.

---

338 EGGERTZ, V. *Fahlun.*—Apparatus for the analysis of iron ores.
339 LYTH, G. W. *Stockholm.*—Goniometer.
340 REYMYRE GLASS WORKS, *Ostgothland.*—Specimens of glass for chemical purposes.
340A JONSSON, A. *Stockholm.*—Spherical compasses.
340B MICHAËLSSON, J. *Stockholm.*—Galactometer.

CLASS 14

341 LINDSTED, P. M. *Gothenburg.*—Photography.
342 UNNA & HÖFFERT, *Gothenburg.* — Photography.

---

343 CARLEMAN, C. G. V. *Stockholm.*—Photographs.

CLASS 15.

346 CEDERGREN, J. T. *Stockholm.*—Transparent clock, enamelled works.
347 LINDEROTH, MRS. B. *Stockholm.*—Watch wheels and movements.
348 LINDEROTH, G. W. *Stockholm.*—Marine chronometers; astronomical pendulum watch.
349 MEURTHIN, J. *Gefle.*—Watch. Parts of a clock.
350 MOLLBERG, L. R. *Stockholm.* —Balances and spiral springs for marine and pocket chronometers.

CLASS 16.

354 HOFSTEDT, C. P. *Stockholm.*—Pianoforte.
355 MALMSJÖ, J. G. *Gothenburg.*—Pianoforte.
356 SÄTHERBERG, A. F. *Norrköping.*—Pianino.

CLASS 17.

362 GEORGII, A. *London.* — Hygienic and medical gymnastic apparatus.
363 LIBERG, A. G. *Stockholm.* —Instruments for dental surgery.

CLASS 18.

367 ALMEDAHL Co. *Gothenburg.*—Cotton yarn, bleached.
368 ANDERSSON, A. *Fritsla Hagen, Westgothland.*—Cotton goods woven in handlooms.
369 ANDERSSON, MRS. C. *Gefle.*—Woven counterpane.
370 ANDERSSON, S. *Kinna Sanden, Westgothland.*—Cotton goods woven in handlooms.
371 BERG COTTON MANUFACTURING Co. *Norrköping.*—Domestic fabrics, bleached and unbleached. Mole-skins.
372 BERG, J. T. *Nääs, Westgothland.* — Cotton goods.
373 EVERS, A. H. *Ahleforss, Westgothland.* —Cotton yarn.
374 HOLMEN Co. *Norrköping.* — Cotton yarns. Domestic fabrics.
375 HÅKANSSON, H. *Fritsla Högen, Westgothland.*—Cotton goods woven in handlooms.
376 JOHANSSON, M. *Källäng, Westgothland.* —Jacquard-woven cotton coverlet.
377 KAMPENHOF Co. *Uddewalla.*—Cotton sailcloth, domestic fabrics, cotton yarns.
378 LARSSON, S. *Stämmemad, Westgothland.*—Cotton goods woven in handlooms.
379 MALMÖ COTTON-WEAVING Co. *Malmö.*—Cotton goods, bleached, unbleached, and dyed.
380 MALMÖ MANUFACTURING Co. *Malmö.* —Cotton yarn.
381 NORRKÖPING COTTON-WEAVING Co. *Norrköping.*—Cotton stuffs.
382 ROSENLUND COTTON-SPINNING Co. *Gothenburg.*—Cotton-yarn and stuffs.

---

383 GIBSON, W. & SONS, *Jonsered, Gothenburg.*—Cotton yarn.

CLASS 19.

386 ALMEDAHL Co. *Gothenburg.*—Samples of flax, linen thread, flax and tow yarns, damask, diaper, and other goods.
387 BREDMAN, S. M. *Sandhem, Smaland.* —Linen yarn, spun by hand.
389 BRODIN, B. C. *Gefle.*—Linen, damask, and diaper table-cloths, napkins, and towels.
388 BRODIN, E. C. *Gefle.*—Linen, damask, and diaper table-cloths and napkins.
390 DAHLQVIST, J. *Själevad, Hernösand.* —Table-cloths and napkins woven in handlooms.
391 FRISK, E. *Hudikswall.*—Flax canvas woven in handlooms.
392 GIBSON, W. & SONS, *Jonsered, Gothenburg.*—Yarns, sail-cloth, cordage, &c.

393 HERNÖSAND LOCAL COMMITTEE. — *Hernösand.*—Flax, linen, thread and yarn.

394 STENBERG, G. WIDOW OF, *Jönköping.* —Table-cloths and napkins.

395 SUNDSTRÖM, F. *Norrköping.*—Cordage and rope-yarns.

396 TROLLE, W. *Gidån, Hernösand.* — Linen goods.

CLASS 20.

401 ALMGREN, K. A. *Stockholm.*—Plain and figured silk, dyed in Sweden.

402 CASPARSON & SCHMIDT, *Stockholm.*— Silk, plain and figured.

403 MEYERSON, L. *Stockholm.*—Silk.

CLASS 21.

407 APPELBERG, C. L. *Norrtelje.*—Waterproof woollen cloth.

408 BEHRLING, P. *Stockholm.* — Woollen shawls and blankets.

409 BOETHIUS & KLING, *Norrköping.* — Woollen goods of various kinds.

410 CARLSVIK MANUFACTURING CO. *Stockholm.*—Woollen, mixed woollen and silk stuffs, yarns, &c., partly dyed.

411 CLAESDOTTER, C. *Sundby, Öland.*—A woollen counterpane.

412 DRAG WOOLLEN MANUFACTURING CO. *Norrköping.*—Woollen cloth of different kinds.

413 ELLIOT, S. & CO. *Stockholm.*—Shawls, woollen and mixed.

414 FÜRSTENBERG, L. & CO. *Gothenburg.* —Shawls and neck-kerchiefs, woollen and mixed.

415 JOHANSSON, M. *Källäng, Westgothland.* —Table-covers of mixed cotton-wool and silk. Shawls of wool, and mixed wool and cotton.

416 LUNDGREN, C. F. *Norrköping.* — Woollen goods.

417 LUNDSTRÖM, C. F. *Jönköping.* — Woollen and mixed stuffs.

418 MEYERSON, L. *Stockholm.*—Stuffs of mixed silk and cotton.

419 NORRKÖPING CORDEROY-WEAVING CO. *Norrköping.*—Woollen goods of various kinds.

420 QVIST, A. *Norrköping.* — Woollen goods.

421 SCHUBERT, J. F. *Norrköping.* — Woollen goods.

422 SMEDJEHOLMEN WOOLLEN MANUFACTURING CO. *Norrköping.*—Woollen cloth of different kinds.

423 VOGELER, J. A. *Norrköping.*—Woven articles of wool and cotton.

424 WAHLQVIST CLOTH MANUFACTURING CO. *Wexiö.* — Woollen cloth used by the Swedish army and navy.

425 WAHREN, H. *Norrköping.*—Woollen goods of various kinds.

426 WELLENIUS, J. F. *Norrköping.* — Woollen cloth.

---

427 ANDERSON, T. W. *Ernsta, Upsal.*— Woollen and linen yarn, homespun.

CLASS 23.

431 LUNDSTRÖM, C. F. *Jönköping.*—Stuffs, dyed by a method invented by the exhibitor.

432 SAHLSTRÖM, C. G. *Jönköping.*—Specimens of stuffs, printed with albumen extracted from fish roe.

CLASS 24.

436 BENSOW & CO. *Stockholm.*—Lace of various kinds.

437 HOLMGREN, W. *Gothenburg.*—Table-cover, netted with silk and embroidered.

---

438 BÄCK, C. MISS, *Gothenburg.*—Embroidered baby linen.

439 ENGLUND, MISS M. E. *Stockholm.* — Lace head-dress.

440 HARTVICK, MRS. A. *Wadstena.*—Bone laces.

CLASS 25.

Sub-Class A.

441 ANDERSSON, J. O. *Gothenburg.*—Fur.

442 BASK, C. J. *Stockholm.*—Hearth-rug of fur, representing the Arms of Sweden.

443 BERGSTRÖM, P. N. *Stockholm.*—Different sorts of skins and furs. Manufactured goods of furs.

444 FORSSELL, D. & CO. *Stockholm.* — Ladies' jackets of martin, otter, and minever; furs, muffs, and collars.

445 NILSSON, P. U. *Jemtland.*—Skins of the bear, elk, and reindeer.

Sub-Class C.

451 BERG, F. J. *Gothenburg.*—Wigs and loose plait.

452 LUNDGREN, F. F. O. *Stockholm.* — Various brushes.

453 LÖFSTRÖM, P. & CO. *Gothenburg.*— Wigs.

454 NIECKELS, T. *Gothenburg.*—Wigs.

455 ZINN, C. M. *Stockholm.*—Wigs of different kinds.

CLASS 26.

Sub-Class A.

461 BERGHOLTZ, J. F. *Wernamo, Smaland.* —Varnished leather.

462 COLLIANDER, J. A. *Gothenburg.* — Leather.

463 LEVISSON, M. SONS, *Gothenburg.*— Leather of different kinds.

Sub-Class B.

466 HOLMBERG, N. P. *Gothenburg.* — Saddlery.

467 KROOK, T. F. *Malmö.* — Harness, saddle, embossed leather straps.

468 SKOGLUND, C. *Stockholm.* — Lady's saddle, with bridle.

CLASS 27.

Sub-Class A.

471 ANGRELIUS, F. R. *Gothenburg.*—Hats.

472 ERICSSON, A. & CO. *Stockholm.*—Hats of various kinds.

Sub-Class C.

476 BERENDT, S. JUN. & Co. *Stockholm.*—Shirts.

477 BROCK, G. F. *Gothenburg.*—Gloves and skins.

478 LÖTHMAN, A. C. *Helsingborg.*—Gloves.

479 MÖLLER, P. D. *Malmö.*—Gloves.

480 SALOMAN, J. *Gothenburg.* — Ready-made linen.

481 SÖDERBERG, J. *Stockholm.*—Hosiery goods.

482 WIECHEL, G. *Norrköping.*—Hosiery goods.

483 LINDBERG, Z. *Gothenburg.*—Shirts.

484 MÖLLER, J. P. *Lund.*—Gloves.

Sub-Class D.

486 FJÄSTAD, P. C. *Stockholm.*—Boots and shoes.

487 HELLMAN, J. *Norrköping.* — Boots, overshoes; fur boots for ladies.

488 SUNDQVIST, L. M. *Gothenburg.*—Boots and shoes.

489 TRANÉ, F. *Stockholm.* — Boots and shoes.

490 TÖRNQVIST, C. W. *Åby, Scania.*—Shooting boots; ankle shoes.

491 LUNDSTEDT, O. *Stockholm.* — Boots and shoes.

CLASS 28.

Sub-Class A.

493 BOCK, C. A. *Klippan, Scania.* — Various kinds of paper.

494 BOMAN, A. *Stockholm.*—Playing cards.

495 BRANDT, J. C. *Stockholm.*—Playing cards.

496 DEUTGEN, E. *Nyqvarn, Stockholm.*—Paper.

497 HEURLIN, G. *Stockholm.* — Playing cards.

498 LITHOGRAPHIC SOCIETY IN NORRKÖPING, *Norrköping.* — Playing cards, stationery, lithography.

499 MOTALA PAPER MANUFACTURING CO. *Ostgothland.*—Paper.

500 MUNKTELL, J. H. HEIRS OF, *Gryksbo, Dalecarlia.* — Filtering paper for chemical purposes.

501 ROSENDAHL MANUFACTURING CO. *Gothenburg.*—Various kinds of paper.

502 STENSHOLM MANUFACTURING CO. *Smaland.*—Specimens of paper, coloured and not coloured.

Sub-Class B.

506 NYBLÆUS, C. G. *Stockholm.*—Samples of ink.

Sub-Class C.

509 BONNIER, A. *Stockholm.*—Specimens of lithography and typography.

510 MANDELGREN, N. M. *Paris.*—Scandinavian monuments, &c. executed in chromo-lithography.

511 MEYER & KÖSTER, *Gothenburg.* — Specimens of lithography.

512 BERLING, F. *Lund.* — Specimen of type.

Sub-Class D.

513 BECK, F. *Stockholm.*—Specimens of bookbinding.

CLASS 29.

Sub-Class A.

516 BONNIER, A. *Stockholm.* — Various maps of Sweden.

517 ERDMANN, A. *Stockholm.*—Geological specimens and maps.

518 LJUNGGREN, G. *Stockholm.*—Economical and statistical maps of different districts of Sweden.

519 HAHR, A. *Stockholm.*—Topographical and statistical maps.

Sub-Class B.

521 CHALMER'S POLYTECHNICAL SCHOOL, *Gothenburg.* — Electro-magnet and dynamometer, made by the pupils of the school.

522 DIRECTION OF THE ROYAL INDUSTRIAL SCHOOL, *Stockholm.*—Drawings, &c. done by the pupils of the school.

523 FAHLGREN, C. J. *Askersund.*—Typograph, by which blind persons can correspond with one another.

524 KLEIN, W. *Malmö.*—Copies for free-hand drawings; executed by the exhibitor's pupils.

525 LYCKEBY IRON-FOUNDRY, *Bleking.*—School furniture.

526 MANILLA INSTITUTE FOR BLIND, DEAF, AND DUMB, *Stockholm.* — Various apparatus for the blind; articles made by them.

527 SILJESTRÖM, P. *Stockholm.*—Various furniture and apparatus for schools. Gymnastic apparatus.

528 BAGGE, J. S. *Fahlun.* — Model of J. Cæsar's Bridge over the Rhine.

CLASS 30.

Sub-Class A.

531 DAHLIN, P. *Stockholm.* — Drawing-room table.

532 PETTERSSON, C. *Gothenburg.*—Writing-table of Jacaranda.

533 CLIFFORD, J. *Stockholm.* — Writing table.

534 GUNDBERG, J. W. *Stockholm.*—Fire-screen.

535 LANGEMEIJER, W. *Stockholm.*—Sofa and chairs

Sub-Class B.

536 GLASELL, C. A. *Gothenburg.*—Looking-glass and table; picture-frame.

537 LINDGREN, P. A. *Söderköping.* — Various articles, carved in wood.

538 MALMBORG, E. *Gothenburg.*—Ornamental figure.

539 MALMSTEN, J. A. *Jönköping.* — Painted blinds.

540 NERPIN, H. *Stockholm.* — Candelabrum, sculptured in wood.

541 NORLING, L. W. *Jönköping.*—Paper-hangings of various kinds.

542 SÖDERSTRÖM, A. J. *Ulricehamn.*— Paper-hangings.

543 ULANDER, M. *Gothenburg.* — Ladies' work-table, the top inlaid with cork, birch-bark, and sponge.

---

544 ISBERG, S., Peasant Woman, *Motala.* —Goblet, carved in wood.

545 TAUTZ, H. *Stockholm.*—Candelabrum of papier maché.

545A DALIN, P. *Stockholm.*—Picture-frame of oak.

CLASS 31.

Sub-Class A.

546 BACKMAN, J. F. *Stockholm.*—Iron safe.

547 BERG, C. J. & TINGBERG, C. G. *Stockholm.*—Economic grate, for the use of coke in porcelain stoves.

548 BOLINDER, J. & C. G. *Stockholm.*— Kitchen ranges; flat-iron warming apparatus.

549 CHRISTIANSSON, G. & SON, *Askersund.* —Hand hammered nails and tacks.

550 FORSGREN, A. *Lindesnæs, Dalecarlia.* —Universal turn-screw.

551 GUNNEBO IRON CO. *Gunnebo, Westerwik.* — Wood screws, and nails of iron and brass; iron wire; iron springs for sofas.

552 GUSTAFSON, C. *Hellstorp, Jönköping.* Specimens of rolled nails.

553 HEDLUND, J. *Eskilstuna.*—Locks of different kinds.

554 HELJESTRAND, C. *Eskilstuna.*—Cork-screws.

555 JOHANSSON, J. M. *Målön, Smaland.*— Iron vices.

556 JÄDER IRON MANUFACTURING CO. *Westmanland.*—Iron manufactures.

557 JÖNKÖPING IRON FOUNDRY, *Jönköping.*—Kitchen ranges; cast-iron pans and kettles; fine castings.

558 KOCKUM, F. H. *Malmö.*—Iron manufactures.

559 LYCKEBY IRON FOUNDRY, *Bleking.*— Garden sofa of iron; flat-iron warming apparatus.

560 MOBERG, A. *Stockholm.* — Kitchen range, with moveable top plate, and copper tank for water.

561 ROBSON, A. *Aspa Iron Works, Nericia.* —Nails.

562 STENMAN, F. *Eskilstuna.*—Locks.

---

563 BERGSTRÖM, J. W. *Stockholm.* — Screwstock on a new principle, invented by O. J. Aqvilon.

564 PETTERSSON, M. *Stockholm.*—Buttons.

Sub-Class B.

566 BILLQVIST, G. W. *Gothenburg.*— Piece of brass-work, embossed by hand.

567 HEDENSTRÖM, SONS, *Westerås.*—Candlesticks; coffee pots.

568 MALMBERG, G. A. & Co. *Stockholm.* —Stew-pans and milk-pans of copper; copper steam-pipe.

569 WESTERBERG, E. *Gusum, Ostgothland.* — Brass-wire gauze, brass-wire, pins, insect-pins.

Sub-Class C.

573 KOCKUM, F. H. *Ronneby, Bleking.*— Pans and kettles of tin-plate; iron and copper nails.

574 LILJA, C. D. *Norrköping.*—Japanned goods.

CLASS 32.

Sub-Class B.

576 ANDERSSON, H. *Malmö.*—Knives and scissors.

577 DAHLBERG, H. *Eskilstuna.* — Steel wares, etched and gilt.

578 GUNNEBO IRON MANUFACTURING CO. *Gunnebo, Westerwik.*—Chaff-knives; scythes.

579 HEDLUND, L. *Eskilstuna.*—Scissors, &c.

580 HELJESTRAND, C. *Eskilstuna.*—Razors.

581 MARÉ, DE A. *Ankarsrum, Westerwik.* —Blades for scythes.

582 STÅHLBERG, L. F. *Eskilstuna.*—Knives and forks.

583 SVENGREN, J. *Eskilstuna.*—Cutlery.

584 OBERG, C. O. *Eskilstuna.*—Files.

CLASS 33.

589 CARLSTRÖM, R. M. *Stockholm.*—Electro-plated vase, candelabra, spoons, forks, &c.

590 FERON, L. C. *Stockholm.* — Various chased silver articles.

591 FYRWALD, C. J. M. & Co. *Stockholm.* Gold-wire articles.

592 HOLM, C. *Norrköping.*—Silver coffee-pot.

593 LARSON, L. & Co. *Gothenburg.* — Punch-bowl, modelled by Qvarnström, property of the Par Bricol Society.

594 MÖLLER, A. W. *Stockholm.*—Electro-plate.

595 PALMGREN, T. *Norrköping.* — Gold bracelet.

596 SJÖGREEN, C. W. *Bruzaholm, Smaland.* —Pearls found in Smaland rivers.

CLASS 34.

Sub-Class A.

599 BROMÖ GLASS WORKS, *Westgothland.* —Specimens of their manufactures.

Sub-Class B.

601 BRUSEWITZ, F. *Limmared, Westgothland.*—Specimens of manufactured glass, and of the materials employed.

602 KOSTA GLASS WORKS, *Smaland.* — Chalice, patin, and other articles of coloured glass and crystal.

CLASS 35.

600 GEIJER, D. R. HEIRS OF, *Rörstrand, Stockholm.*—Dinner services of porcelain, &c.

607 GUSTAFSBERG PORCELAIN MANUFACTURING Co. *Stockholm.*—Porcelain tables, dinner service, &c.

608 HÖGANÄS COLLIERY, *Scania.* — Earthenware goods.

SWITZERLAND.

N.W. Court, No. 3, and N.W. Gallery, No. 2.

CLASS 1.

1 BRUNNER, PROF. C. *Berne.*—Metallic manganese, obtained by a new process.

2 GYPSUM SOCIETY, *Klosters, Grisons.*—Stucco, and casts made with it.

3 LEONI, C. *Coire.* — Marble from the Splugen; a marble monument from the Grisons.

4 LINDENMANN & Co. *Bünzen, Argovie.*—Pressed peat in pipe form.

5 THEOBALD, PROF. G. *Coire.*—Minerals found in Grisons.

6 TUGGINER, WEY, & Co. *Soleure.* — Refractory fire-clay.

CLASS 2.

10 CHAUTEN, *Geneva.*—Vegetable elixir for sea-sickness, &c.

11 COURT, AMI, *Geneva.*—Purple colour for painting on enamel or porcelain.

12 GIMPER, G. *Züric.*—Ethers, oils, and other pharmaceutical products.

13 LACROIX, J. M. *Geneva.*—A specific for the toothache.

14 LAUTERBURG, F. *Berne.*—Waterproof mineral varnish.

15 MATTHEY, A. O. *Locle.* — Diamond powder for polishing steel.

16 MULLER, J. J. & Co. *Basle.*—Extracts of dyewood; aniline and its dyeing derivatives.

17 ST. MORITZ, SOCIETY OF, *Coire.* — Mineral water of St. Moritz.

18 TARASP, SOCIETY OF, *Coire.*—Mineral water of Tarasp.

---

19 DIEDEY, M. *Lausanne.* — "Eau des Alpes," a hair preserve.

CLASS 3.

20 BAUER, C. A. *St. Gall.*—Various objects in gum tragacanth.

21 BECK-LEU, F. *Bekenhof, near Sursee, Lucerne.*—Honey; Alpenschotten sugar, &c.

22 BÉLENOT, F. F. *Mouruz, near Neufchâtel.*—Red Neufchâtel wine of 1857.

23 BÉRAUD, MARC, & Co. *Vevey.*—White wine of Yvorne of 1800 to 1846; wine from a Neufchâtel vine.

24 BERTHOLET & Co. *Vevey.*—Swiss cigars.

25 BILLE, BOREL, & Co. *Thielle, Neufchâtel.*—Havana cigars.

26 BOUVIER, BROS., *Neufchâtel.*—Neufchâtel champagne.

27 BOYMOND, *Geneva.*—Pectoral lozenges. Geneva drops. Cordials.

28 BUNDNERISCHE WEINBAUGESELLSCHAFT, *Coire.*—Wine from the Grisons.

28A CHERVAZ, G. *Vetroz, near Sion.*—Wine from Valais, white and red.

29 CHESSEX, R. *Montreux, Vaud.*—White wine from a Rhénish vine grown at Montreux.

30 CLÉMENT & Co. *Vevey.*—Swiss cigars.

31 CORBOZ-DUBOUX, BROS. *Epesses, Vaud.*—White wine, Lavaux of 1859; white wine of 1857 and 1858.

32 DÉGLON, *Lausanne.*—Helvetian cigars.

33 DELAJOUX, J. *Vevey.*—White wine of 1854, grown at Vevey.

34 BOUVIN, C. JUN. *Sion.*—Wines of the Valais.

35 DESPLANDS & RICHARDET, *Vevey.*—White wine, Lacôte of 1859, and Vevey of 1859.

36 DUBOIS, C. *Vevey.*—White wine from Vevey, 1854 and 1859.

37 DUCHOSAL, DIRECTOR AT *Vernets, Geneva.*—Samples of cereals.

38 DUPLAN & MONNERAT, *Vevey.*—White wine, Yvorne, Clos du Rocher of 1848 and 1858.

39 EICHENBERGER, J. J. *Beinwyl, Argovia.*—Cigars.

40 ELLES, H. *Vevey.*—White Markobrunner, grown at Vevey in 1856 and 1858.

41 FANKHAUSER, J. C. *Lausanne.*—Chocolate.

42 FASSBIND, G. *Arth, Schwytz.*—Cherry-waters.

43 FELDMUHLE, *Rorschach, St. Gall.*—Alimentary pastes.

44 FERT & MOSSU, *Geneva.*—A stomachic made of wine and aromatics.

45 FINAZ, *Geneva.*—Pectoral paste, liquorice juice, cashoo from Bologna, wine of Yvorne, &c.

46 FRÖHLICH, G. *Steffisbourg, Berne.*—Cherry-water.

47 FROSSARD-MULLER & SON, *Payerne, Vaud.*—Gruyère and Emmenthal cheese.

48 GERBER, C. SON, *Steffisbourg, Berne.*—Cherry-water.

49 GILLIARD, ELISE, & Co. *Fleurier, Neufchâtel.*—Green extract of absynthe.

50 KLAUS, C. *Locle.*—Pectoral paste, confectionary, pharmaceutical pastiles, &c.

51 KOEBEL, A. *Sion, Valais.*—Wine from Sion, and cherry-water.

52 KOHLER, AMÉDÉ, & SON, *Lausanne.*—Chocolate.

53 LEGLER, G. *Couvet.*—Extract of absynthe, &c.

54 MASSON, F. *Grandson.*—Cigars.

55 ORMOND & Co. *Vevey.*—Swiss cigars.

56 PASCHOUD & FREYMAN, *Vevey.*—Sparkling wine of Vaud, wine of Yvorne of 1854, Château du Châtelard of 1854, Lavaux of 1854, Lacôte of 1834.

57 LEUBA, A. *Columbier.* — Extract of absinthe.

58 REYMOND & WARNÉRY, *Payerne.*—Cigars and tobacco.

59 DE RIVAZ, A. *Ardon, Valais.*—Wines of Amique.

60 ROSSELET-DUBIED, *Couvet.*—Extract of absynthe, and cherry-water, &c.

61 RUFENACHT, D. *Vevey.* — White wine of Yvorne of 1854.

62 SCHERER, BROS. *Meggen, Lucerne.*—Cherry-water of 1857.

63 SPINTZ, DR. N. *Berzona, Tessin.*—Iodined chocolate.

64 STADELMANN, J. *Escholzmatt, Lucerne.*—Crystallized sugar of milk.

65 SUCHARD, PH. *Neufchâtel.*—Chocolate.

66 VIDOUDEZ & Co. *Lausanne.*—Cigars.

67 VAUTIER, BROS. *Grandson.*—Cigars.

68 VEILLARD, COL. A. *Aigle, Vaud.*—White wine of Yvorne of 1859.

69 WASSALI, F. *Coire.*—Alpine honey, from near the sources of the Rhine..

70 WEBER, F. *Geneva.*—Syrup of lichens.

71 WEBER, S. & SON, *Menzikon.*—Cigars.

72 WICKY-VOGEL, *Schüpfheim.*—Sugar of milk.

73 WIEDMER, M. *Basle.*—Cigars.

74 ZIEGLER-PELLIS, *Winterthur.*—Wine of the Canton of Zürich 1811, 1834, and 1859.

75 MANGE, P. *St. Cergues, Vaud.*—Gentian-water.

76 DELOES, COL. A. *Aigle.*—White wine of Aigle, 1859.

CLASS 4.

80 KLARIR, J. *Appeuzele.* — A box in carved wood.

81 GRUNINGEN, VON, J. G. *Saanen, Berne.*—Wood for sounding boards, ingrained maple, ash, and fir-wood.

82 JAEGER & Co. *Brienz.*—Looking-glass frame, clock-case, &c. in carved wood.

83 MENGOLD, G W. *Coire.* — Fruits from the Grisons in wax.

84 MICHEL, KASPAR, & Co. *Brienz.* — A bouquet-holder, pen-holders, &c. in carved wood.

85 MONNIER & PIGUET, R. *Nyon.*—Toilet soap, pomatum, perfumes.

86 SIEBER, J. H. *Züric.*—Cups, chessmen, brooches, paper knives, walking sticks, &c. in ivory.

87 TAVEL, GROSS, & Co. *Gunten, Berne.*—Wood for sounding boards.

88 WALD, A. H. J. (Bazar Suisse), *Thoune.*—A cabinet and other articles in carved wood.

89 WIRTH, E. BROS. *Brienz.* — Carved wood fancy articles.

CLASS 7.

95 BEUGGER, J. *Winterthur.*—Can-roving frame for cotton.

96 EGLI, J. *Lucerne.* — Boot-trees and lasts.

97 FREULER, M. *Glarus.*—Instruments for the engravers of apparatus for printing cotton cloths.

98 HAAS TYPE-FOUNDRY, *Basle.*—Type-founding machine, cutting machine, German types, and book of proofs.

99 LEDOUX, A. *Geneva.* — Lithographic press; apparatus for endless impression.

100 WAHL & SOCIN, *Basle.*—Silk-ribbon looms.

101 WEINGART, J. *Amerzwyl, Berne.*—Barrels of oak-wood.

102 LANDOLT, EMIL, & MILSTER, J. *St. Georgen, St. Gall.*—Hydraulic press, for alimentary pastes.

103 MASCHINEN-BAUANSTALT, *Frauenfeld.*—A stitching and folding machine, and a newspaper-folding machine.

CLASS 8.

104 ESCHER, WYSS, & Co., *Züric.*—Two compound cylinder marine engines, for shallow river navigation.

105 BELL, T. & F. *Lucerne.* — Paper-machines.

106 COLLADON, PROF. D. *Geneva.*—Floating hydraulic wheel, for the moving principle in a manufactory, or for raising water.

107 DESCHAMPS, ZACHARY, *Geneva.* — An assaying balance.

108 GRABHORN, B. *Geneva.*—An assaying balance.

111 SCHINZ, C. *Züric.*—Model of a glass furnace.

112 SULZER, BROS. *Winterthur.*—Washing-machine for bleachers, dyers, and stuff-printers.

113 THIEMEYER, R. *St. Gall.*—Balances of different qualities.

CLASS 9.

118 BEAUMONT, H. *Bouthillier-de, Geneva.*—Ploughs on a new principle.

119 MARTIN-DUNOYER, *Trélex, Vaud.*—Light plough, for every description of soil.

120 MENZEL, PROF. & GRABERG, *Züric.*—New contrivances to guide the bees in constructing their combs.

CLASS 10.

125 BARGEZI, U. *Soleure.*—Marble slab and table.

126 COLLADON, D. *Geneva.* — Machine for boring stones or the ground.

127 DAPPLES, E. C. E. *Lausanne.*—Sheet iron screw for wooden and cast-iron screw-piles.

128 LEHMANN, J. A. *Sargans, St. Gall.*—A kitchen range, with accessories.

129 PFISTER & SCHIRMER, *St. Gall.* — Cooking stoves, one with, the other without, flues.

130 STAIB, L. F. & Co. *Geneva.*—Caloriferes of cast-iron.

CLASS 11.

133 DEMARTINES, CH. *Lausanne.*—Swiss military head dress.

134 ERLACH, D', R. & Co. *Thoune.*—Rifle and carbine, with barrels of cast steel.

135 KOHLER, L. *Boudry, Neufchâtel.* — The carbine used by the Swiss soldiers.

136 PETER, J. *Geneva.*—Carbines.

137 RIGGENBACH, N. *Olten.*—Carriage of wrought-iron for a 4-pound rifle cannon.

138 SAUERBREY, V. *Basle.*—Double-barrelled guns, one after Le Faucheur; pistols; military carbines.

139 SWISS-BELGIAN INDUSTRIAL SOCIETY, *Schaffhausen.*—The carbine and rifle used by the Swiss soldiers.

141 VERET, COL. J. *Nyon.*—New rifle for infantry.

140 WOLF-BERNHEIM & CO. *Geneva.*—Uniforms of Swiss officers and privates.

141 GOLDSCHMIED, J. J. *Züric.*—Diastimeter for military use.

CLASS 12.

146 SÉNÉCHAUD, L. *Montreux.*—Canoe of sheet iron.

CLASS 13.

150 AMSLER-LAFFON, *Schaffhausen.*—Reversing levels, planimeters with radial movement.

151 ATELIER DES TÉLÉGRAPHES SUISSES, *Berne.*—Several systems of telegraphic apparatus; electric writing chronograph and thermometer.

152 DAGUET, T. *Fribourg.*—Large disk of flint glass, and of crown glass, for optical purposes.

154 GUTKNECHT, J. J. *Landquart.*—Apparatus to measure spirits and water under any pressure. Dry gas-meter.

155 GYSI, F. *Aarau.*—Collection of drawing instruments in German silver; mathematical cases.

156 HIPP, M. *Neufchâtel.*—Telegraphic apparatus, electric clocks, regulators.

157 HOMMEL-ESSER, *Aarau.*—Case of mathematical instruments; pair of compasses, with 29 divisions; pocket case.

158 KERN, J. *Aarau.*—German silver drawing compasses; mathematical cases; engineers' and surveyors' instruments.

159 MONNIER, J. F. *Bevaix, Neufchâtel.*—Barometer and thermometer.

CLASS 14.

164 GEORG, *Basle and Geneva.*—Photographs of pictures in the Bâsle Museum.

166 PONCY, F. *Geneva.*—Photographs.

167 VUAGNAT, *Geneva.*—Photographic visiting cards, &c.

CLASS 15.

171 ASSOCIATION OUVRIÈRE, *Locle.*—Astronomical clock, and watches of different kinds and qualities.

171A AUBERT, BROS. *Sentier Vaud.*—Movements of watches; a chronometer; two repeaters.

172 AUDEMARS, L. *Le Brassus.*—Watches and gold cases.

173 BAUME & LEZARD. *Geneva, and* 21, *Hatton Garden, London.*—Watches.

174 BAUMUL & SON, *Geneva.*—Files and other tools for watchmakers and jewellers.

175 COURVOISIER, F. *Chauxdefonds.*—Gold and silver watches, pocket chronometers.

176 BOREL & COURVOISIER, *Neufchâtel.*—Gold and silver watches.

177 BOURGEAUX & DELAMURE, *Geneva.*—Screw plates and screws for watches, &c.

178 BREITLING - LAEDERICH, *Chauxdefonds.*—Various watches.

179 CAVIN, F. *Couvet.*—Machine for cutting and rounding the teeth of watch and clock wheels.

180 CORNIOLEY & SON, *Geneva.*—Springs for watches.

181 COURVOISIER, A. *Chauxdefonds.*—Horological articles.

182 COURVOISIER, BROS. *Chauxdefonds.*—Gold and silver watches.

183 DARIER, HUGHES, & CO. *Geneva.*—Horological appliances, watch keys, &c.

184 DEVAIN, J. *Locle.*—Jewels for watchmakers.

185 HENRY, J. *Chauxdefonds.*—Dial plates for watches.

186 EHNHUUS, H. 53, *Frith-st. Soho, London.*—Machines and tools for watchmakers; watch-work in various stages of manufacture.

187 GOLAY-LERESCHE, *Geneva.*—Mechanical case with chronometer, thermometer, barometer, watches, &c.

188 GRANDJEAN H. & CO. *Locle.*—Marine and pocket chronometers, watches.

189 GRANGER, J. M. *Geneva.*—Enamelled dial-plates.

190 GROSCLAUDE, C. H. & CO. *Fleurier.*—Marine chronometer, gold and silver watches.

191 GUYE, U. *Fleurier.*—Files, and machine to round the teeth of wheels.

192 HENRI, J. *Chauxdefonds.*—Dial-plate four lines in diameter, with microscopic inscriptions; different dial-plates.

193 HIRSCHI, E. *Chauxdefonds.*—Time-piece springs, &c.

194 HUGUENIN-THIÉBAUD, *Locle.*—Anchors and wheels for the escapement of watches.

195 JACCOTTET, P. E. *Travers.*—Separate parts of watches.

196 JACKY, BROS. *Chauxdefonds.*—Horological articles.

197 JACOT & SANDOZ, *Locle.*—Gold watches.

198 JEANJAQUET, C. *Locle.*—Watch springs.

199 JEANRENAUD, AMI, SEN. *Travers.*—A watch, free escapement, anchor, duplex, dead seconds, lever d'entretien.

200 JEANRENAUD, G. H. *Fleurier.*—Rubies adapted for watch-making, &c.

201 INGOLD, P. F. *Chauxdefonds.*—Files to rectify the teeth of watch wheels.

202 JOHANN, A. *Chauxdefonds.*—Chronometer with diamond crown, &c.

203 JUNOD, EUG. & CO. *Chauxdefonds.*—Chronometer and watches.

204 JUNOD, L. E. *Lucens.*—Jewels for watch-making, sapphires for wire-drawing, &c.

205 KEIGEL & BOREL, *Couvet.*—Machines and tools for watch-repairers.

206 LANG & PADOUX, *Geneva.*—Chronometers and half chronometers with certificates from the Observatory of Geneva.

207 LECOULTRE, U. *Sentier.*—Pinions for watches, chronometers, &c.

208 LEHMANN, C. *Bienne.*—Gold and silver watches.

209 LERESCHE-GOLAY, J. *Vallorbes.*—Tools and materials for watch-making.

210 LESQUEREUX, L. *Locle.* — Watch springs.

211 MAIRET, SYLVAIN, *Locle.*—Keyless clock, minute repeater, pocket chronometer, gold watches.

212 MARCHAND, P. A. *Chauxdefonds.*—A gold repeater and skeleton watch.

213 MATTHEY, A. O. *Locle.*—Polishing apparatus, fancy clock, boxes, electric clock, &c.

214 MATTHEY-DORET, P. *Locle.*—Chronometers and a minute repeater, &c.

215 MONTANDON, C. A. *Locle.*—Chronometers and repeaters.

216 MONTANDON, BROS. *Locle.*—Ladies' and boys' watches.

217 MOULINIÉ & LEGRANDROY, *Geneva.*—Box with artificial singing-birds. Watches, &c.

218 MULLER, A. *Locle.*—Electric chronograph, printing seconds and hundredth parts of seconds.

219 MÜLLER, BROS. *Geneva and London.*—Geneva watches, keys, tools, files, and watch materials.

220 MULLERTZ, J. *Locle.*—Pocket chronometers.

221 NARDIN, U. *Locle.*—Chronometers, remontoirs, repeater, dead seconds, stop, &c.

222 PERREGAUX, H. *Locle.*—Marine chronometer, and chronometer indicating mean temperature.

223 PERRET, A. *Locle.*—Repeaters and other watches, with new movements.

224 PERRET, J. *Chauxdefonds.*—Chronometer and watches.

225 PETITPIERRE, D. L. *Couvet.*—Horological tools. Turn-benches to make, mend, round, and replace pivots.

226 PIAGET, H. *Verrières.*—Gold, silver, and other watches, cylinder and anchor.

227 PIGUET, BROS. *Sentier, Vaud.*—Horological articles, striking clocks, minute repeaters, chronometers, &c.

228 RAUSS, A. *Geneva.*—Enamelled dial-plates for complicated watches.

229 RAYMOND & ROUMIEUX, *Geneva.*—Springs for chronometers, &c.

230 REYMANN, F. SON, *Geneva.*—Springs for watches with horizontal movements. English watches, &c.

231 LECOULTRE, A. *Sentier, Vaud.*—Millionimeter.

232 RINGGIER & Co. *Zofingen.*—Watches.

233 ROBERT-NICOUD, C. A. *Chauxdefonds.*—Chronometers, auxiliary fusees, &c.

234 ROSSEL & SON, *Geneva.*—Horological articles and jewellery.

235 ROULET, G. *Locle.*—Silver watches, anchor and cylinder.

236 SANDOZ BROS. *Ponts-martel, Neufchâtel.*—Various watches.

237 SORDET, H. & SON, *Geneva.*—A watch four lines in diameter, a chronometer, half chronometer with thermometer, &c.

238 VAUTIER, SAMUEL, & SON, *Carouge.*—Files and gravers for watchmakers and jewellers; tools for engravers and chasers.

239 VEILLON-EMERY & Co. *Lausanne.*—Chronometers, plain and complex watches.

240 YONNER, C. A. *Verrières.*—Stones for superior watches, screw plates, &c.

241 PROST, C. *Vevey.*—Pocket chronometer, nickel movement, &c.

CLASS 16.

247 BRUGGER, F. L. *Motier-Travers, Neufchâtel.*—Accordion.

248 GREINER, T. & BREMOND B. *Geneva.*—Musical box.

249 HERTIG, J. *Berne.*—Brass instruments.

250 HUNI & HUBERT, *Züric.*—Pianos.

251 KARRER, S. *Teuffenthal, Argovie.*—Musical box.

252 KÖLLIKER & TROST, *Züric.*—Cottage piano.

253 MULLER, BROS. *Geneva and London.*—Musical boxes.

254 NEEF, G.—*Feuerthalen, Schaffhausen.*—A church harmonium.

255 SPRECHER & Co. *Züric.*—A grand and a cottage piano.

CLASS 17.

259 APPIA, DR. L. *Geneva.*—Apparatus for the transport of the wounded.

260 GOLLIEZ, DR. *Lutry, Vaud.*—Tribulcon: an instrument suited to newly-invented missiles.

261 WINKLER, JOSEF, *Berne.*—Microscopic specimens of teeth and bones, fossil remains from lacustrine formations.

CLASS 18.

266 ANDEREGG, T. *Wattwyl.*—Cotton cloth dyed in grain, ginghams, checks, handkerchiefs.

267 BEUGGER, J. *Winterthur.*—Glass case containing bobbins of cotton.

268 BREITENSTEIN, J. & Co. *Zofingen.*—Cotton fabrics, machine and hand work.

269 BUNTWEBEREI, *Wallenstadt.*—Checks, ginghams, handkerchiefs, and scarfs.

270 EGLI, H. *At the Neuhof, Fischenthal, Züric.*—A piece of cotton cloth.

271 HEINIGER, J. *Burgdorf.*—Canvas for embroidery.

272 HÜSSY, J. R. *Safenwyl.*—Cotton goods.

273 KELLER-STEFFAN, J. G. *Wattwyl.*—Handkerchiefs, gingham, &c.

274 MECHANISCHE WEBEREY, *Altstaetten, St. Gall.*—Handkerchiefs and cotton goods.

275 MÜLLER, J. B. & Co. *Wyl.*—Checks.

276 NEF, J. J. *Herisau.*—Worked muslins.

277 RAMSAUER-AEBLY, J. U. *Herisau.*—White book muslins, white tarlatans.

278 RASCHLE, ABM. *Wattwyl.*—Coloured cotton goods and handkerchiefs.

279 SCHEFER, J. U. *Speicher.*—White, spotted, and figured book muslin.

279A SCHMID, HENGGELER, & Co. *Neu-and Unter-Aegeri.*—Cotton yarns, raw stuff, &c.

280 SPINNEREI, *on the Lorze, Baar, Zug.*—Spun cotton from Surat, &c., with the raw material.

281 STEIGER, SCHOCH, AND EBERHARD, *Herisau.*—Swiss muslins, curtains, dresses, piece goods, &c.

282 SCHMID, H. *Gattikon, Züric.*—Fine yarns.

283 WIEDMER, J. J. *Seon, Argovie.*—Cotton tissues and spring goods made by machine.

CLASS 19.

287 SCHMIED, BROS. *Burgdorf.* — Plain and figured linen ticking, bleached and raw.

CLASS 20.

290 BAER & SPINNER, *Ryfferswyl, Züric.* —Black marceline, black gros du Rhin, &c.

291 BAUMANN, SEN. & CO. *Züric.*—Black and coloured taffeta, velveteen, marceline.

292 BAUMANN & STREULI, *Horgen.*—Gros de Naples and other silk goods.

293 BISCHOFF & SONS, *Basle.*—Plain and figured silk ribbons.

294 BISCHOFF, C. & J. *Basle.*—Black silk, satins, serges, and taffetas.

295 BLEULER & KELLER, *Küssnacht, Züric.* —Poult de soie of different quality.

296 BODMER, H. *Züric.*—Silk gauze.

297 BOELGER, M. *Basle.*—Spun silk for various purposes.

299 BRUNNER, H. *Züric.*—Gros de Naples, satin woven with cotton, &c.

300 BRUPBACHER-MÜLLER, *Züric.* — Silk and satin goods.

302 BARY, DE J. & SON, *Basle.*—Plain and fancy ribbons, &c.

303 DOLDER, A. *Meilen, Züric.*—Gros du Rhin, &c.

304 DREYFUS, I. SONS OF, *Basle.*—Silk and velvet ribbons.

305 DUFOUR & CO. *Thal, St. Gall.*—Silk bolting cloths for mills.

306 EGLI, J. C. *Richterswyl, Züric.*—Gros du Rhin and other silks.

307 FICHTER & SONS, *Basle.*—Silk ribbons.

309 HERZGO & CO. *Aarau.* — Silk ribbons.

310 HÖHN & STÄUBLI, *Horgen, Züric.*— Gros de Naples and other silks.

311 HOTZ & CO. *Oberrieden, Züric.*—Gros du Rhin, &c.

312 HÜNI-STETTLER, *Horgen, Züric.* — Gros de Naples, repps, gros du Rhin, &c.

313 HÜRLIMANN-TRÜMPLER & CO. *Waedenswyl, Züric.*—Poult de soie, checked.

314 MÜLLER-HAUSHEER, *Adlisweil, Züric.* —Black gros du Rhin.

315 NAEF, J. R. & SON, *Cappel.*—Lustring, foulards, gros de Chine, &c.

316 NAEGELI & CO. *Horgen, Züric.*—Gros de Naples, black gros du Rhin.

317 NAEGELI & WILD, *Züric.*—Poult de soie; gros du Rhin, checked.

318 NOTZ & DIGGELMANN, *Züric.*—Checks jaspés, chinés, armures.

319 PREISWERK, DIETRICH, & CO. *Basle.* —Satin ribbons and taffetas, Japan and China silks, &c.

320 REIFF - HUBER, *Wiedikon, Züric.* — Bolting cloth.

322 RUTSCHI & CO. *Züric.*—Satin de Chine, satin de Luxor, gros du Rhin.

323 RYFFEL & CO. *Stäfa, Züric.*—Silks, satins, taffeta, &c.

324 SARASIN & CO. *Basle.*—Silk ribbons.

325 SCHMIED, BROS. *Thalweil, Züric.* — Foulards for handkerchiefs and dresses.

326 SCHNEEBELI, C. SONS OF, *Züric.* — Satins for various purposes.

327 SCHUBIGER & CO. *Utznach, St. Gall.*— Silk goods, taffeta for umbrellas

329 SCHWARZENBACH, J. J. *Kilchberg, Züric.*—Gros du Rhin, &c.

330 SCHWARZENBACH - LANDIS, J. *Thalweil, Züric.*—Poult de soie and taffetas.

331 SIEBER, J. F. *Neumunster, Züric.*— Black gros du Rhin, satin-faced and satin throughout.

332 STAPFER, HÜNI, & CO. *Horgen, Züric.* —Florence, marceline, gros de Naples, gros du Rhin.

333 STAPFER, JOHN, SONS, *Horgen.*—Gros de Naples, poult de soie, checks.

335 STEHLIN & ISELIN, *Niederschönthal, Basle.*—Long spun silk-yarn.

336 STEHLI-HAUSHEER, R. *Lunnern, Zug.* —Marceline, poult de soie, serge de Malaga, &c.

337 STOCKER, J. C. *Züric.*—Marcelines and other silk goods.

338 STÜNZI & SONS, *Horgen.*—Gros d'Alger glacé, black poult de soie, &c.

339 SUREMANN & CO. *Meilen, Züric.* — Satin de Chine, gros du Rhin.

340 TRÜDINGER & CO. *Basle.* — Fancy ribbons.

341 WEBER, FRIEDRICH, *Hausen-on-the-Albis, Züric.*—Gros du Rigi, &c.

342 WIEDMER-HÜNI, J. J. *Horgen.*—Poult de soie, gros du Rhin.

344 WIRZ & CO. *Seefeld, near Züric.* — Black gros du Rhin.

345 ZINGGELER, BROS. *Wädensweil.* — Poult de soie, gros du Rhin.

CLASS 21.

353 BALLY & SCHMITTER, *Aarau.*—Elastic textures for boots.

354 BREITENSTEIN, J. & CO. *Zofingen.*— Plain and figured tissues in colours, wool and cotton.

355 HUBLER & SCHAFROTH, *Burgdorf.*— Samples of unravelled wool, and woollen thread spun from that material.

356 HÜSSY, J. R. *Safenwyl, Argovie.*— Half-woollen stuffs.

357 WILDI & HAUSER, *Langenthal, Berne.* —Half-woollen stuffs.

CLASS 22.

361 GRÄNICHER, S. *Zofingen.*—Oil-cloths for floors, tables, &c.

CLASS 23.

364 HUNERWADEL, A. *Rapperswyl.*—Fabrics in Turkey-red.

366 KAUFMANN, G. *Triengen, Lucerne.*— Skeins of cotton yarn.

367 LUCHSINGER, ELMER, & OERTLI, *Glaris.*—Indiennes, handkerchiefs and sarongs.

368 RIKLI, A. F. *Wangen, Berne.*—Turkey-red, brown, and pink-coloured yarns.

369 SUTER, J. R. *Zofingen, Argovie.* — Turkey-red coloured cotton yarn.

CLASS 24.

374 ALTHERR, J. C. *Speicher, Appenzell.* —Edgings and insertions, machine embroidered muslin curtains and dresses.

375 BAENZIGER, J. *Thal, St. Gall.*—Embroidered muslin, jaconet, batistes, and tulle.

376 EHRENZELLER, F. *St. Gall.* — Embroidered curtains.

377 INDERMUHLE-BUHLER, MRS. *Thoune.* —Embroidered net-work curtains and table-covers.

378 MONTANDON, MISS J. *Motiers-Travers, Neufchâtel.*—Table-covers, collars, cuffs, and pincushion-covers of crochet-work.

379 GORINI, J. B. *St. Gall.*—Embroidered handkerchiefs.

380 NEF, J. J. *Herisau.* — Embroidered curtains, balzorine, shawls, &c.

381 RITTMEYER, B. & Co. *St. Gall.*—Insertions, trimmings, and curtains, embroidered by machine.

382 RAUCH, R. *London* (*St. Gall*).—Embroidered curtains.

383 STÄHELI, WILD C. *St. Gall.* — Fine embroideries.

384 STEIGER-SCHOCH & EBERHARD, *Herisau.*—Embroideries on tulle and muslin.

385 TANNER, B. & H. & KOLLER, *Herisau.* — Embroidered white curtains, silk antimacassars.

CLASS 25.

390 RECHEWERTH, E. *Züric.*—A rug made with the skins of the heads, &c. of animals.

391 ROOS, G. *Lausanne.*—Specimens of the various Swiss furs.

392 ROTH, J. *Wangen, Berne.* — Curled horsehair.

CLASS 26.

396 MERCIER, J. J. & BROS. *Lausanne.*—Leathers and skins, wools, and horse-hair.

397 MERZ, J. *Herisau.*—Dressed skins.

398 PÜNTER, J. J. *Uerikon, Züric.*—Sole-leather, &c.

399 RAICHLEN, L. *Geneva.*—Leather and skins.

401 TESSE. F. *Lausanne.* — White calf-skins.

402 ZÜND, J. U. *Staefa, Züric.* — Hemp hose for fire-engines.

CLASS 27.

405 ANDEREGG, T. *Wattwyl.*—Shirts, and shirt-fronts.

406 BROSSY-PLUMETTAZ, JULIE, *Payerne.* —Flowers and feathers in straw.

407 ECOFFEY, MARIE, *Vevey.*—A crinoline skirt, and stays.

408 FISCHER, BROS. *Meisterschwanden, Argovie.*—Braids of hair, and of manilla and hair; trimmings.

409 FISCHER, J. L. SONS OF, *Dottikon, Argovie.* — Collection of straw and manilla plaits, cigar-cases, &c.

410 HESS-BRUGGER, *Amrisweil, Thurgovie.* —Fancy sleeves, mitts, &c., of English-spun diamond-cotton.

411 ISLER, ALOIS, & Co. *Wildegg, Argovie.* —Plaits, trimmings, and ornaments in straw.

412 ISLER, JACQUES, & Co. *Wohlen, Argovie.*—Straw and horsehair trimmings, and ornaments.

413 LANDERER-ZWILCHENBART, E. *Wohlenschwyl, Argovie.*—Edgings; plaits in straw, hemp, hair, and silk; trimmings, hats, baskets.

414 RUMPF, C. C. *Basle.*—Silk shirts, under-waistcoats, drawers, &c.

415 SCHAERLY-BAERISWYL, J. B. *Fribourg.*—Bonnets of Friboug straw, and plait used in their manufacture.

416 STOLL, J. C. *Schaffhausen.* — Straw hats.

417 SUDAN, MRS. H. & Co. *Fribourg.*—Straw bonnets.

418 THEDY-GREMION, *Enney, near Bulle, Fribourg.*—Straw plaits from Gruyere.

---

419 CHIESA, J. *Loco and Morges.*—Straw plaits.

CLASS 28.

422 DRESCHER, Y. *Lithographical Institute, Züric.* — Lithographs; pictures, in galvano-plastic, relief.

423 GONIN, F. *Nyon.*—Galvanic engraved plates for printing earthenware.

424 RIETER-BIEDERMANN, *Winterthur.* — Printed music, &c.

425 SIEGFRIED, *Wipkingen, Züric.*—Engravings in aqua-tinta.

CLASS 29.

429 BECK, E. *Berne.*—Maps, in relief.

430 BERTSCH, H. *St. Gall.*—Apparatus for natural philosophy.

431 HUTTER, PROF. *Berne.*—Elementary course of drawing.

433 VORUZ, A. *Lausanne.*—Mathematical works.

434 WURSTER, J. & Co. *Topographical Office, Winterthur.*—Geological maps.

CLASS 30.

438 CARRAZ, J. B. *Porrentruy, Berne.* — Buffets of oak, sofa, drawing-room table.

439 GAY, E. & F. *Aigle.*—Brushes, inlaid works of wood, table-mats.

440 LAVANCHY-FARAUDO, *Lausanne.* — Bureau.

441 MONNIER & Co. *Aigle.*—Inlaid and other floorings.

442 WETLI, M. *Berne.* — Bureau with springs.

CLASS 31.

446 BOSSI, G. *Locarno.*—A lock.

447 HARDMEYER, J. *Züric.* — Hammers for dressing mill-stones; articles in cast-steel.

448 KNECHT, J. *Glaris.*—Brass model for calico printing.

449 MADLIGER, J. *Langenthal, Berne.*—A needle perforated lengthwise.

450 MATTHEY, A. O. *Locle.* — Gold and silver electro-plating.

451 NEUHAUS & BLÖSCH, *Bienne.*—Iron wire for wool-cards, and springs.

452 STEINER, A. *Lausanne.*—Iron folding bedstead and spring mattress.

453 KUGLER-DELEIDERRIEZ, *Lausanne.*—Candlesticks, flambeaux, and lamps, of brass and German silver.

CLASS 32.

456 MATTHEY, A. & SONS, *Locle.*—Steel for watch springs, &c.

457 SCHNEIDER, C. F. *Geneva.*—Complicated pocket-knives.

CLASS 33.

459 BERTHOUD, L. *Fleurier.*—Rubies, for wire-drawing, for astronomical purposes.

460 BESSON, J. *Neufchâtel.*—Chasings in steel; statuette, &c.

462 DUBOIS, A. *Chauxdefonds.*—Engravings in gold; castings in silver, enamellings, jewellery, pictures, &c.

463 FAVRE-BULLE, J. C. *Chauxdefonds.*—An oval engraved gold plate.

464 FAVRE-BULLE, L. E. *Locle.*—Decorations for watch-cases.

465 GRANDJEAN-PERRENOUD, H. *Chauxdefonds.* — Engraved decorations, letters, enamel, and diamonds for watchmaking.

467 HUBER, J. J. *Geneva.* — Moveable jewellery.

468 JUNOD, Z. *Locle.*—Jewels, pallets for chronometers, sapphire-gravers and files.

469 KUNDERT, F. *Chauxdefonds.*—Castings and plates in gold and silver; engravings, and precious stones for watchmaking and jewellery.

471 LANG & PADOUX, *Geneva.*—Brooches, bracelets, with landscapes or portraits, from photographs, on enamel.

472 MASSET, L. & SON, *La Motte.*—Frame with perforated rubies for watch-makers; wire-drawing sapphire and diamonds for gold-wire drawers.

CLASS 35.

476 GONIN, F. & BURNAND, A. *Nyon.*—Earthenware; imitations of porcelain; and articles for cooking.

477 LERBER, M. DE, *Romainmotier, Vaud.*—Pipes for wells and drains; bricks.

478 ZIEGLER-PELLIS, *Winterthur.*—Articles in terra-cotta.

---

479 HEIMBERG, BERNE, SOCIETY OF THE POTTERS.—Common pottery.

CLASS 36.

481 CHAMUSSY, J. B. *Geneva.*—Trunks.

482 ISENRING, J. G. *Geneva.*—Articles for travelling.

TURKEY.

N.W. Court and N.W. Gallery, near Horticultural Garden Entrance.

CLASS 1.

1 MEHMED, HADJI, *Constantinople.*—Gold and silver leaf.

2 OHANES, *Constantinople.*—Gold and silver thread, and laces.

3 SIMEON, *Constantinople.*—Diamond for cutting glass.

4 GOVERNOR OF ZANOUN, *Amassia.* — Flint stone.

5 DIMITRI, *Amassia.*—Silver, silver ore, and ingot from the Jolnoon mines.

6 GOVERNOR OF BALIKESSER.—Plaster, stone, mill-stone, and other stones.

7 GOVERNOR OF MONASTIR.—Iron ore, and iron.

8 RESHID, MEHMED, *Aleppo.*—Yellow, black, and red stones, and specimen of their use.

9 GOVERNOR OF PHILIPPOLI. — Lime, limestone, flint, and kufgie stones. Saltpetre.

10 —— OF MOUNT KHANIA.—Stones, and iron ore.

11 —— OF TARASHJA.—Mineral earth.

12 —— GUMULJINA.—Minerals and alum.

13 —— OF CYPRUS.—Copper ore, minerals.

14 —— OF LAPISKY,—Clay, emery, potter's earth.

15 —— OF AVRATHISSAR.—Silver ore.

16 —— OF CARADAGH.—Silver ore, gold dust.

17 —— OF NEVREKOL.—Iron ore.

18 —— OF LESCOFCHA.—Iron ore, and bar iron.

19 —— OF MOSSUL.—Mineral.

20 —— OF KUTAHIEH.—Ecume de mer (mineral).

21 —— OF JANIK.—Iron-stone, and potter's earth.

22 NOOVALLAH, *Conia.*—Saltpetre (rough).

23 GOVERNOR OF TIRNOVA.—Copper.

24 MUSTAPHA, DERVISH, *Bosnia.*—Copper.

25 GOVERNOR OF BOSNIA.—Lead ore and minerals.

26 —— OF DAMASCUS.—Copper and iron ore, and stone.

27 —— OF ADANA.—Lead and copper ore.

28 —— OF LEBANON.—Pit coal, iron ore, and other minerals.

29 —— OF TREBISOND.—Copper and silver ore.

30 —— OF CASTAMOUNI.—Copper ores.

31 —— OF ANGORA.—Copper ore, and mineral salt.

32 —— OF CRETE.—Minerals.

33 APOSTOL OF CRETE.—Silver.

34 GOVERNOR OF BROUSSA.—Marble.

35 KOURSHID, AGA, *Lazistan.*—Minerals.

36 GOVERNOR OF KERESSON.—Iron ore.

37 —— OF KARAHISSARSHARKY.—Alum and flint.

CLASS 3.

38 Kurdogloo, Hafiz, *Amassia.*—Wheat and bamia.

39 Meerlevi, Cheikhi, *Amassia.*—Barley.

40 Hassan, Haji, *Amassia.*—Plums.

41 Akhassan, Hafiz, *Amassia.*—Lentils.

42 Gullekreoglou, *Amassia.*—Peas.

43 Bakal, Kiumil, *Amassia.* — Kidney beans.

44 Bakal, Kehreman, *Amassia.*—Haricot beans.

45 Haji, Efendi, *Amassia.*—Aubergines.

46 Governor of Amassia.—Beans and millet.

47 Osman, Efendi, *Amassia.*—Haricot beans and maize.

48 Kiortash-Ogly, *Amassia.*—Mehleb.

49 Booroon-Ogly, Kirkor, *Amassia.*—Salep.

50 Boornogly, Kerork, *Keupri.*—Wines.

51 Nicholas, *Amassia.*—Litharge and lead.

52 Governor of Carassy.—Wheat, barley, rye, sesame, &c.

53 —— of Kosak.—Pistachio.

54 —— of Balikesser.—Different sorts of corn.

55 —— of Loma.—Raisins.

56 —— of Balikesser. — Olive, and sesame oils.

57 Mekhoory, *Saida.* — Pumpkin and aubergine seeds, and various other seeds.

58 Governor of Latakia.—Various seeds, olives, sesame oil, honey, olive oil, and salt.

59 Hassan, Aga, *Monastir.*—Plums.

60 Governor of Kesree.—Wines.

61 Reshid, Mehmed, *Aleppo.* — Rice, barley, lentils, melon seed, walnuts, junipers, fennel, and liquorice, &c.

62 Reshid, Mehmed, *Aleppo.* — Wines, spices, and dried fruits.

63 Governor of Philippoli. — Rice, wheat, barley, rye, millet, beans, peas, &c.

64 —— of Prasdin.—Raky.

65 Rustem, Usta, *Cozandjik.*—Chestnuts, dried cherries, and wheat.

66 Governor of Djerban.—Wheat and barley.

67 —— of Shaeer.—Butter, olive, and sesame oils.

68 —— of Tripoli.—Olives, walnuts, and almonds.

69 —— of Akkia.—Raky, wine, olives, and olive oil. Various grains and seeds.

70 —— of Beladbechan.—Raky, wine, olives, and olive oil. Various grains and seeds.

71 —— of Drama.—Wheat, barley, maize, and rice.

72 —— of Gumulgina.—Wheat, barley, maize, and rice.

73 —— of Cavala.—Wheat, barley, rye, and maize.

74 —— of Sarishaban.—Wheat, barley, rye, and maize.

75 —— of Curassou.—Wheat, barley, rye, and maize.

76 Governor of Cyprus.—Wines, and various grains and seeds.

77 —— of Dardanelles. — Wheat, sesame, peas.

78 —— of Aivalik.—Olive oil.

79 —— of Lapisky.—Grape molasses.

80 —— of Salonica. — Various grains and seeds.

81 Darbila, *Salonica.*—Samples of flour.

82 Sotin-Oglou, Mical, *Salonica.* — Wines.

83 Governor of Siros.—Various grains and seeds.

84 —— of Nish.—Various grains and seeds.

85 —— of Mossul.—Wheat, millet, and corn.

86 —— of Kutahieh.—Corn, seeds, oils, fruits.

87 —— of Angora.—Wheat.

88 —— of Andrinople.—Bamia, sweetmeats, dried fruit, and oils.

89 —— of Janik.—Corn, seeds, and fruit.

90 Abdullah, Oosta, *Conia.*—Kernel and seed oils.

91 Apostol, *Conia.*—Wine and raky.

92 Governor of Naplouse.—Corn and seeds; oils.

93 —— of Toolcha. — Wheat, barley, millet.

94 —— of Bosnia.—Dried plums.

95 —— of Vidin.—Strawberry wine.

96 —— of Damascus.—Samples of corn and seeds.

97 Beyazid, Efendi, *Damascus.*—Fruits honey, pickles, molasses, sugars, &c.

98 Governor of Adana.—Wheat, barley and sesame.

99 —— of Islimia.—Wines and syrups.

100 —— of Lebanon.—Corn, honey, oils, olives, molasses, fruits, and seeds.

101 Selamouni.—Spirits.

102 Bedros, *Lebanon.*—Wines.

103 Howsy, H. *Lebanon.*—Wines.

104 Reshid, H. *Lebanon.*—Wines.

105 Governor of Scutari of Albania. —Corn and seeds; wines and oils.

106 —— of Mardin.—Dried fruits.

107 —— of Coniah.—Corn and seeds; dried fruits, honey, and molasses.

108 —— of Castamouni.—Rice, saleb, and dried plums.

109 —— of Angora.—Wheat and barley.

110 —— of Crete.—Wines, syrups, and spirits; dried fruits, salep, oils, honey, pickles, and seeds.

111 Aghaco, *Crete.*—Wheat, barley, beans, peas, and almonds.

112 Sami Bey, *Crete.*—Butter, herrings, honey, and olives.

113 Monoulaki, *Crete.*—Olives.

114 Michalaki, *Crete.*—Wines.

115 Tanash, *Broussa.*—Red and white wines.

116 Governor of Broussa.—Olive oil.

117 Kourshid, Aga, *Lazistan.*—Millet.

CLASS 4.

118 Riza, Ali, *Constantinople.* — Wax candles.

119 Papazsgloo, Yanko, *Gueva.*—Silk.

120 Ibrahim, Aga, *Constantinople.* — Soap.

121 Torosoglou, Bogos, *Constantinople.* —Brutia silk.

122 Governor of Broussa.—Silk, silk-worm cods, silk tufts.

123 Brotte, *Constantinople.*—Brutia silk.

124 Zadikoglou, Curubis, *Uskulub.* — Silk.

125 Hussein, *Hadjikioy.*—Hemp.

126 Ismail, Haji, *Medzoza.*—Cotton.

127 Gulbekre-Ogloo, *Amassia.*—Poppy.

128 Boyaji, Meyer, *Amassia.*—Madder-roots.

129 Kishlajickly-Ogly, *Amassia.* — Pérétek.

130 Virmishli, Ahmed, *Amassia.*—Earth used for the manufacture of grape molasses.

131 Booroon-Ogly, Kirkor, *Amassia.*—Opium.

132 Ali, Aga, *Uskulub.*—Yellow berries.

133 Governor of Uskup.—Salt.

134 Hassan, *Keupri.*—Tobacco.

135 Pergama, *Kirkagaj.*—Cotton.

136 Mehmed-Oosta, *Artemid.*—Soap.

137 Governor of Balikesser.—Poppy seed, cotton, nuts, madder roots, madder seed, tobacco, and opium.

138 Khorajan, Madam, *Beyrouth.* — White and yellow silk.

139 Kuri, N. *Beyrouth.*—White silk.

140 Anassir, Hoja, *Beyrouth.*—Yellow silk.

141 Shouan, Beni, *Beyrouth.*—Soap.

142 Mekhoory, *Saïda.*—Cotton in shell.

143 Governor of Latakia.—Tobacco, soap, sponges, and cotton.

144 Mustapha, Haji, *Monartir.*—Fine flax.

145 Governor of Monastir.—Madder roots.

146 Reshid, Mehmed, *Aleppo.*—Gall-nuts, wool, wax, soap, yellow berries, tobacco, cotton, and tallow.

147 Hussein, Haji, *Philippoli.*—Silk.

148 Governor of Djerban.—Opium.

149 —— of Tripoli.—Silk.

150 Manastil, *Shaeer.*—Soap

151 Governor of Shaeer.—Sponges.

152 —— of Tripoli.—Wax.

153 —— of Mount Khania.—Various kinds of grain and seeds.

154 —— of Tripoli.—Various seeds.

155 —— of Belad Bechan.—Tobacco.

156 —— of Drama.—Cotton wool, cotton in shell, tobacco.

157 —— of Gumuljina.—Silk.

158 —— of Cavala. — Tobacco and cotton.

159 —— of Charishaban.—Tobacco and cotton.

160 —— of Curassou.—Tobacco.

161 —— of Cyrus.—Raw silk and cotton.

162 —— of Dardanelles.—Gall-nuts.

163 —— of Aïvalik.—Acorns.

164 Yacofini, *Salonica.*—Raw silk.

165 Governor of Salonica.—Silk cods.

166 —— of Avrathissar.—Silk cods.

167 —— of Siros.—Cotton, tobacco, flax.

168 Halini, Aga, *Smyrna.*—Silk.

169 Governor of Nish.—Yellow dye and dye wood; silk.

170 —— of Caiseria.—Yellow berries.

171 —— of Kutahish.—Tobacco, madder roots, mastic, silk, silk cods, gall-nuts, valonea.

172 —— of Andrinople.—Silk, silk-worm cods, tobacco, wool, hemp, cotton, madder roots.

173 Meledik, Yorgi, *Janik.*—Raw silk.

174 Khoorshid, *Janik.*—Wool.

175 Ahmed, *Janik.*—Tobacco.

176 Governor of Janik.—Hemp.

177 —— of Naplouse. — Cotton, nuts, tobacco.

178 —— of Toolcha.—Wool.

179 Mehmed & Mustapha, *Kurdistan.*—Raw silk and cotton.

180 Governor of Bosnia.—Flax and tobacco.

181 —— of Vidin.—Wheat and maize.

182 Booyana, *Vidin.*—Raw silk.

183 Gourgu, *Vidin.*—Raw silk.

184 Abousaleh, *Damascus.*—Hemp.

185 Halim, El Deval, *Damascus.* — Wax.

186 Governor of Damascus.—Tobacco, shumac, madder roots.

187 Djerdjis, *Damascus.*—Raw silk and cotton.

188 Governor of Magnesia.—Raw silk, tobacco, cotton nuts, madder roots; earth used in the manufacture of grape molasses.

189 —— of Adana.—Cotton, tobacco.

190 —— of Scutari of Albania.—Raw silk, and silk cods.

191 Eddé, M. *Lebanon.*—Raw silk.

192 Governor of Lebanon. — Sugar-cane.

193 Medawar, N. *Lebanon.*—Raw silks.

194 Shahdan, Y. *Lebanon.*—Raw silks.

195 Halil, E. *Lebanon.*—Raw silks.

196 Governor of Lebanon.—Sponges, tobacco, and cocoons.

197 —— of Scutari of Albania.—Tobacco.

198 —— of Conia. — Yellow berries, hemp seed, mastic, madder roots, shumac, opium, wax, valonea, gall nuts, glue.

199 —— of Castamouni.—Saffron, hemp, wax, goats' wool, and linseed.

200 —— of Crete.—Saffron, opium, red and yellow dyeing earth, wax, tobacco, raw silk, cocoons, and dye roots.

201 Aghaco, *Crete.*—Raw silk.

202 Sami Bey, *Crete.*—Various kind of soaps.

203 Manulaki, *Crete.*—Raw silks.

204 Governor of Broussa. — Walnut wood.

205 Constantin, *Broussa.*—Raw silk.

206 Kuleyan, Ovaghim, *Broussa.*—Raw silk.

207 Kirmizian, Garabet, *Broussa.* — Raw silk.

208 Begogloo, Apostol, *Broussa.*—Cocoons.

209 Governor of Broussa. — Valonia and gall nuts.

210 MOOSSA, *Keresson.*—Raw hemp.

211 GOVERNOR OE KARAHISSAR SAHIB. —Opium, poppy seed, and linseed.

CLASS 7.

212 BEYAZID, EFENDI, *Damascus.*—Weaving machine.

CLASS 8.

213 BALYAN, S. *Constantinople.*—Locomotive engine.

CLASS 9.

214 GOVERNOR OF DRAMA.—Rice-cleaning machine.

215 —— OF CARADAGH. — Machine for separating gold dust from sand.

216 HALIM, EL DEVAL, *Damascus.* — Agricultural implements.

CLASS 11.

217 IMPERIAL MANUFACTORY. — Pistols and rifles, for the army and navy.

218 MEHMED BEY, *Tirnova.*—Rifle.

219 ABDULHADI, *Saïda.*—Sword belt.

220 GOVERNOR OF HASKIOY.—Ramrods.

221 SHIMOUN, *Mossoul.*—Musket.

222 HASSAN-OOSTA, *Conia.*—Gun.

223 DERVISH, *Bosnia.*—Cartouch boxes and slippers.

224 SALIH, RAMO, & ABRO, *Bosnia.*—Musket battery.

225 DEVARA, MEHMED, *Damascus.*—Model of gun and carabine.

226 GOVERNOR OF SCUTARI OF ALBANIA. —Cartridge box and musket batteries.

227 GOVERNOR OF TREBISOND. — Gunpowder box.

228 HALIL-OOSTA, *Trebisond.* — Musket battery.

229 AIDIN, AGA, *Uskub.*—Pistols, gun, and rifles.

230 GOVERNOR OF CRETE.—Pistols, with gold ornaments.

231 BEDIGOGLOU, *Karahissar Sahib.* — Musket battery, &c.

Addendum.

217A HETOOM.—Swords, inlaid with gold

CLASS 16.

232 TOPLEE, *Constantinople.* — Turkish musical instruments.

233 AVEDIS, *Constantinople.* — Kettledrums.

CLASS 18.

234 SALIH, *Constantinople.*—Cotton stuff.

235 BATO, *Constantinople.*—Printed calico.

236 ZEITCHA & MARIDRITCHA, *Constantinople.*—Printed calicoes.

237 IMPERIAL MANUFACTORY, *Constantinople.*—Printed calicoes, and calicoes.

238 YORDUM-OOSTA, *Amassia.* — Printed calico.

239 IBRAHIM-OOSTA, *Mersicon, Amassia.* —Cotton stuff for furniture.

240 OSMAN-OOSTA, *Mersicon, Amassia.*—Cotton stuff.

241 DALTABAN-OGLOU, *Mersicon, Amassia.*—Bath towels.

242 HASSAN, HADJ, *Balikesser.*—Printed calico for turban.

243 BEKRI-OOSTA, *Balikesser.* — Printed calico.

244 SONDASH, YONIS, *Saïda.*—Calico and bath towel.

245 GOVERNOR OF SAÏDA.—Cotton edging and button.

246 HICO, *Philippolis.*—Bath, and other towels.

247 GOVERNOR OF PHILIPPOLIS.—Printed calicoes.

248 —— OF SHAEER.—White cotton cap.

249 —— OF TARASHJA.—Printed calico.

250 —— OF BELADBREHAN.—Calico.

251 —— OF DRAMA.—Cotton yarn and cloths.

252 HOFIF, HAJI, *Denisli.*—Cotton stuffs.

253 SELIM, HAJI, *Denisli.*—Cotton bath towel.

254 MUSTAPHA, HAJI, *Denisli.*—Calico.

255 ALIKZADA, MEHMED, *Denisli.*—Calico.

256 GOVERNOR OF CYPRUS. — Printed calico.

257 ELENCO, *Salonica.*—Bath sheets and towels.

258 GOVERNOR OF SIROS. — Cotton fabrics.

259 —— OF SMYRNA.—Cotton yarn.

260 —— OF NISH.—Calicoes.

261 HOOVA, YOUSSUF, *Mossul.*—Cotton stuff.

262 JERGES, SHEMAS, *Mossul.*—Printed calicoes.

263 GOVERNOR OF KUTAHIA. — Cotton goods.

264 GOVERNOR OF ANGORA. — Cotton goods.

265 —— OF ANDRINOPLE.—Cotton goods.

266 —— OF TIRNOVA. — Cotton goods (alaja).

267 COFO, *Bosnia.*—Calico.

268 THEODORA, *Vidin.*—Sheeting.

269 GEBRIL, OSMAN, *Damascus.*—Cotton stuffs.

270 GASSOBY, ABDULLAH, *Damascus.*—Cotton yarn.

271 HAMAS, MUSTAPHA, *Damascus.*—Cotton cloth.

272 ABDELKADER, SEID, *Damascus.* — Cotton towellings.

273 AHMED, SEID, *Damascus.* — Cotton stuff.

274 KLEUHDI, MEHMED, *Damascus.* — Cotton stuff.

275 GOVERNOR OF MAGNEZIA.—Cotton stuff.

276 EDDÉ, M. *Lebanon.*—Cotton stuffs.

277 RESHID HABIB, *Lebanon.* — Cotton stuffs.

278 BOUTROS, ELIAS, *Lebanon.* — Cotton stuffs.

279 HABBAS, A. *Lebanon.* — Cotton sheeting.

280 GOVERNOR OF CRETE.—Cotton socks.

281 KADRI, *Crete.*—Cotton sheeting.

282 HANUSSAKI, MURAD, & RIGEL, *Crete.* —Cotton bagging.

283 KOURSHID, AGA, *Lazistan.*—Cotton yarn.

284 PARSEKT, VARTAN, *Karahissar Sharky.*—Calico and cotton yarn.

285 ANONIA, *Karahissar, Sharky.*—Cotton towels.

CLASS 19.

286 HIDAYET, EFENDI, *Constantinople.*—Embroidered ablution towels.

287 IMPERIAL MANUFACTORY, *Hereké.*—Damasked linen.

288 ——— *Prevesa.*—Linen cloth.

289 BOYAJI, MEGERDICH, *Hajikeny, Amassia.*—Linen cloth.

290 KESHISH, *Hajikeny, Amassia.*—Towels.

291 GOVERNOR OF LATAKIA.—Various linen fabrics.

292 MUSTAPHA, HAJI, *Monastir.*—Flax thread.

293 GOVERNOR OF TARASHJA.—Linen cloth.

294 CHAOUSH, OOSTA, *Denizli.*—Linen for shirting.

295 GOVERNOR OF CYPRUS.—Curtain cloths.

296 —— OF SIROS.—Linen cloth.

297 —— OF NISH.—Hemp cloth, ropes, and pack-thread.

298 AYSHA, *Conia.*—Linen yarn, and cloth.

299 ABOUSALEH, *Damascus.*—Ropes and pack-thread.

300 GOVERNOR OF TREBISOND.—Linen cloths.

301 CHERVISH, JANIK, *Broussa.*—Handkerchiefs and towels.

302 ESKISHERLI, OVANES, *Broussa.*—Bath towels.

303 EMIN, AGA, *Keresson.*—Ropes.

CLASS 20.

304 SALIH, *Constantinople.*—Silk stuff.

305 GOMIDAS, *Constantinople.*—Embroidered gauze for gown.

306 HIDAYET, EFENDI, *Constantinople.*—Embroidered silk covering for coffee board; silk cloth for shirts.

307 SALIH, EFENDI, *Constantinople.*—Silk bath towel.

308 PAPAZOGLOO, YANKO, *Gueva.*—Velvet coffee board.

309 IMPERIAL MANUFACTORY, *Héréké.*—Damasked silks, velvets, silk stuffs for curtains, ribbons, silk cord, satin and silk gauze.

310 BOYHOSS, *Constantinople.*—Silk stuffs.

311 KURDBEKRE, NOURI, *Amassia.*—Silk shirtings.

312 GHANIZADE, MEHMED, *Amassia.*—Brocade.

313 ABDULDJELIL, *Beyrouth.*—Silk shawls, other stuffs.

314 JERDJIS, CONDJI, *Beyrouth.*—Silk stuffs.

315 DJERDJÉSAN, *Beyrouth.*—Silk fabrics.

316 ALY, COOTLY, *Beyrouth.*—Red silk stuff (cootni).

317 DJERDJER, SARÉJI, *Beyrouth.*—Silk stuff (cootni).

318 YAWAB, OUSTA, *Saiola.*—Edging, girdle, and fringe.

319 DJALIN, ILIAS, *Saiola.*—Twisted white silk.

320 HOOKDAR, *Saiola.*—Silk tufts.

321 BILSANAK, ARSLAN, *Saiola.*—Silk ribbon.

322 DANOTO, ANDON, *Aleppo.*—Silk fabrics.

323 KHATIOGLI, ANDON, *Aleppo.*—Various silk fabrics.

324 GOVERNOR OF PHILIPPOLI.—Silk shirting.

325 MOHAMMED, SEID, *Tripoli.*—Silk sheets and trimmings.

326 MOHMED-ALI, OUSTA, *Roustchouk.*—Silk bobbins.

327 GOVERNOR OF SMYRNA.—Silk shirting.

328 HOOVA, HODJA, *Mossul.*—Silk stuffs.

329 GOVERNOR OF ANDRINOPLE.—Silk goods.

330 CHERMEKLY, *Kurdistan.*—Silk stuffs.

331 MUSTAPHA, *Kurdistan.*—Silk stuffs.

332 ADJEM, AHMED, *Damascus.*—Silk shawl.

333 SEMADI, SALIH, *Damascus.*—Silk fabric.

334 BAGDADI, ABDULLAH, *Damascus.*—Silk belts, and silk fabrics.

335 SHEIKUL, HERAKA, *Damascus.*—Silk fabrics.

336 RESHID, SHEIK, *Damascus.*—Silk fabrics.

337 KURISH, HUSSEIN, *Damascus.*—Silk stuffs.

338 GOVERNOR OF MAGHEZIA.—Silk girdles.

339 —— OF SCUTARI OF ALBANIA.—Silk sheeting.

340 EDDÉ, M. *Lebanon.*—Silk stuffs.

341 MESHAKA, J. *Lebanon.*—Silk stuffs.

342 LATIF, M. *Lebanon.*—Silk stuffs.

343 RESHID, ABOUSHEKIR, *Lebanon.*—Silk fabrics.

344 KERKEBESH, *Lebanon.*—Silk millinery.

345 EMIN, EFFENDI, *Trebisond.*—Silk mercery.

346 ALEMDAR, ALI, *Castamouni.*—Silk mercery.

347 GOVERNOR OF CRETE.—Silk cloths and mercery.

348 VASILIKY, *Crete.*—Silk mercery.

349 SOFALAKY, *Crete.*—Silk shirting, &c.

350 ISFAKIANAKY, *Crete.*—Silk sheetings.

351 HASSAN, AGA, *Crete.*—Silk girdle.

352 CHERVISH, JANIK, *Broussa.*—Silk stuffs and mercery.

353 ESKISHEHRLI, OVANES, *Broussa.*—Silk fabrics.

354 HAYRABIT, OVANES, *Broussa.*—Silk fabrics.

355 FOTAGI, AHMED, *Broussa.*—Silk fabrics, and silk and silver fabrics.

356 BELEDIJI, SALEH, *Broussa.*—Silk stuffs.

357 CHERVISH, BEDROS, *Broussa.*—Silk stuffs.

358 SALEH, HAJI, *Broussa.*—Silk and silver cloth.

359 FOOCHIJI, MEHMED, *Broussa.*—Silk and silver cloth.

360 KATINCO, *Broussa*.—Silk purse.

CLASS 21.

361 SALIH, *Constantinople.*—Cotton and silk stuff.

362 RIZA, ALI, *Constantinople.*—Cotton and silk stuff.

363 HIDAYET, EFENDI, *Constantinople.*—Cotton and silk fabrics.

364 AHMED, HAJI, & MEHMED, *Constantinople.*—Silk and cotton fabrics for furniture; woollen stuff called ehram.

365 IMPERIAL MANUFACTORY, *Héréké.*—Damasked woollen.

366 IMPERIAL MANUFACTORY, *Constantinople.*—Socks.

367 GOVERNMENT MANUFACTORY, *Constantinople.*—Woollen cloth for the army; blankets, red caps (fez), and white do.

368 PARSEGH, *Constantinople.* — Stuffs made with goats' wool.

369 HASSAN, EFENDI, *Amassia.*—Cotton and silk shirtings.

370 FATMA, *Amassia.*—Cotton and silk shirtings.

371 KESHISHOGLOU, STEPAN, *Amassia.*—Woollen hosiery.

372 GOVERNOR OF CARASSY.—Woollen aba.

373 —— OF BALIKESSER.—Hosiery.

374 JERDGIS, CONDJI, *Beyrouth.*—Silk and cotton stuff.

375 DJERJES, *Beyrouth.*—Silk and cotton stuff.

376 GOVERNOR OF SOPHIA. — Woollen socks and cloth.

377 —— OF MONASTIR.—Woollen socks.

378 DANOTO, ANTOON, *Aleppo.*—Cotton and silk fabrics.

379 KHATI, ANTOON, *Aleppo.* — Cotton and silk fabrics.

380 SALIM, NAOUM, *Aleppo.*—Cotton and silk fabrics.

381 YOUSSOUF, *Aleppo.*—Cotton and silk fabrics.

382 RISCULAH, *Aleppo.*—Cotton and silk fabrics.

383 GOVERNOR OF PHILIPPOLI.—Woollen stuffs.

384 —— OF PRASDIN.—Woollen shiaks.

385 —— OF PHILIPPOLY.—Cotton and silk stuffs.

386 —— OF HUSKIVY.—Woollen fabrics.

387 IBRAHIM-OGLY, AHMED, *Djerban.* — Woollen ehrams.

388 AHMED-OGLY, MURAD, *Djerban.* — Woollen ehrams.

389 KEHIASGLY, HUSSEIN, *Djerban.* — Woollen ehrams.

390 GOVERNOR OF AKKIA.—Felts.

391 —— OF DRAMA.—Woollen hosiery, and other fabrics.

392 —— OF GUMULGINA. — Woollen fabrics.

393 SOLIMAN, EFENDI, *Denizli.*—Cotton and silk stuff.

394 KEHRIMAN, USTA, *Denizli.*—Cotton and silk stuff.

395 GOVERNOR OF CYPRUS.—Cotton and silk stuff.

396 LUTHFULLAH, *Salonica.* — Silk and cotton bath towels.

397 MARIA, *Salonica.*—Woollen and cotton.

398 CATINA, *Salonica.*—Silk and cotton cloth.

399 AGHOSSOS, *Salonica.*—Woollen aba.

400 GOVERNOR OF AORATHISSAR.—Woollen aba.

401 —— OF SIROS.—Woollen socks.

402 —— OF NISH.—Woollen aba and socks.

403 ABDULLAH, *Mossul.*—Cotton and silk stuff.

404 ABDULLAH & YUSSOUF, *Mossul.* — Woollen stuffs.

405 GOVERNOR OF KUTAHIEH.—Woollen stuffs and hosiery.

406 —— OF ANGORA.—Woollen goods.

407 —— OF ANDRINOPLE. — Woollen goods.

408 SHERIFA, *Conia.*—Hosiery.

409 GOVERNOR OF TIRNOVA.—Silk and woollen stuff.

410 KHOSKOBJI, OGLOU, *Kurdistan.* — Cotton and silk stuffs.

411 TUJAR, OGLOO, *Kurdistan.*—Cotton and silk stuffs.

412 BEDO, OGLOO, *Kurdistan.*—Cotton and silk stuffs.

413 JEVHER, AGHA, *Bosnia.*—Cotton and silk goods.

414 SULEYMAN, *Bosnia.*—Woollen aba.

415 THEODORA, *Vidin.*—Cotton and silk sheeting, and woollen socks.

416 GOVERNOR OF VIDIN.—Hosiery and felt.

417 ADJEIM, AHMED, *Damascus.*—Shawls.

418 KEMALI, SAÏD, *Damascus.* — Cotton and silk goods.

419 HUSSEIN, SHEÏK, *Damascus.*—Cotton and silk goods.

420 MUHYEDDIN, *Damascus.* — Silk and cotton stuffs.

421 GASSOBY, ABDULLAH, *Damascus.*—

422 CURABY, MEHMED, *Damascus.*—Silk and cotton sheeting.

423 KADREEL, HAM, *Damascus.*—Cotton and silk towelling.

424 KASSIM, HAJI, *Damascus.*—Woollen socks.

425 DEVARER, MEHMED, *Damascus.* — Wool yarn.

426 ABDOULLAH, *Damascus.*—Cotton and silk stuff.

427 HAMAS, MUSTAPHA, *Damascus.* — Wool yarn.

428 ABDULKADER, SEID, *Damascus.*—Silk and cotton towelling.

429 GOVERNOR OF ISLIMIA.—Felts.

430 EDDÉ, M. *Lebanon.*—Cotton and silk stuffs.

431 HABIKA, J. *Lebanon.*—Cotton and silk stuffs.

432 NACASH, H. *Lebanon.*—Shawl pattern cloth.
433 HALEBI, J. *Lebanon.*—Printed stuff.
434 KERAM, H. *Lebanon.*—Mixed cloth.
435 RESHID, MICHEL, *Lebanon.*—Silk and cotton shirtings.
436 GOVERNOR OF ANGORA.—Woollen hosiery.
437 VALENDIZO, P. *Crete.* — Woollen hosiery.
438 GOVERNOR OF CRETE.—Cotton and silk sheeting.
439 SOFALAKI, *Crete.*—Silk and cotton shirting.
440 GOVERNOR OF CRETE.—Woollen felt.
441 CHERVISH, JANIK, *Broussa.*—Cotton and silk fabrics.
442 ESKISHEHRLI, OVANES, *Broussa.*—Silk and cotton, and wool and silk fabrics.
443 HAYRABET, OVANES, *Broussa.*—Silk and cotton goods.

CLASS 22.

444 GOVERNOR OF BROUSSA.—Carpets.
445 —— OF SOPHIA.—Carpets.
446 DANOTO, ANDON, *Aleppo.* — Prayer carpet.
447 GOVERNOR OF PHILIPPOLI.—Carpets.
448 —— OF HASKIOY.—Carpets.
449 —— OF AKKIA.—Carpets.
450 TURCOMEN, *Akkia.*—Carpets.
451 GOVERNOR OF BELADBECHAN. — Carpets.
452 ISACHAGHA, *Salonica.*—Carpets.
453 GAVRIL, *Salonica.*—Carpets.
454 GOVERNOR OF SMYRNA.—Carpets.
455 —— OF NISH.—Carpets.
456 —— OF ANGORA.—Carpet.
457 MUSTAPHA & MEHMED, *Conia.* — Carpets.
458 GOVERNOR OF VIDIN.—Carpet.
459 GEBRIL, OSMAN, *Damascus.*—Cotton carpet.
460 HAFIF, HAJI, *Saroukhan.*—Carpet.
461 MUSTAPHA, *Saroukhan.*—Carpet.
462 IBRAHIM, EFENDI, *Saroukhan.* — Carpet.
463 SANDOUKJI, ZADÉ, *Saroukhan.* — Carpet.
464 YAVASH, ZADÉ, *Saroukhan.*—Carpet.
465 SHEVKI, EFFENDI, *Saroukhan.* — Carpet.
466 KADIR, OGLOU, *Saroukhan.*—Carpet.
467 KARASHAHIN, *Saroukhan.*—Carpet.
468 KIATIB, ZADÉ, *Saroukhan.*—Carpet.
469 OSTAOGLOU, *Saroukhan*—Carpet.
470 BAKGHVAN, HASSAN, *Saroukhan.*—Carpet.
471 BEKIR, ZADÉ, *Saroukhan.*—Carpet.
472 HUSSEIN, HAJI, *Saroukhan.*—Carpet.
473 ABDOULLAH, ZADÉ, *Saroukhan.* — Carpet.
474 HEKIM, ZADÉ, *Saroukhan.*—Carpet.
475 SARADJ, MUSTAPHA, *Saroukhan.*—Carpet.
476 MUSTAPHA, NEVAHI, *Saroukhan.*—Carpets.
477 BOURDOURLOU, ZADÉ, *Saroukhan.*—Carpets.
478 PAMOUK, ALI ZADÉ, *Saroukhan.*—Carpets.
479 GOVERNOR OF COOLA.—Carpets.
480 —— OF ADANA.—Carpets.
481 —— OF ANGORA.—Carpets.
482 CATHERINA, *Crete.*—Carpets.
483 MARIA, *Crete.*—Woollen batallia.
484 MURAD, HAJI, *Crete.*—Carpet.
485 HALIL, CHAVOUSH, *Crete.*—Carpet.
486 REGEL, *Crete.*—Carpet.
487 CHERVISH, JANIK, *Broussa.* — Embroidered carpets.
488 NOURI, SULEIMAN, *Broussa.*—Carpet.
489 MEHMED, ALI, *Keresson.*—Carpet.
490 SULEIMAN, OOSTA, *Karahissar Sahib.*—Carpets.
491 GOVERNOR OF KARAHISSAR SAHIB.—Carpets.

CLASS 24.

492 MICHAEL, *Constantinople.*—Embroidered gauze and silk stuffs.
493 HIDAYET, EFENDI, *Constantinople.*—Embroidered handkerchiefs, girdles, &c.
494 YAKUDJIOGLOU, KIRKOR, *Constantinople.*—Embroidered caparison.
495 MERIEM, *Constantinople.* — Embroidered gauze, and cambric for gowns.
496 TAKOUHI, *Constantinople.* — Embroidered handkerchief, silk lace, and purse.
497 YANKO, *Constantinople.* — Lace for hair-dress.
498 MUSTAPHA, EFENDI, *Constantinople.*—Embroidered coffee board covers.
499 MERYEM, *Constantinople.* — Embroidered cambric and scarf.
500 OSMAN, HADJI, *Constantinople.* — Silver lace for head-dress.
501 YELDISYAN, RAPAEL, *Constantinople.*—Silk lace.
502 MOOMKESSER, KESSY, *Beyrouth.*—Silk lace, and silk and silver embroidery.
503 BELSOUK-OGLI, *Saïda.*—Cotton lace.
504 BERZELAY, HAÏM, *Saïda.*—Edgings.
505 GOVERNOR OF LATAKIA.—Silk edging.
506 —— OF PHILIPPOLI.—Lace and embroidery.
507 MEHMED, SEID, *Shaeer.*—Fringes.
508 GOVERNOR OF SIROS.—Lace and edging.
509 AVRAM, *Bosnia.*—Macrama, with gold.
510 KALVA, MAHMOUD. *Damascus.*—Cotton and silver embroidery.
511 GOVERNOR OF MAGNESIA.—Edging and lace.
512 —— OF SCUTARI OF ALBANIA.—Silver lace, silk lace, edgings, &c.
513 NASRANI, H. *Lebanon.*—Embroidered articles.
514 MARIAM, *Trebisond.*—Veil for ladies.
515 GOVERNOR OF CRETE.—Silk and gold lace.
516 EMINÉ, *Crete.*—Handkerchiefs.
517 NASLI, *Crete.*—Handkerchiefs.
518 ESKISHÉHIRLI, OVANE, *Broussa.*—Embroidered upholstery and tapestry.
519 OSMAN, BEY, *Broussa.* — Carriage cushions.
520 FATMA, *Broussa.*—Embroidered handkerchiefs.

521 CHERVISH, JANIK, *Broussa.*—Silver door curtain.

CLASS 25.

522 ARTIN, *Merzikan, Amassia.*—Fitchet skin.

523 KRABELIK, ARTIN, *Merzikan, Amassia.*—Fox, wolf, jackal, cat, hare, and goat-skins.

524 AJABI, MOHIR, *Saïda.*—Hair bags.

525 GOVERNOR OF PHILIPPOLI.—Hair-cloth and hair bags.

526 MEHMED, *Djerban.*—Hair sack.

527 GOVERNOR OF DRAMA.—Fur skins.

528 —— OF GUMULJINA.—Fur skins.

529 —— OF CAVALA.—Fur skins.

530 —— OF SHARISHABAN.—Fur skins.

531 —— OF PRENESHTA.—Hair bags.

532 —— OF SMYRNA.—Hair sacks.

533 —— OF NISH.—Fur skins and goat's hair.

534 —— OF ANGORA.—Goat skins and goat's hair.

535 —— OF JANIK.—Fur skins.

536 EGHIA, OOSTA, *Conia.*—Furs and skins.

537 GURGUI, *Bosnia.*—Fur skins.

538 GOVERNOR OF COOLA.—Morocco leather.

539 —— OF MAGNEZIA.—Morocco leather and fur skins.

540 —— OF ADANA.—Fur skins.

541 —— OF LEBANON.—Hair bags and fur skins.

542 —— OF CASTAMOUNI.—Fur skins.

543 BEKIR, *Keresson.*—Fur skins.

CLASS 26.

544 BOUCHAIN, *Constantinople.*—Harness.

545 IMPERIAL TANNERY, *Constantinople.*—Kid-skins, sole leathers, morocco's, and other leathers.

546 MUSTAPHA, *Constantinople.*—Saddle and harness.

547 YOOSOOF, BABA, *Hajikioy, Amassia.*—Stirrups.

548 HASSAN, HAJI, *Hajikioy, Amassia.*—Morocco leathers.

549 AHMED, HAJI, *Merzikan, Amassia.*—Sole leather.

550 MEHMED, AGA, *Carassy.*—Harness.

551 BEKTASHI-OGLI, SALIH, *Balikesser.*—Morocco leather.

552 HAFIZ, HADJ, *Balikesser.*—Morocco and other leather.

553 MUMKESSER, KESSY, *Beyrouth.*—Morocco and other leather; bridle.

554 GOVERNOR OF BEYROUTH.—Harness.

555 HALIL, SEID, *Saïda.*—Silk bridle.

556 GOVERNOR OF LATAKIA.—Red morocco leather.

557 —— OF SOPHIA.—Halter and whips.

558 DANSTO, ANDON, *Aleppo.*—Harness.

559 RESHID, MEHMED, *Aleppo.*—Morocco leather.

560 AHMED, HAJI, *Philippolis.*—Morocco leather.

561 CALFA, SOLIMAN, *Philippolis.*—Morocco leather.

562 CALFA, HASSAN, *Philippolis.*—Morocco leather.

563 GOVERNOR OF PHILIPPOLIS.—Leathers of various kinds.

564 CARA, HUSSEIN, *Djerban.*—Morocco leather.

565 RAHIM, HAJI, *Djerban.*—Morocco leather.

566 MEHMED, *Djerban.*—Morocco leather.

567 GOVERNOR OF TRIPOLI.—Harness.

568 —— OF SHAEER.—Leathers.

569 AHMED, HAJI, *Shaeer.*—Leathers.

570 GOVERNOR OF DRAMA.—Morocco and other leathers.

571 —— OF GUMULJINA.—Morocco and other leathers.

572 —— OF PRENESHTA.—Sole leather.

573 —— OF CARASSOU.—Morocco and other leathers.

574 ALI & SADIK, *Denizli.*—Morocco and sole leathers.

575 HUSSEIN, USTA, *Salonica.*—Morocco leather.

576 GOVERNOR OF SIROS.—Morocco and other leathers.

577 —— OF NISH.—Morocco and other leathers.

578 —— OF CAISERIA.—Morocco leathers.

579 —— OF KUTAHIEH.—Morocco and other leathers

580 —— OF ANDRINOPLE.—Morocco and other leathers.

581 AHMED, HAJI, *Conia.*—Leathers.

582 EUMER, OOSTA, *Conia.*—Saddle.

583 GOVERNOR OF TIRNOVA.—Horse cloths.

584 —— OF NAPLOUSE.—Morocco leather.

585 ABDOULLAH & IBRAHIM, *Bosnia.*—Horse cloth.

586 BAGDADI, ABDOULLAH, *Damascus.*—Silk harness, horse cloth, and saddle.

587 MELLAH, MUSTAPHA, *Damascus.*—Morocco and other leathers.

588 OMAR, HAJI, *Damascus.*—Horse cloth.

589 DJERDJES, *Damascus.*—Morocco leathers.

590 BADRANI, H. *Lebanon.*—Morocco leathers.

591 MUZIA, EFENDI, *Trebisond.*—Pistol holsters.

592 GOVERNOR OF TREBISOND.—Wallet.

593 HASSAN, ABDI, *Castamouni.*—Morocco leather.

594 ALEMDAR, MEHMED, *Castamouni.*—Morocco leather.

595 SELIM, *Crete.*—Pack saddle.

596 MUSTAPHA, *Crete.*—Harness.

597 AZAMAKI, *Crete.*—Bit, stirrups, &c.

598 AGHACO, *Crete.*—Morocco and other leathers.

599 HALIL, AGA, *Broussa.*—Morocco leather.

600 GOVERNOR OF BROUSSA.—Morocco leather.

601 —— OF KARAHISSAR SAHIB.—Horse cloth.

CLASS 27.

602 MARDIROS, HADJI, *Constantinople.*—Embroidered velvet; ladies' dress.

603 JANIKOGLOU, ARTIN, *Constantinople.* —Velvet, cap, head-dress, tassels, and belt.

604 ZOITCHA & MARDRITCHA, *Constantinople.*—Embroidered slippers.

605 GOVERNOR OF JANINA.—Silver embroidered velvet; Albanian cloths.

606 —— OF HAMA.—Silk mashlah, embroidered with gold.

607 YANKO, *Constantinople.* — Groom's livery, and a suit for pumpers.

608 STEPAN, OOSTA, *Balikesser.*—Boots and shoes.

609 KERKES, ILIAS, *Beyrouth.*—Silk shirt.

610 PALOUSE, ALI, *Beyrouth.*—Shoes and slippers.

611 ZEHAB, SELIM, *Beyrouth.*—Shoes.

612 GOVERNOR OF LAZIKIA.—Boots and shoes.

613 TAHARI, *Monastir.*—Woollen cloaks.

614 GOVERNOR OF MONASTIR.—Cotton, and wool and cotton shirts.

615 DANOTO, ANDON, *Aleppo.*—Neckcloths and gown, with gold thread.

616 DIMITRAKI, *Philippoli.*—Caps.

617 DEJUCIJ, MEHMED, *Tripoli.*—Slippers.

618 GOVERNOR OF TRIPOLI.—Boots and shoes.

619 —— OF SHARISHABAN. — Woollen cloak.

620 —— OF CYPRUS.—Boots.

621 COSTANDI, *Salonica.* — Embroidered bath shirts.

622 GOVERNOR OF NISH.—Red shoes, fur caps, and red cap.

623 AHMED, *Mossul.*—Shoes.

624 GOVERNOR OF ANDRINOPLE.—Boots, shoes, and slippers.

625 IBRAHIM, OOSTA, *Conia.*—Woollen cloak and cap.

626 MEHMED, HAJI, *Conia.*—Boots and shoes.

627 AHMED, HAJI, *Conia.*—Boots and shoes.

628 GOVERNOR OF TIRNOVA.—Silk head stall.

629 —— OF NAPLOUSE.—Cloaks (mashlak).

630 SALIH & ABDULLAH, *Bosnia.*—Shoes.

631 GIBRIL, OSMAN, *Damascus.*—Cloaks (silk and silver).

632 BEDIR, MEHMED, *Damascus.*—Cloak.

633 DJEBRI, ABDELGHANI. *Damascus.*—Wooden shoes, inlaid with mother-of-pearl.

634 MEHMED, ARNAOUD, *Damascus.* — Wooden shoes, inlaid with mother-of-pearl.

635 BOURDJIK, *Damascus.*—Boots, shoes, and slippers.

636 GHANOUM, *Damascus.* — Wool and cotton cloak.

637 DJERDJES, *Damascus.*— Silk cloak.

638 GOVERNOR OF SCUTARI OF ALBANIA. Embroidered cloak.

639 SULEYMAN, *Lebanon.* — Woollen cloaks.

640 ARAB, H. J. *Lebanon.*—Shoes.

641 MUSSA, ALI, *Lebanon.*—Shoes.

642 ISKENDER, *Lebanon.*—Shoes.

643 SHAKIR, H. *Lebanon.*—Embroidered cloaks, &c.

644 NASRANI, H. *Lebanon.*—Slippers (silk), and embroidered caps and bags.

645 HATEM, E. *Lebanon.*—Red slippers and shoes.

646 DAKI, E. *Lebanon.*—Slippers.

647 BEROKES, Y. *Lebanon.*—Jackets.

648 GOVERNOR OF MARDIN. — Woollen cloak.

649 MUZIA, EFENDI, *Trebisond.*—Ladies head-dress.

650 SELIM, *Crete.*—Boots.

651 MANOL, *Crete.*—Boots.

652 LAZOGLOU, AHMED, *Broussa.*—Slippers.

653 MEHMED, ALI, *Lazistan.*—Shoes.

654 AHMED, OOSTA, *Karahissar Sahib.*—Sandals, inlaid with silver.

CLASS 28.

655 KOORSHID, HADJI, *Constantinople.*—Paper and pens.

656 KIAMIL, EFENDI, *Constantinople.*—Inkstand, sand-box, &c.

657 BIDAT, EFENDI, *Constantinople.* — Satchel.

658 MUHENDISYAN, *Constantinople.* — Specimen of printing and engraving.

659 BROGHOLIOS, *Constantinople.* — Alabaster inkstand and sand-box, cistern piece, and fan handle.

660 ELMADJI, MEHMED, *Roustchouk.* — Inkstand.

661 MUZIA, EFENDI, *Trebisond.* — Inkstand.

CLASS 30.

662 HUSSEIN, HADJI, *Constantinople.* — Combs and spoons made of tortoise-shell, ebony, coral, rhinoceros horn, ivory, and mother-of-pearls.

663 RIZA, EFENDI, *Constantinople.*—Perfumery, pastils, and ornaments made with perfumed substances.

664 SHABAN, EFENDI, *Constantinople.*—Cocoa inkstand and coffee-cup.

665 FALINO, *Constantinople.*—Table-cover.

666 MELKIZET, ELEAZAR, *Constantinople.* —Table-cover.

667 RIZA, ALI, *Constantinople.* — Embroidered velvet bolster and amulet case.

668 STAVRI, *Constantinople.*—Drawer and stool of mother-of-pearl.

669 OSTA, ABDULLAH, *Constantinople.*—Chest of mother-of-pearls.

670 SAHAK, *Constantinople.* — Chest of mother of-pearls.

671 OSMAN, HADJI, *Constantinople.*—Embroidered prayer carpet.

672 GUMUSHIAN, BEDROS, *Constantinople.* —Backgammon; ivory chess board.

673 FRANER, OOSTA, *Carassy.* — Olive wood chest.

674 ISMAIL, HADJ, *Balikerser.* — Arm bolster.

675 DONATO, ANDON, *Aleppo.* — Table-covers with gold thread, and coffee board cover.

676 HICO, *Philippolis.*—Table-cover.

677 GOVERNOR OF HASKIOY.—Spoons.

678 —— OF COZANDJIK.—Perfumed oils and waters.

679 GOVERNOR OF TRIPOLI.—Silk and silver tassels, fringes, and others.

680 MELIHA, MICHAEL, *Tripoli.*—Hazel chest.

681 GOVERNOR OF ANDRINOPLE.—Cabinet ware, perfumed soaps and oils.

682 ABDULLAH, OOSTA, *Besnia.*—Spoons with inlaid ornaments.

683 AHMED, *Widin.*—Porte cigar.

684 GEBRI, OSMAN, *Damascus.*—Camp stools.

685 MEHMED, NEDGIM, *Damascus.*—Stool inlaid with mother-of-pearls.

686 ABOU, AHMED, *Damascus.*—Stool inlaid with mother-of-pearls.

687 ABDOULLAH, OOSTA, *Damascus.*—Chest.

688 GOVERNOR OF DAMASCUS.—Trunks inlaid with mother-of-pearls.

689 RESHID, ABOUSHEKIR, *Lebanon.*—Ottoman and cushions, &c.

690 SOLIMAN, HADJI, *Trebisond.*—Trunk.

691 GOVERNOR OF CRETE.—Perfumery.

692 NOURI SULEIMAN, *Broussa.*—Stuff for sofa and chairs.

693 ANDON, HAJI, *Broussa.*—Specimen of sofa furniture.

CLASS 31.

694 SOLIMAN, HAFIS, *Constantinople.*—Yellow tin lantern and lamps.

695 HUSSEIN, AGA, *Constantinople.*—Brass mangal and boore.

696 KEVORK, *Constantinople.*—Lanterns.

697 RAMAZAN, OOSTA, *Balath.*—Coffee-mill.

698 GOVERNOR OF LAZIKIA.—Hardware.

699 —— OF KUTAHIEH.—Horseshoe and nails.

700 —— OF JANIK.—Horseshoeand nails.

701 ALI, OOSTA, *Conia.*—Coffee-mill.

702 IBRAHIM, OOSTA, *Bosnia.*—Copper wares.

703 MUSTAPHA, DERVISH, *Bosnia.*—Copper picknick plate.

704 SULEYMAN & NAZIL, *Bosnia.*—Coffee-mills.

705 BEYAZID, EFFENDI, *Damascus.*—Horseshoe.

706 SEMEK, AGHA, *Trebisond.*—Copper wares.

707 MINASOGLOU, BOGHOS, *Trebisond.*—Copper wares.

708 USENGHIGI, H. *Castamouni.*—Copper wares.

709 APOSTOLOS, *Crete.*—Brass lamp.

710 YARDAN, *Broussa.*—Copper plate and cover.

CLASS 32.

712 KHOORSHID, HADJI, *Constantinople.*—Pen-knife and scissors.

713 SADDEDDIN, *Beyrouth.*—Knives.

714 GOVERNOR OF SOPHIA.—Scissors, with inlaid gold ornaments.

715 SEID, USTA, *Haskioy.*—Knives.

716 GOVERNOR OF SMYRNA.—Knives.

717 —— OF KUTAHIEH.—Knives.

718 —— OF TIRNOVA.—Cutlery.

719 HUSSEIN, HAJI, *Bosnia.*—Daggers and knives.

720 GOVERNOR OF BOSNIA.—Knife.

721 SHERIFF, *Vidin.*—Scissors, knife, &c.

722 OSMAN, *Trebisond.*—Daggers.

723 KILLADDEM, *Trebisond.*—Scissors, knives, forks, &c.

724 MUSTAPHA, *Crete.*—Cutlery.

725 BEKIR, OOSTA, *Broussa.*—Scissors, inlaid with gold.

CLASS 33.

726 MELKIZET, ELIAZAR, *Constantinople.*—Silver ewer and basin, rose-water pot, coffee board, cups for sweetmeats, coffee pot and saucers.

727 LUDFY, *Constantinople.*—Silver case.

728 DJIBOOR, TOOBRE, *Beyrouth.*—Gold and silver jewellery.

729 SHOUBIZI, NOOM, *Beyrouth.*—Silver saucer.

730 ARTUR, YORGI, *Vidin.*—Silver articles.

731 FILAP, ARSENIO, *Vidin.*—Silver cigar-case.

732 GOVERNOR OF TARASHJA.—Silver saucer and cigar-case.

733 ARAKEL, *Conia.*—Jewellery.

734 GOVERNOR OF TIRNOVA.—Watch-case.

735 GORO, *Vidin.*—Silver cups.

736 GOURGI, *Vidin.*—Silver cigar box and glass.

737 DAMAR, HALIL, *Damascus.*—Silver saucers.

738 ABOU, NOUKOUL, *Lebanon.*—Silver wares.

739 SOULMOUN, H. *Lebanon.*—Jewellery.

740 HILMI, EFENDI, *Trebisond.*—Silver jewellery.

741 NICOLA, *Trebisond.*—Silver-plate and jewellery.

742 MEHMED, *Crete.*—Silver watch and chain.

CLASS 35.

743 MOORAD, AGA, *Constantinople.*—Pipe bowls.

744 MOHIS, IBRAHIM, *Beyrouth.*—Pipe and narguilé bowls.

745 MEKHOURI, *Beyrouth.*—Cup, and earthenware pot.

746 GOVERNOR OF BEYROUTH.—Potter's earth.

747 MEKHOORY, *Saïda.*—Earthenware.

748 AJABI, MOHIR, *Saïda.*—Wooden thread.

749 GOVERNOR OF LATAKIA.—Earthenware.

750 MEHMED, SEID, *Shaeer.*—Pipe bowls.

751 ELMADJI, MEHMED, *Roustchouk.*—Pipe bowls, &c.

752 MEHMED ALI, USTI, *Roustchouk.*—Earthenware pottery.

753 MUSTAPHA, USTA, *Roustchouk.*—Water filter.

754 GUMBY, TADEY, *Dardanelles.*—Soup plates.

755 AHMED, MUSTAPHA, *Dardanelles.*—Earthenware.

756 GOVERNOR OF KUTAHIEH.—Earthenware.

757 —— OF ANDRINOPLE.—Water bowls, and other china.

758 MEHMED, *Widin.*—Pipe bowls.

759 AHMED, *Widin.*—Pipe bowls.

760 HASSAN, GHAZI, *Widin.*—Narguila bowls.

761 AHMED, OOSTA, *Widin.*—Bowls.

762 ALI, *Widin.*—Water cup.

763 IBRAHIM, *Widin.*—Cup and cover.

764 BEYAZID, EFENDI, *Damascus.*—Bricks.

CLASS 36.

765 BEDROSS, *Constantinople.* — Amber mouthpiece, and pipes.

766 SHABAN, EFENDI, *Constantinople.*—Ebony pipe.

767 JANIK-OGLOU, ARTIN, *Constantinople.*—Tobacco purse.

768 MAZLOOM-OGLOO, KHATCHADOUR, *Constantinople.*—Silk and shawl purses.

769 RIZA, ALI, *Constantinople.*—Amber mouthpiece.

770 AHMED, OOSTA, *Constantinople.*—Narguila, with inlaid gold ornaments.

771 ZOITCHA & MARIDRITCHA, *Constantinople.*—Amber mouthpiece.

772 IZZET, HAJI, *Constantinople.*—Narguile pipes.

773 DEDA, BESMI, *Constantinople.*—Ebony staves.

774 RUSTEM, USTA, *Haskioy.*—Pipes.

775 GOVERNOR OF ANDRINOPLE.—Pipes and narguilas.

776 —— OF BOSNIA.—Narguile bowls.

777 REMLA, MEHMED, *Damascus.*—Cigar mouthpiece, and pipes.

778 ABOU, ALI, *Damascus.*—Pipes.

779 MEHMED, HAJI, *Damascus.* — Narguilas.

780 ABDEKADER, SHAKIR, *Damascus.*—Narguilas.

781 HAJI, AHMET, *Islimia.*—Ornamental pipe.

782 EMIN, BABA, *Islimia.*—Ornamental pipe.

783 ALI, HAJI, *Islimia.* — Ornamental pipe, and cigaret holder.

784 OSMAN, *Trebisond.*—Ornamental pipes, pipe bowls, and cigar holders.

785 RAZVAN, AGA, *Crete.*—Sticks.

786 SULUK, IBRAHIM, *Broussa.*—Jessamine pipes.

787 AHMET, OOSTA, *Karahissar Sahib.*—Crutchet and club, ornamented with silver.

UNITED STATES.

S.E. Court.

CLASS 1.

1 FEUCHTWANGER, J. W.—42, *Cedar-st. New York.*—1000 specimens of American minerals.

2 MEADS, T.—Cabine of minerals from Lake Superior.

3 NEW JERSEY ZINC CO. *Newark, New Jersey.*—Specimens of zinc ores, with their products; pig and bar iron; steel.

3A MOSHEIMER, J. *Nevada Territory.*—Specimens of gold, silver, quicksilver, copper ores, native sulphur and borax.

3B PRECHT, DR. C. *San Francisco.*—Specimens of crystallised gold, and California marble.

CLASS 2.

4 BAGLEY, M. H. *New York City.*—Crystal carbon oil for lamps.

5 PEASE, S. F.—*Buffalo, New York.*—Samples of carbon and oils for lamps and lubrication.

6 HALE, A. *Lyons, New York.*—Essence of peppermint.

7 PARISH, E. (for the College of Pharmacy of Philadelphia), *Pensylvania.*—Native roots and drugs.

CLASS 3.

8 HOWLAND, O. *Utica, New York.*—Samples of cereals, clover, and timothy seed.

9 HICKER, BROS. *New York City.*—Samples of flour.

10 STEBBINS & CO.—*Rochester, New York.*—Flour.

11 ONONDAGA SALT CO. *Syracuse, New York.*—Samples of table and curing salt.

12 GLENCOVE STARCH CO. *New York City.*—Samples of mazena or corn starch.

13 WADDELL, J. *Springfield, Ohio.*—Indian corn in the ear.

CLASS 4.

14 GLENCOVE STARCH CO. *New York City.*—Samples of starch.

CLASS 5.

15 LAWRENCE & WHITE, *Melrose, New York.*—Lock-nut, and ratchet-washer, giving to nuts and bolts the firmness and safety of rivets.

16 HOLMES, J. E. *New York City.*—Model of improved pneumatic despatch.

17 ROGERS' LOCOMOTIVE WORKS, *Patterson, New York.*—Lithographs and photographs of locomotives.

CLASS 6.

18 BREWSTER & CO. *New York City.*—A phaeton and a road waggon.

19 BLANCHAND & BROWN, *Dayton, Ohio.*—Buggy and waggon spokes.

CLASS 7.

19A WHEELER & WILSON, *New York City.*—American sewing machine.

20 SINGER, J. M. *New York City.*—Sewing machine.

21 WILCOX & GIBBS, *New York City.*—Sewing machine.

21A HOWE SEWING MACHINE CO. *New York.*—Sewing machines.

22 GOODWIN, C. R. *Boston, Mass.*—Ma-

chine for sewing leather, soles of boots and shoes, &c.

23 WRIGHT, H. & Co. 55, *Friday-st. E.C.*—Tape braiding and tape sewing machine.

24 SMITH, A. *West Farms, New York.*—Power loom, for weaving tufted piled fabrics; an entire row of tufts (108 or more) are placed by one operation.

24A RICHARDS, W. D. *Boston, Mass.*—Machinery for sole-cutting and heel-trimming of boots and shoes.

CLASS 8.

25 SANFORD & MALLORY, *New York City.*—Flax and fibre dressing machines.

26 WEMPLE, P. H. *Albany, New York.*—Spacing and boring machine.

27 NEAR, C. *New York City.*—Self-registering dynamometer.

28 WORTHINGTON, R. H. & LEE, W. *New York City.*—Duplex pump.

29 PORTER, C. T. *New York City.*—Stationary engine, and governors.

30 ANDREWS, W. D.—Centrifugal pump, and oscillating engines.

31 LEE & LARNED, *New York City.*—Steam fire engine.

32 DENNISON, C. H. *Rhode Island.*—Wilcox's hot air engine.

33 BLAKE, BROS. *Newhaven, Conn.*—Stone-breaking machine.

34 SANBORN, G. H. *Boston, Mass.*—Rope and cord machine.

34A PORTER, C. T. *New York City.*—Horizontal non-condensing engine indicator for high velocities; governors.

35 SANBORN, G. H. *Boston, Mass.*—Iron refrigerator.

36 GORE, J. C. *Jamaica Plains, Mass.*—Belt shifter.

36A REDSTONES & Co. *Indianapolis, Indiana.*—Model of portable engine.

37 SANBORN, G. H. *Boston, Mass.*—Spindle banding machine; rope and cord machine.

38 ECKELL, J. J. *New York City.*—Combination press and compress.

39 STEELE, H. *Jersey City, New Jersey.*—Pumping engine.

40 HANSBROW, *California.*—Pumps.

41 SWEET, S. *New York.*—Newspaper addressing machine.

42 DEGNER, F. O.—Card and job printing press.

43 SANBORN, G. H. *Boston, Mass.*—Gas regulator.

44 HOLLOWAY & SONS.—American clocks.

44A WARKER & EPPENSTEIN, *New York.*—Apparatus for preparing mineral and soda-water.

44B SANBORN, G. H. *Boston, Mass.*—Gas-pipe tongs; machines used in bookbinding.

44C PACKER, H. H. *Boston, Mass.*—Improved ratchet drill.

44D GIBSON, A. G. *Worcester, Mass.*—Improved carriage coupling.

44E DICKINSON, *New York.*—Model of diamond mill dress.

44F WALCOTT.—Samples of button hole cutters. (Agents, WARNER & Co. 108, *Minories, London.*)

CLASS 9.

45 WOOD, W. A. *Hoosir Falls, New York.*—Self raking reaper, combined reaper and mower, and grass mowing machine.

46 RUSSEL & TREMAIN, *Fayetteville, New York.*—Reaping machine—a new mechanical device.

47 KIRBY & OSBORNE, *Auburn, New York.*—Reaper and mower.

48 REDSTONES, BROS. & Co. *Indianapolis, Indiana.*—Mowing and reaping machine.

49 COE, O. *Ozaukee, Wisconsin.*—Cultivating harrow.

50 DANE, J. F. & Co. *Springfield, Ohio.*—Steel ploughs.

51 BATCHELLOR & SONS, *Wallingford, Vermont.*—Hoes, forks, and rakes.

52 DOUGLAS AXE Co. *Mass.*—American cutlery, axes, &c.

53 BLANCHARD & BROWN, *Dayton, Ohio.*—Cotton planter.

54 WENTWORTH & JARVIS, *Burlington, Iowa.*—Automatic farm gate, and windmill water-elevator.

55 LEVI, A. *Beardsley, New York.*—Hay and earth elevator.

56 PRICE, R. *Albany, New York.*—Churn, and mop.

56A PRINDLE, D. R. *East Bethany, New York.*—Model of corn and bean planter.

CLASS 10.

57 DERROM, A.—Model house, specimens of steam carpentry.

58 SCHOLL, J. G. *Port Washington, Wisconsin.*—Model life-boat.

CLASS 11.

58A DERROM, A.—Pontoons, batteaux, &c.

CLASS 12.

58B WARD, W. H. *Auburn, New York.*—Day and night signal telegraph, telegraph steering and signal lanterns, &c.

CLASS 13.

58C DARLING & SCHWARTZ, *Bangor, Maine.*—Specimens of steel scales and rules.

CLASS 16.

59 STEINWAY & SONS, *New York City.*—Grand and square pianos.

60 DUNHAM, J. *New York City.*—Boudoir piano.

61 HULSKAMP, G. H. *New York City.*—Grand piano, with improved sounding board.

62 DECKER, M. *New York.*—Piano and triolodeons.

CLASS 17.

63 YARD, A. A. *New York City.*—Samples of waterproof adhesive plaster and court plaster.

64 MATTHEWS, PROF. C. *New York City.* —Fumigator.

CLASS 29.

64A CHACE, J. H. & Co. *Portland, Maine.* —Map of the State of Maine.

CLASS 36.

64B McDANIEL. — Specimens of natural flowers.

Addenda.

CLASS 2.

77 RHODES, B. M. *Baltimore, M.* — Barrels super-phosphate lime.

CLASS 3.

78 OSWEGO STARCH CO. *Oswego, New York.*—Samples of prepared corn.

CLASS 4.

79 DUTCHER & ELLERY, *New York.*—American hops.

80 OSWEGO STARCH CO. *Oswego, New York.*—Samples of starch.

81 WILKINS & CO. *New York.*—Bristles and hair.

CLASS 5.

82 HOADLEY, J. C. *Lawrence, Mass.*—Model of trucks for locomotives.

83 TRAIN, G. F. *Boston, Mass.*—Model street tramway carriage.

84 REMINGTON, E. & SONS, *Ilion, New York.*—Revolving stereoscope machine.

85 ROWARD, A. H. *Alleghany City, Pa.* Represented by SCHENLEY, E. W. H. 14, *Princes'-gate.*—Car bumper.

86 WARD, W. H. *Auburn, New York.*—Self-centering railway turn-table.

CLASS 7.

87 CROSBY, C. O. *Boston, Mass. and* 55, *Friday-st. London.*—Machines for preparing tape and joint trimmings, and crimped ruff-lings.

CLASS 8.

88 EDGAR, T. W. *Espy, Pennsylvania.*—Washing-machine.

89 PARKER, D. *Canterbury, New Hampshire.*—Washing-machine.

90 FOOTE, A. M. *New York* (WHEELER & WILSON, 139, *Regent-st.*) — Lock umbrella stand.

91 GODDARD, C. S. *New York.*—Mestizo burring picker.

ERICSSON'S caloric engine, PESANT, BROS. London agents.

SICKLES, E. *New York.*—Steam-steering apparatus.

ROSS, J. *Rochester, New York.*—Conical burr stone mills.

CONROY, E. *Boston, Mass.*—Cork-cutting machine.

CLASS 9.

92 McCORMICK, C. H. *Chicago, Illinois.*—Reaping, mowing, and self-raking machine.

93 PRINDLE, D. R. *East Bethany, New York.*—Agricultural caldron.

94 CENTRAL RAILWAY, *New Jersey.*—Horse-power machine.

CLASS 11.

COLT'S PATENT FIRE-ARMS' MANUFACTURING CO. *New Haven, Conn.*—Samples of guns, pistols, powder flasks, shot pouches, &c.

CLASS 16.

95 HULSKAMP, G. H. *New York City.*—Violins of a new improved construction.

CLASS 18.

96 GARDNER, BREWER, & CO. *Boston, Mass.*—Bleached and brown shirtings.

97 MANCHESTER PRINT WORKS, *Manchester, New Hampshire.*—Cotton prints.

CLASS 21.

98 MANCHESTER PRINT WORKS, *Manchester, New Hampshire.*—De laines and woollen hose.

CLASS 23.

99 MANCHESTER PRINT WORKS, *Manchester, N.H.*—De laines and prints.

CLASS 26.

100 KOHNSTAMM, J. *New York.*—Specimens of leather and leather imitations. (Represented by KOHNSTAMM, H. 33 *Dowgate-hill*).

CLASS 28.

101 DEXTER & CO. *New York.*—Books in the Indian language.

HARVEY, G., portfolios :—

101A Grand golfo for keeping prints, &c.
101B Library portograph on wheels.
101C Sutherland portfolio stand.
101D Vitrifolio for drawings.

101F STEVENS, H. 4, *Trafalgar-sq.*—Specimens of American books, photographs, &c.

101G LOW, S. SON & CO. *Ludgate-hill.*—Specimens of American books.

102 GUN & CO. *Strand, London.*—Specimens of American newspapers.

103 JEWETT, M. P. *Poughkeepsie, New York.* — Catalogues of female seminaries, United States.

CLASS 29.

104 TAINTA, S. & CO. *Philadelphia* (HOLMES, HARRISON, & CO. *London*). — Washington map of the United States.

105 BATES, R. *Philadelphia, Pa.*—Appliances for the cure of stammering.

CLASS 30.

106 PICKHARDT, J. F. C. *New York.*—Extension sofa bedstead.

107 BERTRAM, F. M. B. *New York City, and 4, Gower-st. London.*—Two reception easy rocking, or invalid reclining chairs.

CLASS 32.

108 BLACKWELL, W. 9, *Cranbourne-st. London.*—American (tailors') shears.

CLASS 33.

109 WATTS, A. J. *New York.*—Crystal gold, crystal gold foil, and non-adhesive foil.

110 GREEN, W. H. *Meriden, Conn.*—Revolving castor.

111 BROWN, J. A. & Co. *New York* (Represented by BROWN, B. F. 11, *Cullum-st. London*).—Samples of plated lockets.

CLASS 34.

112 HARTELL & LETCHWORTH *Philadelphia.*—Hermetically sealed glass jars for fruits, and specimens of preserved fruits.

---

113 AMERICAN BANK NOTE CO. *New York.*—Specimens of bank note engraving.

URUGUAY.

N.C. Court, No. 3.

1 MALLMAN & Co. *Soriano.*—Wool.

2 PRANGE, FELZ, & Co. *La Colonia.*—Wool.

3 DRABLE, BROS. *La Colonia.*—Wool.

4 WHITE, W. *La Colonia.*—Wool.

5 HORDEÑANA, SENOR.—Roots.

6 CABAL & WILLIAMS, *Salto.*—Hides; tiger and other skins.

7 HORDENANA, SENOR.—Medicinal roots.

8 CUNHA, S.—Earthen figure.

9 CAMARD, M.—Liqueurs.

10 MAUÁ, BARON DE, *Soriano.*—Wool.

11 BASCÓS & Co.—Flooring tiles, &c.

12 DEPARTMENT OF SAN JOSÉ.—Timber, wheat.

13 DEPARTMENT OF SORIANO.—Flooring tiles, coals, &c.

14 DEPARTMENT OF MALDONALDO.—Coloured marbles, wheat.

15 DEPARTMENT OF SALTO.—Timber, cotton, and harness bridle.

16 DEPARTMENT OF PAYSANDÚ.—Timber, wheat.

17 DEPARTMENT OF MINAS.—Lead iron copper, coals.

18 DEPARTMENT OF TACUAREMBO.—Tobacco, yerba mátte, gold and the stone which produces it.

19 MIGÑONE.—Spirit of Olniz, and wine.

20 IVANICO, *Monte Video.*—Wheat.

21 SINISTRE.—Prepared beef.

22 DIAZ & LIMA, *Soriano.*—Wool.

23 OLIDEN, *Monte Video.*—Prepared beef

24 NIN, P. *Monte Video.*—Cured beef.

25 PROUDFOOT, J. *Glasgow.*—Native saddle and trappings; timber.

26 VARIOUS DEPARTMENTS.—Wheat.

27 THE URUGUAYAN MILL.—Flour.

28 DEPARTMENT OF CANELONES.—Barley.

29 DEPARTMENT OF TACUAREMBO. — Flour, wine.

30 MARTIN, A. *Monte Video.*—Bridles, head-stalls.

31 BURZACO & PIÑEYRUA, *Monte Video.*—Salted cow's-hide.

32 BURZACO & PIÑEYRUA, *Monte Video.*—Hung beef, liquified grease, and fat.

33 GRAÑELY, S. *Monte Video.*—Flour wheat.

34 CRAWFORD, H.—Saddle and trappings.

35 RIVOLTA, *Monte Video.*—Saddle and fittings.

VENEZUELA.

North West Court.

[The following have mostly been collected and forwarded by a Committee appointed by the Venezuelan Government, consisting of Senor Lino J. Revenga, Senor Charles Hahn, and Senor Carlos J. Marxen, and having been delayed on the voyage, were too late to be inserted in the first edition of the Catalogue].

1 THE GOVERNMENT.—The arms of Venezuela, in feathers of the natural colours.

2 SCHEBBYE, *Venezuela.* — Bouquet of flowers, in feathers.

3 ——Roses, in feathers.

4 GUADALUPE NOVELL, *Venezuela.*—A coffee tree, in wax.

5 ——Venezuelan fruits, in wax.

5A ——Branch of cotton tree, in wax.

6 DAVIS, F. L. 13, *Blandford-st.*—Stuffed birds.

7 ——A case of butterflies.

8 ——A totuma carved with a knife.

9 CALCANO, E. *Venezuela.*—Totuma carved with masonic emblems.

10 HEMMING, MRS. F. H. 104, *Gloucester-place.*—Three totumas, one painted.

11 MEYER, F. *Hamburgh.*—A hammock of Margarita cotton, made in that island.

12 ULSTRUP, N. P. B. *Venezuela.*—A hammock made by Indians.

13 ——A hammock.

14 THE COMMITTEE.—A Venezuelan handkerchief, worked at Maracaibo.

15 ——A linen handkerchief.

16 CALCANO, E. *Venezuela.*—A Caraccas shirt.

17 THE COMMITTEE.—A table of various woods.

18 AVILA, *Venezuela.*—A box of various woods.

19 ULSTRUP, N. B. P. *Venezuela.*—A box of Maçanilla palm wood.

20 ——A box.

21 ——A picture frame.

22 MACHADO, *Venezuela.* — Unbleached cotton cloth.

23 THE COMMITTEE.—Cotton wicks.

24 ——Cables of fibres from Rio Negro.

25 ——Rope of cocuiza plant, from Maracaibo.

26 MEYER, F. *Hamburgh.*—Cloth from the cocuiza plant.
27 THE COMMITTEE.—Coffee.
27A NADAL, *Venezuela.*—Coffee.
27B MEYER, F. *Hamburgh.*—Coffee.
28 THE COMMITTEE.—Cocoa.
28A MEYER, F. *Hamburgh.*—Cocoa.
29 THE COMMITTEE.—Beans of various kinds.
30 ——Indian corn.
31 ——Dividivi, from Maracaibo.
32 ——Cochineal, from the Caraccas.
33 STURUP, *Venezuela.*—Cebadilla.
34 ——Vanilla.
35 THE COMMITTEE.—Starch, from the root of the yuca plant.
36 ——Tonquin beans.
37 STURUP, *Venezuela.*—Simarruba.
38 THE COMMITTEE.—Sereipa, from Guayana.
39 ——Secua escandinava: an antidote against poisons, and preservative for iron against rust.
40 CONDE, F. *Venezuela.*—Curara: a cure for hæmorrhage, wounds, and ulcers.
41 ——Espino: a cure for hæmorrhage, wounds, and ulcers.
42 GIL, N. A. *Venezuela.*—Indian balsam.
43 STURUP, G. *Venezuela.*—Root of sarsaparilla.
44 ——Extract of sarsaparilla.
45 ESPINAL, M. *Venezuela.*—Pectoral oil, from sesame.
46 ——Sesame seed.
47 CATALAN & Co. *Venezuela.*—Bitters, from Maracaibo.
48 SYERS, BRAACH, & Co. *Venezuela.*—Bitters, from Angostura.
49 WARBURG & Co. 16, *Devonshire-sq.*—Bitters, from Angostura.
50 THE COMMITTEE.—Preserved oranges.
51 ——Preserved quinces.
52 ——Preserved guayaba.
53 ——Preserved guanàbana.
54 ——Preserved quinces.
55 ——Preserved peaches.
56 ——Imitations of English soap.
57 ——Imitations of Spanish soap.
58 ——Candles of stearine, from Anauco.
59 BOLET, DR. *Venezuela.*—Wax.
60 HAHN, C. *Venezuela.*—White wax, from Carapita.
61 ——Yellow wax.
62 THE COMMITTEE.—Vegetable wax, with fruit and leaf of the tree which produces it.
63 ——Sugar, from Guatire.
64 BARNOLA, J. *Venezuela.*—Chocolate, almonds.
65 ——Vanilla.
66 ——Cinnamon.
67 ——Cocoa.
68 MAYO, C. *Venezuela.*—Chocolates.
69 THE COMMITTEE.—Leaf tobacco, from Guanape.
70 ——Leaf tobacco, from Cumanacoa.
71 ——Cigars, from Cumanacoa.
72 ——Cigars, from Caraccas.
73 ——Cigars, from Carapita.
74 ——Cigars, from Turmero.
75 GARRIDO, F. *Venezuela.*—Snuff.
76 ——Snuff.
77 ——Snuff.
78 THE COMMITTEE.—Tacamahaca.
79 ——Resin of Algarrobo.
80 ——Wool, from Coro.
81 HAHN, C. *Venezuela.*—Goat skins, from Coro.
82 ——Deer skins, from Caraccas.
83 RUETE, RÖHL, & Co. *Venezuela.*—Plaintain leaf, from which paper is made.
84 ——Cottons, from the valleys of Aragua.
85 ——Cotton.
86 STOLTERFOHT & Co. *Liverpool.*—Sea Island cotton, from Maracaibo.
87 BAZLEY, T., M.P. *Manchester.*—Sea Island cotton, from Maracaibo.
88 ——Sea Island cotton, from Maracaibo (manufactured).
89 MEYER, F. *Hamburgh.*—Cotton, from Barquisimeto.
90 ——Cotton.
91 ——Cotton, from Puerto Caballo.
92 HAHN, C. *Venezuela.*—Wild cotton, from Upata Guayana.
92A VARGAS, DR. *Venezuela.*—Eighty-three different kinds of wood.
92B RUETE, BÖHL, &c. *La Guayra.*—Thirty-one different kinds of wood.
92C CONDE, F. *Caraccas.*—Fourteen different kinds of wood.
92D TARTARET & Co. *Venezuela.*—Block of red gateado.
93 MARCANO & Co. & DR. BETANCOURT.—Copper ores from Teques.
94 HEMMING, F. H. 104, *Gloucester-place.*—Copper assayed by Johnson and Matthey.
95 HAHN, C.—Iron ore.
96 ROMERO, J. P.—Silver and lead ore from Carupano.
97 RODRIGUEZ, DR.—Silver ore, mine of "Gran Pobre," Carupano.
98 HAHN, C.—Gold from Caratal, Guayana.
99 MEYER, F. *Hamburgh.*—Gold quartz, from Guayana.
100 HAHN, C.—Gold quartz, from Caratal, Canton of Upata, province of Guayana.
101 REVENGA, L. J.—Green marble from Caracas.
102 ——Red marble.
103 ——Rock crystal.
104 ——Petrified vera wood.

By Authority:
PRINTED BY TRUSCOTT, SON, & SIMMONS,
SUFFOLK LANE, CANNON STREET CITY.

[OVER.

CRYSTAL PALACE.

REFRESHMENT DEPARTMENT.

The period of the International Exhibition affords MR. F. STRANGE the opportunity to announce that every description of Refreshment, from the most simple repast to the most *recherché* banquet, can be obtained at the CRYSTAL PALACE, at prices suitable to the means of all classes of visitors.

He would draw particular attention to his

COLD COLLATION,

at which every description of Cold Meat is supplied *ad libitum*, with Salad, Bread, Cheese, &c., at 1s. 6d. per head; with Chicken, Tongue, Ham, &c., 2s. 6d. per head.

A Room will also be set apart for a

HOT DINNER FROM THE JOINT,

with Vegetables, Bread, Cheese, &c., served *ad libitum*, at 2s. per head.

At the Counters in the Palace and Grounds,

LIGHT REFRESHMENTS,

consisting of Tea, Coffee, Chocolate, Ices, Sandwiches, Meat Pies, Ale, Stout, Lemonade, Soda and Seltzer Water, Perry, Cider, Confectionery, Wines, &c., will be served at the most moderate prices.

An elegant suite of

Private Dining Rooms and Large Coffee Room,

the finest in the world, situated in the South Wing of the Building, has been furnished and decorated in the most costly style. These Rooms command an uninterrupted view of the Grounds and Park, and of the unrivalled scenery of Kent and Surrey. Here visitors may be served with all the delicacies of the season prepared by the most celebrated *chefs* from Paris.

Any of the above Rooms may be engaged for Banquets, Wedding Breakfasts, &c., and will be reserved for parties who may give timely notice of their requirements.

F. STRANGE has collected a STOCK OF WINES which cannot be surpassed in England. It comprises the choicest vintages of France, Germany, Spain, and Portugal.

Whilst arranging for the comforts of the wealthy, F. STRANGE has not forgotten the requirements of visitors of more limited means. In the

THIRD CLASS REFRESHMENT ROOMS

the Public may be supplied with everything of the very best quality at the following prices:—

Plate of Meat		6d.	Cup of Tea	3d.
Bread		1	Roll and Butter	2
Cheese		2	Lemonade, Soda Water, Gingerade . per bottle	3
Porter	per quart	4		
Ale and Stout	do.	6		

The PALE ALE for the Refreshment Department is supplied exclusively by Messrs. Bass & Co., the STOUT, PORTER, and MILD ALE by Messrs. Combe, Delafield, & Co., and the DUBLIN STOUT by Messrs. Guinness & Co., whose names are sufficient to guarantee their being of the best quality.

FREDERICK STRANGE,
Purveyor,
CRYSTAL PALACE, SYDENHAM.

[OVER.

MAPPIN & COMPANY,

OPPOSITE THE PANTHEON,

77 & 78, OXFORD STREET, LONDON.

NOTICE!—All Mappin & Company's celebrated Cutlery is marked with their corporate mark, viz.: M Trustworthy.

Officers' Outfits for India & Colonies.

Military Messes supplied with Cutlery & Plate.

CELEBRATED TABLE KNIVES.

With secure Ivory Handles, which cannot become loose in hot water.

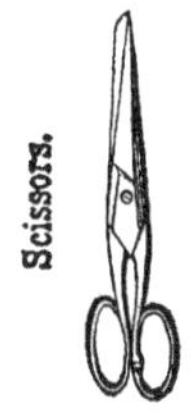

Scissors.

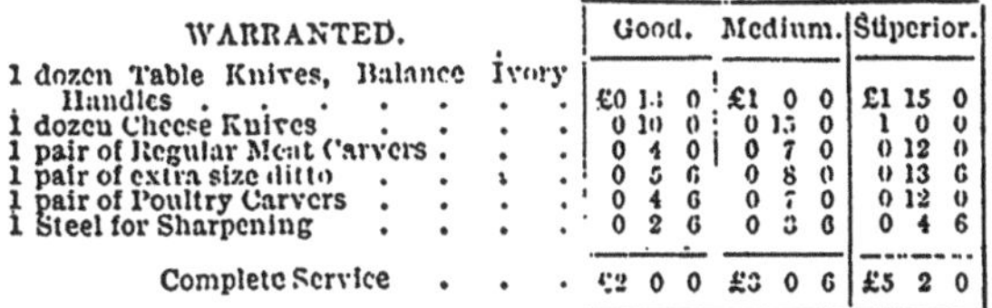

WARRANTED.	Good.	Medium.	Superior.
1 dozen Table Knives, Balance Ivory Handles	£0 13 0	£1 0 0	£1 15 0
1 dozen Cheese Knives	0 10 0	0 15 0	1 0 0
1 pair of Regular Meat Carvers	0 4 0	0 7 0	0 12 0
1 pair of extra size ditto	0 5 6	0 8 0	0 13 6
1 pair of Poultry Carvers	0 4 6	0 7 0	0 12 0
1 Steel for Sharpening	0 2 6	0 3 6	0 4 6
Complete Service	£2 0 0	£3 0 6	£5 2 0

Sporting Knife.

Complete Services of TABLE CUTLERY in Mahogany and Oak Cases for £6 4s. 0d.

Ladies' and Gentlemen's Necessaires, fitted with best cutlery.

RAZORS, SCISSORS, PENKNIVES, in large stock, for exportation, at Sheffield Prices. Razors and Scissors in Russia, Morocco, and other leather cases.

ELECTRO SILVER PLATE.

Guaranteed Quality.

King's Pattern.

Lily Pattern.

MAPPIN & CO.'S SPOONS AND FORKS.

Full Size.	Fiddle.	Thread.	King's.	Lily.
12 Table Forks	£1 7 0	£2 14 0	£2 16 0	£2 14 0
12 Table Spoons	1 7 0	2 14 0	2 16 0	2 14 0
12 Dessert Forks	1 1 0	2 0 0	2 0 0	2 0 0
12 Dessert Spoons	1 1 0	2 0 0	2 0 0	2 0 0
12 Tea Spoons	0 10 6	1 4 0	1 4 0	1 8 0

Each article may be had separately at the same price. The most beautiful and varied assortment to be seen anywhere of tea and dinner services, cruet frames, dish covers, side dishes, waiters, tea-trays, fruit stands, epergnes, &c., the quality excellent, and the prices the lowest possible.

Set of 4 Dish Covers. . £10 0 0	Cruet Frames, 18s. to £10.
" 4 Side Dishes. . 8 10 0	Tea and Coffee Services, complete,
Toast Racks . . . 0 7 0	£1 10s. to £25.

DRESSING BAGS AND CASES.

Celebrated Ladies' Dressing and Travelling Bags, fitted with 14 useful Articles of sterling quality	£1 10 each.
Ditto, ditto, Gentlemen's ditto	2 2 "
Ladies' and Gentlemen's Wood and Leather Dressing Cases, fitted with every article of modern luxury, and of the best quality	1 1 "
Gentlemen's Despatch Boxes, fitted with stationery, wax, &c. &c.	2 2 "

ONLY LONDON ADDRESS—

OPPOSITE THE PANTHEON, OXFORD STREET.

MANUFACTORY—ROYAL CUTLERY WORKS, SHEFFIELD.

Patronized by the late Prince Consort
Great Council Medal Exhibition of 1851
Emperor of Austria
Emperor of the French
Dieu et mon droit
Emperor of Russia
Queen of Spain
1851
And the principal landowners & agriculturalists throughout Europe & the Colonies
Established 1778
View of Leiston Works
Portable Steam Engine
Stone Mill for Grinding Farm Produce
Horizontal Steam Engine
Combined Thrashing Machine & Finishing Dressing Machine
Chaff Cutter
Corn Dressing or Winnowing Machine
Rd. Garrett & Son,
Agricultural Engineers
and Machinists,
Patent Horse Hoe
Taylors Patent Horse Hoe
Economic Seed & Manure Drill
Lever Drag Rake
Portable Steam Engine & Combined Machine at work with arrangement of farm machinery in building.
Leiston Works,
Suffolk, England.
Drill for General Purposes
Suffolk Lever Corn Drill
Illustrated Descriptive Catalogues
In English, French, German, Italian & Spanish,
Sent Postage Free on Application.
Chambers' Patent Manure Distributor.
Horse Power Thrashing Machine Loaded for Travelling
Horse Power Thrashing Machine Ready for Work
Paris 1855

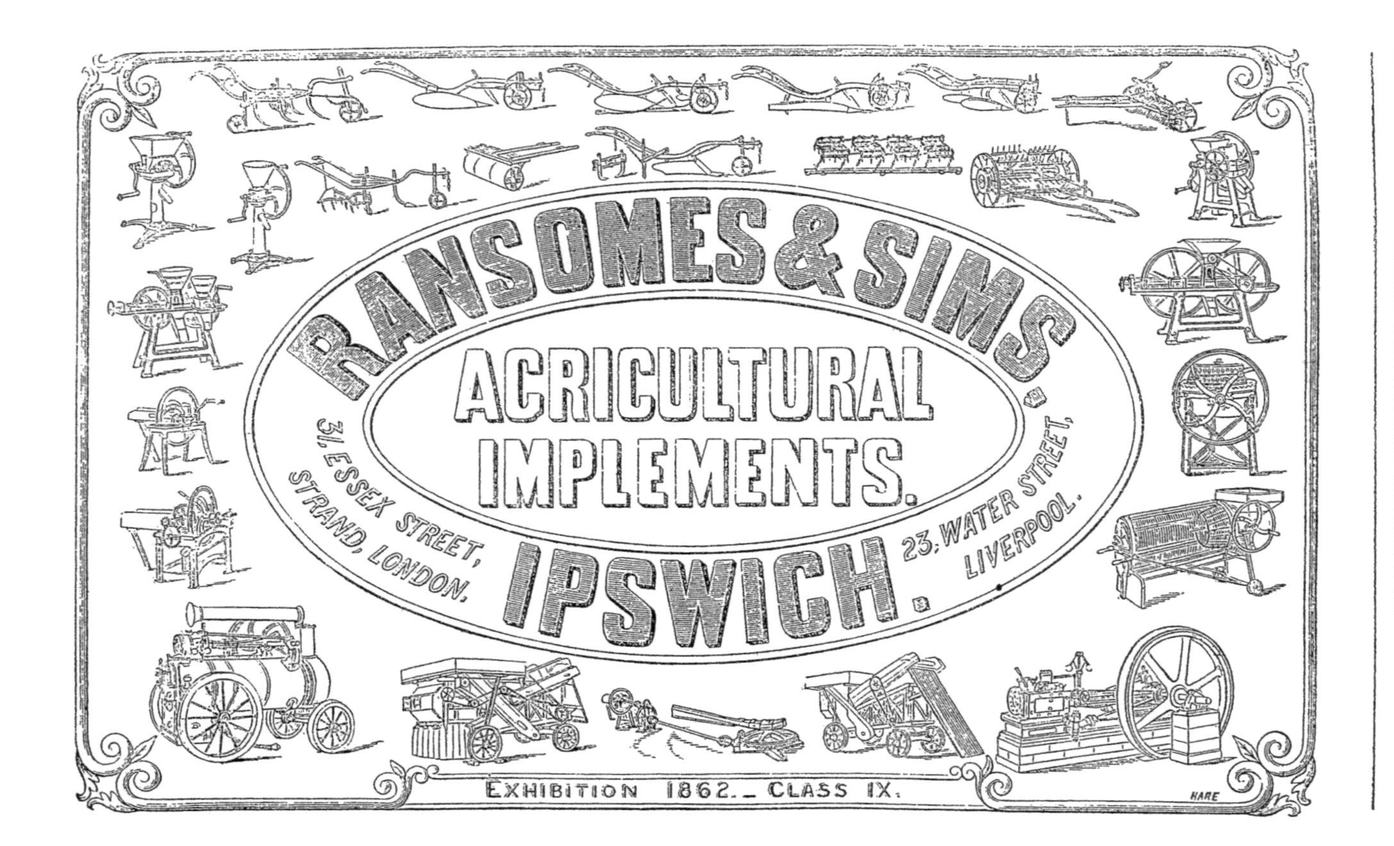
RANSOMES & SIMS.
AGRICULTURAL
IMPLEMENTS.
31, ESSEX STREET,
STRAND, LONDON,
IPSWICH.
23, WATER STREET,
LIVERPOOL.
EXHIBITION 1862.—CLASS IX.
HARE

GRASS MOWING AND REAPING MACHINES.

Class 9, United States Department.

WOOD'S

ROYAL AGRICULTURAL SOCIETY'S FIRST PRIZE

GRASS MOWING MACHINE,

AND WOOD'S PRIZE COMBINED

MOWING AND REAPING MACHINE.

PATRONIZED BY

His late R.H. The Prince Consort.	H.M. The Emperor of the French.	H.M. The King of the Belgians.

Price of Grass Mowing Machines . . . . { For One Horse, £20 0 0; For Two Horses, 22 0 0
Price of Combined Machine, complete in all respects, for Reaping and Mowing . 35 0 0

Awarded the highest Prizes ever offered in England, France, and America, viz.:—

First Prize by the Royal Agricultural Society of England, at Leeds, July, 1861.

First Prize, a Gold Medal and 1000 Francs; also the Grand Gold Medal of Honour by the French Government, at the Great Trial at Vincennes, near Paris, June, 1860.

First Prizes by the United States National Agricultural Society, 1857, 1858, 1859, & 1860.

ALSO, MORE THAN 100 FIRST CLASS COUNTY AND STATE PRIZES.

These Machines are unsurpassed in lightness of draft and quality of work, and will cut any kind of Grass or Grain at the rate of One Acre per hour.

The following Tabular Statement of the Judges of the "Royal Agricultural Society of England," in their Report of the Trial at Leeds, July, 1861 (*vide the Journal of the Society, Vol.* 22, *Part* 2, *No.* 48, *Page* 458), will show the superiority of WOOD'S MACHINES over all others, viz.:—

	Width of Cut.	Speed of Horses in miles per hour.	Quantity Cut in Acres per hour.	Traction Strain on Dynamometer.	Horse Power used continuously.
Cranston's (Wood's) .	4 feet 3 inches	2·523	1·335	132·9	·918
Burgess & Key's . .	4 „ 0 „	2·735	1·326	270·9	1·976
Samuelson's	4 „ 5½ „	2·491	1·333	173·2	1·151

30,000 of these Machines have been made by the Inventor, W. A. WOOD, Esq., Hoosick Falls, N.Y., America, 2200 of which have been sold in Europe during the past three years. Testimonials and further information may be obtained on application to

W. M. CRANSTON,

58, KING WILLIAM STREET, LONDON BRIDGE, LONDON,
AGENT FOR EUROPE.

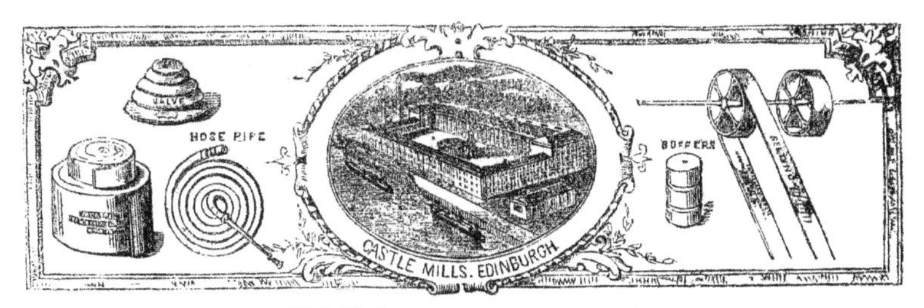

CLASS 8.—WESTERN ANNEXE.

NORTH BRITISH RUBBER COMPANY

(LIMITED),

CASTLE MILLS, EDINBURGH.

Manufacturers of Parmelee's Patent India Rubber Belting, &c. &c.

THE *advantages* which this "*Patent Belting*" possesses over any others are, that when properly used it does not stretch, working equally under water or in warm positions, neither stretching under the former, nor softening or drying hard under the latter circumstances. It does not *Slip*, and from its perfect uniformity of surface, and thickness, its grip on the pulleys is a great gain of power. For *Open Belts* it is unsurpassed, and by numbers of large firms it is used *Crossed* with great success. Its *strength* is one of its chief recommendations. From its being made of Sea Island Cotton of the longest staple, and woven in a peculiar manner, so as to give the greater strength to the warp, when united with layers of India Rubber it becomes a fabric of immense strength, and is capable of doing an amount of work which leather or other fabrics could not stand.

It has quite superseded Leather in Paper Works and other Mills where water is used, or where it is required to work in a damp place. Its superiority over Gutta Percha is in its not being affected by the heat or cold, as it is quite as pliant in the coldest day as in the summer, whereas Gutta Percha becomes stiff when affected by cold.

The India Rubber Belts are easily joined either with lace leather, or, if a heavy belt, by an India Rubber Lap and Screws, made expressly by the Company for the purpose. Endless Belts can be furnished at a small additional cost, which are very convenient for Portable Steam Engines, Thrashing Machines, &c.

Where great speed is required, as for Fans, it is found to work admirably, and is very durable.

For Heavy Belts, say 12 inches to 24 inches wide, and ¾ inch to 1 inch thick, the cost is not more than one-half that of Leather, and its durability much greater. All Belts above 3 inches are cheaper than any other kind in use.

By improvements in the manufacture, the seam or joining is made as strong as any other part of the Belt.

It has been, during the last five years, in use in most of the principal Paper, Flax, and Flour Mills, and Machine Shops, &c., in Great Britain and America.

All *Belts* are stamped with the Company's name, are guaranteed, and may be obtained of all respectable India Rubber Dealers.

The North British Rubber Company also manufacture every article of Vulcanised India Rubber required by Engineers, Mechanics, and Agriculturists, Railway Companies, &c., comprising Hose for conducting Liquids and for Suction, Tubing, Cord, Washers, Valves, Pump Buckets, Piston Packing, Manhole Gaskets, Deckle Straps, and Breast Aprons, for Paper-makers, Wheel Tires for Trucks, Gas Bags, Enamelled Cloth for Carriages, and Sheet Rubber of every thickness, Railway Buffers, Bearing and Draw Springs, &c., and also

INDIA RUBBER OVERSHOES, BOOTS, GOLOSHES, &c.,

For particulars of which see Advertisement under Class 4.

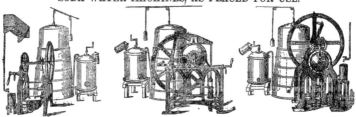

CLAYTON, SHUTTLEWORTH, & CO.,

LINCOLN,

ALSO AT.

78, LOMBARD STREET, LONDON;

125, WEISZGÄRBER, VIENNA;

AND

GEGENÜBER DEM BAHNHOF, PESTH;

MANUFACTURERS OF PORTABLE AND FIXED STEAM-ENGINES;

STEAM CULTIVATING MACHINERY;

Improved Portable Combined Thrashing and Winnowing Machines;

STRAW ELEVATORS;

CORN-GRINDING MILLS; FLOUR-DRESSING MACHINES;

CIRCULAR-SAW BENCHES;

Pumping Machinery for Irrigation; Loam and Mortar Mills; &c. &c.

For the extraordinary increase in their business since the Great Exhibition of 1851, CLAYTON, SHUTTLEWORTH, & CO. avail themselves of the present most fitting opportunity of thanking their friends in all parts of the world. They trust they may still be favoured by their future commands, which they will do all in their power to deserve.

The manufacture of Steam-Engines has been developed in a most extraordinary degree, within the last few years, in England; Engines of the class made by CLAYTON, SHUTTLEWORTH, & CO. have taken the most prominent place in this development.

Commenting on Agricultural Machinery, in the Great Exhibition of 1851, it was considered a remarkable circumstance that one Firm (CLAYTON, SHUTTLEWORTH, & CO.) should have made the *enormous* number of 140 Engines in one year; whereas, in the year just passed, more than four times that number has been constructed and sold, and it has still been found necessary to increase facilities of manufacture, to keep pace with the increased demand.

With every facility for Manufacture, Works the most comprehensive, Plant and Tools of the most modern and best construction, and every other necessary means and appliance at command, CLAYTON, SHUTTLEWORTH, & CO. feel confident they will maintain the position they have long held of the first House of the class.

ILLUSTRATED AND PRICED CATALOGUES

With particulars of the new PATENT SMASHING MACHINE, and STEAM CULTIVATING and PLOUGHING MACHINERY, can be obtained at any of the addresses at the head of this page; and at

Classes 8 and 9 of the Machinery Department of the Great Exhibition of 1862.

MACHINERY PACKED FOR EXPORT IN THE MOST PERFECT MANNER, THIS FIRM HAVING HAD THE GREATEST EXPERIENCE IN SUCH MATTERS.

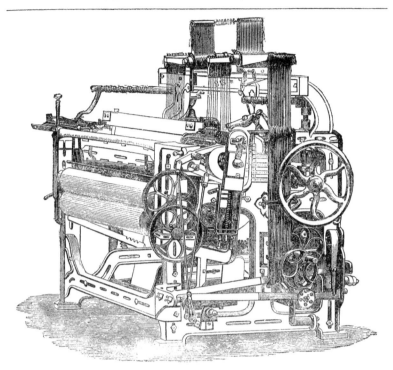

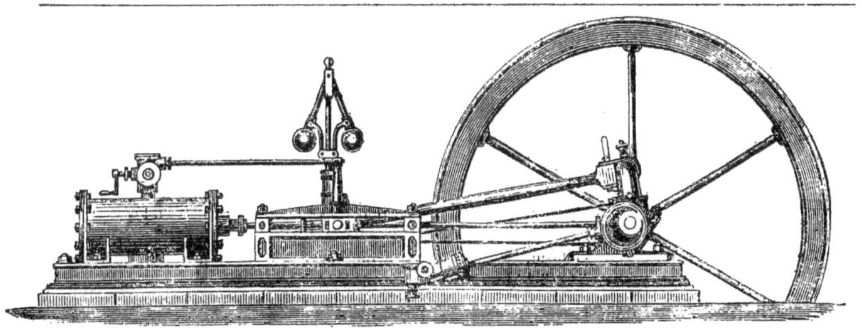

WALTER MACFARLANE & Co.,

ARCHITECTURAL IRONFOUNDERS

AND SANITARY ENGINEERS,

SARACEN FOUNDRY, GLASGOW.

ARCHITECTURAL APPLIANCES.

RAIN WATER PIPES.
RAIN WATER GUTTERS.
RIDGE PLATES.
CRESTING FOR ROOFS.
CRESTING FOR WALLS.
CRESTING FOR BALCONIES.
CRESTING FOR GALLERY FRONTS.
FINIALS AND CROSSES.
BANNERETS AND WEATHER-VANES.
PLUMBERS' CASTINGS.
PLUMBERS' TOOLS.

PUBLIC SANITARY APPLIANCES.

WATER CLOSETS.
ORDURE CLOSETS.
URINALS.
ASH BINS.
WASH-HAND RANGES.
WASH-FOOT RANGES.
BATHS.
BATHING SHADES.
DRINKING FOUNTAINS.
WATER TROUGHS.
BENCH, DESK, & TABLE STANDARDS.

CONTRACTORS BY SPECIAL APPOINTMENT TO HER MAJESTY'S WAR DEPARTMENT.

GEO. BOWER, ENGINEER AND GAS CONTRACTOR, ST. NEOTS, HUNTS.

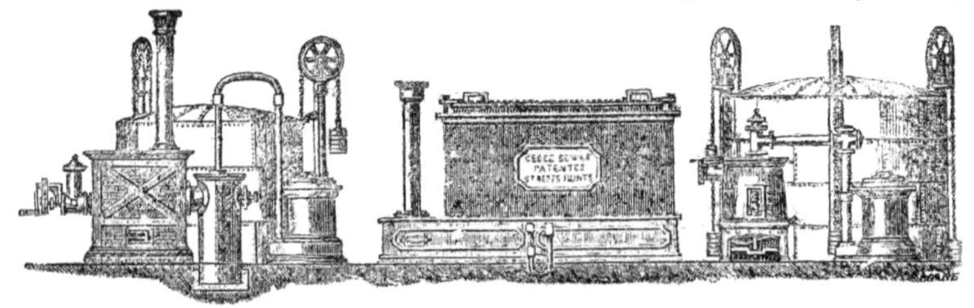

National Gas Apparatus, Patented, 1859.

Combined Gas Purifying Apparatus, to enlarged scale, made both round and rectangular. Patented, 1852.

Vertical Retort Gas Apparatus, Patented, 1860. *Exhibited in Class* 10, *International Exhibition*, 1862.

GAS APPARATUS FOR PRIVATE USE, FACTORIES, VILLAGES & TOWNS.

10 LIGHTS—For small Consumers, requiring no brickwork whatever, perfectly portable, may be placed anywhere, and well adapted for small coal.
20 and up to 80 LIGHTS—With the Retort and purifying portion, portable, easily fixed and managed, and suitable for private Residences.
100 and up to 500 LIGHTS—With Patent Combined Purifier, forming in one the three separate vessels of ordinary construction—simple, and very suitable for Factories, large Mansions, and for Export.

PLANS, SPECIFICATIONS, and ESTIMATES prepared for the Lighting of Villages and Towns.

Gas Works of all sizes, up to 2000 Lights, together with Mains, Syphons, Bends, and Tees, Lamp Columns and Brackets, generally kept in Stock, and for Shipment are delivered at any Port in Great Britain.

Wrought Tube, Fittings, Meters, and every requisite for the entire completion of internal Fittings

References to a selected list of 100 out of a great number of Gas Works of all dimensions, erected by the Advertiser in England and various parts of Europe.

The Combined Purifier in conjunction either with the National or the Vertical Retort Apparatus has obtained a great number of Prizes at various Exhibitions throughout Great Britain.

GEO. BOWER, Engineer and Contractor, St. Neots, Hunts.

N.B.—It is necessary on enquiring for Prices to give the maximum number of Lights, and the longest time they will be burning per day, in Winter.

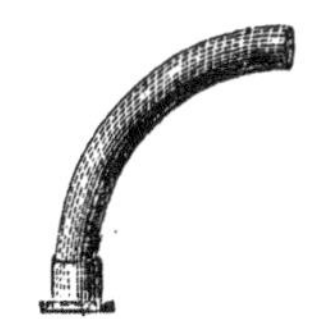

PATENT BITUMENIZED PIPES,

For the Conveyance of Water, Gas, and Drainage; also for Mine Pumps and Air Pipes.

WORKS, BOW, NEAR LONDON.

This important invention for the manufacture of Bitumenized Inoxidable Pipes may confidently be stated to be one of great interest and commercial value, as they possess all the qualities necessary for the conveyance of Water, Gas, Sewage, &c., viz.,—great strength, non-liability to oxidize, lightness, and cheapness. They are guaranteed free from leakage, and resist a pressure of 220 lbs. to the square inch, equal to 506 feet head of water, and can be made equal to any required strength.

They are about one-fourth the weight of iron pipes, and are very considerably cheaper; whilst they are only about one-fourth the price of lead pipes.

They will stand 150 degrees of heat, and will also resist the action of acids and salts, and are not affected by frost or change of temperature; they consequently possess great advantages over all metallic pipes.

The process of manufacture of these pipes consists in causing the material of which they are composed to pass through a reservoir of melted bitumen, after which it is tightly coiled round a mandril to any required thickness, under pressure; a tube of perfect texture is thus formed of great hardness and enormous strength.

The bitumenous composition which unites and covers the pipes, is so well known as an excellent preservative, that it will be unnecessary to demonstrate that when laid underground, the durability of the Patent Pipes may be considered unlimited.

They are made of all sizes, from 2 inches up to 36 inches diameter, in lengths of seven feet, and are connected by socket joints of the same material as the pipes, or with east iron flanges when required.

For gas these pipes are invaluable, the lining effectually resisting the action of gas, and liquid and solid depositations. Wet soils do not oxidize, or affect them injuriously; they are slightly flexible, and therefore *give* in shifting or springy soils, without injuring the joints, and do not sink where iron pipes would, on account of their weight. These pipes are now being used in the Northern Coal-field, near Durham, Newcastle, &c., and pipes are being made up to 2½ inches diameter, for mine pumps, and for the conveyance of *air into pits*, with a simple method of jointing which admits of their being readily moved from place to place as required. The sulphurous waters of mines do not affect them.

The cost of laying these pipes is less than that of iron, and can be done by ordinary workmen, with an entire absence of risk from breakagu.

Although only recently introduced into England, these pipes have been in use on the Continent during the last four years; they have been employed in Paris for the conveyance of gas, and have been ascertained to be as perfect now as when first laid down.

Satisfactory references to engineers and others of the highest standing can be given.

Orders to be addressed to

J. L. NORTON, SOLE AGENT TO THE MANUFACTURERS,

No. 38, La Belle Sauvage Yard, Ludgate Hill, London, E.C.

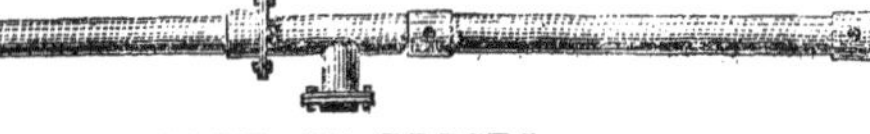

SCALE OF PRICES.

Diameter......	2in.	3in.	4in.	5in.	6in.	7in.	8in.	9in.	10in.	12in.	16in.	20in.	24in.	30in.	36in.
Price ℔ foot with collars	4d.	6d.	8d.	11d.	1/2	1/5	1/3	2/0	2/4	2/8	4/2	6/0	8/6	11/0	13/0
Weight ℔ foot	2lb	3lb	4lb	5½lb	7lb	9lb	12lb	14lb	16lb	20lb	26lb	34lb	45lb	67lb	88lb

Cement for Joining 1d. per lb.

Terms.—Three Months' Bill, or 2½ per Cent. for Cash.

In Class 10—Morewood & Co. Exhibit Galvanized Iron Building Materials.
In Class 31—Morewood & Co. Exhibit General Galvanized Iron Goods.
In Class 1, Open Court—A Shed composed of Continuous Roofing Sheets.

MOREWOOD & Co.

(LATE MOREWOOD & ROGERS),

DOWGATE DOCK, UPPER THAMES STREET, LONDON;

AND LION WORKS, BIRMINGHAM.

PATENT GALVANIZED TINNED IRON, GALVANIZED IRON,

Plain, Corrugated, Curved, or in Tiles. Also

BLACK OR PAINTED CORRUGATED IRON.

Galvanized or Black Cast Gutters, Pipes, &c.

Also Galvanized Water and Gas Tubing, Stamped and Moulded Gutters, Wire, Wire Netting; Hooping, Nails, Rivets, Pails, Bowls, Skips, &c.

Estimates given for every description of Roofs and Buildings, constructed either of Galvanized or Black Iron. Also

MOREWOOD'S PATENT CONTINUOUS ROOFING SHEETS

Of Galvanized Iron, in Lengths of 50 to 500 Feet and upwards, by 2 Feet wide.

CHEAPER THAN FELT WHEN FIXED COMPLETE. For price, instructions as to fixing, &c., apply to the Patentees,

MOREWOOD & CO.,

Dowgate Dock, Upper Thames Street, London, E.C.; and Lion Works, Birmingham Heath, Birmingham.

In Class 1, Open Court, may be seen a Shed, erected by MOREWOOD & Co., and roofed with the Continuous Sheets above mentioned. The sides are also partially made of this metal, in a corrugated form. The Continuous Metal is fixed as easily as canvas, and is confidently recommended as the cheapest building material in use.

CROGGON'S PATENT FELTS,

AS SUPPLIED LARGELY BY THEM TO

HER MAJESTY'S GOVERNMENT.

ASPHALTE ROOFING FELT,

Is perfectly impervious to rain, snow, and frost, and a non-conductor of heat. Price 1d. per square foot, or 8d. per yard of 32 inches wide. In rolls of 25 yards long.

INODOROUS FELT,

Saturated with waterproof materials, free from offensive smell; is very suitable for lining damp walls, &c. Price 1d. per square foot, or 8d. per yard of 32 inches wide. In rolls of 15 yards long.

SHEATHING FELT,

For covering ships' bottoms; insures strength, durability, and an even surface; preserves the timber and caulking of vessels, and prevents the ravages of the worm.

BLACK, price 2d. per sheet—32 by 20 inches. | BROWN, price 2½d. per sheet—32 by 20 inches.

Or in long lengths, at the same ratio.

NON-CONDUCTING FELT,

Is formed entirely of hair, and is used for covering boilers and steam pipes, effecting a saving in fuel of fully 25 per cent., and deadening sound.

SCALE OF PRICES.

	Weight.	Size.				Weight.	Size.	
No. 1.	16 oz. . .	32 in. by 20 in. . .	7d. per sheet.		No. 4.	40 oz. . .	32 in. by 20 in. . .	1s. 1d. per sheet.
" 2.	24 " . .	"	. . 9d. "		" 5.	48 oz. . .	"	. . 1s. 3d. "
" 3.	32 " . .	"	. . 11d. "					

2, DOWGATE HILL, LONDON; 2, GOREE PIAZZAS, LIVERPOOL.

SEWING AND EMBROIDERING MACHINES.

NEWTON WILSON & CO.,

144, HIGH HOLBORN, LONDON, W.C.;

PARIS HOUSE—6, FAUBOURG MONTMARTRE;

PATENTEES AND MANUFACTURERS OF SEWING MACHINES,

AND AGENTS FOR THE "GROVER & BAKER" SEWING MACHINE COMPANY, OF AMERICA.

These machines are both for domestic and manufacturing purposes, and include all the stitches yet produced by the Sewing Machine.

The family machines, unlike all others, do both plain and fancy work, including running, hemming, felling, tucking, gathering, binding, cording, quilting, braiding, and embroidering. They are simple, and light to operate—noiseless in action, perfect in mechanism, and not liable to derangement.

Magnificent specimens of the work of these machines, including almost every garment for man, woman, or child, will be found at the Stand, Class 7B; and patterns of the work, in the different stitches, may be obtained gratis, or will be forwarded post free on application.

The machines in the manufacturing department include special adaptations for different trade purposes—as tailoring, boot-making, harness work, stitching shoe-soles, &c., &c. The chief features of these machines are wide range of application, strength, durability, simplicity, speed, and quietness. Special machines for peculiar embroidery, and stitching two lines of sewing at one time; for herring-bone stitching; for button-hole making; and, finally, for darning stockings.

Every machine is guaranteed, and the possibility of mistake is provided against by the following arrangement, viz.:—that after a month's use of any machine, it may be exchanged, without charge, for any other.

Illustrated Catalogues of the whole of NEWTON WILSON & CO.'s manufacture, in English, French or German, forwarded by post, or obtained free at the Stand in the Process Court.

NEWTON WILSON & Co.'s PATENT CARPET SWEEPERS,

Taking up all the dust and making none, requiring no damping, no kneeling, and no subsequent dusting! Applicable to all kinds of carpets. Prices—12s. 6d., 15s., and 18s.

THE UNRIVALLED

LOCK-STITCH SEWING MACHINE,

MANUFACTURED BY THE

WHEELER & WILSON MANUFACTURING COMPANY,

WITH ALL THE RECENT IMPROVEMENTS,

CRYSTAL CLOTH PRESSER, BINDER, CORDER, HEMMER, &c.,

Will Stitch, Hem, Fell, Bind, Cord, Gather, Embroider, &c.

Is Simple in Design;

Not liable to get out of order;

Elegant in appearance;

Strong & Durable;

The work will not Ravel;

Will make from 500 to 2000 Stitches per Minute;

And proficiency is readily acquired.

They will stitch, with great beauty and regularity, the finest Muslin, or the thickest cloth, and are used extensively by all classes, from the Princess to the Artizan, and also by Dress Makers, Milliners, Mantle Makers, Tailors, Seamstresses; and in the manufacture of Shirts, Collars, Corsets, Lawn Goods, Hats and Caps, Umbrellas, Parasols, Silk Goods, and almost every other description of work which can be done by means of hand sewing.

INSTRUCTIONS GRATIS TO EVERY PURCHASER.

ILLUSTRATED PROSPECTUSES GRATIS AND POST FREE.

Offices and Show Rooms—139, Regent Street, London, W.

SAM^{L} ALLSOPP & SONS

BREWERS,

BURTON-ON-TRENT,

BEG TO ANNOUNCE that the CONTRACTORS for the REFRESHMENT DEPARTMENT of the INTERNATIONAL EXHIBITION obtain their supplies of PALE ALE from Messrs. SAMUEL ALLSOPP & SONS' Brewery, at Burton-on-Trent.

MESSRS. ALLSOPP TAKE THIS OPPORTUNITY OF STATING THAT THEIR LABEL FOR BOTTLED ALE IS AN EXACT FAC-SIMILE OF THE ANNEXED:—

Messrs. ALLSOPP & SONS' PALE MILD and STRONG ALES are supplied (Carriage Free) from .their STORES at

LONDON—61 King William Street
LIVERPOOL—Cook Street
MANCHESTER—Ducie Place
LEEDS—Commercial Buildings
BIRMINGHAM—Upper Temple St.
DERBY—London Road
LEICESTER—Granby Street
BATH—5 Edgar Buildings
DUDLEY—Burnt Tree
CHESTERFIELD—Low Pavement
NOTTINGHAM—Maypole Yard
WOLVERHAMPTON—Exchange St.
WORCESTER—The Cross
STOKE-UPON-TRENT—Wharf St.
SOUTH WALES—13 King St. Bristol
DUBLIN—1 Crampton Quay
CORK—25 Cook Street
GLASGOW—115 St. Vincent Street
EDINBURGH—{ 47 North Bridge / 11 Union Street Lane
PARIS—279 Rue St. Honoré

ON THE SAME TERMS AS FROM THE BREWERY AT BURTON-ON-TRENT.

PHŒNIX
FIRE ASSURANCE COMPANY,
LOMBARD STREET AND CHARING CROSS,
LONDON.

ESTABLISHED IN 1782.

For Insuring every kind of Property, at Home or Abroad, against Loss or Damage by Fire

TRUSTEES AND DIRECTORS.

DECIMUS BURTON, Esq.
TRAVERS BUXTON, Esq.
OCTAVIUS EDWARD COOPE, Esq.
WILLIAM COTTON, Esq.
JOHN DAVIS, Esq.
GEORGE ARTHUR FULLER, Esq.
CHARLES EMANUEL GOODHART, Esq.
JAMES ALEXANDER GORDON, Esq.
EDWARD HAWKINS, Jun., Esq.
KIRKMAN D. HODGSON, Esq., M.P.
WILLIAM JAMES LANCASTER, Esq.
JOHN DORRIEN MAGENS, Esq.
JOHN TIMOTHY OXLEY, Esq.
BENJAMIN SHAW, Esq.
WILLIAM JAMES THOMPSON, Esq.
HENRY HEYMAN TOULMIN, Esq.
MATTHEW WHITING, Esq.

AUDITORS.

JOHN HODGSON, Esq. | PETER MARTINEAU, Esq. | JOSEPH SAMUEL LESCHER, Esq.

GEORGE WILLIAM LOVELL *Secretary.* | JOHN J. BROOMFIELD, *Assistant Secretary.*

Solicitors—Messrs. DAWES & SONS, Angel Court.

The Phœnix Office was established in January, 1782, and has a numerous and wealthy Proprietary, every member of which is responsible in his whole fortune for the engagements of the Company, in addition to a large Capital at all times kept invested in Government and other Securities.

The promptitude and cheerfulness with which the most important liabilities have always been met by this Company are well-known; and the importance of its relations with the public may be estimated by the fact, that since its establishment it has paid more than Eight Millions Sterling in discharge of Claims for Losses by Fire.

PELICAN
LIFE INSURANCE OFFICE
ESTABLISHED IN 1797

No. 70, LOMBARD STREET. E.C., AND 57, CHARING CROSS, S.W.

DIRECTORS.

OCTAVIUS E. COOPE, Esq.
WILLIAM COTTON, Esq., D.C.L., F.R.S.
JOHN DAVIS, Esq.
JAMES A. GORDON, Esq., M.D., F.R.S.
EDWARD HAWKINS, Jun., Esq.
KIRKMAN D. HODGSON, Esq., M.P.
HENRY LANCELOT HOLLAND, Esq.
WILLIAM JAMES LANCASTER, Esq.
JOHN LUBBOCK, Esq., F.R.S.
BENJAMIN SHAW, Esq.
MATTHEW WHITING, Esq.
MARMADUKE WYVILL, Jun., Esq., M.P.

ROBERT TUCKER *Secretary and Actuary.*

EXAMPLES of *the amount of Bonus awarded at the recent division of profits* to Policies of £1000 each, effected for the whole term of life at the undermentioned ages :—

Age when Assured.	Duration of Policy.	Bonus in Cash.	Bonus in Reversion.
20	7 years	£ 29 7 0	£ 66 0 0
	14 years	36 2 0	73 10 0
	21 years	44 8 0	82 0 0
40	7 years	£ 49 13 6	£ 84 10 0
	14 years	61 2 0	95 10 0
	21 years	75 2 6	108 0 0
60	7 years	£ 95 4 6	£127 10 0
	14 years	117 2 6	144 10 0
	21 years	111 1 0	165 10 0

For Prospectuses, Forms of Proposal, &c., apply at the Offices as above, or to any of the Company's Agents.

LONDON & COUNTY BANKING COMPANY.

ESTABLISHED 1836.

Subscribed Capital, £1,500,000, in 30,000 Shares of £50 each.
Paid-up Capital, £500,000.—Reserve Fund, £125,000.

DIRECTORS.

THOS. TYRINGHAM BERNARD, Esq., M.P.
PHILIP PATTON BLYTH, Esq.
JOHN WILLIAM BURMESTER, Esq.
COLES CHILD, Esq.
WILLIAM CORY, Esq.
JOHN FLEMING, Esq.
EDWARD HUGGINS, Esq.
WILLIAM CHAMPION JONES, Esq.
JAMES LAMING, Esq.
WILLIAM LEE, Esq., M.P.
WILLIAM NICOL, Esq., M.P.

GENERAL MANAGER.
WILLIAM MCKEWAN, Esq.

HEAD OFFICE, 21, LOMBARD STREET.

METROPOLITAN BRANCHES.

BAYSWATER . . . 23, Westbourne Grove.
BOROUGH . . . . . 201, High Street, Borough.
COVENT GARDEN 27, James Street.
HANOVER SQUARE 21, Hanover Square.
ISLINGTON . . . . 19, Islington High Street.
KENSINGTON . . . High Street.
KNIGHTSBRIDGE Albert Gate.
OXFORD STREET 441, Oxford Street.
PADDINGTON . 6, Berkeley Place, Edgware Road.
SHOREDITCH . . 187, Shoreditch.

COUNTRY BRANCHES.

Abingdon and Ilsley.
Arundel, Little Hampton, Steyning, and Worthing.
Ashford and Hythe.
Aylesbury, Great Berkhampstead, and Thame.
Banbury.
Basingstoke.
Battle and Robertsbridge.
Bedford.
Bishop's Stortford.
Braintree and Coggeshall.
Brentford and Hounslow.
Brentwood.
Brighton.
Buckingham and Stony Stratford.
Cambridge.
Canterbury.
Chatham, Rochester, and Sheerness.
Chelmsford.
Chichester and Bognor.
Colchester and Sudbury.
Cranbrook and Hawkhurst.
Croydon.
Dartford.
Deptford.
Dorking and Leatherhead.
Dover.
Epsom.
Farnham.
Gravesend.
Greenwich.
Halstead.
Hastings and St. Leonards.
Hertford.
High Wycombe.
Hitchin.
Horsham.
Hungerford.
Huntingdon, St. Ives, and St. Neots.
Kingston-on-Thames.
Leighton Buzzard.
Lewes, Hailsham, and Newhaven.
Luton and Dunstable.
Maidstone and Wrotham.
Maldon.
Newbury.
Newport, Isle of Wight.
Oxford and Witney.
Petersfield.
Petworth, Midhurst, and Pulborough.
Reading & Henley on Thames.
Reigate and Redhill.
Richmond.
Romford.
Rye.
Saffron Walden & Haverhill.
Sandwich.
Sevenoaks.
Sittingbourne.
St. Albans and Hemel Hempsted.
Tenterden.
Tonbridge.
Tonbridge Wells & Ticehurst.
Uxbridge, Rickmansworth, and Watford.
Wallingford and Didcot.
Wantage.
Ware.
Winchester.
Windsor.
Woolwich.

The LONDON AND COUNTY BANK opens—

DRAWING ACCOUNTS with Commercial Houses and Private Individuals, either upon the plan usually adopted by other Bankers, or by charging a small Commission to those persons to whom it may not be convenient to sustain an agreed Permanent Balance.

DEPOSIT ACCOUNTS.—Deposit Receipts are issued for sums of Money placed upon these Accounts, and Interest is allowed for such periods and at such rates as may be agreed upon, reference being had to the state of the Money Market.

CIRCULAR NOTES AND LETTERS OF CREDIT are issued, payable in the principal Cities and Towns of the Continent, in Australia, Canada, India, and China, the United States, and elsewhere.

Great facilities are also afforded to the Customers of the Bank for the receipt of Money from the Towns where the Company has Branches.

The Officers of the Bank are bound not to disclose the transactions of any of its Customers.

By Order of the Directors,
WM. McKEWAN, GENERAL MANAGER.

PRICE'S PATENT CANDLE COMPANY

(LIMITED),

BEG TO CALL SPECIAL ATTENTION TO A FEW OF THEIR MANUFACTURES.

PATENT BELMONTINE CANDLES,

Which surpass Spermaceti in transparency and beauty and burn much longer.

"PRICE'S CANDLES WHICH WILL NOT DROP GREASE"

Are sold in boxes at 1s. and 3s. each. When used as Chamber Candles, prevent damage to dresses and carpets. They should be burnt in the imitation bronze candlesticks, made expressly for them.

GLYCERINE SOAP,

In packets, 1s. each. This is recommended as the best of Toilet Soaps for the Skin, both in cold and hot weather; it contains Price's Patent Glycerine, stirred in after the Soap is made.

SHERWOODOLE,

For removing Grease Stains; like Benzine, but with less smell.

RANGOON OIL,

For Rifles (in Shilling Bottles), and for Artillery, recommended and adopted by the War Office Authorities, "for the preservation of small arms, and articles made of metal, from oxidation."

GISHURST COMPOUND,

For Winter Dressing Fruit Trees and destroying Plant Insects and Mildew.

PATENT SPINDLE AND HEAVY MACHINERY OILS,

Now used in many very great mills in substitution for Sperm Oil. The Mill proprietors report a great saving from its use.

GLYCERINE.

Price's Patent Distilled Glycerine is now well known throughout Europe and America. Every bottle supplied by the Company is tested in their Laboratory, and has its stopper secured by a metallic capsule, lettered "PRICE'S PATENT." Glycerine, besides its important medicinal uses, cures Chapped Hands and Mosquito Bites.

BELMONT, VAUXHALL, LONDON, S.

Lady's Crinoline Riding Boot.

Lady's Crinoline Elastic Ankle Boot.

Gentleman's Crinoline Riding Boot.

CHARLES GODFREY HALL & CO.,

Boot Makers to H.R.H. the Prince Consort and H.R.H. the Prince of Wales

PATENTEES OF THE CRINOLINE BOOTS,

89, REGENT STREET, W.,

SUCCESSORS TO RICHARD HALL,

(The inventor of the Pannus Corium, and of the improved Elastic Enamelled Cloth),

Have recently obtained Her Majesty's Royal Letters Patent for making boots, shoes, and gaiters of various descriptions of *hair*, which will be found to supersede, to a great extent, the use of leather in the upper parts or tops of boots. The Crinoline Boots are made for ladies, gentlemen, and children; they allow freedom of action and circulation to all the tender organs of the foot and ankle, are light, durable and elegant; they retain their position equal to any boot, are applicable to every shape or fashion, and may truly be called *ventilating* boots. The Crinoline Riding Boots are admirably adapted for ladies as well as gentlemen, and especially suited for military gentlemen in the Colonies.

Patented in France and Belgium.—Exhibitors in Class 27 of the International Exhibition.

JONAS BROOK  AND BROTHERS,

MELTHAM MILLS, NEAR HUDDERSFIELD.

20, Cannon Street, West . **London.**
2, Port Street . . . . **Manchester.**
25, Cochrane Street . . **Glasgow.**
76, Castle Street . . . **Bristol.**

117, Boulevard de Sebastopol. **Paris.**
32, Vesey Street . . . . . **New York.**
4, Custom House Square . . **Montreal.**

MANUFACTURERS OF

SEWING COTTON, CROCHET, AND EMBROIDERING.

ONLY PRIZE AT THE LONDON EXHIBITION, 1851.

BROOK'S

Patent Glacé,

SATIN FINISH.

ONLY FIRST-CLASS PRIZE AT THE PARIS EXHIBITION 1855.

BROOK'S

Patent Six Cord

SOFT FINISH.

Brook's Patent Glacé Thread, in White, Black, and Colors.

The extraordinary strength, smoothness, and durability, obtained by this invention, have secured for it great popularity, and it is consequently much imitated in inferior qualities. This Cotton is always labelled BROOK'S PATENT GLACÉ THREAD, and without their name and crest (a GOAT'S HEAD), the words "Glacé" or "Patent Glacé" do not denote that it is of their manufacture.

BROOK'S PATENT NINE AND SIX-CORDS will be found of very superior quality, and is strongly recommended wherever a SOFT COTTON is preferred.

In consequence of the many piracies of trade marks, Messrs. J. BROOK and BROTHERS have determined to rely solely on their *Name and Crest*, a "GOAT'S HEAD," as their Trade Mark.

All BROOK'S Cottons are *guaranteed* to measure the lengths indicated on their TICKETS, in 100, 200, 250, 300, and 500 yards.

SPECIALLY PREPARED FOR LADIES' WORKBOXES,

BROOK'S PATENT GLACÉ, 250 yards; and NINE-CORD, 200 and 300 yards.

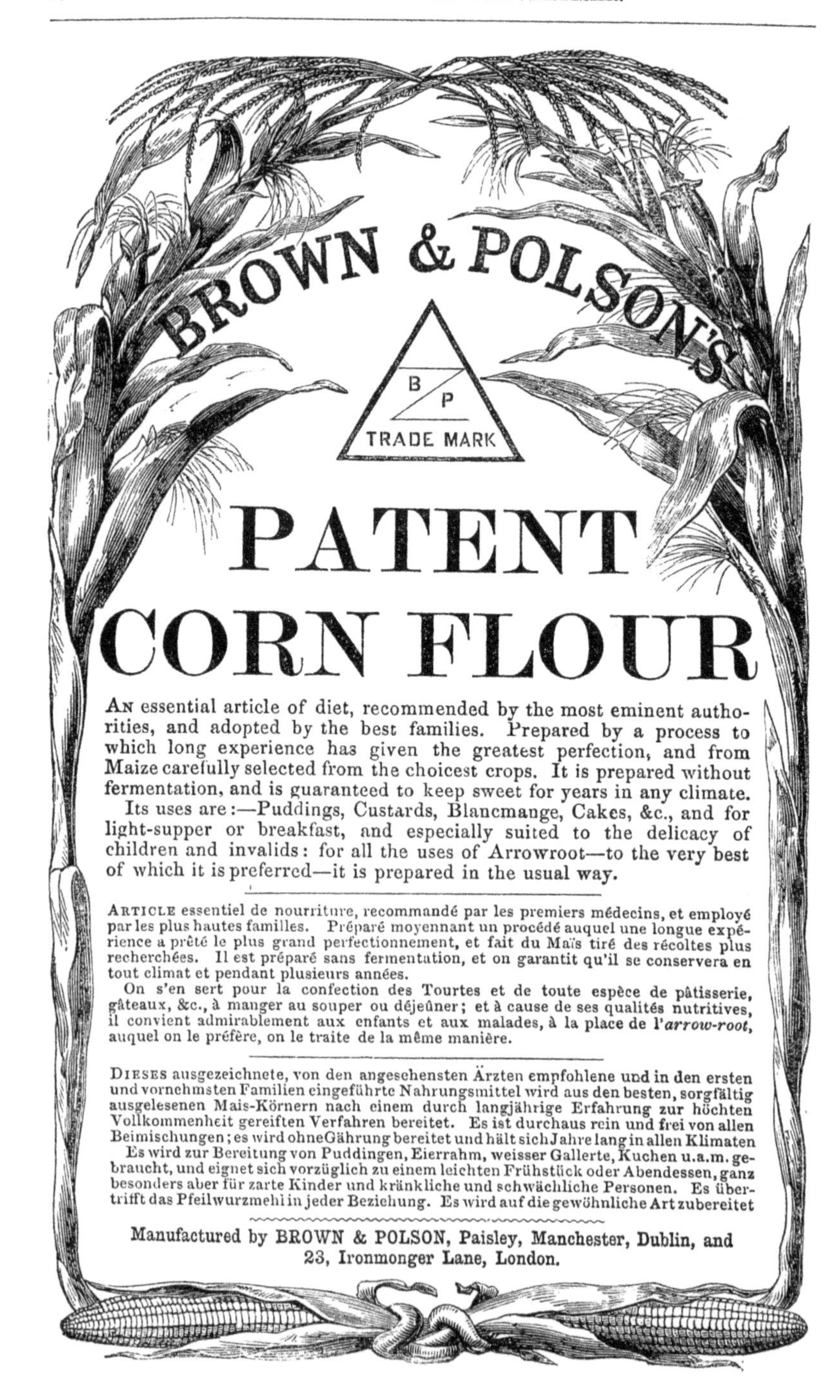
BROWN & POLSON'S
B P
TRADE MARK
PATENT
CORN FLOUR
An essential article of diet, recommended by the most eminent authorities, and adopted by the best families. Prepared by a process to which long experience has given the greatest perfection, and from Maize carefully selected from the choicest crops. It is prepared without fermentation, and is guaranteed to keep sweet for years in any climate.
Its uses are:—Puddings, Custards, Blancmange, Cakes, &c., and for light-supper or breakfast, and especially suited to the delicacy of children and invalids: for all the uses of Arrowroot—to the very best of which it is preferred—it is prepared in the usual way.
Article essentiel de nourriture, recommandé par les premiers médecins, et employé par les plus hautes familles. Préparé moyennant un procédé auquel une longue expérience a prêté le plus grand perfectionnement, et fait du Maïs tiré des récoltes plus recherchées. Il est préparé sans fermentation, et on garantit qu'il se conservera en tout climat et pendant plusieurs années.
On s'en sert pour la confection des Tourtes et de toute espèce de pâtisserie, gâteaux, &c., à manger au souper ou déjeûner; et à cause de ses qualités nutritives, il convient admirablement aux enfants et aux malades, à la place de l'arrow-root, auquel on le préfère, on le traite de la même manière.
Dieses ausgezeichnete, von den angesehensten Ärzten empfohlene und in den ersten und vornehmsten Familien eingeführte Nahrungsmittel wird aus den besten, sorgfältig ausgelesenen Mais-Körnern nach einem durch langjährige Erfahrung zur höchten Vollkommenheit gereiften Verfahren bereitet. Es ist durchaus rein und frei von allen Beimischungen; es wird ohne Gährung bereitet und hält sich Jahre lang in allen Klimaten
Es wird zur Bereitung von Puddingen, Eierrahm, weisser Gallerte, Kuchen u.a.m. gebraucht, und eignet sich vorzüglich zu einem leichten Frühstück oder Abendessen, ganz besonders aber für zarte Kinder und kränkliche und schwächliche Personen. Es übertrifft das Pfeilwurzmehl in jeder Beziehung. Es wird auf die gewöhnliche Art zubereitet
Manufactured by BROWN & POLSON, Paisley, Manchester, Dublin, and 23, Ironmonger Lane, London.

INTERNATIONAL EXHIBITION, 1862.—

IF PLANTED AS DIRECTED THIS WHEAT COSTS

For History and Cultivation of this Wheat, see Essay in the Journal of

BLÉ GÉNÉALOGIQUE.—

HALLETT'S PEDIGREE

"BRED" ON THE SAME PRINCIPLE OF REPEATED SELECTION

1861.—ACTUAL SIZE AFTER FOUR YEARS' REPEATED SELECTION,

DELIVERED AT THE BRIGHTON RAILWAY STATION, PRICE TWO GUINEAS A BUSHEL,

Apply to FREDERIC F. HALLETT, Esq.,

A PRINTED PAPER, "HOW OUR WHEAT CROPS MAY BE DOUBLED,"

Gentlemen wishing to secure this Wheat early next Season, will do well to add their Names at the Wheat is

---

HILL & JONES

(EXHIBITORS)

Wholesale & Export Biscuit Manufacturers, & Confectioners,

JEWRY STREET, LONDON.

HILL & JONES respectfully solicit a trial of their Manufactures by Exporters and Consumers, feeling assured they will be convinced of the superiority of the goods. As a guarantee of quality, most of the following articles may be obtained in bottles and tins, labelled with their name and address. Retail of Grocers, Confectioners, &c., and wholesale only at the above warehouse.

BISCUITS of every variety and the choicest qualities, in 1 lb., 2 lb., 4 lb., and 10 lb. tins, suitable for exportation or home consumption.

LOZENGES of every description, warranted genuine.—Medicated and Proprietary Lozenges prepared with the strictest attention to accuracy.

COMFITS—Almond, Carraway, Coriander, &c., &c., of superior quality, beautifully finished, and tinted with harmless colours by the *new patent steam pans*, by which means a greater delicacy of appearance is imparted.

JUJUBES—Pink, Pine, or Liquorice, rich in flavour, brightness, and colour.

BOILED CONFECTIONERY, manufactured on polished steel plates, thereby dispensing with the necessity for using oil, and avoiding the consequent rancid flavour so generally complained of.

PRESERVED AND CANDIED PEELS—Lemon, Orange, and Citron. These being made by steam process retain their colour and flavour in greater perfection, and for a more lengthened period than when prepared by the old system.

JAMS, MARMALADE, JELLIES, FRENCH SYRUPS, &c., &c., of the choicest descriptions.

Samples can be seen at No. 62, Class 3, where Wholesale Price Lists may be had on application.

MANDER BROTHERS,

VARNISH AND JAPAN MANUFACTURERS,

LONDON—363, Oxford Street (Opposite the Princess's Theatre).
IRELAND—Sackville Place, Dublin.
PARIS—16, Cour des Petites Ecuries.

OBTAINED THE MEDAILLE D'HONNEUR, Paris Exhibition, 1855.

MANDER BROTHERS avail themselves of this medium to announce that they have recently introduced several important improvements into the manufacture of Varnishes, by which they can secure their greater brilliancy, durability, and unvarying excellence.

By the careful observation of sixty years, and by giving their exclusive attention to the production of Varnishes, they have thus succeeded in bringing them to a high state of perfection.

They can refer with satisfaction to the numerous contributions to this Exhibition, which have been finished with their manufactures, particularly to the beautiful decorative and other works of the following well-known firms, to which special reference has been kindly permitted, viz.:—

Carriage Builders—Messrs. R. & J. OFFORD, Wells Street, Oxford Street, London; Messrs. MCNAUGHT & SMITH, Worcester.
Decorator—Mr. T. KERSHAW, 33, Baker Street, Portman Square, London.
Slate Enameller—Mr. G. E. MAGNUS, Pimlico Slate Works, London.
Japanners—Messrs. JOHN BETTRIDGE & CO., Royal Papier Mâché Works, Birmingham.

SAMPLES MAY BE OBTAINED AT THE WAREHOUSE, No. 363, OXFORD STREET.

MANDER FRÈRES,

WOLVERHAMPTON, ET 363, OXFORD STREET, LONDON,

(ETABLIS 1803.)

Manufacturiers de Vernis, pour Carossiers, Décorateurs, et Vernisseurs sur Métaux et Papier Mâché,

Honorés de la Médaille d'Honneur a l'Exposition de Paris, 1855,

Profitent de la présente circonstance pour annoncer qu'ils ont recemment introduit plusieurs améliorations importantes dans la fabrique de leur Vernis, grâce auxquels ils peuvent garantir comme résultat un plus beau lustre, une plus grande durée, et une superiorité invariable de leurs produits.

Par une pratique consécutive de soixante années, et une surveillance incessante dans la fabrication de leurs Vernis, MANDER FRERES ont réussi à les produire à une perfection toute spéciale.

On peut s'en rapporter avec confiance aux nombreuses échantillons de cette Exposition, pour lesquels leurs Vernis ont été employés, notamment aux beaux objets décorés, qui sont exposés par les maisons renommées, dont les noms ci-suivent:—*Carossiers*—Messieurs R. & J. OFFORD, Wells Street, Oxford Street, London; Messieurs MCNAUGHT & SMITH, Worcester. *Peintre et Décorateur*—Monsieur T. KERSHAW, 38, Baker Street, Portman Square, London. *Emailleur d'Ardoises*—Monsieur G. E. MAGNUS, Pimlico Slate Works, London. *Vernisseurs sur Métaux et Papier Mâché*—Messieurs T. BETTRIDGE ET COMPAGNIE, Royal Papier Mâché Works, Birmingham.

Depôt Unique en Paris—Chez Messieurs A. LEVY ET FINGER, 6, Rue de l'Entrepôt, Paris.
Depôt Unique en Belgique—Chez Monsieur J. POISSON, 84, Rue des Palais, Bruxelles.
Agent Principal—Monsieur X. GAGELIN, 16, Cour des Petites Ecuries, Paris.

ON PEUT OBTENIR DES ECHANTILLONS AU DEPOT, 363, OXFORD STREET, LONDON.

GEBRÜDER MANDER,

WOLVERHAMPTON, UND 363, OXFORD STREET, LONDON,

(ETABLIRT SEIT 1803.)

Fabricanten von allerlei Firnissen, fur Wagen Fabricationen, Decorationen Japan und Papier Mache Waaren, u.s.w.

ERHIELTEN DIE EHREN-MEDAILLE IN DER PARISER AUSSTELLUNG, 1855.

GEBRUDER MANDER benutzen diese Gelegenheit dem Publicum anzuzeigen, dass sie kürzlich verschiedene wichtige Verbesserungen in der Firnissfabrication eingeführt haben, wodurch ihren Waaren grösser Brilliance und Dauerhafligkeit gegeben wird.

Durch eine sechzig-jährige genaue Beobachtung in der alleinigen Fabrication dieser Artikel ist es ihnen gelungen dieselben auf einen hohen Grad der Vollkommenheit zu bringen.

GEBRUDER MANDER können mit vollem Vertranen zu solchen Contributionen in der gegenwärtigen Ausstellung hinzeigen die mit ihren Artikeln fabricirt worden sind, und besonders zu den Gegenständen der folgenden wohlbekannten Firmas welche ihnen gütigst erlaubt haben, sich auf ihre Namen zu beziehen, nämlich:—

Wagen Fabrikanten—Messrs. R. & J. OFFORD, Wells Street, Oxford Street, London; Messrs. MCNAUGHT & SMITH, Worcester.
Decorateur—Mr. T. KERSHAW, 38, Baker Street, Portman Square, London.
Slate (Schiefer) Enameller—Mr. G. E. MAGNUS, Pimlico Slate Works, London.
Japanners und Papier Maché Fabricanten—JOHN BETTRIDGE & CO., Royal Papier Maché Works, Birmingham.

MUSTER UND PREISLISTEN KANN MAN IM MAGAZINE, No. 363, OXFORD STREET, LONDON, ERHALTEN.

SUBMARINE TELEGRAPH COMPANY,

IN EXCLUSIVE AND DIRECT COMMUNICATION WITH THE

Continent of Europe, the Channel Islands, Alexandria, and the East.

Via Calais, Boulogne, Dieppe, Coutances, Ostend, Emden, and Tonning.

DIRECT TELEGRAPHIC SERVICE WITH

France, Belgium, Scandinavia, Hanover, Austria, Russia, and the whole South of Europe.

CENTRAL STATION—

58, THREADNEEDLE STREET, CITY, E.C.

(OPPOSITE THE EXCHANGE);

WEST END OFFICE—

43, REGENT'S CIRCUS, PICCADILLY;

CROMWELL ROAD

(OPPOSITE THE INTERNATIONAL EXHIBITION BUILDING);

AND AT

ALL THE LONDON DISTRICT TELEGRAPH STATIONS.

Selection of Charges for Telegrams of 20 words, including address, equally low rates to all other Continental Stations:—

LONDON TO	s.	d.		s.	d.		s.	d.
			Dresden	9	6	Milan	8	6
Aix-la-Chapelle	5	0	Elberfeld	6	6	Moscow	19	0
Alexandria	46	9	Elsinore	8	0	Munich	8	6
Altona	8	0	Epernay	5	0	Nantes	6	0
Amiens	3	6	Florence	9	6	Naples	11	0
Ancona	11	0	Frankfort-on-Main	7	6	Nice	8	6
Antwerp	4	0	Gaeta	11	0	Oporto	13	0
Athens	32	0	Gefle	17	6	Palermo	12	0
Barcelona	9	6	Geneva	7	3	Paris	5	0
Basle	7	3	Genoa	8	6	Prague	9	9
Bayonne	8	6	Ghent	4	0	Ratisbon	8	6
Bergen	19	6	Gothenburg	15	0	Rheims	5	0
Berlin	10	0	Hamburg	8	0	Rome	12	0
Berne	7	3	Hanover	8	0	Rotterdam	5	0
Bilbao	8	6	Havre	3	6	Rouen	3	6
Bologna	9	6	Helsingborg	11	6	St. Petersburg	18	6
Bordeaux	7	3	Honfleur	3	6	Salonica	17	3
Bremen	8	0	Konigsberg	12	6	Santander	9	6
Brest	6	0	Kustendje	17	0	Seville	13	0
Brunswick	8	6	Landscrona	12	6	Smyrna	26	6
Brussels	4	0	Leipsig	9	0	Stockholm	16	3
Bucharest	16	0	Leghorn	8	6	Sundswall	18	6
Cadiz	13	0	Liege	4	0	Taganrog	30	6
Carlscrona	13	9	Lille	3	6	Toulon	8	6
Carthagena	11	6	Lisbon	14	0	Trieste	11	0
Chemnitz	9	0	Livorno—Vercellese	7	3	Turin	7	3
Christiania	17	6	Louvain	4	0	Uddevalla	15	0
Coblentz	6	6	Lyons	7	3	Utrecht	5	0
Cologne	5	0	Madrid	10	6	Venice	8	6
Constantinople	19	6	Malaga	13	0	Verona	9	6
Corfu	16	9	Malta	16	9	Vienna	11	0
Copenhagen	8	0	Mannheim	6	3	Vigo	11	6
Cracow	11	0	Marseilles	8	6	Warsaw	13	6
Dusseldorf	6	6	Messina	12	0	Zurich	7	3

L. W. COURTENAY, Secretary.

Newspaper for the Gardening and Farming Interest.

Every Saturday, price FIVEPENCE, *or* SIXPENCE *Stamped, each Volume complete in itself,*
Thirty-two frequently Thirty-six Folio Pages,

THE GARDENERS' CHRONICLE

AND

AGRICULTURAL GAZETTE;

A WEEKLY RECORD OF RURAL ECONOMY AND GENERAL NEWS.

THE HORTICULTURAL PART EDITED BY PROFESSOR LINDLEY.

As regards the Gardening Part, the principle is to make it a weekly record of everything that bears upon Horticulture, Floriculture, Arboriculture, or Garden Botany, and such Natural History as has a relation to Gardening, with Notices and Criticisms of all Works on such subjects. Connected with this part are

WEEKLY CALENDARS OF GARDENING OPERATIONS,

Given in detail, and adapted to the objects of persons in every station of life; so that the Cottager, with a few rods of ground before his door, the Amateur who has only a Greenhouse, and the Manager of Extensive Gardens, are alike informed of the routine of Operations which the varying seasons render necessary. It moreover contains Reports of Horticultural Exhibitions and Proceedings—Notices of Novelties and Improvements—in fact, everything that can tend to advance the Profession, benefit the condition of the Workman, or conduce to the pleasure of his Employer.

Woodcuts are given whenever the matter treated of requires that mode of illustration.

The Farming Part (under the Editorship of a practical Farmer) treats of—

The Practice of Agriculture.	Results of Experimental Farming.	Foresting.
Agricultural Science.	Growth and Rotation of Crops.	Road-making.
Animal and Vegetable Physiology.	Management of Stock.	Farm-Buildings.
Improvements in Implements described by Woodcuts, whenever requisite.	Veterinary Science.	Labourers.
Improved modes of Husbandry.	Drainage.	Treatment of Poultry.
	Irrigation.	Agricultural Publications.

In short, whatever affects the beneficial employment of capital in land.

Reports are regularly given of the English, Scotch, and Irish Agricultural Societies and Farmers' Clubs—London Markets, Prices of Corn, Hay, Cattle, Seeds, Hops, Potatoes, Butter, Wool, Coal, Timber, Bark, &c., and the Weekly Averages.

An Edition is also published every Monday in time for post, containing Reports on Mark Lane and the Cattle Market.

Replies to Questions connected with the object of the Paper are also furnished weekly.

Lastly, that description of **Domestic and Political News** is introduced which is usually found in a Weekly Newspaper. It is unnecessary to dwell on this head further than to say, that the Proprietors do not range themselves under the banners of any Party; their earnest endeavours are to make THE GARDENERS' CHRONICLE and AGRICULTURAL GAZETTE a full and comprehensive *Record of Facts* only—a Newspaper in the true sense of the word—leaving the Reader to form his own opinions: their object being the elucidation of the laws of Nature, not of Man. The Reader is thus furnished, **in addition to the peculiar features of the Journal,** with such information concerning the events of the day, as supersedes the necessity of his providing himself with any other Weekly Paper.

Office for Advertisements, 41, Wellington Street, Covent Garden, London, W.C.

THE ATHENÆUM.

FROM THE 5TH OF OCTOBER, THE PRICE OF THE ATHENÆUM HAS BEEN THREEPENCE.

Thirty years ago, when the ATHENÆUM came into the hands of its present Proprietors, its price was Eightpence, and its contents, with advertisements, forty-eight columns. Convinced that the circulation of Literary Journals was restricted by high price, and that every advantage offered to the public would bring increase of circulation and authority, the Proprietors reduced the price one-half—to Fourpence. The experiment succeeded, and cheap Literary Journals became the rule.

The Proprietors have always held to the principle then proved. They have given to the public the benefit of every change in the law, increasing the size without increase of price, until the average has become double its former size—above ninety-six columns.

The Proprietors, taking advantage of the abolition of the Paper Duty, therefore resolved that from the 5th of October the price of the ATHENÆUM should be reduced to THREEPENCE.

Office for Advertisements, 20, Wellington Street, Strand, London, W.C.

PERSONS proceeding to India or other Colonies are invited to consult the extensive Catalogues of the Firm of ROBERT COCKS & CO. (supplied gratis and post free), and make their selection before leaving England. Monthly selections of Musical Novelties made up for the Overland Mails. Wholesale Orders executed with accuracy and promptitude. Pianofortes warranted for all climates.

FOR MUSICAL INSTRUMENTS OF EVERY DESCRIPTION, MUSIC OF ALL CLASSES, from the Oratorio and Symphony down to the most simple Instruction Book, for the Fashionable Music of the day, and for everything connected with Music, theoretical and practical, apply to ROBERT COCKS & CO. Pianofortes for Sale or Hire—Price Lists with Drawings *gratis and post free.* Authors' Works printed on their own account.

ROBERT COCKS & CO.'S

NEW DRAWING ROOM

SEMI-COTTAGE PIANOFORTE—£35.

IN ELEGANT WALNUTWOOD CASE,
WITH REGISTERED BLACK KEYS, AND EVERY IMPROVEMENT.

Robert Cocks & Co., 6, New Burlington Street, & 4, Hanover Square, London, W.
Music Publishers to Her Most Gracious Majesty the Queen and the Emperor Napoleon III.

MR. C. F. HANCOCK,

JEWELLER AND SILVERSMITH TO THE PRINCIPAL SOVEREIGNS AND COURTS OF EUROPE,

SOLICITS the honour of a visit to inspect his New Stock of Jewellery and Works of Art and taste in Gold and Silver, now displayed in the New Show Room which he has recently added to his premises.

Miniature Medals of the Order of the Bath, Legion d'Honneur, the Medjidie, with all the different Clasps; also all the Miniature Indian Medals to the present time, with their Clasps, Bars, and Ribbons, including the

CRIMEAN, FRENCH, SARDINIAN, CAFFRE, AND CHINESE.

GREAT VARIETIES OF RACE CUPS AND PRESENTATION PLATE.

N.B.—A LARGE STOCK OF SECOND HAND PLATE.

THE VICTORIA CROSS.

The Miniature Victoria Cross, made from gun-metal taken at Sebastopol, can only be obtained at

C. F. HANCOCK'S,

JEWELLERS' COURT, INTERNATIONAL EXHIBITION, 1862,
38 & 39, BRUTON STREET, BOND STREET, LONDON.

J. R. LOSADA,

CRONOMETRISTA Y RELOJERO DE CÁMARA,

105, REGENT STREET, LONDRES.

(Y Exhibitor, No. 67, Clase 15.)

CONDECORADO TRES VECES POR S. M. POR MERITO EN SU ARTE.

J. R. L. da á su obra toda la perfeccion que permiten los adelantos conocidos en su ramo, y aplica ademas los suyos propios à los volantes de compensacion y á los espirales. Con esta mejora sus Relojes, bien aclimatados, especialmente los de escape Cronométrico ó Duplex, pueden servir para comprobar no solo cálculos de longitud, sinó tambien de la ecuacion del tiempo, y aun notar los retardos y adelantos del paso del sol por el meridiano y de las sombras horarias de un gnomon paralelo al eje terrestre, de cuya correccion se deduce la diferencia entre el tiempo astronómico y el tiempo medio ó civil.

ADVERTENCIA.

J. R. L. hace presente á todos en general, y á sus amigos y favorecedores en particular (que visiten Londres), que sus Rolojes solo se venden en su establecimiento, 105, Regent Street, Londres. Debe hacer presente esto, enconsecuencia de que algunas personas, de malos principios, hacen creer á sus parroquianos, que sus Reloges se pueden obtener en otras partes mas baratos, donde dicen que, J. R. L. los toma, negando que él sea constructor induciéndoles por este medio á comprar lo que no pensaban y pagar el aprendizaje. Se satisfará de esta falsedad á todo el que pase á casa de Losada en donde verá en todos los estados de construccion muchos cientos de Relojes de las diferentes clases superiores que construye.

J. R. L. respectfully informs the public that he introduces into his work every new improvement and discovery, in conjunction with his own, particularly Pendulum Springs and Compensation Balances. By these improvements his Watches, and more especially Chronometers and Duplexes, are fully compensated under all temperatures, serving thereby not only to find the longitude, but also the equation of time, and even to show the sun's retard or advance in passing the meridian, and to correct the indications of the shadows of a gnomon parallel to the earth's axis, by which the difference between astronomical and mean time is deduced.

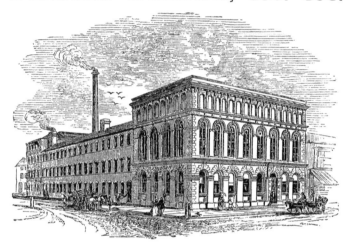

KENT'S PATENTED INVENTIONS,

PROMOTING DOMESTIC ECONOMY.

199, HIGH HOLBORN, AND 329, STRAND, LONDON.

Kent's Patent Rotary Knife Cleaning, Polishing & Sharpening Machine.

A PRIZE MEDAL was awarded to this invention at the Great Exhibition of 1851. Since that period a Second Patent has been granted to G. KENT for certain improvements, which have greatly enhanced its value, not only in general efficiency and durability, but also as a Sharpener of Table Cutlery. G. KENT thinks it necessary to caution intending purchasers against the numerous imitations of his Machine, now being foisted on the public, most of which being made to resemble his are falsely represented to be the "same as Kent's," or "as improvements on Kent's," and although a colour is given to the statements that "parties are at liberty to make them in consequence of Kent's Patent having expired," it is merely an unimportant fact, used to covet a deception, for although true that Kent's *first* patent has lately expired, the Machine, now so well known as "Kent's Rotary Knife Cleaner," is protected by the Second Patent, named above, granted to him for Improvements on the first, and to the vast advantages these improvements give, his Machine owes the reputation it has gained throughout the world.

All are spurious which have not a brass label with the words, "Kent, Patentee and Manufacturer, 199, High Holborn, London."

PRICES AND SIZES.

				£. s.
First Size.—9 Table or Dessert Knives and 1 Carver				14 14
Second	do. 8	,,	,,	12 12
Third	do. 7	,,	,,	10 10
Fourth	do. 6	,,	,,	9 0
Fifth	do. 5	,,	,,	7 10
Sixth	do. 4	,,	,,	6 0
Seventh	do. 4	,,	,,	4 15
Eighth	do. 3	,,	,,	3 3

GRIFFITH'S PATENT WHISK & MIXING MACHINE.

GEO. KENT, Sole Manufacturer.

The character of this article as a Whisk and Mixer, for the use of Cooks, Confectioners, and manufacturers of any liquid or semi-liquid compounds, is now fully established. For beating up Eggs, all kinds of Egg Mixtures, and Batters, it is adopted and highly valued by most of the principal Pastry Cooks, Confectioners, and large private families throughout Great Britain and the Colonies, who testify "that it produces, from a given quantity of ingredients, a larger and better Batter, and in less time than any other process;" whilst manufacturers of Chemicals, Colours, Inks, Soaps, and Candles, as well as Photographers, Wine Merchants, Brewers, &c., find it supply a means of mixing every description of liquid or semi-liquid compound in a manner never before attained. As a Churn it is equally successful, and is acknowledged to bring Butter in less time than any other extant. For rapidity of action, perfection in operation, economy in use, and general efficiency, this Mixer stands unrivalled. They are made of any size or material, according to the purpose for which they are required, from 21s. each.

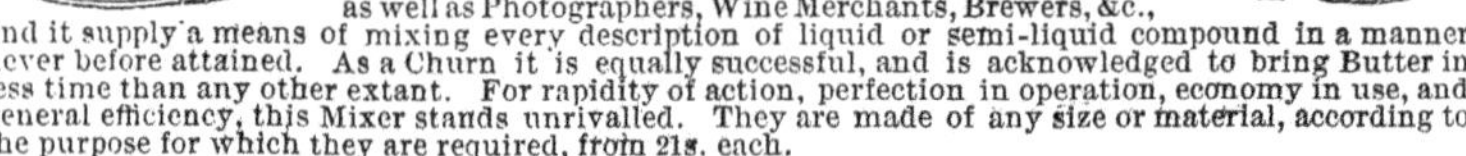

MONROE'S PATENT EGG BEATER.—GEO. KENT, Sole Manufacturer.
With this ingenious little Machine, which is similar in principle to that described above, small quantities of Eggs, all kind of Egg Mixtures and Batters may, in a few minutes, be wrought up to a degree of lightness far superior to anything that can be produced by the ordinary hand whisk. Price 5s.

KENT'S PATENT TRITURATING STRAINER.

The object of KENT'S PATENT TRITURATING STRAINER is to enable Cooks to reduce to pulp, and strain by the same operation, every description of animal and vegetable substance for making Soups, Sauces, Purees, Gravies, Jams, &c.; and that it performs the work in the most perfect manner possible the many hundreds that are annually being sold, and the very high character they bear with every cook using them, will, it is believed, be a sufficient guarantee. By its use the tedious, troublesome, dirty, and wasteful process with the hair sieve and tammy cloth is entirely avoided, whilst the whole of the virtues of the ingredients employed are completely extracted, and brought to the required consistency at a much less cost, and in one-tenth the time usually occupied by those very primitive means.

KENT'S TRITURATOR is in use in most of the Club Houses and respectable Hotels in London and the Provinces. It is also now being adopted in private kitchens, and must in time become an indispensable requisite in every domestic establishment. Price 27s. 6d. and 37s. 6d.

KENT'S PATENT ROTARY CINDER SIFTER, renders unobjectionable a very disagreeable duty, ensures Cinders being properly Sifted, prevents loss, increases cleanliness, and will, in a short time, save its cost.

KENT'S BOX MANGLE, WITH NEW PATENT DOUBLE ACTION. This Mangle reverses its movement by turning the handle always in one direction, but is much lighter than any other description. It is also more easy and rapid in working, and is, in every respect, the best extant. Price, from £9.

KENT'S PATENT WASHING MACHINES. A very simple, economical and effectual mode of cleansing linen, requiring comparatively no hand rubbing, and dispensing with boiling altogether. Price, from £3 10s.

KENT'S PATENT SELF-HEATING BOX IRON is intended for all the purposes to which the old Box and Flat Iron are applied. It will heat in three minutes, without any fire, and remain hot, at a nominal cost, for any length of time. Price, from 5s. 6d.

W. J. BUSH & Co.,

(Exhibitors, Class 2.)

Distillers of Essential Oils & Essences,

Warehouse—30, Liverpool-street, London, E.C.
Works—Ash-grove, Hackney, N.E.
Shippers and large Consumers should buy only of W. J. BUSH & CO.

PAINTS, COLOURS, OILS, &c.—To Exporters.—BLUNDELL, SPENCE, & CO., Hull, and 9, Upper Thames-street, London, prepare at the shortest notice, and put free on board, London and Liverpool, all descriptions of Paints, Colours, and Oils, packed in any form desired, suitable for the Continental, North and South American, East and West Indian, Chinese, Australian, and other markets. Resident Agent, New York, E. HILL, 180, Front-street; Melbourne, R. A. FITCH, 74, Flinders-lane East. Prize Medals—London, 1851; Paris, First Class, 1855.

T. H. FILMER & SONS'

SUPERIOR FURNITURE,

As shown in the

Industrial Exhibition, Classes 22 and 30,

THE CRYSTAL PALACE, SYDENHAM,
And at their extensive
MANUFACTORY AND WAREROOMS,
28, 31, 32, & 34, Berners Street, Oxford Street,
LONDON, W.

FURNITURE, CARPETS, AND BEDDING.—See our Catalogue, elaborately illustrated with 350 Engravings, containing price of every article, and estimates for completely furnishing houses of any class. This very useful guide forwarded, gratis and post-free, on application to LEWIN CRAWCOUR & Co., Cabinet Manufacturers, 22 & 23, Queen's-buildings, Knightsbridge, within five minutes' walk of the Exhibition, and at 12, Sloane-street, Belgravia. Country orders carriage free.

N.B.—An elegant and complete walnut drawing-room suite, 36 guineas; dining-room ditto, in Spanish mahogany, 28 guineas; and two handsome and complete bed-room suites, 30 and 28 guineas each. The whole nearly new, and bargains.

FURNITURE
CARRIAGE FREE.
P. & S. BEYFUS.
144, OXFORD STREET
& 91 TO 95 CITY ROAD.
ILLUSTRATED CATALOGUES GRATIS

SMEE'S SPRING MATTRESS

(TUCKER'S PATENT).

Comfortable, cleanly, simple, portable, and inexpensive. An inspection of this important improvement in Bedding is particularly requested.
Price for the 3-ft. size, 25s.; other sizes in proportion.

WILLIAM SMEE & SONS,

Wholesale Cabinet-makers, Upholsterers, and Bedding Warehousemen,

6, FINSBURY PAVEMENT, LONDON, E.C.

Purchasers are earnestly warned against INFRINGEMENTS and IMITATIONS. Each genuine Mattress bears the label, "Tucker's Patent."

ATKINSON & CO.,

Contractors to Her Majesty's Government,

Cabinet Manufacturers, Upholstery, and Carpet Warehousemen, Drapers, Silk Mercers, &c.,

69, 70, 71, 72, 73, 74, & 75, Westminster Bridge Road, LAMBETH.

N.B.—On the left-hand side after crossing the New Westminster Bridge from the Houses of Parliament.

DALTON & BARTON,

Ribbon, Carriage Lace, & Upholstery Trimming Manufacturers,

173A, ALDERSGATE STREET, LONDON AND COVENTRY.

BRIDGES'S KITCHEN TABLES,

Class 30.

406, OXFORD ST., LONDON, W. (near Soho Square.)

BRIDGES'S PLAIN OAK FURNITURE, &c.

See "Illustrated Catalogue."

Bridges's Prize Butter Prints and Dairy Implements,

To which have been awarded six Prize Medals.

406, Oxford St., London, W. (near Soho Square.)

Silver Medal, First Class, Paris, 1855.

S. NYE & CO.'S PATENT MACHINES,

Of various sizes, for Mincing Meat, Vegetables, &c.; for making Sausages, Mince-Meat, Forced Meat, Potted Meat, and various dishes for Families, Hotel Keepers, Confectioners, Butchers, and also for Hospitals, Lunatic Asylums, and all large Establishments.

Price £1 1*s*., £1 10*s*., £2 2*s*. £3 3*s*., and £7 7*s*.

A SMALL MINCER OR MASTICATOR,

To assist Digestion. Price £1 10*s*.

79, WARDOUR STREET, LONDON.

S. NYE'S PATENT

IMPROVED MILLS,

For Coffee, Pepper, Spice, Rice, Malt, &c., are the best and most convenient made.

Price 8*s*., 10*s*., 14*s*., 15*s*., and 20*s*. each.

79, WARDOUR STREET, LONDON.

SAUSAGE AND MINCING MACHINE.

One Guinea.

THE NEW AMERICAN PATENT.

Sold by the Makers,

BURGESS & KEY,

95, NEWGATE STREET, LONDON,

And all respectable Ironmongers in Town and Country.

WASHING MACHINERY.

THOMAS BRADFORD, Engineer and Patentee, 63, Fleet-street, London, and Manchester.—Comprising Washing Machines, with and without the wringing and mangling apparatus attached, for the use of private family or public institution; Patent Box Mangles, Cottage Mangles, Metal Callendering Machines, Napkin and Linen Presses, Drying Closets, Ironing Stoves and Irons, and every other laundry requisite. Estimates and plans for steam laundries on application, and stating requirements (see Power Machine in motion in *Class* VIII.) The machines may be seen in practical operation daily at the Warehouse, 63, Fleet-street, London.

Descriptive pamphlet post-free or on application.

MODEL INVENTIONS

FOR KNIFE-CLEANING AND SHARPENING.

HILLIARD & CHAPMAN,

PATENTEES, GLASGOW.

1851, Prize Medal. 1862, Class 31.

To Families, Tailors, Shirt Collar, Dress, Boot and Shoe Makers, Saddlers, and others.

W. F. THOMAS & Co.'s
PATENT SEWING MACHINES.

In these Machines, made by the original patentees and introducers of the invention with all the improvements that 14 years' experience has enabled them to effect, stand foremost and unrivalled for all practical purposes. They are guaranteed to perform their work efficiently, with exceedingrapidity, regularity, and durability, and to give no trouble; they are, in fact, the only reliable Machines.

The Stitching produced is alike on both sides of the material, without cord or ridge, and the thread cannot be pulled out.

Hemming, Binding, Gathering, &c., may be accomplished with facility. Illustrated Catalogues and Specimens of the work may be had of W. F. THOMAS & CO., 66, Newgate Street, London.

T. AINSWORTH,
FLAX SPINNER AND LINEN THREAD MANUFACTURER.
Sewing Machine Linen Threads.

CLEATOR MILLS, WHITEHAVEN,
17, LAWRENCE LANE, CHEAPSIDE, LONDON, E.C.
43A, PICCADILLY, MANCHESTER.

GEORGE TOWNSEND & Co.,
MAKERS OF SEWING MACHINE AND OTHER NEEDLES,
HUNT END, NEAR REDDITCH;
AND
12, WALBROOK, LONDON.

Shuttles, Reels, Springs, &c., for every description of Machine.

THE Original Manufacturers of the CROYDON BASKET CARRIAGES.—Messrs. WATERS & SON, Carriage Builders, 5, George-street, Croydon, beg to invite the Public to an inspection of their large stock of light; fashionable Carriages of every description. Basket Carriages in great variety, and of the newest designs. Having extensively enlarged their Premises, they are now enabled to execute any order for Carriages they may be favoured with, upon the shortest notice, and at moderate prices. Messrs. WATERS & SON exhibit at the Exhibition; also, the Crystal Palace, Sydenham.

STEVENSON & ELLIOT,
CARRIAGE BUILDERS,
177, 179, *and* 181, *King Street, Melbourne, Australia. Branch Factory, Stirling, Scotland.*

Barouche, very suitable for the colonies, constructed so as to admit of high wheels, the felloes of which are bent by steam, and have only two pieces on each wheel, altogether combining lightness with strength, and an easy draft for one horse or a pair of ponies. Our Brougham and other carriages are equally light—all built of the very best materials, extra durably varnished, and highly finished in every particular.—Enquiries, &c. to be addressed to the Branch Factory, Stirling, Scotland.

JOHN TIBBITS & SON,
WALSALL,
WHOLESALE SADDLERS AND COACH IRONMONGERS,
Bridle Cutters and Harness Makers,

Manufacturers of all kinds of Harness Mountings, Heraldic Ornaments, Bits, Stirrups, Spurs, Buckles, Whips, &c.
Cavalry Appointments, Axles, Beading, &c.

S. BLACKWELL,
SADDLER, HARNESS MANUFACTURER, &c.
259, OXFORD ST. (near the Marble Arch), W., and No. 42, Class 26, EXHIBITION.

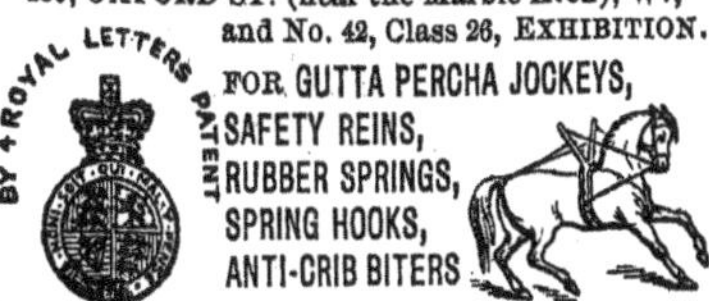
BY 4 ROYAL LETTERS PATENT
FOR GUTTA PERCHA JOCKEYS,
SAFETY REINS,
RUBBER SPRINGS,
SPRING HOOKS,
ANTI-CRIB BITERS

DOERR & REINHART,
WORMS-ON-THE-RHINE, GERMANY,
(Established 1839. *Class* 26, *No.* 531.)

Manufacturers of Patent Calf and Goat Skins, varnished in different colours, Varnished Cow Hides, and Calf Kids. Prize Medals in London, 1851; New York, 1853; Munich, 1854; Silver Medal in Paris, 1855.
Agents in London—Messrs. SCHMITT & DAVID 102, Leadenhall-street, E.C.

GILBY'S
PATENT BREECH-LOADING RIFLES

May be seen in Class 11. They obtained Honourable Mention at the Great French Exhibition. The remaining stock to be sold with right of manufacture. Also, an entirely new system of Breech-loading for Cannon, very rapid and effective, which may be seen on application to JOHN GILBY, Beverley, Inventor and Patentee.

PARKER, FIELD, & SONS,
Gunmakers to Her Majesty, &c., &c., Riflemakers to H. late R. H. the Prince Consort,

233, High Holborn, and Mansell-street, Minories, London, Manufacturers of the ENFIELD RIFLE for the War Department. Rifles, Pistols, and Swords of every description. Small Bore Rifles. New breech-loading Guns, Revolving Pistols, &c.

FREDERICK JOYCE,
Contractor to Her Majesty's War Department,

Patentee and Manufacturer of Sporting and Military Percussion Caps, Breech-loading Cartridge Cases, Gun Wadding, Wire Cartridges, and every description of Sporting Ammunition.
Exhibitor, Class 11.
Office:—57, Upper Thames-street, London.
Established 1820. Wholesale only.

ELEY'S
SPORTING AND MILITARY
AMMUNITION.
Wholesale only,
ELEY BROTHERS
GRAY'S INN ROAD, LONDON.

ROBERT FAUNTLEROY & CO., 99 and 100, Bunhill-row, Finsbury, London, E.C., Foreign Hard-wood, Dye-wood, and Fancy-wood Merchants and Importers. Contractors to Her Majesty's Government. Wholesale and retail. In Class 4 is shown a large Model of the Royal Exchange, London, constructed of a variety of Woods, Corozo, Coquilla, Betel, and Cohoun Nuts, &c. Catalogues gratis. The original Hard-wood business was established in this family by Mr. R. F. in 1732, and his great-grandson's only address is "BUNHILL-ROW."

SHIRTS.
MORNING AND DRESS SHIRTS

Of every description, fitting with precision and ease, at moderate prices.
Measure Papers sent on application.

CAPPER & WATERS,
26, REGENT STREET PICCADILLY, S.W.

STAYS—SKIRTS.

SMITH'S New Patent HARMOZON CORSET.
SMITH'S Patent ROYAL SYMMETRICAL CORSET.
SMITH'S Patent ROYAL RADION CORSET.
CASTLE'S Patent VENTILATING CORSET.
The CARDINIBUS Patent COLLAPSING JUPON.
The IMPERIAL SYMMETRICAL CORSAGE JUPON.
SPIRAL STEEL & BRONZE for SKIRTS (Patented).
ZEPHYR CRINOLINE SKIRTS.

For description of the above see the Exhibition "Illustrated Catalogue," Class 27C. Manufactured by and to be had wholesale only of A. SALOMONS, Old Change, LONDON, of whom may also be obtained every description of STAYS and SKIRTS, ENGLISH, FRENCH, and AMERICAN.

SHIRT MAKERS AND HOSIERS.

CHRISTIAN & RATHBONE,
11, WIGMORE STREET, W.

HATS.—Class 27A.

he attention of the Public is respectfully invited to the HATS, exhibited as above.

The ADVANTAGES offered consist in improved

PLIANCY, VENTILATION, & EXTREME LIGHTNESS.

The lightest weigh 3½ ozs., and afford great relief to those who suffer from the pressure, weight, and general inconvenience of the Hats usually sold; they are well suited for India, China, and the Colonies.

SOLE MANUFACTURERS,

GAIMES, SANDERS, & NICOL,
BIRCHIN LANE, E.C., AND 111, STRAND.

UMBRELLAS AND SUNSHADES, made from French and English Silk, and upon every description of new and patented frame, may be obtained at the Wholesale Manufactory of JOHN MORLAND & SONS, 50, Eastcheap, London-bridge, City. Gingham and Alpaca Umbrellas of all qualities. Silk, Alpaca, Gingham, Steel Frames, Canes, Fittings, and all kinds of Materials for the Trade. Oiled and Japanned Silk and Cambric. Shippers supplied on the most favourable terms.

SAMUEL BROTHERS'

Celebrated Sydenham Trowsers, 17s. 6d.

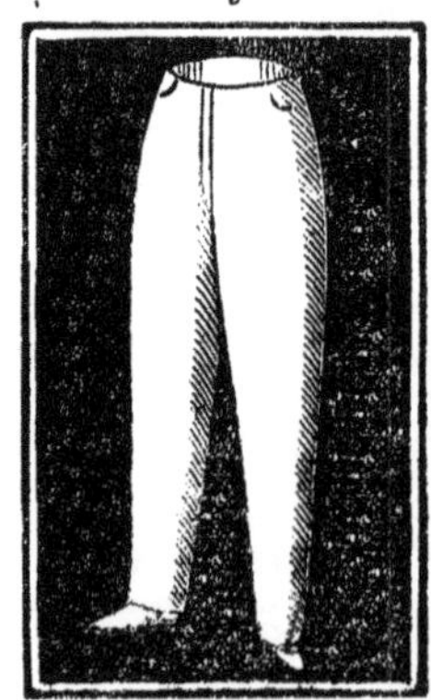

Promenade and Visiting Suits, 42s. to 59s.

Full Dress, 84s.

Boys' Suits, 16s. 6d. to 30s.

Gentlemen and their sons fashionably and perfectly fitted, either to measure, or with superior ready-made Garments.

Complete Outfits.

Book of Fashions and Prices free.

Sole Address—
29 LUDGATE-HILL, LONDON, E.C.

3,148,000
OF
"Sangsters' Alpaca" Umbrellas,

Which obtained a Prize Medal in 1851, have been made under their Patent to January 1st, 1862. To prevent mistakes, their Licensees are now supplied with Labels, bearing the words "Sangsters Alpaca," one of which should be attached to all Umbrellas made of Alpaca. These Umbrellas may be obtained of all Manufacturers and Dealers, and of the Patentees,

W. & J. SANGSTER,

140, Regent-street; 94, Fleet-street; 10, Royal Exchange; and 75, Cheapside.

N.B.—Goods of their own manufacture have a label with the words "Sangsters' Makers."

JOHN PATON, SON, & CO.,

ALLOA, N.B.,

MANUFACTURERS OF WOOLLEN HOSIERY YARNS,

Exhibitors, Class 21.

Samples of Knitted and Loom-wrought Hosiery, Cravats, &c., made by various parties from the yarns, exhibited in case.

SILK MANUFACTURING.

A VALUABLE PATENT

For simplifying the method of doubling silk threads is for Sale. Its importance cannot be over estimated.
Full particulars and specifications can be obtained of Messrs. PATTISON & SON, 57½, Old Broad-street, E.C.

THE GREATEST NOVELTY OF THE AGE.

A CLEAN COLLAR

Every morning without washing or ironing.

Steel Shirt Collars - - - - 1*s*. 6*d*. each.
Steel Wristbands - - - - 2*s*. 6*d*. per pair.
Steel Gauntlets for Ladies - 2*s*. 6*d*. per pair.

These articles exactly resemble the finest linen, and are so manufactured that soils can be removed by the application of a little soap and water, rendering the article ready for use in a moment. Sold by J. H. YEATMAN, Hosier, No. 158, Leadenhall-street, E.C., and 91, Regent-street, W.
Can be obtained by post from the Patentee and Manufacturer, C. F. ATKINSON, Sheffield. Collar sent free by post on receipt of twenty-two stamps; a Pair of Cuffs, thirty-four stamps.

MR. HOMEYER,

OF RANZIN, IN POMERANIA,

Invites Growers and Manufacturers of WOOL to inspect his Fleeces of Long and Short Wools (*vide* Zollverein Department, Catalogue, No. 1200), and to give an opinion on the respective value of these Wools, through medium of the *Mark Lane Express*.
The fleeces are totally free from grease, thereby showing the real quantity of wool which the manufacturer can use.
There are labels attached giving the name of the breeders and the breed of sheep, the weight of fleece and loss of grease in process of washing.
For further particulars apply to Mr. HOMEYER'S Agents in London, Messrs. NICOLSON BROTHERS 5, Jeffrey's-square, City.

BOND'S
PERMANENT MARKING INK

(The Original), for marking on linen, silk, or cotton. Requires no preparation, and is by far the best. NOTICE—The genuine may be distinguished from all spurious imitations by observing that each bottle bears on the label the address of the Inventor, "John Bond, 28, Long-lane, West Smithfield, E.C."
Price 1*s*. per bottle.

RIMMEL'S PATENT PERFUME VAPORIZER,

For diffusing a delightful fragrance, and purifying the air in apartments, ball-rooms, theatres, &c. Price from 6*s*.

Exhibited in Class 4, Eastern Annexe.

E. RIMMEL,
PERFUMER,
96, *Strand, & 24, Cornhill,*
LONDON.

YARDLEY & STATHAM (Established A.D. 1770), Manufacturers of Superior Toilet Soaps and Choice Perfumery, Wholesale and for Exportation, 7, Vine Street, Bloomsbury, London, W.C., and 5, Rue des Vieilles, Haudriettes, Paris.

YARDLEY & STATHAM'S SUN FLOWER OIL SOAP renders the skin beautifully soft, white, and pliant, and emits a refreshing and exquisite odour; it is acknowledged to be the Perfection of Toilet Soaps.

YARDLEY & STATHAM'S HONEY SOAP, invented by them in the year 1845, continues to command the most undeniable appreciation by the public; also their OLD BROWN WINDSOR, GLYCERINE, and other Choice Toilet Soaps, all highly esteemed for their emollient properties and delightful fragrance.

YARDLEY & STATHAM'S COLD CREAM SOAP, prepared expressly for Ladies and Infants, is perfumed with Otto of Roses, and has been justly ranked as the most efficient, yet harmless improver of the complexion.

Specimens exhibited, Class 4D.

YARDLEY & STATHAM'S Fancy Soaps are supplied retail in Tablets and Squares, and 1 lb. Bars, by all Chemists and Dealers in Perfumery; and wholesale at their Manufactory, 7, Vine Street, Bloomsbury, London, and Depôt, 5, Rue des Vieilles, Haudriettes, Paris.

BRECKNELL'S SKIN SOAP, the best for producing a clear and healthy skin, in 1*s*. packets of four tablets or eight squares, may be obtained of all dealers in town and country. N.B. Each tablet and square is stamped with the name of Brecknell. —Brecknell, Turner, & Sons, by appointment to Her Majesty, manufacturers of wax, spermaceti, stearine, composite, and tallow candles; agents to Price's Patent Candle Company, dealers in all other patent candles, all kinds of household and toilet soaps, and in colza, sperm, vegetable, and other lamp oils, &c. Beehive, 31, Haymarket.

J. MARSH, Hair Cutter & Perfumer

To H.R.H. the Prince of Wales, H.M. the King of Saxony, and H.I.H. the Empress of Austria, 175, PICCADILLY, opposite the Burlington Arcade, London, begs most respectfully to call the attention of the Nobility and Gentry to his beautiful specimens of work in Tortoise-shell, Ivory, and Artificial Hair, which for beauty and delicacy of finish cannot be surpassed; also to three New Productions in Perfumery—the Essence of Wood Violets, the Essence of Orange Blossoms, and Extract of Lime Juice and Glycerine, all being Liquid Pomades for the Hair. Class 25, Section 3.

BRUSHES FOR ALL NATIONS.

Manufactured Wholesale and for Exportation (only) by W. S. MOORE, 47, Percival-street, Goswell-street, City, London, E.C. Bone Tooth, Nail, and Shaving Brushes. Ivory Hair, Hat, and Cloth Brushes; also in Rosewood, Satin and all Fancy Woods. Ivory Hand Glasses, Glove Stretchers, Puff Boxes, Paper Knives, and Fancy Ivory Turnery Goods. A very superior Bone Tooth Brush made for exportation, stamped and unstamped, packed nicely in boxes, always in stock. Specimens of the above to be seen in Class 4, and at W. S. MOORE'S, Brush Manufacturer, 47, Percival-street, Goswell-street, City, London, E.C.

BRUSHES! SPONGES!

BARRETT'S very best TOOTH BRUSHES, only 6*d*. each, same quality as those usually charged 1*s*. each; all warranted that the hairs will not come out in the mouth.

BARRETT'S New Penetrating HAIR BRUSH, which thoroughly cleanses the hair, 4*s*. each; and every description of Brush for the TOILET, HOUSE, and STABLE use, of superior quality at half the price usually charged.

SMYRNA SPONGES for the TOILET and BATH at unusually low prices. STABLE SPONGE, 8*s*. per lb. at A. BARRETT'S, 63 & 64, Piccadilly, corner of Albemarle Street, London.

ESTABLISHED 1770.—F. S. CLEAVER'S Prize Medal Honey Soap; Glycerine, Elder Flower, Brown Windsor, &c. &c. F. S. Cleaver's Choice Pomades. F. S. Cleaver's Fashionable Perfumes. F. S. Cleaver's Saponaceous Tooth Powder, and every article connected with the Perfumery Trade. —Manufactory, 32 & 33, Red Lion-street, London, W.C.; 243, Rue St. Denis, Paris; and 31 & 33, Dey-street, New York. Class 4, in the Eastern Annexe of Exhibition Building for 1862.

WILLIAM RIEGER'S TRANSPARENT CRYSTAL SOAP.

The Exhibition Tablet and Shaving Stick,
Wholesale and for Exportation.

26, LAMBETH HILL, E.C.

PHILLIPSON & Co.'s New Perfume "LA DUCHESSE." The most refreshing and durable of the day. Price 2*s*. 6*d*. of all Vendors of Perfumery, or by enclosing a P. O. Order or Stamps, to PHILLIPSON & CO., 1, BUDGE ROW, ST. PAUL'S, LONDON, E.C. The POMADE, 2*s*. 6*d*.; SOAP, 1*s*.; OIL, 2*s*. 6*d*. Perfumery for *every climate*. Catalogues free.

J. & J. STEVENSON,
ARUNDEL COMB WORKS, SHEFFIELD,
AND
9, CRIPPLEGATE BUILDINGS, WOOD STREET, LONDON, E.C.

MANUFACTURERS OF SHELL, HORN, AND INDIA RUBBER COMBS, OF EVERY DESCRIPTION.

Wholesale and for Export only.

NOSOTTI'S LOOKING GLASSES AND GILT DECORATIVE FURNITURE. *Class* 30. Reputed for superior quality and perfection in workmanship. Finished in pure gold, with best French plates. The London and Paris Looking Glass Manufactory, established forty years. Entrance to show-rooms, 398 and 399, Oxford-street. Manufactory, 3 and 4, Great Chapel-street, and 102½, Dean-street, London, W.

UNDER ROYAL PATRONAGE.

GIVE INSTANT RELIEF, & A RAPID CURE OF

ASTHMA, CONSUMPTION,

INFLUENZA, COUGHS, COLDS,

And all Disorders of the Breath, Throat, & Lungs.

Small Books, containing many hundreds of properly-authenticated Cures of Asthma, Consumption, Coughs, Influenza, &c., &c., may be had from every Agent in the Kingdom.

IN COUGHS

The effect of DR. LOCOCK'S PULMONIC WAFERS is truly surprising, as within ten minutes after taking a dose, the most violent cough is subdued. *They have a pleasant taste.*

To Singers and Public Speakers.

These Wafers, by their action on the Throat and Lungs, remove all hoarseness in a few hours, and wonderfully increase the power and flexibility of the Voice.

Prepared only by the Proprietor's Agents, Da Silva and Co., 1, Bride-lane, Fleet-street, London.

Sold at 1s. 1½d., 2s. 9d., and 11s. per box, by all Druggists.

Note.—Full directions are given with every box, in the *English, German, and French languages.*

CAUTION.—The Public is cautioned against purchasing any (so-called) "Locock's Pills," "Locock's Lotions," "Bark," "Tooth Powders," or any other preparation," under the belief that they are connected with "DR. LOCOCK'S WAFERS." The only genuine Medicines are "DR. LOCOCK'S WAFERS," and "DR. LOCOCK'S COSMETIC" (*for the complexion*").

All others are an Imposition upon the public.

DR. HUGO'S MEDICAL ATOMS.

AN ANTIBILIOUS STOMACHIC AND APERIENT MEDICINE

Of the highest reputation, and of so agreeable a taste that it

May be eaten as Confectionary.

In Indigestion and all Disorders of the Stomach, a single dose gives relief. Bad Breath is cured, and Appetite, Strength, and Cheerfulness are rapidly restored.

They are the very best Aperient Medicine, acting gently and effectually; removing Flatulency, Acidity, Heartburn, Bile, Sick Headache, Costiveness, Spasms, Lowness of Spirits, Dimness of Sight, producing refreshing sleep, and restoring tone and vigour to the system.

For Ladies they are a most pleasant remedy.

For Children they are highly recommended, whilst their agreeable taste removes all suspicion of their medicinal character.

ARE SMALL (*though not a Homœopathic Medicine*), AND HAVE A MOST AGREEABLE TASTE.

The full particulars of many most surprising Cures are given with every Packet, together with proper directions for use.

Wholesale Agents, Da Silva and Co., 26, Bride-lane, Fleet-street, London.

Sold at 1s. 1½d., 2s. 9d., and 4s. 6d., by all Druggists.

COMBS. Merchants, Shippers, and the Representatives of Foreign and Colonial Houses are respectfully reminded that the Aberdeen Comb Works are the largest in the world, covering upwards of two acres of ground, and employing 700 hands. Every description of Combs are made from horn, tortoise-shell, india rubber, &c.

Address STEWART R. STEWART & CO., 13, Grocers'-hall-court, Poultry, London, and Aberdeen Comb Works, Aberdeen.

LUMPS OF DELIGHT.—Lovers of the delicious Turkish Confection RAHAT LI KOUM, or Lumps of Delight, are invited to select from one of the largest and most varied stocks in London. 2*s.* per lb.; packages, as imported, 3*s.*, 3*s.* 6*d.*, and 5*s.* each; handsome cartons, for presentation, 3*s.* 6*d.* and upwards. Also Apple and Orange Jelly, Apricot Marmalade, &c. &c. The Standard Tea, black or mixed, 4*s.* per lb. Price lists on application. BARNARD, BRENNAN, & CO., 45, Fleet-street, London, E.C.

JEAN MARIE FARINA,

Rhein, Strasse, Cologne. London: 2, Salter's Hall-court, & 16, St. Swithin's-lane, Cannon-street, City.

Extrait d'Eau de Cologne . 20*s.* per box (6 bottles).
Lavande . . . . 15*s.* " "

As in Farina's Fountains.

Delivered carriage free to all parts of the United Kingdom.

Eau de Cologne and Lavande in bond, for exportation, at 9*s.* and 5*s.* 6*d.* per dozen. Also in wicker bottles.

ÆGLE MARMELOS, OR INDIAN BAEL,

LIQUOR BELÆ,

Prepared from the unripe fruit, so much esteemed in cases of Diarrhœa, Dysentery (whether acute or chronic), and affections of the mucous membrane, &c. &c. Also, the Preserve, or Jelly, from the ripe fruit.

J. G. GOULD, Chemist, 198, Oxford street, corner of Orchard-street, Portman-square, London.

SPIERS & SON, OXFORD,

(102 & 103, HIGH STREET,)

Respectfully invite Tourists to visit their extensive Warehouses for Useful and Ornamental Manufactures, suitable for presents and remembrances of Oxford. Copies of every published guide-book and map of the City and neighbourhood kept in stock.

Exhibitors in the Furniture and Glass Courts.

CLARENCE'S celebrated CAYENNE SAUCE.—This Sauce, by its genuine qualities, has classed itself the first and best of the day, and for its use as a relish to Roast Meat, Game, Poultry, Fish, Steaks, Chops, Cutlets, Gravies, Soups, &c., stands unrivalled. Patronized by the nobility, gentry, and all the club-houses at the West-end. Sold wholesale by the Manufacturer, T. CLARENCE, 2, Church-place, Piccadilly, by St. James's Church; also by Crosse & Blackwell. Retailed by all the principal oil and Italian warehousemen and dealers in Sauces throughout the Kingdom. *Class* 3.

1s. CLOCKS. 1s.

The EVERLASTING, 1*s.*, 1*s.* 6*d.*, 2*s.*, 3*s.*, 4*s.*, 5*s.*, 6*s.*, 7*s.* 6*d.*, 10*s.*, 12*s.* 6*d.*, 15*s.*, &c. Within three miles 2*d.*; booked and packed for rail, 6*d.* extra. The SILENT NIGHT LAMP CLOCK, from 15*s.* 6*d.*—PREECE & CO., 380, Oxford-street, W., Clock Manufactory and Fancy Warehouse. Depôt for DERBYSHIRE SPAR, and English and Foreign ILLUMINATED GLASS. Glass Cobbler Straws, 6*d.* per dozen; 4*s.* per gross.

LANG & PADOUX,

Watchmakers,

GENEVA.

This firm having resolved upon keeping up the old reputation of genuine Geneva watches, its principal object is, not so much to manufacture a large number, as to make them unparalleled for good regulation and solidity, in order to render them perfectly warrantable in every respect.

ROYAL STRAND THEATRE.—Sole Lessee and Manager, Mr. SWANBOROUGH, Sen. Open every Evening for Comedietta, Burlesque, Extravaganza, and Farce. Supported by the most Popular Company in London. Prices—Private Boxes £1 1*s.*, £1 11*s.* 6*d.*, and £2 2*s.*; Stalls 5*s.*; Boxes 3*s.*; Pit 1*s.* 6*d.*; Gallery 6*d.* Second Price at Nine o'Clock to Pit and Boxes only, Boxes 1*s.* 6*d.*, Pit 1*s.* Box Office open from 10 to 5 daily—no charge for Booking.
Acting Manager—Mr. W. H. SWANBOROUGH.

ROYAL COLOSSEUM, Regent's-park and Albany-street, pronounced by the united Press of England and the Continent the most wonderful Shilling Exhibition in the Metropolis. Colossal Dioramas of London, Lisbon, and Paris—Musical and Pictorial Entertainments—Comic Monologue—Oxy-Hydrogen Microscope—Dissolving Views—Swiss Scenery, Mountain Torrents, and Cascades—Waterfalls and Fountains, &c. &c. Open: Morning at Twelve, Evening at Seven. Admission to the whole of the Exhibitions and Entertainments, 1*s.*
Dr. BACHHOFFNER, F.C.S., *Sole Lessee and Manager.*

THE

SOUTH KENSINGTON MUSEUM,

Containing Works of Decorative Art, Modern Pictures, Sculpture and Engravings, Architectural Illustrations, Building Materials, Educational Apparatus and Books, Illustrations of Food and Animal Products, is open FREE on Mondays, Tuesdays, and Saturdays, from 10 a.m. till 10 p.m. The STUDENTS' DAYS are Wednesdays, Thursdays, and Fridays, when the public are admitted on payment of 6*d.* each person; hours from 10 a.m. till 4, 5, or 6 p.m., according to the season.

THE

SCIENCE & ART DEPARTMENT.

For information of the aid afforded by the Science and Art Department of the Committee of Council on Education towards obtaining instruction in Science and Art, bearing on industrial occupations, see the "Illustrated Catalogue" of the International Exhibition, Class 29.

THOMAS CROGER, Exhibitor, 483, Oxford-street, London, Musical Instruments of every description, including the Æolian Harp, for the garden, window-ledge, &c. The Educational Metallic Harmonicon. Pipes Notes, Keys, &c., for Organs and Harmoniums. Illustrated explanatory price lists and testimonials forwarded on application. See also "Official Illustrated Catalogue," Classes 16 and 29.

BY SPECIAL APPOINTMENT.

NUTTING & ADDISON,

PIANOFORTE MAKERS

TO HER MAJESTY,

19, OSNABURGH STREET,

REGENT'S PARK, AND

210 REGENT STREET.

COTTAGE, SEMI-COTTAGE,

BOUDOIR, AND PICCOLO UPRIGHT

PIANOFORTES.

WHEATSTONE'S HARMONIUMS (ENGLISH). —In solid oak cases, manufactured by them, have the full compass of keys, are of the best quality of tone, best workmanship and material, and do not require tuning.

Guineas.

New Patent, five octaves, from CC, double pedals (The best and cheapest Harmonium made).
With one stop, oak case (reduced price) . . . .
Piccolo Piano Model, one stop, polished, with unique wind indicator 10
(With soft and distinct tones, and projecting finger-board.)
With two stops, one set and a-half of vibrators (polished case) 12
(The extra upper half-set of vibrators adds greatly to the effect of the treble, and produces a beautiful diapason-like quality of sound.)
With three stops, large size organ tones (polished case) 15
With five stops, two sets of vibrators (ditto) . 22
With eight stops, two sets of vibrators (ditto) . 24
With ten stops, three sets of vibrators (ditto) . 30
(The best and most effective instrument made.)

For particular description of the above and other Harmoniums, in rosewood and mahogany cases, see Messrs. Wheatstone & Co.'s Illustrated Catalogue, which may be had of them gratis and post-free.

The only Exhibition Prize Medallist for Harmoniums, 1851.

An extensive assortment of French Harmoniums by Alexandrè (including all the latest improvements) at prices from 5 guineas to 150 guineas.

WHEATSTONE & CO.,

Inventors & Patentees of the Concertina,

20, CONDUIT ST., REGENT ST., LONDON.

The original Manufacturers and Importers of Harmoniums.

MUSICAL-BOX DEPOTS,

32, LUDGATE STREET, AND 56, CHEAPSIDE.

NICOLE'S

Celebrated large MUSICAL-BOXES, £1 per Air.
SNUFF-BOXES, from 18s. to 40s.

Catalogues of Tunes and Prices gratis and post-free on application to WALES & MCCULLOCH, as above.

N. WHITEHOUSE,

Practical Optician,

2, CRANBOURN STREET, LEICESTER SQUARE,
Two doors from Burford's Panorama.
ESTABLISHED 1830.

Prices of Dr. Woollaston's Spectacles.

Elastic Blue Steel - - - - - - - - - 6*s.* 6*d.*
Ditto ditto Pebbles - - - - - - - - 15*s.* 6*d.*
Solid Gold Frames and Pebbles - - - - 42*s.* 0*d*

Full particulars on application.
N.B.—Not connected with any other House.

HYDROMETERS, SACCHROMETERS, AND GAUGING RULES,

With all instruments and publications appertaining to Brewers, Distillers, Wine and Spirit Merchants, Wholesale, Retail, and for Exportation.

W. R. LOFTUS, Manufacturer and Publisher to the Revenue of the United Kingdom, 6, Beaufoy-terrace, Edgware-road, London, W.

'None are superior."—*Art Journal.*

ALBUM PORTRAITS, 10 for 10s.; 21 for £1.
ALBUMS, hold 25, gilt-edged, 3s. 6d. each.

LONDON STEREOSCOPIC & PHOTOGRAPHIC COMPANY,

54, CHEAPSIDE (corner of Bow-churchyard),
110, REGENT STREET (opposite Vigo-street).

See extended advertisement, "Fine Arts Catalogue."
GEORGE S. NOTTAGE, *Managing Partner.*

BURROW'S LANDSCAPE GLASSES.

"The best Binoculars yet invented."

In sling cases, 3½ and 6 Guineas each.

TARGET TELESCOPES,

FOR LONG RANGES,

Very light and handy, 25*s.* and 30*s.* each.

W. & J. BURROW,
GREAT MALVERN,

London Agents—B. ARNOLD, 72, Baker-street, W.; WALES & MCCULLOCH, 56, Cheapside, E.C.

HORNE & THORNTHWAITE,

PHILOSOPHICAL AND PHOTOGRAPHIC INSTRUMENT MAKERS TO HER MAJESTY,

121, 122, & 123, *Newgate Street, London, E.C.*

Every requisite for the practice of Photography, and every other branch of Natural Science. Catalogues on application.

JOHN J. GRIFFIN, F.C.S., Manufacturer of Chemical and Philosophical Apparatus, for Educational and Scientific purposes. Specimens may be seen in Classes 13 and 29 of the International Exhibition, in the Educational Department of the South Kensington Museum, and at the Manufactory, No. 119, Bunhill-row, Finsbury, which is open for inspection daily, and where the Apparatus may be examined, and in many cases seen in operation.

THE

LONDON DRAWING ASSOCIATION,

For supplying Engineering, Mechanical, Architectural, and General Drawings, Designs, Maps, Plans, Tracings, &c. &c.

7, DUKE STREET, ADELPHI.

FREDERIC YOUNG, *Manager.*

Mr. CLAUDET,

PHOTOGRAPHER TO THE QUEEN,

BY APPOINTMENT,

107, REGENT STREET, LONDON.

Fourth door from Vigo Street.

Photographic Portraits plain and coloured; Cartes de Visite; Portraits, from Miniature to Life Size; Stereoscopic Portraits.

The following Medals have been awarded to Mr. CLAUDET for the superiority of his Portraits:—

Council Medal, Great Exhibition, 1851.
First-class Silver Medal, Great Exhibition of Paris, 1855.
Silver Medal, Exhibition of Amsterdam, 1855.
Bronze Medal, Exhibition of Brussels, 1856.
Silver Medal, Photographic Exhibition of Scotland, 1860.
Silver Medal, Photographic Exhibition of Birmingham, 1861.

NEWMAN,

MANUFACTURER OF

SUPERFINE ARTIST COLOURS,

VARNISHES, BRUSHES,

And Materials of every description,

24, SOHO SQUARE, LONDON.

PATENTS.

ENGLISH AND FOREIGN PATENT OFFICE.

(*Established* 1830.)

Messrs. DE FONTAINE, MOREAU, & GILBEE, Patent Agents (Chief Offices, 4, South-st., Finsbury, E.C., and 1, Rue de Cyprés, Brussels), undertake the procurement and disposal of Patents in Great Britain and on the Continent. Terms moderate. Circular of information gratis. Agencies in all parts.

NOTICE TO INVENTORS.—Office for Patents, 4, Trafalgar Square, Charing Cross, London.—Printed instructions (gratis) as to the cost of patents for Great Britain or foreign countries. Advice and assistance in disposing of or working inventions. Established 25 years. Full information as to expiring or existing patents at home or abroad.—Apply personally, or by letter, to Messrs. PRINCE & CO., Patent Office, 4, Trafalgar Square, Charing Cross, London, W.C.

LA SURETÉ DU COMMERCE, 11, Rue St. Fiacre, Paris.—The French Society for the Protection of Trade.—Commercial enquiries answered, and Debts recovered in any part of France. The Register contains the names of upwards of 1,200,000 French traders. Addresses and all useful commercial information furnished to subscribers. Annual subscription, Two Guineas.—For further particulars apply to Mr. M. ABRAHAMS, Solicitor to the London Agency, 17, Gresham-street, Bank, E.C.

SHIPPING AGENTS.

WILLIAMS & GEILS,

Forwarding, Shipping, and Insurance Agents for all descriptions of Machinery and all other classes of Merchandize.

166, Fenchurch Street, LONDON, E.C., and Shipping Office Chambers, Junction Place, HULL.

AGENCE MARITIME.

MM. WILLIAMS et GEILS,

Expéditeurs et Agents d'Affaires, et d'Assurances Maritime pour toute espèces de Machines et de Marchandises en général.

à Londres:—166, Fenchurch Street, City, E.C.
à Hull:—Shipping Office Chambers, Junction Place.

SCHIFFS-AGENTUR.

WILLIAMS & GEILS,

Spediteurs, Schiffs-und Assecuranz-Agenten für allerlei Maschinen und Waaren.

Addresse in London:—166, Fenchurch Street, E.C.
Addresse in Hull:—Shipping Office Chambers, Junction Place.

TEETH.

MESSRS. GABRIEL,

THE old-established Dentists, 34, Ludgate Hill, and 27, Harley Street. Exhibitors, Class 17 (see specimens). Patentees and Sole Proprietors of Gabriel's SELF-ADHESIVE TEETH AND SOFT GUMS, without springs and without any operation. One set lasts a life-time. At half the usual prices. Pamphlets gratis. Liverpool, 134, Duke Street; Birmingham, 65, New Street.

ARTIFICIAL TEETH AND PAINLESS DENTISTRY.—MESSRS. MOSELY, Dentists, 30, Berners-street, London, Established 1820, beg to direct attention to a New and Patented improvement in the manufacture of Artificial Teeth, &c. which supersedes all metals and soft or absorbing agents, hitherto the fruitful causes of so many evils to the mouth and gums. A portion of this great improvement consists of a gum-coloured enamelled base for the Artificial Teeth, and as the whole is moulded in a soft state, all inequalities of the gums or roots of teeth are carefully protected, thus insuring a perfect system of painless Dentistry. Neither metals, wires, or unsightly ligatures are required, but a perfectly complete adhesion secured by Mr. Mosely's PATENTED SUCTION PALATE, No 764, Aug. 1855. Consulations, and every information free. Charges unusually moderate. Success guaranteed in all cases by Messrs. MOSELY, 30, Berners-street, Oxford-street.

RUPTURES.—WHITE'S MOC-MAIN PATENT LEVER TRUSS is allowed by 500 medical men to be the best for hernia. It consists of an elastic pad, with a lever, and, instead of the usual spring, a soft band, fitting so closely as to avoid detection. A descriptive circular may be had by post.

JOHN WHITE, Manufacturer, 228, Piccadilly.

Single, 16*s.*, 21*s.*, 26*s.* 6*d.*, and 31*s.* 6*d.*; postage, 1*s.* Double, 31*s.* 6*d.*, 42*s.*, and 52*s.* 6*d.*; postage, 1*s.* 8*d.* P.O.O. to John White, P.O., Piccadilly. [Class 17.

RUPTURE.—'COLES'S TRUSS IS BEST.'—This is the latest invention and the greatest improvement in Trusses. Patronized by Sir Astley Cooper and the most eminent surgeons, worn and recommended by William Cobbett, and which has commanded for thirty years a constantly-increasing reputation. It is what a Truss should be—perfectly efficacious, yet agreeable to the wearer. Read "Cobbett's Legacy to Ruptured Persons," gratis at the Patentees' and Manufactory, 3, Charing-cross. None genuine unless marked with the address. See specimens in Class 17, Exhibition of 1862.

QUININE.

Simplest and best preparation,

WATERS'S QUININE WINE.

Unsurpassed as a tonic—Pleasant to the taste. "Our correspondent may rely upon its purity."—*Lancet.* Sold by all grocers, chemists, and wine merchants, 30*s.* per dozen quart bottles.—ROBERT WATERS, 2, Martin's-lane, Cannon-street, London.

Comfort in Shaving Insured, and Time Saved.

PEARS'S SHAVING STICK

Produces with hot or cold water an instantaneous, unctuous, and consistent lather, which softens the beard, and thereby renders the process of shaving more rapid, easy, and cleanly, than the old mode of using the brush and dish. Pears's Shaving Stick is formed from his Transparent Soap, which has a most agreeable fragrant odour, and its firm consistence makes it more durable than any other soap. Prices of Shaving Sticks, in cases, 1*s.* and 1*s.* 6*d.* each; ditto of Transparent Soap, square tablets, 1*s.* each, and upwards.

Prepared and sold by A. & F. PEARS, 91, Great Russell-street, Bloomsbury, three doors west of the British Museum, and by most respectable perfumers and chemists in town and country. Be sure to ask for "Pears's Shaving Stick." May be had free by post for sixteen postage stamps.

THE GENERAL APOTHECARIES COMPANY (Limited), 49, Berners-street, London, W., and 4, Colquitt-street, Liverpool, was formed for the purpose of supplying the Profession and the public with genuine drugs and pure chemicals. The bonâ fide character of their medicinal preparations is insured by their being respectively ground or manufactured on the premises. All drugs, chemicals, and pharmaceutical preparations are subjected to a careful examination by the Company's analytical chemist.—EDMUND PRATT, *Secretary.*

ANILINE, Nitro Benzol, Oxalic Acid, Citric Acid, Tartaric Acid, Sulphuric Ether, Spirits of Nitre, and all descriptions of pure Chemicals.

FREDERICK ALLEN, Manufacturing Chemist, Bow Common, London, E., invites foreign and home consumers to make trial of any of the above articles direct from the factory. F. A.'s productions have from time to time been tested by analytical chemists eminent for ability, and pronounced of great purity and superior quality. As great care is taken in the manufacture, each article will at all times be found unvarying in quality and character.

GREAT INTERNATIONAL EXHIBITION.

Immediately after the closing will be published, price 5*s.*, a supplementary volume of THE YEAR BOOK OF FACTS IN SCIENCE AND ART. By JOHN TIMBS, F.S.A.,

A compact history of the Exhibition from its first conception to its close describing its origin, progress, and results, its most remarkable objects, official lists, statistical returns, awards, &c. London: Lockwood & Co., 7, Stationers'-hall-court, E.C.

WORTH NOTICE.—What has always been wanted is now published, 5*s.*, post-free, 24th 1,000, enlarged.

THE DICTIONARY APPENDIX and GUIDE to CORRECT SPEAKING and WRITING, with upwards of 7,000 words not found in the Dictionary, comprising the Participles of the Verbs, which perplex all writers. "No person who writes a letter should be without it; those who use it only for one hour cannot fail to appreciate its value."—*Weekly Times.* It is as acceptable as Walker's Dictionary.

JOHN F. SHAW & Co., 48, Paternoster-row, London.

SHAKESPEARE.—THE FAMOUS FOLIO OF 1623.—Now ready, Part I. of the Reprint of this Edition, containing the whole of the Comedies; price 10*s.* 6*d.* To be followed by Part II. The Histories; Part III. The Tragedies. Excepting in regard to size, which is rendered more convenient, it may be truly said,

"One sand another—Not more resembles"

than this Reprint the Original.—Prospectuses and specimen leaves may be had on application.—L. BOOTH, 307, Regent-street, W. Exhibiting, Class 28.

SMITH & SONS' EDUCATIONAL MAPS.—The following are recently added:—

	ft.	in.	ft.	in.	On rllrs.
World on Mercator's Projection	8	6	5	8	30*s.*
World on Globular Projection	7	9	4	2	21*s.*
Africa (with the latest discoveries)	4	4	5	8	16*s.*
Australia (showing the recent explorations)	5	8	4	4	16*s.*

SMITH & SON, 172, Strand (corner of Surrey-street).

MACKAY'S "PERMANENT MANURE"

(So called in contra-distinction to Peruvian guano),

£10. 10*s.* per ton, in 2-cwt bags, delivered free at Agent's Stores.

Manufactured expressly to supply all the ingredients extracted from the soil, by the root crops and cereals, commonly cultivated.

SPECIAL CHARACTERISTICS.

1. Continued effects during successive seasons.
2. Rapidly forwarding the turnip plant in its early stages, and increasing specific gravity of bulbs.
3. Producing potatoes more dry and palatable, and greatly lessening tendency to disease.
4. Increasing yield and weight of grain, and strengthening the straw, thereby enabling it to carry the additional grain without lodging.

Sold by the Manufacturers, GEORGE MACKAY & CO., at Inverness, Scotland (established 1811), and by their Agents at the various towns throughout the Kingdom. (Exhibitor, Class 2.)

THE CONSERVATIVE LAND SOCIETY,

Established 7th Sept., 1852, and enrolled under 6 & 7 William IV., *c.* 32.

TRUSTEES.

The Viscount Ranelagh. | J. C. Cobbold, Esq., M.P.

EXECUTIVE COMMITTEE.

Chairman—Viscount Ranelagh.

Vice-Chairman—Col. Brownlow Knox, M.P.

Bective, Earl of, M.P.	Meyrick, Lt.-Col.
Bourke, Hon. Robert.	Newcomen, C. E., Esq.
Cobbold, J. C., Esq., M.P.	Palk, L., Sir, Bart., M.P.
Currie, H. W., Esq.	Pownall, Henry, Esq.
Holmes, T. Knox, Esq.	Talbot, Hon. & Rev. W. C.
Ingestre, Viscount, M.P.	Winstanley, N., Esq.

Secretary—Charles Lewis Gruneisen, Esq.

Offices—33, Norfolk-street, Strand, London, W.C.

Investments can be made (at the option of the investors) for Savings and Capital, either in the Share, Deposit, or Land Departments, or in all the three divisions of business. Plots, or Houses, or Ground Rents, in eighteen counties.

GLENFIELD PATENT STARCH,

USED IN THE ROYAL LAUNDRY.

THE Ladies are respectfully informed that this Starch is exclusively used in the Royal Laundry; and Her Majesty's Laundress says, that although she has tried Wheaten, Rice, and other Powder Starches, she has found none of them equal to the GLENFIELD, which is the finest Starch she ever used.

In Packets of 1*d.*, 2*d.*, 4*d.*, and 8*d.* each.

WOTHERSPOON & CO., Glasgow and London.

KAMPTULICON, OR ELASTIC FLOORCLOTH. GOUGH & BOYCE, the Original Patentees and Manufacturers of this favourite article, beg to inform the public that they have exclusively supplied it to Windsor Castle, Buckingham Palace, the Tuilleries, the Houses of Parliament, the Bank of England, and other Government offices. It is far superior to oil-cloth, and is especially adapted for churches, offices, halls, passages, stairs, and billiard rooms. Sold by all the best upholsterers in town and country, and also by GOUGH & BOYCE, 12, Bush-lane, Cannon-street, London.

COLT'S

PATENT REPEATING FIRE-ARMS.

CAUTION.

Beware of mongrel imitations and Patent infringements. The genuine weapons are stamped on the barrel, "Address Col. COLT, LONDON," and bear the British Proof-mark on the barrel and between each nipple.

TERMS.

Cash.—The usual discount allowed to the Trade, Army and Navy, East India, Continental, and Colonial Agents. Orders from the country and abroad must be accompanied with cash. Post-office Orders to be made payable at the CHARING CROSS Money Order Office.

Colt's Revolvers.

The favourite sized Pistol for Officers, Army, Navy, and Merchant Service (approved of and adopted universally) is 7½ inch, Rifle bored, Six Shots, weight 2 lbs. 8 oz. Case complete.

Colt's New Pistol Carbine.

The advantages of this Breech Attachment to Colt's 7½ inch Repeating Pistol, and its superiority over all others for Military and Sporting purposes, will be easily understood and appreciated, combining, as it does, *all* the advantages of both Pistol and Carbine, into either of which it can be *instantly* converted.

Four, Five, and Six-inch Rifle Barrel Pocket Pistol,

For Officers, Travellers, and House Protection.

FIVE SHOTS. Weight 23, 24, 26 ozs. (84 Conical and 126 Round Balls to the lb.); with Mould, Wrench, and Cleaning Rod, in Case complete.

New Model (Revolver Shot-and-Ball Gun.

27 inch barrel. FIVE SHOTS. A capital Colonial Gun, for Bird or Beast.

Revolver Rifles.

18, 21, 24, 27, and 30 inch Barrel. FIVE AND SIX SHOTS. For Naval, Military, and Sporting purposes. Large, Medium, and Small Bore, with Sights and all other appurtenances.

Holsters, Belts, Pouches, Ammunition, &c., always ready.

Descriptive and Price Lists Free. Orders, wholesale and retail, promptly executed.

SAMUEL COLT, 14, PALL MALL, S.W.,

Or through any respectable Mercantile, East Indian, or Colonial Agency.

Every Arm is London-proved and bears the Patentee's Trade Marks.

DAY & SON, LITHOGRAPHERS to the QUEEN,

Illustrated, Illuminated, & General Book & Fine-Art Publishers.

Dedicated, by Command, to Her Most Gracious Majesty the Queen,
And by permission of Her Majesty's Commissioners.

MASTERPIECES of INDUSTRIAL ART & SCULPTURE at the INTERNATIONAL EXHIBITION, 1862, by J. B. Waring.

This collection will consist of 300 Plates, containing several hundred Illustrations of the best examples in Sculpture and the Decorative and Industrial Arts; to be executed in the highest style of excellence attainable in Chromo lithography, from coloured Photographs, &c., taken for the purpose, with the express permission of the Exhibitors, by FRANCIS BEDFORD; and will form a complete and valuable epitome of the state of the Industrial Arts throughout the World in the year 1862; a work, as one of reference, calculated to advance the state of these Arts in the future. It will be of such permanent value, and of such elegance and beauty in its production, as to render it necessary for every library in the world, and fit for the drawing-room table.

THE EDITION WILL BE LIMITED TO 2,000 COPIES, and the Stones will then be Destroyed, thus insuring the fullest permanent value for every copy issued.

SUBSCRIBERS' NAMES SHOULD BE SENT TO THE PUBLISHERS AT ONCE. The Work will be published in Parts, each to contain Five Plates and Descriptive Text. The entire work will form Three Volumes. Part I. May 1st. To be completed in 12 months.

Day & Son's Authentic Views of the International Exhibition Building, and its Contents, of all sizes and at all prices.

The Photographs in the East, by Mr. Bedford, who, by command, has accompanied **His Royal Highness the Prince of Wales in his Tour through the Holy Land,** &c. &c. will be published by Messrs. DAY & SON on Mr. Bedford's return. The terms of publication of this highly interesting and beautiful Series may be had on application. A list of Mr. Bedford's English Photographs may also be had.

CAPT. COWPER P. COLES' (R.N.) SHOT-PROOF (Cupola) STEAM RAFT, contrasted with the appropriation of the Invention in the "Monitor." Views, Elevations, Sections, and Plans: with Dimensions, Price 7s. 6d.; also Views of Armour-clad Ships, each 10s. 6d.

Lately Published.

1. The Victoria Psalter. Illuminated by Owen Jones. Dedicated by command to the Queen. Bound in leather, in relief, £12. 12s.
2. Painting in Water-Colour. By Aaron Penley. With Water-Colour Studies, £4. 4s. Proofs, £6. 6s.
3. The Sermon on the Mount. Illuminated by W. & G. Audsley. Magnificently bound, £8. 8s., £10. 10s., and £12. 12s. A splendid and extraordinary work.
4. Manuals for the Practice of Illuminating. By Wyatt & Tymms, 1s. 6d. each.
5. Mr. C. T. Newton's Discoveries at Halicarnassus is out of print, and can only be obtained at the price of £21.
6. Mr. W. Eden Nesfield's work on Mediæval Architecture in France and Italy. Just ready, £4. A list of other Architectural Works.; 7. A List of Illuminated and Illustrated Works; 8. A List of Government Educational Diagrams; 9. A List of Books and Prints, illustrative of all parts of the world; 10. A List of Chromo-lithographs from Drawings, lent for publication by the Queen, may be had on application.

THE DESTROYED PLATE

CHRIST BLESSING LITTLE CHILDREN,

By EASTLAKE and WATT, 22 by 29, on paper 44 by 33.

		published at		price	
Artists' Proofs, India,		£15.	15s.	£5.	5s.
Before Letters,	ditto	12	12	,, 4	4
Inscription Proofs,	ditto	8	8	,, 3	3
After Letters proof, plain,	ditto	5	5	,, 2	2
Prints, plain	ditto	4	4	,, 1	1

Illuminated and Illustrated Works in preparation.

1. Ephesus, and the Temple of Diana. E. Falkener, £2. 2s.
2. The Miniatures and Ornaments of Anglo-Saxon and Irish Manuscripts. J. O. Westwood. 200 copies printed, and the stones destroyed. In 17 parts, at £1. 1s.
3. Anatomy for Artists. By J. Marshall, £1. 1s.
4. The Church's Floral Kalendar. Miss Cuyler, with 38 Illuminated pages by Tymms, £1. 11s. 6d.
5. The Prisoner of Chillon. Illuminated by Audsley, £1. 1s.
6. The Colours of the British Army. By R. F. McNair, in 36 parts at 5s.
7. The History of Joseph and his Brethren. Illuminated and Illustrated by O. Jones and H. Warren, £2. 2s.
8. One Thousand and One Initial Letters. Designed and Illuminated by O. Jones, £4. 4s.
9. Bunyan's Pilgrim's Progress, 30 Water-Colour Drawings. By J. Nash, £3. 3s.
10. Indian Fables, Translated from the Sanscrit and Illustrated in Colours. By Florence Jacomb, £2. 2s.
11. The Art of Decorative Design. By C. Dresser, 200 Illustrations, many Chromo-lithographs.
12. Sketches from Nature in Pencil and Water-Colours. By G. Stubbs, 17 Plates £1. 1s.
13. Passages from English Poets. Illustrated by the Junior Etching Club, 47 plates, proofs £6. 6s. fine copies £3. 3s.

Books nearly out of Print, never to be reproduced, the Stones being destroyed.

1. THE GRAMMAR OF ORNAMENT. By OWEN JONES. Published at £19. 19s., present price £12. 12s.
2. ROBERTS'S SKETCHES IN THE HOLY LAND, &c.

			published at		price	
6 vols.	in parts,		£7.	7s.	£3.	15s.
6 do.	3	ditto.	9	0	,, 4	10
6 do.	3, half mor.,	ditto	10	10	,, 5	0
6 do.	3, morocco,	ditto	11	11	,, 6	0

3. THE ART OF ILLUMINATING. By M. D. WYATT and W. R. TYMMS. Published at £3. 10s., price £2. 2s.

A List of other Works nearly out of print may be had.

COMMISSIONS EXECUTED IN EVERY BRANCH OF THE FINE ARTS.

DAY & SON, LITHOGRAPHERS to the QUEEN, CHROMO-LITHOGRAPHERS, Steel and Copper-plate Engravers and Printers, Draughtsmen and Engravers on Wood, Artistic, Scientific, or Commercial.—Architectural Draughtsmen and Colourists; Letter-press Printers and Bookbinders: in fact, Producers of all Parts and the entirety of Works of every class

BANK NOTE AND CHEQUE ENGRAVERS AND PRINTERS, PHOTOGRAPHERS, ETC.

PATENTEES AND SOLE WORKERS OF A NEW SYSTEM OF AUTOMATIC LITHOGRAPHY AND CHROMO-LITHOGRAPHY, which offers immense advantages to all Consumers of plain, ornamental, or colour Printing. Estimates on application. Picture-Frame Makers, &c.

PRESSES, STONES, AND EVERY MATERIAL FOR THE PRACTICE OF LITHOGRAPHY.

---

4 to 9, GATE STREET, LINCOLN'S-INN FIELDS, LONDON, W.C.

DAY & SON show Specimens of their Productions, and Copies of their Works, at their Stall, North Gallery, near Eastern Dome; and exhibit Colour-printing in action in the Processes Court.

PRINTED BY TRUSCOTT, SON, & SIMMONS, SUFFOLK LANE, CANNON STREET, CITY.

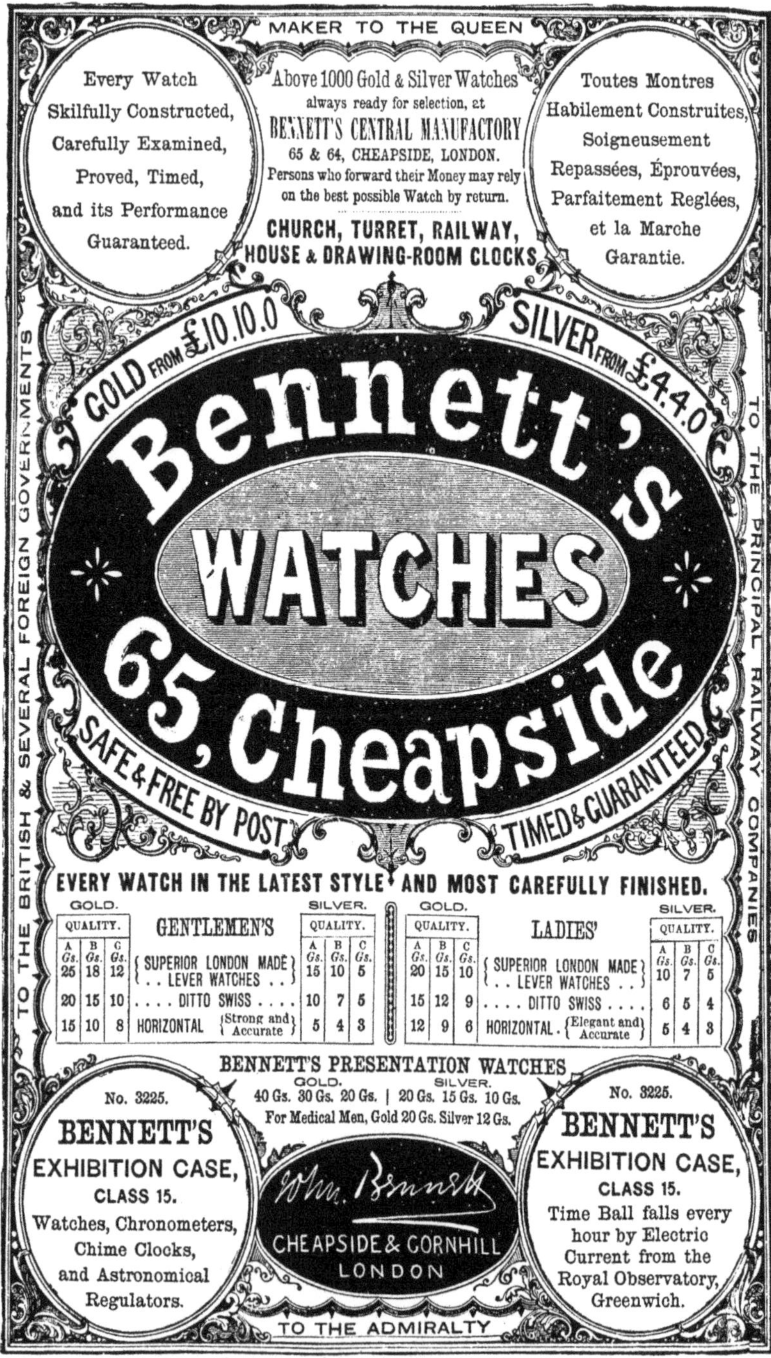

GOLD QUALITY A Gs.	B Gs.	C Gs.	GENTLEMEN'S	SILVER QUALITY A Gs.	B Gs.	C Gs.
25	18	12	SUPERIOR LONDON MADE . . LEVER WATCHES . .	15	10	5
20	15	10	. . . . DITTO SWISS	10	7	5
15	10	8	HORIZONTAL {Strong and Accurate}	5	4	3

GOLD QUALITY A Gs.	B Gs.	C Gs.	LADIES'	SILVER QUALITY A Gs.	B Gs.	C Gs.
20	15	10	SUPERIOR LONDON MADE . . LEVER WATCHES . .	10	7	5
15	12	9	. . . . DITTO SWISS	6	5	4
12	9	6	HORIZONTAL . {Elegant and Accurate}	5	4	3

PRINTED BY TRUSCOTT, SON, & SIMMONS, SUFFOLK LANE, CANNON STREET, CITY.

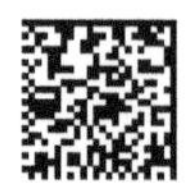

www.ingramcontent.com/pod-product-compliance
Ingram Content Group UK Ltd.
Pitfield, Milton Keynes, MK11 3LW, UK
UKHW040700180125
453697UK00010B/304

9 781108 067157